Lecture Notes in Computer Science 13774

More information about this series at https://link.springer.com/bookseries/558

Takanori Isobe · Santanu Sarkar (Eds.)

Progress in Cryptology – INDOCRYPT 2022

23rd International Conference on Cryptology in India Kolkata, India, December 11–14, 2022 Proceedings

Editors
Takanori Isobe
University of Hyogo
Hyogo, Japan

Santanu Sarkar
Indian Institute of Technology Madras
Chennai, India

ISSN 0302-9743 ISSN 1611-3349 (electronic)
Lecture Notes in Computer Science
ISBN 978-3-031-22911-4 ISBN 978-3-031-22912-1 (eBook)
https://doi.org/10.1007/978-3-031-22912-1

This Springer imprint is published by the registered company Springer Nature Switzerland AG
The registered company address is: Gewerbestrasse 11, 6330 Cham, Switzerland

Preface

With great pleasure, we present the proceedings of INDOCRYPT 2022, the 23rd International Conference on Cryptology in India, organized by The Chatterjee Group - Centers for Research and Education in Science and Technology (TCG CREST), the R. C. Bose Centre for Cryptology and Security, Indian Statistical Institute, Kolkata, and the Bose Institute, Kolkata, under the aegis of the Cryptology Research Society of India (CRSI).

INDOCRYPT began in 2000 under the leadership of Bimal Roy at the Indian Statistical Institute, Kolkata, with an intention to target researchers and academicians in the domain of cryptology. Since its inception, this annual conference has not only been considered as the leading Indian venue on cryptology but also has gained recognition among the prestigious cryptology conferences in the world. Over the last two decades, the conference was held in various cities of India, such as Kolkata (2000, 2006, 2012, 2016), Chennai (2001, 2004, 2007, 2011, 2017), Hyderabad (2002, 2010, 2019), New Delhi (2003, 2009, 2014, 2018), Bangalore (2005, 2015), Kharagpur (2008), Mumbai (2013), and Jaipur (2021). Due to COVID-19 pandemic restrictions, INDOCRYPT went online in 2020. This year was the fifth time the conference was hosted in Kolkata, but in a hybrid mode.

INDOCRYPT 2022 received 88 submissions from 30 different countries in total, among which the papers that were withdrawn before the deadline, or the ones that didn't match the submission policy, were not considered for evaluation. Finally, 74 papers were reviewed by three to four reviewers each. First, the papers went through a double-blind review phase. Next, after a two week discussion phase, with additional comments from the Program Committee members as well as the external reviewers, 31 papers by authors from 17 different countries were finally accepted for presentation in the program and inclusion in this proceedings.

We are immensely thankful to the 52 Program Committee members and the 64 external reviewers, who participated in the process of reviewing and subsequent discussions. Without their tremendous effort, the conference would not have been successful. We would also like to express our gratitude to Springer for their active cooperation and timely production of the conference proceedings. We managed the submissions, reviews, discussions, and proceedings very effectively using the online EasyChair conference management software system and would like to acknowledge this with great regard.

Our program also included three invited talks by V. Kamakoti from IIT Madras, India, Gregor Leander from Ruhr University Bochum, Germany, and Alexander May from Ruhr University Bochum, Germany. Moreover, there were three tutorial talks by Patrick Derbez from University of Rennes 1, France, Mridul Nandi from ISI Kolkata, India, and Santanu Sarkar from IIT Madras, India.

INDOCRYPT 2022 was organized by TCG CREST and the R. C. Bose Centre for Cryptology and Security with the Bose Institute providing the conference venue. We are extremely thankful to the General Co-chairs, Bimal Kumar Roy (ISI Kolkata) and Joydeep Bhattacharya (TCG CREST), for coordinating all the issues related to the organization of the event. We would also like to take this opportunity to thank the Organizing Chair, Organizing Co-chairs, and all members of the Organizing Committee, for their relentless support in successfully hosting the conference.

We are also immensely thankful to the Government of India, the Government of West Bengal, and our sponsors Google, HDFC Bank, Vehere Interactive Pvt. Ltd., AON, KEWAUNEE International Group, TwoPiRadian Infotech Private Limited, and Bosch Global Software Technologies Private Limited, for their generous financial support towards the conference.

Last but not the least, we are extremely thankful to each of the 220 authors who submitted their articles to the conference and those who attended INDOCRYPT 2022.

October 2022

Takanori Isobe
Santanu Sarkar

Organization

General Chairs

Bimal Kumar Roy	ISI Kolkata, India
Joydeep Bhattacharya	TCG CREST, India

Program Co-chairs

Takanori Isobe	University of Hyogo, Japan
Santanu Sarkar	IIT Madras, India

Organizing Chair

Subhamoy Maitra	ISI Kolkata, India

Organizing Co-chairs

Somshubhro Bandyopadhyay	Bose Institute, India
Soumyajit Biswas	TCG CREST, India
Nilanjan Datta	TCG CREST, India

Sponsorship Chair

Rakesh Kumar	ISI Kolkata, India

Accommodation Chair

Bibhas Chandra Das	TCG CREST, India

Organizing Committee

Avik Chakraborti	TCG CREST, India
Shreya Dey	TCG CREST, India
Avijit Dutta	TCG CREST, India
Arpita Maitra	TCG CREST, India
Sougata Mandal	TCG CREST, India
Payel Sadhukhan	TCG CREST, India
Soumya Kanti Saha	TCG CREST, India

Laltu Sardar	TCG CREST, India
Bishakha Sarkar	TCG CREST, India

Program Committee

Avishek Adhakari	Presidency University, India
Shi Bai	Florida Atlantic University, USA
Christof Beierle	Ruhr University Bochum, Bochum, Germany
Rishiraj Bhattacharyya	NISER, India
Christina Boura	University of Versailles, France
Suvradip Chakraborty	ETH Zurich, Switzerland
Anupam Chattopadhyay	NTU, Singapore
Sherman Chow	Chinese University of Hong Kong, Hong Kong
Prem Laxman Das	SETS Chennai, India
Nilanjan Datta	TCG CREST, India
Avijit Dutta	TCG CREST, India
Ratna Dutta	IIT Kharagpur, India
Keita Emura	National Institute of Information and Communications Technology, Japan
Andre Esser	Technology Innovation Institute, Abu Dhabi, UAE
Indivar Gupta	DRDO, Delhi, India
Akinori Hosoyamada	NTT Social Informatics Laboratories, Japan
Mahavir Jhanwar	Ashoka University, India
Selcuk Kavut	Balikesir University, Turkey
Sumit Kumar Pandey	IIT Jammu, India
Jason LeGrow	Verginia Polytechnic Institute and State University, USA
Chaoyun Li	KU Leuven, Belgium
Fukang Liu	University of Hyogo, Japan
Arpita Maitra	TCG CREST, India
Takahiro Matsuda	National Institute of Advanced Industrial Science and Technology, Japan
Willi Meier	FHNW, Brugg-Windisch, Switzerland
Alfred Menezes	University of Waterloo, Canada
Sihem Mesnager	Universities of Paris VIII and XIII, LAGA Lab, France
Kazuhiko Minematsu	NEC, Kawasaki, Japan
Marine Minier	Loria, France
Pratyay Mukherjee	Swirlds Labs/Hedera, USA
Debdeep Mukhopadhyay	IIT Kharagpur, India
Mridul Nandi	ISI, India
David Oswald	University of Birmingham, UK
Saibal Pal	DRDO, Delhi, India

Chester Rebeiro	IIT Madras, Chennai, India
Francesco Regazzoni	University of Amsterdam, Netherlands
Raghavendra Rohit	Technology Innovation Institute, Abu Dhabi, UAE
Sushmita Ruj	University of New South Wales, Sydney, Australia
Somitra Sanadhya	IIT Jodhpur, India
Sourav Sen Gupta	NTU, Singapore
Nicolas Sendrier	Inria, France
Yixin Shen	Royal Holloway, University of London, UK
Bhupendra Singh	DRDO, Bangalore, India
Sujoy Sinha Roy	TU Graz, Austria
Pantelimon Stanica	Naval Postgraduate School, Monterey, USA
Ron Steinfeld	Monash University, Clayton, Australia
Atsushi Takayasu	The University of Tokyo, Japan
Meltem Turan	National Institute of Standards and Technology, USA
Rei Ueno	Tohoku University, Japan
Alexandre Wallet	Inria, France
Yuyu Wang	University of Electronic Science and Technology of China, China
Jun Xu	Institute of Information Engineering, Chinese Academy of Sciences, China

Additional Reviewers

Aikata Aikata
Anubhab Baksi
Pierre Briaud
Bin-Bin Cai
Anirban Chakraborty
Bishwajit Chakraborty
Donghoon Chang
Haokai Changmit Kumar Chauhan
Jorge Chavez-Saab
Pratish Datta
Sabyasachi Dutta
Paul Frixons
David Gerault
Chun Guo
Guifang Huang
Mitsugu Iwamoto
David Jacquemin
Floyd Johnson
Meenakshi Kansal
Hamidreza Khoshakhlagh
Fuyuki Kitagawa
Abhishek Kumar
Kaoru Kurosawa
Virginie Lallemand
Roman Langrehr
Jack P. K. Ma
Gilles Macario-Rat
Monosij Maitra
Siva Kumar Maradana
Subhra Mazumdar
Prasanna Mishra
Girish Mishra
Sayantan Mukherjee
Yusuke Naito
Lucien K. L. Ng
Tran Ngo
Ying-Yu Pan
Tapas Pandit
Amaury Pouly
Mayank Raikwar

Prasanna Ravi
Divya Ravi
Maxime Remaud
Yann Rotella
Debapriya Basu Roy
Partha Sarathi Roy
Rajat Sadhukhan
Yu Sasaki
AndrÃ© Schrottenloher
Jacob Schuldt
Xiangyu Su
Masayuki Tezuka
Toi Tomita
Hikaru Tsuchida
Natarajan Venkatachalam
Javier Verbel
Sulani Kottal Baddhe Vidhanalage
Deepak Vishwakarma
Xiuhua Wang
Harry W.H. Wong
Qianqian Yang
Rui Zhang
Liang Zhao
Lukas Zobernig

Contents

Foundation

Symmetric Key Cryptology

Protocols and Implementation

Cryptanalysis

Boolean Functions

Quantum Cryptography and Cryptanalysis

Post Quantum Cryptography

Foundation

CRS-Updatable Asymmetric Quasi-Adaptive NIZK Arguments

Behzad Abdolmaleki[1(✉)] and Daniel Slamanig[2]

[1] Max Planck Institute for Security and Privacy, Bochum, Germany
behzad.abdolmaleki@mpi-sp.org
[2] AIT Austrian Institute of Technology, Vienna, Austria
daniel.slamanig@ait.ac.at

Abstract. A critical aspect for the practical use of non-interactive zero-knowledge (NIZK) arguments in the common reference string (CRS) model is the demand for a trusted setup, i.e., a trusted generation of the CRS. Recently, motivated by its increased use in real-world applications, there has been a growing interest in concepts that allow to reduce the trust in this setup. In particular one demands that the zero-knowledge and ideally also the soundness property hold even when the CRS generation is subverted. One important line of work in this direction is the so-called updatable CRS for NIZK by Groth *et al.* (CRYPTO'18). The basic idea is that everyone can update a CRS and there is a way to check the correctness of an update. This guarantees that if at least one operation (the generation or one update) have been performed honestly, the zero-knowledge and the soundness properties hold. Later, Lipmaa (SCN'20) adopted this notion of updatable CRS to quasi-adaptive NIZK (QA-NIZK) arguments.

In this work, we continue the study of CRS-updatable QA-NIZK and analyse the most efficient asymmetric QA-NIZKs by González *et al.* (ASIACRYPT'15) in a setting where the CRS is fully subverted and propose an updatable version of it. In contrast to the updatable QA-NIZK by Lipmaa (SCN'20) which represents a *symmetric* QA-NIZK and requires a new non-standard knowledge assumption for the subversion zero-knowledge property, our technique to construct updatable *asymmetric* QA-NIZK is under a well-known standard knowledge assumption, i.e., the Bilinear Diffie-Hellman Knowledge of Exponents assumption. Furthermore, we show the knowledge soundness of the (updatable) asymmetric QA-NIZKs, an open problem posed by Lipmaa, which makes them compatible with modular zk-SNARK frameworks such as LegoSNARK by Campanelli *et al.* (ACM CCS'19).

1 Introduction

Zero-knowledge proofs [24] are a fundamental concept which allows one party (the prover) by interacting with another party (the verifier) to convince the latter that a statement in any NP language is true without revealing any additional information (the zero-knowledge property). At the same time, the prover is not

T. Isobe and S. Sarkar (Eds.): INDOCRYPT 2022, LNCS 13774, pp. 3–25, 2022.
https://doi.org/10.1007/978-3-031-22912-1_1

able to make the verifier accept proofs about false statements (the soundness property). In many of its practical applications it is important to remove interaction in that the prover only needs to compute a single message (a proof), which can then be verified by everyone. These so called non-interactive zero-knowledge (NIZK) proofs, especially for algebraic languages in bilinear groups [26,30,31], play an important role in the design of cryptographic primitives and protocols. The non-interactivity, however, comes at a price and in particular (apart from NIZK secure in the random oracle model) demands a trusted setup that generates a so called common reference string (CRS). This CRS is an input to the prover and all potential verifiers. The critical issue is that if this setup is not performed honestly, i.e., the underlying trapdoor is known to some party, then all security is lost.

A long line of research has focused on obtaining very efficient NIZK proofs in this CRS model [23,27,28,30–35,39], covering efficient pairing-based zero-knowledge Succinct Non-interactive ARguments of Knowledge (zk-SNARKs) for any NP language and succinct Quasi-Adaptive Non-Interactive Zero-Knowledge arguments (QA-NIZKs) for restricted languages, i.e., membership in linear subspaces. QA-NIZKs are a relaxation of NIZK arguments, where the CRS is specialized to the linear space for which membership should be proven [7,8,25,32,33, 35,37,38]. This specialized part is called the language parameter. In this paper our focus will be on QA-NIZK arguments.

1.1 Motivation

For the practical application of NIZK primitives in general, a crucial question is how the CRS generation should be performed. While in theory it is simply assumed that some universally trusted party will perform the CRS generation, such a party is challenging to impossible to find in the real world.

Consequently, this is typically a too strong assumption.

Now there are different approaches to reduce the required trust that needs to be put in the CRS generation. First, the CRS can be generated by a potentially huge set of parties via the use of secure multi-party computation (MPC), so called ceremonies, [1,10,11,36]. And while this approach has seen use in the real world[1], such ceremonies are cumbersome and require significant effort even beyond the technical realisation. Despite the required efforts, however, it can give very strong guarantees, i.e., if at least as one party behaves honest then security is preserved. Second, to remove this additional effort, one can rely on so called subversion NIZKs [9], subversion zk-SNARKS [2,20] and subversion QA-NIZKs [3,6]. In this subversion zero-knowledge model, one introduces a way to check the CRS and the prover does not require to trust the CRS, i.e., the zero-knowledge property (so-called subversion zero-knowledge) is still maintained even if the CRS generation is malicious. Unfortunately, the verifier is still required to trust the CRS generation and it is actually impossible to obtain subversion soundness when at the same time requiring zero-knowledge to hold [9]. Third, an interesting

[1] The "powers of tau" ceremony of Zcash: https://z.cash/technology/paramgen/.

middle ground is the recent technique of a so called *updatable CRS* introduced by Groth *et al.* [29], which is an increasingly popular model [5,12,14,18,22,40–43,46]. In this updatable CRS model, everyone can update a CRS along with providing update proofs such that the correctness of updates can be verified by everyone. This guarantees that zero-knowledge for the prover holds in the presence of an adversarial CRS generator. Moreover, the verifier can trust the CRS, i.e., soundness holds, as long as one operation, either the CRS generation itself or one of the updates of the CRS have been performed honestly. Thus, to be certain that soundness holds, a verifier could do a CRS update on its own and then send the updated CRS to the prover.

Initially, Groth *et al.* [29] defined CRS updates with a focus on zk-SNARKs, and then Lipmaa [40] proposed an updatable CRS version of the QA-NIZK construction of [3,35]. While Lipmaa considers so called symmetric QA-NIZK, i.e., where the language is defined in one of the source groups of a bilinear group, it is not known how this applies to asymmetric QA-NIZK [25], i.e., where the language is defined over both source groups (also called bilateral linear subspaces).[2] Asymmetric QA-NIZKs [17,25,45], however, are useful for many applications where commitments to the same value are available in both source groups of a bilinear group (e.g., proof aggregation, ring signatures, range proofs). As we will discuss soon, despite not being known how to construct it, having what we call an updatable asymmetric QA-NIZK does have interesting implications for concrete applications discussed below.

Applications. zk-SNARKs and QA-NIZKs are appealing as they are succinct, i.e., they allow proving circuits of arbitrary size and linear subspace languages respectively, with a compact proof. They are also concretely very efficient and in particular in bilinear groups we have constructions with proofs represented by three group elements for zk-SNARKs for arithmetic circuits [28], one group element for symmetric QA-NIZK for linear subspace languages [35], and two group elements for asymmetric QA-NIZK for bilateral linear subspace languages [25]. While (asymmetric) QA-NIZKs have many interesting applications (cf. [25,32,33]), our focus will be on their application in the modular design of zk-SNARKs and in particular on LegoSNARK [13].

LegoSNARK is a toolbox for commit-and-prove zk-SNARKs with the aim of constructing a *global* zk-SNARK for some computation C via the linking of *smaller* specialized zk-SNARKs for various subroutines that overall compose to C. The central idea is that by allowing each subroutine of C to be handled by a different proof system, one can select the one that maximizes a metric (e.g., efficiency) that is important for the concrete application. Now LegoSNARK uses succinct QA-NIZKs as efficient zk-SNARKs for linear subspace languages. Abdolmaleki and Slamanig [6] recently showed how one can construct a subversion zero-knowledge variant of symmetric [35] as well as asym-

[2] To avoid confusion we intentionally do not call them QA-NIZK for *symmetric or asymmetric groups* as done in [25], as both types are instantiated in asymmetric, i.e., type-3, bilinear groups.

metric QA-NIZK [25] in a setting where the CRS is subverted but the language parameters are generated honestly. As they mention, the honest language parameters do not represent a problem for practical applications, as they can typically be obtained in a transparent way without trust in their generation (e.g., by deriving them using a random oracle). Furthermore, they show how to integrate a knowledge-sound version of their subversion zero-knowledge symmetric QA-NIZK into LegoSNARK. This represents a step towards a subversion variant of the LegoSNARK toolbox and thus a way to use LegoSNARK with a reduced trust in the required setup.

As most of the recent zk-SNARK constructions focus on the updatable-CRS setting [5,12,14,18,22,41–43,46], it is desirable to enable composable zk-SNARK frameworks such as LegoSNARK also in the updatable CRS setting. If one thereby wants still to take advantage of using QA-NIZK as one of its building blocks, then updatable QA-NIZK are required. While, as mentioned above, there are numerous constructions of zk-SNARKs with an updatable CRS, to date there is only an updatable symmetric QA-NIZK by Lipmaa [40] available. To prove the zero-knowledge property, it requires a new and non-standard knowledge assumption (KW-KE). Adaptive soundness can be shown under a standard assumption, but achieving knowledge soundness, a property that would be required for composable zk-SNARKs, is left as an open problem in [40]. Lipmaa works in a model where the complete CRS including the language parameter (what he calls key) can be generated maliciously. Additionally proofs (what he calls arguments) under previous versions of the CRS can be updated to newer versions of the CRS. While latter extends potential applications, in the context of composable zk-SNARK frameworks, this feature is not required.

Now, apart from the missing knowledge-soundness property in [40], it could be tempting to think that two parallel symmetric QA-NIZK can be used to emulate what is provided by asymmetric QA-NIZK. However, the problem is that one would require an additional "linking proof" that would guarantee that both proofs use the same witness. And exactly this issue, which would increase the proof size and decrease efficiency, is what one can avoid when using asymmetric QA-NIZK in LegoSNARK, whenever the respective commitments are available in both source groups.

Consequently, when having updatable asymmetric QA-NIZKs, which avoid the aforementioned issue, this is another step towards an updatable variant of the LegoSNARK toolbox.

1.2 Our Results

We investigate the most efficient asymmetric QA-NIZK (denoted as Π'_{asy}) by González *et al.* (GHR) [25] in an updatable CRS setting. We show that for Π'_{asy} we can construct updatable asymmetric QA-NIZK arguments (which requires a witness samplable distribution [32]) by extending the CRS suitably and adding two new algorithms for updating the CRS and verify CRS updates. Compared to the recent updatable symmetric QA-NIZK in [40], we consider a variant where the CRS is subverted and can be updated, but the language parameter is chosen honestly. As already mentioned above that latter does not represent a problem for practical applications and in particular composable zk-SNARK frameworks such as LegoSNARK [13].

In contrast to the updatable symmetric QA-NIZK in [40], which relies on a new non-standard knowledge assumption for their subversion zero-knowledge property (KW-KE), our construction of updatable QA-NIZK can be shown to have this property under the Bilinear Diffie-Hellman Knowledge of Exponents (BDH-KE) assumption [2,4] and is asymmetric. Furthermore, under the discrete logarithm assumption in the Algebraic Group Model (AGM) due to Fuchsbauer *et al.* [21], we prove the knowledge soundness property of the proposed updatable asymmetric QA-NIZK. We also show that this also yields the knowledge soundness property of the original GHR asymmetric QA-NIZK.

Technical Overview. In Sect. 3, we give constructions of succinct updatable *asymmetric* QA-NIZK arguments of membership in bilateral linear subspaces. Using implicit notation, we represent the elements of $\mathbb{G}_1$ (respectively of $\mathbb{G}_2$) as $[z]_1 \in \mathbb{G}_1$ (respectively, as $[z]_2 \in \mathbb{G}_2$). Given the language parameters $[\boldsymbol{M}]_1 \in \mathbb{G}_1^{n_1 \times m}$ and $[\boldsymbol{N}]_2 \in \mathbb{G}_2^{n_2 \times m}$, we consider QA-NIZK arguments of membership of the statements $([\boldsymbol{y}]_1, [\boldsymbol{x}]_2)$ in the language

$$\mathcal{L}_{[\boldsymbol{M}]_1,[\boldsymbol{N}]_2} = \left\{([\boldsymbol{y}]_1, [\boldsymbol{x}]_2) \in \mathbb{G}_1^{n_1} \times \mathbb{G}_2^{n_2} : \exists \boldsymbol{w} \in \mathbb{Z}_p^m \text{ s.t. } \boldsymbol{y} = \boldsymbol{M}\boldsymbol{w}, \boldsymbol{x} = \boldsymbol{N}\boldsymbol{w}\right\}.$$

As mentioned, to construct our updatable *asymmetric* QA-NIZK arguments we start from the *asymmetric* QA-NIZK by González *et al.* (GHR) [25] (cf. Fig. 1) and change GHR's QA-NIZK by adding extra elements to the CRS so that the CRS becomes publicly verifiable and trapdoor extractable. Importantly, our aim for the updatable asymmetric QA-NIZK, is to keep the prover and the verifier unchanged compared to GHR's QA-NIZK.

More precisely, the CRS of GHR's QA-NIZK contains $\mathsf{crs} = ([\boldsymbol{A}, \boldsymbol{C}_2, \boldsymbol{P}_2]_2, [\boldsymbol{A}, \boldsymbol{C}_1, \boldsymbol{P}_1]_1)$ where $[\boldsymbol{A}]_i \in \mathbb{G}_i^{k \times k}$, $\boldsymbol{C}_i \in \mathbb{Z}_p^{n_i \times k}$, and $\boldsymbol{P}_i \in \mathbb{Z}_p^{m \times k}$ for $i \in \{1, 2\}$ and integers n_i, m and k. The prover uses $[\boldsymbol{P}_2]_2$ and $[\boldsymbol{P}_1]_1$ to generate a proof and the verifier uses the rest of the CRS to verify the proof. We add two new elements $[\boldsymbol{C}_1]_2$ and $[\boldsymbol{C}_2]_1$ to the CRS of the GHR scheme to make the CRS publicly verifiable. The trapdoor extractability is guaranteed using the new elements $[\boldsymbol{C}_1]_2$ and $[\boldsymbol{C}_2]_1$ and under the Bilinear Diffie-Hellman Knowledge of Exponents assumption (the extracted trapdoor will be used to prove subversion zero-knowledge). To achieve the updatability property, we design two new algorithms Ucrs and Vcrs. The Ucrs algorithm takes the crs and updates it to a new $\mathsf{crs}_{\mathsf{up}}$ so that the update is publicly verifiable. More precisely, given the $\mathsf{crs}_{\mathsf{up}}$, the language parameters $[\boldsymbol{M}]_1$, and $[\boldsymbol{N}]_2$, the Vcrs algorithm checks the well-formedness of $\mathsf{crs}_{\mathsf{up}}$. The latter checking guarantees the existence of a trapdoor $\mathtt{tc}$ for the $\mathsf{crs}_{\mathsf{up}}$, which will be required to prove the zero-knowledge property (cf. Sect. 3.2).

This step is necessary and will be sufficient for subversion zero-knowledge (as the prover can check the well-formedness of the CRS) and updatable soundness (as the verifier can check and update the CRS) in the updatable setting. However, choosing which elements to add to the CRS is not straightforward since the QA-NIZK must remain secure even given this extended CRS as adding too much information into the CRS can easily break the security, i.e., zero-knowledge and/or soundness. For instance, one may achieve the aforementioned properties by adding $[\boldsymbol{P}_1]_2$ and $[\boldsymbol{P}_2]_1$ to the CRS of GHR's QA-NIZK. But adding such elements bring a fundamental issue that under the Bilinear Diffie-Hellman Knowledge of Exponents assumption, the simulator in the zero-knowledge proof can also extract the language parameters $\boldsymbol{M}$ and $\boldsymbol{N}$. Given statements $([\boldsymbol{y}]_1, [\boldsymbol{x}]_2)$, the simulator obtains more information of the witness $\boldsymbol{w}$ of the language $\mathcal{L}_{[\boldsymbol{M}]_1,[\boldsymbol{N}]_2}$, which would violate the zero-knowledge property.

2 Preliminaries

Let PPT denote probabilistic polynomial-time. Let $n \in \mathbb{N}$ be the security parameter. By $x \leftarrow_{\$} \mathcal{D}$ we denote that x is sampled according to distribution $\mathcal{D}$ or uniformly randomly if $\mathcal{D}$ is a set. We denote by $\mathsf{negl}(\lambda)$ an arbitrary negligible function. We write $a \approx_n b$ if $|a - b| \leq \mathsf{negl}(\lambda)$. Algorithm $\mathsf{Pgen}(1^n)$ returns $\mathsf{BG} = (p, \mathbb{G}_1, \mathbb{G}_2, \mathbb{G}_T, \hat{e})$, where $\mathbb{G}_1$, $\mathbb{G}_2$, and $\mathbb{G}_T$ are three additive cyclic groups of prime order p, and $\hat{e} : \mathbb{G}_1 \times \mathbb{G}_2 \to \mathbb{G}_T$ is a non-degenerate efficiently computable bilinear map (pairing). We use the implicit bracket notation of [19], that is, we write $[a]_\iota$ to denote $a g_\iota$ where g_ι is a fixed generator of $\mathbb{G}_\iota$. We denote $\hat{e}([a]_1, [b]_2)$ as $[a]_1[b]_2$. Thus, $[a]_1[b]_2 = [ab]_T$. By $y \leftarrow \mathcal{A}(x; \omega)$ we denote the fact that $\mathcal{A}$, given an input x and random coins ω, outputs y. Let $\mathsf{RND}(\mathcal{A})$ denote the random tape of $\mathcal{A}$, and let $\omega \leftarrow_{\$} \mathsf{RND}(\mathcal{A})$ denote the random choice of the random coins ω from $\mathsf{RND}(\mathcal{A})$.

Computational Assumptions. We require the following assumptions.

Definition 1 (BDH-KE **Assumption** [2,4]). *We say that* BDH-KE *holds relative to* K_0, *if for any PPT adversary* $\mathcal{A}$ *there exists a PPT extractor* $\mathsf{Ext}_{\mathcal{A}}^{\mathsf{BDH\text{-}KE}}$, *such that*

$$\Pr\left[\begin{matrix}\mathsf{p} \leftarrow_{\$} \mathsf{K}_0(1^n); \omega_{\mathcal{A}} \leftarrow_{\$} \mathsf{RND}(\mathcal{A}), \\ ([\alpha_1]_1, [\alpha_2]_2 \| a) \leftarrow (\mathcal{A} \| \mathsf{Ext}_{\mathcal{A}}^{\mathsf{BDH\text{-}KE}})(\mathsf{p}, \omega_{\mathcal{A}})\end{matrix} : [\alpha_1]_1[1]_2 = [1]_1[\alpha_2]_2 \wedge a \neq \alpha_1\right] \approx_n 0 .$$

Here $\mathsf{aux}_{\mathcal{R}}$ is the auxiliary information related to the relation generator of $\mathcal{R}$. Note that the BDH-KE assumption can be considered as a simple case of the PKE assumption of [16]. Also, BDH-KE can be seen as an asymmetric-pairing version of the original KoE assumption [15].
In the following definitions let $\mathcal{D}_k$ be a matrix distribution in $\mathbb{Z}_p^{(k+1)\times k}$.

Definition 2 ($\mathcal{D}_k$**-Matrix Diffie-Hellman** ($\mathcal{D}_k$-MDDH) **Assumption** [44]). *The* $\mathcal{D}_k$-MDDH *assumption for* $\iota \in \{1, 2\}$ *holds relative to* K_0, *if for any PPT adversary* $\mathcal{A}$, $|\mathsf{Exp}_{\mathcal{A}}^{\mathsf{MDDH}}(\mathsf{p}) - 1/2| \approx_n 0$, *where* $\mathsf{Exp}_{\mathcal{A}}^{\mathsf{MDDH}}(\mathsf{p}) :=$

$$\Pr\left[\begin{matrix}\mathsf{p} \leftarrow_{\$} \mathsf{K}_0(1^n); \boldsymbol{A} \leftarrow_{\$} \mathcal{D}_k; \boldsymbol{v} \leftarrow_{\$} \mathbb{Z}_p^k; \\ \boldsymbol{u} \leftarrow_{\$} \mathbb{Z}_p^{k+1}; b \leftarrow_{\$} \{0,1\}; \\ b^* \leftarrow \mathcal{A}(\mathsf{p}, [\boldsymbol{A}]_\iota, [b \cdot \boldsymbol{A}\boldsymbol{v} + (1-b) \cdot \boldsymbol{u}]_\iota)\end{matrix} : b = b^*\right] .$$

Definition 3 (**$\mathcal{D}_k$-SKerMDH Assumption** [25])**.** *The $\mathcal{D}_k$-SKerMDH assumption holds relative to K_0, if for any PPT $\mathcal{A}$,*

$$\Pr\left[\begin{array}{l}\mathsf{p} \leftarrow \mathsf{K}_0(1^n); \boldsymbol{A} \leftarrow_{\$} \mathcal{D}_k; ([\boldsymbol{s}_1]_1, [\boldsymbol{s}_2]_2) \leftarrow \mathcal{A}(\mathsf{p}, [\boldsymbol{A}]_1, [\boldsymbol{A}]_2) : \\ \boldsymbol{s}_1 - \boldsymbol{s}_2 \neq \boldsymbol{0} \wedge \boldsymbol{A}^\top(\boldsymbol{s}_1 - \boldsymbol{s}_2) = \boldsymbol{0}_k\end{array}\right] \approx_n 0 .$$

Let $\mathcal{D}_{\ell k}$ be a probability distribution over matrices in $\mathbb{Z}_p^{\ell \times k}$, where $\ell > k$. Next, we define five commonly used distributions (see [19] for references), where $a, a_i, a_{ij} \leftarrow_{\$} \mathbb{Z}_p^*$: $\mathcal{U}_k$ (uniform), $\mathcal{L}_k$ (linear), $\mathcal{IL}_k$ (incremental linear), $\mathcal{C}_k$ (cascade), $\mathcal{SC}_k$ (symmetric cascade):

$$\mathcal{U}_k\colon \boldsymbol{A} = \begin{pmatrix} a_{11} & \dots & a_{1k} \\ \dots & \dots & \dots \\ a_{k1} & \dots & a_{kk} \\ a_{k+1,1} & \dots & a_{k+1,k} \end{pmatrix}, \quad \mathcal{L}_k\colon \boldsymbol{A} = \begin{pmatrix} a_1 & 0 & \dots & 0 & 0 \\ 0 & a_2 & \dots & 0 & 0 \\ 0 & 0 & \dots & 0 & 0 \\ \dots & \dots & \dots & \dots & \dots \\ 0 & 0 & \dots & 0 & a_k \\ 1 & 1 & \dots & 1 & 1 \end{pmatrix},$$

$$\mathcal{IL}_k\colon \boldsymbol{A} = \begin{pmatrix} a & 0 & \dots & 0 & 0 \\ 0 & a{+}1 & \dots & 0 & 0 \\ 0 & 0 & \dots & 0 & 0 \\ \dots & \dots & \dots & \dots & \dots \\ 0 & 0 & \dots & 0 & a{+}k{-}1 \\ 1 & 1 & \dots & 1 & 1 \end{pmatrix}, \quad \mathcal{C}_k\colon \boldsymbol{A} = \begin{pmatrix} a_1 & 0 & \dots & 0 & 0 \\ 1 & a_2 & \dots & 0 & 0 \\ 0 & 1 & \dots & 0 & 0 \\ \dots & \dots & \dots & \dots & \dots \\ 0 & 0 & \dots & 1 & a_k \\ 0 & 0 & \dots & 0 & 1 \end{pmatrix},$$

$$\mathcal{SC}_k\colon \boldsymbol{A} = \begin{pmatrix} a & 0 & \dots & 0 & 0 \\ 1 & a & \dots & 0 & 0 \\ 0 & 1 & \dots & 0 & 0 \\ \dots & \dots & \dots & \dots & \dots \\ 0 & 0 & \dots & 1 & a \\ 0 & 0 & \dots & 0 & 1 \end{pmatrix}.$$

Assume that $\mathcal{D}_{\ell k}$ outputs matrices $\boldsymbol{A}$ where the upper $k \times k$ submatrix $\bar{\boldsymbol{A}}$ is always invertible, i.e., $\mathcal{D}_{\ell k}$ is robust [32].

QA-NIZK Arguments. Let a language $\mathcal{L}_\varrho$ defined by a relation $\mathcal{R}_\varrho$ which is parametrized by some parameter ϱ, called the language parameter, chosen from a distribution $\mathcal{D}_\mathsf{p}$. We recall the definition of QA-NIZK arguments from Jutla and Roy [32]. A QA-NIZK argument provides a proof for membership of words x with according witnesses $\boldsymbol{w}$ in the language $\mathcal{L}_\varrho$. The distribution $\mathcal{D}_\mathsf{p}$ is witness samplable if there exist an efficient algorithm that samples $(\varrho, \mathsf{tc}_\varrho)$ so that the parameter ϱ is distributed according to $\mathcal{D}_\mathsf{p}$ and membership of the language parameter ϱ can be efficiently verified with tc_ϱ. The CRS of a QA-NIZK depends on a language parameter ϱ and as mentioned in [32], it has to be chosen from a correct distribution $\mathcal{D}_\mathsf{p}$.

Let ϱ be sampled from a distribution $\mathcal{D}_\mathsf{p}$ over associated parameter language $\mathcal{L}_\mathsf{p}$. A QA-NIZK argument in the CRS model contains four PPT algorithms $\Pi = (\mathsf{Pgen}, \mathsf{P}, \mathsf{V}, \mathsf{Sim})$ for a set of witness-relations $\mathcal{R}_\mathsf{p} = \{\mathcal{R}_\varrho\}_{\varrho \in \mathsf{Supp}(\mathcal{D})_\mathsf{p}}$, if the following properties (i-iii) hold. We call the QA-NIZK knowledge sound if instead of (iii) the property (iv) holds. Here, Pgen is the parameter and the CRS generation algorithm, more precisely, Pgen consists of two algorithms K_0 (generates the parameter p) and K (generates the CRS), P is the prover, V is the verifier, and Sim is the simulator.

(i) **Completeness.** For any n, and $(x, \boldsymbol{w}) \in \mathcal{R}_\varrho$,

$$\Pr\left[\begin{array}{l}\mathsf{p} \leftarrow \mathsf{K}_0(1^n); \varrho \leftarrow_{\$} \mathcal{D}_\mathsf{p}; (\mathsf{crs}, \mathsf{tc}) \leftarrow \mathsf{K}(\varrho); \pi \leftarrow \mathsf{P}(\varrho, \mathsf{crs}, x, \boldsymbol{w}) : \\ \mathsf{V}(\varrho, \mathsf{crs}, x, \pi) = 1\end{array}\right] = 1 .$$

(ii) **Statistical Zero-Knowledge.** For any computationally unbounded adversary $\mathcal{A}$, $|\varepsilon_0^{zk} - \varepsilon_1^{zk}| \approx_n 0$, where $\varepsilon_b^{zk} :=$

$$\Pr\left[\mathsf{p} \leftarrow \mathsf{K}_0(1^n); \varrho \leftarrow_\$ \mathcal{D}_\mathsf{p}; (\mathsf{crs}, \mathtt{tc}) \leftarrow \mathsf{K}(\varrho); b \leftarrow_\$ \{0,1\} : \mathcal{A}^{\mathsf{O}_b(\cdot)}(\varrho, \mathsf{crs}) = 1\right].$$

The oracle $\mathsf{O}_0(x, \boldsymbol{w})$ returns $\perp$ (reject) if $(x, \boldsymbol{w}) \notin \mathcal{R}_\varrho$, and otherwise it returns $\mathsf{P}(\varrho, \mathsf{crs}, x, \boldsymbol{w})$. Similarly, $\mathsf{O}_1(x, \boldsymbol{w})$ returns $\perp$ (reject) if $(x, \boldsymbol{w}) \notin \mathcal{R}_\varrho$, and otherwise it returns $\mathsf{Sim}(\varrho, \mathsf{crs}, \mathtt{tc}, x)$.

(iii) **Adaptive Soundness.** For any PPT $\mathcal{A}$,

$$\Pr\left[\begin{matrix}\mathsf{p} \leftarrow \mathsf{K}_0(1^n); \varrho \leftarrow_\$ \mathcal{D}_\mathsf{p}; (\mathsf{crs}, \mathtt{tc}) \leftarrow \mathsf{K}(\varrho); (x, \pi) \leftarrow \mathcal{A}(\varrho, \mathsf{crs}) : \\ \mathsf{V}(\varrho, \mathsf{crs}, x, \pi) = 1 \wedge \neg(\exists \boldsymbol{w} : (x, \boldsymbol{w}) \in \mathcal{R}_\varrho)\end{matrix}\right] \approx_n 0 .$$

$\mathsf{K}([\boldsymbol{M}]_1, [\boldsymbol{N}]_2)$

- $\boldsymbol{A} \leftarrow_\$ \hat{\mathcal{D}}_k; \boldsymbol{K}_1 \leftarrow_\$ \mathbb{Z}_p^{n_1 \times \hat{k}}; \boldsymbol{K}_2 \leftarrow_\$ \mathbb{Z}_p^{n_2 \times \hat{k}}; \boldsymbol{Z} \leftarrow_\$ \mathbb{Z}_p^{m \times \hat{k}}; \boldsymbol{C}_1 \leftarrow \boldsymbol{K}_1 \boldsymbol{A} \in \mathbb{Z}_p^{n_1 \times k}$;
- $\boldsymbol{C}_2 \leftarrow \boldsymbol{K}_2 \boldsymbol{A} \in \mathbb{Z}_p^{n_2 \times k}; [\boldsymbol{P}_1]_1 \leftarrow [\boldsymbol{M}]_1^\top \boldsymbol{K}_2 + [\boldsymbol{Z}]_1 \in \mathbb{Z}_p^{m \times \hat{k}}$;
- $[\boldsymbol{P}_2]_2 \leftarrow [\boldsymbol{N}]_2^\top \boldsymbol{K}_1 + [\boldsymbol{Z}]_2 \in \mathbb{Z}_p^{m \times \hat{k}}$;
- $\mathtt{tc} \leftarrow (\boldsymbol{K}_1, \boldsymbol{K}_2); \mathsf{crs} \leftarrow ([\boldsymbol{A}, \boldsymbol{C}_2, \boldsymbol{P}_2]_2, [\boldsymbol{A}, \boldsymbol{C}_1, \boldsymbol{P}_1]_1)$;
- **return** $(\mathtt{tc}, \mathsf{crs})$;

$\mathsf{P}([\boldsymbol{M}]_1, [\boldsymbol{N}]_2, \mathsf{crs}, [\boldsymbol{y}]_1, [\boldsymbol{x}]_2, \boldsymbol{w})$:

- $\boldsymbol{r} \leftarrow_\$ \mathbb{Z}_p^{\hat{k}}$;
- $[\boldsymbol{\pi}_1]_1 \leftarrow [\boldsymbol{P}_1]_1^\top \boldsymbol{w} + [\boldsymbol{r}]_1 \in \mathbb{G}_1^{\hat{k}}$;
- $[\boldsymbol{\pi}_2]_2 \leftarrow [\boldsymbol{P}_2]_2^\top \boldsymbol{w} + [\boldsymbol{r}]_2 \in \mathbb{G}_2^{\hat{k}}$;
- **return** $([\boldsymbol{\pi}_1]_1, [\boldsymbol{\pi}]_2)$;

$\mathsf{V}([\boldsymbol{M}]_1, [\boldsymbol{N}]_2, \mathsf{crs}, [\boldsymbol{y}]_1, [\boldsymbol{x}]_2, [\boldsymbol{\pi}_1]_1, [\boldsymbol{\pi}_2]_2)$:

- **if** $[\boldsymbol{y}]_1^\top [\boldsymbol{C}_2]_2 - [\boldsymbol{\pi}_1]_1^\top [\boldsymbol{A}]_2 = [\boldsymbol{\pi}_2]_2^\top [\boldsymbol{A}]_1 - [\boldsymbol{x}]_2^\top [\boldsymbol{C}_1]_1$ **return** 1;
- **else** **return** 0;

$\mathsf{Sim}([\boldsymbol{M}]_1, [\boldsymbol{N}]_2, \mathsf{crs}, \mathtt{tc}, [\boldsymbol{y}]_1)$:

- $\boldsymbol{r} \leftarrow_\$ \mathbb{Z}_p^{\hat{k}}$; - $[\boldsymbol{\pi}_1]_1 \leftarrow \boldsymbol{K}_2^\top [\boldsymbol{y}]_1 + [\boldsymbol{r}]_1 \in \mathbb{G}_1^{\hat{k}}$; - $[\boldsymbol{\pi}]_2 \leftarrow \boldsymbol{K}_1^\top [\boldsymbol{x}]_2 + [\boldsymbol{r}]_2 \in \mathbb{G}_1^{\hat{k}}$;
- **return** $([\boldsymbol{\pi}_1]_1, [\boldsymbol{\pi}]_2)$;

Fig. 1. Asymmetric QA-NIZK Π_{asy} ($\hat{\mathcal{D}}_k = \mathcal{D}_k$ and $\hat{k} = k+1$) and Π'_{asy} ($\hat{\mathcal{D}}_k = \bar{\mathcal{D}}_k$ and $\hat{k} = k$) from [25].

(vi) **Adaptive Knowledge Soundness.** For any PPT $\mathcal{A}$ there exists a non-uniform PPT extractor $\mathsf{Ext}_\mathcal{A}$,

$$\Pr\left[\begin{matrix}\mathsf{p} \leftarrow \mathsf{K}_0(1^n); \varrho \leftarrow_\$ \mathcal{D}_\mathsf{p}; (\mathsf{crs}, \mathtt{tc}) \leftarrow \mathsf{K}(\varrho); \omega_\mathcal{A} \leftarrow_\$ \mathsf{RND}(\mathcal{A}); \\ (x, \pi) \leftarrow \mathcal{A}(\omega_\mathcal{A}; \varrho, \mathsf{crs}); \boldsymbol{w} \leftarrow \mathsf{Ext}_\mathcal{A}(\omega_\mathcal{A}; \varrho, \mathsf{crs}) : (x, \boldsymbol{w}) \notin \mathcal{R}_\varrho \\ \wedge \mathsf{V}(\varrho, \mathsf{crs}, x, \pi) = 1\end{matrix}\right] \approx_n 0 .$$

Asymmetric QA-NIZK for Concatenation Languages. We recall the asymmetric QA-NIZK arguments of membership in bilateral linear subspaces of $\mathbb{G}_1^{n_1} \times \mathbb{G}_2^{n_2}$ given by González *et al.* [25] for the language

$$\mathcal{L}_{[\boldsymbol{M}]_1, [\boldsymbol{N}]_2} = \left\{([\boldsymbol{y}]_1, [\boldsymbol{x}]_2) \in \mathbb{G}_1^{n_1} \times \mathbb{G}_2^{n_2} : \exists \boldsymbol{w} \in \mathbb{Z}_p^m \text{ s.t. } \boldsymbol{y} = \boldsymbol{M}\boldsymbol{w}, \boldsymbol{x} = \boldsymbol{N}\boldsymbol{w}\right\} .$$

This language is also known as the concatenation language, since one can define $\boldsymbol{R}$ as a concatenation of language parameters $[\boldsymbol{M}]_1$ and $[\boldsymbol{N}]_2$ so that $\boldsymbol{R} = \begin{pmatrix} [\boldsymbol{M}]_1 \\ [\boldsymbol{N}]_2 \end{pmatrix}$. In other words $([\boldsymbol{y}]_1, [\boldsymbol{x}]_2) \in \mathcal{L}_{[\boldsymbol{M}]_1,[\boldsymbol{N}]_2}$ iff $\begin{pmatrix} [\boldsymbol{y}]_1 \\ [\boldsymbol{x}]_2 \end{pmatrix}$ is in the span of $\boldsymbol{R}$. We recall the full construction of asymmetric QA-NIZK arguments in the CRS model in Fig. 1.

Notice that the QA-NIZK in Fig. 1 for $\mathcal{L}_{[\boldsymbol{M}]_1,[\boldsymbol{N}]_2}$ is a generalization of Π_{as} of [35] in two groups when we set $\hat{\mathcal{D}}_k = \mathcal{D}_k$ and $\hat{k} = k + 1$ (denoted as Π_{asy}). Also it is a generalization of Π'_{as} of [35] in two groups when we set $\hat{\mathcal{D}}_k = \bar{\mathcal{D}}_k$ and $\hat{k} = k$ (denoted as Π'_{asy}).

Theorem 1 *[Theorem 3 of [25]]. If $\hat{\mathcal{D}}_k = \mathcal{D}_k$ and $\hat{k} = k+1$, the QA-NIZK proof system in Fig. 1 is perfect complete, computational adaptive soundness based on the $\mathcal{D}_k$-*SKerMDH *assumption, perfect zero-knowledge.*

Theorem 2 *[Theorem 4 of [25]]. If $\hat{\mathcal{D}}_k = \bar{\mathcal{D}}_k$, $\hat{k} = k$ and $\mathcal{D}_{\mathsf{p}}$ is a witness samplable distribution, Fig. 1 describes a QA-NIZK proof system with perfect completeness, computational adaptive soundness based on the $\mathcal{D}_k$-*KerMDH *assumption, perfect zero-knowledge.*

3 Updatable Asymmetric QA-NIZK

In this section, we investigate asymmetric QA-NIZK arguments when the CRS can be maliciously generated and propose corresponding updatable asymmetric QA-NIZK arguments. Formally, we prove the following theorem:

Theorem 3 *Let $\Pi_{\mathsf{asy\text{-}up}}$ be an updatable asymmetric QA-NIZK argument for linear subspaces from Fig. 4. (i) $\Pi_{\mathsf{asy\text{-}up}}$ is* crs*-update correct,* crs*-update hiding, and complete, (ii) if the* BDH-KE *assumption hold, then $\Pi_{\mathsf{asy\text{-}up}}$ is statistically subversion zero-knowledge, and (iii) if the $\mathcal{D}_k$-*SKerMDH*, (for the case $\hat{\mathcal{D}}_k = \bar{\mathcal{D}}_k$, the distribution $\mathcal{D}_{\mathsf{p}}$ should be witness samplable) then $\Pi_{\mathsf{asy\text{-}up}}$ is computationally updatable sound.*

First, we discuss subversion security of QA-NIZKs in the updatable CRS setting, then propose an updatable version of the most efficient asymetric QA-NIZK construction Π'_{as} in [25] (cf. Fig. 1).

3.1 Security Definitions for Updatable QA-NIZK Arguments

As already mentioned, the notion of *updatability* to achieve subversion security for NIZKs in the CRS model with respect to zero-knowledge and soundness was introduced by Groth *et al.* in [29] with a focus on zk-SNARKs. Later, Lipma [40] applied the underlying ideas to the setting of QA-NIZKs and in particular when both the language parameter ϱ and the CRS can be subverted. More precisely, Lipmaa obtains a version of the Kiltz-Wee QA-NIZK [35] (in the bare public-key (BPK) model) in the aforementioned setting under a new non-falsifiable KW-KE knowledge assumption. In this work, motivated by [6] and their application to composable zk-SNARK frameworks such as LegoSNARK [13], we investigate the security of QA-NIZKs in the CRS model when the CRS is subverted and can be updated but with *honestly chosen* ϱ[3]. Our security definition thus then enables us to construct an updatable asymmetric QA-NIZK that can be used to extend the LegoSANRK [13] with updatable CRS. More precisely such schemes can be used as the updatable zk-SNARKs for bilateral subspace languages as they provide better efficiency than general updatable zk-SNARKs for these types of languages.

Concretely, we define updatable QA-NIZKs security with some changes in the updatable CRS model. A tuple of PPT algorithms $\Pi = (\mathsf{Pgen}, \mathsf{Ucrs}, \mathsf{Vcrs}, \mathsf{P}, \mathsf{V}, \mathsf{Sim})$ is an updatable QA-NIZK if properties (i-v) hold. We call an updatable QA-NIZK updatable knowledge sound if instead of (v) property (vi) holds. Here, $\mathsf{Ucrs}(\varrho, \mathsf{crs})$ is an algorithm to update the CRS that takes the language parameter ϱ and a CRS crs and outputs an updated CRS $\mathsf{crs}_{\mathsf{up}}$ and corresponding trapdoor $\mathsf{tc}_{\mathsf{up}}$. $\mathsf{Vcrs}(\varrho, \mathsf{crs}, \mathsf{crs}_{\mathsf{up}})$ is an algorithm to verify the correctness of a CRS update and takes an old crs to a new CRS $\mathsf{crs}_{\mathsf{up}}$ and checks the well-formedness of the updated CRS.

(i) CRS-update Correctness. For any n,

$$\Pr\left[\begin{array}{l}\mathsf{p} \leftarrow \mathsf{K}_0(1^n); \varrho \leftarrow_{\$} \mathcal{D}_{\mathsf{p}}; (\mathsf{crs}, \mathsf{tc}) \leftarrow \mathsf{K}(\varrho);\\ (\mathsf{crs}_{\mathsf{up}}, \mathsf{tc}_{\mathsf{up}}) \leftarrow \mathsf{Ucrs}(\varrho, \mathsf{crs}) : \ \mathsf{Vcrs}(\varrho, \mathsf{crs}, \mathsf{crs}_{\mathsf{up}}) = 1\end{array}\right] = 1 \ .$$

(ii) CRS-update Hiding. For any n,

$$\Pr\left[\begin{array}{l}\mathsf{p} \leftarrow \mathsf{K}_0(1^n); \varrho \leftarrow_{\$} \mathcal{D}_{\mathsf{p}}; (\mathsf{crs}, \mathsf{tc}) \leftarrow \mathsf{K}(\varrho); (\mathsf{crs}_{\mathsf{up}}, \mathsf{tc}_{\mathsf{up}}) \leftarrow \mathsf{Ucrs}(\varrho, \mathsf{crs})\\ \mathsf{Vcrs}(\varrho, \mathsf{crs}, \mathsf{crs}_{\mathsf{up}}) = 1 : \mathsf{crs}_{\mathsf{up}} \approx_n \mathsf{crs}\end{array}\right] = 1 \ .$$

Note that this property holds the initial crs is maliciously generated and an honest updater Ucrs updates it.

(iii) Completeness. For any n, and $(x, \boldsymbol{w}) \in \mathcal{R}_\varrho$,

$$\Pr\left[\begin{array}{l}\mathsf{p} \leftarrow \mathsf{K}_0(1^n); \varrho \leftarrow_{\$} \mathcal{D}_{\mathsf{p}}; (\mathsf{crs}, \mathsf{tc}) \leftarrow \mathsf{K}(\varrho); (\mathsf{crs}_{\mathsf{up}}, \mathsf{tc}_{\mathsf{up}}) \leftarrow \mathsf{Ucrs}(\varrho, \mathsf{crs});\\ \pi \leftarrow \mathsf{P}(\varrho, \mathsf{crs}_{\mathsf{up}}, x, \boldsymbol{w}) : \mathsf{Vcrs}(\varrho, \mathsf{crs}, \mathsf{crs}_{\mathsf{up}}) = 1 \wedge \mathsf{V}(\varrho, \mathsf{crs}_{\mathsf{up}}, x, \pi) = 1\end{array}\right] = 1 \ .$$

(iv) Statistical Subversion Zero-Knowledge. For any PPT subverter Z there exists a PPT extractor $\mathsf{Ext}_{\mathsf{Z}}$, such that for any computationally unbounded adversary $\mathcal{A}$, $|\varepsilon_0^{zk} - \varepsilon_1^{zk}| \approx_n 0$, where $\varepsilon_b^{zk} :=$

$$\Pr\left[\begin{array}{l}\mathsf{p} \leftarrow \mathsf{K}_0(1^n); \varrho \leftarrow_{\$} \mathcal{D}_{\mathsf{p}}; \omega_{\mathsf{Z}} \leftarrow_{\$} \mathsf{RND}(\mathsf{Z}); (\mathsf{crs}, \mathsf{aux}_{\mathsf{Z}}) \leftarrow \mathsf{Z}(\varrho; \omega_{\mathsf{Z}});\\ \mathsf{tc} \leftarrow \mathsf{Ext}_{\mathsf{Z}}(\varrho; \omega_{\mathsf{Z}}); b \leftarrow_{\$} \{0,1\} : \mathsf{Vcrs}(\varrho, \mathsf{crs}) = 1 \wedge \mathcal{A}^{\mathsf{O}_b(\cdot,\cdot)}(\varrho, \mathsf{crs}, \mathsf{aux}_{\mathsf{Z}}) = 1\end{array}\right].$$

[3] We recall that in such applications ϱ represents public keys of the commitment scheme and can typically derived in a way (e.g., via a random oracle) such that subversion is not possible.

The oracle $\mathsf{O}_0(x, \boldsymbol{w})$ returns $\perp$ (reject) if $(x, \boldsymbol{w}) \notin \mathcal{R}_\varrho$, and otherwise it returns $\mathsf{P}(\varrho, \mathsf{crs}, x, \boldsymbol{w})$. Similarly, $\mathsf{O}_1(x, \boldsymbol{w})$ returns $\perp$ (reject) if $(x, \boldsymbol{w}) \notin \mathcal{R}_\varrho$, and otherwise it returns $\mathsf{Sim}(\varrho, \mathsf{crs}, \mathtt{tc}, x)$.

(v) Updatable Adaptive Soundness. For any PPT $\mathcal{A}$,

$$\Pr\left[\begin{array}{l}\mathsf{p} \leftarrow \mathsf{K}_0(1^n); \varrho \leftarrow_{\$} \mathcal{D}_{\mathsf{p}}; (\mathsf{crs}, \mathtt{tc}) \leftarrow \mathsf{K}(\varrho); (x, \pi, \mathsf{crs}') \leftarrow \mathcal{A}(\varrho, \mathsf{crs}) \\ : (x, \boldsymbol{w}) \notin \mathcal{R}_\varrho \wedge \mathsf{Vcrs}(\varrho, \mathsf{crs}, \mathsf{crs}') = 1 \wedge \mathsf{V}(\varrho, \mathsf{crs}', x, \pi) = 1\end{array}\right] \approx_n 0 .$$

(vi) Updatable Adaptive Knowledge Soundness. For any PPT $\mathcal{A}$ there exists a non-uniform PPT extractor $\mathsf{Ext}_\mathcal{A}$,

$$\Pr\left[\begin{array}{l}\mathsf{p} \leftarrow \mathsf{K}_0(1^n); \varrho \leftarrow_{\$} \mathcal{D}_{\mathsf{p}}; (\mathsf{crs}, \mathtt{tc}) \leftarrow \mathsf{K}(\varrho); \omega_\mathcal{A} \leftarrow_{\$} \mathsf{RND}(\mathcal{A}); \\ (x, \pi, \mathsf{crs}') \leftarrow \mathcal{A}(\omega_\mathcal{A}; \varrho, \mathsf{crs}); \boldsymbol{w} \leftarrow \mathsf{Ext}_\mathcal{A}(\omega_\mathcal{A}; \varrho, \mathsf{crs}, \mathsf{crs}') : (x, \boldsymbol{w}) \notin \mathcal{R}_\varrho \\ \wedge \mathsf{Vcrs}(\varrho, \mathsf{crs}, \mathsf{crs}') = 1 \wedge \mathsf{V}(\varrho, \mathsf{crs}', x, \pi) = 1\end{array}\right] \approx_n 0 .$$

3.2 Construction of Updatable Asymmetric QA-NIZKs

In this section we describe our updatable QA-NIZK for bilateral subspace languages. We first recall some notation and the primitives used in the construction.

Ingredients and Notation. Our updatable *asymmetric* QA-NIZK uses the following assumption and primitives:

- Asymmetric QA-NIZK in the CRS model, i.e., the asymmetric QA-NIZK Π'_{asy} of [25] (cf. Theorem 2).
- The knowledge assumption $\mathsf{BDH\text{-}KE}$ [2,4]. (cf. Definition 1).
- The algorithm $\mathsf{MATV}([\boldsymbol{A}]_2)$ of [3] that checks if a matrix $[\boldsymbol{A}]_2$ from $\mathcal{D}_k \in \{\mathcal{L}_k, \mathcal{IL}_k, \mathcal{C}_k, \mathcal{SC}_k\}$ is efficiently verifiable (cf. Figure 2).
- The algorithm $\mathsf{isinvertible}([\boldsymbol{A}]_2, [\boldsymbol{A}]_1)$ of [3] that checks the invertibility of a square matrix $\boldsymbol{A} \leftarrow_{\$} \mathcal{D}_k = \mathcal{U}_k$ for $k \in \{1, 2\}$ given in both source groups (cf. Figure 3).

Construction. We start with the asymmetric QA-NIZK argument Π'_{asy} from Fig. 1 and show how to obtain an updatable asymmetric QA-NIZK $\Pi'_{\mathsf{asy\text{-}up}}$. To this goal, similar as in previous work on updatable NIZK variants [29,40], we design two new algorithms Ucrs and Vcrs. The Ucrs algorithm takes the original crs and updates it to a new $\mathsf{crs}_{\mathsf{up}}$ such that this update is publicly verifiable. Given the $\mathsf{crs}_{\mathsf{up}}$, the language parameters $[\boldsymbol{M}]_1$, and $[\boldsymbol{N}]_2$, the Vcrs algorithm checks the well-formedness of the $\mathsf{crs}_{\mathsf{up}}$. The latter checking guarantees the existence of a trapdoor $\mathtt{tc}$ for the $\mathsf{crs}_{\mathsf{up}}$, which will be required to prove the zero-knowledge property. Now, we take a closer look at the design of the update procedure.

Updating procedure. The updating phase is tricky as the updated elements need to be publicly verifiable via the Vcrs algorithm. Inspired by [40], we use both multiplicative and additive updating approaches. We let Ucrs adaptively update

$\mathsf{MATV}([\boldsymbol{A}]_2)$ // $\mathcal{D}_k \in \{\mathcal{L}_k, \mathcal{IL}_k, \mathcal{C}_k, \mathcal{SC}_k\}$

check $[a_{11}]_2 \neq [0]_2 \wedge \ldots \wedge [a_{kk}]_2 \neq [0]_2$;
if $\mathcal{D}_k = \mathcal{L}_k$ **then** check $i \neq j \Rightarrow [a_{i,j}]_2 = [0]_2$;
elseif $\mathcal{D}_k = \mathcal{IL}_k$ **then** check $i \neq j \Rightarrow [a_{ij}]_2 = [0]_2$;
$\forall i, [a_{i,i}]_2 = [a_{1,1}]_2 + [i-1]_2$;
elseif $\mathcal{D}_k = \mathcal{C}_k$ **then** check $i \notin \{j, j+1\} \Rightarrow [a_{ij}]_2 = [0]_2$;
$\forall i, [a_{i+1,i}]_2 = [1]_2$;
elseif $\mathcal{D}_k = \mathcal{SC}_k$ **then** check $i \notin \{j, j+1\} \Rightarrow [a_{ij}]_2 = [0]_2$;
$\forall i\,([a_{i+1,i}]_2 = [1]_2 \wedge [a_{ii}]_2 = [a_{11}]_2)$; **fi**
return 1 if all checks pass and 0 otherwise;

Fig. 2. Auxiliary procedure MATV from [3] for $\mathcal{D}_k \in \{\mathcal{L}_k, \mathcal{IL}_k, \mathcal{C}_k, \mathcal{SC}_k\}$.

$\mathsf{isinvertible}([\boldsymbol{A}]_2, [\boldsymbol{A}]_1)$ // $\mathcal{D}_k = \mathcal{U}_k$

if $k = 1$ **then** check $[a_{11}]_2 \neq [0]_2$
else check $[a_{11}, a_{12}]_1 \in \mathbb{G}_1^{1\times 2} \wedge [a_{11}]_1[1]_2 = [1]_1[a_{11}]_2 \wedge$
$[a_{12}]_1[1]_2 = [1]_1[a_{12}]_2 \wedge [a_{11}]_1[a_{22}]_2 - [a_{12}]_1[a_{21}]_2 \neq [0]_T$; **fi**

Fig. 3. Auxiliary procedure isinvertible for $\boldsymbol{A} \in \mathbb{Z}_p^{k\times k}$ and $k \in \{1, 2\}$.

the element $\boldsymbol{P}_i$ for $i \in \{1,2\}$, since due to the structure of crs of our updatable asymmetric QA-NIZK in Fig. 4, by using the crs, $\mathsf{crs}_{\mathsf{int}}$, and $\mathsf{crs}_{\mathsf{up}}$, the Vcrs algorithm can publicly verify them. But updating the element $\boldsymbol{A}$ is more tricky. In particular, if one updates it additively then in order to be able to verify the elements $\boldsymbol{C}_i$, which are needed to make trapdoor extraction possible, one would need to have $[\boldsymbol{K}_i]_i$ for $i \in \{1,2\}$. More precisely, with additively updating $\boldsymbol{A}$, we would have $\boldsymbol{A}_{\mathsf{up}} = \boldsymbol{A} + \boldsymbol{A}_{\mathsf{int}}$ and the updating procedure of elements $\boldsymbol{C}_i$ is as follows:

$$\begin{aligned}[\boldsymbol{C}_{i,\mathsf{up}}]_i = [\boldsymbol{K}_{i,\mathsf{up}}\boldsymbol{A}_{\mathsf{up}}]_i &= [(\boldsymbol{K}_{\mathsf{int}} + \boldsymbol{K}_i)(\boldsymbol{A}_{\mathsf{int}} + \boldsymbol{A})]_i \\ &= [\boldsymbol{C}_{\mathsf{int}}]_i + [\boldsymbol{C}_i]_i + [\boldsymbol{K}_{\mathsf{int}}\boldsymbol{A}]_i + [\boldsymbol{K}_i\boldsymbol{A}_{\mathsf{int}}]_i,\end{aligned}$$

where for verifying $[\boldsymbol{K}_{\mathsf{int}}\boldsymbol{A}]_i$ and $[\boldsymbol{K}_i\boldsymbol{A}_{\mathsf{int}}]_i$ one needs to have $[\boldsymbol{K}_i]_i$. However, having these elements in the crs would leak information about the trapdoor. Thus, we need to update $\boldsymbol{A}$ multiplicatively as $\boldsymbol{A}_{\mathsf{up}} = \boldsymbol{A}\boldsymbol{A}_{\mathsf{int}}$.

Zero-knowledge property. In the zero-knowledge proof, we use the well-known BDH-KE knowledge assumption and show that if the possibly maliciously generated $\mathsf{crs}_{\mathsf{up}}$ passes the Vcrs algorithm, then under the knowledge assumptions there exists an extractor that extracts the trapdoor $\mathtt{tc}$ of $\mathsf{crs}_{\mathsf{up}}$. Using such a trapdoor $\mathtt{tc}$, the simulator can simulate proofs.

Soundness property. Since to achieve publicly verifiability of the $\mathsf{crs}_{\mathsf{up}}$, we add some new elements $[\boldsymbol{C}_2]_1$ and $[\boldsymbol{C}_1]_2$ in the CRS, we need to show that the soundness of the updatable asymmetric QA-NIZKs still holds. We prove the soundness under the standard SKerMDH assumption.

We depict the full construction of the updatable asymmetric QA-NIZK arguments in Fig. 4. Here, the elements with index int are intermediate elements generated by the algorithm Ucrs and can be viewed as update proofs, i.e., enabling to verify consistency of the old and updated crs. The elements with index up are the updated elements, i.e., the new crs. We note that our updatable asymmetric QA-NIZK in Fig. 4, the prover and the verifier are unchanged compared to GHR's QA-NIZK [25].

Remark 1. We note that, one can adapt the updatable asymmetric QA-NIZKs construction in Fig. 4 to other languages like as the *sum in subspace language* and obtain the updatable version of the *argument of sum in subspace* of [25].

3.3 Security Proof for Our Construction

In this section we prove Theorem 3.

Proof. The security properties (i-iii), crs-update correctness and completeness are straightforward from the construction. The crs-update hiding proof is similar [40] (see Appendix A for more details).

(iv: Subversion Zero-Knowledge:) For proving the zero-knowledge property, we need to construct a simulator that can construct proofs without knowing the witness but a trapdoor tc. To this aim, in Lemma 1 (in the extraction phase), we show that from any adversary producing a valid crs from scratch it is possible to extract the trapdoors $(\boldsymbol{K}_1, \boldsymbol{K}_2)$. Then in the simulation phase, given the trapdoor tc, we show how the zero-knowledge simulator can simulate proofs.

Extraction phase. Let the BDH-KE assumption hold. Let $\mathcal{A}$ be an adversary that computes crs so as to break the subversion zero-knowledge property of the updatable asymmetric QA-NIZK in Fig. 4. That is, $\mathcal{A}([\boldsymbol{M}]_1, [\boldsymbol{N}]_2; \omega_{\mathcal{A}})$ outputs $(\mathsf{crs}, \mathsf{aux}_{\mathcal{A}})$. In Lemma 1, based on the BDH-KE assumption, we show how one can construct an extractor to extract the trapdoor tc of a possibly maliciously generated crs.

Lemma 1. *Let the* BDH-KE *assumption hold and let* $[\boldsymbol{M}]_1, [\boldsymbol{N}]_2 \leftarrow_s \mathcal{D}_\mathsf{p}$. *Then for any PPT adversary* $\mathcal{A}$ *there exists extractor* $\mathsf{Ext}_{\mathcal{A}}$ *such that the probability that* $\mathcal{A}$ *on input* $([\boldsymbol{M}]_1, [\boldsymbol{N}]_2)$ *and randomness* ω *outputs* crs *such that* $\mathsf{Vcrs}([\boldsymbol{M}]_1, [\boldsymbol{N}]_2, \mathsf{crs}) = 1$ *and that* $\mathsf{Ext}_{\mathcal{A}}$ *on the same input, outputs* $\mathtt{tc} = (\boldsymbol{K}_1, \boldsymbol{K}_1)$, *is overwhelming.*

Proof. Let adversary $\mathcal{A}$ output crs such that $\mathsf{Vcrs}([\boldsymbol{M}]_1, [\boldsymbol{N}]_2, \mathsf{crs}) = 1$, which guarantees that elements from $\boldsymbol{P}_i$, $\boldsymbol{A}$ and $\boldsymbol{C}_i$ for $i \in \{1, 2\}$ are consistent and in particular that $[\boldsymbol{P}_1]_1[\boldsymbol{A}]_2 - [\boldsymbol{A}]_1[\boldsymbol{P}_2]_2 = [\boldsymbol{M}]_1[\boldsymbol{C}_2]_2 - [\boldsymbol{N}]_2[\boldsymbol{C}_1]_1$and $\boldsymbol{A}$ is invertible. Assume an internal subverter $\mathcal{A}_{\mathsf{BDH\text{-}KE}}$ against the BDH-KE assumption. We

$\mathsf{K}([\boldsymbol{M}]_1, [\boldsymbol{N}]_2)$

- $\boldsymbol{A} \leftarrow_{\$} \hat{\mathcal{D}}_k; \boldsymbol{K}_1 \leftarrow_{\$} \mathbb{Z}_p^{n_2 \times k}; \boldsymbol{K}_2 \leftarrow_{\$} \mathbb{Z}_p^{n_1 \times k}; \boldsymbol{Z} \leftarrow_{\$} \mathbb{Z}_p^{m \times k}; \boldsymbol{C}_1 \leftarrow \boldsymbol{K}_1 \boldsymbol{A} \in \mathbb{Z}_p^{n_2 \times k};$
- $\boldsymbol{C}_2 \leftarrow \boldsymbol{K}_2 \boldsymbol{A} \in \mathbb{Z}_p^{n_1 \times k}; [\boldsymbol{P}_1]_1 \leftarrow [\boldsymbol{M}]_1^\top \boldsymbol{K}_2 + [\boldsymbol{Z}]_1 \in \mathbb{Z}_p^{m \times k};$
- $[\boldsymbol{P}_2]_2 \leftarrow [\boldsymbol{N}]_2^\top \boldsymbol{K}_1 + [\boldsymbol{Z}]_2 \in \mathbb{Z}_p^{m \times k}; \mathsf{crs} \leftarrow ([\boldsymbol{A}, \boldsymbol{C}_1, \boldsymbol{C}_2, \boldsymbol{P}_2]_2, [\boldsymbol{A}, \boldsymbol{C}_1, \boldsymbol{C}_2, \boldsymbol{P}_1]_1);$
- $\mathtt{tc} \leftarrow (\boldsymbol{K}_1, \boldsymbol{K}_2);$
- **return** $(\mathtt{tc}, \mathsf{crs});$

$\mathsf{Ucrs}([\boldsymbol{M}]_1, [\boldsymbol{N}]_2, \mathsf{crs})$:

- $\boldsymbol{A}_{\mathsf{int}} \leftarrow_{\$} \bar{\mathcal{D}}_k; \boldsymbol{K}_{1,\mathsf{int}} \leftarrow_{\$} \mathbb{Z}_p^{n_2 \times k}; \boldsymbol{K}_{2,\mathsf{int}} \leftarrow_{\$} \mathbb{Z}_p^{n_1 \times k}; \boldsymbol{Z}_{\mathsf{int}} \leftarrow_{\$} \mathbb{Z}_p^{m \times k}; \boldsymbol{A}_{\mathsf{up}} \leftarrow \boldsymbol{A}\boldsymbol{A}_{\mathsf{int}};$
- $\boldsymbol{C}_{1,\mathsf{int}} \leftarrow \boldsymbol{K}_{1,\mathsf{int}} \boldsymbol{A}_{\mathsf{up}} \in \mathbb{Z}_p^{n_1 \times k}; \boldsymbol{C}_{2,\mathsf{int}} \leftarrow \boldsymbol{K}_{2,\mathsf{int}} \boldsymbol{A}_{\mathsf{up}} \in \mathbb{Z}_p^{n_2 \times k};$
- $[\boldsymbol{P}_{1,\mathsf{int}}]_1 \leftarrow [\boldsymbol{M}]_1^\top \boldsymbol{K}_{2,\mathsf{int}} + [\boldsymbol{Z}_{\mathsf{int}}]_1 \in \mathbb{Z}_p^{m \times k}; [\boldsymbol{P}_{2,\mathsf{int}}]_2 \leftarrow [\boldsymbol{N}]_2^\top \boldsymbol{K}_{1,\mathsf{int}} + [\boldsymbol{Z}_{\mathsf{int}}]_2 \in \mathbb{Z}_p^{m \times k};$
- $\boldsymbol{C}_{1,\mathsf{up}} \leftarrow \boldsymbol{C}_{1,\mathsf{int}} + \boldsymbol{C}_1 \boldsymbol{A}_{\mathsf{int}}; \boldsymbol{C}_{2,\mathsf{up}} \leftarrow \boldsymbol{C}_{2,\mathsf{int}} + \boldsymbol{C}_2 \boldsymbol{A}_{\mathsf{int}};$
- $[\boldsymbol{P}_{1,\mathsf{up}}]_1 \leftarrow [\boldsymbol{P}_1]_1 + [\boldsymbol{P}_{1,\mathsf{int}}]_1; [\boldsymbol{P}_{2,\mathsf{up}}]_2 \leftarrow [\boldsymbol{P}_2]_2 + [\boldsymbol{P}_{2,\mathsf{int}}]_1;$
- $\mathsf{crs}_{\mathsf{int}} \leftarrow ([\boldsymbol{A}_{\mathsf{int}}, \boldsymbol{C}_{1,\mathsf{int}}, \boldsymbol{C}_{2,\mathsf{int}}.\boldsymbol{P}_{2,\mathsf{int}}]_2, [\boldsymbol{A}_{\mathsf{int}}, \boldsymbol{C}_{1,\mathsf{int}}, \boldsymbol{C}_{2,\mathsf{int}}, \boldsymbol{P}_{1,\mathsf{int}}]_1);$
- **return** $\mathsf{crs}_{\mathsf{up}} \leftarrow ([\boldsymbol{A}_{\mathsf{up}}, \boldsymbol{C}_{1,\mathsf{up}}, \boldsymbol{C}_{2,\mathsf{up}}, \boldsymbol{P}_{2,\mathsf{up}}]_2, [\boldsymbol{A}_{\mathsf{up}}, \boldsymbol{C}_{1,\mathsf{up}}, \boldsymbol{C}_{2,\mathsf{up}}, \boldsymbol{P}_{1,\mathsf{up}}]_1, \mathsf{crs}_{\mathsf{int}});$

$\mathsf{Vcrs}([\boldsymbol{M}]_1, [\boldsymbol{N}]_2, \mathsf{crs}, \mathsf{crs}_{\mathsf{up}})$:

- **if**
  - for $i \in \{1, (1, \mathsf{int}), (1, \mathsf{up})\}$: : **if** $[\boldsymbol{C}_i]_1 \in \mathbb{G}_1^{n_1 \times k} \wedge [\boldsymbol{P}_i]_1 \in \mathbb{G}_1^{m \times k} \wedge [\boldsymbol{A}_i]_1 \in \mathbb{G}_1^{k \times k};$
  - for $i \in \{2, (2, \mathsf{int}), (2, \mathsf{up})\}$: **if** $[\boldsymbol{C}_i]_2 \in \mathbb{G}_2^{n_2 \times k} \wedge [\boldsymbol{P}_i]_2 \in \mathbb{G}_2^{m \times k} \wedge [\boldsymbol{A}_i]_2 \in \mathbb{G}_2^{k \times k};$
  - for $i \in \{1, 2, (1, \mathsf{int}), (1, \mathsf{up}), (2, \mathsf{int}), (2, \mathsf{up})\}$: **if** $[\boldsymbol{A}_i]_1[1]_2 = [1]_1[\boldsymbol{A}_i]_2; \wedge [\boldsymbol{C}_i]_1[1]_2 = [1]_1[\boldsymbol{C}_i]_2;$
  - $[\boldsymbol{A}_{\mathsf{up}}]_1[1]_2 = [\boldsymbol{A}]_1[\boldsymbol{A}_{\mathsf{int}}]_2$
  - $[\boldsymbol{C}_{1,\mathsf{up}}]_1[1]_2 = [\boldsymbol{C}_{1,\mathsf{int}}]_1[1]_2 + [\boldsymbol{C}_1]_1[\boldsymbol{A}_{\mathsf{int}}]_2 \wedge [\boldsymbol{C}_{2,\mathsf{up}}]_1[1]_2 = [\boldsymbol{C}_{2,\mathsf{int}}]_1[1]_2 + [\boldsymbol{C}_2]_1[\boldsymbol{A}_{\mathsf{int}}]_2$
  - $[\boldsymbol{P}_{1,\mathsf{up}}]_1 = [\boldsymbol{P}_{1,\mathsf{int}}]_1 + [\boldsymbol{P}_1]_1 \wedge [\boldsymbol{P}_{2,\mathsf{up}}]_1 = [\boldsymbol{P}_{2,\mathsf{int}}]_1 + [\boldsymbol{P}_2]_1;$
  - $[\boldsymbol{P}_1]_1[\boldsymbol{A}]_2 - [\boldsymbol{A}]_1[\boldsymbol{P}_2]_2 = [\boldsymbol{M}]_1[\boldsymbol{C}_2]_2 - [\boldsymbol{N}]_2[\boldsymbol{C}_1]_1;$
  - $[\boldsymbol{P}_{1,\mathsf{int}}]_1[\boldsymbol{A}_{\mathsf{int}}]_2 - [\boldsymbol{A}_{\mathsf{int}}]_1[\boldsymbol{P}_{2,\mathsf{int}}]_2 = [\boldsymbol{M}]_1[\boldsymbol{C}_{2,\mathsf{int}}]_2 - [\boldsymbol{N}]_2[\boldsymbol{C}_{1,\mathsf{int}}]_1;$
  - $[\boldsymbol{P}_{1,\mathsf{up}}]_1[\boldsymbol{A}_{\mathsf{up}}]_2 - [\boldsymbol{A}_{\mathsf{up}}]_1[\boldsymbol{P}_{2,\mathsf{up}}]_2 = [\boldsymbol{M}]_1[\boldsymbol{C}_{2,\mathsf{up}}]_2 - [\boldsymbol{N}]_2[\boldsymbol{C}_{1,\mathsf{up}}]_1;$
  - for $i \in \mathsf{int}, \mathsf{up}$: **if** $\mathcal{D}_k$ is efficiently verifiable then $\mathsf{MATV}([\boldsymbol{A}_i]_2) = 1 \wedge \mathsf{MATV}([\boldsymbol{A}]_2) = 1$ else check $\mathsf{isinvertible}([\boldsymbol{A}_i]_2, [\boldsymbol{A}_i]_1) = 1;$
- **return** 1;
- **else return** 0;

$\mathsf{P}([\boldsymbol{M}]_1, [\boldsymbol{N}]_2, \mathsf{crs}_{\mathsf{up}}, [\boldsymbol{y}]_1, [\boldsymbol{x}]_2, \boldsymbol{w})$:

- $\boldsymbol{r} \leftarrow_{\$} \mathbb{Z}_p^k; [\boldsymbol{\pi}_1]_1 \leftarrow [\boldsymbol{P}_1]_1^\top \boldsymbol{w} + [\boldsymbol{r}]_1 \in \mathbb{G}_1^k \wedge [\boldsymbol{\pi}_2]_2 \leftarrow [\boldsymbol{P}_2]_2^\top \boldsymbol{w} + [\boldsymbol{r}]_2 \in \mathbb{G}_2^k;$
- **return** $([\boldsymbol{\pi}_1]_1, [\boldsymbol{\pi}]_2);$

$\mathsf{V}([\boldsymbol{M}]_1, [\boldsymbol{N}]_2, \mathsf{crs}_{\mathsf{up}}, [\boldsymbol{y}]_1, [\boldsymbol{x}]_2, [\boldsymbol{\pi}_1]_1, [\boldsymbol{\pi}_2]_2)$:

- **if** $[\boldsymbol{y}]_1^\top [\boldsymbol{C}_2]_2 - [\boldsymbol{\pi}_1]_1^\top [\boldsymbol{A}]_2 = [\boldsymbol{\pi}_2]_2^\top [\boldsymbol{A}]_1 - [\boldsymbol{x}]_2^\top [\boldsymbol{C}_1]_1$ **return** 1;
- **else return** 0;

$\mathsf{Sim}([\boldsymbol{M}]_1, [\boldsymbol{N}]_2, \mathsf{crs}, \mathtt{tc}, [\boldsymbol{y}]_1)$:

- $\boldsymbol{r} \leftarrow_{\$} \mathbb{Z}_p^k$; - $[\boldsymbol{\pi}_1]_1 \leftarrow \boldsymbol{K}_2^\top [\boldsymbol{y}]_1 + [\boldsymbol{r}]_1 \in \mathbb{G}_1^k$; - $[\boldsymbol{\pi}]_2 \leftarrow \boldsymbol{K}_1^\top [\boldsymbol{x}]_2 + [\boldsymbol{r}]_2 \in \mathbb{G}_1^k$;
- **return** $([\boldsymbol{\pi}_1]_1, [\boldsymbol{\pi}]_2);$

Fig. 4. Updatable Asymmetric QA-NIZK $\Pi'_{\mathsf{asy\text{-}up}}$. Here k is an arbitrary value if $\mathcal{D}_k \in \{\mathcal{L}_k, \mathcal{IL}_k, \mathcal{C}_k, \mathcal{SC}_k\}$ and $k \in \{1, 2\}$ in $\mathcal{D}_k = \mathcal{U}_k$.

note that both the subverter and the adversary are in connection and separating them is just for readability of the proof. Let $\omega_{\mathcal{A}} = \omega_{\mathcal{A}_{\mathsf{BDH\text{-}KE}}}$. Let $\mathcal{A}_{\mathsf{BDH\text{-}KE}}$ run $\mathcal{A}$ and output $([\boldsymbol{A}]_1, [\boldsymbol{A}]_2, [\boldsymbol{C}_1, \boldsymbol{C}_2]_1, [\boldsymbol{C}_1, \boldsymbol{C}_2]_2)$. Then under the BDH-KE assumption, there exists an extractor $\mathsf{Ext}^{\mathsf{BDH\text{-}KE}}_{\mathcal{A}_{\mathsf{BDH\text{-}KE}}}$, such that if $\mathsf{Vcrs}([\boldsymbol{M}]_1, [\boldsymbol{N}]_2, \mathsf{crs}) = 1$ then $\mathsf{Ext}^{\mathsf{BDH\text{-}KE}}_{\mathcal{A}_{\mathsf{BDH\text{-}KE}}}([\boldsymbol{M}]_1, [\boldsymbol{N}]_2; \omega_{\mathcal{A}})$ outputs $(\boldsymbol{A}, \boldsymbol{C}_1, \boldsymbol{C}_2)$.

Let $\mathsf{Ext}_{\mathcal{A}}$ be an extractor that with input $([\boldsymbol{M}]_1, [\boldsymbol{N}]_2; \omega_{\mathcal{A}})$ and running $\mathsf{Ext}^{\mathsf{BDH\text{-}KE}}_{\mathcal{A}_{\mathsf{BDH\text{-}KE}}}$ as subroutine, extracts $\mathtt{tc} = (\boldsymbol{K}_1, \boldsymbol{K}_2)$. For the sake of simplicity, the full description of the algorithms is depicted in Fig. 5. More precisely, the extractor $\mathsf{Ext}_{\mathcal{A}}$ first runs $\mathsf{Ext}^{\mathsf{BDH\text{-}KE}}_{\mathcal{A}_{\mathsf{BDH\text{-}KE}}}([\boldsymbol{M}]_1, [\boldsymbol{N}]_2; \omega_{\mathcal{A}})$ which outputs $(\boldsymbol{A}, \boldsymbol{C}_1, \boldsymbol{C}_2)$. Then, $\mathsf{Ext}_{\mathcal{A}}$ computes $(\boldsymbol{K}_1, \boldsymbol{K}_2)$. Indeed, by having $\boldsymbol{A}$, $\boldsymbol{C}_i$, and the fact that $\boldsymbol{A}$ is invertible, the extractor $\mathsf{Ext}_{\mathcal{A}}$ can compute $\boldsymbol{K}_i = \boldsymbol{C}_i \boldsymbol{A}^{-1}$ for $i \in \{1, 2\}$. $\square$

$\mathcal{A}([\boldsymbol{M}]_1, [\boldsymbol{N}]_2; \omega_{\mathcal{A}})$

$(\mathsf{crs}, \mathsf{aux}_{\mathcal{A}}) \leftarrow \mathcal{A}([\boldsymbol{M}]_1, [\boldsymbol{N}]_2; \omega_{\mathcal{A}})$; **return** crs;

$\mathsf{Ext}_{\mathcal{A}}([\boldsymbol{M}]_1; \omega_{\mathcal{A}})$

- $(\boldsymbol{A}, \boldsymbol{C}_1, \boldsymbol{C}_2) \leftarrow \mathsf{Ext}^{\mathsf{BDH\text{-}KE}}_{\mathcal{A}_{\mathsf{BDH\text{-}KE}}}([\boldsymbol{M}]_1, [\boldsymbol{N}]_2; \omega_{\mathcal{A}})$;
Compute $\boldsymbol{K}_1 = \boldsymbol{C}_1\boldsymbol{A}^{-1}$ and $\boldsymbol{K}_2 = \boldsymbol{C}_2\boldsymbol{A}^{-1}$
- **return** $\mathtt{tc} = (\boldsymbol{K}_1, \boldsymbol{K}_2)$;

$\mathcal{A}_{\mathsf{BDH\text{-}KE}}([\boldsymbol{M}]_1, [\boldsymbol{N}]_2; \omega_{\mathcal{A}})$

$(\mathsf{crs}, \mathsf{aux}_{\mathcal{A}}) \leftarrow \mathcal{A}([\boldsymbol{M}]_1, [\boldsymbol{N}]_2; \omega_{\mathcal{A}})$;
return $([\boldsymbol{A}]_1, [\boldsymbol{A}]_2, [\boldsymbol{C}]_1, [\boldsymbol{C}]_2)$;

Fig. 5. The extractors and the constructed adversary $\mathcal{A}$ for Lemma 1.

Simulation phase. In the second step, given the trapdoor $\mathtt{tc}$, we show how a simulator Sim can simulate proofs. Fix concrete values of n, $\mathsf{p} \in \mathrm{im}(\mathsf{Pgen}(1^n))$, $([\boldsymbol{y}]_1, [\boldsymbol{x}]_2, \boldsymbol{w}) \in \mathcal{R}_{[\boldsymbol{M}]_1,[\boldsymbol{N}]_2}$, $\omega_{\mathcal{A}} \in \mathsf{RND}(\mathcal{A})$, and run $\mathsf{Ext}_{\mathcal{A}}([\boldsymbol{M}]_1, [\boldsymbol{N}]_2; \omega_{\mathcal{A}})$ to obtain $(\boldsymbol{K}_1, \boldsymbol{K}_2)$. Thus, it suffices to show that if $\mathsf{Vcrs}([\boldsymbol{M}]_1, [\boldsymbol{N}]_2, \mathsf{crs}) = 1$ and $([\boldsymbol{y}]_1, [\boldsymbol{x}]_2, \boldsymbol{w}) \in \mathcal{R}_{[\boldsymbol{M}]_1,[\boldsymbol{N}]_2}$ then

$$\begin{aligned}\mathsf{O}_0([\boldsymbol{y}]_1, [\boldsymbol{x}]_2, \boldsymbol{w}) &= \mathsf{P}([\boldsymbol{M}]_1, [\boldsymbol{N}]_2, \mathsf{crs}, [\boldsymbol{y}]_1, [\boldsymbol{x}]_2, \boldsymbol{w}) \ ,\\ \mathsf{O}_1([\boldsymbol{y}]_1, [\boldsymbol{x}]_2, \boldsymbol{w}) &= \mathsf{Sim}([\boldsymbol{M}]_1, [\boldsymbol{N}]_2, \mathsf{crs}, [\boldsymbol{y}]_1, [\boldsymbol{x}]_2, \boldsymbol{K}_1, \boldsymbol{K}_2)\end{aligned}$$

have the same distribution. Since O_0 and O_1 have the same distribution, $\Pi'_{\mathsf{asy\text{-}up}}$ is zero-knowledge under the BDH-KE assumption.

(v: Updatable Adaptive Soundness:) The proof is similar to the adaptive soundness proof of Π'_{as} in [25] but with some modifications. Let $m' := n_1 + n_2$ and $\boldsymbol{W} := \left(\begin{smallmatrix}\boldsymbol{M}\\ \boldsymbol{N}\end{smallmatrix}\right)$. Let an adversary $\mathcal{B}$ against $\mathcal{D}_k$-SKerMDH assumption be given a challenge $([\boldsymbol{A}]_1, [\boldsymbol{A}]_2)$, $\boldsymbol{A} \leftarrow \mathcal{D}_k$.

$\mathcal{B}$ samples $([\boldsymbol{M}]_1, [\boldsymbol{N}]_2, \boldsymbol{M}, \boldsymbol{N}) \in \mathcal{R}_{\mathsf{p}}$ and computes $\boldsymbol{W}^{\perp} \in \mathbb{Z}_p^{m' \times (m'-r)}$, where $r = \mathsf{rank}(\boldsymbol{W})$, a basis of the kernel of $\boldsymbol{W}^{\top}$. By definition, $\boldsymbol{W}^{\top} = (\boldsymbol{M}^{\top} || \boldsymbol{N}^{\top})$ and $\boldsymbol{W}^{\top}\boldsymbol{W}^{\perp} = 0$ and thus we can write $\boldsymbol{W}^{\perp} = (\boldsymbol{W}_1, \boldsymbol{W}_2)$, for some matrices such that $\boldsymbol{M}^{\top}\boldsymbol{W}_1 = -\boldsymbol{N}^{\top}\boldsymbol{W}_2$.

Adversary $\mathcal{B}$ samples $\boldsymbol{R} \in \mathbb{Z}_p^{(m'-r-1)\times(k+1)}$ and for $i \in \{1, 2\}$ defines,

$$[\boldsymbol{A}']_i \leftarrow \begin{pmatrix}[\boldsymbol{A}]_i\\ \boldsymbol{R}\cdot[\boldsymbol{A}]_i\end{pmatrix} \in \mathbb{Z}_p^{(k+m'-r)\times k}.$$

Then $\mathcal{B}$ samples $(\boldsymbol{K}'_1, \boldsymbol{K}'_2) \leftarrow \mathbb{Z}_p^{m' \times k}$. Let $\boldsymbol{A}_0$ be the first k rows of $\boldsymbol{A}'$ (or $\boldsymbol{A}$) and $\boldsymbol{A}'$ rows, and $\boldsymbol{T}_{\boldsymbol{A}'} = \boldsymbol{A}'_1\boldsymbol{A}_0^{-1}$. Then $\mathcal{B}$ implicitly sets $(\boldsymbol{K}_1, \boldsymbol{K}_2) := (\boldsymbol{K}'_1, \boldsymbol{K}'_2) + \boldsymbol{T}_{\boldsymbol{A}'}(\boldsymbol{W}_1 || \boldsymbol{W}_2)$, and for $i \in \{1, 2\}$ and computes:

$$[\boldsymbol{C}_1]_i := \boldsymbol{K}_2[\boldsymbol{A}_0]_i = (\boldsymbol{K}'_2 + \boldsymbol{T}_{\boldsymbol{A}'}\boldsymbol{W}_2)[\boldsymbol{A}_0]_i = (\boldsymbol{K}'_2 || \boldsymbol{W}_2)[\boldsymbol{A}']_i,$$

and

$$[\boldsymbol{C}_2]_i := \boldsymbol{K}_1[\boldsymbol{A}_0]_i = (\boldsymbol{K}_1' + \boldsymbol{T}_{\boldsymbol{A}'}\boldsymbol{W}_1)[\boldsymbol{A}_0]_i = (\boldsymbol{K}_1'||\boldsymbol{W}_1)[\boldsymbol{A}']_i.$$

Adversary $\mathcal{B}$ also needs to compute $[\boldsymbol{M}]_1^\top \boldsymbol{K}_2 + [\boldsymbol{Z}]_1$ and $[\boldsymbol{N}]_2^\top \boldsymbol{K}_1 - [\boldsymbol{Z}]_2$. The adversary $\mathcal{B}$ does not know how to compute $\boldsymbol{N}^\top \boldsymbol{K}_1$ or $\boldsymbol{M}^\top \boldsymbol{K}_2$, but she can compute their sum in $\mathbb{Z}_p$ as:

$$\boldsymbol{N}^\top \boldsymbol{K}_1 + \boldsymbol{M}^\top \boldsymbol{K}_2 = \begin{pmatrix} \boldsymbol{M}^\top \\ \boldsymbol{N}^\top \end{pmatrix}(\boldsymbol{K}_1', \boldsymbol{K}_2') + \boldsymbol{T}_{\boldsymbol{A}'}(\boldsymbol{W}_1||\boldsymbol{W}_2) = \boldsymbol{N}^\top \boldsymbol{K}_1' + \boldsymbol{M}^\top \boldsymbol{K}_2' := \boldsymbol{T},$$

due to the fact that $\boldsymbol{M}^\top \boldsymbol{W}_1 = -\boldsymbol{N}^\top \boldsymbol{W}_2$.

Thus, $\mathcal{B}$ picks $\boldsymbol{Z} \leftarrow_{\$} \mathbb{Z}_p^{m \times k}$ and outputs $[\boldsymbol{P}]_2 := [\boldsymbol{T}]_2 - [\boldsymbol{Z}]_2$ and $[\boldsymbol{P}]_1 := [\boldsymbol{Z}]_1$. Now, when the adversary outputs a valid proof for some $([\boldsymbol{y}]_1, [\boldsymbol{x}]_2) \notin \mathcal{L}_{[\boldsymbol{M}]_1,[\boldsymbol{N}]_2}$, it holds that:

$$[\boldsymbol{y}]_1^\top [\boldsymbol{C}_2]_2 - [\boldsymbol{\pi}_1]_1^\top [\boldsymbol{A}_0]_2 = [\boldsymbol{\pi}_2]_2^\top [\boldsymbol{A}_0]_1 - [\boldsymbol{x}]_2^\top [\boldsymbol{C}_1]_1.$$

In which both the RHS and LHS of the last equation are:

$$\text{LHS} = [\boldsymbol{y}]_1^\top (\boldsymbol{K}_2'||\boldsymbol{W}_1)[\boldsymbol{A}']_2 - ([\boldsymbol{\pi}_1]_1^\top || [\boldsymbol{0}_{1\times(m'-r)}]_1)^\top [\boldsymbol{A}']_2 = [\boldsymbol{s}_1^\top]_1 [\boldsymbol{A}']_2,$$

$$\text{RHS} = ([\boldsymbol{\pi}_2]_2^\top || [\boldsymbol{0}_{1\times(m'-r)}]_2)[\boldsymbol{A}']_1 - [\boldsymbol{x}]_2^\top (\boldsymbol{K}_1'||\boldsymbol{W}_2)[\boldsymbol{A}']_1 = [\boldsymbol{s}_2^\top]_2 [\boldsymbol{A}']_1.$$

Here $[\boldsymbol{s}_1^\top]_1 := ([\boldsymbol{y}]_1^\top \boldsymbol{K}_2' - [\boldsymbol{\pi}_1]_1^\top || [\boldsymbol{y}]_1^\top \boldsymbol{W}_1)$ and $[\boldsymbol{s}_2^\top]_2 := ([\boldsymbol{x}]_2^\top \boldsymbol{K}_1' - [\boldsymbol{\pi}_2]_2^\top || - [\boldsymbol{x}]_2^\top \boldsymbol{W}_2)$.

This concludes that $(\boldsymbol{s}_1 - \boldsymbol{s}_2)$ is in the kernel space of $(\boldsymbol{A}')^\top$. In other words, we have that $(\boldsymbol{s}_1 - \boldsymbol{s}_2)^\top \boldsymbol{A}' = \boldsymbol{0}$, and by definition, $\boldsymbol{s}_1 - \boldsymbol{s}_2 = \boldsymbol{c}_1 + \boldsymbol{R}^\top \boldsymbol{c}_2$ and thus

$$(\boldsymbol{s}_1^\top - \boldsymbol{s}_2^\top)\boldsymbol{A} = (\boldsymbol{c}_1^\top + \boldsymbol{c}_2^\top \boldsymbol{R})\boldsymbol{A} = \boldsymbol{c}^\top \boldsymbol{A}' = \boldsymbol{0}_{1\times k} .$$

Since $\boldsymbol{c} \neq \boldsymbol{0}$ and $\boldsymbol{R}$ leaks only through $\boldsymbol{A}'$ (in the definition of $\boldsymbol{C}_i$ for $i \in 1, 2$) as $\boldsymbol{RA}$,

$$\Pr[\boldsymbol{c}_1 + \boldsymbol{R}^\top \boldsymbol{c}_2 = \boldsymbol{0} \mid \boldsymbol{RA}] \leq 1/p ,$$

where the probability is over $\boldsymbol{R}$. This solves the $\mathcal{D}_k$-SKerMDH. □

4 Knowledge Soundness of (Updatable) Asymmetric QA-NIZK Arguments

In the following we investigate a stronger soundness notion, i.e., the knowledge soundness, of (updatable) asymmetric QA-NIZK. We recall that for a proof system to be compatible with modular zk-SNARK frameworks such as LegoSNARK [13], it needs to provide knowledge soundness. Consequently, this guarantees that (updatable) asymmetric QA-NIZK can safely be used within such frameworks.

Similar to as it is done in [6,13], we analyse the knowledge soundness in the Algebraic Group Model (AGM) due to Fuchsbauer *et al.* [21]. In particular, we first directly prove the knowledge soundness property of the updatable asymmetric QA-NIZK in Fig. 4. Moreover, we show that this also yields the knowledge sound property of the original asymmetric QA-NIZK [25] in Fig. 1.

Theorem 4. *Let $\Pi'_{\mathsf{asy\text{-}up}}$ be an asymmetric QA-NIZK argument for linear subspaces from Fig. 1. Assume $\hat{\mathcal{D}_k} = \bar{\mathcal{D}_k}$ and the distribution $\mathcal{D}_{\mathsf{p}}$ is witness samplable matrix distribution. If the discrete logarithm assumption in asymmetric bilinear groups in the AGM holds, then the updatable asymmetric QA-NIZK $\Pi'_{\mathsf{asy\text{-}up}}$ in Fig. 4 is computationally updatable adaptive knowledge sound.*

Proof. We show the theorem under the discrete logarithm assumption in asymmetric bilinear groups in the AGM. Without loss of generality, we consider the updatable asymmetric QA-NIZK scheme $\Pi'_{\mathsf{asy\text{-}up}}$ for $\hat{\mathcal{D}_k} = \bar{\mathcal{D}_k}$, in the MDDH setting where $k = 1$.

Without loss of generality, we assume an algebraic adversary $\mathcal{A}([\boldsymbol{M}]_1, [\boldsymbol{N}]_2, \mathsf{aux})$ against the updatable knowledge soundness of $\Pi'_{\mathsf{asy\text{-}up}}$, where aux is an associated auxiliary input, such that it first generates a $\mathsf{crs}_{\mathcal{A}}$ which verifies under the algorithm Vcrs. Then an honest updator $\mathsf{Ucrs}([\boldsymbol{M}]_1, [\boldsymbol{N}]_2, \mathsf{crs}_{\mathcal{A}})$ outputs an updated $\mathsf{crs} = ([\boldsymbol{A}, \boldsymbol{C}_1, \boldsymbol{C}_2, \boldsymbol{P}_2]_2, [\boldsymbol{A}, \boldsymbol{C}_1, \boldsymbol{C}_2, \boldsymbol{P}_1]_1)$.

Let $[\boldsymbol{\zeta}]_1$ and $[\boldsymbol{\zeta}']_2$ be vectors that contain $\boldsymbol{M}$ (and the portion of aux that has elements from the group $\mathbb{G}_1$) and $\boldsymbol{N}$ (and the portion of aux that has elements from the group $\mathbb{G}_2$). We assume that $[\boldsymbol{\zeta}]_1$ and $[\boldsymbol{\zeta}']_2$ also contains the $\mathsf{crs}_{\mathcal{A}}$'s elements in $\mathbb{G}_1$ and $\mathbb{G}_2$, respectively. Due to the CRS-update hiding property, the $\mathsf{crs}_{\mathcal{A}}$'s elements are indistinguishable from the crs generated by an honest Ucrs and so they will give no advantage to $\mathcal{A}$. Also assume that $[\boldsymbol{\zeta}]_1$ and $[\boldsymbol{\zeta}']_2$ include $[1]_1$ and $[1]_2$, respectively. $\mathcal{A}([\boldsymbol{M}]_1, [\boldsymbol{N}]_2, \mathsf{crs}, \mathsf{aux})$ returns a tuple $([\boldsymbol{y}]_1, [\boldsymbol{x}]_2, [\pi_1]_1, [\pi_2]_2)$ along with coefficients that explain these elements as linear combinations of its input in the groups $\mathbb{G}_1$ and $\mathbb{G}_2$. Let these coefficients be:

$$
\begin{aligned}
[\boldsymbol{y}]_1 &= \boldsymbol{Y}_0[\boldsymbol{P}_1]_1 + \boldsymbol{Y}_1[\boldsymbol{\zeta}]_1 + \boldsymbol{Y}_2[\boldsymbol{A}]_1 + \boldsymbol{Y}_3[\boldsymbol{C}_1]_1 + \boldsymbol{Y}_4[\boldsymbol{C}_2]_1 \\
[\boldsymbol{\pi}]_1 &= \boldsymbol{Z}_0[\boldsymbol{P}_1]_1 + \boldsymbol{Z}_1[\boldsymbol{\zeta}]_1 + \boldsymbol{Z}_2[\boldsymbol{A}]_1 + \boldsymbol{Z}_3[\boldsymbol{C}_1]_1 + \boldsymbol{Z}_4[\boldsymbol{C}_2]_1 \\
[\boldsymbol{x}]_2 &= \boldsymbol{Y}'_0[\boldsymbol{P}_2]_2 + \boldsymbol{Y}'_1[\boldsymbol{\zeta}']_2 + \boldsymbol{Y}'_2[\boldsymbol{A}]_2 + \boldsymbol{Y}'_3[\boldsymbol{C}_1]_1 + \boldsymbol{Y}'_4[\boldsymbol{C}_2]_2 \\
[\boldsymbol{\pi}]_2 &= \boldsymbol{Z}'_0[\boldsymbol{P}_2]_2 + \boldsymbol{Z}'_1[\boldsymbol{\zeta}']_2 + \boldsymbol{Z}'_2[\boldsymbol{A}]_2 + \boldsymbol{Z}'_3[\boldsymbol{C}_1]_2 + \boldsymbol{Z}'_4[\boldsymbol{C}_2]_2
\end{aligned}
$$

Let the extractor $\mathsf{Ext}_{\mathcal{A}}([\boldsymbol{M}]_1, [\boldsymbol{N}]_2, \mathsf{crs}, \mathsf{aux})$ be the algorithm that runs $\mathcal{A}$ and returns $\boldsymbol{w} = \boldsymbol{Z}_0 = \boldsymbol{Z}'_0$. Then, we have to show that the probability that the output of $(\mathcal{A}, \mathsf{Ext}_{\mathcal{A}})$ satisfies verification while $\boldsymbol{y} \neq \boldsymbol{M}\boldsymbol{w}$ and $\boldsymbol{x} \neq \boldsymbol{N}\boldsymbol{w}$ are negligible. In other words, assume that the output of $\mathcal{A}$ is such that $[\boldsymbol{y}]_1 \neq [\boldsymbol{M}]_1\boldsymbol{Z}_0$, $[\boldsymbol{x}]_2 \neq [\boldsymbol{N}]_2\boldsymbol{Z}'_0$, and $[\boldsymbol{y}]_1^\top[\boldsymbol{K}_2 a]_2 - [\boldsymbol{\pi}_1]_1^\top[a]_2 = [\boldsymbol{\pi}_2]_2^\top[a]_1 - [\boldsymbol{x}]_2^\top[\boldsymbol{K}_1 a]_1$; If it happens with non-negligible probability, we can construct an algorithm $\mathcal{B}$ that on input $([\boldsymbol{K}_1, \boldsymbol{K}_2]_1, [\boldsymbol{K}_1, \boldsymbol{K}_2]_2)$ outputs nonzero elements $\boldsymbol{\alpha}, \boldsymbol{\alpha}' \in \mathbb{Z}_p^{\ell\times\ell}$, $\boldsymbol{\beta}, \boldsymbol{\beta}' \in \mathbb{Z}_p^{\ell}$, and $\gamma, \gamma' \in \mathbb{Z}_p$ such that

$$
[\boldsymbol{K}_1^\top \boldsymbol{\alpha} \boldsymbol{K}_1 + \boldsymbol{K}_1^\top \boldsymbol{\beta} + \gamma]_1[1]_2 + [1]_1[\boldsymbol{K}_2^\top \boldsymbol{\alpha}' \boldsymbol{K}_2 + \boldsymbol{K}_2^\top \boldsymbol{\beta}' + \gamma']_2 = [0]_T.
$$

Then we can construct an algorithm $\mathcal{C}$ against the discrete logarithm assumption in asymmetric bilinear groups such that given elements $([t, t']_1, [t, t']_2)$ it returns the exponent $t, t' \in \mathbb{Z}_p$. More precisely, the algorithm $\mathcal{B}([\boldsymbol{K}_1, \boldsymbol{K}_2]_1, [\boldsymbol{K}_1, \boldsymbol{K}_2]_2)$ proceeds as follows:

- Choose $([\boldsymbol{M}]_1, [\boldsymbol{N}]_2, \mathsf{aux})$ from $\mathcal{D}_\mathsf{p}$ along with corresponding elements in $\mathbb{G}_1$ and $\mathbb{G}_2$ (i.e., vectors $\boldsymbol{\zeta}$, $\boldsymbol{\zeta}'$ of entries in $\mathbb{Z}_p$).
- Sample $a \leftarrow_\$ \mathbb{Z}_p$ and run $\mathcal{A}([\boldsymbol{\zeta}', \boldsymbol{C}_1, \boldsymbol{C}_2, \boldsymbol{P}_1, a]_1, [\boldsymbol{\zeta}', a, \boldsymbol{P}_2, \boldsymbol{C}_1, \boldsymbol{C}_2]_2$. We note that $\mathcal{A}$'s input can be efficiently simulated.
- Once received the output of $\mathcal{A}$, it sets $\boldsymbol{\alpha} := \boldsymbol{Y}_0\boldsymbol{M}^\top$, $\boldsymbol{\beta} := \boldsymbol{Y}_1\boldsymbol{\zeta} + \boldsymbol{Y}_2 a + \boldsymbol{Y}_3\boldsymbol{C}_1 + \boldsymbol{Y}_4\boldsymbol{C}_2 - \boldsymbol{M}\boldsymbol{Z}_0$ and $\gamma := -\boldsymbol{Z}_1\boldsymbol{\zeta} - \boldsymbol{Z}_2 a - \boldsymbol{Z}_3\boldsymbol{C}_1 - \boldsymbol{Z}_4\boldsymbol{C}_2$.
- Also it sets $\boldsymbol{\alpha}' := \boldsymbol{Y}'_0\boldsymbol{N}^\top$, $\boldsymbol{\beta}' := \boldsymbol{Y}'_1\boldsymbol{\zeta}' + \boldsymbol{Y}'_2 a + \boldsymbol{Y}'_3\boldsymbol{C}_1 + \boldsymbol{Y}'_4\boldsymbol{C}_2 - \boldsymbol{N}\boldsymbol{Z}'_0$ and $\gamma' := -\boldsymbol{Z}'_1\boldsymbol{\zeta}' - \boldsymbol{Z}'_2 a - \boldsymbol{Z}'_3\boldsymbol{C}_1 - \boldsymbol{Z}'_4\boldsymbol{C}_2$.

Notice that,

$$\begin{aligned}\boldsymbol{K}_1^\top\boldsymbol{\alpha}\boldsymbol{K}_1 + \boldsymbol{K}_1^\top\boldsymbol{\beta} + \gamma = &\boldsymbol{K}_1^\top\boldsymbol{Y}_0\boldsymbol{M}^\top\boldsymbol{K}_1 + \boldsymbol{K}_1^\top\boldsymbol{Y}_1\boldsymbol{\zeta} + \boldsymbol{K}_1^\top\boldsymbol{Y}_2 a + \boldsymbol{K}_1^\top\boldsymbol{Y}_3\boldsymbol{C}_1 + \\ &\boldsymbol{K}_1^\top\boldsymbol{Y}_4\boldsymbol{C}_2 - \boldsymbol{K}_1^\top\boldsymbol{M}\boldsymbol{Z}_0 - \boldsymbol{Z}_1\boldsymbol{\zeta} - \boldsymbol{Z}_2 a - \boldsymbol{Z}_3\boldsymbol{C}_2 - \boldsymbol{Z}_4\boldsymbol{C}_2\\ =&\boldsymbol{K}_1^\top\boldsymbol{Y}_0\boldsymbol{M}^\top\boldsymbol{K}_1 + \boldsymbol{K}_1^\top\boldsymbol{Y}_1\boldsymbol{\zeta} + \boldsymbol{K}_1^\top\boldsymbol{Y}_2 a + \boldsymbol{K}_1^\top\boldsymbol{Y}_3\boldsymbol{C}_1 + \boldsymbol{K}_1^\top\boldsymbol{Y}_4\boldsymbol{C}_2 - \boldsymbol{\pi}_1\\ =&\boldsymbol{K}_1^\top\boldsymbol{y} - \boldsymbol{\pi}_1,\end{aligned}$$

and

$$\begin{aligned}\boldsymbol{K}_2^\top\boldsymbol{\alpha}'\boldsymbol{K}_2 + \boldsymbol{K}_2^\top\boldsymbol{\beta}' + \gamma' = &\boldsymbol{K}_2^\top\boldsymbol{Y}'_0\boldsymbol{N}^\top\boldsymbol{K}_2 + \boldsymbol{K}_2^\top\boldsymbol{Y}'_1\boldsymbol{\zeta}' + \boldsymbol{K}_2^\top\boldsymbol{Y}'_2 a + \boldsymbol{K}_2^\top\boldsymbol{Y}'_3\boldsymbol{C}_1 + \\ &\boldsymbol{K}_2^\top\boldsymbol{Y}'_4\boldsymbol{C}_2 - \boldsymbol{K}_2^\top\boldsymbol{N}\boldsymbol{Z}'_0 - \boldsymbol{Z}'_1\boldsymbol{\zeta}' - \boldsymbol{Z}'_2 a - \boldsymbol{Z}'_3\boldsymbol{C}_2 - \boldsymbol{Z}'_4\boldsymbol{C}_2\\ =&\boldsymbol{K}_2^\top\boldsymbol{Y}'_0\boldsymbol{N}^\top\boldsymbol{K}_2 + \boldsymbol{K}_2^\top\boldsymbol{Y}'_1\boldsymbol{\zeta}' + \boldsymbol{K}_2^\top\boldsymbol{Y}'_2 a + \boldsymbol{K}_2^\top\boldsymbol{Y}'_3\boldsymbol{C}_1 + \boldsymbol{K}_2^\top\boldsymbol{Y}'_4\boldsymbol{C}_2 - \boldsymbol{\pi}_2\\ =&\boldsymbol{K}_2^\top\boldsymbol{x} - \boldsymbol{\pi}_2.\end{aligned}$$

From the verification equation, we have

$$\begin{aligned}&(\boldsymbol{K}_1^\top\boldsymbol{\alpha}\boldsymbol{K}_1 + \boldsymbol{K}_1^\top\boldsymbol{\beta} + \gamma) + (\boldsymbol{K}_2^\top\boldsymbol{\alpha}'\boldsymbol{K}_2 + \boldsymbol{K}_2^\top\boldsymbol{\beta}' + \gamma')\\ &= \boldsymbol{K}_1^\top\boldsymbol{y} - \boldsymbol{\pi}_1 + \boldsymbol{K}_2^\top\boldsymbol{x} - \boldsymbol{\pi}_2 = 0.\end{aligned}$$

Note that, one among $\boldsymbol{\alpha}$, $\boldsymbol{\beta}$, and γ ($\boldsymbol{\alpha}'$, $\boldsymbol{\beta}'$, and γ') must be nonzero. Indeed, if they are all zero then $\boldsymbol{Y}_1\boldsymbol{\zeta} + \boldsymbol{Y}_2 a + \boldsymbol{Y}_3\boldsymbol{C} - \boldsymbol{M}\boldsymbol{Z}_0 = 0$ ($\boldsymbol{Y}'_1\boldsymbol{\zeta}' + \boldsymbol{Y}'_2 a + \boldsymbol{Y}'_3\boldsymbol{C} - \boldsymbol{N}\boldsymbol{Z}'_0 = 0$), that is $\boldsymbol{y} = \boldsymbol{M}\boldsymbol{Z}_0$ ($\boldsymbol{x} = \boldsymbol{N}\boldsymbol{Z}'_0$), which contradicts our assumption on $\mathcal{A}$'s output.

Finally we show how the above problem can be reduced to discrete logarithm problem in asymmetric groups, i.e., the adversary $\mathcal{C}$ on input $([t,t']_1, [t,t']_2)$ returns t' and t'.

Indeed $\mathcal{C}$ samples $\boldsymbol{r}, \boldsymbol{s}, \boldsymbol{r}', \boldsymbol{s}' \in \mathbb{Z}_p^\ell$ and implicitly sets $\boldsymbol{K}_1 = t\boldsymbol{r} + \boldsymbol{s}$ and $\boldsymbol{K}_2 = t\boldsymbol{r}' + \boldsymbol{s}'$. We see that $([\boldsymbol{K}_1, \boldsymbol{K}_2]_1, [\boldsymbol{K}_1, \boldsymbol{K}_2]_2)$ can be efficiently simulated with a distribution identical to the one expected by $\mathcal{B}$. Next, given a solution $(\boldsymbol{\alpha}, \boldsymbol{\beta}, \gamma, \boldsymbol{\alpha}', \boldsymbol{\beta}', \gamma')$ such that $\boldsymbol{K}_1^\top\boldsymbol{\alpha} + \boldsymbol{K}_1^\top\boldsymbol{\beta} + \gamma + \boldsymbol{K}_2^\top\boldsymbol{\alpha}' + \boldsymbol{K}_2^\top\boldsymbol{\beta}' + \gamma' = 0$, one can find $e_1, e_2, e_3 \in \mathbb{Z}_p$ and $e'_1, e'_2, e'_3 \in \mathbb{Z}_p$ such that:

$$\begin{aligned}0 =&(t\boldsymbol{r} + \boldsymbol{s})^\top\boldsymbol{\alpha}(t\boldsymbol{r} + \boldsymbol{s}) + (t\boldsymbol{r} + \boldsymbol{s})^\top\boldsymbol{\beta} + \gamma + (t\boldsymbol{r}' + \boldsymbol{s}')^\top\boldsymbol{\alpha}'(t'\boldsymbol{r}' + \boldsymbol{s}') + \\ &(t'\boldsymbol{r}' + \boldsymbol{s}')^\top\boldsymbol{\beta}' + \gamma'\\ =&t^2(\boldsymbol{r}^\top\boldsymbol{\alpha}\boldsymbol{r}) + t(\boldsymbol{r}^\top\boldsymbol{\alpha}\boldsymbol{s} + \boldsymbol{s}^\top\boldsymbol{\alpha}\boldsymbol{r} + \boldsymbol{r}^\top\boldsymbol{\beta}) + t'^2(\boldsymbol{r}'^\top\boldsymbol{\alpha}'\boldsymbol{r}') + t'(\boldsymbol{r}'^\top\boldsymbol{\alpha}'\boldsymbol{s}' + \\ &\boldsymbol{s}'^\top\boldsymbol{\alpha}'\boldsymbol{r}' + \boldsymbol{r}'^\top\boldsymbol{\beta}') + (\boldsymbol{s}^\top\boldsymbol{\alpha}\boldsymbol{s} + \boldsymbol{s}^\top\boldsymbol{\beta} + \gamma) + (\boldsymbol{s}'^\top\boldsymbol{\alpha}'\boldsymbol{s}' + \boldsymbol{s}'^\top\boldsymbol{\beta}' + \gamma')\\ =&e_1 t^2 + e_2 t + e_3 + e'_1 t'^2 + e'_2 t' + e'_3.\end{aligned}$$

In particular, with overwhelming probability (over the choice of $\boldsymbol{s}$ and $\boldsymbol{s}'$ that are information theoretically hidden from $\mathcal{B}$'s view) $e_3, e_3' \neq 0$. From this solution, $\mathcal{C}$ can solve the system and extract t and t'. □

Theorem 5. *Let $\Pi'_{\mathsf{asy\text{-}up}}$ be an updatable knowledge sound asymmetric QA-NIZK argument for linear subspaces from Theorem 4. Then the asymmetric QA-NIZK Π'_{asy} [25] in Fig. 1 is computationally knowledge sound.*

Proof. We prove it by contradiction in a way that we assume that there is an adversary $\mathcal{A}_{\Pi'_{\mathsf{asy}}}$ that breaks the knowledge soundness of the asymmetric QA-NIZK Π'_{asy} . Then, one can build an adversary $\mathcal{B}_{\Pi'_{\mathsf{asy\text{-}up}}}$ against the updatable asymmetric QA-NIZK $\Pi'_{\mathsf{asy\text{-}up}}$ who runs $\mathcal{A}_{\Pi'_{\mathsf{asy}}}$ as a subroutine algorithm and breaks the updatable knowledge soundness $\Pi'_{\mathsf{asy\text{-}up}}$. More precisely, given $([\boldsymbol{M}]_1, [\boldsymbol{N}]_2, \mathsf{crs})$ where $\mathsf{crs} = ([\boldsymbol{A}, \boldsymbol{C}_1, \boldsymbol{C}_2, \boldsymbol{P}_2]_2, [\boldsymbol{A}, \boldsymbol{C}_1, \boldsymbol{C}_2, \boldsymbol{P}_1]_1)$, the adversary $\mathcal{B}_{\Pi'_{\mathsf{asy\text{-}up}}}$ sets $\mathsf{crs}_{\Pi'_{\mathsf{asy}}} := ([\boldsymbol{A}, \boldsymbol{C}_2, \boldsymbol{P}_2]_2, [\boldsymbol{A}, \boldsymbol{C}_1, \boldsymbol{P}_1]_1)$ and sends it to $\mathcal{A}_{\Pi'_{\mathsf{asy}}}$ and returns back a valid $\pi_{\Pi'_{\mathsf{asy}}}$ for $([\boldsymbol{y}]_1, [\boldsymbol{x}]_2) \notin \mathcal{L}_{[\boldsymbol{M}]_1, [\boldsymbol{N}]_2}$. This concludes the proof. □

5 Discussion and Future Work

In this paper we investigate QA-NIZKs in the full subversion setting via the updatable CRS model. In particular, we analyse the security of the most efficient asymmetric QA-NIZK Π'_{asy} by González et al. [25] for $k = 1, 2$ (when $\mathcal{D}_k = \mathcal{U}_k$) and for arbitrary k (when $\mathcal{D}_k \in \{\mathcal{L}_k, \mathcal{IL}_k, \mathcal{C}_k, \mathcal{SC}_k\}$), when the CRS is full subverted and propose an updatable version of these QA-NIZKs. Since in practice, due to increased efficiency, one is mostly interested in shorter proof size (smaller k as proofs are of size $2k$) and thus even when relying on $\mathcal{D}_k = \mathcal{U}_k$ the focus on schemes for $k = 1, 2$ is most reasonable. Especially as for these values of k one obtains constructions from the most common standard assumptions. But from a theoretical point of view, it is interesting to construct an updatable version of asymmetric QA-NIZK (or even symmetric QA-NIZK) for arbitrary $k > 2$ even in the case of $\mathcal{D}_k = \mathcal{U}_k$. Here, the main obstacle is to design a general (efficient) version of the algorithm $\mathsf{isinvertible}(\cdot, \cdot)$ [3] for checking the invertibility of a matrix of group elements of size k (see Fig. 3).

We recall that our main motivation in this work was to fill the existing gap towards obtaining an updatable version of LegoSNARK [13]. An interesting question is to study how one can combine all the existing updatable CRS building blocks, i.e., updatable CRS SNARKs and QA-NIZKs as well as our construction, to construct an updatable LegoSNARK framework and investigate its efficiency.

Acknowledgements. This work was in part funded by the European Union's Horizon 2020 research and innovation programme under grant agreement no. 871473 (KRAKEN) and no890456 (SLOTMACHINE), and by the Austrian Science Fund (FWF) and netidee SCIENCE under grant agreement P31621-N38 (PROFET). This work has received funding by the German Federal Ministry of Education and Research BMBF (grant 16KISK038, project 6GEM).

A CRS-update Hiding Proof

Lemma 2 (**[40], Lemma 6.**). *Assume that* $\boldsymbol{K}, \boldsymbol{K}_{\mathsf{int}} \in \mathcal{D}_{\boldsymbol{K}}$ *and* $\boldsymbol{A}, \boldsymbol{A}_{\mathsf{int}} \in \mathcal{D}_{\boldsymbol{A}}$, *where* $\mathcal{D}_{\boldsymbol{K}}$ *and* $\mathcal{D}_{\boldsymbol{A}}$ *satisfy the following conditions for random variables* Y_1 *and* Y_2: *(i) if* $Y_1, Y_2 \leftarrow_{\$} \mathcal{D}_{\boldsymbol{K}}$ *then* $Y_1 + Y_2 \in \mathcal{D}_{\boldsymbol{K}}$, *and (ii) if* $Y_1, Y_2 \leftarrow_{\$} \mathcal{D}_{\boldsymbol{A}}$ *then* $Y_1 \cdot Y_2 \in \mathcal{D}_{\boldsymbol{A}}$. *Then,* $\Pi'_{\mathsf{asy\text{-}up}}$ *is key-update hiding.*

Proof. Since $\mathsf{Vcrs}(\mathsf{crs}, \mathsf{lpar}) = 1$, thus, crs is honestly created, $\boldsymbol{C} = \boldsymbol{K}\boldsymbol{A}$. So, $\boldsymbol{C}_{\mathsf{up}} = \boldsymbol{C}\boldsymbol{A}_{\mathsf{int}} + \boldsymbol{K}_{\mathsf{int}}\boldsymbol{A}\boldsymbol{A}_{\mathsf{int}} = (\boldsymbol{K} + \boldsymbol{K}_{\mathsf{int}})\boldsymbol{A}\boldsymbol{A}_{\mathsf{int}} = (\boldsymbol{K} + \boldsymbol{K}_{\mathsf{int}})\boldsymbol{A}_{\mathsf{up}} = \boldsymbol{K}_{\mathsf{up}}\boldsymbol{A}_{\mathsf{up}}$. Similarly holds for $\boldsymbol{P}$. Due to the assumption on $\mathcal{D}_{\boldsymbol{A}}$ and $\mathcal{D}_{\boldsymbol{K}}$, crs and $\mathsf{crs}_{\mathsf{up}}$ come from the same distribution.

References

1. Abdolmaleki, B., Baghery, K., Lipmaa, H., Siim, J., Zajac, M.: UC-secure CRS generation for SNARKs. In: Buchmann, J., Nitaj, A., Eddine Rachidi, T. (eds.) AFRICACRYPT 19. LNCS, vol. 11627, pp. 99–117. Springer, Heidelberg (2019). https://doi.org/10.1007/978-3-030-23696-0_6
2. Abdolmaleki, B., Baghery, K., Lipmaa, H., Zajac, M.: A subversion-resistant SNARK. In: Takagi, T., Peyrin, T. (eds.) ASIACRYPT 2017, Part III. LNCS, vol. 10626, pp. 3–33. Springer, Heidelberg (2017). https://doi.org/10.1007/978-3-319-70700-6_1
3. Abdolmaleki, B., Lipmaa, H., Siim, J., Zajac, M.: On QA-NIZK in the BPK model. In: Kiayias, A., Kohlweiss, M., Wallden, P., Zikas, V. (eds.) PKC 2020, Part I. LNCS, vol. 12110, pp. 590–620. Springer, Heidelberg (2020). https://doi.org/10.1007/978-3-030-45374-9_20
4. Abdolmaleki, B., Lipmaa, H., Siim, J., Zajac, M.: On subversion-resistant snarks. J. Cryptol. **34**(3), 17 (2021)
5. Abdolmaleki, B., Ramacher, S., Slamanig, D.: Lift-and-shift: obtaining simulation extractable subversion and updatable SNARKs generically. In: Ligatti, J., Ou, X., Katz, J., Vigna, G. (eds.) ACM CCS 2020, pp. 1987–2005. ACM Press (2020). https://doi.org/10.1145/3372297.3417228
6. Abdolmaleki, B., Slamanig, D.: Subversion-resistant quasi-adaptive NIZK and applications to modular zk-SNARKs. In: Conti, M., Stevens, M., Krenn, S. (eds.) CANS 2021. LNCS, vol. 13099, pp. 492–512. Springer, Cham (2021). https://doi.org/10.1007/978-3-030-92548-2_26
7. Abe, M., Jutla, C.S., Ohkubo, M., Pan, J., Roy, A., Wang, Y.: Shorter QA-NIZK and SPS with tighter security. In: Galbraith, S.D., Moriai, S. (eds.) ASIACRYPT 2019, Part III. LNCS, vol. 11923, pp. 669–699. Springer, Heidelberg (2019). https://doi.org/10.1007/978-3-030-34618-8_23
8. Abe, M., Jutla, C.S., Ohkubo, M., Roy, A.: Improved (almost) tightly-secure simulation-sound QA-NIZK with applications. In: Peyrin, T., Galbraith, S. (eds.) ASIACRYPT 2018, Part I. LNCS, vol. 11272, pp. 627–656. Springer, Heidelberg (2018). https://doi.org/10.1007/978-3-030-03326-2_21
9. Bellare, M., Fuchsbauer, G., Scafuro, A.: NIZKs with an untrusted CRS: security in the face of parameter subversion. In: Cheon, J.H., Takagi, T. (eds.) ASIACRYPT 2016, Part II. LNCS, vol. 10032, pp. 777–804. Springer, Heidelberg (2016). https://doi.org/10.1007/978-3-662-53890-6_26

10. Ben-Sasson, E., Chiesa, A., Green, M., Tromer, E., Virza, M.: Secure sampling of public parameters for succinct zero knowledge proofs. In: 2015 IEEE Symposium on Security and Privacy, pp. 287–304. IEEE (2015)
11. Bowe, S., Gabizon, A., Green, M.D.: A multi-party protocol for constructing the public parameters of the pinocchio zk-snark. Cryptology ePrint Archive, Report 2017/602 (2017). https://eprint.iacr.org/2017/602
12. Campanelli, M., Faonio, A., Fiore, D., Querol, A., Rodríguez, H.: Lunar: a toolbox for more efficient universal and updatable zkSNARKs and commit-and-prove extensions. In: Tibouchi, M., Wang, H. (eds.) ASIACRYPT 2021, Part III. LNCS, vol. 13092, pp. 3–33. Springer, Heidelberg (2021). https://doi.org/10.1007/978-3-030-92078-4_1
13. Campanelli, M., Fiore, D., Querol, A.: LegoSNARK: Modular design and composition of succinct zero-knowledge proofs. In: Cavallaro, L., Kinder, J., Wang, X., Katz, J. (eds.) ACM CCS 2019, pp. 2075–2092. ACM Press (2019). https://doi.org/10.1145/3319535.3339820
14. Chiesa, A., Hu, Y., Maller, M., Mishra, P., Vesely, N., Ward, N.P.: Marlin: preprocessing zkSNARKs with universal and updatable SRS. In: Canteaut, A., Ishai, Y. (eds.) EUROCRYPT 2020, Part I. LNCS, vol. 12105, pp. 738–768. Springer, Heidelberg (2020). https://doi.org/10.1007/978-3-030-45721-1_26
15. Damgård, I.: Towards practical public key systems secure against chosen ciphertext attacks. In: Feigenbaum, J. (ed.) CRYPTO'91. LNCS, vol. 576, pp. 445–456. Springer, Heidelberg (1992). https://doi.org/10.1007/3-540-46766-1_36
16. Danezis, G., Fournet, C., Groth, J., Kohlweiss, M.: Square span programs with applications to succinct NIZK arguments. In: Sarkar, P., Iwata, T. (eds.) ASIACRYPT 2014, Part I. LNCS, vol. 8873, pp. 532–550. Springer, Heidelberg (2014). https://doi.org/10.1007/978-3-662-45611-8_28
17. Daza, V., González, A., Pindado, Z., Ràfols, C., Silva, J.: Shorter quadratic QA-NIZK proofs. In: Lin, D., Sako, K. (eds.) PKC 2019, Part I. LNCS, vol. 11442, pp. 314–343. Springer, Heidelberg (2019). https://doi.org/10.1007/978-3-030-17253-4_11
18. Daza, V., Ràfols, C., Zacharakis, A.: Updateable inner product argument with logarithmic verifier and applications. In: Kiayias, A., Kohlweiss, M., Wallden, P., Zikas, V. (eds.) PKC 2020, Part I. LNCS, vol. 12110, pp. 527–557. Springer, Heidelberg (2020). https://doi.org/10.1007/978-3-030-45374-9_18
19. Escala, A., Herold, G., Kiltz, E., Ràfols, C., Villar, J.: An algebraic framework for Diffie-Hellman assumptions. In: Canetti, R., Garay, J.A. (eds.) CRYPTO 2013, Part II. LNCS, vol. 8043, pp. 129–147. Springer, Heidelberg (2013). https://doi.org/10.1007/978-3-642-40084-1_8
20. Fuchsbauer, G.: Subversion-zero-knowledge SNARKs. In: Abdalla, M., Dahab, R. (eds.) PKC 2018, Part I. LNCS, vol. 10769, pp. 315–347. Springer, Heidelberg (2018). https://doi.org/10.1007/978-3-319-76578-5_11
21. Fuchsbauer, G., Kiltz, E., Loss, J.: The algebraic group model and its applications. In: Shacham, H., Boldyreva, A. (eds.) CRYPTO 2018, Part II. LNCS, vol. 10992, pp. 33–62. Springer, Heidelberg (2018). https://doi.org/10.1007/978-3-319-96881-0_2
22. Gabizon, A., Williamson, Z.J., Ciobotaru, O.: PLONK: Permutations over Lagrange-bases for oecumenical noninteractive arguments of knowledge. Cryptology ePrint Archive, Report 2019/953 (2019). https://eprint.iacr.org/2019/953

23. Gennaro, R., Gentry, C., Parno, B., Raykova, M.: Quadratic span programs and succinct NIZKs without PCPs. In: Johansson, T., Nguyen, P.Q. (eds.) EUROCRYPT 2013. LNCS, vol. 7881, pp. 626–645. Springer, Heidelberg (2013). https://doi.org/10.1007/978-3-642-38348-9_37
24. Goldwasser, S., Micali, S., Rackoff, C.: The knowledge complexity of interactive proof systems. SIAM J. Comput. **18**(1), 186–208 (1989)
25. González, A., Hevia, A., Ràfols, C.: QA-NIZK arguments in asymmetric groups: new tools and new constructions. In: Iwata, T., Cheon, J.H. (eds.) ASIACRYPT 2015, Part I. LNCS, vol. 9452, pp. 605–629. Springer, Heidelberg (2015). https://doi.org/10.1007/978-3-662-48797-6_25
26. Groth, J.: Simulation-sound NIZK proofs for a practical language and constant size group signatures. In: Lai, X., Chen, K. (eds.) ASIACRYPT 2006. LNCS, vol. 4284, pp. 444–459. Springer, Heidelberg (2006). https://doi.org/10.1007/11935230_29
27. Groth, J.: Short pairing-based non-interactive zero-knowledge arguments. In: Abe, M. (ed.) ASIACRYPT 2010. LNCS, vol. 6477, pp. 321–340. Springer, Heidelberg (2010). https://doi.org/10.1007/978-3-642-17373-8_19
28. Groth, J.: On the size of pairing-based non-interactive arguments. In: Fischlin, M., Coron, J.S. (eds.) EUROCRYPT 2016, Part II. LNCS, vol. 9666, pp. 305–326. Springer, Heidelberg (2016). https://doi.org/10.1007/978-3-662-49896-5_11
29. Groth, J., Kohlweiss, M., Maller, M., Meiklejohn, S., Miers, I.: Updatable and universal common reference strings with applications to zk-SNARKs. In: Shacham, H., Boldyreva, A. (eds.) CRYPTO 2018, Part III. LNCS, vol. 10993, pp. 698–728. Springer, Heidelberg (2018). https://doi.org/10.1007/978-3-319-96878-0_24
30. Groth, J., Ostrovsky, R., Sahai, A.: Perfect non-interactive zero knowledge for NP. In: Vaudenay, S. (ed.) EUROCRYPT 2006. LNCS, vol. 4004, pp. 339–358. Springer, Heidelberg (2006). https://doi.org/10.1007/11761679_21
31. Groth, J., Sahai, A.: Efficient non-interactive proof systems for bilinear groups. In: Smart, N.P. (ed.) EUROCRYPT 2008. LNCS, vol. 4965, pp. 415–432. Springer, Heidelberg (2008). https://doi.org/10.1007/978-3-540-78967-3_24
32. Jutla, C.S., Roy, A.: Shorter quasi-adaptive NIZK proofs for linear subspaces. In: Sako, K., Sarkar, P. (eds.) ASIACRYPT 2013, Part I. LNCS, vol. 8269, pp. 1–20. Springer, Heidelberg (2013). https://doi.org/10.1007/978-3-642-42033-7_1
33. Jutla, C.S., Roy, A.: Switching lemma for bilinear tests and constant-size NIZK proofs for linear subspaces. In: Garay, J.A., Gennaro, R. (eds.) CRYPTO 2014, Part II. LNCS, vol. 8617, pp. 295–312. Springer, Heidelberg (2014). https://doi.org/10.1007/978-3-662-44381-1_17
34. Kilian, J.: A note on efficient zero-knowledge proofs and arguments (extended abstract). In: 24th ACM STOC, pp. 723–732. ACM Press (1992). https://doi.org/10.1145/129712.129782
35. Kiltz, E., Wee, H.: Quasi-adaptive NIZK for linear subspaces revisited. In: Oswald, E., Fischlin, M. (eds.) EUROCRYPT 2015, Part II. LNCS, vol. 9057, pp. 101–128. Springer, Heidelberg (2015). https://doi.org/10.1007/978-3-662-46803-6_4
36. Kohlweiss, M., Maller, M., Siim, J., Volkhov, M.: Snarky ceremonies. In: Tibouchi, M., Wang, H. (eds.) ASIACRYPT 2021, Part III. LNCS, vol. 13092, pp. 98–127. Springer, Heidelberg (2021). https://doi.org/10.1007/978-3-030-92078-4_4
37. Libert, B., Peters, T., Joye, M., Yung, M.: Non-malleability from malleability: Simulation-sound quasi-adaptive NIZK proofs and CCA2-secure encryption from homomorphic signatures. In: Nguyen, P.Q., Oswald, E. (eds.) EUROCRYPT 2014. LNCS, vol. 8441, pp. 514–532. Springer, Heidelberg (2014). https://doi.org/10.1007/978-3-642-55220-5_29

38. Libert, B., Peters, T., Joye, M., Yung, M.: Compactly hiding linear spans - tightly secure constant-size simulation-sound QA-NIZK proofs and applications. In: Iwata, T., Cheon, J.H. (eds.) ASIACRYPT 2015, Part I. LNCS, vol. 9452, pp. 681–707. Springer, Heidelberg (2015). https://doi.org/10.1007/978-3-662-48797-6_28
39. Lipmaa, H.: Progression-free sets and sublinear pairing-based non-interactive zero-knowledge arguments. In: Cramer, R. (ed.) TCC 2012. LNCS, vol. 7194, pp. 169–189. Springer, Heidelberg (2012). https://doi.org/10.1007/978-3-642-28914-9_10
40. Lipmaa, H.: Key-and-argument-updatable QA-NIZKs. In: Galdi, C., Kolesnikov, V. (eds.) SCN 20. LNCS, vol. 12238, pp. 645–669. Springer, Heidelberg (2020). https://doi.org/10.1007/978-3-030-57990-6_32
41. Lipmaa, H.: A unified framework for non-universal snarks. In: Hanaoka, G., Shikata, J., Watanabe, Y. (eds.) Public-Key Cryptography - PKC 2022–25th IACR International Conference on Practice and Theory of Public-Key Cryptography, Virtual Event, Proceedings, Part I. Lecture Notes in Computer Science, vol. 13177, pp. 553–583. Springer (2022)
42. Lipmaa, H., Siim, J., Zajac, M.: Counting vampires: from univariate sumcheck to updatable ZK-SNARK. Cryptology ePrint Archive, Report 2022/406 (2022). https://eprint.iacr.org/2022/406
43. Maller, M., Bowe, S., Kohlweiss, M., Meiklejohn, S.: Sonic: Zero-knowledge SNARKs from linear-size universal and updatable structured reference strings. In: Cavallaro, L., Kinder, J., Wang, X., Katz, J. (eds.) ACM CCS 2019, pp. 2111–2128. ACM Press (2019). https://doi.org/10.1145/3319535.3339817
44. Morillo, P., Ràfols, C., Villar, J.L.: The kernel matrix Diffie-Hellman assumption. In: Cheon, J.H., Takagi, T. (eds.) ASIACRYPT 2016, Part I. LNCS, vol. 10031, pp. 729–758. Springer, Heidelberg (2016). https://doi.org/10.1007/978-3-662-53887-6_27
45. Ràfols, C., Silva, J.: QA-NIZK arguments of same opening for bilateral commitments. In: Nitaj, A., Youssef, A.M. (eds.) AFRICACRYPT 20. LNCS, vol. 12174, pp. 3–23. Springer, Heidelberg (2020). https://doi.org/10.1007/978-3-030-51938-4_1
46. Ràfols, C., Zapico, A.: An algebraic framework for universal and updatable SNARKs. In: Malkin, T., Peikert, C. (eds.) CRYPTO 2021, Part I. LNCS, vol. 12825, pp. 774–804. Springer, Heidelberg, Virtual Event (2021). https://doi.org/10.1007/978-3-030-84242-0_27

ParaDiSE: Efficient Threshold Authenticated Encryption in Fully Malicious Model

Shashank Agrawal[1], Wei Dai[2(✉)], Atul Luykx[3], Pratyay Mukherjee[4], and Peter Rindal[5]

[1] Coinbase, San Jose, USA
[2] Bain Capital Crypto, Foster, USA
me@wdai.us
[3] Google, Mountain View, USA
[4] SupraOracles, San Francisco, USA
[5] Visa Research, San Francisco, USA

Abstract. Threshold cryptographic algorithms achieve robustness against key and access compromise by distributing secret keys among multiple entities. Most prior work focuses on threshold public-key primitives, despite extensive use of authenticated encryption in practice. Though the latter can be deployed in a threshold manner using multi-party computation (MPC), doing so incurs a high communication cost. In contrast, dedicated constructions of threshold authenticated encryption algorithms can achieve high performance. However to date, few such algorithms are known, most notably DiSE (distributed symmetric encryption) by Agrawal et al. (ACM CCS 2018). To achieve *threshold authenticated encryption* (TAE), prior work does not suffice, due to shortcomings in definitions, analysis, and design, allowing for potentially insecure schemes, an undesirable similarity between encryption and decryption, and insufficient understanding of the impact of parameters due to lack of concrete analysis. In response, we revisit the problem of designing secure and efficient TAE schemes. (1) We give new TAE security definitions in the fully malicious setting addressing the aforementioned concerns. (2) We construct efficient schemes satisfying our definitions and perform concrete and more modular security analyses. (3) We conduct an extensive performance evaluation of our constructions, against prior ones.

Keywords: Threshold crypto · Authenticated encryption · Provable security

1 Introduction

Cryptography is increasingly deployed within and across organizations to secure valuable data and enforce authorization. Due to regulations such as GDPR and PCI, or via chip- and cryptocurrency-based payments, a monetary amount can

S. Agrawal, W. Dai and A. Luykx—Work done while at Visa Research.

T. Isobe and S. Sarkar (Eds.): INDOCRYPT 2022, LNCS 13774, pp. 26–51, 2022.
https://doi.org/10.1007/978-3-031-22912-1_2

be attached to the theft, loss, or misuse of cryptographic keys. Securing cryptographic keys is only becoming more important.

Underlining the significance of securing keys is the proliferation of trusted hardware on a wide range of devices, with manufacturers installing *secure enclaves* on mobile devices (Apple [4], Google [5]) and commodity processors (ARM [6], Intel SGX [2]). In addition, organizations increasingly use hardware security modules (HSMs) to generate and secure their cryptographic keys. However, such trusted hardware is expensive to build and deploy, often difficult to use, offers limited flexibility in supported operations, and can be difficult to secure at large scale—for example SGX attacks such as [38,39] or the recent HSM attack [17]. As a result, many seek to reduce reliance on trusted hardware.

Threshold cryptography considers a different, and complementary approach to using trusted hardware: instead of relying on the security of each individual device, a threshold number of devices must be compromised, thereby complicating attacks. Furthermore, as threshold cryptography can be deployed in software, it provides a method of securing keys that is cheaper, faster to deploy, and more flexible. It also provides a way to cryptographically enforce business policies that require a threshold number of stakeholders to sign off on decisions.

The benefits of threshold cryptography have caught the attention of practitioners. For example, the U.S. National Institute of Standards and Technology (NIST) has initiated an effort to standardize threshold cryptography [3]. Furthermore, an increasing number of commercial products use the technology, such as the data protection services offered by Vault [8], and Coinbase Custody [1], and the HSM replacements by Unbound Tech [7] etc.

Threshold symmetric encryption is used in many of the commercial products to protect stored data, generate tokens or randomness. The schemes used vary in sophistication, choosing different trade-offs between security, performance, and other deployment concerns.

1.1 Approaches to Threshold Symmetric Encryption

A naive way of deploying threshold symmetric encryption uses secret sharing. One takes an algorithm, for example the authenticated encryption (AE) scheme AES-GCM [50], and applies a secret sharing scheme to its key. The key shares are sent to different parties so that one has to contact a threshold number of parties to reconstruct the key, to then perform encryption or decryption. However, this approach requires *reconstructing* the key at some point, thereby nearly negating the benefits of splitting the secret among multiple parties in the first place.

Instead, a proper threshold cryptographic implementation of AES-GCM would not require key reconstruction—even while encrypting or decrypting. One could secret share the plaintext instead of the key, and send each share to different parties holding different keys. This avoids key reconstruction, but significantly increases communication and storage if applied to an AE scheme like AES-GCM.

Secure multi-party computation (MPC) also enables implementations of cryptographic algorithms such as AES-GCM in a way that keys remain split during operation. MPC works with any algorithm and therefore is used in applications where backwards compatibility is important, such as the drop-in replacement for HSMs discussed by Archer et al. [15]. However, MPC has a significant performance cost, often requiring multiple rounds of high bandwidth communication among the different parties. Keller et al. [37] present the best-known performance results: per 128 bit AES block, they need anywhere from 2.9 to 8.4 MB of communication and at least 10 communication rounds for two-party computation. For settings where backwards compatibility is not needed, using MPC-friendly ciphers such as MiMC [12] and LowMC [13] instead of AES can improve performance modestly.

Unlike MPC, Agrawal et al.'s threshold AE (TAE) schemes [11]—named *DiSE*—operate in two communication rounds and require anywhere from 32 to 148 bytes per encryption in the two party setting. Furthermore, the DiSE protocols output integrity-protected ciphertext like conventional authenticated encryption schemes and communication complexity does not change with message length. As a result, the DiSE protocols can outperform MPC implementations of authenticated encryption by orders of magnitude. DiSE's efficiency makes it a prime candidate for applications seeking threshold security, which motivates the further study of dedicated TAE schemes.

1.2 Revisiting Threshold Authenticated Encryption

Agrawal et al. [11] (hereafter AMMR) initiate the study of TAE schemes to achieve confidentiality and integrity in a threshold setting while ensuring the underlying master secret key remains distributed during encryption and decryption. Security is defined relative to the threshold t, which is one more than the number of malicious parties the protocols can tolerate. Confidentiality is defined as a CPA-like game where adversaries engage in encryption sessions and the goal is to break semantic security of a challenge ciphertext. Integrity is defined as "one-more ciphertext integrity" where an adversary must provide one more valid ciphertext than its "forgery budget".

We note the following three shortcomings of AMMR's formalization.

Confidentiality Does Not Prevent Key Reconstruction. We show that AMMR's confidentiality definition does not prevent participants from reconstructing master secrets. We give a counter-example scheme that satisfies both confidentiality and integrity as formalized by AMMR, yet allows adversaries to reconstruct the master encryption key, which can then be used to perform encryption without contacting other participants. (See [10] for details.) This is because AMMR's CPA-like confidentiality game does not let the adversary initiate decryption sessions, so schemes can disseminate secret keys during decryption. Therefore, a scheme that is proven secure under AMMR's confidentiality notion may not prevent key reconstruction. Their protocols, however, *do* prevent that.

Loose Notion of Integrity. Participants in the DiSE protocols cannot distinguish whether they are participating in an encryption or a decryption, and adversaries can generate a valid ciphertext while running a decryption session—something that, ideally, should only be possible during encryption. In practice, this can cause difficulties in logging or enforcing permissions and is generally an undesirable property. The fact that participants cannot distinguish encryption from decryption is not just a property of the DiSE protocols but allowed by AMMR's integrity definition (decryption sessions count towards the "forgery budget").

In addition, AMMR's integrity definition allows for "malleable" TAE schemes, where adversaries can participate in an encryption session, make an honest party output an invalid ciphertext, and then "patch" this ciphertext to obtain the correct ciphertext that the honest party should have output (see [10] for more details). As a result, ciphertext integrity is not maintained. In AMMR's integrity game (termed *authenticity* by the authors), ciphertexts generated by an honest party are *not* returned to the adversary which deprives it of having a ciphertext to patch in the first place.

Abstract and Non-concrete Treatment of Security. AMMR's definitional framework captures a wide class of protocols that could have parties arbitrarily interacting with each other. Although general, this approach complicates the formalization of adversarial power. Moreover, the security analyses of the DiSE protocols are asymptotic, providing little guidance on how to securely instantiate parameters in practice.

1.3 Contributions

In light of AMMR's shortcomings, we seek to advance the state-of-the-art in the formalization, design and analysis of TAE schemes. In addition to a concrete security analysis of our schemes, we give the first concrete analysis of distributed pseudorandom functions (DPRFs) and its verifiable extension DVRF, which might be of independent interest. Our analyses use a new modular technique via formalizing a variant of Matrix-DDH assumption [30], called *Tensor DDH*, which helps in a tighter security reduction to DDH than AMMR's.

New Definitions. We introduce new TAE security definitions which fix the aforementioned issues with AMMR's definitions. To do so, we depart from a more abstract description of TAE schemes, and instead only consider TAE schemes which operate in two communication rounds. We present IND-CCA-type definitions capturing confidentiality and integrity, and preventing key reconstruction. Our definitions require that protocols enable participants to distinguish encryption from decryption. Inspired by the definitions of the same name from the public-key literature [25], we present two definitions—CCA and RCCA ("Replayable" CCA)—which capture two different integrity guarantees: RCCA guarantees plaintext integrity, whereas CCA guarantees ciphertext integrity. We believe RCCA to be sufficient for many applications, and propose CCA for settings where ciphertext integrity is important. Note that, as we show below, achieving CCA security comes at a performance cost relative to RCCA security.

New Constructions. We present new constructions satisfying our security definitions. To achieve RCCA we present

1. an approach which departs from DiSE's design (Sect. 5), by using a type of *all-or-nothing transform* [16,46] in combination with forward and inverse block cipher calls during encryption and decryption, respectively, and
2. a new scheme inspired by DiSE combining DPRF and threshold signature.

To achieve CCA, we use a distributed *verifiable* PRF combined with threshold signatures. For a more detailed overview and a comparison among our constructions we refer to Sect. 2.

Performance Evaluation. We present an extensive performance study of our constructions (Appendix A), as well as a comparison with the DiSE protocols. In a three-party setting with threshold set to two, our RCCA-secure random injection-based construction achieves over 777,000 encryptions per second and a latency of 0.1 ms per encryption, and our CCA-secure construction achieves about 350 encryptions per second and a latency of 4 ms per encryption. Although these figures are about 0.7 times those of the comparable DiSE protocols, our constructions guarantee stronger security by satisfying RCCA and CCA notions.

By combining practical considerations, new theoretical design, and concrete analysis, we believe our TAE schemes—collectively named *ParaDiSE*—are sufficiently performant and secure for use in practice, while presenting interesting, novel designs of independent interest.

2 Technical Overview

2.1 Security Definitions

Fully Malicious Security Model. As with AMMR, in our model the adversary obtains the secret keys of corrupt parties and can act on their behalf during encryption or decryption. Moreover, the adversary can initiate the protocols via these corrupt parties and receive the output of the honest parties.

AMMR's message privacy definition does not give the adversary decryption capability, which is what allows for the counter example scheme discussed in [10]. In contrast, our model guarantees that even if the adversary can decrypt honestly generated ciphertext, it still cannot decrypt the challenge ciphertexts. Furthermore, AMMR's authenticity definition assumes the decryption protocol is executed honestly. Instead, we require authenticity even if the adversary deviates from the honest decryption protocol.

Capturing Privacy via the Decryption Criteria. Our threshold IND-RCCA and IND-CCA definitions follow the standard left-or-right indistinguishability model: the adversary submits message pairs (m_0, m_1) to obtain challenge encryptions of message m_b for a hidden bit b, and it breaks privacy if it can guess b. If the adversary asks for a challenge ciphertext and then honestly executes the decryption protocol, it can trivially guess b seeing if decryption returns m_0 or m_1; as with standard AE definitions, we need to prevent such "trivial wins". However, how to prevent such trivial wins in the threshold setting is much less clear. For example, we cannot block the adversary from initiating decryption protocols associated with challenge ciphertext c (the participating parties do not get access to the input of the initiating party). Nevertheless, our security model should capture when the adversary has effectively executed an honest decryption session. This is done via the notion of *decryption criteria*. Informally, the *decryption criteria*, $\mathsf{Eval\text{-}MSet}(c, \mathcal{CR})$, for ciphertext c and corrupt set of parties $\mathcal{CR}$, captures the exact set of messages that needs to be sent (and responded to) from the adversary, in order for the adversary to know the decryption of c.

Capturing Authenticity. In standard AE, authenticity is captured via INT-CTXT, which says that the only valid ciphertext that the adversary can generate is what it receives from the encryption oracle. Similar to privacy, complications arise when moving to our setting. First, the adversary could generate ciphertext by initiating encryption sessions itself. Furthermore, an outside observer (even while seeing all the protocol messages), might have no idea what messages the adversary is trying to encrypt. Hence, to exclude trivial wins, we will give a *forgery* budget to the adversary based on the amount of interactions it has with the honest parties. Second, the notion of valid ciphertext requires the running of a decryption session. But we allow the adversary to deviate arbitrarily in *any* execution of all protocols. This means that an adversary could potentially deviate from the protocol to make the decryption valid for some ciphertext c, while the honest decryption of c would return $\perp$. Note that the previous notion from DiSE completely bypasses this difficulty by *assuming* that decryption is executed honestly. We take a different approach. We first consider a relaxation of authenticity of ciphertext to authenticity of plaintext (analogous to INT-PTXT). We ask that valid decryptions to always decrypt to previous seen messages, even if the adversary deviate arbitrarily during decryption (IND-RCCA). This notion still admits fast symmetric-key based schemes. We also consider providing integrity of ciphertext, while having malicious decryption (IND-CCA). Our model is given in full detail in Sect. 4.3.

2.2 Constructions

We provide an overview for each of our three constructions. First we discuss the random injection-based construction which utilizes a recent construction of indifferentiable authenticated encryption [16]. Then, we describe the generic DPRF-based approach which builds on the DiSE protocol and yields two constructions. The first adds a threshold signature to ensure IND-RCCA. The second adds verifiability to the DPRF to ensure IND-CCA.

Random Injection Based Approach. This approach is based on symmetric-key primitives only and achieves our RCCA security notion. The key is distributed in a t-out-of-n replicated format. In particular, $d = \binom{n}{t-1}$ random keys are shared among the parties such that any t parties together have all d keys, but any strictly smaller subset of them would fall short of at least one key. Let us call this sharing scheme combinatorial secret sharing.

Our construction requires two primitives. A random injection $\mathsf{I} : \mathcal{X} \rightarrow \mathcal{Y}$ and inverse $\mathsf{I}^{-1} : \mathcal{Y} \rightarrow \mathcal{X} \cup \{\perp\}$. Intuitively, each call to $\mathsf{I}(x)$ outputs a uniformly random element from $\mathcal{Y}$ where $|\mathcal{Y}| \gg |\mathcal{X}|$. The inverse function has an authenticity property that it is computationally hard find a y such that $\mathsf{I}^{-1}(y) \neq \perp$ and y was not computed as $y = \mathsf{I}(x)$ for some x. The second primitive is a keyed pseudorandom permutation $\mathsf{PRP} : \mathsf{PRP.kl} \times \{0,1\}^k \rightarrow \{0,1\}^k$ and inverse PRP^{-1}.

Our construction has the encryptor compute $(y_1, y_2, \ldots, y_\ell) \leftarrow y \leftarrow\!\!\$\ \mathsf{I}(m)$ where $y_i \in \{0,1\}^k$ and $\ell \geq d$. The encryptor chooses a set of $t-1$ other parties and computes $e_1 = \mathsf{PRP}_{k_1}(y_1), \ldots, e_d = \mathsf{PRP}_{k_d}(y_d)$ by sending y_i to one of them which possesses k_i. This party returns $e_i = \mathsf{PRP}_{k_1}(y_i)$ back to the encryptor. Importantly, the encryptor sends y_i only to a party that knows k_i. The final ciphertext is defined to be $c = (e_1, \ldots e_d, y_{d+1}, \ldots, y_\ell) \in \{0,1\}^{k\ell}$. During decryption, the decryptor computes all the $y_1 = \mathsf{PRP}^{-1}_{k_1}(e_1), \ldots, y_d = \mathsf{PRP}^{-1}_{k_d}(e_d)$ again by interacting with any other $t-1$ helpers in a similar fashion and then locally computes $m = \mathsf{I}^{-1}(y_1, \ldots, y_\ell)$ which could be $\perp$ if decryption fails.

The security of this scheme crucially builds on the hiding and authenticity properties of random injection. Suppose we have a ciphertext $c = (e_1, \ldots, e_d, y_{d+1}, \ldots, y_\ell)$ of either message m_0 or m_1. Since the adversary only gets to corrupt $t-1$ parties, there must be at least one key k_i that it does not know. Hence, it can only compute PRP_{k_i} via interaction with honest parties that holds this key. Connecting this to our security definition, the decryption criteria for ciphertext c is $\mathsf{Eval\text{-}MSet}(c, \mathcal{CR}) = \{(i, e_i)\}$—meaning that the adversary can trivially decrypt c if it queries e_i to some other honest parties holding key k_i. Hence, assuming that this does not happen, then the adversary have no information about $\mathsf{PRP}^{-1}(e_i)$. And, by the property of I^{-1}, the adversary should gain no information about the original message.

From a high-level, authenticity requires that given (m, y, c) computed as above (recall $y = (y_1, \ldots, y_\ell)$), one can not come up with another triple (m', y', c') without executing a legitimate encryption instance. Let us explore the options for a forgery attack: first if $m' \neq m$ then computing $y' \leftarrow\!\!\$\ \mathsf{I}(m')$ would completely change y' due to the property of I and hence either the attacker needs to predict the PRP outputs, which is hard, or it queries the honest parties for these $\mathsf{PRP}_{k_i}(y'_i)$ values, which we capture via the forgery budget. Also, keeping $m' = m$ and trying with another correctly generated $y^\star$ (since I is randomized) would not help for the same reason. Another forgery strategy could be to mix and match between several y's for which the PRP values are known from prior queries. This is where the property of I comes into play—since the ideal functionality of I generate a uniformly random output each time for any input, there cannot be should not be any collisions among any of the k-bit blocks. In other words, each y_i and e_i value corresponds to at most one message m.

A crucial property of this scheme is a clear distinction between encryption and decryption queries which, in particular, strengthen the security compared to DiSE. During encryption, the (strong) PRP is used in the forward direction while decryption corresponds to an inverse PRP operation. It is then a relatively standard task to prove that mixing these operations would not aid the adversary either in constructing or inverting the I.

DPRF-Based Approach I: Achieving IND-RCCA via Threshold Signature. Our first DPRF-based construction is based on the DiSE construction. We first briefly recall their construction. The DiSE constructions use a *distributed pseudorandom function* (DPRF) to force those encrypting or decrypting to communicate with sufficiently many participants. When encrypting a message m, a commitment $\alpha = \mathsf{Com}(m; r)$ is generated for m with commitment randomness r. The DPRF output, $\beta \leftarrow \mathsf{DP}(j\|\alpha)$, is used as an encryption key to encrypt (m, r) into ciphertext $c \leftarrow \mathsf{enc}_\beta(m\|r)$ with a symmetric-key encryption scheme. When decrypting a ciphertext (α, c), one must recompute $\beta \leftarrow \mathsf{DP}(j\|\alpha)$ and then compute $m \leftarrow \mathsf{dec}_\beta(c)$. $\mathsf{enc}, \mathsf{dec}$ here is a one-time secure encryption scheme. Note that in DiSE, the interactive part of both encryption and decryption protocols are exactly the same, which is just a DPRF query on (j, α).

We want the protocol to reject decryption of ciphertext that was not legitimately generated. We ensure this using a threshold signature scheme. In particular, while computing a $\beta \leftarrow \mathsf{DP}(j\|\alpha)$ during encryption, one also gets a threshold signature σ on $j\|\alpha$. During decryption, each party verifies the signature before responding, and aborts if the verification fails.

Even though adding a signature shouldn't affect privacy, we are proving security against a much stronger model than DiSE. In particular, we need to ensure that privacy is ensured even when the adversary can interact arbitrarily with other honest parties as well as deviate from the honest protocol. Intuitively, we need to ensure that the ability to initiate adversarial decryption sessions and the ability to initiate decryption sessions from honest parties with malicious responses, do not give the adversary extra information about any challenge ciphertext. This is done via a sequence of game hops that "trivialize" the corresponding oracles. Recall that in the standard setting, the decryption oracle can be "trivialized" by simply returning $\perp$ for all ciphertexts that were not seen before. In our setting, we need to be much more careful. First, we move to a game where α corresponds to at most one message m. For adversarially started sessions, we compare the number of valid ciphertexts generated against the *forgery budget*, $\lfloor \mathsf{ct}_{\mathsf{Eval}}/(n - |\mathcal{CR}|)\rfloor$. By the unforgeability of threshold signature, the adversary would need to contact the gap number of parties, $n - |\mathcal{CR}|$, to obtain a fresh signature. Hence, we can ensure integrity for maliciously generated ciphertext. But note that we only offer the integrity of $j\|\alpha$, and in extension, the underlying message.

For ciphertexts generated by sessions initiated by honest parties, we can ensure that the number of valid signatures is the same as the number of sessions that were executed. Moreover, the signature actually offers integrity of the value of $j\|\alpha$, hence we can achieve the notion of IND-RCCA. Using the above, we can "trivialize" the decryption oracle. The rest of the proof is more straight forward. The full detailed proof is given in [10]. Setting up for our next construction, we demonstrate why this scheme cannot achieve IND-CCA. Suppose an adversary starts an encryption session and scrambles its DPRF responses so that the honest party derives some β' to encrypt with. After learning β', the adversary would have gained enough information to know also the *correct* β if everything was executed honestly (this is because of the key-homomorphic properties [23] of the DPRFs). Hence, it can change the ciphertext $c' = \mathsf{enc}_{\beta'}(m\|r)$ to $c = \mathsf{enc}_{\beta}(m\|r)$, which will decrypt correctly. Whereas for IND-CCA, we would need to ensure that the generated ciphertext c' is the valid one.

DPRF Based Approach II: Achieving IND-CCA using DVRF. We take a natural approach to prevent the aforementioned IND-CCA attack by adding a verifiability feature to the above construction. This is achieved by adding a simple and efficient NIZK proof to each partial evaluation, similar to the *strongly-secure* DDH-based DiSE construction. In particular, this prevents an adversary from sending a wrong partial evaluation without breaching the soundness of NIZK proofs, rendering the above attack infeasible. Notably, our formalization of DVRF differs from that of AMMR's and instead follows a recent formalization of Galindo et al. [33]. This allows us to modularize the proof: just by using a DVRF instead of DPRF in the above TAE construction, we upgrade the IND-RCCA secure scheme to an IND-CCA secure one. So, the main task of proving IND-CCA security of the upgraded construction boils down to arguing against CCA *only* attacks like above which are not RCCA.

A New Analysis of DPRF. Finally, we provide a new simpler, modular and tighter analysis of the DDH-based DPRF construction of Naor et al. [45]. Our analysis uses a new variant of Matrix-DDH [30] assumption, that we call Tensor-DDH assumption. This assumption captures the essence of the adversarial view of the DPRF security game (pseudorandomness) into an algebraic framework consisting of tensor products of two secret vectors (provided in the exponent). We show that irrespective of the dimensions of the vectors, this assumption is as hard as DDH with a minimal security loss of a factor 2. In a similar reduction step, AMMR's proof looses a factor that scales with the number of evaluation and challenge queries. Our analysis provides better concrete security.

2.3 Related Work

Threshold Cryptography. Starting with the work of Desmedt [26], most work on threshold cryptography has focused on public-key encryption and digital signatures [22,27]. Starting from Micali and Sidney [42], some work has focused on threshold and distributed PRFs [23,28,45]. However, the pseudo-randomness requirements do not explicitly take into account several avenues of attacks. Agrawal et al. [11] propose stronger notions for distributed DPRFs and build on the constructions of Naor et al. [45] to achieve them.

Threshold Oblivious PRF. Another related notion is distributed/threshold oblivious PRF (TOPRF) [32], which can be thought of as a DPRF, but with an additional requirement of hiding input from the servers. This requirement makes

TOPRF a stronger primitive, which is known to imply oblivious transfer [36]. Furthermore, despite the structural similarities between the DDH-based TOPRF [35] and the DDH-based DVRF [35] we used here, the TOPRF is proven assuming interactive variant of DDH, in contrast to the DVRF which can be reduced to DDH. Therefore, the proof techniques are also quite different.

Authenticated Encryption. Authenticated encryption (AE) has seen a significant amount research since being identified as a primitive worthy of study in its own right [19]. AE research has increasingly addressed practical concerns from a performance, security, and usability point-of-view. AE schemes evolved from a generic composition of encryption and authentication schemes [19] to dedicated schemes like GCM [41]. Initially the description of AE schemes did not match with how they were used in practice, leading to the introduction of nonces [48] and associated data [47]. Further security concerns with how AE schemes are used in practice lead to formalization of different settings and properties, such as varying degrees of nonce-misuse resistance and deterministic AE [49], blockwise adaptive security [31], online authenticated encryption [18], leakage concerns such as unverified plaintext and robustness against it [14,34], and multi-user security [21]. Recently, Barbosa and Farshim proposed indifferentiable authenticated encryption [16], which has many of the properties of the all-or-nothing transform introduced by Rivest [46]. In Sect. 5.1 we discuss how we use these primitives to create one of our TAE schemes.

Multi-party Computation. MPC allows a set of mutually distrustful parties to evaluate a joint function of their inputs without revealing anything more than the function's output. The last 10–15 years have seen tremendous progress in bringing MPC closer to practice.

General-purpose MPC protocols can help parties evaluate any function of their choice but they work with a circuit representation of the function, which leads to a large communication/computation complexity—typically, at least linear in the size of the circuit and the number of parties. Moreover, the parties have to engage in several rounds of communication, with every party talking to every other. (Some recent results reduce the round complexity but have substantially higher computational overhead.) So, general-purpose MPC protocols are not ideal for making a standard AE scheme like AES-GCM distributed. Nonetheless, such a solution would be fully compatible with the standards.

Adaptive DiSE. A recent work [43] defined and constructed stronger TAE schemes that are secure against *adaptive* corruption, in contrast we only focus on static corruption. Nevertheless, the definitions considered in that work is based on the AMMR definitions and hence suffers from similar limitations. Our work can be perceived as orthogonal to that. Augmenting our definitional framework into adaptive setting and constructing secure scheme therein may be an interesting future work.

3 Preliminaries

For a positive integer n, let $[n]$ denote the set $\{1, \ldots, n\}$. For a finite set S, we use $x \leftarrow\!\!\$\; S$ to denote the process of sampling an element uniformly from S and assigning it to x. We $\uplus$ to denote the union of multi-sets.

Let $\mathcal{A}$ be a randomized algorithm. We use $y \leftarrow\!\!\$\; \mathcal{A}(x_1, x_2, \ldots)$ to denote running $\mathcal{A}$ with inputs $x_1, x_2, \ldots$ and assigning its output to variable y. We use the notation $[\![\mathsf{sk}]\!]_n$ to denote the tuple $(\mathsf{sk}_1, \mathsf{sk}_2, \ldots, \mathsf{sk}_n)$. We assume that variables for strings, sets, numbers are initialized to the empty string, the empty set, and zero, respectively. We identify 1 with TRUE and 0 with FALSE.

We borrow some notation from Agrawal et al. [11]. We use $[j : x]$ to denote that the value x is private to party j. For a protocol Π, we write $(c', [j : z']) \leftarrow \Pi(c, [\![\mathbf{k}]\!]_n, [i : (x, y)], [j : z])$ to denote that all parties have a common input c, party ℓ has private input k_ℓ (for all $\ell \in [n]$, this is usually the secret key), party i has two private inputs x and y, and party j has one private input z. After the execution, all parties receive an output c' and j additionally receives z'.

Security Games. Every security game is defined with respect to a protocol Π and an adversary $\mathcal{A}$. Adversary $\mathcal{A}$ gets access to several procedures in the game. When Π is a threshold protocol, we assume that $\mathcal{A}$ consists of two stages $(\mathcal{A}_0, \mathcal{A}_1)$. The first stage adversary $\mathcal{A}_0$ takes input (pp, n, t) and produces some set $C \subset [n]$ with $|C| < t$. The second stage adversary $\mathcal{A}_1$ receives the list of secret keys for parties in C, i.e. $(\mathsf{sk}_i)_{i \in C}$, and access to the procedures defined by the security game. We write $\mathcal{A}_1^{\langle \mathbf{Proc} \rangle}$ to denote the execution of $\mathcal{A}_1$ where $\mathcal{A}_1$ has access to all the available game oracles.

For a game $\mathbf{G}$ with a protocol Π and an adversary $\mathcal{A}$, we use $\Pr *\mathbf{G}_\Pi(\mathcal{A})$ to denote the probability that $\mathbf{G}$ outputs 1. Throughout the paper, n denotes the total number of parties and t the threshold. We define $\Delta_{\mathcal{A}}(O_1\, ;\, O_2) := |\mathbf{P}[\mathcal{A}^{O_1} = 1] - \mathbf{P}[\mathcal{A}^{O_2} = 1]|$, where $\mathcal{A}^O$ denotes $\mathcal{A}$'s output after interacting with oracle O.

Threshold Signature. A threshold signature scheme allows for a signing key to be secret shared among n parties such that any t parties can collectively generate a signature. On a common message m, t parties call $\mathsf{PartSign}(\mathsf{sk}_i, m) \rightarrow \sigma_i$. The σ_i can then be collected and used to produce a signature $\mathsf{CombSig}(\mathsf{vk}, \{(i, \sigma_i)\}_{i \in S}) \rightarrow \sigma$. Signature verification is non-interactive and is performed as $\mathsf{VerSig}(\mathsf{vk}, m, \sigma)$.
For a detailed definition and security properties see [10].

4 Threshold Authenticated Encryption

4.1 Syntax

A threshold authenticated encryption scheme TAE consists of a setup algorithm and protocols for encryption and decryption. Throughout, we let H denote a random oracle that the algorithms can use. Parameter n denotes the number of parties involved in the protocol and parameter t denotes the *threshold* of the protocol. We use the shorthand $\mathcal{P}_i.\mathsf{E}(x, y, \ldots)$ to refer to party i running some algorithm $\mathsf{E}(\mathsf{pp}, \mathsf{sk}_i, \cdot)$ and returning the result.

Setup: takes integers n, t with $1 \leq t \leq n$, and generates n secret key shares $\mathsf{sk}_1, \ldots, \mathsf{sk}_n$ and public parameters pp, denoted $([\![\mathsf{sk}]\!]_n, \mathsf{pp}) \leftarrow\$ \mathsf{Setup}(n, t)$.

Encryption: a 2-round protocol, consisting of three algorithms ($\mathsf{Split}_{\mathsf{enc}}$, $\mathsf{Eval}_{\mathsf{enc}}$, $\mathsf{Combine}_{\mathsf{enc}}$). For an initiating party $j \in [n]$, input m, and set $S \subseteq [n]$, the Enc protocol is executed as follows:

Protocol $\mathsf{Enc}(\mathsf{pp}, [\![\mathsf{sk}]\!]_n, [j : m, S])$

$(L, \mathsf{st}) \leftarrow\$ \mathcal{P}_j.\mathsf{Split}^{\mathrm{H}}_{\mathsf{enc}}(m, S)$
For $(i, x) \in L$ do $r_{i,x} \leftarrow\$ \mathcal{P}_i.\mathsf{Eval}^{\mathrm{H}}_{\mathsf{enc}}(j, x)$
$c \leftarrow\$ \mathsf{Combine}^{\mathrm{H}}_{\mathsf{enc}}(\{r_{i,x}\}_{(i,x)\in L}, \mathsf{st})$
Return c

Protocol $\mathsf{Dec}(\mathsf{pp}, [\![\mathsf{sk}]\!]_n, [j : c, S])$

$(L, \mathsf{st}) \leftarrow \mathcal{P}_j.\mathsf{Split}^{\mathrm{H}}_{\mathsf{dec}}(c, S)$
For $(i, x) \in L$ do $r_{i,x} \leftarrow \mathcal{P}_i.\mathsf{Eval}^{\mathrm{H}}_{\mathsf{dec}}(j, x)$
$m \leftarrow \mathsf{Combine}^{\mathrm{H}}_{\mathsf{dec}}(\{r_{i,x}\}_{(i,x)\in L}, \mathsf{st})$
Return m

The list L (i, x), indicating that message x should be sent to party i for evaluation. Since we assume all parties communicate over authenticated channels, receivers will know the identity of the sender, hence the sending party index j is an input to the evaluation for each receiving party i.

We assume that S always contains the index j. We note that the size of the set L output by $\mathsf{Split}_{\mathsf{enc}}$ (and by $\mathsf{Split}_{\mathsf{dec}}$ below) does not have to match the size of S. This allows multiple messages to be sent to the same party for evaluation.

Decryption: a 2-round protocol, consisting of three deterministic algorithms ($\mathsf{Split}_{\mathsf{dec}}$, $\mathsf{Eval}_{\mathsf{dec}}$, $\mathsf{Combine}_{\mathsf{dec}}$). For an initiating party $j \in [n]$, input c, and set $S \subseteq [n]$, the Dec protocol is executed shown above. We define the finite sets X, Y as $\mathcal{P}_i.\mathsf{Eval}_{\mathsf{dec}} : X \to Y$.

Basic Correctness. We say that TAE satisfies basic correctness if for all positive integers n, t such that $t \leq n$, all $([\![\mathsf{sk}]\!]_n, \mathsf{pp}) \leftarrow\$ \mathsf{Setup}(n, t)$, any $m \in \{0, 1\}^*$, any $S, S' \subseteq [n]$ with $|S|, |S'| \geq t$, and any $i \in S$, $j \in S'$, we have that $m = m'$ where

$$c \leftarrow\$ \mathsf{Enc}(\mathsf{pp}, [\![\mathsf{sk}]\!]_n, [i : m, S]) \,, m' \leftarrow \mathsf{Dec}(\mathsf{pp}, [\![\mathsf{sk}]\!]_n, [j : c, S']) \,.$$

4.2 Decryption Criteria

As is common with CCA-style security games, our games need to prevent the "trivial win" where an adversary decrypts a challenge ciphertext simply by executing decryption as specified by the TAE scheme. What complicates preventing such trivial wins in our setting is the fact that there are many ways to decrypt since basic correctness requires that any group of t out of n parties may decrypt. Furthermore, not only can adversaries corrupt up to $t - 1$ parties and recover their secret keys, but they can also ask honest parties to run $\mathsf{Eval}_{\mathsf{dec}}$.

Given a ciphertext, our goal is to detect when an adversary has collected $\mathsf{Eval}_{\mathsf{dec}}$ output from at least t parties, either through an $\mathsf{Eval}_{\mathsf{dec}}$ query to an honest party or via a corrupted party's key, so that it could decrypt the ciphertext as specified by the protocol. If we can detect such events, then we can rule out trivial decryptions of challenge ciphertexts.

Let $([\![\mathsf{sk}]\!]_n, \mathsf{pp}) \leftarrow\!\!\$\ \mathsf{Setup}(n, t)$, let $S \subseteq [n]$ be a set of parties of size at least t, and $j \in S$ an initiator. When decrypting a ciphertext c, the initiator $\mathcal{P}_j$ first splits c into inputs for the other parties in S: $(L, \mathsf{st}) \leftarrow \mathcal{P}_j.\mathsf{Split}_{\mathsf{dec}}(c, S)$. Then, $\mathcal{P}_j$ sends x to $\mathcal{P}_i$ for evaluation for every $(i, x) \in L$.

Although $\mathsf{Split}_{\mathsf{dec}}(c, S)$ might assign x to $\mathcal{P}_i$, depending on the TAE scheme, there might be other parties with the key material to evaluate x. An adversary $\mathcal{A}$ can ask any party—including corrupt ones—to evaluate x, and is not restricted to the one prescribed in L. Although the corrupt parties can evaluate some of the x in L, a secure TAE scheme will require the adversary to interact with honest parties to evaluate what it cannot. If these x's are queried to honest parties, then $\mathcal{A}$ has enough information to execute the decryption protocol on c.

Without further details about how a given scheme TAE works, the only way to know whether TAE allows $\mathcal{P}_i$ to evaluate x, is to find a set S with $i \in S$ such that $(i, x) \in \mathcal{P}_j.\mathsf{Split}_{\mathsf{dec}}(c, S)$. Instead, we approach this as follows:

1. We require that for all S and $j \in S$, $\mathcal{P}_j.\mathsf{Split}^{\mathrm{H}}_{\mathsf{dec}}(c, S)$ always outputs the same multiset of messages, $\mathsf{Eval\text{-}MSet}^{\mathrm{H}}(c) := \{x\}_{(i,x) \in L}$. In other words, although the assignment to parties could change with S, the set of messages x and their multiplicity stays the same.
2. We assume that a party i can evaluate x if its execution of $\mathsf{Eval}_{\mathsf{dec}}$ produces *any* valid output. Formally, we define a relation $\mathcal{R}^{\mathrm{H}} \subseteq [n] \times X$ where $(i, x) \in \mathcal{R}^{\mathrm{H}} \iff \mathcal{P}_i.\mathsf{Eval}^{\mathrm{H}}_{\mathsf{dec}}(j, x) \neq \bot$.

With the above two assumptions, given a ciphertext c and a party i, we can determine the values that $\mathcal{P}_i$ can evaluate:

$$\left\{x \in \mathsf{Eval\text{-}MSet}^{\mathrm{H}}(c) \mid \mathcal{R}^{\mathrm{H}}(i, x)\right\}. \tag{1}$$

We are now ready to define the *gap* set of messages, i.e. the messages an attacker cannot evaluate on its own. For a ciphertext c and a set of corrupt parties $\mathcal{CR}$,

$$\mathsf{Eval\text{-}MSet}^{\mathrm{H}}(c, \mathcal{CR}) = \mathsf{Eval\text{-}MSet}^{\mathrm{H}}(c) \setminus \left(\biguplus_{i \in \mathcal{CR}} \left\{x \in \mathsf{Eval\text{-}MSet}^{\mathrm{H}}(c) \mid \mathcal{R}^{\mathrm{H}}(i, x)\right\}\right), \tag{2}$$

where $\biguplus$ and $\setminus$ denote union and set-difference over multisets.

Let us take a simple example. Suppose we have a threshold of 3, $S = \{2, 5, 7\}$, and $\mathcal{P}_j.\mathsf{Split}^{\mathrm{H}}_{\mathsf{dec}}(c, S)$ outputs $L = \{(2, \text{“m1”}), (5, \text{“m1”}), (7, \text{“m1”}), (2, \text{“m2”}), (5, \text{“m3”}), (7, \text{“m3”})\}$. Let us consider a rather peculiar $\mathcal{P}_i.\mathsf{Eval}^{\mathrm{H}}_{\mathsf{dec}}(j, x)$ function which has non-$\bot$ output if $x \in \{\text{“m1”}, \text{“m2”}, \ldots, \text{“m}i\text{”}\}$. This in turn similarly defines $\mathcal{R}^{\mathrm{H}}$. Suppose $\mathcal{A}$ corrupts parties $\mathcal{CR} = \{1, 2\}$.

First of all, $\mathsf{Eval\text{-}MSet}^{\mathrm{H}}(c) = \{\text{“m1”}, \text{“m1”}, \text{“m1”}, \text{“m2”}, \text{“m3”}, \text{“m3”}\}$. Party $\mathcal{P}_i$ with $i = 2$ can evaluate $\{x \in \mathsf{Eval\text{-}MSet}^{\mathrm{H}}(c) \mid \mathcal{R}^{\mathrm{H}}(i, x)\} = \{\text{“m1”}\}$ and $i = 2$ can evaluate $\{\text{“m1”}, \text{“m2”}\}$. Thus, the set of messages that can not be evaluated by the corrupt parties is $\mathsf{Eval\text{-}MSet}^{\mathrm{H}}(c, \mathcal{CR}) = \{\text{“m1”}, \text{“m3”}, \text{“m3”}\}$ as per Eq. 2.

$\mathcal{A}$ could get a "m1" message evaluated by $\mathcal{P}_1$ even though L does not prescribe to do so. On the other hand, L indicates that "m1" messages need to be evaluated by three different parties (under their respective keys) while $\mathcal{A}$ has only two under its control. Moreover, none of $\mathcal{P}_1$, $\mathcal{P}_2$ can evaluate "m3". Thus, one can see that $\mathsf{Eval\text{-}MSet}^{\mathrm{H}}(c, \mathcal{CR})$ captures the messages $\mathcal{A}$ cannot process on its own.

4.3 Security

We give two security notions for TAE. First is the IND-RCCA notion, which mirrors the IND-RCCA security for PKE. This notion is relaxed from the standard IND-CCA by targeting the integrity of *plaintext*. Second is the IND-CCA notion, which mirrors the standard IND-CCA notion for standard PKE.

Game $\mathbf{G}^{\mathrm{ind-cca}}_{\mathsf{TAE},n,t}(\mathcal{A})$

$b \leftarrow_\$ \{0,1\}$
$([\![\mathsf{sk}]\!], \mathsf{pp}) \leftarrow_\$ \mathsf{TAE.Setup}(n, t)$
$(\mathcal{CR}, \mathsf{st}_\mathcal{A}) \leftarrow_\$ \mathcal{A}_0(\mathsf{pp})$
$b' \leftarrow_\$ \mathcal{A}_1^{\langle \mathbf{Proc} \rangle}(\mathsf{st}_\mathcal{A}, (\mathsf{sk}_k)_{k \in \mathcal{CR}})$
If $|\mathsf{DecSet}| > \left\lfloor \frac{\mathsf{ct}_{\mathsf{Eval}}}{t - |\mathcal{CR}|} \right\rfloor$ then $\mathsf{forgery} \leftarrow \mathrm{TRUE}$
If $(\exists c \in \mathsf{ChlCtxt} : \mathsf{Eval\text{-}MSet}(c, \mathcal{CR}) \subseteq Q_{\mathsf{dec}})$
  Return $(b'' \leftarrow_\$ \{0,1\}) \vee \mathsf{forgery}$
Return $(b = b') \vee \mathsf{forgery}$

Challenge encryption sessions

Proc $\mathrm{SPLIT}_{\mathsf{enc}}(\mathsf{id}, m_0, m_1, S)$

Require $\mathsf{id} \notin \mathcal{CR}$, $|m_0| = |m_1|$
$u \leftarrow u + 1$; $\mathsf{id}_u \leftarrow \mathsf{id}$
$m_{u,0} \leftarrow m_0$; $m_{u,1} \leftarrow m_1$
$(L_u, \mathsf{st}_u) \leftarrow_\$ \mathsf{Split}_{\mathsf{enc}}(\mathsf{sk}_{\mathsf{id}}, m_{u,b}, S)$
Return $\{(k, x) \in L_u \mid k \in \mathcal{CR}\}$

Proc $\mathrm{COMBINE}_{\mathsf{enc}}(u, \mathsf{rsp})$

For $(k, x) \in L_u$ with $k \notin \mathcal{CR}$ do
  $\mathsf{rsp}[(k, x)] \leftarrow_\$ \mathsf{Eval}_{\mathsf{enc}}(\mathsf{sk}_k, \mathsf{id}_u, x)$
$c_u \leftarrow_\$ \mathsf{Combine}_{\mathsf{enc}}(\mathsf{rsp}[L_u], \mathsf{st}_u)$
If $m_{u,0} = m_{u,1}$ then
  $\mathsf{EncCtxt} \leftarrow \mathsf{EncCtxt} \cup \{c_u\}$
  <u>RCCA</u>: $\mathsf{EncMsg} \leftarrow \mathsf{EncMsg} \cup \{m_{u,0}\}$
Else
  $\mathsf{ChlCtxt} \leftarrow \mathsf{ChlCtxt} \cup \{c_u\}$
  <u>RCCA</u>: $\mathsf{ChlMsg} \leftarrow \mathsf{ChlMsg} \cup \{m_{u,0}, m_{u,1}\}$
Return c_u

Sessions initiated by adversary

Proc $\mathrm{EVAL}_{\mathsf{enc}}(\mathsf{eid}, \mathsf{id}, x)$

Require $\mathsf{eid} \notin \mathcal{CR}$, $\mathsf{id} \in \mathcal{CR}$
$\mathsf{ct}_{\mathsf{Eval}} \leftarrow \mathsf{ct}_{\mathsf{Eval}} + 1$
Return $\mathsf{Eval}_{\mathsf{enc}}(\mathsf{sk}_{\mathsf{eid}}, \mathsf{id}, x)$

Proc $\mathrm{EVAL}_{\mathsf{dec}}(\mathsf{eid}, \mathsf{id}, x)$

Require $\mathsf{eid} \notin \mathcal{CR}$, $\mathsf{id} \in \mathcal{CR}$
$Q_{\mathsf{dec}} \leftarrow Q_{\mathsf{dec}} \uplus \{x\}$
Return $\mathsf{Eval}_{\mathsf{dec}}(\mathsf{sk}_{\mathsf{eid}}, \mathsf{id}, x)$

Decryption sessions

Proc $\mathrm{SPLIT}_{\mathsf{dec}}(\mathsf{id}, c, S)$

Require $\mathsf{id} \notin \mathcal{CR}$
<u>CCA</u>: Require $c \notin \mathsf{ChlCtxt}$
$v \leftarrow v + 1$; $\mathsf{id}_v \leftarrow \mathsf{id}$; $c_v \leftarrow c$
$(L_v, \mathsf{st}_v) \leftarrow \mathsf{Split}_{\mathsf{dec}}(\mathsf{sk}_{\mathsf{id}}, c, S)$
Return $\{(k, x) \in L_v \mid k \in \mathcal{CR}\}$

Proc $\mathrm{COMBINE}_{\mathsf{dec}}(v, \mathsf{rsp})$

For $(k, x) \in L_v$ with $k \notin \mathcal{CR}$ do
  $\mathsf{rsp}[(k, x)] \leftarrow_\$ \mathsf{Eval}_{\mathsf{dec}}(\mathsf{sk}_k, \mathsf{id}_v, x)$
$m_v \leftarrow_\$ \mathsf{Combine}_{\mathsf{dec}}(\mathsf{rsp}[L_v], \mathsf{st}_v)$
<u>RCCA</u>: Require $m_v \notin \mathsf{ChlMsg}$
<u>RCCA</u>: $\mathsf{fresh} \leftarrow (m_v \notin \mathsf{EncMsg})$
<u>CCA</u>: $\mathsf{fresh} \leftarrow (c_v \notin \mathsf{EncCtxt})$
If $m_v \neq \perp$ and fresh then
  <u>RCCA</u>: $\mathsf{DecSet} \leftarrow \mathsf{DecSet} \cup \{m_v\}$
  <u>CCA</u>: $\mathsf{DecSet} \leftarrow \mathsf{DecSet} \cup \{c_v\}$
Return m_v

Fig. 1. IND-RCCA & IND-CCA games for Threshold Authenticated Encryption.

Consider the security game $\mathbf{G}^{\text{ind-rcca}}_{\mathsf{TAE},n,t}$, given in Fig. 1. The goal of the adversary is to either predict the bit b, or generate enough valid decryptions so that the flag $\mathsf{forgery}$ is set to TRUE. Several global variables keep track of the winning condition of the adversary: $\mathsf{ct}_{\mathsf{Eval}}$, $\mathsf{EncCtxt}$, $\mathsf{ChlCtxt}$, $\mathsf{DecCtxt}$, and Q_{dec}.

- The counter $\mathsf{ct}_{\mathsf{Eval}}$ counts the number of times $\mathsf{Eval}_{\mathsf{enc}}$ is called. Note that calling $\mathsf{Eval}_{\mathsf{enc}}$ on honest parties helps the adversary to generate ciphertexts.
- The set $\mathsf{EncCtxt}$ contains the ciphertexts generated by challenge encryption processes where $m_0 = m_1$ which the adversary obtained via GETCTXT.
- The set $\mathsf{ChlCtxt}$ contains the ciphertexts generated by challenge encryption processes where $m_0 \neq m_1$ which the adversary obtained via GETCTXT.
- The set EncMsg contains the set of all encrypted messages (including the challenge messages). This is only used in the RCCA game.
- The set DecSet contains the *valid* and *fresh* decryptions that the adversary can generate. Valid means that the ciphertext c must decrypt correctly. For IND-RCCA, fresh means that ciphertext c decrypts to message $m \neq \perp$ than was not in the set EncMsg. For IND-CCA, fresh means that ciphertext c is not a ciphertext in either $\mathsf{EncCtxt}$ or $\mathsf{ChlCtxt}$.
- Q_{dec} is a *multiset* containing all the queries to $\mathsf{Eval}_{\mathsf{dec}}$ made by the adversary.

We say that the adversary trivially breaks privacy if

$$\exists c \in \mathsf{ChlCtxt} : \mathsf{Eval\text{-}MSet}(c, C) \subseteq Q_{\mathsf{dec}} \,. \tag{3}$$

We also refer to the above check as the *privacy condition*. Intuitively, this captures when the adversary has enough information, through calling $\mathsf{Eval}_{\mathsf{dec}}$, to decrypt a challenge ciphertext c. In this case, the adversary is essentially guaranteed to be able to predict the bit b by correctness of the encryption scheme. Similarly, for authenticity, we need to set aside a budget for the amount of ciphertexts that the adversary can generate herself via invoking $\mathsf{Eval}_{\mathsf{enc}}$ of honest parties. This is captured by the following *authenticity condition*

$$|\mathsf{DecCtxt}| > \left\lfloor \frac{\mathsf{ct}_{\mathsf{Eval}}}{n - |\mathcal{CR}|} \right\rfloor \,. \tag{4}$$

We say that $\mathcal{A}$ breaks authenticity if the above line is true (which will set $\mathsf{forgery}$ to TRUE in the game). We now describe the interfaces exposed to $\mathcal{A}_1$.

Sessions Initiated by Adversary. The adversary can initiate encryption and decryption sessions from corrupt parties. These are achieved via calling $\mathsf{Eval}_{\mathsf{enc}}$ and $\mathsf{Eval}_{\mathsf{dec}}$ on honest parties. Note that the adversary can run Split and $\mathsf{Combine}$ itself. Counters $\mathsf{ct}_{\mathsf{Eval}}$ is incremented every time $\mathsf{Eval}_{\mathsf{enc}}$ is invoked. Multiset Q_{dec} contains the list of all queries made to $\mathsf{Eval}_{\mathsf{dec}}$ (with $\mathsf{id}, \mathsf{eid}$ removed).

Challenge Encryptions. The adversary can initiate challenge encryptions by calling $\text{SPLIT}_{\text{enc}}$ and $\text{COMBINE}_{\text{enc}}$. This is done via keeping track of a session identifier u. The adversary first needs to ask the desired honest party to initiate a session via calling $\text{SPLIT}_{\text{enc}}$ with some message m_0 and m_1 as well as set $S \subseteq [n]$ (we require that $|m_0| = |m_1|$). Oracle $\text{SPLIT}_{\text{enc}}$ will call $\mathsf{Split}_{\mathsf{enc}}(\mathsf{id}, m_b, S)$ and return a session id u, which the adversary needs to supply to $\text{COMBINE}_{\text{enc}}$. Via calling $\text{COMBINE}_{\text{enc}}$ with some session id, the adversary can supply a set of corrupt eval responses, which are used instead of the honest ones for corrupt parties. Any session id u can be input to $\text{COMBINE}_{\text{enc}}$ at most once.

Decryptions. The adversary is able to submit ciphertext to honest parties and initiate decryptions sessions. The adversary's goal here is to submit all the *valid* ciphertext that it can generate. Similar to challenge encryption, the adversary first needs to specify (to $\text{SPLIT}_{\text{dec}}$ oracle) an honest party id, a ciphertext c, and a set S of parties that are chosen to participate in the decryption process. The oracle shall return to the adversary the set of messages (k, x) designated to the corrupt parties k. Next, the adversary can choose to reply with any corrupt responses to these messages via the $\text{COMBINE}_{\text{dec}}$ oracle.

We define the IND-RCCA and IND-CCA advantage of an adversary $\mathcal{A}$ against TAE to be $\mathbf{Adv}^{\text{ind-cca}}_{\mathsf{TAE},n,t}(\mathcal{A}) = 2 \cdot \mathbf{P}\left[\mathbf{G}^{\text{ind-rcca}}_{\mathsf{TAE},n,t}(\mathcal{A})\right] - 1$, and $\mathbf{Adv}^{\text{ind-rcca}}_{\mathsf{TAE},n,t}(\mathcal{A}) = 2 \cdot \mathbf{P}\left[\mathbf{G}^{\text{ind-rcca}}_{\mathsf{TAE},n,t}(\mathcal{A})\right] - 1$, respectively.

Comparison with Security Model of DiSE. First, the DiSE model allows the adversary to generate valid ciphertext by engaging in the decryption protocol. Our model prevents this by not including the queries to $\mathsf{Eval}_{\mathsf{dec}}$ in the definition of the authenticity condition Eq. 4. Second, our security notions allow for malicious adversaries during decryption, whereas DiSE required all parties to behave honestly during decryption. This is a significant strengthening of the model. Third, DiSE targeted privacy and authenticity separately, which allowed for schemes that reconstruct the (master) encryption scheme during decryption. We give such an example (in [10]) which is secure in the model of DiSE but not secure according to our IND-CCA notion.

Finally, there are some subtle differences around what is considered a forgery and how they can be constructed. In DiSE, when an honest party initiates an encryption the resulting ciphertext is not revealed to the adversary in the authenticity game. However, it could be possible for the adversary to take such a ciphertext and generate another which decrypts properly. In fact, [10] discusses this exact scenario. Our definition prevents such attacks by explicitly providing ciphertext generated by honest parties to the adversary.

5 Construction from Indifferential AE

In this section we present our first construction TAE1, based on *random injections.* We first define *random injections* before presenting our construction TAE1 before presenting our construction which we prove secure against the threshold IND-RCCA notion defined previously.

5.1 Random Injection

The core of our first construction is a *random injection.* There are two equivalent ways to view and define this primitive. The first approach is to add authenticity to the notion of All-or-Nothing Transform [29]. The second approach is to view a random injection as an un-keyed indifferentiable Authenticated Encryption [16].

All-or-Nothing Transform. Consider oracles $\mathsf{I} : \mathcal{X} \times \mathcal{R} \rightarrow \mathcal{Y}$ and $\mathsf{I}^{-1} : \mathcal{Y} \rightarrow \mathcal{X}$. We view I as a randomized transform (message space $\mathcal{X}$ and randomness space $\mathcal{R}$), with inverse I^{-1}. Roughly, [29] required indistinguishability of $\mathsf{I}(x_1), \mathsf{I}(x_2)$ given that ℓ bits of the transforms have been erased. An example of such a transform is OAEP [20], which is defined for two random oracles $G : \{0,1\}^k \rightarrow \{0,1\}^{nk}$ and $H : \{0,1\}^{nk} \rightarrow \{0,1\}^k$ as

$$\mathsf{OAEP}(x; r) = (G(r) \oplus x, H(G(r) \oplus x) \oplus r) \ , \tag{5}$$

where $r \leftarrow\!\!\$\ \{0,1\}^k$. The inverse function is defined in the straightforward way. Myers and Shull [44] prove adaptive security of OAEP as defined by [29] (extending prior work by Boyko [24]). The core idea of OAEP is to mask the input x by a random value $G(r)$. In turn, this masked $x \oplus G(r)$ is used to mask r as $r \oplus H(x \oplus G(r))$. Missing any part of the output prevents the function from being inverted efficiently.

Random Injection. We add *authenticity* to the all-or-nothing transform. Specifically, we allow I^{-1} to output $\perp$, meaning $\mathsf{I}^{-1} : \mathcal{Y} \rightarrow \mathcal{X} \cup \{\perp\}$. Intuitively, we would like that all calls to $\mathsf{I}^{-1}(y)$, where y is *not* an output produced by I, to return $\perp$. This is formalized as follows. We define an *ideal random injection*, $\mathsf{I} : \mathcal{X} \times \mathcal{R} \rightarrow \mathcal{Y}$ with associated inverse $\mathsf{I}^{-1} : \mathcal{X} \rightarrow \mathcal{Y} \cup \{\perp\}$, from input domain $\mathcal{X}$ to output domain $\mathcal{Y}$ to be the following.

Proc $\mathsf{I}(x)$	**Proc** $\mathsf{I}^{-1}(y)$
$y \leftarrow\!\!\$\ \mathcal{Y}$; $T[y] \leftarrow x$; Return y	Return $T[y]$

Above, table T is initialized to $\perp$ everywhere. Note that for I to be injective, it should be the case that the number of queries to I, say q, should be much less than $\sqrt{|\mathcal{Y}|}$. Formally, a random injection should be *indifferentiable* [40] from an ideal random injection. Specifically, let $A : \mathcal{X} \rightarrow \mathcal{Y}$ be a randomized transform with inverse A^{-1}, both depending on some random oracle H. We say that A is a (q, ϵ)-RInj if there exists simulator S_H (which has access to $\mathsf{I}, \mathsf{I}^{-1}$) such that the RInj-advantage of any distinguisher $\mathcal{D}$, $\mathbf{Adv}^{\mathrm{rinj}}_{A,A^{-1}}(\mathcal{D}) := \Delta_{\mathcal{D}}\left(A, A^{-1}, H\,;\, \mathsf{I}, \mathsf{I}^{-1}, S_H\right)$, is at most ϵ for distinguisher $\mathcal{D}$ making at most q queries to any oracle.

Indifferentiable Authenticated Encryption. The other approach to view our definition of random injection is through indifferentialble authenticated encryption [16]. Indifferentiable AE is essentially a key-ed version of random injection. [16] present an extension of the OAEP construction which meets this definition. Figure ?? shows the OAEP construction and the (un-keyed) construction of [16] which we denote as Authenticated OAEP (AOAEP). The core difference from OAEP is that the randomness used in OAEP is now chosen as $r' \leftarrow I(x\|r)$ where I is a random oracle. When inverting OAEP, $(x\|r)$ is reconstructed and checked to see if the randomness used to construct the injection is consistent with r'. They show that finding such a consistent OAEP output without querying the oracles in the forward direction is infeasible.

Let OAEP be defined in Eq. 5, with input space $\{0,1\}^{nk+k}$ and I : $\{0,1\}^{nk+k} \rightarrow \{0,1\}^{k}$ be random oracle. $\mathsf{AOAEP} : \{0,1\}^{nk} \rightarrow \{0,1\}^{nk+2k}$ is defined as $\mathsf{AOAEP}(x;r) = \mathsf{OAEP}(x\|r; I(x\|r))$, where r is chosen uniformly at random from $\{0,1\}^{k}$. The inverse $\mathsf{AOAEP}^{-1} : \{0,1\}^{nk+2k} \rightarrow \{0,1\}^{nk}$ is defined as

$$\mathsf{AOAEP}^{-1}(y) = \begin{cases} x & \text{if } y = \mathsf{OAEP}(x\|r; I(x\|r));\ (x\|r) \leftarrow \mathsf{OAEP}^{-1}(y), \\ \bot & \text{otherwise} \end{cases} \tag{6}$$

Theorem 1. *Let* $\mathsf{AOAEP} : \{0,1\}^{nk+2k} \rightarrow \{0,1\}^{nk}$ *be defined above. The proof specifies a simulator* S *such that*

$$\mathbf{Adv}^{\mathrm{rinj}}_{\mathsf{AOAEP}}(\mathcal{D}) \leq \frac{9q^2+q}{2^k} + \frac{3q^2}{2^{nk+k}}\,. \tag{7}$$

where q *is the maximum number of queries to any oracle that* $\mathcal{D}$ *make.*

This follows from [16, Theorem 5]. We also provide a standalone proof in [10] with the bound shown above.

Extension to Variable-Input-Length. Let d be some integer. Consider the following padding function $\mathsf{Pad}_{k,d} : \{0,1\}^* \rightarrow (\{0,1\}^k)^*$. Upon input x, $\mathsf{Pad}_{k,d}$ first append a 1 at the end of x, then it append 0's until the length ℓ is some $m \cdot k$ for some integer $m \geq d$. Consider the variable-input-length transformation $\mathsf{RInj}_d := \mathsf{RInj} \circ \mathsf{Pad}_{k,d}$. It is not hard to show $\mathsf{RInj} \circ \mathsf{Pad}_{k,d}$ is indifferentiable to variable-input-length random injections with the Eq. 7 bound for n set to d.

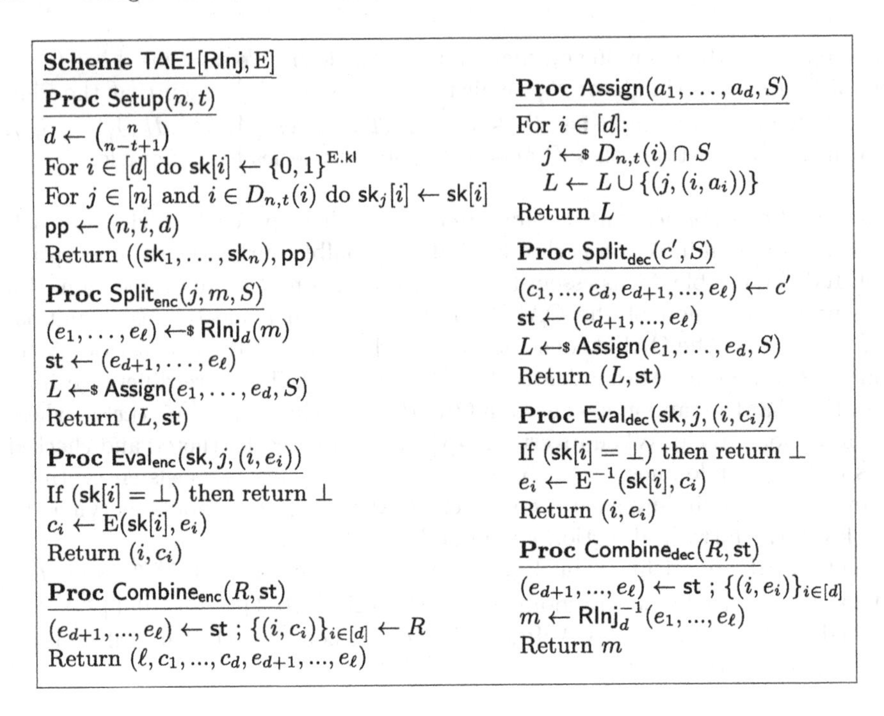

Fig. 2. TAE construction based on random injection RInj and block cipher E : $\{0,1\}^{\mathsf{E.kl}} \times \{0,1\}^k \to \{0,1\}^k$. Recall that $\mathsf{RInj}_d = \mathsf{RInj} \circ \mathsf{Pad}_{k,d}$.

5.2 The Construction

Our TAE construction in Fig. 2 builds on an random injection RInj as defined above. We define the sets $D_{n,t}(i)$ for integers $i \in [d]$ where $d = \binom{n}{n-t+1}$, with the following property: $D_{n,t}(1), \ldots, D_{n,t}(d)$ shall be all the subsets of $[n]$ with size exactly $n-t+1$. Each party $i \in [n]$ will hold secret key sk_j if and only if $i \in D_{n,t}(j)$. Together these secret keys form type of a t-out-of-n replicated secret sharing of the master key $((\mathsf{sk}_1, \ldots, \mathsf{sk}_d))$. To encrypt, the initiating party computes $e \leftarrow\$ \mathsf{RInj}_d(m)$. e is then partitioned into *at least* d block $e_1, \ldots, e_d, \ldots, e_\ell$ with each $e_i \in \{0,1\}^k$. Each e_i is sent to a single party which holds the key sk_i. This party returns $c_i \leftarrow \mathsf{PRP}(\mathsf{sk}_i, e_i)$. The final ciphertext is the comprised of $c_1, \ldots, c_d$ plus any additional blocks of e. An illustration of this is given in Fig. ??. Decryption is defined in the straightforward way where the RInj ensure that the plaintext has not been modified.

5.3 Security

We show that given a secure block cipher E and a secure random injection RInj, scheme TAE = TAE1[RInj, E] is IND-RCCA secure.

Theorem 2. *Let* RInj *be a random injection. Let* $\mathrm{E} : \mathrm{E.kl} \times \{0,1\}^k \to \{0,1\}^k$ *be a block cipher. Let* $\mathsf{TAE} = \mathsf{TAE1}[\mathsf{RInj}, \mathrm{E}]$. *Let* $\mathcal{A}$ *be an adversary against* TAE. *The proof gives adversaries* $\mathcal{B}, \mathcal{C}$, *whose running times are about that of* $\mathcal{A}$ *plus some simulation overhead, such that*

$$\mathbf{Adv}^{\text{ind-cca}}_{\mathsf{TAE},n,t}(\mathcal{A}) \leq 2 \cdot \mathbf{Adv}^{\text{d-prp}}_{\mathrm{E}}(\mathcal{B}) + 2 \cdot \mathbf{Adv}^{\text{rinj}}_{\mathsf{RInj}}(\mathcal{C}) \quad + \frac{(d+2) \cdot q^2}{2^{k-1}}\,, \tag{8}$$

where $q \geq 2$ *is the maximum number of queries to any oracle available to* $\mathcal{A}$, *and* $d = \binom{n}{n-t+1}$ *is assumed to be larger than* 2.

The proof of the above theorem is given in [10].

6 Constructions from Threshold PRF and Signature

In this section, we provide two TAE constructions. One of them achieves IND-RCCA security, whereas the other one achieves IND-CCA security. The first construction is based on a DPRF and a threshold signature scheme. The second construction is achieved simply replacing the DPRF with a DVRF.

A DPRF is a threshold version of a PRF. It consists of an algorithm Setup to generate shares of a key, Eval to generate *partial* PRF values, and Combine to combine them. Informally, a DPRF is considered secure if an attacker under control of less than t parties cannot predict the outcome (from Combine) on any input unless it *collaborates* with enough parties (to reach the threshold). A formal definition has to take several other things into account and is provided in [10]. A DVRF, in addition to DPRF algorithms, has an algorithm called Verify to check if the partial PRF values are *legitimate*, i.e., they can be combined to produce the right output. We also compare with DiSE's DPRF treatment in the Appendix.

We provide our instantiations of DPRF/DVRF which are variants of the DDH-based DPRFs introduced by Naor et al. [45] and used by Agrawal et al. [11] to construct TAE schemes. We provide a new *modular* and *concrete* security analysis for the instantiations [10].

Finally, we provide our TAE constructions that use DPRF/DVRF along with threshold signatures achieving IND-RCCA (Sect. 6.1) and IND-CCA [10].

6.1 IND-RCCA TAE Using DPRF and Threshold Signature

The construction TAE2 is parameterized by a DPRF DP, a threshold signature scheme SIG, and an integer k. The specification of the construction is given in Fig. 3. We explain below the high-level ideas of the scheme.

Keys. Each party holds a key share of the DPRF key and a key share of the threshold signature signing key.

Encryption. When initiator j is encrypting a message m, a commitment $\alpha = \mathrm{H}(m \| r)$ is generated for the message m by using hash function H (modeled as a

Scheme $\mathsf{TAE2}[\mathsf{DP}, \mathsf{SIG}, k]$

Proc $\mathsf{Split}_{\mathsf{enc}}(j, m, S)$

Require $|S| \geq t$
Let S' be a t-sized subset of S
$r \leftarrow \{0,1\}^k$; $\alpha \leftarrow \mathrm{H}(m\|r)$
$\mathsf{st} \leftarrow (j, \alpha, m, r)$
Return $(\{(i, \alpha)\}_{i \in S'}, \mathsf{st})$

Proc $\mathsf{Eval}_{\mathsf{enc}}(\mathsf{sk}_i, j, \alpha)$

$(\mathsf{sk}_{\mathsf{DP},i}, \mathsf{sk}_{\mathsf{SIG},i}) \leftarrow \mathsf{sk}_i$
$y \leftarrow \mathsf{DP.Eval}(\mathsf{sk}_{\mathsf{DP},i}, j\|\alpha)$
$\sigma \leftarrow \mathsf{SIG.PartSign}(\mathsf{sk}_{\mathsf{SIG},i}, j\|\alpha)$
Return (y, σ)

Proc $\mathsf{Combine}_{\mathsf{enc}}(R, \mathsf{st})$

$(j, \alpha, m, r) \leftarrow \mathsf{st}$; $\{(i, (y_i, \sigma_i))\} \leftarrow R$
$\beta \leftarrow \mathsf{DP.Combine}(\{(i, y_i)\})$
$\sigma \leftarrow \mathsf{SIG.CombSig}(\mathsf{vk}, \{(i, \sigma_i)\})$
If $\mathsf{VerSig}(\mathsf{vk}, j\|\alpha, \sigma) \neq 1$ then return $\perp$
$e \leftarrow \mathrm{G}(\beta) \oplus (m\|r)$
Return (j, α, σ, e)

Proc $\mathsf{Setup}(n, t)$

$([\![\mathsf{sk}_{\mathsf{DP}}]\!]_n, \mathsf{pp}_{\mathsf{DP}}) \leftarrow\$ \mathsf{DP.Setup}(n, t)$
$([\![\mathsf{sk}_{\mathsf{SIG}}]\!]_n, \mathsf{vk}) \leftarrow\$ \mathsf{SIG.Setup}(n, t)$
For $i \in [n]$ do $\mathsf{sk}_i \leftarrow (\mathsf{sk}_{\mathsf{DP},i}, \mathsf{sk}_{\mathsf{SIG},i})$
$\mathsf{pp} \leftarrow (\mathsf{pp}_{\mathsf{DP}}, \mathsf{vk})$
Return $((\mathsf{sk}_1, \ldots, \mathsf{sk}_n), \mathsf{pp})$

Proc $\mathsf{Split}_{\mathsf{dec}}(j, c, S)$

Require $|S| \geq t$
Let S' be a t-sized subset of S
$(j, \alpha, \sigma, e) \leftarrow c$; $\mathsf{st} \leftarrow (j, \alpha, e)$
Return $(\{(i, (j\|\alpha, \sigma))\}_{i \in S'}, \mathsf{st})$

Proc $\mathsf{Eval}_{\mathsf{dec}}(\mathsf{sk}_i, j, x)$

$(\mathsf{pp}_{\mathsf{DP}}, \mathsf{vk}) \leftarrow \mathsf{pp}$; $(j\|\alpha, \sigma) \leftarrow x$
$(\mathsf{sk}_{\mathsf{DP},i}, \mathsf{sk}_{\mathsf{SIG},i}) \leftarrow \mathsf{sk}_i$
If $\mathsf{SIG.VerSig}(\mathsf{vk}, j\|\alpha, \sigma)$ then
  Return $\mathsf{DP.Eval}(\mathsf{sk}_{\mathsf{DP},i}, j\|\alpha)$
Else return $\perp$

Proc $\mathsf{Combine}_{\mathsf{dec}}(R, \mathsf{st})$

$(j, \alpha, e) \leftarrow \mathsf{st}$
$\beta \leftarrow \mathsf{DP.Combine}(R)$
$m\|r \leftarrow \mathrm{G}(\beta) \oplus e$
If $(\mathrm{H}(m\|r) \neq \alpha)$ then return $\perp$
Return m

Fig. 3. DPRF & threshold signature based TAE scheme.

random oracle with input space $\{0,1\}^*$ and output space $\mathsf{DP.In}$) and randomly generated r. The DPRF output, $\beta \leftarrow \mathsf{DP}(j\|\alpha)$, is used as an encryption key to encrypt message m and randomness r into $c \leftarrow \mathrm{G}(\beta) \oplus (m\|r)$, where G is a random oracle with input space $\{0,1\}^k$ and output space $\{0,1\}^\infty$. Meanwhile, a threshold signature σ on $j\|\alpha$ is also generated using SIG. The final ciphertext is (j, α, σ, c).

Decryption. When an initiator j' is decrypting a ciphertext (j, α, σ, c) (note that j' does not have to equal j), each party i receives $(j\|\alpha, \sigma)$ and first verifies if σ is a valid signature on $j\|\alpha$ before returning the Eval output of DP. After reconstructing the DPRF output β, the initiator can recover the message m and randomness r. It checks if $\mathrm{H}(m\|r) = \alpha$. If the check succeeds, then plaintext m is returned. Otherwise, decryption fails and $\perp$ is returned.

Capturing the Decryption Criteria. In the scheme TAE2, for a ciphertext $c = (j, \alpha, \sigma, e)$, $\mathsf{Split}_{\mathsf{dec}}(c, S)$ returns a list $(\{(i, (j\|\alpha, \sigma))\}_{i \in S'}$, where S' is a t-sized subset of S. The multiset Eval-MSet(c) just has the element $(j\|\alpha, \sigma)$ repeated t

times. $\mathsf{Eval}_{\mathsf{dec}}$, with inputs sk_i and $(j\|\alpha, \sigma)$, outputs a non-$\perp$ value if σ is a valid signature. Therefore, for a set $\mathcal{CR}$, $\mathsf{Eval\text{-}MSet}(c, \mathcal{CR})$ is either $\mathsf{Eval\text{-}MSet}(c)$ (if σ is invalid) or the set with $(j\|\alpha, \sigma)$ repeated $t - |\mathcal{CR}|$ times.

Theorem 3. *Let* $\mathsf{TAE} = \mathsf{TAE2}[\mathsf{DP}, \mathsf{SIG}, k]$. *Let $\mathcal{A}$ be an adversary in the IND-RCCA game against* TAE *(Fig. 1). Suppose $\mathcal{A}$ makes q_{H} and q_{G} queries to oracles* H *and* G, *respectively. Further, it makes q_{enc}, q_{Eval} and q_{dec} queries to encryption* ($\textsc{Split}_{\mathsf{enc}}$, $\textsc{Combine}_{\mathsf{enc}}$), *evaluation* ($\textsc{Eval}_{\mathsf{enc}}$, $\textsc{Eval}_{\mathsf{dec}}$) *and decryption procedures* ($\textsc{Split}_{\mathsf{dec}}$, $\textsc{Combine}_{\mathsf{dec}}$), *respectively. Then there exist adversaries $\mathcal{B}$ and $\mathcal{C}$ such that*

$$\mathbf{Adv}^{\text{ind-rcca}}_{\mathsf{TAE},n,t}(\mathcal{A}) \leq \frac{(q_{\mathsf{enc}} + q_{\mathsf{dec}})^2}{|\mathsf{DP.In}|} + \frac{q^2_{\mathsf{enc}} + 2 \cdot q_{\mathrm{H}} \cdot q_{\mathsf{enc}}}{2^k} + \frac{2 \cdot q_{\mathrm{G}} \cdot q_{\mathsf{enc}}}{|\mathsf{DP.Out}|} + 2 \cdot \mathbf{Adv}^{\text{sig}}_{\mathsf{SIG},n,t}(\mathcal{B}) + 2 \cdot \mathbf{Adv}^{\text{dprf}}_{\mathsf{DP},n,t}(\mathcal{C}) \,. \tag{9}$$

A proof of the above theorem is given in [10].

Appendix

A Performance Experiments

We implement our protocols TAE1 of Fig. 2 and TAE3 of Fig. ?? and report on their performance. All performance results are obtained on a single laptop with

Table 1. Encryption performance metrics with various number of parties n and threshold t. Throughput is computed by performing many encryptions concurrently (single thread per party). Mbps denotes network bandwidth. Latency is computed by performing sequential encryptions.

t	n	Throughput								Latency			
		(enc/s)				$(Mbps)$				(ms/enc)			
		TAE1	DiSE1	TAE3	DiSE2	TAE1	DiSE1	TAE3	DiSE2	TAE1	DiSE1	TAE3	DiSE2
$n/3$	6	730,627	1,123,555	346	444	88	111	0.3	0.4	0.1	0.1	4.0	3.3
	12	86,588	326,193	145	172	581	97	0.4	0.5	0.4	0.3	7.9	6.7
	18	2,179	13,464	91	105	633	8	0.5	0.5	1.0	0.7	12.1	10.5
$n/2$	4	745,486	1,123,555	346	452	77	111	0.3	0.4	0.1	0.1	4.0	3.3
	6	561,777	722,285	222	259	333	143	0.4	0.5	0.2	0.2	5.6	4.8
	12	24,543	131,324	91	106	988	78	0.5	0.5	0.5	0.5	12.0	10.4
	18	311	3,351	58	68	738	3	0.6	0.5	2.7	1.0	17.9	15.8
$2n/3$	3	777,846	1,123,555	348	445	77	111	0.3	0.4	0.1	0.1	4.0	3.3
	6	421,333	505,600	143	174	500	150	0.4	0.5	0.3	0.3	7.9	6.9
	12	24,845	129,641	65	77	1,400	103	0.6	0.5	0.6	0.6	16.0	14.2
	18	483	6,347	42	49	1,149	8	0.6	0.5	2.3	1.0	24.1	21.4
$n-2$	12	81,548	297,411	51	60	1,637	324	0.6	0.5	0.6	0.6	19.9	17.5
	18	23,905	219,826	30	36	1,983	391	0.6	0.5	1.0	1.0	32.5	28.6
2	12	674,133	1,011,200	337	445	67	100	0.3	0.4	0.2	0.2	4.1	3.4
	18	594,823	919,272	345	441	59	91	0.3	0.4	0.2	0.2	4.1	3.5

an Intel i7 9th Gen (9740H) CPU and 16GB of RAM. Network communication was routed over local host with a theoretical bandwidth of 10Gbps and a measured latency of 0.1 milliseconds. Each party is run on a single thread.

Table 1 contains the results of two experiments. 1) peak encryptions per second each scheme can perform. In particular, 32 byte messages are repeatedly encrypted in an asynchronous manner, where a single party repeatedly initiates 10 batches of 128 encryptions which are processed concurrently. 2) latency of one encryption by running multiple encryptions one at a time in a sequential manner. We report the average time required to perform a single encryption.

We compare with the less secure DiSE schemes [11]. In particular, DiSE was proven secure in an arguably weaker model and does not provide a way to distinguish if the initiating party is performing an decryption or encryption query. We consider the pure symmetric-key based DiSE protocol DiSE1 which utilizes an AES/PRF based DPRF. Like our TAE1 Protocol, DiSE1 does not guarantee that a ciphertext output by encryption is "well formed" if some of the parties behave maliciously. We also consider the DDH-key based DiSE protocol DiSE2 which utilizes ZK-proofs to ensure the correctness of any ciphertext output by the encryption procedure.

Our protocols are very competitive given the added security guarantees. Our symmetric-key based protocol TAE1 achieves a throughput of 778 thousand encryptions per second for $n = 3, t = 2$ while our public-key based protocol TAE3 achieves 346 encryptions per second. This is approximately 0.7 times the throughput of the weaker DiSE protocol. We observe a similar relative performance for other parameter choices when t is close to n or 2. The largest differences occurs for our TAE1 protocol when n is large and $t \approx n/2$. This results in the largest relative communication overhead compared to DiSE1 due to their protocol achieving $O(t)$ communication while ours achieves $O(\binom{n}{t})$ which is maximized for $t = n/2$.

With respect to encryption latency our protocols perform similarly well. Both TAE1 and DiSE1 achieve a latency of 0.1 milliseconds for $n = 3, t = 2$ which is effectively the network latency of just sending the messages. For the public-key based protocol we again observe that the DiSE2 protocol achieves times 0.7 times improvement in latency compared to our TAE3 protocol. This added overhead consists of performing the additional threshold signature.

We argue that the presented performance evaluation shows that our protocols achieve highly practical performance. In particular, the majority of the practical applications of threshold authenticated encryption only require relatively small n, e.g. $n \in \{3, 4, 5\}$. For this range of parameters both of our protocols are highly competitive with the DiSE protocols while providing stronger security guarantees. Our schemes also preserve the property that the network communication overhead is independent of the length of the message being encryption. This property is not enjoyed by generic MPC based approaches, e.g. [37].

References

1. Coinbase custody. custody.coinbase.com/. Use of secret sharing described in [9]
2. Intel Software Guard Extensions. software.intel.com/en-us/sgx
3. NIST tcg. csrc.nist.gov/Projects/threshold-cryptography
4. Secure Enclave overview - Apple Support.support.apple.com/guide/security/secure-enclave-overview-sec59b0b31ff/1/web/1
5. Titan M makes Pixel 3 our most secure phone yet. www.blog.google/products/pixel/titan-m-makes-pixel-3-our-most-secure-phone-yet/
6. TrustZone. developer.arm.com/ip-products/security-ip/trustzone
7. Unbound Tech. www.unboundtech.com/. Use of MPC mentioned in [15]
8. Vault Seal. www.vaultproject.io/docs/concepts/seal.html
9. [Podcast] Institutional Cryptoasset Custody w/Sam McIngvale of Coinbase Custody - (Eps. 0028–0029), July 2019. blog.nomics.com/flippening/coinbase-custody-sam-mcingvale/
10. Agrawal, S., Dai, W., Luykx, A., Mukerjee, P., Rindal., P.: ParaDiSE: efficient threshold authenticated encryption in fully malicious model. Cryptology ePrint Archive, Report 2022/1449 (2022). https://eprint.iacr.org/2022/1449
11. Agrawal, S., Mohassel, P., Mukherjee, P., Rindal, P.: DiSE: distributed symmetric-key encryption. In: Lie, D., Mannan, M., Backes, M., Wang, X. (eds.) ACM CCS 2018, pp. 1993–2010. ACM Press, October 2018
12. Albrecht, M.R., Grassi, L., Rechberger, C., Roy, A., Tiessen, T.: MiMC: efficient encryption and cryptographic hashing with minimal multiplicative complexity. In: Cheon, J.H., Takagi, T. (eds.) ASIACRYPT 2016. Part I, volume 10031 of LNCS, pp. 191–219. Springer, Heidelberg (2016)
13. Albrecht, M.R., Rechberger, C., Schneider, T., Tiessen, T., Zohner, M.: Ciphers for MPC and FHE. In: Oswald, E., Fischlin, M. (eds.) EUROCRYPT 2015. Part I, volume 9056 of LNCS, pp. 430–454. Springer, Heidelberg (2015)
14. Andreeva, E., Bogdanov, A., Luykx, A., Mennink, B., Mouha, N., Yasuda, K.: How to securely release unverified plaintext in authenticated encryption. In: Sarkar, P., Iwata, T. (eds.) ASIACRYPT 2014. LNCS, vol. 8873, pp. 105–125. Springer, Heidelberg (2014). https://doi.org/10.1007/978-3-662-45611-8_6
15. Archer, D.W., et al.: From keys to databases - real-world applications of secure multi-party computation. Comput. J. **61**(12), 1749–1771 (2018)
16. Barbosa, M., Farshim, P.: Indifferentiable authenticated encryption. In: Shacham, H., Boldyreva, A. (eds.) CRYPTO 2018. Part I, volume 10991 of LNCS, pp. 187–220. Springer, Heidelberg (2018)
17. Bedrune, J.-B., Campana, G.: Everybody be cool, this is a robbery! www.sstic.org/media/SSTIC2019/SSTIC-actes/hsm/SSTIC2019-Article-hsm-campana_bedrune_neNSDyL.pdf
18. Bellare, M., Boldyreva, A., Knudsen, L.R., Namprempre, C.: On-line ciphers and the hash-CBC constructions. J. Cryptol. **25**(4), 640–679 (2012)
19. Bellare, M., Namprempre, C.: Authenticated encryption: relations among notions and analysis of the generic composition paradigm. In: Okamoto, T. (ed.) ASIACRYPT 2000. LNCS, vol. 1976, pp. 531–545. Springer, Heidelberg (2000). https://doi.org/10.1007/3-540-44448-3_41
20. Bellare, M., Rogaway, P.: Optimal asymmetric encryption. In: De Santis, A. (ed.) EUROCRYPT 1994. LNCS, vol. 950, pp. 92–111. Springer, Heidelberg (1995). https://doi.org/10.1007/BFb0053428

21. Bellare, M., Tackmann, B.: The multi-user security of authenticated encryption: AES-GCM in TLS 1.3. In: Robshaw, M., Katz, J. (eds.) CRYPTO 2016. LNCS, vol. 9814, pp. 247–276. Springer, Heidelberg (2016). https://doi.org/10.1007/978-3-662-53018-4_10
22. Bendlin, R., Damgård, I.: Threshold decryption and zero-knowledge proofs for lattice-based cryptosystems. In: Micciancio, D. (ed.) TCC 2010. LNCS, vol. 5978, pp. 201–218. Springer, Heidelberg (2010)
23. Boneh, D., Lewi, K., Montgomery, H., Raghunathan, A.: Key homomorphic PRFs and their applications. In: Canetti, R., Garay, J.A. (eds.) CRYPTO 2013. LNCS, vol. 8042, pp. 410–428. Springer, Heidelberg (2013). https://doi.org/10.1007/978-3-642-40041-4_23
24. Boyko, V.: On the security properties of OAEP as an all-or-nothing transform. In: Wiener, M. (ed.) CRYPTO 1999. LNCS, vol. 1666, pp. 503–518. Springer, Heidelberg (1999). https://doi.org/10.1007/3-540-48405-1_32
25. Canetti, R., Krawczyk, H., Nielsen, J.B.: Relaxing chosen-ciphertext security. In: Boneh, D. (ed.) CRYPTO 2003. LNCS, vol. 2729, pp. 565–582. Springer, Heidelberg (2003). https://doi.org/10.1007/978-3-540-45146-4_33
26. Desmedt, Y.: Society and group oriented cryptography: a new concept. In: Pomerance, C. (ed.) CRYPTO 1987. LNCS, vol. 293, pp. 120–127. Springer, Heidelberg (1988). https://doi.org/10.1007/3-540-48184-2_8
27. Desmedt, Y., Frankel, Y.: Threshold cryptosystems. In: Brassard, G. (ed.) CRYPTO 1989. LNCS, vol. 435, pp. 307–315. Springer, New York (1990). https://doi.org/10.1007/0-387-34805-0_28
28. Dodis, Y.: Efficient construction of (distributed) verifiable random functions. In: Desmedt, Y.G. (ed.) PKC 2003. LNCS, vol. 2567, pp. 1–17. Springer, Heidelberg (2003). https://doi.org/10.1007/3-540-36288-6_1
29. Dodis, Y., Sahai, A., Smith, A.: On perfect and adaptive security in exposure-resilient cryptography. In: Pfitzmann, B. (ed.) EUROCRYPT 2001. LNCS, vol. 2045, pp. 301–324. Springer, Heidelberg (2001). https://doi.org/10.1007/3-540-44987-6_19
30. Escala, A., Herold, G., Kiltz, E., Ràfols, C., Villar, J.: An algebraic framework for diffie-hellman assumptions. In: Canetti, R., Garay, J.A. (eds.) CRYPTO 2013. LNCS, vol. 8043, pp. 129–147. Springer, Heidelberg (2013). https://doi.org/10.1007/978-3-642-40084-1_8
31. Fouque, P.-A., Joux, A., Martinet, G., Valette, F.: Authenticated on-line encryption. In: Matsui, M., Zuccherato, R.J. (eds.) SAC 2003. LNCS, vol. 3006, pp. 145–159. Springer, Heidelberg (2004). https://doi.org/10.1007/978-3-540-24654-1_11
32. Freedman, M.J., Ishai, Y., Pinkas, B., Reingold, O.: Keyword search and oblivious pseudorandom functions. In: Kilian, J. (ed.) TCC 2005. LNCS, vol. 3378, pp. 303–324. Springer, Heidelberg (2005). https://doi.org/10.1007/978-3-540-30576-7_17
33. Galindo, D., Liu, J., Ordean, M., Wong, J.-M.: Fully distributed verifiable random functions and their application to decentralised random beacons. Cryptology ePrint Archive, Report 2020/096 (2020). https://eprint.iacr.org/2020/096
34. Hoang, V.T., Krovetz, T., Rogaway, P.: Robust authenticated-encryption AEZ and the problem that it solves. In: Oswald, E., Fischlin, M. (eds.) EUROCRYPT 2015. LNCS, vol. 9056, pp. 15–44. Springer, Heidelberg (2015). https://doi.org/10.1007/978-3-662-46800-5_2

35. Jarecki, S., Kiayias, A., Krawczyk, H., Xu, J.: TOPPSS: cost-minimal password-protected secret sharing based on threshold OPRF. In: Gollmann, D., Miyaji, A., Kikuchi, H. (eds.) ACNS 2017. LNCS, vol. 10355, pp. 39–58. Springer, Cham (2017). https://doi.org/10.1007/978-3-319-61204-1_3
36. Jarecki, S., Liu, X.: Efficient oblivious pseudorandom function with applications to adaptive OT and secure computation of set intersection. In: Reingold, O. (ed.) TCC 2009. LNCS, vol. 5444, pp. 577–594. Springer, Heidelberg (2009). https://doi.org/10.1007/978-3-642-00457-5_34
37. Keller, M., Orsini, E., Rotaru, D., Scholl, P., Soria-Vazquez, E., Vivek, S.: Faster secure multi-party computation of AES and DES using lookup tables. In: ACNS (2017)
38. Kocher, P., et al.: Spectre attacks: exploiting speculative execution. In: 40th IEEE Symposium on Security and Privacy (S&P'19) (2019)
39. Lipp, M., et al.: Meltdown: reading kernel memory from user space. In: 27th USENIX Security Symposium (USENIX Security 18) (2018)
40. Maurer, U., Renner, R., Holenstein, C.: Indifferentiability, impossibility results on reductions, and applications to the random oracle methodology. In: Naor, M. (ed.) TCC 2004. LNCS, vol. 2951, pp. 21–39. Springer, Heidelberg (2004). https://doi.org/10.1007/978-3-540-24638-1_2
41. McGrew, D.A., Viega, J.: The security and performance of the galois/counter mode (GCM) of operation. In: INDOCRYPT (2004)
42. Micali, S., Sidney, R.: A simple method for generating and sharing pseudo-random functions, with applications to clipper-like key escrow systems. In: Coppersmith, D. (ed.) CRYPTO 1995. LNCS, vol. 963, pp. 185–196. Springer, Heidelberg (1995). https://doi.org/10.1007/3-540-44750-4_15
43. Mukherjee, P.: Adaptively secure threshold symmetric-key encryption. In: Bhargavan, K., Oswald, E., Prabhakaran, M. (eds.) INDOCRYPT 2020. LNCS, vol. 12578, pp. 465–487. Springer, Cham (2020). https://doi.org/10.1007/978-3-030-65277-7_21
44. Myers, S., Shull, A.: Practical revocation and key rotation. In: Smart, N.P. (ed.) CT-RSA 2018. LNCS, vol. 10808, pp. 157–178. Springer, Cham (2018). https://doi.org/10.1007/978-3-319-76953-0_9
45. Naor, M., Pinkas, B., Reingold, O.: Distributed pseudo-random functions and KDCs. In: Stern, J. (ed.) EUROCRYPT 1999. LNCS, vol. 1592, pp. 327–346. Springer, Heidelberg (1999). https://doi.org/10.1007/3-540-48910-X_23
46. Rivest, R.L.: All-or-nothing encryption and the package transform. In: Biham, E. (ed.) FSE 1997. LNCS, vol. 1267, pp. 210–218. Springer, Heidelberg (1997). https://doi.org/10.1007/BFb0052348
47. Rogaway, P.: Authenticated-encryption with associated-data. In: Atluri, V. (ed.) ACM CCS 2002, pp. 98–107. ACM Press, November 2002
48. Rogaway, P.: Nonce-based symmetric encryption. In: Roy, B., Meier, W. (eds.) FSE 2004. LNCS, vol. 3017, pp. 348–358. Springer, Heidelberg (2004). https://doi.org/10.1007/978-3-540-25937-4_22
49. Rogaway, P., Shrimpton, T.: A provable-security treatment of the key-wrap problem. In: Vaudenay, S. (ed.) EUROCRYPT 2006. LNCS, vol. 4004, pp. 373–390. Springer, Heidelberg (2006). https://doi.org/10.1007/11761679_23
50. Salowey, J.A., McGrew, D., Choudhury, A.: AES Galois Counter Mode (GCM) Cipher Suites for TLS. RFC 5288, August 2008

Stronger Security and Generic Constructions for Adaptor Signatures

Wei Dai[1(✉)], Tatsuaki Okamoto[2], and Go Yamamoto[3]

[1] Bain Capital Crypto, Boston, USA
me@wdai.us
[2] NTT, Koto City, Japan
[3] NTT Research, Palo Alto, USA

Abstract. Adaptor signatures have seen wide applications in layer-2 and peer-to-peer blockchain applications such as atomic swaps and payment channels. We first identify two shortcomings of previous literature on adaptor signatures. (1) Current aim of "script-less" adaptor signatures restricts instantiability, limiting designs based on BLS or current NIST PQC candidates. (2) We identify gaps in current formulations of security. In particular, we show that current notions do not rule out a class of insecure schemes. Moreover, a natural property concerning the on-chain *unlinkability* of adaptor signatures has not been formalized. We then address these shortcomings by providing new and stronger security notions, as well as new generic constructions from *any* signature scheme and hard relation. On definitions:

1. We develop security notions that strictly imply previous notions.
2. We formalize the notion of unlinkability for adaptor signatures.
3. We give modular proof frameworks that facilitate simpler proofs.

On constructions:

1. We give a generic construction of adaptor signature from *any* signature scheme and *any* hard relation, showing that theoretically, (linkable) adaptor signatures can be constructed from any one-way function.
2. We also give an *unlinkable* adaptor signature construction from any signature scheme and any strongly random-self reducible relation, which we show instantiations of using DL, RSA, and LWE.

1 Introduction

1.1 Background and Problems

The invention of Bitcoin has ignited a vast amount of research in the area of distributed ledger technologies in the past decade. Scalability and interoperability are two crucial challenges for the mass adoption of blockchain technology. One fruitful direction towards solving these challenges is the study of layer-2 and peer-to-peer (P2P) protocols. For example, atomic swaps allow users to swap asset across different chains without a trusted third party, payment channels

W. Dai and T. Okamoto—Work done while at NTT Research.

T. Isobe and S. Sarkar (Eds.): INDOCRYPT 2022, LNCS 13774, pp. 52–77, 2022.
https://doi.org/10.1007/978-3-031-22912-1_3

allow a group of users to conduct many off-chain payments while only posting a small number of transactions to the blockchain, while payment channel networks (PCNs) generalize payment channels to enable large-scale P2P payments. One important functionality underlying these applications is that of *atomicity*. Roughly, *atomicity* guarantees that a set of transactions should either all be posted (to their respective ledgers) or none are. We briefly review the two currently known techniques to achieve atomicity.

Atomicity in blockchains: scripting vs. adaptor signatures. One seminal technique put forth by the work of the Bitcoin Lightning Network is the so-called "Hash Time Lock Contracts" (HTLC) [21]. Roughly, it is a Bitcoin spending script that, in addition to verification of a signature, also checks the release of a hash pre-image. Atomicity for a set of transactions is achieved by requiring the release of the same pre-image. Hence, if one transaction is processed, then the same pre-image can be used in other transactions. Such techniques are used to build applications such as atomic swaps, payment channels, and payment channel networks.

Aiming to eliminate the use of special scripts for the above applications, Poelstra proposed a technique called "adaptor signatures" [19,20], which replaces the role of "hash-locks". Roughly, an adaptor signature for some underlying signature scheme achieves the following: If Alice gives to Bob a "pre-signature" $\hat{\sigma}$ on some message m and instance Y, then any signature on message m, that is valid against public key of Alice, given by Bob will allow Alice to learn the witness y for instance Y. Hence, adaptor signature is a "script-less" hash-lock, in the sense that the release of the signature by Bob also releases the witness to Alice. Following the work of Poelstra, there have been numerous work proposing various forms of adaptor signatures [2,9,15,24] and applying them in applications to payment channels [1,15].

Do we need script-less adaptor signatures? It is understood [8,19,20] that eliminating the need for scripts allows "applications" to blockchains that do not support it. Indeed, adaptor signatures allow atomic swap between say Bitcoin and Monero [11]. However, we point out that it is not clear if constructions of more complex applications, e.g. payment channels, can be realized for script-less blockchains such as Monero. Indeed, currently known techniques for payment channels crucially rely on the use of scripts [1,2,15]. Hence, we reconsider the requirement of "script-less"—we consider adaptor signatures whose verification requires minimal additions to the underlying signature supported by the blockchain. This leads to the first shortcoming of the current approach.

Problem 1: Restricted design space and limited compatibility with post-quantum signature candidates. Previous works on adaptor signatures have one common theme—the adaptor signature scheme *must* work with a known underlying signature scheme, that is supported directly by the blockchain, e.g. ECDSA [14] and Schnorr [23]. This is due to the desire to eliminate the use of more complex spending scripts such as HTLC. However, one significant downside to such an approach is that adaptor signatures are not possible for all signature schemes. Indeed, it is known [8] that adaptor signature schemes are *impossible* for unique signatures such as BLS [5].

The problem of aiming for "script-less" adaptor signatures is even more pronounced when considering post-quantum (PQ) adaptor signatures. Neither of the two currently known post-quantum constructions, i.e. LAS [9] (based on LWE), and IAS [24] (based on isogenies), are compatible with the NIST PQC Round 3 PQ signature candidates [16]. The underlying signature scheme of LAS is a (strict) variant of Dilithium [7]. The underlying signature scheme of IAS is CSI-FiSh [4], which was proposed after the NIST PQC submission deadline. This means that even if the community arrives at a consensus on PQ standard signatures, further standardization and selection efforts might be required for extensions to adaptor signatures.

Prior works also leave open some natural theoretical questions, which we raise and answer in our work: What are the minimal assumptions required to construct adaptor signatures? In particular, are adaptor signatures in Minicrypt (i.e. can be constructed from one-way functions)?

Problem 2: Definitional gaps. The security definition for adaptor signatures, first proposed by [2], is later adopted by all subsequent follow-up works [1,8,9,24]. However, there is a serious gap in this security definition. In particular, we demonstrate that such notions do not guarantee security against *multiple* queries to the pre-signature oracle with the same message and instance pair (m, Y). Moreover, we show that such gap is not just theoretical—we give a counterexample scheme that is secure against previous notions but renders many applications insecure if they are used in practice. The root cause of this gap is the weak forms of attacks considered. Specifically, the security game (for notions called aEUF-CMA and aWitExt) only allows a single challenge query to pre-sign, and hence does not rule out attacks that access pre-sign more than once.[1] Secondly, previous definitions fail to capture a natural property of adaptor signature schemes—on-chain unlinkability. For example, it is understood that atomic swaps based on Schnorr signatures give more privacy guarantees than HTLC-based solutions. Indeed, we show that such a property does not follow from prior formalization.

1.2 Our Contributions

Our work addresses the problems outlined above. First, we give stronger definitions for adaptor signatures, as well as new definitions of unlinkability. We also give a modular proof framework to facilitate simpler proofs. Second, we give generic constructions of adaptor signatures from *any* secure signature scheme and *any* hard relation. The construction is unlinkable if the relation is assumed to be strongly random-self-reducible (SRSR). Our constructions are compatible

[1] We remark that this is *not* a weakness of previous constructions, but rather a gap in the formal security guarantees and the security is expected for applications.

with any of the current NIST PQC candidates. Answering the theoretical questions, we show (linkable) adaptor signatures can be constructed from one-way functions, and unlinkability can be achieved assuming additionally the existence of SRSR relations.

New security notions and modular proof framework. A significant portion of our technical contribution is regarding the security definition of adaptor signature. First, we close the definitional gap by giving two security notions, called (strong) full extractability, abbreviated as (S)FExt, that exposes rich sets of attack interfaces. We show that full extractability *strictly* implies previous notions. Next, we formalize the notion of unlinkability for adaptor signatures. Lastly, we present a modular proof framework, allowing the security of adaptor signatures to be proved against much simpler notions than full extractability, which we call simple and unique extractability.

Generic construction and unifying perspective. We revisit the main design constraint so far considered for adaptor signatures, namely that they should be "script-less scripts." We allow the use of minimal scripts and define "augmented" signature scheme SigR, which is defined against any signature Sig and any relation $\mathbb{R}$. Augmented scheme SigR additionally verifies the release of a witness alongside a valid signature. We then give an adaptor signature scheme GAS_1 for SigR and prove it secure (Theorem 5). We remark that adaptor signature GAS_1 is generic in the sense that it is constructed from *any* signature scheme and *any* hard relation. However, we remark that implementing augmented signature schemes for existing blockchains such as Bitcoin amounts to using scripts. Our work can be seen as a formalization of HTLCs as adaptor signatures.

Achieving on-chain unlinkability. We formally define the notion of unlinkability, which has not been formally studied previously. Unlinkability asks adapted signatures to be indistinguishable from standard ones, even if one knows the instance and witness pair used to derive the adapted signature. We show how to add unlinkability to GAS_1, assuming that relation R is strongly random-self-reducible (SRSR), obtaining a new construction which we name GAS_2, whose security proofs are given in Theorem 7. We show how to instantiate SRSR from standard number theoretical problems such as DL and RSA, as well as the learning with errors problem (LWE).

Flexible instantiations and minimal assumptions. We remark that our generic constructions can work with *any* signature scheme and *any* hard relation. In particular, our constructions can be instantiated with any of the NIST PQC Round 3 candidate schemes. For example, we could use Rainbow, which is NIST PQC round 3 [16] finalist based on multivariate polynomials, and with any post-quantum-secure SRSR relation, which could be based on lattice problems. More generally, our work shows that linkable adaptor signature is in Minicrypt (Theorem 6), meaning it can be constructed assuming the existence of one-way functions (GAS_1). Our construction of GAS_2 shows that the existence of strong random-self reducible relations implies the existence of unlinkable adaptor signatures (Theorem 8). Moreover, we remark that the overhead of GAS_1 and GAS_2 are also *minimal* in terms of computational overhead and signature size. GAS_1 has virtually no computational overhead and GAS_2 requires re-randomization of instances from the hard relation. An adapted signature for both schemes contains a standard signature of the underlying signature scheme as well as an instance and witness pair for the hard relation.

2 Preliminary

We use $[n]$ for a positive integer n to denote the set $\{1,\ldots,n\}$. Let S be a finite set. We use $x \twoheadleftarrow S$ to denote sampling from set S uniformly at random and assigning the result to variable x. "PT" denotes polynomial-time, also referred to as efficient. We use $\lambda \in \mathbb{N}$ to denote the security parameter. We recall that a function $f : \mathbb{N} \to \mathbb{R}$ is negligible if for every $c \in \mathbb{N}$, there exists $n_c \in \mathbb{N}$ such that $|f(n)| < n^{-c}$ for all $n > n_c$. Algorithms are probabilistic unless specified otherwise. Suppose $\mathcal{A}$ is an algorithm expecting oracles $\mathsf{O}_1,\ldots,$ we use $x \twoheadleftarrow \mathcal{A}^{\mathsf{O}_1,\cdots}(\cdots)$ to denote an execution of algorithm $\mathcal{A}$ and assigning its output to variable x. We use $[\mathcal{A}^{\mathsf{O}_1,\cdots}(\cdots)]$ to denote the set of all possible outputs of algorithm $\mathcal{A}$. We use $S \xleftarrow{\cup} x$ to denote adding an element x to the set S. Integer variables are initialized to 0 and set variables are initialized to the empty set. We adopt the code-based game-playing framework of [3]. A game $\mathbf{G}$ is usually parameterized by some cryptographic scheme S and an adversary $\mathcal{A}$. A game consists of list of named oracles. The execution of a game is the execution of the Main procedure. Variables in game oracles are global by default. An "Assert" statement inside an oracle call will immediately terminate the execution of the oracle call and return False if the given expression evaluates to False. For an example of a game, see Fig. 3.

Relations and random self-reducibility. We recall that a relation $\mathsf{R} \subseteq \{0,1\}^* \times \{0,1\}^*$ is an NP relation if there exists PT algorithm $\mathsf{R.Vf}$ such that $\mathsf{R.Vf}(Y, y) = \mathsf{True}$ if and only if $(Y, y) \in \mathsf{R}$. The language L_R for relation R is defined as the set $\{Y \in \{0,1\}^* \mid \exists y \in \{0,1\}^* : (Y, y) \in \mathsf{R}\}$. A generator for R is a PT algorithm $\mathsf{R.Gen}$ that on input 1^λ returns a pair $(Y, y) \in \mathsf{R}$. For any positive integer q, we define the q-one-wayness advantage of an adversary $\mathcal{A}$ against R to be $\mathbf{Adv}^{\text{q-ow}}_{\mathsf{R},\mathcal{A}}(\lambda) := \Pr[\mathbf{G}^{\text{q-ow}}_{R,q,\mathcal{A}}(\lambda)]$, where the one-wayness game is given below.

Game $\mathbf{G}^{\text{q-ow}}_{\mathsf{R},\mathcal{A}}(\lambda)$:

```
1 For i ∈ [q] do (Y_i, ·) ↞ R.Gen(1^λ)
2 (I, y') ↞ A(Y_1, ..., Y_q) ; Return (Y_I, y') ∈ R
```

Above, adversary $\mathcal{A}$ returns an index $I \in \{1,\ldots,q\}$ and a guess y'. We say that R is a hard relation if for polynomial q and any PT $\mathcal{A}$, $\mathbf{Adv}^{\text{q-ow}}_{R,\mathcal{A}}(\lambda)$ is negligible. We define the following unique-witness advantage of an adversary $\mathcal{A}$ to be the $\mathbf{Adv}^{\text{uwit}}_{\mathsf{R},\mathcal{A}}(\lambda) := \Pr[(Y, y) \in \mathsf{R}, (Y, y') \in \mathsf{R}, y \neq y' \mid (Y, y, y') \twoheadleftarrow \mathcal{A}(1^\lambda)]$.

We say that a hard relation R is random self-reducible (RSR) if there exists sets $\mathsf{R}.R_\lambda$ and efficient deterministic algorithms $\mathsf{R}.A, \mathsf{R}.B, \mathsf{R}.C$ such that the following holds for any $(Y, y) \in [\mathsf{R}.\mathsf{Gen}(1^\lambda)]$ and $r \twoheadleftarrow \mathsf{R}.R_\lambda$: (1) $Y' \leftarrow \mathsf{R}.A(Y, r)$ is distributed identically to $\mathsf{R}.\mathsf{Gen}(1^\lambda)$. (2) For $y' \leftarrow \mathsf{R}.B(y, r)$ it holds that $(Y', y') \in R$ where $Y' = \mathsf{R}.A(Y, r)$. (3) For $y \leftarrow \mathsf{R}.C(y', r)$, where $y' = \mathsf{R}.B(y, r)$, it holds that $(Y, y) \in \mathsf{R}$.

Pictorial depiction			Game $\mathbf{G}^{\mathsf{correct}}_{\mathsf{aSig},m}(\lambda)$
Alice		**Bob**	1 $(\mathsf{pk}, \mathsf{sk}) \twoheadleftarrow \mathsf{KeyGen}(1^\lambda)$
Public input: pk, Y, m		Public input: pk, Y, m	2 $(Y, y) \twoheadleftarrow R.\mathsf{Gen}(1^\lambda)$
Private input: sk		Private input: y	3 $\hat{\sigma} \twoheadleftarrow \mathsf{pSign}(\mathsf{sk}, m, Y)$
3 $\hat{\sigma} \twoheadleftarrow \mathsf{pSign}(\mathsf{sk}, m, Y)$	$\xrightarrow{\hat{\sigma}}$	4 $b_1 \leftarrow \mathsf{pVrf}(\mathsf{pk}, m, \hat{\sigma})$	4 $b_1 \leftarrow \mathsf{pVrf}(\mathsf{pk}, m, \hat{\sigma}, Y)$
		5 If not b_1 then Abort	5 If not b_1 then Abort
8 $y \leftarrow \mathsf{Ext}(\hat{\sigma}, \sigma, Y)$	$\xleftarrow{\sigma}$	6 $\sigma \leftarrow \mathsf{Adapt}(\mathsf{pk}, \hat{\sigma}, y)$	6 $\sigma \leftarrow \mathsf{Adapt}(\mathsf{pk}, \hat{\sigma}, y)$
9 $b_3 \leftarrow (Y, y) \in R$		7 $b_2 \leftarrow \mathsf{Vrf}(\mathsf{pk}, m, \sigma)$	7 $b_2 \leftarrow \mathsf{Vrf}(\mathsf{pk}, m, \sigma)$
			8 $y \leftarrow \mathsf{Ext}(Y, \hat{\sigma}, \sigma)$
			9 $b_3 \leftarrow (Y, y) \in R$
			10 Return $(b_1 \wedge b_2 \wedge b_3)$

Fig. 1. Left: Pictorial depiction of an honest execution of adaptor signing protocol between Alice and Bob. Alice holds the secret key sk corresponding to her public key pk. Bob holds the witness y corresponding to his public instance Y. We assume that Alice and Bob have agreed on the message m to be signed before the execution of the protocol. In this depiction, we have simplified the release of σ from Bob. In practical settings, Alice could learn σ from an indirect channel, e.g. a public ledger. We also remark that Bob does not need his private input y for lines 4 and 5, and his private input y is only needed for line 6. **Right**: game defining the correctness of an adaptor signature scheme.

Signature schemes. A signature scheme Sig consists of PT algorithms KeyGen, Sign, and Vrf. Via $(\mathsf{pk}, \mathsf{sk}) \twoheadleftarrow \mathsf{KeyGen}(1^\lambda)$, the key generation algorithm generates a public key pk and a secret key sk. Via $\sigma \twoheadleftarrow \mathsf{Sign}(\mathsf{sk}, m)$, the signing algorithms generates a signature σ. Via $b \twoheadleftarrow \mathsf{Vrf}(\mathsf{pk}, m, \sigma)$, the verification algorithm returns a boolean value $b \in \{\mathsf{True}, \mathsf{False}\}$, indicating the validity of the message signature pair. Consider the game $\mathbf{G}^{\text{uf-cma}}_{\mathsf{Sig}}$ given below.

Game $\mathbf{G}^{\text{uf-cma}}_{\mathsf{Sig},\mathcal{A}}(\lambda)$, $\mathbf{G}^{\text{suf-cma}}_{\mathsf{Sig},\mathcal{A}}(\lambda)$

1 $(\mathsf{pk}, \mathsf{sk}) \twoheadleftarrow \mathsf{KeyGen}(1^\lambda)$; $(m, \sigma) \twoheadleftarrow \mathcal{A}^{\mathsf{Sign}}(\mathsf{pk})$; Assert $\mathsf{Vrf}(\mathsf{pk}, m, \sigma)$
2 $\mathbf{G}^{\text{uf-cma}}_{\mathsf{Sig},\mathcal{A}}$: Return $(m \notin S)$
3 $\mathbf{G}^{\text{suf-cma}}_{\mathsf{Sig},\mathcal{A}}$: Return $((m, \sigma) \notin U)$

$\mathsf{Sign}(m)$:
4 $\sigma \twoheadleftarrow \mathsf{Sign}(\mathsf{sk}, m)$; $S \leftarrow S \cup \{m\}$; $U \leftarrow U \cup \{(m, \sigma)\}$; Return σ

We define the (S)UF-CMA advantage of an adversary $\mathcal{A}$ against Sig to be $\mathbf{Adv}^{\text{uf-cma}}_{\mathsf{Sig},\mathcal{A}}(\lambda)$ ($\mathbf{Adv}^{\text{suf-cma}}_{\mathsf{Sig},\mathcal{A}}(\lambda)$). We say that scheme Sig is (S)UF-CMA-secure if $\mathbf{Adv}^{\text{uf-cma}}_{\mathsf{Sig},\mathcal{A}}(\lambda)$ ($\mathbf{Adv}^{\text{suf-cma}}_{\mathsf{Sig},\mathcal{A}}(\lambda)$) is negligible for any PT adversary $\mathcal{A}$.

3 Definitions and Relations

The inception of Adaptor signatures was due to the idea of "script-less scripts" by Polstra, who proposed ways to construct applications of atomic swaps [20] and atomic multi-hop payments [19] without the use of special spending scripts. In a follow-up work, Malavolta el al. implicitly gave constructions of Schnorr and ECDSA adaptor signatures in their work on anonymous multi-hop locks for payment channel networks [15]. The notion of adaptor signature was formally defined and studied by [2] and independently by [10]. Follow-up works have constructed two-party adaptor signatures [8] and post-quantum adaptor signatures [9,24], as well as building other applications on top of adaptor signatures [1]. All follow-up works follow the security definitions of [2]. Our definitions align with that of [2] in terms of syntax, correctness, and basic security (called pre-signature adaptability). Our framework deviates from and significantly improves upon the main security definitions of [2].

Notion(s)	Secret key holder Alice	Witness holder Bob	Observer Carol
Correctness	Honest	Honest	-
Adaptability	Malicious	Honest	-
Extractability	Honest	Malicious	-
Unlink (new)	Honest	Honest	Malicious

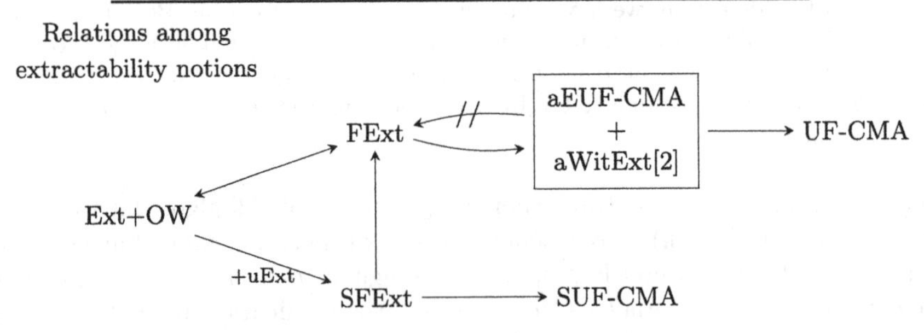

Fig. 2. Top: Guide to correctness and security notions. **Bottom**: Relations among security notions for the secret-key holder (Alice). An arrow A → B means that if an adaptor signature scheme is A-secure, then it is B-secure. Additional assumptions are marked on top of arrows. All implications are tight, meaning reductions preserve running time and success advantage up to small additive constants.

Syntax and correctness. Let $\mathsf{Sig} = (\mathsf{KeyGen}, \mathsf{Sign}, \mathsf{Vrf})$ be a signature scheme. Let R be a hard relation with generator Gen. An adaptor signature scheme aSig for signature scheme Sig and relation R specifies (probabilistic) algorithm pSign, and deterministic algorithms pVrf, Adapt, and Ext. We assume that algorithms of Sig, i.e. $\mathsf{KeyGen}, \mathsf{Sign}, \mathsf{Vrf}$, are additionally defined to be algorithms of aSig. Let $(\mathrm{pk}, \mathrm{sk}) \in [\mathsf{Sig}.\mathsf{KeyGen}(1^\lambda)]$ and $(Y, y) \in [\mathsf{Gen}(1^\lambda)]$. The adaptor signature algorithms behave as follows.

- Via $\hat{\sigma} \twoheadleftarrow \mathsf{pSign}(\mathrm{sk}, m, Y)$ the pre-sign algorithm generates a pre-signature.
- Via $b \leftarrow \mathsf{pVrf}(\mathrm{pk}, m, \hat{\sigma}, Y)$, the pre-signature verification returns a boolean.
- Via $\sigma \leftarrow \mathsf{Adapt}(\mathrm{pk}, \hat{\sigma}, y)$, the adapt algorithm returns a signature σ.
- Via $y \leftarrow \mathsf{Ext}(\sigma, \hat{\sigma}, Y)$, the extract algorithm extracts a witness y.

Game $\mathbf{G}^{\mathrm{fext}}_{\mathsf{aSig},\mathcal{A}}(\lambda)$, $\mathbf{G}^{\mathrm{sfext}}_{\mathsf{aSig},\mathcal{A}}(\lambda)$

1 $(\mathrm{pk}, \mathrm{sk}) \twoheadleftarrow \mathsf{KeyGen}(1^\lambda)$
2 $(m^*, \sigma^*) \twoheadleftarrow \mathcal{A}^{\mathrm{NewY,Sign,pSign}}(\mathrm{pk})$
3 Assert $\mathsf{Vrf}(\mathrm{pk}, m^*, \sigma^*)$ // Forgery must be valid
4 $\mathbf{G}^{\mathrm{fext}}_{\mathsf{aSig},\mathcal{A}}$: Assert $(m^* \notin S)$
5 $\mathbf{G}^{\mathrm{sfext}}_{\mathsf{aSig},\mathcal{A}}$: Assert $((m^*, \sigma^*) \notin U)$
6 Ret $\big(\forall (Y, \hat{\sigma}) \in T[m^*]$ st $Y \notin C$ // All adversarial $Y, \hat{\sigma}$
$: (Y, \mathsf{Ext}(Y, \hat{\sigma}, \sigma^*)) \notin R\big)$ // Extraction fails

NewY:
7 $(Y, \cdot) \twoheadleftarrow \mathsf{R.Gen}(1^\lambda)$; $C \overset{\cup}{\leftarrow} Y$; Return Y

Sign(m):
8 $\sigma \twoheadleftarrow \mathsf{Sign}(\mathrm{sk}, m)$
9 $S \overset{\cup}{\leftarrow} m$
10 $U \overset{\cup}{\leftarrow} (m, \sigma)$
11 Return σ

pSign(m, Y):
12 $\hat{\sigma} \twoheadleftarrow \mathsf{pSign}(\mathrm{sk}, m, Y)$
13 $T[m] \overset{\cup}{\leftarrow} (Y, \hat{\sigma})$
14 Return $\hat{\sigma}$

Fig. 3. Games defining (strong) full extractability notions.

To introduce correctness, let us first consider an example of honest execution of adaptor signature between two parties Alice and Bob. Consider the protocol given in Fig. 1. In typical usage of adaptor signatures, algorithms pSign and Ext are executed by some party (Alice) holding sk, and pVrf and Adapt are executed by a party (Bob) holding secret witness y. We note that Alice and Bob can execute the protocol up to step 5 even if Bob does not know secret witness y; on the other hand, execution of steps 6–9 requires Bob's knowledge of the witness y. We say that aSig is *correctness* if for all message m, $\Pr[\mathbf{G}^{\mathrm{correct}}_{\mathsf{aSig},m}(\lambda)] = 1$. We remark that inputs to extraction are all public, meaning that any external observer can extract out a witness y. This is an intended property of adaptor signatures which allows Alice to delegate witness extraction to third parties Figs 2 and 7.

Adaptor signatures with canonical signing. Any adaptor signature scheme aSig gives an alternative way to generating signatures via pSign and Adapt. Specifically, we consider the following signing algorithm.

Algorithm $\mathsf{Sign}'(m)$

1 $(Y, y) \twoheadleftarrow \mathsf{R.Gen}(1^\lambda)$; $\hat{\sigma} \twoheadleftarrow \mathsf{pSign}(\mathrm{sk}, m, Y)$; Return $\sigma \leftarrow \mathsf{Adapt}(\mathrm{pk}, \hat{\sigma}, y)$

We say that an adaptor signature scheme has *canonical* signing if the above signing algorithm gives signatures that are *identically distributed* to those given by Sign for any secret-key sk and message m. Any adaptor signature scheme can be turned into a canonical one by simply replacing the signing algorithm with the one defined above. All schemes considered in this work are canonical without modifications.

Security of Bob: Adaptability. The first security notion we introduce is called *adaptability*. Intuitively, it guarantees that if Bob is convinced of the validity of the pre-signature $\hat{\sigma}$ and knows a corresponding witness y, then Bob can generate a valid signature σ. More specifically, we ask that for any pk, m, and $\hat{\sigma}$, if $(Y, y) \in R$ and $\mathsf{pVrf}(\mathrm{pk}, m, \hat{\sigma}, Y)$ returns true, it must be that $\mathsf{Adapt}(\mathrm{pk}, \sigma, y)$ returns a valid signature σ on message m wrt to public key pk. Referring back to Fig. 1, pre-signature adaptability guarantees the safety of Bob—he can always turn a valid pre-signature into a signature if he learns a corresponding witness. Formally, consider the following game.

Game $\mathbf{G}^{\mathrm{adapt}}_{\mathsf{aSig},\mathcal{A}}(\lambda)$

1 $(\mathrm{pk}, \mathrm{sk}) \leftarrow\!\!\leftarrow \mathsf{KeyGen}(1^\lambda)$; $(m, \hat{\sigma}, (Y, y)) \leftarrow\!\!\leftarrow \mathcal{A}(\mathrm{pk})$
2 Assert $((Y, y) \in R \wedge \mathsf{pVrf}(\mathrm{pk}, m, \hat{\sigma}, Y))$
3 Return $\mathsf{Vrf}(\mathrm{pk}, m, \mathsf{Adapt}(\mathrm{pk}, \hat{\sigma}, y))$

We say that aSig satisfies *pre-signature adaptability* if $1 - \Pr[\mathbf{G}^{\mathrm{adapt}}_{\mathsf{aSig},\mathcal{A}}(\lambda)]$ is negligible for all PT adversary $\mathcal{A}$. We say that aSig has *perfect pre-signature adaptability* if for all adversary $\mathcal{A}$, $\Pr[\mathbf{G}^{\mathrm{adapt}}_{\mathsf{aSig},\mathcal{A}}(\lambda)] = 1$. Our definition of pre-signature adaptability aligns with that of [2].

Security of Alice: Full Extractability (FExt). The most involved part of the security definition is for the secret-key holder of the signature scheme (Alice). This is where previous definitions fall short. We give a unified definition of security for the safety of the secret-key holder. Before we give the formal definition, we give a high-level description of the available attack surfaces. A secret-key holder, Alice, could potentially expose the following interfaces.

- $\mathsf{Sign}(\mathrm{sk}, m)$ for adversarially chosen m. Exposing such signing oracle models the applications in which honestly generated signatures of Alice are released. Note that we cannot prevent previous honestly generated signatures from being valid. This is similar to the signing oracle that is in UF-CMA and SUF-CMA notions. In fact, we will consider two variants of security for Alice as well.
- $\mathsf{pSign}(\mathrm{sk}, m, Y)$ for adversarially chosen m and instance Y. This models all interactions that Alice could have with external parties where Alice gives out pre-signatures. Each query generates a tuple $(m, Y, \hat{\sigma})$, and we store them in a table indexed by message m, i.e. each query adds the pair $(Y, \hat{\sigma})$ to the set $T[m]$, which is initialized to the empty set.
- Forgery guarantee: For Alice, after given out many signatures and pre-signatures, the following guarantee is desired: if some forgery (m^*, σ^*) is given by an adversary, then one of the following must hold: (1) (m^*, σ^*) must have come from a signing query (2) There must be a corresponding tuple $(Y, \hat{\sigma})$ in table $T[m]$ such that σ^* gives a valid extraction, i.e. $\mathsf{Ext}(Y, \hat{\sigma}, \sigma^*)$ gives some y such that $(Y, y) \in \mathsf{R}$.

Additionally, in the above scenario where Bob (the adversary) returns a signature that extracts, we would like to additionally restrict the *instances* with respect to which extraction could happen. Specifically, we would like to separate instances given to pSign into two categories: (1) those for which Bob knows a witness and (2) those for which Bob does not know a witness. This is to ensure that even if Bob learns a pre-signature on some instance Y for which it does not know the witness, it cannot adapt it into a valid signature. This is achieved in the formal security notion by introducing an oracle NewY that samples honest instances for Bob.

To summarize, extractability guarantees that if Alice gives out signatures and pre-signatures, then the only forgery that some adversary Bob can give is (1) those already given by Alice as signatures (2) some forgery that leads to a valid extraction of a witness. Formally, consider the games $\mathbf{G}^{\mathsf{fext}}_{\mathsf{aSig},\mathcal{A}}$ and $\mathbf{G}^{\mathsf{sfext}}_{\mathsf{aSig},\mathcal{A}}$ given in Fig. 3. In either game, the adversary is given some honestly sampled public key pk and has access to oracles NewY, Sign, pSign (line 2). Each query to Sign is recorded as the allowed forgery budget (set S for the normal case and U for the strong case). Each query m, Y to pSign derives some $\hat{\sigma}$ (line 15) and these values are recorded in table $T[m]$ (line 14). In the end, to win the game, the adversary needs to produce a valid forgery (line 2–3) which is fresh (line 4 or 5) and does not produce any valid extraction (line 6) against instances chosen by Bob (i.e. excluding those Y generated by NewY). Note that the full extractability game only requires the forgery to be on a fresh message (line 4), and the strong full extractability game requires the entire forgery to be fresh (line 5). This differentiation is consistent with the difference between UF-CMA and SUF-CMA security of signatures. We define the (S)FExt advantage of an adversary $\mathcal{A}$ to be $\mathbf{Adv}^{\mathsf{fext}}_{\mathsf{aSig},\mathcal{A}}(\lambda)$ ($\mathbf{Adv}^{\mathsf{sfext}}_{\mathsf{aSig},\mathcal{A}}(\lambda)$), and we say that scheme aSig is (S)FExt-secure if the corresponding advantage function if negligible for efficient adversaries.

Intuitively, if messages to be signed and pre-signed has high entropy and do not repeat, then it suffices to only assume FExt. On the other hand, if applications expect messages to repeat, then it is crucial to additionally aim for SFExt.

Strict implications to notions given by [2]. We show that FExt implies previous notions of unforgeability and extractability (Theorem 1), formally aEUF-CMA and aWitExt as defined in [2]. Their notions have since been adopted in following works [1,8,9,24]. Roughly, their security notion guarantees "unforgeability" and "extractability" against a single challenge pre-sign query. Unlike our full extractability notion, their notions specify an *explicit* phase for the challenge message selection and forgery generation. This results in the adversary only learning *exactly one* pre-signature $\hat{\sigma}^* = \mathsf{pSign}(\mathrm{sk}, m^*, Y^*)$ on challenge message m^* and instance Y^*. It is not hard to see that our notion of FExt implies aEUF-CMA and aWitExt. On the other hand, we give a counterexample scheme to show that such implication is *strict* (Theorem 2), meaning there are schemes which are aEUF-CMA- and aWitExt-secure that are not FExt-secure. We give a sketch of the counterexample below and give the full analysis in Sect. 3.1.

Counterexample. We will modify a secure (in the sense of aEUF-CMA and aWitExt) adaptor signature scheme aSig so that pSign leaks an additional signature on message m if and only if pSign is called with the same message m and instance Y more than once. In more detail, pSign will do the following: upon a query $\mathsf{pSign}(\mathrm{sk}, m, Y)$, we first compute a pre-signature $\hat{\sigma}$ exactly as in aSig. Additional to $\hat{\sigma}$, pSign will return some string C_b for a random bit b where (1) C_0 is a random encryption pad and (2) $C_1 = C_0 \oplus \sigma$ is a one-time pad encryption of a signature $\sigma = \mathsf{Sign}(\mathrm{sk}, m; r)$ under randomness r. To make these values consistent across different runs of pSign, we derive values of (C_0, r) via a PRF F (with secret key sk) applied to the input (m, Y), i.e. $(C_0, r) \leftarrow \mathsf{F}(\mathrm{sk}, (m, Y))$. Hence, any single call to $\mathsf{pSign}(m, Y)$ does not reveal any information. Indeed, We will show that the resulting scheme is secure in the sense of aEUF-CMA and aWitExt. However, even two calls $\mathsf{pSign}(m, Y)$ reveals a fresh signature with half probability, hence breaking FExt security.

Our counterexample scheme demonstrates that security guaranteed by aEUF-CMA and aWitExt is weaker than expected for applications where many protocols might be executed concurrently since both notions only guarantee security against a single challenge instance. Moreover, aEUF-CMA is also a *selective* notion, in that the honest instance Y_1 is not known to the adversary until *after* the adversary selects a challenge message m^*. This further weakens the security guaranteed. In contrast, our security notion FE gives the challenge instance Y_1 to the adversary and allows any number of challenge queries to pSign.

Modular proofs from simpler notions. Full extractability and strong full extractability are fairly complex notions, where the adversary is given many attack interfaces. To facilitate simpler proofs and better intuitive understanding, we give a framework in Sect. 3.2 for proving FExt and SFExt security. In particular, we show that proofs can be modularized if a simplified notion called *simple extractability* is achieved (Theorem 3). Roughly, simple extractability removes the interfaces of NewY and Sign. Furthermore, we show that if the adaptor signature scheme is also *uniquely extractable* then it also satisfies strong full extractability (Theorem 4). Roughly, unique extractability says that with access to an oracle pSign, an adversary cannot find two valid signatures that both extracts.

New privacy notion: unlinkability. Unlinkability requires adapted signatures, using pSign and Adapt, to be indistinguishable from honestly generated signatures from Sign, even with adversarial access to pre-signatures and signatures. Formally, consider the following game $\mathbf{G}^{\mathrm{unlink}}_{\mathsf{aSig},\mathcal{A}}$.

Game $\mathbf{G}^{\mathrm{unlink}}_{\mathsf{aSig},\mathcal{A}}(\lambda)$	SignChl$(m, (Y, y))$:	Sign(m):
1 $b \twoheadleftarrow \{0,1\}$	5 Assert $((Y, y) \in R)$	10 $\sigma \twoheadleftarrow \mathsf{Sign}(\mathrm{sk}, m)$
2 $(\mathrm{pk}, \mathrm{sk}) \twoheadleftarrow \mathsf{KeyGen}(1^\lambda)$	6 $\hat{\sigma} \twoheadleftarrow \mathsf{pSign}(\mathrm{sk}, m, Y)$	11 Return σ
3 $b' \twoheadleftarrow \mathcal{A}^{\mathrm{SignChl,Sign,pSign}}(\mathrm{pk})$	7 $\sigma_0 \twoheadleftarrow \mathsf{Adapt}(\mathrm{pk}, \hat{\sigma}, y)$	pSign(m, Y):
4 Return $(b = b')$	8 $\sigma_1 \twoheadleftarrow \mathsf{Sign}(\mathrm{sk}, m)$	12 $\hat{\sigma} \twoheadleftarrow \mathsf{pSign}(\mathrm{sk}, m, Y)$
	9 Return σ_b	13 Return $\hat{\sigma}$

We define the unlink-advantage of an adversary $\mathcal{A}$ against adaptor signature scheme aSig to be $\mathbf{Adv}^{\mathrm{unlink}}_{\mathsf{aSig},\mathcal{A}}(\lambda) = 2\Pr[\mathbf{G}^{\mathrm{unlink}}_{\mathsf{aSig},\mathcal{A}}(\lambda)] - 1$. We say that scheme aSig is (1) unlinkable if $\mathbf{Adv}^{\text{unlink}}_{\mathsf{aSig},\mathcal{A}}(\lambda)$ is negligible for efficient adversaries (2) perfectly unlinkable if the unlink advantage is 0 for any adversary.

We briefly explain how unlinkability guarantees on-chain privacy for the application of atomic swaps. In atomic swaps, Alice and Bob aim to atomically post signature σ_A (for some message m_A) and σ_B (for some message m_B) to a public ledger. To do this, Bob would first generate a pair (Y, y) and gives Alice his pre-signature $\hat{\sigma}_B \twoheadleftarrow \mathsf{pSign}(\mathrm{sk}_B, m_B, Y)$. Alice will verify the validity of such pre-signature before giving her pre-signature $\hat{\sigma}_A \twoheadleftarrow \mathsf{pSign}(\mathrm{sk}_A, m_A, Y)$ to Bob. Now, Bob can adapt the pre-signature of Alice to a valid signature, via $\sigma_A \leftarrow \mathsf{Adapt}(\mathrm{pk}_A, \hat{\sigma}_A, y)$, and post to the ledger. But if Bob does so, then Alice can extract witness y via $y \leftarrow \mathsf{Ext}(Y, \hat{\sigma}_A, \sigma_A)$ and adapt the pre-signature of Bob to a valid signature using witness y. Unlinkability of the adaptor signature scheme ensures that the adapted signatures σ_A and σ_B to be indistinguishable from honestly generated signatures. Hence, an outside observer cannot deduce that σ_A and σ_B are "linked."

Game $\mathbf{G}^{\text{aEUFCMA}}_{\mathsf{aSig},\mathcal{A}}(\lambda)$

1 $(\mathrm{pk}, \mathrm{sk}) \twoheadleftarrow \mathsf{KeyGen}(1^\lambda)$
2 $(Y^*, y^*) \twoheadleftarrow \mathsf{Gen}(1^\lambda)$
3 $\sigma^* \twoheadleftarrow \mathcal{A}^{\mathrm{Sign},\mathrm{pSign},\mathrm{pSignChl}}(\mathrm{pk})$
4 Return $(\mathsf{Vrf}(\mathrm{pk}, m^*, \sigma^*) \wedge m^* \notin S)$

pSignChl(m^*): // Exactly once
5 $\hat{\sigma} \twoheadleftarrow \mathsf{pSign}(\mathrm{sk}, m, Y^*)$; Return $(\hat{\sigma}, Y^*)$

Sign(m):
6 $S \leftarrow S \cup \{m\}$
7 Return $\sigma \twoheadleftarrow \mathsf{Sign}(\mathrm{sk}, m)$

pSign(m, Y):
8 $S \leftarrow S \cup \{m\}$
9 Return $\hat{\sigma} \twoheadleftarrow \mathsf{pSign}(\mathrm{sk}, m, Y)$

Game $\mathbf{G}^{\text{aWitExt}}_{\mathsf{aSig},\mathcal{A}}(\lambda)$

1 $(\mathrm{pk}, \mathrm{sk}) \twoheadleftarrow \mathsf{KeyGen}(1^\lambda)$
2 $\sigma^* \twoheadleftarrow \mathcal{A}^{\mathrm{Sign},\mathrm{pSign},\mathrm{pSignChl}}(\mathrm{pk})$
3 Assert $(\mathsf{Vrf}(\mathrm{pk}, m^*, \sigma^*) \wedge m^* \notin S)$
4 Return $((Y^*, \mathsf{Ext}(Y^*, \hat{\sigma}^*, \sigma^*)) \notin R)$

pSignChl(m^*, Y^*): // Exactly once
5 Return $\hat{\sigma}^* \twoheadleftarrow \mathsf{pSign}(\mathrm{sk}, m, Y^*)$

Sign(m):
6 $S \leftarrow S \cup \{m\}$
7 Return $\sigma \twoheadleftarrow \mathsf{Sign}(\mathrm{sk}, m)$

pSign(m, Y):
8 $S \leftarrow S \cup \{m\}$
9 Return $\hat{\sigma} \twoheadleftarrow \mathsf{pSign}(\mathrm{sk}, m, Y)$

Fig. 4. Games defining aEUF-CMA and aWitExt notions [2] for adaptor signature scheme aSig.

3.1 Relations with Previous Notions

Restating aEUF-CMA and aWitExt notions of [2]. Their security notions are defined as games where the adversary is run in two stages. We note that it is not clear (in their pseudocode) if the first and second stage of the adversaries are allowed to share any state. Hence, we take the stronger interpretation that implicit state-sharing is allowed. To keep the presentation consistent, we slightly rewrite their security games (while preserving the semantics, assuming the state of first stage adversary is passed to the second stage) so that there is only a single stage. The functionality of the two stages is instead realized via a challenge oracle that can only be called once[2]. Formally, we introduce the notions of [2] by considering games $\mathbf{G}^{\text{aEUFCMA}}$ and $\mathbf{G}^{\text{aWitExt}}$ given in Fig. 4.

Full extractability implies previous notions. We first show that our notion of full extractability implies both aEUF-CMA and aWitExt. Specifically, we show that given any adversary attacking aEUF-CMA or aWitExt, then we can give an adversary attacking FExt. Notice that an aEUF-CMA adversary or aWitExt adversary specifies a message m^* and eventually returns σ^*. We will construct a FExt adversary that (1) forwards all Sign and pSign queries (2) simulates pSignChl oracle with pSign while recording m^* and (3) returns exactly m^*, σ^* at the end. The proof below checks that these adversaries win the FExt game whenever the starting adversary wins aEUF-CMA game or aWitExt game.

Theorem 1 (FExt $\implies$ aEUF-CMA + aWitExt). *Let $\mathcal{A}_{\text{aEUFCMA}}$ and $\mathcal{A}_{\text{aWitExt}}$ be aEUF-CMA and aWitExt adversaries. The proof constructs adversaries $\mathcal{A}_{\text{fext},1}$ and $\mathcal{A}_{\text{fext},2}$, which have the same running time as the given adversaries, such that*

$$\Pr[\mathbf{G}^{\text{aEUFCMA}}_{\mathsf{aSig},\mathcal{A}_{\text{aEUFCMA}}}(\lambda)] = \mathbf{Adv}^{\text{fext}}_{\mathsf{aSig},\mathcal{A}_{\text{fext},1}}(\lambda)\,, \tag{1}$$

$$\Pr[\mathbf{G}^{\text{aWitExt}}_{\mathsf{aSig},\mathcal{A}_{\text{aWitExt}}}(\lambda)] = \mathbf{Adv}^{\text{fext}}_{\mathsf{aSig},\mathcal{A}_{\text{fext},2}}(\lambda)\,. \tag{2}$$

Proof (of Theorem 1). Consider the following adversaries $\mathcal{A}_{\text{fext},1}$ and $\mathcal{A}_{\text{fext},2}$.

Adversary $\mathcal{A}^{\text{NewY,Sign,pSign}}_{\text{fext},1}(\text{pk})$:	Adversary $\mathcal{A}^{\text{NewY,Sign,pSign}}_{\text{fext},2}(\text{pk})$:
1 $\sigma^* \twoheadleftarrow \mathcal{A}^{\text{Sign,pSign,pSignChl}}_{\text{aEUFCMA}}(\text{pk})$	1 $\sigma^* \twoheadleftarrow \mathcal{A}^{\text{Sign,pSign,pSignChl}}_{\text{aWitExt}}(\text{pk})$
2 Return (m^*, σ^*)	2 Return (m^*, σ^*)
pSignChl(m):	pSignChl(m, Y):
3 $m^* \leftarrow m$; $Y^* \leftarrow \text{NewY}()$	3 $m^* \leftarrow m$; $Y^* \leftarrow Y$
4 Return $\hat{\sigma}^* \twoheadleftarrow \text{pSign}(m^*, Y^*)$	4 Return $\hat{\sigma}^* \twoheadleftarrow \text{pSign}(m^*, Y^*)$

We check that the given adversaries wins if the given reduction adversary does. For aEUF-CMA, $\mathcal{A}_{\text{aEUFCMA}}$ wins if m^* has not been queried previously to either Sign or pSign oracles of game $\mathbf{G}^{\text{aEUFCMA}}$. Furthermore, for the single pSignChl query that $\mathcal{A}_{\text{aEUFCMA}}$ makes, our reduction adversary $\mathcal{A}_{\text{fext},1}$ uses an instance Y^* from NewY to derive a presignature σ^*. Therefore, for $\mathcal{A}_{\text{fext},1}$ the table at m^* is a singleton set, i.e. $T[m^*] = \{(Y^*, \hat{\sigma}^*)\}$ for which Y^* is in the challenge set of instances C. Hence, our adversary $\mathcal{A}_{\text{fext},1}$ wins as long as signature σ^* is valid on m^*, which is exactly the condition that $\mathcal{A}_{\text{aEUFCMA}}$ needs to satisfy to win as well. This justifies (1).

A similar analysis holds for $\mathcal{A}_{\text{aWitExt}}$, namely that $T[m^*] = \{(Y^*, \hat{\sigma}^*)\}$ at the end of its execution. Notice that $\mathcal{A}_{\text{aWitExt}}$ wins if $\mathsf{Ext}(Y^*, \hat{\sigma}^*, \sigma^*)$ is not a valid

[2] One can think of the "first stage" as everything leading up to the challenge oracle call and the "second stage" being everything following the challenge oracle call.

extraction, which is exactly what $\mathcal{A}_{\mathrm{fext},2}$ needs to satisfy as well to win. This justifies (2). □

Insufficiency of previous notions. To show that previous notions are insufficient, we give a scheme that is secure against previous notions, namely aEUF-CMA and aWitExt, but not FExt-secure. The informal intuition on our scheme is as follows. We will modify a secure adaptor signature scheme aSig so that pSign leaks (depending on a random coin flip) either (1) a one-time encryption pad C_0 or (2) a one-time encryption $C_1 = C_0 \oplus \sigma$, where σ is a signature on m. We use a PRF to derive C_0 and the signing randomness for σ so that they are consistent across different runs of pSign.

Formally, let aSig be any FE-secure adaptor signature for underlying signature scheme Sig and relation R. Suppose that $\mathsf{Sig.KeyGen}(1^\lambda)$ returns secret keys that are uniformly distributed over some set S_λ. Let $L_\lambda = \{Y \mid \exists y : (Y, y) \in [\mathsf{Gen}(1^\lambda)]\}$. Let $(\mathrm{pk}, \cdot) \in [\mathsf{Sig.KeyGen}(1^\lambda)]$. Suppose $\mathsf{Sig.Sign}(\mathrm{pk}, \cdot)$ uses random coins of length at most r_λ and that $[\mathsf{Sig.Sign}(\mathrm{pk}, \cdot)] \subseteq \{0,1\}^{n_\lambda}$. Let F be a pseudo-random function of the form $F_\lambda : S_\lambda \times (\{0,1\}^* \times L_\lambda) \to \{0,1\}^{n_\lambda} \times \{0,1\}^{r_\lambda}$, where S_λ is the key space and $\{0,1\}^* \times L_\lambda$ is the input space. Consider adaptor signature scheme AS_0 given below.

```
Scheme AS_0
pSign(sk, m, Y):                          Adapt(pk, σ̂, y):
 1  b ←$ {0,1}                             9  (σ̂', C) ← σ̂
 2  σ̂' ←$ aSig.pSign(sk, m, Y)            10  σ ← Adapt(pk, σ̂', y)
 3  (C_0, r) ← F_sk(m, Y)                 11  Return σ
 4  σ ← Sig.Sign(sk, m; r)                Ext(σ̂, σ, Y):
 5  C_1 ← C_0 ⊕ σ                         12  (σ̂', C) ← σ̂
 6  Return (σ̂, C_b)                       13  y ← aSig.Ext(σ̂', σ, Y)
pVrf(pk, m, σ̂, Y):                        14  Return y
 7  (σ̂', C) ← σ̂
 8  Return aSig.pVrf(pk, m, σ̂', Y)
```

We claim that AS_0 satisfies aEUF-CMA and aWitExt (if aSig is FExt-secure and F is a secure PRF), but AS_0 is not FE-secure even if aSig is FE-secure. Intuitively, any single run of $\mathsf{pSign}(\mathrm{sk}, m, Y)$ only leaks a presignature, as the second part of the output C_b is random. However, given any two evaluations of $(\hat{\sigma}_i, C_i) \leftarrow\!\!\$\ \mathsf{pSign}(\mathrm{sk}, m, Y)$ for $i \in \{1,2\}$, it holds with probability $1/2$ that $\sigma = C_1 \oplus C_2$ is a valid signature on m. This breaks extractability. Since in the FE game, adversaries are allowed to query pSign any number of times for any given Y, even for the challenge instance Y_1. However, for aEUF-CMA and aWitExt, the adversary is only allowed to call pSignChl exactly once. This means that pSignChl does not leak an extra signature and the scheme can be shown to satisfy aEUF-CMA and aWitExt.

Theorem 2 (aEUF-CMA+aWitExt $\not\Rightarrow$ FExt). *Scheme* AS_0 *satisfies aEUF-CMA and aWitExt if* aSig *does and F is a secure PRF. In particular, for any adversary* $\mathcal{A}_{\mathrm{aEUFCMA}}$ *and* $\mathcal{A}_{\mathrm{aWitExt}}$, *the proof gives reduction adversaries*

$\mathcal{A}'_{\mathrm{aEUFCMA}}$, $\mathcal{A}'_{\mathrm{aWitExt}}$, $\mathcal{A}_{\mathrm{prf},1}$, *and* $\mathcal{A}_{\mathrm{prf},2}$, *all about as efficient as the starting adversaries, such that*

$$\Pr[\mathbf{G}^{\mathrm{aEUFCMA}}_{\mathsf{AS}_0,\mathcal{A}_{\mathrm{aEUFCMA}}}(\lambda)] \leq \Pr[\mathbf{G}^{\mathrm{aEUFCMA}}_{\mathsf{aSig},\mathcal{A}'_{\mathrm{aEUFCMA}}}(\lambda)] + \mathbf{Adv}^{\mathrm{prf}}_{F,\mathcal{A}_{\mathrm{prf},1}}(\lambda)\,, \tag{3}$$

$$\Pr[\mathbf{G}^{\mathrm{aWitExt}}_{\mathsf{AS}_0,\mathcal{A}_{\mathrm{aWitExt}}}(\lambda)] \leq \Pr[\mathbf{G}^{\mathrm{aWitExt}}_{\mathsf{aSig},\mathcal{A}'_{\mathrm{aWitExt}}}(\lambda)] + \mathbf{Adv}^{\mathrm{prf}}_{F,\mathcal{A}_{\mathrm{prf},2}}(\lambda)\,. \tag{4}$$

However, scheme AS_0 *is not Ext-secure even if* aSig *is. In particular, the proof gives efficient adversary* $\mathcal{A}_{\mathrm{ext}}$, *that makes two queries to* pSign, *such that*

$$\mathbf{Adv}^{\mathrm{fext}}_{\mathsf{AS}_0,\mathcal{A}_{\mathrm{fext}}}(\lambda) = \frac{1}{2}\,. \tag{5}$$

3.2 Modular Proofs from Simple Notions

Full extractability and strong full extractability are fairly complex notions, where the adversary is given many attack interfaces. To facilitate simpler proofs and better intuitive understanding, we give a framework for proving FExt and SFExt security. In particular, we show that proofs can be modularized if a simpler extractability notion (Ext) is achieved.

Simple Extractability (Ext). We first define simple extractability (Ext). The notion eliminates some of the attack surfaces considered in FExt *without* weakening the security guaranteed. The formal definition is given in Fig. 5. The adversary is given access to a pre-signature oracle pSign and, to win, must produce a valid signature that does not extract against *any* previous queries to pSign. We show that, assuming R is a hard relation, Ext implies FExt.

Theorem 3 (Ext + OW $\implies$ FExt). *Let* aSig *be an adaptor signature scheme for a hard relation* R. *Suppose it satisfies Ext then it also satisfies FExt. Formally, given any FExt-adversary* $\mathcal{A}_{\mathrm{fext}}$, *we can construct* $\mathcal{A}_{\mathrm{ext}}$ *and* $\mathcal{A}_{\mathrm{ow}}$ *such that*

$$\mathbf{Adv}^{\mathrm{fext}}_{\mathsf{aSig},\mathcal{A}_{\mathrm{fext}}}(\lambda) \leq \mathbf{Adv}^{\mathrm{ext}}_{\mathsf{aSig},\mathcal{A}_{\mathrm{ext}}}(\lambda) + \mathbf{Adv}^{\mathrm{q\text{-}ow}}_{\mathsf{R},\mathcal{A}_{\mathrm{ow}}}(\lambda)\,. \tag{6}$$

If $\mathcal{A}_{\mathrm{fext}}$ *makes* q_{Sign} *and* q_{pSign} *queries to* Sign *and* pSign, *respectively. Then* $\mathcal{A}_{\mathrm{ext}}$ *makes* $q_{\mathrm{Sign}} + q_{\mathrm{pSign}}$ *queries to the* pSign *oracles. Furthermore,* $\mathcal{A}_{\mathrm{ext}}$ *and* $\mathcal{A}_{\mathrm{ow}}$ *are about as efficient as* $\mathcal{A}_{\mathrm{fext}}$.

We give a high-level proof sketch here and a full proof in [6]. The reduction adversary $\mathcal{A}_{\mathrm{ext}}$ will need to simulate NewY and Sign, since it only has access to a pre-signature oracle. The adversary will simulate NewY itself and use pSign to simulate queries to Sign. The latter is possible due to the requirement of aSig to be canonical. In particular, each query to Sign is simulated by first sampling a fresh pair $(Y, y) \in \mathsf{R}$ and a signature is then derived using oracle pSign and Adapt algorithm. In doing so, table T for adversary $\mathcal{A}_{\mathrm{ext}}$ becomes larger than that for $\mathcal{A}_{\mathrm{fext}}$. However, it is not hard to see that for the forgery message m^*, the set $T[m^*]$ is the same for both $\mathcal{A}_{\mathrm{fext}}$ and $\mathcal{A}_{\mathrm{ext}}$, assuming $\mathcal{A}_{\mathrm{fext}}$ wins. This is because $\mathcal{A}_{\mathrm{fext}}$ can only win if the returned message m^* is not in the set S, which means that there were no previous queries of the form Sign(m^*). Finally, we need to make sure that the forgery (m^*, σ^*) does not extract for a challenge instance $Y \in C$ (those Y that was returned by NewY). This event should not happen with high probability since we have assumed that R is hard. Indeed, it is not hard to give a OW adversary whose OW-advantage can be used to upper-bound the probability of this event.

Next, we show that if aSig is shown to satisfy Ext, then we can also show that it satisfies SFExt security if an additional security notion, which is conceptually simple and easy to verify, is satisfied.

SFExt from unique extractability (uExt). Unique extractability requires that an adversary, with access to a pre-signature oracle, cannot find $(m, Y, \hat{\sigma}, \sigma, \sigma')$ where σ, σ' are two distinct valid signatures on message m that also both extracts against instance Y and pre-signature $\hat{\sigma}$. Formally, consider the game $\mathbf{G}^{\text{uext}}$ given in Fig. 5. The adversary has access to a pre-signature oracle returning honestly generated pre-signatures. The adversary wins if it successfully finds two distinct valid signatures σ, σ' that both extracts. We define the uExt advantage of an adversary $\mathcal{A}$ to be $\mathbf{Adv}^{\text{uext}}_{\text{aSig},\mathcal{A}}(\lambda)$. We say that aSig satisfies uExt if the advantage of any efficient adversary is negligible. We show that if an adaptor signature for a hard relation R satisfies Ext, and uExt, then it must also satisfy SFExt.

Game $\mathbf{G}^{\text{ext}}_{\text{aSig},\mathcal{A}}(\lambda)$

1 $(\text{pk}, \text{sk}) \twoheadleftarrow \mathsf{KeyGen}(1^\lambda)$; $(m^*, \sigma) \twoheadleftarrow \mathcal{A}^{\text{pSign}}(\text{pk})$; Assert $\mathsf{Vrf}(\text{pk}, m^*, \sigma)$
2 Return $(\forall (Y, \hat{\sigma}) \in T[m^*] : (Y, \mathsf{Ext}(Y, \hat{\sigma}, \sigma)) \notin R)$ // Extraction fails for all $Y, \hat{\sigma}$

pSign(m, Y):
3 $\hat{\sigma} \twoheadleftarrow \mathsf{pSign}(\text{sk}, m, Y)$; $T[m] \xleftarrow{\cup} (Y, \hat{\sigma})$; Return $\hat{\sigma}$

Game $\mathbf{G}^{\text{uext}}_{\text{aSig},\mathcal{A}}(\lambda)$

1 $(\text{pk}, \text{sk}) \twoheadleftarrow \mathsf{KeyGen}(1^\lambda)$; $(m, Y, \hat{\sigma}, \sigma, \sigma') \twoheadleftarrow \mathcal{A}^{\text{pSign}}(\text{pk})$
2 Assert $(\sigma \neq \sigma' \wedge \mathsf{pVrf}(\text{pk}, m, \hat{\sigma}, Y) \wedge \mathsf{Vrf}(\text{pk}, m, \sigma) \wedge \mathsf{Vrf}(\text{pk}, m, \sigma'))$
3 $y \leftarrow \mathsf{Ext}(Y, \hat{\sigma}, \sigma)$; $y' \leftarrow \mathsf{Ext}(Y, \hat{\sigma}, \sigma')$
4 Return $((Y, y) \in R \wedge (Y, y') \in R)$ // Both extraction succeeds

pSign(m, Y):
5 Return $\hat{\sigma} \twoheadleftarrow \mathsf{pSign}(\text{sk}, m, Y)$

Fig. 5. Games defining extractability and unique extractability.

Theorem 4 (Ext + OW + uExt $\implies$ SFExt). *Let* aSig *be a canonical adaptor signature scheme for a hard relation* R. *Suppose it satisfies Ext and uExt, then it also satisfies SFExt. Formally, given any SFExt-adversary* $\mathcal{A}_{\text{sfext}}$, *we can construct* $\mathcal{A}_{\text{ow}}, \mathcal{A}_{\text{ext}}, \mathcal{A}_{\text{uext}}$ *such that*

$$\mathbf{Adv}^{\text{sfext}}_{\text{aSig},\mathcal{A}_{\text{sfext}}}(\lambda) \leq \mathbf{Adv}^{\text{ext}}_{\text{aSig},\mathcal{A}_{\text{ext}}}(\lambda) + \mathbf{Adv}^{\text{q-ow}}_{\text{R},\mathcal{A}_{\text{ow}}}(\lambda) + \mathbf{Adv}^{\text{uext}}_{\text{aSig},\mathcal{A}_{\text{uext}}}(\lambda) . \quad (7)$$

If $\mathcal{A}_{\mathrm{sfext}}$ makes q_{Sign} and q_{pSign} queries to Sign *and* pSign, *respectively. Then $\mathcal{A}_{\mathrm{ext}}$ and $\mathcal{A}_{\mathrm{uext}}$ makes $q_{\mathrm{Sign}} + q_{\mathrm{pSign}}$ queries to their* pSign *oracles. Furthermore, all adversaries are about as efficient as $\mathcal{A}_{\mathrm{sfext}}$.*

We give a high-level proof sketch here and the full proof in [6]. Similar to before, the reduction adversary $\mathcal{A}_{\mathrm{ext}}$ will need to simulate oracles NewY and Sign, which is done exactly as in the proof of Theorem 3. However, for SFExt, we can no longer assume that table $T[m^*]$ is the same for our reduction adversary. This is because $\mathcal{A}_{\mathrm{sfext}}$ can return a valid forgery message m^* for which there were previous queries of the form Sign(m^*). However, since for all simulated Sign queries, the reduction adversary $\mathcal{A}_{\mathrm{ext}}$ already knows a valid signature that extracts (the signature that is returned at the end of the Sign oracle call), we can use the notion of unique extractability to upper bound the probability that the forgery signature σ^* also extracts. On the other hand, if the forgery does not extract, then our reduction adversary $\mathcal{A}_{\mathrm{fext}}$ will win by simply forwarding the forgery (m^*, σ^*). Finally, a reduction to one-wayness of R is done similarly as before to bound the probability that extraction succeeds for some $Y \in C$.

Scheme	Assumptions		Security		
	on Sig	on R	FExt	SFExt	Unlink
GAS_1	UF-CMA	OW	✓	×	×
GAS_1	SUF-CMA	OW, uWit	✓	✓	×
GAS_2	UF-CMA	OW, SRSR	✓	×	✓
GAS_2	SUF-CMA	OW, SRSR, uWit	✓	✓	✓

Fig. 6. Table comparing constructions and their instantiations.

In upshot, to prove FExt or SFExt, we can first show that a scheme aSig satisfies Ext, which implies FExt security. If aSig is shown to additionally satisfy uExt, then we know that aSig is also SFExt-secure.

4 Generic Constructions

In this section, we give generic constructions of adaptor signatures from *any* signature scheme Sig and *any* hard relation R. The reason we use the word "from" but not "for" is that the generated signature σ' is not exactly a standard signature σ for Sig. However, the verification of σ' will require *minimal* modification to verification of σ, in that it only additionally perform a membership check of relation R.

In more detail, our adaptor signature schemes generate signatures σ' that are a combination of a standard signature σ from Sig and a pair (Y, y) from R, i.e. $\sigma' = (\sigma, Y, y)$. Additionally, the verification must perform checks of the standard signature σ as well as a membership check that $(Y, y) \in$ R. Hence, in terms of applications, our adaptor signatures can be supported by the blockchain as only as it supports signature Sig and relation $\mathbb{R}$ as well as basic scripting capabilities. For example, if Sig is taken to be ECDSA over secp256k1 and R is taken to be the relation induced by hashing 256-bit inputs with sha256, then SigR can be realized via a Bitcoin script, which coincide with the construction of a "hash-lock" contract. We first formalize such an "augmented" signature scheme.

Augmented signature schemes SigR. Let Sig be any signature scheme and R be any hard relation. Roughly, the augmented signature scheme SigR is a signature scheme whose signatures additionally (1) attest to an instance Y alongside a message m and (2) releases a valid witness y of instance Y. Formally, the signing and verification algorithms are given in Fig. 7 (key generation is unchanged).

Our constructions can be seen as a generalization to "hash-lock" contracts (coined and used by the Bitcoin Lightning network [21]). We study the notion of adaptor signatures in a general setting where there is no restriction on the underlying signatures schemes. We remark that implementing augmented signature schemes for existing blockchains such as Bitcoin require usage of "scripts."

Scheme SigR

Sign(sk, m):
1 $(Y, y) \twoheadleftarrow R.\mathsf{Gen}(1^\lambda)$
2 $\sigma \twoheadleftarrow \mathsf{Sig.Sign}(\mathrm{sk}, (m, Y))$
3 Return (σ, Y, y)

Vrf(pk, m, σ'):
4 $(\sigma, Y, y) \leftarrow \sigma'$
5 Return $(\mathsf{Sig.Vrf}(\mathrm{pk}, (m, Y), \sigma) \wedge (Y, y) \in R)$

Scheme GAS_1

pSign(sk, m, Y):
1 $\sigma' \twoheadleftarrow \mathsf{Sig.Sign}(\mathrm{sk}, (m, Y))$
2 Return (σ', Y)

pVrf(pk, m, $\hat{\sigma}$, Y):
3 Return $\mathsf{Sig.Vrf}(\mathrm{pk}, (m, Y), \hat{\sigma})$

Adapt(pk, $\hat{\sigma}$, y):
4 $(\sigma', Y) \leftarrow \hat{\sigma}$; $\sigma \leftarrow (\sigma', Y, y)$
5 Return σ

Ext($\hat{\sigma}$, σ, Y):
6 $(\sigma', Y, y) \leftarrow \sigma$
7 Return y

Scheme GAS_2

pSign(sk, m, Y):
1 $r \twoheadleftarrow R.R_\lambda$; $Y' \leftarrow R.A(Y, r)$
2 $\sigma' \twoheadleftarrow \mathsf{Sig.Sign}(\mathrm{sk}, (m, Y'))$
3 Return (σ', Y, r)

pVrf(pk, m, $\hat{\sigma}$, Y):
4 $(\sigma', \cdot, r) \leftarrow \hat{\sigma}$; $Y' \leftarrow R.A(Y, r)$
5 Return $\mathsf{Sig.Vrf}(\mathrm{pk}, (m, Y'), \sigma')$

Adapt(pk, $\hat{\sigma}$, y):
6 $(\sigma', Y, r) \leftarrow \hat{\sigma}$; $y' \leftarrow R.B(y, r)$
7 $Y' \leftarrow \mathrm{R}.A(Y, r)$
8 Return (σ', Y', y')

Ext($\hat{\sigma}$, σ, Y):
9 $(\sigma', Y, r) \leftarrow \hat{\sigma}$; $(\sigma', Y', y') \leftarrow \sigma$
10 $y \leftarrow R.C(y', r)$
11 Return y

Fig. 7. Top: Augmented signature scheme SigR for any signature scheme Sig and relation R. Bottom: adaptor signature schemes GAS_1 and GAS_2 for signature scheme SigR.

We first present a construction GAS_1 that achieves all security properties, but unlinkability.

Generic Adaptor Signature (GAS) 1. Let Sig be any signature scheme and any hard relation R. Consider construction GAS_1 given in Fig. 7. We show that GAS_1 satisfies FExt as long as Sig is Unforgeable, and additionally SFExt if Sig is strongly unforgeable and R has unique witnesses.

Theorem 5. *Adaptor signature scheme* GAS_1 *satisfies correctness and pre-signature adaptability. If* Sig *is UF-CMA-secure and R is one-way then* GAS_1 *is FExt-secure. Given adversary* $\mathcal{A}_{\mathrm{fext}}$*, we can construct adversaries* $\mathcal{A}_{\mathrm{uf\text{-}cma}}$ *and* $\mathcal{A}_{\mathrm{ow}}$*, with running times similar to that of* $\mathcal{A}_{\mathrm{fext}}$ *such that*

$$\mathbf{Adv}^{\mathrm{fext}}_{\mathsf{GAS}_1,\mathcal{A}_{\mathrm{fext}}}(\lambda) \leq \mathbf{Adv}^{\mathrm{uf\text{-}cma}}_{\mathsf{Sig},\mathcal{A}_{\mathrm{uf\text{-}cma}}}(\lambda) + \mathbf{Adv}^{\mathrm{q\text{-}ow}}_{R,\mathcal{A}_{\mathrm{ow}}}(\lambda)\,. \tag{8}$$

Furthermore, if Sig *is SUF-CMA-secure and* R *has unique witnesses, then* GAS_1 *is SFExt-secure. Formally, given adversary* $\mathcal{A}_{\mathrm{sfext}}$*, we can construct adversaries* $\mathcal{A}_{\mathrm{suf\text{-}cma}}$*,* $\mathcal{A}_{\mathrm{ow}}$*, and* $\mathcal{A}_{\mathrm{uwit}}$ *with running times similar to that of* $\mathcal{A}_{\mathrm{sfext}}$ *such that*

$$\mathbf{Adv}^{\mathrm{sfext}}_{\mathsf{GAS}_1,\mathcal{A}_{\mathrm{sfext}}}(\lambda) \leq 2\mathbf{Adv}^{\mathrm{suf\text{-}cma}}_{\mathsf{Sig},\mathcal{A}_{\mathrm{uf\text{-}cma}}}(\lambda) + \mathbf{Adv}^{\mathrm{q\text{-}ow}}_{R,\mathcal{A}_{\mathrm{ow}}}(\lambda) + \mathbf{Adv}^{\mathrm{uwit}}_{\mathsf{R},\mathcal{A}_{\mathrm{uwit}}}\,. \tag{9}$$

We give a rough proof intuition here and full proof in Appendix A.1. Correctness and adaptability are straightforward to check. We rely on Theorem 3 and Theorem 4 so that we only need to verify Ext- and uExt-security of GAS_1. It is not hard to verify that extractability follows from the unforgeability of Sig. For unique extractability, it is not hard to see that we at least need to assume strong extractability of Sig and that R has unique witnesses, since otherwise SigR is not strongly unforgeable. It turns out that these assumptions are also sufficient to show unique extractability.

We observe the following theorem.

Theorem 6. *If one-way functions exist then SFExt-secure adaptor signatures exist.*

The proof follows from the fact that one-way functions imply SUF-CMA-secure signatures [12,17,22] as well as (e.g. length doubling) pseudo-random generators (PRG) [13], which in turn imply hard relations with computationally unique witnesses consisting of pairs $(\mathsf{PRG}(x), x)$.

Our second construction adds *unlinkability* to GAS_1. To achieve this, we additionally need to assume that relation R is random-self-reducible. For example, hash-preimage relation is *not* RSR while discrete-logarithm relation is.

Generic Adaptor Signature (GAS) 2. First, we assume that R is random-self-reducible. The idea for adding unlinkability is simple: in pSign, we first derive a random instance Y' from the input instance Y and then only use Y' in Sig.Sign; furthermore, we need to return the randomness r for the derivation of Y' as part of the pre-signature. To keep the scheme well specified, other parts of the scheme are modified accordingly. Formally, consider construction GAS_2 given in Fig. 7.

We will show that in addition to all the properties of GAS_1, scheme GAS_2 also achieves perfect unlinkability.

We will however need a slightly stronger form of RSR called strong RSR, which is captured via the following security game.

Game $\mathbf{G}^{\mathrm{srsr}}_{\mathsf{R},\mathcal{A}}(\lambda)$:

1 $b \twoheadleftarrow \{0,1\}$; $b' \twoheadleftarrow \mathcal{A}^{\mathrm{New}}()$

New(Y, y):

2 Assert $(Y, y \in \mathsf{R})$; $(Y', y') \twoheadleftarrow \mathsf{R.Gen}(1^\lambda)$; $r \twoheadleftarrow \mathsf{R}.R_\lambda$

3 $Y_0 \leftarrow \mathsf{R}.A(Y', r)$; $y_0 \leftarrow \mathsf{R}.B(y', r)$

4 $Y_1 \leftarrow \mathsf{R}.A(Y, r)$; $y_1 \leftarrow \mathsf{R}.B(y, r)$

5 Return (Y_b, y_b)

We say that R is strongly random self-reducible (SRSR) if the advantage of any PT adversary $\mathcal{A}$, defined to be $\mathbf{Adv}^{\mathrm{srsr}}_{\mathsf{R},\mathcal{A}}(\lambda) := 2\Pr[\mathbf{G}^{\mathrm{srsr}}_{\mathsf{R},\mathcal{A}}(\lambda)] - 1$, is negligible.

Theorem 7. *Adaptor signature scheme* GAS_2 *satisfies correctness and pre-signature adaptability. If* R *is strongly random-self reducible, then* GAS_2 *is unlinkable. Specifically, given any adversary* $\mathcal{A}_{\mathrm{unlink}}$*, the proof gives an adversary* $\mathcal{A}_{\mathrm{srsr}}$*, as efficient as* $\mathcal{A}_{\mathrm{unlink}}$*, such that*

$$\mathbf{Adv}^{\mathrm{unlink}}_{\mathsf{GAS}_2,\mathcal{A}_{\mathrm{unlink}}}(\lambda) \leq \mathbf{Adv}^{\mathrm{srsr}}_{\mathsf{R},\mathcal{A}_{\mathrm{srsr}}}(\lambda) . \tag{10}$$

Furthermore, if Sig *is UF-CMA-secure and R is one-way then* GAS_2 *is FExt-secure. Given adversary* $\mathcal{A}_{\mathrm{fext}}$*, we can construct adversaries* $\mathcal{A}_{\mathrm{uf\text{-}cma}}$ *and* $\mathcal{A}_{\mathrm{ow}}$*, with running times similar to that of* $\mathcal{A}_{\mathrm{uf}}$*, such that*

$$\mathbf{Adv}^{\mathrm{fext}}_{\mathsf{GAS}_2,\mathcal{A}_{\mathrm{fext}}}(\lambda) \leq \mathbf{Adv}^{\mathrm{uf\text{-}cma}}_{\mathsf{Sig},\mathcal{A}_{\mathrm{uf\text{-}cma}}}(\lambda) + \mathbf{Adv}^{\mathrm{q\text{-}ow}}_{R,\mathcal{A}_{\mathrm{ow}}}(\lambda) . \tag{11}$$

Lastly, if Sig *is SUF-CMA-secure and* R *has unique witnesses, then* GAS_1 *is SFExt-secure. Formally, given adversary* $\mathcal{A}_{\mathrm{sfext}}$*, we can construct adversaries* $\mathcal{A}_{\mathrm{suf\text{-}cma}}$*,* $\mathcal{A}_{\mathrm{ow}}$*, and* $\mathcal{A}_{\mathrm{uwit}}$ *with running times similar to that of* $\mathcal{A}_{\mathrm{sfext}}$ *such that*

$$\mathbf{Adv}^{\mathrm{sfext}}_{\mathsf{GAS}_1,\mathcal{A}_{\mathrm{sfext}}}(\lambda) \leq 2\mathbf{Adv}^{\mathrm{suf\text{-}cma}}_{\mathsf{Sig},\mathcal{A}_{\mathrm{uf\text{-}cma}}}(\lambda) + \mathbf{Adv}^{\mathrm{q\text{-}ow}}_{R,\mathcal{A}_{\mathrm{ow}}}(\lambda) + \mathbf{Adv}^{\mathrm{uwit}}_{\mathsf{R},\mathcal{A}_{\mathrm{uwit}}} . \tag{12}$$

Unlinkability follows from SRSR property in a straightforward manner and the rest of the proofs are very similar to those for Theorem 5 and are given in Appendix A.2. Note that any hard relation trivially implies one-way functions: for example, the mapping $f_\lambda(r) := Y$, where $(Y, y) \leftarrow \mathsf{R.Gen}(1^\lambda; r)$ and r is any element of the randomness space of $\mathsf{R.Gen}(1^\lambda)$, is one-way. Hence, similar to Theorem 6, we observe the following theorem.

Theorem 8. *If SRSR relations exist then SFExt-secure and unlinkable adaptor signatures exist.*

Strong RSR relations from any epimorphic (homomorphic and onto) one-way function. Suppose $f_\lambda : D_\lambda \to R_\lambda$ is a homomorphic one-way function, where D_λ and R_λ are both abelian groups (where D_λ has group operation $+$ and R_λ

has group operation $\cdot$). For example, two instantiations are $f_{\mathbb{G},g}(y) = g^y$ and $f_{(N,e)}(y) = y^e \mod N$, where $(\mathbb{G}, g)$ is a group instance and (N, e) is an RSA public key. Then it is clear that the relation containing pairs $(f(y), y)$ can be made strongly RSR. In particular, we consider $\mathsf{R}.A(Y, r) := Y \cdot f(r)$, where r is sampled uniformly randomly from D_λ. The corresponding algorithms B and C are defined as $\mathsf{R}.B(y, r) = y + r$ and $\mathsf{R}.C(y', r) = y' - r$. It is easy to check that the relation is SRSR since $(f(y+r), y+r)$ is uniformly random regardless of the value of y, as long as f is homomorphic and onto.

Strong RSR relations from LWE. We sketch how LWE gives rise to a SRSR relation. Recall that for LWE, the dimension n is the LWE security parameter. Take any m that is polynomial in n. We fix[3] a random matrix $A_\lambda \in \mathbb{Z}_q^{m\times n}$ for each security parameter λ. (We can take $n = \mathcal{O}(\lambda)$.) Let q be the modulus and α_e and α_t be parameters that we shall fix at the end. Consider the following relation R consisting of pairs (Y, y) with $y \twoheadleftarrow \mathbb{Z}_q^n$ and $Y = Ay + e$, where each component of e is sampled from a discrete Gaussian of width $\alpha_e q$. We define the rerandomize algorithm $\mathsf{R}.A(Y, (r, t)) := Y + A_\lambda r + t$, where r is uniformly random in $\mathbb{Z}_q^n$ and $t \in \mathbb{Z}_q^m$ is such that each component of t is sampled from a discrete Gaussian of width $\alpha_t q$. Lastly, we define $\mathsf{R}.B(y, (r, t)) := y + r$ and $\mathsf{R}.B(y, (r, t)) := y - r$. Above, components of e and t are from (discretized) Gaussian distributions of parameter $\alpha_e q$ and $\alpha_t q$, respectively, where $\alpha_e = \mathcal{O}(1/f^2(n))$, $\alpha_t = \mathcal{O}(1/f(n))$, and $q = \mathcal{O}(f^3(n))$ for a super-polynomial function $f(n)$. For example, the parameter of distribution e is $\mathcal{O}(n^{\log n})$, that of t is $\mathcal{O}(n^{2\log n})$, and modulus q is $\mathcal{O}(n^{3\log n})$. Note that given some value of error e, one cannot distinguish between $e + t$ and $e' + t$ for freshly sampled e' (i.i.d to e) and t. This is because t is "wider" than e by a factor of $n^{\log(n)}$, which is super-polynomial. Finally, by [18, Theorem 4.2.4], the relation R is SRSR if GapSVP_γ and SIVP_γ are hard against quantum adversaries, where $\gamma = \widetilde{\mathcal{O}}(n^{1-\log(n)})$.

A Omitted Proofs

A.1 Proof of Theorem 5

Proof (of Theorem 5). First, correctness holds by construction. Next, we check adaptability. Let $(\mathrm{pk}, \mathrm{sk}) \in [\mathsf{KeyGen}(1^\lambda)]$ and $m \in \{0,1\}^*$. Let $\hat{\sigma}, (Y, y)$ be such that $(Y, y) \in R$ and $\mathsf{pVrf}(\mathrm{pk}, m, \hat{\sigma}, Y) = \mathsf{True}$. This means that $\hat{\sigma} = (\sigma, Y)$ and $\mathsf{Sig.Vrf}(\mathrm{pk}, (m, Y), \sigma) = \mathsf{True}$. Hence, by the verification of SigR, it must be that $\mathsf{SigR.Vrf}(\mathrm{pk}, m, (\sigma, Y, y)) = \mathsf{True}$.

We move on to FExt and SFExt. With the help of Theorem 3 and Theorem 4, we simply need to show that for any adversary $\mathcal{A}_{\mathrm{ext}}$ and $\mathcal{A}_{\mathrm{uext}}$,

$$\mathbf{Adv}^{\mathrm{ext}}_{\mathsf{GAS}_1,\mathcal{A}_{\mathrm{ext}}}(\lambda) \le \mathbf{Adv}^{\mathrm{uf\text{-}cma}}_{\mathsf{Sig},\mathcal{A}_{\mathrm{uf\text{-}cma}}}(\lambda)\,, \tag{13}$$

$$\mathbf{Adv}^{\mathrm{uext}}_{\mathsf{GAS}_1,\mathcal{A}_{\mathrm{uext}}}(\lambda) \le \mathbf{Adv}^{\mathrm{suf\text{-}cma}}_{\mathsf{Sig},\mathcal{A}_{\mathrm{suf\text{-}cma}}}(\lambda) + \mathbf{Adv}^{\mathrm{uwit}}_{\mathsf{R},\mathcal{A}_{\mathrm{uwit}}}(\lambda)\,, \tag{14}$$

[3] More formally, A should be sampled as a parameter for each security parameter, but we fix such A here for simplicity.

where $\mathcal{A}_{\text{uf-cma}}, \mathcal{A}_{\text{suf-cma}}, \mathcal{A}_{\text{uwit}}$ are reduction adversaries to be constructed.

We first show (13). Consider the following game $\mathbf{G}_0$ and adversary $\mathcal{A}_{\text{uf-cma}}$.

Game $\mathbf{G}_0$	Adversary $\mathcal{A}_{\text{uf-cma}}^{\text{Sign}}(\text{pk})$:
1 $(\text{pk}, \text{sk}) \twoheadleftarrow \mathsf{Sig.KeyGen}(1^\lambda)$	1 $(m^*, \sigma^*) \twoheadleftarrow \mathcal{A}_{\text{ext}}^{\text{pSign}}(\text{pk})$
2 $(m^*, \sigma^*) \twoheadleftarrow \mathcal{A}_{\text{ext}}^{\text{Sign},\text{pSign}}(\text{pk})$	2 $(\sigma', Y^*, y^*) \leftarrow \sigma^*$
3 $(\sigma', Y^*, y^*) \leftarrow \sigma^*$	3 Return $((m^*, Y^*), \sigma')$
4 Assert $(\mathsf{Sig.Vrf}(\text{pk}, (m^*, Y^*), \sigma') \wedge (Y^*, y^*) \in R)$	pSign(m, Y):
5 Return $(\forall Y \in T[m^*] : (Y, y^*) \notin R)$	4 $\sigma \twoheadleftarrow \text{Sign}((m, Y))$
pSign(m, Y):	5 Return (σ, Y)
6 $\sigma \twoheadleftarrow \mathsf{Sig.Sign}(\text{sk}, (m, Y))$; $T[m] \xleftarrow{\cup} Y$	
7 Return (σ, Y)	

We claim that

$$\mathbf{Adv}^{\text{ext}}_{\mathsf{Sig},\mathcal{A}_{\text{ext}}}(\lambda) = \Pr[\mathbf{G}_0] \leq \mathbf{Adv}^{\text{uf-cma}}_{\mathsf{Sig},\mathcal{A}_{\text{uf-cma}}}(\lambda) . \tag{15}$$

This is straightforward, because if $\mathbf{G}_0$ returns true, then it must be that Y^* returned by the adversary is fresh, meaning it has not queried $\mathsf{pSign}(m^*, Y^*)$ previously. Finally, we note that adversary $\mathcal{A}_{\text{uf-cma}}$ also wins exactly when (m^*, Y^*) is fresh.

Next, we bound (14). Consider the following games $\mathbf{G}_1, \mathbf{G}_2, \mathbf{G}_3$. Game $\mathbf{G}_1$ is $\mathbf{G}^{\text{uext}}_{\mathsf{GAS}_1,\mathcal{A}_{\text{uext}}}$. Games $\mathbf{G}_2$ and $\mathbf{G}_3$ rewrites the winning condition of $\mathbf{G}_1$ depending on disjoint events b_1 and b_2.

Game $\mathbf{G}_1$, $\mathbf{G}_2$, $\mathbf{G}_3$

1 $(\text{pk}, \text{sk}) \twoheadleftarrow \mathsf{Sig.KeyGen}(1^\lambda)$
2 $(m, Y, \hat{\sigma}, \sigma, \sigma') \twoheadleftarrow \mathcal{A}_{\text{uext}}^{\text{pSign}}(\text{pk})$
3 $(\sigma_0, Y, y) \leftarrow \sigma$; $(\sigma_1, Y', y') \leftarrow \sigma'$
4 $b_1 \leftarrow (Y \neq Y')$
5 $b_2 \leftarrow (Y = Y') \wedge (y \neq y')$
6 $b_3 \leftarrow \mathsf{Vrf}(\text{pk}, (m, Y), \sigma_0) \wedge \mathsf{Vrf}(\text{pk}, (m, Y'), \sigma_1) \wedge (Y, y) \in \mathsf{R} \wedge (Y, y') \in \mathsf{R}$
7 $\mathbf{G}_1$: Return $(b_1 \vee b_2) \wedge b_3$
8 $\mathbf{G}_2$: Return $b_1 \wedge b_3$
9 $\mathbf{G}_3$: Return $b_2 \wedge b_3$

pSign(m, Y):
10 $\sigma \twoheadleftarrow \mathsf{Sig.Sign}(\text{sk}, (m, Y))$; Return (σ, Y)

Clearly, we have

$$\mathbf{Adv}^{\text{uext}}_{\mathsf{GAS}_1,\mathcal{A}_{\text{uext}}}(\lambda) = \Pr[\mathbf{G}_1] = \Pr[\mathbf{G}_2] + \Pr[\mathbf{G}_3] . \tag{16}$$

Next, we construct adversaries $\mathcal{A}_{\text{suf-cma}}$ and $\mathcal{A}_{\text{uwit}}$, such that

$$\Pr[\mathbf{G}_2] \leq \mathbf{Adv}^{\text{suf-cma}}_{\mathsf{Sig},\mathcal{A}_{\text{suf-cma}}}(\lambda) , \tag{17}$$

$$\Pr[\mathbf{G}_3] \leq \mathbf{Adv}^{\text{uwit}}_{\mathsf{R},\mathcal{A}_{\text{uwit}}}(\lambda) . \tag{18}$$

This is straightforward, $\mathcal{A}_{\text{suf-cma}}$ can simulate pSign with its Sign oracle, and $\mathcal{A}_{\text{uwit}}$ can sample its own key pair to simulate game $\mathbf{G}_3$. The specifications of these adversaries are given below.

Adversary $\mathcal{A}^{\text{Sign}}_{\text{suf-cma}}(\text{pk})$:	Adversary $\mathcal{A}_{\text{uwit}}()$:
1 $(m, Y, \hat{\sigma}, \sigma, \sigma') \twoheadleftarrow \mathcal{A}^{\text{pSign}}_{\text{uext}}(\text{pk})$	1 $(\text{pk}, \text{sk}) \twoheadleftarrow \mathsf{Sig.KeyGen}(1^\lambda)$
2 $(\sigma_0, Y_0, y_0) \leftarrow \sigma$; $(\sigma_1, Y_1, y_1) \leftarrow \sigma'$	2 $(m, Y, \hat{\sigma}, \sigma, \sigma') \twoheadleftarrow \mathcal{A}^{\text{pSign}}_{\text{uext}}(\text{pk})$
3 If $\exists i \in \{1,2\} : (m, Y_i, \sigma_i) \notin U$ then	3 $(\sigma_0, Y_0, y_0) \leftarrow \sigma$; $(\sigma_1, Y_1, y_1) \leftarrow \sigma'$
4 Return $((m, Y_i), \sigma_i)$	4 Return (Y_0, y_0, y_1)
pSign(m, Y):	pSign(m, Y):
5 $\sigma \twoheadleftarrow \text{Sign}((m, Y))$	5 $\sigma \twoheadleftarrow \mathsf{Sign}(\text{sk}, (m, Y))$
6 $U \xleftarrow{\cup} (m, Y, \sigma)$	6 Return (σ, Y)
7 Return (σ, Y)	

This concludes the proof of Theorem 5. □

A.2 Proof of Theorem 7

Proof (of Theorem 7). First, correctness and adaptability holds similar to GAS_1. We give a reduction that turns any unlink adversary to a strong RSR adversary for R. The reduction is very straightforward and we keep the descript at a high-level here. The SRSR adversary sample a key pair $(\text{pk}, \text{sk}) \twoheadleftarrow \mathsf{Sig.KeyGen}(1^\lambda)$, using which it can run pSign and Sign algorithms. It can simulate oracles Sign and pSign honestly. It uses the New oracle given to it from the strong RSR game to simulate SignChl, the pair (Y, y) that is in the input of SignChl is simply forwarded to New.

We check extractability. Similar to GAS_1, we need to show that for any adversary $\mathcal{A}_{\text{ext}}$ and $\mathcal{A}_{\text{uext}}$,

$$\mathbf{Adv}^{\text{ext}}_{\mathsf{GAS}_2, \mathcal{A}_{\text{ext}}}(\lambda) \leq \mathbf{Adv}^{\text{uf-cma}}_{\mathsf{Sig}, \mathcal{A}_{\text{uf-cma}}}(\lambda) , \tag{19}$$

$$\mathbf{Adv}^{\text{uext}}_{\mathsf{GAS}_2, \mathcal{A}_{\text{uext}}}(\lambda) \leq \mathbf{Adv}^{\text{suf-cma}}_{\mathsf{Sig}, \mathcal{A}_{\text{suf-cma}}}(\lambda) + \mathbf{Adv}^{\text{uwit}}_{\mathsf{R}, \mathcal{A}_{\text{uwit}}}(\lambda) , \tag{20}$$

where $\mathcal{A}_{\text{uf-cma}}, \mathcal{A}_{\text{suf-cma}}, \mathcal{A}_{\text{uwit}}$ are reduction adversaries to be constructed.

We first show (19). Consider the following game $\mathbf{G}_0$ and adversary $\mathcal{A}_{\text{uf-cma}}$.

Game $\mathbf{G}_0$	Adversary $\mathcal{A}^{\text{Sign}}_{\text{uf-cma}}(\text{pk})$:
1 $(\text{pk}, \text{sk}) \twoheadleftarrow \mathsf{Sig.KeyGen}(1^\lambda)$	1 $(m^*, \sigma^*) \twoheadleftarrow \mathcal{A}^{\text{pSign}}_{\text{ext}}(\text{pk})$
2 $(m^*, \sigma^*) \twoheadleftarrow \mathcal{A}^{\text{pSign}}_{\text{ext}}(\text{pk})$	2 $(\sigma', Y^*, y^*) \leftarrow \sigma^*$
3 $(\sigma', Y^*, y^*) \leftarrow \sigma^*$	3 Return $((m^*, Y^*), \sigma')$
4 Assert $(\mathsf{Sig.Vrf}(\text{pk}, (m^*, Y^*), \sigma') \wedge (Y^*, y^*) \in R)$	pSign(m, Y):
5 Return $(\forall (Y, r) \in T[m^*] : (Y, R.C(y^*, r)) \notin R)$	4 $r \twoheadleftarrow R.R_\lambda$
pSign(m, Y):	5 $Y' \leftarrow R.A(Y, r)$
6 $r \twoheadleftarrow \mathsf{R}.R_\lambda$; $Y' \leftarrow R.A(Y, r)$; $T[m] \xleftarrow{\cup} (Y, r)$	6 $\sigma' \twoheadleftarrow \text{Sign}((m, Y'))$
7 $\sigma' \twoheadleftarrow \mathsf{Sig.Sign}(\text{sk}, (m, Y'))$; Return (σ', Y, r)	7 Return (σ', Y, r)

We claim that

$$\mathbf{Adv}^{\text{ext}}_{\mathsf{Sig},\mathcal{A}_{\text{ext}}}(\lambda) = \Pr[\mathbf{G}_0] \leq \mathbf{Adv}^{\text{uf-cma}}_{\mathsf{Sig},\mathcal{A}_{\text{uf-cma}}}(\lambda) . \tag{21}$$

We claim that if $\mathbf{G}_0$ returns true, it must be that Y^* returned by adversary is fresh, meaning the adversary has not queried $\mathsf{Sign}((m^*, Y^*))$ previously. Seeking a contradiction, suppose that adversary has incurred a query $\mathsf{Sign}(m^*, Y^*)$, then this query must have come from some query $\mathsf{pSign}(m^*, Y_0)$, where the game has sampled some r_0 such that $\mathsf{R}.A(Y_0, r_0) = Y^*$. By line 4, $(Y^*, y^*) \in R$. So, $\mathsf{R}.C(y^*, r)$ must be a witness of Y_0. This means that the game must return False at line 5. Therefore, there was no signature on message (m^*, Y^*) if the game returns True. We note that adversary $\mathcal{A}_{\text{uf-cma}}$ also wins exactly when (m^*, Y^*) is fresh. This verifies (19).

Next, we bound (20). Consider the following games $\mathbf{G}_1, \mathbf{G}_2, \mathbf{G}_3$. Game $\mathbf{G}_1$ is $\mathbf{G}^{\text{uext}}_{\mathsf{GAS}_2,\mathcal{A}_{\text{uext}}}$. Games $\mathbf{G}_2$ and $\mathbf{G}_3$ rewrites the winning condition of $\mathbf{G}_1$ depending on disjoint events b_1 and b_2.

```
Game G1, G2, G3
1  (pk, sk) <<- Sig.KeyGen(1^λ)
2  (m, Y, σ̂, σ, σ') <<- A_uext^pSign(pk)
3  (σ0, Y, y) <- σ ; (σ1, Y', y') <- σ'
4  b1 <- (Y ≠ Y')
5  b2 <- (Y = Y') ∧ (y ≠ y')
6  b3 <- Vrf(pk, (m, Y), σ0) ∧ Vrf(pk, (m, Y'), σ1) ∧ (Y, y) ∈ R ∧ (Y, y') ∈ R
7  G1: Return (b1 ∨ b2) ∧ b3
8  G2: Return b1 ∧ b3
9  G3: Return b2 ∧ b3

pSign(m, Y):
10 r <<- R.R_λ ; Y' <- R.A(Y, r) ; σ' <<- Sign(sk, (m, Y')) ; Return (σ', Y, r)
```

Clearly, we have

$$\mathbf{Adv}^{\text{uext}}_{\mathsf{GAS}_1,\mathcal{A}_{\text{uext}}}(\lambda) = \Pr[\mathbf{G}_1] = \Pr[\mathbf{G}_2] + \Pr[\mathbf{G}_3] . \tag{22}$$

Next, we construct adversaries $\mathcal{A}_{\text{suf-cma}}$ and $\mathcal{A}_{\text{uwit}}$, such that

$$\Pr[\mathbf{G}_2] \leq \mathbf{Adv}^{\text{suf-cma}}_{\mathsf{Sig},\mathcal{A}_{\text{suf-cma}}}(\lambda) , \tag{23}$$

$$\Pr[\mathbf{G}_3] \leq \mathbf{Adv}^{\text{uwit}}_{\mathsf{R},\mathcal{A}_{\text{uwit}}}(\lambda) . \tag{24}$$

This is straightforward, $\mathcal{A}_{\text{suf-cma}}$ can simulate pSign with its Sign oracle, and $\mathcal{A}_{\text{uwit}}$ can sample its own key pair to simulate game $\mathbf{G}_3$. The specifications of these adversaries are given below.

Adversary $\mathcal{A}^{\text{Sign}}_{\text{suf-cma}}(\text{pk})$:	Adversary $\mathcal{A}_{\text{uwit}}()$:
1 $(m, Y, \hat{\sigma}, \sigma, \sigma') \twoheadleftarrow \mathcal{A}^{\text{pSign}}_{\text{uext}}(\text{pk})$	1 $(\text{pk}, \text{sk}) \twoheadleftarrow \mathsf{Sig.KeyGen}(1^\lambda)$
2 $(\sigma_0, Y_0, y_0) \leftarrow \sigma$; $(\sigma_1, Y_1, y_1) \leftarrow \sigma'$	2 $(m, Y, \hat{\sigma}, \sigma, \sigma') \twoheadleftarrow \mathcal{A}^{\text{pSign}}_{\text{uext}}(\text{pk})$
3 If $\exists i \in \{1,2\} : (m, Y_i, \sigma_i) \notin U$ then	3 $(\sigma_0, Y_0, y_0) \leftarrow \sigma$; $(\sigma_1, Y_1, y_1) \leftarrow \sigma'$
4 Return $((m, Y_i), \sigma_i)$	4 Return (Y_0, y_0, y_1)
pSign(m, Y):	pSign(m, Y):
5 $r \twoheadleftarrow \mathsf{R}.R_\lambda$; $Y' \leftarrow \mathsf{R}.A(Y, r)$	5 $r \twoheadleftarrow \mathsf{R}.R_\lambda$; $Y' \leftarrow \mathsf{R}.A(Y, r)$
6 $\sigma' \twoheadleftarrow \text{Sign}((m, Y'))$	6 $\sigma' \twoheadleftarrow \mathsf{Sign}(\text{sk}, (m, Y'))$; Return (σ', Y, r)
7 $U \overset{\cup}{\leftarrow} (m, Y', \sigma')$; Return (σ', Y, r)	

This concludes the proof of Theorem 7. □

References

1. Aumayr, L., et al.: Bitcoin-compatible virtual channels. Cryptology ePrint Archive, Report 2020/554 (2020). https://eprint.iacr.org/2020/554
2. Aumayr, L., et al.: Generalized bitcoin-compatible channels. Cryptology ePrint Archive, Report 2020/476 (2020). https://eprint.iacr.org/2020/476
3. Bellare, M., Rogaway, P.: The security of triple encryption and a framework for code-based game-playing proofs. In: Vaudenay, S. (ed.) Advances in Cryptology - EUROCRYPT 2006. LNCS, vol. 4004, pp. 409–426. Springer, Heidelberg (2006). https://doi.org/10.1007/11761679_25
4. Beullens, W., Kleinjung, T., Vercauteren, F.: CSI-FiSh: Efficient isogeny based signatures through class group computations. In: Galbraith, S.D., Moriai, S. (eds.) ASIACRYPT 2019. Part I, volume 11921 of LNCS, pp. 227–247. Springer, Heidelberg (2019). https://doi.org/10.1007/978-3-030-34578-5_9
5. Boneh, D., Lynn, B., Shacham, H.: Short signatures from the Weil pairing. In: Boyd, C. (ed.) ASIACRYPT 2001. LNCS, vol. 2248, pp. 514–532. Springer, Heidelberg (2001). https://doi.org/10.1007/s00145-004-0314-9
6. Dai, W., Okamoto, T., Yamamoto, G.: Stronger security and generic constructions for adaptor signatures. Cryptology ePrint Archive (2022)
7. Ducas, L., Lepoint, T., Lyubashevsky, V., Schwabe, P., Seiler, G., Stehle, D.: CRYSTALS - Dilithium: digital signatures from module lattices. Cryptology ePrint Archive, Report 2017/633 (2017). https://eprint.iacr.org/2017/633
8. Erwig, A., Faust, S., Hostáková, K., Maitra, M., Riahi, S.: Two-party adaptor signatures from identification schemes. In: Garay, J. (ed.) PKC 2021. Part I, volume 12710 of LNCS, pp. 451–480. Springer, Heidelberg (2021). https://doi.org/10.1007/978-3-030-75245-3_17
9. Esgin, M.F., Ersoy, O., Erkin, Z.: Post-quantum adaptor signatures and payment channel networks. Cryptology ePrint Archive, Report 2020/845 (2020). https://eprint.iacr.org/2020/845
10. Fournier, L.: One-time verifiably encrypted signatures aka adaptor signatures (2019)
11. Gugger, J.: Bitcoin-monero cross-chain atomic swap. Cryptology ePrint Archive, Report 2020/1126 (2020). https://eprint.iacr.org/2020/1126

12. Huang, Q., Wong, D.S., Zhao, Y.: Generic transformation to strongly unforgeable signatures. In: Katz, J., Yung, M. (eds.) ACNS 07. LNCS, vol. 4521, pp. 1–17. Springer, Heidelberg (2007). https://doi.org/10.1007/978-3-540-72738-5_1
13. Impagliazzo, R., Levin, L.A., Luby, M.: Pseudo-random generation from one-way functions (extended abstracts). In: 21st ACM STOC, pp. 12–24. ACM Press (1989)
14. Johnson, D., Menezes, A., Vanstone, S.: The elliptic curve digital signature algorithm (ECDSA). Int. J. Inf. Secur. **1**(1), 36–63 (2001)
15. Malavolta, G., Moreno-Sanchez, P., Schneidewind, C., Kate, A., Maffei, M.: Anonymous multi-hop locks for blockchain scalability and interoperability. In: NDSS 2019. The Internet Society (2019)
16. Moody, D., et al.: Status report on the second round of the NIST post-quantum cryptography standardization process (2020)
17. Naor, M., Yung, M.: Universal one-way hash functions and their cryptographic applications. In: 21st ACM STOC, pp. 33–43. ACM Press (1989)
18. Peikert, C.: How (not) to instantiate ring-LWE. Cryptology ePrint Archive, Report 2016/351 (2016). https://eprint.iacr.org/2016/351
19. Poelstra, A.: Lightning in scriptless scripts (2017). https://lists.launchpad.net/mimblewimble/msg00086.html. Accessed Aug 2021
20. Poelstra, A.: Scriptless scripts (2017). https://download.wpsoftware.net/bitcoin/wizardry/mw-slides/2017-03-mit-bitcoin-expo/slides.pdf. Accessed Aug 2021
21. Poon, J., Dryja, T.: The bitcoin lightning network: scalable off-chain instant payments (2016). https://lightning.network/lightning-network-paper.pdf. Accessed: Aug 2021
22. Rompel, J.: One-way functions are necessary and sufficient for secure signatures. In: 22nd ACM STOC, pp. 387–394. ACM Press (1990)
23. Schnorr, C.-P.: Efficient signature generation by smart cards. J. Crypt. **4**(3), 161–174 (1991)
24. Tairi, E., Moreno-Sanchez, P., Maffei, M.: Post-quantum adaptor signature for privacy-preserving off-chain payments. Cryptology ePrint Archive, Report 2020/1345 (2020). https://eprint.iacr.org/2020/1345

Entropic Hardness of Module-LWE from Module-NTRU

Katharina Boudgoust[1], Corentin Jeudy[2,3](✉), Adeline Roux-Langlois[4], and Weiqiang Wen[5]

[1] Department Computer Science, Aarhus University, Aarhus, Denmark
katharina.boudgoust@cs.au.dk

[2] Univ Rennes, CNRS, IRISA, Rennes, France

[3] Orange Labs, Applied Crypto Group, Cesson-Sévigné, France
corentin.jeudy@irisa.fr

[4] Normandie Univ, UNICAEN, ENSICAEN, CNRS, GREYC, 14000 Caen, France
adeline.roux-langlois@cnrs.fr

[5] LTCI, Telecom Paris, Institut Polytechnique de Paris, Paris, France
weiqiang.wen@telecom-paris.fr

Abstract. The Module Learning With Errors problem (M-LWE) has gained popularity in recent years for its security-efficiency balance,and its hardness has been established for a number of variants. In this paper, we focus on proving the hardness of (search) M-LWE for general secret distributions, provided they carry sufficient min-entropy. This is called entropic hardness of M-LWE. First, we adapt the line of proof of Brakerski and Döttling on R-LWE (TCC'20) to prove that the existence of certain distributions implies the entropic hardness of M-LWE. Then, we provide one such distribution whose required properties rely on the hardness of the decisional Module-NTRU problem.

Keywords: Lattice-based cryptography · Module learning with errors · Entropic hardness · Module-NTRU

1 Introduction

The *Learning With Errors* (LWE) [27] and NTRU [17] problems are the most widespread computational assumptions for designing lattice-based cryptosystems. The LWE problem asks to find a secret $\mathbf{s} \in \mathbb{Z}_q^d$ given the noisy system $(\mathbf{A}, \mathbf{b} = \mathbf{A}\mathbf{s} + \mathbf{e} \bmod q)$ for $\mathbf{A} \in \mathbb{Z}_q^{m \times d}$ uniformly random and $\mathbf{e}$ drawn from ψ^m, where ψ is an error distribution over $\mathbb{Z}$ (or $\mathbb{R}$). In the decisional variant, one has to distinguish such $\mathbf{b}$ from a uniform vector $\mathbf{u}$ over $\mathbb{Z}_q$ (or $\mathbb{T}_q = \mathbb{R}/q\mathbb{Z}$). Although the error distribution can be arbitrary, most theoretical proofs use a Gaussian distribution. LWE benefits from strong hardness guarantees, as its average-case formulation is proven to be at least as hard as worst-case lattice problems. The parameter d is known as the LWE dimension and is often seen as the security parameter of the problem as it is linked to the dimension of the

T. Isobe and S. Sarkar (Eds.): INDOCRYPT 2022, LNCS 13774, pp. 78–99, 2022.
https://doi.org/10.1007/978-3-031-22912-1_4

underlying lattice. The NTRU problem is defined over a more algebraic setting, using algebraic integers instead of rational ones. In a number field K, consider R to be the ring of algebraic integers in K. (Decisional) NTRU asks to distinguish between $gf_q^{-1} \bmod q$ and u, where f and g are short elements of $R_q = R/qR$, f_q^{-1} is the R_q-inverse of f and u is uniform over R_q. The search version, consisting in finding f and g given $gf_q^{-1} \bmod q$, has recently been linked to standard ideal lattice problems [26]. Although very few theoretical hardness results are known for NTRU, it has been widely studied for more than two decades from a cryptanalytic standpoint. Unless for overstretched parameter sets, e.g. [15], it is believed to be a reliable hardness assumption to design public-key cryptosystems, e.g., [2,11].

Although LWE allows for designing provably secure cryptosystems, the resulting schemes are usually not practical enough to be used in real-world systems. With this perspective in mind, several algebraically structured variants of LWE [9,18, 20,29] were introduced to improve efficiency in terms of computation and storage, while maintaining strong enough hardness guarantees from problems over structured lattices. The underlying algebraic framework is the same as the one in the NTRU problem. In this paper, we focus on the *Module Learning With Errors* (M-LWE) problem which is similar to the original LWE formulation. The set of integers $\mathbb{Z}$ is replaced by a ring of algebraic integers R, as above. The problem is formulated over the free R-module R^d, where d is called the rank of the module. Then, for a modulus q, and an error distribution ψ over R (or $K_{\mathbb{R}} = K \otimes_{\mathbb{Q}} \mathbb{R}$), and given $(\mathbf{A}, \mathbf{b} = \mathbf{A}\mathbf{s} + \mathbf{e} \bmod q)$ with $\mathbf{A}$ uniform in $R_q^{m \times d}$, $\mathbf{s} \in R_q^d$ and $\mathbf{e}$ drawn from ψ^m, the goal is to recover $\mathbf{s}$. The decisional version asks to distinguish such a $\mathbf{b}$ from a uniformly distributed vector $\mathbf{u}$. When $d = 1$, we call it *Ring* LWE (R-LWE). A module version of NTRU was also recently studied by Chuengsatiansup et al. [12]. It consists in distinguishing $\mathbf{G}\mathbf{F}_q^{-1} \bmod q$, with $(\mathbf{F}, \mathbf{G})$ short matrices in $R^{d\times d} \times R^{m\times d}$ and $\mathbf{F}_q^{-1}$ the R_q-inverse[1] of $\mathbf{F}$, from a uniform matrix $\mathbf{U} \in R_q^{m\times d}$. The module versions M-LWE and M-NTRU give more flexibility to adjust the balance between efficiency and security.

The average-case formulation of M-LWE is parameterized by a distribution $\mathcal{S}$ on the secret $\mathbf{s}$. The original definition uses $\mathcal{S}$ to be the uniform distribution over R_q^d. Even though this choice is the most natural one for theoretical results, practical schemes vary from it. For efficiency reasons, it is advantageous to choose secret distributions that lead to small-norm secrets, as is done in the M-LWE-based schemes Kyber [3] and Dilithium [13] which were chosen for standardization by NIST. The study of LWE with small-norm secrets [10,16,21] has recently been extended to the module setting by Boudgoust et al. [4,5]. The second reason to deviate from the uniform distribution stems from the situation where the key is not sampled from the prescribed distribution but from an imperfect one. Braskerski and Döttling conducted a study of LWE with general entropic secret distributions [7], which they afterward extended to the ring setting ($d = 1$) [8]. It was left open to thoroughly generalize this proof method to

[1] As we use both the R_q-inverse and the K-inverse, we insist on differentiating them as $\mathbf{F}_q^{-1}$ and $\mathbf{F}^{-1}$ respectively.

larger ranks $d > 1$. Hardness results encapsulating imperfect distributions provide theoretical insights on the resistance to key leakage, like cold boot attacks for instance. These attacks leverage the physical properties of the hardware to recover remanent information from the memory. In the representative cold boot attack on M-LWE based schemes [1], the adversary can manage to obtain a faulty version of the (long-term) secret key $\mathbf{s}$ stored in memory. From the faulty key $\widetilde{\mathbf{s}} = \mathbf{s} + \Delta\mathbf{s}$, the adversary can now recover the full secret $\mathbf{s}$ by targeting a new M-LWE instance $(\mathbf{A}, \mathbf{A}\widetilde{\mathbf{s}} - \mathbf{b} \bmod q) = (\mathbf{A}, \mathbf{A}\Delta\mathbf{s} - \mathbf{e} \bmod q)$, where the new secret $\Delta\mathbf{s}$ is only promised to have certain entropy. The motivation is therefore to prove the hardness of M-LWE when some secret key information is leaked to the attacker, i.e., the remaining entropy in $\Delta\mathbf{s}$ is smaller than that of $\mathbf{s}$.

Our Contributions. In this paper, we extend the line of work of Brakerski and Döttling on the entropic hardness to the module setting, i.e., $d \geq 1$, which has gained popularity over its preceding variants. Our main contribution is given in the following informal theorem. For a complete statement, refer to Theorem 5.

Theorem 1 (Informal). *Assuming decisional* M-NTRU *is hard, the problem search* M-LWE *with secret distribution $\mathcal{S}$ is hard for a sufficiently large Gaussian noise and provided that $\mathcal{S}$ has large enough min-entropy.*

The first step of our proof is to translate the M-NTRU problem from its algebraic form to a non-algebraic one, which we call Structured NTRU (S-NTRU). This simply consists in embedding the algebraic setting into vectors and matrices over the integers or the reals. We similarly define Structured LWE (S-LWE), generalizing the notion given in [8]. The core of our work is then to prove the hardness of S-LWE from that of S-NTRU.

To do so, we construct a *sometimes lossy pseudorandom distribution*, as introduced in [8]. The pseudorandomness property, proven under the hardness of S-NTRU, essentially allows us to replace the uniform matrix $\mathbf{A}$ from the S-LWE instance by the S-NTRU matrix $\mathbf{G}\mathbf{F}_q^{-1} \bmod q$, where $\mathbf{F}_q^{-1}$ is the R_q-inverse of $\mathbf{F}$. Then, the sometimes lossiness property translates the fact that going from $\mathbf{s}$ to $\mathbf{G}\mathbf{F}_q^{-1}\mathbf{s} + \mathbf{e} \bmod q$ loses enough information on $\mathbf{s}$ so that it is hard to recover it. The sometimes lossiness thus entails the hardness of S-LWE when $\mathbf{A} = \mathbf{G}\mathbf{F}_q^{-1} \bmod q$. By the pseudorandomness, it then yields the hardness of S-LWE for a uniform $\mathbf{A}$. Section 4 gives explicit conditions on the distributions of $\mathbf{F}$ and $\mathbf{G}$ for the sometimes lossiness property to be satisfied. It requires $\mathbf{F}$ and $\mathbf{G}$ to be somewhat well-behaved in terms of invertibility and spectral bounds. More precisely, $\mathbf{F}$ must be invertible in R_q, the distributions on $\mathbf{F}$ and $\mathbf{G}$ should both minimize their largest singular value, and the distribution of $\mathbf{F}$ should also maximize its smallest singular value. These conditions are obtained by generalizing the approach from [8] in the case of rings ($d = 1$), by introducing the dimension d associated to the module rank.

We then exhibit in Sect. 5 a distribution on $\mathbf{F}$ and $\mathbf{G}$ that satisfies the required conditions, which is our main contribution. The invertibility in R_q has been studied in [12], which naturally implies invertibility in K. The spectral analysis is more tricky and is the object of Sects. 5.1 and 5.2. We give technical results on the smallest (and largest) singular values of discrete Gaussian matrices over the ring of integers of a number field, which might be of independent interest. In addition to dealing with matrices, our techniques differ from that of [8] as we adopt a different, and inherently simpler, distribution on $\mathbf{F}$ and $\mathbf{G}$. This provides another valid way to construct sometimes lossy pseudorandom distributions from module-based assumptions. For our distribution, the pseudorandomness property is obtained by assuming that $\mathbf{G}\mathbf{F}_q^{-1} \bmod q \in R_q^{m\times d}$ is indistinguishable from uniform, hence the M-NTRU assumption for a rectangular matrix $\mathbf{G}$. The approach in [8] requires the HNF-R-LWE assumption in addition to NTRU. Also, our simpler approach leads to improved parameters both in our module generalization but also in the ring setting when setting $d = 1$, using a *multiple public key* version of NTRU instead of HNF-R-LWE and NTRU. The high-level picture of the entire proof is summarized in Fig. 1.

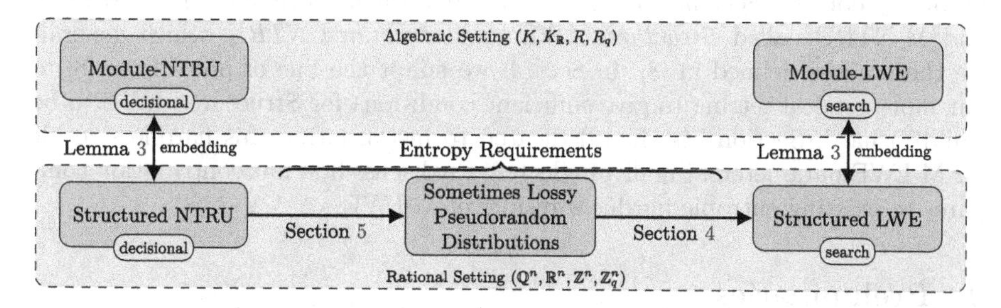

Fig. 1. High-level summary of the contributions of this work, proving Theorem 1.

One advantage of removing the HNF-M-LWE assumption is that our proof can be seen as a reduction from decisional M-NTRU to search M-LWE. This generalizes the observation of Peikert [25] that decisional NTRU reduces to search R-LWE. Another upside in only assuming M-NTRU is that we directly recover the expected statistical hardness of M-LWE from that of M-NTRU whenever the parameters are large enough. We discuss briefly this statistical result in Sect. 5.3 but note that it seems hard to build cryptosystems purely based on the statistical hardness of M-LWE in these parameter regimes. Our reduction is also rank-preserving between M-NTRU and M-LWE. Although it prevents us from reaching unusually short secrets, it can still lead to better parameters compared to (close to) rank-preserving reductions based on M-LWE assumptions, as discussed in Sect. 6. As explained, we rely on a rectangular formulation of M-NTRU which is similar to the multiple public keys version in the case of NTRU. For non-overstretched parameters, we have no reason to believe that this multiple public keys version, or rectangular M-NTRU in our case, is a substantially weaker assumption than square M-NTRU.

Lin et al. [19] recently adapted the proof method from [7] on LWE to modules, which uses a sensibly different approach from the one we described above. Their proof is based on an M-LWE hardness assumption, while ours is based on M-NTRU. This assumption allows them to tweak the starting rank k of the module to reach smaller or larger secrets and noise, depending on which one to optimize. However, their result does not provide a rank-preserving reduction as $k < d$. Additionally, when k is close to d, our proof can lead to better parameters, at the expense of trading the underlying assumption. We discuss these differences in Sect. 6.

Open Questions. Our proof only shows the entropic hardness of search M-LWE, and we leave it as an open problem to extend it to the decisional variant. One possibility (of more general interest) would be to find a search-to-decision reduction for M-LWE that preserves the (non-uniform) secret distribution. Additionally, it would be interesting to have a worst-case to average-case reduction from module lattice problems to decision M-NTRU, which would be of independent interest.

Organization. In Sect. 2, we introduce the notions and results that are needed in our proof. In Sect. 3, we provide an equivalent formulation of M-LWE and M-NTRU called *Structured LWE* and *Structured NTRU*, which generalize the notions defined in [8]. In Sect. 4, we adapt the line of proof from [8] to our more general setting to give sufficient conditions for Structured LWE to be (mildly) hard. Section 5 is then dedicated to instantiating this hardness result for M-LWE parameters. Finally, in Sect. 6, we discuss how our contribution compares to existing entropic hardness results of M-LWE.

2 Preliminaries

In this paper, q denotes a positive integer, $\mathbb{Z}$ the set of integers, and $\mathbb{Z}_q$ the quotient ring $\mathbb{Z}/q\mathbb{Z}$. We use $\mathbb{Q}$ and $\mathbb{R}$ to denote the fields of rationals and reals respectively. For a positive integer n, we use $[n]$ to denote $\{1, \ldots, n\}$. The vectors are written in bold lowercase letters $\mathbf{a}$ while the matrices are in bold uppercase letters $\mathbf{A}$. The transpose and Hermitian operators for vectors and matrices are denoted with the superscript T and $\dagger$ respectively. The identity matrix of size $n \times n$ is denoted by $\mathbf{I}_n$. For a vector $\mathbf{a} \in \mathbb{C}^n$, we define its Euclidean norm as $\|\mathbf{a}\|_2 = (\sum_{i \in [n]} |a_i|^2)^{1/2}$, and its infinity norm as $\|\mathbf{a}\|_\infty = \max_{i \in [n]} |a_i|$. Further, we denote by $\operatorname{diag}(\mathbf{a})$ the diagonal matrix whose diagonal entries are the entries of $\mathbf{a}$. For a matrix $\mathbf{A} = [a_{ij}]_{i \in [n], j \in [m]} \in \mathbb{C}^{n \times m}$, we define its spectral norm as $\|\mathbf{A}\|_2 = \max_{\mathbf{x} \in \mathbb{C}^m \setminus \{\mathbf{0}\}} \|\mathbf{A}\mathbf{x}\|_2 / \|\mathbf{x}\|_2$, which corresponds to its *largest singular value*. The *smallest singular value* of $\mathbf{A}$ is denoted by $s_{\min}(\mathbf{A})$. For a ring R and integers k, q, we denote $GL_k(R, q)$ the set of matrices of $R^{k \times k}$ that are invertible in $R_q = R/qR$. The uniform distribution over a finite set S is denoted by $U(S)$, and the statistical distance between two discrete distributions P and Q over a countable set S is defined as $\Delta(P, Q) = \frac{1}{2}\sum_{x \in S} |P(x) - Q(x)|$. Finally, the action of sampling $x \in S$ from a distribution P is denoted by $x \hookleftarrow P$.

2.1 Algebraic Number Theory

Since we focus on the problem once embedded into the integers, we only give the necessary notations related to algebraic number theory. The complete background is deferred to the full version [6, Sec. 2.1].

In this paper, we consider a number field $K = \mathbb{Q}(\zeta)$ of degree n and its ring of integers, which we denote by R. We also define the field tensor product as $K_{\mathbb{R}} = K \otimes_{\mathbb{Q}} \mathbb{R}$ and the torus $\mathbb{T}_q = K_{\mathbb{R}}/qR$ for an integer q. The canonical embedding is denoted by σ, while the coefficient embedding is denoted by τ. We also consider the embedding σ_H defined by $\mathbf{U}_H^\dagger \sigma$ where

$$\mathbf{U}_H = \frac{1}{\sqrt{2}} \begin{bmatrix} \sqrt{2}\mathbf{I}_{t_1} & \mathbf{0} & \mathbf{0} \\ \mathbf{0} & \mathbf{I}_{t_2} & i\mathbf{I}_{t_2} \\ \mathbf{0} & \mathbf{I}_{t_2} & -i\mathbf{I}_{t_2} \end{bmatrix},$$

with t_1 the number of real field embeddings of K, and t_2 the number of pairs of complex field embeddings. The embeddings σ and τ are also linked by the Vandermonde matrix of the field denoted by $\mathbf{V}$, i.e., $\sigma = \mathbf{V}\tau$. Finally, the multiplication of two elements x, y of K translate into a matrix multiplication of the embeddings. More precisely, for an embedding $\theta \in \{\sigma, \tau, \sigma_H\}$, there exists a ring homomorphism M_θ from K to $\mathbb{C}^{n \times n}$ such that for any $x, y \in K$, it holds $\theta(xy) = M_\theta(x) \cdot \theta(y)$. We have $M_\sigma(x) = \mathrm{diag}(\sigma(x))$, $M_{\sigma_H}(x) = \mathbf{U}_H^\dagger M_\sigma(x) \mathbf{U}_H$, and $M_\tau(x) = \sum_{k=0}^{n-1} \langle \tau(x), \mathbf{e}_k \rangle \mathbf{C}^k$, where $(\mathbf{e}_k)_k$ is the canonical basis of $\mathbb{C}^n$, and $\mathbf{C}$ is the companion matrix of the defining polynomial of K. All the embeddings and multiplication matrix maps extend to vectors and matrices over K in the natural way. For an integer η, we define $S_\eta = \tau^{-1}(\{-\eta, \ldots, \eta\}^n)$.

2.2 Lattices

A full-rank lattice Λ of rank n can be defined by a basis $\mathbf{B} = [\mathbf{b}_i]_{i \in [n]} \in \mathbb{R}^{n \times n}$ as $\Lambda = \sum_{i \in [n]} \mathbb{Z} \cdot \mathbf{b}_i$. We sometimes use $\Lambda(\mathbf{B})$ to specify which basis of the lattice we consider. The *dual lattice* of a lattice Λ is defined by $\Lambda^* = \{\mathbf{x} \in \mathrm{Span}(\Lambda) : \forall \mathbf{y} \in \Lambda, \langle \mathbf{x}, \mathbf{y} \rangle \in \mathbb{Z}\}$.

Structured Lattices. Any ideal $\mathcal{I}$ of R (resp. R-module $M \subseteq K^d$) embeds into a lattice via τ or σ_H called *ideal lattice* (resp. *module lattice*). In the rest of the paper, we denote by $\mathbf{L}_R$ a basis of the lattice $\sigma_H(R)$ and $\mathbf{B}_R = \mathbf{L}_R^{-1}$. For every integer d, we also define $\mathbf{L}_{R^d} = \mathbf{I}_d \otimes \mathbf{L}_R$, which is a basis of $\sigma_H(R^d)$, and $\mathbf{B}_{R^d} = \mathbf{I}_d \otimes \mathbf{L}_R = \mathbf{L}_{R^d}^{-1}$. Since each element of the lattice $\sigma_H(R)$ can be represented as $\mathbf{L}_R \mathbf{x}$ for some $\mathbf{x} \in \mathbb{Z}^n$, we can map R to $\mathbb{Z}^n$ by $\mathbf{B}_R \sigma_H$. We denote by $M_R(x)$ the associated multiplication matrix, i.e.,

$$M_R(x) = \mathbf{B}_R M_{\sigma_H}(x) \mathbf{B}_R^{-1} = (\mathbf{B}_R \mathbf{U}_H^\dagger) \mathrm{diag}(\sigma(x)) (\mathbf{B}_R \mathbf{U}_H^\dagger)^{-1}.$$

Notice that when the power basis is a $\mathbb{Z}$-basis of R, e.g. cyclotomics, we can choose $\mathbf{L}_R = [\sigma_H(1) \mid \ldots \mid \sigma_H(\zeta^{n-1})] = \mathbf{U}_H^\dagger \mathbf{V}$. In that case, we have $\mathbf{B}_R \sigma_H = \tau$ and also $M_R(\cdot) = M_\tau(\cdot)$. We keep the notations without assuming how the basis $\mathbf{L}_R$ is chosen. We only use this specific choice of basis in Sect. 5. The map $M_R(\cdot)$ is extended coefficient-wise to vectors and matrices. It maps (matrices of) ring elements to (block matrices of) structured matrices.

2.3 Probabilities

We first recall a trivial lemma that bounds the maximal singular value of a random matrix $\mathbf{Z}$ by bounding the singular values of its *blocks*. The result can be made deterministic for $\delta = 0$. The proof is quite standard and provided in the full version [6, Lem. 2.1] for completeness.

Lemma 1. *Let a, b, c, d be integers. Let $\mathbf{Z} = [\mathbf{Z}_{ij}]_{(i,j)\in[a]\times[b]} \in \mathbb{R}^{ac\times bd}$ be a random block matrix where each $\mathbf{Z}_{ij} \in \mathbb{R}^{c\times d}$. Assume it holds for all (i,j) that $\mathbb{P}[\|\mathbf{Z}_{ij}\|_2 \geq \gamma] \leq \delta$, for $0 \leq \delta \leq 1$. Then, we have $\mathbb{P}[\|\mathbf{Z}\|_2 \geq \sqrt{ab}\cdot\gamma] \leq ab\cdot\delta$.*

Gaussians. For a full-rank matrix $\mathbf{S} \in \mathbb{R}^{m\times n}$ $(m \geq n)$, and a center $\mathbf{c} \in \mathbb{R}^n$ we define the Gaussian function by $\rho_{\mathbf{c},\mathbf{S}}(\mathbf{x}) = \exp(-\pi(\mathbf{x}-\mathbf{c})^T(\mathbf{S}^T\mathbf{S})^{-1}(\mathbf{x}-\mathbf{c}))$. We can then define the continuous Gaussian distribution $D_{\mathbf{c},\mathbf{S}}$ whose density is $\det(\mathbf{S})^{-1}\rho_{\mathbf{c},\mathbf{S}}$. If $\mathbf{S} = s\mathbf{I}_n$, then we simply write $D_{\mathbf{c},s}$, and we omit $\mathbf{c}$ if it is $\mathbf{0}$. We then define the discrete Gaussian distribution $\mathcal{D}_{\Lambda,\mathbf{c},\mathbf{S}}$ over a lattice Λ by conditioning on $\mathbf{x}$ being in the lattice, i.e., for all $\mathbf{x} \in \Lambda$, $\mathcal{D}_{\Lambda,\mathbf{c},\mathbf{S}}(\mathbf{x}) = D_{\mathbf{c},\mathbf{S}}(\mathbf{x})/D_{\mathbf{c},\mathbf{S}}(\Lambda)$, where $D_{\mathbf{c},\mathbf{S}}(\Lambda) = \sum_{\mathbf{y}\in\Lambda} D_{\mathbf{c},\mathbf{S}}(\mathbf{y})$. We use the simplified notation $\mathcal{D}_{R,\mathbf{c},\mathbf{S}}$ to denote $\mathcal{D}_{\sigma_H(R),\mathbf{c},\mathbf{S}}$. Additionally, by abuse of notation, we also use $D_{\mathbf{c},\mathbf{S}}$ to denote the distribution obtained by sampling $\mathbf{x}$ from $D_{\mathbf{c},\mathbf{S}}$ and outputting $\sigma_H^{-1}(\mathbf{x}) \in K_{\mathbb{R}}$. For $\varepsilon > 0$, we denote by $\eta_\varepsilon(\Lambda)$ the smoothing parameter of a lattice Λ [23], which is defined by $\eta_\varepsilon(\Lambda) = \min\{s > 0 : \rho_{1/s}(\Lambda^*) \leq 1+\varepsilon\}$.

Min-Entropy. Let $\mathbf{x}$ follow a discrete distribution on a set X, and $\mathbf{z}$ follow a (possibly continuous) distribution on a (measurable) set Z. The *average conditional min-entropy*[2] of $\mathbf{x}$ given $\mathbf{z}$ is $\widetilde{H}_\infty(\mathbf{x}|\mathbf{z}) = -\log_2(\mathbb{E}_{\mathbf{z}'}[\max_{\mathbf{x}'\in X}\mathbb{P}[\mathbf{x}=\mathbf{x}'|\mathbf{z}=\mathbf{z}']])$. If $\mathbf{z}$ is deterministic, we obtain the definition of the *min-entropy* of $\mathbf{x}$ as $H_\infty(\mathbf{x}) = -\log_2(\max_{\mathbf{x}'\in X}\mathbb{P}[\mathbf{x}=\mathbf{x}'])$. For $\varepsilon > 0$, we also define the ε-smooth average conditional min-entropy by $\widetilde{H}_\infty^\varepsilon(\mathbf{x}|\mathbf{z}) = \max\{\widetilde{H}_\infty(\mathbf{x}'|\mathbf{z}') : \Delta((\mathbf{x}',\mathbf{z}'),(\mathbf{x},\mathbf{z})) \leq \varepsilon\}$. For convenience, we simply refer to all these notions as min-entropy instead of their full name when it is clear from the context or notations. But it should be noted that the notions are distinct.

2.4 Noise Lossiness

We recall the notion of *noise lossiness* of a distribution $\mathcal{S}$ of secrets as introduced in [7,8]. It quantifies how much information is lost about a secret from $\mathcal{S}$ when perturbed by a Gaussian noise. As we are in the module setting, we highlight the dimension as nd where n is the ring degree, and d the module rank.

Definition 1 (Noise Lossiness). *Let n, d, q be integers and $s > 0$ be a Gaussian parameter. Let $\mathbf{B}$ be a non-singular matrix in $\mathbb{R}^{nd\times nd}$. Let $\mathcal{S}$ be a distribution of secrets over $\mathbb{Z}_q^{nd}$. The noise lossiness $\nu_{s\mathbf{B}}(\mathcal{S})$ is defined by $\nu_{s\mathbf{B}}(\mathcal{S}) = \widetilde{H}_\infty(\mathbf{s}|\mathbf{s}+\mathbf{e})$, where $\mathbf{s} \hookleftarrow \mathcal{S}$, and $\mathbf{e} \hookleftarrow D_{s\mathbf{B}}$.*

[2] The (non-average) conditional min-entropy of $\mathbf{x}$ given $\mathbf{z}$ is denoted by $H_\infty(\mathbf{x}|\mathbf{z})$ instead of $\widetilde{H}_\infty(\mathbf{x}|\mathbf{z})$, and given by $H_\infty(\mathbf{x}|\mathbf{z}) = -\log_2\left(\max_{\mathbf{z}'\in Z}\max_{\mathbf{x}'\in X}\mathbb{P}[\mathbf{x}=\mathbf{x}'|\mathbf{z}=\mathbf{z}']\right)$.

We also recall the bounds on the noise lossiness derived in [7] in the case of general distributions, as well as that of distributions over bounded secrets.

Lemma 2 **([7], Lem. 5.2 & 5.4).** *Let n, d, q be integers, and a Gaussian parameter $s > 0$. Let $\mathcal{S}$ be any distribution over $\mathbb{Z}_q^{nd}$. If $s \leq q\sqrt{\pi/\ln(4nd)}$, then it holds that $\nu_s(\mathcal{S}) \geq H_\infty(\mathcal{S}) - nd \cdot \log_2(q/s) - 1$. Alternatively, if $\mathcal{S}$ is r-bounded (for the Euclidean norm), then it holds that $\nu_s(\mathcal{S}) \geq H_\infty(\mathcal{S}) - \sqrt{2\pi nd}\log_2(e) \cdot \frac{r}{s}$, with no restriction on s.*

2.5 Module Learning with Errors and Module NTRU

In this work, we deal with *Module Learning With Errors* (M-LWE) over the (primal) ring of integers R of a number field K. Additionally, we do not limit the secret distribution to be uniform, and we thus define M-LWE for an arbitrary distribution of secrets $\mathcal{S}$.

Definition 2. *Let K be a number field of degree n, and R its ring of integers. Let d, q, m be positive integers. Finally, let $\mathcal{S}$ be a secret distribution supported on R_q^d, and ψ a distribution over $K_\mathbb{R}$. The search* $\text{M-LWE}_{n,d,q,m,\psi,\mathcal{S}}$ *problem is to find the secret* $\mathbf{s}$ *given* $(\mathbf{A}, \mathbf{b}) = (\mathbf{A}, \mathbf{A}\mathbf{s} + \mathbf{e} \bmod qR)$ *for* $\mathbf{A} \hookleftarrow U(R_q^{m\times d})$, $\mathbf{s} \hookleftarrow \mathcal{S}$, *and* $\mathbf{e} \hookleftarrow \psi^m$*. The decisional version consists in deciding whether such* $(\mathbf{A}, \mathbf{b})$ *is distributed as above or if it is uniform over* $R_q^{m\times d} \times \mathbb{T}_q^m$.

We sometimes use a discrete error distribution ψ over R instead of $K_\mathbb{R}$. The standard form of M-LWE corresponds to $\mathcal{S} = U(R_q^d)$, for which we omit the $\mathcal{S}$ in the notation. For arbitrary secret distributions $\mathcal{S}$, we analyze the hardness of the problem based on some requirement on the entropy of $\mathcal{S}$. This is why these cases are also referred to as *entropic* M-LWE or *entropic hardness* of M-LWE. We also recall the Module-NTRU (M-NTRU) problem defined in [12].

Definition 3 **(Module-NTRU (M-NTRU)).** *Let R be the ring of integers of a number field K and let q be a modulus. Let m, d be positive integers, and ψ be a distribution on R. Let* $\mathbf{G} \hookleftarrow \psi^{m\times d}$ *and* $\mathbf{F} \hookleftarrow \psi^{d\times d}$ *conditioned on* $\mathbf{F} \in GL_d(R, q)$*. Let* $\mathbf{F}_q^{-1}$ *be the R_q-inverse of* $\mathbf{F}$*. The* $\text{M-NTRU}_{n,d,q,m,\psi}$ *problem asks to distinguish between* $\mathbf{G}\mathbf{F}_q^{-1} \in R_q^{m\times d}$ *and a uniformly random* $\mathbf{A} \hookleftarrow U(R_q^{m\times d})$.

In this paper, we consider $\psi = \mathcal{D}_{R,\gamma}$ for some Gaussian parameter $\gamma > 0$. In this case, we simply denote it as $\text{M-NTRU}_{n,d,q,m,\gamma}$.

3 Structured LWE

In this section, we provide another formulation of LWE (resp. NTRU), which is equivalent to M-LWE (resp. M-NTRU) in a specific setting. Finally, we detail the notion of *mild hardness* presented in [8], and how it relates to the standard definition of hardness.

3.1 Structured LWE and Structured NTRU

In the following, we define a version of LWE that generalizes the *Structured LWE* problem from [8]. We also define the *Structured NTRU* problem which is a generalization of the DSR problem from [8]. We use the name Structured NTRU only because it is to M-NTRU what Structured LWE is to M-LWE. NTRU already being a structured problem, it should not be interpreted as if there exists an *unstructured* version of NTRU.

The difference stems in introducing the extra dimension of the module rank d. Instead of considering vectors of m matrices of size $n \times n$, we consider block matrices of size $m \times d$ with $n \times n$ blocks. We only define the search variant of Structured LWE and the decisional variant of Structured NTRU as they are the only one needed in this paper, but one could define the other versions in the natural way. The main motivation for working with these problems is the simpler analysis due to its formulation over $\mathbb{Z}$ instead of R. Furthermore, both S-LWE and S-NTRU can be instantiated with distributions that are not directly linked to M-LWE and M-NTRU, and are therefore more general.

The technical difficulties of considering arbitrary ranks d, as opposed to just $d = 1$ in [8], arise in Sects. 4 and 5 when dealing with matrices over R rather than single ring elements.

Definition 4 (Structured LWE, [8, Def. 3.1] adapted). *Let n, d, q, and m be positive integers. Let $\mathcal{M}$ be a distribution of matrices on $\mathbb{Z}_q^{n\times n}$, and Υ be a distribution of error-distributions on $\mathbb{R}^n$. Furthermore, let $\mathcal{S}$ be a distribution on $\mathbb{Z}_q^{nd}$. The goal of the* $\text{S-LWE}_{n,d,q,m,\mathcal{M},\Upsilon,\mathcal{S}}$ *problem is to find the secret* $\mathbf{s} \hookleftarrow \mathcal{S}$ *given* $(\mathbf{A}, \mathbf{y}) = (\mathbf{A}, \mathbf{A}\mathbf{s} + \mathbf{e} \bmod q)$, *with* $\mathbf{A} \hookleftarrow \mathcal{M}^{m\times d}$, *and* $\mathbf{e} \hookleftarrow \psi^m$ *where* $\psi \hookleftarrow \Upsilon$.

If m is not specified, it means we consider the samples one by one. A single sample follows the same definition for $m = 1$. When Υ is the deterministic distribution outputting some error distribution ψ, we simply use $\text{S-LWE}_{n,d,q,m,\mathcal{M},\psi,\mathcal{S}}$.

Definition 5 (Structured NTRU [8, Def. 5.1] adapted). *Let n, d, q, and m be positive integers. Let $\mathcal{M}$ be a distribution of matrices on $\mathbb{Z}_q^{n\times n}$, and Ψ a distribution on $GL_{nd}(\mathbb{Z}, q) \times \mathbb{Z}^{nm\times nd}$. The* $\text{S-NTRU}_{n,d,q,m,\mathcal{M},\Psi}$ *problem is to distinguish the two following distributions (1)* $\mathbf{G}\cdot\mathbf{F}_q^{-1} \bmod q$, *where* $(\mathbf{F}, \mathbf{G}) \hookleftarrow \Psi$, *and* $\mathbf{F}_q^{-1}$ *is the $\mathbb{Z}_q$-inverse of* $\mathbf{F} \bmod q$, *and (2)* $\mathbf{U} \hookleftarrow \mathcal{M}^{m\times d}$.

Notice that these formulations are very similar to Definitions 2 and 3 but where the ring R is embedded as $\mathbb{Z}^n$.

The following states that they are equivalent up to carefully chosen mappings between the different distributions, as done in [8] for the case of R-LWE. In particular, the transformation consists in embedding the ring elements using M_R and $\mathbf{B}_{R}\sigma_H$. M_R maps a ring element to a structured matrix, thus motivating the names Structured LWE and Structured NTRU.

We provide the proof in the full version [6, Lem. 3.1] for completeness.

Lemma 3. *Let $K = \mathbb{Q}(\zeta)$ be a number field of degree n, and R its ring of integers. Let d, q, m be positive integers. We set $\mathcal{M} = M_R(U(R_q))$ the distribution over $\mathbb{Z}_q^{n\times n}$, where M_R is defined in Sect. 2.2.*

LWE. *Let ψ' be a distribution over $K_\mathbb{R}$, and $\mathcal{S}'$ a distribution over R_q^d. Define $\psi = (\mathbf{B}_R\sigma_H)(\psi')$, and $\mathcal{S} = (\mathbf{B}_R\sigma_H)(\mathcal{S}')$, where $\mathbf{B}_R$ is defined in Sect. 2.2. Then, the two problems* M-LWE$_{n,d,q,m,\psi',\mathcal{S}'}$ *and* S-LWE$_{n,d,q,m,\mathcal{M},\psi,\mathcal{S}}$ *are equivalent.*

NTRU. *Now, let ψ' be a distribution over R. We define Ψ to be the distribution over $GL_{nd}(\mathbb{Z}, q) \times \mathbb{Z}^{nm\times nd}$ obtained by drawing $\mathbf{F}$ from $(\psi')^{d\times d}$ conditioned on being in $GL_d(R, q)$, $\mathbf{G}$ from $(\psi')^{m\times d}$ and outputting $(M_R(\mathbf{F}), M_R(\mathbf{G}))$. Then, the two problems* M-NTRU$_{n,d,q,m,\psi'}$ *and* S-NTRU$_{n,d,q,m,\mathcal{M},\Psi}$ *are equivalent.*

Recall that in cyclotomic fields for example, we can choose $\mathbf{B}_R = (\mathbf{U}_H^\dagger \mathbf{V})^{-1}$, which leads to $M_R = M_\tau$ and $\mathbf{B}_R\sigma_H = \tau$. In this case, it simply uses the coefficient embedding to embed M-LWE (resp. M-NTRU) into S-LWE (resp. S-NTRU). Note that the hardness of S-NTRU can be established for other distributions Ψ, but based on different assumptions than M-NTRU. We discuss it in Sect. 5.2. The distribution of the blocks is chosen to be $\mathcal{M} = M_R(U(R_q))$. The reader can keep this choice in mind, but we point out that the results of Sects. 3.2 and 4 hold for arbitrary distributions $\mathcal{M}$, $\mathcal{S}$ and ψ.

3.2 (Mild) Hardness

We consider the two notions of hardness for S-LWE as is done in [8], namely standard hardness and the weaker notion of *mild hardness*. We show that standard hardness naturally implies mild hardness, while the converse requires an a priori unbounded number of samples in order to use a success amplification argument. All the proofs are provided in the full version [6, Lem. 3.2 & 3.3].

Definition 6 (Standard and Mild Hardness). *Let n, d, and q be positive integers. Let $\mathcal{M}$ be a distribution over $\mathbb{Z}_q^{n\times n}$, and Υ a distribution of distributions over $\mathbb{R}^n$. Finally, let $\mathcal{S}$ be a distribution on $\mathbb{Z}_q^{nd}$. For any $(\mathbf{s}, \psi)$ sampled from $(\mathcal{S}, \Upsilon)$, we denote by $\mathcal{O}_{\mathbf{s},\psi}$ the (randomized) oracle that, when called, returns $(\mathbf{A}_i, \mathbf{A}_i\mathbf{s} + \mathbf{e}_i \bmod q)$, where $\mathbf{A}_i \hookleftarrow \mathcal{M}^{1\times d}$ and $\mathbf{e}_i \hookleftarrow \psi$.*

The S-LWE$_{n,d,q,\mathcal{M},\Upsilon,\mathcal{S}}$ *problem is* standard *hard, if for every* PPT *adversary $\mathcal{A}$ and every non-negligible function ε, there exists a negligible function ν such that $\mathbb{P}_{\substack{\mathbf{s}\hookleftarrow\mathcal{S}\\ \psi\hookleftarrow\Upsilon}}[\mathbb{P}_{\mathcal{A},\mathcal{O}_{\mathbf{s},\psi}}[\mathcal{A}^{\mathcal{O}_{\mathbf{s},\psi}}(1^\lambda) = \mathbf{s}] \geq \varepsilon(\lambda)] \leq \nu(\lambda)$, where $\mathcal{A}^{\mathcal{O}_{\mathbf{s},\psi}}$ means that the adversary has access to $\mathcal{O}_{\mathbf{s},\psi}$ as a black-box and can thus query it as many times as they want. The internal probability is over the random coins of $\mathcal{A}$ and the random coins of $\mathcal{O}_{\mathbf{s},\psi}$ (meaning over the randomness of the $(\mathbf{A}_i, \mathbf{e}_i)$).*

We now say that the S-LWE$_{n,d,q,\mathcal{M},\Upsilon,\mathcal{S}}$ *problem is* mildly *hard, if for every* PPT *adversary $\mathcal{A}$ and every negligible function μ, there exists a negligible function ν such that $\mathbb{P}_{\substack{\mathbf{s}\hookleftarrow\mathcal{S}\\ \psi\hookleftarrow\Upsilon}}[\mathbb{P}_{\mathcal{A},\mathcal{O}_{\mathbf{s},\psi}}[\mathcal{A}^{\mathcal{O}_{\mathbf{s},\psi}}(1^\lambda) = \mathbf{s}] \geq 1 - \mu(\lambda)] \leq \nu(\lambda)$. When the number of available samples m is fixed a priori, we use the same definitions except that $\mathcal{A}$ is only allowed at most m queries to the oracle. The samples can be written in matrix form as $(\mathbf{A}, \mathbf{A}\mathbf{s} + \mathbf{e} \bmod q)$ with $\mathbf{A} \hookleftarrow \mathcal{M}^{m\times d}$ and $\mathbf{e} \hookleftarrow \psi^m$.*

Lemma 4. *Let n, d, q be positive integers. Let $\mathcal{M}$ be a distribution over $\mathbb{Z}_q^{n\times n}$, and Υ a distribution of distributions over $\mathbb{R}^n$. Finally, let $\mathcal{S}$ be a distribution on $\mathbb{Z}_q^{nd}$. If* S-LWE$_{n,d,q,\mathcal{M},\Upsilon,\mathcal{S}}$ *is standard hard, then it is also mildly hard. The same result holds when the number of available samples m is fixed.*

Lemma 5 ([8, Lem. 3.4] adapted). *Let n, d, q be positive integers. Let $\mathcal{M}$ be a distribution over $\mathbb{Z}_q^{n\times n}$, and Υ a distribution of distributions over $\mathbb{R}^n$. Finally, let $\mathcal{S}$ be a distribution on $\mathbb{Z}_q^{nd}$. If* S-LWE$_{n,d,q,\mathcal{M},\Upsilon,\mathcal{S}}$ *is mildly hard, then it is also standard hard.*

Lemma 5 only holds for an unbounded number of samples, which is not always realistic. In order to generate new samples from a fixed number, one can use the rerandomization lemma from [8, Lem. 3.5] which straightforwardly adapt to our more general setting.

We defer it to the full version [6, Sec. 3.4] due to lack of space.

4 Entropic Hardness of Structured LWE

In this section, we adapt the notion of *sometimes lossy pseudorandom distribution* from [8] to our more general version of Structured LWE. They gather two main properties, *pseudorandomness* and *sometimes lossiness*, which are essential in proving the entropic hardness of S-LWE. Section 4.1 formalizes this idea that if there exists a sometimes lossy pseudorandom distribution, then S-LWE is mildly hard. Then, Sect. 4.2 gives sufficient conditions to construct such distributions. We defer the proofs of this section to the full version [6]. The matrices in this section are over $\mathbb{Z}$, $\mathbb{Z}_q$, $\mathbb{Q}$ or $\mathbb{R}$, and not R, R_q, K or $K_\mathbb{R}$.

Definition 7 ([8, Def. 4.1] adapted). *Let n, d, q, and m be positive integers. Let $\mathcal{X}$ be a distribution on $\mathbb{Z}_q^{nm\times nd}$, $\mathcal{M}$ a distribution on $\mathbb{Z}_q^{n\times n}$, $\mathcal{S}$ a distribution on $\mathbb{Z}_q^{nd}$ and ψ an error distribution on $\mathbb{R}^n$. We say that $\mathcal{X}$ is a* sometimes lossy pseudorandom distribution *for $(\mathcal{S}, \mathcal{M}, \psi)$ if there exists a negligible function ε, a $\kappa = \omega(\log_2 \lambda)$ and a $\delta \geq 1/poly(\lambda)$ such that the following properties hold.*

Pseudorandomness: *$\mathcal{X}$ is computationally indistinguishable from $\mathcal{M}^{m\times d}$*
Sometimes Lossiness: $\mathbb{P}_{\mathbf{A}\hookleftarrow\mathcal{X}}[\widetilde{H}_\infty^\varepsilon(\mathbf{s}|\mathbf{A}, \mathbf{A}\mathbf{s}+\mathbf{e} \bmod q) \geq \kappa] \geq \delta$, *where $\mathbf{s} \hookleftarrow \mathcal{S}$ and $\mathbf{e} \hookleftarrow \psi^m$.*

4.1 From Sometimes Lossiness to the Entropic Hardness of Structured LWE

The following theorem adapted from [8] states that the existence of a sometimes lossy pseudorandom distribution implies the mild hardness of Structured LWE. The proof can be found in the full version [6, Thm. 4.1]. The pseudorandomness property essentially allows us to trade the uniform matrix $\mathbf{A}$ from the S-LWE instance for the matrix $\mathbf{G}\mathbf{F}_q^{-1} \bmod q$ (where $\mathbf{F}_q^{-1}$ is the $\mathbb{Z}_q$-inverse of a short matrix $\mathbf{F}$) as in Definition 3. Then, the sometimes lossiness property translates the fact that going from $\mathbf{s}$ to $\mathbf{G}\mathbf{F}_q^{-1}\mathbf{s} + \mathbf{e} \bmod q$ loses enough information on $\mathbf{s}$ (with non-negligible probability over the choice of $\mathbf{F}$ and $\mathbf{G}$) so that it is hard to recover $\mathbf{s}$. The sometimes lossiness thus entails the hardness of S-LWE when $\mathbf{A} = \mathbf{G}\mathbf{F}_q^{-1}$ instead of being uniform. By the pseudorandomness, it then yields the hardness of S-LWE.

Theorem 2 **([8, Thm. 4.2] adapted).** *Let n, d, q, m be positive integers. Let $\mathcal{X}$ be a distribution on $\mathbb{Z}_q^{nm \times nd}$, $\mathcal{M}$ a distribution on $\mathbb{Z}_q^{n \times n}$, $\mathcal{S}$ a distribution on $\mathbb{Z}_q^{nd}$ and ψ an error distribution on $\mathbb{R}^n$. We assume that all the distributions are efficiently sampleable. If the distribution $\mathcal{X}$ is a sometimes lossy pseudorandom distribution for $(\mathcal{M}, \mathcal{S}, \psi)$, then $\text{S-LWE}_{n,d,q,m,\mathcal{M},\psi,\mathcal{S}}$ is mildly hard.*

4.2 Construction of Sometimes Lossy Pseudorandom Distributions

We now provide the generalization of [8] to our new problem S-NTRU in order to give sufficient conditions to construct sometimes lossy pseudorandom distributions, and thus get the mild hardness of $\text{S-LWE}_{n,d,q,m,\mathcal{M},\psi,\mathcal{S}}$ by Theorem 2.

Albeit more general, the proof follows the same structure as the one from [8], which is why we defer it to the full version [6, Thm. 4.3]. The goal is to prove that $\mathbf{s}$ has sufficient min-entropy left, even if $\mathbf{G}\mathbf{F}_q^{-1}\mathbf{s} + \mathbf{e}_0$ is known. Recall that $\mathbf{B}_R \in \mathbb{R}^{n \times n}$ and for $\ell \geq 1$, $\mathbf{B}_{R^\ell} = \mathbf{I}_\ell \otimes \mathbf{B}_R$, as defined in Sect. 2.2. Note that in the following, we need both the inverse modulo q ($\mathbf{F}_q^{-1}$) and the rational inverse ($\mathbf{F}^{-1}$) of a matrix $\mathbf{F} \in \mathbb{Z}_q^{nd \times nd}$. The invertibility modulo q implies that the determinant of $\mathbf{F}$ is a unit in $\mathbb{Z}_q$, and is therefore non-zero when seen as an element of $\mathbb{Q}$, which in turns implies the rational invertibility.

Theorem 3 **([8, Thm. 5.8] adapted).** *Let n, d, m, and q be positive integers, and $\beta_1, \beta_2 > 0$. Let Ψ be a distribution on $GL_{nd}(\mathbb{Z}, q) \times \mathbb{Z}^{nm \times nd}$, $\mathcal{M}$ a distribution on $\mathbb{Z}_q^{n \times n}$, and $\mathcal{S}$ a distribution on $\mathbb{Z}_q^{nd}$. Assume $\text{S-NTRU}_{n,d,q,m,\mathcal{M},\Psi}$ is hard. Additionally, assume that if $(\mathbf{F}, \mathbf{G}) \hookleftarrow \Psi$ then*

- $\left\|\mathbf{B}_{R^m}^{-1}\mathbf{G}\mathbf{F}^{-1}\mathbf{B}_{R^d}\right\|_2 \leq \beta_1$ *where $\mathbf{F}^{-1}$ is the rational inverse of $\mathbf{F}$.*
- $\left\|\mathbf{B}_{R^d}^{-1}\mathbf{F}\mathbf{B}_{R^d}\right\|_2 \leq \beta_2$

with probability at least $\delta \geq 1/\mathsf{poly}(\lambda)$ over the choice of $(\mathbf{F}, \mathbf{G})$. Define the distribution $\mathcal{X}$ on $\mathbb{Z}_q^{nm \times nd}$ by $\mathbf{G}\mathbf{F}_q^{-1}$, where $(\mathbf{F}, \mathbf{G}) \hookleftarrow \Psi$ and $\mathbf{F}_q^{-1} \in \mathbb{Z}_q^{nd \times nd}$ is the $\mathbb{Z}_q$-inverse of $\mathbf{F}$. Let $s > \beta_2 \eta_\varepsilon(\Lambda(\mathbf{B}_{R^d}^{-1}))$ and $s_0 > 2^{3/2}\beta_1 s$. Further assume that $\nu_{s\mathbf{B}_{R^d}}(\mathcal{S}) \geq nd\log_2(\beta_2) + \omega(\log_2(\lambda))$.

Then $\mathcal{X}$ is a sometimes lossy pseudorandom distribution for $(\mathcal{S}, \mathcal{M}, D_{s_0\mathbf{B}_{R^m}})$.

Therefore, Theorems 3 and 2 together yield the following immediate corollary.

Corollary 1. *Assume that the conditions of Theorem 3 are satisfied. Then the problem $\text{S-LWE}_{n,d,q,m,\mathcal{M},\psi,\mathcal{S}}$ is mildly hard.*

5 Instantiation for M-LWE

This section constitute our main contribution, which consists in concretely exhibiting a sometimes lossy pseudorandom distribution that implies the entropic hardness of M-LWE. We thus set the parameters so that it fits the requirements of both Sects. 3.1 and 4. As seen in the latter in Theorem 3, the S-NTRU problem must be hard for this distribution, and the distribution also needs to be somewhat well behaved in terms of its spectral properties. Lemma 3 gives the equivalence between M-NTRU and S-NTRU, which allows for expressing the entire result in the more algebraic module setting. The more technical aspect of this section comes from Sect. 5.1, in which we study the spectral properties that we need. In particular, we derive a lower bound on the smallest singular value of discrete Gaussian matrices over R (once embedded via M_{σ_H}). Then, in Sect. 5.2, we define this distribution and verify that it indeed leads to a sometimes lossy pseudorandom distribution. Combining it with Corollary 1 and the equivalence between M-LWE and S-LWE of Lemma 3, we then obtain the entropic (mild) hardness of M-LWE. All the results in this section hold for arbitrary number fields at the exception of Corollary 3 which is stated for cyclotomic fields.

5.1 Invertibility and Singular Values of Discrete Gaussian Matrices

We now recall [12, Thm. A.5] that gives the density of square discrete Gaussian matrices over R_q that are invertible modulo qR. This theorem gives concrete conditions so that $\mathcal{D}_{GL_d(R,q),\gamma}$ is efficiently sampleable. The proofs of [12, Thm. A.5, Lem. A.6] depend on the embedding that is chosen to represent R as a lattice. The paper uses Gaussian distributions in the coefficient embedding over the lattice $\tau(R)$ which differ from our context. As such we need to adapt the proofs for Gaussian distributions in the canonical embedding over the lattice $\sigma_H(R)$. The changes are mostly limited to volume arguments as the volume of the lattice R depends on the embedding. Also, note that the proofs of [12, Thm. A.5, Lem. A.6] still hold in any number field and for any splitting behaviour of q provided that it is unramified, no matter the size of the norm of its prime ideal factors.

Theorem 4 ([12, Thm. A.5] adapted). *Let K be a number field of degree n and R its ring of integers. Let $d \geq 1$ and $q > 2n$ be an unramified prime. We define $N_{\max} = \max_{\mathfrak{p}|qR,\mathfrak{p}\text{ prime}} N(\mathfrak{p})$ and $N_{\min} = \min_{\mathfrak{p}|qR,\mathfrak{p}\text{ prime}} N(\mathfrak{p})$. Assume that $\gamma \geq 2^{1/(2d-1)} \cdot (|\Delta_K| \cdot N_{\max}^{(d-1)/(2d-1)})^{1/n}$. Then*

$$\rho_\gamma(R^{d\times d} \setminus GL_d(R,q)) \leq \frac{2r}{N_{\min}} \cdot \frac{\gamma^{nd^2}}{|\Delta_K|^{d^2/2}} \cdot (1 + 8d^2 2^{-n}) \leq \frac{2r}{N_{\min}} \cdot (1 + 8d^2 2^{-n}) \cdot \rho_\gamma(R^{d\times d}),$$

where r is the number of prime factors of qR, and Δ_K the discriminant of K.

Remark 1. Consider $q = 1 \bmod \nu$ over the ν-th cyclotomic field. Then, qR fully splits into n distinct prime ideals, each of norm q ($N_{\min} = N_{\max}$). Thus, if $\gamma \geq 2^{1/(2d-1)}(|\Delta_K| \cdot q^{(d-1)/(2d-1)})^{1/n}$ which is roughly $\Omega(n)$, then we have that

$$\mathbb{P}_{\mathbf{F} \hookleftarrow \mathcal{D}_{R,\gamma}^{d\times d}}[\mathbf{F} \notin GL_d(R,q)] \leq \frac{2n}{q} + \mathsf{negl}(n) \tag{1}$$

Note that if $q \leq 2n$, the inequality is vacuous. However, in practice q is usually much larger. For example, Kyber [3] uses $q \geq 13 \cdot n$ while the signature Dilithium [13] uses $q \geq n^{5/2} \gg 2n$. This yields a probability of invertibility that is sufficient for this work, while allowing for reducing the parameter γ as much as possible. More precisely, it allows for taking $\gamma = \Omega(n)$ with a constant close to 1. Also, as mentioned before, the invertibility in R_q implies that the determinant is a unit of R_q. As such, it is a non-zero element when seen in K which implies the K-invertibility.

Although it seems folklore, we weren't able to find a Gaussian tail bound on $\sigma(x)$ in the infinity norm for $x \hookleftarrow \mathcal{D}_{R,\gamma}$. We therefore provide the following lemma, whose proof is mostly based on [24, Cor. 5.3], and which proves that $\|\sigma(x)\|_\infty \leq \gamma \log_2 n$ with overwhelming probability. Most of the tail bounds are with respect to the Euclidean norm and thus require an extra $\sqrt{n}$ factor. Here, we are only interested in the infinity norm. The proof is given in the full version [6, Sec. B.5].

Lemma 6. *Let R be a ring of integers of degree n. Then for any $\gamma > 0$ and any $t \geq 0$, it holds that $\mathbb{P}_{f \hookleftarrow \mathcal{D}_{R,\gamma}}[\|\sigma(f)\|_\infty \leq \gamma t] \geq 1 - 2ne^{-\pi t^2}$. Choosing $t = \log_2 n$ gives $\|\sigma(f)\|_\infty \leq \gamma \log_2 n$ with overwhelming probability.*

The main challenge in instantiating Theorem 3 is to provide a decent bound for $\|\mathbf{G}'(\mathbf{F}')^{-1}\|_2$. It seems to require knowledge on the smallest singular value of $\mathbf{F}'$, which in our case is taken from a discrete Gaussian distribution. We now provide a lower bound on the smallest singular value of discrete Gaussian matrices. This automatically gives an upper bound on $\|(\mathbf{F}')^{-1}\|_2$, as $\|(\mathbf{F}')^{-1}\|_2 = 1/s_{\min}(\mathbf{F}')$.

Lemma 7. *Let $K = \mathbb{Q}(\zeta)$ be a number field of degree n, and R its ring of integers. Let $\mathcal{I}$ be any fractional ideal of R. Let $\gamma > 0$ be a Gaussian parameter. Then, for all $\delta \geq 0$, it holds that*

$$\mathbb{P}_{\mathbf{F} \hookleftarrow \mathcal{D}_{\mathcal{I},\gamma}^{d\times d}}\left[s_{\min}(M_{\sigma_H}(\mathbf{F})) \leq \frac{\delta}{\sqrt{d}}\right] \leq nC_\gamma \delta + nc_\gamma^d,$$

with $C_\gamma > 0$ and $c_\gamma \in (0,1)$ parameters depending on γ.

Proof. **Spectral analysis.** For convenience, we define $S(\mathbf{A})$ to be the set of singular values of any complex matrix $\mathbf{A}$. First of all, note that $M_{\sigma_H}(\mathbf{F}) = (\mathbf{I}_d \otimes \mathbf{U}_H^\dagger) M_\sigma(\mathbf{F})(\mathbf{I}_d \otimes \mathbf{U}_H)$. Since $\mathbf{U}_H$ is unitary, we have $S(M_{\sigma_H}(\mathbf{F})) = S(M_\sigma(\mathbf{F}))$. Recall that $M_\sigma(\mathbf{F})$ is the block matrix of size $nd \times nd$ whose block $(i,j) \in [d]^2$

is $\text{diag}(\sigma(f_{ij}))$. The matrix can therefore be seen as a $d \times d$ matrix with blocks of size $n \times n$. The idea is now to permute the rows and columns of $M_\sigma(\mathbf{F})$ to end up with a matrix of size $n \times n$ with blocks of size $d \times d$ only on the diagonal, as noticed in e.g. [14]. For that, we define the following permutation π of $[nd]$. For all $i \in [nd]$, write $i - 1 = k_1^{(i)} + nk_2^{(i)}$, with $k_1^{(i)} \in \{0, \dots, n-1\}$ and $k_2^{(i)} \in \{0, \dots, d-1\}$. Then, define $\pi(i) = 1 + k_2^{(i)} + dk_1^{(i)}$. This is a well-defined permutation based on the uniqueness of the Euclidean division. We can then define the associated permutation matrix $\mathbf{P}_\pi = [\delta_{i,\pi(j)}]_{(i,j)\in[nd]^2} \in \mathbb{R}^{nd \times nd}$. Then, it holds that $\mathbf{P}_\pi M_\sigma(\mathbf{F})\mathbf{P}_\pi^T = \text{diag}(\sigma_1(\mathbf{F}), \dots, \sigma_n(\mathbf{F}))$. Since $\mathbf{P}_\pi$ is a permutation matrix, it is unitary and therefore $S(M_\sigma(\mathbf{F})) = S(\mathbf{P}_\pi M_\sigma(\mathbf{F})\mathbf{P}_\pi^T)$. As $\mathbf{P}_\pi M_\sigma(\mathbf{F})\mathbf{P}_\pi^T$ is block-diagonal, we directly get the singular values by $S(\mathbf{P}_\pi M_\sigma(\mathbf{F})\mathbf{P}_\pi^T) = \cup_{k\in[n]} S(\sigma_k(\mathbf{F}))$. This proves that $S(M_{\sigma_H}(\mathbf{F})) = \cup_{k\in[n]} S(\sigma_k(\mathbf{F}))$. In particular, taking the minimum of the sets yields $s_{\min}(M_{\sigma_H}(\mathbf{F})) = \min_{k\in[n]} s_{\min}(\sigma_k(\mathbf{F}))$.

Random matrix theory. By [22, Lem. 2.8], for all $i, j \in [d]$ and unit vector $\mathbf{u} \in \mathbb{C}^n$, $\langle \sigma_H(f_{ij}), \mathbf{u} \rangle$ is sub-Gaussian with sub-Gaussian moment γ. Hence, since the rows of $\mathbf{U}_H$ are unit vectors of $\mathbb{C}^n$, it holds that for all $k \in [n]$ and $i, j \in [d]$, $\sigma_k(f_{ij})$ is sub-Gaussian with moment γ. Thus, for all $k \in [n]$, $\sigma_k(\mathbf{F})$ has independent and identically distributed sub-Gaussian entries. A result from random matrix theory by Rudelson and Vershynin [28, Thm 1.2] yields

$$\mathbb{P}\left[s_{\min}(\sigma_k(\mathbf{F})) \leq \frac{\delta}{\sqrt{d}}\right] \leq C_\gamma \cdot \delta + c_\gamma^d, \tag{2}$$

for some parameters $C_\gamma > 0$ and $c_\gamma \in (0,1)$ that only depend on the sub-Gaussian moment γ, for all $k \in [n]$. A union-bound gives that $\mathbb{P}[s_{\min}(M_{\sigma_H}(\mathbf{F})) \leq \delta/\sqrt{d}] \leq nC_\gamma\delta + nc_\gamma^d$, thus concluding the proof.

□

An immediate application of this lemma is for $\mathcal{I} = R$, which gives a probabilistic bound depending on the constants C_γ and c_γ, which in turns depend on γ. Aside from experimentally noticing that C_γ decreases polynomially with γ and that c_γ seems to decrease exponentially with γ, we do not have closed-form expression of these constants. By scaling the distribution by γ and applying Lemma 7 for $\mathcal{I} = \gamma^{-1}R$, we obtain

$$\mathbb{P}_{\mathbf{F}\hookleftarrow\mathcal{D}_{R,\gamma}^{d\times d}}\left[s_{\min}(M_{\sigma_H}(\mathbf{F})) \leq \frac{\gamma\delta}{\sqrt{d}}\right] \leq nC\delta + nc^d,$$

with $C > 0$ and $c \in (0,1)$ no longer depending on γ. In this case, we can mitigate the union bound blow-up by choosing $\delta = n^{-3/2}$ for example, yielding $s_{\min}(M_{\sigma_H}(\mathbf{F})) \geq \gamma/n\sqrt{nd}$ with non-negligible probability. In what follows, we thus use the following corollary. We discuss in Sect. 6 how we can experimentally expect a better bound.

Corollary 2. *Let $K = \mathbb{Q}(\zeta)$ be a number field of degree n, and R its ring of integers. Let $\gamma > 0$ be a Gaussian parameter. It holds that*

$$\mathbb{P}_{\mathbf{F}\hookleftarrow\mathcal{D}_{R,\gamma}^{d\times d}}\left[s_{\min}(M_{\sigma_H}(\mathbf{F})) \leq \frac{\gamma}{n\sqrt{nd}}\right] \leq O(n^{-1/2}).$$

5.2 Instantiation

We now define a distribution Ψ over $GL_{nd}(\mathbb{Z}, q) \times \mathbb{Z}^{nm \times nd}$ and prove it verifies the conditions of Theorem 3 under a careful choice of parameters. This distribution is actually a direct application of Definition 3 for a Gaussian distribution ψ. As opposed to what is done in [8], we no longer need to assume the hardness of Hermite Normal Form M-LWE and we solely rely on the M-NTRU assumption for *rectangular* matrices. Nonetheless, the distribution proposed in [8, Sec. 6] can be adapted to the module setting. We defer the analysis to the full version [6, App. A] due to lack of space. In this case, the hardness of S-NTRU is proven under that of M-NTRU for *square* matrices and of HNF-M-LWE. It thus uses a more standard formulation of M-NTRU but at the expense of an additional assumption, and also slightly worse parameters. This highlights a trade-off between the underlying hardness assumptions and the parameters.

Definition 8. *Let K be a number field of degree n, and R its ring of integers. Let d, q, m be positive integers. Let $\gamma > 0$ be a Gaussian parameter. We define the distribution Ψ as follows:*

- *Choose $\mathbf{F} \hookleftarrow \mathcal{D}_{R,\gamma}^{d\times d}$ such that $\mathbf{F} \in GL_d(R, q)$;*
- *Choose $\mathbf{G} \hookleftarrow \mathcal{D}_{R,\gamma}^{m\times d}$.*
- *Output $(M_R(\mathbf{F}), M_R(\mathbf{G}))$.*

Note that by Theorem 4, Ψ is efficiently sampleable if γ is sufficiently large, depending on the splitting behaviour of q. In particular, as stated in Remark 1, one can choose $\gamma = [\Theta(n)$ and have a non-negligible probability that a sample from $\mathcal{D}_{R^{d\times d},\gamma}$ is also in $GL_d(R, q)$ for a fully splitted prime q in a cyclotomic field. Also, since M_R is a ring homomorphism, if $\mathbf{F}$ is invertible modulo q, then so is $M_R(\mathbf{F})$, and vice-versa. By Lemma 3, M-NTRU and S-NTRU, are equivalent for a specific connection between the distributions. We specifically defined Ψ so that it matches with M-NTRU$_{n,d,q,m,\gamma}$. Hence, assuming the hardness of S-NTRU for this distribution Ψ is equivalent to assuming the hardness of M-NTRU$_{n,d,q,m,\gamma}$. We now show that the distribution Ψ leads to a sometimes lossy pseudorandom distribution. By Theorem 3 it suffices to bound the maximal singular values of $\mathbf{B}_{R^d}^{-1} M_R(\mathbf{F}) \mathbf{B}_{R^d}$, and $\mathbf{B}_{R^m}^{-1} M_R(\mathbf{G}) M_R(\mathbf{F})^{-1} \mathbf{B}_{R^d}$, which is the object of the following lemma.

Lemma 8. *Let K be a number field of degree n, and R its ring of integers. Let d, q, m be positive integers, and $\gamma > 0$. Let $(M_R(\mathbf{F}), M_R(\mathbf{G})) \hookleftarrow \Psi$. Then, (1) $\left\|\mathbf{B}_{R^d}^{-1} M_R(\mathbf{F}) \mathbf{B}_{R^d}\right\|_2 \leq d\gamma \log_2 n$ and (2) $\left\|\mathbf{B}_{R^m}^{-1} M_R(\mathbf{G}) M_R(\mathbf{F})^{-1} \mathbf{B}_{R^d}\right\|_2 \leq nd\sqrt{nm} \log_2 n$ except with probability at most $O(n^{-1/2}) + 2d(d+m)ne^{-\pi \log_2^2 n}$ over the choice of $(M_R(\mathbf{F}), M_R(\mathbf{G}))$. When $d, m =$ poly(n), this probability is simply $O(n^{-1/2})$.*

Proof. **1.** It holds that $\mathbf{B}_{R^d}^{-1} M_R(\mathbf{F}) \mathbf{B}_{R^d} = M_{\sigma_H}(\mathbf{F})$. Let (i, j) be in $[d]^2$. As $\mathbf{U}_H$ is unitary, we have $\|M_{\sigma_H}(f_{ij})\|_2 = \|M_\sigma(f_{ij})\|_2 = \|\sigma(f_{ij})\|_\infty$. As $f_{ij} \hookleftarrow \mathcal{D}_{R,\gamma}$, Lemma 6 gives that $\|\sigma(f_{ij})\|_\infty \leq \gamma \log_2 n$ except with a probability of at most

$2ne^{-\pi \log_2^2 n}$. Lemma 1 yields $\left\|\mathbf{B}_{R^d}^{-1} M_R(\mathbf{F})\mathbf{B}_{R^d}\right\|_2 \leq d\gamma \log_2 n$ except with a probability of at most $d^2 \cdot 2ne^{-\pi \log_2^2 n}$ which is negligible for $d = \mathsf{poly}(n)$.

2. We now bound $\left\|\mathbf{B}_{R^m}^{-1} M_R(\mathbf{G}) M_R(\mathbf{F})^{-1}\mathbf{B}_{R^d}\right\|_2$ from above. Note that we have

$$\mathbf{B}_{R^m}^{-1} M_R(\mathbf{G}) M_R(\mathbf{F})^{-1}\mathbf{B}_{R^d} = M_{\sigma_H}(\mathbf{G}) M_{\sigma_H}(\mathbf{F})^{-1}.$$

As $\|.\|_2$ is sub-multiplicative, we simply bound $\|M_{\sigma_H}(\mathbf{G})\|_2$ and $\left\|M_{\sigma_H}(\mathbf{F})^{-1}\right\|_2$ from above separately. Lemma 6 and 1 again yield $\|M_{\sigma_H}(\mathbf{G})\|_2 \leq \sqrt{md}\gamma \log_2 n$, except with probability at most $md \cdot 2ne^{-\pi \log_2^2 n}$. Finally, by Corollary 2 we have that $\left\|M_{\sigma_H}(\mathbf{F})^{-1}\right\|_2 \leq n\sqrt{nd}/\gamma$ except with a probability of at most $O(n^{-1/2})$.

Hence $\left\|\mathbf{B}_{R^m}^{-1} M_R(\mathbf{G}) M_R(\mathbf{F})^{-1}\mathbf{B}_{R^d}\right\|_2 \leq nd\sqrt{nm} \log_2 n$,
except with probability at most $O(n^{-1/2}) + md \cdot 2ne^{-\pi \log_2^2 n}$.

□

We can now summarize the results of this section in our main theorem by combining Lemma 8 with Corollary 1. Using the equivalence of Lemma 3 between the module and structured formulations, we have the following.

Theorem 5. *Let K be a number field of degree n, and R its ring of integers. Let m, d be positive integers. Let q be a positive integer and $\gamma > 0$ be such that $\mathcal{D}_{GL_d(R,q),\gamma}$ is efficiently sampleable. Let $\mathcal{S}$ be a distribution on R_q^d and $\mathcal{S}' = \mathbf{B}_R \sigma_H(\mathcal{S})$. Assume the hardness of* M-NTRU$_{n,d,q,m,\mathcal{D}_{R,\gamma}}$, *and that $\nu_s(\mathcal{S}') \geq nd \log_2(d\gamma \log_2 n) + \omega(\log_2 \lambda)$ for some $s > \gamma d \log_2(n) \cdot \eta_\varepsilon(R)$. Then, for $s_0 \geq 2^{3/2} \cdot s \cdot n\sqrt{nmd} \log_2 n$, we have that* M-LWE$_{n,d,q,m,D_{s_0},\mathcal{S}}$ *is mildly hard.*

What characterizes the secret distribution is the noise lossiness condition of Theorem 5. By using Lemma 2, we can have concrete conditions on the entropy of the secret distribution.

Corollary 3. *Let K be the ν-th cyclotomic field of degree $n = \varphi(\nu)$, and R its ring of integers. Let m, d be positive integers. Let $q > 2n$ be a prime such that $q = 1 \bmod \nu$, and $\gamma > 2^{1/(2d-1)} nq^{(d-1)/(n(2d-1))}$. Let $\mathcal{S}$ be a distribution on R_q^d and $\mathcal{S}' = \tau(\mathcal{S})$ (see Sect. 3.1). Assume the hardness of* M-NTRU$_{n,d,q,m,\mathcal{D}_{R,\gamma}}$. *Also, assume that for some $\gamma d \log_2(n) \cdot \eta_\varepsilon(R) < s < q\sqrt{\pi \ln(4nd)}$ it holds that*

$$H_\infty(\mathcal{S}) \geq nd \log_2(d\gamma \log_2 n) + nd \log_2(q/s) + \omega(\log_2 \lambda).$$

Then, for $s_0 \geq 2^{3/2} \cdot s \cdot n\sqrt{nmd} \log_2 n$, we have that M-LWE$_{n,d,q,m,D_{s_0},\mathcal{S}}$ *is mildly hard.*

If $\mathcal{S}$ is supported on S_η^d for some positive integer $\eta \geq 2$, and for some $s > \gamma d \log_2(n) \cdot \eta_\varepsilon(R)$ it holds that

$$H_\infty(\mathcal{S}) \geq nd \log_2(d\gamma \log_2 n) + \sqrt{2\pi} \log_2 e \cdot nd \cdot \eta/s + \omega(\log_2 \lambda),$$

then the conclusion still holds.

Proof (of Corollary 3). Note that as τ is a bijection, $\mathcal{S}$ and $\mathcal{S}'$ have the same entropy. Hence, the first statement is obtained simply by combining Theorem 5 with Lemma 2. For the second statement, we take a bounded distribution. In practical uses of M-LWE, the bounds are considered on the coefficients of the polynomials, which is why we consider a bound on the infinity norm of the coefficient embedding of the secrets. To use Lemma 2, we simply have to translate this bound into a bound on the secrets from $\mathcal{S}'$ in the Euclidean norm. Since K is a cyclotomic field, we can choose $\mathbf{B}_R = (\mathbf{U}_H^\dagger \mathbf{V})^{-1}$ as discussed in Sect. 2.2, which yields $\mathbf{B}_R \sigma_H = \tau$. Hence, we indeed have $\mathcal{S}' = \mathbf{B}_R \sigma_H(\mathcal{S})$ as required by Theorem 5. Now, as $\tau(\mathcal{S}) = \mathcal{S}'$ is supported on $\{-\eta, \ldots, \eta\}^{nd}$, for $\mathbf{s} \hookleftarrow \mathcal{S}$ it holds that $\|\mathbf{s}'\|_2 = \|\mathbf{B}_R \sigma_H(\mathbf{s})\|_2 = \|\tau(\mathbf{s})\|_2 \leq \eta\sqrt{nd}$. Applying Lemma 2 thus yields the second statement.

□

5.3 On the Statistical Entropic Hardness of M-LWE

Chuengsatiansup et al. [12] show that if the Gaussian $\psi = \mathcal{D}_{R,\gamma}$ is sufficiently wide, and where q does not split too much, then the M-NTRU$_{n,d,q,m,\gamma}$ problem is statistically hard, as we restate below. Note that the original statement requires $d \geq m$, which is actually not needed in the proof. The final proof of Theorem 6 uses Theorem 4, and also the volume of R but in a ratio, which does not affect the result when changing the embedding.

Theorem 6 (**[12, Thm. A.1] adapted**)**.** *Let K be a number field of degree n and R its ring of integers. Let $m, d \geq 1$ and q be an unramified prime such that $\min_{\mathfrak{p}|qR, \mathfrak{p} \text{ prime}} N(\mathfrak{p}) = 2^{\Omega(n)}$. We define $N_{\max} = \max_{\mathfrak{p}|qR, \mathfrak{p} \text{ prime}} N(\mathfrak{p})$. Let $\gamma \geq \max\left(2nq^{m/(d+m)+2/(n(d+m))}, 2^{1/(2d-1)} |\Delta_K|^{1/n} N_{\max}^{(d-1)/(n(2d-1))}\right)$. Then, let $\mathcal{X}_\gamma$ be the distribution of $\mathbf{G}\mathbf{F}_q^{-1} \bmod qR$ where $(\mathbf{F}, \mathbf{G}) \hookleftarrow \mathcal{D}_{R,\gamma}^{d\times d} \times \mathcal{D}_{R,\gamma}^{m\times d}$ with $\mathbf{F} \in GL_d(R, q)$, $\mathbf{F}_q^{-1}$ the R_q-inverse of $\mathbf{F}$. Then, $\Delta(\mathcal{X}_\gamma, \mathcal{U}(R_q^{m\times d})) \leq 2^{-\Omega(n)}$.*

Our hardness assumption for the mild hardness of Entropic M-LWE is statistically thus proven by Theorem 6 for wide Gaussian distributions and a modulus q that does not split into too many factors. In this case, our result introduces non-trivial lower bounds on the entropy of the secret and the size of the noise such that the M-LWE becomes statistically hard. By Theorem 5, this provides the mild hardness of M-LWE$_{n,d,q,m,D_{s_0},\mathcal{S}}$ with no computational assumption whatsoever. Nevertheless, the parameter γ required by Theorem 6 is roughly $2n\sqrt{q}$ and hence makes the parameters of M-LWE not usable in practice. In particular, the entropy of the secret distribution must be very large. It then requires that both the size of the secret and the size of the masking noise s must be of the order of at least $\sqrt{q}$, making it hard to build usable cryptosystems purely based on it. As such, the statistical result should only be seen as an interesting byproduct of the theoretical proof, but not as a groundbreaking result for practical applications.

6 Related Work

In this section, we detail our main result of Theorem 5 and how it places with respect to existing hardness results for entropic M-LWE. To simplify the concrete comparison of parameters, we use the case of power-of-two cyclotomic fields and use the formulation of Corollary 3.

First, note that our reduction is rank-preserving in the sense that the module rank from our M-NTRU assumption equals the final module rank for entropic M-LWE. It can be advantageous for the concrete hardness analysis, but it also gives less room to tweak the parameters in order to achieve small secrets. In particular, it cannot achieve uniform binary secrets, a variant studied by Boudgoust et al. [4,5], as it does not carry enough entropy when the rank is preserved. Although the case of binary secret is not encompassed, it is still interesting to see what we can achieve in terms of minimal bound η on the secret coefficients (i.e., coefficients in $\{-\eta, \ldots, \eta\}$). When looking at bounded secret distributions we take the second statement of Corollary 3, and for a uniform distribution to have maximal entropy.

Our reduction provides theoretical insights and gives confidence in the fact that M-LWE-based cryptosystems are resilient even if the secret distribution present a certain amount of leakage. We insist on the fact that our proof does not encompass practical parameters, but can still be instantiated with concrete parameters that satisfy all the conditions. We give for example one such set of parameters in Table 1.

Table 1. Example parameter sets verifying the conditions of Corollary 3.

n	d	m	q	γ	ε	s	η	s_0
256	4	4	1105625551361	297	2^{-100}	11894537	4780	34450257913

The final noise s_0 can be improved by a factor of roughly n experimentally. Indeed, the bound on the smallest singular value of Lemma 7 can be applied for $\mathcal{I} = R$ directly which introduces the constant C_γ. Experimentally, it seems like $C_\gamma = O(1/\gamma^\delta)$ for $\delta \in (3/2, 2)$. Hence, for $\gamma = \Omega(n)$, it would yield $s_{\min}(M_\sigma(\mathbf{F})) \geq \gamma/\sqrt{nd}$ with non-negligible probability. This would thus save a factor of n in the expression of s_0. We have studied the lower bound with a heuristics for cyclotomic fields, and with γ under the condition of Theorem 4. It yields a lower bound of $\gamma/\sqrt{nd}$ on the smallest singular value with (experimental) probability at least $3/4$ (and going to 1 with polynomial speed as n grows). A bound of $\gamma/(10\sqrt{nd})$ is verified with (experimental) probability of at least $99/100$. The bound seems coherent with the extensive research around spectral estimations of random matrices. The smallest singular value of random matrices has been widely studied in order to prove the Von Neumann & Goldstine conjecture [30] that for a centered unit-variance random matrix $\mathbf{A}$ of size $N \times N$, $s_{\min}(\mathbf{A})$ is asymptotically equivalent to $1/\sqrt{N}$ with high probability.

This conjecture has been proven for specific distributions satisfying various conditions on the entries, which our matrix $M_{\sigma_H}(\mathbf{F})$ for $\mathbf{F} \hookleftarrow \mathcal{D}_{R,\gamma}^{d\times d}$ unfortunately does not verify. The bound seems however to hold heuristically.

As mentioned in the introduction, Lin et al. [19][3] adapted the lossy argument approach of [7] to the module setting. In order to compare with our hardness result, we need to make some modifications to their result to adapt it to the primal ring. We also need to adjust it to our definition of M-LWE which differs by a factor of q, i.e., considering $\mathbf{As} + \mathbf{e} \bmod qR$ instead of $q^{-1}\mathbf{As} + \mathbf{e}' \bmod R$ with $\mathbf{e}'$ having a Gaussian parameter in $(0, 1)$.

We only summarize the comparison in Table 2 due to lack of space.

A detailed description of this table is given in the full version [6, Sec. 6]. It essentially shows that we trade the underlying hardness assumption for differences in the parameters. In particular, the main difference comes from comparing $k \log_2 q$ with $d \log_2(nd \log_2 n)$, as γ can be as low as n. In some parameter regimes, our proof method thus leads to slightly improved parameters. For example, for $n = 256$, $q = n^3$, and k close to d (close to rank-preserving reduction), we achieve better parameters in terms of noise and secret size.

Table 2. Comparison of [19] in the primal ring and Corollary 3 for the hardness of M-LWE$_{n,d,q,m,D_{s_0},\mathcal{S}}$ over power-of-two cyclotomic fields of degree n, with module rank d and secret distribution $\mathcal{S}$. For clarity, we have $C' = \sqrt{2\pi}\log_2 e$. The $\eta_{\min}$ and s_0 are obtained by fixing a slack $\alpha = 1/n$, which adds mild conditions on s.
[a]: The result for arbitrary number fields is that of Theorem 5, but Corollary 3 is stated for cyclotomic fields.

	[19]	**Corollary** 3
Number fields	Arbitrary	Arbitrary[a]
Constant secrets	Yes	No
Rank-preserving	No	Yes
Hardness assumption	M-LWE$_{k,\mathcal{D}_{R,s_1}}$	M-NTRU$_{d,\gamma=\Omega(n)}$
General Distribution $\mathcal{S}$		
Minimal Entropy $H_\infty(\mathcal{S}) - \omega(\log_2 \lambda)$	$nk\log_2 q + nd\log_2(\frac{q\sqrt{n}}{s})$	$nd\log_2(d\gamma\log_2 n) + nd\log_2(\frac{q}{s})$
Maximal Masking Noise s	$s < q\sqrt{n}\sqrt{\frac{\pi}{\ln(4nd)}}$	$d\gamma\log_2(n)\eta_\varepsilon(R) < s < q\sqrt{\frac{\pi}{\ln(4nd)}}$
Bounded Distribution $\mathcal{S}$ (over S_η^d)		
Minimal Entropy $H_\infty(\mathcal{S}) - \omega(\log_2 \lambda)$	$nk\log_2 q + \frac{C' nd \cdot n\eta}{s}$	$nd\log_2(d\gamma\log_2 n) + \frac{C' nd\eta}{s}$
Minimal secret bound $\eta_{\min}$	$\frac{1}{2}q^{\frac{k+1/n}{d}}$	$\frac{1}{2}(nd\log_2 n)^{1+\frac{1}{nd}}$
Minimal noise s_0	$s_1 \cdot O\left(\frac{\sqrt{m}n^2 d}{\log_2 q} q^{\frac{k+1/n}{d}}\right)$	$\sqrt{8}(nd\log_2 n)^2\sqrt{nm}\cdot\eta_\varepsilon(R)$

[3] Note that at the time of writing, the paper by Lin et al. is only accessible on ePrint and has not yet been peer-reviewed.

Acknowledgments. This work was supported by the European Union PROMETHEUS project (Horizon 2020 Research and Innovation Program, grant 780701), by the PEPR quantique France 2030 programme (ANR-22-PETQ-0008), and further supported by the Danish Independent Research Council under project number 0165-00107B (C3PO). We thank Alexandre Wallet and Damien Stehlé for helpful discussions. We also thank our anonymous referees of Eurocrypt 2022 and Indocrypt 2022 for their thorough proof reading and constructive feedback.

References

1. Albrecht, M.R., Deo, A., Paterson, K.G.: Cold boot attacks on ring and module LWE keys under the NTT. IACR Trans. Cryptogr. Hardw. Embed. Syst. **2018**(3), 173–213 (2018)
2. Bernstein, D. J., et al.: NTRU prime round-3 candidate to the NIST post-quantum cryptography standardisation project (2020)
3. Bos, J. W., et al.: CRYSTALS - kyber: a CCA-secure module-lattice-based KEM. In: Euro S and P, pp. 353–367. IEEE (2018)
4. Boudgoust, K., Jeudy, C., Roux-Langlois, A., Wen, W.: Towards classical hardness of module-LWE: the linear rank case. In: Moriai, S., Wang, H. (eds.) ASIACRYPT 2020. LNCS, vol. 12492, pp. 289–317. Springer, Cham (2020). https://doi.org/10.1007/978-3-030-64834-3_10
5. Boudgoust, K., Jeudy, C., Roux-Langlois, A., Wen, W.: On the hardness of module-LWE with binary secret. In: Paterson, K.G. (ed.) CT-RSA 2021. LNCS, vol. 12704, pp. 503–526. Springer, Cham (2021). https://doi.org/10.1007/978-3-030-75539-3_21
6. Boudgoust, K., Jeudy, C., Roux-Langlois, A., Wen, W.: Entropic hardness of module-LWE from module-NTRU. IACR Cryptol. ePrint Arch, p. 245 (2022)
7. Brakerski, Z., Döttling, N.: Hardness of LWE on general entropic distributions. In: Canteaut, A., Ishai, Y. (eds.) EUROCRYPT 2020. LNCS, vol. 12106, pp. 551–575. Springer, Cham (2020). https://doi.org/10.1007/978-3-030-45724-2_19
8. Brakerski, Z., Döttling, N.: Lossiness and entropic hardness for ring-LWE. In: Pass, R., Pietrzak, K. (eds.) TCC 2020. LNCS, vol. 12550, pp. 1–27. Springer, Cham (2020). https://doi.org/10.1007/978-3-030-64375-1_1
9. Brakerski, Z., Gentry, C., Vaikuntanathan, V.: (leveled) fully homomorphic encryption without bootstrapping. In: ITCS, pp. 309–325. ACM (2012)
10. Brakerski, Z., Langlois, A., Peikert, C., Regev, O., Stehlé, D.: Classical hardness of learning with errors. In: STOC, pp. 575–584. ACM (2013)
11. Chen, C., et al.: NTRU round-3 candidate to the NIST post-quantum cryptography standardisation project (2020)
12. Chuengsatiansup, C., Prest, T., Stehlé, D., Wallet, A., Xagawa, K.: Modfalcon: compact signatures based on module-NTRU lattices. In: AsiaCCS, pp. 853–866. ACM (2020)
13. Ducas, L., et al.: Crystals-dilithium: a lattice-based digital signature scheme. IACR Trans. Cryptogr. Hardw. Embed. Syst. **2018**(1), 238–268 (2018)
14. Ducas, L., Micciancio, D.: Improved short lattice signatures in the standard model. In: Garay, J.A., Gennaro, R. (eds.) CRYPTO 2014. LNCS, vol. 8616, pp. 335–352. Springer, Heidelberg (2014). https://doi.org/10.1007/978-3-662-44371-2_19
15. Ducas, L., van Woerden, W.: NTRU fatigue: how stretched is overstretched? In: Tibouchi, M., Wang, H. (eds.) ASIACRYPT 2021. LNCS, vol. 13093, pp. 3–32. Springer, Cham (2021). https://doi.org/10.1007/978-3-030-92068-5_1

16. Goldwasser, S., Tauman Kalai, Y., Peikert, C., Vaikuntanathan, V.: Robustness of the learning with errors assumption. In: ICS, pp. 230–240. Tsinghua University Press (2010)
17. Hoffstein, J., Pipher, J., Silverman, J.H.: NTRU: a ring-based public key cryptosystem. In: Buhler, J.P. (ed.) ANTS 1998. LNCS, vol. 1423, pp. 267–288. Springer, Heidelberg (1998). https://doi.org/10.1007/BFb0054868
18. Langlois, A., Stehlé, D.: Worst-case to average-case reductions for module lattices. Des. Codes Cryptogr. **75**(3), 565–599 (2015)
19. Lin, H., Wang, Y., Wang, M.: Hardness of module-LWE and ring-LWE on general entropic distributions. IACR Cryptol. ePrint Arch, p. 1238 (2020)
20. Lyubashevsky, V., Peikert, C., Regev, O.: On ideal lattices and learning with errors over rings. In: Gilbert, H. (ed.) EUROCRYPT 2010. LNCS, vol. 6110, pp. 1–23. Springer, Heidelberg (2010). https://doi.org/10.1007/978-3-642-13190-5_1
21. Micciancio, D.: On the hardness of learning with errors with binary secrets. Theory Comput. **14**(1), 1–17 (2018)
22. Micciancio, D., Peikert, C.: Trapdoors for lattices: simpler, tighter, faster, smaller. In: Pointcheval, D., Johansson, T. (eds.) EUROCRYPT 2012. LNCS, vol. 7237, pp. 700–718. Springer, Heidelberg (2012). https://doi.org/10.1007/978-3-642-29011-4_41
23. Micciancio, D., Regev, O.: Worst-case to average-case reductions based on gaussian measures. SIAM J. Comput. **37**(1), 267–302 (2007)
24. Peikert, C.: Limits on the hardness of lattice problems in phl_{php} norms. Comput. Complex. **17**(2), 300–351 (2008)
25. Peikert, C.: A decade of lattice cryptography. Found. Trends Theor. Comput. Sci. **10**(4), 283–424 (2016)
26. Pellet-Mary, A., Stehlé, D.: On the hardness of the NTRU problem. In: Tibouchi, M., Wang, H. (eds.) ASIACRYPT 2021. LNCS, vol. 13090, pp. 3–35. Springer, Cham (2021). https://doi.org/10.1007/978-3-030-92062-3_1
27. Regev, O.: On lattices, learning with errors, random linear codes, and cryptography. In: STOC, pp. 84–93. ACM (2005)
28. Rudelson, M., Vershynin, R.: The littlewood-offord problem and invertibility of random matrices. Adv. Math. **218**, 600–633 (2008)
29. Stehlé, D., Steinfeld, R., Tanaka, K., Xagawa, K.: Efficient public key encryption based on ideal lattices. In: Matsui, M. (ed.) ASIACRYPT 2009. LNCS, vol. 5912, pp. 617–635. Springer, Heidelberg (2009). https://doi.org/10.1007/978-3-642-10366-7_36
30. von Neumann, J., Goldstine, H.H.: Numerical inverting of matrices of high order. Bull. Amer. Math. Soc. **53**, 1021–1099 (1947)

18. [illegible] (2016)
19. [illegible] https://doi.org/10.1007/[illegible]
[illegible] (2011)
20. [illegible] (2020)
21. [illegible] https://doi.org/10.1007/978-3-[illegible]
22. [illegible]
23. [illegible]
24. [illegible]

1. [illegible]
2. [illegible] https://doi.org/10.1007/978-3-030-[illegible]
3. [illegible]
4. [illegible]
5. [illegible]
6. [illegible]

Symmetric Key Cryptology

New Algorithm for Exhausting Optimal Permutations for Generalized Feistel Networks

Stéphanie Delaune[1], Patrick Derbez[1], Arthur Gontier[1](✉), and Charles Prud'homme[2]

[1] Univ Rennes, CNRS, IRISA, Rennes, France
{stephanie.delaune,patrick.derbez,arthur.gontier}@irisa.fr
[2] TASC, IMT-Atlantique, LS2N-CNRS, 44307 Nantes, France
charles.prudhomme@imt-atlantique.fr

Abstract. The Feistel construction is one of the most studied ways of building block ciphers. Several generalizations were proposed in the literature, leading to the Generalized Feistel Network (GFN) construction, in which the round function operates on each pair of blocks in parallel until all branches are permuted. At FSE'10, Suzaki and Minematsu studied the diffusion of such construction, raising the question of how many rounds are required so that each block of the ciphertext depends on all blocks of the plaintext. Exhausting all possible permutations up to 16 blocks, they observed that there were always optimal permutations mapping even-number input blocks to odd-number output blocks and vice versa. Recently, both Cauchois *et al.* and Derbez *et al.* proposed new algorithms to build optimal even-odd permutations for up to 36 blocks. In this paper, we present a new algorithm based on iterative path building to search for optimal Feistel permutation. This algorithm is much faster in exhausting optimal non-even-odd permutations than all the previous approaches. Our first result is a computational proof that no non-even-odd permutation reaches a better diffusion round than optimal even-odd permutations up to 32 blocks. Furthermore, it is well known that permutations with an optimal diffusion round do not always lead to optimal permutations against differential cryptanalysis. We investigate several new criteria to build permutations leading to more secure GFN.

Keywords: Block cipher · Feistel network · Differential analysis

1 Introduction

The Feistel Network is a classical design of modern block ciphers, used for many primitives as DES [6], TWINE [11] and SIMON [2]. The core idea of such a construction is to split the plaintext into two halves of equal length called blocks. At each round, the second block is duplicated and one side goes through a function F and is then xored to the first block. The two resulting blocks are then inverted.

The work presented in this article was funded by the French National Research Agency as part of the DeCrypt project (ANR- 18-CE39-0007).

T. Isobe and S. Sarkar (Eds.): INDOCRYPT 2022, LNCS 13774, pp. 103–124, 2022.
https://doi.org/10.1007/978-3-031-22912-1_5

One big advantage of this scheme is that the function F has not to be invertible since the decryption function is the same as the encryption one in reverse order. Since its introduction, several improvements have been proposed to the original design. In particular, at ASIACRYPT'96, Nyberg defined the Generalized Feistel Network (GFN) which splits the message into $2k$ blocks and uses a round function of the form:

$$(x_0, x_1, \ldots, x_{2k-1}) \to \pi(x_0 \oplus F_{i,0}, x_1, \ldots, x_{2k-2} \oplus F_{i,k-1}, x_{2k-1})$$

where each $F_{i,j}$ is a pseudorandom function, and π is a permutation of the blocks [7]. This design was for instance used in both the block ciphers `TWINE` [11] and `PICCOLO` [9]. It is a generalization of the more classical *Type-2 Feistel* construction proposed by Zheng *et al.* at CRYPTO'89 [12], in which the permutation π is always the cyclic shift.

Cryptographic properties of GFN highly depend on the permutation used for blocks. For instance, if the identity function was chosen as the permutation, the resulting block cipher would be very weak as the parallel application of weak ciphers. Thus, selecting the *optimal* permutation is an interesting task for designers. At FSE'10, Suzaki and Minematsu focused on finding the permutations reaching the *lowest diffusion rounds* [10]. More precisely, they searched for the permutations minimizing the number of rounds required to achieve full block diffusion: each block of the ciphertext depends on all blocks of the plaintext and vice-versa. This criterion is tied to the resistance of the resulting cipher against impossible differential attacks, a powerful cryptanalysis technique. Along with a lower bound on the diffusion round of a GFN of $2k$ blocks, Suzaki and Minematsu gave optimal permutations (w.r.t. the diffusion round) for $2 \leq 2k \leq 16$. It is worthy to note that such an optimal permutation was then used to design block ciphers such as `TWINE` [11]. At FSE'19 Cauchois *et al.* identified new equivalence classes regarding the diffusion rounds and, together with new algorithms, were able to give optimal permutations up to $2k = 20$ [3]. Furthermore, restricting the search to even-odd permutations (i.e. permutations sending blocks of even index to blocks of odd ones and vice-versa), they were able to find the best even-odd permutations up to $2k = 24$. Finally, few months later, Derbez *et al.* proposed a new characterization of the problem restricted to even-odd permutations as well as a clever algorithm to exhaust the search space. As a result they found the best even-odd permutations up to $2k = 36$ [5]. In particular, they solved the problem opened by Suzaki and Minematsu regarding the case $2k = 32$.

It is also possible to optimize GFN for other criteria than the diffusion round. For instance in [8], Shi *et al.* searched for the permutations offering the best resistance against Demirci-Selçuk meet-in-the-middle attacks [4].

Our contribution. Since the original work of Suzaki and Minematsu [10], most of the new algorithms to find the permutations lowering the diffusion round were dedicated to the even-odd case. There are two main reasons for that. First, considering even-odd permutations only does highly reduce the search space, making it possible to exhaust it. Second, it was shown that up to $2k = 20$ at least one of the optimal permutations is an even-odd permutation.

In this paper, we focus on non-even-odd permutations and propose a new algorithm to solve the general case. In previous approaches, the core part of algorithms was somehow dedicated to answering the question: does block i diffuse into all blocks after R rounds? In our new algorithm, we answer the question: does block i diffuse to block j after R rounds? This more precise question allows us to cut the search earlier than previous algorithms while exhausting the permutations. Thus, our first result is a computational proof that, up to $2k = 32$, there is always at least one even-odd permutation which is optimal regarding the diffusion round. The best known diffusion rounds for even-odd and non-even-odd permutations are given in Table 1.

In the second part of the paper, we investigate more sophisticated criteria than the diffusion round and study whether the optimal permutations lead to optimal GFN regarding differential cryptanalysis.

Table 1. State of the art regarding optimal Diffusion Round. k is the number of Feistel pairs and the references are : Suzaki et al. [10], Cauchois et al. [3], Derbez et al. [5]

$2k$	Fibonacci bound	Even-odd		Non-even-odd	
		DR	Ref	DR	Ref
6	5	5	[10]	6	[3,10]
8	6	6		6	
10	6	7		7	
12	7	8		8	
14	7	8		8	
16	7	8		8	
18	8	8	[3]	9	[3]
20	8	9		9	
22	8	8		9	new
24	8	9		≥ 9	
26	8	9	[5]	≥ 9	
28	9	9		≥ 9	
30	9	9		≥ 9	
32	9	9		≥ 9	

2 Preliminaries

We recall in this section some notions and useful results that will be used throughout this paper.

2.1 Generalized Feistel Networks

Generalized Feistel Networks have been introduced by [7] as a generalization of Type-2 Feistel construction [12]. Roughly, the cycle shift performed at each round in [12] is replaced by an arbitrary permutation leading to stronger schemes with better diffusion if the permutation is chosen wisely.

Definition 1. *A* Generalised Feistel Network *(GFN) is defined by a number k of Feistel pairs, a word size n, a number of rounds r, a permutation π over $2k$ elements (named* blocks*), and $r \cdot k$ cryptographic keyed functions F_j^i from $\mathbb{F}_2^n$ to $\mathbb{F}_2^n$ (with $1 \leq i \leq r$, and $1 \leq j \leq k$). The ciphertext of a message of size $2k \cdot n$ is given by $\mathcal{R}_r \circ \ldots \circ \mathcal{R}_1$, where $\mathcal{R}_i$ is the round function:*

$$\mathcal{R}_i : (X_0, \ldots, X_{2k-1}) \rightarrow \pi(X_0 \oplus F_1^i(X_1), X_1, \ldots, X_{2k-2} \oplus F_k^i(x_{2k-1}), X_{2k-1})$$

In this paper, neither the word size n, nor the exact definition of the keyed functions F_i^j are relevant. Hence, we simply use F hereafter, and we denote GFN_π^k a GFN with k Feistel pairs using permutation π.

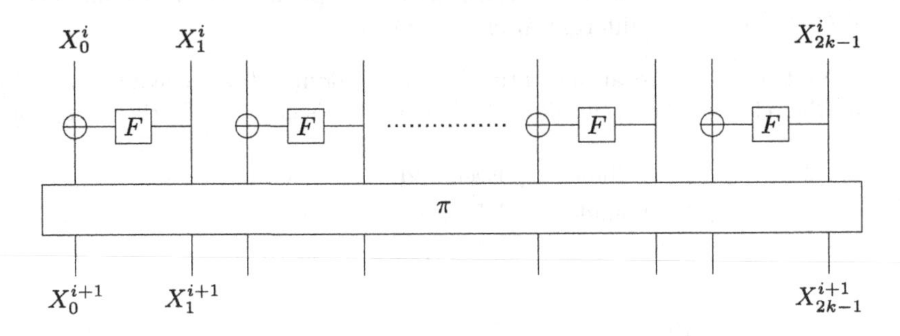

Fig. 1. Round function $\mathcal{R}_i$ of a GFN with k Feistel pairs

In the following, we denote by $X^i = (X_0^i, X_1^i, \ldots, X_{2k-1}^i)$ the input data of the $i+1^{\text{th}}$ round for $i \geq 0$. We say that X_j^i is an even block when j is even, and an odd one otherwise. An illustration for round $\mathcal{R}_i$ is given in Fig. 1.

2.2 Diffusion Round

In [10], it has been observed that the *diffusion round* of a permutation π (denoted $DR(\pi)$) is closely related to the security of the corresponding GFN against some of the attacks mentioned above. Intuitively, the diffusion round is the round at which full diffusion is achieved. In other words, assuming good enough functions $F_{i,j}$, the diffusion round is the round from which every bit of the ciphertext depends on every bit of the plaintext. We now formally recall the definition of this notion.

Given $r > 0$ and $i, j \in \{0, \ldots, 2k-1\}$, if X_i^r is expressed by a formal expression containing a non-zero term in X_j^0, we say that X_j^0 *diffuses* to X_i^r, and we say that X_j^0 fully diffuses after r rounds when X_j^0 diffuses to X_i^r for all $i \in \{0, \ldots, 2k-1\}$. For instance, we have that X_0^0 diffuses to $X_{\pi(0)}^1$ whereas X_1^0 diffuses to both $X_{\pi(0)}^1$ and $X_{\pi(1)}^1$. In general, an even block X_i^r will only diffuse to its successor $X_{\pi(i)}^{r+1}$, whereas an odd block X_i^r will diffuse to its successor $X_{\pi(i)}^{r+1}$ and the successor of its even neighbour $X_{\pi(i-1)}^{r+1}$.

Definition 2. *Given a permutation π over $2k$ elements, we denote $DR_i(\pi)$ as the minimum number of rounds r such that X_i^0* fully diffuses *after r rounds. Then, the* diffusion round *of a permutation π is given by*
$DR(\pi) = \max_{0 \le i < 2k}\{DR_i(\pi)\}$.

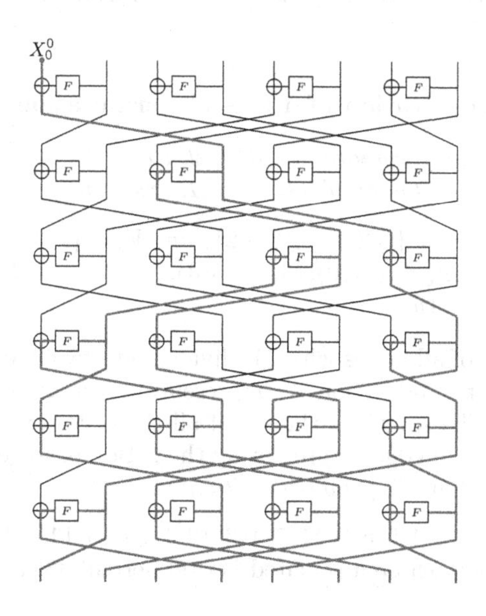

Fig. 2. Diffusion of X_0^0 after $r = 6$ successive rounds

Example 1. Let π = (3,0,5,6,1,2,7,4). This is an even-odd permutation. Figure 2 illustrates the diffusion of X_0^0 after successive rounds. For instance, we have that X_0^0 diffuses to X_5^2 and X_6^2, and full diffusion regarding X_0^0 is reached after 6 rounds, thus $DR_0(\pi) = 6$.

In GFN, decryption is made using π^{-1} and thus we want full diffusion to be effective for π and π^{-1}. We denote $DR^*(\pi) = \max(DR(\pi), DR(\pi^{-1}))$.

As recalled in introduction, finding permutations minimizing the diffusion round has deserved a lot of attention during the past few years. To ease the problem of finding optimal permutations, the focus has been made on even-odd permutations as they seem to achieve better diffusion [3,5]. The belief that even-odd permutations are better has only been formally established by exhausting all the optimal permutations up to $2k = 20$ [3]. In this paper, relying on a novel algorithm based on iterative path building, we will show that this is true up to $2k = 32$.

3 Path Building Algorithm

In this section, we first explain how to represent a permutation π over $2k$ elements as a graph before describing our algorithm. This representation fits well the understanding of our algorithm since its core idea is to build paths. This graph will also be of great help to propose a new characterization of the notion of diffusion round.

3.1 Graph Representation of a Feistel Permutation

Definition 3. *Given a permutation π over $2k$ elements, the* Feistel permutation graph *associated to π is the graph $G_\pi = (V, E)$ where:*

- $V = V_\mathsf{e} \cup V_\mathsf{o}$ *with* $V_\mathsf{e} = \{0, 2, \ldots, 2k-2\}$, *and* $V_\mathsf{o} = \{1, 3, \ldots, 2k-1\}$;
- $E = E_\epsilon \cup E_\pi$ *with* $E_\epsilon = \{(1,0), (3,2), (5,4), \ldots, (2k-1, 2k-2)\}$, *and* $E_\pi = \{(u, v) \mid u, v \in V \wedge \pi(u) = v\}$.

The set V is the set of all nodes which is divided into two halves, the set of even nodes V_e and the set of odd nodes V_o representing respectively the even blocks and the odd ones. The set E_π is the set of all the edges of the permutation π, whereas E_ϵ is the set of edges representing the S-Box passages from the odd to the even nodes (also called *epsilon-transitions*).

Example 2. Let $\pi = (2, 4, 5, 6, 9, 11, 7, 1, 3, 12, 15, 0, 13, 14, 8, 10)$. This is a non-even-odd permutation whose associated Feistel permutation graph is as follows:

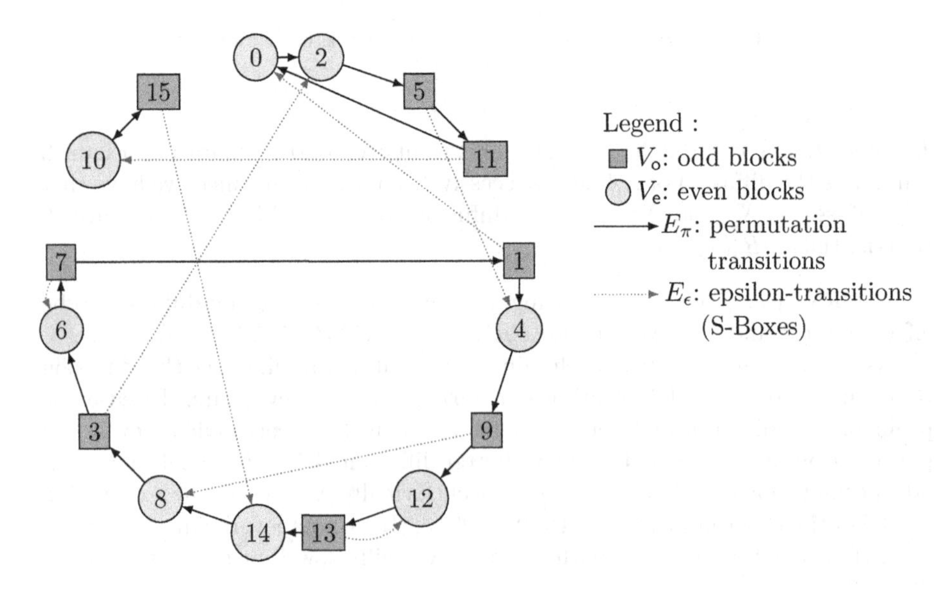

In the following, we will often refer to the Feistel permutation graph G_π of a permutation π. The sets $V_\mathsf{e}, V_\mathsf{o}, E_\pi, E_\epsilon$ will be used to represent the even blocks, the odd blocks, the permutation transitions and the ϵ-transitions.

Definition 4. *A* path $p = (e_1, \ldots, e_n)$ *is a finite sequence of edges from E which joins two nodes from V. Moreover, when* $e_n \in E_\pi$*, such a path is called a* diffusable path *(or* d-path *for short).*

We say that a path p is of *length* ℓ if there are exactly ℓ edges from E_π in p. Note that there can be multiple occurrences of the same edge in a path. We sometimes need to consider d-paths since a Feistel round is composed of one edge in E_ϵ followed by one edge in E_π. Based on this graph representation, we propose a new characterization of $DR(\pi)$.

Corollary 1. *$DR(\pi)$ is the smallest integer R such that:*
$\forall\ u, v \in V$, there exists a d-path of length R from u to v in G_π.

In order to compute the diffusion round of a permutation π, we can consider the d-paths of a certain length between all pairs of nodes in the graph G_π. As already noticed in [5], in the specific setting of even-odd permutations, it is actually sufficient to consider some specific sets of nodes, and only paths of length $R-1$ to establish that the diffusion round is equal to R. In the following, we formally define these specific paths for the general case (Proposition 1) and the even-odd case (Proposition 2).

Proposition 1. *Let π be a permutation, $DR(\pi)$ is the smallest integer R such that: $\forall a \in V_\mathsf{e}$, $\forall b \in V_\mathsf{o}$, there exists a path of length $R-1$ from a to b in G_π.*

Proof. Let $a \in V_\mathsf{e}$ and $b \in V_\mathsf{o}$ we have that $(a+1, a), (b, b-1) \in E_\epsilon$ with $a+1 \in V_\mathsf{o}$ and $b-1 \in V_\mathsf{e}$. Furthermore, we have $g, h \in V$ such that $(b, g), (b-1, h) \in E_\pi$ (see the graph below with $i = a+1$ and $j = b-1$).

1) From Corollary 1, we know that there is a d-path of length R from a to g, thus there is a path of length $R-1$ from a to b.

2) Now, suppose that there is a $R' < DR(\pi)$ such that $\forall\ a \in V_\mathsf{e},\ b \in V_\mathsf{o}$ there is a path of length $R'-1$ from a to b. We then have a d-path of length R' from i to g, from i to h and from a to h. Since we have these d-paths for each pair $a \in V_\mathsf{e}, b \in V_\mathsf{o}$, we have full diffusion with R' leading to a contradiction. □

For any permutation π, the Proposition 1 reduces the number of paths we have to consider when studying diffusion. In the case of an even-odd permutation, the length of these paths can be further reduced.

Proposition 2. *Let π be an even-odd permutation, $DR(\pi)$ is the smallest integer R such that: $\forall c \in V_\mathsf{o}$, $\forall d \in V_\mathsf{e}$, there exists a path of length $R-3$ from c to d in G_π.*

Proof. Let $b, c \in V_o$ and $a, d \in V_e$ with $(a, c), (d, b) \in E_\pi$. We have that $(a + 1, a), (b, b - 1) \in E_\epsilon$ with $a + 1 \in V_o$ and $b - 1 \in V_e$. Furthermore, we have $g, h \in V$ such that $(b, g), (b - 1, h) \in E_\pi$ (see the graph below with $i = a + 1$ and $j = b - 1$).

1) From Proposition 1, we know that there is a path of length $R - 1$ from a to b, thus there is a path of length $R - 3$ from c to d.

2) Now suppose that there is $R' < DR(\pi)$ such that $\forall\, c \in V_o,\ d \in V_e$ there is a path of length $R' - 3$ from c to d. We then have a d-path of length R' from i to g, from i to h and from a to h. Since we have these d-paths for all pairs $a \in V_e, b \in V_o$ then we have full diffusion with R' leading to a contradiction. □

3.2 The MakePath Algorithm

We present a new algorithm to search for permutations with optimal diffusion round. Our algorithm is based on path building to efficiently enumerate permutations with full diffusion or any other path-based property. Thanks to Propositions 1 and 2, we will only consider paths of length $R - 3$ from odd to even nodes in the even-odd case and paths of length $R - 1$ from even to odd nodes in the general case. To obtain effective procedures, we enumerate the paths while building a Feistel permutation graph. With this method, the more paths we add to the graph, the fewer possibilities remain for the following ones. Thanks to this, we can also define a strategy to cut the search as soon as possible by trying the paths with the least possibilities first. Our algorithm is composed of the three following functions:

- MakePath builds all the possible paths from a node a to a node b and is described in Algorithm 1. Starting from node a, the function calls itself on each possible next node for the path until all paths reach b with the length R. More precisely, on a node x, there is only three possibilities. If x is odd, there is one call to the even node $x - 1$. In this call, the length l does not decrease because ϵ-transitions are not counted in the path length (line 2–3). If $\pi[x]$ has already been fixed, we have no choice, and thus we follow it (line 4–5). If $\pi[x]$ is free, we have to try all the remaining free nodes (line 7–9). On each valid path, the function calls NextPath that will choose the next path to build (line 13).
- HasProperty checks whether the property of interest is satisfied between two nodes. For example, when considering the full diffusion property, we have to check whether a path of length R exists between 2 nodes (more details in Sect. 4).

- NEXTPATH chooses two nodes a and b that does not have the property described in HASPROPERTY. If such a pair of nodes exists, it calls MAKEPATH on it to link them with the next path. It is described in Algorithm 2. For the choice of a and b, the strategy consists of starting by the paths with the least possibilities. To do so, we can either count the remaining possible paths during the search, or we can set a static path priority (more details in Sect. 4.1).

Algorithm 1: MAKEPATH(x, π, b, l)

```
   Data: x: current node, π: partial permutation, b: target node, l: remaining
         length to reach R
 1 if l > 0 then
 2 |   if x is odd then
 3 |   |   MAKEPATH(x − 1, π, b, l);
 4 |   if π[x] is fixed then
 5 |   |   MAKEPATH(π[x], π, b, l − 1);
 6 |   else
 7 |   |   for all y not used in π do
 8 |   |   |   π[x] ← y;
 9 |   |   |   MAKEPATH(y, π, b, l − 1);
10 |   |   end
11 |   |   free π[x];
12 |   end
13 else if x = b then NEXTPATH(π) ;
```

Algorithm 2: NEXTPATH(π)

```
   Data: π: partial permutation
 1 for all (a, b) given by STRATEGY() do
 2 |   if ¬HASPROPERTY(a, π, b, R) then
 3 |   |   MAKEPATH(a, π, b, R);
 4 |   |   return;
 5 end
 6 Add π to solution pool
```

Our algorithm starts by a call to NEXTPATH with an undefined permutation and a given global parameter R. It stops when one of the following conditions holds:

1. There is no possible path from a to b, and thus there is no solution.
2. The permutation is complete, i.e. fully defined: it is a solution if HASPROPERTY is true for each pair of nodes.
3. The algorithm ends without fixing the whole permutation. In this case, any completion of the permutation lead to a valid solution.

Once all the recursive branches of our algorithm have been explored, all the paths of length R have been exhausted. Thus, at the end of the algorithm, we find all the permutations achieving full diffusion at round R if any. The algorithm can build these permutations from scratch, but it will find a lot of similar solutions. Indeed, starting by a graph like the one given in Example 2, a similar graph can be obtained by simply relabelling the Feistel pairs. To avoid these redundancies, we need to break some symmetries before running the search. To do so, we will rely on the notion of *skeleton* defined in the following section.

3.3 Skeletons

As explained in [3], in the even-odd case, the permutation can be split in two parts, the odd to even edges and the even to odd edges. This makes the search easier $(k!)^2$. Moreover, half of the permutation can be further reduced to all its possible cycle decompositions to break some symmetries. This reduces the search to $\mathcal{N}_k k!$ where $\mathcal{N}_k$ is the number of partitions of k. In the following, we propose a generalization of the cycle decompositions to consider non-even-odd permutations as well, and we rely for that on our graph representation.

Definition 5 (ϵ-cycle). *An ϵ-cycle is a path $c = (e_1, \ldots, e_{2l})$ in which the first and last nodes are equal and edges alternate between E_π and E_ϵ one by one.*

We note a l-ϵ-cycle an ϵ-cycle of size l i.e. with l ϵ-transitions. Moreover, we will only use one representative of $c = (e_1, \ldots, e_{2l})$ and we will not consider all the equivalent ϵ-cycles like $(e_{2l}, e_1, \ldots, e_{2l-1})$ or $(e_1, \ldots, e_{2l}, e_1, \ldots, e_{2l})$. Some examples are given in Fig. 3.

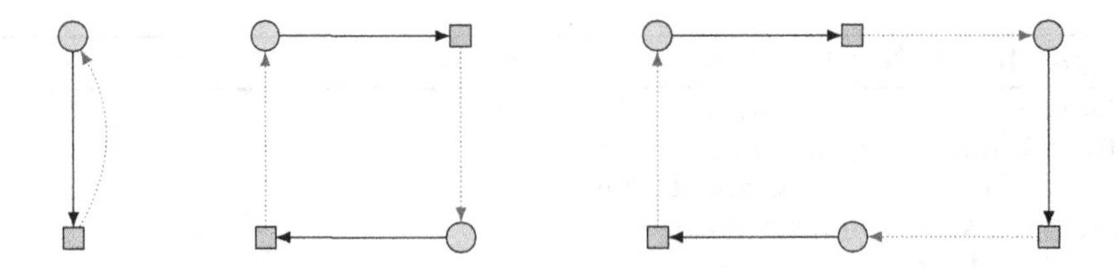

Fig. 3. 1-ϵ-cycle, 2-ϵ-cycle, and 3-ϵ-cycle

Let P be a partition of the integer k. For each $i \in P$, we fix one representative ϵ-cycle of the corresponding size. For example, there are three possible decompositions in ϵ-cycle for $k = 3$, i.e. $\{3\}$, $\{2, 1\}$, and $\{1, 1, 1\}$. This corresponds to one 3-ϵ-cycle, or one 2-ϵ-cycle with one 1-ϵ-cycle, or three 1-ϵ-cycles. This holds only for the even-odd case. To have a similar method in the general case, we rely on ϵ-chains to handle the non-even-odd parts of the permutation.

Definition 6 (ϵ-chain). *An ϵ-chain is a path $ch = (e_1, \ldots, e_{2l+1})$ in which the two first nodes are in V_{o} and the two last nodes are in V_{e}. The edges alternate between E_π and E_ϵ one by one.*

We note an l-ϵ-chain an ϵ-chain of size l, i.e. with l ϵ-transitions. Except for the first and the last node, all the nodes in an ϵ-chain are pairwise distinct. Indeed, if a node appears two times in an ϵ-chain, then it is not an ϵ-chain but an ϵ-cycle. However, the first and last node can be in an other structure, like an other ϵ-cycle or an other ϵ-chain, or in the same ϵ-chain, making the ϵ-chain loops on itself. This loop may occur at the beginning of the ϵ-chain, at its end, or on both sides. Some examples of free and looping chains are given in respectively Figs. 4 and 5.

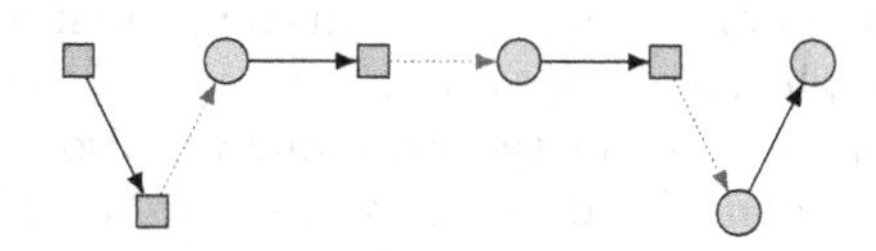

Fig. 4. A 3-ϵ-chain.

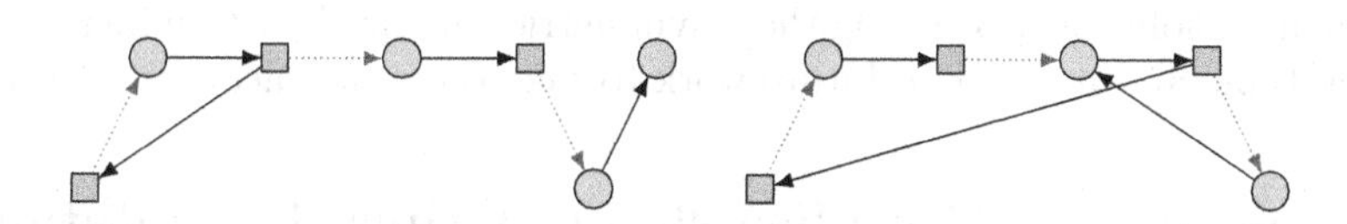

Fig. 5. Two 3-ϵ-chains looping on themselves.

Definition 7. *A* skeleton *of size k is a set of ϵ-cycles and ϵ-chains whose sum of sizes is k.*

Example 3. The skeleton of the graph given in Example 2 is depicted below (see Fig. 6). It is composed of three ϵ-cycles of size 3, 1, and 1, as well as two ϵ-chains of size 2 and 1.

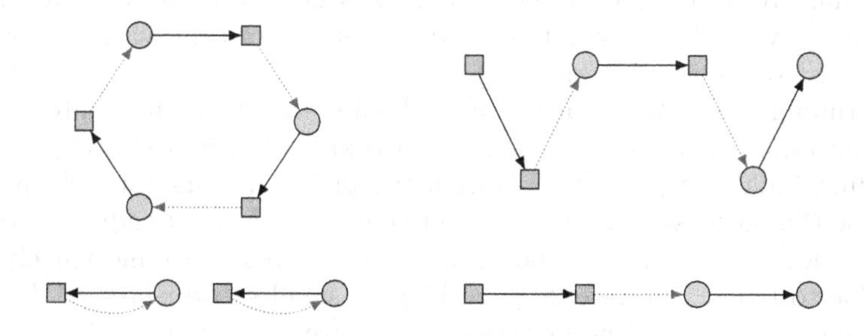

Fig. 6. Skeleton of Example 2

The skeleton of Fig. 6 is also valid for graphs similar to Example 2 but with different node numbers. In fact we can permute two pairs of nodes to find a different permutation having the same skeleton. This is why we will only use one representative of each skeleton. The number of skeletons is given by the formula $\sum_{i=0}^{k} \mathcal{N}_i \times \mathcal{N}_{k-i}$ with $\mathcal{N}_i$ the number of partitions of the integer i. The formula has two parts, one for the ϵ-cycles with $\mathcal{N}_i$ and one for the ϵ-chains with $\mathcal{N}_{k-i}$. The formula then sums the skeletons with each possible division into ϵ-cycles and ϵ-chains. For $2k = 16$ there are 22 even-odd skeletons and 163 skeletons with at least one ϵ-chain. For $2k = 32$ there are 231 even-odd skeletons and 5591 non-even-odd ones. Starting from a skeleton, we can complete it with edges to make a Feistel permutation graph. These edges are $\{(a, b) \mid a \in V_o, b \in V_e\}$. Furthermore, if there is one or more ϵ-chain in the skeleton, we also need to fix the first and last node of the ϵ-chains. To do this, we use the MAKEPATH algorithm on each partial solution (skeleton). It is much faster than building the permutation from scratch because some symmetries are broken. The algorithm can be run on each skeleton independently to facilitate parallelization. Nevertheless, there are some symmetries left in our algorithm. Indeed, a l-ϵ-cycle will produce l similar solutions. Moreover, if there are m times the same ϵ-cycle or ϵ-chain, there will be $m!$ similar solutions. Breaking these symmetries in our algorithm increases its running time, and it is left to future work to take them into account effectively.

4 Non-even-odd Case: Search for Optimal Permutations

The search for optimal permutation has been focused on even-odd permutation because in practice, the non-even-odd ones where never better up to $2k = 20$. In this section, we first use our algorithm to show that this is true for up to 32 blocks. We then give a useful example that we found while looking for a general proof.

4.1 Up to $2k = 32$

To test whether a non-even-odd permutation can have a better diffusion round than the even-odd ones, we used Algorithm 1 on all the skeletons having at least one ϵ-chain. We fixed R to be one round less than the diffusion round known for the best even-odd permutation, and ran our algorithm with the property HASPATH (described in Algorithm 3).

The running time of our algorithm is highly related to the strategy implemented into the NEXTPATH function (Algorithm 2). The best strategy we found was to first build the paths that start and end on the smallest ϵ-chains. This is because the paths starting by consecutive even nodes and ending by consecutive odd nodes have the least possibilities and therefore are most likely to be impossible to build. The case $2k = 22$ is quite small so we increased R to find the optimal non-even-odd permutations. They are given in Table 2. These optimal permutations have a diffusion round of 9 which is one round more than the optimal even-odd permutations.

Algorithm 3: HasPath(x, π, b, l)

```
Data: x: current node, π: partial permutation, b: target node, l: remaining
      length
1 if l > 0 then
2   return (x is odd ∧ HasPath(x − 1, π, b, l))
3          ∨ (π[x] is fixed ∧ HasPath(π[x], π, b, l − 1));
4 else return x = b ;
```

Table 2. Optimal Non-even-odd permutations for 2k=22

π =(3, 18, 5, 16, 7, 12, 9, 10, 1, 14, 13, 2, 15, 8, 11, 21, 17, 4, 19, 6, 0, 20)
π =(3, 6, 5, 12, 7, 10, 9, 18, 1, 2, 13, 4, 15, 16, 17, 8, 11, 21, 19, 14, 0, 20)
π =(3, 12, 5, 0, 7, 10, 9, 18, 1, 2, 13, 4, 15, 16, 17, 21, 11, 8, 19, 14, 6, 20)
π =(3, 8, 5, 16, 7, 21, 9, 14, 1, 2, 13, 18, 15, 0, 17, 6, 11, 12, 19, 4, 10, 20)
π =(3, 21, 5, 10, 7, 0, 9, 14, 1, 2, 13, 18, 15, 8, 17, 6, 11, 12, 19, 4, 16, 20)
π =(3, 8, 5, 6, 7, 4, 1, 12, 11, 2, 9, 21, 15, 19, 13, 17, 10, 16, 14, 20, 0, 18)
π =(3, 4, 5, 14, 7, 0, 9, 16, 11, 2, 1, 12, 15, 21, 13, 6, 19, 10, 17, 8, 18, 20)
π =(3, 6, 5, 10, 7, 16, 9, 18, 11, 14, 1, 2, 15, 4, 13, 0, 19, 8, 17, 21, 12, 20)

For $2k = 24$ to $2k = 32$, our algorithm ended without finding any non-even-odd permutations with a better diffusion round than the optimal even-odd ones. As a result, we establish that the non-even-odd permutations do not achieve a better diffusion round than the even-odd permutations up to $2k = 32$. All results are summarized in Table 1 and have been obtained on a 128 core CPU. The hardest instance with 32 blocks and $R = 8$ takes around 8 h of computing time. In Cauchois et al [3], it is mentioned that "$2^{46.4}$ tests of diffusion rounds" are needed when considering 20 blocks. Actually, our algorithm is faster and tackles this instance in around 8 s on our supercomputer. The source code is publicly available at https://gitlab.inria.fr/agontier/ANewAlgoForGFN.

4.2 Towards an Impossibility Result

Intuitively, a non-even-odd permutation should not reach a better diffusion round than the optimal even-odd one. Indeed, every time there are two consecutive odd nodes $u, v \in V_o$ such that $(u, v) \in E_\pi$, there are also somewhere in the graph G_π two consecutive even nodes $x, y \in V_e$ such that $(x, y) \in E_\pi$. We recall that each odd node has two outgoing edges (one in E_π and one in E_ϵ) whereas each even node has only one. Therefore, all the paths starting from the node x have one edge less to achieve full diffusion and any path that passes through (u, v) will gain one edge. Since the number of even to even edges is the same as the number of odd to odd edges, one could think that they compensate.

One of our objective during this work was to provide a formal proof that the diffusion round of the non-even-odd permutations are also bounded by the

Fibonacci bound as for even-odd permutations. Thus we made the conjecture that the total number of paths in a permutation graph and its inverse permutation graph could not exceed the sum of the Fibonacci bounds. However, we found a non-even-odd permutation for which the number of odd nodes reached from the even nodes was in total, and with redundancy, greater than the even-odd Fibonacci bound, which suggests that an improvement of the diffusion round is possible by considering non-even-odd permutations.

Example 4. We consider the permutation π =(3,2,1,5,0,6,7,4) depicted in the leftmost graph of Fig. 7. The rightmost one represents π^{-1}.

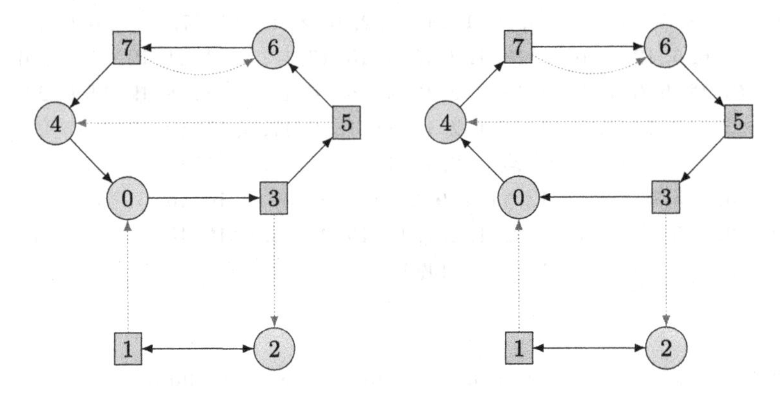

Fig. 7. Permutation graph of π and π^{-1}

On these two graphs, we give in Table 3 the number of paths of length $R = 5$ that ends on an odd node from each even node. There are 22 paths for π, and 21 paths for π^{-1}.

Table 3. Number of paths in π and π^{-1}

start node	0	2	4	6
number of paths	5	8	5	4

start node	0	2	4	6
number of paths	4	5	5	7

When considering only the even-odd permutations, the maximum number of paths given by the Fibonacci suite is 5 for each node and thus $4 \times 5 = 20$ in total. This example shows that the diffusion round in the general case (i.e. considering both even-odd and non-even-odd permutations) cannot be bounded by the Fibonacci suite if we consider the sum of all paths on π and π^{-1}. However, we may note that there is one node (e.g. node 6 for π) having less paths than the Fibonacci suite. We always observe this phenomenon on the permutations we considered. We think that to establish an impossibility result (a non-even-odd permutation can not be better than the optimal even-odd one), we should focus on these nodes.

5 Even-odd Case: Search for New Properties

As studied in the literature, the diffusion round is a property that can be used to find *good* Feistel permutations. This criteria is tied to the resistance of the resulting ciphertext against e.g. impossible differentials, saturation attacks and pseudorandomness analysis [10]. However, permutations with optimal diffusion round can also be weak against other cryptanalysis techniques. For instance, the designers of `WARP` [1] selected a permutation achieving full diffusion in 10 rounds while permutations with a diffusion round of 9 actually exist. The main reason is that all optimal permutations for the diffusion round are much weaker regarding truncated differential cryptanalysis than the one they selected. These permutations require at least 32 rounds to reach 64 active S-Boxes, while the permutation used in `WARP` (which is non optional w.r.t. the diffusion round) only requires 19 rounds to reach the same resistance.

Therefore, it would be interesting to look for other properties which might lead to stronger ciphers. With our algorithm it is quite simple to change the property we are looking for as we only need to provide a new HasProperty function. In this section, we thus propose several properties derived from the diffusion round and study the quality of their solutions against truncated differential cryptanalysis. We consider two properties, the first one is a generalization of the diffusion round where we consider not one but X paths between each pair of blocks. The second one consists of counting the S-Boxes on each path instead of the paths themselves.

5.1 Number of Paths

The diffusion round property ensures that each solution has at least one d-path of length R between each pair of blocks. We propose a new property parameterized by an integer X, namely X-DR, which extends the diffusion round to at least X d-paths of length R between each pair of blocks.

Definition 8. *X-$DR(\pi)$ is the smallest integer R such that:*
$\forall u, v \in V$, there are X d-paths of length R from u to v in G_π.

This new property introduces the parameter X denoting the minimum number of paths we want between each pair of nodes. When $X = 1$, this corresponds to the full diffusion property. To use this new property in our algorithm, the call to HasProperty line 2 of Algorithm 2 is replaced by a call to NumberOfPaths with the slight modification that this number of paths must be greater or equal to the parameter X. This function counts the number of paths between two nodes, it is given in Algorithm 4.

Since we want more than one path between two nodes, the function MakePath may need to create multiple paths. Due to these multiple paths, we must set an order between paths to prevent introducing new symmetries. For example, we should not build a path p after a path q if we already tried to build them in the other order. Proposition 2, stated and proved for the diffusion

Algorithm 4: NumberOfPath(x, π, b, l)

```
Data: x: current node, π: partial permutation, b: target node, l: remaining
      length
1 if l > 0 then
2   if π[x] is fixed then
3     if x is odd then
4       return
          NumberOfPath(x − 1, π, b, l) + NumberOfPath(π[x], π, b, l − 1);
5     else return NumberOfPath(π[x], π, b, l − 1) ;
6   else return 0 ;
7 else
8   if x = b then return 1 ;
9   else return 0 ;
```

round, is still valid when considering *X-DR*. It is stated in Proposition 3, and for sake of completeness the proof is given in Appendix.

Proposition 3. *Let π be an even-odd permutation, X-$DR(\pi)$ is the smallest integer R such that: $\forall c \in V_o, d \in V_e$, there are X paths of length $R-3$ from c to d in G_π.*

To compare this criterion w.r.t. truncated differential analysis, we computed the minimal number of active S-Boxes for each possible permutation for $k=6$, $k=7$, and $k=8$. We give in Table 4 the best number (i.e. the minimum one) we obtained from round 1 to round 16 :

Table 4. Best minimal number of active S-Boxes for each round

k	Round															
	1	2	3	4	5	6	7	8	9	10	11	12	13	14	15	16
6	0	1	2	3	4	6	8	11	14	16	19	22	24	26	28	29
7	0	1	2	3	4	6	8	11	14	19	23	26	28	30	33	35
8	0	1	2	3	4	6	8	11	14	19	22	26	29	31	34	37

Then, we took the 500 first solutions given by our algorithm for the criterion. We computed the minimal number of active S-Boxes for each of these solutions, and we counted the number of solutions that reached the optimal value for each round from 10 to 16. The results are given in Table 5. Note that to get 500 solutions, we sometimes needed to consider the criterion to a higher round than the optimal one. For example the diffusion round for $k=6$ is $R=8$. However, there are only 245 solutions with these parameters. Thus, we had to increase R until we reached 500 solutions. This is summarized in the range column of Table 5.

Table 5. Number of solutions with an optimal number of active S-Boxes from round 10 to round 16 in the 500 first solutions considering $k = 8$

X-DR	Round							
	10	11	12	13	14	15	16	Range
1 path	9	16	0	0	0	0	0	8
2 paths	24	37	0	0	0	0	0	9
3 paths	0	4	0	0	0	0	0	10
4 paths	15	15	0	0	0	0	0	10–11
5 paths	0	1	0	0	0	0	0	11
6 paths	9	9	0	0	0	0	0	11–12
7 paths	0	0	0	0	0	0	0	12
8 paths	0	0	0	0	0	0	0	12

For $k = 8$, we do not see a trend and we have similar results for $k = 7$ and $k = 6$. In fact, the property seems uncorrelated to the optimal number of active S-Boxes. We can see that increasing the parameter X increases the round R we need to go to find 500 solutions. Indeed when we search for two paths instead of one, the property is so strict that there are no solutions for $R = 8$. We also see that few to none of the 500 solutions are optimal in general.

5.2 Number of S-Boxes

Having X paths between each pair of blocks does not ensure that these paths are "good" from the differential analysis point of view. Instead of constraining the number of paths, we propose to ensure that a minimum number of S-Boxes are present in the d-paths between each pair of blocks.

Definition 9. *X-$SB(\pi)$ is the smallest integer R such that: $\forall u, v \in V$, there are X S-Boxes traversed by d-paths of length R from u to v in G_π. A S-Box reached by two paths of the same length will be counted only once.*

For example, in the two paths of length 5 from a to d depicted below, the S-Box corresponding to the red edge (b, b') will be counted twice (as it occurs at two different lengths), whereas the S-Box corresponding to the red edges (a', c) will be counted only once (even if it occurs on both paths).

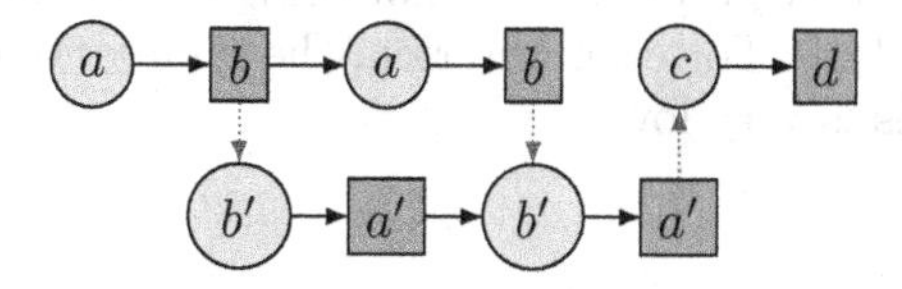

To use this new property in our algorithm, the call to HasProperty line 2 of Algorithm 2 is replaced by a call to DetectSBoxes with the slight modification that the sum of detected S-Boxes must be greater or equal to the parameter X.

DetectSBoxes is described in Algorithm 5. Unlike paths, we cannot simply count the S-Boxes because of the redundancy described in the previous example. We have to use a Boolean matrix of dimension 2 or an equivalent structure to remember at which path length l we encounter each S-Box.

Algorithm 5: DetectSBoxes(x, π, b, l, M)

```
Data: x: current node, π: partial permutation, b: target node, l: remaining
      length, M: Boolean matrix of dimension 2
1  M0 ← Matrix filled with false values;
2  if l > 0 then
3  |   if π[x] is fixed then
4  |   |   if x is odd then
5  |   |   |   M2 ← copy(M);
6  |   |   |   M2[x, l] ← true;
7  |   |   |   M3 ← DetectSBoxes(π[x], π, b, l − 1, M);
8  |   |   |   M4 ← DetectSBoxes(x − 1, π, b, l, M2);
9  |   |   |   return Bit-wise OR(M3,M4);
10 |   |   else return DetectSBoxes(π[x], π, b, l − 1, M) ;
11 |   else return M0 ;
12 else
13 |   if x = b then return M ;
14 |   else return M0 ;
```

Proposition 2, stated and proved for the diffusion round, is also valid when considering X-SB. It is stated in Proposition 4, and for sake of completeness, the proof is given in Appendix.

Proposition 4. *Let π be an even-odd permutation, X-$SB(\pi)$ is the smallest integer R such that: $\forall c \in V_o, d \in V_e$, there are X S-Boxes traversed by paths of length $R-3$ from c to d in G_π. A S-Box reached by two paths of the same length will be counted only once.*

As for the X-DR criteria, we looked at the quality of optimal permutations for the X-SB criteria regarding truncated differential cryptanalysis for $k = 6$, $k = 7$, and $k = 8$. The results are summarized in Table 6 for $k = 8$ and are similar for lower k.

Overall, these two new properties did not bring better solutions for the truncated differential analysis. For each criterion, the number of optimal solution in the 500 first solutions is very low.

5.3 TWINE

Finally, we studied our criteria on the permutation used in `TWINE` [11]. The values of our criteria for `TWINE` are given in Table 7. To see if these are good values, we

Table 6. Number of solutions with an optimal number of active S-Boxes from round 10 to round 16 in the 500 first solutions considering $k = 8$

X-SB	Round							
	10	11	12	13	14	15	16	Range
1 S-Box	25	44	0	0	0	0	0	8
2 S-Boxes	25	44	0	0	0	0	0	8
3 S-Boxes	0	1	0	0	0	0	0	9
4 S-Boxes	18	30	0	0	0	0	0	9
5 S-Boxes	4	12	0	0	0	0	0	9–10
6 S-Boxes	4	9	0	2	2	2	0	9–10
7 S-Boxes	0	6	0	0	0	0	0	10
8 S-Boxes	0	9	0	0	0	0	0	10–11
9 S-Boxes	0	1	0	1	15	1	0	11
10 S-Boxes	0	4	0	0	0	0	0	11
11 S-Boxes	0	6	0	0	0	0	0	11–12
12 S-Boxes	0	0	0	0	0	0	0	11–12

used our algorithm to enumerate permutations with strictly greater values for our criteria. The algorithm concluded that there is no permutation with a better X-SB than `TWINE` up to $X = 22$. The experimentation was not done beyond due to its computational cost.

Table 7. X-DR and X-SB values for `TWINE`

1 to 2-SB	3 to 6-SB	7 to 8-SB	9 to 14-SB	15 to 22-SB
8	9	10	11	12
1-DR	2-DR	3-DR	4 to 5-DR	6 to 9-DR
8	9	10	11	12

However, `TWINE` is not optimal for 4-DR and 6-DR. There is only one permutation that is optimal on 4-DR and 6-DR at the same time. This permutation is $\pi =(3, 4, 5, 8, 1, 12, 9, 10, 11, 2, 7, 14, 13, 6, 15, 0)$. To compare it with `TWINE`, we computed the truncated differentials on both permutations in Table 8.

This new permutation π is better than `TWINE` and optimal at round 10. However, it is worse for rounds 13 to 16. In fact, in all the $k = 8$ permutations, none can reach the optimal number of active S-Boxes at every round.

Table 8. Truncated Differentials for TWINE and π

Round	8	9	10	11	12	13	14	15	16
TWINE	**11**	**14**	18	**22**	24	27	30	32	35
π	**11**	**14**	**19**	**22**	24	26	28	30	32

6 Conclusion

In this paper, we proposed a new generic algorithm based on path building to enumerate permutations regarding a chosen property for Generalized Feistel Networks. The main advantage of our algorithm is that it is not restricted to the even-odd permutations nor the diffusion round property. Furthermore, it was fast enough to prove that no non-even-odd permutation reaches a strictly better diffusion round than optimal even-odd permutations up to 32 blocks. Thus we fully solved the problem opened by Suzaki and Minematsu in [10] and partially solved by Derbez *et al.* in [5].

However, in both [5] and [1], it was highlighted that optimal permutations regarding the diffusion round might still lead to ciphers far from offering an optimal resistance against differential cryptanalysis. We thus tried two more complex properties derived from the diffusion round and studied the quality of the solutions they provide against truncated differential cryptanalysis.

Future work. We believe that providing a formal proof that there is always at least one even-odd permutation optimal with respect to the diffusion round would be a great result which should lead to a better understanding of GFN. We are confident that obtaining such a proof is possible and the particular example described Sect. 4.2 seems to be a good starting point. Another interesting problem concerns properties that would ensure some level of resistance against differential cryptanalysis. Indeed, our work clearly shows that permutations ensuring fast and strong diffusion are rarely optimal regarding this type of distinguishers.

Appendix A Proofs of Proposition 3 and 4

Proposition 3. *Let π be an even-odd permutation π, X-$DR(\pi)$ is the smallest integer R such that: $\forall c \in V_o, d \in V_e$, there are X paths of length $R-3$ from c to d in G_π.*

Proof. Let $b, c \in V_o$ and $a, d \in V_e$ with $(a,c),(d,b) \in E_\pi$. We have that $(a+1,a),(b,b-1) \in E_\epsilon$ with $a+1 \in V_o$ and $b-1 \in V_e$. Furthermore, we have $g,h \in V$ such that $(b,g),(b-1,h) \in E_\pi$ (see the graph below with $i = a+1$ and $j = b-1$).

1) From Definition 8, we know that there is X d-paths of length R from a to g, thus there is X paths of length $R-3$ from c to d.

2) Now suppose that there is $R' < X\text{-}DR(\pi)$ such that $\forall\ c \in V_o,\ d \in V_e$ there is X paths of length $R'-3$ from c to d. We then have X d-paths of length R' from i to g, from i to h and from a to h. Since we have these d-paths for all pairs $a \in V_e, b \in V_o$ then we have full diffusion with $X\text{-}DR(\pi) = R'$ and thus the contradiction $X\text{-}DR(\pi) < X\text{-}DR(\pi)$. □

Proposition 4. *Let π be an even-odd permutation π, $X\text{-}SB(\pi)$ is the smallest integer R such that: $\forall c \in V_o, d \in V_e$, there are X S-Boxes traversed by paths of length $R-3$ from c to d in G_π. A S-Box reached by two paths at the same time will be counted only once.*

Proof. Let $b, c \in V_o$ and $a, d \in V_e$ with $(a,c),(d,b) \in E_\pi$. We have that $(a+1, a),(b, b-1) \in E_\epsilon$ with $a+1 \in V_o$ and $b-1 \in V_e$. Furthermore, we have $g, h \in V$ such that $(b,g),(b-1,h) \in E_\pi$ (see the graph below with $i = a+1$ and $j = b-1$).

1) From Definition 9, we know that there is X S-Boxes in all the d-paths of length R from a to g, thus there is X S-Boxes in all paths of length $R-3$ from c to d.

2) Now suppose that there is $R' < X\text{-}SB(\pi)$ such that $\forall\ c \in V_o,\ d \in V_e$ there is X S-Boxes in all the paths of length $R'-3$ from c to d. We then have X S-Boxes in all the d-paths of length R' from i to g, from i to h and from a to h. Since we have these d-paths for all pairs $a \in V_e, b \in V_o$ then we have full diffusion with $X\text{-}SB(\pi) = R'$ and thus the contradiction $X\text{-}SB(\pi) < X\text{-}SB(\pi)$. □

References

1. Banik, S.: WARP?: revisiting GFN for lightweight 128-bit block cipher. In: Dunkelman, O., Jacobson, Jr., M.J., O'Flynn, C. (eds.) SAC 2020. LNCS, vol. 12804, pp. 535–564. Springer, Cham (2021). https://doi.org/10.1007/978-3-030-81652-0_21
2. Beaulieu, R., Shors, D., Smith, J., Treatman-Clark, S., Weeks, B., Wingers, L.: The SIMON and SPECK families of lightweight block ciphers. IACR Cryptology ePrint Archive **2013**, 404 (2013). http://eprint.iacr.org/2013/404
3. Cauchois, V., Gomez, C., Thomas, G.: General diffusion analysis: How to find optimal permutations for generalized type-ii feistel schemes. IACR Trans. Symmetric Cryptol. **2019**(1), 264–301 (2019). https://doi.org/10.13154/tosc.v2019.i1.264-301, https://doi.org/10.13154/tosc.v2019.i1.264-301
4. Derbez, P., Fouque, P.-A.: Automatic search of meet-in-the-middle and impossible differential attacks. In: Robshaw, M., Katz, J. (eds.) CRYPTO 2016. LNCS, vol. 9815, pp. 157–184. Springer, Heidelberg (2016). https://doi.org/10.1007/978-3-662-53008-5_6

5. Derbez, P., Fouque, P., Lambin, B., Mollimard, V.: Efficient search for optimal diffusion layers of generalized feistel networks. IACR Trans. Symmetric Cryptol. **2019**(2), 218–240 (2019). https://doi.org/10.13154/tosc.v2019.i2.218-240 https://doi.org/10.13154/tosc.v2019.i2.218-240 https://doi.org/10.13154/tosc.v2019.i2.218-240
6. DES: Data Encryption Standard. FIPS PUB 46, Federal information processing standards publication 46 (1977)
7. Nyberg, K.: Generalized feistel networks. In: Kim, K., Matsumoto, T. (eds.) ASIACRYPT 1996. LNCS, vol. 1163, pp. 91–104. Springer, Heidelberg (1996). https://doi.org/10.1007/BFb0034838
8. Shi, D., Sun, S., Derbez, P., Todo, Y., Sun, B., Hu, L.: Programming the demirci-selçuk meet-in-the-middle attack with constraints. In: Peyrin, T., Galbraith, S. (eds.) ASIACRYPT 2018. LNCS, vol. 11273, pp. 3–34. Springer, Cham (2018). https://doi.org/10.1007/978-3-030-03329-3_1
9. Shibutani, K., Isobe, T., Hiwatari, H., Mitsuda, A., Akishita, T., Shirai, T.: Piccolo: an ultra-lightweight blockcipher. In: Preneel, B., Takagi, T. (eds.) CHES 2011. LNCS, vol. 6917, pp. 342–357. Springer, Heidelberg (2011). https://doi.org/10.1007/978-3-642-23951-9_23
10. Suzaki, T., Minematsu, K.: Improving the generalized feistel. In: Hong, S., Iwata, T. (eds.) FSE 2010. LNCS, vol. 6147, pp. 19–39. Springer, Heidelberg (2010). https://doi.org/10.1007/978-3-642-13858-4_2
11. Suzaki, T., Minematsu, K., Morioka, S., Kobayashi, E.: TWINE: a lightweight block cipher for multiple platforms. In: Knudsen, L.R., Wu, H. (eds.) SAC 2012. LNCS, vol. 7707, pp. 339–354. Springer, Heidelberg (2013). https://doi.org/10.1007/978-3-642-35999-6_22
12. Zheng, Y., Matsumoto, T., Imai, H.: On the construction of block ciphers provably secure and not relying on any unproved hypotheses. In: Brassard, G. (ed.) CRYPTO 1989. LNCS, vol. 435, pp. 461–480. Springer, New York (1990). https://doi.org/10.1007/0-387-34805-0_42

Minimizing Even-Mansour Ciphers for Sequential Indifferentiability (Without Key Schedules)

Shanjie Xu[1,2], Qi Da[1,2], and Chun Guo[1,2,3](✉)

[1] Key Laboratory of Cryptologic Technology and Information Security of Ministry of Education, Shandong University, Qingdao, Shandong 266237, China
{shanjie1997,daqi}@mail.sdu.edu.cn, chun.guo@sdu.edu.cn
[2] School of Cyber Science and Technology, Shandong University, Qingdao, Shandong, China
[3] Shandong Research Institute of Industrial Technology, Jinan, Shandong, China

Abstract. Iterated Even-Mansour (IEM) schemes consist of a small number of fixed permutations separated by round key additions. They enjoy provable security, assuming the permutations are *public and random.* In particular, regarding chosen-key security in the sense of *sequential indifferentiability (seq-indifferentiability)*, Cogliati and Seurin (EUROCRYPT 2015) showed that without key schedule functions, the 4-round *Even-Mansour with Independent Permutations and no key schedule* $\mathrm{EMIP}_4(k,u) = k \oplus \mathbf{p}_4\big(k \oplus \mathbf{p}_3\big(k \oplus \mathbf{p}_2(k \oplus \mathbf{p}_1(k \oplus u))\big)\big)$ is sequentially indifferentiable.

Minimizing IEM variants for classical strong (tweakable) pseudorandom security has stimulated an attractive line of research. In this paper, we seek for minimizing the EMIP_4 construction while retaining seq-indifferentiability. We first consider EMSP, a natural variant of EMIP using *a single round permutation.* Unfortunately, we exhibit a slide attack against EMSP with *any number of rounds.* In light of this, we show that the 4-round $\mathrm{EM2P}_4^{\mathbf{p}_1,\mathbf{p}_2}(k,u) = k \oplus \mathbf{p}_1\big(k \oplus \mathbf{p}_2\big(k \oplus \mathbf{p}_2(k \oplus \mathbf{p}_1(k \oplus u))\big)\big)$ using 2 independent random permutations $\mathbf{p}_1, \mathbf{p}_2$ is seq-indifferentiable. This provides the *minimal seq-indifferentiable IEM without key schedule.*

Keywords: Blockcipher · Sequential indifferentiability · Key-alternating cipher · Iterated even-mansour cipher

1 Introduction

A fundamental cryptographic problem is to construct secure blockciphers from keyless permutations. A natural solution is the Iterated Even-Mansour (IEM) scheme (a.k.a. key-alternating cipher) initiated in [19] and extended and popularized in a series of works [1,4,17,24]. Given t permutations $\mathbf{p}_1, ..., \mathbf{p}_t : \{0,1\}^n \to \{0,1\}^n$ and a *key schedule* $\overrightarrow{\varphi} = (\varphi_0, ..., \varphi_t)$, $\varphi_i : \{0,1\}^\kappa \to \{0,1\}^n$, and for $(k,u) \in \{0,1\}^\kappa \times \{0,1\}^n$, the scheme is defined as

$$\mathrm{EM}[\overrightarrow{\varphi}]_t(k,u) := \varphi_t(k) \oplus \mathbf{p}_t\big(...\varphi_2(k) \oplus \mathbf{p}_2\big(\varphi_1(k) \oplus \mathbf{p}_1(\varphi_0(k) \oplus u)\big)...\big).$$

T. Isobe and S. Sarkar (Eds.): INDOCRYPT 2022, LNCS 13774, pp. 125–145, 2022.
https://doi.org/10.1007/978-3-031-22912-1_6

It abstracts *substitution-permutation network* that has been used by a number of standards [26,27,33]. Modeling $\mathbf{p}_1, ..., \mathbf{p}_t$ as public random permutations, variants of this scheme provably achieve various security notions, including indistinguishability [4,6,7,19,25,28,32,36,37], related-key security [8,20], known-key security [2,9], chosen-key security in the sense of correlation intractability [8,23], and indifferentiability [1,13,29]. Despite the theoretical uninstantiatability of the random oracle model [5], such arguments dismiss generic attacks and are typically viewed as evidences of the soundness of the design approaches.

Indifferentiability of IEM. The classical security definition for a blockcipher is *indistinguishability from a (secret) random permutation.* Though, reliable blockciphers are broadly used as *ideal ciphers*, i.e., randomly chosen blockciphers. Motivated by this, the notion of *indifferentiability* [31] *from ideal ciphers* was proposed [1,11,29] as the strongest security for blockcipher structures built upon (public) random functions and random permutations. Briefly speaking, for the IEM cipher $\mathrm{EM}^{\mathcal{P}}$ built upon random permutations $\mathcal{P}$, if there exists an efficient simulator $\mathcal{S}^E$ that queries an ideal cipher E to mimic its (non-existent) underlying permutations, such that $(E, \mathcal{S}^E)$ is indistinguishable from $(\mathrm{EM}^{\mathcal{P}}, \mathcal{P})$, then $\mathrm{EM}^{\mathcal{P}}$ is indifferentiable from E [31]. This property implies that the cipher $\mathrm{EM}^{\mathcal{P}}$ inherits all ideal cipher-properties defined by single-stage security games, including security against (various forms of) related-key and chosen-key attacks.

As results, Andreeva et al. [1] proposed the IEM variant $\mathrm{EMKD}_t(k, u) = \mathbf{h}(k) \oplus \mathbf{p}_t(...\mathbf{h}(k) \oplus \mathbf{p}_2(\mathbf{h}(k) \oplus \mathbf{p}_1(\mathbf{h}(k) \oplus u))...)$ using a random oracle $\mathbf{h} : \{0,1\}^\kappa \to \{0,1\}^n$ to derive the round key $\mathbf{h}(k)$, and proved indifferentiability at 5 rounds. Concurrently, Lampe and Seurin [29] proposed to consider the *"single-key" Even-Mansour* variant $\mathrm{EMIP}_t(k, u) = k \oplus \mathbf{p}_t(...k \oplus \mathbf{p}_2(k \oplus \mathbf{p}_1(k \oplus u))...)$ without any non-trivial key schedule, and proved indifferentiability at 12 rounds. Both results are tightened in subsequent works [13,22], showing that 3-round EMKD and 5-round EMIP achieve indifferentiability.

Sequential Indifferentiability. Indifferentiable blockciphers [1,11,13,22,29] typically require unnecessarily complicated constructions [35], and their practical influences are not as notable as the analogues for hash function [10,15]. To remedy, weaker security definitions have been proposed [2,9,30,34]. In particular, to formalize *chosen-key security*, Mandal et al. [30] and subsequently Cogliati and Seurin [8] advocated the notion of *sequential-indifferentiability (seq-indifferentiability)*, which is a variant of indifferentiability concentrating on distinguishers that follow a strict restriction on the order of queries. The usefulness of seq-indifferentiability lies in its implication towards *correlation intractability* [5], meaning that no (chosen-key) adversary can find inputs/outputs of the blockcipher that satisfies evasive relations. For the aforementioned Even-Mansour variants, seq-indifferentiability (and CI) have been established for 3-round EMKD [23] and 4-round EMIP [8], both of which are tight. The fact that 4-round EMIP is seq-indifferentiable/CI but not "fully" indifferentiable also separated the two security notions [13].

Our Question. Besides initial positive results on the general $\text{EM}[\overrightarrow{\varphi}]_t$ model, another attractive line of work has been set to seek for *minimizing IEM cipher* for certain security properties. In detail, Dunkelman [17] was the first to minimize the 1-round Even-Mansour cipher by halving the key size without affecting its SPRP security. Following this and with significant technical novelty, Chen et al. [6] proposed minimal 2-round IEM variants with beyond-birthday SPRP security. Subsequently, Dutta [18] extended the discussion to tweakable Even-Mansour (TEM) ciphers and proposed minimal 2-round and 4-round IEM variants, depending on the assumptions on tweak schedule functions.

Regarding (seq-)indifferentiability, we stress that all the aforementioned results on IEM [1,8,13,22,23,29] requires *using t independent random permutations in the t rounds*. As will be elaborated, this independence is crucial for their (seq-)indifferentiability simulators. A natural next step is to investigate whether (weaker) indifferentiability is achievable using a single permutation. In particular, without key schedule, does the *single-permutation Even-Mansour* variant $\text{EMSP}_t(k, u) = k \oplus \mathbf{p}(...k \oplus \mathbf{p}(k \oplus \mathbf{p}(k \oplus u))...)$ suffice?

1.1 Our Contributions

We make the first step towards answering our question and analyze the IEM cipher with identical permutation w.r.t. the seq-indifferentiability.

New Attack Against Seq-Indifferentiability. Our first observation is that, even in the weaker model of seq-indifferentiability, the aforementioned *"single-key", single-permutation Even-Mansour* variant EMSP remains *insecure*, regardless of the number of rounds. Concretely, we exhibit a chosen-key attack that makes just 1 permutation query and 1 encryption query. Our attack utilized a sort of weakness that is related to slide attacks [3]. In detail, in the EMSP construction, a single input/output pair $\mathbf{p}(x) = y$ of the permutation already yields a full t-round EMSP_t evaluation $y \to \underbrace{(x, y) \to ... \to (x, y)}_{t \text{ times}} \to x$ with $k = x \oplus y$, by acting as the involved evaluations in all the t rounds.

Minimal and Secure Construction. Given our negative result on EMSP, to achieve security, one has to enhance 4-round EMSP by using at least 2 independent random permutations. This consideration yields a minimal IEM solution scheme $\text{EM2P}_4^{\mathbf{p}_1,\mathbf{p}_2} : \{0,1\}^n \times \{0,1\}^n \to \{0,1\}^n$ uses two random permutations $\mathbf{p}_1, \mathbf{p}_2$ though no key schedule:

$$\text{EM2P}_4^{\mathbf{p}_1,\mathbf{p}_2}(k, u) := k \oplus \mathbf{p}_1\big(k \oplus \mathbf{p}_2\big(k \oplus \mathbf{p}_2(k \oplus \mathbf{p}_1(k \oplus u))\big)\big).$$

See Fig. 1 for an illustration. We established seq-indifferentiability for $\text{EM2P}_4^{\mathbf{p}_1,\mathbf{p}_2}$ with $O(q^2)$ simulator complexity and $O(q^4/2^n)$ security which are comparable with EMIP_4 [8]. For ease of comparison, we summarize our results and the existing in Table 1.

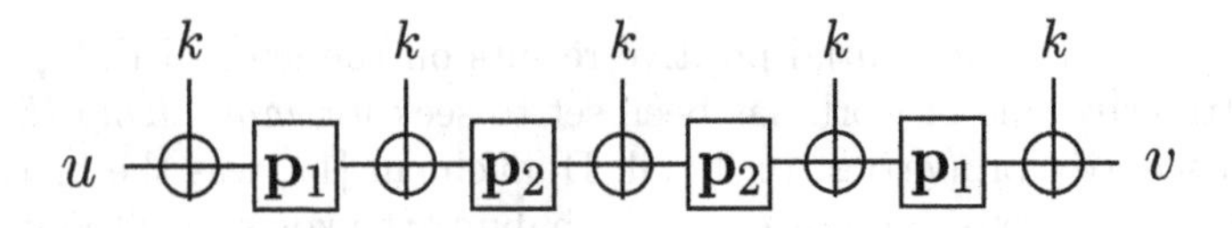

Fig. 1. The minimal construction $\text{EM2P}_4^{\mathbf{p}_1,\mathbf{p}_2}$ using two independent random permutations $\mathbf{p}_1, \mathbf{p}_2 : \{0,1\}^n \to \{0,1\}^n$ and no key schedule.

Table 1. Comparison of ours with existing seq-indifferentiable/CI IEM results. The column **Key sch.** indicates the key schedule functions in the schemes. The column **Complex.** indicates the simulator complexities.

Scheme	♯Rounds	♯Primitives	Key sch.	Complex.	Bounds	Ref.
$\text{EMIP}_4^{\mathbf{p}_1,\mathbf{p}_2,\mathbf{p}_3,\mathbf{p}_4}$	4	4	no	q^2	$q^4/2^n$	[8]
$\text{EMKD}_3^{\mathbf{h},\mathbf{p}_1,\mathbf{p}_2,\mathbf{p}_3}$	3	4	random oracle **h**	q^2	$q^4/2^n$	[23]
$\text{EMSP}^{\mathbf{p}}$	t	1	no	insecure	insecure	Sect. 3
$\text{EM2P}_4^{\mathbf{p}_1,\mathbf{p}_2}$	4	**2**	no	q^2	$q^4/2^n$	Sect. 4

Proof Approach. Our proof for the seq-indifferentiability of $\text{EM2P}_4^{\mathbf{p}_1,\mathbf{p}_2}$ is an extension of [8], with subtle changes addressing new collision events due to permutation-reusing.

In general, to establish indifferentiability-type security, the first step is to construct a simulator that resists obvious attack. Then, it remains to argue:

- The simulator is efficient, i.e., its complexity can be bounded;
- The simulator gives rise to an ideal world $(E, \mathcal{S}^E)$ that is indistinguishable from the real world $(\text{EM}^{\mathcal{P}}, \mathcal{P})$.

To design a simulator, we mostly follow the simulator strategy for EMIP_4 (which uses *independent* permutations) [8], taking queries to the middle (2nd and 3rd) rounds as "signals" for chain detection and the outer (1st and 4th) rounds for adaptations.

For example, a distinguisher D may arbitrarily pick $k, u \in \{0,1\}^n$ and evaluate $x_1 \leftarrow k \oplus u$, $\mathbf{p}_1(x_1) \to y_1$, $x_2 \leftarrow k \oplus y_1$, $\mathbf{p}_2(x_2) \to y_2$, $x_3 \leftarrow k \oplus y_2$, $\mathbf{p}_2(x_3) \to y_3$, $x_4 \leftarrow k \oplus y_4$, $\mathbf{p}_1(x_4) \to y_4$, $x_5 \leftarrow k \oplus y_4$. This creates a sequence of four *(query) records* $\big((1, x_1, y_1), (2, x_2, y_2), (2, x_3, y_3), (1, x_4, y_4)\big)$ that will be called a *computation chain* (the number 1 or 2 indicates the index of the permutation). When D is in the real world $(\text{EM2P}_4^{\mathbf{p}_1,\mathbf{p}_2}, (\mathbf{p}_1, \mathbf{p}_2))$, it necessarily holds $\text{EM2P}_4^{\mathbf{p}_1,\mathbf{p}_2}(k, u) = x_5$. To be consistent with this in the ideal world $(E, \mathcal{S}^E)$, $\mathcal{S}$ should "detect" such actions of D, "run ahead" of D and define some simulated (query) records to "complete" a similar computation chain.

The crucial observation on $EM2P_4$ is that permutations used in the middle (2nd and 3rd) rounds and the outer (1st and 4th) rounds remain independent. Consequently, upon D querying the permutation, the simulator can identify in clear if D is evaluating in the middle (when D queries P_2) or in the outer rounds (when D queries P_1). With these ideas, every time D queries P_2 or P_2^{-1}, our simulator completes all new pairs of records $\big((2,x,y),(2,x',y')\big)$ of P_2.[1]

Concretely, facing the aforementioned attack, $\mathcal{S}$ pinpoints the key $k = y_2 \oplus x_3$ and recognize the "partial chain" $\big((1,x_1,y_1),(2,x_2,y_2),(2,x_3,y_3)\big)$ upon the third permutation query $P_2(x_3) \to y_3$. $\mathcal{S}$ then queries the ideal cipher $E(k, k \oplus x_1) \to x_5$ and *adapts* the simulated P_1 by enforcing $P_1(k \oplus y_3) := k \oplus x_5$. As such, a simulated computation chain $\big((1,x_1,y_1),(2,x_2,y_2),(2,x_3,y_3),(1,k \oplus y_3, k \oplus x_5)\big)$ with $E(k, k \oplus x_1) = x_5$ is completed. Worth noting, queries to P_2 only function as "signals" for detection, while adaptations only create records on P_1 (such "adapted" records thus won't trigger new detection). This idea of assigning a unique role to every round/simulated primitive was initiated in [11], and it indeed significantly simplifies arguments.

Of course, D may pick $k', y_4' \in \{0,1\}^n$ and evaluate "conversely". In this case, our simulator detects the "partial chain" $\big((2,x_2',y_2'),(2,x_3',y_3'),(1,x_4',y_4')\big)$ after D's third query $P_2^{-1}(y_2') \to x_2'$, queries $E^{-1}(k', k' \oplus y_4') \to x_0'$ and pre-enforces $P_1(k' \oplus x_0') := k' \oplus x_5'$ to reach $\big((1,k' \oplus x_0', k' \oplus x_5'),(2,x_2',y_2'),(2,x_3',y_3'),(1,x_4',y_4')\big)$ with $E(k', k' \oplus x_1') = x_5'$. In the seq-indifferentiability setting, these have covered all adversarial possibilities. In particular, the distinguisher D cannot pick k', y_1' and evaluate $P_1^{-1}(y_1') \to x_1'$, $u' \leftarrow k' \oplus x_1'$, $E(k', u') \to v'$, and $P_1^{-1}(k' \oplus v') \to x_4'$, since this violates the query restriction. This greatly simplifies simulation [8,21,23,30] compared with the "full" indifferentiability setting.

Compared with [8], our novelty lies in handling new collision events that are harmless in the setting of $EMIP_4$. E.g., consider the previous example of enforcing $P_1(k \oplus y_3) := k \oplus x_5$ to complete $\big((1,x_1,y_1),(2,x_2,y_2),(2,x_3,y_3)\big)$. Since the 1st and 4th rounds are using the same permutation P_1, the collisions $k \oplus y_3 = x_1$ and $k \oplus x_5 = y_1$ also incur inconsistency in the simulated P_1 and prevent adaptation. But we do not need a paradigm-level shift: with all such events characterized, the proof follows that for $EMIP_4$. Clearly, the simulator detects and completes $O(q^2)$ chains, and indistinguishability of $(E, \mathcal{S}^E)$ and $(EM2P_4^{\mathbf{P}_1,\mathbf{P}_2}, \mathcal{P})$ follows a randomness mapping argument similar to [8].

1.2 Organization

Section 2 serves notations and definitions. Then, in Sect. 3 and 4, we provide our attack on $EMSP_t^{\mathbf{P}}$ and sequential indifferentiability of 4-round $EM2P_4^{\mathbf{P}_1,\mathbf{P}_2}$ respectively. We finally conclude in Sect. 5.

[1] In comparison, Cogliati and Seurin's simulator for $EMIP_4$ completes all newly constituted pairs $\big((2,x_2,y_2),(3,x_3,y_3)\big)$ of records of P_2 *and* P_3.

2 Preliminaries

Notation. An n-bit random permutation $\mathbf{p} : \{0,1\}^n \to \{0,1\}^n$ is a permutation that is uniformly chosen from all $(2^n)!$ possible choices, and its inverse is denoted by $\mathbf{p}^{-1}$. Denote by $\mathcal{P}$ a tuple of independent random permutations $(\mathbf{p}_1, ..., \mathbf{p}_r)$, where the number t depends on the concrete context (and will be made concrete later). For integers κ and n, an ideal blockcipher $E[\kappa, n] : \{0,1\}^\kappa \times \{0,1\}^n \to \{0,1\}^n$ is chosen randomly from the set of all blockciphers with key space $\{0,1\}^\kappa$ and message and ciphertext space $\{0,1\}^n$. For each key $k \in \{0,1\}^\kappa$, the map $E(k, \cdot)$ is a random permutation with inversion oracle $E^{-1}(k, \cdot)$. Since we focus on the case of $\kappa = n$, we will simply use E instead of $E[n, n]$.

Sequential Indifferentiability. The notion of sequential indifferentiability (seq-indifferentiability), introduced by Mandal et al. [30], is a weakened variant of (full) indifferentiability of Maurer et al. [31] tailored to *sequential distinguishers* [30], a class of restricted distinguishers. For concreteness, our formalism concentrates on blockciphers. Consider the blockcipher construction $\mathcal{C}^{\mathcal{P}}$ built upon several random permutations $\mathcal{P}$. A distinguisher $D^{\mathcal{C}^{\mathcal{P}}, \mathcal{P}}$ with oracle access to both the cipher and the underlying permutations is trying to distinguish $\mathcal{C}^{\mathcal{P}}$ from the ideal cipher E. Then, D is *sequential*, if it proceeds in the following steps in a strict order: (1) queries the underlying permutations $\mathcal{P}$ in arbitrary; (2) queries the cipher $\mathcal{C}^{\mathcal{P}}$ in arbitrary; (3) outputs, and cannot query $\mathcal{P}$ again in this phase. This order of queries is illustrated by the numbers in Fig. 2.

In this setting, if there is a simulator $\mathcal{S}^E$ that has access to E and can mimic $\mathcal{P}$ such that in the view of any sequential distinguisher D, the system $(E, \mathcal{S}^E)$ is indistinguishable from the system $(\mathcal{C}^{\mathcal{P}}, \mathcal{P})$, then $\mathcal{C}^{\mathcal{P}}$ is *sequentially indifferentiable* (seq-indifferentiable) from E.

To characterize the adversarial power, we define a notion *total oracle query cost* of D, which refers to the total number of queries received by $\mathcal{P}$ (from D or $\mathcal{C}^{\mathcal{P}}$) when D interacts with $(\mathcal{C}^{\mathcal{P}}, \mathcal{P})$ [30]. Then, the definition of seq-indifferentiability due to Cogliati and Seurin [8] is as follows.

Definition 1 (Seq-indifferentiability). *A blockcipher construction $\mathcal{C}^{\mathcal{P}}$ with oracle access to a tuple of random permutations $\mathcal{P}$ is statistically and strongly $(q, \sigma, t, \varepsilon)$-seq-indifferentiable from an ideal cipher E, if there exists a simulator $\mathcal{S}^E$ such that for any sequential distinguisher D of total oracle query cost at most q, $\mathcal{S}^E$ issues at most σ queries to E and runs in time at most t, and it holds*

$$\left|\Pr_{\mathcal{P}}[D^{\mathcal{C}^{\mathcal{P}}, \mathcal{P}} = 1] - \Pr_E[D^{E, \mathcal{S}^E} = 1]\right| \leq \varepsilon.$$

If D makes q queries, then its total oracle query cost is poly(q). As a concrete example, the t-round EM cipher $\mathrm{EM}_t^{\mathcal{P}}$ makes t queries to $\mathcal{P}$ to answer any query it receives, and if D makes q_e queries to $\mathrm{EM}_t^{\mathcal{P}}$ and q_p queries to $\mathcal{P}$, then the total oracle query cost of D is $q_p + tq_e = \mathrm{poly}(q_p + q_e) = \mathrm{poly}(q)$.

Albeit being weaker than "full" indifferentiability [31] (which can be viewed as seq-indifferentiability without restricting distinguishers to sequential), seq-indifferentiability already implies *correlation intractability* in the ideal model [8,

30]. The notion of correlation intractability was introduced by Canetti et al. [5] and adapted to ideal models by Mandal et al. [30] to formalize the hardness of finding exploitable relation between the inputs and outputs of function ensembles. For simplicity, we only present asymptotic definitions. Consider a relation $\mathcal{R}$ over pairs of binary sequences.

- $\mathcal{R}$ is *evasive with respect to an ideal cipher* E, if no efficient oracle Turing machine $\mathcal{M}^E$ can output an m-tuple $(x_1, \ldots, x_m)$ such that $\big((x_1, \ldots, x_m), (E(x_1), \ldots, E(x_m))\big) \in \mathcal{R}$ with a significant success probability;
- An idealized blockcipher $\mathrm{EM}^{\mathcal{P}}$ is *correlation intractable with respect to* $\mathcal{R}$, if no efficient oracle Turing machine $\mathcal{M}^{\mathcal{P}}$ can output an m-tuple $(x_1, \ldots, x_m)$ such that $\big((x_1, \ldots, x_m), (\mathrm{EM}^{\mathcal{P}}(x_1), \ldots, \mathrm{EM}^{\mathcal{P}}(x_m))\big) \in \mathcal{R}$ with a significant success probability.

With these, the implication [8,30] states that if $\mathrm{EM}^{\mathcal{P}}$ is seq-indifferentiable from E, then for any m-ary relation $\mathcal{R}$ which is evasive with respect to E, $\mathrm{EM}^{\mathcal{P}}$ is correlation intractable with respect to $\mathcal{R}$.

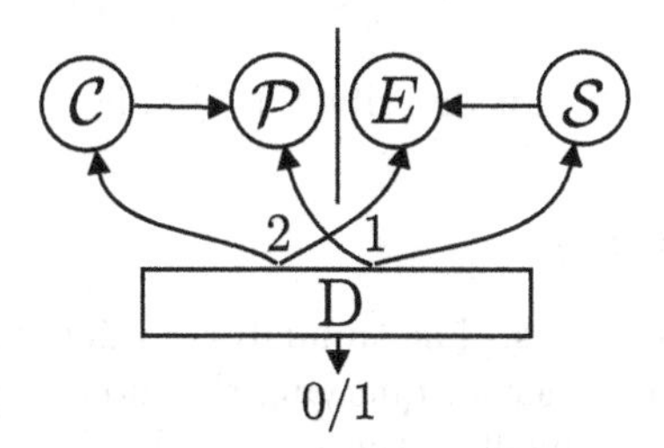

Fig. 2. Setting for seq-indifferentiability. The numbers 1 and 2 indicate the query order that D has to follow.

3 Slide Attack on the Single-Key, Single-Permutation EMSP

The t-round $\mathrm{EMSP}_t^{\mathbf{P}}$ uses the same permutation in every round, and is defined as

$$\mathrm{EMSP}_t^{\mathbf{P}}(k, u) := k \oplus \mathbf{p}\big(...k \oplus \mathbf{p}\big(k \oplus \mathbf{p}(k \oplus \mathbf{p}(k \oplus u))\big)...\big).$$

Our attack proceeds as follows.

1. Picks $x \in \{0,1\}^n$ in arbitrary and query $\mathbf{p}(x) \to y$.
2. Computes $k \leftarrow x \oplus y$. Outputs 1 if and only if $E(k, y) = x$.

Clearly, it always outputs 1 when interacting with $(\mathrm{EMSP}_t^{\mathbf{P}}, \mathbf{p})$ with *any rounds* t. In the ideal world, the simulator has to find a triple $(x \oplus y, y, x) \in (\{0,1\}^n)^3$ such that $E(x \oplus y, y) = x$ for the ideal cipher E. When the simulator makes q_S queries, it is easy to see: the probability that a forward ideal cipher query $E(x \oplus y, y)$

responds with x is at most $1/(2^n - q_S)$; the probability that a backward query $E^{-1}(x \oplus y, y)$ responds with x is at most $1/(2^n - q_S)$. Thus, the probability that the simulator pinpoints $E(x \oplus y, y) = x$ is at most $q_S/(2^n - q_S)$, and the attack advantage is at least $1 - q_S/(2^n - q_S)$.

It is also easy to see that, the above attack essentially leverages a relation that is evasive [8] w.r.t. an ideal cipher.

4 Seq-Indifferentiability of EM2P$_4$

This section proves seq-indifferentiability for the 4-round $\text{EM2P}_4^{\mathbf{p}_1,\mathbf{p}_2}$, the variant of single-key IEM using two permutations $\mathbf{p}_1, \mathbf{p}_2$, as shown in Fig. 1.

Theorem 1. *Assume that* $\mathbf{p}_1$ *and* $\mathbf{p}_2$ *are two independent random permutations. Then, the 4-round single-key Even-Mansour scheme* $EM2P_4^{\mathbf{p}_1,\mathbf{p}_2}$ *defined as*

$$EM2P_4^{\mathbf{p}_1,\mathbf{p}_2}(k,u) := k \oplus \mathbf{p}_1(k \oplus \mathbf{p}_2(k \oplus \mathbf{p}_2(k \oplus \mathbf{p}_1(k \oplus u))))$$

is strongly and statistically $(q, \sigma, t, \varepsilon)$*-seq-indifferentiable from an ideal cipher* E*, where* $\sigma = q^2$, $t = O(q^2)$, *and* $\varepsilon \le \frac{20q^3+29q^4}{2^n} = O(\frac{q^4}{2^n})$ *(assuming* $q+2q^2 \le 2^n/2$*).*

To prove Theorem 1, we first describe our simulator in Sect. 4.1.

4.1 Simulator of EM2P$_4$

Randomness and Interfaces. The simulator $\mathcal{S}$ offers four interfaces P_1, P_1^{-1}, P_2 and P_2^{-1} to the distinguisher for querying the internal permutations, and the input of the query is any element in the set $\{0,1\}^n$.

To handily describe lazying sampling during simulation, we follow previous works [1,11–14,16,21,29] and make the randomness used by $\mathcal{S}$ explicit through two random permutations p_1 and p_2. Namely, $\mathcal{S}$ queries $\mathbf{p}_1$ and $\mathbf{p}_2$ (see below for concreteness) to have a random value z rather than straightforwardly sampling $z \xleftarrow{\$} \{0,1\}^n$. Let $\mathcal{P} = (\mathbf{p}_1, \mathbf{p}_2)$. We denote by $\mathcal{S}^{E,\mathcal{P}}$ the simulator that emulates the primitives for E and queries $\mathbf{p}_1$ and $\mathbf{p}_2$ for necessary random values. As argued in [1], explicit randomness is merely an equivalent formalism of lazying sampling.

Maintaining Query Records. To keep track of previously answered permutation queries, $\mathcal{S}$ internally maintains two sets Π_1 and Π_2 that have entries in the form of $(i, x, y) \in \{1,2\} \times \{0,1\}^n \times \{0,1\}^n$. $\mathcal{S}$ will ensure that for any $x \in \{0,1\}^n$ and $i \in \{1,2\}$, there is at most one $y \in \{0,1\}^n$ such that $(i, x, y) \in \Pi_i$, and vice versa. As will be elaborated later, $\mathcal{S}$ aborts whenever it fails to ensure such consistency. By this, the sets Π_1 and Π_2 will define two partial permutations, and we denote by $domain(\Pi_i)$ ($range(\Pi_i)$, resp.) the (time-dependent) set of all n-bit values x (y, resp.) satisfying $\exists z \in \{0,1\}^n$ s.t. $(i, x, z) \in \Pi_i$ ($(i, z, y) \in \Pi_i$, resp.). We further denote by $\Pi_i(x)$ ($\Pi_i^{-1}(y)$, resp.) the corresponding value of z.

Simulation Strategy. Upon the distinguisher D querying $P_i(x)$ ($P_i^{-1}(y)$, resp.), $\mathcal{S}$ checks if $x \in \Pi_1$ ($y \in \Pi_1^{-1}$, resp.), and answers with $\Pi_1(x)$ ($\Pi_1^{-1}(y)$, resp.) when it is the case. Otherwise, the query is new, and $\mathcal{S}$ queries $\mathbf{p}_i$ for $y \leftarrow \mathbf{p}_i(x)$ ($x \leftarrow \mathbf{p}_i^{-1}(y)$, resp.). If $y \notin range(\Pi_i)$, $\mathcal{S}$ adds the record (i, x, y) to the set Π_i; otherwise, $\mathcal{S}$ aborts to avoid inconsistency in Π_i (as mentioned). Then, when $i = 1$, $\mathcal{S}$ simply answers with x (y, resp.); when $i = 2$, $\mathcal{S}$ completes the partial chains formed by this new record $(2, x, y)$ and previously created records in Π_2 (as mentioned in the Introduction).

In detail, when the new adversarial query is to $P_2(x)$ and $\mathcal{S}$ adds a new record $(2, x, y)$ to Π_2, $\mathcal{S}$ considers all pairs of triples $\big((2, x, y), (2, x', y')\big) \in (\Pi_2)^2$ (including the pair $\big((2, x, y), (2, x, y)\big)$) *and* all $\big((2, x', y'), (2, x, y)\big) \in (\Pi_2)^2$ (with $x' \neq x$ for distinction). Then,

- For every pair $\big((2, x, y), (2, x', y')\big) \in (\Pi_2)^2$, $\mathcal{S}$ computes $k \leftarrow y \oplus x'$ and $x_4 \leftarrow y' \oplus k$. $\mathcal{S}$ then internally invokes P_1 to have $y_4 \leftarrow P_1(x_4)$ and $v \leftarrow y_4 \oplus k$. $\mathcal{S}$ then queries the ideal cipher to have $u \leftarrow E^{-1}(k, v)$, and further computes $x_1 \leftarrow u \oplus k$ and $y_1 \leftarrow x \oplus k$. Finally, if $x_1 \notin domain(\Pi_1)$ and $y_1 \notin range(\Pi_1)$, $\mathcal{S}$ adds the record (i, x, y) to the set Π_i, to complete the 4-chain $\big((1, x_1, y_1), (2, x, y), (2, x', y'), (1, x_4, y_4)\big)$; otherwise, $\mathcal{S}$ aborts to avoid inconsistency. The record $(1, x_1, y_1)$ is called *adapted*, since it is created to "link" the simulated computation. In our pseudocode, this process is implemented as a procedure $Complete^-$;
- For every pair $\big((2, x', y'), (2, x, y)\big) \in (\Pi_2)^2$, $\mathcal{S}$ computes $k \leftarrow y' \oplus x$, $y_1 \leftarrow x' \oplus k$, $x_1 \leftarrow P_1^{-1}(y_1)$, $u \leftarrow x_1 \oplus k$; $v \leftarrow E(k, u)$, $y_4 \leftarrow v \oplus k$ and $x_4 \leftarrow y \oplus k$. $\mathcal{S}$ finally adds the adapted record $(1, x_4, y_4)$ to Π_1 when $x_4 \notin domain(\Pi_1)$ and $y_4 \notin range(\Pi_1)$, to complete $\big((1, x_1, y_1), (2, x', y'), (2, x, y), (1, x_4, y_4)\big)$, or aborts otherwise. In our pseudocode, this process is implemented as a procedure $Complete^+$.

Upon D querying $P_2^{-1}(y)$, the simulator actions are similar to $P_2(x)$ by symmetry. Our strategy is formally described via pseudocode in the next paragraph.

Simulator in Pseudocode

1: **Simulator** $\mathcal{S}^{E,\mathcal{P}}$
2: **Variables:** Sets Π_1, Π_2, X_{Dom}, and X_{Rng}, all initially empty

3: **public procedure** $P_1(x)$
4:   **if** $x \notin domain(\Pi_1)$ **then**
5:     $y \leftarrow \mathbf{p}_1(x)$
6:     **if** $\Pi_1^{-1}(y) \neq \perp$ **then abort**
7:     **if** $y \in X_{Rng}$ **then abort**
8:     $\Pi_1 \leftarrow \Pi_1 \cup \{(1, x, y)\}$
9:   **return** $\Pi_1(x)$

10: **public procedure** $P_1^{-1}(y)$
11:   **if** $y \notin range(\Pi_1)$ **then**
12:     $x \leftarrow \mathbf{p}_1^{-1}(y)$
13:     **if** $\Pi_1(x) \neq \perp$ **then abort**
14:     **if** $x \in X_{Dom}$ **then abort**
15:     $\Pi_1 \leftarrow \Pi_1 \cup \{(1, x, y)\}$
16:   **return** $\Pi_1^{-1}(y)$

```
17: public procedure P_2(x)
18:   if x ∉ domain(Π_2) then
19:     y ← p_2(x)
20:     Π_2 ← Π_2 ∪ {(2, x, y)}
21:     forall (2, x', y') ∈ Π_2 do
22:       // 3^+ chain
23:       k ← y' ⊕ x
24:       if y ⊕ k ∈ domain(Π_1)
25:         then abort
26:       X_Dom ← X_Dom ∪ {y ⊕ k}
27:       X_Rng ← X_Rng ∪ {x' ⊕ k}
28:       // 2^+ chain
29:       k ← y ⊕ x'
30:       if x ⊕ k ∈ range(Π_1)
31:         then abort
32:       if ∃(2, x'', y'') ∈ Π_2 :
            x' ⊕ y' ⊕ x = x ⊕ y ⊕ x''
          then abort
33:       X_Dom ← X_Dom ∪ {y' ⊕ k}
34:       X_Rng ← X_Rng ∪ {x ⊕ k}
35:     forall (2, x', y') ∈ Π_2
          s.t. x' ≠ x do
36:       k ← x ⊕ y'
37:       Complete^+(x', k)
38:     forall (2, x', y') ∈ Π_2 do
39:       k ← y ⊕ x'
40:       Complete^-(y', k)
41:   // Clear the pending sets
42:   X_Dom ← ∅, X_Rng ← ∅
43:   return Π_2(x)

44: public procedure P_2^{-1}(y)
45:   if y ∉ range(Π_2) then
46:     x ← p_2^{-1}(y)
47:     Π_2 ← Π_2 ∪ {(2, x, y)}
48:     forall (2, x', y') ∈ Π_2 do
49:       // 2^- chain
50:       k ← y ⊕ x'
51:       if x ⊕ k ∈ range(Π_1)
52:         then abort
53:       X_Dom ← X_Dom ∪ {y' ⊕ k}
54:       X_Rng ← X_Rng ∪ {x ⊕ k}
55:       // 3^- chain
56:       k ← y' ⊕ x
57:       if y ⊕ k ∈ domain(Π_1)
58:         then abort
59:       if ∃(2, x'', y'') ∈ Π_2 :
            y' ⊕ x' ⊕ y = y'' ⊕ x ⊕ y
          then abort
60:       X_Dom ← X_Dom ∪ {y ⊕ k}
61:       X_Rng ← X_Rng ∪ {x' ⊕ k}
62:     forall (2, x', y') ∈ Π_2
          s.t. x' ≠ x do
63:       k ← y ⊕ x'
64:       Complete^-(y', k)
65:     forall (2, x', y') ∈ Π_2 do
66:       k ← x ⊕ y'
67:       Complete^+(x', k)
68:   // Clear the pending sets
69:   X_Dom ← ∅, X_Rng ← ∅
70:   return Π_2^{-1}(y)

71: private procedure Complete^+(x_2, k)
72:   y_1 ← x_2 ⊕ k, x_1 ← P_1^{-1}(y_1)
73:   u ← x_1 ⊕ k, v ← E(k, u)
74:   y_4 ← v ⊕ k
75:   y_2 ← P_2(x)
76:   x_3 ← y_2 ⊕ k, y_3 ← P_2(x_3)
77:   x_4 ← y_3 ⊕ k
78:   if x_4 ∈ domain(Π_1) then abort
79:   if y_4 ∈ range(Π_1) then abort
80:   if y_4 ∈ X_Rng then abort
81:   Π_1 ← Π_1 ∪ {(1, x_4, y_4)}

82: private procedure Complete^-(y_3, k)
83:   x_4 ← y_3 ⊕ k, y_4 ← P_1(x_4)
84:   v ← y_4 ⊕ k, u ← E^{-1}(k, v)
85:   x_1 ← u ⊕ k
86:   x_3 ← P_2^{-1}(y_3)
87:   y_2 ← x_3 ⊕ k, x_2 ← P_2^{-1}(y_2)
88:   y_1 ← x_2 ⊕ k
89:   if x_1 ∈ domain(Π_1) then abort
90:   if y_1 ∈ range(Π_1) then abort
91:   if x_1 ∈ X_Dom then abort
92:   Π_1 ← Π_1 ∪ {(1, x_1, y_1)}
```

We identify a number of bad events during the simulation and coded them in $\mathcal{S}$. The occurrence of such events indicates potential abortions due to adaptations in future. In detail, before calling $Complete^+$ and $Complete^-$, $\mathcal{S}$ creates two sets X_{Rng} and X_{Dom} for the values that will be used in subsequent adaptations: for every $x \in X_{Dom}$, $\mathcal{S}$ will create an adapted record of the form $(1, x, \star)$; for every

$y \in X_{Rng}$, $\mathcal{S}$ will create an adapted record of the form $(1, \star, y)$. Therefore, collisions among values in X_{Dom} and $domain(\Pi_1)$ (resp., X_{Rng} and $range(\Pi_1)$) already indicate the failure of some future adaptations. Thus, once such events occur, $\mathcal{S}$ also aborts to terminate the doomed execution.

4.2 The Indistinguishability Proof

It remains to establish two claims for any distinguisher D: (a) the simulator $\mathcal{S}^{E,\mathcal{P}}$ has bounded complexity; (b) the real and ideal worlds are indistinguishable. To this end, we introduce a helper intermediate system in the next paragraph. Then, subsequent paragraphs establish claims (a) and (b) in turn.

Intermediate System. As shown in Fig. 3, we use three systems for the proof. In detail, let $\Sigma_1(E, \mathcal{S}^{E,\mathcal{P}})$ be the system capturing the ideal world, where E is an ideal cipher and $\mathbf{p}_1$, $\mathbf{p}_2$ are independent random permutations; and let $\Sigma_3(\mathrm{EM2P}_4^{\mathcal{P}}, \mathcal{P})$ be the real world.

We follow [8,30] and introduce $\Sigma_2(\mathrm{EM2P}_4^{\mathcal{S}^{E,\mathcal{P}}}, \mathcal{S}^{E,\mathcal{P}})$ as an intermediate system, which is modified from Σ_1 by replacing E with an $\mathrm{EM2P}_4$ instance that queries the simulator to evaluate.

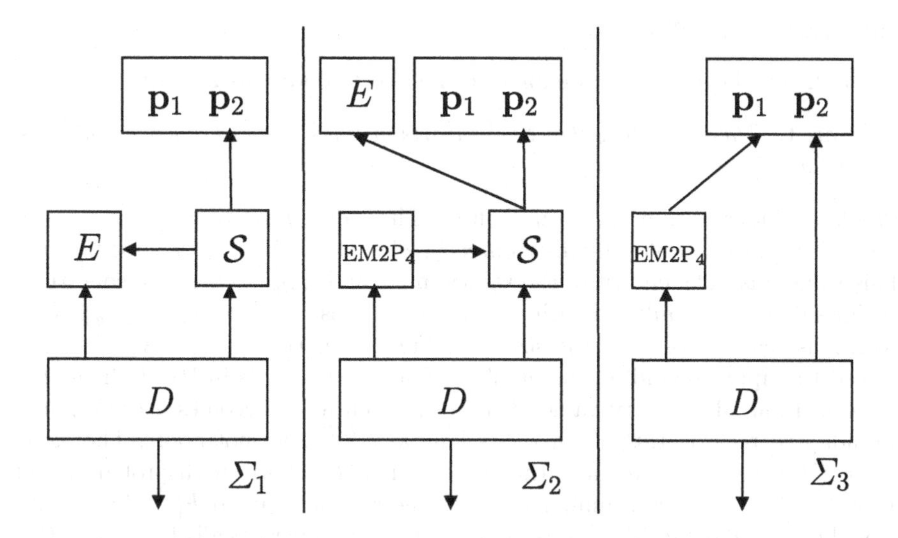

Fig. 3. Systems used in the proof.

Then, consider a fixed sequential distinguisher D of total oracle query cost at most q. The remaining key points are as follows.

Complexity of $\mathcal{S}^{E,\mathcal{P}}$. As the key observation, $\mathcal{S}^{E,\mathcal{P}}$ never adds records to Π_2 internally. Thus, $|\Pi_2|$ increases by 1 after each adversarial query, and thus $|\Pi_2| \le q$. By this, the number of detected chains $\big((2, x_2, y_2), (2, x_2', y_2')\big) \in (\Pi_2)^2$

is at most q^2. This also means $\mathcal{S}^{E,\mathcal{P}}$ makes at most q^2 queries to E, since such a query only appears during completing a detected chain. For each detected chain, $\mathcal{S}^{E,\mathcal{P}}$ adds at most 2 records to Π_1. Moreover, $|\Pi_1|$ may also increase by q due to D straightforwardly querying P_1 or P_1^{-1}. It thus holds $|\Pi_1| \leq q+2q^2$. Finally, the running time is dominated by completing chains, and is thus $O(q^2)$.

Indistinguishability of Σ_1, Σ_2 and Σ_3. First, we need to show that the two simulated permutations are consistent, which is of course necessary for indistinguishability. Note that the occurrence of such inconsistency would particularly render $\mathcal{S}^{E,\mathcal{P}}$ abort. Therefore, via a fine-grained analysis of the various involved values, we establish an upper bound on the probability that $\mathcal{S}^{E,\mathcal{P}}$ aborts.

4.3 Abort Probability of $\mathcal{S}^{E,\mathcal{P}}$

As discussed in Sect. 4.2, when the total oracle query cost of D does not exceed q, it holds $|\Pi_2| \leq q$, and the total number of detected chains $((2, x_2, y_2), (2, x_2', y_2')) \in (\Pi_2)^2$ is at most q^2. The latter means:

(i) the number of adapted records in Π_1 is at most q;
(ii) the number of calls to P_1 and P_1^{-1} is at most $q+q^2$ in total (which is the number of detected chains plus the number of adversarial queries to P_1 and P_1^{-1});
(iii) $|X_{Dom}| \leq q^2$, $|X_{Rng}| \leq q^2$.

With the above bounds, we analyze the abort conditions in turn.

Lemma 1. *The probability that $\mathcal{S}^{E,\mathcal{P}}$ aborts at lines 6, 7, 13 and 14 is at most $(2q^3 + 2q^4)/2^n$.*

Proof. Consider lines 6 and 7 in P_1 first. The value $y \leftarrow \mathbf{p}_1(x)$ newly "downloaded" from $\mathbf{p}_1$ is uniformly distributed in $2^n - |\Pi_1| \geq 2^n - q - 2q^2$ possibilities. This value y is independent of the values in Π_1 and X_{Rng}. Thus, the conditions for lines 6 and 7 are fulfilled with probability at most $|range(\Pi_1) \cup X_{Rng}|$. However, it is easy to see that, the size of the union set $range(\Pi_1) \cup X_{Rng}$ cannot exceed the upper bound on the number of adapted records in Π_1 at the end of the execution, since every value y' in X_{Rng} eventually becomes a corresponding adapted record $(1, x', y')$ in Π_1 as long as $\mathcal{S}^{E,\mathcal{P}}$ does not abort. Therefore, $|range(\Pi_1) \cup X_{Rng}| \leq q^2$, and thus each call to P_1 aborts with probability at most $q^2/(2^n - q - 2q^2)$. Similarly by symmetry, each call to P_1^{-1} aborts with probability at most $q^2/(2^n - q - 2q^2)$. Since the number of calls to P_1 and P_1^{-1} is at most $q + q^2$ in total, the probability that $\mathcal{S}^{E,\mathcal{P}}$ aborts at lines 6, 7, 13 and 14 is at most

$$(q+q^2) \cdot \frac{q^2}{2^n - (q+2q^2)} \leq \frac{2q^3 + 2q^4}{2^n},$$

assuming $q + 2q^2 \leq 2^n/2$. □

Next, we analyze the probability of the "early abort" conditions in P_2 and P_2^{-1}.

Lemma 2. *The probability that $\mathcal{S}^{E,\mathcal{P}}$ aborts at lines 25, 31 and 32 in the procedure P_2 (resp., lines 52, 58 and 59 in the procedure P_2^{-1}) is at most $(6q^3 + 8q^4)/2^n$.*

Proof. Consider the conditions in P_2 first. The value $y \leftarrow \mathbf{p}_1(x)$ newly "downloaded" from $\mathbf{p}_1$ is uniformly distributed in $2^n - |\Pi_1| \geq 2^n - q - 2q^2$ possibilities. Moreover, this value y is independent of the values in Π_1, Π_2 and X_{Rng}.

With the above in mind, we analyze the conditions in turn. First, consider line 25. For every detected partial chain $\big((2, x', y'), (2, x, y)\big)$, the condition $y \oplus k \in domain(\Pi_1)$ translates into $y \oplus y' \oplus x \in domain(\Pi_1)$, which holds with probability at most $|domain(\Pi_1)|/(2^n - q - 2q^2) \leq (q + 2q^2)/(2^n - q - 2q^2)$ (since $|\Pi_1| \leq q + 2q^2$).

The arguments for the remaining conditions are similar: since y is uniform,

- for every detected partial chain $\big((2, x, y), (2, x', y')\big)$, the condition $x \oplus k \in range(\Pi_1) \Leftrightarrow x \oplus y \oplus x' \in range(\Pi_1)$ is fulfilled with probability at most $(q + 2q^2)/(2^n - q)$ (again using $|\Pi_1| \leq q + 2q^2$);
- for every detected partial chain $\big((2, x', y'), (2, x, y)\big)$, the probability to have $x' \oplus y' \oplus x = x \oplus y \oplus x''$ for some $(2, x'', y'') \in \Pi_2$ is at most $q/(2^n - q - 2q^2)$(since $|\Pi_2| \leq q$).

Since the number of detected partial chains $\big((2, x', y'), (2, x, y)\big)$ is at most $|\Pi_2| \leq q$, the probability that a single query or call to $P_2(x)$ aborts at lines 25, 31 and 32 is at most

$$q \times \Big(\frac{q + 2q^2}{2^n - q - 2q^2} + \frac{q + 2q^2}{2^n - q - 2q^2} + \frac{q}{2^n - q - 2q^2}\Big) \leq \frac{3q^2 + 4q^3}{2^n - q - 2q^2} \leq \frac{6q^2 + 8q^3}{2^n},$$

assuming $q + 2q^2 \leq 2^n/2$.

The above complete the analysis for P_2. The analysis for lines 52, 58 and 59 in P_2^{-1} is similar by symmetry, yielding the same bound. Summing over the at most q queries or calls to P_2 and P_2^{-1}, we reach the claimed bound $q(6q^2 + 8q^3)/2^n \leq 6q^3 + 8q^4/2^n$. □

For the subsequent argument, we introduce a bad event BadE_ℓ regarding the ideal cipher queries made during $\mathcal{S}$ processing the ℓ-th adversarial query to $P_2(x^{(\ell)})$ or $P_2^{-1}(y^{(\ell)})$. Formally, BadE_ℓ occurs if:

- In this period, during a call to $Complete^+(x_2, k)$ in this period, a query to $v \leftarrow E(k, u)$ is made, and the response satisfies $v \oplus k \in range(\Pi_1)$ or $v \oplus k \in X_{Rng}$; or
- In this period, during a call to $Complete^-(y_3, k)$ in this period, a query to $u \leftarrow E^{-1}(k, v)$ is made, and the response satisfies $u \oplus k \in domain(\Pi_1)$ or $u \oplus k \in X_{Dom}$.

To analyze BadE_ℓ, we need a helper lemma as follows.

Lemma 3. *Inside every call to $Complete^+$, resp. $Complete^-$, the ideal cipher query $E(k, u)$, resp. $E^{-1}(k, v)$, is fresh. Namely, the simulator $\mathcal{S}^{E,\mathcal{P}}$ never made this query $E(k, u)$, resp. $E^{-1}(k, v)$, before.*

Proof. Assume that this does not hold. Then this means that such a query previously occurred when completing another chain. By the construction of $EM2P_4$ and our simulator, this means right after the call to $Complete^+$ or $Complete^-$ that queried $E(k,u)$, all the four corresponding round inputs/outputs $(1,x_1,y_1)$, $(2,x_2,y_2)$, $(2,x_3,y_3)$ and $(1,x_4,y_4)$ with $k = u \oplus x_1 = y_1 \oplus x_2 = ... = y_4 \oplus E(k,u)$ have been in Π_1 and Π_2. This in particular includes the query to P_2/P_2^{-1} that was purported to incur the current call to $Complete^+/Complete^-$. But since the query to P_2/P_2^{-1} is not new, this contradicts the construction of our simulator. Therefore, the ideal cipher query must be fresh. □

The probability of BadE_ℓ is then bounded as follows.

Lemma 4. *In each call to $Complete^+$ or $Complete^-$, the probability that BadE_ℓ occurs is at most $2(q+2q^2)/2^n$.*

Proof. We first analyze the abort probabilities of calls to $Complete^+$ and $Complete^-$. Consider a call to $Complete^+(x_2,k)$ first. By Lemma 3, the ideal cipher query $E(k,u) \to v$ made inside this call is new. Since $\mathcal{S}^{E,\mathcal{P}}$ makes at most q^2 queries to E, the value v is uniform in at least $2^n - q^2$ possibilities. Furthermore, v is independent of the values in X_{Rng} and $range(\Pi_1)$. Therefore,

$$\Pr[v \oplus k \in (X_{Rng} \cup range(\Pi_1))] \leq \frac{|X_{Rng} \cup range(\Pi_1)|}{2^n - q^2}.$$

It is easy to see that $|X_{Rng} \cup range(\Pi_1)|$ cannot exceed the upper bound $q+2q^2$ on $|\Pi_1|$ at the end of the execution. Therefore, the probability to have BadE_ℓ in a call to $Complete^+(x_2,k)$ is at most $(q+2q^2)/(2^n - q^2)$.

The analysis of $Complete^-(y_3,k)$ is similar by symmetry, yielding the same bound $(q+2q^2)/(2^n - q^2)$. Assuming $q^2 \leq 2^n/2$, we obtain the claim. □

Then, we address the abort probability due to adaptations in $Complete^+$ and $Complete^-$ call.

Lemma 5. *The probability that $\mathcal{S}^{E,\mathcal{P}}$ aborts at lines 78, 79, and 80; 89, 90, and 91 is at most $(2q^3 + 4q^4)/2^n$.*

Proof. Noting that $Complete^+$ and $Complete^-$ are only called during processing adversarial queries to $P_2(x)/P_2^{-1}(y)$, we quickly sketch the process of the latter. Wlog we focus on processing a query $P_2(x)$, as the case of $P_2^{-1}(y)$ is similar by symmetry.

Upon D making the ℓ-th query to $P_2(x^{(\ell)})$, $\mathcal{S}^{E,\mathcal{P}}$ first "downloads" the response $y^{(\ell)} \leftarrow \mathbf{p}_2(x)$ from $\mathbf{p}_2$ and then detects a number of partial chains as follows:

$$\begin{aligned}
2^+\text{chains}: &\ ((2,x^{(1)},y^{(1)}),(2,x^{(\ell)},y^{(\ell)})), ..., ((2,x^{(\ell-1)},y^{(\ell-1)}),(2,x^{(\ell)},y^{(\ell)})),\\
3^+\text{chains}: &\ ((2,x^{(\ell)},y^{(\ell)}),(2,x^{(1)},y^{(1)})), ..., ((2,x^{(\ell)},y^{(\ell)}),(2,x^{(\ell-1)},y^{(\ell-1)})),\\
&\ ((2,x^{(\ell)},y^{(\ell)}),(2,x^{(\ell)},y^{(\ell)})),
\end{aligned}$$

where $(2, x^{(1)}, y^{(1)}), ..., (2, x^{(\ell-1)}, y^{(\ell-1)}) \in \Pi_2$ are the triples created due to the earlier $\ell - 1$ adversarial queries to P_2 or P_2^{-1}. For conceptual convenience we refer to the former type of chains as 2^+*chains* and the latter as 3^+*chains*. $\mathcal{S}$ then proceeds in two steps:

- First, completes the 3^+chains in turn, making a number of calls to $Complete^+$;
- Second, completes the 2^+chains in turn, making a number of calls to $Complete^-$.

We proceed to argue that, during processing the ℓ-th query to $P_2(x^{(\ell)})$, the above calls to $Complete^+/Complete^-$ abort with probability at most $\big(2(2\ell - 1)(q + 2q^2)\big)/2^n$ in total.

To this end, consider the j-th 3^+chain $\big((2, x^{(j)}, y^{(j)}), (2, x^{(\ell)}, y^{(\ell)})\big)$. Let $k^{(j)} = y^{(j)} \oplus x^{(\ell)}$ and $x_4^{(j)} = k^{(j)} \oplus y^{(\ell)}$. Since $\mathcal{S}$ did not abort at line 25, it holds $x_4^{(j)} \notin domain(\Pi_1)$ right after $\mathcal{S}$ "downloads" $y^{(\ell)} \leftarrow \mathbf{p}_2(x)$. We then show that $x_4^{(j)} \notin domain(\Pi_1)$ is kept till the call to $Complete^+(x^{(j)}, k^{(j)})$ adapts by adding $(1, x_4^{(j)}, y_4^{(j)})$ to Π_1, so that lines 78, 79 and 80 won't cause abort.

- First, for any 3^+chain $\big((2, x^{(j')}, y^{(j')}), (2, x^{(\ell)}, y^{(\ell)})\big)$ completed before the chain $\big((2, x^{(j)}, y^{(j)}), (2, x^{(\ell)}, y^{(\ell)})\big)$, its adaptation cannot add $(1, x_4^{(j)}, \star)$ to Π_1, since its adapted pair is of the form $x_4^{(j')} = y^{(j')} \oplus x^{(\ell)} \oplus y^{(\ell)} \neq x_4^{(j)}$;
- Second, internal queries to $P_1^{-1}(y_1) \to x_1$ (with $x_1 \leftarrow \mathbf{p}_1^{-1}(y_1)$) during this period cannot add $(1, x_4^{(j)}, \star)$ to Π_1, since $x_4^{(j)}$ was added to X_{Dom} and since $x_1 \notin X_{Dom}$ (otherwise $\mathcal{S}$ has aborted at line 7).

Thus, line 78 won't cause abort at all. On the other hand, with $\neg\mathsf{BadE}_\ell$ as the condition, $y_4^{(j)} \notin (range(\Pi_1) \cup X_{Rng})$ necessarily holds. Therefore, in the call to $Complete^+(x^{(j)}, k^{(j)})$ adapts, lines 79 and 80 will not cause abort. The above completes the argument for $Complete^+$ calls due to 3^+chains.

We then address 2^+chains by considering the j-th $\big((2, x^{(\ell)}, y^{(\ell)}), (2, x^{(j)}, y^{(j)})\big)$. Let $k^{(j)} = y^{(j)} \oplus x^{(\ell)}$ and $x_4^{(j)} = k^{(j)} \oplus y^{(\ell)}$. Since $\mathcal{S}$ did not abort at line 25, it holds $x_4^{(j)} \notin domain(\Pi_1)$ right after $\mathcal{S}$ downloads $y^{(\ell)} \leftarrow \mathbf{p}_2(x)$. We then show that $x_4^{(j)} \notin domain(\Pi_1)$ is kept till the call to $Complete^+(x^{(j)}, k^{(j)})$ adapts by adding $(1, x_4^{(j)}, y_4^{(j)})$ to Π_1, so that lines 78, 79 and 80 won't cause abort.

Therefore, during processing the ℓ-th query to $P_2(x^{(\ell)})$ or $P_2^{-1}(y^{(\ell)})$, the probability that $\mathcal{S}$ aborts in each call to $Complete^+$ or $Complete^-$ is equal to $\Pr[\mathsf{BadE}_\ell]$, which does not exceed $2(q + 2q^2)/2^n$ by Lemma 4.

To summarize, recall that the total number of detected partial chains/calls to $Complete^+$ or $Complete^-$ is bounded by $|\Pi_2|^2 \le q^2$. Therefore, the probability that $\mathcal{S}^{E,\mathcal{P}}$ aborts at lines 78, 79, and 80; 89, 90, and 91 is bounded by

$$q^2 \times \left(\frac{2(q + 2q^2)}{2^n}\right) \le \frac{2q^3 + 4q^4}{2^n},$$

as claimed. □

Lemma 6. *The probability that $\mathcal{S}^{E,\mathcal{P}}$ aborts in D^{Σ_2} is at most $(10q^3+14q^4)/2^n$.*

Proof. Gathering Lemmas 1, 2 and 5 yields the bound.

4.4 Indistinguishability of Σ_1 and Σ_3

A random tuple $(E,\mathcal{P})$ is *good*, if $\mathcal{S}^{E,\mathcal{P}}$ does not abort in $D^{\Sigma_2(E,\mathcal{P})}$. It can be proved that, for any good tuple $(E,\mathcal{P})$, the transcript of the interaction of D with $\Sigma_1(E,\mathcal{P})$ and $\Sigma_2(E,\mathcal{P})$ is exactly the same. This means the gap between Σ_1 and Σ_2 is the abort probability.

Σ_1 to Σ_2.

Lemma 7. *For any distinguisher D of total oracle query cost at most q, it holds*

$$\left|\Pr[D^{\Sigma_1(E,\mathcal{S}^{E,\mathcal{P}})}=1]-\Pr[D^{\Sigma_2(EM2P_4^{\mathcal{S}^{E,\mathcal{P}}},\mathcal{S}^{E,\mathcal{P}})}=1]\right|\leq\frac{10q^3+14q^4}{2^n}.$$

Proof. Note that in Σ_1 and Σ_2, the sequential distinguisher D necessarily first queries $\mathcal{S}$ and then E (in Σ_1) or $\mathrm{EM2P}_4$ (in Σ_2) only. Thus, the transcript of the first phase of the interaction (i.e., for the queries of D to $\mathcal{S}^{E,\mathcal{P}}$) are clearly the same, since in both cases they are answered by $\mathcal{S}$ using the same randomness $(E,\mathcal{P})$. For the second phase of the interaction (i.e., queries of D to its left oracle), it directly follows from the adaptation mechanism. Hence, the transcripts of the interaction of D with $\Sigma_1(E,\mathcal{P})$ and $\Sigma_2(E,\mathcal{P})$ are the same for any good tuple $(E,\mathcal{P})$. Further using Lemma 6 yields

$$\begin{aligned}&\left|\Pr[D^{\Sigma_1(E,\mathcal{S}^{E,\mathcal{P}})}=1]-\Pr[D^{\Sigma_2(\mathrm{EM2P}_4^{\mathcal{S}^{E,\mathcal{P}}},\mathcal{S}^{E,\mathcal{P}})}=1]\right|\\ \leq\ &\Pr[(E,\mathcal{P})\ \textit{is bad}]\leq\frac{10q^3+14q^4}{2^n},\end{aligned}$$

as claimed. □

Σ_2 to Σ_3: Randomness Mapping. We now bound the gap between Σ_2 and Σ_3. Following [8,11], the technique is the randomness mapping argument.

We define a map Λ mapping pairs $(E,\mathcal{P})$ either to the special symbol $\perp$ when $(E,\mathcal{P})$ is bad, or to a pair of *partial permutations* $\mathcal{P}'=(\mathbf{p}'_1,\mathbf{p}'_2)$ when $(E,\mathcal{P})$ is good. A partial permutation is functions $\mathbf{p}'_i\colon\{0,1\}^n\to\{0,1\}^n\cup\{*\}$ and $\mathbf{p}_i'^{-1}\colon\{0,1\}^n\to\{0,1\}^n\cup\{*\}$, such that for all $x,y\in\{0,1\}^n$, $\mathbf{p}'_i(x)=y\neq *\Leftrightarrow\mathbf{p}_i'^{-1}(y)=x\neq *$.

Then map Λ is defined for good pairs $(E,\mathcal{P})$ as follows: run $D^{\Sigma_2(E,\mathcal{P})}$, and consider the tables Π_i of the simulator at the end of the execution: then fill all undefined entries of the Π_i's with the special symbol $*$. The result is exactly $\Lambda(E,\mathcal{P})$. Since for a good pair $(E,\mathcal{P})$, the simulator never overwrite an entry in its tables, it follows that $\Lambda(E,\mathcal{P})$ is a pair of partial permutations as just defined above. We say that a pair of partial permutations $\mathcal{P}'=(\mathbf{p}'_1,\mathbf{p}'_2)$ is good if it has a good preimage by Λ. Then, we say that a pair of permutations $\mathcal{P}$ extends a pair of partial permutations $\mathcal{P}' = (\mathbf{p}'_1, \mathbf{p}'_2)$, denoted $\mathcal{P} \vdash \mathcal{P}'$, if for each $i = 1, 2$, $\mathbf{p}_i$ and $\mathbf{p}'_i$ agree on all entries such that $\mathbf{p}'_i(x) \neq *$ and $\mathbf{p}'^{-1}_i(y) \neq *$.

By definition of the randomness mapping, for any good tuple of partial permutations $\mathcal{P}'$, the outputs of $D^{\Sigma_2(E,\mathcal{P})}$ and $D^{\Sigma_3(\mathcal{P})}$ are equal for any pair $(E, \mathcal{P})$ such that $\Lambda(E, \mathcal{P}) = \mathcal{P}'$ and any tuple of permutations $\mathcal{P}$ such that $\mathcal{P} \vdash \mathcal{P}'$. Let Ω_1 be the set of partial permutations $\mathcal{P}'$ such that $D^{\Sigma_2(E,\mathcal{P})}$ output 1 for any pair $(E, \mathcal{P})$ such that $\Lambda(E, \mathcal{P}) = \mathcal{P}'$. Then, we have the following ratio.

Lemma 8. *Consider a fixed distinguisher D with total oracle query cost at most q. Then, for any $\mathcal{P}' = (\mathbf{p}'_1, \mathbf{p}'_2) \in \Omega_1$, it holds*

$$\frac{\Pr[\mathcal{P} \vdash \mathcal{P}']}{\Pr[\Lambda(E, \mathcal{P}) = \mathcal{P}']} \geq 1 - \frac{q^4}{2^n}.$$

Proof. Since the number of "non-empty" entries $\mathbf{p}'_1(x) \neq *$ and $\mathbf{p}'_2(x) \neq *$ are $|\Pi_1|$ and $|\Pi_2|$ respectively, we have

$$\Pr[\mathcal{P} \vdash \mathcal{P}'] = \Big(\prod_{j=0}^{|\Pi_1|-1} \frac{1}{2^n - j}\Big)\Big(\prod_{j=0}^{|\Pi_2|-1} \frac{1}{2^n - j}\Big).$$

Fix any good preimage $(\widetilde{E}, \widetilde{\mathcal{P}})$ of $\mathcal{P}'$. One can check that for any tuple $(E, \mathcal{P})$, $\Lambda(E, \mathcal{P}) = \mathcal{P}'$ *iff* the transcript of the interaction of $\mathcal{S}$ with $(E, \mathcal{P})$ in $D^{\Sigma_2(E,\mathcal{P})}$ is the same as the transcript of the interaction of $\mathcal{S}$ with $(\widetilde{E}, \widetilde{\mathcal{P}})$ in $D^{\Sigma_2(\widetilde{E},\widetilde{\mathcal{P}})}$.

Assume that during the Σ_2 execution $D^{\Sigma_2(\mathrm{EM2P}_4^{\mathcal{S}^{E,\mathcal{P}}}, \mathcal{S}^{E,\mathcal{P}})}$, $\mathcal{S}$ makes q_e, q_1 and q_2 queries to E, $\mathbf{p}_1$ and $\mathbf{p}_2$ respectively. Then,

$$\Pr[\Lambda(E, \mathcal{P}) = \mathcal{P}'] \leq \Big(\prod_{j=0}^{q_e-1} \frac{1}{2^n - j}\Big)\Big(\prod_{j=0}^{q_1-1} \frac{1}{2^n - j}\Big)\Big(\prod_{j=0}^{q_2-1} \frac{1}{2^n - j}\Big).$$

It is easy to see that, $q_e + q_1 + q_2 = |\Pi_1| + |\Pi_2|$: because $q_1 + q_2$ equal the number of lazily sampled records in Π_1 and Π_2, while q_e equal the number of adapted records in Π_1.

Furthermore, $q_e \leq q^2$ by Sect. 4.2. It thus holds

$$\begin{aligned}
\frac{\Pr[\mathcal{P} \vdash \mathcal{P}']}{\Pr[\Lambda(E, \mathcal{P}) = \mathcal{P}']} &\geq \frac{\Big(\prod_{j=0}^{|\Pi_1|-1} \frac{1}{2^n-j}\Big)\Big(\prod_{j=0}^{|\Pi_2|-1} \frac{1}{2^n-j}\Big)}{\Big(\prod_{j=0}^{q_e-1} \frac{1}{2^n-j}\Big)\Big(\prod_{j=0}^{q_1-1} \frac{1}{2^n-j}\Big)\Big(\prod_{j=0}^{q_2-1} \frac{1}{2^n-j}\Big)} \\
&\geq \prod_{j=0}^{q^2-1} \Big(1 - \frac{j}{2^n}\Big) \\
&\geq 1 - \frac{(q^2)^2}{2^n} \geq 1 - \frac{q^4}{2^n},
\end{aligned}$$

as claimed. □

Lemma 9. *For any distinguisher D with total oracle query cost at most q, it holds*

$$\left|\Pr\left[D^{\Sigma_2(EM2P_4^{\mathcal{S}^{E,\mathcal{P}}},\mathcal{S}^{E,\mathcal{P}})}=1\right]-\Pr\left[D^{\Sigma_3(EM2P_4^{\mathcal{P}},\mathcal{P})}=1\right]\right|\le\frac{10q^3+15q^4}{2^n}.$$

Proof. Gathering Lemmas 6 and 8 yields

$$\begin{aligned}
&\left|\Pr\left[D^{\Sigma_2(\mathrm{EM2P}_4^{\mathcal{S}^{E,\mathcal{P}}},\mathcal{S}^{E,\mathcal{P}})}=1\right]-\Pr\left[D^{\Sigma_3(\mathrm{EM2P}_4^{\mathcal{P}},\mathcal{P})}=1\right]\right|\\
\le&\Pr\left[(E,\mathcal{P})\ is\ bad\right]+\sum_{\mathcal{P}'\in\Omega_1}\Pr\left[\Lambda(E,\mathcal{P})=\mathcal{P}'\right]-\sum_{\mathcal{P}'\in\Omega_1}\Pr\left[\mathcal{P}\vdash\mathcal{P}'\right]\\
\le&\Pr\left[(E,\mathcal{P})\ is\ bad\right]+\sum_{\mathcal{P}'\in\Omega_1}\Pr\left[\Lambda(E,\mathcal{P})=\mathcal{P}'\right]\left(1-\frac{\Pr\left[\mathcal{P}\vdash\mathcal{P}'\right]}{\Pr\left[\Lambda(E,\mathcal{P})=\mathcal{P}'\right]}\right)\\
\le&\frac{10q^3+14q^4}{2^n}+\frac{q^4}{2^n}\sum_{\mathcal{P}'\in\Omega_1}\Pr\left[\Lambda(E,\mathcal{P})=\mathcal{P}'\right]\\
\le&\frac{10q^3+15q^4}{2^n},
\end{aligned}$$

as claimed. □

Gathering Lemmas 7 and 9 yields the bound in Theorem 1.

5 Conclusion

We make a step towards minimizing the 4-round iterated Even-Mansour ciphers while retaining sequential indifferentiability. On the negative side, we exhibit an attack against single-key, single-permutation Even-Mansour with any rounds; on the positive side, we prove sequential indifferentiability for 4-round single-key Even-Mansour using 2 permutations. These provide the minimal Even-Mansour variant that achieve sequential indifferentiability without key schedule functions.

Acknowledgements. We sincerely thank the anonymous reviewers for their helpful comments. Chun Guo was partly supported by the National Natural Science Foundation of China (Grant No. 62002202).

References

1. Andreeva, E., Bogdanov, A., Dodis, Y., Mennink, B., Steinberger, J.P.: On the indifferentiability of key-alternating ciphers. In: Canetti, R., Garay, J.A. (eds.) CRYPTO 2013. LNCS, vol. 8042, pp. 531–550. Springer, Heidelberg (2013). https://doi.org/10.1007/978-3-642-40041-4_29
2. Andreeva, E., Bogdanov, A., Mennink, B.: Towards understanding the known-key security of block ciphers. In: Moriai, S. (ed.) FSE 2013. LNCS, vol. 8424, pp. 348–366. Springer, Heidelberg (2014). https://doi.org/10.1007/978-3-662-43933-3_18

3. Biryukov, A., Wagner, D.: Slide attacks. In: Knudsen, L. (ed.) FSE 1999. LNCS, vol. 1636, pp. 245–259. Springer, Heidelberg (1999). https://doi.org/10.1007/3-540-48519-8_18
4. Bogdanov, A., Knudsen, L.R., Leander, G., Standaert, F.-X., Steinberger, J., Tischhauser, E.: Key-alternating ciphers in a provable setting: encryption using a small number of public permutations. In: Pointcheval, D., Johansson, T. (eds.) EUROCRYPT 2012. LNCS, vol. 7237, pp. 45–62. Springer, Heidelberg (2012). https://doi.org/10.1007/978-3-642-29011-4_5
5. Canetti, R., Goldreich, O., Halevi, S.: The random oracle methodology, revisited. J. ACM **51**(4), 557–594 (2004). https://doi.org/10.1145/1008731.1008734
6. Chen, S., Lampe, R., Lee, J., Seurin, Y., Steinberger, J.: Minimizing the two-round even–mansour cipher. J. Cryptol. **31**(4), 1064–1119 (2018). https://doi.org/10.1007/s00145-018-9295-y
7. Chen, S., Steinberger, J.P.: Tight security bounds for key-alternating ciphers. In: Nguyen, P.Q., Oswald, E. (eds.) EUROCRYPT 2014. LNCS, vol. 8441, pp. 327–350. Springer, Heidelberg (May 2014). https://doi.org/10.1007/978-3-642-55220-5_19
8. Cogliati, B., Seurin, Y.: On the provable security of the iterated even-mansour cipher against related-key and chosen-key attacks. In: Oswald, E., Fischlin, M. (eds.) EUROCRYPT 2015. LNCS, vol. 9056, pp. 584–613. Springer, Heidelberg (2015). https://doi.org/10.1007/978-3-662-46800-5_23
9. Cogliati, B., Seurin, Y.: Strengthening the known-key security notion for block ciphers. In: Peyrin, T. (ed.) FSE 2016. LNCS, vol. 9783, pp. 494–513. Springer, Heidelberg (2016). https://doi.org/10.1007/978-3-662-52993-5_25
10. Coron, J.-S., Dodis, Y., Malinaud, C., Puniya, P.: Merkle-Damgård revisited: how to construct a hash function. In: Shoup, V. (ed.) CRYPTO 2005. LNCS, vol. 3621, pp. 430–448. Springer, Heidelberg (2005). https://doi.org/10.1007/11535218_26
11. Coron, J.-S., Holenstein, T., Künzler, R., Patarin, J., Seurin, Y., Tessaro, S.: How to build an ideal cipher: the indifferentiability of the feistel construction. J. Cryptol. **29**(1), 61–114 (2014). https://doi.org/10.1007/s00145-014-9189-6
12. Dachman-Soled, D., Katz, J., Thiruvengadam, A.: 10-round feistel is indifferentiable from an ideal cipher. In: Fischlin, M., Coron, J.-S. (eds.) EUROCRYPT 2016. LNCS, vol. 9666, pp. 649–678. Springer, Heidelberg (2016). https://doi.org/10.1007/978-3-662-49896-5_23
13. Dai, Y., Seurin, Y., Steinberger, J., Thiruvengadam, A.: Indifferentiability of iterated even-mansour ciphers with non-idealized key-schedules: five rounds are necessary and sufficient. In: Katz, J., Shacham, H. (eds.) CRYPTO 2017. LNCS, vol. 10403, pp. 524–555. Springer, Cham (2017). https://doi.org/10.1007/978-3-319-63697-9_18
14. Dai, Y., Steinberger, J.: Indifferentiability of 8-round feistel networks. In: Robshaw, M., Katz, J. (eds.) CRYPTO 2016. LNCS, vol. 9814, pp. 95–120. Springer, Heidelberg (2016). https://doi.org/10.1007/978-3-662-53018-4_4
15. Dodis, Y., Ristenpart, T., Steinberger, J., Tessaro, S.: To hash or not to hash again? (in)differentiability results for H^2 and HMAC. In: Safavi-Naini, R., Canetti, R. (eds.) CRYPTO 2012. LNCS, vol. 7417, pp. 348–366. Springer, Heidelberg (2012). https://doi.org/10.1007/978-3-642-32009-5_21
16. Dodis, Y., Stam, M., Steinberger, J., Liu, T.: Indifferentiability of confusion-diffusion networks. In: Fischlin, M., Coron, J.-S. (eds.) EUROCRYPT 2016. LNCS, vol. 9666, pp. 679–704. Springer, Heidelberg (2016). https://doi.org/10.1007/978-3-662-49896-5_24

17. Dunkelman, O., Keller, N., Shamir, A.: Minimalism in cryptography: the even-mansour scheme revisited. In: Pointcheval, D., Johansson, T. (eds.) EUROCRYPT 2012. LNCS, vol. 7237, pp. 336–354. Springer, Heidelberg (2012). https://doi.org/10.1007/978-3-642-29011-4_21
18. Dutta, A.: Minimizing the two-round tweakable even-mansour cipher. In: Moriai, S., Wang, H. (eds.) ASIACRYPT 2020. LNCS, vol. 12491, pp. 601–629. Springer, Cham (2020). https://doi.org/10.1007/978-3-030-64837-4_20
19. Even, S., Mansour, Y.: A construction of a cipher from a single pseudorandom permutation. J. Cryptol. **10**(3), 151–161 (1997). https://doi.org/10.1007/s001459900025
20. Farshim, P., Procter, G.: The related-key security of iterated even–mansour ciphers. In: Leander, G. (ed.) FSE 2015. LNCS, vol. 9054, pp. 342–363. Springer, Heidelberg (2015). https://doi.org/10.1007/978-3-662-48116-5_17
21. Guo, C., Lin, D.: A synthetic indifferentiability analysis of interleaved double-key even-mansour ciphers. In: Iwata, T., Cheon, J.H. (eds.) ASIACRYPT 2015. LNCS, vol. 9453, pp. 389–410. Springer, Heidelberg (2015). https://doi.org/10.1007/978-3-662-48800-3_16
22. Guo, C., Lin, D.: Indifferentiability of 3-round even-mansour with random oracle key derivation. IACR Cryptol. ePrint Arch, p. 894 (2016). http://eprint.iacr.org/2016/894
23. Guo, C., Lin, D.: Separating invertible key derivations from non-invertible ones: sequential indifferentiability of 3-round even-mansour. Des. Codes Crypt. **81**(1), 109–129 (2016)
24. Guo, J., Peyrin, T., Poschmann, A., Robshaw, M.: The LED block cipher. In: Preneel, B., Takagi, T. (eds.) CHES 2011. LNCS, vol. 6917, pp. 326–341. Springer, Heidelberg (2011). https://doi.org/10.1007/978-3-642-23951-9_22
25. Hoang, V.T., Tessaro, S.: Key-alternating ciphers and key-length extension: exact bounds and multi-user security. In: Robshaw, M., Katz, J. (eds.) CRYPTO 2016. LNCS, vol. 9814, pp. 3–32. Springer, Heidelberg (2016). https://doi.org/10.1007/978-3-662-53018-4_1
26. ISO/IEC: Information technology — security techniques – lightweight cryptography – part 2: Block ciphers. ISO/IEC 29192-2:2012 (2012). https://www.iso.org/standard/56552.html
27. ISO/IEC: Information security – encryption algorithms – part 7: Tweakable block ciphers. ISO/IEC FDIS 18033-7 (2021). https://www.iso.org/standard/80505.html
28. Lampe, R., Patarin, J., Seurin, Y.: An asymptotically tight security analysis of the iterated even-mansour cipher. In: Wang, X., Sako, K. (eds.) ASIACRYPT 2012. LNCS, vol. 7658, pp. 278–295. Springer, Heidelberg (2012). https://doi.org/10.1007/978-3-642-34961-4_18
29. Lampe, R., Seurin, Y.: How to construct an ideal cipher from a small set of public permutations. In: Sako, K., Sarkar, P. (eds.) ASIACRYPT 2013. LNCS, vol. 8269, pp. 444–463. Springer, Heidelberg (2013). https://doi.org/10.1007/978-3-642-42033-7_23
30. Mandal, A., Patarin, J., Seurin, Y.: On the public indifferentiability and correlation intractability of the 6-round feistel construction. In: Cramer, R. (ed.) TCC 2012. LNCS, vol. 7194, pp. 285–302. Springer, Heidelberg (2012). https://doi.org/10.1007/978-3-642-28914-9_16
31. Maurer, U., Renner, R., Holenstein, C.: Indifferentiability, impossibility results on reductions, and applications to the random oracle methodology. In: Naor, M. (ed.) TCC 2004. LNCS, vol. 2951, pp. 21–39. Springer, Heidelberg (2004). https://doi.org/10.1007/978-3-540-24638-1_2

32. Mouha, N., Luykx, A.: Multi-key security: the even-mansour construction revisited. In: Gennaro, R., Robshaw, M. (eds.) CRYPTO 2015. LNCS, vol. 9215, pp. 209–223. Springer, Heidelberg (2015). https://doi.org/10.1007/978-3-662-47989-6_10
33. Pub, N.F.: 197: Advanced encryption standard (aes). Federal Inf. Process. Stand. Publication **197**(441), 0311 (2001)
34. Soni, P., Tessaro, S.: Public-seed pseudorandom permutations. In: Coron, J.-S., Nielsen, J.B. (eds.) EUROCRYPT 2017. LNCS, vol. 10211, pp. 412–441. Springer, Cham (2017). https://doi.org/10.1007/978-3-319-56614-6_14
35. Soni, P., Tessaro, S.: Naor-reingold goes public: the complexity of known-key security. In: Nielsen, J.B., Rijmen, V. (eds.) EUROCRYPT 2018. LNCS, vol. 10822, pp. 653–684. Springer, Cham (2018). https://doi.org/10.1007/978-3-319-78372-7_21
36. Tessaro, S., Zhang, X.: Tight security for key-alternating ciphers with correlated sub-keys. In: Tibouchi, M., Wang, H. (eds.) Advances in Cryptology - ASIACRYPT 2021, Part III. Lecture Notes in Computer Science, vol. 13092, pp. 435–464. Springer (2021). https://doi.org/10.1007/978-3-030-92078-4
37. Wu, Y., Yu, L., Cao, Z., Dong, X.: Tight security analysis of 3-round key-alternating cipher with a single permutation. In: Moriai, S., Wang, H. (eds.) ASIACRYPT 2020. LNCS, vol. 12491, pp. 662–693. Springer, Cham (2020). https://doi.org/10.1007/978-3-030-64837-4_22

INT-RUP Security of SAEB and TinyJAMBU

Nilanjan Datta[1], Avijit Dutta[1], and Shibam Ghosh[2(✉)]

[1] TCG Centres for Research and Education in Science and Technology, Kolkata, India
{nilanjan.datta,avijit.dutta}@tcgcrest.org

[2] Department of Computer Science, University of Haifa, Haifa, Israel
sghosh03@campus.haifa.ac.il

Abstract. The INT-RUP security of an authenticated encryption (AE) scheme is a well studied problem which deals with the integrity security of an AE scheme in the setting of releasing unverified plaintext model. Popular INT-RUP secure constructions either require a large state (e.g. GCM-RUP, LOCUS, Oribatida) or employ a two-pass mode (e.g. MON-DAE) that does not allow on-the-fly data processing. This motivates us to turn our attention to feedback type AE constructions that allow small state implementation as well as on-the-fly computation capability. In CT-RSA 2016, Chakraborti et al. have demonstrated a generic INT-RUP attack on rate-1 block cipher based feedback type AE schemes. Their results inspire us to study about feedback type AE constructions at a reduced rate. In this paper, we consider two such recent designs, SAEB and TinyJAMBU and we analyze their integrity security in the setting of releasing unverified plaintext model. We found an INT-RUP attack on SAEB with roughly 2^{32} decryption queries. However, the concrete analysis shows that if we reduce its rate to 32 bits, SAEB achieves the desired INT-RUP security bound without any additional overhead. Moreover, we have also analyzed TinyJAMBU, one of the finalists of the NIST LwC, and found it to be INT-RUP secure. To the best of our knowledge, this is the first work reporting the INT-RUP security analysis of the block cipher based single state, single pass, on-the-fly, inverse-free authenticated ciphers.

1 Introduction

In the last few years, the increasing growth of the Internet of Things (IoT) comes with high demands on and constrictive conditions for cryptographic schemes. Such constraints may come in various types, as these small interconnected devices may have to operate with low power, low area, low memory, or otherwise. Lightweight cryptography is about developing cryptographic solutions for such constrained environments and partly ignited by the CAESAR [14] and the ongoing NIST Lightweight Competition (LwC) [29]. As a result of these competitions, the cryptographic community has witnessed the rise of various

T. Isobe and S. Sarkar (Eds.): INDOCRYPT 2022, LNCS 13774, pp. 146–170, 2022.
https://doi.org/10.1007/978-3-031-22912-1_7

lightweight authenticated encryption schemes in recent years. These include block cipher based constructions such as CLOC [27], JAMBU [38], COFB [19], SAEB [33], SUNDAE [10], permutation based constructions such as ASCON [24], ACORN [37], Beetle [18] and tweakable block cipher based constructions such as Deoxys AEAD [28], Romulus and Remus [26], Skinny AEAD [11] etc. However, in the paper, we confine our discussion only to block cipher and permutation based AE schemes.

1.1 Designing Area-Efficient Authenticated Ciphers

Often Lightweight Authenticated Encryption (AE) schemes are sequential in nature. This is primarily due to the fact that sequential modes consume less state size (the memory needed for storing internal values), implying smaller hardware footprint. In addition, sequential modes usually offer inverse-free property (except a few construction such as CBC and its variants[1]) of the underlying primitive, which also suits one of the very basic needs of implementing ciphers in lightweight environments.

Block Cipher Based Designs. The design of almost every block cipher based sequential AE schemes starts with a fixed initial state and processes each input block in sequence by applying a feedback function on the previous block cipher output, some secret auxiliary state, and the current input (message or associated data). This feedback function derives the next block cipher input, updated secret auxiliary state, and the current output (in case of message blocks). Thus, any block cipher based sequential AE scheme can be described by the underlying block cipher, the secret auxiliary state and the feedback function. Consequently, the AE scheme's efficiency and hardware footprint largely depend on the efficiency and the hardware footprint of the underlying block cipher, feedback function, and the secret auxiliary state. Zhang et al. [40] have proposed one such block cipher based sequential AE scheme called iFEED that uses plaintext feedback and achieves optimal rate[2] (i.e. rate-1). However, it requires a state size of $(3n + k)$-bits, where n and k are the block size and the key size of the underlying block cipher respectively. CPFB, proposed by Montes and Penazzi [31], is a notable scheme that reduces the state size to $(2n + k)$-bits, at the cost of reducing the rate to 3/4. In CHES'17, Chakraborti et al. proposed COFB [19], the first feedback type AE scheme, achieves rate-1 with a state size of $1.5n + k$-bits. Recently, it has been proven in [16] that the state size for any feedback type rate-1 AE scheme cannot be less than $1.5n + k$-bits. In the same paper, authors have also proposed a hybrid feedback type AE scheme called HyENA [16] that achieves rate-1 with the state size $1.5n + k$-bits but with a reduced XOR count.

[1] Some inverse-free modes are not sequential, e.g., CTR, OTR, GCM etc.

[2] rate is defined as the inverse of the number of block cipher calls required to process a single block of message, where a block refers to the block size of the block cipher.

Sponge Type Designs. An alternative way to avoid the generation of the auxiliary states from the block cipher based designs is to use a public permutation based sponge mode of operations. Since the selection of Ascon [24] in the final portfolio of the CAESAR [14] competition, sponge based designs have gained a huge momentum. In the ongoing NIST LwC competition [29], out of 57 submissions, 25 submissions are based on sponge type designs and out of 10 finalists, 5 candidates are sponge type designs [29]. The primary feature that one can get out of sponge type designs is that unlike block cipher based constructions, it does not require any key scheduling algorithm to invoke. This feature proves to be beneficial from the storage point of view when the data size of the underlying permutation for any sponge type design is less than the combination of the block size and key size of the underlying block cipher for any block cipher based designs. In such cases, sponge type mode becomes an excellent choice for area-efficient designs. Moreover, the additional feature of having no inverse call to the underlying permutation at the time of executing verified decryption algorithm, ensures an extremely low hardware footprint in a combined encryption-decryption implementation of the mode. By leveraging the advantages of sponge-type structure in block cipher based designs (albeit block cipher based schemes are required to store extra k-bit state for storing the keys), a few block cipher based sponge-type designs have recently been proposed. This includes CAESAR candidate JAMBU [38], and two NIST LwC candidates SAEAES [32] (which is an instantiation of SAEB [33] with AES-128 block cipher) and TinyJAMBU [39], where all the three AE schemes use a block size of 128-bits along with a block cipher key of 128-bits, employing an extremely small overall state of size 256-bits.

1.2 Authenticated Ciphers Under Release of Unverified Plaintext (RUP) Setting

In traditional authenticated ciphers, the verification must be done prior to release of plaintexts to the user. However, in resource constrained environments with limited memory, it may not be feasible to store the whole plaintext and one might be forced to release the plaintext before verification. As a result, an attacker can get hold of the insecure memory of the IoT devices to get access of the unverified plaintext, which can leak significant informations about the cipher to break its security. Moreover, real-time streaming protocols (e.g., SRTP, SSH, and SRTCP) and Optical Transport Networks sometimes need to release *plaintexts in segments with intermediate tags* on the fly to reduce end-to-end latency and storage. Owing to these real-time applications, we require a security notion which should ensure that when a cryptographic scheme satisfies the security notion, releasing the unverified plaintext will not lead to any threat to the system. *Releasing Unverified Plaintext* (RUP) security fulfills the above demand.

In [6], Andreeva et al. formalized the security notion of an authenticated encryption scheme under the release of unverified plaintext setting. In this model, the encryption functionality $\mathcal{E}$ remains, but it separates the decryption/verification functionality $\mathcal{DV}$ into a decryption functionality $\mathcal{D}$ and a verification functionality $\mathcal{V}$. Likewise the usual security notion of any AE scheme that ensures both confidentiality and integrity, Andreeva et al. [6] have suggested to achieve the confidentiality and integrity security for any AE scheme in the RUP model using IND-CPA + PA1/PA2 notion and INT-RUP notion respectively. For the confidentiality model in PA1 setting, i.e., IND-CPA + PA1, the adversary is given access to $\mathcal{E}$ and to either $\mathcal{D}$ or a simulator. The purpose of this notion is to complement the conventional confidentiality in the sense that it measures the advantage an adversary can gain from actually having access to $\mathcal{D}$. The integrity notion of an authenticated encryption scheme under this model, i.e., INT-RUP, allows an adversary to interact with $\mathcal{E}, \mathcal{D}, \mathcal{V}$; and asks it to forge a valid ciphertext, i.e., make a new, valid query to $\mathcal{V}$ oracle. The adversary potentially possesess significantly more power in this model due to the access to the decryption oracle $\mathcal{D}$.

Andreeva et al. [6] have shown that OCB [35], COPA [7] are insecure in the RUP security model. In [23], Datta et al. mounted an INT-RUP attack on any Encrypt-Linear mix-Encrypt type authenticated ciphers that includes the CAESAR standard COLM [5]. In another direction, Chakraborti et al. [17] mounted an INT-RUP attack on iFeed [40]. Adopting a similar attack strategy, they have shown a generic INT-RUP attack [17] on rate-1 block cipher based feedback type AEAD constructions. At the same time, they also proposed a scheme called mCPFB [17] and claimed that the INT-RUP security could be achieved at the cost of the rate of the construction. Similar approach have been used in OCBIC [41] and LOCUS [15], which builds upon OCB [35], and LOTUS [15], which builds upon OTR [30]. For both the constructions LOCUS and LOTUS, IND-CPA + PA1 and INT-RUP both security notions have been achieved at the cost of additional block cipher invocations, which halves the rate of the construction. Note that these modes are parallel in structure, and all of them require a state of size at least $3n + k$-bits, where n is the block size and k is the key size of the underlying block cipher. Ashur et al. [9] proposed an alternative notion of RUP security, called RUPAE. This notion focuses on nonce-based authenticated encryption, and proposed a RUP-variant of GCM [1], dubbed GCM-RUP [9], in the described nonce-based model. On the other extreme, Chang et al. [20] introduced the notion of AERUP which unifies the notions of RUP privacy (i.e., IND-CPA + PA1) and integrity (i.e., INT-RUP) for deterministic authenticated ciphers. They also proposed a simple variant of SUNDAE [10], dubbed MONDAE [20], that achieves confidentiality and integrity security in RUP setting under this newly introduced AERUP model. However, MONDAE is a two-pass authenticated encryption mode. Hence, it does not have the on-the-fly decryption feature. In a nutshell, while looking at the ciphers with RUP security, either the constructions lose on state size (e.g., LOCUS [15], mCPFB [17], GCM-RUP [9], requires at least $3n + k$-bits) or the construction does not have the on-the-fly decryption feature (e.g., MONDAE [20]). The above discussion makes us raise the question:

Can we have a block cipher based INT-RUP secure design with on-the-fly decryption feature with a total of $n + k$-bits state?

1.3 Towards RUP-Secure Single-state On-the-Fly Authenticated Encryption

The above question turns our attention to study the INT-RUP security of sponge based modes. In [13], Bhattacharjee et al. have studied the INT-RUP security of permutation based sponge type designs. They have presented an INT-RUP attack on generic duplex constructions, with attack complexity $O(q_d q_p/2^c)$, where q_d is the number of decryption queries, q_p is the number of primitive queries to the permutation, and c is the capacity part of the construction. They have also shown that such attacks can be extended to other Sponge variants such as Beetle [18] and SPoC [2]. The main idea of the attack is to exploit a collision between an inner state of the construction and a primitive query. To resist such attacks, the authors used the concept of masking the previous state and proposed a new cipher called Oribatida [13] that achieves INT-RUP security of $O(q_d^2/2^c)$. However, this comes at the cost of an additional state.

When we move to block cipher based AEAD constructions, Chakraborti et al. [17] have shown that any feedback type rate-1 block cipher based AEAD construction is not INT-RUP secure. Adopting the idea used in iFeed, they have shown a generic INT-RUP attack on rate-1 block cipher based feedback type AEAD constructions. This result immediately rules out the popular area-efficient block cipher based designs such as COFB and HyENA to have INT-RUP security. Therefore, the focus goes to feedback type constructions with a lower rate. This makes us look into block cipher based sponge type constructions such as SAEB [33] and TinyJAMBU [39]. Due to the inverse-free implementation with $n+k$-bits state, these constructions are incredibly lightweight and ideally suited for resource constraint applications. At the same time, these constructions have the capability of on-the-fly computation of plaintext/ciphertext blocks. Thus, they are ideally suited for applications where RUP security would be of extreme relevance. However, the current literature does not say anything about these block cipher based constructions, and hence investigating their RUP security seems an exciting research direction. In this regard, we would like to mention that in a recent work, Andreeva et al. [4] have shown $2^{n/2}$ INT-RUP security bound on a forkcipher [8] based construction, called SAEF [3]. The structure of SAEF resembles to the CBC mode of operation, where one of the output blocks of the forkcipher is used to XOR-mask the input and sometimes output of the next primitive call.

1.4 Our Contribution and Significance of the Result

In this paper, we study the INT-RUP security of two constructions, namely SAEB [33] by Naito et al. and TinyJAMBU [39] by Wu et al. Our contribution is threefold:

(i) We have shown an INT-RUP attack on SAEB that uses a single encryption query, and roughly $2^{c/2}$ decryption queries, where c is the capacity part of the construction. The attack is applicable for any choices of rate and capacity.

(ii) We have investigated the INT-RUP security bound of SAEB. We have shown that it offers roughly $q_d^2/2^c$ INT-RUP security, where q_d is the number of decryption queries and c is the capacity part of the construction. Combining the proven security bound of SAEB with its attack complexity establishes the tightness of the security bound of SAEB. This result signifies that if we instantiate SAEB with a standard 128-bit block cipher and put a restriction that at a time 32-bits of the message will be injected to SAEB, the mode achieves INT-RUP security up to 2^{48} blocks, which satisfies the NIST criteria of having 2^{50} bytes of data complexity. However, for SAEAES, where we inject 64-bits of message at a time to the construction, achieves INT-RUP security up to 2^{32} blocks.

(iii) Finally, we consider the INT-RUP security of TinyJAMBU, one of the finalists of the NIST LwC. Interestingly TinyJAMBU has a unique structure, where message injection and ciphertext release occur from different parts of the state. We have proved that TinyJAMBU offers roughly $q_v \sigma_d / 2^{n-r}$ INT-RUP security, where σ_d is the total number of blocks in all the decryption queries, r-bits of the message is injected at a time to the construction and n is the block size of the underlying block cipher.

Thus, in this work, we have obtained the INT-RUP security bounds of SAEB and TinyJAMBU. To the best of our knowledge, this is the first work that reports two single-pass[3], inverse-free AEAD constructions, achieving INT-RUP security, while keeping the on-the-fly decryption property intact. We would like to point here that both of these construction will not preserve confidentiality in the RUP setting as these are not two-pass modes. This paper solely focuses on studying the integrity security of the two constructions in the RUP setting.

2 Preliminaries

For $n \in \mathbb{N}$, $[n]$ denotes the set $\{1, 2, \ldots, n\}$ and for $a < b \in \mathbb{N}$, we write $[a, b]$ to denote the set $\{a, a+1, \ldots, b\}$. We write $(a, b) < (a', b')$ to denote that either $a < a'$ or ($a = a'$ and $b < b'$). For a finite set $\mathcal{X}$, $\mathsf{X} \leftarrow_{\$} \mathcal{X}$ denotes the uniform at random sampling of X from $\mathcal{X}$. For $n \in \mathbb{N}$, we write $\{0,1\}^+$ and $\{0,1\}^n$ to denote the set of all non-empty binary strings, and the set of all n-bit binary strings, respectively. We write $\emptyset$ to denote the empty string, and $\{0,1\}^* = \{0,1\}^+ \cup \{\emptyset\}$. For any two strings $X, Y \in \{0,1\}^*$, we write $X \| Y$ to denote the concatenation of the string X followed by the string Y. For $X \in \{0,1\}^*$, $|X|$ denotes the length (number of the bits) of X. For any non-empty binary string $X \in \{0,1\}^+$, $(X[1], \ldots, X[k]) \xleftarrow{n} X$ denotes the n-bit parsing of X, where $|X[i]| = n$ for $1 \leq i \leq k-1$, and $1 \leq |X[k]| \leq n$. We use the notation $X[a \ldots b]$ to denote bit string $X[a] \| X[a+1] \| \cdots \| X[b]$. If $a = 1$, then we write $X[\ldots b]$ to denote $X[1 \ldots b]$. In this paper, we fix a positive integer n and define the function ozs over the set of any binary string, as $\mathsf{ozs}(X) := X \| 1 \| 0^{n-(|X| \bmod n)-1}$. Note that the function is

[3] GCM-RUP [9] also achieves inverse-free, INT-RUP security and on-the-fly decryption property but it requires two pass.

injective and maps all m-bit binary strings to a multiple of n-bit binary strings by appropriately padding the string with 10^*. For any real number X, $\lceil X \rceil$ denotes the smallest integer X' such that $X' \geq X$. For any $X \in \{0,1\}^+$ and an integer $i \leq |X|$, $\lfloor X \rfloor_i$ ($\lceil X \rceil_i$) returns the least significant (most significant, resp.) i-bits of X. For any integer i, we denote the n-bit unsigned representation of i as $\langle i \rangle_n$.

2.1 Authenticated Encryption

An authenticated encryption (AE) is an integrated scheme that provides both privacy and integrity of a plaintext $M \in \{0,1\}^*$ and integrity of an associated data $A \in \{0,1\}^*$. Taking a nonce $N \in \mathcal{N}$ (which is a unique value for each encryption), where $\mathcal{N}$ is the nonce space, together with the associated data A and the plaintext M, the encryption function of AE, enc_K, produces a tagged-ciphertext $(C,T) \in \{0,1\}^* \times \{0,1\}^\tau$ with $|C| = |M|$. We denote the length in blocks of the associated data, message and ciphertext with a, m and c, respectively. The corresponding decryption function, dec_K, takes $(N,A,C,T) \in \mathcal{N} \times \{0,1\}^* \times \{0,1\}^* \times \{0,1\}^\tau$ and returns a decrypted plaintext $M \in \{0,1\}^*$ when the authentication on (N,A,C,T) is successful; otherwise, it returns the atomic error symbol denoted by $\perp$. Following Andreeva et al. [6], we separate the decryption algorithm into plaintext computation and tag verification. Formally, the decryption interface provides two algorithms, a decryption function dec_K that takes (N,A,C) and returns a decrypted plaintext M irrespective of the authentication result (hence we drop the tag value), and a verification function ver_K that takes (N,A,C,T) and returns $\top$ when the authentication succeeds; otherwise, it returns $\perp$.

2.2 Integrity Security in RUP Setting

Following the definition of Andreeva et al. [6], we define the integrity security of an authenticated encryption scheme in the RUP setting. We consider an information theoretic adversary $\mathscr{A}$ with access to a triplet of oracles of an authenticated encryption scheme Θ - for a uniformly sampled secret key K, the encryption oracle $\Theta.\mathsf{enc}_K$, decryption oracle $\Theta.\mathsf{dec}_K$ and the verification oracle $\Theta.\mathsf{ver}_K$. We say that $\mathscr{A}$ forges Θ under the RUP setting if $\mathscr{A}$ can compute a tuple (N,A,C,T) satisfying $\Theta.\mathsf{ver}_K(N,A,C,T) \neq \perp$, without querying (N,A,M) to $\Theta.\mathsf{enc}_\mathsf{K}$ and receiving (C,T) as a response, i.e. (N,A,C,T) is a non-trivial forgery. We assume that $\mathscr{A}$ can make decryption queries of the form (N,A,C) to the oracle $\Theta.\mathsf{dec}_K$, with no restriction on nonce repetitions, and receive the corresponding response M, whereas nonces should be distinct for every encryption queries to $\Theta.enc_K$. Then, the integrity security or equivalently the forging advantage of Θ for the adversary $\mathscr{A}$ in the RUP setting is defined as

$$\mathsf{Pr}[K \leftarrow_{\$} \{0,1\}^k : \mathscr{A}^{\Theta.\mathsf{enc}_K, \Theta.\mathsf{dec}_K, \Theta.\mathsf{ver}_K} \text{ forges }].$$

We assume that $\mathscr{A}$ does not make any query to the oracles for which it can compute the corresponding response on its own. We call such an adversary a *non-trivial* adversary. Following [6], we view the forging advantage of Θ in the RUP setting as an equivalent distinguishing game between two worlds. The real world consists of $(\Theta.\mathsf{enc}_K, \Theta.\mathsf{dec}_K, \Theta.\mathsf{ver}_K)$ for a uniformly chosen key K, whereas the ideal world consists of $(\Theta.\mathsf{enc}_K, \Theta.\mathsf{dec}_K, \perp)$, i.e., the verification oracle in the real world is replaced by the reject symbol. This means all verification attempts in the ideal world will lead to a rejection. Under this equivalent setting, the integrity advantage for any distinguisher $\mathscr{A}$ is defined as

$$\mathbf{Adv}_{\Theta}^{\mathsf{int\text{-}rup}}(\mathscr{A}) := \left|\mathsf{Pr}[\mathscr{A}^{(\Theta.\mathsf{enc}_K, \Theta.\mathsf{dec}_K, \Theta.\mathsf{ver}_K)} = 1] - \mathsf{Pr}[\mathscr{A}^{(\Theta.\mathsf{enc}_K, \Theta.\mathsf{dec}_K, \perp)} = 1]\right|,$$

where $\perp$ denotes the degenerate oracle that always returns $\perp$ symbol and the probability is defined over the randomness of K. The integrity under RUP advantage of Θ is defined as

$$\mathbf{Adv}_{\Theta}^{\mathsf{int\text{-}rup}}(q_e, q_d, q_v, \sigma_e, \sigma_d, \sigma_v) := \max_{\mathscr{A}} \mathbf{Adv}_{\Theta}^{\mathsf{int\text{-}rup}}(\mathscr{A}),$$

where the maximum is taken over all distinguishers making q_e encryption queries, q_d decryption queries and q_v verification queries. Here σ_e, σ_d and σ_v denotes the total number of block cipher calls with distinct inputs in encryption, decryption, and verification queries, respectively. Throughout this paper, we write $(q_e, q_d, q_v, \sigma_e, \sigma_d, \sigma_v)$-distinguisher to represent a distinguisher that makes q_e encryption queries with a total of σ_e many primitive calls with distinct inputs in encryption queries, q_d decryption queries with a total of σ_d many primitive calls with distinct inputs in decryption queries and q_v verification queries with a total of σ_v many primitive calls with distinct inputs in verification queries.

Let Λ_1 and Λ_0 denote the random variable induced by the interaction of $\mathscr{A}$ with the real oracle and the ideal oracle, respectively. The probability of realizing a transcript ω in the ideal oracle (i.e., $\mathsf{Pr}[\Lambda_0 = \omega]$) is called the *ideal interpolation probability*. Similarly, one can define the *real interpolation probability*. A transcript ω is said to be *attainable* with respect to $\mathscr{A}$ if the ideal interpolation probability is non-zero (i.e., $\mathsf{Pr}[\Lambda_0 = \omega] > 0$). We denote the set of all attainable transcripts by Ω. Following these notations, we state the main result of the H-Coefficient Technique in Theorem 1. The proof of this theorem can be found in [34].

Theorem 1 (H-Coefficient Technique). *Suppose for some* $\Omega_{\mathsf{bad}} \subseteq \Omega$, *which we call the* bad *set of transcripts, the following conditions hold:*

1. $\mathsf{Pr}[\Lambda_0 \in \Omega_{\mathsf{bad}}] \leq \epsilon_1$,
2. *For any good transcript* $\omega \in \Omega \setminus \Omega_{\mathsf{bad}}$, $\mathsf{Pr}[\Lambda_1 = \omega] \geq (1 - \epsilon_2) \cdot \mathsf{Pr}[\Lambda_0 = \omega]$.

Then, we have

$$\mathbf{Adv}_{\Theta}^{\mathsf{int\text{-}rup}}(q_e, q_d, q_v, \sigma_e, \sigma_d, \sigma_v) \leq \epsilon_1 + \epsilon_2. \quad (1)$$

We will apply the H-Coefficient technique to bound the integrity security of the two block cipher based authenticated ciphers SAEB and TinyJAMBU in the RUP model. To do this, we first replace the underlying primitive of the construction, which is a block cipher, with a random permutation at the cost of the PRP advantage of the block cipher. Then, we bound the distinguishing advantage of the resulting construction (whose underlying primitive is a random permutation) from the ideal one. We bound this advantage against an adversary $\mathscr{A}$ that is computationally unbounded (i.e., no bound on the time complexity, but bounded on the number of queries that it can ask to the oracle) and hence deterministic. We call them *information-theoretic* adversary. Therefore, from now onwards, we skip the time parameter from their corresponding advantage definitions.

3 SAEB AEAD Mode and Its INT-RUP Security

SAEB [33] is a block cipher based AEAD scheme, proposed by Naito et al. in TCHES'18. The design principle of SAEB follows the sponge duplex mode based on block ciphers. Similar to permutation based sponge constructions, SAEB injects r-bits of the message at a time to the construction, called the message injection rate, c-bits capacity, and the overall block size is $n = r + c$-bits. The algorithmic description of the encryption function of SAEB is presented in Fig. 1, and its schematic diagram is depicted in Fig. 2. An instantiation of SAEB with AES-128, called SAEAES [32], was submitted in NIST LwC and is one of the second round candidates of the competition. In the original proposal of the scheme, the recommended parameters of SAEAES are $r = 64$ and $c = 64$. The pictorial representation of the encryption algorithm of SAEB is given in Fig. 2. Naito et al. [33] have shown that SAEB achieves birthday bound security with the dominating term being $(\sigma_a + \sigma_d)/2^c + (\sigma_e + \sigma_d)^2/2^n$, where σ_a is the number of associated data blocks across all the queries, σ_e is the total number of block cipher calls with distinct inputs in encryption queries, and σ_d is the total number of block cipher calls with distinct inputs in decryption queries. However,

$\mathsf{SAEB}(K, N, A, M)$

1: $A[1]\|\ldots\|A[a] \xleftarrow{r} A\|10^*$;
2: $M[1]\|\ldots\|M[m] \xleftarrow{r} M\|10^*$;
3: $s[1] \leftarrow A[1]\|0^c$;
4: **for** $i = 2$ **to** $a - 1$;
5: $\quad s[i] \leftarrow E_K(s[i-1]) \oplus A[i]\|0^\star$;
6: $s[a] \leftarrow E_K(s[a-1]) \oplus A[a]\|\mathsf{const}_1$;
7: $s[a+1] \leftarrow E_K(s[a]) \oplus N\|\mathsf{const}_2$;
8: **for** $i = 2$ **to** $m - 1$;
9: $\quad s[a+i] \leftarrow E_K(s[a+i-1]) \oplus M[i]\|0^\star$;
10: $\quad C[i] \leftarrow \lceil s[a+i] \rceil_r$;
11: $s[a+m+1] \leftarrow E_K(s[a+m]) \oplus M[m]\|\mathsf{const}_3$;
12: $T \leftarrow \lceil E_K(s[a+m+1]) \rceil_r$;
return (C, T);

Fig. 1. Encryption algorithm of SAEB authenticated encryption mode.

the designers have not analyzed the construction in the RUP setting, and to the best of our knowledge, no prior work has addressed the issue of analyzing the security of this construction in the RUP setting. In the subsequent sections, we analyze the INT-RUP security of SAEB. In particular, we show in Sect. 3.1 that an adversary $\mathscr{A}$ with roughly $2^{c/2}$ decryption queries, can forge SAEB in the release of unverified plaintext setting and in Sect. 4, we give an upper bound of the order $2^{c/2}$ on the INT-RUP security of SAEB.

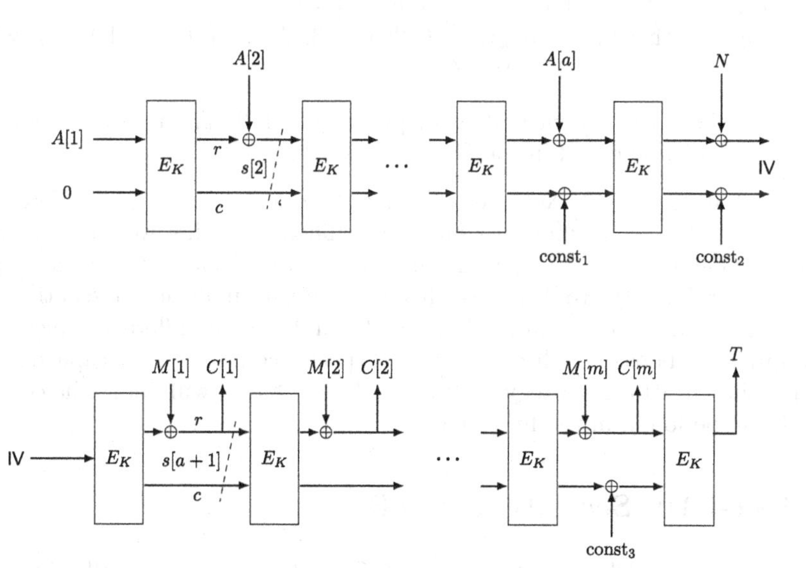

Fig. 2. Encryption algorithm of SAEB Authenticated Encryption for a block associated data, and m block message. const_1, const_2, const_3 are two-bit distinct constants used for domain separation. $\mathsf{const}_1 = 01/10$ based on whether $A[a]$ is a full or partial block. $\mathsf{const}_2 = 11$ and $\mathsf{const}_3 = 01/10$ based on whether $M[m]$ is a full or partial block.

3.1 INT-RUP Attack on SAEB

In this section, we show a forging attack on SAEB in the INT-RUP setting with $2^{c/2}$ decryption queries and a single encryption query. Here we describe the adversary $\mathcal{A}$ that primarily exploits the fact that during a decryption call, an adversary can control the rate part of the input of the block cipher by directly injecting the ciphertext into the rate part of the block cipher to mount the attack: First, we describe the attack when $r \geq c/2$, then extend it for $r < c/2$. Let $r \geq c/2$. We describe an adversary $\mathscr{A}$ that mounts an INT-RUP attack against SAEB with roughly $2^{c/2}$ decryption queries and a single encryption query as follows:

1. $\mathscr{A}$ chooses an arbitrary r-bit nonce N, an arbitrary r-bit associated data A and an arbitrary r-bit ciphertext data C, and then makes $2^{c/2+1}$ decryption queries of the form $(N,\ A,\ C_i[1] \parallel C[2] \parallel C[3] \parallel \ \dots \ \parallel C[\ell+1])_{i=1,\dots,2^{c/2}}$,

with distinct r-bit $C_i[1]$ values such that $C[2] = C[3] = \ldots = C[\ell + 1] = C$ and $\ell = \lceil c/r \rceil + 1$. Let the unverified released plaintext be $M_i[1] \parallel M_i[2] \parallel M_i[3] \parallel \ldots \parallel M_i[\ell + 1]$.
2. Assume there exist two indices $j, k \in [2^{c/2}]$ for which $M_j[a] = M_k[a]$ for all $a \in [3, \ell + 1]$.
3. $\mathscr{A}$ makes an encryption query with $(N,\ A,\ M_j[1] \parallel M_j[2] \parallel M_j[3] \parallel \ldots \parallel M_j[\ell + 1])$. Let the tagged ciphertext be $(C_j[1] \parallel C[2] \parallel C[3] \parallel \ldots \parallel C[\ell + 1],\ T)$, where $C[2] = C[3] = \ldots = C[\ell + 1] = C$.
4. $\mathscr{A}$ forges with $(N,\ A,\ (C_k[1] \parallel C[2] \parallel C[3] \parallel \ldots \parallel C[\ell + 1]),\ T)$, where $C[2] = C[3] = \ldots = C[\ell + 1] = C$.

It is easy to see that $\mathcal{A}$ succeeds with probability 1/2. The technical details of the analysis can be found in the Full version [22].

ATTACK WHEN $r < c/2$. Now, we consider the case when $r < c/2$. Note that, when $r < c/2$, then varying just one r-bit ciphertext string would result in at most 2^r different values. This would not ensure a collision in the capacity part with high probability. To deal with this, one can vary multiple consecutive r-bit ciphertext strings, say s many, which results in 2^{rs} many different values. If we appropriately choose s with $rs \geq c/2$, we expect a collision in the capacity part. Then a similar attack strategy, as described for $r \geq c/2$, will hold. The concrete attack can be found in the full version.

4 INT-RUP Security of SAEB

In this section, we show that SAEB is INT-RUP secure against all adversaries that make roughly $2^{c/2}$ decryption queries, where c is the capacity of the construction. We prove the security of the construction in the information-theoretic setting, where a uniform random n-bit permutation P replaces the underlying block cipher of the construction at the cost of the prp advantage of the block cipher E and denote the resulting construction as SAEB*[P]. In the following, we state and prove the int-rup security result of SAEB*[P].

Theorem 2. *Let $\mathsf{P} \leftarrow_{\$} \mathsf{Perm}(n)$ be an uniformly sampled n-bit random permutation. The* INT-*RUP advantage for any $(q_e, q_d, q_v, \sigma_e, \sigma_d, \sigma_v)$-distinguisher $\mathscr{A}$ against the construction* SAEB*[P] *that makes at most q_e encryption, q_d decryption and q_v verification queries with at most σ_e primitive calls with distinct inputs in encryption queries, σ_d primitive calls with distinct inputs in decryption queries and σ_v primitive calls with distinct inputs in verification queries having a total of $\sigma = \sigma_e + \sigma_d + \sigma_v$ primitive calls with distinct inputs such that $\rho \leq \sigma_e$, where ρ is a parameter, is given by*

$$\mathbf{Adv}^{\mathsf{int\text{-}rup}}_{\mathsf{SAEB^*[P]}}(\mathscr{A}) \leq \frac{\sigma^2}{2^{n+1}} + \frac{\sigma_e^{\rho}}{(2^r)^{\rho-1}} + \frac{q_v(2\rho + \sigma_d)}{2^c} + \frac{q_v}{2^{\tau}} + \frac{q_v(\sigma_e + \sigma_d)}{2^c}.$$

Proof. As the first step of the proof, we slightly modify the construction by replacing the random permutation with a random function $R \leftarrow_{\$} \mathsf{Funcs}(\{0,1\}^n)$. We denote the resulting construction as Θ. This modification comes at the cost of birthday bound complexity due to the PRF-PRP switching lemma [12,21]. We consider a computationally unbounded non-trivial deterministic distinguisher $\mathscr{A}$ that interacts with a triplet of oracles in either of the two worlds: in the real world, it interacts with $(\Theta.\mathsf{enc}_R, \Theta.\mathsf{dec}_R, \Theta.\mathsf{ver}_R)$, and in the ideal world, it interacts with $(\Theta.\mathsf{enc}_R, \Theta.\mathsf{dec}_R, \perp)$, where $\perp$ denotes the oracle that always rejects the verification attempts. We summarize $\mathscr{A}$'s query-response in a transcript ω which is segregated into a transcript of encryption queries, decryption queries, verification queries. Basically, we segregate the transcript ω into three parts ω^+, ω^-, and $\omega^\times$, where $\omega^+ = \{(N_1^+, A_1^+, M_1^+, C_1^+, T_1^+), \ldots, (N_{q_e}^+, A_{q_e}^+, M_{q_e}^+, C_{q_e}^+, T_{q_e}^+)\}$ is a tuple of encryption queries, $\omega^- = \{(N_1^-, A_1^-, C_1^-, M_1^-), (N_2^-, A_2^-, C_2^-, M_2^-), \ldots, (N_{q_d}^-, A_{q_d}^-, C_{q_d}^-, M_{q_d}^-)\}$ is a tuple of decryption queries, and a tuple of verification queries $\omega^\times = \{(N_1^\times, A_1^\times, C_1^\times, T_1^\times, \perp_1), \ldots, (N_{q_v}^\times, A_{q_v}^\times, C_{q_v}^\times, T_{q_v}^\times, \perp_{q_v})\}$ such that $\omega = \omega^+ \cup \omega^- \cup \omega^\times$. We modify the experiment by releasing all internal state values to adversary $\mathscr{A}$ after it makes all the encryption, decryption and verification queries, and before it outputs the decision bit b. We denote the j-th internal state value in the i-th encryption, decryption and verification query as $s_i^+[j], s_i^-[j]$ and $s_i^\times[j]$ respectively. In general, we denote it as $s_i^\star[j]$, where $\star \in \{+, -, \times\}$. The length of associated data, message and ciphertext in the i-th query is denoted as $a_i^\star, m_i^\star$ and $c_i^\star$ respectively where $\star \in \{+, -, \times\}$. In the real world, these internal state variables for every encryption, decryption and verification queries are computed by the corresponding oracles that faithfully evaluate SAEB^*. Note that the sequence of internal state values, in the real world for the i-th encryption query of length $a_i^+ + m_i^+ + 1$, are defined as follows:

$$\begin{cases} s_i^+[1] = A_i^+[1] \| \langle 0 \rangle_{c_1} \\ s_i^+[k] = A_i^+[k] \| \langle 0 \rangle_{c_1} \oplus R(s_i^+[k-1]), \text{ for } 2 \le k < a_i^+ \\ s_i^+[a_i^+] = \mathsf{ozs}(A_i^+[a_i^+]) \| \langle \mathsf{const}_1 \rangle_c \oplus R(s_i^+[a_i^+ - 1]) \\ s_i^+[a_i^+ + 1] = N_i^+ \| \langle \mathsf{const}_2 \rangle_c \oplus R(s_i^+[a_i^+]) \\ s_i^+[a_i^+ + 1 + k] = M_i^+[k] \| \langle 0 \rangle_c \oplus R(s_i^+[k-1]), \text{ for } 1 \le k < m_i^+ \\ s_i^+[a_i^+ + 1 + m_i^+] = \mathsf{ozs}(M_i^+[m_i^+]) \| \langle \mathsf{const}_3 \rangle_c \oplus R(s_i^+[a_i^+ + m_i^+]) \end{cases} \tag{2}$$

Similarly, we define the internal state values in the real world for decryption and verification queries. In the ideal world, as the encryption and decryption oracles are identical to that of the real world, the intermediate state variables for every encryption and decryption queries are faithfully evaluated by the corresponding oracles, and hence the sequence of state values for i-th encryption query is identically defined to Eq. (2). Similar to the real world, we also define the internal state values in the ideal world for decryption queries. As the verification oracle $\perp$ in the ideal world always returns rejects and does not compute anything, the internal state variables are not defined. Therefore, we have to define the sampling of the intermediate state variables for every verification query in the ideal world. To achieve this, for every verification query (N, A, C, T), the verification oracle for the ideal world invokes the decryption oracle $\Theta.\mathsf{dec}_R$ with input (N, A, C) ignoring the output of r-bit plaintext strings. Finally, the verification oracle ignores the checking of the computed tag T^* with the given tag T. Thus, the sequence of internal state values is defined for verification queries. Let the modified attack transcripts be $\omega_{\mathtt{new}} = \omega^+_{\mathtt{new}} \cup \omega^-_{\mathtt{new}} \cup \omega^\times_{\mathtt{new}}$, where

$$\begin{cases} \omega^+_{\mathtt{new}} = \omega^+ \cup \{s^+_i[j] : i \in [q_e], j \in [a_i + m_i + 1]\} \\ \omega^-_{\mathtt{new}} = \omega^- \cup \{s^-_i[j] : i \in [q_d], j \in [a_i + m_i + 1]\} \\ \omega^\times_{\mathtt{new}} = \omega^\times \cup \{s^\times_i[j] : i \in [q_v], j \in [a_i + m_i + 1]\}. \end{cases}$$

For a given transcript $\omega_{\mathtt{new}}$, we reorder the transcript so that all the encryption queries appear first, followed by all the decryption queries and finally, all the verification queries. It is easy to see that a state collision occurs in $s^\star_i[j]$ and $s^\circledast_{i'}[j]$ with probability 1, where $\star, \circledast \in \{+, -, \times\}$, if

- $A^\star_i[\ldots j] = A^\circledast_{i'}[\ldots j]$, when $j \leq a^\star_i$,
- $A^\star_i = A^\circledast_{i'}$, $N^\star_i = N^\circledast_{i'}$, when $j = a^\star_i + 1$,
- $N^\star_i = N^\circledast_{i'}$, $A^\star_i = A^\circledast_i$, $M^\star_i[\ldots(j - a^\star_i - 1)] = M^\circledast_{i'}[\ldots(j - a^\star_i - 1)]$, when $a^\star_i + 1 < j \leq a^\star_i + 1 + m^\star_i$.

A state collision that happens with probability 1 is called a *trivial collision*, and any other state collision is *non-trivial.* For $j > 1$, we write $\mathsf{ancestor}(s^\star_i[j])$ to denote the sequence of state values $(s^\star_i[1], \ldots, s^\star_i[j-1])$ that leads to $s^\star_i[j]$ in the i-th query and $\mathsf{ancestor}(s^\star_i[1]) = \phi$, where $\star \in \{+, -, \times\}$. Using this notion, we say that a trivial state collision between $s^\star_i[j]$ and $s^\circledast_{i'}[j]$ occurs if and only if $\mathsf{ancestor}(s^\star_i[j]) = \mathsf{ancestor}(s^\circledast_{i'}[j])$ for some j. Let $D^\star_i := A^\star_i \| N^\star_i$ and $d^\star_i := a^\star_i + 1$ where $\star \in \{+, -, \times\}$. Let us consider two queries $\star, \circledast \in \{+, -, \times\}$ with distinct query indices i and i', where $i, i' \in [q_e + q_d + q_v]$. We define the *longest common prefix* of (i, i'), denoted as $\mathsf{LCP}(i, i')$

$$\begin{cases} j, \text{ if } j \in [d^\star_i], D^\star_i[1..j] = D^\circledast_{i'}[1..j], D^\star_i[j+1] \neq D^\circledast_{i'}[j+1] \\ d^\star_i, \text{ if } D^\star_i = D^\circledast_{i'}, M^\star_i[1] \neq M^\circledast_{i'}[1] \\ j, \text{ if } j \in [d^\star_i, d^\star_i + m^\star_i], D^\star_i = D^\circledast_{i'}, M^\star_i[1..j] = M^\circledast_{i'}[1..j], M^\star_i[j+1] \neq M^\circledast_{i'}[j+1] \\ d^\star_i + m^\star_i, \text{ if } D^\star_i = D^\circledast_{i'}, M^\star_i = M^\circledast_{i'}[1..m^\star_i] \end{cases}$$

Consequently, we define $\mathsf{LLCP}(i) \triangleq \max_{i'<i}\{\mathsf{LCP}(i, i')\}$, that denotes the longest common prefix of query index $i \in [q_e + q_d + q_v]$.

4.1 Definition and Probability of Bad Transcripts

In this section, we define and bound the probability of bad transcripts in the ideal world. Let Ω be the set of all attainable transcripts and $\omega_{\mathtt{new}} \in \Omega$ be one such attainable transcript. We say that transcript $\omega_{\mathtt{new}}$ is **bad**, i.e., $\omega_{\mathtt{new}} \in \Omega_{\mathtt{bad}}$, if at least one of the following holds:

1. Coll: there exists i, j, i', j' with $(i', j') < (i, j)$ with $\mathsf{LLCP}(i) < j \leq a_i + m_i + 1$ and $i \leq [q_e + q_d + q_v]$ such that $s^\star_i[j] = s^\circledast_{i'}[j']$, where $\star, \circledast \in \{+, -, \times\}$.

2. mColl: $\exists i_1, j_1, \ldots, i_\rho, j_\rho$ with $\{i_1, \ldots, i_\rho\} \subseteq [q_e]$ and for all $1 \le k \le \rho$, $j_k \in [m_{i_k}]$, such that $C^+_{i_1}[j_1] = \cdots = C^+_{i_\rho}[j_\rho]$.
3. Forge: This event happens if for some verification query, all its intermediate states prior to the final state matches with intermediate state from some encryption or decryption queries; and the final state of the verification query matches with the final state of an encryption query for which the tag matches. In other words, $\exists i \in [q_v]$ such that for the i-th verification query $(N^\times_i, A^\times_i, C^\times_i, T^+_i)$, the following two events hold:

$$\begin{cases} \forall j \in [a^\times_i + 1, (a^\times_i + c^\times_i)], \exists i', j' \text{ such that } s^\times_i[j] = s^+_{i'}[j'] \text{ or } s^\times_i[j] = s^-_{i'}[j'], \\ \exists f \in [q_e] \text{ such that } s^\times_i[a^\times_i + c^\times_i + 1] = s^+_f[a^+_f + c^+_f + 1] \text{ with } T^+_i = T^+_f. \end{cases}$$

We now compute the probability of a transcript being bad in the ideal world. Using the union bound, we have

$$\Pr[\Lambda_0 \in \Omega_{\mathsf{bad}}] = \Pr[\mathsf{Coll} \vee \mathsf{mColl} \vee \mathsf{Forge}]. \tag{3}$$

Bounding Coll**:** For this event to happen, we know that there exists at least one pair of indices $(i', j') < (i, j)$ such that $\mathsf{LLCP}(i) < j \le a_i + m_i + 1$ and $s^\star_i[j] = s^\circledast_{i'}[j']$. For any value of $j \in [1, a_i + m_i + 1]$, we have,

$$s^\star_i[j] = s^\star_{i'}[j] \Leftrightarrow R(s^\star_i[j-1]) \oplus R(s^\star_{i'}[j-1]) = x^\star_i[j] \oplus x^\circledast_{i'}[j] \tag{4}$$

where $x^\star_i[j]$ and $x^\circledast_{i'}[j]$ are two n-bit strings. Note that if $j > a_i + 1$, the first r-bits of $x^\star_i[j]$ is xored with message block if $\star = +$ or the first r-bits of $x^\star_i[j]$ is replaced by the ciphertext block if $\star \in \{-, \times\}$. Similarly, if $j > a_i + 1$ the first r-bits of $x^\circledast_{i'}[j]$ is xored with message block if $\circledast = +$ or the first r-bits of $x^\circledast_{i'}[j]$ is replaced by a ciphertext block if $\circledast \in \{-, \times\}$. On the other hand, if $j \le a_i + 1$, then first r-bits of $x^\star_i[j]$ and $x^\circledast_{i'}[j]$ is xored with associated data or nonce for $\star, \circledast \in \{+, -, \times\}$. First, let us consider the case when $j = j' = \mathsf{LLCP}(i) + 1$ and $j > 1$. Then there exists some i' such that $i' < i$ and $\mathsf{LLCP}(i)$ holds for i'. Now we consider the values of $s^\star_i[j-1]$ and $s^\circledast_{i'}[j'-1]$. Based on the values of j, we get following cases.

1. <u>Case $j < a^\star_i + 1$.</u> In this case, we have $D^\star_i[1..j-1] = D^\circledast_{i'}[1..j-1]$ i.e., $A^\star_i[1..j] = A^\circledast_{i'}[1..j]$ Then from the Equation (2), we have $s^\star_i[j-1] = s^\circledast_{i'}[j'-1]$.
2. <u>Case $j = a^\star_i + 1$.</u> In this case, $\mathsf{LLCP}(i) = a^\star_i$. Thus, $D^\star_i[1..a^\star_i] = D^\circledast_{i'}[1..a^\star_i]$ i.e., $A^\star_i = A^\circledast_{i'}$. Then from the Eq. (2), we have $s^\star_i[j-1] = s^\circledast_{i'}[j'-1]$.
3. <u>Case $j = a^\star_i + 2$.</u> In this case, the associated date and nonce in i-th and i'-th query matches. So, we have $s^\star_i[j-1] = s^\circledast_{i'}[j'-1]$.
4. <u>Case $a^\star_i + 2 \le j \le a^\star_i + m^\star_i + 1$.</u> In this case, the nonce and associated data in the i-th and i'-th query matches. Also, the message/ciphertext in i-th and i'-th query matches up to $(j - a^\star_i - 2)$-th block. Thus from the Eq. (10), we have $s^\star_i[j-1] = s^\circledast_{i'}[j'-1]$.

Therefore, for any value of $j \in [2, a_i + m_i + 1]$, $s_i^\star[j-1] = s_{i'}^{\circledast}[j'-1]$. Thus, the probability of the event,

$$s_i^\star[j] = s_{i'}^\star[j] \Leftrightarrow R(s_i^\star[j-1]) \oplus R(s_{i'}^\star[j-1]) = x_i^\star[j] \oplus x_{i'}^{\circledast}[j] \Leftrightarrow x_i^\star[j] \oplus x_{i'}^{\circledast}[j] = 0$$

is zero. On the other hand, for all $i' \leq i$ and $j' \neq j$ or $j \neq \mathsf{LLCP}(i) + 1$, $R(s_i^\star[j-1]) \oplus R(s_{i'}^\star[j-1]) = x_i^\star[j] \oplus x_{i'}^{\circledast}[j]$ holds with probability at most 2^{-n}. Thus, the probability that two states collide is 2^{-n}. Note that there are σ possible values of (i, j) in a transcript, each having no more than σ possible values of (i', j'), where σ is the total number of permutation calls including all encryption, decryption and verification queries, such that $s_i^\star[j] = s_{i'}^{\circledast}[j']$ holds. Therefore, we have

$$\Pr[\mathsf{Coll}] \leq \binom{\sigma}{2} \frac{1}{2^n} \leq \frac{\sigma^2}{2^{n+1}}. \tag{5}$$

Bounding mColl**:** We bound this event by conditioning the event that Coll does not occur. As there are no non-trivial collisions, $\mathsf{ancestor}(s_{i_1}^+[j_1])$, $\mathsf{ancestor}(s_{i_2}^+[j_2])$, ..., $\mathsf{ancestor}(s_{i_\rho}^+[j_\rho])$ are all distinct and fresh. Therefore, all the outputs $R(s_{i_1}^+[j_1])$, $R(s_{i_2}^+[j_2]), ..., R(s_{i_\rho}^+[j_\rho])$ are all uniformly sampled over $\{0,1\}^n$. Thus, from the randomness of R, we can view this event as throwing σ_e balls into 2^r bins (as we are seeking collisions in the rate part) uniformly at random, where σ_e denotes the total number of primitive calls including all encryption queries and we want to find the probability that there is a bin that contains ρ or more balls. In other words, ρ or more outputs take some constant value c. This event occurs with probability at most $(\frac{1}{2^r})^\rho$. Again, we have 2^r choices for the constant value c. Therefore, by varying the choices of all encryption queries, we have

$$\Pr[\mathsf{mColl} \mid \overline{\mathsf{Coll}}] \leq \binom{\sigma_e}{\rho} \times 2^r (\frac{1}{2^r})^\rho \leq \frac{\sigma_e^\rho}{(2^r)^{\rho-1}}, \tag{6}$$

where the last inequality follows from Stirling's approximation ignoring the constant term.

Bounding Forge**:** We fix a verification query $(N_i^\times, A_i^\times, C_i^\times, T_i^\times)$ with associated data length $a_i^\times$ and ciphertext length $c_i^\times$ such that $\forall a_i^\times + 1 \leq j \leq (a_i^\times + c_i^\times), \exists i', j'$ s.t $s_i^\times[j] = s_{i'}^\star[j']$ where $\star \in \{+,-\}$. Let $\overline{j}$ be the largest index for which $s_i^\times[\overline{j}]$ does have a trivial collision with $s_{i'}^\star[j]$. We bound the probability of Forge when Coll and mColl do not occur. Now, we consider the following two cases based on the values of $\overline{j}$.

(a) Consider $a_i^\times + 1 \leq \overline{j} < a_i^\times + c_i^\times$. In this case the associated data, nonce and some parts of the message match with some previous query. Here, the adversary can control the rate part and so $s_i^\times[\overline{j}+1]$ matches with some encryption or decryption query with probability at most $\frac{\rho+\sigma_d}{2^c}$.

(b) Finally, consider the case $\overline{j} = a_i^\times + c_i^\times$. So the final state matches with some previous encryption query with probability at most $\frac{\rho}{2^c}$.

Combining everything together and by varying over the choices of all the verification queries,

$$\Pr[\mathsf{Forge} \mid \overline{\mathsf{Coll}} \wedge \overline{\mathsf{mColl}}] \leq \frac{q_v(2\rho + \sigma_d)}{2^c}. \tag{7}$$

From Eq. (3)–Eq. (7), we obtain the probability of a transcript being bad as,

$$\begin{aligned}\Pr[\Lambda_0 \in \Omega_{\mathsf{bad}}] &\leq \Pr[\mathsf{Coll}] + \Pr[\mathsf{mColl} \mid \overline{\mathsf{Coll}}] + \Pr[\mathsf{Forge} \mid \overline{\mathsf{Coll}} \wedge \overline{\mathsf{mColl}}] \\ &\leq \frac{\sigma^2}{2^{n+1}} + \frac{\sigma_e^\rho}{(2^r)^{\rho-1}} + \frac{q_v(2\rho + \sigma_d)}{2^c}. \end{aligned} \tag{8}$$

4.2 Analysis of the Good Transcripts

In this section, we show that for a good transcript $\omega_{\mathtt{new}} \in \Omega_{\mathtt{new}}$, the probability of realizing $\omega_{\mathtt{new}}$ in the real world is as likely as in the ideal world. It is easy to see that for a good transcript $\omega_{\mathtt{new}}$, we have that $\Pr[\Lambda_1^+ = \omega_{\mathtt{new}}^+, \Lambda_1^- = \omega_{\mathtt{new}}^-] = \Pr[\Lambda_0^+ = \omega_{\mathtt{new}}^+, \Lambda_0^- = \omega_{\mathtt{new}}^-]$. Thus, the ratio of interpolation probabilities is given by

$$\begin{aligned}\frac{\Pr[\Lambda_1 = \omega_{\mathtt{new}}]}{\Pr[\Lambda_0 = \omega_{\mathtt{new}}]} &= \Pr[\Lambda_1^\times = \omega_{\mathtt{new}}^\times | \Lambda_1^+ = \omega_{\mathtt{new}}^+, \Lambda_1^+ = \omega_{\mathtt{new}}^+] \\ &\geq 1 - \Pr[\Lambda_1^\times \neq \omega_{\mathtt{new}}^\times \mid \Lambda_1^+ = \omega_{\mathtt{new}}^+, \Lambda_1^- = \omega_{\mathtt{new}}^-],\end{aligned}$$

where we have used the fact that $\Pr[\Lambda_0^\times = \omega_{\mathtt{new}}^\times \mid \Lambda_0^+ = \omega_{\mathtt{new}}^+, \Lambda_0^- = \omega_{\mathtt{new}}^-] = 1$, because in the ideal world, the response to any verification query is $\perp$. For $i \in [q_v]$, let $\mathbf{E}_i$ denote the event $\mathbf{E}_i \triangleq \left(\bar{\lambda}^i \neq \perp \mid \Lambda_1^+ = \omega_{\mathtt{new}}^+, \Lambda_1^- = \omega_{\mathtt{new}}^-\right)$, where $\bar{\lambda}^i$ be the random variable that denotes the response to the i-th verification query in the real world. Then, we have

$$\Pr[\Lambda_1^\times \neq \omega_{\mathtt{new}}^\times \mid \Lambda_1^+ = \omega_{\mathtt{new}}^+, \Lambda_1^- = \omega_{\mathtt{new}}^-] \leq \sum_{i \in [q_v]} \Pr[\mathbf{E}_i].$$

We fix $i \in [q_v]$ and let the i-th verification query be $(N_i^\times, A_i^\times, C_i^\times, T_i^\times)$, of associated data length $a_i^\times$ and ciphertext length $c_i^\times$. We want to bound $\Pr[\mathbf{E}_i]$. This probability is non-zero only if $\Theta.\mathsf{ver}_R$ returns anything other than $\perp$. If $s_i^\times[a_i^\times + c_i^\times + 1]$ does not match with the final state of any encryption query, $\Pr[\mathbf{E}_i] \leq \frac{1}{2^\tau}$ holds trivially. Suppose, there exists some encryption query such that the final state matches with $s_i^\times[a_i^\times + c_i^\times + 1]$. In this case, there must exist some j such that $s_i^\times[j]$ is fresh, otherwise Forge is true. Let j^* be the maximum of such j. Then, $s_i^\times[j^* + 1]$ matches with some previous encryption or decryption state with probability at most $\frac{(\sigma_e + \sigma_d)}{2^c}$, as the adversary can control only the rate-part of these states. Putting everything together we get:

$$\Pr[\mathbf{E}_i] \leq \frac{1}{2^\tau} + \frac{(\sigma_e + \sigma_d)}{2^c}.$$

Therefore, we have

$$\Pr[\Lambda_1^{\times} \neq \omega_{\mathtt{new}}^{\times} \mid \Lambda_1^{+} = \omega_{\mathtt{new}}^{+}, \Lambda_1^{-} = \omega_{\mathtt{new}}^{-}] \leq \frac{q_v}{2^{\tau}} + \frac{q_v(\sigma_e + \sigma_d)}{2^c}. \quad (9)$$

The result follows as we combine Eq. (8), Eq. (9) and Theorem 1. □

SIGNIFICANCE OF INT-RUP SECURITY OF SAEB: If we instantiate SAEB with AES-128 block cipher and restrict its message *injection* rate to $r = 32$-bits, then the capacity c will be of 96-bits, and hence SAEB-32 will provide INT-RUP security up to 2^{48} decryption/verification blocks[4], which satisfies the NIST criteria of having 2^{50} byte of data-complexity. However, for SAEAES, where the message injection rate is $r = 64$-bits, we only achieve 2^{32} block of INT-RUP security. This result signifies that if we wish to have an INT-RUP secure variant of SAEB, we can simply use the same construction as SAEAES but with a lower message injection rate.

5 TinyJAMBU and Its INT-RUP Security

TinyJAMBU is one of the finalists of the NIST lightweight competition. The design principle of TinyJAMBU follows the sponge duplex mode based on keyed permutations derived from lightweight LFSRs. Unlike SAEB, TinyJAMBU injects the message in a specific part of the state and squeeze from a different part of the state to output the ciphertext. Therefore, we have a r-bit message injection part, r-bit squeezing part, and a c-bit unaltered capacity part, which is xored with the frame constants. Together, we have a total state size of $n = c + 2r$-bits. TinyJAMBU uses two different keyed permutations with the same key K. These two keyed permutations are similar in structure but differs only in the number of rounds. One permutation consists of 384 rounds of a LFSR, which we denote as $P_K^{(1)}$ and the another permutation consists of 1024 rounds of the same LFSR, which we denote as $P_K^{(2)}$. The encryption and the decryption algorithm of TinyJAMBU starts with the Init function that mixes the key K and the nonce N to produce a pseudorandom state. In particular, the Init function consists of two steps: key set up phase and nonce set up phase. In the key setup phase, an 128-bit register is initialized with all 0 and update the state by the keyed permutation $P_K^{(2)}$. In the nonce setup phase, an 96-bits nonce N is splitted up into three 32-bits nonces $N[1] \| N[2] \| N[3]$. Followed by it, for each $i \in \{1, 2, 3\}$, it updates the intermediate state s by xoring $0 \| \mathsf{const}_i \| 0$ with the current value of s. Then, invoke $P_K^{(1)}$ on s and finally xor the output with $N[i]$ to update the intermediate state s.

The algorithmic description of the encryption function of TinyJAMBU is given in Fig. 3, and its schematic diagram is depicted in Fig. 4. In the original proposal of the scheme [39], the recommended parameters of TinyJAMBU are $r = 32$, $c = 64$. const_i denotes 3-bits frame constants for $i \in \{1, 2, 3, 4\}$, where $\mathsf{const}_1 = 001, \mathsf{const}_2 = 011, \mathsf{const}_3 = 101, \mathsf{const}_4 = 111$.

[4] Security bound of the SAEB is moot if the number of encryption blocks exceeds 2^{32}.

TinyJAMBU(K, N, A, M)

1 : $A[1]\|\ldots\|A[a] \xleftarrow{r} A;\ M[1]\|\ldots\|M[m] \xleftarrow{r} M;$
2 : $s[0] \leftarrow \mathsf{Init}(K, N);$
3 : **for** $i = 1$ **to** $a - 1;$
4 : $\quad s[i-1] \leftarrow s[i-1] \oplus (0\|\mathsf{const}_2\|0);$
5 : $\quad s[i] \leftarrow P_K^{(1)}(s[i-1]) \oplus A[i];$
6 : $s[a-1] \leftarrow s[a-1] \oplus (0\|\mathsf{const}_2\|0);$
7 : $s[a] \leftarrow P_K^{(1)}(s[a-1]) \oplus A[a]\|10^{r-|A[a]|-1}\|lp(A[a]);$
8 : **for** $i = 1$ **to** $m - 1;$
9 : $\quad s[a+i-1] \leftarrow s[a+i-1] \oplus (0\|\mathsf{const}_3\|0);$
10 : $\quad s[a+i] \leftarrow P_K^{(2)}(s[a+i-1]) \oplus M[i];$
11 : $\quad C[i] \leftarrow M[i] \oplus s[a+i][64...(64+r-1)];$
12 : $s[a+m-1] \leftarrow s[a+m-1] \oplus (0\|\mathsf{const}_3\|0);$
13 : $s[a+m] \leftarrow P_K^{(2)}(s[a+m-1]) \oplus M[m]\|10^{r-|M[m]|-1}\|0^r\|lp(M[m]);$
14 : $C[m] \leftarrow M[m] \oplus s[a+m][64...(64+|M[m]|-1)];$
15 : $s[a+m] \leftarrow P_K^{(2)}(s[a+m] \oplus (0\|\mathsf{const}_4\|0));$
16 : $T_1 \leftarrow s[a+m][64...(64+\tau/2-1)];$
17 : $s[a+m] \leftarrow P_K^{(1)}(s[a+m] \oplus (0\|\mathsf{const}_4\|0));$
18 : $T_2 \leftarrow s[a+m][64...(64+\tau/2-1)];$
19 : $T \leftarrow T_1\|T_2;$
return $(C, T);$

Fig. 3. Formal specification of TinyJAMBU authenticated encryption mode. The function $lp(X) := \lfloor |X|/8 \rfloor$ denotes the binary representation of the number of bytes present in a binary string X.

TinyJAMBU achieves birthday bound security with dominant terms being $(e\sigma_e/\rho 2^r)^\rho(2^r/\sqrt{\rho}) + (\sigma_a + \sigma_d)(\rho-1)/2^{c+(r/2)+1}$, where ρ is a properly chosen constant. In the following, we state and prove the INT-RUP security result of TinyJAMBU*[P].

Theorem 3. *Let* $\mathsf{P} \leftarrow_{\$} \mathsf{Perm}(n)$ *be an* n*-bit uniform random permutation. Let* r, c *and* ρ *be three parameters such that* $n = c + 2r$ *and let* τ *be the bit size of the tag output by* TinyJAMBU. *The* INT-*RUP advantage for any* $(q_e, q_d, q_v, \sigma_e, \sigma_d, \sigma_v)$*-distinguisher* $\mathscr{A}$ *against the construction* TinyJAMBU*[P] *that makes at most* q_e *encryption,* q_d *decryption and* q_v *verification queries with at most* σ_e *primitive calls with distinct inputs in encryption queries,* σ_d *primitive calls with distinct inputs in decryption queries and* σ_v *primitive calls with distinct inputs in verification queries having a total of* $\sigma = \sigma_e + \sigma_d + \sigma_v$ *primitive calls with distinct inputs such that* $\rho \le \sigma_e$ *is given by*

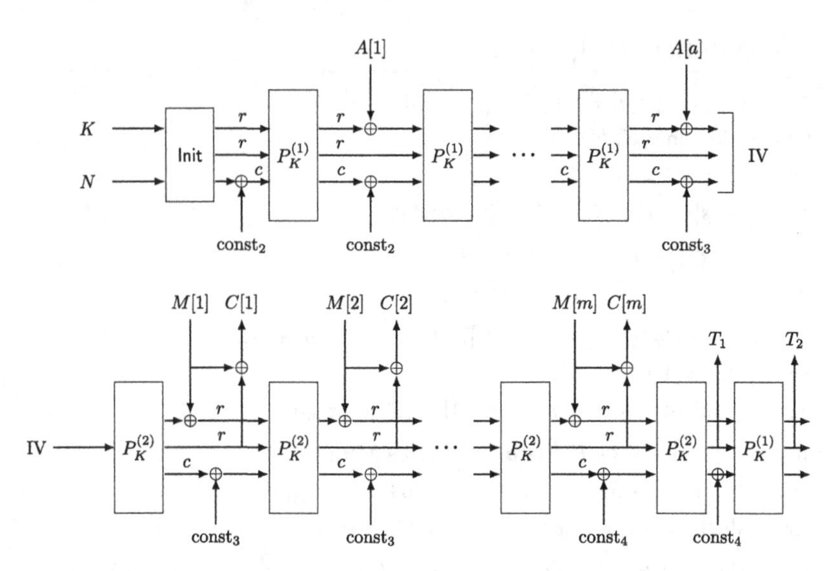

Fig. 4. TinyJAMBU Authenticated Encryption for a block associated data, and m block message. $\mathsf{const}_2 = 011$, $\mathsf{const}_3 = 101$, $\mathsf{const}_3 = 111$ are small distinct constants used for domain separation. 32-bits of the message and associated data are injected to the construction.

$$\mathbf{Adv}^{\text{int-rup}}_{\mathsf{TinyJAMBU}^*}(\mathscr{A}) \leq \frac{\sigma^2}{2^{n+1}} + \frac{\sigma_e^\rho}{(2^r)^{\rho-1}} + \frac{q_v(2\rho + \sigma_d)}{2^{n-r}} + \frac{q_v}{2^\tau} + \frac{q_v(\sigma_e + \sigma_d)}{2^{n-r}}.$$

Remark 1. Recently two independent works by Sibleyras et al. [36] and Dunkelman et al. [25] have shown some vulnerabilities on the underlying permutation of TinyJAMBU. Note that these results do not imply any insecurity of the mode TinyJAMBU per se. In this paper, we prove the INT-RUP security of TinyJAMBU by viewing it as a mode which is build on top of some secure keyed permutations. Note that the construction uses two different keyed permutations with the same key but differs only in the number of rounds, we model the security proof of the construction in the standard setting, where we replace these two keyed permutations with an *n-bit uniform random permutation* and denote the resulting construction as TinyJAMBU*[P].

Proof. We proceed similar to the proof of Theorem 2. We modify the construction by replacing the permutations with random function $R \leftarrow_\$ \mathsf{Funcs}(\{0,1\}^n)$ and denoting the resulting construction as Θ. We consider a distinguisher $\mathscr{A}$, that interacts with a triplet of oracles in either of the worlds: in the real world, it interacts with $(\Theta.\mathsf{enc}_R, \Theta.\mathsf{dec}_R, \Theta.\mathsf{ver}_R)$ and in the ideal world it interacts with $(\Theta.\mathsf{enc}_R, \Theta.\mathsf{dec}_R, \perp)$. We summarize the queries in a transcript ω and segregate the transcript ω into three parts ω^+, ω^-, and $\omega^\times$ as described for the analysis of SAEB and we have $\omega = \omega^+ \cup \omega^- \cup \omega^\times$. The length of associated data, message and ciphertext in the i-th query is denoted as $a_i^\star, m_i^\star$ and $c_i^\star$ respectively where $\star \in \{+, -, \times\}$. We denote the j-th input to R in the i-th encryption, decryption and verification query as $s_i^+[j], s_i^-[j]$ and $s_i^\times[j]$ respectively and in general $s_i^\star[j]$ where $\star \in \{+,-,\times\}$. The sequence of internal state values in the real world for i-th encryption query of length a_i+m_i are defined according to the construction. as follows:

$$\begin{cases} s_i^+[0] = R(0) \\ s_i^+[1] = R(s_i^+[0] \oplus 0\|\langle \mathsf{const}_1\rangle_2\|0) \oplus N_i^+[1] \\ s_i^+[2] = R(s_i^+[1] \oplus 0\|\langle \mathsf{const}_1\rangle_2\|0) \oplus N_i^+[2] \\ s_i^+[3] = R(s_i^+[2] \oplus 0\|\langle \mathsf{const}_1\rangle_2\|0) \oplus N_i^+[3] \\ s_i^+[k+3] = R(s_i^+[k-1] \oplus 0\|\langle \mathsf{const}_2\rangle_2\|0) \oplus A_i^+[k], \text{ for } 1 \le k < a_i^+ \\ s_i^+[a_i+3] = R(s_i^+[a_i^+-1] \oplus 0\|\langle \mathsf{const}_2\rangle_2\|0) \oplus lp(A_i^+[a_i]) \\ s_i^+[a_i+3+k] = R(s_i^+[k-1] \oplus 0\|\langle \mathsf{const}_3\rangle_2\|0) \oplus M_i^+[k], \text{ for } 1 \le k < m_i \\ s_i^+[a_i+3+m_i] = R(s_i^+[a_i^+ + m_i^+] \oplus 0\|\langle \mathsf{const}_3\rangle_2\|0) \oplus lp(M_i^+[m_i]) \oplus 0\|\langle \mathsf{const}_4\rangle_2\|0 \end{cases} \tag{10}$$

In the above description, the nonce in the i-th encryption query is $N_i^+ = N_i^+[1]\|N_i^+[2]\|N_i^+[3]$ is processed in three steps. We modify the experiment by releasing all internal state values to the adversary $\mathscr{A}$ before it outputs the decision bit b, but after it makes all the queries. As the verification oracle $\perp$ in the ideal world always returns reject and does not compute anything, the internal state variables are not defined. Thus, similar to the proof of Theorem 2, we have to define the sampling of the intermediate state variables for every verification query in the ideal world. To achieve this, for every verification query (N, A, C, T) made by adversary $\mathcal{A}$, the verification oracle for the ideal world invokes the decryption oracle $\Theta.\mathsf{dec}_R$ with input (N, A, C) ignoring the output of r-bit plaintext strings. Finally, the verification oracle ignores the checking of the computed tag T^* with the given tag T. Hence, the sequence of internal state values is defined for verification queries. Let the modified attack transcripts be $\omega_{\texttt{new}} = \omega_{\texttt{new}}^+ \cup \omega_{\texttt{new}}^- \cup \omega_{\texttt{new}}^\times$. For a given transcript $\omega_{\texttt{new}}$, we reorder the transcript in such a way that all the encryption queries appear first, followed by all the decryption queries and finally, all the verification queries. Now, it is easy to see that a state collision in $s_i^\star[j]$ and $s_{i'}^\circledast[j]$ occurs with probability 1 if

- $N_i^\star = N_{i'}^\circledast$, $A_i^\star[\ldots j] = A_{i'}^\circledast[\ldots j]$, when $j \le a_i^\star$,
- $N_i^\star = N_{i'}^\circledast$, $A_i^\star = A_{i'}^\circledast$, $M_i^\star[\ldots(j-a_i^\star)] = M_{i'}^\circledast[\ldots(j-a_i^\star)]$, when $a_i^\star < j \le a_i^\star + m_i^\star$.

A state collision with probability 1 is called trivial, and any other state collision is called non-trivial. We denote $D_i^\star = N_i^\star\|A_i^\star$ and $d_i^\star = a_i^\star+3$ where $\star \in \{+,-,\times\}$. Let us consider two query $\star, \circledast \in \{+,-,\times\}$ with distinct query indices i and i', where $i, i' \in [q_e+q_d+q_v]$. Similar to the proof of Theorem 2, we use the notations $\mathsf{ancestor}$, $\mathsf{LCP}(i,i')$ and $\mathsf{LLCP}(i)$.

5.1 Definition and Probability of Bad Transcripts

In this section, we define and bound the probability of bad transcripts in the ideal world. The idea of this proof is almost similar to that of Theorem 2. However, the bounds are different because the adversary cannot control the rate part from where the squeezing occurs. Let Ω be the set of all attainable transcripts and $\omega_{\mathtt{new}} \in \Omega$ be one such attainable transcript. We say that transcript $\omega_{\mathtt{new}}$ is **bad**, i.e., $\omega_{\mathtt{new}} \in \Omega_{\mathtt{bad}}$, if at least one of the following holds:

1. Coll: there exists i, j, i', j' with $(i', j') < (i, j)$ with $\mathsf{LLCP}(i) < j \leq a_i + m_i + 3$ and $i \leq [q_e + q_d + q_v]$ such that $s_i^\star[j] = s_{i'}^{\circledast}[j']$, where $\star, \circledast \in \{+, -, \times\}$.
2. mColl: $\exists i_1, j_1, \ldots, i_\rho, j_\rho$ with $\{i_1, \ldots, i_\rho\} \subseteq [q_e]$ and for all $1 \leq k \leq \rho$, $j_k \in [m_{i_k}]$, such that $C_{i_1}^+[j_1] = \cdots = C_{i_\rho}^+[j_\rho]$.
3. Forge: This event happens, if for some verification query, all its intermediate states prior to the final state match with the intermediate state from some encryption or decryption queries, and the final state of the verification query matches with the final state of the encryption query for which the tag matches. In other words, $\exists i \in [q_v]$ such that for the i-th verification query $(N_i^\times, A_i^\times, C_i^\times, T_i^+)$, the following two events hold:

$$\begin{cases} \forall j \in [a_i^\times, (a_i^\times + c_i^\times - 1)], \exists i', j' \text{ such that } s_i^\times[j] = s_{i'}^+[j'] \text{ or } s_i^\times[j] = s_{i'}^-[j'], \\ \exists f \in [q_e] \text{ such that } s_i^\times[a_i^\times + c_i^\times] = s_f^+[a_f^+ + c_f^+] \text{ with } T_i^+ = T_f^+. \end{cases}$$

We now compute the probability of a transcript being bad in the ideal world. If $\omega_{\mathtt{new}}$ is a transcript observed in the ideal world, we want to calculate the probability of $\omega_{\mathtt{new}}$ to satisfy one of the above conditions. By applying the union bound, we have

$$\begin{aligned} \Pr[\Lambda_0 \in \Omega_{\mathtt{bad}}] &= \Pr[\mathsf{Coll} \vee \mathsf{mColl} \vee \mathsf{Forge}] \\ &\leq \Pr[\mathsf{Coll}] + \Pr[\mathsf{mColl} \mid \overline{\mathsf{Coll}}] + \Pr[\mathsf{Forge} \mid \overline{\mathsf{Coll}} \wedge \overline{\mathsf{mColl}}] \\ &\leq \frac{\sigma^2}{2^{n+1}} + \frac{\sigma_e^\rho}{(2^r)^{\rho-1}} + \frac{q_v(2\rho + \sigma_d)}{2^{n-r}}. \end{aligned} \tag{11}$$

The calculation for the bounds for each of the above terms are similar to the one used for SAEB, and the details can be found in the full version [22].

5.2 Analysis of the Good Transcripts

In this section, we show that for a good transcript $\omega_{\mathtt{new}} \in \Omega_{\mathtt{new}}$, the probability of realizing $\omega_{\mathtt{new}}$ in the real world is as likely as in the ideal world. It is easy to see that for a good transcript $\omega_{\mathtt{new}}$, we have $\Pr[\Lambda_1^+ = \omega_{\mathtt{new}}^+, \Lambda_1^- = \omega_{\mathtt{new}}^-] = \Pr[\Lambda_0^+ = \omega_{\mathtt{new}}^+, \Lambda_0^- = \omega_{\mathtt{new}}^-]$. Thus, the ratio of interpolation probabilities is given by

$$\begin{aligned} \frac{\Pr[\Lambda_1 = \omega_{\mathtt{new}}]}{\Pr[\Lambda_0 = \omega_{\mathtt{new}}]} &= \Pr[\Lambda_1^\times = \omega_{\mathtt{new}}^\times | \Lambda_1^+ = \omega_{\mathtt{new}}^+, \Lambda_1^+ = \omega_{\mathtt{new}}^+] \\ &\geq 1 - \Pr[\Lambda_1^\times \neq \omega_{\mathtt{new}}^\times \mid \Lambda_1^+ = \omega_{\mathtt{new}}^+, \Lambda_1^- = \omega_{\mathtt{new}}^-]. \end{aligned} \tag{12}$$

Now let us bound the term in the right hand side of the equation. For $i \in [q_v]$, let $\mathbf{E}_i$ denote the following event

$$\mathbf{E}_i \triangleq \left(\bar{\lambda}^i \neq \perp \mid \Lambda_1^+ = \omega_{\mathtt{new}}^+, \Lambda_1^- = \omega_{\mathtt{new}}^-\right),$$

where $\bar{\lambda}^i$ be the random variable that denotes the response to the i-th verification query in the real world. Then, we have

$$\Pr[\Lambda_1^\times \neq \omega_{\mathtt{new}}^\times \mid \Lambda_1^+ = \omega_{\mathtt{new}}^+, \Lambda_1^- = \omega_{\mathtt{new}}^-] \leq \sum_{i \in [q_v]} \Pr[\mathbf{E}_i].$$

We fix $i \in [q_v]$ and let the i-th verification query be $(N_i^\times, A_i^\times, C_i^\times, T_i^\times)$. Assume that the associated data length is $a_i^\times$, and the ciphertext length is $c_i^\times$. We want to bound $\Pr[\mathbf{E}_i]$. This probability is non-zero only if $\Theta.\mathsf{ver}_{R_1,R_2}$ returns anything other than $\perp$. If $s_i^\times[a_i^\times + c_i^\times]$ does not match with the final state of any encryption query, we have $\Pr[\mathbf{E}_i] \leq \frac{1}{2^\tau}$. Suppose there is some encryption query such that the final state matches with $s_i^\times[a_i^\times + c_i^\times]$. In this case, there must be some j such that $s_i^\times[j]$ is fresh; otherwise Forge is true. Let j^* be the maximum of such j. Then, $s_i^\times[j^* + 1]$ matches with some previous encryption or decryption states with probability at most $\frac{(\sigma_e + \sigma_d)}{2^{n-r}}$. Note that the adversary can control only the first r-bits of the state. Putting everything together we get:

$$\Pr[\mathbf{E}_i] \leq \frac{1}{2^\tau} + \frac{(\sigma_e + \sigma_d)}{2^{n-r}}.$$

Therefore, we have

$$\begin{aligned} \frac{\Pr[\Lambda_1 = \omega_{\mathtt{new}}]}{\Pr[\Lambda_0 = \omega_{\mathtt{new}}]} &\geq 1 - (\sum_{i \in [q_v]} \frac{1}{2^\tau} + \frac{(\sigma_e + \sigma_d)}{2^{n-r}}) \\ &\geq 1 - (\frac{q_v}{2^\tau} + \frac{q_v(\sigma_e + \sigma_d)}{2^{n-r}}). \end{aligned} \tag{13}$$

Finally, we obtain the result combining Eq. (11), Eq. (13) and using Theorem 1. □

6 Conclusion and Future Works

In this paper, we have analyzed the INT-RUP security of SAEB and TinyJAMBU. Our analysis on TinyJAMBU is particularly relevant from the NIST lightweight competition perspective, and we believe that our result may positively impact TinyJAMBU during the final portfolio selection. Our result on SAEB depicts that the INT-RUP security can be achieved by controlling the message injection rate without incurring any additional overheads. A similar analysis for the permutation based constructions may be considered as a future research direction. However, we would like to mention that the trick, similar to SAEB, cannot be applied on TinyJAMBU as message injection and ciphertext release occur from different parts of its state. It would be interesting to come up with a matching INT-RUP attack on TinyJAMBU. Note that Oribatida achieves INT-RUP security, but at the cost of maintaining an additional state. Investigating INT-RUP secure permutation based sponge constructions without any additional overhead seems a challenging open problem.

References

1. Recommendation for Block Cipher Modes of Operation: Galois/Counter Mode (GCM) and GMAC. NIST Special Publication 800-38D. National Institute of Standards and Technology (2007)
2. AlTawy, R., Gong, G., He, M., Jha, A., Mandal, K., Nandi, M., Rohit, R.; SpoC: an authenticated cipher submission to the NIST LWC competition (2019). https://csrc.nist.gov/projects/lightweight-cryptography/round-2-candidates
3. Andreeva, E., Bhati, A.S., Vizar, D.: Nonce-misuse security of the SAEF authenticated encryption mode. Cryptology ePrint Archive, Report 2020/1524 (2020)
4. Andreeva, E., Bhati, A.S., Vizar, D.: Rup security of the SAEF authenticated encryption mode. Cryptology ePrint Archive, Report 2021/103 (2021)
5. Andreeva, E., et al.: COLM v1. CAESAR Competition
6. Andreeva, E., Bogdanov, A., Luykx, A., Mennink, B., Mouha, N., Yasuda, K.: How to securely release unverified plaintext in authenticated encryption. In: Sarkar, P., Iwata, T. (eds.) ASIACRYPT 2014. LNCS, vol. 8873, pp. 105–125. Springer, Heidelberg (2014). https://doi.org/10.1007/978-3-662-45611-8_6
7. Andreeva, E., Bogdanov, A., Luykx, A., Mennink, B., Tischhauser, E., Yasuda, K.: AES-COPA, vol 2. Submission to CAESAR (2015). https://competitions.cr.yp.to/round2/aescopav2.pdf
8. Andreeva, E., Lallemand, V., Purnal, A., Reyhanitabar, R., Roy, A., Vizár, D.: Forkcipher: a new primitive for authenticated encryption of very short messages. In: Galbraith, S.D., Moriai, S. (eds.) ASIACRYPT 2019. LNCS, vol. 11922, pp. 153–182. Springer, Cham (2019). https://doi.org/10.1007/978-3-030-34621-8_6
9. Ashur, T., Dunkelman, O., Luykx, A.: Boosting authenticated encryption robustness with minimal modifications. In: Katz, J., Shacham, H. (eds.) CRYPTO 2017. LNCS, vol. 10403, pp. 3–33. Springer, Cham (2017). https://doi.org/10.1007/978-3-319-63697-9_1
10. Banik, S., Bogdanov, A., Luykx, A., Tischhauser, E.: Sundae: small universal deterministic authenticated encryption for the internet of things. IACR Trans. Symmetric Cryptol. **3**, 2018 (2018)
11. Beierle, C., et al.: SKINNY-AEAD and skinny-hash. IACR Trans. Symmetric Cryptol. **2020**(S1), 88–131 (2020)
12. Bellare, M., Rogaway, P.: The security of triple encryption and a framework for code-based game-playing proofs. In: Vaudenay, S. (ed.) EUROCRYPT 2006. LNCS, vol. 4004, pp. 409–426. Springer, Heidelberg (2006). https://doi.org/10.1007/11761679_25
13. Bhattacharjee, A., López, C.M., List, E., Nandi, M.: The oribatida v1.3 family of lightweight authenticated encryption schemes. J. Math. Cryptol. **15**(1) (2021)
14. CAESAR: Competition for Authenticated Encryption: Security, Applicability, and Robustness (2014). http://competitions.cr.yp.to/caesar.html
15. Chakraborti, A., Datta, N., Jha, A., Mancillas-López, C., Nandi, M., Sasaki, Yu.: INT-RUP secure lightweight parallel AE modes. IACR Trans. Symmetric Cryptol. **2019**(4), 81–118 (2019)
16. Chakraborti, A., Datta, N., Jha, A., Mitragotri, S., Nandi, M.: From combined to hybrid: Making feedback-based AE even smaller. IACR Trans. Symmetric Cryptol. **2020**(S1), 417–445 (2020)
17. Chakraborti, A., Datta, N., Nandi, M.: INT-RUP analysis of block-cipher based authenticated encryption schemes. In: Sako, K. (ed.) CT-RSA 2016. LNCS, vol. 9610, pp. 39–54. Springer, Cham (2016). https://doi.org/10.1007/978-3-319-29485-8_3

18. Chakraborti, A., Datta, N., Nandi, M., Yasuda, K.: Beetle family of lightweight and secure authenticated encryption ciphers. IACR Trans. Cryptogr. Hardw. Embed. Syst. **2018**(2), 218–241 (2018)
19. Chakraborti, A., Iwata, T., Minematsu, K., Nandi, M.: Blockcipher-based authenticated encryption: how small can we go? In: CHES 2017, Proceedings, pp. 277–298 (2017)
20. Chang, D., et al.: Release of unverified plaintext: tight unified model and application to ANYDAE. IACR Trans. Symmetric Cryptol. **2019**(4), 119–146 (2019)
21. Chang, D., Nandi, M.: A short proof of the PRP/PRF switching lemma. IACR Cryptol. ePrint Arch. **2008**, 78 (2008)
22. Datta, N., Dutta, A., Ghosh, S.: INT-RUP security of SAEB and tinyjambu. Cryptology ePrint Archive, Paper 2022/1414 (2022). https://eprint.iacr.org/2022/1414
23. Datta, N., Luykx, A., Mennink, B., Nandi, M.: Understanding RUP integrity of COLM. IACR Trans. Symmetric Cryptol. **2017**(2), 143–161 (2017)
24. Dobraunig, C., Eichlseder, M., Mendel, F., Schläffer, M.: Ascon v1.2. Submission to CAESAR (2016). https://competitions.cr.yp.to/round3/asconv12.pdf
25. Dunkelman, O., Lambooij, E., Ghosh, S.: Practical related-key forgery attacks on the full tinyjambu-192/256. Cryptology ePrint Archive, Paper 2022/1122 (2022)
26. Iwata, T., Khairallah, M., Minematsu, K., Peyrin, T.: Duel of the titans: the romulus and remus families of lightweight AEAD algorithms. IACR Trans. Symmetric Cryptol. **2020**(1), 43–120 (2020)
27. Iwata, T., Minematsu, K., Guo, J., Morioka, S., Kobayashi, E.: CAESAR Candidate CLOC. DIAC (2014)
28. Jean, J., Nikolic, I., Peyrin, T., Seurin, Y.: The deoxys AEAD family. J. Cryptol. **34**(3), 31 (2021)
29. McKay, K.A., Bassham, L., Turan, M.S., Mouha, N.: Report on lightweight cryptography (2017). http://nvlpubs.nist.gov/nistpubs/ir/2017/NIST.IR.8114.pdf
30. Minematsu, K.: AES-OTR v3.1. Submission to CAESAR (2016). https://competitions.cr.yp.to/round3/aesotrv31.pdf
31. Montes, M., Penazzi, D.: AES-CPFB v1. Submission to CAESAR (2015). https://competitions.cr.yp.to/round1/aescpfbv1.pdf
32. Naito, Y., Matsui, M., Sakai, Y., Suzuki, D., Sakiyama, K., Sugawara, T.: SAEAES: submission to NIST LwC (2019). https://csrc.nist.gov/CSRC/media/Projects/lightweight-cryptography/documents/round-2/spec-doc-rnd2/SAEAES-spec-round2.pdf
33. Naito, Y., Matsui, M., Sugawara, T., Suzuki, D.: SAEB: A lightweight blockcipher-based AEAD mode of operation. IACR Trans. Cryptogr. Hardw. Embed. Syst. **2018**(2), 192–217 (2018)
34. Patarin, J.: The "coefficients h" technique. In: Selected Areas in Cryptography, pp. 328–345 (2008)
35. Rogaway, P.: Efficient instantiations of tweakable blockciphers and refinements to modes OCB and PMAC. In: Lee, P.J. (ed.) ASIACRYPT 2004. LNCS, vol. 3329, pp. 16–31. Springer, Heidelberg (2004). https://doi.org/10.1007/978-3-540-30539-2_2
36. Sibleyras, F., Sasaki, Y., Todo, Y., Hosoyamada, A., Yasuda, K.: Birthday-bound slide attacks on TinyJAMBU's keyed-permutations for all key sizes. In: Cheng, C.M., Akiyama, M. (eds.) IWSEC 2022. LNCS, vol. 13504, pp. 107–127. Springer, Cham (2022). https://doi.org/10.1007/978-3-031-15255-9_6
37. Wu, H.: ACORN: a lightweight authenticated cipher (v3). Submission to CAESAR (2016). https://competitions.cr.yp.to/round3/acornv3.pdf

38. Wu, H., Huang, T.: The JAMBU lightweight authentication encryption mode (v2.1). Submission to CAESAR (2016). https://competitions.cr.yp.to/round3/jambuv21.pdf
39. Wu, H., Huang, T.: TinyJAMBU: a family of lightweight authenticated encryption algorithms: submission to NIST LwC (2019). https://csrc.nist.gov/CSRC/media/Projects/lightweight-cryptography/documents/finalist-round/updated-spec-doc/tinyjambu-spec-final.pdf
40. Zhang, L., Wu, W., Sui, H., Wang, P.: iFeed[AES] v1. Submission to CAESAR (2014). https://competitions.cr.yp.to/round1/ifeedaesv1.pdf
41. Zhang, P., Wang, P., Hu, H.: The INT-RUP security of OCB with intermediate (parity) checksum. IACR Cryptology ePrint Archive (2017). https://eprint.iacr.org/2016/1059.pdf

Offset-Based BBB-Secure Tweakable Block-ciphers with Updatable Caches

Arghya Bhattacharjee[1], Ritam Bhaumik[2(✉)], and Mridul Nandi[1]

[1] Indian Statistical Institute, Kolkata, India
[2] Inria, Paris, France
bhaumik.ritam@gmail.com

Abstract. A nonce-respecting tweakable blockcipher is the building-block for the OCB authenticated encryption mode. An XEX-based TBC is used to process each block in OCB. However, XEX can provide at most birthday bound privacy security, whereas in Asiacrypt 2017, beyond-birthday-bound (BBB) forging security of OCB3 was shown in [14]. In this paper we study how at a small cost we can construct a nonce-respecting BBB-secure tweakable blockcipher. We propose the OTBC-3 construction, which maintains a cache that can be easily updated when used in an OCB-like mode. We show how this can be used in a BBB-secure variant of OCB with some additional keys and a few extra blockcipher calls but roughly the same amortised rate.

Keywords: OCB · tweakable block-cipher · Authenticated encryption · Updatable offsets · Beyond-birthday-bound security

1 Introduction

Authenticated encryption (AE) is a symmetric-key cryptographic function for providing a combined guarantee of privacy (or confidentiality) and authenticity (or integrity) of plaintexts. Beginning with the formalisation by Katz and Yung [35] and Bellare and Namprempre [11,12], and the constructions by Jutla [33,34], the practical significance of AE has been accepted in the community, and over the last decade or so the design and analysis of AE modes has been a very active area of research in symmetric-key cryptography.

Associated data (AD) is the data that is not confidential but contributes to the authentication of the message, and AE with associated data (AEAD), formalised by Rogaway [44], takes both a plaintext and some AD as input. AEAD ensures confidentiality of plaintexts and authenticity of both plaintexts and AD. The most popular form of AEAD is based on a nonce, and is called nonce-based AEAD (NAEAD). A nonce is a non-repeating value for each encryption, and can

R. Bhaumik—This project has received funding from the European Research Council (ERC) under the European Union's Horizon 2020 research and innovation programme (grant agreement no. 714294 - acronym QUASYModo).

T. Isobe and S. Sarkar (Eds.): INDOCRYPT 2022, LNCS 13774, pp. 171–194, 2022.
https://doi.org/10.1007/978-3-031-22912-1_8

be realised for instance with a counter. NAEAD is commonly built as a mode of operation of a blockcipher. However, there is often an inherent limitation on the security caused by the birthday paradox on the input or output of a blockcipher, which ensures only $(n/2)$-bit security of NAEAD if a blockcipher with n-bit blocks is used. The $(n/2)$-bit security is commonly referred to as birthday-bound (BB) security. Possible solutions to break this barrier exist, i.e., NAEAD with beyond-birthday-bound (BBB) security. However, they come with an extra computational cost.

One way to get around this obstacle is to use a tweakable blockcipher (TBC) as the underlying primitive instead of classical blockciphers. A TBC was formalised by Liskov, Rivest and Wagner [38,39], and it has an extra t-bit tweak input to provide variability, i.e., it provides a family of 2^t independent blockciphers indexed by the tweak. Starting from the early Hasty Pudding Cipher [49], many TBC designs have been proposed, including Threefish (in Skein [20]), Deoxys-BC [32], Joltik-BC [31], and KIASU-BC from the TWEAKEY framework [30], and Scream [22], where the last four schemes were submitted to CAESAR (Competition for Authenticated Encryption: Security, Applicability, and Robustness) [1]. We also see other examples including SKINNY [8,9], QARMA [6], CRAFT [10], the TBCs in the proposals for the NIST Lightweight Cryptography project [3], OPP [21] for permutation-based instantiations of OCB3 that uses a (tweakable) Even-Mansour construction, and a construction by Naito [41].

One of the most popular TBC-based NAEAD schemes is OCB. There are three main variants of OCB. The first, now called OCB1 (2001) [47], was motivated by Charanjit Jutla's IAPM [33,34]. A second version, now called OCB2 (2004) [2,45], added support for associated data (AD) and redeveloped the mode using the idea of a tweakable blockcipher. Later OCB2 was found to have a disastrous bug [26,27]. The final version of OCB, called OCB3 (2011) [37], corrected some missteps taken with OCB2 and achieved the best performance yet. OCB3 is simple, parallelisable, efficient, provably secure with BB security, and its security is well analysed [4,5,46]. It is specified in RFC 7253 [36] and was selected for the CAESAR final portfolio.

In recent times, OCB has been analysed in much detail from various perspectives. A blockcipher-based NAEAD scheme OTR and its TBC-based counterpart $\mathbb{OTR}$ were designed by Minematsu [40] which improve OCB by removing the necessity of the decryption routine of the underlying blockcipher or TBC (this property is often called as the inverse-freeness). Bhaumik and Nandi [14] showed that when the number of encryption query blocks is not more than birthday-bound (an assumption without which the privacy guarantee of OCB3 disappears), even an adversary making forging attempts with the number of blocks in the order of $2^n/\ell_{\mathrm{MAX}}$ (n being the block-size and ℓ_{MAX} being the length of the longest block) may fail to break the integrity of OCB3. Zhang et al. [51,52] described a new notion, called plaintext or ciphertext checksum (PCC), which is a generalisation of plaintext checksum (used to generate the tag of OCB), and proved that all authenticated encryption schemes with PCC are insecure in the INT-RUP security model. Then they fixed the weakness of PCC, and described a new approach called intermediate (parity) checksum (I(P)C for short). Based on the I(P)C approach, they provided two modified schemes OCB-IC and OCB-IPC to settle the INT-RUP of OCB in the nonce-misuse setting. They proved that OCB-IC and OCB-IPC are INT-RUP up to the birthday bound in the nonce-misuse setting if the underlying tweakable blockcipher is a secure mixed tweakable pseudorandom permutation (MTPRP). The security bound of OCB-IPC is proved to be tighter than OCB-IC. To improve their speed, they utilised a "prove-then-prune" approach: prove security and instantiate with a scaled-down primitive (e.g., reducing rounds for the underlying primitive invocations). Bao et al. [7] introduced a scheme called XTX*, based on previous tweak extension schemes for TBCs, and defined ZOCB and ZOTR for nonce-based authenticated encryption with associated data. While ΘCB and $\mathbb{OTR}$ have an independent part to process AD, their schemes integrated this process into the encryption part of a plaintext by using the tweak input of the TBC, and thus achieved full absorption and full parallelisability simultaneously.

OCB has also found its place in other domains of cryptology like lightweight cryptology and quantum cryptology. Chakraborti et al. [15] proposed a lightweight authenticated encryption (AE) scheme, called Light-OCB, which can be viewed as a lighter variant of OCB as well as a faster variant of LOCUS-AEAD [16] which has been a Round 2 candidate of the NIST Lightweight Cryptography project. Bhaumik et al. [13] proposed a new rate-one parallelisable mode named QCB inspired by TAE and OCB and prove its security against quantum superposition queries.

There are two limitations on OCB that we would like to emphasise. The first is that OCB's security crucially depends on the encrypting party not repeating a nonce. The mode should never be used in situations where that can't be assured; one should instead employ a misuse-resistant AE scheme [48]. These include AES-GCM-SIV [23,24], COLM, and Deoxys-II. A second limitation of OCB is its birthday-bound degradation in provable security. This limitation implies that, given OCB's 128-bit block-size, one must avoid operating on anything near 2^{64} blocks of data. The RFC on OCB [36] asserts that a given key should be used to encrypt at most 2^{48} blocks (4 petabytes), including the associated data. Practical AE modes that avoid the birthday-bound degradation in security are now known [1,24,28,29,43].

1.1 Our Contributions

In this paper we explore ways of designing an offset-based tweakable blockcipher that can be used to obtain an OCB-like authenticated encryption mode with better security guarantees. First we show that when using an n-bit nonce (where n is the width of the block-cipher) it is difficult to go beyond the birthday-bound if we use the same offset to mask the input and the output (OTBC-0). Next we show that if we take fully independent offsets for masking inputs and outputs for each message, we get full security in the nonce-respecting scenario (OTBC-1); however, this does not fit well in the OCB-like mode, because new additional random-function calls are needed to process each message block.

We proceed to introduce the notion of *updatable offsets*, and explain why TBCs with updatable offsets are well-suited to build an OCB-like mode. Then we build a simple TBC with updatable offsets (OTBC-2), and give a birthday-attack on it that demonstrates that such a construction is not sufficient to get beyond-birthday security for the OCB. Finally, we introduce the notion of offsets that are not updatable by themselves, but are efficiently computable from updatable *caches*. As the most important technical contribution of the paper, we instantiate a TBC with this property (OTBC-3) and show that it achieves a beyond-birthday TPRP security in the number of nonces queried, as long as the maximum length of each message (i.e., the maximum number of times each block is used) is not very high. Additionally, we also show that OTBC-3 achieves at least security up to the birthday-bound even when nonce is misused and inverse queries are allowed.

Finally, we use OTBC-3 to design an authenticated encryption mode called OCB+, which is beyond-birthday secure in both privacy and authenticity. We argue how the privacy bound follows from our security proof of OTBC-3, while the authenticity can be proved in the exact same way as in [14]. OCB+ uses nine random function calls for processing each nonce, so its rate is approximately $\sigma/(\sigma + 9q)$, where σ is the total number of blocks including messages and associated data, and q is the number of distinct nonces. When the messages are sufficiently long, this rate comes close to 1, making this as efficient as OCB3, but with a BBB security guarantee.

2 Preliminaries

Throughout the paper N will mean 2^n. For any positive integer m, $[m]$ will denote the set $\{1, \ldots, m\}$. Matrices will be denoted with boldface letters, and for a matrix $\mathbf{H}$, $|\mathbf{H}|$ will denote its determinant. We'll use the Pochhammer falling factorial power notation

$$(a)_b := a(a-1)\ldots(a-b+1).$$

For ease of notation we write $+$ to denote field addition (bitwise XOR) when used between two or more field elements. Field multiplication in $\mathbb{GF}(2^n)$ is denoted with a bold dot $(\cdot)$.

2.1 Distinguishing Advantage

For two oracles $\mathcal{O}_0$ and $\mathcal{O}_1$, an algorithm $\mathcal{A}$ which tries to distinguish between $\mathcal{O}_0$ and $\mathcal{O}_1$ is called a distinguishing adversary. $\mathcal{A}$ plays an interactive game with $\mathcal{O}_b$ where b is unknown to $\mathcal{A}$, and then outputs a guess for b; $\mathcal{A}$ wins when the guessed bit matches b. The distinguishing advantage of $\mathcal{A}$ is defined as

$$\mathbf{Adv}^{\mathcal{O}_1,\mathcal{O}_0}(\mathcal{A}) := \left|\Pr_{\mathcal{O}_0}[\mathcal{A} \Rightarrow 1] - \Pr_{\mathcal{O}_1}[\mathcal{A} \Rightarrow 1]\right|,$$

where the subscript of Pr denotes the oracle with which $\mathcal{A}$ is playing.

$\mathcal{O}_0$ conventionally represents an ideal primitive, while $\mathcal{O}_1$ represents either an actual construction or a mode of operation built using some other ideal primitives. We use the standard terms real oracle and ideal oracle for $\mathcal{O}_1$ and $\mathcal{O}_0$ respectively. Typically the goal of the function F represented by $\mathcal{O}_1$ is to emulate the ideal primitive F^* represented by $\mathcal{O}_0$. A security game is a distinguishing game with an optional set of additional restrictions, chosen to reflect the desired security goal. When we talk of distinguishing advantage between F and F^* with a specific security game $\mathcal{G}$ in mind, we include $\mathcal{G}$ in the subscript, e.g., $\mathbf{Adv}_{\mathcal{G}}^{F,F^*}(\mathcal{A})$. (We note that this notation is general enough to capture games where each oracle implements multiple functions, e.g., F can handle both encryption and decryption queries by accepting an extra bit to indicate the direction of queries.) Also we sometimes drop the ideal primitive and simply write $\mathbf{Adv}_{\mathcal{G}}^{F}(\mathcal{A})$ when the ideal primitive is clear from the context.

2.2 TPRP, TPRP* and TSPRP Security Notions

Given a tweak-space $\mathcal{W}$, let $\mathsf{Perm}(\mathcal{W}, n)$ be the set of all functions $\tilde{\pi} : \mathcal{W} \times \{0,1\}^n \to \{0,1\}^n$ such that for any tweak $W \in \mathcal{W}$, $\tilde{\pi}(W, \cdot)$ is a permutation over $\{0,1\}^n$. Then a $\tilde{\pi}^*$ distributed uniformly at random over $\mathsf{Perm}(\mathcal{W}, n)$ will be called a *tweakable random permutation* (TRP).

Let $\mathcal{K}$ denote a key-space. Then $\tilde{E} : \mathcal{K} \times \mathcal{W} \times \{0,1\}^n \to \{0,1\}^n$ will be called a *tweakable pseudorandom permutation* (TPRP) if for a key K distributed uniformly at random over $\mathcal{K}$ and for any adversary $\mathcal{A}$ trying to distinguish $\tilde{E}_K := \tilde{E}(K, \cdot, \cdot)$ from $\tilde{\pi}^*$, $\mathbf{Adv}^{\tilde{E}_K, \tilde{\pi}^*}(\mathcal{A})$ is small. We call this game the TPRP game and denote the advantage of $\mathcal{A}$ as $\mathbf{Adv}_{\text{TPRP}}^{\tilde{E}}(\mathcal{A})$ in short.

We will be more interested in a modified version of the TPRP game, where $\mathcal{A}$ is under the added restriction that no two queries can be made with the same tweak. We call this the *tweak respecting pseudorandom permutation* (TPRP*) game, and denote the corresponding advantage of $\mathcal{A}$ as $\mathbf{Adv}_{\text{TPRP*}}^{\tilde{E}}(\mathcal{A})$.

Finally, the *tweakable strong pseudorandom permutation* (TSPRP) game allows $\mathcal{A}$ to make both encryption and decryption queries to the oracle. The advantage term of $\mathcal{A}$ in a TSPRP game will be denoted $\mathbf{Adv}_{\text{TSPRP}}^{\tilde{E}}(\mathcal{A})$.

2.3 Authenticated Encryption and Its Security Notion

A *nonce-based Authenticated Encryption with associated data* (NAEAD) involves a key-space $\mathcal{K}$, a nonce-space $\mathcal{N}$, an associated-data-space $\mathcal{AD}$, a message space $\mathcal{M}$ and a tag space $\mathcal{T}$ along with two functions $\mathsf{Enc} : \mathcal{K} \times \mathcal{N} \times \mathcal{AD} \times \mathcal{M} \to \mathcal{M} \times \mathcal{T}$ (called the Encryption Function) and $\mathsf{Dec} : \mathcal{K} \times \mathcal{N} \times \mathcal{AD} \times \mathcal{M} \times \mathcal{T} \to \mathcal{M} \cup \{\bot\}$ (called the Decryption Function) with the correctness condition that for any $K \in \mathcal{K}, N \in \mathcal{N}, A \in \mathcal{AD}$ and $M \in \mathcal{M}$, it holds that

$$\mathsf{Dec}(K, N, A, \mathsf{Enc}(K, N, A, M)) = M\,.$$

The NAEAD security game is played between the $(\mathsf{Enc}, \mathsf{Dec})$ scheme described above and an ideal oracle $(\mathsf{Enc}^*, \mathsf{Dec}^*)$ where $\mathsf{Enc}^* : \mathcal{K} \times \mathcal{N} \times \mathcal{AD} \times \mathcal{M} \to$

$\mathcal{M} \times \mathcal{T}$ is an ideal random function and $\mathsf{Dec}^* : \mathcal{K} \times \mathcal{N} \times \mathcal{AD} \times \mathcal{M} \times \mathcal{T} \to \{\perp\}$ is a constant function. The adversary $\mathcal{A}$ can make encryption or decryption queries to the oracle. In addition we assume the following restrictions:

1. $\mathcal{A}$ should be once-respecting, i.e., should not repeat a nonce in more than one encryption queries; and
2. $\mathcal{A}$ should not make *pointless* queries, i.e., should not repeat the same query multiple times or should not make the decryption query (N, A, C, T) if it has already made an encryption query (N, A, M) and received (C, T) in response.

The distinguishing advantage of $\mathcal{A}$ for an NAEAD scheme $\mathcal{E}$ will be denoted by $\mathbf{Adv}^{\mathcal{E}}_{\mathrm{NAEAD}}(\mathcal{A})$. The following two security notions are captured in this advantage.

1. Privacy or Confidentiality, i.e., $\mathcal{A}$ should not be able to distinguish the real oracle from the ideal oracle.
2. Authenticity or Integrity, i.e., $\mathcal{A}$ should not be able to forge the real oracle. In other words, $\mathcal{A}$ should not be able to make a decryption query to the real oracle to which the response isn't $\perp$.

2.4 Coefficients H Technique

The H-coefficient technique is a proof method by Patarin [42] that was modernized by Chen and Steinberger [17,50]. A distinguisher $\mathcal{A}$ interacts with oracles $\mathcal{O}$ (The oracle $\mathcal{O}$ could be a sequence of multiple oracles.) and obtains outputs from a real world $\mathcal{O}_1$ or an ideal world $\mathcal{O}_0$. The results of its interaction are collected in a transcript τ. The oracles can sample random coins before the experiment (often a key or an ideal primitive that is sampled beforehand) and are then deterministic. A transcript τ is attainable if $\mathcal{A}$ can observe τ with non-zero probability in the ideal world.

The Fundamental Theorem of the H-coefficients technique, whose proof can be found, e.g., in [17,42,50], states the following:

Theorem 1 ([42]). Assume, there exist $\epsilon_1, \epsilon_2 \geq 0$ such that

$$\Pr_{\mathcal{O}_0}[\mathsf{bad}] \leq \epsilon_1,$$

and for any attainable transcript τ obtained without encountering bad,

$$\frac{\Pr_{\mathcal{O}_1}[\tau]}{\Pr_{\mathcal{O}_0}[\tau]} \geq 1 - \epsilon_2.$$

Then, for all adversaries $\mathcal{A}$, it holds that $\mathbf{Adv}^{\mathcal{O}_0,\mathcal{O}_1}(\mathcal{A}) \leq \epsilon_1 + \epsilon_2$.

The technique has been generalized by Hoang and Tessaro [25] in their expectation method, which allowed them to derive the Fundamental Theorem as a corollary. Since we only consider bad events in the ideal world, we will write $\Pr_{\mathcal{O}_0}[\mathsf{bad}]$ simply as $\Pr[\mathsf{bad}]$ when there is no scope for confusion; the same notation is used when the event bad is broken down into further sub-events.

2.5 Mirror Theory

Consider a sequence of n-bit variables $W_1, \ldots, W_t$, subject to r bi-variate equations of the form

$$W_i + W_j = \delta_{ij}.$$

Consider the graph with $W_1, \ldots, W_t$ as vertices and the bi-variate equations as weighted edges with δ_{ij} the weight between W_i and W_j. Suppose we can show that the graph is cycle-free, and that each path has a non-zero sum of weights. Let $\xi_{\max}$ be the size of the largest component of this graph. Then Mirror Theory tells us that as long as $\xi_{\max}^2 \leq \sqrt{N}/\log_2 N$ and $t \leq N/12\xi_{\max}^2$, the number of solutions to the system of equations such that W_i's are all distinct is at least $(N)_t/N^r$. [18,19]

3 Finding a Suitable Tweakable Block-cipher

We set out to find an offset-based Tweakable Block-cipher that could give us a beyond-birthday security bound for OCB+. The general structure of this is as follows:

$$C = \pi(M + T) + \widehat{T},$$

where the offsets T and $\widehat{T}$ are functions of the nonce $\mathcal{N}$ and the block-number i.

3.1 Attempt with Same Offset

The first question we asked is whether it is possible to achieve this by having $T = \widehat{T}$, i.e., adding the same offset before and after the blockcipher call, like in OCB. The most powerful version of this is to have

$$T = \widehat{T} = f(\mathcal{N}, i)$$

for some $2n$-bit-to-n-bit random function f. This we call OTBC-0, defined as

$$\mathsf{OTBC\text{-}0}(\mathcal{N}, i, M) := \pi(M + f(\mathcal{N}, i)) + f(\mathcal{N}, i).$$

This construction is shown in Fig. 1.

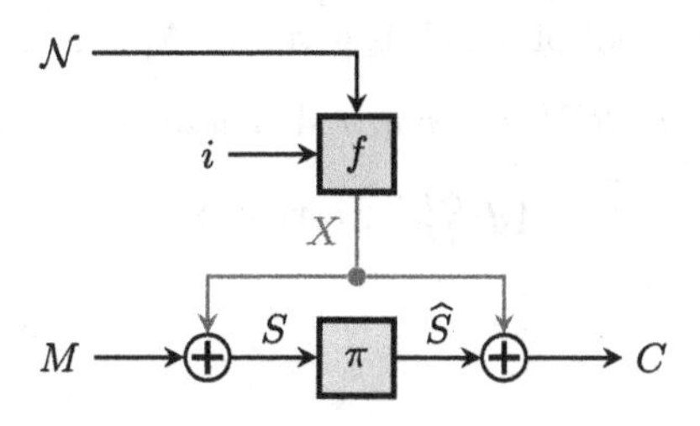

Fig. 1. OTBC-0: Same offset.

Birthday Attack on OTBC-0. Unfortunately, OTBC-0 fails to give us beyond birthday-bound security. This is because for two queries with the same message, there is a collision in the ciphertext whenever there is a collision in the output of f; in addition the ciphertext-collision can also happen if the sum of the outputs of π and f collide. This shows that the collision probability at C is roughly double the collision probability in an ideal tweakable block-cipher, which can be detected in the birthday-bound. A more formal description of the attack is given in Appendix ??.

3.2 Independent Offsets

We deduce from the preceding subsection that using the same offset above and below can never give us beyond-birthday TPRP* security for the tweakable block-cipher. We next examine the most powerful version of this possible, where the two offsets on either side of π come from two completely independent $2n$-bit-to-n-bit random functions f_1 and f_2. This we call OTBC-1, defined as

$$\mathsf{OTBC\text{-}1}(\mathcal{N}, i, M) := \pi(M + f_1(\mathcal{N}, i)) + f_2(\mathcal{N}, i).$$

This construction is shown in Fig. 2.

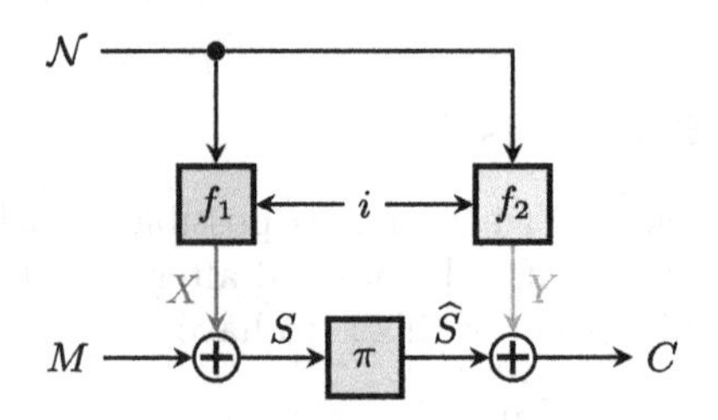

Fig. 2. OTBC-1: Different random offsets.

Security of OTBC-1. As it turns out, OTBC-1 trivially achieves full TPRP* security. This is because in a tweak-respecting game, the offsets are always random and independent of all other offsets in the game, making it impossible to glean any information from the oracle responses. We formally state this as the following theorem, the proof of which is given in Appendix ??.

Theorem 2. *For any TPRP* adversary $\mathcal{A}$ making q queries, we have*

$$\boldsymbol{Adv}^{\mathsf{OTBC\text{-}1}}_{TPRP^*}(\mathcal{A}) = 0.$$

3.3 Updatable Offsets

While OTBC-1 is a fully secure tweakable blockcipher, it's not very interesting to us in the context of OCB+. This is because when the same nonce is used with different block-numbers (as we need for OCB+), new calls to f_1 and f_2 are needed for each new block-number. Thus we need three primitive calls to process every block of message, which robs us of the main advantage of an OCB-like design.

This points us to the next desirable feature we need in the offsets: they should be *efficiently updatable* when we keep the nonce same and increment the block-number. We call a $2n$-bit-to-n-bit function h efficiently updatable on the second input if there is an efficiently computable function g (called the *update* function) such that for each i we have

$$h(\mathcal{N}, i+1) = g(i, h(\mathcal{N}, i)).$$

In other words, given $h(\mathcal{N}, i)$ has already been computed, $h(\mathcal{N}, i+1)$ can be computed through the update function g while bypassing a fresh call to h. (For this to make sense, of course, h should be computationally heavy and g should be much faster than h.) Note that the update function may or may not use i as an additional argument; while in this work we'll only consider update functions that are *stationary* (i.e., ignore the block-number i, and apply the same function at each block to get the offset for the next block), it is possible to have an update function that varies with i but still satisfies the above-discussed criteria.

The Simplest Updatable Design. The simplest way to design an updatable function is to call a random function f on the nonce $\mathcal{N}$ once, and then use a stationary update function to obtain the offset for each successive block-number. This can be formally defined as follows:

$$\begin{aligned} h(\mathcal{N}, 1) &= g(f(\mathcal{N})), \\ h(\mathcal{N}, i) &= g(h(\mathcal{N}, i-1)) = g^i(f(\mathcal{N})), \qquad i \geq 2. \end{aligned}$$

Using these updatable offsets with two independent random functions f_1 and f_2 for input-masking and output-masking respectively, we can define a tweakable block-cipher OTBC-g as

$$\mathsf{OTBC\text{-}g}(\mathcal{N}, i, M) = \pi\left(M + g^i(f_1(\mathcal{N}))\right) + g^i(f_2(\mathcal{N})).$$

Instantiating OTBC-g. In commonly used finite fields, there generally exist primitive elements that allow very fast multiplication. As an instantiation of g, we use multiplication with one such fixed primitive α. Concretely, we define the update function as

$$g(f(\mathcal{N})) = \alpha \cdot f(\mathcal{N}).$$

Thus, we use as the updatable offsets

$$T = \alpha^i \cdot f_1(\mathcal{N}), \qquad \widehat{T} = \alpha^i \cdot f_2(\mathcal{N}).$$

This gives us the construction OTBC-2, defined as

$$\mathsf{OTBC\text{-}2}(\mathcal{N}, i, M) = \pi\left(M + \alpha^i \cdot f_1(\mathcal{N})\right) + \alpha^i \cdot f_2(\mathcal{N}).$$

This construction is shown in Fig. 3.

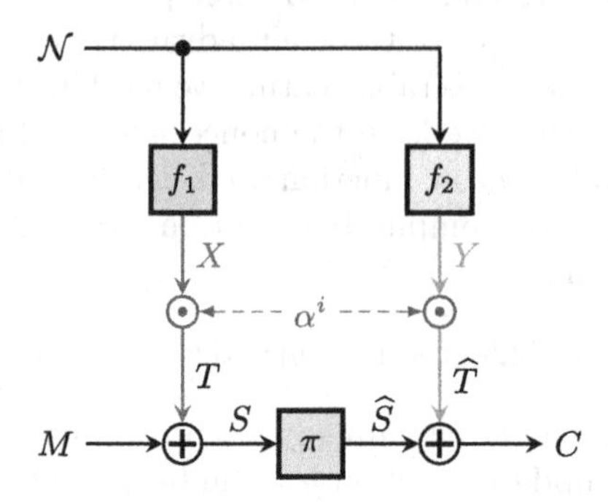

Fig. 3. OTBC-2: Updatable offsets with two independent random-function calls.

Attack on OTBC-2. Unfortunately, this simple updatable function is not sufficient to give us beyond-birthday-bound security. This is because since the update function is linear and publicly known, we can make queries such that successive message blocks under the same nonce follow the update relation, which forces the successive S blocks to also conform to the update relation. Thus, one collision on S between two different nonces ensures that successive blocks also see an S-collision, which can be exploited in a distinguishing attack. This we state as the following theorem, the proof of which is given in Appendix ??.

Theorem 3. *There exists a distinguisher $\mathcal{A}$ querying with q nonces and L blocks under each nonce with $L \geq 12$ in a TPRP* game against OTBC-2 such that*

$$\mathbf{Adv}^{OTBC\text{-}2}_{TPRP^*}(\mathcal{A}) \geq \Omega\left(\frac{q^2 L^2}{N}\right).$$

3.4 Offsets with Updatable Caches

To get around this problem, we observe that in order to use an offset-based tweakable block-cipher in OCB+, we don't really need it to be updatable; it is enough for it to maintain a small and updatable hidden state or *cache*, such that the offsets are efficiently computable from the cache. Letting ψ denote the *caching function*, g the update function as before, h the offset-generating function, and φ the cache-to-offset function, we have

$$\psi(\mathcal{N}, i+1) = g(i, \psi(\mathcal{N}, i)), \qquad h(\mathcal{N}, i) = \varphi(\psi(\mathcal{N}, i)).$$

Again, for this to make sense, g and ψ should be computationally heavy when computed from scratch, while g and φ should be much faster.

Updatable Caches, Non-updatable Offsets. To avoid the kind of attack that we found on OTBC-2, we want to design a tweakable block-cipher with offsets which are not themselves updatable, but are efficiently computable from updatable caches. This makes the offsets more independent, while still giving us a means of updating them efficiently at a small additional cost.

One simple way to achieve this is to use two independent random functions f_1 and f_2 on the nonce, put the outputs in the cache as two different *branches*, and use two different update functions g and g' on the two branches; the offset can then be generated as the sum of the two branches. This can be formally defined as follows:

$$\begin{aligned}
\psi(\mathcal{N}, 1) &= (g(f_1(\mathcal{N})), g'(f_2(\mathcal{N}))), \\
\psi(\mathcal{N}, i) &= [g, g'](\psi(\mathcal{N}, i-1)) = (g^i(f_1(\mathcal{N})), g'^i(f_2(\mathcal{N}))), \quad i \geq 2, \\
\varphi(x, y) &= x + y, \\
h(\mathcal{N}, i) &= g^i(f_1(\mathcal{N})) + g'^i(f_2(\mathcal{N})) = \varphi(\psi(\mathcal{N}, i)),
\end{aligned}$$

where $[g, g']$ denotes the two-input function that applies g to the first input and g' to the second input. Note that $h(\mathcal{N}, i)$ is not efficiently computable from $h(\mathcal{N}, i-1)$ without accessing the cache $\psi(\mathcal{N}, i-1)$, which makes the offsets themselves non-updatable in the absence of the cache. Using these offsets we can define a tweakable block-cipher OTBC-gg' as

$$\mathsf{OTBC\text{-}gg'}(\mathcal{N}, i, M) = \pi\left(M + g^i(f_1(\mathcal{N})) + g'^i(f_2(\mathcal{N}))\right) + f_3(\mathcal{N}) + g^i(\pi(0^n)).$$

where f_3 is a third independent random-function. Note that we do not bother to use the non-updatable updates for masking the output, because $\mathcal{A}$ can make only encryption queries, and thus cannot exploit the same weakness in the output-masking.

Instantiating OTBC-gg'. As the main contribution of this section, we propose a concrete instantiation of OTBC-gg' and analyse its security. As before we keep the field-multiplication by α as g, and for g' we use field-multiplication by α^2. The resulting tweakable block-cipher, called OTBC-3, is defined as

$$\mathsf{OTBC\text{-}3}(\mathcal{N}, i, M) = \pi\left(M + \alpha^i \cdot f_1(\mathcal{N}) + \alpha^{2i} \cdot f_2(\mathcal{N})\right) + f_3(\mathcal{N}) + \alpha^i \cdot \pi(0^n).$$

This construction is shown in Fig. 4.

3.5 TPRP* Security Analysis of OTBC-3

Consider a distinguisher $\mathcal{A}$ making σ encryption queries to OTBC-3 with q distinct nonces and $\ell^{(j)} \leq L$ block-numbers $1, \ldots, \ell^{(j)}$ for the j-th nonce for each $j \in [q]$. Then we have the following result.

Theorem 4. *As long as $\sigma \leq N/n^2L^2$, we have*

$$\boldsymbol{Adv}^{\mathsf{OTBC\text{-}3}}_{\mathsf{TPRP^*}}(\mathcal{A}) \leq \frac{n\sigma L}{N}.$$

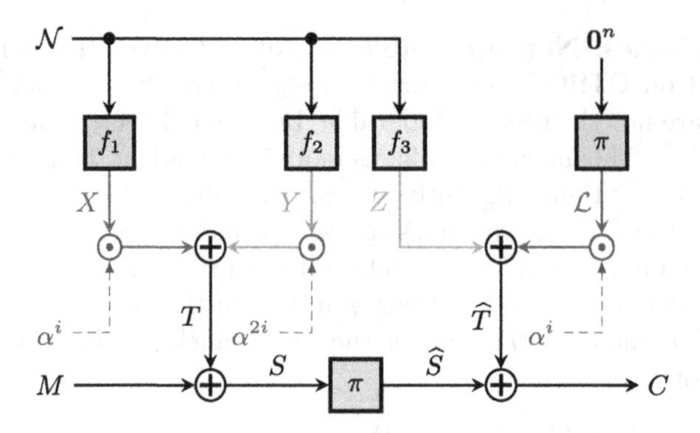

Fig. 4. OTBC-3: Offsets with updatable caches using three independent random-function calls.

Algorithm 1. OTBC-$3^{f_1,f_2,f_3,\pi}(\mathcal{N}, i, M)$

1: $T \leftarrow \alpha^i f_1(\mathcal{N}) \oplus \alpha^{2i} f_2(\mathcal{N})$
2: $\widehat{T} \leftarrow f_3(\mathcal{N}) \oplus \alpha^i \pi(0^n)$
3: $S \leftarrow M \oplus T$
4: $\widehat{S} \leftarrow \pi(S)$
5: $C \leftarrow \widehat{S} \oplus \widehat{T}$
6: **return** C

Proof. In this proof, we'll use the following lemma, the proof of which is deferred to Appendix ??.

Lemma 1. *For some $r \geq 2$ and $2r$ numbers $i_1, i'_1, \ldots, i_r, i'_r < N$ such that $i_j \neq i'_j$ for each $j \in [r]$, define*

$$\mathbf{B}_r = \begin{bmatrix} \alpha^{i_1} & \alpha^{2i_1} & \alpha^{i'_2} & \alpha^{2i'_2} & 0 & 0 & 0 & 0 & \cdots & 0 & 0 & 0 & 0 \\ 0 & 0 & \alpha^{i_2} & \alpha^{2i_2} & \alpha^{i'_3} & \alpha^{2i'_3} & 0 & 0 & \cdots & 0 & 0 & 0 & 0 \\ 0 & 0 & 0 & 0 & \alpha^{i_3} & \alpha^{2i_3} & \alpha^{i'_4} & \alpha^{2i'_4} & \cdots & 0 & 0 & 0 & 0 \\ \vdots & \vdots & \vdots & \vdots & \vdots & \vdots & \vdots & \vdots & \ddots & \vdots & \vdots & \vdots & \vdots \\ 0 & 0 & 0 & 0 & 0 & 0 & 0 & 0 & \cdots & \alpha^{i_{r-1}} & \alpha^{2i_{r-1}} & \alpha^{i'_r} & \alpha^{2i'_r} \\ \alpha^{i'_1} & \alpha^{2i'_1} & 0 & 0 & 0 & 0 & 0 & 0 & \cdots & 0 & 0 & \alpha^{i_r} & \alpha^{2i_r} \end{bmatrix}.$$

Then $\mathbf{B}_r$ is at least of rank r.

Label the q nonces $\mathcal{N}^{(1)}, \ldots, \mathcal{N}^{(q)}$. For the j-th nonce, there are $\ell^{(j)}$ queries $(\mathcal{N}^{(j)}, 1, M_1^{(j)}), \ldots, (\mathcal{N}^{(j)}, \ell^{(j)}, M_{\ell^{(j)}}^{(j)})$, with outputs $(C_1^{(j)}, \ldots C_{\ell^{(j)}}^{(j)})$ respectively. For the internal transcript, we have $\mathcal{L}$, the encryption of 0 with π, and for the j-th nonce, we have the three random-function outputs $X^{(j)}, Y^{(j)}, Z^{(j)}$; finally, we have the (input, output) pairs $(S_1^{(j)}, \widehat{S}_1^{(j)}), \ldots, (S_{\ell^{(j)}}^{(j)}, \widehat{S}_{\ell^{(j)}}^{(j)})$ to π, and the (input-offset, output-offset) pairs $(T_1^{(j)}, \widehat{T}_1^{(j)}), \ldots, (T_{\ell^{(j)}}^{(j)}, \widehat{T}_{\ell^{(j)}}^{(j)})$. Then this extended transcript satisfies the following equations for each $j \in [q]$ and each $i \in [\ell^{(j)}]$:

$$S_i^{(j)} = M_i^{(j)} + T_i^{(j)}, \qquad \widehat{S}_i^{(j)} = C_i^{(j)} + \widehat{T}_i^{(j)},$$
$$T_i^{(j)} = \alpha^i \cdot X^{(j)} + \alpha^{2i} \cdot Y^{(j)}, \qquad \widehat{T}_i^{(j)} = Z^{(j)} + \alpha_i \cdot \mathcal{L}.$$

Internal Sampling. Following the query phase of the game, in the ideal world we sample the internal transcript as follows (subject to certain bad events to be defined subsequently):

- Sample $X^{(j)}, Y^{(j)}$ uniformly at random for each $j \in [q]$;
- Check for bad1, bad2, bad3, bad4;
- Sample $\mathcal{L}$ uniformly at random;
- Check for bad5, bad6;
- Let $S_1, \ldots, S_t$ be a labeling of the unique values in $\{S_i^{(j)} \mid j \in [q], i \in [\ell^{(j)}]\}$;
- Sample $\{\widehat{S}_k \mid k \in [t]\}$ directly from good set, subject to the equations $\widehat{S}_i^{(j)} + \widehat{S}_{i'}^{(j)} = C_i^{(j)} + C_{i'}^{(j)} + (\alpha^i + \alpha^{i'}) \cdot \mathcal{L}$ for each $j \in [q]$ and each $i, i' \in [\ell^{(j)}]$.

Before describing the bad events bad1, ..., bad6, we define two graphs on the extended transcript.

Transcript Graph. For distinct $j_1, j_2 \in [q]$, there is an edge (j_1, j_2) in G if we have some $i_1 \in [\ell^{(j_1)}]$ and some $i_2 \in [\ell^{(j_2)}]$ such that $S_{i_1}^{(j_1)} = S_{i_2}^{(j_2)}$.

We will refer to paths of length 2 in G as *links*. A link (j_1, j_2, j_3) formed with the collisions $S_{i_1}^{(j_1)} = S_{i_2}^{(j_2)}$ and $S_{i_2'}^{(j_2)} = S_{i_3}^{(j_3)}$ for some $i_1 \in [\ell^{(j_1)}]$, $i_2, i_2' \in [\ell^{(j_2)}]$ and $i_3 \in [\ell^{(j_3)}]$ is called *degenerate* if $i_2 = i_2'$ and *non-degenerate* otherwise. We observe that the above link being degenerate implies $S_{i_1}^{(j_1)} = S_{i_3}^{(j_3)}$, so (j_1, j_3) is also an edge in G. By *short-circuiting* a degenerate link (j_1, j_2, j_3) we will refer to the operation of replacing it with the edge (j_1, j_3).

A path of length ≥ 3 is called non-degenerate if at least one of its sublinks is non-degenerate. When a non-degenerate path contains a degenerate sublink, we can short-circuit it to obtain a shorter non-degenerate path. We can repeat this operation as long as the path contains degenerate sublinks to end up with a *minimal* non-degenerate path. When the initial path is a cycle, we end up with either a minimal non-degenerate cycle or a *double-collision* edge, i.e., an edge (j_1, j_2) in G such that for distinct $i_1, i_1' \in [\ell^{(j_1)}]$ and distinct $i_2, i_2' \in [\ell^{(j_2)}]$ we have $S_{i_1}^{(j_1)} = S_{i_2}^{(j_2)}$, and $S_{i_1'}^{(j_1)} = S_{i_2'}^{(j_2)}$.

Dual Graph (for Mirror Theory). We also define a second graph H on the transcript, which is something of a dual of the first. This is the graph we need to check for the conditions necessary to apply mirror theory. First consider the graph H' such that the vertices of H' are the distinct values $S_1, \ldots, S_t$, and there is an edge between S_i and $S_{i'}$ in H if they appear in the same nonce, i.e., if there is some $j \in [q]$, $i, i' \in [\ell^{(j)}]$ such that $\widehat{S}_i^{(j)} + \widehat{S}_{i'}^{(j)} = C_i^{(j)} + C_{i'}^{(j)} + (\alpha^i + \alpha^{i'}) \cdot \mathcal{L}$; further, the weight of this edge is then $C_i^{(j)} + C_{i'}^{(j)} + (\alpha^i + \alpha^{i'}) \cdot \mathcal{L}$.

From H' we get H by dropping all *redundant edges*—for each $j \in [\ell^{(j)}]$, out of the fully connected subgraph of G with $\binom{\ell^{(j)}}{2}$ edges, we only keep a spanning tree of $\ell^{(j)} - 1$ edges, and drop the rest. For instance, one way of choosing H could be to just keep the edge between $\widehat{S}_i^{(j)}$ and $\widehat{S}_{i+1}^{(j)}$ for each $i \in [\ell^{(j)} - 1]$. (Note that we assume here that all $\widehat{S}_i^{(j)}$ are distinct within any j, because that is the only use-case we'll need; the notions however easily generalise to graphs with intra-nonce collisions.)

We observe that H is cycle-free as long as G is cycle-free, and that the size $\xi_{\max}$ of the largest component of H is at most LM when M is the size of the largest component of G.

Bad Events. Based on the graphs G and H defined above, we can describe our bad events.

- bad1: *We have $j \in [q]$ and distinct $i, i' \in [\ell^{(j)}]$ such that $S_i^{(j)} = S_{i'}^{(j)}$.*
- bad2: *There is a double-collision edge in G.*
- bad3: *There is a minimal non-degenerate cycle in G.*
- bad4: *G has a component of size $> n$.*
- bad5: *We have $j \in [q]$ and distinct $i, i' \in [\ell^{(j)}]$ such that $C_i^{(j)} + C_{i'}^{(j)} = (\alpha^i + \alpha^{i'}) \cdot \mathcal{L}$.*
- bad6: *We have a path in H on which the edge-weights sum to 0.*

Next we give an upper bound on the probability of at least one bad event happening in the ideal world. Define

$$\mathsf{bad} := \bigcup_{p=1}^{6} \mathsf{bad[p]}.$$

Then we have the following lemma.

Lemma 2. *In the ideal world,*

$$\Pr[\mathit{bad}] \leq \frac{n\sigma L}{N}.$$

Proof (of Lemma 1). We bound the probability of each of the six bad events one by one below.

bad1: *We have $j \in [q]$ and distinct $i, i' \in [\ell^{(j)}]$ such that $S_i^{(j)} = S_{i'}^{(j)}$.*
For a fixed choice of indices j, i and i', the probability of the event comes out to be $1/N$ due to the randomness of $T_i^{(j)}$ or $T_{i'}^{(j)}$. From union bound over all possible choices of indices, we obtain

$$\Pr[\mathsf{bad1}] \leq \frac{1}{N} \sum_{j=1}^{q} \ell^{(j)2} \leq \frac{L}{N} \sum_{j=1}^{q} \ell^{(j)} \leq \frac{\sigma L}{N}.$$

bad2: *There is a double-collision edge in G.*
This implies that we have distinct $j_1, j_2 \in [q]$, distinct $i_1, i'_1 \in [\ell^{(j_1)}]$, and distinct $i_2, i'_2 \in [\ell^{(j_2)}]$ such that $S^{(j_1)}_{i_1} = S^{(j_2)}_{i_2}$, and $S^{(j_1)}_{i'_1} = S^{(j_2)}_{i'_2}$. This can be written as $\mathbf{B}_2\mathbf{v} = \mathbf{c}$, where

$$\mathbf{B}_2 = \begin{bmatrix} \alpha^{i_1} & \alpha^{2i_1} & \alpha^{i_2} & \alpha^{2i_2} \\ \alpha^{i'_1} & \alpha^{2i'_1} & \alpha^{i'_2} & \alpha^{2i'_2} \end{bmatrix}, \mathbf{v} = \begin{bmatrix} X^{(j_1)} \\ Y^{(j_1)} \\ X^{(j_2)} \\ Y^{(j_2)} \end{bmatrix}, \mathbf{c} = \begin{bmatrix} M^{(j_1)}_{i_1} + M^{(j_2)}_{i_2} \\ M^{(j_1)}_{i'_1} + M^{(j_2)}_{i'_2} \end{bmatrix}.$$

$\mathbf{B}_2$ is of rank 2 by Lemma 1. Thus, when we fix $j_1, j_2, i_1, i'_1, i_2, i'_2$, we have

$$\Pr[\mathbf{B}_2\mathbf{v} = \mathbf{c}] \leq \frac{1}{N^2}.$$

Thus,

$$\Pr[\mathsf{bad2}] \leq \frac{1}{N^2} \sum_{j_1=1}^{q} \sum_{j_2=1}^{q} \ell^{(j_1)2} \ell^{(j_2)2} \leq \frac{L^2}{N^2} \sum_{j_1=1}^{q} \sum_{j_2=1}^{q} \ell^{(j_1)} \ell^{(j_2)} \leq \frac{\sigma^2 L^2}{N^2}.$$

bad3: *There is a minimal non-degenerate cycle in the transcript graph.*
First, suppose there is a minimal non-degenerate cycle of length 3. Thus, we have distinct $j_1, j_2, j_3 \in [q]$, distinct $i_1, i'_1 \in [\ell^{(j_1)}]$, distinct $i_2, i'_2 \in [\ell^{(j_2)}]$, and distinct $i_3, i'_3 \in [\ell^{(j_3)}]$ such that $S^{(j_1)}_{i_1} = S^{(j_2)}_{i'_2}$, $S^{(j_2)}_{i_2} = S^{(j_3)}_{i'_3}$, and $S^{(j_3)}_{i_3} = S^{(j_1)}_{i'_1}$. (We name the indices like this for symmetry.) As before, this can be written as $\mathbf{B}_3\mathbf{v} = \mathbf{c}$, where

$$\mathbf{B}_3 = \begin{bmatrix} \alpha^{i_1} & \alpha^{2i_1} & \alpha^{i'_2} & \alpha^{2i'_2} & 0 & 0 \\ 0 & 0 & \alpha^{i_2} & \alpha^{2i_2} & \alpha^{i'_3} & \alpha^{2i'_3} \\ \alpha^{i'_1} & \alpha^{2i'_1} & 0 & 0 & \alpha^{i_3} & \alpha^{2i_3} \end{bmatrix}, \mathbf{v} = \begin{bmatrix} X^{(j_1)} \\ Y^{(j_1)} \\ X^{(j_2)} \\ Y^{(j_2)} \\ X^{(j_3)} \\ Y^{(j_3)} \end{bmatrix}, \mathbf{c} = \begin{bmatrix} M^{(j_1)}_{i_1} + M^{(j_2)}_{i'_2} \\ M^{(j_2)}_{i_2} + M^{(j_3)}_{i'_3} \\ M^{(j_3)}_{i_3} + M^{(j_1)}_{i'_1} \end{bmatrix}.$$

$\mathbf{B}_3$ is of rank 3 by Lemma 1. Thus, when we fix $j_1, j_2, j_3, i_1, i'_1, i_2, i'_2, i_3, i'_3$, we have

$$\Pr[\mathbf{B}_3\mathbf{v} = \mathbf{c}] \leq \frac{1}{N^3}.$$

Next, suppose there is a minimal non-degenerate cycle of length $r \geq 4$. Thus we have distinct $j_1, \ldots, j_r \in [q]$; for $u \in [r-1]$ we have $i_u \in [\ell^{(j_u)}]$ and $i'_{u+1} \in [\ell^{(j_{u+1})}]$ such that $S^{(j_u)}_{i_u} = S^{(j_{u+1})}_{i'_{u+1}}$; and finally, we have $i_r \in [\ell^{(j_r)}]$ and $i'_1 \in [\ell^{(j_1)}]$ such that $S^{(j_r)}_{i_r} = S^{(j_1)}_{i'_1}$; the cycle being minimal non-degenerate implies that for each $u \in [r]$, $i_u \neq i'_u$. This can be written as $\mathbf{B}_r\mathbf{v} = \mathbf{c}$, where

$$\mathbf{B}_r = \begin{bmatrix} \alpha^{i_1} & \alpha^{2i_1} & \alpha^{i'_2} & \alpha^{2i'_2} & 0 & 0 & 0 & 0 & \cdots & 0 & 0 & 0 & 0 \\ 0 & 0 & \alpha^{i_2} & \alpha^{2i_2} & \alpha^{i'_3} & \alpha^{2i'_3} & 0 & 0 & \cdots & 0 & 0 & 0 & 0 \\ 0 & 0 & 0 & 0 & \alpha^{i_3} & \alpha^{2i_3} & \alpha^{i'_4} & \alpha^{2i'_4} & \cdots & 0 & 0 & 0 & 0 \\ \vdots & \vdots & \vdots & \vdots & \vdots & \vdots & \vdots & \vdots & \ddots & \vdots & \vdots & \vdots & \vdots \\ 0 & 0 & 0 & 0 & 0 & 0 & 0 & 0 & \cdots & \alpha^{i_{r-1}} & \alpha^{2i_{r-1}} & \alpha^{i'_r} & \alpha^{2i'_r} \\ \alpha^{i'_1} & \alpha^{2i'_1} & 0 & 0 & 0 & 0 & 0 & 0 & \cdots & 0 & 0 & \alpha^{i_r} & \alpha^{2i_r} \end{bmatrix},$$

$$\mathbf{v} = \begin{bmatrix} X^{(j_1)} \\ Y^{(j_1)} \\ X^{(j_2)} \\ Y^{(j_2)} \\ \vdots \\ X^{(j_r)} \\ Y^{(j_r)} \end{bmatrix}, \mathbf{c} = \begin{bmatrix} M_{i_1}^{(j_1)} + M_{i'_2}^{(j_2)} \\ M_{i_2}^{(j_2)} + M_{i'_3}^{(j_3)} \\ M_{i_3}^{(j_3)} + M_{i'_4}^{(j_4)} \\ \vdots \\ M_{i_{r-1}}^{(j_{r-1})} + M_{i'_r}^{(j_r)} \\ M_{i_r}^{(j_r)} + M_{i'_1}^{(j_1)} \end{bmatrix}.$$

$\mathbf{B}_r$ is of rank r by Lemma 1. Thus, for each $r \geq 3$, when we fix $j_1, \ldots, j_r, i_1, i'_1, \ldots, i_r, i'_r$, we have

$$\Pr[\mathbf{B}_r \mathbf{v} = \mathbf{c}] \leq \frac{1}{N^r}.$$

Assuming $2\sigma L \leq N$, we have

$$\Pr[\mathsf{bad3}] \leq \sum_{r=3}^{q} \frac{\prod_{u=1}^{r} \ell^{(j_u)2}}{N^r}$$
$$\leq \sum_{r=3}^{q} \left(\left(\frac{L}{N}\right)^r \prod_{u=1}^{r} \ell^{(j_u)} \right) \leq \sum_{r=3}^{q} \left(\frac{\sigma L}{N}\right)^r \leq \frac{2\sigma^3 L^3}{N^3}.$$

bad4: *G has a component of size $> n$.*
For a component of size M, the minimum number of nonces in that component should be $p+1$ where $p = \lceil M/L \rceil - 1$ with p collisions among themselves. In other words, $\exists$ distinct $j_1, j_2, \cdots, j_{p+1} \in [q]$ and $i_1 \in \ell^{(j_1)}$, $i_2, i'_2 \in \ell^{(j_2)}$, $i_3, i'_3 \in \ell^{(j_3)}$, $\cdots$, $i_p, i'_p \in \ell^{(j_p)}$, $i_{p+1} \in \ell^{(j_{p+1})}$ such that

$$S_{i_1}^{(j_1)} = S_{i_2}^{(j_2)}, S_{i'_2}^{(j_2)} = S_{i_3}^{(j_3)}, \ldots, S_{i'_p}^{(j_p)} = S_{i_{p+1}}^{(j_{p+1})}.$$

For a fixed choice of indices, the probability of the event comes out to be $1/N^p$. The independence assumption comes from the fact that every equation from the system of equations mentioned above introduces a fresh nonce. From union bound over all the possible choices of indices, we obtain

$$\begin{aligned}\Pr[\mathsf{bad4}] &\leq \frac{1}{N^p}\sum_{j_1=1}^{q}\sum_{j_2=1}^{q}\cdots\sum_{j_{p+1}=1}^{q}\ell^{(j_1)2}\ell^{(j_2)2}\ldots\ell^{(j_{p+1})2}\\ &\leq \frac{L^{p+1}}{N^p}\sum_{j_1=1}^{q}\sum_{j_2=1}^{q}\cdots\sum_{j_{p+1}=1}^{q}\ell^{(j_1)}\ell^{(j_2)}\ldots\ell^{(j_{p+1})}\\ &\leq \frac{\sigma^{p+1}L^{p+1}}{N^p}=\frac{\sigma L}{N}\left(\frac{\sigma^p L^p}{N^{p-1}}\right).\end{aligned}$$

Assuming $\sigma L \leq N/2$ and $p = n$, we get

$$\Pr[\mathsf{bad4}] \leq \frac{\sigma L}{N}.$$

bad5: *We have $j \in [q]$ and distinct $i, i' \in [\ell_j]$ such that $C_i^{(j)} + C_{i'}^{(j)} = (\alpha^i + \alpha^{i'})\cdot\mathcal{L}$.* For a fixed choice of indices j, i and i', the probability of the event comes out to be $1/N$ due to the randomness of $\mathcal{L}$. From union bound over all possible choices of indices, we obtain

$$\Pr[\mathsf{bad5}] \leq \frac{1}{N}\sum_{j=1}^{q}\ell^{(j)2} \leq \frac{L}{N}\sum_{j=1}^{q}\ell^{(j)} \leq \frac{\sigma L}{N}.$$

bad6: Suppose the first and last vertices on a path inside some component are $\widehat{S}_i^{(j)}$ and $\widehat{S}_{i'}^{(j')}$. Also suppose that the path goes through $x_1, x_2, \cdots, x_y$ vertices of position $i_1, i_2, \cdots, i_y$ respectively. Then this bad event implies

$$C_i^{(j)} + C_{i'}^{(j')} + (\alpha^i + x_1\alpha^{i_1} + \cdots + x_y\alpha^{i_y} + \alpha^{i'})\cdot\mathcal{L} = 0.$$

For a fixed choice of the vertex pair $(\widehat{S}_i^{(j)}, \widehat{S}_{i'}^{(j')})$, the probability of the event comes out to be $1/N$ due to the randomness of $\mathcal{L}$. Applying union bound over all possible vertex pairs, and summing over all components $\mathcal{C}$ of G, we get

$$\begin{aligned}\Pr[\mathsf{bad6}] &\leq \sum_{\mathcal{C}}\frac{1}{2N}\cdot\left(\sum_{j\in\mathcal{C}}\ell^{(j)}\right)^2\\ &\leq \sum_{\mathcal{C}}\frac{1}{2N}\cdot\xi_{\max}\cdot\sum_{j\in\mathcal{C}}\ell^{(j)} = \frac{\xi_{\max}\sigma}{2N} \leq \frac{n\sigma L}{2N}.\end{aligned}$$

Thus, by union-bound, we have

$$\Pr[\mathsf{bad}] \leq \frac{4\sigma L}{N} + \frac{\sigma^2 L^2}{N^2} + \frac{2\sigma^3 L^3}{N^3} + \frac{n\sigma L}{2N} \leq \frac{n\sigma L}{N},$$

which completes the proof of the lemma. □

Bounding the Ratio of Good Probabilities. Let τ be a good transcript. In the real world, there are q distinct inputs to f_1, q distinct inputs to f_2, and t distinct inputs to π. Thus,

$$\Pr_{\mathcal{O}_1}[\tau] = \frac{1}{N^{2q}(N)_t}.$$

In the ideal world, in the online stage, there are σ outputs that are sampled uniformly at random. In the offline stage, q more values are sampled uniformly, and finally t variables are sampled from the good set subject to r non-redundant equations (we calculate r later). Since $\sigma < N/n^2L^2$, and none of the bad events has happened, the conditions for applying mirror theory are fulfilled. Thus, using mirror theory,

$$\Pr_{\mathcal{O}_0}[\tau] \leq \frac{1}{N^{\sigma+q}} \cdot \frac{N^r}{(N)_t} \leq \frac{1}{N^{\sigma+q-r}(N)_t}.$$

To calculate r, we note that every repeated use of a nonce adds a non-redundant equation to the system. Thus, $r = \sigma - q$, giving us

$$\Pr_{\mathcal{O}_0}[\tau] \leq \frac{1}{N^{2q}(N)_t}.$$

Thus, we have

$$\frac{\Pr_{\mathcal{O}_1}[\tau]}{\Pr_{\mathcal{O}_0}[\tau]} \geq 1,$$

Applying the H-Coefficient Technique with $\epsilon_1 = n\sigma L/N$ and $\epsilon_2 = 0$ completes the proof. □

Appendix ?? gives a birthday-bound TSPRP proof for OTBC-3.

4 An Application of OTBC-3

Using the tweakable block-cipher OTBC-3, we define an authenticated encryption scheme OCB+ that is about as efficient as OCB3 while providing a higher degree of privacy guarantee without affecting the authenticity guarantee of OCB3. This is shown in Fig. 5.

4.1 Nonce Handling

OCB+ uses a nonce $\mathcal{N}$ of $n-2$ bits, with the final two bits reserved for domain separation. $\mathcal{N}\|00$ is used for processing the message blocks, $\mathcal{N}\|01$ is used for processing the tag, and $\mathcal{N}\|10$ is used for handling the associated data.

4.2 Handling Incomplete Blocks

Incomplete blocks can be handled in the same way as in OCB3, modifying the masking constants for the incomplete blocks. This does not affect the privacy bound significantly, and since the focus of this work is to improve the privacy guarantee of OCB3, we skip giving specific details on how to handle incomplete blocks in OCB+.

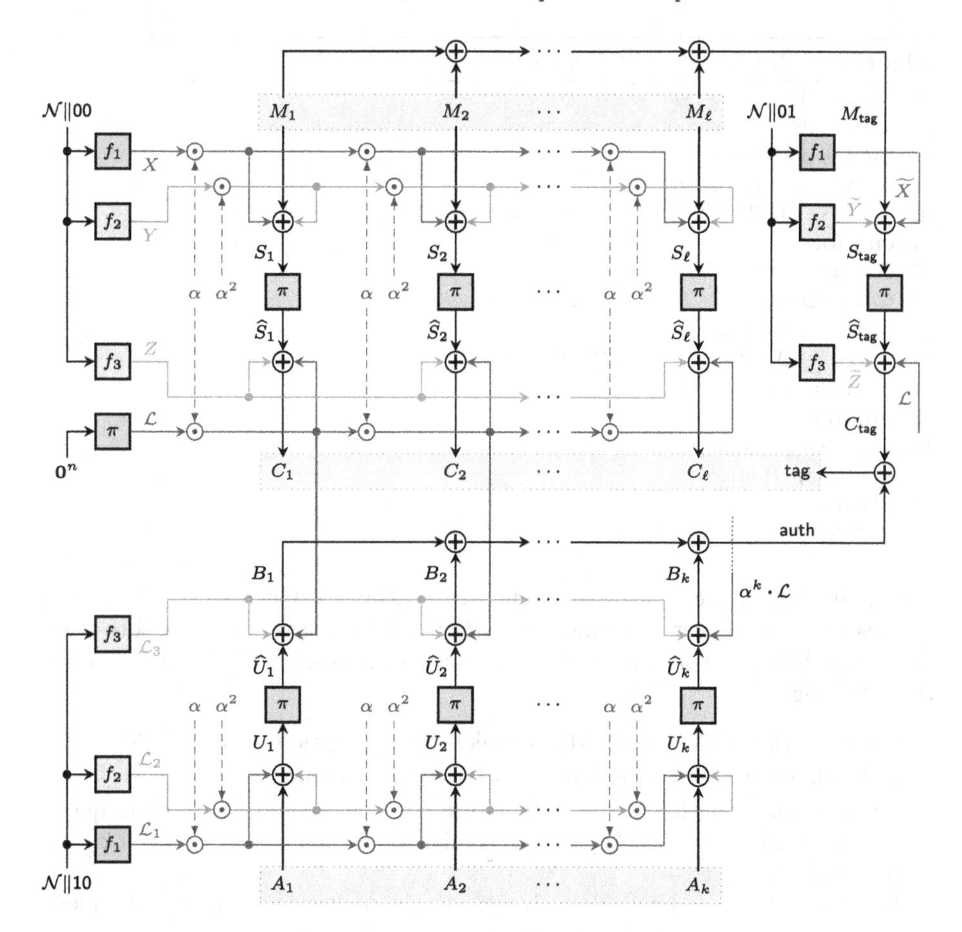

Fig. 5. The OCB+ construction. α is a primitive field-element that allows efficient multiplication.

4.3 Security Claims

We claim that as long as the maximum length L permitted for each message (i.e., the maximum number of blocks encrypted using the same nonce) is small, OCB+ provides both beyond-birthday privacy and beyond-birthday authenticity. Formally we claim the following.

Theorem 5. *Consider a distinguisher $\mathcal{A}$ of OCB+ which can make q encryption queries with distinct nonces with σ blocks and q' decryption queries to its challenger. Suppose the length of the i-th message and the i-th associated data are ℓ_i and k_i respectively, where $\ell_i, k_i \leq L \forall i \in [q_e]$. As long as $\sigma \leq N/n^2L^2$, we have*

$$\boldsymbol{Adv}^{\mathsf{OCB+}}_{\mathsf{NAEAD}}(\mathcal{A}) \leq \frac{n\sigma L}{N} + O\left(\frac{q'L}{N}\right).$$

Algorithm 2. $\mathsf{OCB+}^{f_1,f_2,f_3,\pi}(\mathcal{N}, A, M)$

1: $M_{\mathsf{tag}} \leftarrow 0^n$
2: $\mathsf{auth} \leftarrow 0^n$
3: **for** $i \leftarrow 1$ to ℓ **do**
4: $\quad M_{\mathsf{tag}} \leftarrow M_{\mathsf{tag}} \oplus M_i$
5: $\quad C_i \leftarrow \mathsf{OTBC\text{-}3}^{f_1,f_2,f_3,\pi}(\mathcal{N}\|00, i, M_i)$
6: **end for**
7: $C \leftarrow C_1\| \cdots \|C_\ell$
8: $C_{\mathsf{tag}} \leftarrow \mathsf{OTBC\text{-}3}^{f_1,f_2,f_3,\pi}(\mathcal{N}\|01, 0, M_{\mathsf{tag}})$
9: **for** $i \leftarrow 1$ to k **do**
10: $\quad B_i \leftarrow \mathsf{OTBC\text{-}3}^{f_1,f_2,f_3,\pi}(\mathcal{N}\|10, i, A_i)$
11: $\quad \mathsf{auth} \leftarrow \mathsf{auth} \oplus B_i$
12: **end for**
13: $\mathsf{tag} \leftarrow C_{\mathsf{tag}} \oplus \mathsf{auth}$
14: $T \leftarrow \mathsf{chop}_\tau(\mathsf{tag})$
15: **return** (C, T)

Proof. Suppose there is a distinguisher $\mathcal{B}$ of OTBC-3 which can make $\sigma + q$ queries to its challenger and which works in the following way. It runs $\mathcal{A}$ to start the game. Whenever $\mathcal{A}$ makes the i-th encryption query $(\mathcal{N}^i, A^i, M^i)$, $\mathcal{B}$ does the following.

- For the j-th message block M_j^i, it makes the encryption query$(\mathcal{N}^i\|00, j, M_j^i)$ to its challenger. Suppose it receives C_j^i as the response.
- Suppose the length of M^i is ℓ_i blocks. It makes and encryption query $(\mathcal{N}^i\|01, 0, M_1^i + \cdots + M_{\ell_i}^i)$ to its challenger. Suppose it receives C_{tag}^i as response.
- For the j-th associated data block A_j^i, it makes the encryption query $(\mathcal{N}^i\|10, j, A_j^i)$ to its challenger. Suppose it receives B_j^i as response.
- Suppose the length of A^i is k_i blocks. It calculates $\mathsf{auth}^i = B_1^i + \cdots + B_{k_i}^i$.
- Finally it returns $(C_1^i\| \cdots \|C_{\ell_i}^i, \mathsf{chop}_\tau(C_{\mathsf{tag}}^i + \mathsf{auth}^i))$ to $\mathcal{A}$.

Once $\mathcal{A}$ submits its decision bit, $\mathcal{B}$ carries it forward to its challenger as its own decision bit as well. Then we obtain the following privacy advantage of $\mathcal{A}$:

$$\mathbf{Adv}_{\mathsf{priv}}^{\mathsf{OCB+}}(\mathcal{A}) = \mathbf{Adv}_{\mathrm{TPRP*}}^{\mathsf{OTBC\text{-}3}}(\mathcal{B}).$$

Combining this result with Theorem 4, we obtain

$$\mathbf{Adv}_{\mathsf{priv}}^{\mathsf{OCB+}}(\mathcal{A}) \leq \frac{n\sigma L}{N}. \tag{1}$$

From the security analysis in Section 4 of [14], we obtain the following authenticity advantage of $\mathcal{A}$.

$$\mathbf{Adv}_{\mathsf{auth}}^{\mathsf{OCB+}}(\mathcal{A}) \leq O\left(\frac{q'L}{N}\right). \tag{2}$$

The result of Theorem 5 follows directly from (1) and (2). □

References

1. CAESAR: Competition for Authenticated Encryption: Security, Applicability, and Robustness. https://competitions.cr.yp.to/caesar-submissions.html
2. Information technology - Security techniques - Authenticated encryption. ISO/IEC 19772:2009 (2009)
3. NIST Lightweight Cryptography. https://csrc.nist.gov/Projects/lightweight-cryptography
4. Aoki, K., Yasuda, K.: The security of the OCB mode of operation without the SPRP assumption. In: Susilo, W., Reyhanitabar, R. (eds.) ProvSec 2013. LNCS, vol. 8209, pp. 202–220. Springer, Heidelberg (2013). https://doi.org/10.1007/978-3-642-41227-1_12
5. Ashur, T., Dunkelman, O., Luykx, A.: Boosting authenticated encryption robustness with minimal modifications. In: Katz, J., Shacham, H. (eds.) CRYPTO 2017. LNCS, vol. 10403, pp. 3–33. Springer, Cham (2017). https://doi.org/10.1007/978-3-319-63697-9_1
6. Avanzi, R.: The QARMA block cipher family. Almost MDS matrices over rings with zero divisors, nearly symmetric even-mansour constructions with non-involutory central rounds, and search heuristics for low-latency s-boxes. IACR Trans. Symmetric Cryptol. **2017**(1), 4–44 (2017)
7. Bao, Z., Guo, J., Iwata, T., Minematsu, K.: ZOCB and ZOTR: tweakable blockcipher modes for authenticated encryption with full absorption. IACR Trans. Symmetric Cryptol. **2019**(2), 1–54 (2019)
8. Beierle, C., et al.: The SKINNY family of block ciphers and its low-latency variant MANTIS. In: Robshaw, M., Katz, J. (eds.) CRYPTO 2016. LNCS, vol. 9815, pp. 123–153. Springer, Heidelberg (2016). https://doi.org/10.1007/978-3-662-53008-5_5
9. Beierle, C., et al.: SKINNY-AEAD and SKINNY-Hash. IACR Trans. Symmetric Cryptol. **2020**(S1), 88–131 (2020)
10. Beierle, C., Leander, G., Moradi, A., Rasoolzadeh, S.: Craft: lightweight tweakable block cipher with efficient protection against DFA attacks. IACR Trans. Symmetric Cryptol. **2019**(1), 5–45 (2019)
11. Bellare, M., Namprempre, C.: Authenticated encryption: relations among notions and analysis of the generic composition paradigm. J. Cryptol. **21**(4), 469–491 (2008). https://doi.org/10.1007/s00145-008-9026-x
12. Bellare, M., Namprempre, C.: Authenticated encryption: relations among notions and analysis of the generic composition paradigm. In: Okamoto, T. (ed.) ASIACRYPT 2000. LNCS, vol. 1976, pp. 531–545. Springer, Heidelberg (2000). https://doi.org/10.1007/3-540-44448-3_41
13. Bhaumik, R., et al.: QCB: efficient quantum-secure authenticated encryption. In: Tibouchi, M., Wang, H. (eds.) ASIACRYPT 2021. LNCS, vol. 13090, pp. 668–698. Springer, Cham (2021). https://doi.org/10.1007/978-3-030-92062-3_23
14. Bhaumik, R., Nandi, M.: Improved security for OCB3. In: Takagi, T., Peyrin, T. (eds.) ASIACRYPT 2017. LNCS, vol. 10625, pp. 638–666. Springer, Cham (2017). https://doi.org/10.1007/978-3-319-70697-9_22
15. Chakraborti, A., Datta, N., Jha, A., Mancillas-López, C., Nandi, M.: Light-OCB: parallel lightweight authenticated cipher with full security. In: Batina, L., Picek, S., Mondal, M. (eds.) SPACE 2021. LNCS, vol. 13162, pp. 22–41. Springer, Cham (2022). https://doi.org/10.1007/978-3-030-95085-9_2

16. Chakraborti, A., Datta, N., Jha, A., Mancillas-López, C., Nandi, M., Sasaki, Yu.: Int-rup secure lightweight parallel ae modes. IACR Trans. Symmetric Cryptol. **2019**(4), 81–118 (2020)
17. Chen, S., Steinberger, J.: Tight security bounds for key-alternating ciphers. In: Nguyen, P.Q., Oswald, E. (eds.) EUROCRYPT 2014. LNCS, vol. 8441, pp. 327–350. Springer, Heidelberg (2014). https://doi.org/10.1007/978-3-642-55220-5_19
18. Cogliati, B., Dutta, A., Nandi, M., Patarin, J., Saha, A.: Proof of mirror theory for any $\xi_{\max}$. IACR Cryptol. ePrint Arch., 686 (2022)
19. Dutta, A., Nandi, M., Saha, A.: Proof of mirror theory for $\xi_{\max} = 2$. IEEE Trans. Inf. Theory **68**(9), 6218–6232 (2022)
20. Ferguson, N., et al.: The Skein Hash Function Family. SHA3 submission to NIST (Round 3) (2010)
21. Granger, R., Jovanovic, P., Mennink, B., Neves, S.: Improved masking for tweakable blockciphers with applications to authenticated encryption. In: Fischlin, M., Coron, J.-S. (eds.) EUROCRYPT 2016. LNCS, vol. 9665, pp. 263–293. Springer, Heidelberg (2016). https://doi.org/10.1007/978-3-662-49890-3_11
22. Grosso, V., et al.: SCREAM v3. Submission to CAESAR competition (2015)
23. Gueron, S., Langley, A., Lindell, Y.: AES-GCM-SIV: nonce misuse-resistant authenticated encryption. RFC 8452 April 2019. https://doi.org/10.17487/RFC8452, https://www.rfc-editor.org/info/rfc8452
24. Gueron, S., Lindell, Y.: GCM-SIV: full nonce misuse-resistant authenticated encryption at under one cycle per byte. In: Proceedings of the 22nd ACM SIGSAC Conference on Computer and Communications Security, CCS '15, New York, NY, USA, pp. 109–119. Association for Computing Machinery (2015)
25. Hoang, V.T., Tessaro, S.: Key-alternating ciphers and key-length extension: exact bounds and multi-user security. In: Robshaw, M., Katz, J. (eds.) CRYPTO 2016. LNCS, vol. 9814, pp. 3–32. Springer, Heidelberg (2016). https://doi.org/10.1007/978-3-662-53018-4_1
26. Inoue, A., Iwata, T., Minematsu, K., Poettering, B.: Cryptanalysis of OCB2: attacks on authenticity and confidentiality. J. Cryptol. **33**(4), 1871–1913 (2020). https://doi.org/10.1007/s00145-020-09359-8
27. Inoue, A., Iwata, T., Minematsu, K., Poettering, B.: Cryptanalysis of OCB2: attacks on authenticity and confidentiality. In: Boldyreva, A., Micciancio, D. (eds.) CRYPTO 2019. LNCS, vol. 11692, pp. 3–31. Springer, Cham (2019). https://doi.org/10.1007/978-3-030-26948-7_1
28. Iwata, T.: New blockcipher modes of operation with beyond the birthday bound security. In: Robshaw, M. (ed.) FSE 2006. LNCS, vol. 4047, pp. 310–327. Springer, Heidelberg (2006). https://doi.org/10.1007/11799313_20
29. Iwata, T.: Authenticated encryption mode for beyond the birthday bound security. In: Vaudenay, S. (ed.) AFRICACRYPT 2008. LNCS, vol. 5023, pp. 125–142. Springer, Heidelberg (2008). https://doi.org/10.1007/978-3-540-68164-9_9
30. Jean, J., Nikolić, I., Peyrin, T.: Tweaks and keys for block ciphers: the `TWEAKEY` framework. In: Sarkar, P., Iwata, T. (eds.) ASIACRYPT 2014. LNCS, vol. 8874, pp. 274–288. Springer, Heidelberg (2014). https://doi.org/10.1007/978-3-662-45608-8_15
31. Jean, J., Nikolic, I., Peyrin, T.: Joltik v1.3. CAESAR Round, 2 (2015)
32. Jean, J., Nikolic, I., Peyrin, T., Seurin, Y.: The Deoxys AEAD family. J. Cryptol. **34**(3), 31 (2021)
33. Jutla, C.S.: Encryption modes with almost free message integrity. In: Pfitzmann, B. (ed.) EUROCRYPT 2001. LNCS, vol. 2045, pp. 529–544. Springer, Heidelberg (2001). https://doi.org/10.1007/3-540-44987-6_32

34. Jutla, C.S.: Encryption modes with almost free message integrity. J. Cryptol. **21**(4), 547–578 (2008). https://doi.org/10.1007/s00145-008-9024-z
35. Katz, J., Yung, M.: Unforgeable encryption and chosen ciphertext secure modes of operation. In: Goos, G., Hartmanis, J., van Leeuwen, J., Schneier, B. (eds.) FSE 2000. LNCS, vol. 1978, pp. 284–299. Springer, Heidelberg (2001). https://doi.org/10.1007/3-540-44706-7_20
36. Krovetz, T., Rogaway, P.: The OCB Authenticated-Encryption Algorithm. RFC 7253, May 2014. https://doi.org/10.17487/RFC7253, https://www.rfc-editor.org/info/rfc7253
37. Krovetz, T., Rogaway, P.: The software performance of authenticated-encryption modes. In: Joux, A. (ed.) FSE 2011. LNCS, vol. 6733, pp. 306–327. Springer, Heidelberg (2011). https://doi.org/10.1007/978-3-642-21702-9_18
38. Liskov, M., Rivest, R.L., Wagner, D.: Tweakable block ciphers. J. Cryptol. **24**, 588–613 (2011). https://doi.org/10.1007/s00145-010-9073-y
39. Liskov, M., Rivest, R.L., Wagner, D.: Tweakable block ciphers. In: Yung, M. (ed.) CRYPTO 2002. LNCS, vol. 2442, pp. 31–46. Springer, Heidelberg (2002). https://doi.org/10.1007/3-540-45708-9_3
40. Minematsu, K.: Parallelizable rate-1 authenticated encryption from pseudorandom functions. In: Nguyen, P.Q., Oswald, E. (eds.) EUROCRYPT 2014. LNCS, vol. 8441, pp. 275–292. Springer, Heidelberg (2014). https://doi.org/10.1007/978-3-642-55220-5_16
41. Naito, Y.: Tweakable blockciphers for efficient authenticated encryptions with beyond the birthday-bound security. IACR Trans. Symmetric Cryptol. **2017**(2), 1–26 (2017)
42. Patarin, J.: The "Coefficients H" technique. In: Avanzi, R.M., Keliher, L., Sica, F. (eds.) SAC 2008. LNCS, vol. 5381, pp. 328–345. Springer, Heidelberg (2009). https://doi.org/10.1007/978-3-642-04159-4_21
43. Peyrin, T., Seurin, Y.: Counter-in-tweak: authenticated encryption modes for tweakable block ciphers. In: Robshaw, M., Katz, J. (eds.) CRYPTO 2016. LNCS, vol. 9814, pp. 33–63. Springer, Heidelberg (2016). https://doi.org/10.1007/978-3-662-53018-4_2
44. Rogaway, P.: Authenticated-encryption with associated-data. In: Proceedings of the 9th ACM Conference on Computer and Communications Security, CCS 2002, New York, NY, USA, pp. 98–107. Association for Computing Machinery (2002)
45. Rogaway, P.: Efficient instantiations of tweakable blockciphers and refinements to modes OCB and PMAC. In: Lee, P.J. (ed.) ASIACRYPT 2004. LNCS, vol. 3329, pp. 16–31. Springer, Heidelberg (2004). https://doi.org/10.1007/978-3-540-30539-2_2
46. Rogaway, P., Bellare, M., Black, J.: OCB: a block-cipher mode of operation for efficient authenticated encryption. ACM Trans. Inf. Syst. Secur. **6**(3), 365–403 (2003)
47. Rogaway, P., Bellare, M., Black, J., Krovetz, T.: OCB: a block-cipher mode of operation for efficient authenticated encryption. In: Proceedings of the 8th ACM Conference on Computer and Communications Security, CCS 2001, New York, NY, USA, pp. 196–205. Association for Computing Machinery (2001)
48. Rogaway, P., Shrimpton, T.: A provable-security treatment of the key-wrap problem. In: Vaudenay, S. (ed.) EUROCRYPT 2006. LNCS, vol. 4004, pp. 373–390. Springer, Heidelberg (2006). https://doi.org/10.1007/11761679_23
49. Schroeppel, R.: The Hasty Pudding Cipher. AES submission to NIST (1998)
50. John Steinberger Shan Chen. Tight security bounds for key-alternating ciphers. Cryptology ePrint Archive, Report 2013/222 (2013). https://ia.cr/2013/222

51. Ping Zhang, Peng Wang, and Honggang Hu. The int-rup security of ocb with intermediate (parity) checksum. Cryptology ePrint Archive, Report 2016/1059, 2016. https://ia.cr/2016/1059
52. Zhang, P., Wang, P., Hu, H., Cheng, C., Kuai, W.: INT-RUP security of checksum-based authenticated encryption. In: Okamoto, T., Yu, Y., Au, M.H., Li, Y. (eds.) ProvSec 2017. LNCS, vol. 10592, pp. 147–166. Springer, Cham (2017). https://doi.org/10.1007/978-3-319-68637-0_9

ISAP+: ISAP with Fast Authentication

Arghya Bhattacharjee[1], Avik Chakraborti[2], Nilanjan Datta[2(✉)], Cuauhtemoc Mancillas-López[3], and Mridul Nandi[1]

[1] Indian Statistical Institute, Kolkata, India
[2] TCG Centres for Research and Education in Science and Technology, Kolkata, India
nilanjan.datta@tcgcrest.org
[3] Computer Science Department, CINVESTAV-IPN, Mexico City, Mexico
cuauhtemoc.mancillas@cinvestav.mx

Abstract. This paper analyses the lightweight, sponge-based NAEAD mode ISAP, one of the finalists of the NIST Lightweight Cryptography (LWC) standardisation project, that achieves high-throughput with inherent protection against differential power analysis (DPA). We observe that ISAP requires 256-bit capacity in the authentication module to satisfy the NIST LWC security criteria. In this paper, we study the analysis carefully and observe that this is primarily due to the collision in the associated data part of the hash function which can be used in the forgery of the mode. However, the same is not applicable to the ciphertext part of the hash function because a collision in the ciphertext part does not always lead to a forgery. In this context, we define a new security notion, named 2PI+ security, which is a strictly stronger notion than the collision security, and show that the security of a class of encrypt-then-hash based MAC type of authenticated encryptions, that includes ISAP, reduces to the 2PI+ security of the underlying hash function used in the authentication module. Next we investigate and observe that a feed-forward variant of the generic sponge hash achieves better 2PI+ security as compared to the generic sponge hash. We use this fact to present a close variant of ISAP, named ISAP+, which is structurally similar to ISAP, except that it uses the feed-forward variant of the generic sponge hash in the authentication module. This improves the overall security of the mode, and hence we can set the capacity of the ciphertext part to 192 bits (to achieve a higher throughput) and yet satisfy the NIST LWC security criteria.

Keywords: Authenticated encryption · ISAP · ISAP+ · Re-keying · Side channel resistant · 2PI+ · Sponge

1 Introduction

The emergence of side-channel and fault attacks [10,11,27,28] has made it clear that cryptographic implementations may not always behave like a black box. Instead, they might behave like a grey box where the attacker has physical access

T. Isobe and S. Sarkar (Eds.): INDOCRYPT 2022, LNCS 13774, pp. 195–219, 2022.
https://doi.org/10.1007/978-3-031-22912-1_9

to the device executing a cryptographic task. As a result, designers have started to design side-channel countermeasures such as masking [12,23]. However, cryptographic primitives like block ciphers (for example, AES [16] and ARX-based designs [6]) are costly to be mask-protected against side-channel attacks. Consequently, designing primitives or modes with inherent side-channel protection is becoming an essential and popular design goal. In this line of design, several block ciphers (e.g., Noekeon [26], PICARO [33], Zorro [21]) and permutations (e.g., ASCON-p [20], KECCAK-p [1,25]) have been proposed with dedicated structures to reduce the resource requirements for masking. In addition, a few NAEAD (Nonce based Authenticated Encryption with Associated Data) modes, such as ASCON [13,20], PRIMATES [2], SCREAM [24], KETJE/KEYAK [9] have been proposed and submitted to the CAESAR [15] competition with the same goal in mind. However, they still lead to significant overheads.

In [30], Medwed et al. have proposed a new technique called fresh *re-keying* to have inherent side-channel protection. Following their work, a series of works [4,4,29] have been proposed that use this novel concept. This technique requires a side-channel resistant fresh key computation function with the nonce and the master key as the inputs. The main idea behind these designs is to ensure that different session keys are used for different nonces. Hence, a new nonce should be used to generate a fresh session key.

1.1 ISAP and Its Variants

In [19], Dobraunig et al. proposed a new authenticated encryption, dubbed ISAP v1, following the re-keying strategy. It is a sponge-based design [7,8] that follows the Encrypt-then-MAC paradigm [5]. We'll traditionally use the terms *rate* and *capacity* to represent the exposed part and the hidden part of the state of the sponge construction respectively. ISAP v1 claims to offer higher-order differential power analysis (DPA) protection provided by an inherent design strategy that combines a sponge-based stream cipher for the encryption module with a sponge-based MAC (suffix keyed) for the authentication module. Both the modules compute fresh session keys using a GGM [22] tree-like function to strengthen the key computation against side-channel attacks.

Later, ISAP v1 was improved to ISAP v2 [14,18], and was submitted to the NIST Lightweight Cryptography (LWC) standardization project [31] and currently one of the finalists. ISAP v2 is equipped with several promising features and currently is considered to be a strong candidate for the competition. It recommends two variants, instantiated with the lightweight permutations ASCON-p and KECCAK-p[400]. Precisely, ISAP v2 retains all the inherent DPA resistance properties of ISAP v1 along with a better resistance against other implementation based attacks. In addition, ISAP v2 is even more efficient in hardware resources than the first version. ISAP v2 has been highly praised by the cryptography community, and to the best of our knowledge, it is the only inherently DPA protected NAEAD mode that aims to be implemented on lightweight platforms. The ISAP mode is flexible and can be instantiated with any sufficiently large permutation. Precisely, the security claims made by the designers depict that ISAP needs around 256-bit capacity to satisfy the NIST security criteria and hence needs a permutation with the state size larger than 256-bits. Thus it is highly desirable to analyze the mode further to understand whether it can be designed with a smaller capacity and hence a higher rate, that directly impacts the throughput.

1.2 Improving the Throughput of ISAP

ISAP v2 proposes four instances with the ASCON-p and the KECCAK-p permutations. ISAP v2 with ASCON-p (a 320-bit permutation) is designed with a 64 bit rate. However, it is better to achieve a higher rate design for a higher throughput. The observation is similar for the other instances as well. A potential direction can be to design an algorithm with an improved security bound over the capacity to increase the rate without compromise in the security level. An increased security bound can also help the designers to achieve the same security with a lower state size. This in turn may reduce the register size by using a permutation with a smaller state. A potential choice can be to analyze ISAP with a focus on the BBB (Beyond Birthday Bound) NAEAD security.

We observe that ISAP adopts an efficient approach of applying an unkeyed hash function on the nonce, the associated data, and the ciphertext, and then uses a PRF (Pseudo Random Function) on the hash value. In this case, the security of the NAEAD mode boils down to the collision security of the underlying hash function. This mode can bypass the requirement of storing the master key and can get rid of the key register. However, the hash collision results in a relatively low security bound that may not always be acceptable in ultra-lightweight applications as low security bound forces the designs to adopt primitives with high state size. Thus, an increase in the security bound has the full potential to increase the hardware performance of the design significantly. Motivated by this issue, we aim to study the tightness of the security bound to optimize the throughput and the hardware footprint. Note that, ISAP is already an efficient construction and has been reviewed rigorously by various research groups. Hence, a more detailed mode analysis and any possible mode optimization can further strengthen the construction.

The security proof in [19] by Dobraunig et al. showed that ISAP achieves security up to the birthday bound on the capacity, i.e. of $O(T^2/2^c)$, where T is the time complexity, and c is the capacity size (in bits). We observe that this factor arises due to the simple sponge-type hash applied on the nonce, the associated data, and the random ciphertext. It is obvious to get a collision in the nonce and the associated data that can be trivially used by an adversary to mount a forgery. However, it is not evident how a collision in the random ciphertexts can lead to such an attack. In this regard we investigate the amount of the ciphertext-bit that can be injected per permutation call during the hash and ask the question:

"Can we increase the rate of absorption of the ciphertext blocks in the hash?"

We believe that a positive answer to this question will not only result in a more efficient construction, but more importantly contribute significantly in the direction of NAEAD mode analysis.

1.3 Our Contributions

In this paper, we study a simple variant of ISAP that achieves higher throughput keeping all the primary features of ISAP intact. Our contribution is three-fold:

1. First, in Sect. 3, we propose a permutation-based generic EtHM (Encrypt then Hash based MAC) type NAEAD mode using a PRF and an unkeyed hash function. This is essentially a generalisation of ISAP type constructions. Note that this generic mode does not guarantee any side-channel resistance; only proper instantiation of the PRF ensures that. In Sect. 3.2, we have shown that the NAEAD security of EtHM can be expressed in terms of the PRF security of a fixed input length, variable output length keyed function F, the 2PI+ security of H. Intuitively, the 2PI+ notion demands that given a challenge random message of some length chosen by the adversary, it is difficult for an adversary to compute the second pre-image of the random message. We have introduced the notion in Sect. 2.6. This is in contrast with the traditional collision security that was used for the analysis of ISAP.
2. Next, in Sect. 4, we first show that for generic sponge hash, a collision attack can be extended to a 2PI+attack. Thus, the generic sponge hash achieves 2PI+ security of $\Omega(T^2/2^c)$, where c is the capacity of the sponge hash. Next, we consider a feed-forward variant of the sponge hash that uses (i) generic sponge hash to process the nonce and the associated data and (ii) a feed-forward variant of the sponge hash to process the message. We show that the feed-forward property ensures that a collsion attack can not be extended to mount a 2PI+ attack. In fact, we prove that this variant of sponge hash obtains an improved security of $O(DT/2^c)$. Note that D and T are data and time complexity respectively, and we typically allow $T \approx D^2$. Hence, feed-forward based sponge archives a better 2PI+security as we consider security in terms of D and T instead of traditional one parameter security.
3. Finally, in Sect. 5, we have considered a simple variant of ISAP with minimal changes, named ISAP+, which is a particular instantiation of the generic EtHM mode. To be specific, the differences between ISAP+ and ISAP are as follows:
   (a) Instead of using the generic sponge hash as used in ISAP, we use the feed-forward variant of sponge hash as discussed above.
   (b) In the authentication module of ISAP+, we use the capacity of c' bits for nonce, associated data and first block of ciphertext processing. For rest of the ciphertext blocks, we use capacity of c bits.
   (c) We make a separation among the messages depending on whether its length is less than r' bits or not. The domain separation is performed by adding 0 or 1 to the capacity part before the final permutation call.

This modification ensures that ISAP+ achieves improved security of $O(T^2/2^{c'} +DT/2^c)$, where n is the state-size or the size of the permutation (in bits), $c' = n - r$, $c' = n - r'$). This security boost allows the designer to choose c' and c ($< c'$) effectively to obtain better throughput.

1.4 Relevance of the Work

To understand the relevance of the improved security, let us consider the instantiation of ISAP with ASCON and KECCAK and compare them with ISAP+. To satisfy the NIST requirements, ISAP+ can use $c = 192$, $c' = 256$, and hence, it has a injection rate of $r = 128$-bits for the ciphertexts and an injection rate of $r = 64$ bits for associated data. Table 1 demonstrate a comparative study of ISAP and ISAP+ in terms of the number of permutation calls required for the authentication module.

Table 1. Comparative Study of ISAP+ and ISAP on the no. of permutation calls in the authentication module for associated data of length a bits, message of length m bits.

Mode	Permutation	Parameters	# permutation calls
ISAP	ASCON	$r = 64$	$\lceil \frac{a+m+1}{64} \rceil$
ISAP+	ASCON	$r = 128,\ r' = 64$	$\lceil \frac{a+1}{64} \rceil + \lceil \frac{m}{128} \rceil$
ISAP	KECCAK	$r = 144$	$\lceil \frac{a+m+1}{144} \rceil$
ISAP+	KECCAK	$r = 208,\ r' = 144$	$\lceil \frac{a+1}{144} \rceil + \lceil \frac{m}{208} \rceil$

This result demonstrates that for applications that require long message processing, ISAP+ performs better than ISAP in terms of throughput and speed. Let us consider a concrete example. Consider encrypting a message of length 1 MB with an associate data of length 1 KB using ASCON permutation. With ISAP, the authentication module requires around $1,31,201$ many primitive calls. On the other hand, with ISAP+ this requires only $65,665$ many primitive calls, which is almost half as compared to ISAP.

This paper depicts the robustness of the mode ISAP, and how one can increase the throughtput of the mode at the cost of some hardware area preserving all the inherent security features. This result seems relevant to the cryptography community in the sense that ISAP is also a finalist of the NIST LWC project for standardization.

2 Preliminaries

2.1 Notations

We'll usually use lowercase letters (e.g., x, y) for integers and indices, uppercase letters (e.g., X, Y) for binary strings and functions, and calligraphic uppercase letters (e.g., $\mathcal{X}$, $\mathcal{Y}$) for sets and spaces. $\mathbb{N}$ and $\mathbb{Z}$ will be used to denote the set of natural numbers and the set of integers respectively. 0^x and 1^y will denote the sequence of x 0's and y 1's respectively. $\{0,1\}^x$ and $\{0,1\}^*$ will denote the set of binary strings of length x and the set of all binary strings respectively. For any $X \in \{0,1\}^*$, $|X|$ and $\|X\|$ will denote the number of bits, and the number of blocks of the binary string X respectively, where the size of the blocks should be clear from the context. For two binary strings X and Y, $X\|Y$ will denote the concatenation of X and Y. For any $X \in \{0,1\}^*$, we define the parsing of X into r-bit blocks as $X_1 \cdots X_x \leftarrow_r X$, where $|X_i| = r$ for all $i < x$ and $1 \leq |X_x| \leq r$ such that $X = X_1\| \cdots \|X_x$. For any $X \in \{0,1\}^*$, $X_1 \cdots X_x \twoheadleftarrow_r X$ does the work of $X_1 \cdots X_x \leftarrow_r X$, and follows it by the compulsory 10^* padding. Given any sequence $X = X_1 \cdots X_x$ and $1 \leq a \leq b \leq x$, we'll represent the subsequence $X_a \ldots X_b$ by $X[a \cdots b]$. For integers $a \leq b$, we'll write $[a \cdots b]$ for the set $\{a, \ldots, b\}$, and for integers $1 \leq a$, we'll write $[a]$ for the set $\{1, \ldots, a\}$. We'll use the notations $\lceil x \rceil_r$ and $\lfloor x \rfloor_r$ to denote the decimal ceiling and floor function on the integer x respectively, and similarly, $\lceil X \rceil_r$ and $\lfloor X \rfloor_r$, to denote the most significant x bits and the least significant x bits of the binary string X respectively. By $X \xleftarrow{\$} \mathcal{X}$, we'll denote that X is chosen uniformly at random from the set $\mathcal{X}$.

2.2 Distinguishing Advantage

For two oracles $\mathcal{O}_0$ and $\mathcal{O}_1$, an algorithm $\mathcal{A}$ which tries to distinguish between $\mathcal{O}_0$ and $\mathcal{O}_1$ is called a distinguishing adversary. $\mathcal{A}$ plays an interactive game with $\mathcal{O}_b$ where b is unknown to $\mathcal{A}$, and then outputs a bit $b_{\mathcal{A}}$. The winning event is $[b_{\mathcal{A}} = b]$. The distinguishing advantage of $\mathcal{A}$ is defined as

$$\mathbf{Adv}_{\mathcal{O}_1,\mathcal{O}_0}(\mathcal{A}) := |\Pr[b_{\mathcal{A}} = 1|b = 1] - \Pr[b_{\mathcal{A}} = 1|b = 0]| .$$

Let $\mathbf{A}[q,t]$ be the class of all distinguishing adversaries limited to q oracle queries and t computations. We define

$$\mathbf{Adv}_{\mathcal{O}_1,\mathcal{O}_0}[q,t] := \max_{\mathbf{A}[q,t]} \mathbf{Adv}_{\mathcal{O}_1,\mathcal{O}_0}(\mathcal{A}) .$$

When the adversaries in $\mathbf{A}[q,t]$ are allowed to make both encryption queries and decryption queries to the oracle, this is written as $\mathbf{Adv}_{\pm\mathcal{O}_1,\pm\mathcal{O}_0}[q,q',t]$, where q is the maximum number of encryption queries allowed and q' is the maximum number of decryption queries allowed. Enc_b and Dec_b denote the encryption and the decryption function associated with $\mathcal{O}_b$ respectively. $\mathcal{O}_0$ conventionally represents an ideal primitive, while $\mathcal{O}_1$ represents either an actual construction or a mode of operation built of some other ideal primitives. Typically the goal of the function represented by $\mathcal{O}_1$ is to emulate the ideal primitive represented by $\mathcal{O}_0$. We use the standard terms real oracle and ideal oracle for $\mathcal{O}_1$ and $\mathcal{O}_0$ respectively. A security game is a distinguishing game with an optional set of additional restrictions, chosen to reflect the desired security goal. When we talk of distinguishing advantage with a specific security game $\mathcal{G}$ in mind, we include $\mathcal{G}$ in the superscript, e.g., $\mathbf{Adv}^{\mathcal{G}}_{\mathcal{O}_1,\mathcal{O}_0}(\mathcal{A})$. Also we sometimes drop the ideal oracle and simply write $\mathbf{Adv}^{\mathcal{G}}_{\mathcal{O}_1}(\mathcal{A})$ when the ideal oracle is clear from the context.

2.3 Authenticated Encryption and Its Security Notion

A **N**once based **A**uthenticated **E**ncryption with **A**ssociated **D**ata (**NAEAD**) involves a key space $\mathcal{K}$, a nonce space $\mathcal{N}$, an associated data space $\mathcal{AD}$, a message space $\mathcal{M}$ and a tag space $\mathcal{T}$ along with two functions $\mathsf{Enc} : \mathcal{K} \times \mathcal{N} \times \mathcal{AD} \times \mathcal{M} \to \mathcal{M} \times \mathcal{T}$ (called the Encryption Function) and $\mathsf{Dec} : \mathcal{K} \times \mathcal{N} \times \mathcal{AD} \times \mathcal{M} \times \mathcal{T} \to \mathcal{M} \cup \{\perp\}$ (called the Decryption Function) with the correctness condition that for any $K \in \mathcal{K}, N \in \mathcal{N}, A \in \mathcal{AD}$ and $M \in \mathcal{M}$, it holds that

$$\mathsf{Dec}(K, N, A, \mathsf{Enc}(K, N, A, M)) = M\,.$$

In the NAEAD security game, the real oracle involves such a pair of functions Enc_1 and Dec_1 with $K \xleftarrow{\$} \mathcal{K}$. On the other hand, the ideal oracle involves an ideal random function $\mathsf{Enc}_0 : \mathcal{K} \times \mathcal{N} \times \mathcal{AD} \times \mathcal{M} \to \mathcal{M} \times \mathcal{T}$ and a constant function $\mathsf{Dec}_0 : \mathcal{K} \times \mathcal{N} \times \mathcal{AD} \times \mathcal{M} \times \mathcal{T} \to \{\perp\}$. The adversary ($\mathcal{A}$) which interacts with one of the two oracles is supposed to be:

1. Nonce-respecting, i.e., $\mathcal{A}$ should not repeat a nonce in more than one encryption queries, and
2. Non-repeating, i.e., $\mathcal{A}$ should not make the decryption query (N, A, C, T) if it has already made the encryption query (N, A, M) and received (C, T) in response.

The distinguishing advantage of $\mathcal{A}$ will be denoted by $\mathbf{Adv}^{\mathrm{NAEAD}}_{(\mathsf{Enc}_1,\mathsf{Dec}_1)}(\mathcal{A})$. The following two security notions are captured in this advantage.

1. Privacy or Confidentiality, i.e., $\mathcal{A}$ should not be able to distinguish the real oracle from the ideal oracle.
2. Authenticity or Integrity, i.e., $\mathcal{A}$ should not be able to forge the real oracle. In other words, $\mathcal{A}$ should not be able to make a decryption query to the real oracle to which the response isn't $\perp$.

2.4 The Coefficients H Technique

The Coefficients H Technique is a proof method by Patarin [32]. Consider two oracles $\mathcal{O}_0$ (the ideal oracle) and $\mathcal{O}_1$ (the real oracle). Let $\mathcal{T}$ denote the set of all possible transcripts (i.e., the set of all query-response pairs) an adversary can obtain. For any transcript $\tau \in \mathcal{T}$, we will denote the probability to realize the transcript as $\mathsf{ip}_{\mathsf{real}}(\tau)$ or $\mathsf{ip}_{\mathsf{ideal}}(\tau)$ when it is interacting with the real or the ideal oracle respectively. We call them the interpolation probabilities. W.l.o.g., we assume that the adversary is deterministic. Hence, the interpolation probabilities are the properties of the oracles only. As we deal with stateless oracles, these probabilities are independent of the order of the query-response pairs in the transcript.

Theorem 1. *Suppose for a set $\mathcal{T}_{good} \subseteq \mathcal{T}$ of transcripts (called the* **good transcripts***) the following holds:*

1. *For any adversary $\mathcal{A}$ interacting with $\mathcal{O}_0$ (the ideal oracle), the probability of getting a transcript in $\mathcal{T}_{good}$ is at least $1 - \epsilon_{bad}$. We may denote the set $\mathcal{T} \setminus \mathcal{T}_{good}$ by $\mathcal{T}_{bad}$. Hence, the probability of getting a transcript in $\mathcal{T}_{bad}$ is at most ϵ_{bad}.*
2. *For any adversary $\mathcal{A}$ and for any transcript $\tau \in \mathcal{T}_{good}$,*

$$\mathsf{ip}_{\mathsf{real}}(\tau) \geq (1 - \epsilon_{ratio}) \cdot \mathsf{ip}_{\mathsf{ideal}}(\tau).$$

For an oracle $\mathcal{O}_1$ (the real oracle) satisfying (1) and (2), we have

$$\mathbf{Adv}_{\mathcal{O}_0,\mathcal{O}_1}(\mathcal{A}) \leq \epsilon_{bad} + \epsilon_{ratio}.$$

2.5 Fixed Input - Variable Output PRFs with Prefix Property

A fixed input variable output function (FIL-VOL) is a keyed function $F_K : \{0,1\}^\star \times \{0,1\} \times \mathbb{N} \rightarrow \{0,1\}^\star$ that takes as input an input a string $I \in \{0,1\}^\star$, a flag $b \in \{0,1\}$ as input, a positive integer $\ell \in \mathbb{N}$, and outputs a string $O \in \{0,1\}^\ell$, i.e., $O := F_K(I, b; \ell)$. We call such a keyed function a FIL-VOL pseudo random function maintaining the prefix-property if

- for all inputs $(I, b; \ell), (I', b'; \ell')$ with $(I, b) \neq (I', b')$, $F_K(I, b; \ell)$, $F_K(I', b'; \ell)$ are distributed uniformly at random, and
- for all inputs $(I, b; \ell), (I, b; \ell')$, with $\ell' > \ell$, $\lceil F_K(I', b'; \ell')\rceil_\ell = F_K(I', b'; \ell)$.

More formally,

$$\mathbf{Adv}_F^{\mathsf{PRF}}(\mathcal{A}) := |\Pr[\mathcal{A}^{F_K} = 1] - \Pr[\mathcal{A}^f = 1]|,$$

where f is a random function from same domain and range maintaining the prefix property.

2.6 Multi-target 2nd Pre-image with Associated Data

In this section, we discuss the notion of multi-target 2nd pre-image security of permutation-based hash functions.

In this setting, an adversary (say $\mathcal{A}$) chooses q (nonce, associated data, length)-tuples to the challenger $\mathcal{C}$, say $(N_i, A_i, \ell_i)_{i=1..q}$. The challenger in turn returns q uniformly random messages of specified lengths respectively, say

$C_1, \ldots, C_q$. The queries $(N_i, A_i, C_i)_{i=1..q}$ are called *challenge queries*. The goal of $\mathcal{A}$ is to return q' many $(N'_j, A'_j, C'_j)_{j=1..q'}$ (called response queries) tuples such that at least one of the hash values of (N'_j, A'_j, C'_j) matches with the hash value of any one of the (N_i, A_i, C_i). Note that the adversaries are allowed to set some challenge queries as response queries: $(N'_j, A'_j, C'_j) = (N_i, A_i, C_i)$, for some i, j. However, for the winning event the challenge and response queries should be distinct. The adversary can make up to q_p queries to p or p^{-1}. Formally, the advantage of $\mathcal{A}$ is defined as

$$\mathbf{Adv}_{\mathrm{H}}^{\mathsf{2PI+}}(\mathcal{A}) := \Pr[\exists i, j,\ H^p(N'_j, A'_j, C'_j) = H^p(N_i, A_i, C_i),\\ (N'_j, A'_j, C'_j) \neq (N_i, A_i, C_i)].$$

where H is an IV-based hash function. Note that the adversary is allowed to make hash queries before, after, or in between its interaction with the challenger to obtain the challenge message(s). Also, note that the 2PI+ security does not depend on the message length. The fact that the adversary submits a length ℓ_i to the challenger to obtain each message before the submission of the challenge message is merely because the 2PI+ security notion enables its adversary to obtain messages of whatever lengths it pleases.

3 An EtHM Paradigm for NAEAD

This section introduces an efficient generalized Encrypt-then-Hash based MAC (EtHM) paradigm for NAEAD modes. This is a generalized paradigm for constructing side-channel resilient modes such as ISAP.

3.1 Specification

Let n, k and τ be positive integers such that $n > \tau$. The construction takes as input a plaintext M, a nonce N, an associated data A, and outputs a ciphertext C and a tag T. Given a permutation based FIL-VOL keyed-function with prefix property F_K^p, and a permutation based un-keyed hash function $H^p : \{0,1\}^* \rightarrow \{0,1\}^n$ the mode works as follows.

$$C = M \oplus F_K^p(N, 0; |M|),$$
$$T \| D = p(F_K^p(X, 1; |X|) \| Z), \text{ where } X \| Z = H^p(N, A, C).$$

The authenticated encryption module is pictorially depicted in Fig. 1. Note that T denotes the most significant τ bits of the output of the permutation call. The least significant $(n - \tau)$ bits is denoted by D. Note that we do not need D from the construction point of view, however, we require it during the security analysis. Notations F, H and F_K^p, H^p have been used in this paper interchangeably for convenience, and aren't supposed to create any confusion. From time to time, we'll address this paradigm as EtHM only.

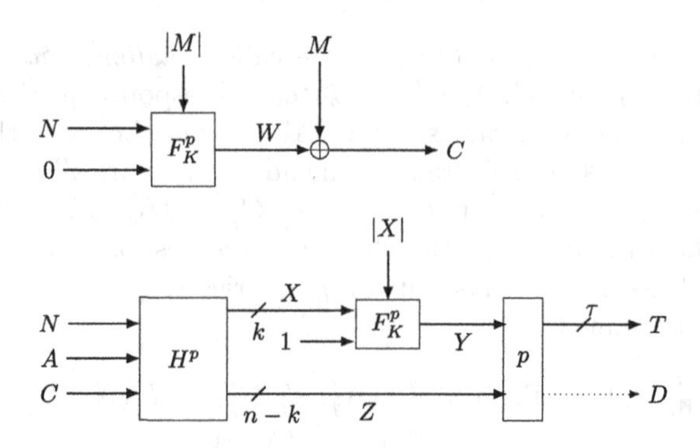

Fig. 1. Authenticated encryption module of the EtHM paradigm.

3.2 Security of EtHM

In this subsection, we analyse the NAEAD security of EtHM with F as the underlying function and H as the multi-target IV-respecting second pre-image resistant hash function. Formally, we prove the following theorem.

Theorem 2 (NAEAD Security of EtHM). *Consider EtHM based on a function F and a hash function H. For all deterministic nonce-respecting non-repeating query making adversary $\mathcal{A}$ which can make at most q_e encryption queries, q_v decryption queries and q_p primitive queries to p and its inverse and assuming $q = q_e + q_v$, there exists two adversaries $\mathcal{B}_1$ and $\mathcal{B}_2$ such that the NAEAD advantage of $\mathcal{A}$ can be bounded by*

$$\mathbf{Adv}_{\mathsf{EtHM}}^{NAEAD}(\mathcal{A}) \leq \mathbf{Adv}_{F}^{\mathsf{PRF}}(\mathcal{B}_1) + 2\mathbf{Adv}_{\lfloor H \rfloor_{n-k}}^{\mathsf{2PI+}}(\mathcal{B}_2) + \frac{qq_p}{2^n} + \frac{2kq_v}{2^k} + \frac{q_v}{2^\tau} + \frac{\binom{q_p}{k}}{2^{\tau(k-1)}} + \frac{\binom{q_p}{k}}{2^{(n-k)(k-1)}},$$

where $\mathcal{B}_1$ can make $2q$ PRF queries and $\mathcal{B}_2$ can make q challenge queries, q response queries and q_p primitive queries to p and its inverse.

Proof. Let $\mathbf{Enc}^{F_K^p,p}$ and $\mathbf{Dec}^{F_K^p,p}$ be the encryption and the decryption function of EtHM respectively. Let us call its oracle $\mathcal{O}_1 = (\mathbf{Enc}^{F_K^p,p}, \mathbf{Dec}^{F_K^p,p}, p)$. We have to upper-bound the distinguishing advantage of $\mathcal{A}$ interacting with $\mathcal{O}_1$ or $\mathcal{O}_3 = (\$, \perp, p)$. For our purpose, we define an intermediate oracle by replacing F_K^p in $\mathcal{O}_1$ by a random functions \$. Let us call this new intermediate oracle $\mathcal{O}_2 = (\mathbf{Enc}^{\$,p}, \mathbf{Dec}^{\$,p}, p)$. We will employ a standard reduction proof. We break down the distinguishing game of $\mathcal{A}$ using the triangle inequality as follows.

$$\begin{aligned} \mathbf{Adv}_{\mathsf{EtHM}}^{\mathrm{NAEAD}}(\mathcal{A}) &= \left|\Pr[\mathcal{A}^{\mathcal{O}_1} = 1] - \Pr[\mathcal{A}^{\mathcal{O}_3} = 1]\right| \\ &\leq \left|\Pr[\mathcal{A}^{\mathcal{O}_1} = 1] - \Pr[\mathcal{A}^{\mathcal{O}_2} = 1]\right| \\ &\quad + \left|\Pr[\mathcal{A}^{\mathcal{O}_2} = 1] - \Pr[\mathcal{A}^{\mathcal{O}_3} = 1]\right|. \end{aligned} \quad (1)$$

Now, we bound each of the two terms.

□ Bounding $|\Pr[\mathcal{A}^{\mathcal{O}_1} = 1] - \Pr[\mathcal{A}^{\mathcal{O}_2} = 1]|$. We bound this term by the PRF advantage of F. For that, let us consider the following adversary $\mathcal{B}_1$ that runs $\mathcal{A}$ (any distinguisher of $\mathcal{O}_1$ and $\mathcal{O}_2$) as follows.

- Whenever $\mathcal{A}$ submits an encryption query (N, A, M), $\mathcal{B}_1$ submits $(N, |M|, 0)$ to its challenger. Suppose the challenger returns C. $\mathcal{B}_1$ calculates $X\|Z = H^p(N, A, C)$ with $|X| = k$ and $|Z| = n - k$ and submits $(X, 1; k)$ to its challenger. Suppose the challenger returns Y. $\mathcal{B}_1$ returns $(C, p(Y\|Z))$ to $\mathcal{A}$.
- Similarly, whenever $\mathcal{A}$ submits a decryption query $(\hat{N}, \hat{A}, \hat{C}, \hat{T})$, $\mathcal{B}_1$ submits $(\hat{N}, 0; |\hat{C}|)$ to its challenger. Suppose the challenger returns $\hat{M}$. $\mathcal{B}_1$ calculates $\hat{X}\|\hat{Z} = H^p(\hat{N}, \hat{A}, \hat{C})$ with $|\hat{X}| = k$ and $|\hat{Z}| = n - k$ and submits $(\hat{X}, 1; k)$ to its challenger. Suppose the challenger returns $\hat{Y}$. $\mathcal{B}_1$ calculates $\lceil p(\hat{Y}\|\hat{Z})\rceil_\tau$ If $T = \hat{T}$, then $\mathcal{B}_1$ returns $\hat{M}$ to $\mathcal{A}$. Otherwise it returns $\perp$.
- At the end of the game, $\mathcal{A}$ submits the decision bit to $\mathcal{B}_1$ which it forwards to its challenger. Note that when $\mathcal{A}$ supposedly interacts with $\mathcal{O}_1$ or $\mathcal{B}_1$ supposedly interacts with F_K^p, they submit $b = 1$. Otherwise, they submit $b = 0$.

It is easy to see $\Pr[\mathcal{A}^{\mathcal{O}_1} = 1] = \Pr[\mathcal{B}_1^{F_K^p} = 1]$ and $\Pr[\mathcal{A}^{\mathcal{O}_2} = 1] = \Pr[\mathcal{B}_1^{\$} = 1]$, and hence we obtain the following.

$$|\Pr[\mathcal{A}^{\mathcal{O}_1} = 1] - \Pr[\mathcal{A}^{\mathcal{O}_2} = 1]| = \mathbf{Adv}_F^{\mathsf{PRF}}(\mathcal{B}_1). \tag{2}$$

□ Bounding $|\Pr[\mathcal{A}^{\mathcal{O}_2} = 1] - \Pr[\mathcal{A}^{\mathcal{O}_3} = 1]|$. This bound follows from the lemma given below, the proof of which is deferred to the next section.

Lemma 1. *Let $\mathcal{A}$ be a deterministic nonce-respecting non-repeating query making adversary interacting with oracle $\mathcal{O}_2$ or $\mathcal{O}_3$ which can make at most q_e encryption queries, q_v decryption queries and q_p primitive queries to p and its inverse. Assuming $q = q_e + q_v$, there exists an adversary $\mathcal{B}_2$ such that the NAEAD advantage of $\mathcal{A}$ can be bounded by*

$$|\Pr[\mathcal{A}^{\mathcal{O}_2} = 1] - \Pr[\mathcal{A}^{\mathcal{O}_3} = 1]| \leq 2\mathbf{Adv}_{\lfloor H\rfloor_{n-k}}^{2PI+}(\mathcal{B}_2) + \frac{qq_p}{2^n} + \frac{2kq_v}{2^k} + \frac{q_v}{2^\tau} + \frac{\binom{q_p}{k}}{2^{\tau(k-1)}} + \frac{\binom{q_p}{k}}{2^{(n-k)(k-1)}},$$

where $\mathcal{B}_2$ can make q challenge queries, q response queries and q_p primitive queries to p and its inverse.

The proof of the theorem follows from Eq. 1, Eq. 2 and Lemma 1.

3.3 Proof of Lemma 1

Now we'll prove Lemma 1 using Coefficients H Technique step by step.

Step I: Sampling of the Ideal Oracle and Defining the Bad Events. We start with sampling of the ideal oracle and go on mentioning the bad events whenever they occur. Note that whenever we mention a bad event, even if it's not explicitly mentioned, it's implicitly understood that the previous bad events haven't occurred.

In the online phase, the adversary interacts with the oracles and receives the corresponding responses. In this phase, it can make any construction or permutation query. The i-th encryption query is (N^i, A^i, M^i), the i-th decryption query is $(\hat{N}^i, \hat{A}^i, \hat{C}^i, \hat{T}^i)$, $H(\hat{N}^i, \hat{A}^i, \hat{C}^i) = \hat{X}^i \| \hat{Z}^i$ with $|\hat{X}^i| = k$ and $|\hat{Z}^i| = n-k$, the i-th permutation query is U^i if it's a forward query (i.e., p query), and V^i if it's a backward query (i.e., p^{-1} query).

1. Return (C^i, T^i), $\forall i \in [q_e]$, where $C^i \xleftarrow{\$} \{0,1\}^{|M^i|}, T^i \xleftarrow{\$} \{0,1\}^{\tau}$.
2. Return $\perp, \forall i \in [q_v]$.
3. Return the true output values of the permutation queries.
4. Set $X^i := \lceil H(N^i, A^i, C^i) \rceil_k, \hat{X}^i := \lceil H(\hat{N}^i, \hat{A}^i, \hat{C}^i) \rceil_k$,
   $Z^i := \lfloor H(N^i, A^i, C^i) \rfloor_{n-k}, \hat{Z}^i := \lfloor H(\hat{N}^i, \hat{A}^i, \hat{C}^i) \rfloor_{n-k}$

The adversary aborts if the following (bad) event occurs.

- bad1: $\exists i \in [q_e]$ and $j \in [q_v]$ with $i \neq j$ and $(N^i, A^i, C^i) \neq (\hat{N}^j, \hat{A}^j, \hat{C}^j)$ such that $Z^i = \hat{Z}^j$.
- bad2: $\exists i, j \in [q_e]$ with $i \neq j$ such that $Z^i = Z^j$.

In the offline phase, the adversary can no longer interact with any oracle, but the challenger may release some additional information to the adversary before it submits its decision.

1. $Y^i \xleftarrow{\$} \{0,1\}^k$, $\forall i \in [q_e]$ and $j \in [i-1]$ with $X^i \neq X^j$.
2. $\hat{Y}^i \xleftarrow{\$} \{0,1\}^k$, $\forall i \in [q_v], j \in [i-1]$ and $\ell \in [q_e]$ with $\hat{X}^i \neq \hat{X}^j$ and $\hat{X}^i \neq X^\ell$.

Again, the adversary aborts if any of the following (bad) events occur.

- bad3: $\exists i \in [q_e]$ and $j \in [q_p]$ such that $Y^i \| Z^i = U^j$.
- bad4: There is a k-multi-collision at the τ most significant bits of the output of the forward permutation queries.
- bad5: There is a k-multi-collision at the $(n-k)$ least significant bits of the output of the backward permutation queries.
- bad6: $\exists i \in [q_v]$ and $j \in [q_p]$ such that $\hat{Y}^i \| \hat{Z}^i = U^j$.

If none of the bad events occur, then

1. $D^i \xleftarrow{\$} \{0,1\}^{n-\tau}$, $\forall i \in [q_e]$,
2. $\hat{T}'^i \| \hat{D}^i \xleftarrow{\$} \{0,1\}^n$, $\forall i \in [q_v]$.

Again, the adversary aborts if any of the following (bad) event occurs.

- bad7: $\exists i \in [q_v]$ such that $\hat{T}^i = \hat{T}'^i$.

Step II: Bounding the Probability of the Bad Events. Now we'll upper bound the probabilities of the bad events.

- bad1: This event says that the capacity part of the hash of an encryption query matches with the capacity part of the hash of a forging query. This is nothing but computing a second pre-image corresponding to a challenge (N, A, C), where C is chosen uniformly at random. Thus, the probability of this event is bounded by the 2PI+ security of H.

$$\Pr[\mathsf{bad1}] \leq \mathbf{Adv}^{\mathsf{2PI+}}_{\lfloor H \rfloor_{n-k}}(\mathcal{B}_2),$$

  where $\mathcal{B}_2$ can make q challenge queries, q response queries and q_p primitive queries to p and its inverse.
- bad2: This event says that the capacity part of the hash of an encryption query matches with the capacity part of the hash of another encryption query. This is again nothing but computing a second pre-image corresponding to a challenge (N, A, C), where C is chosen uniformly at random. Thus, the probability of this event is bounded by the 2PI+ security of H.

$$\Pr[\mathsf{bad2}] \leq \mathbf{Adv}^{\mathsf{2PI+}}_{\lfloor H \rfloor_{n-k}}(\mathcal{B}_2),$$

  where $\mathcal{B}_2$ can make q challenge queries, q response queries and q_p primitive queries to p and its inverse.
- bad3: For a fixed encryption query and a fixed permutation query, the probability of this event comes out to be equal to $1/2^n$ due to the randomness of U^j. Applying union bound over all possible choices, we obtain

$$\Pr[\mathsf{bad3}] \leq \frac{q_e q_p}{2^n}.$$

- bad4: For a fixed k-tuple of forward permutation queries, the probability of this event comes out to be equal to $1/2^{\tau(k-1)}$ due to the randomness of the permutation output. Applying union bound over all possible choices, we obtain

$$\Pr[\mathsf{bad4}] \leq \frac{\binom{q_p}{k}}{2^{\tau(k-1)}}.$$

- bad5: For a fixed k-tuple of backward permutation queries, the probability of this event comes out to be equal to $1/2^{(n-k)(k-1)}$ due to the randomness of the permutation output. Applying union bound over all possible choices, we obtain

$$\Pr[\mathsf{bad5}] \leq \frac{\binom{q_p}{k}}{2^{(n-k)(k-1)}}.$$

- bad6: We analyse this bad event in the three following sub-cases.
  - In this case, the number of multi-collision at the τ most significant bits of the output of the forward permutation queries is at most k. So the adversary can make a hash query (N, A, C) to obtain $X \| Z$, fix Z as the

least significant bits and vary the rest of the bits to obtain the multi-collision. Suppose the multi-collision happens at the value T. In that case, if the adversary makes the decryption query (N, A, C, T), then the probability of $\mathsf{bad6}$ comes out to be equal to $k/2^k$. For q_v decryption queries, this probability comes out to be equal to $kq_v/2^k$.

- In this case, the number of multi-collision at the $(n-k)$ least significant bits of the output of the backward permutation queries is at most k. So the adversary can fix the τ most significant bits (say T) and vary the rest of the bits to obtain the multi-collisions. Suppose the multi collisions happen at the values $Z_1, Z_2, \cdots, Z_m$. Also suppose that the adversary has q_1 hash pre-images of Z_1, q_2 hash pre-images of Z_2, $\cdots$, q_m hash pre-images of Z_m, where $q_1 + q_2 + \cdots + q_m = q_v$. For $i \in [r]$, suppose the adversary has a pre-image (N, A, C) of Z_i. In that case, if the adversary makes the decryption query (N, A, C, T), then the probability of $\mathsf{bad6}$ comes out to be equal to $k/2^k$. For q_v pre-images, this probability comes out to be equal to $kq_v/2^k$.
- If the previous two cases don't occur, i.e., there is no multi-collision, then for a fixed decryption query and a fixed permutation query, he probability of $\mathsf{bad6}$ comes out to be equal to $1/2^n$ due to the randomness of U^j. For q_v decryption queries and q_p permutation queries, this probability comes out to be equal to $q_v q_p/2^n$.

Combining all the three cases, we obtain

$$\Pr[\mathsf{bad6} | (\overline{\mathsf{bad3}} \wedge \overline{\mathsf{bad4}} \wedge \overline{\mathsf{bad5}})] \leq \frac{2kq_v}{2^k} + \frac{q_v q_p}{2^n}.$$

- $\mathsf{bad7}$: For a fixed decryption query, the probability of this event comes out to be equal to $1/2^\tau$ due to the randomness of $\hat{T}'^i$. Applying union bound over all possible choices, we obtain

$$\Pr[\mathsf{bad7}] \leq \frac{q_v}{2^\tau}.$$

Combining everything, we obtain

$$\begin{aligned} \epsilon_{\mathsf{bad}} := \Pr[\mathsf{bad}] &\leq \Pr[\mathsf{bad1} \vee \mathsf{bad2} \vee \cdots \vee \mathsf{bad7}] \\ &\leq 2\mathbf{Adv}^{\mathsf{2PI+}}_{\lfloor H \rfloor_{n-k}}(\mathcal{B}_2) + \frac{qq_p}{2^n} + \frac{2kq_v}{2^k} + \frac{q_v}{2^\tau} \\ &+ \frac{\binom{q_p}{k}}{2^{\tau(k-1)}} + \frac{\binom{q_p}{k}}{2^{(n-k)(k-1)}}. \end{aligned} \quad (3)$$

Step III: Ratio of Good Interpolation Probabilities. We recall that to obtain oracle $\mathcal{O}_2$, we replace the function F of $\mathcal{O}_1$ with a random function \$. All the remaining specification of $\mathcal{O}_2$ are similar to $\mathcal{O}_1$ (see Sect. 3.1). Let q_x be the number of construction queries with distinct X^i's and $\hat{X}^i$'s and q' be the number of construction queries with distinct (N^i, A^i, C^i)'s and $(\hat{N}^i, \hat{A}^i, \hat{C}^i)$'s. For any good transcript τ, we get

$$\Pr_{\mathcal{O}_2}[\tau] = \frac{1}{2^{n\sigma_e}} \frac{1}{2^{kq_x}} \frac{1}{(2^n)_{q'+q_p}}.$$

The first term corresponds to the number of choices for W^i. The second term corresponds to the number of choices for Y^i. The third term corresponds to the number of choices for the outputs of the distinct permutation calls. We also get

$$\Pr_{\mathcal{O}_3}[\tau] = \frac{1}{2^{n\sigma_e}} \frac{1}{2^{nq'}} \frac{1}{2^{kq_x}} \frac{1}{(2^n)_{q_p}}.$$

The first term corresponds to the number of choices for C^i. The second term corresponds to the number of choices for $T^i \| D^i$. The third term corresponds to the number of choices for Y^i. The fourth term corresponds to the number of choices for the outputs of the distinct permutation calls. Thus we finally obtain

$$\frac{\Pr_{\mathcal{O}_2}[\tau]}{\Pr_{\mathcal{O}_3}[\tau]} \geq 1 \text{, i.e., } \epsilon_{\mathsf{good}} = 0. \tag{4}$$

Step IV: Final Calculation. The Lemma follows as we use Eq. 3 and Eq. 4 in Theorem 1.

4 Multi-target 2nd Pre-image Security of Sponge Based Hashes

This section analyses the 2PI+ security of the sponge hash and some of it's variants.

4.1 Sponge Hash and Its 2PI+ Security

First we briefly revisit the sponge hash. Consider the initial state to be $N\|IV$ for some fixed IV. Let $p \in \mathsf{Perm}$ where Perm is the set of all permutations on $\{0,1\}^n$. We call the r most significant bits of the state as rate and the c' least significant bits of the state as capacity. The associated data A and the message C is absorbed in r'-bit blocks by subsequent p-calls, and the output of the last p-call is the hash output T. Figure 2 illustrates the sponge hash. Now let us look at its 2PI+ security.

The following attack demonstrates that the sponge hash is vulnerable to a meet-in-the-middle attack as follows.

- Suppose an adversary (say $\mathcal{A}$) submits $(N, A, 2)$ and receives the random message $C_1\|C_2$ from its challenger where $|C_1| = |C_2| = r'$.
- $\mathcal{A}$ computes the hash as $H = p(p(S_1 \oplus C_1\|S_2 \oplus 0^\star 1) \oplus (C_2\|0^c))$. Suppose $H = p(Y_2\|Z_2)$ where $|Y_2| = r'$ and $|Z_2| = c$.
- $\mathcal{A}$ makes some p-queries of the form $\star\|IV$ and some p^{-1}-queries of the form $\star\|Z_2$, and stores the p-query outputs in the list $\mathcal{L}_1$ and the p^{-1}-query outputs in the list $\mathcal{L}_2$.

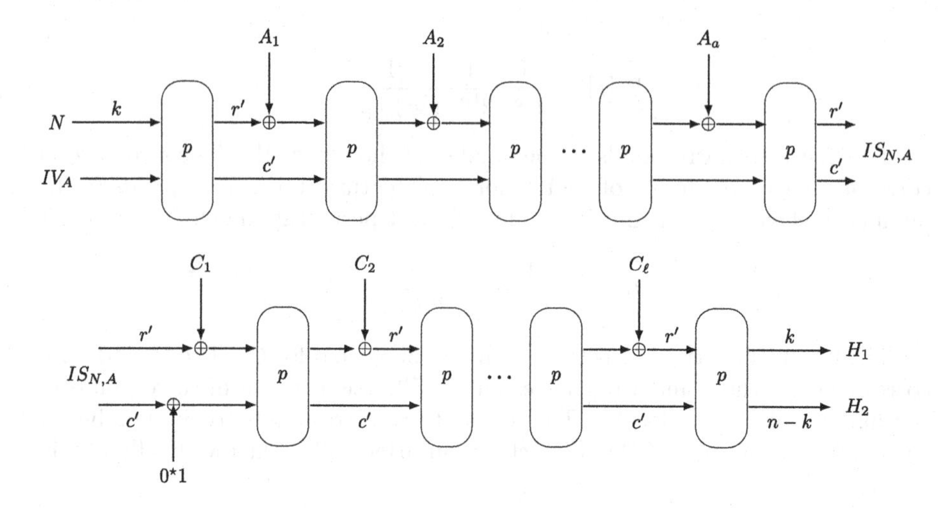

Fig. 2. Sponge hash with ℓ message blocks.

- Suppose the capacity of one entry in $\mathcal{L}_1$ (say $Y_1\|Z_1$ where $|Y_1| = r'$ and $|Z_1| = c'$) matches with the capacity of one entry in $\mathcal{L}_2$ (say $Y_1^\star\|Z_1$). Suppose $p(N'\|IV) = Y_1\|Z_1$ and $p^{-1}(Y_2^\star\|Z_2) = Y_1^\star\|Z_1$.
- $\mathcal{A}$ returns $(N', \epsilon, (Y_1 \oplus Y_1^\star)\|(Y_2 \oplus Y_2^\star))$ to its challenger as the second pre-image of the random message $(N, C_1\|C_2)$.

It is easy to see that the attack succeeds with probability $\frac{|\mathcal{L}_1\|\mathcal{L}_2|}{2^{c'}}$. In other words, if the adversary is able to make around $2^{c'/2}$ p-queries and p^{-1}-queries each, it would be able to mount this 2PI+ attack with very high probability. Thus, for the sponge hash, the 2PI+ security reduces to the collision security due to the above meet-in-the-middle attack, and the 2PI+ security for sponge hash is $\Omega(q_p^2/2^{c'})$. Now, we are more interested in some other hash functions where a such collision attack doesn't induce a 2PI+ attack.

4.2 Feed Forward Based Sponge Hash and Its 2PI+ Security

Now, we consider a feed forward variant of the sponge hash. The nonce and associated data processing remains as it is. However, the following modifications during the random message processing:

- The capacity part of the output of the i-th permutation is xored with the previous state capacity to obtain the updated i-th state capacity.
- The message injection rate for the first block of random ciphertext remains r' bits, and for all successive blocks the rate is r bits, where $r \geq r'$. To make things compatible, the capacity part before the first p-call is chopped to the least significant c-bits while feed-forwarding.
- We use a domain separation before the final permutation call depending on the size of the random message. If the size is less than or equal to r', we xor 1 in the capacity.

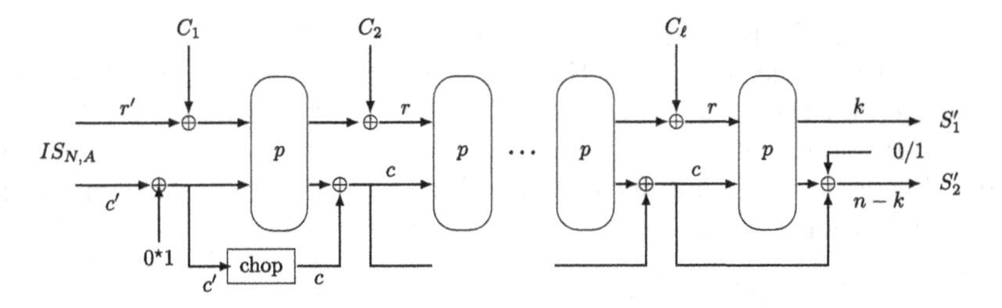

Fig. 3. The feed forward variant of the sponge hash. The initial state $IS_{N,A}$ is generated identically as in the sponge hash, depicted in Fig. 2.

Figure 3 illustrates the feed forward variant of the sponge hash with n-bit hash value $H_1 \| H_2$. It is easy to see that the attack on the sponge hash can not be extended to this hash due to the feed-forward functionality of this hash. Now let us look at its 2PI+ security. Formally, we state the following lemma:

Lemma 2. *Let H be the feed-forward based sponge hash as defined as above. The 2PI+ security of the construction is given by*

$$\mathbf{Adv}^{\mathsf{2PI+}}_{\lfloor H \rfloor_{n-k}}(\mathcal{A}) \leq \frac{q_p^2}{2^{c'}} + \frac{(q_p + \sigma_v)\sigma_e}{2^c} + \frac{\sigma_e^2}{2^c} + \frac{q_v}{2^{n-k}},$$

where $\mathcal{A}$ makes at most q_e challenge queries with an aggregate of σ_e blocks, q_v forging attempt queries with an aggregate of σ_v blocks, and q_p many permutation queries.

Proof. First let us consider the scenario for all challenge queries with $\ell \geq r'$. Suppose at the i-th step, the adversary (say $\mathcal{A}$) submits the i-th message length and receives the random message C^i from its challenger. $\mathcal{A}$ makes successive queries to p to derive the hash value corresponding to the fixed IV and C^i. Moreover, the adversary makes several additional queries to p or p^{-1}.

Graph Based Representation. Now we draw a graph corresponding to all the challenge, permutation queries and forging attempts made by the adversary $\mathcal{A}$. A node of the graph is an n-bit state value. For a challenge or response query, we consider all the permutation inputs as nodes. Suppose the $(i-1)^{th}$ and i^{th} permutation inputs are X_{i-1}, and X_i respectively, then we draw an edge from node X_{i-1} to X_i with edge labelled as C_i, where C_i is i^{th} message injected. The starting vertex for each query (N, A, C) is defined as $IS_{N,A} \oplus (C_1 \| 0)$. Now we consider the direct permutation queries. suppose $\mathcal{A}$ makes a p query with the input X, and the output is Y (i.e., $Y = p(X)$), then we draw an edge from vertex X to vertex $Y \oplus (0 \| \lfloor X \rfloor_c)$. Similarly, if $\mathcal{A}$ makes a p^{-1} query with input $Y^\star$, and the output is $X^\star$ (i.e., $p^{-1}(Y^\star) = X^\star$), we draw an edge from $X^\star$ to $Y^\star \oplus (0 \| \lfloor X^\star \rfloor_c)$ with label 0. Essentially, the p^{-1} queries behave similar to the p queries, and we obtain a directed edge-labelled graph. This is depicted in Fig. 4. Thus, overall we have a graph corresponding to all the queries. All the nodes computed during the hash computation (corresponding to the challenge queries) are called "H"-nodes and all the other nodes are called "P"-nodes. So, by definition, the number of H-nodes is σ_e, the total number of primitive calls required for the hash computation of all the challenge messages. The total number of P-nodes are bounded by $(q_p + \sigma_v)$, q_p being the total number of direct p and p^{-1} calls, and σ_v being the number of p calls used in the hash computation for the verification queries.

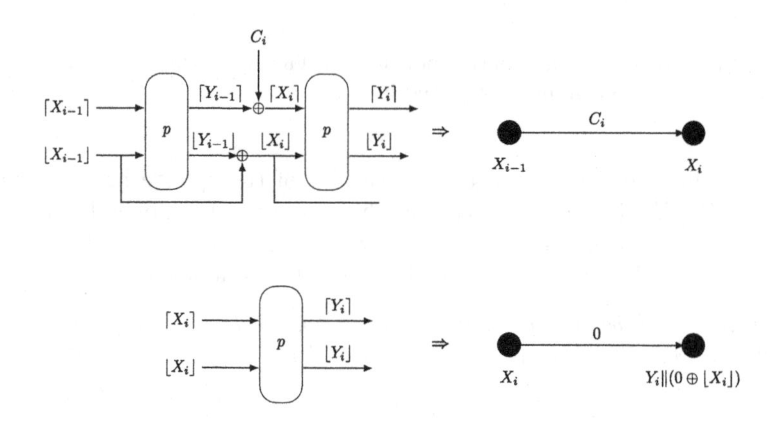

Fig. 4. The graph representation: challenge and forging queries (top), direct permutation queries (bottom).

Definition and Bounding the Probability of a Bad Graph. We call a collision occurs in two nodes if there capacity values are same. Now we call such a graph *bad* if there is a collision (i) among two starting "H" nodes, or (ii) due to a "H" node and a "P" node, (iii) between two "H" nodes. Now let us try to bound the probability that a graph is bad. For the first case, the initial state collision will reduce to a simple collision attack. This is due to the fact that the nonce and associated data are chosen by the adversary. Hence, this probability can be bounded by $q_p^2/2^{c'}$. For case (ii) and (iii), such a collision will occur with probability at most $\frac{1}{2^c - q_p}$, and the number of possible choice of H nodes and P nodes are σ_e and $(q_p + \sigma_v)$ respectively. Thus, the probability that a graph is *bad* can be bounded by $(\frac{q_p^2}{2^{c'}} + \frac{(q_p+\sigma_v)\sigma_e}{2^c} + \frac{\sigma_e^2}{2^c})$.

Bounding 2PI+ Security for A Good Graph. It is easy to see that if a graph is not *bad*, then we do not have any forgeries, except for random hash value matching, which can be bounded by $\frac{q_v}{2^{n-k}}$.

Combining everything together, the lemma follows. □

Note that to extend the analysis for shorter challenge queries with $\ell \geq r'$ we need a domain separator at the end (adding 1 at the capacity). This is to resist an attack by guessing the random ciphertext and transferring a collision attack into a 2PI+attack.

5 ISAP+: A Throughput-Efficient Variant of ISAP

In this section, we describe the ISAP+ family of NAEAD mode by instantiating EtHM with a sponge based PRF and the hybrid sponge hash and ultimately come up with the complete specification details of ISAP+.

5.1 Specification of ISAP+

Let n, k, r, r' and r_0 be five positive integers satisfying $1 < r, r', r_0 < n$, and IV_{KE}, IV_{KA} and IV_A be three $(n-k)$-bit binary numbers. We call the last three numbers as the initialization vectors. Let $c = n - r, c' = n - r'$ and $c_0 = n - r_0$. Let p be an n-bit public permutation. The authenticated encryption module of ISAP+ uses a secret key $K \in \{0,1\}^k$, receives a nonce $N \in \{0,1\}^k$, an associated data $A \in \{0,1\}^*$ and a message $M \in \{0,1\}^*$ as inputs, and returns a ciphertext $C \in \{0,1\}^{|M|}$ and a tag $T \in \{0,1\}^k$. The verified decryption module uses the same secret key K and receives a nonce $N \in \{0,1\}^k$, an associated data $A \in \{0,1\}^*$, a ciphertext $C \in \{0,1\}^*$ and a tag $T \in \{0,1\}^\tau$ as inputs. In case of successful verification, it returns a message $M \in \{0,1\}^{|C|}$. In case the verification fails, it returns $\perp$. Both the modules use a sub-module named rekeying. The complete specification of ISAP+ is provided in Fig. 5. The pictorial representation of the same is provided in Figs. 6, 7, and 8.

Viewing ISAP+ as an Instantiation of EtHM: It is easy to that ISAP+ can be viewed as an instantiation of EtHM where the hash function H^p is given by the feed-forward variant of sponge hash as depicted in Fig. 8 and the FIL-VOL keyed function F_k^p is described as follows:

- When $flag = 1$ (i.e., inside encryption module), F_k^p involves the rekeying function with $(n-k)$-bit output followed by p calls as depicted in Figs. 6 and 7. The inputs are the nonce N, $flag = 1$ and a parameter ℓ which represents the message length. The number of p calls is equal to the number of r-bit message blocks.
- When $flag = 0$ (i.e., inside authentication module), F_k^p involves only the rekeying function with k-bit output as depicted in Fig. 6. The inputs are the k most significant bits of the hash output, $flag = 0$ and a parameter $\ell = k$.

5.2 Design Rationale

In this section, we'll try to highlight and explain the main points regarding what motivated the design of EtHM, and in particular, ISAP+.

□ **Improved Rate for Ciphertext Processing in the Hash:** As we move from collision security to 2PI+ security at the ciphertext absorption phase of the authentication module, we achieve the same security with a smaller capacity size, which allows us to use a larger rate size for ciphertext absorption.
□ **Last Domain Separator:** The last domain separator is crucial to domain separate the short and long messages. Without this domain separator, we can

Algorithm ISAP+.AE$_K(N, A, M)$

1. $C \leftarrow$ ISAP+.Enc/Dec(K, N, M)
2. $T \leftarrow$ ISAP+.Auth(K, N, A, C)
3. **return** (C, T)

Algorithm ISAP+.Auth(K, N, A, C)

1. $A_1 \cdots A_a \twoheadleftarrow_{r'} A$
2. **if** $|C| < r'$ **then**
3. $\quad C_1 \leftarrow C \| 10^{r'-|C|}$
4. **else if** $|C| = r'$ **then**
5. $\quad C_1 \leftarrow C$
6. $\quad C_2 \leftarrow 10^r$
7. **else**
8. $\quad C_1 \leftarrow \lceil C \rceil_{r'}$
9. $\quad C_2 \cdots C_\ell \twoheadleftarrow_r \lfloor C \rfloor_{|C|-r'}$
10. $S \leftarrow N \| IV_A$
11. **for** $i = 1$ **to** a
12. $\quad S \leftarrow p(S) \oplus (A_i \| 0^{c'})$
13. $S \leftarrow p(S) \oplus (C_1 \| 0^{c'}) \oplus 0^{n-1}1$
14. **for** $i = 2$ **to** ℓ
15. $\quad S \leftarrow p(S) \oplus (C_i \| 0^c) \oplus 0^r \| \lfloor S \rfloor_c$
16. $S \leftarrow p(S) \oplus 0^r \| \lfloor S \rfloor_c$
17. $S \leftarrow$ (ISAP+.RK$(K, \lceil S \rceil_k, 0, k)) \| \lfloor S \rfloor_{n-k}$
18. **if** $|C| < r'$ **then**
19. $\quad S \leftarrow S \oplus (0^{n-1} \| 1)$
20. **return** $T \leftarrow \lceil p(S) \rceil_k$

Algorithm ISAP+.VD$_K(N, A, C, T)$

1. $T' \leftarrow$ ISAP+.Auth(K, N, A, C)
2. **if** $T = T'$ **then**
3. $\quad$ **return** ISAP+.Enc/Dec(K, N, C)
4. **else**
5. $\quad$ **return** $\perp$

Algorithm ISAP+.Enc/Dec(K, N, X)

1. $X_1 \cdots X_\ell \twoheadleftarrow_{r'} X$
2. $S \leftarrow N \|$(ISAP+.RK$(K, N, 1, n-k))$
3. **for** $i = 1$ **to** ℓ
4. $\quad S \leftarrow p(S)$
5. $\quad Y_i \leftarrow \lceil S \rceil_{r'} \oplus X_i$
6. $Y \leftarrow \lceil Y_1 \| \cdots \| Y_\ell \rceil_{|X|}$
7. **return** Y

Algorithm ISAP+.RK$(K, X, flag, z)$

1. $IV \leftarrow (flag = 1)?\ IV_{KE} : IV_{KA}$
2. $X_1 \cdots X_w \leftarrow_{r_0} X$
3. $S \leftarrow p(K \| IV)$
4. **for** $i = 1$ **to** $(w-1)$
5. $\quad S \leftarrow p((\lceil S \rceil_{r_0} \oplus X_i) \| \lfloor S \rfloor_{n-r_0})$
6. $S \leftarrow p((\lceil S \rceil_{|X_w|} \oplus X_w) \| \lfloor S \rfloor_{n-|X_w|})$
7. **return** $\lceil S \rceil_z$

Fig. 5. Formal specification of the authenticated encryption and the verified decryption algorithms of ISAP+.

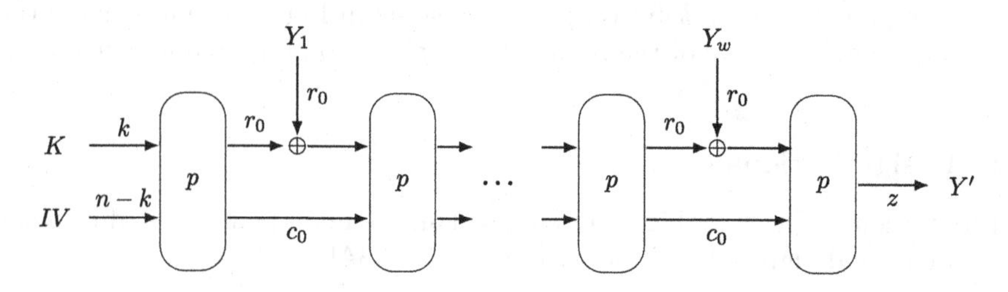

Fig. 6. Re-keying module of ISAP+ on a w-bit input Y.

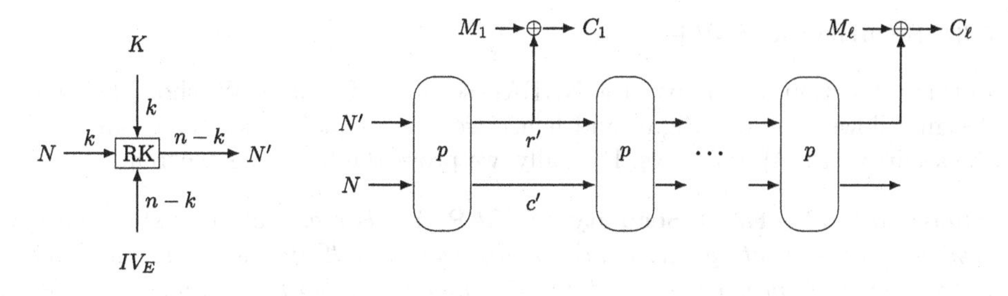

Fig. 7. Encryption module of ISAP+ for ℓ block message.

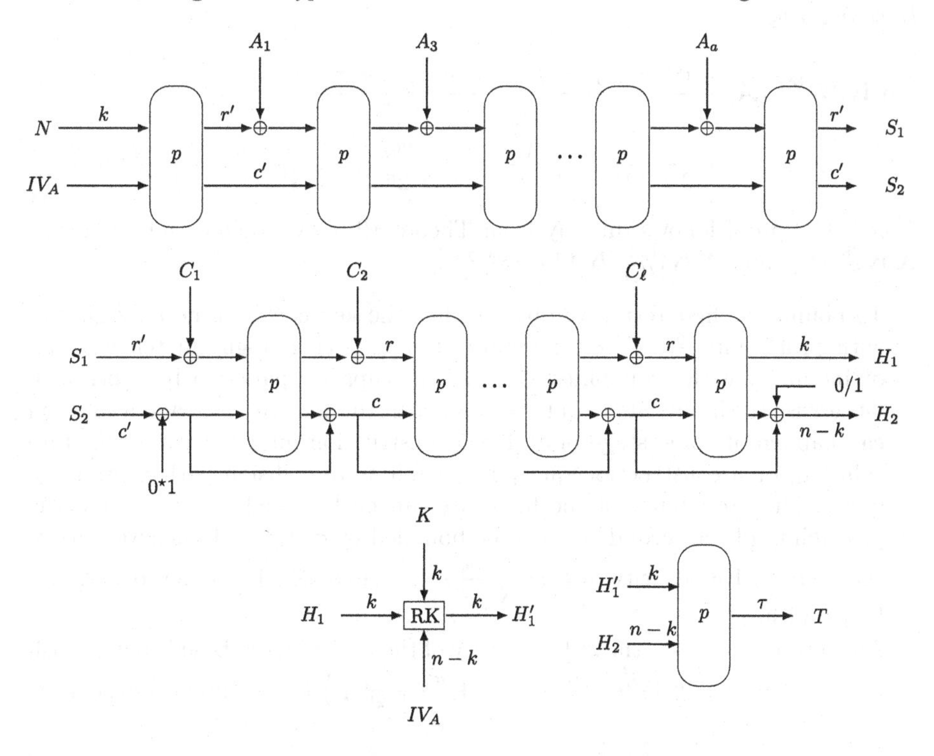

Fig. 8. Authentication module of ISAP+ for a block associated data and ℓ block message.

have a forgery with one encryption query which consists of a message that is less than one block in length and the corresponding forging attempt which consists of more than one ciphertext blocks. As a result, a separator bit, applied to the capacity just before the last permutation call, allows us to differentiate these two cases, and ensure that the input to the last permutation is distinct for each of the two queries, which in turn prevents the attack.

5.3 Security of ISAP+

In this subsection, we analyse the NAEAD security of ISAP+. We show that our design follows that paradigm, and hence we can adapt its security result, and the security of ISAP+ follows. Formally, we prove the following theorem.

Theorem 3 (NAEAD Security of ISAP+). *For all deterministic nonce-respecting non-repeating query making adversary $\mathcal{A}$ of ISAP+ which can make at most q_e encryption queries of a total of maximum σ_e blocks, q_v forging queries and q_p primitive queries to p and its inverse, the NAEAD advantage of $\mathcal{A}$ can be bounded by*

$$\begin{aligned}\mathbf{Adv}_{\mathsf{ISAP+}}^{NAEAD}(\mathcal{A}) \leq &\frac{\sigma_e^2 + \sigma_e q_p + q_p^2}{2^{c'}} + \frac{\sigma_e^2 + \sigma_e q_p + \sigma_e \sigma_v}{2^c} \\ &+ \frac{\binom{q_p}{k}}{2^{\tau(k-1)}} + \frac{\binom{q_p}{k}}{2^{(n-k)(k-1)}} + \frac{q q_p}{2^n} + \frac{2kq_v + q_p}{2^k} + \frac{q_p + q_v}{2^{n-k}} + \frac{q_v}{2^\tau}.\end{aligned}$$

Proof. The proof follows directly from Theorem 2, as we bound the two terms $\mathbf{Adv}_F^{\mathsf{PRF}}(\mathcal{B}_1)$ and $\mathbf{Adv}_{\lfloor H \rfloor_c}^{\mathsf{2PI+}}(\mathcal{B}_2)$ for ISAP+.

- To bound the first term, we observe that the key can be randomly guesses with probability $\frac{q_p}{2^k}$. Also, the state after re-keying might match with an offline query with probability $\frac{q_p}{2^{n-k}}$, as the capacity part can be controlled. Otherwise the inputs of the outer sponge construction are fresh, and a collision can happen at some stage only if two construction queries have a full state collision, or a construction query has a full state collision with a primitive query. The probability of the first case can be bounded by $\sigma_e^2/2^{c'}$ and the probability of the second case can be bounded by $\sigma_e q_p/2^{c'}$. Hence we achieve the overall PRF security of F as $\left(\frac{\sigma_e q_p + \sigma_e^2}{2^{c'}} + \frac{q_p}{2^k} + \frac{q_p}{2^{n-k}}\right)$. Further details can be found in [3].
- The second term, i.e., the 2PI+ security of the feed-forward based sponge hash can be bounded by $\left(\frac{q_p^2}{2^{c'}} + \frac{(q_p+\sigma_v)\sigma_e}{2^c} + \frac{\sigma_e^2}{2^c} + \frac{q_v}{2^{n-k}}\right)$ (See Lemma 2, Sect. 4.2).

□

Side-Channel Resistance. ISAP+ inherits its security against side-channel leakage directly from ISAP. In [17], the author have clearly mentioned that "There are no requirements on the implementation of the hash function H, since it processes only publicly known data." Following their argument, ISAP+ achieves same leakage resilience as it modifies only the hash function of ISAP and retains the rest of the design as it is. Accordingly, ISAP+ will provide a similar result on the leakage resilience bound as given in [18, Theorem 1].

6 Conclusion

In this paper, we have proposed a generic framework for a permutation-based EtHM type NAEAD mode using a PRF and an unkeyed hash function with 2PI+ security. We have shown that ISAP follows the framework EtHM and hence it's security boils down to the 2PI+ security of the underlying hash function. We propose a feed-forward variant of the sponge hash function with improved security and use it to design a new variant of ISAP that achieves improved security, and that in turn improves the throughput of the construction. Designing some new hash functions with better 2PI+ security and improving the security or throughput of the mode instantiated with the newly designed hash seems to be a challenging open problem.

Acknowledgement. The authors would like to thank all the anonymous reviewers for their valuable comments and suggestions. Cuauhtemoc Mancillas López is partially supported by the Cryptography Research Center of the Technology Innovation Institute (TII), Abu Dhabi (UAE), under the TII- Cuauhtemoc project.

References

1. National Institute of Standards and Technology: FIPS PUB 202: SHA-3Standard: Permutation-based hash and extendable-output functions. Federal Information Processing Standards Publication 202, U.S. Department of Commerce, August 2015
2. Andreeva, E., et al.: PRIMATEs v1.02. Submission to CAESAR (2016). https://competitions.cr.yp.to/round2/primatesv102.pdf
3. Andreeva, E., Daemen, J., Mennink, B., Van Assche, G.: Security of keyed sponge constructions using a modular proof approach. In: Leander, G. (ed.) FSE 2015. LNCS, vol. 9054, pp. 364–384. Springer, Heidelberg (2015). https://doi.org/10.1007/978-3-662-48116-5_18
4. Belaïd, S., et al.: Towards fresh re-keying with leakage-resilient PRFs: cipher design principles and analysis. J. Cryptogr. Eng. **4**(3), 157–171 (2014)
5. Bellare, M., Namprempre, C.: Authenticated encryption: relations among notions and analysis of the generic composition paradigm. In: Okamoto, T. (ed.) ASIACRYPT 2000. LNCS, vol. 1976, pp. 531–545. Springer, Heidelberg (2000). https://doi.org/10.1007/3-540-44448-3_41
6. Bernstein, D.J.: ChaCha, a variant of Salsa20 (2008)
7. Bertoni, G., Daemen, J., Peeters, M., Van Assche, G.: Sponge functions, 2007. In: Ecrypt Hash Workshop (2007)
8. Bertoni, G., Daemen, J., Peeters, M., Van Assche, G.: On the indifferentiability of the sponge construction. In: Smart, N. (ed.) EUROCRYPT 2008. LNCS, vol. 4965, pp. 181–197. Springer, Heidelberg (2008). https://doi.org/10.1007/978-3-540-78967-3_11
9. Bertoni, G., Daemen, M.P.J., Van Assche, G., Van Keer, R.: Ketje v2. Submission to CAESAR (2016). https://competitions.cr.yp.to/round3/ketjev2.pdf
10. Biham, E., Shamir, A.: Differential fault analysis of secret key cryptosystems. In: Kaliski, B.S. (ed.) CRYPTO 1997. LNCS, vol. 1294, pp. 513–525. Springer, Heidelberg (1997). https://doi.org/10.1007/BFb0052259

11. Boneh, D., DeMillo, R.A., Lipton, R.J.: On the importance of checking cryptographic protocols for faults. In: Fumy, W. (ed.) EUROCRYPT 1997. LNCS, vol. 1233, pp. 37–51. Springer, Heidelberg (1997). https://doi.org/10.1007/3-540-69053-0_4
12. Chari, S., Jutla, C.S., Rao, J.R., Rohatgi, P.: Towards sound approaches to counteract power-analysis attacks. In: Wiener, M. (ed.) CRYPTO 1999. LNCS, vol. 1666, pp. 398–412. Springer, Heidelberg (1999). https://doi.org/10.1007/3-540-48405-1_26
13. Dobraunig, C., Eichlseder, M., Mendel, F., Schläffer, M.: Ascon v1.2. Submission to NIST Lightweight Cryptography, 2019 (2019). https://csrc.nist.gov/CSRC/media/Projects/lightweight-cryptography/documents/finalist-round/updated-spec-doc/ascon-spec-final.pdf
14. Dobraunig, C., et al.: ISAP v2.0. Submission to NIST (2019). https://csrc.nist.gov/CSRC/media/Projects/Lightweight-Cryptography/documents/round-1/spec-doc/ISAP-spec.pdf
15. CAESAR Committee: CAESAR: Competition for Authenticated Encryption: Security, Applicability, and Robustness. http://competitions.cr.yp.to/caesar.html/
16. Daemen, J., Rijmen, V.: The Design of Rijndael: AES - The Advanced Encryption Standard. Information Security and Cryptography, Springer, Heidelberg (2002). https://doi.org/10.1007/978-3-662-04722-4
17. Dobraunig, C., et al.: ISAP v2.0. https://csrc.nist.gov/CSRC/media/Projects/lightweight-cryptography/documents/finalist-round/updated-spec-doc/isap-spec-final.pdf
18. Dobraunig, C., et al.: ISAP v2.0. IACR Trans. Symmetric Cryptol. **2020**(S1), 390–416 (2020)
19. Dobraunig, C., Eichlseder, M., Mangard, S., Mendel, F., Unterluggauer, T.: ISAP - towards side-channel secure authenticated encryption. IACR Trans. Symmetric Cryptol. **2017**(1), 80–105 (2017)
20. Dobraunig, C., Eichlseder, M., Mendel, F., Schläffer, M., Ascon v1.2. Submission to CAESAR (2016). https://competitions.cr.yp.to/round3/asconv12.pdf
21. Gérard, B., Grosso, V., Naya-Plasencia, M., Standaert, F.-X.: Block ciphers that are easier to mask: how far can we go? In: Bertoni, G., Coron, J.-S. (eds.) CHES 2013. LNCS, vol. 8086, pp. 383–399. Springer, Heidelberg (2013). https://doi.org/10.1007/978-3-642-40349-1_22
22. Goldreich, O., Goldwasser, S., Micali, S.: How to construct random functions. J. ACM **33**(4), 792–807 (1986)
23. Goubin, L., Patarin, J.: DES and differential power analysis the "duplication" method. In: Koç, Ç.K., Paar, C. (eds.) CHES 1999. LNCS, vol. 1717, pp. 158–172. Springer, Heidelberg (1999). https://doi.org/10.1007/3-540-48059-5_15
24. Grosso, V., et al.: SCREAM side-channel resistant authenticated encryption with masking. Submission to CAESAR (2015). https://competitions.cr.yp.to/round2/screamv3.pdf
25. Bertoni, G., Daemen, J., Peeters, M., Van Assche, G.: The Keccak reference (version 3.0) (2011). https://keccak.team/files/Keccak-reference-3.0.pdf
26. Daemen, J., Peeters, M., Van Assche, G., Rijmen, V.: The NOEKEON block cipher, 2000. Nessie Proposal (2020). https://competitions.cr.yp.to/round3/acornv3.pdf
27. Kocher, P.C.: Timing Attacks on implementations of Diffie-Hellman, RSA, DSS, and other systems. In: Koblitz, N. (ed.) CRYPTO 1996. LNCS, vol. 1109, pp. 104–113. Springer, Heidelberg (1996). https://doi.org/10.1007/3-540-68697-5_9

28. Kocher, P., Jaffe, J., Jun, B.: Differential power analysis. In: Wiener, M. (ed.) CRYPTO 1999. LNCS, vol. 1666, pp. 388–397. Springer, Heidelberg (1999). https://doi.org/10.1007/3-540-48405-1_25
29. Medwed, M., Petit, C., Regazzoni, F., Renauld, M., Standaert, F.-X.: Fresh re-keying II: securing multiple parties against side-channel and fault attacks. In: Prouff, E. (ed.) CARDIS 2011. LNCS, vol. 7079, pp. 115–132. Springer, Heidelberg (2011). https://doi.org/10.1007/978-3-642-27257-8_8
30. Medwed, M., Standaert, F.-X., Großschädl, J., Regazzoni, F.: Fresh re-keying: security against side-channel and fault attacks for low-cost devices. In: Bernstein, D.J., Lange, T. (eds.) AFRICACRYPT 2010. LNCS, vol. 6055, pp. 279–296. Springer, Heidelberg (2010). https://doi.org/10.1007/978-3-642-12678-9_17
31. NIST: Lightweight cryptography. https://csrc.nist.gov/Projects/Lightweight-Cryptography
32. Patarin, J.: The "coefficients H" technique. In: Avanzi, R.M., Keliher, L., Sica, F. (eds.) SAC 2008. LNCS, vol. 5381, pp. 328–345. Springer, Heidelberg (2009). https://doi.org/10.1007/978-3-642-04159-4_21
33. Piret, G., Roche, T., Carlet, C.: PICARO – a block cipher allowing efficient higher-order side-channel resistance. In: Bao, F., Samarati, P., Zhou, J. (eds.) ACNS 2012. LNCS, vol. 7341, pp. 311–328. Springer, Heidelberg (2012). https://doi.org/10.1007/978-3-642-31284-7_19

Protocols and Implementation

Revisiting the Efficiency of Perfectly Secure Asynchronous Multi-party Computation Against General Adversaries

Ananya Appan, Anirudh Chandramouli, and Ashish Choudhury(✉)

International Institute of Information Technology, Bangalore, India
ashish.choudhury@iiitb.ac.in

Abstract. In this paper, we present a *perfectly-secure* multi-party computation (MPC) protocol in the *asynchronous* communication setting with *optimal* resilience. Our protocol is secure against a computationally-unbounded *malicious* adversary characterized by an *adversary structure* $\mathcal{Z}$, which enumerates all possible subsets of potentially corrupt parties. The protocol incurs an *amortized* communication of $\mathcal{O}(|\mathcal{Z}|^2)$ bits per multiplication. This improves upon the previous best protocol of Choudhury and Pappu (INDOCRYPT 2020), which requires an *amortized* communication of $\mathcal{O}(|\mathcal{Z}|^3)$ bits per multiplication. Previously, perfectly-secure MPC with amortized communication of $\mathcal{O}(|\mathcal{Z}|^2)$ bits per multiplication was known *only* in the relatively simpler *synchronous* communication setting (Hirt and Tschudi, ASIACRYPT 2013).

Keywords: Byzantine faults · Secret-sharing · Unconditional-security · Byzantine agreement · Privacy · Multi-party computation · Non-threshold adversary

1 Introduction

Secure *multi-party computation* (MPC) [5,17,26,28] is a fundamental problem in secure distributed computing. Consider a set of n mutually-distrusting parties $\mathcal{P} = \{P_1, \ldots, P_n\}$, where a subset of parties can be corrupted by a *computationally-unbounded malicious* (Byzantine) adversary Adv. Informally, an

A. Appan and A. Chandramouli—Work done when the author was a student at International Institute of Information Technology, Bangalore
The full version of the article is available at [1]
A. Choudhury—This research is an outcome of the R&D work undertaken in the project under the Visvesvaraya PhD Scheme of Ministry of Electronics & Information Technology, Government of India, being implemented by Digital India Corporation (formerly Media Lab Asia). The author is also thankful to the Electronics, IT & BT Government of Karnataka for supporting this work under the CIET project.

T. Isobe and S. Sarkar (Eds.): INDOCRYPT 2022, LNCS 13774, pp. 223–248, 2022.
https://doi.org/10.1007/978-3-031-22912-1_10

MPC protocol allows the parties to securely compute any function f of their private inputs, while ensuring that their respective inputs remain private. The most popular way of characterizing Adv is through a *threshold*, by assuming that it can corrupt *any* subset of up to t parties. In this setting, MPC with *perfect security* (where no error is allowed in the outcome) is achievable iff $t < n/3$ [5]. Hirt and Maurer [18] generalized the threshold model by introducing the general-adversary model (also known as the *non-threshold* setting). In this setting, Adv is characterized by a monotone *adversary structure* $\mathcal{Z} = \{Z_1, \ldots, Z_h\} \subset 2^{\mathcal{P}}$ which enumerates all possible subsets of potentially corrupt parties, where Adv can select any subset of parties $Z^\star \in \mathcal{Z}$ for corruption. Modelling the distrust in the system through $\mathcal{Z}$ allows for more flexibility (compared to the threshold model), especially when $\mathcal{P}$ is not too large. In the general-adversary model, MPC with perfect security is achievable iff $\mathcal{Z}$ satisfies the $\mathbb{Q}^{(3)}(\mathcal{P}, \mathcal{Z})$ condition.[1]

Following the seminal work of [5] all generic perfectly-secure MPC protocols follow the paradigm of *shared circuit-evaluation*. In this paradigm, it is assumed that f is abstracted as a publicly-known arithmetic circuit ckt over some finite field $\mathbb{F}$. The problem of securely computing f then boils down to "securely evaluating" the circuit ckt. To achieve this goal, the parties jointly and securely evaluate each gate in ckt in a secret-shared fashion, where each value during the circuit-evaluation remains secret-shared. In more detail, each party first secret-shares its input for f, with every party holding a share for each input, such that the shares of the corrupt parties reveal no additional information about the underlying shared values. The parties then maintain the following *gate-invariant* for each gate in ckt: given the gate-inputs in a secret-shared fashion, the parties get the gate-output in a secret-shared fashion without revealing any additional information about the gate-inputs and gate-output. Finally, the function-output (which is secret-shared) is publicly reconstructed. Intuitively, security follows because the adversary does not learn any additional information beyond the inputs of the corrupt parties and the function output, since the shares learnt by Adv are independent of the underlying values.

How the above gate-invariant is maintained depends on the type of gate and the type of secret-sharing deployed. Typically, the underlying secret-sharing is *linear*, such as Shamir's [27] for the case of *threshold adversaries*, and the replicated secret-sharing [20,24] for the case of general adversaries.[2] Consequently, maintaining the invariant for linear gates in ckt is "free" (completely *non-interactive*). However, to maintain the gate-invariant for non-linear (multiplication) gates, the parties need to interact. Consequently, the communication complexity (namely, the total number of bits communicated by uncorrupted parties) of any generic MPC protocol is dominated by the communication complexity of evaluating the multiplication gates in ckt. Hence, the focus is to improve the *amortized* communication complexity per multiplication gate. The amortized

[1] $\mathcal{Z}$ satisfies the $\mathbb{Q}^{(k)}(\mathcal{P}, \mathcal{Z})$ condition [18], if the union of no k sets from $\mathcal{Z}$ covers $\mathcal{P}$.

[2] A secret-sharing scheme is called *linear*, if the shares are computed as a linear function of the secret and the underlying randomness used in the scheme.

complexity is derived under the assumption that the circuit is "large enough", so that the terms that are *independent* of the circuit size can be ignored.

In terms of communication efficiency, MPC protocols against general adversaries are *inherently less efficient* than those against threshold adversaries, by several orders of magnitude. Protocols against threshold adversaries typically incur an *amortized* communication of $n^{\mathcal{O}(1)}$ bits per multiplication, compared to $|\mathcal{Z}|^{\mathcal{O}(1)}$ bits per multiplication required against general adversaries. Since $|\mathcal{Z}|$ could be exponentially large in n, the *exact* exponent is very important. For instance, as noted in [19], if $n = 25$, then $|\mathcal{Z}|$ is around one million. Consequently, a protocol with an amortized communication complexity of $\mathcal{O}(|\mathcal{Z}|^2)$ bits per multiplication is preferred over a protocol with an amortized communication complexity of $\mathcal{O}(|\mathcal{Z}|^3)$ bits. The most efficient perfectly-secure MPC protocol against general adversaries is due to [19], which incurs an *amortized* communication of $\mathcal{O}(|\mathcal{Z}|^2 \cdot (n^5 \log|\mathbb{F}| + n^6) + |\mathcal{Z}| \cdot (n^7 \log|\mathbb{F}| + n^8))$ bits per multiplication. The complexity is derived by substituting the broadcasts done in their protocol through the reliable broadcast protocol of [15], as referred in [19].

Our Motivation and Results: All the above results hold in the *synchronous* setting, where the parties are assumed to be globally synchronized, with strict upper bounds on the message delay. Hence, any "late" message is attributed to a corrupt sender party. Such strict time-outs are, however, extremely difficult to maintain in real-world networks like the Internet, which are better modelled by the *asynchronous* communication setting [7]. Here, no timing assumptions are made and messages can be arbitrarily (but finitely) delayed, with every message sent being delivered *eventually*, but need not be in the same order in which they were sent. Asynchronous protocols are inherently more complex and less efficient by several orders of magnitude when compared to their synchronous counterparts. This is because in any asynchronous protocol, a *slow* (but uncorrupted) sender party *cannot* be distinguished from a *corrupt* sender party who does not send any message. Consequently, to avoid an endless wait, the parties *cannot* afford to wait to receive messages from all the parties, which results in unknowingly ignoring messages from a subset of potentially honest parties. The *resilience* (fault-tolerance) of *asynchronous* MPC (AMPC) protocols is poor compared to synchronous MPC protocols. For instance, perfectly-secure AMPC against threshold adversaries is achievable iff $t < n/4$. Against general adversaries, perfectly secure AMPC requires $\mathcal{Z}$ to satisfy the $\mathbb{Q}^{(4)}(\mathcal{P}, \mathcal{Z})$ [22].

Compared to synchronous MPC protocols, AMPC protocols are not very well-studied [3,4,6,11,25], especially against general adversaries. The most efficient perfectly-secure AMPC protocol against general adversaries is due to [10]. In this work, we design a new communication efficient perfectly-secure AMPC protocol against general adversaries. The *amortized* efficiency of our protocol is comparable with the most efficient perfectly-secure MPC protocol against general adversaries in the *synchronous* communication setting [19], especially if we focus on the exponent of $|\mathcal{Z}|$. Our result compared with relevant existing results is presented in Table 1.

Table 1. Amortized communication complexity per multiplication of different perfectly-secure MPC protocols against general adversaries

Setting	Reference	Condition	Communication complexity
Synchronous	[19]	$\mathbb{Q}^{(3)}(\mathcal{P}, \mathcal{Z})$	$\mathcal{O}(\lvert\mathcal{Z}\rvert^2 \cdot (n^5 \log \lvert\mathbb{F}\rvert + n^6) + \lvert\mathcal{Z}\rvert \cdot (n^7 \log \lvert\mathbb{F}\rvert + n^8))$
Asynchronous	[10]	$\mathbb{Q}^{(4)}(\mathcal{P}, \mathcal{Z})$	$\mathcal{O}(\lvert\mathcal{Z}\rvert^3 \cdot (n^7 \log \lvert\mathbb{F}\rvert + n^9 \cdot (\log n + \log \lvert\mathcal{Z}\rvert)))$
Asynchronous	This work	$\mathbb{Q}^{(4)}(\mathcal{P}, \mathcal{Z})$	$\mathcal{O}(\lvert\mathcal{Z}\rvert^2 \cdot n^7 \log \lvert\mathbb{F}\rvert + \lvert\mathcal{Z}\rvert \cdot n^9 \log n)$

1.1 Technical Overview

Our protocol is designed in the pre-processing model, where the parties first generate secret-shared random *multiplication-triples* of the form (a, b, c), where $c = ab$. These are used later to efficiently evaluate the multiplication gates using Beaver's method [2]. At the heart of our pre-processing phase protocol lies an efficient *asynchronous* multiplication protocol to securely multiply two secret-shared values. The protocol closely follows the *synchronous* multiplication protocol of [19]. However, there are several non-trivial challenges (discussed in the sequel) while adapting the protocol to the *asynchronous* setting.

The MPC protocol of [19] as well as ours are based on the secret-sharing used in [24], which considers a *sharing specification* $\mathbb{S} \stackrel{def}{=} \{\mathcal{P} \backslash Z | Z \in \mathcal{Z}\}$. A value $s \in \mathbb{F}$ is said to be secret-shared with respect to $\mathbb{S}$ if there exist shares $s_1, \ldots, s_{|\mathbb{S}|}$ such that $s = s_1 + \ldots + s_{|\mathbb{S}|}$. Additionally, for each $q = 1, \ldots, |\mathbb{S}|$, the share s_q is known to every (honest) party in S_q. A sharing of s is denoted by $[s]$, where $[s]_q$ denotes the qth share. The *synchronous* multiplication protocol of [19], which assumes the $\mathbb{Q}^{(3)}(\mathcal{P}, \mathcal{Z})$ condition takes as input $[a], [b]$ and securely generates a random sharing $[ab]$. Note that the following hold: $ab = \sum_{(p,q) \in \{1,\ldots,|\mathbb{S}|\} \times \{1,\ldots,|\mathbb{S}|\}} [a]_p [b]_q$. The main idea is that since $S_p \cap S_q \neq \emptyset$, a *publicly-known* party from $S_p \cap S_q$ can be *designated* to secret-share the summand $[a]_p [b]_q$. For efficiency, every designated "summand-sharing party" can sum up all the summands assigned to it and share the sum instead. If *no* summand-sharing party behaves *maliciously*, then the sum of all secret-shared sums leads to a secret-sharing of ab.

To deal with cheating, [19] first designed an *optimistic* multiplication protocol Π_{OptMult}, which takes an additional parameter $Z \in \mathcal{Z}$ and generates a secret-sharing of ab, *provided* Adv corrupts a set of parties $Z^\star \subseteq Z$. The idea used in Π_{OptMult} is the same as above, except that the summand-sharing parties are now "restricted" to the subset $\mathcal{P} \setminus Z$. Since $(S_p \cap S_q) \setminus Z$ will be non-empty (as otherwise $\mathcal{Z}$ *does not* satisfy the $\mathbb{Q}^{(3)}(\mathcal{P}, \mathcal{Z})$ condition), it is guaranteed that each summand $[a]_p [b]_q$ can be assigned to a designated party in $\mathcal{P} \setminus Z$. Since the parties will *not* know the exact set of corrupt parties, they run an instance of Π_{OptMult}, once for each $Z \in \mathcal{Z}$. This guarantees that at least one of these instances generates a secret-sharing of ab. By comparing the output sharings generated in all the instances of Π_{OptMult}, the parties can *detect* whether any cheating has occurred. If no cheating is detected, then any of the output sharings can serve as the sharing of ab. Else, the parties consider a pair of *conflicting*

Π_{OptMult} instances (whose resultant output sharings are different) and proceed to a *cheater-identification* phase. In this phase, based on the values shared by the summand-sharing parties in the conflicting Π_{OptMult} instances, the parties identify at least one corrupt summand-sharing party. This phase *necessarily* requires the participation of *all* the summand-sharing parties from the conflicting Π_{OptMult} instances. Once a corrupt summand-sharing party is identified, the parties disregard all output sharings of Π_{OptMult} instances involving that party. This process of comparing the output sharings of Π_{OptMult} instances and identifying corrupt parties continues, until all the remaining output sharings are for the same value.

Challenges in the Asynchronous Setting: There are *two* main non-trivial challenges while applying the above ideas in an *asynchronous* setting. First, in Π_{OptMult}, a potentially *corrupt* party may *never* share the sum of the summands *designated* to that party, leading to an *indefinite* wait. To deal with this, we notice that since $\mathcal{Z}$ satisfies the $\mathbb{Q}^{(4)}(\mathcal{P}, \mathcal{Z})$ condition [22] (as we consider the *asynchronous* setting), each $(S_p \cap S_q) \setminus Z$ contains at least one *honest* party. So instead of designating a *single* party for the summand $[a]_p[b]_q$, *each* party in $\mathcal{P} \setminus Z$ shares the sum of *all* the summands it is "capable" of, thus guaranteeing that each $[a]_p[b]_q$ is considered for sharing by at least one (honest) party. However, we have to ensure that a summand is *not* shared multiple times.

The *second* challenge is that if there is a pair of conflicting Π_{OptMult} instances, the potentially *corrupt* summand-sharing parties from these instances *may not* participate further in the cheater-identification phase. To get around this, the multiplication protocol now proceeds in *iterations*, where in each iteration, the parties run an instance of the *asynchronous* Π_{OptMult} (outlined above) for each $Z \in \mathcal{Z}$ and compare the outputs from each instance. They then proceed to the respective cheater-identification phase if the outputs are not the same. However, the summand-sharing parties from previous iterations are *not* allowed to participate in future iterations until they participate in the cheater-identification phase of all the previous iterations. This prevents the *corrupt* summand-sharing parties in previous iterations from acting as summand-sharing parties in future iterations, until they clear their "pending tasks", in which case they are caught and discarded. We stress that the *honest* parties are eventually "released" to act as summand-sharing parties in future iterations. Once the parties reach an iteration where the outputs of all the Π_{OptMult} instances are the same (which happens eventually), the protocol stops.

2 Preliminaries and Existing Asynchronous Primitives

We assume that the parties are connected by pair-wise secure channels. The adversary Adv is *malicious* and *static*, and decides the set of corrupt parties at the beginning of the protocol execution. Parties not under the control of Adv are called *honest*. Given $\mathcal{P}' \subseteq \mathcal{P}$, we say that $\mathcal{Z}$ satisfies the $\mathbb{Q}^{(k)}(\mathcal{P}', \mathcal{Z})$ condition if, for every $Z_{i_1}, \ldots, Z_{i_k} \in \mathcal{Z}$, the condition $\mathcal{P}' \not\subseteq Z_{i_1} \cup \ldots \cup Z_{i_k}$ holds.

We assume that the parties want to compute a function f represented by a *publicly* known arithmetic circuit ckt over a finite field $\mathbb{F}$ consisting of linear and non-linear gates, with M being the number of multiplication gates. Without loss of generality, we assume that each $P_i \in \mathcal{P}$ has an input $x^{(i)}$ for f, and that all the parties want to learn the single output $y = f(x^{(1)}, \ldots, x^{(n)})$.

In our protocols, we use a secret-sharing based on the one from [24], defined with respect to a *sharing specification* $\mathbb{S} = \{S_1, \ldots, S_h\} \stackrel{def}{=} \{\mathcal{P} \setminus Z | Z \in \mathcal{Z}\}$, where $h = |\mathbb{S}| = |\mathcal{Z}|$. This sharing specification $\mathbb{S}$ is $\mathcal{Z}$-*private*, meaning that for every $Z \in \mathcal{Z}$, there is an $S \in \mathbb{S}$ such that $Z \cap S = \emptyset$. Moreover, if $\mathcal{Z}$ satisfies the $\mathbb{Q}^{(4)}(\mathcal{P}, \mathcal{Z})$ condition, then $\mathbb{S}$ satisfies the $\mathbb{Q}^{(3)}(\mathbb{S}, \mathcal{Z})$ condition, meaning that for every $Z_{i_1}, Z_{i_2}, Z_{i_3} \in \mathcal{Z}$ and every $S \in \mathbb{S}$, the condition $S \not\subseteq Z_{i_1} \cup Z_{i_2} \cup Z_{i_3}$ holds. In general, we say that $\mathbb{S}$ satisfies the $\mathbb{Q}^{(k)}(\mathbb{S}, \mathcal{Z})$ condition if for every $Z_{i_1}, \ldots, Z_{i_k} \in \mathcal{Z}$ and every $S \in \mathbb{S}$, the condition $S \not\subseteq Z_{i_1} \cup \ldots \cup Z_{i_k}$ holds.

Definition 1 ([24]). *A value $s \in \mathbb{F}$ is said to be secret-shared with respect to $\mathbb{S}$, if there exist shares $s_1, \ldots, s_{|\mathbb{S}|}$ such that $s = s_1 + \ldots + s_{|\mathbb{S}|}$, and for $q = 1, \ldots, |\mathbb{S}|$, the share s_q is known to every (honest) party in S_q.*

A sharing of s is denoted by $[s]$, where $[s]_q$ denotes the qth share. Note that P_i will hold the shares $\{[s]_q\}_{P_i \in S_q}$. The above secret-sharing is *linear*, as $[c_1 s_1 + c_2 s_2] = c_1[s_1] + c_2[s_2]$ for any publicly-known $c_1, c_2 \in \mathbb{F}$.

2.1 The Asynchronous Universal Composability (UC) Framework

Unlike the previous unconditionally-secure AMPC protocols [3,6,10,11,25], we prove the security of our protocols using the UC framework [8,9,16], based on the real-world/ideal-world paradigm, which we discuss next. The discussion is based on the description of the framework against threshold adversaries as provided in [12] (which is further based on [13,21]). We adapt the framework for the case of general adversaries. Informally, the security of a protocol is argued by "comparing" the capabilities of the adversary in two separate worlds. In the *real-world*, the parties exchange messages among themselves, computed as per a given protocol. In the *ideal-world*, the parties do not interact with *each other*, but with a *trusted* third-party (an ideal functionality), which enables the parties to obtain the result of the computation based on the inputs provided by the parties. Informally, a protocol is considered to be secure if whatever an adversary can do in the real protocol can be also done in the ideal-world.

The Asynchronous Real-World: An execution of a protocol Π in the real-world consists of n *interactive Turing machines* (ITMs) representing the parties in $\mathcal{P}$. Additionally, there is an ITM for representing the adversary Adv. Each ITM is initialized with its random coins and possible inputs. Additionally, Adv may have some auxiliary input z. Following the convention of [7], the protocol operates *asynchronously* by a sequence of *activations*, where at each point, a single ITM is active. Once activated, a party can perform some local computation, write on its output tape, or send messages to other parties. On the other hand, if the adversary is activated, it can send messages on the behalf of corrupt parties. The protocol execution is complete once all honest parties obtain their respective outputs. We let $\mathsf{REAL}_{\Pi,\mathsf{Adv}(z),Z^\star}(\vec{x})$ denote the random variable consisting of the output of the honest parties and the view of the adversary Adv during the execution of a protocol Π. Here, Adv controls parties in $Z^\star$ during the execution of protocol Π with inputs $\vec{x} = (x^{(1)}, \ldots, x^{(n)})$ for the parties (where party P_i has input $x^{(i)}$), and auxiliary input z for Adv.

The Asynchronous Ideal-World: A protocol in the ideal-world consists of *dummy* parties $P_1, \ldots, P_n$, an ideal-world adversary $\mathcal{S}$ (also called *simulator*) and an ideal functionality $\mathcal{F}_{\mathsf{AMPC}}$. We consider *static* corruptions such that the set of corrupt parties $Z^\star$ is fixed at the beginning of the computation and is known to $\mathcal{S}$. $\mathcal{F}_{\mathsf{AMPC}}$ receives the inputs from the respective dummy parties, performs the desired computation f on the received inputs, and sends the outputs to the respective parties. The ideal-world adversary *does not* see and *cannot* delay the communication between the parties and $\mathcal{F}_{\mathsf{AMPC}}$. However, it can communicate with $\mathcal{F}_{\mathsf{AMPC}}$ on the behalf of corrupt parties.

Since $\mathcal{F}_{\mathsf{AMPC}}$ models a real-world protocol *asynchronous* protocol, ideal functionalities must consider some inherent limitations to model the asynchronous communication model with eventual delivery. For example, in a real-world protocol, the adversary can decide when each honest party learns the output since it has full control over message scheduling. To model the notion of time in the ideal-world, [21] uses the concept of *number of activations*. Namely, once $\mathcal{F}_{\mathsf{AMPC}}$ has computed the output for some party, it *does not* ask "permission" from $\mathcal{S}$ to deliver it to the respective party. Instead, the corresponding party must "request" $\mathcal{F}_{\mathsf{AMPC}}$ for the output, which can be done only when the concerned party is active. Moreover, the adversary can "instruct" $\mathcal{F}_{\mathsf{AMPC}}$ to delay the output for each party by ignoring the corresponding requests, but only for a polynomial number of activations. If a party is activated sufficiently many times, the party will eventually receive the output from $\mathcal{F}_{\mathsf{AMPC}}$ and hence, ideal computation eventually completes. That is, each honest party eventually obtains its desired output. As in [12], we use the term "*$\mathcal{F}_{\mathsf{AMPC}}$ sends a request-based delayed output to P_i*", to describe the above interaction.

Another limitation is that in a real-world AMPC protocol, the (honest) parties *cannot* afford for all the parties to provide their input for the computation to avoid an endless wait, as the corrupt parties may decide not to provide their inputs. Hence, *every* AMPC protocol suffers from *input deprivation*, where inputs of a subset of potentially honest parties (which is decided by the choice of adversarial message scheduling) may get ignored during computation. Consequently, once a "core set" of parties $\mathcal{CS}$ provide their inputs for the computation, where $\mathcal{P} \setminus \mathcal{CS} \in \mathcal{Z}$, the parties have to start computing the function by assuming some default input for the left-over parties. To model this in the ideal-world, $\mathcal{S}$ is given the provision to decide the set $\mathcal{CS}$ of parties whose inputs should be taken into consideration by $\mathcal{F}_{\mathsf{AMPC}}$. We stress that $\mathcal{S}$ *cannot* delay sending $\mathcal{CS}$ to $\mathcal{F}_{\mathsf{AMPC}}$ indefinitely. This is because in the real-world protocol, Adv *cannot* prevent the honest parties from providing their inputs indefinitely (Fig. 1).

Functionality $\mathcal{F}_{\mathsf{AMPC}}$

$\mathcal{F}_{\mathsf{AMPC}}$ proceeds as follows, running with the parties $\mathcal{P} = \{P_1, \ldots, P_n\}$ and an adversary $\mathcal{S}$, and is parametrized by an n-party function $f : \mathbb{F}^n \rightarrow \mathbb{F}$ and an adversary structure $\mathcal{Z} \subset 2^{\mathcal{P}}$.

1. For each party $P_i \in \mathcal{P}$, initialize an input value $x^{(i)} = \perp$.
2. Upon receiving a message $(\mathsf{inp}, \mathsf{sid}, v)$ from some $P_i \in \mathcal{P}$ (or from $\mathcal{S}$ if P_i is *corrupt*), do the following:
   - Ignore the message if the output has already been computed;
   - Else, set $x^{(i)} = v$ and send $(\mathsf{inp}, \mathsf{sid}, P_i)$ to $\mathcal{S}$.[a]
3. Upon receiving a message $(\mathsf{coreset}, \mathsf{sid}, \mathcal{CS})$ from $\mathcal{S}$, do the following:[b]
   - Ignore the message if $(\mathcal{P} \setminus \mathcal{CS}) \notin \mathcal{Z}$ or if output has already been computed;
   - Else, record $\mathcal{CS}$ and set $x^{(i)} = 0$ for every $P_i \notin \mathcal{CS}$.[c]
4. If $\mathcal{CS}$ has been recorded and the value $x^{(i)}$ has been set to a value different from $\perp$ for every $P_i \in \mathcal{CS}$, then compute $y \stackrel{def}{=} f(x^{(1)}, \ldots, x^{(n)})$ and generate a request-based delayed output $(\mathsf{out}, \mathsf{sid}, (\mathcal{CS}, y))$ for every $P_i \in \mathcal{P}$.

[a] If P_i is corrupt, then no need to send $(\mathsf{inp}, \mathsf{sid}, P_i)$ to $\mathcal{S}$ as the input has been provided by $\mathcal{S}$ only.

[b] $\mathcal{S}$ cannot delay sending $\mathcal{CS}$ indefinitely; see the discussion before the description of the functionality.

[c] It is possible that for some $P_i \notin \mathcal{CS}$, the input has been set to a value different from 0 during step 1 and $x^{(i)}$ is now reset to 0. This models the scenario that in the real-world protocol, even if P_i is able to provide its input, P_i's inclusion to $\mathcal{CS}$ finally depends upon message scheduling, which is under adversarial control.

Fig. 1. Ideal functionality for asynchronous MPC for session id sid.

We let $\mathsf{IDEAL}_{\mathcal{F}_{\mathsf{AMPC}}, \mathcal{S}(z), Z^\star}(\vec{x})$ denote the random variable consisting of the output of the honest parties and the view of the adversary $\mathcal{S}$, controlling the parties in $Z^\star$, with the parties having inputs $\vec{x} = (x^{(1)}, \ldots, x^{(n)})$ (where party P_i has input x_i), and auxiliary input z for $\mathcal{S}$.

We say that a real-world protocol Π *securely realizes* $\mathcal{F}_{\mathsf{AMPC}}$ *with perfectly-security* if and only if for every Adv, there exists an $\mathcal{S}$ whose running time is polynomial in the running time of Adv, such that for every possible $Z^\star$, every possible $\vec{x} \in \mathbb{F}^n$ and every possible $z \in \{0,1\}^\star$, it holds that the random variables

$$\left\{\mathsf{REAL}_{\Pi, \mathsf{Adv}(z), Z^\star}(\vec{x})\right\} \quad \text{and} \quad \left\{\mathsf{IDEAL}_{\mathcal{F}_{\mathsf{AMPC}}, \mathcal{S}(z), Z^\star}(\vec{x})\right\}$$

are identically distributed.

The Universal-Composability (UC) Framework: While the real-world/ideal-world based security paradigm is used to define the security of a protocol in the "stand-alone" setting, the more powerful UC framework [8,9] is used to define the security of a protocol when multiple instances of the protocol might be running in parallel, possibly along with other protocols. Informally, the security in the UC-framework is still argued by comparing the real-world and the ideal-world. However, now, in both worlds, the computation takes place in the presence of an additional interactive process (modelled as an ITM) called the *environment* and denoted by Env. Roughly speaking, Env models the "external environment" in which protocol execution takes place. The interaction between Env and the various entities takes place as follows in the two worlds.

In the real-world, the environment gives inputs to the honest parties, receives their outputs, and can communicate with the adversary at any point during the execution. During the protocol execution, the environment gets activated first. Once activated, the environment can either activate one of the parties by providing some input or activate Adv by sending it a message. Once a party completes its operations upon getting activated, the control is returned to the environment. Once Adv gets activated, it can communicate with the environment (apart from sending messages to the honest parties). The environment also fully controls the corrupt parties that send all the messages they receive to Env, and follow the orders of Env. The protocol execution is completed once Env stops activating other parties, and outputs a single bit.

In the ideal-model, Env gives inputs to the (dummy) honest parties, receives their outputs, and can communicate with $\mathcal{S}$ at any point during the execution. The dummy parties act as channels between Env and the functionality. That is, they send the inputs received from Env to functionality and transfer the output they receive from the functionality to Env. The activation sequence in this world is similar to the one in the real-world. The protocol execution is completed once Env stops activating other parties and outputs a single bit.

A protocol is said to be UC-secure with *perfect-security*, if for every real-world adversary Adv there exists a simulator $\mathcal{S}$, such that for any environment Env, the environment cannot distinguish the real-world from the ideal-world.

The Hybrid Model: In a $\mathcal{G}$-hybrid model, a protocol execution proceeds as in the real-world. However, the parties have access to an ideal functionality $\mathcal{G}$ for some specific task. During the protocol execution, the parties communicate with $\mathcal{G}$ as in the ideal-world. The UC framework guarantees that an ideal functionality in a hybrid model can be replaced with a protocol that UC-securely realizes $\mathcal{G}$. This is specifically due to the following composition theorem from [8,9].

Theorem 1 ([8,9]). Let Π be a protocol that UC-securely realizes some functionality $\mathcal{F}$ in the $\mathcal{G}$-hybrid model and let ρ be a protocol that UC-securely realizes $\mathcal{G}$. Moreover, let Π^ρ denote the protocol that is obtained from Π by replacing every ideal call to $\mathcal{G}$ with the protocol ρ. Then Π^ρ UC-securely realizes $\mathcal{F}$ in the model where the parties do not have access to the functionality $\mathcal{G}$.

2.2 Existing Asynchronous Primitives

Asynchronous Reliable Broadcast (Acast): Acast allows a designated *sender* $P_S \in \mathcal{P}$ to identically send a message $m \in \{0,1\}^\ell$ to all the parties. If P_S is *honest*, then all honest parties eventually output m. If P_S is *corrupt* and some honest party outputs $m^\star$, then every other honest party eventually outputs $m^\star$. The above requirements are formalized by an ideal functionality $\mathcal{F}_{\mathsf{Acast}}$ (see

[1]). In [23], a perfectly-secure Acast protocol is presented with a communication complexity of $\mathcal{O}(n^2\ell)$ bits, provided $\mathcal{Z}$ satisfies the $\mathbb{Q}^{(3)}(\mathcal{P}, \mathcal{Z})$ condition.

Asynchronous Byzantine Agreement (ABA): The ideal ABA functionality is presented in Fig. 2, which is a generalization of the corresponding functionality against *threshold* adversaries, as presented in [13]. It can be considered as a special case of the AMPC functionality, which looks at the set of inputs provided by the set of parties in $\mathcal{CS}$, where $\mathcal{CS}$ is decided by the ideal-world adversary. If the input bits provided by all the honest parties in $\mathcal{CS}$ are the same, then it is set as the output bit. Else, the output bit is set to be the input bit provided by some corrupt party in $\mathcal{CS}$ (for example, the first corrupt party in $\mathcal{CS}$ according to lexicographic ordering). In the functionality, the input bits provided by various parties are considered to be the "votes" of the respective parties.

Functionality $\mathcal{F}_{\mathsf{ABA}}$

$\mathcal{F}_{\mathsf{ABA}}$ proceeds as follows, running with $\mathcal{P} = \{P_1, \ldots, P_n\}$ and adversary $\mathcal{S}$, and is parametrized by an adversary-structure $\mathcal{Z}$. Let $Z^\star$ denote the set of corrupt parties, where $Z^\star \in \mathcal{Z}$ and let $\mathcal{H} = \mathcal{P} \setminus Z^\star$. For each P_i, initialize an input value $x^{(i)} = \bot$.

1. Upon receiving a message $(\mathsf{vote}, \mathsf{sid}, b)$ from some $P_i \in \mathcal{P}$ (or from $\mathcal{S}$ if P_i is *corrupt*) where $b \in \{0, 1\}$, do the following:
   - Ignore the message if the output has been already computed;
   - Else, set $x^{(i)} = b$ and send $(\mathsf{vote}, \mathsf{sid}, P_i, b)$ to $\mathcal{S}$.[a]
2. Upon receiving a message $(\mathsf{coreset}, \mathsf{sid}, \mathcal{CS})$ from $\mathcal{S}$, do the following:[b]
   - Ignore the message if $(\mathcal{P} \setminus \mathcal{CS}) \notin \mathcal{Z}$ or if output has been already computed;
   - Else, record $\mathcal{CS}$.
3. If the set $\mathcal{CS}$ has been recorded and the value $x^{(i)}$ has been set to a value different from $\bot$ for every $P_i \in \mathcal{CS}$, then compute the output y as follows and generate a request-based delayed output $(\mathsf{decide}, \mathsf{sid}, (\mathcal{CS}, y))$ for every $P_i \in \mathcal{P}$.
   - If $x^{(i)} = b$ holds for all $P_i \in (\mathcal{H} \cap \mathcal{CS})$, then set $y = b$.
   - Else, set $y = x^{(i)}$, where P_i is the party with the smallest index in $\mathcal{CS} \cap Z^\star$.

[a] If $P_i \in Z^\star$, then no need to send $(\mathsf{vote}, \mathsf{sid}, P_i, b)$ to $\mathcal{S}$ as the input has been provided by $\mathcal{S}$ only.
[b] As in the case of $\mathcal{F}_{\mathsf{AMPC}}$, $\mathcal{S}$ cannot delay sending $\mathcal{CS}$ indefinitely.

Fig. 2. The ideal functionality for asynchronous Byzantine agreement for session id sid.

In [10], a perfectly-secure ABA protocol Π_{ABA} is presented, provided $\mathcal{Z}$ satisfies the $\mathbb{Q}^{(4)}(\mathcal{P}, \mathcal{Z})$ condition, which holds for our case. In the protocol, all honest parties eventually compute their output with probability 1 and the protocol incurs an expected communication of $\mathcal{O}(|\mathcal{Z}| \cdot (n^6 \log |\mathbb{F}| + n^8(\log n + \log |\mathcal{Z}|)))$ bits.[3]

Verifiable Secret-Sharing (VSS): A VSS protocol allows a designated *dealer* $P_D \in \mathcal{P}$ to verifiably secret-share its input $s \in \mathbb{F}$. If P_D is *honest*, then the honest parties eventually complete the protocol with $[s]$. The verifiability property guarantees that if P_D is *corrupt* and some honest party completes the protocol, then all honest parties eventually complete the protocol with a secret-sharing of some value. These requirements are formalized through the functionality $\mathcal{F}_{\mathsf{VSS}}$ (Fig. 3). The functionality, upon receiving a vector of shares from P_{D}, distributes the appropriate shares to the respective parties. The dealer's input is defined *implicitly* as the sum of provided shares. We will use $\mathcal{F}_{\mathsf{VSS}}$ in our protocols as follows: P_D on having the input s, sends a random vector of shares $(s_1, \ldots, s_{|\mathbb{S}|})$ to $\mathcal{F}_{\mathsf{VSS}}$ where $s_1 + \ldots + s_{|\mathbb{S}|} = s$. If P_D is *honest*, then the view of Adv will be independent of s and the probability distribution of the shares learnt by Adv will be *independent* of the dealer's input, since $\mathbb{S}$ is $\mathcal{Z}$*-private.*

Functionality $\mathcal{F}_{\mathsf{VSS}}$

$\mathcal{F}_{\mathsf{VSS}}$ proceeds as follows for each $P_i \in \mathcal{P}$ and an adversary $\mathcal{S}$, and is parametrized by the adversary structure $\mathcal{Z}$, sharing specification $\mathbb{S} = \{S_1, \ldots . S_h\} = \{\mathcal{P} \setminus Z | Z \in \mathcal{Z}\}$ and a dealer P_D. Let $Z^\star \in \mathcal{Z}$ be the set of corrupt parties.

- On receiving $(\mathsf{dealer}, \mathsf{sid}, P_{\mathsf{D}}, (s_1, \ldots, s_h))$ from P_{D} (or from $\mathcal{S}$ if $P_{\mathsf{D}} \in Z^\star$), set $s = \sum_{q=1,\ldots,h} s_q$ and for $q = 1, \ldots, h$, set $[s]_q = s_q$. Generate a request-based delayed output $(\mathsf{share}, \mathsf{sid}, P_D, \{[s]_q\}_{P_i \in S_q})$ for each $P_i \notin Z^\star$.[a]

[a] If P_D is *corrupt*, then $\mathcal{S}$ may not send any input to $\mathcal{F}_{\mathsf{VSS}}$, in which case the functionality will not generate any output.

Fig. 3. The ideal functionality for VSS for session id sid.

In [10], a *perfectly-secure* VSS protocol Π_{PVSS} is presented, to securely realize $\mathcal{F}_{\mathsf{VSS}}$, provided $\mathcal{Z}$ satisfies the $\mathbb{Q}^{(4)}(\mathcal{P}, \mathcal{Z})$ condition. The protocol incurs a communication of $\mathcal{O}(|\mathcal{Z}| \cdot n^2 \log |\mathbb{F}| + n^4 \log n)$ bits (see [1] for the full details).

Default Secret-Sharing: The perfectly-secure protocol Π_{PerDefSh} takes a *public* input $s \in \mathbb{F}$ and $\mathbb{S}$ to *non-interactively* generate $[s]$, where the parties collectively set $[s]_1 = s$ and $[s]_2 = \ldots = [s]_{|\mathbb{S}|} = 0$.

Reconstruction Protocols: Let the parties hold $[s]$. Then, [10] presents a perfectly-secure protocol $\Pi_{\mathsf{PerRecShare}}$ to reconstruct $[s]_q$ for any given $q \in \{1, \ldots, |\mathbb{S}|\}$ and a perfectly-secure protocol Π_{PerRec} to reconstruct s, provided $\mathbb{S}$ satisfies the $\mathbb{Q}^{(2)}(\mathbb{S}, \mathcal{Z})$ condition. The protocols incur a communication of $\mathcal{O}(n^2 \log |\mathbb{F}|)$ and $\mathcal{O}(|\mathcal{Z}| \cdot n^2 \log |\mathbb{F}|)$ bits respectively.

[3] From [14], every *deterministic* ABA protocol must have non-terminating runs, where the parties may run the protocol forever, *without* obtaining any output. To circumvent this result, randomized ABA protocols are considered and the best we can hope for from such protocols is that the parties eventually obtain an output, asymptotically with probability 1 (this property is called *almost-surely termination* property).

3 Perfectly-Secure Pre-processing Phase Protocol

We present a perfectly-secure protocol which generates a secret-sharing of M random multiplication-triples unknown to the adversary. The protocol realizes the ideal functionality $\mathcal{F}_{\mathsf{Triples}}$ (Fig. 4) which allows the ideal-world adversary to specify the shares for each of the output triples on the behalf of corrupt parties. The functionality then "completes" the sharing of all the triples randomly, while keeping them "consistent" with the shares specified by the adversary.

Functionality $\mathcal{F}_{\mathsf{Triples}}$

$\mathcal{F}_{\mathsf{Triples}}$ proceeds as follows, running with the parties $\mathcal{P}$ and an adversary $\mathcal{S}$, and is parametrized by an adversary-structure $\mathcal{Z}$ and $\mathcal{Z}$-*private* sharing specification $\mathbb{S} = \{S_1, \ldots, S_h\} = \{\mathcal{P} \setminus Z | Z \in \mathcal{Z}\}$. Let $Z^\star$ denote the set of corrupt parties.

- If there exists a set of parties $\mathcal{A}$ such that $\mathcal{P} \setminus \mathcal{A} \in \mathcal{Z}$ and every $P_i \in \mathcal{A}$ has sent the message $(\mathsf{triples}, \mathsf{sid}, P_i)$, then send $(\mathsf{triples}, \mathsf{sid}, \mathcal{A})$ to $\mathcal{S}$ and prepare the output as follows.
  - Generate secret-sharing of M random multiplication-triples. To generate one such sharing, randomly select $a, b \in \mathbb{F}$, compute $c = ab$ and execute the steps labelled **Single Sharing Generation** for a, b and c.
  - Let $\{([a^{(\ell)}], [b^{(\ell)}], [c^{(\ell)}])\}_{\ell \in \{1,\ldots,M\}}$ be the resultant secret-sharing of the multiplication-triples. Send a request-based delayed output $(\mathsf{tripleshares}, \mathsf{sid}, \{[a^{(\ell)}]_q, [b^{(\ell)}]_q, [c^{(\ell)}]_q\}_{\ell \in \{1,\ldots,M\}, P_i \in S_q})$ to each $P_i \in \mathcal{P} \setminus Z^\star$ (no need to send the respective shares to the parties in $Z^\star$, as $\mathcal{S}$ already has the shares of all the corrupt parties).

  Single Sharing Generation: Do the following to generate a secret-sharing of a given value s.
  - Upon receiving $(\mathsf{shares}, \mathsf{sid}, \{s_q\}_{S_q \cap Z^\star \neq \emptyset})$ from $\mathcal{S}$, randomly select $s_q \in \mathbb{F}$ corresponding to each $S_q \in \mathbb{S}$ for which $S_q \cap Z^\star = \emptyset$, such that $\sum_{S_q \cap Z^\star \neq \emptyset} s_q + \sum_{S_q \cap Z^\star = \emptyset} s_q = s$ holds. [a] For $q = 1, \ldots, h$, set $[s]_q = s_q$.

[a] $\mathcal{S}$ *cannot* delay sending the shares on the behalf of the corrupt parties indefinitely as, in our real-world protocol, the adversary *cannot* indefinitely delay the generation of secret-shared multiplication-triples.

Fig. 4. Ideal functionality for asynchronous pre-processing phase with session id sid.

This provision is made because in our pre-processing phase protocol, the real-world adversary will have full control over the shares of the corrupt parties corresponding to the random multiplication-triples generated in the protocol. We present a protocol for securely realizing $\mathcal{F}_{\mathsf{Triples}}$. For this, we need a multiplication protocol which takes as input $\{([a^{(\ell)}], [b^{(\ell)}])\}_{\ell=1,\ldots,M}$ and outputs $\{[c^{(\ell)}]\}_{\ell=1,\ldots,M}$, where $c^{(\ell)} = a^{(\ell)} b^{(\ell)}$, without revealing any additional information about $a^{(\ell)}, b^{(\ell)}$. We first explain and present the protocol assuming $M = 1$, where the inputs are $[a]$ and $[b]$. The goal is to securely generate a random sharing $[ab]$. The modifications to handle M pairs of inputs are straightforward.

We briefly recall the high-level idea behind our multiplication protocol which had been discussed in detail in Sect. 1.1. We first design an *asynchronous* optimistic multiplication protocol Π_{OptMult} which takes as input a set $Z \in \mathcal{Z}$ and generates a secret-sharing of ab, *provided* Adv corrupts a set of parties $Z^\star \subseteq Z$. Using protocol Π_{OptMult}, the parties then proceed in *iterations*, where in each iteration, the parties run an instance of the *asynchronous* Π_{OptMult} for each $Z \in \mathcal{Z}$. They then compare the outputs from each instance to detect if the corrupt parties cheated in any of the instances, and proceed to the respective cheater-identification phase, if any cheating is detected. If an iteration "fails" (meaning that cheating is detected in the form of a pair of "conflicting" Π_{OptMult} instances where the resultant secret-shared outputs are *different*), then the parties temporarily "wait-list" all the parties who have shared any summand during the conflicting instances of Π_{OptMult} for that iteration. The summand-sharing parties stay on the waiting-list till they complete all their supposed tasks in the corresponding cheater-identification phase, after which they are "released" to participate in instances of Π_{OptMult} in future iterations. This mechanism ensures that if an iteration fails, then the cheating parties from that iteration cannot participate in future iterations till they participate in the pending cheater identification phase of the failed iteration, in which case they are eventually discarded by all honest parties. This process is repeated till the parties reach a "successful" iteration where no cheating is detected (where the outputs of all the Π_{OptMult} instances are the same). We will show that there will be *at most* $t(tn + 1) + 1$ iterations within which a successful iteration is reached, where t is the cardinality of the maximum-sized subset in $\mathcal{Z}$.

Based on the above discussion, we next present protocols $\Pi_{\mathsf{OptMult}}, \Pi_{\mathsf{MultCI}}$ and Π_{Mult}. Protocol Π_{MultCI} (multiplication with cheater-identification) represents an iteration as discussed above. In the protocol, the parties run an instance of Π_{OptMult} for each $Z \in \mathcal{Z}$. If a pair of conflicting Π_{OptMult} instances with different outputs are identified, then the parties proceed to execute the corresponding cheater-identification phase. Protocol Π_{Mult} iteratively calls Π_{MultCI} multiple times till it reaches a "successful" instance of Π_{MultCI} (where the outputs of all the instances of Π_{OptMult} are the same). Across all the instances of these protocols, the parties maintain the following *dynamic* sets:

- $\mathcal{W}^{(i)}_{\mathsf{iter}}$: Denotes the parties *wait-listed* by P_i corresponding to instance number iter of Π_{MultCI} during Π_{Mult}. If P_i detects any cheating during the instance number iter of Π_{MultCI} with a pair of conflicting Π_{OptMult} instances, then all the summand-sharing parties from the conflicting instances are included in $\mathcal{W}^{(i)}_{\mathsf{iter}}$. These parties are removed from $\mathcal{W}^{(i)}_{\mathsf{iter}}$ only if they execute their respective steps of the corresponding cheater-identification phase.

- $\mathcal{LD}^{(i)}_{\mathsf{iter}}$: The set of parties from $\mathcal{W}^{(i)}_{\mathsf{iter}}$ which are *locally discarded* by P_i during the cheater-identification phase of instance number iter of Π_{MultCI} in Π_{Mult}.
- $\mathcal{GD}$: Denotes the set of parties, *globally discarded* by *all* (honest) parties across various instances of Π_{MultCI} in protocol Π_{Mult}.[4]

Looking ahead, these sets will be maintained such that no honest party is ever included in the $\mathcal{GD}$ and $\mathcal{LD}^{(i)}_{\mathsf{iter}}$ sets of any honest P_i. Moreover, any honest party which is included in the $\mathcal{W}^{(i)}_{\mathsf{iter}}$ set of any honest P_i will eventually be removed.

3.1 Optimistic Multiplication Protocol

Protocol Π_{OptMult} is executed with respect to a given $Z \in \mathcal{Z}$ and iteration number iter. The inputs of the protocol are $[a]$ and $[b]$. The protocol is guaranteed to eventually generate an output, which will be $[ab]$ if no party *outside* the set Z behaves maliciously. The idea behind the protocol is as follows. Since $ab = \sum_{(p,q) \in \mathbb{S} \times \mathbb{S}} [a]_p[b]_q$, a secret-sharing of ab can be computed *locally* from the shares of the summands $[a]_p[b]_q$, owing to the linearity property of the secret-sharing. As $\mathcal{Z}$ satisfies the $\mathbb{Q}^{(4)}(\mathcal{P}, \mathcal{Z})$ condition, each $(S_p \cap S_q) \setminus Z$ contains at least one *honest* party. Since the parties may not know the identity of the honest parties in the set $(S_p \cap S_q) \setminus Z$, *every* party in $(S_p \cap S_q) \setminus Z$ tries to secret-share the summand $[a]_p[b]_q$. For the sake of efficiency, instead of sharing a single summand, each party in $\mathcal{P} \setminus Z$ tries to act as a summand-sharing party and shares the sum of all the summands it is "capable" of. To avoid "repetition" of summands, the parties select *distinct* summand-sharing parties in hops. The summands which the selected party is capable of sharing are "marked" as shared, ensuring that they are not considered in future hops. To agree on the summand-sharing party of each hop, the parties execute an instance of the *agreement on common subset* (ACS) primitive [4], where one instance of ABA is invoked on the behalf of each candidate summand-sharing party. While voting for a candidate party in $\mathcal{P} \setminus Z$ during a hop, the parties ensure that the candidate has indeed secret-shared some sum and satisfies the following conditions:

- The candidate party has not been selected in an earlier hop.
- The candidate party does not belong to the waiting list or the list of locally-discarded parties of any previous iteration.
- The candidate does not belong to the list of globally-discarded parties (Fig. 5).

We next formally prove the properties of the protocol Π_{OptMult}. While proving these properties, we assume that for every iter, no honest party is ever included in the set $\mathcal{GD}$ and all honest parties are eventually removed from the $\mathcal{W}^{(i)}_{\mathsf{iter}'}, \mathcal{LD}^{(i)}_{\mathsf{iter}'}$ sets of every honest P_i for every $\mathsf{iter}' < \mathsf{iter}$. Looking ahead, these conditions are guaranteed in the protocols Π_{MultCI} and Π_{Mult} (where these sets are constructed and managed), where Π_{OptMult} is used as a subprotocol.

[4] The reason for two different discarded sets is that the various instances of cheater-identification are executed *asynchronously*, thus resulting in a corrupt party to be identified by different honest parties during different iterations.

Protocol $\Pi_{\mathsf{OptMult}}(\mathcal{P}, \mathcal{Z}, \mathbb{S}, [a], [b], Z, \mathsf{iter})$

- **Initialization**
  - Initialize the set of ordered pair of indices of *all* summands : $\mathsf{Summands}_{(Z,\mathsf{iter})} = \{(p,q)\}_{p,q=1,\ldots,|\mathbb{S}|}$.
  - Initialize the summand indices corresponding to $P_j \in \mathcal{P} \setminus Z$: $\mathsf{Summands}^{(j)}_{(Z,\mathsf{iter})} = \{(p,q)\}_{P_j \in S_p \cap S_q}$.
  - Initialize the set of summands-sharing parties : $\mathsf{Selected}_{(Z,\mathsf{iter})} = \emptyset$. Initialize the hop number $\mathsf{hop} = 1$.
- Do the following till $\mathsf{Summands}_{(Z,\mathsf{iter})} \neq \emptyset$:
  - **Sharing Summands**:
    1. If $P_i \notin Z$ and $P_i \notin \mathsf{Selected}_{(Z,\mathsf{iter})}$, then compute $c^{(i)}_{(Z,\mathsf{iter})} = \sum_{(p,q)\in \mathsf{Summands}^{(i)}_{(Z,\mathsf{iter})}} [a]_p[b]_q$. Randomly select the shares $c^{(i)}_{(Z,\mathsf{iter})_1}, \ldots, c^{(i)}_{(Z,\mathsf{iter})_h}$, such that $c^{(i)}_{(Z,\mathsf{iter})_1} + \ldots + c^{(i)}_{(Z,\mathsf{iter})_h} = c^{(i)}_{(Z,\mathsf{iter})}$. Call $\mathcal{F}_{\mathsf{VSS}}$ with message $(\mathsf{dealer}, \mathsf{sid}_{\mathsf{hop},i,\mathsf{iter},Z}, (c^{(i)}_{(Z,\mathsf{iter})_1}, \ldots, c^{(i)}_{(Z,\mathsf{iter})_h}))$, where $\mathsf{sid}_{\mathsf{hop},i,\mathsf{iter},Z} = \mathsf{hop}||\mathsf{sid}||i||\mathsf{iter}||Z$.[a]
    2. Keep requesting for an output from $\mathcal{F}_{\mathsf{VSS}}$ with $\mathsf{sid}_{\mathsf{hop},j,\mathsf{iter},Z}$, for $j = 1, \ldots, n$, till an output is received.
  - **Selecting Summand-Sharing Party Through ACS**:
    1. For $j = 1, \ldots, n$, send $(\mathsf{vote}, \mathsf{sid}_{\mathsf{hop},j,\mathsf{iter},Z}, 1)$ to $\mathcal{F}_{\mathsf{ABA}}$, if *all* the following conditions hold:
       - $P_j \notin \mathcal{GD}$, $P_j \notin Z$ and $P_j \notin \mathsf{Selected}_{(Z,\mathsf{iter})}$. Moreover, $\forall \mathsf{iter}' < \mathsf{iter}$, $P_j \notin \mathcal{W}^{(i)}_{\mathsf{iter}'}$ and $P_j \notin \mathcal{LD}^{(i)}_{\mathsf{iter}'}$.
       - An output $(\mathsf{share}, \mathsf{sid}_{\mathsf{hop},j,\mathsf{iter},Z}, P_j, \{[c^{(j)}_{(Z,\mathsf{iter})}]_q\}_{P_i \in S_q})$ is received from $\mathcal{F}_{\mathsf{VSS}}$, with $\mathsf{sid}_{\mathsf{hop},j,\mathsf{iter},Z}$.
    2. For $j = 1, \ldots, n$, request for an output from $\mathcal{F}_{\mathsf{ABA}}$ with $\mathsf{sid}_{\mathsf{hop},j,\mathsf{iter},Z}$, until an output is received.
    3. If $\exists P_j \in \mathcal{P}$ such that $(\mathsf{decide}, \mathsf{sid}_{\mathsf{hop},j,\mathsf{iter},Z}, 1)$ is received from $\mathcal{F}_{\mathsf{ABA}}$ with $\mathsf{sid}_{\mathsf{hop},j,\mathsf{iter},Z}$, then for each $P_k \in \mathcal{P}$ for which no vote message has been sent yet, send $(\mathsf{vote}, \mathsf{sid}_{\mathsf{hop},k,\mathsf{iter},Z}, 0)$ to $\mathcal{F}_{\mathsf{ABA}}$ with $\mathsf{sid}_{\mathsf{hop},k,\mathsf{iter},Z}$.
    4. Once an output $(\mathsf{decide}, \mathsf{sid}_{\mathsf{hop},j,\mathsf{iter},Z}, v_j)$ is received from $\mathcal{F}_{\mathsf{ABA}}$ with $\mathsf{sid}_{\mathsf{hop},j,\mathsf{iter},Z}$ for all $j \in \{1, \ldots, n\}$, select the least indexed P_j, such that $v_j = 1$. Then set $\mathsf{hop} = \mathsf{hop} + 1$ and update the following.
       - $\mathsf{Selected}_{(Z,\mathsf{iter})} = \mathsf{Selected}_{(Z,\mathsf{iter})} \cup \{P_j\}$. $\mathsf{Summands}_{(Z,\mathsf{iter})} = \mathsf{Summands}_{(Z,\mathsf{iter})} \setminus \mathsf{Summands}^{(j)}_{(Z,\mathsf{iter})}$.
       - $\forall P_k \in \mathcal{P} \setminus \{Z \cup \mathsf{Selected}_{(Z,\mathsf{iter})}\}$: $\mathsf{Summands}^{(k)}_{(Z,\mathsf{iter})} = \mathsf{Summands}^{(k)}_{(Z,\mathsf{iter})} \setminus \mathsf{Summands}^{(j)}_{(Z,\mathsf{iter})}$.
- $\forall P_j \in \mathcal{P} \setminus \mathsf{Selected}_{(Z,\mathsf{iter})}$, participate in an instance of Π_{PerDefSh} with input $c^{(j)}_{(Z,\mathsf{iter})} = 0$. Output $\{[c^{(1)}_{(Z,\mathsf{iter})}]_q, \ldots, [c^{(n)}_{(Z,\mathsf{iter})}]_q, [c_{(Z,\mathsf{iter})}]_q\}_{P_i \in S_q}$, where $c_{(Z,\mathsf{iter})} \stackrel{def}{=} c^{(1)}_{(Z,\mathsf{iter})} + \ldots + c^{(n)}_{(Z,\mathsf{iter})}$.

[a] The notation $\mathsf{sid}_{\mathsf{hop},i,\mathsf{iter},Z}$ is used to distinguish among the different calls to $\mathcal{F}_{\mathsf{VSS}}$ and $\mathcal{F}_{\mathsf{ABA}}$ within each hop.

Fig. 5. Optimistic multiplication in $(\mathcal{F}_{\mathsf{VSS}}, \mathcal{F}_{\mathsf{ABA}})$-hybrid for iteration iter and session id sid, assuming Z to be corrupt. The above code is executed by each P_i, who implicitly uses the dynamic sets $\mathcal{GD}$, $\mathcal{W}^{(i)}_{\mathsf{iter}'}$ and $\mathcal{LD}^{(i)}_{\mathsf{iter}'}$ for $\mathsf{iter}' < \mathsf{iter}$

Claim 1. For every $Z \in \mathcal{Z}$ and every ordered pair $(p, q) \in \{1, \ldots, |\mathbb{S}|\} \times \{1, \ldots, |\mathbb{S}|\}$, the set $(S_p \cap S_q) \setminus Z$ contains at least one honest party.

Proof. From the definition of $\mathbb{S}$, we have $S_p = \mathcal{P} \setminus Z_p$ and $S_q = \mathcal{P} \setminus Z_q$, where $Z_p, Z_q \in \mathcal{Z}$. Let $Z^\star \in \mathcal{Z}$ be the set of corrupt parties during the protocol Π_{OptMult}. If $(S_p \cap S_q) \setminus Z$ *does not* contain any honest party, then it implies that $((S_p \cap S_q) \setminus Z) \subseteq Z^\star$. This further implies that $\mathcal{P} \subseteq Z_p \cup Z_q \cup Z \cup Z^\star$, implying that $\mathcal{Z}$ *does not* satisfy the $\mathbb{Q}^{(4)}(\mathcal{P}, \mathcal{Z})$ condition, which is a contradiction.

Claim 2. For every $Z \in \mathcal{Z}$, if all honest parties participate during the hop number hop in the protocol Π_{OptMult}, then all honest parties eventually obtain a common summand-sharing party, say P_j, for this hop, such that the honest parties will eventually hold $[c^{(j)}_{(Z,\mathsf{iter})}]$. Moreover, party P_j will be distinct from the summand-sharing party selected for any hop number $\mathsf{hop}' < \mathsf{hop}$.

Proof. Since all honest parties participate in hop number hop, it follows that $\mathsf{Summands}_{(Z,\mathsf{iter})} \neq \emptyset$ at the beginning of hop number hop. This implies that there exists at least one ordered pair $(p, q) \in \mathsf{Summands}_{(Z,\mathsf{iter})}$. From Claim 1, there exists at least one *honest* party in $(S_p \cap S_q) \setminus Z$, say P_k, who will have both the shares $[a]_p$ as well as $[b]_q$ (and hence the summand $[a]_p[b]_q$). We also note that P_k *would not* have been selected as the common summand-sharing party in any previous $\mathsf{hop}' < \mathsf{hop}$, as otherwise, P_k would have already included the summand $[a]_p[b]_q$ in the sum $c^{(k)}_{(Z,\mathsf{iter})}$ shared by P_k during hop hop', implying that $(p, q) \notin \mathsf{Summands}_{(Z,\mathsf{iter})}$. Now, during the hop number hop, party P_k will randomly secret-share the sum $c^{(k)}_{(Z,\mathsf{iter})}$ by making a call to $\mathcal{F}_{\mathsf{VSS}}$, and every honest P_i will eventually receive an output $(\mathsf{share}, \mathsf{sid}_{\mathsf{hop},k,\mathsf{iter},Z}, P_k, \{[c^{(k)}_{(Z,\mathsf{iter})}]_q\}_{P_i \in S_q})$ from $\mathcal{F}_{\mathsf{VSS}}$ with $\mathsf{sid}_{\mathsf{hop},k,\mathsf{iter},Z}$. Moreover, P_k *will not* be present in the set $\mathcal{GD}$ and if P_k is present in the sets $\mathcal{W}^{(i)}_{\mathsf{iter}'}, \mathcal{LD}^{(i)}_{\mathsf{iter}'}$ of any honest P_i for any $\mathsf{iter}' < \mathsf{iter}$, then will eventually be removed from these sets.

We next claim that during the hop number hop, there will be at least one instance of $\mathcal{F}_{\mathsf{ABA}}$ corresponding to which *all* honest parties eventually receive the output 1. For this, we consider two possible cases:

- *At least one honest party participates with input* 0 *in the* $\mathcal{F}_{\mathsf{ABA}}$ *instance corresponding to* P_k: Let P_i be an *honest* party, who sends $(\mathsf{vote}, \mathsf{sid}_{\mathsf{hop},k,\mathsf{iter},Z}, 0)$ to $\mathcal{F}_{\mathsf{ABA}}$ with $\mathsf{sid}_{\mathsf{hop},k,\mathsf{iter},Z}$. Then it follows that there exists some $P_j \in \mathcal{P}$, such that P_i has received $(\mathsf{decide}, \mathsf{sid}_{\mathsf{hop},j,\mathsf{iter},Z}, 1)$ as the output from $\mathcal{F}_{\mathsf{ABA}}$ with $\mathsf{sid}_{\mathsf{hop},j,\mathsf{iter},Z}$. Hence, *every* honest party will eventually receive the output $(\mathsf{decide}, \mathsf{sid}_{\mathsf{hop},j,\mathsf{iter},Z}, 1)$ as the output from $\mathcal{F}_{\mathsf{ABA}}$ with $\mathsf{sid}_{\mathsf{hop},j,\mathsf{iter},Z}$.
- *No honest party participates with input* 0 *in the* $\mathcal{F}_{\mathsf{ABA}}$ *instance corresponding to* P_k: In this case, *every* honest party will eventually send $(\mathsf{vote}, \mathsf{sid}_{\mathsf{hop},k,\mathsf{iter},Z}, 1)$ to $\mathcal{F}_{\mathsf{ABA}}$ with $\mathsf{sid}_{\mathsf{hop},k,\mathsf{iter},Z}$ and eventually receives the output $(\mathsf{decide}, \mathsf{sid}_{\mathsf{hop},k,\mathsf{iter},Z}, 1)$ from $\mathcal{F}_{\mathsf{ABA}}$.

Now, based on the above claim, we can further claim that all honest parties will eventually participate with some input in all the n instances of $\mathcal{F}_{\mathsf{ABA}}$ invoked during the hop number hop and hence, all the n instances of $\mathcal{F}_{\mathsf{ABA}}$ during the hop number hop will eventually produce an output. Since the summand-sharing party for hop number hop corresponds to the least indexed $\mathcal{F}_{\mathsf{ABA}}$ instance in which all the honest parties obtain 1 as the output, it follows that eventually, the honest parties will select a summand-sharing party. Moreover, this summand-sharing party will be common, as it is based on the outcome of $\mathcal{F}_{\mathsf{ABA}}$ instances.

Let P_j be the summand-sharing party for the hop number hop. We next show that the honest parties will eventually hold $[c^{(j)}_{(Z,\mathsf{iter})}]$. For this, we note that since P_j has been selected as the summand-sharing party, *at least* one *honest* party, say P_i, must have sent $(\mathsf{vote}, \mathsf{sid}_{\mathsf{hop},j,\mathsf{iter},Z}, 1)$ to $\mathcal{F}_{\mathsf{ABA}}$ with $\mathsf{sid}_{\mathsf{hop},j,\mathsf{iter},Z}$. If not, then $\mathcal{F}_{\mathsf{ABA}}$ with $\mathsf{sid}_{\mathsf{hop},j,\mathsf{iter},Z}$ will never produce the output $(\mathsf{decide}, \mathsf{sid}_{\mathsf{hop},j,\mathsf{iter},Z}, 1)$ and hence, P_j will not be the summand-sharing party for the hop number hop. Now since P_i sent $(\mathsf{vote}, \mathsf{sid}_{\mathsf{hop},j,\mathsf{iter},Z}, 1)$ to $\mathcal{F}_{\mathsf{ABA}}$, it follows that P_i has received an output $(\mathsf{share}, \mathsf{sid}_{\mathsf{hop},j,\mathsf{iter},Z}, P_j, \{[c^{(j)}_{(Z,\mathsf{iter})}]_q\}_{P_i \in S_q})$ from $\mathcal{F}_{\mathsf{VSS}}$ with $\mathsf{sid}_{\mathsf{hop},j,\mathsf{iter},Z}$. This implies that P_j must have sent the message $(\mathsf{dealer}, \mathsf{sid}_{\mathsf{hop},j,\mathsf{iter},Z}, (c^{(j)}_{(\mathsf{iter},Z)_1}, \ldots, c^{(j)}_{(\mathsf{iter},Z)_h}))$ to $\mathcal{F}_{\mathsf{VSS}}$ with $\mathsf{sid}_{\mathsf{hop},j,\mathsf{iter},Z}$. Consequently, every honest party will eventually receive their respective outputs from $\mathcal{F}_{\mathsf{VSS}}$ with $\mathsf{sid}_{\mathsf{hop},j,\mathsf{iter},Z}$ and hence, the honest parties will eventually hold $[c^{(j)}_{(Z,\mathsf{iter})}]$.

Finally, to complete the proof of the claim, we need to show that party P_j is different from the summand-sharing parties selected during the hops $1, \ldots, \mathsf{hop}-1$. If P_j has been selected as a summand-sharing party for any hop number $\mathsf{hop}' < \mathsf{hop}$, then no honest party ever sends $(\mathsf{vote}, \mathsf{sid}_{\mathsf{hop},j,\mathsf{iter},Z}, 1)$ to $\mathcal{F}_{\mathsf{ABA}}$ with $\mathsf{sid}_{\mathsf{hop},j,\mathsf{iter},Z}$. Consequently, $\mathcal{F}_{\mathsf{ABA}}$ with $\mathsf{sid}_{\mathsf{hop},j,\mathsf{iter},Z}$ will never send the output $(\mathsf{decide}, \mathsf{sid}_{\mathsf{hop},j,\mathsf{iter},Z}, 1)$ to any honest party and hence P_j will not be selected as the summand-sharing party for hop number hop, which is a contradiction.

Claim 3. In protocol Π_{OptMult}, all honest parties eventually obtain an output. The protocol makes $\mathcal{O}(n^2)$ calls to $\mathcal{F}_{\mathsf{VSS}}$ and $\mathcal{F}_{\mathsf{ABA}}$.

Proof. From Claim 1 and 2, it follows that the number of hops in the protocol is $\mathcal{O}(n)$, as in each hop a new summand-sharing party is selected and if all honest parties are included in the set of summand-sharing parties $\mathsf{Selected}_{(Z,\mathsf{iter})}$, then $\mathsf{Summands}_{(Z,\mathsf{iter})}$ becomes $\emptyset$. The proof now follows from the fact that in each hop, there are $\mathcal{O}(n)$ calls to $\mathcal{F}_{\mathsf{VSS}}$ and $\mathcal{F}_{\mathsf{ABA}}$.

Claim 4. In Π_{OptMult}, if no party in $\mathcal{P} \setminus Z$ behaves maliciously, then for each $P_i \in \mathsf{Selected}_{(Z,\mathsf{iter})}$, $c^{(i)}_{(Z,\mathsf{iter})} = \sum_{(p,q) \in \mathsf{Summands}^{(i)}_{(Z,\mathsf{iter})}} [a]_p[b]_q$ holds and $c_{(Z,\mathsf{iter})} = ab$.

Proof. From the protocol steps, it follows that $\mathsf{Selected}_{(Z,\mathsf{iter})} \cap Z = \emptyset$, as no honest part ever votes for any party from Z as a candidate summand-sharing party during any hop in the protocol. Now since $\mathsf{Selected}_{(Z,\mathsf{iter})} \subseteq (\mathcal{P} \setminus Z)$, if no party in $\mathcal{P} \setminus Z$ behaves maliciously, then it implies that every party $P_i \in \mathsf{Selected}_{(Z,\mathsf{iter})}$ behaves honestly and secret-shares $c^{(i)}_{(Z,\mathsf{iter})}$ by calling $\mathcal{F}_{\mathsf{VSS}}$, where $c^{(i)}_{(Z,\mathsf{iter})} = \sum_{(p,q) \in \mathsf{Summands}^{(i)}_{(Z,\mathsf{iter})}} [a]_p[b]_q$. Moreover, from the protocol steps, it follows that for every $P_j, P_k \in \mathsf{Selected}_{(Z,\mathsf{iter})}$:

$$\mathsf{Summands}^{(j)}_{(Z,\mathsf{iter})} \cap \mathsf{Summands}^{(k)}_{(Z,\mathsf{iter})} = \emptyset.$$

To prove this, suppose P_j and P_k are included in $\mathsf{Selected}_{(Z,\mathsf{iter})}$ during hop number hop_j and hop_k respectively, where without loss of generality, $\mathsf{hop}_j < \mathsf{hop}_k$. Then from the protocol steps, during hop_j, the parties would set $\mathsf{Summands}^{(k)}_{(Z,\mathsf{iter})} = \mathsf{Summands}^{(k)}_{(Z,\mathsf{iter})} \setminus \mathsf{Summands}^{(j)}_{(Z,\mathsf{iter})}$. This ensures that during hop_k, there exists no ordered pair $(p,q) \in \{1,\ldots,|\mathbb{S}|\} \times \{1,\ldots,|\mathbb{S}|\}$, such that $(p,q) \in \mathsf{Summands}^{(j)}_{(Z,\mathsf{iter})} \cap \mathsf{Summands}^{(k)}_{(Z,\mathsf{iter})}$.

Since all the parties $P_i \in \mathsf{Selected}_{(Z,\mathsf{iter})}$ have behaved honestly, from the protocol steps, it also follows that :

$$\bigcup_{P_i \in \mathsf{Selected}_{(Z,\mathsf{iter})}} \mathsf{Summands}^{(i)}_{(Z,\mathsf{iter})} = \{(p,q)\}_{p,q=1,\ldots,|\mathbb{S}|}.$$

Finally, from the protocol steps, it follows that $\forall P_j \in \mathcal{P} \setminus \mathsf{Selected}_{(Z,\mathsf{iter})}$, the condition $c^{(j)}_{(Z,\mathsf{iter})} = 0$ holds. Now since $c_{(Z,\mathsf{iter})} = c^{(1)}_{(Z,\mathsf{iter})} + \ldots + c^{(n)}_{(Z,\mathsf{iter})}$, it follows that if no party in $\mathcal{P} \setminus Z$ behaves maliciously, then $c_{(Z,\mathsf{iter})} = ab$ holds.

Claim 5. In Π_{OptMult}, no additional information about a and b is leaked to Adv.

Proof. Let $Z^\star \in \mathcal{Z}$ be the set of corrupt parties. To prove the claim, we argue that in the protocol, Adv does not learn any additional information about the shares $\{[a]_p, [b]_p\}_{S_p \cap Z^\star = \emptyset}$. For this, consider an arbitrary summand $[a]_p[b]_q$ where $S_p \cap Z^\star = \emptyset$ and where $q \in \{1,\ldots,h\}$. Clearly, the summand $[a]_p[b]_q$ will not be available with any party in $Z^\star$. Let P_j be the party from $\mathsf{Selected}_{(Z,\mathsf{iter})}$, such that $(p,q) \in \mathsf{Summands}^{(j)}_{(Z,\mathsf{iter})}$; i.e. the summand $[a]_p[b]_q$ is included by P_j while computing the summand-sum $c^{(j)}_{(Z,\mathsf{iter})}$. Clearly P_j is *honest*, since $P_j \notin Z^\star$. In the protocol, party P_j randomly secret-shares the summand-sum $c^{(j)}_{(Z,\mathsf{iter})}$ by supplying a random vector of shares for $c^{(j)}_{(Z,\mathsf{iter})}$ to the corresponding $\mathcal{F}_{\mathsf{VSS}}$. Now, since $\mathbb{S}$ is $\mathcal{Z}$-*private*, it follows that the shares $\{[c^{(j)}_{(Z,\mathsf{iter})}]_r\}_{S_r \cap Z^\star \neq \emptyset}$ learnt by Adv in the protocol will be independent of the summand $[a]_p[b]_q$ and hence, independent of $[a]_p$. Using a similar argument, we can conclude that the shares learnt by Adv in the protocol will be independent of the summands $[a]_q[b]_p$ (and hence independent of $[b]_p$), where $S_p \cap Z^\star = \emptyset$, and where $q \in \{1,\ldots,h\}$.

The proof of Lemma 1 now follows from Claims 1–5.

Lemma 1. Let $\mathcal{Z}$ satisfy the $\mathbb{Q}^{(4)}(\mathcal{P},\mathcal{Z})$ condition, $\mathbb{S} = \{\mathcal{P} \setminus Z | Z \in \mathcal{Z}\}$ and let all honest parties participate in the instance $\Pi_{\mathsf{OptMult}}(\mathcal{P},\mathcal{Z},\mathbb{S},[a],[b],Z,\mathsf{iter})$. Then all honest parties eventually compute $[c_{(Z,\mathsf{iter})}], [c^{(1)}_{(Z,\mathsf{iter})}], \ldots, [c^{(n)}_{(Z,\mathsf{iter})}]$ where $c_{(Z,\mathsf{iter})} = c^{(1)}_{(Z,\mathsf{iter})} + \ldots + c^{(n)}_{(Z,\mathsf{iter})}$, provided no honest party is included in the $\mathcal{GD}$ and $\mathcal{LD}^{(i)}_{\mathsf{iter}}$, sets and each honest party in the $\mathcal{W}^{(i)}_{\mathsf{iter}}$, sets of every honest P_i is eventually removed, for all $\text{iter}' < \text{iter}$. If no party in $\mathcal{P} \setminus Z$ acts maliciously, then $c_{(Z,\text{iter})} = ab$. In the protocol, Adv does not learn anything additional about a and b. The protocol makes $\mathcal{O}(n^2)$ calls to $\mathcal{F}_{\text{VSS}}$ and $\mathcal{F}_{\text{ABA}}$.

Protocol Π_{OptMult} for M Pairs of Inputs: Now in each hop, every P_i calls $\mathcal{F}_{\text{VSS}}$ M times to share M summations. While voting for a candidate summand-sharing party in a hop, the parties check whether it has shared M values. Hence, there will be $\mathcal{O}(n^2 M)$ calls to $\mathcal{F}_{\text{VSS}}$, but *only* $\mathcal{O}(n^2)$ calls to $\mathcal{F}_{\text{ABA}}$.

3.2 Multiplication Protocol with Cheater Identification

Based on protocol Π_{OptMult}, we next present the protocol Π_{MultCI} with cheater identification (Fig. 6). The protocol takes as input an iteration number iter and $[a], [b]$. If *no* party behaves maliciously, then the protocol securely outputs $[ab]$. In the protocol, parties execute an instance of Π_{OptMult} for each $Z \in \mathcal{Z}$ and compare the outputs. Since at least one of the Π_{OptMult} instances is guaranteed to output $[ab]$, if all the outputs are the same, then no cheating has occurred. Otherwise, the parties identify a pair of *conflicting* Π_{OptMult} instances with different outputs, executed with respect to Z and Z', and flag the iteration as "failed". Let $\text{Selected}_{(Z,\text{iter})}$ and $\text{Selected}_{(Z',\text{iter})}$ be the summand-sharing parties in the conflicting Π_{OptMult} instances. The parties next proceed to a cheater-identification phase to identify at least one corrupt party in $\text{Selected}_{(Z,\text{iter})} \cup \text{Selected}_{(Z',\text{iter})}$.

Each $P_j \in \text{Selected}_{(Z,\text{iter})}$ is made to share the sum of the summands from its summand-list[5] overlapping with the summand-list of each $P_k \in \text{Selected}_{(Z',\text{iter})}$ and vice-versa. Next, these "partitions" are compared, based on which at least one corrupt party in $\text{Selected}_{(Z,\text{iter})} \cup \text{Selected}_{(Z',\text{iter})}$ is guaranteed to be identified provided *all* the parties in $\text{Selected}_{(Z,\text{iter})} \cup \text{Selected}_{(Z',\text{iter})}$ secret-share the

Protocol $\Pi_{\text{MultCI}}(\mathcal{P}, \mathcal{Z}, \mathbb{S}, [a], [b], \text{iter})$

- **Initialization**: Initialize $\mathcal{W}^{(i)}_{\text{iter}} = \mathcal{LD}^{(i)}_{\text{iter}} = \emptyset$ and $\text{flag}^{(i)}_{\text{iter}} = \perp$. Fix some $Z' \in \mathcal{Z}$.
- **Running Optimistic Multiplication**:
  - For each $Z \in \mathcal{Z}$, participate in the instance $\Pi_{\text{OptMult}}(\mathcal{P}, \mathcal{Z}, \mathbb{S}, [a], [b], Z, \text{iter})$ with session id sid. Let $\{[c^{(1)}_{(Z,\text{iter})}]_q, \ldots, [c^{(n)}_{(Z,\text{iter})}]_q, [c_{(Z,\text{iter})}]_q\}_{P_i \in S_q}$ be the output obtained. Moreover, let $\text{Selected}_{(Z,\text{iter})}$ be the set of summand-sharing parties and for each $P_j \in \text{Selected}_{(Z,\text{iter})}$, let $\text{Summands}^{(j)}_{(Z,\text{iter})}$ be the set of ordered pairs of indices corresponding to the summands whose sum has been shared by P_j, during this instance of Π_{OptMult}.
  - Corresponding to every $Z \in \mathcal{Z}$, participate in an instance of Π_{PerRec} to reconstruct $c_{(Z,\text{iter})} - c_{(Z',\text{iter})}$.

Fig. 6. Code for P_i for multiplication with cheater identification for iteration iter and session id sid, in the $\mathcal{F}_{\text{VSS}}$-hybrid

[5] Here, the summand-list of a selected party refers to the summands it was supposed to share during the respective Π_{OptMult} instance of that iteration.

- **Output in Case of Success**: If $\forall Z \in \mathcal{Z}$, $c_{(Z,\mathsf{iter})} - c_{(Z',\mathsf{iter})} = 0$, then set $\mathsf{flag}^{(i)}_{\mathsf{iter}} = 0$ and output $\{[c_{(Z',\mathsf{iter})}]_q\}_{P_i \in \mathcal{S}_q}$.
- **Waiting-List and Cheater Identification in Case of Failure**: If $\exists Z \in \mathcal{Z}$: $c_{(Z,\mathsf{iter})} - c_{(Z',\mathsf{iter})} \neq 0$, then let Z be the first set such that $c_{(Z,\mathsf{iter})} - c_{(Z',\mathsf{iter})} \neq 0$. Set the *conflicting-sets* to be Z, Z', $\mathsf{flag}^{(i)}_{\mathsf{iter}} = 1$ and proceed as follows.
  - **Wait-listing Parties**: Set $\mathcal{W}^{(i)}_{\mathsf{iter}} = \mathsf{Selected}_{(Z,\mathsf{iter})} \cup \mathsf{Selected}_{(Z',\mathsf{iter})}$.
  - **Sharing Partition of the Summand-Sums**:
    1. If $P_i \in \mathsf{Selected}_{(Z,\mathsf{iter})}$, compute $d^{(ij)}_{(Z,\mathsf{iter})} = \sum_{(p,q) \in \mathsf{Summands}^{(i)}_{(Z,\mathsf{iter})} \cap \mathsf{Summands}^{(j)}_{(Z',\mathsf{iter})}} [a]_p[b]_q$, for every $P_j \in \mathsf{Selected}_{(Z',\mathsf{iter})}$. Randomly pick $d^{(ij)}_{(Z,\mathsf{iter})_1}, \ldots, d^{(ij)}_{(Z,\mathsf{iter})_h}$ such that $d^{(ij)}_{(Z,\mathsf{iter})_1} + \ldots + d^{(ij)}_{(Z,\mathsf{iter})_h} = d^{(ij)}_{(Z,\mathsf{iter})}$. Send $(\mathsf{dealer}, \mathsf{sid}_{i,j,\mathsf{iter},Z}, (d^{(ij)}_{(Z,\mathsf{iter})_1}, \ldots, d^{(ij)}_{(Z,\mathsf{iter})_h})$ to $\mathcal{F}_{\mathsf{VSS}}$, where $\mathsf{sid}_{i,j,\mathsf{iter},Z} = \mathsf{sid}||i||j||\mathsf{iter}||Z$.
    2. If $P_i \in \mathsf{Selected}_{(Z',\mathsf{iter})}$, compute $e^{(ij)}_{(Z',\mathsf{iter})} = \sum_{(p,q) \in \mathsf{Summands}^{(i)}_{(Z',\mathsf{iter})} \cap \mathsf{Summands}^{(j)}_{(Z,\mathsf{iter})}} [a]_p[b]_q$, for all $P_j \in \mathsf{Selected}_{(Z,\mathsf{iter})}$. Randomly pick $e^{(ij)}_{(Z',\mathsf{iter})_1}, \ldots, e^{(ij)}_{(Z',\mathsf{iter})_h}$ which sum up to $e^{(ij)}_{(Z',\mathsf{iter})}$ and then send $(\mathsf{dealer}, \mathsf{sid}_{i,j,\mathsf{iter},Z'}, (e^{(ij)}_{(Z',\mathsf{iter})_1}, \ldots, e^{(ij)}_{(Z',\mathsf{iter})_h})$ to $\mathcal{F}_{\mathsf{VSS}}$, where $\mathsf{sid}_{i,j,\mathsf{iter},Z'} = \mathsf{sid}||i||j||\mathsf{iter}||Z'$.
    3. Corresponding to every $P_j \in \mathsf{Selected}_{(Z,\mathsf{iter})}$ and every $P_k \in \mathsf{Selected}_{(Z',\mathsf{iter})}$, keep requesting for output from $\mathcal{F}_{\mathsf{VSS}}$ with session id $\mathsf{sid}_{j,k,\mathsf{iter},Z}$, till an output is obtained.
    4. Corresponding to every $P_j \in \mathsf{Selected}_{(Z',\mathsf{iter})}$ and every $P_k \in \mathsf{Selected}_{(Z,\mathsf{iter})}$, keep requesting for output from $\mathcal{F}_{\mathsf{VSS}}$ with session id $\mathsf{sid}_{j,k,\mathsf{iter},Z'}$, till an output is obtained.
  - **Removing Parties from Wait List**: Set $\mathcal{W}^{(i)}_{\mathsf{iter}} = \mathcal{W}^{(i)}_{\mathsf{iter}} \setminus \{P_j\}$, if *all* the following criteria pertaining to P_j hold:
    1. $P_j \in \mathsf{Selected}_{(Z,\mathsf{iter})}$: if an output $(\mathsf{share}, \mathsf{sid}_{j,k,\mathsf{iter},Z}, P_j, \{[d^{(jk)}_{(Z,\mathsf{iter})}]_q\}_{P_i \in \mathcal{S}_q})$ is received from $\mathcal{F}_{\mathsf{VSS}}$ with session id $\mathsf{sid}_{j,k,\mathsf{iter},Z}$, corresponding to each $P_k \in \mathsf{Selected}_{(Z',\mathsf{iter})}$,
    2. $P_j \in \mathsf{Selected}_{(Z',\mathsf{iter})}$: if an output $(\mathsf{share}, \mathsf{sid}_{j,k,\mathsf{iter},Z'}, P_j, \{[e^{(jk)}_{(Z',\mathsf{iter})}]_q\}_{P_i \in \mathcal{S}_q})$ is received from $\mathcal{F}_{\mathsf{VSS}}$ with session id $\mathsf{sid}_{j,k,\mathsf{iter},Z'}$, corresponding to every $P_k \in \mathsf{Selected}_{(Z,\mathsf{iter})}$.
  - **Verifying the Summand-Sum Partitions and Locally Identifying Corrupt Parties**:
    1. For every $P_j \in \mathsf{Selected}_{(Z,\mathsf{iter})}$, participate in an instance of Π_{PerRec} to reconstruct the difference value $c^{(j)}_{(Z,\mathsf{iter})} - \sum_{P_k \in \mathsf{Selected}_{(Z',\mathsf{iter})}} d^{(jk)}_{(Z,\mathsf{iter})}$. If the difference is not 0, then set $\mathcal{LD}^{(i)}_{\mathsf{iter}} = \mathcal{LD}^{(i)}_{\mathsf{iter}} \cup \{P_j\}$.
    2. For every $P_j \in \mathsf{Selected}_{(Z',\mathsf{iter})}$, participate in an instance of Π_{PerRec} to reconstruct the difference value $c^{(j)}_{(Z',\mathsf{iter})} - \sum_{P_k \in \mathsf{Selected}_{(Z,\mathsf{iter})}} e^{(jk)}_{(Z',\mathsf{iter})}$. If the difference is not 0, then set $\mathcal{LD}^{(i)}_{\mathsf{iter}} = \mathcal{LD}^{(i)}_{\mathsf{iter}} \cup \{P_j\}$.

Fig. 6. (*continued*)

3. For each ordered pair (P_j, P_k) where $P_j \in \mathsf{Selected}_{(Z,\mathsf{iter})}$ and $P_k \in \mathsf{Selected}_{(Z',\mathsf{iter})}$, participate in an instance of Π_{PerRec} to reconstruct $d^{(jk)}_{(Z,\mathsf{iter})} - e^{(kj)}_{(Z',\mathsf{iter})}$. If the value is not 0, then do the following:
   i. Participate in instances of Π_{PerRec} to reconstruct $d^{(jk)}_{(Z,\mathsf{iter})}$ and $e^{(kj)}_{(Z',\mathsf{iter})}$. Participate in instances of $\Pi_{\mathsf{PerRecShare}}$ to reconstruct $[a]_p$ and $[b]_q$, such that $(p,q) \in \mathsf{Summands}^{(j)}_{(Z,\mathsf{iter})} \cap \mathsf{Summands}^{(k)}_{(Z',\mathsf{iter})}$.
   ii. Compare $\sum_{(p,q)\in\mathsf{Summands}^{(j)}_{(Z,\mathsf{iter})} \cap \mathsf{Summands}^{(k)}_{(Z',\mathsf{iter})}} [a]_p[b]_q$ with $d^{(jk)}_{(Z,\mathsf{iter})}$ and $e^{(kj)}_{(Z',\mathsf{iter})}$ and identify the corrupt party $P_c \in \{P_j, P_k\}$. Set $\mathcal{LD}^{(i)}_{\mathsf{iter}} = \mathcal{LD}^{(i)}_{\mathsf{iter}} \cup \{P_c\}$.

Fig. 6. (*continued*)

required partitions. The cheater-identification phase will be "stuck" if the *corrupt* parties in $\mathsf{Selected}_{(Z,\mathsf{iter})} \cup \mathsf{Selected}_{(Z',\mathsf{iter})}$ do not participate. To prevent such corrupt parties from causing future instances of Π_{MultCI} to fail, the parties wait-list all the parties in $\mathsf{Selected}_{(Z,\mathsf{iter})} \cup \mathsf{Selected}_{(Z',\mathsf{iter})}$. A party is then "released" only after it has shared all the required values as part of the cheater-identification phase. Every honest party is eventually released from the waiting-list. This wait-listing guarantees that corrupt parties will be barred from acting as summand-sharing parties as part of the Π_{OptMult} instances of future invocations of Π_{MultCI}, until they participate in the cheater-identification phase of previously failed instances of Π_{MultCI}. Since the cheater-identification phase is executed asynchronously, each party maintains its own set of *locally-discarded* parties, where corrupt parties are included as and when identified.

In Lemma 2 (see [1] for the proof), we say that an instance of Π_{MultCI} is *successful*, if $c_{(Z,\mathsf{curr})} - c_{(Z',\mathsf{curr})} = 0$ for all $Z \in \mathcal{Z}$ with respect to the publicly-known $Z' \in \mathcal{Z}$ fixed in the protocol, else the instance *fails*.

Lemma 2. Let $\mathcal{Z}$ satisfy the $\mathbb{Q}^{(4)}(\mathcal{P}, \mathcal{Z})$ condition and let all honest parties participate in $\Pi_{\mathsf{MultCI}}(\mathcal{P}, \mathcal{Z}, \mathbb{S}, [a], [b], \mathsf{iter})$. Then, Adv does not learn any additional information about a and b. Moreover, the following hold.

- The instance will eventually be deemed to succeed or fail by the honest parties, where for a successful instance, the parties output a sharing of ab.
- If the instance is not successful, then the honest parties will agree on a pair $Z, Z' \in \mathcal{Z}$ such that $c_{(Z,\mathsf{iter})} - c_{(Z',\mathsf{iter})} \neq 0$. Moreover, all honest parties present in the $\mathcal{W}^{(i)}_{\mathsf{iter}}$ set of any honest party P_i will eventually be removed and no honest party is ever included in the $\mathcal{LD}^{(i)}_{\mathsf{iter}}$ set of any honest P_i. Furthermore, there will be a pair of parties P_j, P_k from $\mathsf{Selected}_{(Z,\mathsf{iter})} \cup \mathsf{Selected}_{(Z',\mathsf{iter})}$, with at least one of them being maliciously-corrupt, such that if both P_j and P_k are removed from the set $\mathcal{W}^{(h)}_{\mathsf{iter}}$ of any honest party P_h, then eventually the corrupt party(ies) among P_j, P_k will be included in the set $\mathcal{LD}^{(i)}_{\mathsf{iter}}$ of every honest P_i.

- The protocol makes $\mathcal{O}(|\mathcal{Z}|n^2)$ calls to $\mathcal{F}_{\mathsf{VSS}}$ and $\mathcal{F}_{\mathsf{ABA}}$ and communicates $\mathcal{O}((|\mathcal{Z}|^2n^2 + |\mathcal{Z}|n^4)\log|\mathbb{F}|)$ bits.

Protocol Π_{MultCI} for M Pairs of Inputs: The parties now run instances of Π_{OptMult} with M pairs of inputs (see [1] for the details). The protocol will make $\mathcal{O}(M\cdot|\mathcal{Z}|\cdot n^2)$ calls to $\mathcal{F}_{\mathsf{VSS}}$, $\mathcal{O}(|\mathcal{Z}|\cdot n^2)$ calls to $\mathcal{F}_{\mathsf{ABA}}$ and incurs a communication of $\mathcal{O}((M\cdot|\mathcal{Z}|^2\cdot n^2 + |\mathcal{Z}|\cdot n^4)\log|\mathbb{F}|)$ bits.

3.3 Multiplication Protocol

Protocol Π_{Mult} (Fig. 7) takes inputs $[a], [b]$ and securely generates $[ab]$. The protocol proceeds in iterations, where in each iteration, an instance of Π_{MultCI} is invoked. If the iteration is successful, then the parties take the output of the corresponding Π_{MultCI} instance. Else, they proceed to the next iteration, with the cheater-identification phase of failed Π_{MultCI} instances running in the background. Let t be the cardinality of maximum-sized subset from $\mathcal{Z}$. To upper

Protocol $\Pi_{\mathsf{Mult}}(\mathcal{P}, \mathcal{Z}, \mathbb{S}, [a], [b])$

- **Initialization**: Set $t = \max\{|Z| : Z \in \mathcal{Z}\}$, initialize $\mathcal{GD} = \emptyset$ and $\mathsf{iter} = 1$.
- **Multiplication with Cheater Identification**: Participate in the instance $\Pi_{\mathsf{MultCI}}(\mathcal{P}, \mathcal{Z}, \mathbb{S}, [a], [b], \mathsf{iter})$ with sid.
  - **Positive Output**: If $\mathsf{flag}^{(i)}_{\mathsf{iter}}$ is set to 0, then output the shares obtained during the Π_{MultCI} instance.
  - **Negative Output**: If $\mathsf{flag}^{(i)}_{\mathsf{iter}}$ is set to 1, then proceed as follows.
    - **Identifying a Cheater Party Through ACS**: If $\mathsf{iter} = k\cdot(tn+1)$ for some $k \geq 1$, then do the following.
      1. Let $\mathcal{LD}^{(i)}_r$ be the set of locally-discarded parties for the instance $\Pi_{\mathsf{MultCI}}(\mathcal{P}, \mathcal{Z}, \mathbb{S}, [a], [b], r)$, for $r = 1, \dots, \mathsf{iter}$. For $j = 1, \dots, n$, send $(\mathsf{vote}, \mathsf{sid}_{j,\mathsf{iter},k}, 1)$ to $\mathcal{F}_{\mathsf{ABA}}$ where $\mathsf{sid}_{j,\mathsf{iter},k} = \mathsf{sid}||j||\mathsf{iter}||k$, if for any $r \in \{1, \dots, \mathsf{iter}\}$, party P_j is present in $\mathcal{LD}^{(i)}_r$ and $P_j \notin \mathcal{GD}$.
      2. For $j = 1, \dots, n$, keep requesting for an output from $\mathcal{F}_{\mathsf{ABA}}$ with $\mathsf{sid}_{j,\mathsf{iter},k}$, until an output is received.
      3. If $\exists P_j \in \mathcal{P}$ such that $(\mathsf{decide}, \mathsf{sid}_{j,\mathsf{iter},k}, 1)$ is received from $\mathcal{F}_{\mathsf{ABA}}$ with $\mathsf{sid}_{j,\mathsf{iter},k}$, then for each $P_\ell \in \mathcal{P}$, for which no vote message has been sent yet, send $(\mathsf{vote}, \mathsf{sid}_{\ell,\mathsf{iter},k}, 0)$ to $\mathcal{F}_{\mathsf{ABA}}$ with $\mathsf{sid}_{\ell,\mathsf{iter},k}$.
      4. Once an output $(\mathsf{decide}, \mathsf{sid}_{\ell,\mathsf{iter},k}, v_\ell)$ is received from $\mathcal{F}_{\mathsf{ABA}}$ with $\mathsf{sid}_{\ell,\mathsf{iter},k}$ for every $\ell \in \{1, \dots, n\}$, select the minimum indexed party P_j from $\mathcal{P}$, such that $v_j = 1$. Then set $\mathcal{GD} = \mathcal{GD} \cup \{P_j\}$, set $\mathsf{iter} = \mathsf{iter} + 1$ and go to the step labelled **Multiplication with Cheater Identification.**
    - Else set $\mathsf{iter} = \mathsf{iter} + 1$ and go to the step **Multiplication with Cheater Identification.**

Fig. 7. Multiplication protocol in the $(\mathcal{F}_{\mathsf{VSS}}, \mathcal{F}_{\mathsf{ABA}})$-hybrid for sid. The above code is executed by every party P_i

bound the number of failed iterations, the parties run ACS after every $tn + 1$ failed iterations to "globally" discard a new corrupt party. This is done through calls to $\mathcal{F}_{\mathsf{ABA}}$, where the parties vote for a candidate corrupt party, based on the $\mathcal{LD}$ sets of *all* failed iterations. The idea is that during these $tn + 1$ failed iterations, there will be at least one *corrupt* party who is eventually included in the $\mathcal{LD}$ set of *every* honest party. This is because there can be at most tn distinct pairs of "conflicting-parties" across the $tn + 1$ failed iterations (follows from Lemma 2). At least one conflicting pair, say (P_j, P_k), is guaranteed to repeat among the $tn + 1$ failed instances, with *both* P_j and P_k being removed from the previous waiting-lists. Thus, the corrupt party(ies) among P_j, P_k is eventually included to the $\mathcal{LD}$ sets. There can be at most $t(tn + 1)$ failed iterations after which *all* the corrupt parties will be discarded and the next iteration is guaranteed to be successful, with only *honest* parties acting as the candidate summand-sharing parties in the underlying instances of Π_{OptMult}.

Lemma 3. Let $\mathcal{Z}$ satisfy the $\mathbb{Q}^{(4)}(\mathcal{P}, \mathcal{Z})$ condition and let $\mathbb{S} = \{\mathcal{P} \setminus Z | Z \in \mathcal{Z}\}$. Then Π_{Mult} takes at most $t(tn + 1)$ iterations and all honest parties eventually output a secret-sharing of $[ab]$, where $t = \max\{|Z| : Z \in \mathcal{Z}\}$. In the protocol, Adv does not learn anything additional about a and b. The protocol makes $\mathcal{O}(|\mathcal{Z}| \cdot n^5)$ calls to $\mathcal{F}_{\mathsf{VSS}}$ and $\mathcal{F}_{\mathsf{ABA}}$ and additionally incurs a communication of $\mathcal{O}(|\mathcal{Z}|^2 \cdot n^5 \log |\mathbb{F}| + |\mathcal{Z}| \cdot n^7 \log |\mathbb{F}|)$ bits (see [1] for the proof).

Protocol Π_{Mult} for M Pairs of Inputs: To handle M pairs of inputs, the instances of Π_{MultCI} are now executed with M pairs of inputs in each iteration. This requires $\mathcal{O}(M \cdot |\mathcal{Z}| \cdot n^5)$ calls to $\mathcal{F}_{\mathsf{VSS}}$, $\mathcal{O}(|\mathcal{Z}| \cdot n^5)$ calls to $\mathcal{F}_{\mathsf{ABA}}$ and communication of $\mathcal{O}((M \cdot |\mathcal{Z}|^2 \cdot n^5 + |\mathcal{Z}| \cdot n^7) \log |\mathbb{F}|)$ bits.

3.4 The Pre-processing Phase Protocol

In protocol $\Pi_{\mathsf{PerTriples}}$, the parties first jointly generate secret-sharing of M random pairs of values, followed by running an instance of Π_{Mult} to securely compute the product of these pairs; see [1] for the full details and proof of Theorem 2.

Theorem 2. If $\mathcal{Z}$ satisfies $\mathbb{Q}^{(4)}(\mathcal{P}, \mathcal{Z})$ condition, then $\Pi_{\mathsf{PerTriples}}$ is a perfectly-secure protocol for realizing $\mathcal{F}_{\mathsf{Triples}}$ with UC-security in the $(\mathcal{F}_{\mathsf{VSS}}, \mathcal{F}_{\mathsf{ABA}})$-hybrid model. The protocol makes $\mathcal{O}(M \cdot |\mathcal{Z}| \cdot n^5)$ calls to $\mathcal{F}_{\mathsf{VSS}}$, $\mathcal{O}(|\mathcal{Z}| \cdot n^5)$ calls to $\mathcal{F}_{\mathsf{ABA}}$ and incurs a communication of $\mathcal{O}(M \cdot |\mathcal{Z}|^2 \cdot n^5 \log |\mathbb{F}| + |\mathcal{Z}| \cdot n^7 \log |\mathbb{F}|)$ bits.

4 MPC Protocol in the Pre-processing Model

In the MPC protocol Π_{AMPC}, the parties first generate secret-shared random multiplication-triples through $\mathcal{F}_{\mathsf{Triples}}$. Each party then randomly secret-shares its input for ckt through $\mathcal{F}_{\mathsf{VSS}}$. To avoid an indefinite wait, the parties agree on a common subset of parties, whose inputs are eventually secret-shared, through ACS. The parties then jointly evaluate each gate in ckt in a secret-shared fashion by generating a secret-sharing of the gate-output from a secret-sharing of the gate-input(s). Linear gates are evaluated non-interactively due to the linearity of secret-sharing. To evaluate multiplication gates, the parties deploy Beaver's method [2], using the secret-shared multiplication-triples generated by $\mathcal{F}_{\mathsf{Triples}}$. Finally, the parties publicly reconstruct the secret-shared function output (see [1] for the description of the protocol and proof of Theorem 3).

Theorem 3. Protocol Π_{AMPC} UC-securely realizes the functionality $\mathcal{F}_{\mathsf{AMPC}}$ for securely computing f with perfect security in the $(\mathcal{F}_{\mathsf{Triples}}, \mathcal{F}_{\mathsf{VSS}}, \mathcal{F}_{\mathsf{ABA}})$-hybrid model, in the presence of a static malicious adversary characterized by an adversary-structure $\mathcal{Z}$ satisfying the $\mathbb{Q}^{(3)}(\mathcal{P}, \mathcal{Z})$ condition. The protocol makes one call to $\mathcal{F}_{\mathsf{Triples}}$ and $\mathcal{O}(n)$ calls to $\mathcal{F}_{\mathsf{VSS}}$ and $\mathcal{F}_{\mathsf{ABA}}$ and additionally incurs a communication of $\mathcal{O}(M \cdot |\mathcal{Z}| \cdot n^2 \log |\mathbb{F}|)$ bits, where M is the number of multiplication gates in the circuit ckt representing f.

By replacing the calls to $\mathcal{F}_{\mathsf{Triples}}$ and $\mathcal{F}_{\mathsf{VSS}}$ in Π_{AMPC} with perfectly-secure protocol $\Pi_{\mathsf{PerTriples}}$ and Π_{PVSS} respectively and by replacing the calls to $\mathcal{F}_{\mathsf{ABA}}$ in Π_{AMPC} with the almost-surely terminating ABA protocol Π_{ABA} of [10], we get the following corollary of Theorem 3.

Corollary 3. *If $\mathcal{Z}$ satisfies the $\mathbb{Q}^{(4)}(\mathcal{P}, \mathcal{Z})$ condition, then Π_{AMPC} UC-securely realizes $\mathcal{F}_{\mathsf{AMPC}}$ in plain model. The protocol incurs a total communication of $\mathcal{O}(M \cdot (|\mathcal{Z}|^2 \cdot n^7 \log |\mathbb{F}| + |\mathcal{Z}| \cdot n^9 \log n) + |\mathcal{Z}|^2 (n^{11} \log |\mathbb{F}| + n^{13}(\log n + \log |\mathcal{Z}|)))$ bits, where M is the number of multiplication gates in* ckt.

References

1. Appan, A., Chandramouli, A., Choudhury, A.: Revisiting the efficiency of asynchronous multi party computation against general adversaries. IACR Cryptology ePrint Archive, p. 651 (2022)
2. Beaver, D.: Efficient multiparty protocols using circuit randomization. In: Feigenbaum, J. (ed.) CRYPTO 1991. LNCS, vol. 576, pp. 420–432. Springer, Heidelberg (1992). https://doi.org/10.1007/3-540-46766-1_34
3. Beerliová-Trubíniová, Z., Hirt, M.: Simple and efficient perfectly-secure asynchronous MPC. In: Kurosawa, K. (ed.) ASIACRYPT 2007. LNCS, vol. 4833, pp. 376–392. Springer, Heidelberg (2007). https://doi.org/10.1007/978-3-540-76900-2_23
4. Ben-Or, M., Canetti, R., Goldreich, O.: Asynchronous secure computation. In: STOC, pp. 52–61. ACM (1993). https://doi.org/10.1145/167088.167109
5. Ben-Or, M., Goldwasser, S., Wigderson, A.: Completeness theorems for non-cryptographic fault-tolerant distributed computation (extended abstract). In: STOC, pp. 1–10. ACM (1988). https://doi.org/10.1145/62212.62213
6. Ben-Or, M., Kelmer, B., Rabin, T.: Asynchronous secure computations with optimal resilience (extended abstract). In: PODC, pp. 183–192. ACM (1994). https://doi.org/10.1145/197917.198088
7. Canetti, R.: Studies in secure multiparty computation and applications. Ph.D. thesis, Weizmann Institute, Israel (1995)
8. Canetti, R.: Universally composable security: a new paradigm for cryptographic protocols. In: FOCS, pp. 136–145. IEEE Computer Society (2001). https://doi.org/10.1109/SFCS.2001.959888

9. Canetti, R.: Universally composable security. J. ACM **67**(5), 28:1–28:94 (2020). https://doi.org/10.1145/3402457
10. Choudhury, A., Pappu, N.: Perfectly-secure asynchronous MPC for general adversaries (extended abstract). In: Bhargavan, K., Oswald, E., Prabhakaran, M. (eds.) INDOCRYPT 2020. LNCS, vol. 12578, pp. 786–809. Springer, Cham (2020). https://doi.org/10.1007/978-3-030-65277-7_35
11. Choudhury, A., Patra, A.: An efficient framework for unconditionally secure multiparty computation. IEEE Trans. Inf. Theory **63**(1), 428–468 (2017). https://doi.org/10.1109/TIT.2016.2614685
12. Cohen, R.: Asynchronous secure multiparty computation in constant time. In: Cheng, C.-M., Chung, K.-M., Persiano, G., Yang, B.-Y. (eds.) PKC 2016. LNCS, vol. 9615, pp. 183–207. Springer, Heidelberg (2016). https://doi.org/10.1007/978-3-662-49387-8_8
13. Coretti, S., Garay, J., Hirt, M., Zikas, V.: Constant-round asynchronous multiparty computation based on one-way functions. In: Cheon, J.H., Takagi, T. (eds.) ASIACRYPT 2016. LNCS, vol. 10032, pp. 998–1021. Springer, Heidelberg (2016). https://doi.org/10.1007/978-3-662-53890-6_33
14. Fischer, M.J., Lynch, N.A., Paterson, M.: Impossibility of Distributed Consensus with One Faulty Process. J. ACM **32**(2), 374–382 (1985). https://doi.org/10.1145/3149.214121
15. Fitzi, M., Maurer, U.: Efficient Byzantine agreement secure against general adversaries. In: Kutten, S. (ed.) DISC 1998. LNCS, vol. 1499, pp. 134–148. Springer, Heidelberg (1998). https://doi.org/10.1007/BFb0056479
16. Goldreich, O.: The Foundations of Cryptography - Volume 2, Basic Applications. Cambridge University Press, Cambridge (2004). https://doi.org/10.5555/1804390
17. Goldreich, O., Micali, S., Wigderson, A.: How to play any mental game or a completeness theorem for protocols with honest majority. In: STOC, pp. 218–229. ACM (1987). https://doi.org/10.1145/28395.28420
18. Hirt, M., Maurer, U.: Player simulation and general adversary structures in perfect multiparty computation. J. Cryptol. **13**(1), 31–60 (2000). https://doi.org/10.1007/s001459910003
19. Hirt, M., Tschudi, D.: Efficient general-adversary multi-party computation. In: Sako, K., Sarkar, P. (eds.) ASIACRYPT 2013. LNCS, vol. 8270, pp. 181–200. Springer, Heidelberg (2013). https://doi.org/10.1007/978-3-642-42045-0_10
20. Ito, M., Saito, A., Nishizeki, T.: Secret sharing schemes realizing general access structures). In: Global Telecommunication Conference, Globecom, pp. 99–102. IEEE Computer Society (1987). https://doi.org/10.1002/ecjc.4430720906
21. Katz, J., Maurer, U., Tackmann, B., Zikas, V.: Universally composable synchronous computation. In: Sahai, A. (ed.) TCC 2013. LNCS, vol. 7785, pp. 477–498. Springer, Heidelberg (2013). https://doi.org/10.1007/978-3-642-36594-2_27
22. Kumar, M.V.N.A., Srinathan, K., Rangan, C.P.: Asynchronous perfectly secure computation tolerating generalized adversaries. In: Batten, L., Seberry, J. (eds.) ACISP 2002. LNCS, vol. 2384, pp. 497–511. Springer, Heidelberg (2002). https://doi.org/10.1007/3-540-45450-0_37
23. Kursawe, K., Freiling, F.C.: Byzantine fault tolerance on general hybrid adversary structures. Technical report, RWTH Aachen (2005)
24. Maurer, U.: Secure multi-party computation made simple. In: Cimato, S., Persiano, G., Galdi, C. (eds.) SCN 2002. LNCS, vol. 2576, pp. 14–28. Springer, Heidelberg (2003). https://doi.org/10.1007/3-540-36413-7_2

25. Patra, A., Choudhury, A., Pandu Rangan, C.: Efficient asynchronous verifiable secret sharing and multiparty computation. J. Cryptol. **28**(1), 49–109 (2013). https://doi.org/10.1007/s00145-013-9172-7
26. Rabin, T., Ben-Or, M.: Verifiable secret sharing and multiparty protocols with honest majority (extended abstract). In: STOC, pp. 73–85. ACM (1989). https://doi.org/10.1145/73007.73014
27. Shamir, A.: How to share a secret. Commun. ACM **22**(11), 612–613 (1979). https://doi.org/10.1145/359168.359176
28. Yao, A.C.: Protocols for secure computations (extended abstract). In: FOCS, pp. 160–164. IEEE Computer Society (1982). https://doi.org/10.1109/SFCS.1982.38

Protego: Efficient, Revocable and Auditable Anonymous Credentials with Applications to Hyperledger Fabric

Aisling Connolly[1], Jérôme Deschamps[2], Pascal Lafourcade[2], and Octavio Perez Kempner[3,4](✉)

[1] DFINITY, Zürich, Switzerland
[2] LIMOS, University Clermont Auvergne, Aubière, France
jerome.deschamps@etu.uca.fr, pascal.lafourcade@uca.fr
[3] DIENS, École normale supérieure, CNRS, PSL University, Paris, France
octavio.perez.kempner@ens.fr
[4] be-ys Research, Clermont-Ferrand, France

Abstract. Recent works to improve privacy in permissioned blockchains like Hyperledger Fabric rely on Idemix, the only anonymous credential system that has been integrated to date. The current Idemix implementation in Hyperledger Fabric (v2.4) only supports a fixed set of attributes; it does not support revocation features, nor does it support anonymous endorsement of transactions (in Fabric, transactions need to be approved by a subset of peers before consensus). A prototype Idemix extension by Bogatov *et al.* (CANS, 2021) was proposed to include revocation, auditability, and to gain privacy for users. In this work, we explore how to gain efficiency, functionality, and further privacy, departing from recent works on anonymous credentials based on Structure-Preserving Signatures on Equivalence Classes. As a result, we extend previous works to build a new anonymous credential scheme called Protego. We also present a variant of it (Protego Duo) based on a different approach to hiding the identity of an issuer during showings. We also discuss how both can be integrated into Hyperledger Fabric and provide a prototype implementation. Finally, our results show that Protego and Protego Duo are at least twice as fast as state-of-the-art approaches based on Idemix.

Keywords: Anonymous credentials · Auditability · Hyperledger fabric · Mercurial signatures · Permissioned blockchains

1 Introduction

When first introduced, the core use of blockchains was in the *permissionless* setting; anyone could join and participate. Over the years, blockchains have

A. Connolly—Work done while the author was at Wordline Global.

T. Isobe and S. Sarkar (Eds.): INDOCRYPT 2022, LNCS 13774, pp. 249–271, 2022.
https://doi.org/10.1007/978-3-031-22912-1_11

also found use within consortiums, where several authorized organizations wish to share information among the group, but not necessarily to the public as a whole. This need gave rise to *permissioned* blockchains whereby authorities are established to define a set of participants. When a federation of authorities (*consortium*), each in control of a subset of participants, shares the blockchain's governance, the term *federated* is also used to describe such blockchains.

The use of federated blockchains increased to address the need to run a common business logic within a closed environment. As an example, one can consider pharmaceutical companies that would like to trade sensitive information about product developments and agree on supplies or prices in a consortium with partial trust. A recurrent problem in such scenarios is that of privacy while being compliant with regulations and *Know Your Customer* practices. Agreeing with other entities to run a shared business logic should not imply that everything needs be public within the consortium. Privacy still needs to be provided without affecting existing regulations, *e.g.*, when considering bilateral agreements.

The most developed permissioned platform is Hyperledger Fabric (or simply Fabric). In Fabric, users submit transaction proposals to a subset of peers (called *endorsers*) that vouch for their execution. By default, it provides no privacy features as everything (users and transactions) is public. Reading the blockchain anyone can know (1) who triggered a smart contract using which arguments (transaction proposals are signed by the clients), (2) who vouched for its execution (endorsers also sign their responses) concerning reading and writing sets; and (3) why a given transaction was marked as invalid (either because of invalid read/write sets or because the endorsement policy check failed). Furthermore, checking access control and endorsement policies links different organizations, users and their attributes to concrete actions on the system.

Such limitations severely restrict the use of Fabric. From the user perspective this impacts the enforcement of different regulations. For organizations, the case is similar. Consider a consortium of pharmaceuticals that run a common business logic to exchange information on medical research. If the entity behind a request is known, other organizations can infer (based on the request) which drug the entity in question is trying to develop. If the endorsers are known, information about who executes what can disclose business relations.

Motivated by the need to protect business interests and to meet regulatory requirements, some privacy features were integrated using the Identity Mixer [11,21] (or Idemix for short). This anonymous attribute-based credential (ABC) scheme gave the first glimpse of privacy for users within a consortium. Idemix allows a Membership Service Provider (MSP) to issue credentials enabling users to sign transactions anonymously. In brief, users generate a zero-knowledge proof attesting that the MSP issued them a credential on its attributes to sign a transaction. Fabric's support for Idemix was added in v1.3, providing the first solution to tackle the problem of *participant* privacy. Unfortunately, as for v2.4 the Idemix implementation still suffers severe limitations:

1. It supports a fixed set of only four attributes.
2. It does not support revocation features.

3. Credentials leak the MSP ID, meaning that anonymity is local to users within an organization. For this reason, current deployments can only use a single MSP for the whole network, introducing a single point of failure.
4. It does not support the issuance of Idemix credentials for the endorsing peers, meaning that the identity of endorsers is always leaked.

The most promising effort to extend the functionality of Idemix appeared in [7]. Their aim was to extend the original credential system to support *delegatable credentials* [9], while integrating revocation and auditability features (solving three of the four limitations). Below we outline the main ideas introduced in [7].

Delegatable Credentials. In a bid to overcome the issue of Idemix credentials leaking the MSP ID and thus the affiliation of the user, a trusted root authority provides credentials to intermediate authorities. This way users can obtain credentials from intermediate authorities. To sign a transaction, the user must generate a zero-knowledge proof attesting that (1) the signer owns the credential; (2) the signature is valid; (3) all adjacent delegation levels are legitimate; and (4) that the top-level public key belongs to the root authority.

Revocation and Auditability. To generate efficient proofs of non-revocation, the system timeline is divided into *epochs*. Issued credentials are only valid for a given epoch, and must be reissued as the timeline advances. For each epoch, a user requests a revocation handle that binds their public key to the epoch. When presenting a credential, the user also provides a proof of non-revocation. To enable auditing of a transaction, users verifiably encrypt their public key under an authorized auditor's public key.

To date, some functionalities remain limited. (1) There is still no notion of privacy for endorsers. (2) Delegatable credentials require proving knowledge of a list of keys. (3) The root authority is still a single point of failure. (4) Selective disclosure of attributes requires computation linear in the size of all the attributes encoded in the credential. (5) Many zero-knowledge proofs need to be generated for each transaction. (6) Many pairings need to be computed for verification.

Recent results [12,14,15] introduced newer models based on Structure-Preserving Signatures on Equivalence Classes (SPS-EQ) to build ABC's, providing a host of extra functionalities and more efficient constructions. The main goal of this work is to leverage such results, position them in the blockchain scenario and provide an alternative to Idemix (and its extension) to overcome existing privacy and functional limitations while also improving efficiency.

Contributions. To build an ABC scheme that overcomes the inherent limitations from Idemix and its extension, we argue that changing some of the underlying building blocks is necessary. Therefore, we take the framework from [15] as our starting point, incorporate the recent improvements from [12], and include the revocation extension originally proposed in [14]. From there, we extend the ABC model to support auditability features and adapt it to non-interactive showings. To do so, we rely on the random oracle model (already present in the blockchain setting). We also present and discuss two alternatives to the use of delegatable credentials (as used in [7]) to hide the identity of credential issuers, following the formalizations from [12] and [6] but using new approaches.

When compared to [12], the modifications are: (1) we adapt the ABC model to non-interactive showings, (2) we extend the model defining a revocation authority as in [14], and an auditing authority (not considered in the previous works), (3) we keep the SCDS scheme from [12] as it is but replace the signature scheme with the one given in [13], and (4) we build a *malleable* NIZK argument that can be pre-computed to obtain a more efficient issuer-hiding feature.

As a result, we build Protego, an ABC suitable for permissioned blockchains. We also present Protego Duo, a variant based on a different approach to hide the identity of credential issuers. Both support revocation and auditing features, which are important to enable a broader variety of use cases for permissioned blockchains. We discuss how to integrate our work with Fabric, compare it with Idemix and its recent extensions, and provide a prototype implementation showing that Protego and Protego Duo are faster than the most recent Idemix extension (see Sect. 5 for a detailed evaluation and benchmarks). Furthermore, a showing proof in Protego Duo is constant-size (8.3 kB), surpassing [7] in which the proof size grows linearly with the number of *attributes and delegation levels*.

Related Work. We describe the related work following two main streams; the results addressing privacy concerns in Fabric, and parallel research developments.

Privacy Concerns in Fabric. The most closely related work appears with the introduction of Idemix [11,21] and its extension to include revocation and auditability [7]. Adding auditability is crucial for permissioned blockchains as they are often used in heavily regulated industries. Privacy-preserving auditing for distributed ledgers was introduced in [18] under the guise of zkLedger. This general solution offered great functionality in that it provided confidentiality of transactions, and privacy of the users within the transaction. However, it assumed low transaction volume between few participants and as such is quite limited in scalability. Fabric-friendly auditable token payments were introduced in [2] and were based on threshold blind signatures. The core idea to achieve auditability was to encrypt the user's public key under the public key of an auditor. This is the same approach in [7], which we also use in this work. Although the auditing ideas are similar, the construction pertains solely to transaction privacy and offers no identity privacy for a user. Following the approach of gaining auditability of transactions, auditable smart contracts were captured by FabZK [16] which is based on Pedersen commitments and zero-knowledge proofs. To achieve auditability, the structure of the ledger is modified, and as such, would need to make considerable changes to existing used permissioned blockchain platforms.

One of the limitations in Idemix and its extension is the lack of privacy or anonymity for endorsing peers. A potential solution to this was proposed in [17], where the endorsement policy is based on a ring signature scheme such that the endorsement set itself is not revealed, but only that sufficiently many endorsement signatures were obtained. Another approach to obtain privacy-preserving endorsements was described in [3], leveraging Idemix credentials to gain endorser-privacy, and as such, inherits the limitations (notably leaking the endorser's organization) that come with Idemix.

Attribute-Based Credentials. Early anonymous credential schemes were built from blind signatures, whereby a user obtained a blind signature from an issuer on the user's commitment to its attributes. When the user later authenticates, they provide the signature, the shown attributes, and a proof of knowledge of all unshown attributes. These schemes are limited as they can only be shown once. Subsequent work like the one underlying Idemix [10] allowed for an arbitrary number of unlinkable showings. A user obtains a signature on a commitment on attributes, randomizes the signature, and proves in zero-knowledge that the randomized signature corresponds to the shown and unshown attributes.

Recent work from [15] circumvented inefficiencies in the above ideas by coining two new primitives: set-commitment schemes, and SPS-EQ. As a result, authors obtained a scheme allowing to randomize both the signature and the commitment on a set of attributes. Furthermore, a subset-opening of the set-commitment yielded constant-size selective showing of attributes.

New work from [12] extended [15], improving the expressivity, efficiency trade-offs and introducing the notion of *signer-hiding* (also known as *issuer-hiding* [6]) to allow users to easily randomize the public key used to generate a signature to hide the identity of credential issuers. Authors achieve the previous points using a *Set-Commitment scheme supporting Disjoint Sets* (SCDS) and mercurial signatures [13]. The latter primitive extends SPS-EQ to consider equivalence classes not only on the message space but also on the key space.

We build on top of the above-mentioned works but unlike [12], we work with the generic group model [20] as our main motivation is the proposal of efficient alternatives. For this reason, we use the mercurial signature scheme from [13].

2 Cryptographic Background

Below, we walk through the different building blocks mentioning how and why these components yield greater functionality and efficiency for a credential system in the permissioned blockchain setting like Fabric. Subsequently, we introduce the necessary notation and syntax for the following sections.

SCDS. Using commitment schemes that allow to commit to *sets* of attributes enables constant-size openings of subsets (selective disclosure) of the committed sets. These schemes support commitment randomization without the need to rely on zero-knowledge proofs of correct randomization, as the corresponding witness for openings can be adapted accordingly with respect to the randomization of the committed set. The set-commitment scheme presented in [12] extends [15] to support openings of attribute sets *disjoint* from the committed set. This is particularly useful in the permissioned blockchain setting, *e.g.*, to model access control policies. Furthermore, the scheme from [12] also supports the use of proof of exponentiations to outsource some of the computational cost from the verifier to the prover. In the case of Fabric, this is a particularly interesting feature to make the endorser's verification faster when validating a transaction proposal.

Mercurial Signatures. The introduction of SPS-EQ in [15] allowed to adapt a signature on a representative message to a signature on a different representative (in a given equivalence class) without knowledge of the secret key. If the adapted signature is indistinguishable from a fresh signature on a random message, the scheme satisfies the notion of *perfect adaption*. This, together with the randomizability of the set-commitment scheme, allows to consistently and efficiently update the signature of a credential, bypassing the need to generate and keep account of pseudonyms and NIZK proofs that are required in all previous works based on Idemix. Using mercurial signatures as in [12] allows to easily randomize the corresponding public keys while consistently adapting the signatures.

Issuer-Hiding. In [12], since users can consistently randomize the signature on their credential and the issuer's public key (as previously mentioned), a fully adaptive NIZK argument is used to prove that a randomized issuer key belongs to the equivalence class of one of the keys contained in a list of issuers keys. This way, the randomized issuer key can be used to verify the credential while hiding the issuer's identity (like in a ring signature). In permissioned blockchains where there are multiple organizations that issue credentials, such a NIZK allows users holding valid signatures to pick any subset of issuer's public keys to generate a proof. Another approach following the work from [6] (briefly discussed in [12]) is to consider issuer-policies. An issuer-policy is a set $\{(\sigma_i, \mathsf{opk}_i)_{i\in[n]}\}$ of signatures on issuer's public keys generated by some verification secret key vsk. To hide the identity of an issuer j, a user consistently randomizes the pair $(\sigma_j, \mathsf{opk}_j)$ to obtain a randomized public key opk'_j. It then adapts the signature σ on its credential the same way, and presents opk'_j to the verifier. If the verifier accepts the signature σ_j on opk'_j (using vpk), it proceeds to verify σ using opk'_j. Issuer-policies can be specified by the entity that created the smart contract and defined within using the entity's verification key pair. Unlike the first approach where users choose the issuer's anonymity set, here it is determined by the policy maker.

2.1 Notation

Let BGGen be a p.p.t algorithm that on input 1^λ with λ the security parameter, returns a description $\mathsf{BG} = (p, \mathbb{G}_1, \mathbb{G}_2, \mathbb{G}_T, P_1, P_2, e)$ of an asymmetric (Type-3) bilinear group where $\mathbb{G}_1, \mathbb{G}_2, \mathbb{G}_T$ are cyclic groups of prime order p with $\lceil \log_2 p \rceil = \lambda$, P_1 and P_2 are generators of $\mathbb{G}_1$ and $\mathbb{G}_2$, and $e : \mathbb{G}_1 \times \mathbb{G}_2 \to \mathbb{G}_T$ is an efficiently computable (non-degenerate) bilinear map. For all $a \in \mathbb{Z}_p$, $[a]_s = aP_s \in \mathbb{G}_s$ denotes the implicit representation of a in $\mathbb{G}_s$ for $s \in \{1, 2\}$. For vectors $\mathbf{a}, \mathbf{b}$ we extend the pairing notation to $e([\mathbf{a}]_1, [\mathbf{b}]_2) := [\mathbf{ab}]_T \in \mathbb{G}_T$. $r \xleftarrow{\$} \mathcal{S}$ denotes sampling r from set $\mathcal{S}$ uniformly at random. $\mathcal{A}(x; y)$ indicates that y is passed directly to $\mathcal{A}$ on input x. Hash functions are denoted by $\mathcal{H}$.

2.2 Set-Commitment Scheme Supporting Disjoint Sets [12]

Syntax. A *set-commitment scheme supporting disjoint sets* (SCDS) consists of the following p.p.t algorithms:

$\mathsf{Setup}(1^\lambda, 1^q)$ is a probabilistic algorithm which takes as input a security parameter λ and an upper bound q for the cardinality of committed sets, both in unary form. It outputs public parameters pp (including an evaluation key ek), and discards the trapdoor key s used to generate them. $\mathbb{Z}_p^* \setminus \{s\}$ defines the domain of set elements for sets of maximum cardinality q.

$\mathsf{TSetup}(1^\lambda, 1^q)$ is equivalent to Setup but also returns the trapdoor key s.

$\mathsf{Commit}(\mathsf{pp}, \mathcal{X})$ is a probabilistic algorithm which takes as input pp and a set $\mathcal{X}$ with $1 \leq |\mathcal{X}| \leq q$. It outputs a commitment C on set $\mathcal{X}$ and opening information O.

$\mathsf{Open}(\mathsf{pp}, C, \mathcal{X}, O)$ is a deterministic algorithm which takes as input pp, a commitment C, a set $\mathcal{X}$, and opening information O. It outputs 1 if and only if O is a valid opening of C on $\mathcal{X}$.

$\mathsf{OpenSS}(\mathsf{pp}, C, \mathcal{X}, O, \mathcal{S})$ is a deterministic algorithm which takes as input pp, a commitment C, a set $\mathcal{X}$, opening information O, and a non-empty set $\mathcal{S}$. If $\mathcal{S}$ is a subset of $\mathcal{X}$ committed to in C, OpenSS outputs a witness wss that attests to it. Otherwise, outputs $\perp$.

$\mathsf{OpenDS}(\mathsf{pp}, C, \mathcal{X}, O, \mathcal{D})$ is a deterministic algorithm which takes as input pp, a commitment C, a set $\mathcal{X}$, opening information O, and a non-empty set $\mathcal{D}$. If $\mathcal{D}$ is disjoint from $\mathcal{X}$ committed to in C, OpenDS outputs a witness wds that attests to it. Otherwise, outputs $\perp$.

$\mathsf{VerifySS}(\mathsf{pp}, C, \mathcal{S}, \mathsf{wss})$ is a deterministic algorithm which takes as input pp, a commitment C, a non-empty set $\mathcal{S}$, and a witness wss. If wss is a valid witness for $\mathcal{S}$ a subset of the set committed to in C, it outputs 1 and otherwise $\perp$.

$\mathsf{VerifyDS}(\mathsf{pp}, C, \mathcal{D}, \mathsf{wds})$ takes as input pp, a commitment C, a non-empty set $\mathcal{D}$, and a witness wss. If wds is a valid witness for $\mathcal{D}$ being disjoint from the set committed to in C, it outputs 1 and otherwise $\perp$.

$\mathsf{PoE}(\mathsf{pp}, \mathcal{X}, \alpha)$ takes as input pp, a non-empty set $\mathcal{X}$, and a randomly-chosen value α. It computes a proof of exponentiation for the characteristic polynomial of $\mathcal{X}$ and outputs a proof π_Q and a witness Q.

Security Properties. Correctness requires that (1) for a set $\mathcal{X}$, $\mathsf{Open}(\mathsf{Commit}\ (\mathsf{pp},\mathcal{X}),\mathcal{X}) = 1$, (2) for $\mathcal{S}$ a subset of $\mathcal{X}$, $\mathsf{VerifySS}(\mathcal{S}, \mathsf{OpenSS}(\mathsf{Commit}(\mathsf{pp}, \mathcal{X}))) = 1$, and (3) for all possible sets $\mathcal{D}$ disjoint from $\mathcal{X}$, $\mathsf{VerifyDS}(\mathcal{D}, \mathsf{OpenDS}(\mathsf{Commit}(\mathsf{pp}, \mathcal{X}))) = 1$. The scheme should also be (1) *binding* whereby each commitment overwhelmingly pertains to one particular set of attributes, (2) *hiding* whereby an adversary, given access to opening oracles, should not be able to distinguish which of two sets a commitment was generated on, and (3) *sound* in that sets which are not subsets of the committed set do not verify under $\mathsf{VerifySS}$, and sets that are not disjoint from the committed set do not verify under $\mathsf{VerifyDS}$.

2.3 Structure-Preserving Signatures on Equivalence Classes [12]

Syntax. A Structure-Preserving Signature on Equivalence Classes (SPS-EQ) consists of the following algorithms:

$\mathsf{ParGen}(1^\lambda)$ is a p.p.t algorithm that, given a security parameter λ, outputs public parameters pp.

$\mathsf{TParGen}(1^\lambda)$ is like ParGen but it also returns a trapdoor td (if any).

$\mathsf{KGen}(\mathsf{pp}, \ell)$ is a p.p.t algorithm that, given pp and a vector length $\ell > 1$, outputs a key pair $(\mathsf{sk}, \mathsf{pk})$.

$\mathsf{Sign}(\mathsf{pp}, \mathsf{sk}, \mathbf{m})$ is a p.p.t algorithm that, given pp, a representative $\mathbf{m} \in (\mathbb{G}_i^*)^\ell$ for class $[\mathbf{m}]_\mathcal{R}$, a secret key sk, outputs a signature $\sigma' = (\sigma, \tau)$ (potentially including a tag τ) on the message $\mathbf{m}$.

$\mathsf{ChgRep}(\mathsf{pp}, \mathbf{m}, (\sigma, \tau), \mu, \mathsf{pk})$ is a p.p.t algorithm that takes as input pp, a representative message $\mathbf{m} \in (\mathbb{G}_i^*)^\ell$, a signature σ (and potentially a tag τ), a scalar μ, and a public key pk. It computes an updated signature σ' on new representative $\mathbf{m}^* = \mu\mathbf{m}$ and outputs $(\mathbf{m}^*, \sigma')$.

$\mathsf{Verify}(\mathsf{pp}, \mathbf{m}, (\sigma, \tau), \mathsf{pk})$ is a deterministic algorithm that takes as input pp, a representative message $\mathbf{m}$, a signature σ (potentially including a tag τ), and public key pk. If σ is a valid signature on $\mathbf{m}$ it outputs 1 and 0 otherwise.

For mercurial signatures, the algorithms $\mathsf{ConvertPK}(\mathsf{pk}, \rho)$ and $\mathsf{ConvertSK}(\mathsf{sk}, \rho)$ are included to compute new representatives for public and secret keys. ChgRep is extended to adapt signatures with respect to new key representatives as well.

Security Properties. SPS-EQ should be correct, existentially unforgeable against chosen-message attacks and have perfect adaption (in the vein of [12]).

3 Our ABC Model

We can rely on the random oracle model and apply the Fiat-Shamir transform to the ABC scheme from [12] (the showing protocol is a three move public coin one). However, in the previous ABC, interaction is required in the showing protocol to provide freshness (*i.e.*, to avoid replay attacks). To overcome this issue, we require the user to send the transaction proposal during the first move. Thus, applying the Fiat-Shamir transform to the first move bounds the credential showing to that particular transaction so that it cannot be replayed. Security is defined following the usual properties from [12,14,15]. In addition, we also consider the issuer-hiding notion from [12] and introduce a new one for auditability. However, we do not consider replay-attacks as in the previous models since they can be trivially detected for the same transaction.

ABC Syntax. An ABC consists of the following p.p.t algorithms:

$\mathsf{Setup}(1^\lambda, \mathsf{aux})$ takes a security parameter λ and some optional auxiliary information aux (which may fix an universe of attributes, attribute values and other parameters) and outputs public parameters pp discarding any trapdoor.

$\mathsf{TSetup}(1^\lambda, \mathsf{aux})$ like Setup but returns a trapdoor.

$\mathsf{OKGen}(\mathsf{pp})$ takes pp and outputs an organization key pair $(\mathsf{osk}, \mathsf{opk})$.

$\mathsf{UKGen}(\mathsf{pp})$ takes pp and outputs a user key pair $(\mathsf{usk}, \mathsf{upk})$.

$\mathsf{AAKGen}(\mathsf{pp})$ takes pp and outputs an auditor key pair $(\mathsf{ask}, \mathsf{apk})$.

$\mathsf{RAKGen}(\mathsf{pp})$ takes pp and outputs a revocation key pair $(\mathsf{rsk}, \mathsf{rpk})$.

$\mathsf{Obtain}(\mathsf{pp}, \mathsf{usk}, \mathsf{opk}, \mathsf{apk}, \mathcal{X}, \mathsf{nym})$and $\mathsf{Issue}(\mathsf{pp},\mathsf{upk},\mathsf{osk},\mathsf{apk},\mathcal{X},\mathsf{nym})$ are run by a user and the organization respectively, who interact during execution. Obtain takes pp, the user's secret key usk, an organization's public key opk, an auditor's public key apk, an attribute set $\mathcal{X}$ of size $|\mathcal{X}| < q$, and a pseudonym nym used for revocation. Issue takes pp, a public key upk, a secret key osk, an auditor's public key apk, an attribute set $\mathcal{X}$ of size $|\mathcal{X}| < q$, and a pseudonym nym. At the end of this protocol, Obtain outputs a credential cred on $\mathcal{X}$ for the user or $\perp$ if the execution failed.

$\mathsf{Show}(\mathsf{pp}, \mathsf{opk}, \mathsf{upk}, \mathsf{usk}, \mathsf{cred}, \mathcal{X}, \mathcal{S}, \mathcal{D}, \mathsf{aux})$ takes pp, a public key opk, a key pair (usk, upk), a credential cred for the attribute set $\mathcal{X}$, potentially non-empty sets $\mathcal{S} \subseteq \mathcal{X}$, $\mathcal{D} \nsubseteq \mathcal{X}$ representing attributes sets being a subset ($\mathcal{S}$) or disjoint ($\mathcal{D}$) to the attribute set ($\mathcal{X}$) committed in the credential, and auxiliary information aux. It outputs a proof π.

$\mathsf{Verify}(\mathsf{pp}, \mathsf{opk}, \mathcal{X}, \mathcal{S}, \mathcal{D}, \pi, \mathsf{aux})$ takes pp, the (potentially empty) sets $\mathcal{S}$ and $\mathcal{D}$, a proof π and auxiliary information aux. It outputs 1 or 0 indicating whether the credential showing proof π was accepted or not.

$\mathsf{RSetup}(\mathsf{pp}, (\mathsf{rsk}, \mathsf{rpk}), \mathsf{NYM}, \mathsf{RNYM})$ takes pp, a revocation key pair (rsk, rpk) and two disjoint lists NYM and RNYM (holding valid and revoked pseudonyms). It outputs auxiliary information $\mathsf{aux}_{\mathsf{rev}}$ for the revocation authority and revocation information $\mathbb{R} = (\mathbb{R}_V, \mathbb{R}_S)$. $\mathbb{R}_V$ is needed for verifying the revocation status and $\mathbb{R}_S$ is a list holding the revocation information per nym.

$\mathsf{Revoke}(\mathsf{pp}, (\mathsf{rsk}, \mathsf{rpk}), \mathsf{aux}_{\mathsf{rev}}, \mathbb{R}, b)$ takes pp, (rsk, rpk), $\mathsf{aux}_{\mathsf{rev}}$, $\mathbb{R}$ and a bit b indicating revoked/unrevoked. It outputs information $\mathbb{R}'$ and $\mathsf{aux}'_{\mathsf{rev}}$.

$\mathsf{AuditEnc}(\mathsf{upk}, \mathsf{apk})$ takes upk and apk. It outputs an encryption enc of upk under apk and auxiliary information α.

$\mathsf{AuditDec}(\mathsf{enc}, \mathsf{ask})$ takes enc and ask. It outputs a decryption of enc using ask.

$\mathsf{AuditPrv}(\mathsf{enc}, \alpha, \mathsf{usk}, \mathsf{apk})$ takes enc, α, usk, and apk. It generates a proof for enc being the encryption of upk under apk and outputs a proof π.

$\mathsf{AuditVerify}(\mathsf{apk}, \pi)$ takes apk and a proof π for the correct encryption of a user's public key under apk and outputs 1 if and only if the proof verifies.

To introduce the formal security model, we consider a single revocation, issuing and auditability authority. Extension to the multi-issuing and multi-auditing setting is straightforward as each key can be generated independently. For multiple revocation authorities, one needs to consider multiple revocation accumulators and thus adapt the scheme accordingly. Issuer-hiding and auditability properties are considered independently as extensions. Let us denote by Tx the universe of transactions tx represented as bitstrings. We also use the following auxiliary lists, sets and global variables in oracles and formal definitions. N represents the set of all pseudonyms nym while the sets NYM and RNYM represent the subsets of unrevoked and revoked pseudonyms respectively. Therefore, we have that $\mathsf{NYM} \cap \mathsf{RNYM} = \emptyset \wedge \mathsf{NYM} \cup \mathsf{RNYM} = \mathsf{N}$. NYM, HU and CU are lists that keep track of which nym is assigned to which user, honest users and corrupt users, respectively. The global variables RI and $\mathsf{NYM}_{\mathsf{LoR}}$ (initially set to $\perp$) store the revocation information $(\mathbb{R}_S, \mathbb{R}_V)$ and the pseudonyms used in $\mathcal{O}_{\mathsf{LoR}}$ respectively. The oracles are defined as follows:

$\mathcal{O}_{\mathtt{HU}}(i)$ takes as input a user identity i. If $i \in \mathtt{HU} \cup \mathtt{CU}$, it returns $\perp$. Otherwise, it creates a new honest user i by running $(\mathtt{USK}[i], \mathtt{UPK}[i]) \xleftarrow{\$} \mathsf{UKGen}(\mathsf{opk})$, adding i to the honest user list HU and returning $\mathtt{UPK}[i]$.

$\mathcal{O}_{\mathtt{CU}}(i, \mathsf{upk})$ takes as input a user identity i and (optionally) upk; if user i does not exist, a new corrupt user with public key upk is registered, while if i is honest, its secret key and all credentials are leaked. If $i \in \mathtt{CU}$, $i \in I_{\mathsf{LoR}}$ (that is, i is a challenge user in the anonymity game) or if $\mathsf{NYM}_{\mathsf{LoR}} \cap \mathsf{N}[i] \neq \emptyset$ then the oracle returns $\perp$. If $i \in \mathtt{HU}$ then the oracle removes i from HU and adds it to CU; it returns $\mathtt{USK}[i]$ and $\mathtt{CRED}[j]$ for all j with $\mathtt{OWNR}[j] = i$. Otherwise (*i.e.*, $i \notin \mathtt{HU} \cup \mathtt{CU}$), it adds i to CU and sets $\mathtt{UPK}[i] \leftarrow \mathsf{upk}$.

$\mathcal{O}_{\mathtt{RN}}(\mathsf{rsk}, \mathsf{rpk}, \mathsf{REV})$ takes as input the revocation secret key rsk, the revocation public key rpk and a list REV of pseudonyms to be revoked. If $\mathsf{REV} \cap \mathsf{RNYM} \neq \emptyset$ or $\mathsf{REV} \not\subseteq \mathsf{N}$ return $\perp$. Otherwise, set $\mathsf{RNYM} \leftarrow \mathsf{RNYM} \cup \mathsf{REV}$ and $\mathsf{RI} \leftarrow \mathsf{Revoke}(\mathsf{pp}, (\mathsf{rsk}, \mathsf{rpk}), \mathsf{RNYM}, \mathsf{RI}, 1)$.

$\mathcal{O}_{\mathsf{ObtIss}}(i, \mathcal{X})$ takes as input a user identity i, a pseudonym nym and a set of attributes $\mathcal{X}$. If $i \notin \mathtt{HU}$ or $\exists\, j : \mathtt{NYM}[j] = \mathsf{nym}$, it returns $\perp$. Otherwise, it issues a credential to i by running $(\mathsf{cred}, \top) \xleftarrow{\$} \mathsf{Obtain}(\mathsf{pp}, \mathtt{USK}[i], \mathsf{opk}, \mathsf{apk}, \mathcal{X}, \mathsf{nym})$, $\mathsf{Issue}(\mathsf{pp}, \mathtt{UPK}[i], \mathsf{osk}, \mathsf{apk}, \mathcal{X}, \mathsf{nym})$. If $\mathsf{cred} = \perp$, it returns $\perp$. Else, it appends $(i, \mathsf{cred}, \mathcal{X}, \mathsf{nym})$ to $(\mathtt{OWNR}, \mathtt{CRED}, \mathtt{ATTR}, \mathtt{NYM})$ and returns $\top$.

$\mathcal{O}_{\mathsf{Obtain}}(i, \mathcal{X})$ lets the adversary $\mathcal{A}$, who impersonates a malicious organization, issue a credential to an honest user. It takes as input a user identity i, a pseudonym nym and a set of attributes $\mathcal{X}$. If $i \notin \mathtt{HU}$, it returns $\perp$. Otherwise, it runs $(\mathsf{cred}, \cdot) \xleftarrow{\$} \mathsf{Obtain}(\mathsf{pp}, \mathtt{USK}[i], \mathsf{opk}, \mathsf{apk}, \mathcal{X}, \mathsf{nym}), \cdot)$, where the Issue part is executed by $\mathcal{A}$. If $\mathsf{cred} = \perp$, it returns $\perp$. Else, it appends $(i, \mathsf{cred}, \mathcal{X}, \mathsf{nym})$ to $(\mathtt{OWNR}, \mathtt{CRED}, \mathtt{ATTR}, \mathtt{NYM})$ and returns $\top$.

$\mathcal{O}_{\mathsf{Issue}}(i, \mathcal{X})$ lets the adversary $\mathcal{A}$, who impersonates a malicious user, obtain a credential from an honest organization. It takes as input a user identity i, a pseudonym nym and a set of attributes $\mathcal{X}$. If $i \notin \mathtt{CU}$, it returns $\perp$. Otherwise, it runs $(\cdot, I) \xleftarrow{\$} (\cdot, \mathsf{Issue}(\mathsf{pp}, \mathtt{UPK}[i], \mathsf{osk}, \mathsf{apk}, \mathcal{X}, \mathsf{nym}))$, where the Obtain part is executed by $\mathcal{A}$. If $\mathrm{I} = \perp$, it returns $\perp$. Else, it appends $(i, \perp, \mathcal{X}, \mathsf{nym})$ to $(\mathtt{OWNR}, \mathtt{CRED}, \mathtt{ATTR}, \mathtt{NYM})$ and returns $\top$.

$\mathcal{O}_{\mathsf{Show}}(j, \mathcal{S}, \mathcal{D})$ lets the adversary $\mathcal{A}$ play a dishonest verifier during a showing by an honest user. It takes as input an index of an issuance j and attributes sets $\mathcal{S}$ and $\mathcal{D}$. Let $i \xleftarrow{\$} \mathtt{OWNR}[j]$. If $i \notin \mathtt{HU}$, it returns $\perp$. Otherwise, it runs $(S, \cdot) \xleftarrow{\$} \mathsf{Show}(\mathsf{pp}, \mathtt{USK}[i], \mathtt{UPK}[i], \mathsf{opk}, \mathtt{ATTR}[j], \mathcal{S}, \mathcal{D}, \mathtt{CRED}[j], \mathsf{RI}, \mathsf{apk}, \mathsf{tx}), \cdot)$

$\mathcal{O}_{\mathsf{LoR}}(j_0, j_1, \mathcal{S}, \mathcal{D})$ is the challenge oracle in the anonymity game where $\mathcal{A}$ runs Verify and must distinguish (multiple) showings of two credentials $\mathtt{CRED}[j_0]$ and $\mathtt{CRED}[j_1]$. The oracle takes two issuance indices j_0 and j_1 and attribute sets $\mathcal{S}$ and $\mathcal{D}$. If $J_{\mathsf{LoR}} \neq \emptyset$ and $J_{\mathsf{LoR}} \neq \{j_0, j_1\}$, it returns $\perp$. Let $i_0 \xleftarrow{\$} \mathtt{OWNR}[j_0]$ and $i_1 \xleftarrow{\$} \mathtt{OWNR}[j_1]$. If $J_{\mathsf{LoR}} \neq \emptyset$ then it sets $J_{\mathsf{LoR}} \xleftarrow{\$} \{j_0, j_1\}$ and $I_{\mathsf{LoR}} \xleftarrow{\$} \{i_0, i_1\}$. If $i_0, i_1 \neq \mathtt{HU} \vee \mathsf{N}[i_0] = \perp \vee \mathsf{N}[i_1] = \perp \vee \mathsf{N}[i_0] \in \mathsf{RNYM} \vee \mathsf{N}[i_1] \in \mathsf{RNYM} \vee \mathcal{S} \not\subseteq \mathtt{ATTR}[j_0] \cap \mathtt{ATTR}[j_1] \vee \mathcal{D} \cap \{\mathtt{ATTR}[j_0] \cup \mathtt{ATTR}[j_1]\} \neq \emptyset$, it returns $\perp$. Else, it adds $\mathsf{N}[i_b]$ to $\mathsf{NYM}_{\mathsf{LoR}}$ and runs $(S, \cdot) \xleftarrow{\$} (\mathsf{Show}(\mathsf{pp}, \mathtt{USK}[j_b], \mathtt{UPK}[j_b], \mathsf{opk}, \mathtt{ATTR}[j_b], \mathcal{S}, \mathcal{D}, \mathtt{CRED}[j_b], \mathsf{RI}, \mathsf{apk}, \mathsf{tx}), \cdot)$ (with b set by the experiment)

Intuitively, *correctness* requires that a credential showing with respect to a non-empty sets $\mathcal{S}$ and $\mathcal{D}$ of attributes always verifies if it was issued honestly on some attribute set $\mathcal{X}$ with $\mathcal{S} \subset \mathcal{X}$ and $\mathcal{D} \cap \mathcal{X} \neq \emptyset$.

Definition 1 (Correctness). *An ABC system is correct if* $\forall\ \lambda > 0$, $\forall\ q, q' > 0$, $\forall\ \mathcal{X} : 0 < |\mathcal{X}| \leq q$, $\forall\ \emptyset \neq \mathcal{S} \subset \mathcal{X}$, $\forall\ \emptyset \neq \mathcal{D} \not\subseteq \mathcal{X} : 0 < |\mathcal{D}| \leq q$, $\forall$ $\mathsf{NYM}, \mathsf{RNYM} \subseteq \mathsf{N} : 0 < |\mathsf{N}| \leq q' \wedge \mathsf{NYM} \cap \mathsf{RNYM} = \emptyset$, $\forall\ \mathsf{nym} \in \mathsf{NYM}$, $\forall\ \mathsf{nym}' \in \mathsf{RNYM}$ *it holds that:*
$\mathsf{pp} \xleftarrow{\$} \mathsf{Setup}(1^\lambda, (1^q, 1^{q'}))$; $(\mathsf{rsk}, \mathsf{rpk}) \xleftarrow{\$} \mathsf{RAKGen}(\mathsf{pp})$; $(\mathsf{ask}, \mathsf{apk}) \xleftarrow{\$} \mathsf{AAKGen}(\mathsf{pp})$; $(\mathbb{R}, \mathsf{aux}_{\mathsf{rev}}) \leftarrow \mathsf{RSetup}(\mathsf{pp}, \mathsf{rpk}, \mathsf{NYM}, \mathsf{RNYM})$; $(\mathsf{osk}, \mathsf{opk}) \xleftarrow{\$} \mathsf{OKGen}(\mathsf{pp})$; $(\mathsf{usk}, \mathsf{upk}) \xleftarrow{\$} \mathsf{UKGen}(\mathsf{pp})$; $(\mathsf{cred}, \top) \xleftarrow{\$} (\mathsf{Obtain}(\mathsf{pp}, \mathsf{usk}, \mathsf{opk}, \mathsf{apk}, \mathcal{X}, \mathsf{nym}), \mathsf{Issue}(\mathsf{pp}, \mathsf{upk}, \mathsf{osk}, \mathsf{apk}, \mathcal{X}, \mathsf{nym}))$; $(\mathbb{R}_S, \mathbb{R}_V) \leftarrow \mathsf{Revoke}(\mathsf{pp}, \mathbb{R}, \mathsf{aux}_{\mathsf{rev}}, \mathsf{nym}', 1)$; $\Omega \leftarrow \mathsf{Show}(\mathsf{pp}, \mathsf{usk}, \mathsf{upk}, \mathsf{opk}, \mathsf{cred}, \mathcal{S}, \mathcal{D}, \mathbb{R}, \mathsf{apk}, \mathsf{tx})$; $1 \leftarrow \mathsf{Verify}(\mathsf{pp}, \mathcal{S}, \mathcal{D}, \mathsf{opk}, \mathbb{R}_V, \mathsf{rpk}, \mathsf{apk}, \mathsf{tx}, \Omega)$

We now provide a formal definition for *unforgeablility*. Given at least one non-empty set $\mathcal{S} \subset \mathcal{X}$ or $\mathcal{D} \not\subseteq \mathcal{X}$, a user in possession of a credential for the attribute set $\mathcal{X}$ cannot perform a valid showing for $\mathcal{D} \subset \mathcal{X}$ nor for $\mathcal{S} \not\subseteq \mathcal{X}$. Moreover, revocated users cannot perform valid showings and no coalition of malicious users can combine their credentials and prove possession of a set of attributes which no single member has. This holds even after seeing showings of arbitrary credentials by honest users.

Definition 2 (Unforgeability). *An ABC system is unforgeable, if* $\forall\ \lambda, q, q' > 0$ *and* p.p.t *adversaries* $\mathcal{A}$ *having oracle access to* $\mathcal{O} := \{\mathcal{O}_{HU}, \mathcal{O}_{CU}, \mathcal{O}_{RN}, \mathcal{O}_{\mathsf{ObtIss}}, \mathcal{O}_{\mathsf{Issue}}, \mathcal{O}_{\mathsf{Show}}\}$ *the following probability is negligible.*

$$\Pr\left[\begin{array}{l}\mathsf{pp} \xleftarrow{\$} \mathsf{Setup}(1^\lambda, (1^q, 1^{q'})); (\mathsf{rsk}, \mathsf{rpk}) \xleftarrow{\$} \mathsf{RAKGen}(\mathsf{pp}); (\mathsf{ask}, \mathsf{apk}) \xleftarrow{\$} \mathsf{AAKGen}(\mathsf{pp});\\ (\mathsf{osk}, \mathsf{opk}) \xleftarrow{\$} \mathsf{OKGen}(\mathsf{pp}); (\mathcal{S}, \mathcal{D}, \mathsf{st}) \xleftarrow{\$} \mathcal{A}^{\mathcal{O}}(\mathsf{pp}, \mathsf{opk}, \mathsf{rpk}, \mathsf{apk});\\ (\cdot, b^*) \xleftarrow{\$} (\mathcal{A}(\mathsf{st}), \mathsf{Verify}(\mathsf{pp}, \mathcal{S}, \mathcal{D}, \mathsf{opk}, \mathsf{rpk}, \mathsf{apk}, \mathsf{RI}, \mathsf{tx}, \Omega)) :\\ b^* = 1 \wedge \forall\ j : \mathit{OWNR}[j] \in \mathit{CU} \implies (\mathsf{N}[j] = \perp\ \vee\ (\mathsf{N}[j] \neq \perp\ \wedge (\mathcal{S} \not\subseteq \mathit{ATTR}[j]\ \vee\\ \mathcal{D} \subseteq \mathit{ATTR}[j] \vee \mathsf{N}[j] \in \mathsf{RNYM}))\end{array}\right]$$

For *anonymity*, during a showing, no verifier and no (malicious) organization (even if they collude) should be able to identify the user or learn anything about the user, except that she owns a valid credential for the shown attributes. Furthermore, different showings of the same credential are unlinkable.

Definition 3 (Anonymity). *An ABC system is anonymous, if* $\forall\ \lambda, q, q' > 0$ *and all* p.p.t *adversaries* $\mathcal{A}$ *having oracle access to* $\mathcal{O} := \{\mathcal{O}_{HU}, \mathcal{O}_{CU}, \mathcal{O}_{RN}, \mathcal{O}_{\mathsf{Obtain}}, \mathcal{O}_{\mathsf{Show}}, \mathcal{O}_{\mathsf{LoR}}\}$ *the following probability is negligible.*

$$\Pr\left[\begin{array}{l}\mathsf{pp} \xleftarrow{\$} \mathsf{Setup}(1^\lambda, (1^q, 1^{q'})); (\mathsf{ask}, \mathsf{apk}) \xleftarrow{\$} \mathsf{AAKGen}(\mathsf{pp});\\ b \xleftarrow{\$} \{0,1\}; (\mathsf{opk}, \mathsf{rpk}, \mathsf{st}) \xleftarrow{\$} \mathcal{A}(\mathsf{pp}); b^* \xleftarrow{\$} \mathcal{A}^{\mathcal{O}}(\mathsf{st})\end{array} : b^* = b\right] - \frac{1}{2}$$

Issuer-hiding states that no adversary (*i.e.*, a malicious verifier) cannot tell with high probability who is the issuer of a credential issued to an honest user.

Definition 4 (Issuer-hiding). *An ABC system supports issuer-hiding if for all* $\lambda > 0$, *all* $q > 0$, *all* $n > 0$, *all* $t > 0$, *all* $\mathcal{X}$ *with* $0 < |\mathcal{X}| \leq t$, *all* $\emptyset \neq \mathcal{S} \subset \mathcal{X}$ *and* $\emptyset \neq \mathcal{D} \not\subseteq \mathcal{X}$ *with* $0 < |\mathcal{D}| \leq t$, *and* p.p.t *adversaries* $\mathcal{A}$, *the following holds.*

$$\Pr\left[\begin{array}{l}\mathsf{pp} \xleftarrow{\$} \mathsf{Setup}(1^\lambda, 1^q); \forall\, i \in [n] : (\mathsf{osk}_i, \mathsf{opk}_i) \xleftarrow{\$} \mathsf{OKGen}(\mathsf{pp});\\ (\mathsf{usk}, \mathsf{upk}) \xleftarrow{\$} \mathsf{UKGen}(\mathsf{pp}); j \xleftarrow{\$} [n];\\ (\mathsf{cred}, \top) \xleftarrow{\$} (\mathsf{Obtain}(\mathsf{usk}, \mathsf{opk}_j, \mathcal{X}), \mathsf{Issue}(\mathsf{upk}, \mathsf{osk}_j, \mathcal{X}));\\ j^* \xleftarrow{\$} \mathcal{A}^{\mathcal{O}_{\mathsf{Show}}}(\mathsf{pp}, \mathcal{S}, \mathcal{D}, (\mathsf{opk}_i)_{i \in [n]})\end{array} : j^* = j\right] \leq \frac{1}{n}$$

Finally, *auditability* requires that showings correctly encrypt users' keys.

Definition 5 (Auditability). *An ABC scheme is auditable, if* $\forall\ \lambda, q, q' > 0, \mathsf{tx}, \mathsf{nym} \in \mathsf{NYM}, \mathbb{R}, \mathbb{R}_V$ *and all* p.p.t *adversaries* $\mathcal{A}$ *having oracle access to* $\mathcal{O}_{\mathsf{Issue}}$, *the following probability is negligible.*

$$\Pr\left[\begin{array}{l}\mathsf{pp} \xleftarrow{\$} \mathsf{Setup}(1^\lambda, (1^q, 1^{q'})); (\mathsf{rsk}, \mathsf{rpk}) \xleftarrow{\$} \mathsf{RAKGen}(\mathsf{pp});\\ (\mathsf{ask}, \mathsf{apk}) \xleftarrow{\$} \mathsf{AAKGen}(\mathsf{pp}); (\mathsf{osk}, \mathsf{opk}) \xleftarrow{\$} \mathsf{OKGen}(\mathsf{pp});\\ (\mathsf{usk}, \mathsf{upk}) \xleftarrow{\$} \mathsf{UKGen}(\mathsf{pp});\\ (\mathcal{S}, \mathcal{D}, \mathsf{enc}, \Omega, \mathsf{st}) \xleftarrow{\$} \mathcal{A}^{\mathcal{O}}(\mathsf{pp}, \mathsf{opk}, \mathsf{rpk}, \mathsf{apk}, \mathsf{usk}, \mathsf{upk}, \mathsf{nym});\\ (\cdot, b^*) \xleftarrow{\$} (\mathcal{A}(\mathsf{st}), \mathsf{Verify}(\mathsf{pp}, \mathcal{S}, \mathcal{D}, \mathsf{opk}, \mathbb{R}_V, \mathsf{rpk}, \mathsf{apk}, \mathsf{RI}, \mathsf{tx}, \Omega)) :\\ b^* = 1 \wedge \mathsf{upk} \neq \mathsf{AuditDec}(\mathsf{enc}, \mathsf{ask})\end{array}\right]$$

4 Protego

We elaborate on the decisions that led to the design of our ABC scheme Protego. Subsequently, we discuss our construction and the integration with Fabric.

Revocation. We opt to integrate the work from [14] as pointed out in [12]. The revocation system from [14] defines a revocation authority responsible for managing an allow and deny list of revocation handlers. The authority publishes an accumulator RevAcc representing the deny list, and maintains a public list of non-membership witnesses for unrevoked users. During the issuing protocol, users are given a revocation handle that is encoded in the credential. To prove that they are not revoked during a showing, the user consistently randomizes its credential with the accumulator and the corresponding non-membership witness. Then the verifier checks that the (randomized) witness is valid for the revocation handle (encoded in the user credential), and with respect to the (randomized) accumulator. To work, the user must compute a Zero-Knowledge Proof of Knowledge (ZKPoK) on the correct randomization of the non-membership witness and the accumulator. As explained in [14], the revocation handle encoded in the user's credential is of the form $\mathsf{usk}_2(b + \mathsf{nym})P_1$, where usk_2 is an additional user secret key required for anonymity and nym is the pseudonym used for revocation. For this reason, users are required to manage augmented keys of the form $\mathsf{upk} = (\mathsf{upk}_1, \mathsf{upk}_2)$, $\mathsf{usk} = (\mathsf{usk}_1, \mathsf{usk}_2)$. Furthermore, for technical reasons, another component $\mathsf{usk}_2 Q$, where Q is a random element in $\mathbb{G}_1$ with unknown discrete logarithm, must be included in the credential.

Auditability. A credential in [12,15] and [14] contains a tuple (C, rC, P_1) where C is the set commitment on the user attributes, r is a random value used for technical purposes and P_1 is used to compute a ZKPoK of the randomizer μ in $(\mu C, \mu rC, \mu P_1)$ during a showing. We borrow the idea of using a verifiable variant of ElGamal from [7] to prove the well-formedness of a ciphertext (encrypting the user's public key) with respect to the auditor's key. Therefore, we add the user's public key upk_1 and the auditor's public key apk as the sixth and seventh components to the credential. Thus, we now have revocable credentials of the form $(C, rC, P_1, \mathsf{usk}_2(b + \mathsf{nym})P_1, \mathsf{usk}_2 Q, \mathsf{upk}_1, \mathsf{apk})$, which can be randomized to obtain a tuple $(\mu C, \mu rC, \mu P_1, \mu \mathsf{usk}_2(b + \mathsf{nym})P_1, \mu \mathsf{usk}_2 Q, \mu \mathsf{usk}_1 P_1, \mu \mathsf{apk})$. We exploit this fact to allow the user to generate an *audit* proof that can be publicly verified without leaking information about the user's public key. This way, verifiers can check a proof using the sixth and seventh component in the credential to be sure that (1) the user encrypted a public key for which it has the corresponding secret key, and (2) using the correct one. Since the issuing authority signs the credential, the randomization needs to be consistent. Modifications required to implement our auditability approach are as follows:

1. The user randomizes its credential as usual to obtain a new one of the form $(C_1', C_2', C_3', C_4', C_5', C_6', C_7') = (\mu C_1, \mu C_2, \mu P_1, \mu C_4, \mu C_5, \mu \mathsf{upk}_1, \mu \mathsf{apk})$. Since only the user knows the randomizer μ, its public key remains hidden.
2. The user picks $\alpha \in \mathbb{Z}_p$ and encrypts its own public key using ElGamal encryption with auditor's public key apk and randomness α to obtain a ciphertext $\mathsf{enc} = (\mathsf{enc}_1, \mathsf{enc}_2) = (\mathsf{upk}_1 + \alpha \mathsf{apk}, \alpha P_1)$.
3. The user runs the algorithm $\mathsf{AuditPrv}$ (Fig. 1) with input $(\mathsf{enc}, \alpha, \mathsf{usk}_1, \mathsf{apk})$ to obtain c, z_1 and z_2.
4. Then, the user picks $\beta \leftarrow \mathbb{Z}_p$, computes $t_1 = \beta P_2$, $t_2 = \beta\mu P_2$, $t_3 = \alpha\beta P_2$ and sends $(\mathsf{enc}, c, z_1, z_2, t_1, t_2, t_3)$ to the verifier alongside the randomized credential from step 1.
5. The verifier checks the well-formedness of the ElGamal encryption pair running the algorithm $\mathsf{AuditVerify}$ (Fig. 1) with input $(c, \mathsf{enc}, z_1, z_2)$. If the check succeeds, it checks the following pairing equations to verify that the encrypted public key is the one in the credential:
$e(\mathsf{enc}_2, t_2) = e(C_3', t_3) \wedge e(\mathsf{enc}_2, t_1) = e(P_1, t_3) \wedge e(\mathsf{enc}_1, t_2) = e(C_6', t_1) + e(C_7', t_3)$

Observe that the verifier knows $\mu P_1 = C_3'$, $\mu \mathsf{usk}_1 P_1 = C_6'$, $\mu \mathsf{ask} P_1 = C_7'$, $(\mathsf{usk}_1 + \alpha \mathsf{ask})P_1 = \mathsf{enc}_1$, $\alpha P_1 = \mathsf{enc}_2$, $\beta P_2 = t_1$, $\beta\mu P_2 = t_2$ and $\alpha\beta P_2 = t_3$. With β the user is able to randomize the other values so that the pairing equation can be checked to verify the relation between the ElGamal ciphertext and the randomized public key in C_6', without leaking information about the user's public key. Furthermore, the first two pairing equations verify the well-formedness of t_1, t_2 and t_3 with respect to the user's credential and the ciphertext. Hence, the verifier will not be able to recover the user's public key nor the user cheat.

The proposed solution only adds two elements to the credential, while requiring the user to send two more elements in $\mathbb{G}_1$, three in $\mathbb{Z}_p$ and three in $\mathbb{G}_2$, for a total of eight. Computational cost remains low as it just involves the computation of seven pairings, the ElGamal encryption and two Schnorr proofs [19].

Issuer-Hiding. We incorporate the issuer-hiding approaches discussed in Sect. 2. The work from [12] relies on a NIZK argument to prove that a randomized public key belongs to the equivalence class of one of the public keys contained in a list of issuer keys. We adapt the proof system to the signature used in this work, and extend it to make it *malleable* (see Appendix A) so that users can compute the proof once and then adapt it during showings with little computational cost (instead of having to compute it from scratch). This efficiency improvement is very useful in permissioned blockchains as the set of authorities tends to stay the same over time. For both approaches we observe that the mercurial signature used in this work only provides a weak form of issuer-hiding. Given a signature that has been adapted to verify under a randomized public key pk' in the equivalence class of pk, the owner of pk can recognize it. Thus, issuers can know which transactions belong to users from their organizations (but not to which particular user) and which ones don't by reading the non-interactive showing proof (it contains the issuer's randomized public key). However, we argue that in the permissioned blockchain setting this provides a fair trade-off as a minimum traceability level is important for compliance and auditability purposes.

Our Construction. Compared to [12], we make use of a hash function to apply the (strong) Fiat-Shamir transform while adding the previously discussed auditability and revocation features. Therefore we implement the ZKPoK's as Schnorr proofs (unlike [12] which followed Remark 1 from [15]).

In Fig. 1 we present the setup, key generation, revocation and auditing algorithms. The setup algorithm also takes a bound q' on the maximun number of revocated pseudonyms for the revocation accumulator. The revocation authority is responsible for running the Revoke algorithm and updating the accumulator.

Obtain and Issue have constant-size communication and are given in Fig. 2. For Show and Verify we present *Protego* and *Protego Duo*, depending on the issuer-hiding approach. Protego is given in Fig. 3 and produces a variable-length proof as it relies on the (mallable) NIZK proof. Protego Duo produces a constant-size proof and is depicted in Fig. 4. The differences are highlighted with grey. For both, after the credential is updated, the user randomizes the revocation accumulator, witnesses, and generates the Schnorr proofs. Following the auditing proof, the Fiat-Shamir transform is applied, the ZKPoK's and PoE's are computed, returning the showing proof. Verify takes a proof (depending on the case), computes the challenge and verifies each of the statements.

Integration with Fabric. A multi-party computation ceremony can be run for the CRS generation of the Setup algorithm to ensure that no organization knows the trapdoors of the different components. As we are in the permissioned setting it is plausible to assume that at least one of the organizations is honest. By allowing *users and endorsers* to obtain credentials, both can produce showing proofs. Users can generate showing proofs to prove that they satisfy the access policy for the execution of a particular transaction proposal. Furthermore, by computing the PoE's, the verification time for endorsers improves substantially. Similarly, endorsers can prove that they satisfy a given endorsement policy

$\underline{\mathsf{Setup}(1^\lambda, \mathsf{aux}):}$
$(q, q') \leftarrow \mathsf{aux}$; **pick** $\mathcal{H} : \{0,1\}^* \rightarrow \mathbb{Z}_p^*$; $Q \xleftarrow{\$} \mathbb{G}_1$; $(\mathsf{rev_{pp}}, \mathsf{rev_{td}}) \xleftarrow{\$} \mathsf{RevAcc.Setup}(1^\lambda, q')$
$(\mathsf{BG}, \mathsf{scds_{pp}}, \mathsf{scds_{td}}) \xleftarrow{\$} \mathsf{SCDS.Setup}(1^\lambda, q)$; $(\mathsf{sps_{pp}}, \mathsf{sps_{td}}) \xleftarrow{\$} \mathsf{SPS\text{-}EQ.ParGen}(1^\lambda; \mathsf{BG})$
return $(\mathcal{H}, \mathsf{BG}, \mathsf{rev_{pp}}, Q, \mathsf{scds_{pp}}, \mathsf{sps_{pp}})$
$\underline{\mathsf{TSetup}(1^\lambda, \mathsf{aux}):}$
$(q, q') \leftarrow \mathsf{aux}$; **pick** $\mathcal{H} : \{0,1\}^* \rightarrow \mathbb{Z}_p^*$; $Q \xleftarrow{\$} \mathbb{G}_1$; $(\mathsf{rev_{pp}}, \mathsf{rev_{td}}) \xleftarrow{\$} \mathsf{RevAcc.Setup}(1^\lambda, q')$
$(\mathsf{BG}, \mathsf{scds_{pp}}, \mathsf{scds_{td}}) \xleftarrow{\$} \mathsf{SCDS.Setup}(1^\lambda, q)$; $(\mathsf{sps_{pp}}, \mathsf{sps_{td}}) \xleftarrow{\$} \mathsf{SPS\text{-}EQ.ParGen}(1^\lambda; \mathsf{BG})$
$\mathsf{td} = (\mathsf{rev_{td}}, \mathsf{scds_{td}}, \mathsf{sps_{td}})$; **return** $(\mathcal{H}, \mathsf{BG}, \mathsf{rev_{pp}}, Q, \mathsf{scds_{pp}}, \mathsf{sps_{pp}}, \mathsf{td})$
$\underline{\mathsf{RevAcc.Setup}(1^\lambda, 1^q):}$ $\mathsf{BG} \xleftarrow{\$} \mathsf{BGGen}(1^\lambda)$; $b \xleftarrow{\$} \mathbb{Z}_p^*$; **return** $(\mathsf{BG}, (b^i P_1, b^i P_2)_{i \in [q]})$
$\underline{\mathsf{AAKGen}(\mathsf{pp}):}$ $\mathsf{ask} \xleftarrow{\$} Z_p^*$; $\mathsf{apk} \leftarrow \mathsf{ask} P_1$; **return** $(\mathsf{ask}, \mathsf{apk})$
$\underline{\mathsf{RAKGen}(\mathsf{pp}):}$ $\mathsf{rsk} \xleftarrow{\$} Z_p^*$; $\mathsf{rpk} \leftarrow \mathsf{rsk} P_2$; **return** $(\mathsf{rpk}, \mathsf{rsk})$
$\underline{\mathsf{OKGen}(\mathsf{pp}):}$ **return** $\mathsf{SPS\text{-}EQ.KGen}(\mathsf{BG}, \mathsf{sps_{pp}}, 7)$
$\underline{\mathsf{UKGen}(\mathsf{pp}):}$ $\mathsf{usk}_1, \mathsf{usk}_2 \xleftarrow{\$} Z_p^*$; $(\mathsf{upk}_1, \mathsf{upk}_2) \leftarrow (\mathsf{usk}_1 P_1, \mathsf{usk}_2 P_1)$
return $((\mathsf{usk}_1, \mathsf{usk}_2), (\mathsf{upk}_1, \mathsf{upk}_2))$
$\underline{\mathsf{RSetup}(\mathsf{pp}, (\mathsf{rsk}, \mathsf{rpk}), \mathsf{NYM}, \mathsf{RNYM}):}$
$(\Pi_{\mathsf{rev}}, \mathsf{aux_{rev}}) \leftarrow \mathsf{RevAcc.Commit}(\mathsf{rev_{pp}}, \mathsf{RNYM})$
foreach $\mathsf{nym} \in \mathsf{NYM}$ **do** $\mathsf{WIT}[\mathsf{nym}] \leftarrow \mathsf{RevAcc.NonMemWit}(\mathsf{pp}, \Pi_{\mathsf{rev}}, \mathsf{aux_{rev}}, \mathsf{nym})$
return $((\Pi_{\mathsf{rev}}, \mathsf{WIT}), \mathsf{aux_{rev}})$
$\underline{\mathsf{RevAcc.Commit}(\mathsf{pp}, \mathcal{X}; \mathsf{rsk}):}$
check $|\mathcal{X}| \leq q \wedge \nexists\, b' \in \mathcal{X} : b' P_1 = b P_1$; $\Pi_{\mathsf{rev}} \leftarrow \mathsf{rsk}^{-1} \cdot \mathsf{Ch}_{\mathcal{X}}(s) P_1$; $\mathsf{aux_{rev}} \leftarrow \mathcal{X}$
return $(\Pi_{\mathsf{rev}}, \mathsf{aux_{rev}})$
$\underline{\mathsf{Revoke}(\mathsf{pp}, \mathbb{R}, \mathsf{aux_{rev}}, \mathsf{nym}, b):}$
parse $\mathbb{R} = (\Pi_{\mathsf{rev}}, \mathsf{WIT})$; **parse** $\mathsf{aux_{rev}} = \mathsf{RNYM}$
if $b = 1$
  $\mathsf{NYM} \leftarrow \mathsf{NYM} \setminus \{\mathsf{nym}\}$; $\mathsf{RNYM} \leftarrow \mathsf{RNYM} \cup \{\mathsf{nym}\}$
  $(\Pi'_{\mathsf{rev}}, \mathsf{aux}'_{\mathsf{rev}}) \leftarrow \mathsf{RevAcc.Add}(\mathsf{pp}, \Pi_{\mathsf{rev}}, \mathsf{RNYM}, \mathsf{nym})$
else
  $\mathsf{NYM} \leftarrow \mathsf{NYM} \cup \{\mathsf{nym}\}$; $\mathsf{RNYM} \leftarrow \mathsf{RNYM} \setminus \{\mathsf{nym}\}$
  $(\Pi'_{\mathsf{rev}}, \mathsf{aux}'_{\mathsf{rev}}) \leftarrow \mathsf{RevAcc.Del}(\mathsf{pp}, \Pi_{\mathsf{rev}}, \mathsf{RNYM}, \mathsf{nym})$
foreach $\mathsf{nym}' \in \mathsf{NYM}$ **do** $\mathsf{WIT}[\mathsf{nym}'] \leftarrow \mathsf{RevAcc.NonMemWit}(\mathsf{pp}, \Pi'_{\mathsf{rev}}, \mathsf{aux}'_{\mathsf{rev}}, \mathsf{nym}')$
return $((\Pi'_{\mathsf{rev}}, \mathsf{WIT}), \mathsf{aux}'_{\mathsf{rev}})$
$\underline{\mathsf{RevAcc.Add}(\mathsf{pp}, \mathsf{rsk}, \Pi_{\mathsf{rev}}, \mathsf{aux_{rev}}, \mathsf{nym}):}$
parse $\mathsf{aux_{rev}} = \mathcal{X}$; $\mathcal{X} \leftarrow \mathcal{X} \cup \{\mathsf{nym}\}$; **return** $\mathsf{RevAcc.Commit}(\mathsf{pp}, \mathcal{X}; \mathsf{rsk})$
$\underline{\mathsf{RevAcc.Del}(\mathsf{pp}, \mathsf{rsk}, \Pi_{\mathsf{rev}} \mathsf{aux_{rev}}, \mathsf{nym}):}$
parse $\mathsf{aux_{rev}} = \mathcal{X}$; $\mathcal{X} \leftarrow \mathcal{X} \setminus \{\mathsf{nym}\}$; **return** $\mathsf{RevAcc.Commit}(\mathsf{pp}, \mathcal{X}; \mathsf{rsk})$
$\underline{\mathsf{RevAcc.NonMemWit}(\mathsf{pp}, \Pi_{\mathsf{rev}}, \mathsf{aux_{rev}}, \mathsf{nym}):}$
$\mathcal{X} \leftarrow \mathsf{aux_{rev}}$; **check** $\mathsf{nym} \notin \mathcal{X}$; Let $q(X)$ and $d \in \mathbb{Z}_p^*$ s.t. $\mathsf{Ch}_{\mathcal{X}}(X) = q(X)(X + \mathsf{nym}) + d$
return $(q(b) P_2, d)$
$\underline{\mathsf{RevAcc.VerifyWit}(\mathsf{pp}, \Pi_{\mathsf{rev}}, \mathsf{nym}, \mathsf{wss_{rev}}):}$
$(\mathsf{wss}^1_{\mathsf{rev}}, \mathsf{wss}^2_{\mathsf{rev}}) \leftarrow \mathsf{wss_{rev}}$; **return** $e(\Pi_{\mathsf{rev}}, \mathsf{rpk}) = e((b + \mathsf{nym}) P_1, \mathsf{wss}^1_{\mathsf{rev}}) e(\mathsf{wss}^2_{\mathsf{rev}} P_1, P_2)$
$\underline{\mathsf{AuditEnc}(\mathsf{upk}, \mathsf{apk}):}$ $\alpha \leftarrow \mathbb{Z}_p$; $\mathsf{enc} \leftarrow (\mathsf{upk} + \alpha \mathsf{apk}, \alpha P_1)$; **return** (enc, α)
$\underline{\mathsf{AuditDec}(\mathsf{enc}, \mathsf{ask}):}$ $(\mathsf{enc}_1, \mathsf{enc}_2) \leftarrow \mathsf{enc}$; **return** $(\mathsf{enc}_1 - \mathsf{ask} \cdot \mathsf{enc}_2)$
$\underline{\mathsf{AuditPrv}(\mathsf{enc}, \alpha, \mathsf{usk}, \mathsf{apk}):}$
$r_1, r_2 \leftarrow \mathbb{Z}_p$; $\mathsf{com}_1 \leftarrow r_1 P_1 + r_2 \mathsf{apk}$; $\mathsf{com}_2 \leftarrow r_2 P_1$; $c \leftarrow \mathcal{H}(\mathsf{com}_1, \mathsf{com}_2, \mathsf{enc})$
$z_1 \leftarrow r_1 + c \cdot \mathsf{usk}$; $z_2 \leftarrow r_2 + c \cdot \alpha$; **return** (c, z_1, z_2)
$\underline{\mathsf{AuditVerify}(\mathsf{apk}, c, \mathsf{enc}, z_1, z_2):}$
$\mathsf{com}_1 \leftarrow z_1 P_1 + z_2 \mathsf{apk} - c\mathsf{enc}_1$; $\mathsf{com}_2 \leftarrow z_2 P_1 - c\mathsf{enc}_2$; $c' \leftarrow \mathcal{H}(\mathsf{com}_1, \mathsf{com}_2, \mathsf{enc})$
return $c' = c$

Fig. 1. Protego: setup, key generation, revocation and auditing algorithms.

$\underline{\mathsf{Obtain}(\mathsf{pp}, \mathsf{usk}, \mathsf{opk}, \mathsf{apk}, \mathcal{X}, \mathsf{nym})}$		$\underline{\mathsf{Issue}(\mathsf{pp}, \mathsf{upk}, \mathsf{osk}, \mathsf{apk}, \mathcal{X}, \mathsf{nym})}$
$r_1, r_2, r_3 \xleftarrow{\$} \mathbb{Z}_p^*;\ a_1 \leftarrow r_1 P_1;\ a_2 \leftarrow r_2 P_1$		
$a_3 \leftarrow r_3 Q;\ C_4 \leftarrow \mathsf{usk}_2(b + \mathsf{nym})P_1$		
$C_5 \leftarrow \mathsf{usk}_2 Q;\ e \leftarrow \mathcal{H}(\mathsf{upk}_1, \mathsf{upk}_2, C_5, a_1, a_2, a_3)$		
$z_1 \leftarrow r_1 + e \cdot \mathsf{usk}_1$		
$z_2 \leftarrow r_2 + e \cdot \mathsf{usk}_2;\ z_3 \leftarrow r_3 + e \cdot \mathsf{usk}_2$		
$(C_1, O) \leftarrow \mathsf{SCDS.Commit}(\mathsf{scds_{pp}}, \mathcal{X}; \mathsf{usk}_1)$		
$r_4 \xleftarrow{\$} \mathbb{Z}_p^*;\ C_2 \leftarrow r_4 \cdot C_1$		
$\Sigma \leftarrow (C_1, C_2, C_4, C_5, (a_i, z_i)_{i \in [3]})$	$\xrightarrow{\Sigma}$	$e \leftarrow \mathcal{H}(\mathsf{upk}_1, \mathsf{upk}_2, C_5, a_1, a_2, a_3)$
		check
		$z_1 P_1 = a_1 + e \cdot \mathsf{upk}_1$
		$z_2 P_1 = a_2 + e \cdot \mathsf{upk}_2;\ z_3 Q = a_3 + e \cdot C_5$
		$e(C_1, P_2) \neq e(\mathsf{upk}_1, \mathsf{Ch}_{\mathcal{X}}(s)P_2)$
		$\forall\, x \in \mathcal{X} : xP_1 \neq \mathsf{ek}_1^0$
check		$e(C_4, P_2) = e(\mathsf{upk}_2, (b + \mathsf{nym})P_2)$
$\mathsf{SPS\text{-}EQ.Verify}(\mathsf{sps_{pp}},$	$\xleftarrow{\sigma}$	$\sigma \leftarrow \mathsf{SPS\text{-}EQ.Sign}(\mathsf{sps_{pp}},$
$(C_1, C_2, P_1, C_4, C_5, \mathsf{upk}_1, \mathsf{apk}), \sigma, \mathsf{opk})$		$(C_1, C_2, P_1, C_4, C_5, \mathsf{upk}_1, \mathsf{apk}), \mathsf{osk})$
return $\mathsf{cred} \leftarrow (C_1, C_4, C_5, \sigma, r_4, \mathsf{nym}, O)$		

Fig. 2. Protego: obtain and issue algorithms.

attaching a showing proof to their endorements. Even if the endorsement policies are defined in a privacy-preserving way as suggested in [3], endorsers can still compute selective AND and NAND clauses for the respective pseudonyms defined by the policy using their credentials. Endorsers should also use the read and write sets to from the transaction proposals to generate showing proofs.

Security Proofs. We present the main theorems and proofs for Protego (which are analogous for Protego Duo). Correctness and issuer-hiding follow from [12].

Theorem 1. *If the q-co-DL assumption holds, the ZKPoK's have perfect ZK,* SCDS *is sound,* SPS-EQ *is* EUF-CMA *secure, and* RevAcc *is collision-free then Protego is unforgeable.*

Proof Sketch. The proof follows from [12] (Th. 6) and [14] (Th. 3) whereby we assume there is an efficient adversary $\mathcal{A}$ winning the unforgeability game with non-negligible probability. We use $\mathcal{A}$ considering the following types of attacks:

Type 1. Adversary $\mathcal{A}$ conducts a valid showing so that $\mathsf{nym}^* = \perp$. Then we construct an adversary $\mathcal{B}$ that uses $\mathcal{A}$ to break the EUF-CMA security.

Type 2. Adversary $\mathcal{A}$ manages to conduct a showing accepted by the verifier using the credential of user i^* under nym^* with respect to $\mathcal{S}^*$ such that $\mathcal{S}^* \not\subseteq \mathtt{ATTR}[\mathsf{nym}]$ or with respect to $\mathcal{D}^*$ such that $\mathcal{D}^* \subseteq \mathtt{ATTR}[\mathsf{nym}]$ holds. Then we construct an adversary $\mathcal{B}$ that uses $\mathcal{A}$ to break the soundness of the set-commitment scheme SCDS.

Type 3. Adversary $\mathcal{A}$ manages to conduct a showing accepted by the verifier reusing a showing based on the credential of a user i^* under nym^* with $i^* \in \mathtt{HU}$, whose secret usk_{i^*} and credentials it does not know.

$\underline{\mathsf{Show}(\mathsf{pp}, \mathsf{usk}, \mathsf{upk}, \mathsf{opk}_j, \mathsf{cred}, \mathcal{S}, \mathcal{D}, (\mathsf{opk}_i)_{i\in[n]}, (\mathsf{opk}_j^i, w_j^i)_{i\in[2]}, \Omega, \mathbb{R}, \mathsf{apk}, \mathsf{tx})}$

$(C_1, C_4, C_5, \sigma, r, \mathsf{nym}, O) \leftarrow \mathsf{cred};\ (\Pi_{\mathsf{rev}}, \mathsf{WIT}) \leftarrow \mathbb{R}; \beta, \mu, \rho, \gamma, \tau, (r_i)_{i\in[5]} \xleftarrow{\$} \mathbb{Z}_p^*$
if $O = (1, (o_1, o_2))$ **then** $O' = (1, (\mu \cdot o_1, o_2))$ **else** $O' = \mu \cdot O$
$\mathsf{opk}_j' \leftarrow \mathsf{ConvertPK}(\mathsf{opk}_j, \rho w_j^1 + \gamma w_j^2); \Omega' \leftarrow \mathsf{SH.ZKEval}(\mathsf{opk}_j^1, \mathsf{opk}_j^2, \Omega; \rho, \gamma)$
$\sigma' \xleftarrow{\$} \mathsf{SPS\text{-}EQ.ChgRep}(\mathsf{sps}_{\mathsf{pp}}, (C_1, rC_1, P_1, C_4, C_5, \mathsf{upk}_1, \mathsf{apk}), \sigma, \mu, \rho w_j^1 + \gamma w_j^2, \mathsf{opk}_j)$
$\mathsf{cred}' \leftarrow ((C_i)_{i\in[7]} = \mu \cdot (C_1, rC_1, P_1, C_4, C_5, \mathsf{upk}_1, \mathsf{apk}), \sigma')$
$\mathsf{wss} \leftarrow \mathsf{SCDS.OpenSS}(\mathsf{scds}_{\mathsf{pp}}, C_1, \mathcal{S}, O'); \mathsf{wds} \leftarrow \mathsf{SCDS.OpenDS}(\mathsf{scds}_{\mathsf{pp}}, C_1, \mathcal{D}, O')$
$\mathsf{wss}_{\mathsf{rev}} \leftarrow \mathsf{WIT}[\mathsf{nym}]; \mathsf{wss}_{\mathsf{rev}}' \leftarrow (\tau \mathsf{wss}_{\mathsf{rev}}^1, \mathsf{usk}_2 \mu\tau \mathsf{wss}_{\mathsf{rev}}^2 P_1)$
$a_1 \leftarrow r_1 C_1; a_2 \leftarrow r_2 P_1; a_3 \leftarrow r_3 \Pi_{\mathsf{rev}}; a_4 \leftarrow r_4 Q; a_5 \leftarrow r_5 P_1; \Pi_{\mathsf{rev}}' \leftarrow (\mathsf{usk}_2 \mu\tau) \Pi_{\mathsf{rev}}$
$(\mathsf{enc}, \alpha) \leftarrow \mathsf{AuditEnc}(\mathsf{apk}, \mathsf{upk}_1); t_1 = \beta P_2; t_2 = \beta\mu P_2; t_3 = \alpha\beta P_2$
$\Pi \leftarrow \mathsf{AuditPrv}(\mathsf{enc}, \alpha, \mathsf{usk}, \mathsf{apk})$
$e \leftarrow \mathcal{H}(\mathcal{S}, \mathcal{D}, \mathsf{apk}, \mathsf{tx}, \mathsf{enc}, \Pi, \mathsf{opk}_j', (\mathsf{opk}_i)_{i\in[n]}, \Omega', (a_i)_{i\in[5]}, (t_i)_{i\in[3]}, \mathsf{cred}', \mathsf{wss}, \mathsf{wds}, \mathsf{tx},$
$C_2, C_3, \Pi_{\mathsf{rev}}', C_5, \mathsf{wss}_{\mathsf{rev}}')$
$z_1 \leftarrow r_1 + e \cdot r;\ z_2 \leftarrow r_2 + e \cdot \mu;\ z_3 \leftarrow r_3 + e \cdot (\mathsf{usk}_2 \mu\tau);\ z_4 \leftarrow r_4 + e \cdot (\mathsf{usk}_2 \mu)$
$z_5 \leftarrow r_5 + e \cdot (\mathsf{usk}_2 \mu\tau \mathsf{wss}_{\mathsf{rev}}^2)$
$\pi_1 \leftarrow \mathsf{SCDS.PoE}(\mathsf{scds}_{\mathsf{pp}}, \mathcal{S}, e); \pi_2 \leftarrow \mathsf{SCDS.PoE}(\mathsf{scds}_{\mathsf{pp}}, \mathcal{D}, e)$
return $(\mathsf{enc}, (t_i)_{i\in[3]}, \mathsf{opk}', (\mathsf{opk}_i)_{i\in[n]}, \Omega', \mathsf{cred}', \mathsf{wss}, \mathsf{wds}, \mathsf{wss}_{\mathsf{rev}}', \Pi_{\mathsf{rev}}', \Pi, \pi_1, \pi_2, (a_i, z_i)_{i\in[5]})$

$\underline{\mathsf{Verify}(\mathsf{pp}, \mathcal{S}, \mathcal{D}, \Pi_{\mathsf{rev}}, \mathsf{rpk}, \mathsf{apk}, \mathsf{tx}, \Omega)}$

$(\mathsf{enc}, (t_i)_{i\in[3]}, \mathsf{opk}', (\mathsf{opk}_i)_{i\in[n]}, \Omega', \mathsf{cred}', \mathsf{wss}, \mathsf{wds}, \mathsf{wss}_{\mathsf{rev}}', \Pi_{\mathsf{rev}}', \Pi, \pi_1, \pi_2, (a_i, z_i)_{i\in[5]}) \leftarrow \Omega$
$(C_1, C_2, C_3, C_4, C_5, C_6, C_7, \sigma) \leftarrow \mathsf{cred}'$
$e \leftarrow \mathcal{H}(\mathcal{S}, \mathcal{D}, \mathsf{apk}, \mathsf{tx}, \mathsf{enc}, \Pi, \mathsf{opk}', (\mathsf{opk}_i)_{i\in[n]}, \Omega', (a_i)_{i\in[5]}, (t_i)_{i\in[3]}, \mathsf{cred}', \mathsf{wss}, \mathsf{wds}, \mathsf{tx},$
$C_2, C_3, \Pi_{\mathsf{rev}}', C_5, \mathsf{wss}_{\mathsf{rev}}')$
check
$z_1 C_1 = a_1 + eC_2;\ z_2 P_1 = a_2 + eC_3;\ z_3 \Pi_{\mathsf{rev}} = a_3 + e\Pi_{\mathsf{rev}}';\ z_4 Q = a_4 + eC_5$
$z_5 P_1 = a_5 + e\mathsf{wss}_{\mathsf{rev}}';\ \mathsf{RevAcc.VerifyWit}(\Pi_{\mathsf{rev}}', C_4, \mathsf{wss}_{\mathsf{rev}}');\ \mathsf{AuditVerify}(\mathsf{enc}, \Pi_2)$
$e(\mathsf{enc}_1, t_2) = e(C_6, t_1) + e(C_7, t_3);\ e(\mathsf{enc}_2, t_2) = e(C_3, t_3);\ e(\mathsf{enc}_2, t_1) = e(P_1, t_3)$
$\mathsf{SCDS.VerifySS}(C_1, \mathcal{S}, \mathsf{wss}; \pi_1, e);\ \mathsf{SCDS.VerifyDS}(C_1, \mathcal{D}, \mathsf{wds}; \pi_2, e)$
$\mathsf{SH.Verify}((\mathsf{opk}_i)_{i\in[n]}, \mathsf{opk}', \Omega');\ \mathsf{SPS\text{-}EQ.Verify}(\mathsf{cred}', \mathsf{opk}')$

Fig. 3. Protego: show and verify algorithms.

Type 4. Adversary $\mathcal{A}$ manages to conduct a showing accepted by the verifier using some credential corresponding to a revoked pseudonym $\mathsf{nym}^* \in \mathsf{RNYM}$. Then, we construct an adversary $\mathcal{B}$ that uses $\mathcal{A}$ to break the binding property of the revocation accumulator RevAcc.

Types 1 and 2 follow the proofs of [12] (Th. 6) as the underlying primitives remain unchanged. For Type 3, we leverage the fact that reusing a showing would only allow the adversary to generate a valid showing for *the same* original transaction tx (that is timestamped), and hence, we do not consider it as an attack. Observe that any modification done to the original tx will lead to a different challenge and thus the rest of the proofs (showing, revocation and auditing) will not pass. Finally, Type 4 follows from [14] (Th. 3). □

Theorem 2. *If the DDH assumption holds, the* SPS-EQ *perfectly adapts signatures, and* $\mathcal{H}$ *is assumed to be a random oracle, then Protego is anonymous.*

Proof Sketch. The proof follows from [12] (Th. 7) and [14] (Th. 4). However, we must also to take into account the RO model and the addition of the auditing

$\underline{\mathsf{Show}(\mathsf{pp}, \mathsf{usk}, \mathsf{upk}, \mathsf{opk}_j, \mathsf{cred}, \mathcal{S}, \mathcal{D}, \mathsf{opk}_j, \sigma_j, \mathbb{R}, \mathsf{apk}, \mathsf{tx})}$

$(C_1, C_4, C_5, \sigma, r, \mathsf{nym}, O) \leftarrow \mathsf{cred}$; $(\Pi_{\mathsf{rev}}, \mathsf{WIT}) \leftarrow \mathbb{R}$; $\beta, \mu, \rho, \gamma, \tau, (r_i)_{i\in[5]} \overset{\$}{\leftarrow} \mathbb{Z}_p^*$
if $O = (1, (o_1, o_2))$ **then** $O' = (1, (\mu \cdot o_1, o_2))$ **else** $O' = \mu \cdot O$
$\mathsf{opk}'_j \leftarrow \mathsf{ConvertPK}(\mathsf{opk}_j, \rho)$; $\sigma'_j \overset{\$}{\leftarrow} \mathsf{SPS\text{-}EQ.ChgRep}(\mathsf{sps_{pp}}, \mathsf{opk}_j, \sigma_j, \rho)$
$\sigma' \overset{\$}{\leftarrow} \mathsf{SPS\text{-}EQ.ChgRep}(\mathsf{sps_{pp}}, (C_1, rC_1, P_1, C_4, C_5, \mathsf{upk}_1, \mathsf{apk}), \sigma, \mu, \rho, \mathsf{opk}_j)$
$\mathsf{cred}' \leftarrow ((C_i)_{i\in[7]} = \mu \cdot (C_1, rC_1, P_1, C_4, C_5, \mathsf{upk}_1, \mathsf{apk}), \sigma')$
$\mathsf{wss} \leftarrow \mathsf{SCDS.OpenSS}(\mathsf{scds_{pp}}, C_1, \mathcal{S}, O')$; $\mathsf{wds} \leftarrow \mathsf{SCDS.OpenDS}(\mathsf{scds_{pp}}, C_1, \mathcal{D}, O')$
$\mathsf{wss_{rev}} \leftarrow \mathsf{WIT}[\mathsf{nym}]$; $\mathsf{wss}'_{\mathsf{rev}} \leftarrow (\tau \mathsf{wss}^1_{\mathsf{rev}}, \mathsf{usk}_2\mu\tau\mathsf{wss}^2_{\mathsf{rev}}P_1)$
$a_1 \leftarrow r_1C_1; a_2 \leftarrow r_2P_1; a_3 \leftarrow r_3\Pi_{\mathsf{rev}}; a_4 \leftarrow r_4Q; a_5 \leftarrow r_5P_1; \Pi'_{\mathsf{rev}} \leftarrow (\mathsf{usk}_2\mu\tau)\Pi_{\mathsf{rev}}$
$(\mathsf{enc}, \alpha) \leftarrow \mathsf{AuditEnc}(\mathsf{apk}, \mathsf{upk}_1)$; $t_1 = \beta P_2; t_2 = \beta\mu P_2; t_3 = \alpha\beta P_2$
$\Pi \leftarrow \mathsf{AuditPrv}(\mathsf{enc}, \alpha, \mathsf{usk}, \mathsf{apk})$
$e \leftarrow \mathcal{H}(\mathcal{S}, \mathcal{D}, \mathsf{apk}, \mathsf{tx}, \mathsf{enc}, \Pi, \mathsf{opk}'_j, \sigma'_j, (a_i)_{i\in[5]}, (t_i)_{i\in[3]}, \mathsf{cred}', \mathsf{wss}, \mathsf{wds}, \mathsf{tx},$
$C_2, C_3, \Pi'_{\mathsf{rev}}, C_5, \mathsf{wss}'_{\mathsf{rev}})$
$z_1 \leftarrow r_1 + e \cdot r;\ z_2 \leftarrow r_2 + e \cdot \mu;\ z_3 \leftarrow r_3 + e \cdot (\mathsf{usk}_2\mu\tau);\ z_4 \leftarrow r_4 + e \cdot (\mathsf{usk}_2\mu)$
$z_5 \leftarrow r_5 + e \cdot (\mathsf{usk}_2\mu\tau\mathsf{wss}^2_{\mathsf{rev}})$
$\pi_1 \leftarrow \mathsf{SCDS.PoE}(\mathsf{scds_{pp}}, \mathcal{S}, e)$; $\pi_2 \leftarrow \mathsf{SCDS.PoE}(\mathsf{scds_{pp}}, \mathcal{D}, e)$
return $(\mathsf{enc}, (t_i)_{i\in[3]}, \mathsf{opk}'_j, \sigma'_j, \mathsf{cred}', \mathsf{wss}, \mathsf{wds}, \mathsf{wss}'_{\mathsf{rev}}, \Pi'_{\mathsf{rev}}, \Pi, \pi_1, \pi_2, (a_i, z_i)_{i\in[5]})$

$\underline{\mathsf{Verify}(\mathsf{pp}, \mathcal{S}, \mathcal{D}, \Pi_{\mathsf{rev}}, \mathsf{rpk}, \mathsf{apk}, \mathsf{vpk}, \mathsf{tx}, \Omega)}$

$(\mathsf{enc}, (t_i)_{i\in[3]}, \mathsf{opk}', \sigma', \mathsf{cred}', \mathsf{wss}, \mathsf{wds}, \mathsf{wss}'_{\mathsf{rev}}, \Pi'_{\mathsf{rev}}, \Pi, \pi_1, \pi_2, (a_i, z_i)_{i\in[5]}) \leftarrow \Omega$
$(C_1, C_2, C_3, C_4, C_5, C_6, C_7, \sigma) \leftarrow \mathsf{cred}'$
$e \leftarrow \mathcal{H}(\mathcal{S}, \mathcal{D}, \mathsf{apk}, \mathsf{tx}, \mathsf{enc}, \Pi, \mathsf{opk}', \sigma', (a_i)_{i\in[5]}, (t_i)_{i\in[3]}, \mathsf{cred}', \mathsf{wss}, \mathsf{wds}, \mathsf{tx},$
$C_2, C_3, \Pi'_{\mathsf{rev}}, C_5, \mathsf{wss}'_{\mathsf{rev}})$
check
$z_1C_1 = a_1 + eC_2;\ z_2P_1 = a_2 + eC_3;\ z_3\Pi_{\mathsf{rev}} = a_3 + e\Pi'_{\mathsf{rev}};\ z_4Q = a_4 + eC_5$
$z_5P_1 = a_5 + e\mathsf{wss}'_{\mathsf{rev}}$; $\mathsf{RevAcc.VerifyWit}(\Pi'_{\mathsf{rev}}, C_4, \mathsf{wss}'_{\mathsf{rev}})$; $\mathsf{AuditVerify}(\mathsf{enc}, \Pi_2)$
$e(\mathsf{enc}_1, t_2) = e(C_6, t_1) + e(C_7, t_3);\ e(\mathsf{enc}_2, t_2) = e(C_3, t_3);\ e(\mathsf{enc}_2, t_1) = e(P_1, t_3)$
$\mathsf{SCDS.VerifySS}(C_1, \mathcal{S}, \mathsf{wss}; \pi_1, e)$; $\mathsf{SCDS.VerifyDS}(C_1, \mathcal{D}, \mathsf{wds}; \pi_2, e)$
$\mathsf{SPS\text{-}EQ.Verify}(\mathsf{opk}', \mathsf{vpk})$; $\mathsf{SPS\text{-}EQ.Verify}(\mathsf{cred}', \mathsf{opk}')$

Fig. 4. Protego Duo: show and verify algorithms.

features. The extra credential components for the auditing are randomized during every credential showing like the rest of the components. Similarly, the user generates a new encryption of the auditor's public key with a fresh α, while a fresh β is used to randomize the values t_i. Since ElGamal encryption is IND-CPA secure and key-private [5], the ciphertexts produced by the user are indistinguishable and do not leak information about the user's public key nor the auditor's. □

Theorem 3. *If the algorithms* AuditPrv *and* AuditVerify *are a NIZK proof system and the* SPS-EQ *is* EUF-CMA *secure then Protego is auditable.*

Proof. If the verification returns true, we have that $\exists\ (\mathsf{enc}_1^*, \mathsf{enc}_2^*) = ((\delta^* + \alpha^*\mathsf{ask})P_1, \alpha^*P_1)$ for some δ^* and α^* chosen by the adversary. Moreover, because of the unforgeability of the signature scheme, the verification implies that $C_3 = \mu^*P_1$, $C_6 = \mu^*\mathsf{usk}_1P_1$ and $C_7 = \mu^*\mathsf{ask}P_1$ for some μ^* chosen by the adversary.

Table 1. Running time for the different algorithms in milliseconds.

	Revocation				Signature			Issuer-hiding NIZK		
	$n = 10$		$n = 100$		$\ell = 7$ (for Protego)			$n = 5$		
Scheme	Prove	Verify	Prove	Verify	Sign	Verify	ChgRep	Prove	Verify	ZKEval
[7]	28	64	**28**	64	23	59	N/A	N/A	N/A	N/A
Protego	**7.7**	**4.2**	77.4	**4.2**	**3.4**	**11**	8.8	103	118	59

As a result, we can re-write the pairing equations for the audit proof as:

$$e(\alpha^* P_1, t_2^*) = e(\mu^* P_1, t_3^*)$$
$$e(\alpha^* P_1, t_1^*) = e(P_1, t_3^*)$$
$$e((\delta^* + \alpha^* \mathsf{ask}) P_1, t_2^*) = e(\mu^* \mathsf{usk}_1 P_1, t_1^*) + e(\mu^* \mathsf{ask} P_1, t_3^*)$$

where t_1^*, t_2^* and t_3^* are also chosen by the adversary. We show that $\delta^* = \mathsf{usk}_1$, which implies that $\mathsf{upk}_1 = \mathsf{AuditDec}(\mathsf{enc}, \mathsf{ask})$. Looking at the first two equations in the target group, we have that $\alpha^* t_2^* = \mu^* t_3^*$ and $\alpha^* t_1^* = t_3^*$, concluding that $t_2^* = \mu^* t_1^*$. Replacing t_2^* and t_3^* in third one and simplyfing we obtain $(\delta^* + \alpha^* \mathsf{ask})\mu^* t_1^* = \mu^* \mathsf{usk}_1 t_1^* + \mu \mathsf{ask} \alpha^* t_1^*$. Therefore, we have $\mu^* \delta^* t_1^* + \mu^* \alpha^* \mathsf{ask} t_1^* = \mu^* \mathsf{usk}_1 t_1^* + \mu^* \alpha^* \mathsf{ask} t_1^*$, deducing that $\delta^* = \mathsf{usk}_1$. □

5 Evaluation

We implemented a prototype of Protego and Protego Duo (available in [1]), using Rust with the bls12-381 curve and the BLAKE3 hash function. Our signature implementation is based on the one from [8] but using the bls12-381 curve instead of BN curves [4]. As a result, we obtain times up to 67% faster when compared to [8]. To run the benchmarks a Macbook Air (Chip M2 & 16GB RAM) was used with no extra optimizations, using the nightly compiler, and the *Criterion* library. For all values, the standard deviation was at most 1 millisecond.

Issue and Obtain take roughly 20 ms each when issuing a credential for 10 attributes. Both scale linearly on the number of attributes. To evaluate showing and verification, we considered the PoE in the showing protocol. Therefore, verification running time remains (almost) constant[1] regardless the number of shown attributes, credential size, and issuer-hiding approach. If the PoE is disabled, showing running time would be smaller while verification would increase linearly with the number of shown attributes. An auditing proof in Protego takes roughly 1 and 1.5 ms for proof generation and verification, surpassing the values from [7]. In Table 1 we report the revocation and signing algorithms, including our issuer-hiding NIZK with $n = 5$. For Protego, we consider a signature for vectors of length seven (the size of a credential). In our case, the revocation

[1] Asymptotic complexity is O(1) (considering exponentiations and pairings) but some multiplications depending on the shown attributes are required, hence the difference.

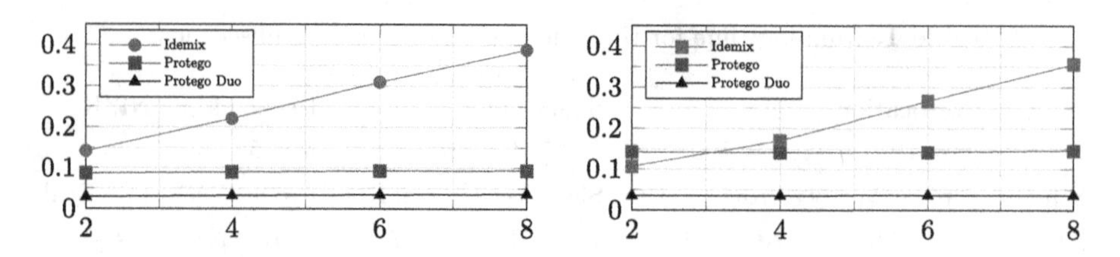

Fig. 5. From left to right, showing and verification times (in seconds) for the different schemes considering credential showings for 2, 4, 6 and 8 attributes.

Table 2. Protocols' comparison showing the running times in milliseconds.

	$k=2$		$k=4$		$k=6$		$k=8$		$k=10$	
Scheme	Show	Verify	Show	Verify	Show	Verify	Show	Verify	Show	Verify
[7]	141	106	220	170	309	266	388	356	–	–
Protego	86	142	89	140	92	141	93	145	96	145
Protego Duo	**29**	**35**	**32**	**35**	**34**	**36**	**37**	**36**	**39**	**38**

witnesses are computed by the authority (in linear time) and then randomized by the users (in constant time). For this reason we consider the generation of a single witness for a revocation lists of 10 and a hundred elements (although in practice one would expect it to be closer to 10). For [7], we consider the total time to generate and verify a signature in a user level $L = 2$ (involving two delegations), with revocation times in $\mathbb{G}_2$.

Comparison with the Idemix Extension from [7]**.** The computational cost for the prover and verifier grows linearly with the number of attributes in the credential and delegation levels for [7]. In Protego Duo, the prover computational cost is $O(n-k)$ for showings involving k-attributes out of n, which in practice is much better. Verification cost in Protego and Protego Duo is almost constant (or $\mathrm{O}(k)$ if the PoE is disabled). The two works are compared in Fig. 5 using the same hardware (exact times are also given in Table 2). For [7], we consider a delegation level $L = 2$, which corresponds to a user level given that the root authority is at $L = 0$ and organizations start at $L = 1$. Regarding the attributes, [7] we could only retrieve information considering proofs for credential possesion below ten attributes (assuming a minimal overhead when all attributes are shown as authors suggest). Therefore, we report credential possesions for [7] considering up to 8 attributes, and selective disclosures of k-out-of-10 attributes in ours. For Protego, we consider five authorities for the NIZK proof, which would suffice for practical scenarios like a consortium of pharmaceuticals.

6 Conclusions and Future Work

We presented here the first SPS-EQ credential scheme modified to work with permissioned blockchains. The versatility of Protego alongside the efficiency gains (at least twice as fast as the most recent Idemix extension), enables a broader scope of applications in such a setting. Depending on the context, the PoE's can be computed or not, the credential issuer can be hidden or not, and one can select only subsets or disjoint sets to generate the proofs. Similarly, auditability and revocation features can be considered as optional, showing its flexibility.

As future directions to explore, we consider the following points: (1) adding confidentiality of transactions to a Protego-like credential scheme, (2) adding more power to the users (*i.e.*, how to define precise notions of user-invoked regulatory measures), and (3) extend our results to the multi-authority setting, where users can get attributes from multiple authorities instead of a single one.

Acknowledgements. We thank the anonymous reviewers for their valuable feedback. The European Commission partially supported Octavio Perez Kempner's work as part of the CUREX project (H2020-SC1-FA-DTS-2018-1 under grant agreement No 826404).

A Our NIZK Argument for Issuer-hiding

We refer the reader to [12] (Sect. 3.1) for the basic syntax and security properties of malleable NIZK proof systems. In Fig. 6 we build a fully adaptive malleable NIZK argument following the construction from [12]. The main idea is that given two proofs π_1 and π_2 for statements $\mathbf{x}_1 = w_1\mathbf{v}_i$ and $\mathbf{x}_2 = w_2\mathbf{v}_i$, one can compute

$\underline{\mathsf{SH.PGen}(1^\lambda)}$:
$\mathsf{BG} \xleftarrow{\$} \mathsf{BGGen}(1^\lambda);\ z \xleftarrow{\$} \mathbb{Z}_p$
return $(\mathsf{BG}, [z]_1)$

$\underline{\mathsf{SH.PSim}(\mathsf{crs}, \mathsf{td}, (\mathbf{v}_i)_{i\in[n]}, [\mathbf{x}_1]_2, [\mathbf{x}_2]_2)}$:
$\delta, z_1, \ldots, z_{n-1} \xleftarrow{\$} \mathbb{Z}_p^*$
$z_n \leftarrow \delta\mathsf{td} - \sum_{i=1}^{i=n-1} z_j$
for all $\mathbf{i} \in [n]$ **do**
  $d_i \xleftarrow{\$} \mathbb{Z}_p;\ [\mathbf{a}_i]_2 \leftarrow d_i \cdot \mathbf{v}_i - z_i \cdot \mathbf{x}$
return $(([\mathbf{a}_n]_2, [d_n]_1, [z_n]_1)_{n\in[n]}, \delta P_2)$

$\underline{\mathsf{SH.ZKEval}(\mathsf{crs}, [\mathbf{x}_1]_2, [\mathbf{x}_2]_2, \pi; \alpha, \beta)}$:
// $[\mathbf{x}']_2 = (\alpha w_1 + \beta w_2)[\mathbf{v}_i]_2$
$(([\mathbf{a}_n^j]_2, [d_n^j]_1, [z_n]_1)_{n\in[n]}^{j\in[2]}, Z_2) \leftarrow \pi$
$\delta \xleftarrow{\$} \mathbb{Z}_p^*;\ Z_2' \leftarrow \delta Z_2$
for all $i \in [n]$ **do**
  $[z_i']_1 \leftarrow \delta[z_i]_1;$
  $[d_i']_2 \leftarrow \delta\alpha[d_i^1]_2 + \delta\beta[d_i^2]_2;$
  $[\mathbf{a}_i']_2 \leftarrow \delta\alpha[\mathbf{a}_i^1]_2 + \delta\beta[\mathbf{a}_i^2]_2;$
return $(([\mathbf{a}_n']_2, [d_n']_1, [z_n']_1)_{n\in[n]}, Z_2')$

$\underline{\mathsf{SH.TPGen}(1^\lambda)}$:
$\mathsf{BG} \xleftarrow{\$} \mathsf{BGGen}(1^\lambda);\ z \xleftarrow{\$} \mathbb{Z}_p;\ \mathsf{td} \leftarrow z$
return $(\mathsf{BG}, [z]_1, \mathsf{td})$

$\underline{\mathsf{SH.Prove}(\mathsf{crs}, ([\mathbf{v}_i]_2)_{i\in[n]}, ([\mathbf{x}_j]_2, w_j)_{j\in[2]})}$:
// $[\mathbf{x}_1]_2 = w_1[\mathbf{v}_i]_2,\ [\mathbf{x}_2]_2 = w_2[\mathbf{v}_i]_2$
$\delta, r_1, r_2, z_1, \ldots, z_{n-1} \xleftarrow{\$} \mathbb{Z}_p^*$
$[z_n]_1 \leftarrow \delta[z]_1 - \sum_{i=1}^{i=n-1}[z_i]_1$
$([\mathbf{a}_i^j]_2, [d_i^j]_1) \leftarrow (r_j[\mathbf{v}_i]_2, w_j[z_i]_1 + [r_j]_1)$
for all $k \neq i \in [n],\ j \in [2]$ **do**
  $d_k^j \xleftarrow{\$} \mathbb{Z}_p;\ [\mathbf{a}_k^j]_2 \leftarrow d_k^j[\mathbf{v}_k]_2 - z_k[\mathbf{x}_j]_2$
return $(([\mathbf{a}_n^j]_2, [d_n^j]_1, [z_n]_1)_{n\in[n]}^{j\in[2]}, \delta P_2)$

$\underline{\mathsf{SH.Verify}(\mathsf{crs}, ([\mathbf{v}_i]_2)_{i\in[n]}, [\mathbf{x}]_2, \pi)}$:
$(([\mathbf{a}_n]_2, [d_n]_1, [z_n]_1)_{n\in[n]}, Z_2) \leftarrow \pi$
check $e([z]_1, Z_2) = e(\sum_{i=1}^{i=n}[z_i]_1, [1]_2)$
for all $\mathbf{i} \in [n]$ **do**
  check $e([d_i]_1, [\mathbf{v}_i]_2) = e([z_i]_1, [\mathbf{x}]_2) + e([1]_1, [\mathbf{a}_i]_2)$

Fig. 6. Our fully adaptive malleable NIZK argument

a valid proof π for the statement $\mathbf{x} = (\alpha w_1 + \beta w_2)\mathbf{v}_i$ with fresh α and β. The derivation privacy property of the proof system ensures that π looks like a freshly computed proof. Security follows from theorems 2 and 8 from [12].

References

1. Implementation. https://github.com/octaviopk9/indocrypt_protego
2. Androulaki, E., Camenisch, J., Caro, A.D., Dubovitskaya, M., Elkhiyaoui, K., Tackmann, B.: Privacy-preserving auditable token payments in a permissioned blockchain system. In: Proceedings of the 2nd ACM Conference on Advances in Financial Technologies. AFT 2020, pp. 255–267. Association for Computing Machinery, New York (2020)
3. Androulaki, E., De Caro, A., Neugschwandtner, M., Sorniotti, A.: Endorsement in hyperledger fabric. In: 2019 IEEE International Conference on Blockchain (Blockchain), pp. 510–519 (2019)
4. Barreto, P.S.L.M., Naehrig, M.: Pairing-friendly elliptic curves of prime order, pp. 319–331 (2006). https://doi.org/10.1007/11693383_22
5. Bellare, M., Boldyreva, A., Desai, A., Pointcheval, D.: Key-privacy in public-key encryption, pp. 566–582 (2001). https://doi.org/10.1007/3-540-45682-1_33
6. Bobolz, J., Eidens, F., Krenn, S., Ramacher, S., Samelin, K.: Issuer-hiding attribute-based credentials. In: Conti, M., Stevens, M., Krenn, S. (eds.) CANS 2021. LNCS, vol. 13099, pp. 158–178. Springer, Cham (2021). https://doi.org/10.1007/978-3-030-92548-2_9
7. Bogatov, D., De Caro, A., Elkhiyaoui, K., Tackmann, B.: Anonymous transactions with revocation and auditing in hyperledger fabric. In: Conti, M., Stevens, M., Krenn, S. (eds.) CANS 2021. LNCS, vol. 13099, pp. 435–459. Springer, Cham (2021). https://doi.org/10.1007/978-3-030-92548-2_23
8. Burkhart, M.: Mercurial signatures implementation. Github (2020). https://github.com/burkh4rt/Mercurial-Signatures
9. Camenisch, J., Drijvers, M., Dubovitskaya, M.: Practical UC-secure delegatable credentials with attributes and their application to blockchain, pp. 683–699 (2017). https://doi.org/10.1145/3133956.3134025
10. Camenisch, J., Lysyanskaya, A.: Signature schemes and anonymous credentials from bilinear maps, pp. 56–72 (2004). https://doi.org/10.1007/978-3-540-28628-8_4
11. Camenisch, J., Van Herreweghen, E.: Design and implementation of the idemix anonymous credential system, pp. 21–30 (2002). https://doi.org/10.1145/586110.586114
12. Connolly, A., Lafourcade, P., Perez Kempner, O.: Improved Constructions of Anonymous Credentials from Structure-Preserving Signatures on Equivalence Classes. In: Hanaoka, G., Shikata, J., Watanabe, Y. (eds.) PKC 2022. LNCS, vol. 13177, pp. 409–438. Springer, Cham (2022). https://doi.org/10.1007/978-3-030-97121-2_15
13. Crites, E.C., Lysyanskaya, A.: Delegatable anonymous credentials from mercurial signatures, pp. 535–555 (2019). https://doi.org/10.1007/978-3-030-12612-4_27
14. Derler, D., Hanser, C., Slamanig, D.: A new approach to efficient revocable attribute-based anonymous credentials, pp. 57–74 (2015). https://doi.org/10.1007/978-3-319-27239-9_4

15. Fuchsbauer, G., Hanser, C., Slamanig, D.: Structure-preserving signatures on equivalence classes and constant-size anonymous credentials. J. Cryptol. **32**(2), 498–546 (2018). https://doi.org/10.1007/s00145-018-9281-4
16. Kang, H., Dai, T., Jean-Louis, N., Tao, S., Gu, X.: Fabzk: supporting privacy-preserving, auditable smart contracts in hyperledger fabric. In: 2019 49th Annual IEEE/IFIP International Conference on Dependable Systems and Networks (DSN), pp. 543–555 (2019)
17. Mazumdar, S., Ruj, S.: Design of anonymous endorsement system in hyperledger fabric. IEEE Trans. Emerg. Top. Comput. 1 (2019)
18. Narula, N., Vasquez, W., Virza, M.: Zkledger: privacy-preserving auditing for distributed ledgers. In: Proceedings of the 15th USENIX Conference on Networked Systems Design and Implementation. NSDI 2018, pp. 65–80. USENIX Association, USA (2018)
19. Schnorr, C.P.: Efficient identification and signatures for smart cards, pp. 239–252 (1990). https://doi.org/10.1007/0-387-34805-0_22
20. Shoup, V.: Lower bounds for discrete logarithms and related problems, pp. 256–266 (1997). https://doi.org/10.1007/3-540-69053-0_18
21. Zurich, I.R.: Specification of the identity mixer cryptographic library v2.3.0 (2013)

Hybrid Scalar/Vector Implementations of Keccak and SPHINCS^{+} on AArch64

Hanno Becker[1(✉)] and Matthias J. Kannwischer[2(✉)]

[1] Arm Limited, Cambridge, UK
hanno.becker@arm.com
[2] Academia Sinica, Taipei, Taiwan
matthias@kannwischer.eu

Abstract. This paper presents two new techniques for the fast implementation of the Keccak permutation on the A-profile of the Armarchitecture: First, the elimination of explicit rotations in the Keccak permutation through Barrel shifting, applicable to scalar AArch64 implementations of `Keccak-f1600`. Second, the construction of hybrid implementations concurrently leveraging both the scalar and the Neon instruction sets of AArch64. The resulting performance improvements are demonstrated in the example of the hash-based signature scheme SPHINCS^{+}, one of the recently announced winners of the NIST post-quantum cryptography project: We achieve up to 1.89× performance improvements compared to the state of the art. Our implementations target the Arm Cortex-{A55,A510,A78,A710,X1,X2} processors common in client devices such as mobile phones.

Keywords: Arm · AArch64 · Armv8-A · Keccak · SIMD · Neon · Post-quantum cryptography · SPHINCS+

1 Introduction

Hash functions and extendable-output functions based on the Keccak-ppermutations have gained popularity since their standardization as SHA-3 and SHAKE in FIPS202 [Dwo15] through the US National Institute for Standards and Technology (NIST) in 2012. Post-quantum cryptography (PQC) in particular makes extensive use of SHA-3 and SHAKE as building blocks: In July 2022, NIST announced [ACD+22] the four schemes it intends to include in its first PQC standard – updating the standards for key-establishment [Nat18, Nat19] and digital signatures [Nat13] – and all four selected schemes make use of SHA-3. Among them is the hash-based signature scheme SPHINCS^{+} [HBD+22], and three lattice-based schemes: the key-encapsulation scheme Kyber [SAB+22], and the digital signature schemes Dilithium [LDK+22] and Falcon [PFH+22].

While the selected lattice-based schemes provide very good performance and often outperform classical public-key cryptography, hash-based signatures come at a much higher cost. For example, pqm4 [KPR+] – a benchmarking project

T. Isobe and S. Sarkar (Eds.): INDOCRYPT 2022, LNCS 13774, pp. 272–293, 2022.
https://doi.org/10.1007/978-3-031-22912-1_12

for post-quantum cryptography on the Arm Cortex-M4 – reports 4 million clock cycles for signing of `dilithium2`, 18 million clock cycles for `falcon512-tree`, and 400 million clock cycles for `sphincs-sha256-128f-simple` – the *fastest* SPHINCS$^+$ parameter set. While signing performance appears to favour lattice-based signatures, hash-based signatures come with two important advantages: Firstly, they only rely on the collision-resistance and pre-image resistance of the underlying hash function, while lattice-based signatures rely on computation problems (M-LWE, M-SIS, and NTRU). Secondly, hash-based signatures have much smaller public keys of just 32 to 64 bytes, while Dilithium requires at least 1 312 bytes and Falcon requires at least 897 bytes.

Due to the expected upcoming deployment of SPHINCS$^+$, it is essential to understand the performance of SPHINCS$^+$ on a variety of platforms. Unsurprisingly, the computational bottleneck of hash-based signatures are the invocations of the used hash function and, consequently, having a fast hash implementation results in a fast SPHINCS$^+$ implementation. Furthermore, SPHINCS$^+$ can make use of parallel hash implementations which is particularly useful on platforms providing SIMD instructions allowing to compute multiple hashes at once.

In this work, we study scalar and SIMD implementations of `Keccak-f1600` on the AArch64 instruction set of the Arm architecture, and showcase their performance by integrating them into implementations of SPHINCS$^+$. We target the Arm Cortex-{A55,A510,A78,A710,X1,X2} processors common in client devices such as mobile phones, and which are representative of the breadth of implementations of the A-profile of the Arm architecture across the performance/power/area spectrum.

Contributions. We make the following contributions:

1. We shorten scalar AArch64 implementations of `Keccak-f1600` by trading standalone rotations for extensive use of the Barrel shifter. On our target CPUs, this technique leads to a significant performance improvement.
2. We show that 2-way parallel implementations of `Keccak-f1600` using the Armv8.4-A SHA-3 Neon instructions can sometimes be sped up by also mixing in regular Neon instructions, leading to better hardware utilization.
3. We present Scalar/Neon hybrid implementations for 3-, 4- and 5-way parallel `Keccak-f1600`. They compute a 2-way parallel Keccak on the Neon units in parallel with further permutation(s) on the scalar execution units. We investigate such Scalar/Neon hybrids with and without the SHA-3 instructions.
4. We showcase our `Keccak-f1600` implementations by plugging them into SPHINCS$^+$ and achieve signing speed-ups of up to 1.89× over the state of the art.
5. We present detailed analyses of the relation between our target microarchitectures and optimization potentials for our `Keccak-f1600` implementations.

Source Code. Our implementations are available at https://gitlab.com/arm-research/security/pqax.

Related Work. Kölbl [Köl18] studies the implementations of SPHINCS (the predecessor of SPHINCS$^+$) for AArch64, targeting the Cortex-A57 and Cortex-A72 CPUs. For Keccak, Kölbl makes use of a two-way parallel Neon implementation from the eXtended Keccak code package (XKCP) [DHP+]. Westerbaan [Wes] presents a two-way parallel Neon implementation of Keccak using the Armv8.4-A SHA-3 instructions. This implementation is also used in the SPHINCS$^+$ NIST PQC submission [HBD+22]. Lattice-based cryptography on AArch64 has been studied by Nguyen and Gaj [NG21] as well as Becker, Hwang, Kannwischer, Yang, and Yang [BHK+21]. Hybrid implementations have previously been applied in other contexts: Bernstein and Schwabe [BS12] present a scalar/vector hybrid implementation of the Salsa20 cipher for Armv7-A, and Lenngren [Len19] presents a scalar/vector hybrid implementation of the key-exchange mechanism X25519 for Armv8-A.

Applicability Beyond This Work. Our work has application beyond what is presented in this paper. In particular, it can be useful for the following:

- **Stateful hash-based signatures.** Stateful hash-based signature schemes like XMSS [HBG+18] or LMS [MCF19] can also be implemented in a parallel fashion. Hence, our implementations can be integrated into XMSS or LMS.
- **Other post-quantum candidates.** Other post-quantum schemes also benefit from faster hashing. Notably, Kyber and Dilithium are designed to leverage fast parallel hashing. We therefore believe that our implementations will enable speed-ups for those schemes, but leave a detailed evaluation to future work.
- **KangarooTwelve.** Closely related to SHA-3 is KangarooTwelve [BDP+18] which also builds on the Keccak-ppermutation but uses 12 rounds instead of 24. The techniques presented here apply to KangorooTwelve as well.

Structure. Section 2 provides background on Keccak, SPHINCS$^+$ and the Arm architecture. Section 3 and Sect. 4 present our novel implementation techniques for `Keccak-f1600`, including improvements to scalar implementations, parallel Neon implementations, and as the main novelty, hybrid implementations. Finally, in Sect. 5 we present the performance results for our `Keccak-f1600` and SPHINCS$^+$ implementations on the Cortex-{A55,A510,A78,A710,X1,X2}.

2 Preliminaries

2.1 Keccak

Keccak [BDH+] is a family of permutations, instances of which form the basis of the SHA-3 standard [Dwo15] including the SHA-3 hash functions and the SHAKE extendable output functions (XOF); the reader unfamiliar with the notion of a XOF may think of it as a cryptographic hash function with flexible output size, generalizing the classical use case of hashing arbitrary-size inputs

```
1  // r[x,y], RC[i] are constants fixed in the specification
2  keccak-f1600(A)
3    for i in 0..23:
4      // Θ step
5      C[x]   = A[x,0] xor ... xor A[x,4],          for   x=0..4
6      D[x]   = C[x-1] xor rot(C[x+1],1),          for   x=0..4
7      A[x,y] = A[x,y] xor D[x],                   for x,y=0..4
8      // ρ + π steps
9      B[y,2*x+3*y] = rot(A[x,y], r[x,y]),         for x,y=0..4
10     // χ step
11     A[x,y] = B[x,y] xor
12              ((not B[x+1,y]) and B[x+2,y]),     for x,y=0..4
13     // ι step
14     A[0,0] = A[0,0] xor RC[i]
```

Fig. 1. Pseudocode for Keccak-f1600

into a fixed-size output. The reverse use case – expanding a fixed-sized input into a variable-size output – is useful, for example, for randomness expansion, and is being used for that purpose in the various NIST PQC schemes.

The core of Keccak within SHA-3 is the Keccak permutation Keccak-f1600, operating on a 1600-bit state viewed as a 5×5 matrix $\mathtt{A}[x, y]$ of 64-bit values. It consists of 24 rounds of 5 operations $(\theta, \rho, \pi, \chi, \iota)$ each. χ is the only non-linear operation, while ρ and π are mere bit-permutations, and θ and ι are linear operations. The pseudocode specification of Keccak-f1600 is given in Fig. 1.

2.2 SPHINCS+

Based on SPHINCS [BHH+15], SPHINCS+ [HBD+22] is a stateless hash-based signature scheme that was selected as one of the winners of the NIST PQC project [NIS16]. At the core, SPHINCS+ relies on three building blocks: An improved version of Winternitz One-Time Signatures (WOTS+), the multi-tree version of the eXtended Merkle Signature Scheme (XMSS^{MT}), and the Forest Of Random Subsets (FORS) few time signature scheme. We briefly recall the main concepts and refer to the SPHINCS+ specification [HBD+22] for details.

WOTS+. WOTS+ [Hül13] is a hash-based one-time signature scheme, working roughly as follows: The secret key is a tuple of random values $s_0, s_1, \ldots, s_{\ell-1}$ in the domain of an underlying hash function h, and the public key consists of the repeated hash $h^{2^k-1}(s_i)$, where k is a fixed parameter. Signing works by splitting a message in k-bit blocks $m_i < 2^k$ and revealing the partial preimages $h^{m_i}(s_i)$ of the public keys. Verification checks that they are, in fact, preimages. As stated, this is flawed since knowing the signature for a block m_i allows forging a signature for any $m_i' > m_i$, but this can be fixed through a checksum. We refer the interested reader to [Hül13] for further details.

XMSS^{MT}. The idea of XMSS [BDH11] is to combine multiple one-time public keys into a single many-time public key by means of a hash tree. The leaves of the hash tree correspond to hashes of the one-time public keys, and the root of the hash tree is the XMSS public key. Signing means signing with one of the one-time keys at the leaves, and providing an authentication path to the root of the hash tree. The signer must carefully track which leaf keys have already been used, and never use the same leaf key twice. XMSS^{MT} builds on XMSS, replacing the single hash tree by a hyper-tree, i.e., multiple layers of XMSSwhere the WOTS$^+$ keys on upper layers are used to sign the XMSSroots of the lower layers. By doing so, key generation is limited to the upmost tree and signing only needs to compute relatively small trees. However, this comes at the cost of inflated signature sizes as one XMSS^{MT} consists of multiple XMSSsignatures.

Eliminating the State. SPHINCS$^+$ eliminates the state from XMSS^{MT} by using a very large hyper-tree and pseudo-randomly selecting leaves for signing. As collisions may still occur, it uses FORS on the lowest layer.

SPHINCS$^+$ Parameters. SPHINCS$^+$ specifies 36 parameter sets consisting of all possible combinations of (a) a hash function (SHAKE, SHA-2, or Haraka), (b) a security level (128, 192 or 256 bits), (c) an optimization target (**s** for small signature, or **f** for fast signing), and (d) a tweakable hash function ("simple", comparable to LMS [MCF19], or "robust", comparable to XMSS [BDH11, HBG+18]). Parameter sets are named accordingly, e.g., `sphincs-shake-128f-simple`. In this work, we focus on the SHAKE parameter sets.

(Parallel) Hashing in SPHINCS$^+$. Key generation, signing, and verification in SPHINCS$^+$ are dominated by hashing and benefit from parallelization.

We begin with WOTS$^+$-based XMSS, which offers three independent potentials for parallelization: First, it is straightforward to extend WOTS$^+$ key generation to compute multiple hash chains in parallel. This works for any parallelization degree and benefits XMSS key generation, signing and verification. Second, XMSS key generation and signing require the computation of a large hash tree where the leaves are the hashes of freshly generated WOTS$^+$ public keys. This is dominated by the leaf computations and can be sped up by parallelizing multiple WOTS$^+$ key generations. Again, the approach works for any parallelization degree. Third, for 2/4-fold parallelization, a single hash-tree computation may be further parallelized as demonstrated in [HBD+22], though for a WOTS$^+$-based hash tree, this offers only a negligible performance improvement.

We parallelize XMSS verification via the first approach for parallelization, and XMSS key generation and signing via the second. For 2/4-fold parallelization, we also apply the third approach, but mainly for uniformity with FORS: FORS also relies on tree hashing and benefits from the second and third parallelization approaches – moreover, parallelized tree hashing is much more impactful for FORS due to the cheaper leaf computations. For FORS only, which involves *multiple* hash trees, there is also the potential of performing an N-way parallel hash tree computation, but we leave exploring this for future work.

It should be noted that for degrees of parallelization which are not aligned to the total number of invocations, an overhead occurs. For example, the hypertrees in `sphincs-shake-128f-simple` have only 8 leaves, which does not suit 5-way parallelization. We believe that further study in the best use of the parallelization potentials would be beneficial, and encourage research.

2.3 ArmArchitecture

Computing based on the Armarchitecture is ubiquitous. At the coarsest level, one can distinguish three *profiles*: The application (A) profile targeting high performance markets, such as PC, mobile, and server; the real-time (R) profile for timing-sensitive and safety-critical systems; and the embedded (M) profile for secure, low-power, and low-cost microprocessors.

In this article, we focus on the A-profile of the Arm architecture. Numerous iterations of the A-profile exists, such as Armv7-A, Armv8-A and, as of late, Armv9-A, each including a respective set of extensions. We specifically focus on the AArch64 64-bit instruction set architecture introduced with Armv8-A, as well as the SHA-3 extension which is part of Armv8.4-A.

Implementations of the A-profile of the Armarchitecture form a spectrum in itself: To name some, it includes power-efficient CPUs like the Cortex-A7 for Linux-based embedded IoT devices, cores like Cortex-A710 and Cortex-X2 for client devices such as desktops or mobile phones, as well as the *Arm*® Neoverse™ IP for infrastructure applications. In this article, we focus on two recent generations of cores for the client market, Cortex-{A55,A510,A78,A710,X1,X2}. However, we expect that our optimizations do also apply to the Neoverse N1 and Neoverse V1 infrastructure cores.

It is common for Arm-based SoCs, particularly in the client market, to host multiple CPUs targeting different power/performance profiles, and to dynamically switch between them depending on demand. Originally, this was known as *Arm*®big.LITTLE™, distinguishing between a high-efficiency "LITTLE" CPU and a high-performance "big" CPU. Nowadays, *Arm*®DynamIQ™ allows for more flexibility in the configuration of CPUs on a SoC, and DynamIQ-based SoCs often host *three* different ArmCPUs targeting different performance/power profiles. Two such triples are Cortex-{A55,A78,X1} and Cortex-{A510,A710,X2}.

On a microarchitectural level, "LITTLE" cores are typically based on an *in-order* pipeline with some support for superscalar execution. For example, the Cortex-A53 and Cortex-A55 CPUs support dual-issuing of scalar instructions, while the Cortex-A510 CPU is even capable of *triple*-issuing scalar instructions. In terms of SIMD capabilities, Cortex-A53 and Cortex-A55 can single-issue 128-bit Neon instructions, while Cortex-A510 offers an interesting novelty: Pairs of Cortex-A510 CPUs are joined to a Cortex-A510 complex, sharing up to two 128-bit SIMD units. That is, if only one of the cores uses the SIMD units, dual-issuing of 128-bit Neon instructions is possible.

The "medium" Cortex-A7x and "big" Cortex-X cores are based on *out-of-order* pipelines with multiple scalar and SIMD execution units. For example, all

```
1  // θ step
2  eor C0, A00, A10
3  eor C0, C0,  A20
4  eor C0, C0,  A30
5  eor C0, C0,  A40
6  eor C1, A01, A11
7  eor C1, C1,  A21
8  eor C1, C1,  A31
9  eor C1, C1,  A41
10 eor C2, A02, A12
11 eor C2, C2,  A22
12 eor C2, C2,  A32
13 eor C2, C2,  A42
14 eor C3, A03, A13
15 eor C3, C3,  A23
16 eor C3, C3,  A33
17 eor C3, C3,  A43
18 eor C4, A04, A14
19 eor C4, C4,  A24
20 eor C4, C4,  A34
21 eor C4, C4,  A44
22 eor D1, C0, C2, ROR #63
23 eor D3, C2, C4, ROR #63
24 eor D0, C4, C1, ROR #63
25 eor D2, C1, C3, ROR #63
26 eor D4, C3, C0, ROR #63
27 // ρ, π, rest of θ steps
28 eor B00, A00, D0
29 eor B40, A02, D2
30 ror B40, B40, #2
31 eor B02, A22, D2
32 ror B02, B02, #21
33 eor B22, A23, D3
34 ror B22, B22, #39
35 eor B23, A34, D4
36 ror B23, B23, #56
37 eor B34, A43, D3
38 ror B34, B34, #8
39 ...
40 // χ step
41 bic tmp, B12, B11
42 eor A10, tmp,  B10
43 bic tmp, B13, B12
44 eor A11, tmp,  B11
45 bic tmp, B14, B13
46 eor A12, tmp,  B12
47 bic tmp, B10, B14
48 eor A13, tmp,  B13
49 ...
50
51 // ι step
52 eor A00, A00, RC
```

Fig. 2. 'Canonical' scalar AArch64 implementation of one `Keccak-f1600` round.

of Cortex-{A78,A710,X1,X2} have four scalar execution units. In terms of their SIMD capabilities, Cortex-A7x cores typically have two Neon execution units, while Cortex-X CPUs have four. Such information, as well as detailed listings of latencies and throughput per instructions, are provided in the publicly available software optimization guides [Armb, Arma, Armd, Armc, Arme, Armf].

3 Keccak on AArch64 –Architecture

This is the first of two section presenting our implementations of `Keccak-f1600` on the AArch64 instruction set architecture, the second being Sect. 4. Here, we focus on *architectural* considerations: We exhibit ways to express `Keccak-f1600` through the AArch64 instruction set and its extensions. We discuss three approaches: A scalar implementation, an Armv8-ANeonSIMD implementations, and an Armv8.4-A Neon SIMD implementation leveraging the SHA-3 extension.

It is difficult to define meaningful metrics for performance at the architectural level: The number of instructions, the depth and the width (i.e., the amount of instruction level parallelism) of a computation are first approximations, but the actual performance will typically also heavily depend on the target microarchitecture – which is to be expected considering wide range of implementations of the Armarchitecture across the performance/power spectrum.

In light of the above, the goal of this section is merely to provide us with a 'pool' of implementation approaches for `Keccak-f1600`. The study of their suitability for our target microarchitectures, as well as further microarchitecture specific optimizations, are the subject of Sect. 4.

3.1 Scalar Implementation

The description of `Keccak-f1600` from Sect. 2.1 admits a straightforward mapping to the AArch64 instruction set architecture: The 1600-bit state can be maintained in 25 general purpose registers of 64 bits each, and the bitwise operations performed in the θ, ρ, χ and τ steps can be implemented using the `XOR`, `ROR`, `BIC` instructions. This 'canonical' implementation is presented in Fig. 2.

Eliminating Rotations. The canonical implementation can be shortened by eliminating explicit rotations as follows. For any bitwise operation `OP` such as

```
1  // θ step
2  eor C2, A42, A02, ROR #52
3  eor C0, A00, A10, ROR #61
4  eor C4, A24, A14, ROR #50
5  eor C1, A21, A31, ROR #57
6  eor C3, A03, A23, ROR #63
7  ...
8  eor C2, C2, A12, ROR #5
9  eor C0, C0, A40, ROR #25
10 eor C4, C4, A44, ROR #15
11 eor C1, C1, A11, ROR #27
12 eor C3, C3, A43, ROR #2
13 eor D1, C0, C2, ROR #61
14 ror C2, C2, 62
15 eor D3, C2, C4, ROR #57
16 ror C4, C4, 58
17 eor D0, C4, C1, ROR #55
18 ror C1, C1, 56
19 eor D2, C1, C3, ROR #63
20 eor D4, C3, C0, ROR #63
```

```
21 // ρ, π, rest of θ steps
22 eor B00, D0, A00
23 eor B40, D2, A02, ROR #50
24 eor B02, D2, A22, ROR #46
25 eor B22, D3, A23, ROR #63
26 eor B23, D4, A34, ROR #28
27 eor B34, D3, A43, ROR #2
28 eor B43, D0, A30, ROR #54
29 eor B20, D1, A01, ROR #43
30 eor B41, D3, A13, ROR #36
31 eor B13, D1, A31, ROR #49
32 eor B21, D2, A12, ROR #3
33 eor B12, D0, A20, ROR #39
34 eor B10, D3, A03
35 eor B03, D3, A33, ROR #37
36 eor B33, D2, A32, ROR #8
37 eor B32, D1, A21, ROR #56
38 eor B11, D4, A14, ROR #44
39 eor B14, D2, A42, ROR #62
40 ...
```

```
41 // χ step
42 bic tmp0, B12,  B11, ROR #47
43 bic tmp1, B13,  B12, ROR #42
44 eor A10,  tmp0, B10, ROR #39
45 bic tmp0, B14,  B13, ROR #16
46 eor A11,  tmp1, B11, ROR #25
47 bic tmp1, B10,  B14, ROR #31
48 eor A12,  tmp0, B12, ROR #58
49 bic tmp0, B11,  B10, ROR #56
50 eor A13,  tmp1, B13, ROR #47
51 bic tmp1, B22,  B21, ROR #19
52 eor A14,  tmp0, B14, ROR #23
53 bic tmp0, B23,  B22, ROR #47
54 eor A20,  tmp1, B20, ROR #24
55 bic tmp1, B24,  B23, ROR #10
56 eor A21,  tmp0, B21, ROR #2
57 bic tmp0, B20,  B24, ROR #47
58 ...
59 // ι step
60 eor A00, A00, RC
```

Fig. 3. Keccak-f1600 round without explicit rotations.

XOR or BIC, it holds that $(x \texttt{ OP } y) \lll \texttt{imm} = (x \lll \texttt{imm}) \texttt{ OP } (y \lll \texttt{imm})$, so

$$(x \lll \texttt{imm0}) \texttt{ OP } (y \lll \texttt{imm1}) = (x \texttt{ OP } (y \lll \texttt{imm1} - \texttt{imm0})) \lll \texttt{imm0} \quad (1)$$

This trivial identity replaces the explicit rotations $x \lll \texttt{imm0}$ and $y \lll \texttt{imm1}$ with the combination of a shifted application of OP and an explicit rotation of its result. Since AArch64 offers shifted variants of logical operations as discussed in Sect. 2.3, this eliminates one explicit rotation. Moreover, if $\texttt{imm0} = 0$ or $\texttt{imm1} = 0$, no explicit rotation remains. Finally, if the result is used in another bitwise operation, the process can be repeated, deferring all explicit rotations to the very end of the computation, where only one rotation per output has to be performed. We call this process "lazy rotation" in the following.

The idea of lazy rotations can be applied to Keccak-f1600 in order to defer the explicit rotations in the π-step. At first, however, it would seem that the entire Keccak-f1600 loop would need to be unrolled to benefit from the idea, as performing rotations at a later stage in the loop is still as expensive as performing them at the π-step. Luckily, this is not the case, as we explain now.

Assume we have deferred explicit rotations in the π-step to the end of the first Keccak-f1600 iteration, so that the true state $\texttt{A}[x, y]$ can be obtained form the software state $\texttt{A}'[x, y]$ via $\texttt{A}[x, y] = \texttt{A}'[x, y] \lll \texttt{s}[x, y]$ for some constants $\texttt{s}[x, y]$. In the θ-step for the next iteration, we can then compute $\texttt{D}[x]$ via lazy rotations, obtaining a value $\texttt{D}'[x]$ so that the true $\texttt{D}[x]$ is again given by $\texttt{D}[x] = \texttt{D}'[x] \oplus \texttt{t}[x]$ for suitable constants $\texttt{t}[x]$. If we then *explicitly* rotate $\texttt{D}'[x]$ to obtain the true $\texttt{D}[x]$, the final part $\texttt{A}[x, y] \leftarrow \texttt{A}[x, y] \oplus \texttt{D}[x] = (\texttt{A}'[x, y] \lll \texttt{s}[x, y]) \oplus \texttt{D}[x]$ can be computed using a shifted XOR without any deferred rotation. By breaking the chain of deferred rotations at $\texttt{D}[x]$, we prevent an accumulation of deferred rotations which would otherwise force us to unroll the loop.

The above explains how to trade 25 explicit rotations in the π-step for 5 explicit rotations in the θ-step. In fact, it turns out that 2 of the 5 deferred rotations for $\texttt{D}[x]$ are 0, so that only 3 explicit rotations are necessary. The final result is presented in Fig. 3.

Register Allocation. We aim to keep most operations in-place to reduce the number of MOV operations. In the notation of Fig. 1, we'd like $\texttt{loc}\,\texttt{B}[x,y] = \texttt{loc}\,\texttt{A}[x,y]$ for most (x,y), where $\texttt{loc}\,\texttt{X}$ is the register location used by X. Without backup MOVs, however, we cannot have $\texttt{loc}\,\texttt{B}[x,y] = \texttt{loc}\,\texttt{A}[x,y]$ for all x,y: Otherwise, there'd be cyclic dependencies in the computation of both

$$\texttt{B}[x',y'] = \texttt{A}[x,y] \oplus \texttt{D}[x] \quad \text{and} \qquad (\theta\pi)$$
$$\texttt{A}[x,y] = \texttt{B}[x,y] \oplus (\neg\texttt{B}[x+1,y]\ \&\ \texttt{B}[x+2,y]) \qquad (\chi)$$

preventing in-place computation – we use the shorthand $(x',y') := (y, 2x+3y)$ here and below. The goal is to slightly offset $\{\texttt{loc}\,\texttt{B}[]\}$ from $\{\texttt{loc}\,\texttt{A}[]\}$ for the computation of $(\theta\pi)$, and to move entries back to their original place in (χ). Concretely, we set $\texttt{loc}\,\texttt{B}[x,y] = \texttt{loc}\,\texttt{A}[x,y]$ for $x \notin \{0,1\}$ and $\texttt{loc}\,\texttt{B}[x,y] = \texttt{loc}\,\texttt{A}[x,(y+1)\%5]$ for $x \in \{0,1\}$ and $y \in \{1,2,3,4\}$, while using fresh registers for $\texttt{B}[0,0]$ and $\texttt{B}[1,0]$ – this choice will become clear soon.

The computation of $(\theta\pi)$ then proceeds in a chained fashion: After computing $\texttt{B}[x_0',y_0']$ from $\texttt{A}[x_0,y_0]$, we continue with $\texttt{B}[x_1',y_1']$ s.t. $\texttt{loc}\,\texttt{B}[x_1',y_1'] = \texttt{loc}\,\texttt{A}[x_0,y_0]$ – that is, once we have used some $\texttt{A}[]$ to compute the corresponding $\texttt{B}[]$, we overwrite its location next. Starting with $\texttt{B}[0,0]$ or $\texttt{B}[1,0]$ (which use fresh registers), it terminates once we reach the computation of $\texttt{B}[0',0']$ or $\texttt{B}[1',0']$ from $\texttt{A}[0,1]$ or $\texttt{A}[1,1]$, because $\texttt{loc}\,\texttt{A}[0,1]$ and $\texttt{loc}\,\texttt{A}[1,1]$ aren't used by $\texttt{B}[]$.

In principle, the chained computation of $(\theta\pi)$ just described does not depend on the particular choice of $\texttt{loc}\,\texttt{B}[]$, but the lengths of the resulting chains do: Our specific choice leads to a length-24 chain from $\texttt{B}[0,0]$ to $\texttt{A}[0,1]$, and a length-1 chain from $\texttt{B}[1,0]$ to $\texttt{A}[1,1]$. This matters for register allocation: At the time of $(\theta\pi)$, we are already using 30 registers – 25 for the state $\texttt{A}[]$ and 5 for $\texttt{D}[]$ – so only *one* remains, yet we need *two* fresh locations for $\texttt{B}[0,0]$ and $\texttt{B}[1,0]$. We solve this by using the single free location for $\texttt{B}[0,0]$, while *after* computing its length-24 chain, all but one $\texttt{D}[]$ are no longer needed, so $\texttt{B}[1,0]$ can use any of those.

Finally, we compute (χ), where the special role of $x = 0,1$ in the definition of $\texttt{loc}\,\texttt{B}[]$ becomes important: Namely, when we compute $\texttt{A}[0,y], \texttt{A}[1,y]$ from $\texttt{B}[*,y]$, we cannot yet overwrite any $\texttt{loc}\,\texttt{B}[*,y]$ as they're still needed for subsequent (χ) steps. We thus require $\texttt{loc}\,\texttt{A}[0,y], \texttt{loc}\,\texttt{A}[1,y] \notin \{\texttt{loc}\,\texttt{B}[*,y]\}$. On the other hand, after computing $\texttt{A}[0,y], \texttt{A}[1,y]$ out-of-place, we may compute $\texttt{A}[2,y], \texttt{A}[3,y], \texttt{A}[4,y]$ in-place since they're no longer used as input for (χ). This motivates our choice of $\texttt{loc}\,\texttt{B}[x,y] = \texttt{loc}\,\texttt{A}[x,y]$ for $x \neq 0,1$, while offsetting $\texttt{loc}\,\texttt{B}[0,y], \texttt{loc}\,\texttt{B}[1,y]$.

Overall, the above yields an in-place implementation of a single Keccak-f1600 round using 31 registers, and without using any MOV instructions or stack spilling.

Statistics. Each round in our scalar Keccak-f1600 implementation uses 76× EOR, 25× BIC and 3× ROR instructions, totalling 104 arithmetic instructions. In fact, the first round does not require RORs, but we need 23 RORs after the last round. Overall, we get to $24 \times 104 - 3 + 23 = 2516$ arithmetic instructions for the core of Keccak-f1600. Taking into account function preamble and postamble, we get to 2747 instructions executed per Keccak-f1600 invocation.

3.2 Armv8.4-A Neon Implementation

The Armv8.4-A architecture introduces the SHA-3 extension adding the following instructions: `EOR3` for the computation of three-fold `XOR`s; `RAX1` for the combination of a rotation-by-1 and a `XOR`; `XAR` for the combination of a `XOR` and a rotation; and finally `BCAX` for the combination of a `BIC` and `XOR`. Those instructions enable a straightforward implementation of `Keccak-f1600` on Armv8.4-A, with `EOR3` and `RAX1` handling the first part of the θ-step, `XAR` handling the second part of the θ-step merged with the ρ-step, and `BCAX` handling the τ-step. The first public implementation along those lines was [Wes]. We slightly refine it by removing explicit `MOV` instructions as detailed in Sect. 3.1.

Statistics. Each round requires $10\times$ `EOR3` instructions, $5\times$ `RAX1` instructions, $24\times$ `XAR`, $2\times$ `EOR` and $25\times$ `BCAX`. Overall, it thus uses $24 \times 66 = 1584$ vector instructions, $24 \times 64 = 1536$ of which from the Armv8.4-A SHA-3 extension.

3.3 Armv8-A Neon Implementation

To implement `Keccak-f1600` on Armv8-A Neon instructions, the structure of the Armv8.4-A code can be retained, while implementing `EOR3`, `RAX1`, `XAR`, and `BCAX` as macros based on Armv8-A Neon instructions. Rotations are constructed from a left shift (`SHL`) and a right shift with insert (`SRI`). An implementation along those lines was first developed in [Ngu] based on intrinsics; here, we use a version in handwritten assembly, minimizing vector moves and stack spills.

Statistics. When implementing `EOR3` via $2\times$ `EOR`, `RAX1` and `XAR` via `EOR+SHL+SRI`, and `BCAX` via `BIC+EOR`, each `Keccak-f1600` round consists of $76\times$ `EOR`, $29\times$ `SRI`, $30\times$ `SHL` and $25\times$ `BIC` instructions totalling 160 Neon instructions per round, and $24 \times 160 = 3840$ Neon instructions for all of `Keccak-f1600`.

4 `Keccak-f1600` on AArch64 –Microarchitecture

Here, we study the implementations presented in Sect. 3 from a microarchitectural perspective. We first comment on each approach separately, and then present Scalar/Neon hybrid implementations, the main novelty of this paper.

4.1 Scalar Implementation

Recall from Sect. 3.1 the main ideas of our scalar `Keccak-f1600` implementation: Eliminating explicit rotations through extensive use of shifted instructions, and eliminating explicit `MOV`s through careful register management.

```
1  // Naive
2  eor C0, A20, A40, ROR #50
3  eor C0, A30, C0, ROR #49
4  eor C0, A10, C0, ROR #57
5  eor C0, A00, C0, ROR #61
6  eor C1, A41, A11, ROR #60
7  eor C1, A01, C1, ROR #44
8  eor C1, A31, C1, ROR #58
9  eor C1, A21, C1, ROR #57
10 ...
11 eor C4, A04, A44, ROR #53
12 eor C4, A34, C4, ROR #56
13 eor C4, A14, C4, ROR #48
14 eor C4, A24, C4, ROR #50
15
16 ror C1, C1, 56
17 ror C4, C4, 58
18 ror C2, C2, 62
19
20 eor D1, C0, C2, ROR #63
21 eor D3, C2, C4, ROR #63
22 eor D0, C4, C1, ROR #63
23 eor D2, C1, C3, ROR #63
24 eor D4, C3, C0, ROR #63
```

```
1  // Better (fine for A55)
2  eor C0, A20, A40, ROR #50
3  eor C1, A41, A11, ROR #60
4  eor C2, A32, A12, ROR #59
5  eor C3, A13, A43, ROR #30
6  eor C4, A04, A44, ROR #53
7  eor C0, A30, C0,  ROR #49
8  eor C1, A01, C1,  ROR #44
9  eor C2, A22, C2,  ROR #26
10 eor C3, A33, C3,  ROR #63
11 eor C4, A34, C4,  ROR #56
12 ...
13 eor C0, A00, C0,  ROR #61
14 eor C1, A21, C1,  ROR #57
15 eor C2, A42, C2,  ROR #52
16 eor C3, A03, C3,  ROR #63
17 eor C4, A24, C4,  ROR #50
18
19 ror C1, C1, 56
20 ror C4, C4, 58
21 ror C2, C2, 62
22
23 eor D1, C0, C2, ROR #63
24 eor D3, C2, C4, ROR #63
25 eor D0, C4, C1, ROR #63
26 eor D2, C1, C3, ROR #63
27 eor D4, C3, C0, ROR #63
```

```
1  // Even better (for A510)
2  eor C2, A42, A02, ROR #52
3  eor C0, A00, A10, ROR #61
4  eor C4, A24, A14, ROR #50
5  eor C1, A21, A31, ROR #57
6  eor C3, A03, A23, ROR #63
7  eor C2, C2,  A22, ROR #48
8  eor C0, C0,  A30, ROR #54
9  eor C4, C4,  A34, ROR #34
10 eor C1, C1,  A01, ROR #51
11 eor C3, C3,  A33, ROR #37
12 ...
13 eor C2, C2,  A12, ROR #5
14 eor C0, C0,  A40, ROR #25
15 eor C4, C4,  A44, ROR #15
16 eor C1, C1,  A11, ROR #27
17 eor C3, C3,  A43, ROR #2
18
19 eor D1, C0, C2, ROR #61
20 ror C2, C2, 62
21 eor D3, C2, C4, ROR #57
22 ror C4, C4, 58
23 eor D0, C4, C1, ROR #55
24 ror C1, C1, 56
25 eor D2, C1, C3, ROR #63
26 eor D4, C3, C0, ROR #63
```

Fig. 4. θ-step optimized for dual-issuing capability of the A55 (middle) and triple-issuing capability of the A510 (right) compared to the naïve approach (left)

The Cost of Shifted Instructions. Our rotation-elimination implementation is only useful if shifted instructions have the same throughput as unshifted instructions, which is the case for all our targets Cortex-{A55,A510,A78,A710,X1,X2}. However, there *are* exceptions, such as the Cortex-A72, and for such CPUs, rotation-elimination may lead to worse performance despite having a lower instruction count. However, as we see below for Cortex-A55 and Cortex-A510, an increased *latency* for shifted instructions need not be problematic.

The Cost of `MOV`s. Eliminating `MOV`s for general purpose registers is a microoptimization primarily useful for LITTLE CPUs. High-end out-of-order CPUs, in turn, can sometimes implement such `MOV`s with zero latency (e.g. [Arme, Section 4.14]) and therefore show little benefit from reduced register movement.

Optimization for In-Order Execution. Optimizing code for in-order execution requires careful scheduling of instructions for latency, throughput, and the width of the superscalar pipeline. We now make this concrete for `Keccak-f1600` and our in-order target microarchitectures Cortex-A55 and Cortex-A510.

We begin by discussing Cortex-A55. As detailed in the Software Optimization Guide [Armb], Cortex-A55 is capable of issuing logical instructions with shift at a rate of 2 IPC and a latency of 2 cycles. This is sufficient for a stall-free execution of the column-wise 5-fold `XOR`s in the θ-step, *provided* one alternates between the columns; Fig. 4 shows both the naïve and slow approach (left column), as well as an interleaved implementation suitable for Cortex-A55 (middle column).

We next consider the χ-step: Looking at the naïve implementation in Fig. 5 (left column), it would seem that with a dual-issuing core and a latency of 2-cycles per shifted instruction, it should stall. However, as explained in [Armb, Figure 1 and Sect. 3.1.1], the execution of shifted instructions is pipelined, and

```
1 // Naive
2 bic tmp, A12_, A11_, ROR #47
3 eor A10, tmp,  A10_, ROR #39
4 bic tmp, A13_, A12_, ROR #42
5 eor A11, tmp,  A11_, ROR #25
6 bic tmp, A14_, A13_, ROR #16
7 eor A12, tmp,  A12_, ROR #58
8 bic tmp, A10_, A14_, ROR #31
9 eor A13, tmp,  A13_, ROR #47
10 bic tmp, A11_, A10_, ROR #56
11 eor A14, tmp,  A14_, ROR #23
12 bic tmp, A22_, A21_, ROR #19
13 eor A20, tmp,  A20_, ROR #24
14 bic tmp, A23_, A22_, ROR #47
15 eor A21, tmp,  A21_, ROR #2
16 ...
```

```
1 // Improved for triple-issuing
2 bic tmp0, A12_, A11_, ROR #47
3 bic tmp1, A13_, A12_, ROR #42
4 eor A10, tmp0,  A10_, ROR #39
5 bic tmp0, A14_, A13_, ROR #16
6 eor A11, tmp1,  A11_, ROR #25
7 bic tmp1, A10_, A14_, ROR #31
8 eor A12, tmp0,  A12_, ROR #58
9 bic tmp0, A11_, A10_, ROR #56
10 eor A13, tmp1,  A13_, ROR #47
11 bic tmp1, A22_, A21_, ROR #19
12 eor A14, tmp0,  A14_, ROR #23
13 bic tmp0, A23_, A22_, ROR #47
14 eor A20, tmp1,  A20_, ROR #24
15 bic tmp1, A24_, A23_, ROR #10
16 eor A21, tmp0,  A21_, ROR #2
17 ...
```

Fig. 5. χ-step optimized for triple-issuing on the A510 (right) compared to the naïve implementation (left)

appropriate forwarding paths provide an effective 1-cycle latency between shifted instructions in case the output of a shifted instruction is used as an *unshifted* input in the consuming instruction; luckily, we are in such a special case.

We now turn to Cortex-A510. As can be seen in the software optimization guide [Arma], Cortex-A510 can issue shifted instructions at a rate of *three* instructions per cycle (the first "LITTLE" core with such capabilities) and 2-cycle latency. Moreover, our experiments suggest that we again have a 1-cycle effective latency for outputs of shifted instructions being used as non-shifted inputs.

To leverage the triple-issuing capability of Cortex-A510, the following adjustments have to be made: Firstly, for the columnwise `XOR`s in the θ-step, use the accumulator as a *non-shifted* input only. The right column in Fig. 4 shows an adjusted version suitable for triple-issuing on Cortex-A510. Secondly, the χ-step cannot be triple-issued when written as in Fig. 5 (left column); instead, one has to manually interleave the computation as in Fig. 5 (right column). With those adjustments in place, the `Keccak-f1600` code is mostly triple-issuable, as the performance numbers in Sect. 5 will confirm.

4.2 Armv8-A Neon Implementation

Recall that our Armv8-A implementation replaces the SHA-3 instructions `RAX1`, `XAR`, `BCAX`, and `EOR3` by macros based on Armv8-A Neon instructions.

Suitability for In-Order Microarchitectures. Generally, implementations based on defining high-level operations such as the SHA-3 operations as assembly macros tend to be unsuitable for in-order execution, as they cannot exploit parallelism at the macro-level and are thus unlikely to obey instruction latencies. For example, on Cortex-A510, `EOR`, `SRI`, and `SHL` have a latency of 3 cycles, so an implementation of `XAR` as `EOR+SHL+SRI` will have a total latency of 9 cycles.

On Cortex-A55, however, we are lucky that logical SIMD instructions have a 1-cycle latency. Moreover, we observe experimentally that `SHL+SRI` pairs synthesizing a rotation do also run without stalls – the implementations of the SHA-3 macros `EOR3`, `BCAX`, `RAX1`, `XAR` can therefore run stall-free. We should expect a performance not far off the optimum of 1 Neon instruction per cycle, and we do not see significant further optimization potential in this approach.

Improving performance of our Armv8-A Neon code on Cortex-A510 requires instruction level parallelism through the interleaving of the SHA-3 macros. Due to the very high register pressure, however, this requires a lot of stack spilling, which is tedious to do by hand. Instead, it makes more sense in this case to work with intrinsics as done in [Ngu], and leave register allocation and stack spilling to the compiler. However, as we shall see in Sect. 5, both scalar and Armv8.4-A-based `Keccak-f1600` implementations perform better than the Armv8-A based implementation anyway, so the point is moot and we do not explore it further.

Suitability for Out-of-Order Microarchitectures. Generally speaking, the fact that the SHA-3 macros are not scheduled for latency is less problematic for out-of-order cores than for in-order cores, as the microarchitecture will leverage out-of-order execution and register renaming to create the required instruction level parallelism. Still, there is room for further optimization, as we now explain.

The first optimization concerns the availability of functionality on the different SIMD units. For our out-of-order target microarchitectures, the `EOR` and `BIC` instructions can run on all SIMD units. However, the `SHL` and `SRI` instructions, which we use heavily to synthesize 64-bit rotations, are only supported by 50% of the SIMD units – one in the case of Cortex-A78 and Cortex-A710, and two in the case of Cortex-X1 and Cortex-X2. This limits the maximum throughput of the `XAR` and `RAX` blocks, at least when looked at in isolation. In the context of an entire `Keccak-f1600` round, however, `SHL+SRI` make up for *less* than 50% of SIMD instructions, so that manual interleaving of the `XAR` and `RAX` blocks with surrounding code mitigates the throughput loss. Additionally, we replace instances of `SHL X, A, #1` by `ADD X, A, A` (this applies to all `RAX1` blocks and one `XAR` invocation), reducing the pressure on the `SHL/SRI`-capable SIMD units, since (like `EOR` and `BIC`) `ADD` can run on all SIMD units.

The second optimization concerns the θ step: We found that by moving the 5-fold `EOR`s into the previous iteration, we can alleviate a performance bottleneck at the θ step resulting from the lack of instruction level parallelism. For example, with `EOR` having a latency of 2 cycles, one would need at least 8 independent data streams to keep all 4 SIMD units on the Cortex-X1 and Cortex-X2 busy.

4.3 Armv8.4-A Neon Implementation

Suitability for In-Order Cores. As for the scalar implementation, we schedule code for latency to ensure fast execution on Cortex-A510, the basis being the latencies of the SHA-3 instructions as documented in the SWOG [Arma], and the fact each core in a Cortex-A510 complex has up to two SIMD units, depending on whether the other core in the complex is also performing SIMD operations. It is noteworthy that in such a configuration, Cortex-A510 has more throughput for SHA-3 operations than Cortex-A710 and Cortex-X2.

We found that scheduling the code for latency was mostly straightforward, one exception being the `RAX1` instruction, which on Cortex-A510 has a latency of 8 cycles: Here, it seems preferable to express the operation through other Neon instructions of lower latencies.

Suitability for Out-of-Order Cores. For our out-of-order Armv8.4-A targets Cortex-A710 and Cortex-X2, we believe that a "standard" Armv8.4-A implementation along the lines of [Wes] does not have significant microarchitecture-specific optimization potential: As explained in [Armc, Armf], both cores have a single SIMD unit supporting the SHA-3 instructions, limiting a pure Armv8.4-A implementation to 1536 cycles at best, which our implementations already come very close to both for Cortex-A710 and Cortex-X2 – see Sect. 5.

4.4 Hybrid Implementations

The idea for hybrid implementations is simple and general: Given code paths A and B exercising different execution resources, interleave them to facilitate parallel execution by the underlying microarchitecture. Ideally, if the runtimes of A and B are t_A and t_B, respectively, one hopes to achieve the *joint* functionality of A, B in runtime $\max\{t_A, t_B\}$, instead of the sequential $t_A + t_B$.

When constructing a hybrid, one has to consider the individual performance of the code paths to be interleaved, and balance them accordingly to maximize the gain $(t_A + t_B) - \max\{t_A, t_B\} = \min\{t_A, t_B\}$: For example, if path A is $2\times$ as fast as path B, one should interleave 2 copies of A with a single copy of B.

Hybrid implementations have previously been applied in other contexts: Bernstein and Schwabe [BS12] present a scalar/Neon hybrid implementation of the Salsa20 cipher for Armv7-A, and Lenngren [Len19] presents a scalar/Neon hybrid implementation of the key-exchange mechanism X25519 for Armv8-A.

Suitability for Different Microarchitectures. A hybrid can reach ideal performance $\max\{t_A, t_B\}$ only if the target has the bandwidth to process A and B in parallel. Otherwise, there will be arbitration, with full arbitration leading to sequential performance $t_A + t_B$. It is therefore important to understand the target's maximum $w_{\max}$ of instructions per cycle (IPC), as well as the IPCs w_A ad w_B targetted by A and B. Only if $w_A + w_B \leq w_{\max}$ there is a chance to unlock performance $\max\{t_A, t_B\}$ through a hybrid.

For example, Lenngren [Len19] constructs a Scalar/Neon hybrid for X25519 on Cortex-A53, leveraging that (a) generally, Cortex-A53 can achieve up to 2 IPC, but (b) scalar multiplication and SIMD instructions are limited to 1 IPC. Manual interleaving of scalar and SIMD implementations unlocks an IPC of $\approx$2.

For out-of-order CPUs, the necessity for manual interleaving depends on the target microarchitecture: If the paths to be interleaved are loops of the same size and the out-of-order execution window exceeds the loop body, an alternation of iterations from the two paths may eventually execute in parallel even without manual interleaving. For `Keccak-f1600`, however, each round is large, so we manually interleave scalar and Neon iterations to facilitate parallel execution.

Scalar/Neon Hybrid. We apply the idea of hybrid implementations to our scalar and Neon implementations of `Keccak-f1600`: Concretely, we construct implementations of N-way parallel `Keccak-f1600` by interleaving $N-2$ scalar computations of `Keccak-f1600` with a Neon-based computation of `Keccak-f1600-x2`.

Interleaving the scalar and Neon `Keccak-f1600` implementations was straightforward since the only shared architectural resource is the loop counter. Practically, we wrote code side by side to facilitate readability, as shown in Fig. 6.

The choice of N depends on the relative speed of the scalar and Neon code. For example, on Cortex-X1 and Cortex-X2, we chose $N = 3$ and $N = 4$, implementing `Keccak-f1600-xN` from one or two scalar `Keccak-f1600` and one Neon `Keccak-f1600-x2`. On Cortex-A78, we found that $N = 5$ was more suitable.

We comment on the feasibility of hybrids on our targets: For Cortex-A55 and Cortex-A510, our scalar code come close to the issue limit of 2 and 3 IPC, while the SIMD code reaches less than 1 IPC on Cortex-A55 and close to 2 IPC on Cortex-A510. We don't see meaningful speedup through hybrids.

For Cortex-A78 and Cortex-A710, the scalar and Neon implementations target an IPC of 4 and 2, respectively. Since Cortex-A710 has a maximum IPC of 5, they cannot be interleaved without penalty. Cortex-A78, in turn, has a maximum IPC of 6, so a scalar/Neon appears feasible. However, our initial attempt of constructing `Keccak-f1600-x5` on Cortex-A78 fell $> 20\%$ short of our expectations, and only after a significant code-size reduction, we achieved the desired performance. We explain this as follows: While Cortex-A78 has a maximum IPC of 6, the instruction decoder has a maximum IPC of 4. An IPC > 4 can only be unlocked through the use of the "MOP-cache", hosting decoded instructions, but our unrolled code failed to achieve a good hitrate. Once the code was shortened to fit in the MOP-cache, performance reached the expected level.

Neon/Neon Hybrid. An implementation based purely on the Armv8.4-A SHA-3 instructions will only exercise those Neon units implementing the SHA-3 extension. In the case of our targets Cortex-A710 and Cortex-X2, these are 50% and 25% of all Neon units, respectively – the remaining units stay idle.

We have therefore developed hybrid Armv8-A/Armv8.4-A implementations of `Keccak-f1600-x2`, mixing SHA-3 instructions with regular Neon instructions, to achieve better utilization of the SIMD units. This is a different kind of hybrid than the Scalar/Neon one, as we're alternating between different implementation strategies rather than interleaving them. The balance between SHA-3 and regular Neon instructions depends on the share of SIMD execution units implementing the SHA-3 instructions. For example, on Cortex-X2, we strive for 3 regular Neon

```
1 eor sC0, sA30, sA40   ;   eor3 vC0.16b, vA00.16b, vA10.16b, vA20.16b
2 eor sC1, sA31, sA41   ;
3 eor sC2, sA32, sA42   ;
4 eor sC3, sA33, sA43   ;   eor3 vC0.16b, vC0.16b, vA30.16b, vA40.16b
5 eor sC4, sA34, sA44   ;
6 eor sC0, sA20, sC0    ;
7 eor sC1, sA21, sC1    ;   eor3 vC1.16b, vA01.16b, vA11.16b, vA21.16b
8 eor sC2, sA22, sC2    ;
9 eor sC3, sA23, sC3    ;
10 eor sC4, sA24, sC4   ;   eor3 vC1.16b, vC1.16b, vA31.16b, vA41.16b
11 eor sC0, sA10, sC0   ;
12 eor sC1, sA11, sC1   ;
13 eor sC2, sA12, sC2   ;   eor3 vC2.16b, vA02.16b, vA12.16b, vA22.16b
14 ...
```

Fig. 6. Interleaving of scalar and Armv8.4-A Keccak-f1600 code

instructions for 1× SHA-3 instruction, keeping all four SIMD units busy, while on Cortex-A710, the balance should be 1/1.

Scalar/Neon/Neon Hybrid. Finally, we have also experimented with "triple" hybrid implementations interleaving a scalar implementation with the Neon/Neon hybrid described in the previous section. In addition to Keccak-f1600-x4, we also considered an implementation Keccak-f1600-x3 interleaving one scalar computation with one hybrid Neon/Neon implementation of Keccak-f1600-x2.

5 Results

5.1 Benchmarking Environments

Cortex-{X1,A78,A55}. Our first benchmarking platform is a Lantronix Snapdragon 888 hardware development kit with a Qualcomm Snapdragon SM8350P SoC. It is an Arm DynamIQ SoC featuring one high-performance Arm Cortex-X1 core, three Arm Cortex-A78 cores, and four energy-efficient in-order Cortex-A55 cores. The SoC implements the Armv8.2-A instruction set. It also implements the Armv8.4-A dot product instructions, but not the Armv8.4-A SHA-3 instructions. The hardware development kit comes with a rooted Android 11 which allows us to run cross-compiled static executables.

Cortex-{X2,A710,A510}. Our second benchmarking platform is a Samsung S22 smartphone with a Samsung Exynos 2200 (S5E9925) SoC. It is an Arm DynamIQ SoC consisting of one high-performance Cortex-X2 core, three Cortex-A710 cores, and 4 energy-efficient in-order Cortex-A510 cores – the first generation of cores implementing the Armv9-A architecture. The Armv8.4-A SHA-3 extension is also implemented. The SoC is running a rooted Android 12. Our benchmarks suggest that the four Cortex-A510 cores are paired in two Cortex-A510 complexes with shared SIMD units; our benchmarks only use one Cortex-A510 a time, therefore allowing it to leverage 2 SIMD units.

Compiler and Benchmarking. We cross-compile our software using the Arm GNU toolchain[1] version 11.3.Rel1. We then copy the executable to the device and run it on a specific core via `taskset`. If we find the desired core disabled for power-saving, we first create artifical load on the system to re-enable it. We use the `perf_events` syscalls for cycle counting. For benchmarking our individual `Keccak-f1600` functions, we warm the cache by running the function 1 000 times, and then report the median of 100 samples of the average performance of 100 function invocations (the averaging amortizes the cost of the `perf` syscalls).

5.2 `Keccak-f1600` Performance

The results of our performance measurements for `Keccak-f1600` are shown in Table 1. As reference points, we use the `crypto_hash/keccakc512/simple` scalar C implementation from SUPERCOP [Kee], the Armv8-A implementation from [Ngu], and the Armv8.4-A implementation from [Wes]. We will now comment and interpret results for each CPU separately.

Table 1. Cycle counts for various implementations of `Keccak-f1600`. "Neon+SHA-3" indicates implementations using the SHA-3 instructions. Numbers in brackets are normalized with respect to the amount of parallelization.

Approach			Cortex-X1		Cortex-A78		Cortex-A55	
C	[Kee]	1x	811	(811)	819	(819)	1 935	(1935)
Scalar	Ours	1x	690	(690)	709	(709)	1 418	(1418)
Neon	[Ngu]	2x	1 370	(685)	2 409	(1204)	5 222	(2611)
Neon	Ours	2x	1 317	(658)	2 197	(1098)	4 560	(2280)
Scalar/Neon	Ours	4x	1 524	(381)	2 201	(550)	7 288	(1822)
Scalar/Neon	Ours	5x	2 161	(432)	2 191	(438)	8 960	(1792)
Approach			Cortex-X2		Cortex-A710		Cortex-A510	
C	[Kee]	1x	817	(817)	820	(820)	1 375	(1375)
Scalar	Ours	1x	687	(687)	701	(701)	968	(968)
Neon	[Ngu]	2x	1 325	(662)	2 391	(1195)	3 397	(1698)
Neon	Ours	2x	1 274	(637)	2 044	(1022)	6 970	(3485)
Neon+SHA-3	[Wes]	2x	1 547	(773)	1 550	(775)	2 268	(1134)
Neon+SHA-3	Ours	2x	1 547	(773)	1 549	(774)	1 144	(572)
Neon/Neon+SHA-3	Ours	2x	944	(472)	1 502	(751)	4 449	(2224)
Scalar/Neon/Neon+SHA-3	Ours	3x	985	(328)	1 532	(510)	4 534	(1511)
Scalar/Neon	Ours	4x	1 469	(367)	2 229	(557)	7 384	(1846)
Scalar/Neon+SHA-3	Ours	4x	1 551	(387)	1 608	(402)	3 545	(886)
Scalar/Neon	Ours	5x	2 152	(430)	2 535	(507)	7 169	(1433)
Scalar/Neon/Neon+SHA-3	Ours	4x	1 439	(359)	1 755	(438)	4 487	(1121)

[1] https://developer.arm.com/downloads/-/arm-gnu-toolchain-downloads.

Cortex-A55 and Cortex-A510. We observe a significant speedup from the C scalar implementation to our hand-optimized assembly implementation: 1.36× for Cortex-A55 and 1.42× for Cortex-A510. We further note that the scalar performance is close to the theoretical optimum: With ≈ 2750 instructions in total (see Sect. 3.1) and a maximum issue rate of 2 instructions per cycle on Cortex-A55, the theoretical performance limits on Cortex-A55 are ≈ 1375 cycles. Similar, the maximum issue rate of 3 instructions per cycle on Cortex-A510 leads to a theoretical performance limit of ≈ 917 cycles.

As expected (see Sect. 4.2), the pure Neon implementation is not competitive for neither Cortex-A55 nor Cortex-A510. In particular, we confirm that the macro-based implementation performs very poorly on Cortex-A510 since latencies are not obeyed, while the intrinics-based implementation from [Ngu] does better at scheduling the code for latency.

For Cortex-A510, we observe a significant speedup from the Armv8.4-A implementation, explained by the presence of 2 SIMD units capable of executing the SHA-3 Neon instructions. The very large performance gap between our implementation and that of [Wes] is largely due to the high latency of `RAX1`, which we have circumvented as described in Sect. 4.3.

Finally, we observe that hybrid implementations are not beneficial on in-order cores, as we expected in Sect. 4.4.

We take away that Cortex-A55 and Cortex-A510 have very efficient scalar implementations which fully leverage the potential for superscalar execution. On Cortex-A55, the scalar implementation should even be used for batched applications of `Keccak-f1600`. On Cortex-A510, batched applications of `Keccak-f1600` should use the Armv8.4-A based implementation.

Cortex-A78 and Cortex-A710. We observe a speedup of 1.15× for our scalar implementation compared to the baseline C implementation. We don't gain as much as for Cortex-A55 and Cortex-A510, which is expected since scheduling for latency is less important for out-of-order cores. Moreover, our scalar implementation is close to the theoretical optimum: With 2516 arithmetic instructions in the core of `Keccak-f1600`, and 4 scalar units, performance is bounded by ≈ 629 cycles, *ignoring* preamble and postamble.

Next, we comment at the Armv8-A Neon performance. Recalling that the core of the implementation performs 3840 Neon arithmetic instructions, and Cortex-A78 and Cortex-A710 have maximum Neon IPC of 2, our implementations are reasonably close to the theoretical optimum, yet around 1.5× slower than the scalar implementation. For Cortex-A78, the `Keccak-f1600-x5` hybrid achieves near optimal performance, leveraging up to 6 IPC on Cortex-A78. For Cortex-A710 in turn, we confirm that the 5-way hybrid cannot work due to the maximum of 5 IPC on Cortex-A710.

Finally, we look at the Armv8.4-A Neon performance. With a single Neon unit implementing the SHA-3 instructions, we cannot do better than 1536 cycles, and our implementation comes very close to that, providing a meaningful speedup of 1.32× over the Armv8-A Neon implementation. Yet, it is still slightly slower than the scalar implementation, but a `Keccak-f1600-x4` scalar/Armv8.4-A Neon hybrid gets the best of the fast scalar implementation and the SHA-3 instructions. This implementation leverages the maximum throughput of 5 IPC for Cortex-A710: 4 IPC for the scalar implementation, and 1 IPC for the Neon implementation. This also explains why the Scalar/Neon/Neon hybrid is worse than the Scalar/Neon hybrid: There is no bandwidth to leverage all four scalar units and both Neon units in every cycle.

Cortex-X1 and Cortex-X2. For scalar `Keccak-f1600`, we get essentially the same performance as for Cortex-A78 and Cortex-A710, and the same comments apply – this is unsurprising given that Cortex-{A78,A710,X1,X2} all have the same throughput and latency for the relevant scalar instructions.

Next, we look at the performance of the Armv8-A Neon implementations. We observe that it is 5%–10% faster than 2× the scalar implementation – i.e., for batched computations of `Keccak-f1600`, scalar and Armv8-A Neon implementation are roughly on par. We also note that the performance is lower than what the theoretical maximum of 4 Neon IPC for Cortex-X1 and Cortex-X2 would suggest: With 3840 Neon arithmetic instructions, one could hope for $\approx$ 1000 cycles. We believe that the difficulty in going significantly beyond 3 IPC lies in the `Keccak-f1600` computation "narrowing" at the θ step, and in the `SHL+SRI`-based rotations having a maximum IPC of 2 (see Sect. 4.2). Nonetheless, we cannot exclude further optimization potential, and encourage research.

The roughly equal performance of scalar and Armv8-A Neon implementation motivates why we pair 2× and 1× `Keccak-f1600-x2` when constructing the Scalar/Neon-Armv8-A hybrid for `Keccak-f1600-x4`. We observe that the hybrid is only slightly above the theoretical optimum, confirming that the frontends of Cortex-X1 and Cortex-X2 are wide enough to process both implementations.

Next, we comment on the performance of the Armv8.4-A Neon implementation on Cortex-X2. First, one observes that the pure Armv8.4-A implementation is slower than the Armv8-A implementation. While this may come as a surprise, the reason is clear: The SHA-3 instructions are implemented on 1 out of 4 Neon units, while the logical operations underlying the Armv8-A implementation are available on all units. Accordingly, we observe a significant speedup for the Neon/Neon hybrid, since it puts all Neon units to use. In fact, this hybrid is sufficiently fast to make a 3-way batched Scalar/Neon/Neon hybrid useful, and this implementation yields the best batched performance. A 4-way batched Scalar/Neon/Neon implementation brings little benefit compared to a Scalar/Armv8-A Neon hybrid: that's because the bottleneck is the scalar code anyway.

5.3 SPHINCS$^+$ Performance

Table 2 shows the performance of SPHINCS$^+$ (v3.1) based on our `Keccak-f1600` implementations, in comparison to previous implementations. We only display

Table 2. Performance results for SPHINCS$^+$. For each platform, we pick the `Keccak-f1600` implementation that achieves the best performance.

Parameter set	Impl.	Key generation		Signing		Verification	
Cortex-X1							
128f-robust	C [Kee]	7 358k		170 826k		11 503k	
	[Ngu]	6 112k	(1.00 ×)	141 857k	(1.00 ×)	9 835k	(1.00 ×)
	Ours	3 491k	(1.75 ×)	81 198k	(1.75 ×)	5 881k	(1.67 ×)
128s-robust	C [Kee]	470 976k		3 546 272k		4 168k	
	[Ngu]	391 075k	(1.00 ×)	2 937 624k	(1.00 ×)	3 634k	(1.00 ×)
	Ours	223 778k	(1.75 ×)	1 681 496k	(1.75 ×)	2 139k	(1.70 ×)
Cortex-A78							
128f-robust	C [Kee]	7 507k	(1.00 ×)	174 285k	(1.00 ×)	11 912k	(1.00 ×)
	[Ngu]	10 731k		249 061k		16 939k	
	Ours	5 043k	(1.49 ×)	117 280k	(1.49 ×)	7 949k	(1.50 ×)
128s-robust	C [Kee]	479 608k	(1.00 ×)	3 603 102k	(1.00 ×)	4 277k	(1.00 ×)
	[Ngu]	686 059k		5 153 452k		6 359k	
	Ours	262 264k	(1.83 ×)	2 029 133k	(1.78 ×)	2 534k	(1.69 ×)
Cortex-A55							
128f-robust	C [Kee]	18 035k	(1.00 ×)	418 555k	(1.00 ×)	27 322k	(1.00 ×)
	[Ngu]	23 444k		544 203k		37 017k	
	Ours	13 078k	(1.38 ×)	304 188k	(1.38 ×)	21 855k	(1.25 ×)
128s-robust	C [Kee]	1 153 927k	(1.00 ×)	8 667 372k	(1.00 ×)	10 415k	(1.00 ×)
	[Ngu]	1 500 186k		11 269 260k		13 301k	
	Ours	835 847k	(1.38 ×)	6 278 826k	(1.38 ×)	6 916k	(1.51 ×)
Cortex-X2							
128f-robust	C [Kee]	7 481k		173 680k		11 409k	
	[Ngu]	5 946k	(1.00 ×)	138 094k	(1.00 ×)	9 400k	(1.00 ×)
	[Wes]	6 930k		160 942k		11 298k	
	Ours	3 315k	(1.79 ×)	77 038k	(1.79 ×)	5 544k	(1.70 ×)
128s-robust	C [Kee]	479 373k		3 601 405k		4 374k	
	[Ngu]	381 170k	(1.00 ×)	2 863 365k	(1.00 ×)	3 312k	(1.00 ×)
	[Wes]	443 343k		3 330 902k		3 937k	
	Ours	194 295k	(1.96 ×)	1 517 988k	(1.89 ×)	1 849k	(1.79 ×)
Cortex-A710							
128f-robust	C [Kee]	7 571k		175 706k		11 796k	
	[Ngu]	10 641k		247 082k		17 210k	
	[Wes]	6 980k	(1.00 ×)	162 090k	(1.00 ×)	11 338k	(1.00 ×)
	Ours	3 743k	(1.86 ×)	87 052k	(1.86 ×)	6 071k	(1.87 ×)
128s-robust	C [Kee]	483 664k		3 633 790k		4 194k	
	[Ngu]	681 006k		5 118 302k		6 188k	
	[Wes]	446 644k	(1.00 ×)	3 356 044k	(1.00 ×)	3 850k	(1.00 ×)
	Ours	239 634k	(1.86 ×)	1 800 720k	(1.86 ×)	2 147k	(1.79 ×)
Cortex-A510							
128f-robust	C [Kee]	13 787k		315 780k		21 640k	
	[Ngu]	15 270k		354 191k		24 771k	
	[Wes]	10 600k	(1.00 ×)	245 623k	(1.00 ×)	16 866k	(1.00 ×)
	Ours	5 428k	(1.95 ×)	125 818k	(1.95 ×)	8 920k	(1.89 ×)
128s-robust	C [Kee]	871 396k		6 548 093k		7 969k	
	[Ngu]	974 307k		7 322 458k		8 397k	
	[Wes]	661 699k	(1.00 ×)	4 991 715k	(1.00 ×)	5 791k	(1.00 ×)
	Ours	347 614k	(1.90 ×)	2 610 123k	(1.91 ×)	3 322k	(1.74 ×)

results for the "robust" 128-bit parameter sets, but note that our implementations work for all other parameter sets, too, and show similar speedups. Full results are available alongside the code. We see significant performance improvements of up to 1.89× compared to the state of the art.

Acknowledgments. Matthias J. Kannwischer was supported by the Taiwan Ministry of Science and Technology through Academia Sinica Investigator Award AS-IA-109-M01 and the Executive Yuan Data Safety and Talent Cultivation Project (AS-KPQ-109-DSTCP).

References

[ACD+22] Alagic, G., et al.: Status report on the third round of the NIST post-quantum cryptography standardization process, 2022-07-05 04:07:00 (2022)

[Arma] Arm Limited: Cortex-A510 Software Optimization Guide

[Armb] Arm Limited: Cortex-A55 Software Optimization Guide

[Armc] Arm Limited: Cortex-A710 Software Optimization Guide

[Armd] Arm Limited: Cortex-A78 Software Optimization Guide

[Arme] Arm Limited: Cortex-X1 Software Optimization Guide

[Armf] Arm Limited: Cortex-X2 Software Optimization Guide

[BDH+] Bertoni, G., Daemen, J., Hoffert, S., Peeters, M., Van Assche, G., Van Keer, R.: Keccak. https://keccak.team/keccak.html

[BDH11] Buchmann, J., Dahmen, E., Hülsing, A.: XMSS - a practical forward secure signature scheme based on minimal security assumptions. In: Yang, B.-Y. (ed.) PQCrypto 2011. LNCS, vol. 7071, pp. 117–129. Springer, Heidelberg (2011). https://doi.org/10.1007/978-3-642-25405-5_8

[BDP+18] Bertoni, G., Daemen, J., Peeters, M., Van Assche, G., Van Keer, R., Viguier, B.: KangarooTwelve: Fast Hashing Based on Keccak-p. In: Preneel, B., Vercauteren, F. (eds.) ACNS 2018. LNCS, vol. 10892, pp. 400–418. Springer, Cham (2018). https://doi.org/10.1007/978-3-319-93387-0_21

[BHH+15] Bernstein, D.J., et al.: SPHINCS: practical stateless hash-based signatures. In: Oswald, E., Fischlin, M. (eds.) EUROCRYPT 2015. LNCS, vol. 9056, pp. 368–397. Springer, Heidelberg (2015). https://doi.org/10.1007/978-3-662-46800-5_15

[BHK+21] Becker, H., Hwang, V., Kannwischer, M.J., Yang, B.-Y., Yang, S.-Y.: Neon NTT: faster Dilithium, Kyber, and saber on Cortex-A72 and Apple M1. IACR Trans. Cryptogr. Hardw. Embed. Syst. **2022**(1), 221–244 (2021)

[BS12] Bernstein, D.J., Schwabe, P.: NEON crypto. In: Prouff, E., Schaumont, P. (eds.) CHES 2012. LNCS, vol. 7428, pp. 320–339. Springer, Heidelberg (2012). https://doi.org/10.1007/978-3-642-33027-8_19

[DHP+] Daemen, J., Hoffert, S., Peeters, M., Van Assche, G., Van Keer, R.: eXtended Keccak Code Package. https://github.com/XKCP/XKCP

[Dwo15] Dworkin, M.: SHA-3 standard: permutation-based hash and extendable-output functions, 2015-08-04 (2015)

[HBD+22] Hülsing, A., et al.: SPHINCS+. Technical report (2022). https://sphincs.org/

[HBG+18] Hülsing, A., Butin, D., Gazdag, S.-L., Rijneveld, J., Mohaisen, A.: XMSS: eXtended merkle signature scheme. RFC 8391, May 2018. https://rfc-editor.org/rfc/rfc8391.txt

[Hül13] Hülsing, A.: W-OTS+ – shorter signatures for hash-based signature schemes. In: Youssef, A., Nitaj, A., Hassanien, A.E. (eds.) AFRICACRYPT 2013. LNCS, vol. 7918, pp. 173–188. Springer, Heidelberg (2013). https://doi.org/10.1007/978-3-642-38553-7_10

[Kee] Van Keer, R.: "simple" Keccak C implementation in SUPERCOP (crypto_hash/keccakc512/simple). https://bench.cr.yp.to/supercop.html

[Köl18] Kölbl, S.: Putting wings on SPHINCS. In: Lange, T., Steinwandt, R. (eds.) PQCrypto 2018. LNCS, vol. 10786, pp. 205–226. Springer, Cham (2018). https://doi.org/10.1007/978-3-319-79063-3_10

[KPR+] Kannwischer, M.J., Petri, R., Rijneveld, J., Schwabe, P., Stoffelen, K.: PQM4: post-quantum crypto library for the ARM Cortex-M4. https://github.com/mupq/pqm4

[LDK+22] Lyubashevsky, V., et al.: CRYSTALS-DILITHIUM. Technical report (2022). https://pq-crystals.org/dilithium/

[Len19] Lenngren, E.: AArch64 optimized implementation for X25519 (2019). https://github.com/Emill/X25519-AArch64

[MCF19] McGrew, D., Curcio, M., Fluhrer, S.: Leighton-Micali hash-based signatures. RFC 8554, April 2019. https://rfc-editor.org/rfc/rfc8554.txt

[Nat13] National Institute of Standards and Technology: FIPS186-4: Digital Signature Standard (DSS) (2013). https://doi.org/10.6028/NIST.FIPS.186-4

[Nat18] National Institute of Standards and Technology: NIST SP 800-56A Rev. 3: Recommendation for Pair-Wise Key-Establishment Schemes Using Discrete Logarithm Cryptography (2018). https://doi.org/10.6028/NIST.SP.800-56Ar3

[Nat19] National Institute of Standards and Technology: NIST SP 800-56B Rev. 2: Recommendation for Pair-Wise Key-Establishment Using Integer Factorization Cryptography (2019). https://doi.org/10.6028/NIST.SP.800-56Br2

[NG21] Nguyen, D.T., Gaj, K.: Fast NEON-based multiplication for lattice-based NIST post-quantum cryptography finalists. In: Cheon, J.H., Tillich, J.-P. (eds.) PQCrypto 2021 2021. LNCS, vol. 12841, pp. 234–254. Springer, Cham (2021). https://doi.org/10.1007/978-3-030-81293-5_13

[Ngu] Nguyen, D.T.: Armv8-A Neon implementation for Keccak-f1600. https://github.com/cothan/NEON-SHA3_2x

[NIS16] NIST Computer Security Division. Post-Quantum Cryptography Standardization (2016). https://csrc.nist.gov/Projects/Post-Quantum-Cryptography

[PFH+22] Prest, T., et al.: FALCON. Technical report (2022). https://falcon-sign.info/

[SAB+22] Schwabe, P., et al.: CRYSTALS-KYBER. Technical report (2022). https://pq-crystals.org/kyber/

[Wes] Westerbaan, B.: ARMV8.4-A implementation for Keccak-f1600. https://github.com/bwesterb/armed-keccak

Parallel Isogeny Path Finding with Limited Memory

Emanuele Bellini[1], Jorge Chavez-Saab[1,3](✉), Jesús-Javier Chi-Domínguez[1], Andre Esser[1], Sorina Ionica[2], Luis Rivera-Zamarripa[1], Francisco Rodríguez-Henríquez[1,3], Monika Trimoska[4], and Floyd Zweydinger[5]

[1] Technology Innovation Institute, Abu Dhabi, United Arab Emirates
{emanuele.bellini,jorge.saab,jesus.dominguez,andre.esser, luis.zamarripa,francisco.rodriguez}@tii.ae
[2] Université de Picardie Jules Verne, Amiens, France
sorina.ionica@u-picardie.fr
[3] CINVESTAV-IPN, Mexico City, Mexico
francisco@cs.cinvestav.mx
[4] Radboud University, Nijmegen, The Netherlands
monika.trimoska@ru.nl
[5] Ruhr University Bochum, Bochum, Germany
floyd.zweydinger@rub.de

Abstract. The security guarantees of most isogeny-based protocols rely on the computational hardness of finding an isogeny between two supersingular isogenous curves defined over a prime field $\mathbb{F}_q$ with q a power of a large prime p. In most scenarios, the isogeny is known to be of degree ℓ^e for some small prime ℓ. We call this problem the Supersingular Fixed-Degree Isogeny Path (*SIPFD*) problem. It is believed that the most general version of *SIPFD* is not solvable faster than in exponential time by classical as well as quantum attackers.

In a classical setting, a meet-in-the-middle algorithm is the fastest known strategy for solving the *SIPFD*. However, due to its stringent memory requirements, it quickly becomes infeasible for moderately large *SIPFD* instances. In a practical setting, one has therefore to resort to time-memory trade-offs to instantiate attacks on the *SIPFD*. This is particularly true for GPU platforms, which are inherently more memory-constrained than CPU architectures. In such a setting, a van Oorschot-Wiener-based collision finding algorithm offers a better asymptotic scaling. Finding the best algorithmic choice for solving instances under a fixed prime size, memory budget and computational platform remains so far an open problem.

To answer this question, we present a precise estimation of the costs of both strategies considering most recent algorithmic improvements. As a second main contribution, we substantiate our estimations via optimized software implementations of both algorithms. In this context, we provide the first optimized GPU implementation of the van Oorschot-Wiener approach for solving the *SIPFD*. Based on practical measurements we extrapolate the running times for solving different-sized instances. Finally, we give estimates of the costs of computing

T. Isobe and S. Sarkar (Eds.): INDOCRYPT 2022, LNCS 13774, pp. 294–316, 2022.
https://doi.org/10.1007/978-3-031-22912-1_13

a degree-2^{88} isogeny using our CUDA software library running on an NVIDIA A100 GPU server.

Keywords: Isogenies · Cryptanalysis · GPU · Golden collision search · Meet-in-the-middle · Time-memory trade-offs · Efficient implementation

1 Introduction

Let E_0 and E_1 be two supersingular isogenous elliptic curves defined over a finite field $\mathbb{F}_q$, with q a power of a large prime p. Computing an isogeny $\phi\colon E_0 \to E_1$ is believed to be hard in the classical as well as the quantum setting and is known as the Supersingular Isogeny Path (*SIP*) problem. In many scenarios, the isogeny is of known degree ℓ^e for some small prime ℓ and we refer to this variant as the Supersingular Fixed-Degree Isogeny Path (*SIPFD*) problem. Investigating the concrete computational hardness of *SIPFD* and the best approaches to tackle it in multi- and many-core CPU and GPU platforms, is the main focus of this work.

In the context of cryptographic protocols, *SIP* was first studied by Charles, Goren and Lauter [6]. They reduced the collision resistance of a provably secure hash function to the problem of finding two isogenies of equal degree ℓ^n for a small prime ℓ and $n \in \mathbb{Z}$ between any two supersingular elliptic curves. This in turn may also be tackled as a *SIPFD* problem.

Variants of the *SIPFD* problem form the basis of several isogeny-based signatures [17,33]. Further, *SIPFD* has been used as foundation of recently proposed cryptographic primitives such as Verifiable Delay Functions [7,11]. Based on the intractability of the *SIPFD* problem, Jao and De Feo proposed the Supersingular Isogeny-based Diffie-Hellman key exchange protocol (SIDH) [10,18]. Apart from revealing the isogeny degree, SIDH also reveals the evaluation of its secret isogenies at a large torsion subgroup. This weaker variant of *SIPFD* was dubbed as the *Computational Supersingular Isogeny* (CSSI) problem [10]. SIKE [2], a variant of SIDH equipped with a key encapsulation mechanism, was one of the few schemes that made it to the fourth round of the NIST standardization effort as a KEM candidate [25]. Until recently, the best-known algorithms for breaking SIDH or SIKE had an exponential time complexity in both, classical and quantum settings.

However, in July 2022, Castryck and Decru [5] proposed a surprising attack that (heuristically) solves the CSSI problem in polynomial-time. This attack relies on the knowledge of three crucial pieces of information, namely, (i) the degree of the isogeny ϕ; (ii) the endomorphism ring of the starting curve E_0; and (iii) the images $\phi(P_0), \phi(Q_0)$ of Alice's generator points $\langle P_0, Q_0 \rangle = E[2^a]$, where the prime $p = 2^a 3^b - 1$ is the underlying prime used by SIKE instantiations. Recall that (ii) and (iii) are only known in the specific case of the CSSI problem, but not in the more general case of the *SIPFD* problem. Furthermore, another attack by Maino and Martindale [21] and yet another one by Robert [27] quickly followed. Maino and Martindale's attack relies on several crucial steps used in [5], but does not require knowledge of the endomorphism ring associated to the base curve. Robert's attack can also break SIDH for any random starting supersingular elliptic curve.

Despite the short time elapsed since the publication of Castryck and Decru's attack, several countermeasures have already been proposed by trying to hide the degree of the isogeny [24], the endomorphism ring of the base curve [4], or the images of the torsion points [15]. At this point, only time will tell if SIDH/SIKE will ever recover from the attacks on the CSSI problem. But even if this never happens, the theoretical and practical importance of the *SIPFD* problem still stands. For instance, the constructions from [17, Section 4], [7], and [20, Section 5.3] do not append images of auxiliary points to their public keys. In turn the Castryck-Decru family of attacks does not apply, making the security of those applications entirely based on the *SIPFD* problem.

Known Attacks on the *SIPFD* Problem. Even before the publication of the attack in [5], it was wildly believed that the best approaches for solving the CSSI problem are classical and not quantum [19]. Here we present a brief summary of the different assumptions made across the last decade about the cost of solving the *SIPFD* problem. We stress that while all these advances were made with SIKE as main motivation, the fact that they did not make use of the torsion point images means that they still represent the state-of-the-art for attacks against the general *SIPFD* problem. The fastest known algorithm for solving *SIP* has computational complexity $\tilde{O}(\sqrt{p})$ [13,16,22]. However, if the secret isogeny is of known degree ℓ^e, there might exist more efficient algorithms for solving the *SIPFD*. Indeed, in their NIST first round submission, the SIKE team [2] argued that the best classical attack against the CSSI problem was to treat it as an *SIPFD* problem and use a MitM approach with a time and memory cost of $O(\ell^{\frac{e}{2}})$, which is more efficient for SIKE and all instantiations of the *SIPFD* where $\ell^e \leq p$.

By assuming an unlimited memory budget and memory queries with zero time cost, the MitM attack is indeed the best attack against the *SIPFD* problem. Nevertheless, in [1], the authors argued that the van Oorschot-Wiener (vOW) golden collision search, which yields a better time-memory trade-off curve, is the best classical approach for large instances. The rationale used is that the $O(\ell^{\frac{e}{2}})$ memory requirement for launching the MitM attack is infeasible for the cryptographic parameter sizes. Since the best known generic attacks against AES use a negligible amount of memory, it is just natural to set an upper bound on the available classical memory when evaluating the cost of solving *SIPFD* instantiations in the context of NIST security levels 1 to 5.

To increase interest in studying the CSSI problem Costello published in [8] two Microsoft $IKE challenges, a small and a large one using a 182- and a 217-bit prime number, respectively. These two CSSI instances are known as $IKEp182 and $IKEp217 challenges.[1] A few months later, the solution of $IKEp182 was

[1] The precise specifications can be found in https://github.com/microsoft/SIKE-challenges.

announced by Udovenko and Vitto in [30]. The authors treated this challenge as an instance of *SIPFD*, and then used a MitM approach largely following the description given in [9] along with several clever sorting and sieving tricks for optimizing data queries for their disk-based storage solution. The authors reported that their attack had a timing cost of less than 10 core-years, but at the price of using 256 TiB of high-performance network storage memory.

It is obvious that this memory requirements quickly render the strategy unfeasible for larger non-toy instances. As mentioned in [1], there exists a time-memory trade-off variant of the MitM algorithm (*cf.* Sect. 2.2), which was adopted by Udovenko and Vitto to bring the storage requirements of their attack down to about 70 TiB.

However, determining the best algorithmic choice for solving instances of given size under a certain memory budget and computational platform remains so far an open problem. In this work we present a framework predicting that both MitM variants are outperformed by the vOW golden collision approach even for moderately large *SIPFD* instances. We then substantiate our claims by extrapolating results of our implementations, accounting for practical effects such as memory access costs.

Our Contributions. In [1] it was found that vOW is a better approach than MitM to tackle large *SIPFD* instances. However, the small Microsoft challenge $IKEp182 was broken, before the Castryck-Decru attack was known, using a MitM strategy [30]. As discussed in [30], it remains unclear for which instance sizes and memory availability, vOW outperforms MitM. In this work we answer this question from a theoretical and practical perspective. Theoretically, we give a precise estimation of the costs of both strategies including most recent algorithmic improvements. Practically we substantiate our estimations via optimized implementations and extensive benchmarking performed in CPU and GPU platforms.

Moreover, in the case of CPU platforms, we present a detailed framework that for a fixed memory budget and prime size, predicts when a pure MitM approach, batched (limited memory) MitM or vOW approach becomes the optimal design choice for attacking *SIPFD* (see Sect. 3 and Fig. 2). The predictions of our model are backed up by practical experiments on small *SIPFD* instances and extrapolations based on the obtained practical timings of our implementations.

We additionally provide the first optimized GPU implementation of the vOW attack on *SIPFD*, outperforming a CPU based implementation by a factor of almost two magnitudes. We provide medium sized experimental data points using our GPU implementation including extrapolations to larger instances. More concretely, our implementation solves *SIPFD* instances with isogeny degree 2^{88} with primes of bit size 180 (comparable to the instance solved in [30]) using 16 GPUs each equipped with only 80 GiB of memory in about 4 months. Based on our experimental results we conclude that vOW is the preferred choice for any larger *SIPFD* instances on reasonable hardware.

Our CPU and GPU software libraries are open-source and available at https://github.com/TheSIPFDTeam/SIPFD.

Outline. The remainder of this work is organized as follows. In Sect. 2 we present a formal definition of *SIPFD* and relevant mathematical background. We also give a detailed explanation of the MitM and vOW strategies. In Sect. 3 we present a careful estimation of the cost of the MitM and vOW strategies and their corresponding trade-offs in the context of the *SIPFD*. In Sect. 4 we present our implementation of the CPU-based MitM attack and the vOW strategy on a multi-core GPU platform.

2 Preliminaries

2.1 Elliptic Curves and Isogenies

Let $\mathbb{F}_p$ be the prime field with p elements and let E be a supersingular elliptic curve defined over $\mathbb{F}_{p^2}$. A common choice, convenient for implementations, is to choose p such that $p \equiv 3 \bmod 4$, and take $\mathbb{F}_{p^2} = \mathbb{F}_p[i]/(i^2+1)$ the quadratic extension of $\mathbb{F}_p$. Moreover, we will assume that E is given by a Montgomery equation:

$$E\colon y^2 = x^3 + Ax^2 + x, \quad A \in \mathbb{F}_{p^2} \setminus \{\pm 2\}.$$

The set of points satisfying this equation along with a point at infinity $\mathcal{O}_E$ form an abelian group. The point $\mathcal{O}_E$ plays the role of the neutral element. In general, we write the sum of d copies of P as $[d]P$ and if k is the smallest scalar such that $[k]P = \mathcal{O}_E$, we say that P is an order-k elliptic curve point. The d-torsion subgroup, denoted by $E[d]$, is the set of points $\{P \in E(\overline{\mathbb{F}}_p) \mid [d]P = \mathcal{O}_E\}$. If $\gcd(p, d) = 1$, then $E[d]$, as a subgroup of E, is isomorphic to $\mathbb{Z}/n\mathbb{Z} \times \mathbb{Z}/n\mathbb{Z}$. The j-invariant of the curve E is given by $j(E) := \frac{256(A^2-3)^3}{A^2-4}$. It has the useful property that two curves are isomorphic if and only if they have the same j-invariant.

An isogeny $\phi\colon E \to E'$ is a rational map (roughly speaking a pair of quotient of polynomials) such that $\phi(\mathcal{O}_E) = \mathcal{O}_{E'}$. This implies that ϕ is a group homomorphism (see for instance [32]). By a theorem of Tate [28], an isogeny defined over $\mathbb{F}_{p^2}$ exists if and only if $\#E(\mathbb{F}_{p^2}) = \#E'(\mathbb{F}_{p^2})$. If there is an isogeny ϕ between E and E', then we say that the two curves are *isogenous.* The kernel $\ker\phi$ is the set of points in the domain curve E which are mapped to the identity point $\mathcal{O}_{E'}$. If we restrict to separable maps only, any isogeny ϕ is uniquely determined by its kernel up to an isomorphism. We say ϕ is a d-isogeny or an isogeny of degree d whenever $\#\ker\phi = d$. Any isogeny can be written as a composition of prime-degree isogenies and the degree is multiplicative, in the sense that $\deg(\phi_1 \circ \phi_2) = \deg(\phi_1)\deg(\phi_2)$.

For each isogeny $\phi\colon E \to E'$ there also exists a dual d-isogeny $\hat{\phi} : E' \to E$ satisfying $\phi \circ \hat{\phi} = [d]$ and $\hat{\phi} \circ \phi = [d]$, where $[d]$ is the isogeny $P \to [d]P$ on E and E' respectively. Isogenies of degree d are computed in time $O(d)$ by using Vélu's formulas, or for sufficiently large d in $O(\sqrt{d})$, using the more recent $\sqrt{\text{é}}$lu's formulas [3]. In practice, if the degree is d^e with d small, one splits a d^e-isogeny as the composition of e d-isogenies each computed with Vélu's formula.

We are now ready to state a formal definition of the *SIPFD* problem.

Definition 1 (*SIPFD* **problem).** *Let p, ℓ be two prime numbers. Consider E and E' two supersingular elliptic curves defined over $\mathbb{F}_{p^2}$ such that $\#E(\mathbb{F}_{p^2}) = \#E'(\mathbb{F}_{p^2})$. Given $e \in \mathbb{N}$ find an isogeny of degree ℓ^e from E to E', if it exists.*

Concretely, in the remainder of this work we assume that the secret isogeny is of degree 2^e i.e., we fix $\ell = 2$, and define it over $\mathbb{F}_{p^2}$. Moreover, we assume $p = f \cdot 2^e - 1$ for some odd cofactor f and that $\#E(\mathbb{F}_{p^2}) = (p+1)^2$. This allows us to have the 2^e-torsion subgroup defined over $\mathbb{F}_{p^2}$. Finally, we work with instances where e is half the bitlength of the prime p. While all these conditions are more specific than the general *SIPFD* problem, they are efficiency-oriented decisions that are common practice in isogeny-based protocols and we do not exploit them beyond that.

Let P, Q be a basis of $E[2^e]$. Then the kernel of any 2^e-isogeny can be written as either $\langle P + [\mathrm{sk}]Q \rangle$, with $\mathrm{sk} \in \{0, \ldots, 2^e - 1\}$ or $\langle [\mathrm{sk}][2]P + Q \rangle$, with $\mathrm{sk} \in \{0, \ldots, 2^{e-1} - 1\}$. For simplicity, in our implementation we work with isogeny kernels are always of the form $\langle P + [\mathrm{sk}]Q \rangle$. There is little loss of generality, since attacking the remaining kernels would only require re-labelling the basis and re-running the algorithm.

An isogeny $\phi : E_0 \to E_1$ of degree ℓ^e can be written as a composition $\phi = \phi_1 \circ \phi_0$ of two isogenies of degree $\ell^{e/2}$ (assuming an even e for simplicity), where $\phi_0 : E_0 \to E_m$ and $\phi_1 : E_m \to E_1$ for some middle curve E_m. Since there exists a dual isogeny $\hat{\phi}_1 : E_1 \to E_m$, one can conduct a Meet in the Middle (MitM) attack by exploring all the possible $\ell^{e/2}$-isogenies emanating from E_0 and E_1, and finding the pair of isogenies that arrive to the same curve E_m (up to isomorphism). The largest attack recorded on the *SIPFD* problem, conducted by Udovenko and Vitto[2] [30], used this strategy to break an instance with $\ell = 2$ and $e = 88$.

2.2 Meet in the Middle (MitM)

Let us briefly recall the MitM procedure to solve the *SIPFD* for $\ell = 2$. We first compute and store all $2^{e/2}$-isogenous curves to E_0 in a table T (identified via their j-invariants). Then we proceed by computing each $2^{e/2}$-isogenous curve to E_1 and check if its j-invariant is present in table T. Any matching pair then allows to recover the secret isogeny as outlined in the previous section.

Complexity. Let $N := 2^{e/2}$. The worst-case time complexity of the MitM attack is $2N$ evaluations of degree-$2^{e/2}$ isogenies, while in the average case $1.5N$ such evaluations are necessary. The space complexity is dominated by the size of the table to store the N j-invariants and scalars.

In a memory restricted setting, where the table size is limited to W entries, the MitM attack is performed in batches. In each batch, we compute and store the output of W isogenies from E_0, then compute and compare against each

[2] This work was realized as an attack on SIKE, but does not exploit the torsion point images and can be regarded as an attack on *SIPFD* in general.

of the N isogenies from E_1 without storing them. The number of batches is N/W where each batch performs W isogenies from E_0 and N isogenies from E_1, yielding a total of $\frac{N}{W}(N+W)$ evaluations of $2^{e/2}$-isogenies.

Depth-First Search Methodology. In 2018, Adj *et al.* [1, §3.2] showed that computing the isogenies from each side in a depth-first tree fashion yields performance improvements. The improvement stems from the iterative construction of the $2^{e/2}$-isogenies as $e/2$ degree-2 isogenies. Here, whenever two isogenies share the same initial path, the depth-first approach avoids re-computation of those steps.

In order to adapt to the limited-memory scenario, let us assume that the available memory can hold $W = 2^{\omega}$ entries. Then each batch of isogenies from E_0 can be obtained by following a fixed path for the first $e/2-\omega$ steps, and then computing the whole subtree of depth ω from this node.

Also, the attack is easy to parallelize. Assuming 2^c threads are used, all trees can be branched sequentially for c steps to obtain 2^c subtrees, each of which is assigned to a different core. This methodology for evaluating trees in batches and with multiple cores is summarized in Fig. 1.

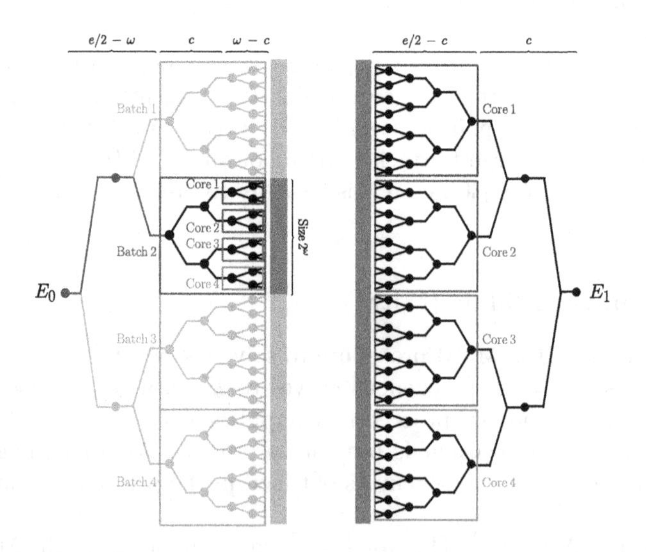

Fig. 1. The batched Meet-in-the-Middle depth-first approach: finding a 2^e-isogeny between E_0 and E_1 with 2^c cores and 2^{ω} memory. In this example, $e = 12$, $c = 2$, and $\omega = 4$.

Since a binary tree of depth ω has $2^{\omega+1}-2$ edges, each batch is computing $e/2-\omega+2^{\omega+1}-2$ isogenies of degree 2 for the side corresponding to E_0, and the whole tree with $2^{e/2+1}-2$ isogenies for the side corresponding to E_1. The expected cost corresponding to half of the batches is then

$$\frac{1}{2}2^{e/2-\omega}\left(2^{e/2+1}+2^{\omega+1}+e/2-\omega-4\right)\approx 2^{e-\omega}$$

computations of 2-isogenies.

2.3 Parallel Collision Search

Given a random function $f: S \to S$, van Oorschot and Wiener's method [26] is a parallel procedure to find collisions in f. The main idea of the algorithm is to construct in parallel several chains of evaluations $x_i = f(x_{i-1})$, starting from random seeds x_0. Further, a small fraction of the points in S is called *distinguished* based on an arbitrary criterion (e.g. that the binary representation of $x \in S$ ends with a certain number of zeros). A chain continues until it reaches a distinguished point. Then this point is compared against a hash table including all previously found distinguished points. Further, to avoid infinite loops, chains are aborted after their length exceeds a specified threshold.

Two chains ending in the same distinguished point indicate a collision between those chains. This collision can be efficiently reconstructed if the seeds x_0, x'_0 and the lengths d, d' of the colliding chains are known. Therefore, assuming $d > d'$ we take $d-d'$ steps on the longer chain (starting from x_0). From there on we take simultaneous steps on both chains, checking after each step if the collision has occurred. Hence, the hash table stores for each found distinguished point the triplet (x_0, x_d, d) indexed by x_d.

Complexity. Let N be the size of the set S, θ the proportion of points that are distinguished, and W the amount of distinguished triplets that we can store. Since each chain has an average length of $1/\theta$, the chains represented by the stored triplets (once the hash table is completely filled) include an average of W/θ points. Therefore the probability that a given evaluation of f collides with any of these points is $W/N\theta$. After a collision takes place, the chain needs to continue for an additional $1/\theta$ steps on average before it reaches a distinguished point and the collision is detected. At this point, the two involved chains must be reconstructed from the start to find the exact step at which the collision occurred, yielding a total of $N\theta/W + 3/\theta$ evaluations of f to find a collision. The optimal choice for θ is $\sqrt{3W/N}$ yielding a cost of $2\sqrt{3N/W}$ per collision. Note, however that this analysis assumes a table that already contains W triplets. To capture the transition effect of the table filling up, van Oorschot and Wiener [26] model $\theta = \alpha\sqrt{W/N}$ for a parameter α that is experimentally measured to be optimal at $\alpha = 2.25$. The resulting cost per collision is found to be linear in $\sqrt{N/W}$ as long as $2^{10} < W < N/2^{10}$.

Note that any random function from S to itself is expected to have $N/2$ collisions, however, many applications, including the *SIPFD*, require looking for one specific collision that we refer to as the "golden collision" [12,14,23,31]. This means that the attack has to find $N/4$ different collisions on average before stumbling upon the golden collision, bringing the total cost to $\mathcal{O}(\sqrt{N^3/W})$ function evaluations.

Application to the *SIPFD* Problem. To attack the *SIPFD* problem and find the kernel of a degree-2^e isogeny between E_0 and E_1, we assume for simplicity that e is even and define $S = \{0,1\} \times \{0, \dots, 2^{e/2} - 1\}$ so that $N = 2^{e/2+1}$. We also define the map $g : S \to \mathbb{F}_{p^2}$, $(c, k) \mapsto j(E_c/\langle P_c + [k]Q_c\rangle)$, where (P_c, Q_c) are a predefined basis of the $2^{e/2}$-torsion on either side as before. As explained in Sect. 2.1, the function g yields a bijection between S and the set of $2^{e/2}$-isogenies with kernel $\langle P_c + [k]Q_c\rangle$ from the curves on either side. A collision $g(c, k) = g(c', k')$ with $c \neq c'$ implies two isogenous curves starting on opposite sides and meeting at a middle curve (up to isomorphism).

To apply the parallel collision search, we need a function f that maps S back to itself. Hence, we have to work with the composition $f = h \circ g$ where h is an arbitrary function mapping j-invariants back to S. This composition introduces several fake collisions that are produced by the underlying hash function while there is still only one (golden) collision that leads to the secret isogeny.

Note that for a certain (unlucky) choice of hash function h the golden collision might not be detectable.[3] Therefore, we have to periodically switch the hash function h. More precisely, we switch the function whenever we found a certain amount C of distinguished points. If we model $C = \beta \cdot W$ for some constant β, then each hash function will have a probability of $2\beta W/N$ for finding the golden collision. Experimentally, van Oorschot and Wiener [26] found $\beta = 10$ to perform best, and the average running time of the attack is measured to be $(2.5\sqrt{N^3/W})/m$, where m is the number of processors computing paths in parallel.

3 Accurate Formulas for vOW and MitM

So far, we have provided theoretical cost functions for the golden collision search in terms of the number of evaluations of the function f, and for the batched depth-first MitM in terms of the number of 2-isogeny evaluations. We now provide a more detailed cost model in terms of elliptic curve operations to make these costs directly comparable. These formulas give a first indication of which memory regime favors which algorithm and, further, they form the starting point for parameter selection in our implementation.

3.1 Meet in the Middle

For the depth-first MitM, we have counted only the 2-isogeny evaluations but the total cost involves also obtaining the kernel points of each isogeny and pushing the basis points through the isogeny. As described in [1], the total cost of processing a node at depth d can be summarized as:

- $2^{e/2-d}$ point doublings to compute the kernel points
- 2 isogeny constructions to compute the children nodes

[3] For instance, one of the points that leads to the golden collision might be part of a cycle that does not reach a distinguished point.

- 1 point doubling, 1 point addition, and 6 isogeny evaluations to push the basis through the isogenies.

Nodes at the second-to-last level represent an exception since once we obtain the leaves, we no longer require pushing the bases and instead we need to compute the j-invariant.

Let us refer by ADD, DBL, ISOG, EVAL, JINV to the cost of a point addition, point doubling, 2-isogeny construction, 2-isogeny evaluation at a point, and j-invariant computation, respectively. The total cost of computing a tree of depth $e/2$ is then

$$\begin{aligned}\text{DFS}(e/2) &= \sum_{d=0}^{e/2-2} 2^d \left((2^{e/2-d}+1)\texttt{DBL} + 2\texttt{ISOG} + 1\texttt{ADD} + 6\texttt{EVAL}\right) \\ &\quad + 2^{e/2-1}\left(2\texttt{DBL} + 2\texttt{ISOG} + 2\texttt{JINV}\right) \\ &= 2^{e/2-1}\left((e+1)\texttt{DBL} + 4\texttt{ISOG} + 1\texttt{ADD} + 6\texttt{EVAL} + 2\texttt{JINV}\right) + O(1).\end{aligned}$$

The expected time of the whole MitM attack using 2^ω memory entries, which computes a tree of depth ω on one side and a tree of depth $e/2$ on the other side for each batch, is then

$$\begin{aligned}\text{MitM}(e,\omega) &= \frac{2^{e/2}}{2\cdot 2^\omega}(DFS(\omega) + DFS(e/2)) \\ &\approx (2^{e-\omega-2} + 2^{e/2-2})\left(\texttt{DBL} + 4\texttt{ISOG} + 1\texttt{ADD} + 6\texttt{EVAL} + 2\texttt{JINV}\right) \\ &\quad + (2^{e-\omega-2}e/2 + 2^{e/2-2}\omega)\texttt{DBL}.\end{aligned}$$

3.2 Golden Collision Search

For the golden collision search, the cost of an evaluation of the random function, given a scalar $k \in \mathbb{Z}_{2^{e/2}}$ and a bit $c \in \{0,1\}$, consists of

- computing the kernel point $P_i + [k]Q_i$,
- constructing a single $2^{e/2}$-isogeny with said kernel and
- computing the j-invariant of the output curve.

The first step is usually done with a three-point Montgomery ladder which has an average cost of $\frac{e}{2}(\texttt{DBL}+\texttt{ADD})$. For the second step, it is shown in [10] that a "balanced" strategy for computing a $2^{e/2}$-isogeny costs about $\frac{e}{4}\log(e/2)\texttt{DBL} + \frac{e}{4}\log(e/2)\texttt{EVAL} + \frac{e}{2}\texttt{ISOG}$. Hence, the total expected sequential time of the golden collision search is

$$\begin{aligned}\text{GCS}(e,\omega) &= 2.5\cdot 2^{3(e/2+1)/2-\omega/2} \\ &\quad \times \left(\frac{e}{4}\log(e/2)(\texttt{DBL}+\texttt{EVAL}) + \frac{e}{2}\texttt{ISOG} + \texttt{JINV} + \frac{e}{2}(\texttt{DBL}+\texttt{ADD})\right).\end{aligned}$$

3.3 Simplified Cost Models for Montgomery Curves

Assuming that we use Montgomery curve arithmetic, then the cost of curve operations can be expressed in terms of field additions, multiplications, squares and inverses ($\mathtt{A}$, $\mathtt{M}$, $\mathtt{S}$, $\mathtt{I}$, respectively) as follows (compare to [2])

$$\mathtt{DBL} = 4\mathtt{A} + 4\mathtt{M} + 2\mathtt{S}, \quad \mathtt{ADD} = 6\mathtt{A} + 4\mathtt{M} + 2\mathtt{S}, \quad \mathtt{ISOG} = \mathtt{A} + 2\mathtt{S} \quad \text{and}$$
$$\mathtt{EVAL} = 6\mathtt{A} + 4\mathtt{M}, \quad \mathtt{JINV} = 8\mathtt{A} + 3\mathtt{M} + 4\mathtt{S} + \mathtt{I}.$$

Moreover, we assume $\mathtt{M} = 1.5\mathtt{S} = 100\mathtt{A} = 0.02\mathtt{I}$ which we have obtained experimentally from our quadratic field arithmetic implementation. The cost models can then be written in units of $\mathtt{M}$ as

$$\text{MitM}(e, \omega)/\mathtt{M} \approx \frac{22799}{600}\left(2^{e-\omega} + 2^{e/2}\right) + \frac{403}{300}\left(2^{e-\omega} \cdot e/2 + 2^{e/2} \cdot \omega\right) \quad (1)$$

and

$$\text{GCS}(e, \omega)/\mathtt{M} \approx 2.5 \cdot 2^{3(e/2+1)/2-\omega/2}\left(\frac{4181}{75} + \frac{1211}{200}e + \frac{283}{120}e\log(e/2)\right) \quad (2)$$

For a given value of e and a memory budget ω, we can now determine which algorithm is favorable. Figure 2 visualizes three different regions. For $\omega \geq e/2$ the full MitM attack without batching can be applied. The batched MitM attack is found to have a narrow area of application at the border of the region where the golden collision search is optimal, which dominates the largest part of the limited-memory area.

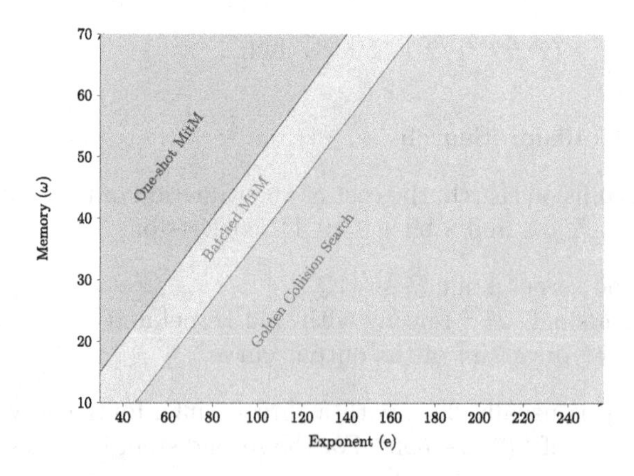

Fig. 2. Regions in the (e, ω) space where each attack is optimal for solving a SIPFD problem of size 2^e with memory limited to 2^w entries.

We would like to stress, that this comparison is based on idealized models involving only underlying field arithmetic operations. They do not take into account any practical effects, as e.g. memory access timings or parallelization issues. Nevertheless, it gives a first indication of the superiority of the golden collision search in the limited memory setting.

4 Practical Results on Solving the *SIPFD*

In this section we present our experimental results with a focus on our GPU implementation of the van Oorschot and Wiener golden collision search. But first, let us start with an experimental validation of our theoretical estimates of the MitM algorithm and its batched version from Sect. 3.1.

4.1 Practical Results of Our MitM CPU Implementation

We have implemented the batched depth-first MitM attack and run experiments on an AMD EPYC 7763 64-Core processor at 2.45 GHz, running 32 threads in parallel.

The j-invariants in each batch are stored in RAM, along with the corresponding scalar sk. Each processor maintains an array with j-invariants that have been calculated and sort lexicographically, to reduce the number of memory accesses when searching for the collision.

For measuring the performance of the batched depth-first MitM with the memory parameter ω, we fix a small instance with exponent $e = 50$ and benchmark the attack for ω with $18 \leq \omega \leq 25$. These timings are compared to Eq. 1, using a separate benchmark for the cost of M, i.e. a multiplication operation of our implementation. As shown in Fig. 3, the experimental measurements are found to adhere to the model up to an overhead factor of about 2, which is explained by the memory access times and sorting overheads that are not accounted for in Eq. 1.

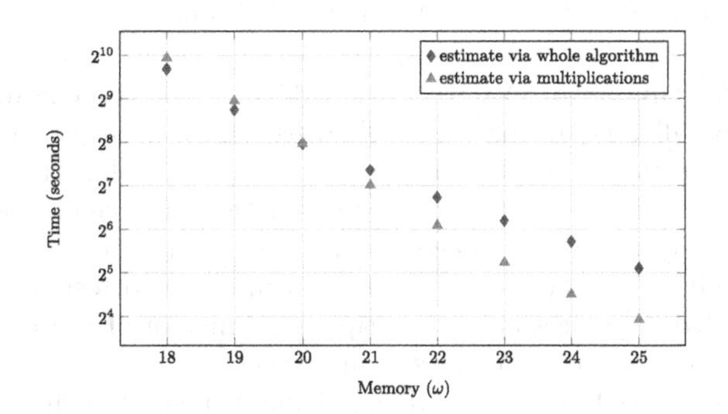

Fig. 3. Completion time of the MitM attack for an exponent $e = 50$ using 32 physical processors and different memory bounds compared to the prediction in Eq. 1.

We then tested the attack for increasing values of e while limiting the memory to $\omega \leq 28$. For $e > 56$, the batched MitM must be used and we have estimated the complexity of the whole attack by completing a single batch. As expected, Fig. 4 shows that the slope of the cost changes drastically once we enter the limited-memory region. The overhead factor between the experimental results and the theoretical model is always found to be less than 2.6. We conclude that

Eq. 1 can be used to estimate the cost of the attack for larger parameters without significant overhead.

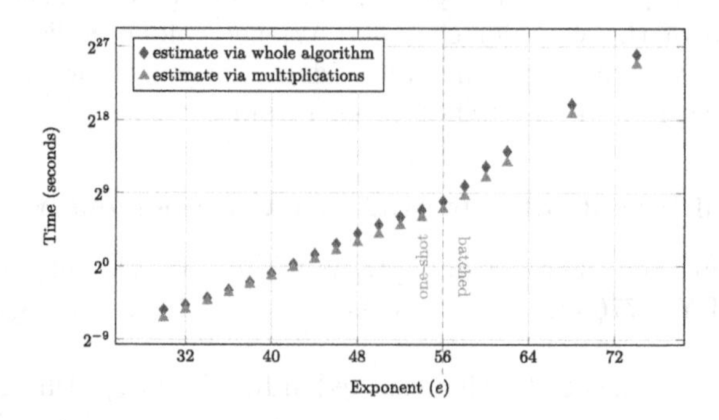

Fig. 4. Completion time of the MitM attack for various exponent sizes.

For comparison, the instance solved by Udovenko and Vitto in [30] was in the unlimited-memory setting using $e = 88$ and $\omega = 44$. Based on our model and adjusting to their clock frequency, we obtain an estimate of 9.47 core-years for the attack. This is close to Udovenko and Vitto's experimental result of 8.5 core-years, despite the fact that they used network storage.

4.2 Practical Considerations for Our vOW GPU Implementation

Let us give a brief explanation of the GPU architecture we used, followed by a summary of practical features of our implementation.

GPU Architecture. An NVIDIA CUDA device allows to execute thousands of threads in parallel. Following the Single Instructions Multiple Thread (SIMT) paradigm, a collection of 32 threads is bundled in *warps* that can only perform the same instruction on different data. One of the main challenges when programming CUDA devices is to decrease the memory latency, i.e., the time the threads are waiting for the data to be loaded into the corresponding registers. Therefore all CUDA devices have a multiple-level memory hierarchy incorporating memory and caches of different size and speed.

The NVIDIA A100 has an 80 GB sized main memory, connected to other GPUs in the same cluster via a high throughput bus called *NVLINK*. However, for performing computations, data must be propagated through the two levels of caches down to the registers. Each thread has only a very limited amount of these registers. Whenever more registers are addressed than physically available, the memory must be outsourced to other memory levels, causing latency and stalls. Further, whenever more threads are requested than the hardware can handle concurrently, a scheduling is performed, by swapping active threads against queued ones. As a consequence, caches must be invalided, which leads to further memory latency. However, there is usually an optimal number of concurrent threads such that memory latency can be minimized by an optimal scheduling.

GPU Potential of vOW. Note that the major task performed inside the vOW algorithm is the computation of chains of evaluations of the given function on different inputs. Therefore, it fits into the SIMT paradigm and can effectively be parallelized on the GPU. Further, since the devices are inherently memory-constrained, they profit from the good asymptotic trade-off curve of the vOW collision search.

Practical Features

Hash Function. For performance improvements, we heuristically model hash functions with ℓ-bit output as the projection to the first ℓ bits of the input. To obtain a randomized version we xor a fixed random nonce to the output. That is, for a given nonce $\mathbf{r} \in \mathbb{F}_2^\ell$ the hash function $h_\mathbf{r}\colon \mathbb{F}_2^* \mapsto \mathbb{F}_2^\ell$ is defined as $h_\mathbf{r}(\mathbf{x}) := (x_1, \ldots, x_\ell) + \mathbf{r}$. This is justified by the fact that the inputs usually inherit already enough randomness, which is confirmed in our experiments.

Memory Optimizations. The bit-size of every triplet (x_0, x_d, d) is roughly $e + \log(20/\theta)$, since x_0, x_d encode $2^{e/2}$-isogenies and the length of each chain is $d < 20/\theta$. However, due to our hash function choice, we can omit $\log W$ bits of x_d referring to its address in the table, plus another $\log(1/\theta)$ bits from the fact that it is a distinguished point, giving a size of roughly $e + \log(20) - \log W$ bits per triplet.

PTX Assembly. We provide core functionalities of our GPU implementation in PTX (Parallel Thread eXecution) assembly, which is the low level instruction set of NVIDIA CUDA GPUs. This includes our own optimized $\mathbb{F}_p$ arithmetic. In this context, we provide optimized version of both the schoolbook and the Karatsuba algorithm for integer multiplication, as well as the Montgomery reduction.

Data Structure. For storing distinguished points we compare the performance of a standard hash table against the Packed Radix-Tree-List (PRTL) proposed in [29]. The PRTL is a hash table that stores a linked list at each address, instead of single elements. This avoids the need for element replacement in case of hash collisions. Further it identifies the address of an element via its prefix (*radix*) and stores only the prefix-truncated element. The *packed* property of the PRTL relates to distinguished point triplets being stored as a single bit-vector, thus, avoiding the waste of space due to alignment. We ran CPU experiments with both data structures to identify the optimal choice prior to translating the code to the GPU setting. Eventually, we adopted the packed property and the use of prefixes, while we found no improvement in performance from using linked lists.

Precomputation. As discussed in [9], the time T_f required for a function evaluation can be decreased via precomputation. For a depth parameter d, one can precompute the 2^d curves corresponding to all the 2^d–isogenies from E_0 and E_1. When computing a $2^{e/2}$-isogeny, the initial d steps are replaced by a table lookup and we end up computing only a $2^{e/2-d}$-isogeny. Note that the memory needed for precomputation grows exponentially with d and so asymptotically it does not play a relevant role. However, for relatively small parameters it can provide valuable savings and speed up our experiments without affecting metrics such as the number of calls to f.

4.3 Practical Results of Our vOW GPU Implementation

In the following we use the practical performance of our implementation together with the known theoretical behavior to extrapolate the time to solve larger instances. In the original work of van Oorschodt-Wiener the time complexity of the procedure was found to be well approximated by

$$\frac{1}{m}(2.5\sqrt{N^3/W}) \cdot T_f, \tag{3}$$

where T_f is the cost per function evaluation. Therefore, we measure the cost T_f of our implementation which then allows us to derive an estimate for arbitrary instances. Further, we compare this estimate against the theoretical estimate via Eq. 2 and an estimate based on collecting a certain amount of distinguished points.

Additionally, we verify that our GPU implementation using the functions specified in Sect. 2.3 has a similar behavior as the CPU implementation using random functions of [26]. This increases the reliability in our estimates, as it shows that the time complexity of our implementation is still well approximated by Eq. 3. Let us start with this verification.

Verifying the Theoretical Behavior. In [26] van Oorschot and Wiener find that on average it takes $\frac{0.45N}{W}$ randomized versions of the function to find the solution, which in our case corresponds to random choices of the hash function (compare to Sect. 2.3). In their experiments, the function is changed after $\beta \cdot W$ distinguished points have been discovered, where a value of $\beta = 10$ is found to be optimal. Further, chains are aborted after they reach a length of $20\theta^{-1}$, i.e., 20 times their expected length.

Optimal Value of β. Let us first verify that an amount of $10 \cdot W$ distinguished points until we abort the collision search for the current version of the function is still a suitable choice for our implementation. Table 1 shows the average running time of our vOW implementation using different values of β. We conclude that the values around $\beta = 10$ give comparable performance, with $\beta = 10$ being optimal in most of the experiments. The results are averaged over 100 ($e = 34$) and 50 ($e = 36$) runs respectively.

Table 1. Running time in seconds for different values of β.

e	ω	$\beta = 5$	$\beta = 10$	$\beta = 15$	$\beta = 20$
34	8	405.08	384.74	371.67	**335.88**
	9	244.30	**198.86**	238.60	285.97
	10	173.73	207.37	**136.80**	179.93
36	9	704.65	**567.89**	654.15	599.61
	10	419.87	**373.16**	489.71	542.00
	11	398.72	365.62	**314.26**	290.49

Expected Number of Randomized Versions of the Function. Now that we confirmed the optimal choice of β, we expect that the required amount of random functions until success also matches the one from [26]. In this case, the number of required randomizations of the function until the golden collision is found should follow a geometric distribution with parameter close to $\frac{W}{0.45N}$.

We confirm this distribution in an experiment for $e = 30$, in which case we have $N = 2^{e/2+1} = 2^{16}$ and use a hash table that can store up to $W = 2^7$ distinguished points. We then solved 1000 such instances and recorded for each the number of randomized versions of the function until the solution was found. On average, it took 208.28 versions compared to the approximation of $\frac{0.45N}{W} = 230.4$, despite slightly surpassing the $W \leq N/2^{10}$ limit where the vOW experiments took place. In Fig. 5 we visualize the obtained frequencies (triangles) and give as comparison the probabilities of the geometric distribution with parameter $\frac{1}{208.27}$ (diamonds). In this figure we accumulated the frequencies in each interval of size 20 to allow for a better visualization.

Measuring the Time per Function Evaluation. Next we measured the time per function evaluation that the GPU implementation requires on our hardware for different values of e. To pick our parameters, we first set W to the largest power of 2 such that the memory would not surpass our GPU's 80 GB budget,

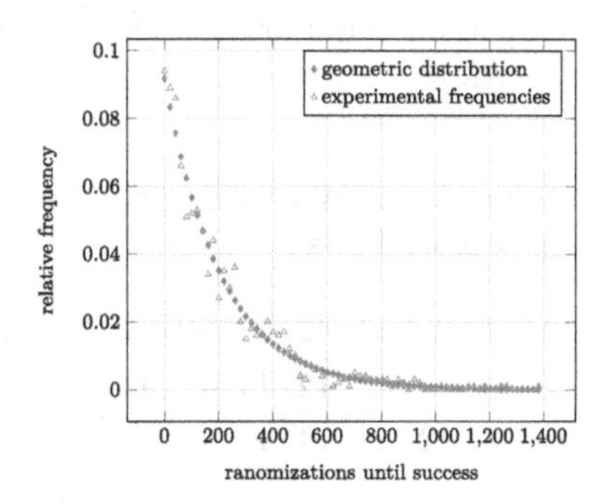

Fig. 5. Number of used randomizations to find the solution for $e = 30$, $W = 2^7$.

then chose the largest precomputation depth that would fit in the remaining memory. In the smaller instances, the memory and precomputation depth were additionally subject to a cap of $W \leq N/2^8$ and $d \leq e/4$ in the smaller instances. After performing the precomputation, we measured the time per function evaluation as illustrated in Fig. 6. The jumps in the graph indicate that the bitsize of the used prime, which is roughly $2e$, exceeds the next 64-bit boundary. In those cases the prime occupies an additional register, which leads to a slowdown of the $\mathbb{F}_{p^2}$-arithmetic.

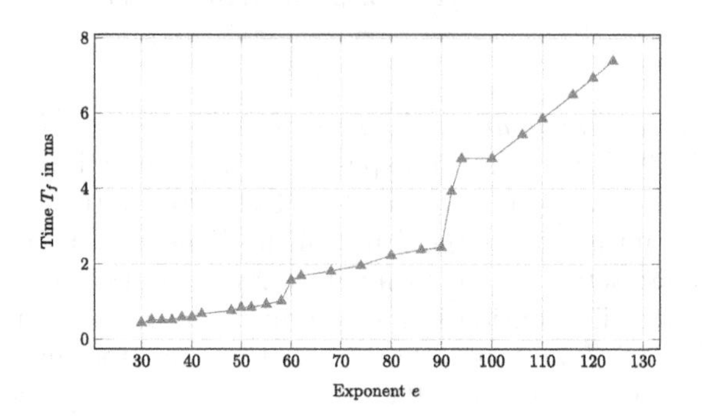

Fig. 6. Cost per function evaluation using 6912 threads in parallel. Each data point is averaged over 4096 evaluations.

Performance Estimation Using a Single GPU. Now, the measured timings allow us to estimate the time required by our implementation to solve larger instances. To compute this estimate we use Eq. 3 with the measured value for T_f and the number of concurrent threads m used on the GPU. The resulting estimate is shown in Fig. 7 (diamonds).

Note that the steeper incline in the estimation for $e > 62$ stems from the fact that for $e = 62$ we reach the maximum number of concurrent threads for our implementation, which we find to be 27,648 threads. Further, from $e = 80$ onwards we additionally hit our hash table memory limit of $W = 2^{33}$ elements. We summarize in Table 2 optimal configurations for the *SIPFD* instances executed on our single GPU platform.

Table 2. Optimal configurations for vOW on single GPU with 80 GB memory. Configurations for $e > 80$ match the one of $e = 80$.

e	30	32	34	36	38	40	42	48	50	52	56	62	68	74	80
d	9	9	10	11	11	12	12	14	14	15	16	17	19	21	22
$\log W$	8	9	10	11	12	13	14	17	18	19	21	24	27	30	33

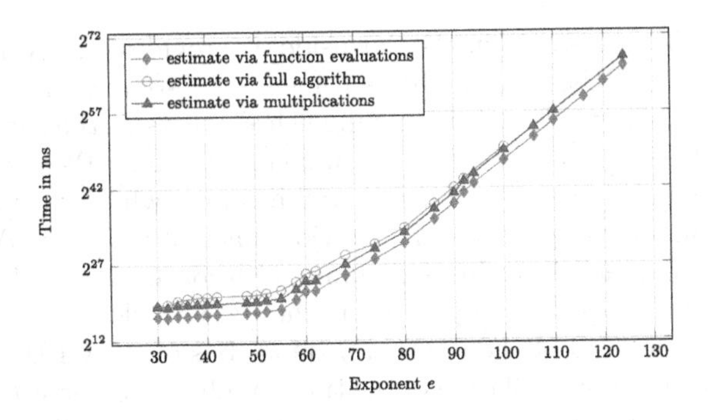

Fig. 7. Estimated time to solve instances of *SIPFD* on a single GPU.

We also obtain an alternative estimate based on the time to finish one version of the random function in the full implementation of the attack. That is, we measure the time to obtain $10 \cdot W$ distinguished points and then multiply by the average number $\frac{0.45N}{W}$ of random functions needed. This method should capture the performance more accurately as it includes practical effects such as the memory access costs. For $e \leq 62$ we averaged 100 experiments of completing a random function, while for larger instances we decreased the number of experiments and for $e \geq 76$ we only computed a $1/2^{10}$ fraction of the needed points and scaled the resulting time accordingly. The results of this second estimation are also shown in Fig. 7 (circles) and present an overhead factor of about 8. This overhead is likely the result of imperfect parallelization speedups in GPUs, as well as the cost of memory accesses, but it is observed to decrease towards larger instances.

Finally, we benchmarked the average cost of field multiplications in our GPU setup to obtain a third estimate based on Eq. 2, which is also presented in Fig. 7 (triangles). This estimate closely matches the estimate via the full algorithm, especially for larger instances where distinguished points are rare and memory accesses are more sporadic.

Overall, our measurements support the use of any of the three methods described to obtain accurate extrapolations of the algorithm's running time. For a concrete example, we estimate that a problem with $e = 88$ which corresponds to the instance solved by Udovenko and Vitto in [30], would take about 44 years on a single GPU with 80 GB memory limit. While this is not yet very impressive, compared to the 10 CPU years reported in [30], a single GPU is far less expensive and powerful than the 128 TB network storage cluster used for that record. Therefore in the following section we give an estimate of the attack when scaling to a multiple GPU architecture.

Multiple GPU Estimation. We explored different strategies for parallelizing the vOW algorithm across multiple GPUs. In the first strategy, every GPU independently runs its own instantiation of the algorithm. The advantage of this approach lies its simplicity, which minimizes overhead since no communication between GPUs is necessary. On the downside, it provides only a linear speedup in the number of GPUs, since additional memory resources are not shared. In our second approach, GPUs report distinguished points to the same hash table, which is stored distributed over the global memory of all GPUs. The advantage here clearly lies in the increase of the overall memory, which allows to make use of the good time-memory trade-off behavior inherent to the vOW algorithm. However, this approach introduces a communication overhead due to the distributed memory access. On top of that, the data needs to be send over the slower *NVLINK* instead of the internal memory bus of the GPU.

We performed an extrapolation of the time to solve different sized instances in the distributed setting, similar to the extrapolation via the full algorithm in the single GPU setting. In this experiment, we allocated a hash table able to store up to $W = 2^{34}$ distinguished triplets, which for large instances corresponds to about 200 GB, across the memory of four GPUs connected via an *NVLINK* bus. We then measured the time to collect and store a certain amount X of distinguished points. Multiplying this time by $\frac{10 \cdot W}{X} \cdot 0.45 \cdot \frac{2^{e/2+1}}{W} = 4.5 \cdot 2^{e/2+1}/X$, gives an extrapolation of the running time of completing the whole attack.

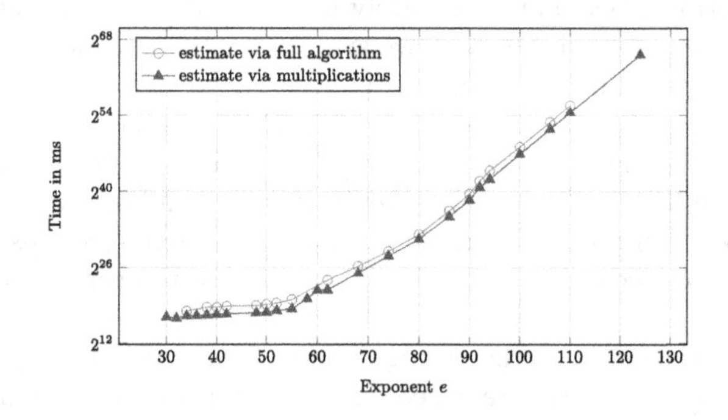

Fig. 8. Estimated time to solve instances of *SIPFD* on 4 GPUs connected via an *NVLINK* bus.

Figure 8 visualizes the obtained extrapolations (circles) in comparison to the estimate via the multiplication benchmark (triangles), i.e., using Eq. 2. We observe, similar to the single GPU case, a slight underestimation by using Eq. 2, which for larger instances vanishes. For the larger instances we obtain an underestimation by a factor of roughly two, which corresponds to the performance difference of the *NVLINK* bus in comparison to the internal memory bus. However, since for larger instances with fixed memory budget the time to compute distinguished points dominates, the factor is expected to vanish. Hence, we finally conclude that using the distributed memory architecture does not lead to unexpected performance slowdowns.

Comparing Both Strategies. Let us determine, which of the parallelization strategies is preferable for a specific amount of GPUs. For large instances, the computational cost of the multi-GPU as well as the single-GPU setting, are well approximated by Eq. 2. Therefore the speedup when parallelizing via distributed memory using X GPUs is

$$\frac{\mathrm{GCS}(e,\omega)}{\mathrm{GCS}(e,\omega+\log X)/X}=X^{3/2},$$

and, hence, preferable over the strategy via independent executions with a speedup of only X. Also, if comparing the exact numbers obtained from the estimate via the full algorithm in the distributed memory setting and the single GPU setting, we find that the distributed setting offers a better practical performance already for $e \geq 62$.

Extrapolating $e = 88$ and the Way Forward. Based on our practical timings we estimate the time to solve an instance with $e = 88$ on 4 GPUs to about 32 GPU years in comparison to roughly 44 GPU years in the single GPU setting. Moreover, if we scale the attack to 16 GPUs, which is the maximum that the *NVLINK* bus currently supports, we estimate the time to only 5.6 GPU years, which means the experiment would finish in about 4 months. We therefore conclude from our experiments that for larger instances, with a memory budget of 128 TB in the MitM case and 80 GB per device in the GPU case, the vOW algorithm is the preferred choice.

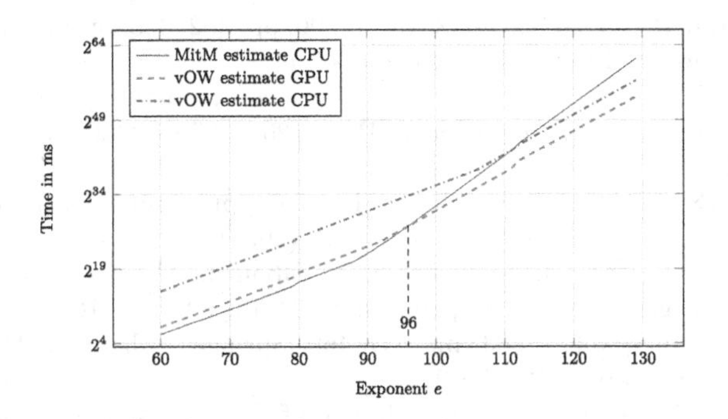

Fig. 9. Estimated time to solve instances of *SIPFD* on 16 NVIDIA Ampere GPUs with 80 GB each connected via an *NVLINK* bus in comparison to a cluster with 128 TB storage and 256 cores.

In Fig. 9 we visualize the result of the estimation via Eq. 1 and 2 in both settings assuming 256 cores with 128 TB of memory in the CPU case and 16 NVIDIA A100s connected via an *NVLINK* bus in the GPU case. This figure illustrates the estimate for running the MitM on the CPU (solid line) and the vOW on the GPU system (dashed line). We find that under these fixed resources, the break-even point from where vOW offers a better performance lies at $e = 96$. Additionally, we provide the estimate if we instead execute vOW on the corresponding CPU system (dash dotted line). Observe, that even under the unrealistic assumption that the 128 TB of memory would allow for efficient random access (for the vOW hash table), it does not outperform the GPU based approach for any instance size. Moreover, even under this memory advantage in case of $e = 96$, the GPU implementation offers a speedup of almost two magnitudes (82x). We conclude that the way forward when tackling larger instances of the *SIPFD* clearly favors vOW implementations on GPU platforms.

Acknowledgements. This project has received funding from the European Commission through the ERC Starting Grant 805031 (EPOQUE) and from BMBF Industrial Blockchain - iBlockchain.

References

1. Adj, G., Cervantes-Vázquez, D., Chi-Domínguez, J.J., Menezes, A., Rodríguez-Henríquez, F.: On the cost of computing isogenies between supersingular elliptic curves. In: Cid, C., Jacobson, M., Jr. (eds.) SAC 2018. LNCS, vol. 11349, pp. 322–343. Springer, Cham (2019). https://doi.org/10.1007/978-3-030-10970-7_15
2. Azarderakhsh, R., et al.: Supersingular Isogeny Key Encapsulation. Third Round Candidate of the NIST's post-quantum cryptography standardization process (2020), available at: https://sike.org/
3. Bernstein, D.J., Feo, L.D., Leroux, A., Smith, B.: Faster computation of isogenies of large prime degree. ANTS XIV, Open Book Ser. **4**, 39–55 (2020)
4. Burdges, J., De Feo, L.: Delay encryption. In: Canteaut, A., Standaert, F.-X. (eds.) EUROCRYPT 2021. LNCS, vol. 12696, pp. 302–326. Springer, Cham (2021). https://doi.org/10.1007/978-3-030-77870-5_11
5. Castryck, W., Decru, T.: An efficient key recovery attack on sidh (preliminary version). Cryptology ePrint Archive, Paper 2022/975 (2022). https://eprint.iacr.org/2022/975
6. Charles, D.X., Lauter, K.E., Goren, E.Z.: Cryptographic hash functions from expander graphs. J. Cryptol. **22**(1), 93–113 (2009)
7. Chavez-Saab, J., Rodríguez-Henríquez, F., Tibouchi, M.: Verifiable isogeny walks: towards an isogeny-based postquantum VDF. In: AlTawy, R., Hülsing, A. (eds.) SAC 2021. LNCS, vol. 13203, pp. 441–460. Springer, Cham (2022). https://doi.org/10.1007/978-3-030-99277-4_21
8. Costello, C.: The case for SIKE: a decade of the supersingular isogeny problem. IACR Cryptology ePrint Archive, p. 543 (2021). https://eprint.iacr.org/2021/543
9. Costello, C., Longa, P., Naehrig, M., Renes, J., Virdia, F.: Improved classical cryptanalysis of SIKE in practice. In: Kiayias, A., Kohlweiss, M., Wallden, P., Zikas, V. (eds.) PKC 2020. LNCS, vol. 12111, pp. 505–534. Springer, Cham (2020). https://doi.org/10.1007/978-3-030-45388-6_18
10. De Feo, L., Jao, D., Plût, J.: Towards quantum-resistant cryptosystems from supersingular elliptic curve isogenies. J. Math. Cryptol. **8**(3), 209–247 (2014)
11. De Feo, L., Masson, S., Petit, C., Sanso, A.: Verifiable delay functions from supersingular isogenies and pairings. In: Galbraith, S.D., Moriai, S. (eds.) ASIACRYPT 2019. LNCS, vol. 11921, pp. 248–277. Springer, Cham (2019). https://doi.org/10.1007/978-3-030-34578-5_10
12. Delaplace, C., Esser, A., May, A.: Improved low-memory subset sum and LPN algorithms via multiple collisions. In: Albrecht, M. (ed.) IMACC 2019. LNCS, vol. 11929, pp. 178–199. Springer, Cham (2019). https://doi.org/10.1007/978-3-030-35199-1_9

13. Delfs, C., Galbraith, S.D.: Computing isogenies between supersingular elliptic curves over $\mathbb{F}_p$. Des. Codes Cryptogr. **78**(2), 425–440 (2016)
14. Esser, A., May, A.: Low weight discrete logarithm and subset sum in $2^{0.65n}$ with polynomial memory. In: Canteaut, A., Ishai, Y. (eds.) EUROCRYPT 2020. LNCS, vol. 12107, pp. 94–122. Springer, Cham (2020). https://doi.org/10.1007/978-3-030-45727-3_4
15. Fouotsa, T.B.: SIDH with masked torsion point images. Cryptology ePrint Archive, Paper 2022/1054 (2022). https://eprint.iacr.org/2022/1054
16. Galbraith, S.D.: Constructing isogenies between elliptic curves over finite fields. LMS J. Comput. Math. **2**, 118–138 (1999)
17. Galbraith, S.D., Petit, C., Silva, J.: Identification protocols and signature schemes based on supersingular isogeny problems. In: Takagi, T., Peyrin, T. (eds.) ASIACRYPT 2017. LNCS, vol. 10624, pp. 3–33. Springer, Cham (2017). https://doi.org/10.1007/978-3-319-70694-8_1
18. Jao, D., De Feo, L.: Towards quantum-resistant cryptosystems from supersingular elliptic curve isogenies. In: Yang, B.-Y. (ed.) PQCrypto 2011. LNCS, vol. 7071, pp. 19–34. Springer, Heidelberg (2011). https://doi.org/10.1007/978-3-642-25405-5_2
19. Jaques, S., Schanck, J.M.: Quantum cryptanalysis in the RAM model: claw-finding attacks on SIKE. In: Boldyreva, A., Micciancio, D. (eds.) CRYPTO 2019. LNCS, vol. 11692, pp. 32–61. Springer, Cham (2019). https://doi.org/10.1007/978-3-030-26948-7_2
20. De Feo, L., Dobson, S., Galbraith, S.D., Zobernig, L.: SIDH proof of knowledge. IACR Cryptology ePrint Archive, p. 1023 (2021), https://eprint.iacr.org/2021/1023, to appear in ASIACRYPT 2022
21. Maino, L., Martindale, C.: An attack on SIDH with arbitrary starting curve. Cryptology ePrint Archive, Paper 2022/1026 (2022). https://eprint.iacr.org/2022/1026
22. Corte-Real Santos, M., Costello, C., Shi, J.: Accelerating the Delfs-Galbraith algorithm with fast subfield root detection. In: Dodis, Y., Shrimpton, T. (eds.) CRYPTO 2022. LNCS, vol. 13509, pp. 285–314. Springer, Cham (2022). https://doi.org/10.1007/978-3-031-15982-4_10
23. May, A.: How to meet ternary LWE keys. In: Malkin, T., Peikert, C. (eds.) CRYPTO 2021. LNCS, vol. 12826, pp. 701–731. Springer, Cham (2021). https://doi.org/10.1007/978-3-030-84245-1_24
24. Moriya, T.: Masked-degree SIDH. Cryptology ePrint Archive, Paper 2022/1019 (2022). https://eprint.iacr.org/2022/1019
25. NIST: NIST Post-Quantum Cryptography Standardization Process. Second Round Candidates (2017). https://csrc.nist.gov/projects/post-quantum-cryptography/round-2-submissions
26. van Oorschot, P.C., Wiener, M.J.: Parallel collision search with cryptanalytic applications. J. Cryptol. **12**(1), 1–28 (1999)
27. Robert, D.: Breaking SIDH in polynomial time. Cryptology ePrint Archive, Paper 2022/1038 (2022). https://eprint.iacr.org/2022/1038
28. TATE, J.: Endomorphisms of abelian varieties over finite fields. Inventiones Mathematicae **2**, 134–144 (1966)
29. Trimoska, M., Ionica, S., Dequen, G.: Time-memory analysis of parallel collision search algorithms. IACR Trans. Cryptogr. Hardw. Embed. Syst. **2021**(2), 254–274 (2021)
30. Udovenko, A., Vitto, G.: Breaking the SIKEp182 challenge. Cryptology ePrint Archive, Paper 2021/1421. Accepted to the SAC 2022 Conference (2021). https://eprint.iacr.org/2021/1421

31. van Vredendaal, C.: Reduced memory meet-in-the-middle attack against the NTRU private key. LMS J. Comput. Math. **19**(A), 43–57 (2016)
32. Washington, L.C.: Elliptic Curves: Number Theory and Cryptography, 2nd edn. Chapman & Hall/CRC, Hoboken (2008)
33. Yoo, Y., Azarderakhsh, R., Jalali, A., Jao, D., Soukharev, V.: A post-quantum digital signature scheme based on supersingular isogenies. In: Kiayias, A. (ed.) FC 2017. LNCS, vol. 10322, pp. 163–181. Springer, Cham (2017). https://doi.org/10.1007/978-3-319-70972-7_9

Cryptanalysis

Distinguishing Error of Nonlinear Invariant Attacks

Subhabrata Samajder(✉) and Palash Sarkar

Applied Statistics Unit, Indian Statistical Institute, 203, B.T.Road, Kolkata 700108, India
subhabrata.samajder@gmail.com, palash@isical.ac.in

Abstract. Todo et al. (2018) have proposed nonlinear invariant attacks which consider correlations between nonlinear input and output combiners for a key alternating block cipher. In its basic form, a nonlinear invariant attack is a distinguishing attack. In this paper we obtain precise expressions for the errors of nonlinear invariant attacks in distinguishing a key alternating cipher from either a uniform random function or a uniform random permutation.

Keywords: Block cipher · Nonlinear invariant attack · Distinguishing error

Mathematics Subject Classification (2010): 94A60 · 68P25 · 62P99

1 Introduction

Let $E : \mathcal{K} \times \{0,1\}^n \to \{0,1\}^n$ be an r-round block cipher, where $\mathcal{K}$ is the key space of the block cipher. For $K \in \mathcal{K}$, we write $E_K(\cdot)$ to denote $E(K, \cdot)$. A distinguishing attack based on correlation between input and output combiners of a block cipher proceeds as follows. Let $g_0 : \{0,1\}^n \to \{0,1\}$ and $g_r : \{0,1\}^n \to \{0,1\}$ be two n-variable Boolean functions. The function g_0 serves as a combiner of the input of E_K while the function g_r serves as a combiner of the output of E_K. The correlation between the input and output combiners is the correlation between g_0 and $g_r \circ E_K$. This correlation is captured by considering the weight of the function $f_{E_K} : \{0,1\}^n \to \{0,1\}$ defined by $f_{E_K}(\alpha) = g_0(\alpha) \oplus g_r(E_K(\alpha))$. Suppose it is possible to find some property of E such that f_{E_K} has a nature which is different from that of f_π (resp. f_ρ) where π is a uniform random permutation of $\{0,1\}^n$ (resp. ρ is a uniform random function from $\{0,1\}^n$ to $\{0,1\}^n$) for a large class of keys, called weak keys. Then for any weak key K, this property forms the basis of distinguishing E_K from π (resp. ρ).

Distinguishing attacks based on linear cryptanalysis [15] is the classical example of the above scenario. For such an attack, the functions g_0 and g_r are linear functions. Linear cryptanalysis has an extensive history and has been successfully applied to both block and stream ciphers. Though linear cryptanalysis has been studied as a distinguishing attack, in its most potent form, linear cryptanalysis can be a key recovery attack [15].

T. Isobe and S. Sarkar (Eds.): INDOCRYPT 2022, LNCS 13774, pp. 319–335, 2022.
https://doi.org/10.1007/978-3-031-22912-1_14

Considering g_0 and g_r to be nonlinear functions makes the analysis complicated. This was first discussed by Harpes et al. [7] and later by Knudsen and Robshaw [13]. Todo et al. [16] introduced an interesting nonlinear attack, i.e., g_0 and g_r are nonlinear functions. Consider E_K to be a key alternating block cipher. Suppose there are Boolean functions g_0 and g_r and a weak key K such that $g_0(P) \oplus g_r(E_K(P))$ takes the same value for all possible choices of the plaintext P. This property can be exploited to obtain a distinguishing attack on the block cipher E_K. Such an attack is called a nonlinear invariant attack and the functions g_0 and g_r are called nonlinear invariants[1]. For the practical block ciphers SCREAM, iSCREAM and Midori64, the existence of nonlinear invariants (with $g_0 = g_r$) and related weak keys were described in [16].

Formally, let $\mathcal{D}^{\mathcal{O}}$ be a distinguisher which has access to an oracle $\mathcal{O}$. The distinguisher chooses distinct plaintexts $P_1, \ldots, P_N$ and queries the oracle, receiving in response the values $\mathcal{O}(P_1), \ldots, \mathcal{O}(P_N)$. $\mathcal{D}$ outputs 1 if $g_0(P_1) \oplus g_r(\mathcal{O}(P_1)) = \cdots = g_0(P_N) \oplus g_r(\mathcal{O}(P_N))$, otherwise, it outputs 0. Let $\mathcal{D}^{\mathcal{O}} \Rightarrow 1$ denote the event that $\mathcal{D}$ outputs 1 after interacting with the oracle $\mathcal{O}$. Then the advantage of the distinguisher $\mathcal{D}$ in distinguishing E_K from a uniform random permutation π is

$$\mathsf{Adv}(\mathcal{D}) = \left|\Pr[\mathcal{D}^{E_K} \Rightarrow 1] - \Pr[\mathcal{D}^{\pi} \Rightarrow 1]\right|. \tag{1}$$

If g_0 and g_r are non-linear invariants and K is a weak key, then $\Pr[\mathcal{D}^{E_K} \Rightarrow 1] = 1$ and so $\mathsf{Adv}(\mathcal{D}) = 1 - \varepsilon$, where

$$\varepsilon = \Pr[g_0(P_1) \oplus g_r(\pi(P_1)) = \cdots = g_0(P_N) \oplus g_r(\pi(P_N))]. \tag{2}$$

If $\varepsilon = 0$, then $\mathcal{D}$ has the maximum possible advantage which is 1. The quantity ε denotes the distinguishing error. A higher value of ε lowers the advantage of $\mathcal{D}$. The efficacy of the distinguisher $\mathcal{D}$ is determined by its advantage which in turn is determined by the value of ε. So, to know how good $\mathcal{D}$ is at distinguishing E_K from π, it is required to determine the value of ε.

Our Contributions

This work performs an analysis of the distinguishing error of nonlinear invariant attacks. We determine the probability that $g_0(P_1) \oplus g_r(\pi(P_1)) = \cdots = g_0(P_N) \oplus g_r(\pi(P_N))$. This is done in two cases, namely, when $P_1, \ldots, P_N$ are chosen under uniform random sampling without replacement and when $P_1, \ldots, P_N$ are some fixed distinct n-bit values. Further, these probabilities are also computed when π is replaced by a uniform random function ρ. Our analysis provides expressions for the error probabilities of the corresponding distinguishers. Such an analysis was not performed in [16]. Some of the consequences of our analysis are as follows.

1. It turns out that the error probability considered in [16] is that of distinguishing E_K from a uniform random function. The error probability of distinguishing E_K from a uniform random permutation is obtained here for the first time.

[1] While g_0 and g_r can be linear, the scenario that is relevant for the present work is when they are nonlinear.

2. The general form of the error probabilities are derived without any restriction on g_0 and g_r. When g_0 and g_r are balanced functions, we prove the following two results.
   (a) The error in distinguishing from a uniform random function is $1/2^{N-1}$.
   (b) The error in distinguishing from a uniform random permutation is at least as large as the error in distinguishing from a uniform random function. This is a consequence of Jensen's inequality. For moderate values of N, the error in distinguishing from a uniform random permutation is almost the same as the error in distinguishing from a uniform random function.

Distinguishers in Symmetric Key Cryptanalysis

There is a long history of distinguishing attacks on various kinds of symmetric key ciphers. A comprehensive discussion of this literature is beyond the scope of this work. To highlight the enduring interest in distinguishing attacks, we mention a few works. The notion of optimal distinguishers for linear cryptanalysis of block ciphers has been studied in [2]. A number of works have explored distinguishers for the AES [4–6,10,12]. Distinguishing attacks are also applicable to stream ciphers. We refer to [8] for an overview. This line of work continue to be of interest as is evidenced by [9,11]. Distinguishing attacks have also been proposed on hash functions [1,3] and T-functions [14]. In general, whenever a distinguisher is proposed, its effectiveness needs to be justified by a mathematical analysis of its advantage. The present work provides such mathematical justification of the effectiveness of distinguishers for nonlinear invariant attacks.

2 Nonlinear Invariant Attack

We provide a brief description of the nonlinear invariant attack for key alternating ciphers. Our description follows the suggestion in Sect. 7 of [16] where the nonlinear invariants are allowed to be different for the different rounds.

Let $E : \mathcal{K} \times \{0,1\}^n \rightarrow \{0,1\}^n$ be a key alternating block cipher which iterates a round function $R : \{0,1\}^n \rightarrow \{0,1\}^n$ over r rounds. For an n-bit string L, define $R_L : \{0,1\}^n \rightarrow \{0,1\}^n$ as $R_L(\alpha) = R(\alpha \oplus L)$. Let $K \in \mathcal{K}$. For a plaintext P, let the ciphertext C be $C = E_K(P)$ which is obtained in the following manner. The secret key K is used to obtain the round keys $K_0, \ldots, K_{r-1}$. Then

$$C = (R_{K_{r-1}} \circ R_{K_{r-2}} \circ \cdots \circ R_{K_0})(P).$$

Suppose there are functions $g_0, \ldots, g_r : \{0,1\}^n \rightarrow \{0,1\}$ and constants $c_0, \ldots, c_{r-1} \in \{0,1\}$, such that there are round keys $K_0, \ldots, K_{r-1}$ for which

$$g_{i+1}(R(\alpha \oplus K_i)) = g_i(\alpha \oplus K_i) \oplus c_i = g_i(\alpha) \oplus g_i(K_i) \oplus c_i \qquad (3)$$

for all $\alpha \in \{0,1\}^n$. Then $g_0, \ldots, g_r$ are called nonlinear invariants with associated constants $c_0, \ldots, c_{r-1}$. The key K as well as the round keys $K_0, \ldots, K_{r-1}$ obtained from K are called weak keys.

Before proceeding, we introduce a notation. Let $S : \{0,1\}^n \rightarrow \{0,1\}^n$ and $\phi, \psi : \{0,1\}^n \rightarrow \{0,1\}$. We define the function $f_S[\phi, \psi] : \{0,1\}^n \rightarrow \{0,1\}$ as follows:

$$f_S[\phi, \psi](\alpha) = \phi(\alpha) \oplus \psi(S(\alpha)). \tag{4}$$

The primary requirement in a weak key nonlinear invariant attack is the property given in the following proposition. This property has been derived in [16] for the case where the functions $g_0, \ldots, g_r$ are all equal.

Proposition 1. *Let $E : \mathcal{K} \times \{0,1\}^n \rightarrow \{0,1\}^n$ be an r-round key alternating cipher. Suppose $g_0, \ldots, g_r$ are nonlinear invariants with associated constants $c_0, \ldots, c_{r-1}$ such that there are weak round keys $K_0, \ldots, K_{r-1}$ obtained from a key K. Then for any $\alpha \in \{0,1\}^n$, $f_{E_K}[g_0, g_r](\alpha)$ is a constant which is independent of α.*

Proof. There are n-bit strings $\alpha_1, \ldots, \alpha_{r-1}$ such that $\alpha_0 = \alpha$; $\alpha_{i+1} = R_{K_i}(\alpha_i) = R(\alpha_i \oplus K_i)$ for $i = 0, \ldots, r-1$; and $\beta = \alpha_r = E_K(\alpha)$. The following holds.

$$\begin{aligned}
g_r(\beta) &= g_r(R(\alpha_{r-1} \oplus K_{r-1})) \\
&= g_{r-1}(\alpha_{r-1}) \oplus g_{r-1}(K_{r-1}) \oplus c_{r-1} \\
&= (g_{r-1}(R(\alpha_{r-2} \oplus K_{r-2}))) \oplus g_{r-1}(K_{r-1}) \oplus c_{r-1} \\
&= g_{r-2}(\alpha_{r-2}) \oplus (g_{r-2}(K_{r-2}) \oplus g_{r-1}(K_{r-1})) \oplus (c_{r-2} \oplus c_{r-1}) \\
&\vdots \\
&= g_0(\alpha) \oplus \left(\bigoplus_{i=0}^{r-1} g_i(K_i) \right) \oplus \left(\bigoplus_{i=0}^{r-1} c_i \right).
\end{aligned}$$

So,

$$g_0(\alpha) \oplus g_r(\beta) = \left(\bigoplus_{i=0}^{r-1} g_i(K_i) \right) \oplus \left(\bigoplus_{i=0}^{r-1} c_i \right). \tag{5}$$

The right hand side of (5) is determined by the functions $g_0, \ldots, g_{r-1}$, the constants $c_0, \ldots, c_{r-1}$ and the round keys $K_0, \ldots, K_{r-1}$. In particular, it is independent of α (and also $\beta = E_K(\alpha)$). □

Proposition 1 shows that if $g_0, \ldots, g_r$ are nonlinear invariants for some weak keys $K_0, \ldots, K_{r-1}$, then for all 2^n n-bit strings α, $g_0(\alpha) \oplus g_r(E_K(\alpha))$ is a constant. We next consider the following question. Suppose g_0 and g_r are any two n-variable Boolean functions, $\alpha_1, \ldots, \alpha_N$ and $\beta_1, \ldots, \beta_N$ are arbitrary n-bit strings, what is the maximum value of N such that $g_0(\alpha_1) \oplus g_r(\beta_1) = \cdots = g_0(\alpha_N) \oplus g_r(\beta_N)$ holds? This question is answered in the next proposition. For a Boolean function f, by $\mathsf{wt}(f)$ we denote the Hamming weight of f.

Proposition 2. *Let $g_0, g_r : \{0,1\}^n \rightarrow \{0,1\}$. Let $\alpha_1, \ldots, \alpha_N$ and $\beta_1, \ldots, \beta_N$ be arbitrary n-bit strings such that*

$$g_0(\alpha_1) \oplus g_r(\beta_1) = \cdots = g_0(\alpha_N) \oplus g_r(\beta_N).$$

Then $N \leq \mathfrak{N}$, where

$$\mathfrak{N} = \max\left(\min(2^n + w_0 - w_r, 2^n - w_0 + w_r), \min(w_0 + w_r, 2^{n+1} - w_0 - w_r)\right). \quad (6)$$

Here $w_0 = \mathsf{wt}(g_0)$ and $w_r = \mathsf{wt}(g_r)$. If $w_0 = w_r$, then the right hand side of (6) is equal to 2^n.

Proof. The condition $g_0(\alpha_1) \oplus g_r(\beta_1) = \cdots = g_0(\alpha_N) \oplus g_r(\beta_N)$ can occur in two ways, namely that all of the individual expressions are equal to 0 or, all of these are equal to 1.

Consider the maximum possible value of N such that $g_0(\alpha_1) \oplus g_r(\beta_1) = \cdots = g_0(\alpha_N) \oplus g_r(\beta_N) = 0$. An individual relation $g_0(\alpha_i) \oplus g_r(\beta_i)$ can be 0 in two possible ways, either $g_0(\alpha_i) = g_r(\beta_i) = 0$ or $g_0(\alpha_i) = g_r(\beta_i) = 1$. Suppose there are N_0 α_i's such that $g_0(\alpha_i) = g_r(\beta_i) = 0$ and there are N_1 α_i's such that $g_0(\alpha_i) = g_r(\beta_i) = 1$. Since $g_0(\alpha_i) = 1$ for N_1 i's, it follows that $N_1 \leq w_0$ and similarly, $N_1 \leq w_r$ so that $N_1 \leq \min(w_0, w_r)$. A similar argument shows that $N_0 \leq \min(2^n - w_0, 2^n - w_r)$. Since $N = N_0 + N_1$, we have $N \leq \min(w_0, w_r) + \min(2^n - w_0, 2^n - w_r)$.

Now consider the maximum possible value of N such that $g_0(\alpha_1) \oplus g_r(\beta_1) = \cdots = g_0(\alpha_N) \oplus g_r(\beta_N) = 1$. An argument similar to the above shows that $N \leq \min(2^n - w_0, w_r) + \min(w_0, 2^n - w_r)$.

The maximum value of N such that $g_0(\alpha_1) \oplus g_r(\beta_1) = \cdots = g_0(\alpha_N) \oplus g_r(\beta_N)$ is either the maximum value of N such that $g_0(\alpha_1) \oplus g_r(\beta_1) = \cdots = g_0(\alpha_N) \oplus g_r(\beta_N) = 0$ or the maximum value of N such that $g_0(\alpha_1) \oplus g_r(\beta_1) = \cdots = g_0(\alpha_N) \oplus g_r(\beta_N) = 1$. This shows that

$$N \leq \max\left(\min(w_0, w_r) + \min(2^n - w_0, 2^n - w_r), \min(w_0, 2^n - w_r) + \min(2^n - w_0, w_r)\right). \quad (7)$$

A simple and routine argument shows that the right hand side of (7) is equal to the right hand side of (6). There are four cases to consider, namely ($w_0 \leq w_r$ and $w_0 \leq 2^n - w_r$), ($w_0 \leq w_r$ and $w_0 > 2^n - w_r$), ($w_0 > w_r$ and $w_0 \leq 2^n - w_r$), and ($w_0 > w_r$ and $w_0 > 2^n - w_r$), and in each case it is required to argue that the right hand sides of (7) and (6) are equal. We provide the argument for the case $w_0 \leq w_r$ and $w_0 > 2^n - w_r$, the other cases being similar. The inequality $w_0 > 2^n - w_r$ is equivalent to $w_0 + w_r > 2^{n+1} - (w_0 + w_r)$. So under the condition $w_0 \leq w_r$ and $w_0 > 2^n - w_r$, the right hand sides of both (7) and (6) are equal to $\max(w_0 + 2^n - w_r, 2^{n+1} - w_0 - w_r)$. □

Remark: Consider Propositions 1 and 2 together. If $g_0, \ldots, g_r$ are nonlinear invariants, then for all 2^n n-bit strings α, $g_0(\alpha) \oplus g_r(E_K(\alpha))$ is a constant. So, if $\mathfrak{N} < 2^n$, then there are no choices of Boolean functions $g_1, \ldots, g_{r-1}$, such that $g_0, g_1, \ldots, g_{r-1}, g_r$ are nonlinear invariants.

Notation: For the convenience of the ensuing description, we introduce some notation.

- For a Boolean function f and $\overline{\alpha} = (\alpha_1, \ldots, \alpha_N)$ where $\alpha_i \in \{0,1\}^n$ for $i = 1, \ldots, N$, define $\Psi(f, \overline{\alpha}) = (f(\alpha_1), \ldots, f(\alpha_N))$.
- For $0 \leq w \leq 2^n$, let $\mathcal{F}_w$ be the set of all n-variable Boolean functions having weight w.
- Given g_0, for $0 \leq \ell \leq N$, let $\mathcal{P}_\ell[g_0]$ be the set of all $\overline{\alpha} = (\alpha_1, \ldots, \alpha_N)$, where $\alpha_1, \ldots, \alpha_N$ are distinct n-bit strings, such that $g_0(\alpha_i) = 1$ for exactly ℓ of the α_i's, i.e., $\mathcal{P}_\ell[g_0] = \{\overline{\alpha} = (\alpha_1, \ldots, \alpha_N) : \#\{i : g_0(\alpha_i) = 1\} = \ell\}$. When g_0 is clear from the context we will simply write $\mathcal{P}_\ell$ instead of $\mathcal{P}_\ell[g_0]$.

Lemma 1. *Let $\overline{P} = (P_1, \ldots, P_N)$ where $P_1, \ldots, P_N$ are chosen from $\{0,1\}^n$ under uniform random sampling without replacement. Then*

$$\Pr[\overline{P} \in \mathcal{P}_\ell[g_0]] = \frac{\binom{w_0}{\ell}\binom{2^n - w_0}{N-\ell}}{\binom{2^n}{N}}, \tag{8}$$

where $w_0 = \mathsf{wt}(g_0)$.

Proof. The event $\overline{P} \in \mathcal{P}_\ell[g_0]$ occurs if exactly ℓ of the P_i's fall in the support of g_0 while the other $N - \ell$ of the P_i's fall outside the support of g_0. Let us call strings in the support of g_0 to be red and the strings outside the support of g_0 to be black. So, there are w_0 red strings and $2^n - w_0$ black strings. The random experiment consists of choosing N distinct strings from 2^n strings such that ℓ are red and $N - \ell$ are black. This is the setting of hypergeometric distribution and the required probability is given by the right hand side of (8). □

2.1 Building Distinguishers

Proposition 1 provides a structural property for a key alternating cipher E. Suppose $g_0, \ldots, g_r$ are nonlinear invariants (with associated constants $c_0, \ldots, c_{r-1}$) and K is such that $K_0, \ldots, K_{r-1}$ are weak keys, then for any plaintext P, $g_0(P) \oplus g_r(E_K(P))$ is a constant. To be able to distinguish E_K from a uniform random permutation π (resp. a uniform random function ρ), it is required to obtain the probability that $g_0(P) \oplus g_r(\pi(P))$ (resp. $g_0(P) \oplus g_r(\rho(P))$) is a constant.

The availability of a single plaintext is not sufficient to construct a meaningful distinguisher. So, suppose plaintexts $P_1, \ldots, P_N$ are used by the distinguishing algorithm. Since it is not useful to repeat plaintexts, without loss of generality, we may assume $P_1, \ldots, P_N$ to be distinct. From Proposition 1, we have that

$$f_{E_K}[g_0, g_r](P_1) = f_{E_K}[g_0, g_r](P_2) = \cdots = f_{E_K}[g_0, g_r](P_N). \tag{9}$$

Distinguishing from a Uniform Random Permutation: Since for each K, E_K is a bijective map, the appropriate goal would be to distinguish E_K from a uniform random permutation π of $\{0,1\}^n$. To build a distinguisher, it is required to know the probability of the following event.

$$\mathcal{E}^\pi : f_\pi[g_0, g_r](P_1) = f_\pi[g_0, g_r](P_2) = \cdots = f_\pi[g_0, g_r](P_N).$$

The event $\mathcal{E}^\pi$ can be written as the disjoint union of two events $\mathcal{E}_0^\pi$ and $\mathcal{E}_1^\pi$, i.e., $\mathcal{E}^\pi = \mathcal{E}_0^\pi \cup \mathcal{E}_1^\pi$, where

$$\begin{array}{l} \mathcal{E}_0^\pi : f_\pi[g_0, g_r](P_1) = 0, f_\pi[g_0, g_r](P_2) = 0, \ldots, f_\pi[g_0, g_r](P_N) = 0; \\ \mathcal{E}_1^\pi : f_\pi[g_0, g_r](P_1) = 1, f_\pi[g_0, g_r](P_2) = 1, \ldots, f_\pi[g_0, g_r](P_N) = 1. \end{array} \quad (10)$$

So,

$$\Pr[\mathcal{E}^\pi] = \Pr[\mathcal{E}_0^\pi] + \Pr[\mathcal{E}_1^\pi]. \quad (11)$$

The distinguisher $\mathcal{D}^\mathcal{O}$ which distinguishes E_K from π using a nonlinear invariant attack has been defined in Sect. 1. Note that the distinguishing error ε defined in (2) is exactly $\Pr[\mathcal{E}^\pi]$.

Uniform Random Function: Considering a block cipher to be a map from n-bit strings to n-bit strings, a weaker goal would be to distinguish E_K from a uniform random function ρ from $\{0,1\}^n$ to $\{0,1\}^n$. The events $\mathcal{E}^\rho, \mathcal{E}_0^\rho$ and $\mathcal{E}_1^\rho$ are defined in a manner similar to $\mathcal{E}^\pi, \mathcal{E}_0^\pi$ and $\mathcal{E}_1^\pi$ respectively with π replaced by ρ. To build a distinguisher, it is required to obtain the probability of $\mathcal{E}^\rho$. As in the case of uniform random permutation, a distinguisher can make only one-sided error when the oracle uses ρ and the probability of this error is $\Pr[\mathcal{E}^\rho]$.

Choice of Plaintexts: On being provided with distinct plaintexts $P_1, \ldots, P_N$, the distinguisher can make an error. The error probability depends on the manner in which $P_1, \ldots, P_N$ are chosen. We will analyse the error probability under the following two possible scenarios.

Uniform random sampling without replacement: In this analysis, we assume that $P_1, \ldots, P_N$ are chosen from $\{0,1\}^n$ using uniform random sampling without replacement.

Fixed values: In this analysis, it is assumed that $P_1, \ldots, P_N$ are distinct fixed n-bit strings, i.e., there is no randomness in the plaintexts. Suppose $(P_1, \ldots, P_N) \in \mathcal{P}_\ell[g_0]$, i.e., there are exactly ℓ P_i's such that $g_0(P_i) = 1$. We show that the probability of error depends on ℓ.

We introduce the following notation to denote the four different kinds of error probabilities that can occur.

- $\varepsilon_{\pi,\$}$ is the error probability of distinguishing E_K from a uniform random permutation π when $P_1, \ldots, P_N$ are chosen under uniform random sampling without replacement, i.e., $\varepsilon_{\pi,\$} = \Pr[\mathcal{E}^\pi]$ when $P_1, \ldots, P_N$ are chosen under uniform random sampling without replacement.
- $\varepsilon_{\pi,\ell}$ is the error probability of distinguishing E_K from a uniform random permutation π when $(P_1, \ldots, P_N) \in \mathcal{P}_\ell$, i.e., $\varepsilon_{\pi,\ell} = \Pr[\mathcal{E}^\pi]$ when $(P_1, \ldots, P_N) \in \mathcal{P}_\ell$.
- $\varepsilon_{\rho,\$}$ is the error probability of distinguishing E_K from a uniform random function ρ when $P_1, \ldots, P_N$ are chosen under uniform random sampling without replacement, i.e., $\varepsilon_{\rho,\$} = \Pr[\mathcal{E}^\rho]$ when $P_1, \ldots, P_N$ are chosen under uniform random sampling without replacement.

– $\varepsilon_{\rho,\ell}$ is the error probability of distinguishing E_K from a uniform random function ρ when $(P_1,\ldots,P_N) \in \mathcal{P}_\ell$, i.e., $\varepsilon_{\rho,\ell} = \Pr[\mathcal{E}^\rho]$ when $(P_1,\ldots,P_N) \in \mathcal{P}_\ell$.

3 Error Probability for Uniform Random Function

In this section, we obtain expressions for $\varepsilon_{\rho,\$}$ and $\varepsilon_{\rho,\ell}$. The expression for $\varepsilon_{\rho,\$}$ is given in Theorem 1. Corollary 1 to Lemma 2 provides the expression for $\varepsilon_{\rho,\ell}$.

Lemma 2. *Let g_0 and g_r be two n-variable Boolean functions. Let ρ be a uniform random function and $F = f_\rho[g_0, g_r] = g_0 \oplus (g_r \circ \rho)$. Let $\overline{\alpha} = (\alpha_1,\ldots,\alpha_N)$ where $\alpha_1,\ldots,\alpha_N$ are distinct n-bit strings. Then*

$$\begin{aligned}\Pr[\Psi(F,\overline{\alpha}) = (0,\ldots,0)] &= \prod_{i=1}^{N}\left(\frac{w_r}{2^n}g_0(\alpha_i) + \frac{2^n - w_r}{2^n}(1 - g_0(\alpha_i))\right)\\ &= \left(\frac{w_r}{2^n}\right)^{\ell}\left(\frac{2^n - w_r}{2^n}\right)^{N-\ell}; \qquad (12)\\ \Pr[\Psi(F,\overline{\alpha}) = (1,\ldots,1)] &= \prod_{i=1}^{N}\left(\frac{2^n - w_r}{2^n}g_0(\alpha_i) + \frac{w_r}{2^n}(1 - g_0(\alpha_i))\right)\\ &= \left(\frac{w_r}{2^n}\right)^{N-\ell}\left(\frac{2^n - w_r}{2^n}\right)^{\ell}. \qquad (13)\end{aligned}$$

where $w_r = \mathsf{wt}(g_r)$ and ℓ is such that $\overline{\alpha} \in \mathcal{P}_\ell$.

Further, if g_r is balanced, then $\Pr[\Psi(F,\overline{\alpha}) = (0,\ldots,0)] = \Pr[\Psi(F,\overline{\alpha}) = (1,\ldots,1)] = 1/2^N$.

Proof. Consider $\Psi(F,\overline{\alpha}) = (0,\ldots,0)$ which is the following event:

$$g_r(\rho(\alpha_1)) = g_0(\alpha_1),\ldots, g_r(\rho(\alpha_N)) = g_0(\alpha_N).$$

Since $\alpha_1,\ldots,\alpha_N$ are distinct and ρ is a uniform random function, the n-bit strings $X_1 = \rho(\alpha_1),\ldots,X_N = \rho(\alpha_N)$ are independent and uniformly distributed over $\{0,1\}^n$. Let $p_i = \Pr[g_r(\rho(\alpha_i)) = g_0(\alpha_i)] = \Pr[g_r(X_i) = g_0(\alpha_i)]$ for $i = 1,\ldots,N$. Since $X_1,\ldots,X_N$ are independent, so are the events $g_r(X_1) = g_0(\alpha_1)$, $\ldots, g_r(X_N) = g_0(\alpha_N)$. Consequently,

$$\begin{aligned}\Pr[\Psi(F,\overline{\alpha}) = (0,\ldots,0)] &= \Pr[g_r(X_1) = g_0(\alpha_1),\ldots, g_r(X_N) = g_0(\alpha_N)]\\ &= p_1\cdots p_N.\end{aligned}$$

Since X_i is uniformly distributed over $\{0,1\}^n$, the event $g_r(X_i) = 1$ occurs if and only if X_i falls within the support of g_r and the probability of this is $w_r/2^n$. Similarly, the event $g_r(X_i) = 0$ occurs with probability $(2^n - w_r)/2^n$.

$$p_i = \Pr[g_r(X_i) = g_0(\alpha_i)] = \begin{cases}\Pr[g_r(X_i) = 1] = \frac{w_r}{2^n} & \text{if } g_0(\alpha_i) = 1;\\ \Pr[g_r(X_i) = 0] = \frac{2^n - w_r}{2^n} & \text{if } g_0(\alpha_i) = 0.\end{cases}$$

This can be compactly written as

$$p_i = \frac{w_r}{2^n} g_0(\alpha_i) + \frac{2^n - w_r}{2^n}(1 - g_0(\alpha_i)).$$

Let $\overline{\alpha} \in \mathcal{P}_\ell$. Then for exactly ℓ of the α_i's we have $g_0(\alpha_i) = 1$ while for the other $N - \ell$ of the α_i's, we have $g_0(\alpha_i) = 0$. This consideration leads to (12).

The proof for (13) is similar. If g_r is balanced, then $w_r = 2^{n-1}$ which shows the last part of the theorem. □

Corollary 1. *Let g_0 and g_r be two n-variable Boolean functions. Let ρ be a uniform random function and $F = f_\rho[g_0, g_r] = g_0 \oplus (g_r \circ \rho)$. Let $\overline{P} = (P_1, \ldots, P_N) \in \mathcal{P}_\ell$. Then*

$$\varepsilon_{\rho,\ell} = \Pr[\mathcal{E}^\rho] = \left(\frac{w_r}{2^n}\right)^\ell \left(\frac{2^n - w_r}{2^n}\right)^{N-\ell} + \left(\frac{w_r}{2^n}\right)^{N-\ell} \left(\frac{2^n - w_r}{2^n}\right)^\ell. \quad (14)$$

Further, if g_r is balanced, then $\varepsilon_{\rho,\ell} = 1/2^{N-1}$.

Corollary 1 raises the relevant question that when g_r is not balanced (i.e., $w_r \neq 2^{n-1}$) how should ℓ be chosen such that $\varepsilon_{\rho,\ell}$ is minimum. To answer this, consider $\varepsilon_{\rho,\ell}$ as a function of ℓ. Let $p = \frac{w_r}{2^n}$. Since $w_r \neq 2^{n-1}$, $p \neq 1/2$. Then,

$$\begin{aligned}
\varepsilon_{\rho,\ell} &= p^\ell(1-p)^{N-\ell} + p^{N-\ell}(1-p)^\ell \\
\Rightarrow \frac{d\varepsilon_{\rho,\ell}}{d\ell} &= p^\ell(1-p)^{N-\ell}\{\ln p - \ln(1-p)\} \\
&\quad + p^{N-\ell}(1-p)^\ell\{-\ln p + \ln(1-p)\} \\
&= \{\ln p - \ln(1-p)\}\left\{p^\ell(1-p)^{N-\ell} - p^{N-\ell}(1-p)^\ell\right\}.
\end{aligned}$$

If $\ell = N/2$, then $d\varepsilon_{\rho,\ell}/d\ell = 0$. A routine analysis now shows that $d\varepsilon_{\rho,\ell}/d\ell < 0$ for $\ell < N/2$ and $d\varepsilon_{\rho,\ell}/d\ell > 0$ for $\ell > N/2$ irrespective of whether $p > 1/2$ or $p < 1/2$. So, for $p \neq 1/2$, $\varepsilon_{\rho,\ell}$ takes the minimum value for $\ell = N/2$.

Theorem 1. *Let g_0 and g_r be two n-variable Boolean functions. Let ρ be a uniform random function from $\{0,1\}^n$ to $\{0,1\}^n$ and $F = f_\rho[g_0, g_r] = g_0 \oplus (g_r \circ \rho)$. Let $\overline{P} = (P_1, \ldots, P_N)$ where $P_1, \ldots, P_N$ are chosen from $\{0,1\}^n$ under uniform random sampling without replacement and independently of F (i.e. $P_1, \ldots, P_N$ and F are statistically independent). Then*

$$\left.\begin{aligned}
&\Pr[\mathcal{E}_0^\rho] = \Pr[\Psi(F, \overline{P}) = (0, \ldots, 0)] \\
&= \sum_{\ell=0}^{N} \left(\frac{w_r}{2^n}\right)^\ell \left(\frac{2^n - w_r}{2^n}\right)^{N-\ell} \cdot \frac{\binom{w_0}{\ell}\binom{2^n - w_0}{N-\ell}}{\binom{2^n}{N}}; \\
&\Pr[\mathcal{E}_1^\rho] = \Pr[\Psi(F, \overline{P}) = (1, \ldots, 1)] \\
&= \sum_{\ell=0}^{N} \left(\frac{2^n - w_r}{2^n}\right)^\ell \left(\frac{w_r}{2^n}\right)^{N-\ell} \cdot \frac{\binom{w_0}{\ell}\binom{2^n - w_0}{N-\ell}}{\binom{2^n}{N}}.
\end{aligned}\right\} \quad (15)$$

Here $w_0 = \mathsf{wt}(g_0)$ and $w_r = \mathsf{wt}(g_r)$.

Consequently,

$$\begin{aligned}\varepsilon_{\rho,\$} &= \Pr[\mathcal{E}^{\rho}]\\ &= \sum_{\ell=0}^{N}\left(\left(\frac{w_r}{2^n}\right)^{\ell}\left(\frac{2^n-w_r}{2^n}\right)^{N-\ell}+\left(\frac{2^n-w_r}{2^n}\right)^{\ell}\left(\frac{w_r}{2^n}\right)^{N-\ell}\right)\cdot\frac{\binom{w_0}{\ell}\binom{2^n-w_0}{N-\ell}}{\binom{2^n}{N}}.\end{aligned} \tag{16}$$

Further, if g_r is balanced, then $\varepsilon_{\rho,\$} = 1/2^{N-1}$.

Proof. Consider the event $\mathcal{E}_0^{\rho}$.

$$\begin{aligned}\Pr[\mathcal{E}_0^{\rho}] &= \Pr[\Psi(F,\overline{P}) = (0,\ldots,0)]\\ &= \sum_{\ell=0}^{N}\Pr[\Psi(F,\overline{P}) = (0,\ldots,0), \overline{P}\in\mathcal{P}_{\ell}]\\ &= \sum_{\ell=0}^{N}\sum_{\overline{\alpha}\in\mathcal{P}_{\ell}}\Pr[\Psi(F,\overline{P}) = (0,\ldots,0), \overline{P}=\overline{\alpha}]\\ &= \sum_{\ell=0}^{N}\sum_{\overline{\alpha}\in\mathcal{P}_{\ell}}\Pr[\Psi(F,\overline{\alpha}) = (0,\ldots,0), \overline{P}=\overline{\alpha}]\\ &= \sum_{\ell=0}^{N}\sum_{\overline{\alpha}\in\mathcal{P}_{\ell}}\Pr[\Psi(F,\overline{\alpha}) = (0,\ldots,0)]\cdot\Pr[\overline{P}=\overline{\alpha}]\\ &\quad\text{(since } F \text{ and } \overline{P} \text{ are independent)}\\ &= \sum_{\ell=0}^{N}\sum_{\overline{\alpha}\in\mathcal{P}_{\ell}}\left(\frac{w_r}{2^n}\right)^{\ell}\left(\frac{2^n-w_r}{2^n}\right)^{N-\ell}\cdot\Pr[\overline{P}=\overline{\alpha}] \quad \text{(from Lemma 2)}\\ &= \sum_{\ell=0}^{N}\left(\frac{w_r}{2^n}\right)^{\ell}\left(\frac{2^n-w_r}{2^n}\right)^{N-\ell}\sum_{\overline{\alpha}\in\mathcal{P}_{\ell}}\Pr[\overline{P}=\overline{\alpha}]\\ &= \sum_{\ell=0}^{N}\left(\frac{w_r}{2^n}\right)^{\ell}\left(\frac{2^n-w_r}{2^n}\right)^{N-\ell}\Pr[\overline{P}\in\mathcal{P}_{\ell}]\\ &= \sum_{\ell=0}^{N}\left(\frac{w_r}{2^n}\right)^{\ell}\left(\frac{2^n-w_r}{2^n}\right)^{N-\ell}\cdot\frac{\binom{w_0}{\ell}\binom{2^n-w_0}{N-\ell}}{\binom{2^n}{N}} \quad \text{(from Lemma 1).}\end{aligned}$$

The probability of the event $\mathcal{E}_1^{\rho}$ is similarly obtained. Since $\mathcal{E}^{\rho}$ is the disjoint union of $\mathcal{E}_0^{\rho}$ and $\mathcal{E}_1^{\rho}$, we obtain (16).

If g_r is balanced, $w_r = 2^{n-1}$ and we have

$$\varepsilon_{\rho,\$} = \frac{1}{2^{N-1}}\sum_{\ell=0}^{N}\frac{\binom{w_0}{\ell}\binom{2^n-w_0}{N-\ell}}{\binom{2^n}{N}} = \frac{1}{2^{N-1}}.$$

The last equality holds since $\binom{w_0}{\ell}\binom{2^n-w_0}{N-\ell}/\binom{2^n}{N}$ is the probability that a random variable X equals ℓ where X follows the hypergeometric distribution $\mathsf{HG}(2^n, w_0, N)$ and so $\sum_{\ell=0}^{N} \Pr[X = \ell] = 1$. □

Remarks:

1. From Corollary 1 and Theorem 1, we have that if g_r is balanced, then $\varepsilon_{\rho,\ell} = \varepsilon_{\rho,\$} = 1/2^{N-1}$, i.e., the error probability of the distinguisher is determined only by the number of distinct plaintexts that are used and not on whether these are fixed or chosen using uniform random sampling without replacement.
2. It has been mentioned in [16] that the distinguishing error of a nonlinear invariant attack is $1/2^{N-1}$. The above analysis shows that this is the error in distinguishing from a uniform random function.

4 Error Probability for Uniform Random Permutation

In this section, we obtain expressions for $\varepsilon_{\pi,\$}$ and $\varepsilon_{\pi,\ell}$. The expression for $\varepsilon_{\pi,\$}$ is given in Theorem 2. Lemmas 3 and 1 are intermediate steps to proving the theorem. Corollary 2 provides the expression for $\varepsilon_{\pi,\ell}$.

Lemma 3. *Let g_0 and g_r be n-variable Boolean functions. Let π be a uniform random permutation and $F = f_\pi[g_0, g_r] = g_0 \oplus (g_r \circ \pi)$. Let $\alpha_1, \ldots, \alpha_N$ be distinct n-bit strings such that $\#\{i : g_0(\alpha_i) = 1\} = \ell$. Denote $\overline{\alpha} = (\alpha_1, \ldots, \alpha_N)$. Then*

$$\begin{aligned} \Pr[\Psi(F, \overline{\alpha}) = (0, \ldots, 0)] &= \frac{\binom{2^n-w_r}{N-\ell}\binom{w_r}{\ell}}{\binom{2^n}{N}\binom{N}{\ell}} \quad \text{and} \\ \Pr[\Psi(F, \overline{\alpha}) = (1, \ldots, 1)] &= \frac{\binom{w_r}{N-\ell}\binom{2^n-w_r}{\ell}}{\binom{2^n}{N}\binom{N}{\ell}}, \end{aligned} \tag{17}$$

where $w_r = \mathsf{wt}(g_r)$.

Proof. Consider the first statement. It is given that $g_0(\alpha_i) = 1$ for exactly ℓ of the α_i's.

Let us start with the special case where $g_0(\alpha_1) = \cdots = g_0(\alpha_\ell) = 1$ and $g_0(\alpha_{\ell+1}) = \cdots = g_0(\alpha_N) = 0$. Then the event $\Psi(F, \overline{\alpha}) = (0, \ldots, 0)$ holds if and only if $g_r(\pi(\alpha_1)) = \cdots = g_r(\pi(\alpha_\ell)) = 1$ and $g_r(\pi(\alpha_{\ell+1})) = \cdots = g_r(\pi(\alpha_N)) = 0$. Since $\alpha_1, \ldots, \alpha_N$ are distinct n-bit strings and π is a uniform random permutation of $\{0,1\}^n$, the random quantities $\pi(\alpha_1), \ldots, \pi(\alpha_N)$ can be thought of as being chosen from $\{0,1\}^n$ using uniform random sampling without replacement. Further, $g_r(\pi(\alpha_i)) = 1$ (resp. 0) if and only if $\pi(\alpha_i)$ falls within (resp. outside) the support of g_r.

From the above considerations, the probability that $g_r(\pi(\alpha_1)) = 1$ is $w_r/2^n$; the probability that $g_r(\pi(\alpha_2)) = 1$ given that $g_r(\pi(\alpha_1)) = 1$ is $(w_r - 1)/(2^n - 1)$; continuing, the probability that $g_r(\pi(\alpha_\ell)) = 1$ given that $g_r(\pi(\alpha_1)) = 1, \ldots,$

$g_r(\pi(\alpha_{\ell-1})) = 1$ is $(w_r - \ell + 1)/(2^n - \ell + 1)$; the probability that $g_r(\pi(\alpha_{\ell+1})) = 0$ given that $g_r(\pi(\alpha_1)) = 1, \ldots, g_r(\pi(\alpha_\ell)) = 1$ is $(2^n - w_r)/(2^n - \ell)$; the probability that $g_r(\pi(\alpha_{\ell+2})) = 0$ given that $g_r(\pi(\alpha_1)) = 1, \ldots, g_r(\pi(\alpha_\ell)) = 1$ and $g_r(\pi(\alpha_{\ell+1})) = 0$ is $(2^n - w_r - 1)/(2^n - \ell - 1)$; continuing, the probability that $g_r(\pi(\alpha_N)) = 0$ given that $g_r(\pi(\alpha_1)) = 1, \ldots, g_r(\pi(\alpha_\ell)) = 1$ and $g_r(\pi(\alpha_{\ell+1})) = 0, \ldots, g_r(\pi(\alpha_{N-1})) = 0$ is $(2^n - w_r - (N - \ell) + 1)/(2^n - (N - 1))$. So,

$$\begin{aligned} &\Pr[\Psi(F, \overline{\alpha}) = (0, \ldots, 0)] \\ &= \frac{w_r(w_r - 1) \cdots (w_r - \ell + 1)(2^n - w_r)(2^n - w_r - 1) \cdots (2^n - w_r - (N - \ell) + 1)}{2^n(2^n - 1) \cdots (2^n - N + 1)}. \end{aligned} \tag{18}$$

Consider now the general case where there are exactly ℓ values of i such that $g_0(\alpha_i) = 1$ and these are not necessarily the first ℓ α_i's. Following the argument given above for the special case, it is not difficult to see that the probability of $\Psi(F, \overline{\alpha}) = (0, \ldots, 0)$ in the general case is also given by (18). In particular, the argument shows that the numerator of the probability in the general case is a reordering of the numerator of (18) while the denominator remains the same. So, in all cases the probability of $\Psi(F, \overline{\alpha}) = (0, \ldots, 0)$ is given by (18). The right hand side of (18) simplifies to the following expression

$$\frac{w_r!(2^n - w_r)!}{2^n!} \cdot \frac{(2^n - N)!}{(w_r - \ell)!(2^n - w_r - (N - \ell))!}. \tag{19}$$

Multiplying the numerator and denominator of (19) by $N!\ell!(N - \ell)!$, we obtain

$$\frac{w_r!}{\ell!(w_r - \ell)!} \cdot \frac{(2^n - w_r)!}{(N - \ell)!(2^n - w_r - (N - \ell))!} \cdot \frac{N!(2^n - N)!}{2^n!} \cdot \frac{\ell!(N - \ell)!}{N!}. \tag{20}$$

So finally we obtain (18) to be equal to

$$\Pr[\Psi(F, \overline{\alpha}) = (0, \ldots, 0)] = \frac{\binom{2^n - w_r}{N - \ell}\binom{w_r}{\ell}}{\binom{2^n}{N}\binom{N}{\ell}}.$$

This shows the first statement. The other statement is obtained similarly. □

Corollary 2. *Let g_0 and g_r be two n-variable Boolean functions. Let π be a uniform random permutation and $F = f_\pi[g_0, g_r] = g_0 \oplus (g_r \circ \pi)$. Let $\overline{P} = (P_1, \ldots, P_N) \in \mathcal{P}_\ell$. Then*

$$\varepsilon_{\pi,\ell} = \Pr[\mathcal{E}^\pi] = \frac{\binom{2^n - w_r}{N - \ell}\binom{w_r}{\ell}}{\binom{2^n}{N}\binom{N}{\ell}} + \frac{\binom{w_r}{N - \ell}\binom{2^n - w_r}{\ell}}{\binom{2^n}{N}\binom{N}{\ell}}, \tag{21}$$

where $w_r = \mathsf{wt}(g_r)$.

As in the case for uniform random function, a relevant issue is the choice of ℓ such that $\varepsilon_{\pi,\ell}$ is minimum. Unlike the case for uniform random function, it is difficult to analytically tackle this point for $\varepsilon_{\pi,\ell}$. We have run numerical experiments to explore the nature of ℓ for which $\varepsilon_{\pi,\ell}$ is the minimum. There does not seem to be any clear indication of what should be the value of ℓ. Also, we have observed that the minimum value of $\varepsilon_{\pi,\ell}$ is not much lesser than that obtained by evaluating the expression for $\varepsilon_{\pi,\$}$ as obtained in the following result.

Theorem 2. *Let g_0 and g_r be two n-variable Boolean functions. Let π be a uniform random permutation of $\{0,1\}^n$ and $F = f_\pi[g_0, g_r] = g_0 \oplus (g_r \circ \pi)$. Let $\overline{P} = (P_1, \ldots, P_N)$ where $P_1, \ldots, P_N$ are chosen from $\{0,1\}^n$ under uniform random sampling without replacement and independently of F (i.e. $P_1, \ldots, P_N$ and F are statistically independent). Then*

$$\begin{aligned}\Pr[\mathcal{E}_0^\pi] = \Pr[\Psi(F, \overline{P}) = (0, \ldots, 0)] &= \sum_{\ell=0}^{N} \frac{\binom{2^n - w_r}{N-\ell}\binom{w_r}{\ell}}{\binom{2^n}{N}\binom{N}{\ell}} \cdot \frac{\binom{w_0}{\ell}\binom{2^n - w_0}{N-\ell}}{\binom{2^n}{N}}; \\ \Pr[\mathcal{E}_1^\pi] = \Pr[\Psi(F, \overline{P}) = (1, \ldots, 1)] &= \sum_{\ell=0}^{N} \frac{\binom{w_r}{N-\ell}\binom{2^n - w_r}{\ell}}{\binom{2^n}{N}\binom{N}{\ell}} \cdot \frac{\binom{w_0}{\ell}\binom{2^n - w_0}{N-\ell}}{\binom{2^n}{N}}.\end{aligned} \tag{22}$$

Here $w_0 = \mathsf{wt}(g_0)$ and $w_r = \mathsf{wt}(g_r)$. Consequently,

$$\varepsilon_{\pi,\$} = \Pr[\mathcal{E}^\pi] = \sum_{\ell=0}^{N} \frac{\binom{2^n - w_r}{N-\ell}\binom{w_r}{\ell} + \binom{w_r}{N-\ell}\binom{2^n - w_r}{\ell}}{\binom{2^n}{N}\binom{N}{\ell}} \cdot \frac{\binom{w_0}{\ell}\binom{2^n - w_0}{N-\ell}}{\binom{2^n}{N}}. \tag{23}$$

If both g_0 and g_r are balanced, then $\varepsilon_{\pi,\$}$ is the expectation of $2p(X)/\binom{N}{X}$, i.e.,

$$\varepsilon_{\pi,\$} = \mathbf{E}\left[\frac{2p(X)}{\binom{N}{X}}\right] \tag{24}$$

where X follows $\mathsf{HG}(2^n, 2^{n-1}, N)$ and for $\ell = 0, \ldots, N$, $p(\ell)$ is the probability that $X = \ell$.

Proof. Consider $\Pr[\mathcal{E}_0^\pi]$.

$$\begin{aligned}&\Pr[\Psi(F, \overline{P}) = (0, \ldots, 0)] \\ &= \sum_{\ell=0}^{N} \sum_{\overline{\alpha} \in \mathcal{P}_\ell} \Pr[\Psi(F, \overline{P}) = (0, \ldots, 0), \overline{P} = \overline{\alpha}] \\ &= \sum_{\ell=0}^{N} \sum_{\overline{\alpha} \in \mathcal{P}_\ell} \Pr[\Psi(F, \overline{\alpha}) = (0, \ldots, 0), \overline{P} = \overline{\alpha}] \\ &= \sum_{\ell=0}^{N} \sum_{\overline{\alpha} \in \mathcal{P}_\ell} \Pr[\Psi(F, \overline{\alpha}) = (0, \ldots, 0)] \cdot \Pr[\overline{P} = \overline{\alpha}] \\ &\quad \text{(since } F \text{ and } \overline{P} \text{ are independent)} \\ &= \sum_{\ell=0}^{N} \sum_{\overline{\alpha} \in \mathcal{P}_\ell} \frac{\binom{2^n - w_r}{N-\ell}\binom{w_r}{\ell}}{\binom{2^n}{N}\binom{N}{\ell}} \cdot \Pr[\overline{P} = \overline{\alpha}] \quad \text{(from Lemma 3)}\end{aligned}$$

$$= \sum_{\ell=0}^{N} \frac{\binom{2^n - w_r}{N-\ell}\binom{w_r}{\ell}}{\binom{2^n}{N}\binom{N}{\ell}} \sum_{\overline{\alpha} \in \mathcal{P}_\ell} \Pr[\overline{P} = \overline{\alpha}]$$

$$= \sum_{\ell=0}^{N} \frac{\binom{2^n - w_r}{N-\ell}\binom{w_r}{\ell}}{\binom{2^n}{N}\binom{N}{\ell}} \Pr[\overline{P} \in \mathcal{P}_\ell]$$

$$= \sum_{\ell=0}^{N} \frac{\binom{2^n - w_r}{N-\ell}\binom{w_r}{\ell}}{\binom{2^n}{N}\binom{N}{\ell}} \cdot \frac{\binom{w_0}{\ell}\binom{2^n - w_0}{N-\ell}}{\binom{2^n}{N}} \quad \text{(from Lemma 1).}$$

$\Pr[\mathcal{E}_1^\pi]$ is obtained similarly. Further, the probability of $\mathcal{E}^\pi$ is obtained from (11).

If both g_0 and g_r are balanced, then $w_0 = w_r = 2^{n-1}$ and we have

$$\varepsilon_{\pi,\$} = \sum_{\ell=0}^{N} \frac{2\binom{2^{n-1}}{N-\ell}\binom{2^{n-1}}{\ell}}{\binom{2^n}{N}\binom{N}{\ell}} \cdot \frac{\binom{2^{n-1}}{\ell}\binom{2^{n-1}}{N-\ell}}{\binom{2^n}{N}} = \sum_{\ell=0}^{N} \frac{2p(\ell)}{\binom{N}{\ell}} \cdot \frac{\binom{2^{n-1}}{\ell}\binom{2^{n-1}}{N-\ell}}{\binom{2^n}{N}}$$

$$= \mathbf{E}\left[\frac{2p(X)}{\binom{N}{X}}\right].$$

□

The next result shows that when g_0 and g_r are balanced, the distinguishing error for uniform random permutations is at least as large as that for uniform random functions.

Theorem 3. *Let g_0 and g_r be two balanced n-variable Boolean functions. Let π be a uniform random permutation of $\{0,1\}^n$ and ρ be a uniform random function from $\{0,1\}^n$ to $\{0,1\}^n$. Define $F_\pi = f_\pi[g_0, g_r] = g_0 \oplus (g_r \circ \pi)$ and $F_\rho = f_\rho[g_0, g_r] = g_0 \oplus (g_r \circ \rho)$. Let $\overline{P} = (P_1, \ldots, P_N)$ where $P_1, \ldots, P_N$ are chosen from $\{0,1\}^n$ under uniform random sampling without replacement and these are independent of F_ρ or F_π. Let*

$$\begin{aligned} \varepsilon_{\pi,\$} &= \Pr[\mathcal{E}^\pi] = \Pr[\mathcal{E}^\pi] + \Pr[\mathcal{E}_0^\pi] \\ &= \Pr[\Psi(F_\pi, \overline{P}) = (0, \ldots, 0)] + \Pr[\Psi(F_\pi, \overline{P}) = (1, \ldots, 1)]; \\ \varepsilon_{\rho,\$} &= \Pr[\mathcal{E}^\rho] = \Pr[\mathcal{E}^\rho] + \Pr[\mathcal{E}_0^\rho] \\ &= \Pr[\Psi(F_\rho, \overline{P}) = (0, \ldots, 0)] + \Pr[\Psi(F_\rho, \overline{P}) = (1, \ldots, 1)]. \end{aligned}$$

Then $\varepsilon_{\pi,\$} \geq \varepsilon_{\rho,\$}$.

Proof. It is given that g_0 and g_r are both balanced. From Theorem 1, it follows that $\varepsilon_{\rho,\$} = 1/2^{N-1}$. From Theorem 2, we have that $\varepsilon_{\pi,\$}$ is the expectation of $2p(X)/\binom{N}{X}$, i.e., $\varepsilon_{\pi,\$} = \mathbf{E}[2p(X)/\binom{N}{X}]$, where X follows $\mathsf{HG}(2^n, 2^{n-1}, N)$ and for $\ell = 0, \ldots, N$, $p(\ell)$ is the probability that $X = \ell$.

Let $Y = 2p(X)/\binom{N}{X}$. Using Jensen's inequality, we obtain

$$\begin{aligned}
\frac{1}{\mathbf{E}[Y]} &\leq \mathbf{E}\left[\frac{1}{Y}\right] \\
&= \mathbf{E}\left[\frac{\binom{N}{X}}{2p(X)}\right] \\
&= \sum_{\ell=0}^{N} \frac{\binom{N}{\ell}}{2p(\ell)} \cdot \Pr[X = \ell] \\
&= \sum_{\ell=0}^{N} \frac{\binom{N}{\ell}}{2p(\ell)} \cdot p(\ell) \\
&= 2^{N-1}.
\end{aligned}$$

Noting $\varepsilon_{\pi,\$} = \mathbf{E}[Y]$ and $\varepsilon_{\rho,\$} = 1/2^{N-1}$ gives the desired result. □

5 Computational Results

This section gives a summary of the computations done with the expressions of the error probabilities of nonlinear invariant attack presented in Sects. 3 and 4.

In our computations we have used the following Stirling's approximation to compute the binomial coefficients.

$$\binom{\mathfrak{k}}{i} \approx \frac{1}{\sqrt{2\pi\mathfrak{k}}(i/\mathfrak{k})^{i+\frac{1}{2}}(1-i/\mathfrak{k})^{\mathfrak{k}-i+\frac{1}{2}}}.$$

The computations were done for $n = 16, 32, 48$ and 64; and $N = 2^{\mathfrak{n}}$ for $\mathfrak{n} = 2, 4, 8$ and 16, except that the case $N = 2^{16}$ was not considered when $n = 16$. Further, we have considered balanced g_0 and g_r, i.e., $\mathsf{wt}(g_0) = \mathsf{wt}(g_r) = 2^{n-1}$. As a result, $\varepsilon_{\rho,\$}$, which is the error probability of distinguishing from a uniform random function, is equal to $1/2^{N-1}$.

Comparison Between $\varepsilon_{\pi,\$}$ and $\varepsilon_{\rho,\$}$. Table 1 gives the value of $\varepsilon_{\pi,\$}$ and the ratio $\varepsilon_{\pi,\$}/\varepsilon_{\rho,\$} = 2^{N-1}\varepsilon_{\pi,\$}$ for different values of n and $\mathfrak{n}$. It may be noted that the last column of the table confirms Theorem 3 which shows that for balanced g_0 and g_r, $\varepsilon_{\pi,\$} \geq \varepsilon_{\rho,\$} = 1/2^{N-1}$. Further, the ratio is close to 1. This may be explained by referring to the proof of Theorem 3. The result $\varepsilon_{\pi,\$} \geq 1/2^{N-1}$ is obtained using Jensen's inequality to the convex function $f(x) = 1/x$. In the case that the convex function is a straight line Jensen's inequality is tight. In the range of x where Jensen's inequality is applied, it turns out that $f(x)$ behaves almost like a straight line. Consequently, the inequality is almost tight in this range of applicability.

Table 1. Comparison between $\varepsilon_{\pi,\$}$ and $\varepsilon_{\rho,\$} = 2^{-(N-1)}$.

n	$\mathfrak{n}$	$\varepsilon_{\pi,\$}$	$2^{N-1} \times \varepsilon_{\pi,\$}$
16	2	0.133739	1.069910
	4	0.000031	1.017414
	8	1.728943×10^{-77}	1.000990
32	2	0.133739	1.069908
	4	0.000031	1.017415
	8	1.728930×10^{-77}	1.000982
	16	$9.982420\times10^{-19729}$	1.000004
48	2	0.133739	1.069908
	4	0.000031	1.017415
	8	1.728930×10^{-77}	1.000982
	16	$9.982420\times10^{-19729}$	1.000004
64	2	0.133739	1.069908
	4	0.000031	1.017415
	8	1.728930×10^{-77}	1.000982
	16	$9.982420\times10^{-19729}$	1.000004

6 Conclusion

In this paper, we have performed a detailed analysis of the distinguishing error of nonlinear invariant attacks. We have obtained precise expressions for the error of nonlinear invariant attacks in distinguishing a key alternating cipher from either a uniform random function or a uniform random permutation. It has been theoretically proven that the distinguishing error probability in the case of uniform random permutation is greater than the distinguishing error probability in the case of uniform random function.

Acknowledgement. We thank the reviewers for their helpful comments which have helped in improving the paper.

References

1. Aumasson, J.-P., Meier, W.: Zero-sum distinguishers for reduced Keccak-f and for the core functions of Luffa and Hamsi (2009). https://131002.net/data/papers/AM09.pdf. Accessed on 30 Jun 2020
2. Baignères, T., Junod, P., Vaudenay, S.: How far can we go beyond linear cryptanalysis? In: Lee, P.J. (ed.) ASIACRYPT 2004. LNCS, vol. 3329, pp. 432–450. Springer, Heidelberg (2004). https://doi.org/10.1007/978-3-540-30539-2_31
3. Boura, C., Canteaut, A.: Zero-sum distinguishers for iterated permutations and application to Keccak-f and Hamsi-256. In: Biryukov, A., Gong, G., Stinson, D.R. (eds.) SAC 2010. LNCS, vol. 6544, pp. 1–17. Springer, Heidelberg (2011). https://doi.org/10.1007/978-3-642-19574-7_1

4. Boura, C., Canteaut, A., Coggia, D.: A general proof framework for recent AES distinguishers. IACR Trans. Symmetric Cryptol. **2019**(1), 170–191 (2019)
5. Gilbert, H.: A simplified representation of AES. In: Sarkar, P., Iwata, T. (eds.) ASIACRYPT 2014. LNCS, vol. 8873, pp. 200–222. Springer, Heidelberg (2014). https://doi.org/10.1007/978-3-662-45611-8_11
6. Grassi, L., Rechberger, C.: Revisiting gilbert's known-key distinguisher. Des. Codes Cryptogr. **88**(7), 1401–1445 (2020)
7. Harpes, C., Kramer, G.G., Massey, J.L.: A generalization of linear cryptanalysis and the applicability of Matsui's Piling-up lemma. In: Guillou, L.C., Quisquater, J.-J. (eds.) EUROCRYPT 1995. LNCS, vol. 921, pp. 24–38. Springer, Heidelberg (1995). https://doi.org/10.1007/3-540-49264-X_3
8. Hell, M., Johansson, T., Brynielsson, L.: An overview of distinguishing attacks on stream ciphers. Cryptogr. Commun. **1**(1), 71–94 (2009)
9. Hell, M., Johansson, T., Brynielsson, L., Englund, H.: Improved distinguishers on stream ciphers with certain weak feedback polynomials. IEEE Trans. Inf. Theor. **58**(9), 6183–6193 (2012)
10. Jean, J., Naya-Plasencia, M., Peyrin, T.: Multiple limited-birthday distinguishers and applications. In: Lange, T., Lauter, K., Lisoněk, P. (eds.) SAC 2013. LNCS, vol. 8282, pp. 533–550. Springer, Heidelberg (2014). https://doi.org/10.1007/978-3-662-43414-7_27
11. Kesarwani, A., Roy, D., Sarkar, S., Meier, W.: New cube distinguishers on NFSR-based stream ciphers. Des. Codes Cryptogr. **88**(1), 173–199 (2020)
12. Knudsen, L.R., Rijmen, V.: Known-key distinguishers for some block ciphers. In: Kurosawa, K. (ed.) ASIACRYPT 2007. LNCS, vol. 4833, pp. 315–324. Springer, Heidelberg (2007). https://doi.org/10.1007/978-3-540-76900-2_19
13. Knudsen, L.R., Robshaw, M.J.B.: Non-linear approximations in linear cryptanalysis. In: Maurer, U. (ed.) EUROCRYPT 1996. LNCS, vol. 1070, pp. 224–236. Springer, Heidelberg (1996). https://doi.org/10.1007/3-540-68339-9_20
14. Künzli, S., Junod, P., Meier, W.: Distinguishing attacks on T-functions. In: Dawson, E., Vaudenay, S. (eds.) Mycrypt 2005. LNCS, vol. 3715, pp. 2–15. Springer, Heidelberg (2005). https://doi.org/10.1007/11554868_2
15. Matsui, M.: Linear cryptanalysis method for DES cipher. In: Helleseth, T. (ed.) EUROCRYPT 1993. LNCS, vol. 765, pp. 386–397. Springer, Heidelberg (1994). https://doi.org/10.1007/3-540-48285-7_33
16. Todo, Y., Leander, G., Sasaki, Yu.: Nonlinear invariant attack: practical attack on full SCREAM, iSCREAM, and Midori64. J. Cryptol. **32**(4), 1383–1422 (2018). https://doi.org/10.1007/s00145-018-9285-0

Weak Subtweakeys in SKINNY

Daniël Kuijsters(✉), Denise Verbakel, and Joan Daemen

Radboud University, Nijmegen, The Netherlands
{Daniel.Kuijsters,denise.verbakel,joan.daemen}@ru.nl

Abstract. Lightweight cryptography is characterized by the need for low implementation cost, while still providing sufficient security. This requires careful analysis of building blocks and their composition.

SKINNY is an ISO/IEC standardized family of tweakable block ciphers and is used in the NIST lightweight cryptography standardization process finalist ROMULUS. We present non-trivial linear approximations of two-round SKINNY that have correlation one or minus one and that hold for a large fraction of all round tweakeys. Moreover, we show how these could have been avoided.

Keywords: Cryptanalysis · Lightweight symmetric cryptography · Block ciphers

1 Introduction

In 2018, NIST initiated a process for the standardization of *lightweight cryptography* [14], i.e., cryptography that is suitable for use in constrained environments. A typical cryptographic primitive is built by composing a relatively simple round function with itself a number of times. To choose this number of rounds, a trade-off is made between the security margin and the performance.

One of the finalists in this standardization process is the ROMULUS [8] scheme for authenticated encryption with associated data. This scheme is based on the ISO/IEC 18033-7:2022 [1] standardized lightweight tweakable block cipher SKINNY [2].

Two of the most important techniques for the analysis of symmetric primitives are differential [3] and linear cryptanalysis [12]. To reason about the security against these attacks, the designers of SKINNY have computed lower bounds on the number of *active* S-boxes in linear and differential trails. However, at the end of Sect. 4.1 of [2] they write:

> The above bounds are for single characteristic, thus it will be interesting to take a look at differentials and linear hulls. Being a rather complex task, we leave this as future work.

Building on the work of [4,15] investigated clustering of two-round trails in SKINNY and in this paper we report and explain its most striking finding.

T. Isobe and S. Sarkar (Eds.): INDOCRYPT 2022, LNCS 13774, pp. 336–348, 2022.
https://doi.org/10.1007/978-3-031-22912-1_15

By examination of two rounds, we argue why it is sensible to look at the substructure that consists of a double S-box with a subtweakey addition in between. We study this double S-box structure both from an algebraic point of view and a statistical point of view. We found that for some subtweakeys there are non-trivial *perfect* linear approximations, i.e., that have correlation one or minus one. We present them in this paper together with their constituent linear trails. For both the version of SKINNY that uses the 4-bit S-box and the version that uses the 8-bit S-box, we present one non-trivial perfect linear approximation of the double S-box structure that holds for 1/4 of all subtweakeys and four non-trivial perfect linear approximations that each hold for 1/16 of all subtweakeys. In total, 1/4 of the subtweakeys is *weak*, i.e., it has an associated non-trivial perfect linear approximation. The linear approximations of the double S-box structure can be extended to linear approximations of the full two rounds of SKINNY. From the fact that the double S-box structure appears in four different locations, it follows that $1 - (3/4)^4 \approx 68\%$ of the round tweakeys is weak, i.e., two rounds have a non-trivial perfect linear approximation.

Despite requiring more resources to compute, this shows that for many round tweakeys two rounds are weaker than a single round. Moreover, this also shows that the bounds on the squared correlations of linear approximations that are based on counting the number of active S-boxes in linear trails may not be readily assumed.

We conclude by showing how this undesired property could have easily been avoided by composing the S-box with a permutation of its output bits, which has a negligible impact on the implementation cost.

1.1 Outline and Contributions

In Sect. 2 we remind the reader of the parts of the SKINNY block cipher specification that are relevant to our analysis. We argue why it is reasonable to study the double S-box structure and explore its algebraic properties. Section 3 serves as a reminder for the reader of the relevant statistical analysis tools of linear cryptanalysis. Section 4 presents our findings from the study of the linear trails of the double S-box structure. We show how the problem could have been avoided in Sect. 5. Finally, we state the main message behind our findings in Sect. 6.

2 The SKINNY Family of Block Ciphers

SKINNY [2] is a family of tweakable block ciphers. A member of the SKINNY family is denoted by SKINNY-b-t, where b denotes the block size and t denotes the size of the tweakey [10]. The block size b is equal to 64 bits or 128 bits. The tweakey t is b, $2b$, or $3b$ bits.

The AES-like [7] data path of the SKINNY block cipher is the repeated application of a round function on a representation of the state as a four by four array of m-bit vectors, where m is either four or eight.

Pairs (i, j) comprising a row index i and column index j with $0 \leq i, j \leq 3$ are used to index into the state array. For example, $(0, 0)$ refers to the entry in the top left and $(3, 3)$ to the entry in the bottom right. The m-bit entries $x^{(i,j)}$ are of the form $(x_{m-1}^{(i,j)}, \ldots, x_0^{(i,j)})$.

The round function consists of the following steps in sequence: `SubCells`, `AddConstants`, `AddRoundTweakey`, `ShiftRows`, and `MixColumns`.

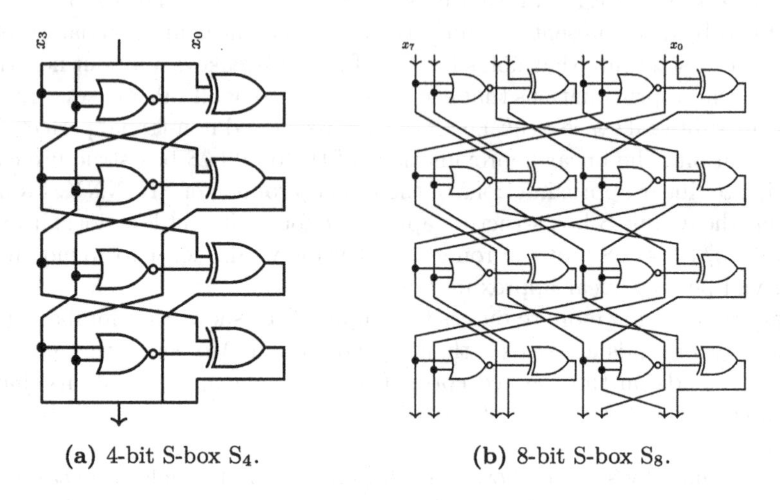

(a) 4-bit S-box S_4. **(b)** 8-bit S-box S_8.

Fig. 1. Circuit-level representation of S_4 and S_8. (Figure adapted from [9].)

Figure 1 shows the circuit-level view of the S-boxes that are used in the `SubCells` step of SKINNY.

The block matrix that is used in the `MixColumns` step is equal to

$$M = \begin{pmatrix} 1 & 0 & 1 & 1 \\ 1 & 0 & 0 & 0 \\ 0 & 1 & 1 & 0 \\ 1 & 0 & 1 & 0 \end{pmatrix},$$

where 0 denotes the zero matrix of size $m \times m$ and 1 denotes the identity matrix of size m. Each of the four columns of the state is multiplied by M in parallel.

The composition of two rounds is depicted in Fig. 2.

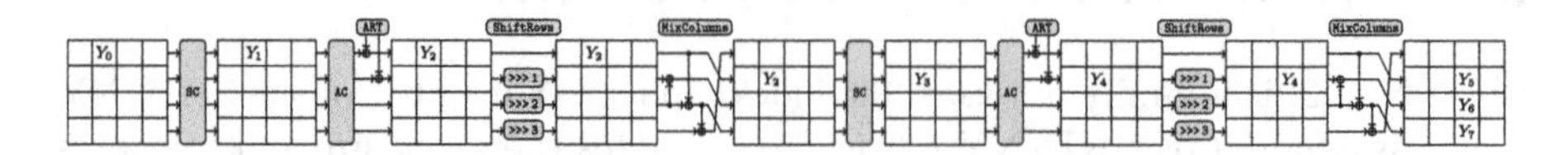

Fig. 2. Two-round SKINNY. (Figure adapted from [9].)

Consider the entry of the state at position $(0, 1)$ in Fig. 2. It is of the form $Y_0 = x^{(0,1)}$. This expression propagates through the step functions of two rounds and leads to the following intermediate expressions:

$$\begin{aligned} Y_1 &= \mathrm{S}_m(x^{(0,1)}) \\ Y_2 &= \mathrm{S}_m(x^{(0,1)}) + k^{(0)} \\ Y_3 &= \mathrm{S}_m(\mathrm{S}_m(x^{(0,1)}) + k^{(0)}) \\ Y_4 &= \mathrm{S}_m(\mathrm{S}_m(x^{(0,1)}) + k^{(0)}) + k^{(1)} \\ Y_5 + Y_6 + Y_7 &= \mathrm{S}_m(\mathrm{S}_m(x^{(0,1)}) + k^{(0)}) + k^{(1)}, \end{aligned}$$

Here, $k^{(0)}$ and $k^{(1)}$ are subtweakeys, which are linear expressions in the cipher key and tweak bits (assuming that the tweakey does not consist entirely of cipher key bits). These linear expressions depend on the round number, but they are known to the attacker. The tweak can be chosen by the attacker and the cipher key is unknown to the attacker. By choosing the tweak, the attacker can attain all values of $k^{(0)}$ and $k^{(1)}$ for a given cipher key.

The final expression shows that the sum of certain triples of state entries at the output of the second round is equal to the application of two S-boxes and subtweakey additions to a single entry of the input to the first round. The second subtweakey addition does not have an important influence on the statistical properties of this expression, so we remove it and turn our attention to the properties of the function

$$\mathrm{D}_{m,k} = \mathrm{S}_m \circ \mathrm{T}_{m,k} \circ \mathrm{S}_m ,$$

where $\mathrm{T}_{m,k}$ is defined by $x \mapsto x + k$ for $x \in \mathbb{F}_2^m$. We will refer to $\mathrm{D}_{m,k}$ as the *double S-box structure.*

For reasons of simplicity, we study SKINNY-64-t, i.e., the version with 4-bit S-boxes. However, our results can be extended to the case of 8-bit S-boxes as well.

By concatenating two copies of the 4-bit S-box circuit with a subtweakey addition layer in between we obtain the circuit-level view of $\mathrm{D}_{4,k}$ that is depicted in Fig. 3. Consider the input x_1. It passes through an XOR gate, the subtweakey addition layer, and finally through a second XOR gate before being routed to the third component of the output of $\mathrm{D}_{4,k}$. If $k_3 = k_2 = 0$, then the XOR gates cancel each other out and the third component of $\mathrm{D}_{4,k}$ is equal to $x_1 + k_0$. This observation does not depend on the value of k_1.

Let us now derive this same result in an algebraic way. Of course, we could compute the algebraic expression for $\mathrm{D}_{4,k}$ directly, but it is more insightful to study the S-box and its inverse.

The 4-bit S-box is of the form

$$\mathrm{S}_4 = \mathrm{N}_4 \circ \mathrm{L}_4 \circ \mathrm{N}_4 \circ \mathrm{L}_4 \circ \mathrm{N}_4 \circ \mathrm{L}_4 \circ \mathrm{N}_4$$

where

$$\begin{aligned} \mathrm{N}_4(x_3, x_2, x_1, x_0) &= (x_3, x_2, x_1, x_2x_3 + x_0 + x_2 + x_3 + 1) \quad \text{and} \\ \mathrm{L}_4(x_3, x_2, x_1, x_0) &= (x_2, x_1, x_0, x_3) . \end{aligned}$$

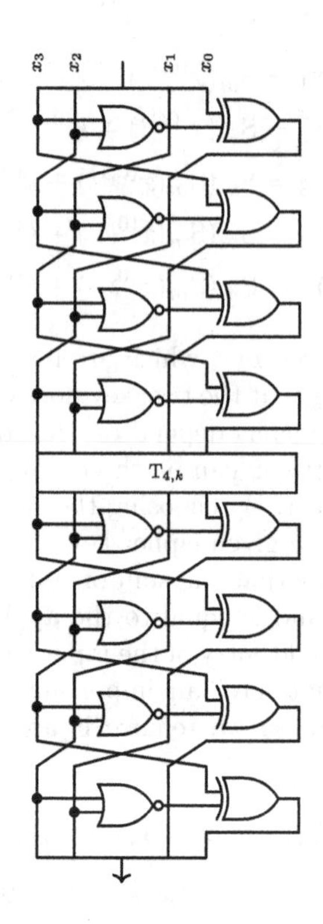

Fig. 3. Circuit-level representation of $\mathrm{D}_{4,k}$. (Figure adapted from [9].)

It follows that $\mathrm{S}_4 = (\mathrm{S}_4^{(3)}, \mathrm{S}_4^{(2)}, \mathrm{S}_4^{(1)}, \mathrm{S}_4^{(0)})$ where

$$\begin{aligned}
\mathrm{S}_4^{(3)} &= x_2x_3 + x_0 + x_2 + x_3 + 1 \\
\mathrm{S}_4^{(2)} &= x_1x_2 + x_1 + x_2 + x_3 + 1 \\
\mathrm{S}_4^{(1)} &= x_1x_2x_3 + x_0x_1 + x_1x_2 + x_1x_3 + x_2x_3 + x_0 + x_3 \\
\mathrm{S}_4^{(0)} &= x_0x_1x_2 + x_1x_2x_3 + x_0x_1 + x_0x_2 + x_0x_3 + x_1x_3 + x_1 + x_2 + x_3
\end{aligned}$$

The S-box has a generalized Feistel structure [13]. Therefore, it is not difficult to deduce that the inverse of $\mathrm{T}_{4,k} \circ \mathrm{S}_4$ is of the form

$$I_{4,k} = (\mathrm{T}_{4,k} \circ \mathrm{S}_4)^{-1} = \mathrm{N}_4 \circ \mathrm{R}_4 \circ \mathrm{N}_4 \circ \mathrm{R}_4 \circ \mathrm{N}_4 \circ \mathrm{R}_4 \circ \mathrm{N}_4 \circ \mathrm{T}_{4,k}\,,$$

where $\mathrm{R}_4(x_3, x_2, x_1, x_0) = (x_0, x_3, x_2, x_1)$. It follows that $I_{4,k}$ is of the form $(I_{4,k}^{(3)}, I_{4,k}^{(2)}, I_{4,k}^{(1)}, I_{4,k}^{(0)})$ where

$$
\begin{aligned}
I_{4,k}^{(3)} &= x_1x_2x_3 + x_0x_1 + x_0x_3 + x_1x_2(k_3+1) + x_1x_3(k_2+1) + x_2x_3k_1 \\
&\quad + x_1(k_2k_3 + k_0 + k_2 + k_3) + x_2(k_1k_3 + k_1 + 1) + x_3(k_1k_2 + k_0 + k_1 + 1) \\
&\quad + x_0(k_1 + k_3) + k_1k_2k_3 + k_0k_1 + k_0k_3 + k_1k_2 + k_1k_3 + k_2 + k_3\,, \\
I_{4,k}^{(2)} &= x_0x_3 + x_2x_3 + x_0(k_3+1) + x_2(k_3+1) + x_3(k_0+k_2) + x_1 + k_0k_3 + k_2k_3 \\
&\quad + k_0 + k_1 + k_2\,, \\
I_{4,k}^{(1)} &= x_2x_3 + x_2(k_3+1) + x_3(k_2+1) + x_0 + k_2k_3 + k_0 + k_2 + k_3 + 1\,, \\
I_{4,k}^{(0)} &= x_0x_2x_3 + x_1x_2x_3 + x_0x_2(k_3+1) + x_0x_3k_2 + x_1x_2k_3 + x_1x_3(k_2+1) \\
&\quad + x_2x_3(k_0+k_1) + x_0x_1 + x_0(k_2k_3 + k_1 + k_2 + 1) + x_1(k_2k_3 + k_0 + k_3 + 1) \\
&\quad + x_2(k_0k_3 + k_1k_3 + k_0 + 1) + x_3(k_0k_2 + k_1k_2 + k_1) + k_0k_2k_3 + k_1k_2k_3 \\
&\quad + k_0k_1 + k_0k_2 + k_1k_3 + k_0 + k_1 + k_2 + 1\,.
\end{aligned}
$$

We observe that if $k_3 = k_2 = 0$, then the component $I_{4,k}^{(1)}$ differs from $\mathrm{S}_4^{(3)}$ by the constant k_0 for any value of k_1. This implies that $\mathrm{D}_{4,(0,0,k_1,k_0)}^{(3)} = x_1 + k_0$.

3 Linear Cryptanalysis

To analyze $\mathrm{D}_{m,k}$ in more detail, we use the statistical framework of linear cryptanalysis [6,12].

The important concept here is a *linear approximation*, i.e., an ordered pair of linear masks $(u, v) \in \mathbb{F}_2^m \times \mathbb{F}_2^m$ that determine linear combinations of output and input bits, respectively. A mask u defines a *linear functional*

$$
x \mapsto u^\top x = u_0x_0 + \cdots + u_{m-1}x_{m-1}\,.
$$

We measure the quality of a linear approximation with the correlation between the linear functionals defined by the masks.

Definition 1. *The (signed)* correlation *between the linear functional defined by the mask $u \in \mathbb{F}_2^m$ at the output of a function $G\colon \mathbb{F}_2^m \to \mathbb{F}_2^m$ and the linear functional defined by the mask $v \in \mathbb{F}_2^m$ at its input is defined as*

$$
\mathrm{C_G}(u, v) = \frac{1}{2^m} \sum_{x \in \mathbb{F}_2^m} (-1)^{u^\top \mathrm{G}(x) + v^\top x}\,.
$$

The $2^m \times 2^m$ matrix $\mathrm{C_G}$ with entries $\mathrm{C_G}(u, v)$ is called the *correlation matrix* of the function G. We call a linear approximation with a correlation of one or minus one *perfect.*

In addition to specifying masks at the input and output of $\mathrm{D}_{m,k}$, we may also specify intermediate masks.

Definition 2. *A sequence $(u, v, w) \in \mathbb{F}_2^m \times \mathbb{F}_2^m \times \mathbb{F}_2^m$ is called a* linear trail *of $\mathrm{D}_{m,k}$ if it satisfies the following conditions:*

1. $\mathrm{C}_{\mathrm{S}_m}(u,v) \neq 0$;
2. $\mathrm{C}_{\mathrm{S}_m}(v,w) \neq 0$.

Each of the trails contributes to the correlation of the linear approximation.

Definition 3. *The* correlation contribution *of a linear trail* (u,v,w) *over* $\mathrm{D}_{m,k}$ *equals*

$$\mathrm{C}_{\mathrm{D}_{m,k}}(u,v,w) = (-1)^{v^\top k}\, \mathrm{C}_{\mathrm{S}_m}(u,v)\, \mathrm{C}_{\mathrm{S}_m}(v,w)\,.$$

From the theory of correlation matrices [6], it follows that

$$\begin{aligned}\mathrm{C}_{\mathrm{D}_{m,k}}(u,v) &= \sum_{v \in \mathbb{F}_2^m} \mathrm{C}_{\mathrm{D}_{m,k}}(u,v,w) \\ &= \sum_{v \in \mathbb{F}_2^m} (-1)^{v^\top k}\, \mathrm{C}_{\mathrm{S}_m}(u,v)\, \mathrm{C}_{\mathrm{S}_m}(v,w)\,.\end{aligned}$$

4 Linear Trails of $\mathbf{S}_m \circ \mathbf{T}_{m,k} \circ \mathbf{S}_m$

We can now translate the observations from Sect. 2 into the language of linear cryptanalysis. The observations state that the linear approximation $(\mathtt{1000}, \mathtt{0010})$ of $\mathrm{D}_{4,(0,0,k_1,k_0)}$ is perfect for all $k_0, k_1 \in \mathbb{F}_2$.

One way of seeing this is directly from the fact that

$$\begin{aligned}(\mathtt{1000})^\top \mathrm{D}_{4,(0,0,k_1,k_0)} &= \mathrm{D}^{(3)}_{4,(0,0,k_1,k_0)} \\ &= x_1 + k_0 \\ &= (\mathtt{0010})^\top x + k_0\,.\end{aligned}$$

Hence, the correlation is one if k_0 is zero and minus one otherwise.

An alternative view is the following. Due to the equivalence of vectorial Boolean functions and their correlation matrices [6], equality of $\mathrm{S}_4^{(3)}$ and $I_{4,k}^{(1)}$ implies equality of row $\mathtt{1000}$ of $\mathrm{C}_{\mathrm{S}_4}$ and row $\mathtt{0010}$ of $\mathrm{C}_{I_{4,k}}$. The latter corresponds to column $\mathtt{0010}$ of $C_{\mathrm{T}_{4,k} \circ \mathrm{S}_4}$. These are exactly the two vectors that we need to multiply in order to compute $\mathrm{C}_{\mathrm{D}_{4,k}}(\mathtt{1000}, \mathtt{0010})$. Using the orthogonality relations [11], it is not difficult to show that this correlation is either one or minus one, depending on the constant difference between $\mathrm{S}_4^{(3)}$ and $I_{4,k}^{(1)}$, which only influences the sign.

In general, we have computed all the non-trivial perfect linear approximations for each of the 2^m subtweakeys. This was accomplished by considering all the possible linear trails over $\mathrm{D}_{4,k}$. The results are found in Table 1 for the case $m = 4$, i.e., for the 4-bit S-box, and in Table 2 for the case $m = 8$, i.e., for the 8-bit S-box. The first column lists the output masks and the third column lists the input masks. An asterisk denotes that the linear approximation holds for any subtweakey bit in that position. It turns out that in both cases such linear approximations exist for a quarter of the subtweakeys. We call subtweakeys for which this property holds *weak*.

Consider a fixed subtweakey. If (u_1, w_1) and (u_2, w_2) are two perfect linear approximations, then their sum $(u_1 + u_2, w_1 + w_2)$ is again a perfect linear approximation, as evidenced by the tables. Moreover, the pair $(0, 0)$ is always a perfect linear approximation. It follows that the perfect linear approximations for a fixed subtweakey form a linear subspace of $\mathbb{F}_2^m \times \mathbb{F}_2^m$.

5 Patching the Problem

To patch the problem, we search within a specific subset of S-boxes that are *permutation equivalent* [5] to the original.

Definition 4. *Two functions* $\mathrm{F}\colon \mathbb{F}_2^m \to \mathbb{F}_2^m$ *and* $\mathrm{G}\colon \mathbb{F}_2^m \to \mathbb{F}_2^m$ *are called* permutation equivalent *if there exist bit permutations* σ *and* τ *such that*

$$\mathrm{F} = \tau \circ \mathrm{G} \circ \sigma\,.$$

A bit permutation τ *is a permutation of* $\{0, \ldots, m-1\}$ *that has been extended to* $\mathbb{F}_2^m$ *by*

$$(x_{m-1}, \ldots, x_0) \mapsto (x_{\tau(m-1)}, \ldots, x_{\tau(0)})\,.$$

Many of the cryptographic properties of an S-box are preserved by permutation equivalence, e.g., the algebraic degree, the differential uniformity, the linearity, and the branch number. Moreover, the impact of a bit permutation on the implementation cost is negligible. For example, in hardware it amounts to rewiring of the signals. We have restricted our search to those permutation equivalent S-boxes for which σ is the identity.

Any bit permutation applied to the output bits of S_4 permutes the columns of its correlation matrix. Indeed, we have

$$\mathrm{C}_{\mathrm{G}}(u, v) = \mathrm{C}_{\mathrm{S}_4}(u, \tau^{-1}(v))\,.$$

Table 3 lists the bit permutations τ and the ratio of subtweakeys for which there exist non-trivial perfect linear approximations. For example, the row "(x_2, x_1, x_0, x_3) 0" corresponds to the bit permutation $\tau = \mathrm{L}_4$ for which no subtweakeys are weak. It turns out that there exist many permutation equivalent S-boxes for which the double S-box structure does not have non-trivial perfect linear approximations for any subtweakey.

Similarly, for the 8-bit S-box we found that there exist many permutation equivalent S-boxes for which there exist no non-trivial perfect linear approximations. An example of such an S-box is obtained by applying the bit permutation $\tau(x_7, x_6, x_5, x_4, x_3, x_2, x_1, x_0) = (x_7, x_5, x_6, x_4, x_3, x_2, x_1, x_0)$. Because the number of possible bit permutations is large, we did not include them all here.

Table 1. Perfect linear approximations of $S_4 \circ T_{4,k} \circ S_4$ and their constituent linear trails.

Output Mask u	Intermediate Mask v	Input Mask w	Subtweakey k	$C_{D_{4,k}}(u,w)$	$C_{T_{4,k}}(v,v)$	$C_S(u,v)$	$C_S(v,w)$
1000	0001	0010	00*k_0	$(-1)^{k_0}$	$(-1)^{k_0}$	1/2	1/2
	0101				$(-1)^{k_0}$	−1/2	−1/2
	1001				$(-1)^{k_0}$	−1/2	−1/2
	1101				$(-1)^{k_0}$	−1/2	−1/2
1010	0001	1110	0001	1	−1	−1/4	1/4
	0011				−1	1/4	−1/4
	0100				1	−1/2	−1/2
	0101				−1	1/4	−1/4
	0110				1	−1/2	−1/2
	0111				−1	−1/4	1/4
	1001				−1	−1/4	1/4
	1011				−1	1/4	−1/4
	1101				−1	−1/4	1/4
	1111				−1	1/4	−1/4
0010	0001	1100	0001	−1	−1	1/4	1/4
	0011				−1	1/4	1/4
	0100				1	1/2	−1/2
	0101				−1	−1/4	−1/4
	0110				1	−1/2	1/2
	0111				−1	−1/4	−1/4
	1001				−1	1/4	1/4
	1011				−1	1/4	1/4
	1101				−1	1/4	1/4
	1111				−1	1/4	1/4
0010	0001	1110	0011	−1	−1	1/4	1/4
	0011				1	1/4	−1/4
	0100				1	1/2	−1/2
	0101				−1	−1/4	−1/4
	0110				−1	−1/2	−1/2
	0111				1	−1/4	1/4
	1001				−1	1/4	1/4
	1011				1	1/4	−1/4
	1101				−1	1/4	1/4
	1111				1	1/4	−1/4
1010	0001	1100	0011	1	−1	−1/4	1/4
	0011				1	1/4	1/4
	0100				1	−1/2	−1/2
	0101				−1	1/4	−1/4
	0110				−1	−1/2	1/2
	0111				1	−1/4	−1/4
	1001				−1	−1/4	1/4
	1011				1	1/4	1/4
	1101				−1	−1/4	1/4
	1111				1	1/4	1/4

Table 2. Perfect linear approximations of $S_8 \circ T_{8,k} \circ S_8$ and their constituent linear trails.

Output Mask u	Intermediate Mask v	Input Mask w	Subtweakey k	$C_{D_{8,k}}(u,w)$	$C_{T_{8,k}}(v,v)$	$C_S(u,v)$	$C_S(v,w)$
01000000	00010000	00001000	$00{*}k_4{*}{*}{*}{*}$	$(-1)^{k_4}$	$(-1)^{k_4}$	1/2	1/2
	01010000				$(-1)^{k_4}$	−1/2	−1/2
	10010000				$(-1)^{k_4}$	−1/2	−1/2
	11010000				$(-1)^{k_4}$	−1/2	−1/2
10010000	00001000	00000010	0001****	−1	1	−1/2	1/2
	00011000				−1	−1/4	−1/4
	00101000				1	1/2	−1/2
	00111000				−1	−1/4	−1/4
	01011000				−1	1/4	1/4
	01111000				−1	1/4	1/4
	10011000				−1	1/4	1/4
	10111000				−1	1/4	1/4
	11011000				−1	1/4	1/4
	11111000				−1	1/4	1/4
11010000	00001000	00001010	0001****	1	1	−1/2	−1/2
	00011000				−1	−1/4	1/4
	00101000				1	−1/2	−1/2
	00111000				−1	1/4	−1/4
	01011000				−1	1/4	−1/4
	01111000				−1	−1/4	1/4
	10011000				−1	1/4	−1/4
	10111000				−1	−1/4	1/4
	11011000				−1	1/4	−1/4
	11111000				−1	−1/4	1/4
10010000	00001000	00001010	0011****	1	1	−1/2	−1/2
	00011000				−1	−1/4	1/4
	00101000				−1	1/2	−1/2
	00111000				1	−1/4	−1/4
	01011000				−1	1/4	−1/4
	01111000				1	1/4	1/4
	10011000				−1	1/4	−1/4
	10111000				1	1/4	1/4
	11011000				−1	1/4	−1/4
	11111000				1	1/4	1/4
11010000	00001000	00000010	0011****	−1	1	−1/2	1/2
	00011000				−1	−1/4	−1/4
	00101000				−1	−1/2	−1/2
	00111000				1	1/4	−1/4
	01011000				−1	1/4	1/4
	01111000				1	−1/4	1/4
	10011000				−1	1/4	1/4
	10111000				1	−1/4	1/4
	11011000				−1	1/4	1/4
	11111000				1	−1/4	1/4

Table 3. Permutation equivalent S-boxes and their ratio of weak subtweakeys.

$\tau(x_3, x_2, x_1, x_0)$	Ratio of weak subtweakeys
(x_3, x_2, x_1, x_0)	4/16
(x_2, x_3, x_1, x_0)	6/16
(x_3, x_1, x_2, x_0)	0
(x_2, x_1, x_3, x_0)	0
(x_1, x_3, x_2, x_0)	0
(x_1, x_2, x_3, x_0)	2/16
(x_3, x_2, x_0, x_1)	0
(x_2, x_3, x_0, x_1)	0
(x_3, x_1, x_0, x_2)	0
(x_2, x_1, x_0, x_3)	0
(x_1, x_3, x_0, x_2)	5/16
(x_1, x_2, x_0, x_3)	0
(x_3, x_0, x_2, x_1)	7/16
(x_2, x_0, x_3, x_1)	0
(x_3, x_0, x_1, x_2)	0
(x_2, x_0, x_1, x_3)	0
(x_1, x_0, x_3, x_2)	6/16
(x_1, x_0, x_2, x_3)	0
(x_0, x_3, x_2, x_1)	10/16
(x_0, x_2, x_3, x_1)	8/16
(x_0, x_3, x_1, x_2)	0
(x_0, x_2, x_1, x_3)	0
(x_0, x_1, x_3, x_2)	0
(x_0, x_1, x_2, x_3)	0

6 Conclusion

The main message that we want to communicate is that the composition of individually strong cryptographic functions may produce a weaker function for a large subset of the round tweakey space. In SKINNY, this weakness holds for *any* cipher key, because the subtweakeys are computed from the both the cipher key and the tweak, the latter of which is chosen by the user. In small structures, such undesired properties can be practically revealed through a combination of algebraic and statistical analysis. This shows that counting the number of active S-boxes in trails may have little meaning. Such properties could have been avoided by moving to a slightly different function at a negligible implementation cost.

We did not expect this kind of problem to exist for the 8-bit version of the SKINNY S-box. However, like the 4-bit S-box, in the composition of the two 8-bit S-boxes, the first stage of the second S-box and the final stage of the first S-box are the same, leading to cancellation. If the matrix that is used in the MixColumns step did not have a row with a single one, then this double S-box structure would not exist. As a result, this particular problem would not be there.

Acknowledgements. Joan Daemen and Daniël Kuijsters are supported by the European Research Council under the ERC advanced grant agreement under grant ERC-2017-ADG Nr. 788980 ESCADA.

References

1. 27, I.J.S.: Information Security "Encryption Algorithms" Part 7: Tweakable Block Ciphers, 1st edn. International Organization for Standardization, Vernier, Geneva, Switzerland (2022). https://www.iso.org/standard/80505.html
2. Beierle, C., et al.: The SKINNY family of block ciphers and its low-latency Variant MANTIS. In: Robshaw, M., Katz, J. (eds.) CRYPTO 2016. LNCS, vol. 9815, pp. 123–153. Springer, Heidelberg (2016). https://doi.org/10.1007/978-3-662-53008-5_5
3. Biham, E., Shamir, A.: Differential cryptanalysis of des-like cryptosystems. In: CRYPTO 1990 (1990). https://doi.org/10.1007/3-540-38424-3_1
4. Bordes, N., Daemen, J., Kuijsters, D., Assche, G.V.: Thinking outside the superbox. In: Malkin, T., Peikert, C. (eds.) Advances in Cryptology - CRYPTO 2021–41st Annual International Cryptology Conference, CRYPTO 2021, Virtual Event, 16–20 August 2021, Proceedings, Part III. LNCS, vol. 12827, pp. 337–367. Springer, Cham (2021). https://doi.org/10.1007/978-3-030-84252-9_12
5. Carlet, C.: Boolean Functions for Cryptography and Coding Theory. Cambridge University Press, Cambridge (2021)
6. Daemen, J.: Cipher and hash function design, strategies based on linear and differential cryptanalysis, Ph.D. Thesis. K.U.Leuven (1995)
7. Daemen, J., Rijmen, V.: The Design of Rijndael - The Advanced Encryption Standard (AES), Information Security and Cryptography. 2nd edn. Springer, Heidelberg (2020). https://doi.org/10.1007/978-3-662-04722-4
8. Iwata, T., Khairallah, M., Minematsu, K., Peyrin, T.: Duel of the titans: The romulus and remus families of lightweight AEAD algorithms. IACR Trans. Symmet. Cryptol. **2020**(1), 43–120 (2020), https://doi.org/10.13154/tosc.v2020.i1.43-120
9. Jean, J.: TikZ for Cryptographers(2016). https://www.iacr.org/authors/tikz/
10. Jean, J., Nikolić, I., Peyrin, T.: Tweaks and keys for block ciphers: The TWEAKEY framework. In: Sarkar, P., Iwata, T. (eds.) ASIACRYPT 2014. LNCS, vol. 8874, pp. 274–288. Springer, Heidelberg (2014). https://doi.org/10.1007/978-3-662-45608-8_15
11. Lidl, R., Niederreiter, H.: Finite fields, Encyclopedia of Mathematics and its Applications, vol. 20, 2nd edn. Cambridge University Press, Cambridge (1997)
12. Matsui, M.: Linear cryptanalysis method for DES cipher. In: Helleseth, T. (ed.) Proceedings of Advances in Cryptology - EUROCRYPT 1999 (1993)

13. Nyberg, K.: Generalized feistel networks. In: Kim, K., Matsumoto, T. (eds.) ASIACRYPT 1996. LNCS, vol. 1163, pp. 91–104. Springer, Heidelberg (1996). https://doi.org/10.1007/BFb0034838
14. Turan, M.S., et al.: Status report on the second round of the NIST lightweight cryptography standardization process (2021–07-20 04:07:00 2021). https://tsapps.nist.gov/publication/get_pdf.cfm?pub_id=932630
15. Verbakel, D.: Influence of design on differential and linear propagation properties of block cipher family skinny. Bachelor's thesis, Radboud University, Nijmegen, The Netherlands (2021)

Full Round Zero-Sum Distinguishers on TinyJAMBU-128 and TinyJAMBU-192 Keyed-Permutation in the Known-Key Setting

Orr Dunkelman, Shibam Ghosh(✉), and Eran Lambooij

Department of Computer Science, University of Haifa, Haifa, Israel
orrd@cs.haifa.ac.il, sghosh03@campus.haifa.ac.il, eran@hideinplainsight.io

Abstract. TinyJAMBU is one of the finalists in the NIST lightweight standardization competition. This paper presents full round practical zero-sum distinguishers on the keyed permutation used in TinyJAMBU.

We propose a full round zero-sum distinguisher on the 128- and 192-bit key variants and a reduced round zero-sum distinguisher for the 256-bit key variant in the known-key settings. Our best known-key distinguisher works with 2^{16} data/time complexity on the full 128-bit version and with 2^{23} data/time complexity on the full 192-bit version. For the 256-bit version, we can distinguish 1152 rounds (out of 1280 rounds) in the known-key settings. In addition, we present the best zero-sum distinguishers in the secret-key settings: with complexity 2^{23} we can distinguish 544 rounds in the forward direction or 576 rounds in the backward direction.

For finding the zero-sum distinguisher, we bound the algebraic degree of the TinyJAMBU permutation using the monomial prediction technique proposed by Hu *et al.* at ASIACRYPT 2020. We model the monomial prediction rule on TinyJAMBU in MILP and find upper bounds on the degree by computing the parity of the number of solutions.

1 Introduction

Lightweight cryptographic primitives are essential for providing security for highly resource-constrained devices that transmit sensitive information. Thus, recent years have seen a substantial increase in the development of lightweight symmetric cryptographic primitives. As a response, NIST started the lightweight cryptography competition [25] in 2018. The competition aims to standardize lightweight authenticated encryption algorithms and lightweight hash functions. In 2021, NIST announced ten finalists out of the initial 56 candidates.

TinyJAMBU [32,33], proposed by Wu *et al.*, is one of the finalists of the NIST lightweight competition. The design principle of TinyJAMBU follows the sponge duplex mode using a keyed permutation. The core component of TinyJAMBU is a keyed permutation derived from a lightweight NFSR that contains a single NAND gate as the non-linear component.

T. Isobe and S. Sarkar (Eds.): INDOCRYPT 2022, LNCS 13774, pp. 349–372, 2022.
https://doi.org/10.1007/978-3-031-22912-1_16

TinyJAMBU uses two different keyed permutations with the same key K. These are similar in structure but differ only in the number of rounds. We denote these two permutations $\mathcal{P}^a$ and $\mathcal{P}^b$. The permutation $\mathcal{P}^a$ is used in steps where no output is observed. $\mathcal{P}^b$ is used in the first initialization step and when an internal state value is partially leaked. In the original submission of TinyJAMBU, the round number of $\mathcal{P}^a$ was 384 for all variants. The round number of $\mathcal{P}^b$ was 1024, 1152, and 1280 for 128-, 192-, and 256-bit key variants, respectively.

1.1 Existing Analysis on TinyJAMBU Permutation

The designers of TinyJAMBU [32,33] provide a rigorous security analysis of the underlying permutation against various attacks. From the differential and linear attack perspectives, the designers [32] count the least number of active NAND gates to claim security against those attacks. Later it was improved by Saha *et al.* in [23] using (first order) correlated NAND gates. The authors in [23] propose differential characteristics through 338 and 384 rounds of the keyed permutation with probabilities of $2^{-62.68}$ and $2^{-70.68}$, respectively. These attacks lead to a forgery attack in the TinyJAMBU mode. In response to this result, the designers of TinyJAMBU increased the number of rounds of $\mathcal{P}^a$ from 384 to 640 [33]. More recently Dunkelman *et al.* reported related-key forgery attacks on the full TinyJAMBU-192 and TinyJAMBU-256 (after the tweak) [11].

Sibleyras *et al.* [24] discuss slide attack on the full round TinyJAMBU permutations. These reported attacks are key recovery attacks on TinyJAMBU permutation with data/time complexities of $2^{65}, 2^{66}$, and $2^{69.5}$ for 128-, 192-, and 256-bit key variants, respectively.

Regarding algebraic attacks, the designers of TinyJAMBU [32,33] claim that all the ciphertext bits are affected by the input bits after 598 rounds. This statement was supported by a recent work by Dutta *et al.* [12], where the authors bound the degree using the monomial prediction technique [17] and showed that TinyJAMBU is secure against 32-sized cube attacks after 445 rounds. In addition, they propose cube distinguishers in the weak-key setting for 451 rounds and 476 rounds of TinyJAMBU permutation and the size of the weak-key set is 2^{108}. Another cryptanalysis from an algebraic perspective was reported recently by Teng *et al.* [26], where the authors propose cube attacks against TinyJAMBU. The authors propose basic cube distinguisher on 438 rounds TinyJAMBU by considering TinyJAMBU mode as a black-box.

1.2 Our Contributions

This paper aims to study the algebraic properties of TinyJAMBU permutation as a standalone primitive. Notably, we construct zero-sum [2,3,5] distinguishers that distinguish the TinyJAMBU keyed permutation in practical complexity. Zero-sum distinguishers [2,3,5] can be used to distinguish permutations from a random permutation by suggesting a subset of inputs whose corresponding outputs are summed to zero. One way to do so is to bound the algebraic degree of the output bits, as a bit of degree d is balanced over any cube of degree $d+1$.

The most precise technique for upper bounding the degree of a Boolean function is the *Monomial prediction technique*, proposed at ASIACRYPT 2020 [17]. The *Monomial prediction technique* recursively predicts the existence of a monomial in the polynomial representation of the output bits. One can model the rules of monomial prediction using automatic tools like MILP [17] or CP/SAT [14].

We use zero-sum distinguishers both in the known-key [20] and the secret-key settings [19,21]. We show that one can bound the degree of an output bit after 544 rounds of the TinyJAMBU permutation by 22 by carefully choosing the cube variables. Furthermore, we show that the degree of the inverse TinyJAMBU permutation increases slower than in the forward direction. Using this, we show that in the backwards direction, a bit can be represented as a degree 22 polynomial after 576 rounds. Combining these results, we then show full round zero-sum distinguishers on 128- and 192-bit key versions and reduced round distinguisher on 256-bit key version in the known key settings. Our best zero-sum distinguishers are summarized in Table 1, which are the best-known algebraic distinguishers on TinyJAMBU permutation till date.

Table 1. Zero-sum (ZS) distinguishers on the full round TinyJAMBU permutation.

Key size	#Rounds	Complexity	Model	Type	Section
All	480(/1024)	2^{16}	Secret Key	ZS	5
All	544(/1024)	2^{23}	Secret Key	ZS	5.1
128	1024	2^{16}	Known Key	ZS	5.1
192	1152	2^{23}	Known Key	ZS	5.2
256	1152(/1280)	2^{23}	Known Key	ZS	5.2

1.3 Paper Structure

In Sect. 2, we define the notation for the paper as well as giving a short introduction of the monomial prediction technique. We give a short overview of the TinyJAMBU mode and keyed primitive in Sect. 3. Then, in Sect. 4, we show how to build a zero-sum distinguisher on the full keyed permutation of TinyJAMBU-128 and TinyJAMBU-192. In Sect. 5 we show how improve data/time complexity of the zero-sum distinguisher on the full keyed permutation of TinyJAMBU-128 and TinyJAMBU-192 Finally, we conclude the paper in Sect. 6.

2 Preliminaries

2.1 Notations

The size of a set $\mathbb{S}$ is denoted as $||\mathbb{S}||$. The Hamming weight of $a \in \mathbb{F}_2^n$ is defined as $wt(a) = \sum_{i=1}^{i=n} a_i$. We use bold lowercase letters to represent vectors in a binary field. For any n-bit vector $\mathbf{s} \in \mathbb{F}_2^n$, its i-th coordinate is denoted by s_i, thus we have $\mathbf{s} = (s_{n-1}, ..., s_0)$. $\mathbf{0}$ represents the binary vector with all elements being 0. For any vector $\mathbf{u} \in \mathbb{F}_2^n$ and $\mathbf{x} \in \mathbb{F}_2^n$, we define the *bit product* as $\mathbf{x}^{\mathbf{u}} = \prod_{i=1}^{n} {x_i}^{u_i}$.

Let $Y \subseteq \mathbb{F}_2^n$ be a multi-set of vectors. A coordinate position $0 \leq i < n$ is called a *balanced position* if $\bigoplus_{\mathbf{y} \in Y} y_i = 0$.

2.2 Boolean Functions and Upper Bounds on the Degree

An n-variable Boolean function f is a function from $\mathbb{F}_2^n$ to $\mathbb{F}_2$. If f is a Boolean function then there exists a unique multivariate polynomial in

$$\mathbb{F}_2[x_0, x_1, ..., x_{n-1}]/(x_0^2 + x_0, x_1^2 + x_1, ..., x_{n-1}^2 + x_{n-1})$$

such that

$$f(x_0, x_1, ..., x_{n-1}) = \bigoplus_{\mathbf{u} \in \mathbb{F}_2^n} a_{\mathbf{u}}^f \mathbf{x}^{\mathbf{u}}.$$

This multivariate polynomial is called the algebraic normal form (ANF) of f. In this paper, we primarily look at the algebraic degree of a Boolean function. The definition of the algebraic degree of a Boolean function is as follows

Definition 1. *The algebraic degree of a function $f : \mathbb{F}_2^n \rightarrow \mathbb{F}_2$ is d if d is the degree of the monomial with the highest algebraic degree in the ANF of f, i.e.,*

$$d = \max_{\mathbf{u} \in \mathbb{F}_2^n, a_{\mathbf{u}}^f \neq 0} wt(\mathbf{u}).$$

Many well-known attacks such as integral attacks [18], higher-order differential attacks [21], cube attacks [9], and zero sum distinguishers [4] exploit a low degree of a Boolean function. Hence, the problem of finding the degree of a cryptographic function is important in cryptanalysis. Numerous methods for computing (or bounding) the degree of cryptographic Boolean functions have been proposed in the literature. A study on various degree evaluation methods can be found in [7].

Canteaut *et al.* proposed a method for upper bounding the algebraic degree of composite functions at EUROCRYPT 2002 [6], which was improved by Boura et al. [3] at FSE 2011 with applications to the Keccak-f function.

In CRYPTO 2017, Liu [22] proposed the *numeric mapping* technique, which is an approach for upper bounding the algebraic degree of a non-linear feedback shift register based stream ciphers. The main idea of this technique is to estimate the degree of a monomial by computing the sum of degrees of all the variables contained in this monomial. Thus, one can use the degree of previous states to estimate the degree of the current state. Using the *numeric mapping* technique, the author found several zero-sum distinguishers for round-reduced Trivium, Kreyvium, and TriviA-SC.

Another approach for the degree evaluation is based on the division property which was proposed as a generalization of the integral property by Todo at Eurocrypt 2015 [27]. It is used in [28] to offer the first attack on the full

MISTY1 cipher. The division property proposed by Todo is word-based, i.e., the propagation of the division property captures information only at the word level.

In FSE 2016, Todo and Morii introduced the bit-based division property [29]. However, the accuracy of this approach is determined by the accuracy of the "propagation rules" of the underlying detection algorithms for division properties. The *monomial prediction* technique, proposed at ASIACRYPT 2020 [17], offers an exact method to do so. We use the monomial prediction technique in this paper to find an upper bound on the degree of a Boolean function. In the following subsection we briefly discuss how to use it to compute the algebraic degree.

2.3 Monomial Prediction

Finding algebraic properties, such as the degree of the vectorial Boolean functions corresponding to cryptographic primitives, is usually very hard regarding computational complexity. One way to determine various algebraic properties of a Boolean function is to determine the exact algebraic structure, i.e., predict the presence of particular monomials in its ANF. Monomial prediction technique [17] is an indirect way to determine the presence of a particular monomial in the bit product function of the output bits. In this section, we discuss an overview of monomial prediction technique and how to find the degree of a Boolean function with this technique.

Let $f : \mathbb{F}_2^n \rightarrow \mathbb{F}_2^m$ be a vectorial Boolean function that maps $\mathbf{x} = (x_{n-1}, ..., x_0) \in \mathbb{F}_2^n$ to $\mathbf{y} = (y_{m-1}, ..., y_0)$ with $y_i = f_i(\mathbf{x})$, where $f_i : \mathbb{F}_2^n \rightarrow \mathbb{F}_2$ is a Boolean function called the i-th coordinate of f. The monomial prediction problem is to identify if the monomial $\mathbf{x}^{\mathbf{u}}$ is present or absent as a monomial in the polynomial representation of $\mathbf{y}^{\mathbf{v}}$ for some $\mathbf{u} \in \mathbb{F}_2^n$ and $\mathbf{v} \in \mathbb{F}_2^m$. We denote that the monomial $\mathbf{x}^{\mathbf{u}}$ is present in $\mathbf{y}^{\mathbf{v}}$ by $\mathbf{x}^{\mathbf{u}} \rightarrow \mathbf{y}^{\mathbf{v}}$, likewise, $\mathbf{x}^{\mathbf{u}} \nrightarrow \mathbf{y}^{\mathbf{v}}$ denotes that $\mathbf{x}^{\mathbf{u}}$ is not present in $\mathbf{y}^{\mathbf{v}}$.

As most of the Boolean functions that appear in symmetric cryptographic primitives are built as a composition of a sequence of vectorial Boolean functions, the authors of [17] proposed a recursive prediction model. Naturally, the function f can be written as a composition of round functions:

$$\mathbf{y} = f(\mathbf{x}) = f^{(r-1)} \circ f^{(r-2)} \circ \cdots \circ f^{(0)}(\mathbf{x})$$

and $\mathbf{x}^{(i+1)} = f^{(i)}(\mathbf{x}^{(i)})$ for all $0 \leq i < r$. We can represent $\mathbf{x}^{(i)}$ as a function of $\mathbf{x}^{(j)}$ for $j < i$ as $\mathbf{x}^{(i)} = f^{(i-1)} \circ \cdots \circ f^{(j)}(\mathbf{x}^{(j)})$.

Definition 2. *(Monomial Trail [17]) Let* $\mathbf{x}^{(i+1)} = f^{(i)}(\mathbf{x}^{(i)})$ *for* $0 \leq i < r$. *We call a sequence of monomials* $((\mathbf{x}^{(0)})^{\mathbf{u}^{(0)}}, (\mathbf{x}^{(1)})^{\mathbf{u}^{(1)}}, ..., (\mathbf{x}^{(r)})^{\mathbf{u}^{(r)}})$ *an* r*-round* ***monomial trail*** *connecting* $(\mathbf{x}^{(0)})^{\mathbf{u}^{(0)}}$ *to* $(\mathbf{x}^{(r)})^{\mathbf{u}^{(r)}}$ *with respect to the composite function* $f(\mathbf{x}) = f^{(r-1)} \circ f^{(r-2)} \circ \cdots \circ f^{(0)}(\mathbf{x})$ *if*

$$(\mathbf{x}^{(0)})^{\mathbf{u}^{(0)}} \rightarrow (\mathbf{x}^{(1)})^{\mathbf{u}^{(1)}} \rightarrow \cdots \rightarrow (\mathbf{x}^{(i)})^{\mathbf{u}^{(i)}} \rightarrow \cdots \rightarrow (\mathbf{x}^{(r)})^{\mathbf{u}^{(r)}}$$

If there is at least one monomial trail connecting $(\mathbf{x}^{(0)})^{\mathbf{u}^{(0)}}$ *to* $(\mathbf{x}^{(r)})^{\mathbf{u}^{(r)}}$*, we denote it as* $(\mathbf{x}^{(0)})^{\mathbf{u}^{(0)}} \rightsquigarrow (\mathbf{x}^{(r)})^{\mathbf{u}^{(r)}}$.

It is important to note that there can be multiple monomial trails connecting $(\mathbf{x}^{(0)})^{\mathbf{u}^{(0)}}$ to $(\mathbf{x}^{(r)})^{\mathbf{u}^{(r)}}$. This leads to the definition of a *monomial hull* which is the set of all trails connecting $(\mathbf{x}^{(0)})^{\mathbf{u}^{(0)}}$ and $(\mathbf{x}^{(r)})^{\mathbf{u}^{(r)}}$, which is denoted as $\{(\mathbf{x}^{(0)})^{\mathbf{u}^{(0)}} \nrightarrow (\mathbf{x}^{(r)})^{\mathbf{u}^{(r)}}\}$. The size of a *monomial hull* can be found recursively as given in the following Lemma 1 from [17].

Lemma 1. *([17]) For $r \geq 1$, $(\mathbf{x}^{(0)})^{\mathbf{u}^{(0)}} \rightsquigarrow (\mathbf{x}^{(\mathbf{r})})^{\mathbf{u}^{(r)}}$, then*

$$\|\{(\mathbf{x}^{(0)})^{\mathbf{u}^{(0)}} \nrightarrow (\mathbf{x}^{(\mathbf{r})})^{\mathbf{u}^{(r)}}\}\| = \begin{cases} 1, & \text{if } r = 1 \\ \displaystyle\sum_{(\mathbf{x}^{(r-1)})^{\mathbf{u}^{(r-1)}} \rightarrow (\mathbf{x}^{(r)})^{\mathbf{u}^{(r)}}} \|\{(\mathbf{x}^{(0)})^{\mathbf{u}^{(0)}} \nrightarrow (\mathbf{x}^{(r-1)})^{\mathbf{u}^{(r-1)}}\}\|, & \text{otherwise} \end{cases}$$

It is easy to observe that the presence or absence of a monomial in $(\mathbf{x}^{(r)})^{\mathbf{u}^{(r)}}$, depends on the parity of the size of the *monomial hull*, which is captured in Proposition 1.

Proposition 1. *([17]) $(\mathbf{x}^{(0)})^{\mathbf{u}^{(0)}} \rightarrow (\mathbf{x}^{(r)})^{\mathbf{u}^{(r)}}$ if and only if $\|\{(\mathbf{x}^{(0)})^{\mathbf{u}^{(0)}} \nrightarrow (\mathbf{x}^{(r)})^{\mathbf{u}^{(r)}}\}\|$ is odd.*

To conclude, the problem of finding the degree of a polynomial is reduced to finding the parity of the size of the monomial hull of the monomial with the highest degree.

2.4 Computing the Algebraic Degree Using Monomial Prediction

To determine the algebraic degree d, of a Boolean function f, one can use the *monomial prediction technique* [17]. To compute an upper bound on the degree we only need to prove the existence of a monomial $\mathbf{x}^{\mathbf{u}}$ with $wt(\mathbf{u}) = d$ such that $\mathbf{x}^{\mathbf{u}} \rightarrow f$ and $\mathbf{x}^{\mathbf{u}'} \nrightarrow f$ for all $\mathbf{u}'$ with $wt(\mathbf{u}') > wt(\mathbf{u})$.

To automate this task, the authors in [17] proposed a Mixed Integer Linear Programming (MILP) approach. The core idea is to model the monomial trails of a function with linear inequalities so that only the valid trails satisfy the system. For more information on the MILP modeling, we refer to [15–17].

To find the existence of a specific monomial, one can choose a monomial (i.e., the exponent vector $\mathbf{u}$) and check for the feasibility of the MILP model. On the other hand, one can find the maximum degree by setting the objective function of the MILP model to maximize $wt(\mathbf{u})$ according to $\max_{\mathbf{x}^{\mathbf{u}} \rightarrow f} wt(\mathbf{u})$. Finally, to confirm the presence of a monomial, we need to check if the number of solutions is even or odd. To do this, *PoolSearchMode* of the Gurobi[1] solver is used in [17]. The *PoolSearchMode* is implemented by Gurobi solver to systematically search for multiple solutions.

[1] https://www.gurobi.com.

3 The Specification of TinyJAMBU

The TinyJAMBU [32] family of authenticated encryption with associated data (AEAD), is a small variant of JAMBU [31]. It is one of the finalists of the NIST lightweight competition. The design principle of TinyJAMBU is based on the sponge duplex mode using keyed permutations derived from a lightweight NLFSR.

In this paper we call the internal keyed permutation the TinyJAMBU permutation. The round function of the TinyJAMBU permutation is defined as follows: $\mathcal{P}_k : \mathbb{F}_2^{128} \to \mathbb{F}_2^{128}$ such that for a key bit $k \in \mathbb{F}_2$ and input $\mathbf{x} = (x_{127}, ..., x_0)$, $\mathcal{P}_k(\mathbf{x}) = (z_{127}, ..., z_0)$ where

$$\begin{cases} z_{127} = 1 \oplus k \oplus x_{91} \oplus x_{85}x_{70} \oplus x_{47} \oplus x_0 \\ z_i = x_{i+1}, \ \text{for } 0 \le i \le 126. \end{cases} \tag{1}$$

We denote the r-round permutation by $\mathcal{P}_{\mathsf{K}}^r$. Given a key $\mathsf{K} = (k_{\kappa-1}, ..., k_0)$ of size κ and a 128-bit input, $\mathcal{P}_{\mathsf{K}}^r$ outputs a 128-bit state by calling the round function r times as follows

$$\mathcal{P}_{\mathsf{K}}^r(\mathbf{x}) = \mathcal{P}_{k_{\kappa-1}} \cdots \circ \mathcal{P}_{k_1} \circ \mathcal{P}_{k_0}(\mathbf{x}),$$

where each subscript i of the key bits are computed as $k_i = k_{i (\mathsf{mod}\ \kappa)}$. The round function of the TinyJAMBU permutation is depicted in Fig. 1.

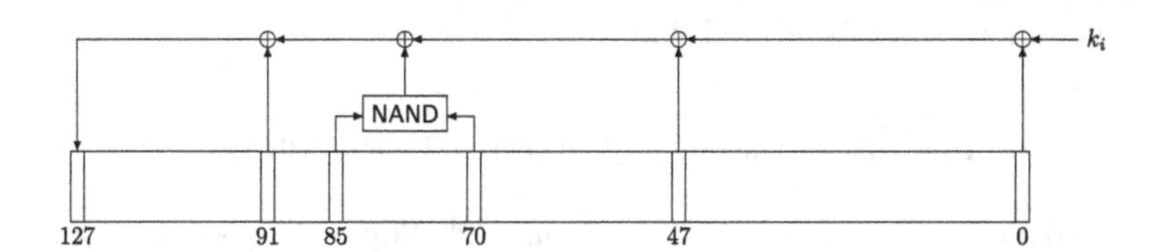

Fig. 1. Round function of the TinyJAMBU permutation.

We define the inverse of $\mathcal{P}_{\mathsf{K}}^r$ as $\mathcal{P}_{\mathsf{K}}^{-r}$, such that for any K and $\mathbf{x}$, $\mathcal{P}_{\mathsf{K}}^{-r}(\mathcal{P}_{\mathsf{K}}^r)(\mathbf{x}) = \mathbf{x}$. We follow the convention that the input to the $\mathcal{P}_{\mathsf{K}}^{-r}$ is rotated, i.e., the i-th input bit becomes $127 - i$ for all $0 \le i < 128$. The round function of the TinyJAMBU inverse permutation is defined as follows: $\mathcal{P}_k^{-1} : \mathbb{F}_2^{128} \to \mathbb{F}_2^{128}$ such that for a key bit $k \in \mathbb{F}_2$ and input $\mathbf{z} = (z_{127}, ..., z_0)$, $\mathcal{P}_k(\mathbf{z}) = (x_{127}, ..., x_0)$ where

$$\begin{cases} x_{127} = 1 \oplus k \oplus x_{37} \oplus x_{43}x_{58} \oplus x_{81} \oplus x_0 \\ z_i = x_{i+1}, \ \text{for } 0 \le i \le 126. \end{cases} \tag{2}$$

Given a key $\mathsf{K} = (k_{\kappa-1}, ..., k_0)$, the TinyJAMBU inverse function is computed as follows

$$\mathcal{P}_{\mathsf{K}}^{-r}(\mathbf{x}) = \mathcal{P}_{k_{\kappa-1}}^{-1} \cdots \circ \mathcal{P}_{k_1}^{-1} \circ \mathcal{P}_{k_0}^{-1}(\mathbf{x}),$$

where each subscript i of the key bits are computed as $k_i = k_{(\kappa-i)(\mathsf{mod}\ \kappa)}$. The round function of the inverse TinyJAMBU permutation is depicted in Fig. 2.

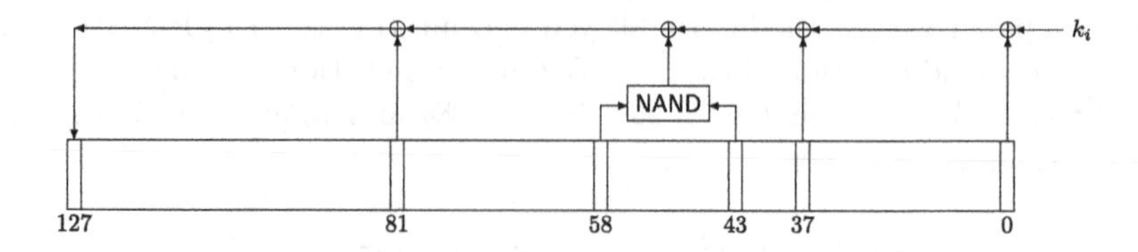

Fig. 2. Round function of the inverse TinyJAMBU permutation.

TinyJAMBU has a 32-bit message injection part (rate), a 32-bit squeezing part, and a 96-bit unaltered capacity part. The capacity part of the state is XORed with the 3-bit frame constants denoted by const_i. TinyJAMBU uses two different keyed permutations using the same key K in different phases of the encryption. We denote them as $\mathcal{P}^a$ and $\mathcal{P}^b$. The only differene between these two permutation is the number of rounds.

TinyJAMBU has three variants based on key size used in the permutation, whose parameters are listed in Table 2.

Table 2. Parameters of the tweaked TinyJAMBU [33].

Variant	#Rounds $\mathcal{P}^a$	#Rounds $\mathcal{P}^b$	State	Key	Nonce	Tag
TinyJAMBU-128	640	1024	128	128	96	64
TinyJAMBU-192	640	1152	128	192	96	64
TinyJAMBU-256	640	1280	128	256	96	64

We denote internal state as s. The encryption algorithm of TinyJAMBU can be divided into the following four phases.

Initialization. In this step $\mathcal{P}^b$ is applied to the 128-bit initial state $s = (0, 0, ..., 0)$ to inject the key into the state. After that, in the nonce setup phase, a 96-bit nonce N is split up into three 32-bit nonce parts $N_0 \| N_1 \| N_2$ and for each part of the nonce the state is updated with $\mathcal{P}^a$ after which the nonce is added to the most significant part of the state. A depiction of the initialization is given in Fig. 3.

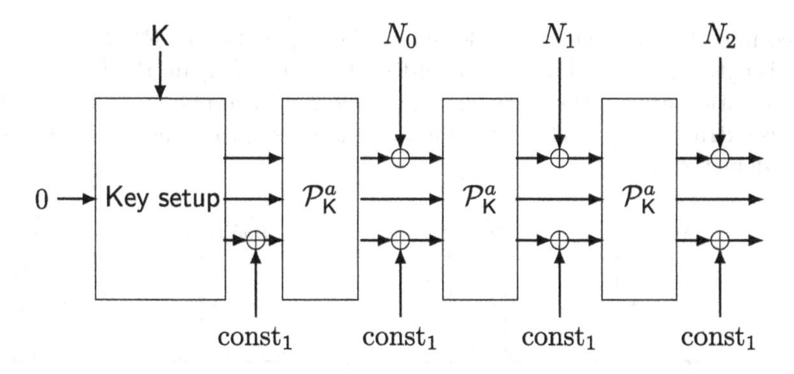

Fig. 3. TinyJAMBU initialization.

Associated Data Processing. The associated data is divided into 32-bit blocks. For each block, the 3-bit frame constant of the associated data phase is XORed with the state and then the state is updated with $\mathcal{P}^a$, after which the 32 bits of the associated data part is XORed with the state. The schematic diagram associated data processing and finalization step is depicted in Fig. 4.

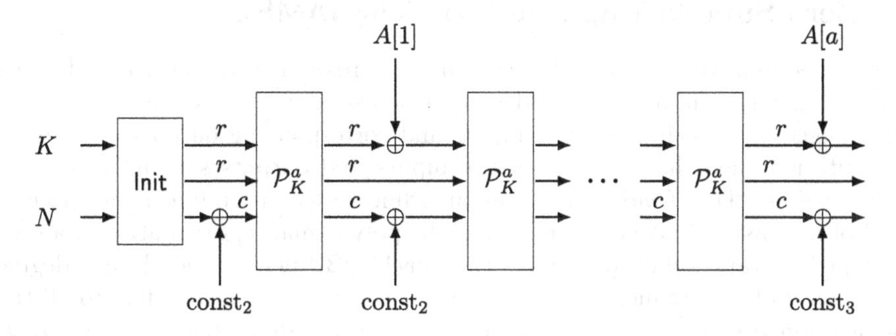

Fig. 4. TinyJAMBU Authenticated Encryption for a blocks of associated data, and m blocks of message.

Encryption. The plaintext is divided into 32-bit blocks. For each block, the frame bits for encryption are XORed into the state. Then the state is updated with $\mathcal{P}^b$, after which the plaintext block is XORed into the most significant part of the state. Finally, we obtain the 32-bit ciphertext block by XORing bits $95\ldots64$ of the state with the plaintext block. Note that, the plaintext and nonce are added to the 32 most significant bits of the state which are $127\ldots96$ and the key stream used for encryption, is obtained from bits $95\ldots64$.

Finalization. After encrypting the plaintext, the 64-bit authentication tag $T_0\|T_1$ is generated in two steps. First, the frame bits for the finalization are

XORed into the state which is followed by the application of $\mathcal{P}^a$ after which 32-bit T_0 is extracted from bit 95...64 of the state. Then, again the frame bits for the finalization are XORed with the state followed by application of $\mathcal{P}^a$ and 32-bit T_1 is extracted. The schematic diagram plaintext processing and finalization step is depicted in Fig. 5.

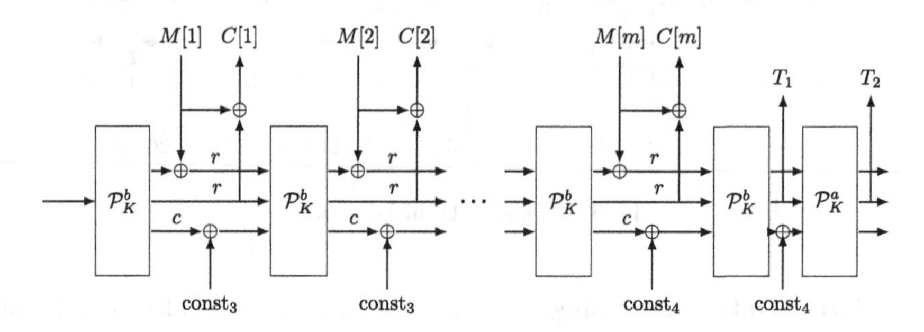

Fig. 5. TinyJAMBU Authenticated Encryption for a blocks of associated data, and m blocks of message.

4 Zero-Sum Distinguishers on TinyJAMBU

In this section, we show how to construct a zero-sum distinguisher [2] for the TinyJAMBU permutation. The idea of this distinguisher, as the name 'zero-sum' suggests, is to find a set of inputs and corresponding outputs of an n-bit permutation such that the bits in the inputs and outputs sum to 0 over $\mathbb{F}_2$. In the case of TinyJAMBU we choose an affine vector space ν as a subspace of $\mathbb{F}_2^{128}$ of dimension d. We then show that the polynomial representation of all (or some) of the targeted output bits of the TinyJAMBU permutation have a degree less than d after r rounds. In such a case, the outputs corresponding to all the elements in ν sum to 0 for the targeted bits as the dimension is greater than the degree. In other words, if the algebraic degree of a Boolean function f is less than d, there exists an affine vector space ν of dimension at least d for which $\bigoplus_{\mathbf{x}\in\nu} f(\mathbf{x}) = 0$. The time and data complexity of the attack is $O(2^d)$ and the memory complexity is $O(1)$.

The input set ν usually consists of inputs taking all possible combinations in d input bits and the remaining bits take a fixed value. Thus the input set forms an affine vector space of dimension d. So the resulting output sets are the d-th derivative of the corresponding vectorial Boolean function with respect to ν. Recall that this idea was first proposed as the higher order differential attack [19,21].

In the following we discuss various zero-sum distinguishers on the TinyJAMBU permutation. The basis for all the distinguishers is finding an upper bound on the degree of the polynomial representation of input/output bits. This is done using a MILP model of the monomial prediction trail of the TinyJAMBU permutation.

4.1 MILP Modeling

Let us consider a function $f : \mathbb{F}_2^m \rightarrow \mathbb{F}_2^n$ such that $\mathbf{y} = f(\mathbf{x})$. Every pair of $(\mathbf{u}, \mathbf{v})$ is a valid monomial trail through f if and only if $\mathbf{x}^{\mathbf{u}} \rightarrow \mathbf{y}^{\mathbf{v}}$. The main motivation of MILP modeling is that it is easy to test validity of a monomial trail using the MILP. Let us consider the following monomial trail

$$(\mathbf{x}^{(0)})^{\mathbf{u}^{(0)}} \rightarrow (\mathbf{x}^{(1)})^{\mathbf{u}^{(1)}} \rightarrow \cdots \rightarrow (\mathbf{x}^{(i)})^{\mathbf{u}^{(i)}} \rightarrow \cdots \rightarrow (\mathbf{x}^{(r)})^{\mathbf{u}^{(r)}}$$

We consider the transition of the exponents $(\mathbf{u}^{(0)}), \mathbf{u}^{(1)}), ..., \mathbf{u}^{(r)}))$ through the round functions and construct a MILP model by modeling the propagation of monomial trails. Any function can be decomposed into smaller operations, namely, COPY, XOR, AND and NOT. We recall the MILP model from [15,16] that supports the following operations.

COPY [15]. Consider the function $\mathsf{COPY} : \mathbb{F}_2 \rightarrow \mathbb{F}_2^m$ such that $\mathsf{COPY}(a) = (a, a, ..., a)$, i.e., one bit is copied to m bits. Let $(u, (v_1, v_2, ..., v_m))$ denote the monomial trail through the COPY function, then, it can be represented using the following MILP constraints:

$$\begin{cases} v_1 + v_2 + \cdots + v_m \geq u \\ u \geq v_i, \forall i \in \{1, 2, ..., m\} \\ u, v_1, ..., v_m \text{ are all binary variables.} \end{cases}$$

XOR [15]. Consider the function $\mathsf{XOR} : \mathbb{F}_2^m \rightarrow \mathbb{F}_2$ such that $\mathsf{XOR}(a_1, ..., a_m) = a_1 \oplus \cdots \oplus a_m$. Let $((u_1, u_2, ..., u_m), v)$ denote the monomial trail through the XOR function, then, it can be represented using the following MILP constraints:

$$\begin{cases} u_1 + u_2 + \cdots + u_m - v = 0 \\ u_1, ..., u_m, v \text{ are all binary variables.} \end{cases}$$

AND [15]. Consider the function $\mathsf{AND} : \mathbb{F}_2^m \rightarrow \mathbb{F}_2$ such that $\mathsf{AND}(a_1, ..., a_m) = a_1 a_2 \cdots a_m$. Let $((u_1, u_2, ..., u_m), v)$ denote the monomial trail through the AND function, then, it can be represented using the following MILP constraints:

$$\begin{cases} v = u_i, \forall i \in \{1, 2, ..., m\} \\ u_1, ..., u_m, v \text{ are all binary variables.} \end{cases}$$

NOT [16]. Consider the function $\mathsf{NOT} : \mathbb{F}_2 \rightarrow \mathbb{F}_2$ such that $\mathsf{NOT}(a) = 1 \oplus a$. Let (u, v) denote the monomial trail through the NOT function, then, it can be represented using the following MILP constraint:

$$\begin{cases} v \geq u \\ u, v \text{ are binary variables.} \end{cases}$$

4.2 MILP Model for the Monomial Trails of TinyJAMBU

We now discuss a MILP model to capture the valid monomial trails through the TinyJAMBU permutation. We denote the i-th intermediate state of the TinyJAMBU permutation and the TinyJAMBU inverse permutation as $\mathbf{x}^{(i)} = (x^i_{127}, \ldots, x^i_0)$ and $\bar{\mathbf{x}}^{(i)} = (\bar{x}^i_{127}, \ldots, \bar{x}^i_0)$ for $0 \leq i \leq r$, respectively. Thus the input (state) variables are $\mathbf{x}^{(0)} = (x^0_{127}, \ldots, x^0_0)$, and the output (state) variables after r-rounds are $\mathbf{x}^{(r)} = (x^r_{127}, \ldots, x^r_0)$. We let $\mathbf{u}^{(i)} = (u^i_{127}, \ldots, u^i_0)$ and $\bar{\mathbf{u}}^{(i)} = (\bar{u}^i_{127}, \ldots, \bar{u}^i_0)$ denote the exponents of the i-th intermediate state of the TinyJAMBU permutation and the TinyJAMBU inverse permutation, respectively. Also the exponents corresponding to the key variables are denoted by $(u^0_{128+\kappa}, \ldots, u^0_{128})$.

To represent the feedback polynomial of TinyJAMBU permutation and its inverse, we define the function $\mathsf{CORE} : \mathbb{F}_2^5 \rightarrow \mathbb{F}_2^5$ such that

$$\mathsf{CORE}(x_5, x_4, x_3, x_2, x_1, x_0) = (y_5, y_4, y_3, y_2, y_1, y_0)$$

where

$$(y_5, y_4, y_3, y_2, y_1, y_0) = \begin{cases} y_5 = x_5 \\ y_4 = x_4 \\ y_3 = x_3 \\ y_2 = x_2 \\ y_1 = x_1 \\ y_0 = 1 \oplus x_0 \oplus x_1 \oplus x_2 x_3 \oplus x_4 \oplus x_5. \end{cases} \tag{3}$$

The way we model the i-th round works in two steps. At the first step the CORE function is applied to $(k^i_i, x^i_{91}, x^i_{85}, x^i_{70}, x^i_{47}, x^i_0)$, i.e.,

$$(k^{i+1}_i, x^{i+1}_{91}, x^{i+1}_{85}, x^{i+1}_{70}, x^{i+1}_{47}, x^{i+1}_0) = \mathsf{CORE}(k^i_i, x^i_{91}, x^i_{85}, x^i_{70}, x^i_{47}, x^i_0).$$

In the second step all the state variables and key variables are rotated, i.e., $x^{i+1}_j = x^{i+1}_{j+1 \bmod 128}$ and $k^{i+1}_j = k^{i+1}_{j+1 \bmod 128}$. To construct a monomial trail of the round function of TinyJAMBU permutation, we decompose the CORE function given in Eq. 3 into COPY, XOR, AND and NOT operations. Then we can model the monomial trails through the CORE function as shown in Table 3, by introducing 7 intermediate variables w_i for $i = 0, 1, \ldots, 7$. In Table 3, u_i's and v_i's represent the MILP variables to denote exponents of x_i's and y_i's respectively. For the second step we just rotate the indices. In Algorithm 1, we discuss how to generate a complete MILP model for TinyJAMBU permutation, where $\mathcal{L}$ represents the inequalities for the CORE function as given in Table 3.

The model for the inverse TinyJAMBU permutation works almost similarly. The only difference is the input to the CORE function. In the case of the inverse round, the CORE function is applied as follows

$$(k^{i+1}_i, x^{i+1}_{81}, x^{i+1}_{58}, x^{i+1}_{43}, x^{i+1}_{37}, x^{i+1}_0) = \mathsf{CORE}(k^i_i, x^i_{81}, x^i_{58}, x^i_{43}, x^i_{37}, x^i_0).$$

Table 3. Inequalities to represent the CORE function.

Operation	Trail	MILP constraints
$x_1 \xrightarrow{\text{COPY}} (z_1, y_1)$	$(u_1, (w_1, v_1))$	$u_1 \leq w_1,\ u_1 \leq v_1,\ w_1 + v_1 \geq u_1$
$x_2 \xrightarrow{\text{COPY}} (z_2, y_2)$	$(u_2, (w_2, v_2))$	$u_2 \leq w_2,\ u_2 \leq v_2,\ w_2 + v_2 \geq u_2$
$x_3 \xrightarrow{\text{COPY}} (z_3, y_3)$	$(u_3, (w_3, v_3))$	$u_3 \leq w_3,\ u_3 \leq v_3,\ w_3 + v_3 \geq u_3$
$x_4 \xrightarrow{\text{COPY}} (z_4, y_4)$	$(u_4, (w_4, v_4))$	$u_4 \leq w_4,\ u_4 \leq v_4,\ w_4 + v_4 \geq u_4$
$x_5 \xrightarrow{\text{COPY}} (z_5, y_5)$	$(u_5, (w_5, v_5))$	$u_5 \leq w_5,\ u_5 \leq v_5,\ w_5 + v_5 \geq u_5$
$(z_2, z_3) \xrightarrow{\text{AND}} z_6$	$((w_2, w_3), w_6)$	$w_6 == z_2,\ w_6 == z_3$
$(x_0, z_1, z_6, z_4, z_5) \xrightarrow{\text{XOR}} z_7$	$((u_0, w_1, w_6, w_4, w_5), w_7)$	$w_7 == u_0 + w_1 + w_6 + w_4 + w_5$
$z_7 \xrightarrow{\text{NOT}} y_0$	(w_7, v_0)	$v_0 \geq w_7$

4.3 Degree Estimation of the TinyJAMBU Permutation

We now bound the degree of TinyJAMBU permutation using the MILP model generated in Algorithm 1. In the following discussion, we consider the key $\mathsf{K} = (k_{\kappa-1}, ..., k_0)$ as a constant. Thus, the j-th state bit after i rounds of $\mathcal{P}_{\mathsf{K}}$, x_j^i, is a polynomial on $\mathbf{x}^{(0)} = (x_{127}^0, \ldots, x_0^0)$, and we need to determine the degree of the polynomial over $\mathbf{x}^{(0)} = (x_{127}^0, \ldots, x_0^0)$. In all of our attacks, we fix our target output (state) bit to the 127-th bit and thus for an r-round permutation $\mathbf{u}^{(r)} = (u_{127}^r, ..., u_0^r)$ is

$$u_j^r = \begin{cases} 1, & \text{if } j = 127 \\ 0, & \text{otherwise.} \end{cases} \tag{4}$$

We discuss the detailed procedures to find the degree of the TinyJAMBU permutation in Algorithm 2, where we call Algorithm 1 to generate a MILP model for the TinyJAMBU. Algorithm 2 works as follows: we first generate a MILP model $\mathcal{M}$ and solve the model to get a possible monomial $\mathbf{m}$ in the polynomial representation of the 127-th bit of the output state. Our next task is to confirm the presence of the monomial $\mathbf{m}$ in the polynomial representation of the 127-th bit of the output state. To do so we generate another MILP model $\mathcal{M}'$ (line 11 in Algorithm 2). Here we set the initial variables $(\mathbf{u}'^{(0)})$ of $\mathcal{M}'$ according to the solution of $\mathcal{M}$ and solve it. In this case we count the number of solutions of $\mathcal{M}'$. If the number of solutions is odd, then we confirm that the obtained monomial $\mathbf{m}$ exists (line 19 in Algorithm 2) in the polynomial representation of the 127-th bit of the output state. Otherwise, if the number of solution is even, then the number of trails from the monomial $\mathbf{m}$ is even. So, we remove the solution corresponding to $\mathbf{m}$ (line 22-26 in Algorithm 2) and solve the model $\mathcal{M}$ again. We do this until we get a monomial with an odd number of trails.

Algorithm 1. TinyJAMBU$_{MILP}$ (Generating a MILP model)

1: Input: r, the targeted number of rounds
2: Output: The MILP model $\mathcal{M}$ for r-round TinyJAMBU permutation, MILP variable representing initial and final states
3: Declare an empty MILP model $\mathcal{M}$;
4: $\mathcal{M}.var \leftarrow \mathbf{u}^{(0)} = (u^0_{127+\kappa}, \ldots, u^0_{128}, u^0_{127}, \ldots, u^0_0)$;
5: $\mathcal{M}.var \leftarrow \mathbf{s} = (s_{127+\kappa}, \ldots, s_0)$;
6: $\mathbf{s} \leftarrow \mathbf{u}^{(0)}$;
7: **for** $i = 0; i < r; i = i + 1$ **do**
8: $\quad\mathcal{M}.var \leftarrow (w_7, \ldots, w_0)$;
9: $\quad\mathcal{M}.var \leftarrow (v_{128}, v_{91}, v_{85}, v_{70}, v_{47}, v_0)$;
10: $\quad\mathcal{M}.con \leftarrow \mathcal{L}(s_{128}, s_{91}, s_{85}, s_{70}, s_{47}, s_0, v_{128}, v_{91}, v_{85}, v_{70}, v_{47}, v_0, w_7, \ldots, w_0)$;
11: $\quad s_i = v_i, i \in \{0, 47, 70, 85, 91, 128\}$;
12: $\quad$**for** $j = 0; j < 128; j = j + 1$ **do**
13: $\qquad s_j \leftarrow s_{j+1 \mod 128}$;
14: $\quad$**for** $j = 128; j < \kappa; j = j + 1$ **do**
15: $\qquad s_j \leftarrow s_{128+\{(j+1) \mod \kappa\}}$;
16: $\mathbf{u}^{\mathrm{r}} \leftarrow \mathbf{s}$
17: **return** $\mathcal{M}, \mathbf{u}^0, \mathbf{u}^r$;

We have listed our findings of degree for TinyJAMBU-128 permutation and its inverse permutation in Table 7 and Table 8, respectively, in Appendix A. The source code for MILP modeling can be found in

https://github.com/ShibamCrS/zeroSumDistinguishersOnTinyJambu.git.

4.4 Basic Zero-Sum Distinguisher

We now discuss a basic zero-sum distinguisher on TinyJAMBU-128. To find this we upper bound the degree of TinyJAMBU permutation using Algorithm 2. We set the objective function to maximize the sum of initial variables, i.e., $\sum_{j=0}^{127} u_j^0$ and check for the parity of the number of solutions as in Algorithm 2. With the help of Algorithm 2 we are able to evaluate an upper bound on the algebraic degree of the TinyJAMBU-128 up to 333 rounds. The degree of all the bits in $\mathbf{x}^{(333)}$ is upper-bounded by 38 for 333 rounds. Thus, if we consider an affine subspace ν of dimension 39 and consider the sum of the output states over all the elements of ν, we get $\mathbf{0}$, i.e.,

$$\bigoplus_{\mathbf{x}^{(0)} \in \nu} \mathcal{P}_{\mathsf{K}}^{333}(\mathbf{x}^{(0)}) = \mathbf{0}.$$

Which gives us that all the state bits are balanced after 333 rounds.

Algorithm 2. Degree estimation of the TinyJAMBU permutation

```
1: Input: r, the targeted number of rounds
2: Output: The degree d of r-round TinyJAMBU permutation
3: (M, u^(0), u^(r)) ← TinyJAMBU_MILP(r);
4: M.constraint ← {u^r_127 = 1};
5: for i = 0; i < 127; i = i + 1 do
6:     M.constraint ← {u^r_i = 0};
7: M.objective ← max(u^0_127 + ... + u^0_0);
8: while true do
9:     M.update();
10:    M.optimize();
11:    if M.status is OPTIMAL then
12:        d ← M.objvalue
13:        (M', u'^(0), u'^(r)) ← TinyJAMBU_MILP(r);
14:        u^r_127 ← 1;
15:        for i = 0; i < 127; i = i + 1 do
16:            u^r_i ← 0;
17:        for i = 0; i < 128; i = i + 1 do
18:            u'^0_i ← u^0_i.value;
19:        M'.optimize();
20:        if M'.status is OPTIMAL then
21:            if Number of solution in M' is odd then
22:                return d;
23:            else
24:                for i = 0; i < 127 + κ; i = i + 1 do
25:                    if u^0_i.value = 0 then
26:                        M.constraint ← {u^0_i = 1};
27:                    else
28:                        M.constraint ← {u^0_i = 0};
```

We can extend this distinguisher to cover more rounds due to the basic nature of the shift register. From the design of the round function, we can observe that each bit is shifted to its previous bit, i.e., $x^i_j = x^i_{j+1}$. Thus after 448 rounds the following 12 bits are still balanced

$$(x_{11}^{(448)}, x_{10}^{(448)}, \ldots x_0^{(448)}).$$

The complexity of this distinguisher is 2^{39}. When the same affine subspace passes through a random permutation, the outputs satisfy the above condition with probability 2^{-12}. Thus, the distinguishing advantage of this distinguisher $1 - 2^{-12}$.

We can increase the distinguishing advantage by considering ℓ affine subspace of dimension 39 instead of taking one affine subspace. This comes at the cost of complexity. In this case the distinguishing advantage of this distinguisher is $1 - 2^{-12\ell}$ and the complexity is $\ell 2^{39}$. Let us take $\ell = 4$. Then the distinguishing advantage increases to $1 - 2^{-48}$ and complexity of this distinguisher is 2^{41}.

We can increase the number of rounds up to 460 and still can use this zero-sum property. However, the distinguishing advantage of the attack decreases. If $\mathcal{A}$ is the distinguishing advantage, then the number of rounds $r \propto \log(1 - \mathcal{A})$, where $333 \leq r \leq 460$.

Zero-Sum Distinguisher of the Inverse Permutation. We now discuss a basic zero-sum distinguisher on the inverse permutation. Similar to the forward direction, here also we set the objective function to maximize $\sum_{j=0}^{127} \bar{u}_j^0$ and to check for the parity of the number of solutions. We noticed in our experiments that in the inverse direction the degree increases significantly slower compared to the forward direction. This is because the update function has only one non-linear operation on bit number 70 and 85 and for the inverse permutation, the first non-linear operation is after 70 rounds, while in the forward direction the first non-linear operation occurs after 42 rounds.

With the help of monomial prediction, we are able to evaluate an upper bound on the algebraic degree of the inverse TinyJAMBU-128 up to 502 rounds. The degree of all the bits in $\bar{\mathbf{x}}^{(502)}$ is upper-bounded by 41 for 502 rounds. Thus if we consider an affine subspace ν of dimension 42 and consider the sum of the output states over all the elements of ν, we get $\mathbf{0}$, i.e.,

$$\bigoplus_{\mathbf{x}^{(0)} \in \nu} \mathcal{P}_{\mathsf{K}}^{-502}(\bar{\mathbf{x}}^{(0)}) = \mathbf{0}.$$

As in the forward direction, we can extend this distinguisher for more rounds using shifting property of the register. After 608 rounds the following 21 bits are still balanced

$$(\bar{x}_{20}^{(608)}, \bar{x}_{19}^{(608)}, ..., \bar{x}_{0}^{(608)}).$$

The data and time complexity of this distinguisher is 2^{42} and the distinguishing advantage of this distinguisher $1 - 2^{-21}$. Similar to the forward direction, we can increase the distinguishing advantage by considering few more subspaces of dimension 41. If we consider 4 subspaces of dimension 41, the distinguishing advantage increases to $1 - 2^{-84}$ and complexity of this distinguisher is 2^{44}.

4.5 Extending to Full Rounds Using Inside-Out Approach

We extend the above zero-sum distinguishers to a full round distinguisher using the inside-out approach. We start in the middle of the permutation and compute the degree outwards. We consider an affine subspace ν of dimension d and for all 2^d possible intermediate states we compute the outputs. Suppose that we consider the state after r_1 rounds for an r-round permutation where $r_1 + r_2 = r$. For all these 2^d intermediate states $\mathbf{x}^{(\mathbf{r_1})} \in \mathbb{F}_2^{128}$, we compute $\mathcal{P}^{r_2}(\mathbf{x}^{(\mathbf{r_1})})$ and $\mathcal{P}^{-r_1}(\mathbf{x}^{(\mathbf{r_1})})$. If the degree of both functions $\mathcal{P}^{r_2}$ and $\mathcal{P}^{-r_1}$ are less than d, we get zero-sum on both outputs, i.e.,

$$\bigoplus_{\mathbf{x}^{(\mathbf{r_1})} \in \nu} \mathcal{P}^{r_2}(\mathbf{x}^{(\mathbf{r_1})}) = \bigoplus_{\mathbf{x}^{(\mathbf{r_1})} \in \nu} \mathcal{P}^{-r_1}(\mathbf{x}^{(\mathbf{r_1})}) = \mathbf{0}.$$

The idea is depicted in Fig. 6.

$$\text{zero-sum} \xleftarrow[deg(\mathcal{P}^{-r_1}) < d]{} \nu \xrightarrow[deg(\mathcal{P}^{r_2}) < d]{} \text{zero-sum}$$

Fig. 6. Zero-sum distinguisher from an intermediate vector space ν.

However, to use this distinguisher to distinguish the TinyJAMBU permutation from a random permutation, we need to know the secret key K, i.e., this is a *known-key zero-sum* distinguisher. This cryptographic model, distinguishing a cryptographic permutation from a random permutation with the knowledge of the key, was introduced by Knudsen *et al.* in [20] at ASIACRYPT 2007. The authors in [20] found a distinguisher on the 7-round AES [8] using this model. There are other examples of this approach in the literature, for example on KECCAK and Luffa [3,5], PHOTON [30], Ascon [10] or MiMC [13].

To apply the technique to the TinyJAMBU-128 permutation, we bound the degree of $\mathcal{P}^{r_2}$ and $\mathcal{P}^{-r_1}$ using a MILP model of the monomial prediction rules. We find that the degree of all the bits of $\mathbf{x}^{(332)}$ is upper-bounded by 37 after applying 332 rounds of $\mathcal{P}$. On the other hand, the degree of all the bits of $\bar{\mathbf{x}}^{(482)}$ is upper bounded by 37 for round number 482 of $\mathcal{P}^{-1}$. We set $r_1 = 448$ and $r_2 = 576$ giving $r = r_1 + r_2 = 1024$. If we take $\mathcal{P}^{448}(\mathbf{x}^{(\mathbf{576})}) = (z_{127}, ..., z_0)$, then we get the zero-sum at the bit positions $(z_{10}, z_9, ..., z_0)$. Similarly, if $\mathcal{P}^{-576}(\mathbf{x}^{(\mathbf{576})}) = (y_{127}, ..., y_0)$, then we get the zero-sum at the following 33 bit positions $(y_{32}, y_{31}, ..., y_0)$. Thus we consider an affine vector space ν of dimension 38 and compute outwards by setting the intermediate state to all possible 2^{38} vectors from ν. We collect the outputs of $\mathcal{P}^{448}$ and the outputs of $\mathcal{P}^{-576}$. Finally, we have a set of inputs $X = \{\mathbf{x} = (x_{127}, ..., x_0)\}$ with $\bigoplus_{\mathbf{x} \in X} x_i = 0$ for all $0 \le i \le 32$. Also, a set of outputs $Z = \{\mathcal{P}^{1024}(\mathbf{x}) : \mathbf{x} \in X\}$ such that $\bigoplus_{\mathbf{x} \in X} z_i = \bigoplus_{\mathbf{x} \in X} (\mathcal{P}^{1024}(\mathbf{x}))_i = 0$ for all $0 \le i \le 10$. Combining the two results we have a zero-sum distinguisher of full round TinyJAMBU-128 permutation with time and data complexity $O(2^{38})$. See Fig. 7 for a depiction of the attack.

$$\text{zero-sum} \xleftarrow[deg(\mathcal{P}^{-576}) < 38]{} \nu \xrightarrow[deg(\mathcal{P}^{448}) < 38]{} \text{zero-sum}$$

Fig. 7. Zero-sum distinguisher on full TinyJAMBU-128.

5 Improved Zero-Sum Distinguisher

In the previous section, we suggested a distinguisher on the full round TinyJAMBU-128 permutation by considering degree of the TinyJAMBU permutation and inverse permutation over all 128 state variables. In this section we improve the time and data complexity of the distinguisher by choosing the proper variables (or in other words a proper affine subspace ν), i.e., we want to find the degree of the TinyJAMBU permutation over some selected variables. This scenario is also similar to the cube attack [9] or a cube tester [1].

Given a Boolean polynomial f in m variables, we choose a set of variables (also called the cube variables in a cube attack) of size n and set the rest of the $m-n$ variables to a constant (typically they are set to zero). We reduced the function f to a function on n-variables. If we can show that the reduced function has degree lower than some value d, we have a zero-sum property.

Proper Choice of Variables. Let us consider the following subspace ν of dimension 23 from $\mathbb{F}_2^{128}$

$$\mathbf{x} \in \nu \iff \begin{cases} x_{i+47} = x_i, \text{ if } 0 \leq i \leq 22 \\ 0, \text{ if } 23 \leq i \leq 46 \\ 0, \text{ if } 70 \leq i \leq 127. \end{cases} \tag{5}$$

From the algebraic description of the round permutation we get that after one round, the feedback polynomial is $\mathbf{x}_{127}^{(1)} = 1 \oplus x_0^{(0)} \oplus x_{47}^{(0)} \oplus x_{70}^{(0)} x_{85}^{(0)} \oplus x_{91}^{(0)} \oplus k_0$. If we set $\mathbf{x}^{(0)}$ to any vector from ν, we have $\mathbf{x}_{127}^{(1)} = 1 \oplus k_0$. This is also true for first 23 rounds. Therefore, for the first 23 rounds the degree of the feedback polynomial is 0 in the cube variables. Then, from round 24 to 47, the degree of the feedback polynomials is also 0 in the cube variables. It is now clear how the locations of the fixed 0 bits were chosen - to ensure that as many forward rounds with degree 0 as possible. We conclude the results with the following Lemma 2.

Lemma 2. *If we choose the input state $\mathbf{x}^{(0)}$ of the TinyJAMBU permutation from the subspace ν as described in* (5)*, we get $deg(x_{127}^{(i)}) = 0$ for all $0 \leq i \leq 47$. More precisely, if $\mathcal{P}_{\mathsf{K}}^{47}(\mathbf{x}^{(0)}) = \mathbf{z}$, then the coordinates of $\mathbf{z}$ satisfies the following conditions:*

$$\begin{aligned} z_i &= x_i^{(0)} \text{ for } 0 \leq i \leq 22 \\ z_i &= 0 \text{ for } 22 \leq i \leq 80 \\ z_i &= 1 \oplus k_{81-i}, \text{ for } 81 \leq i \leq 117 \\ z_i &= k_{155-i} \oplus k_{118-i}, \text{ for } 118 \leq i \leq 127 \end{aligned}$$

After 47 rounds, the degree is upper bounded by 1 and state bits 23 to 128 do not contain any cube variables. So we do not consider these MILP variables in the objective function.

After 388 rounds the degree in all bits of $\mathbf{x}^{(388)}$ is upper bounded by 22. Consequently, as we have started after 47 rounds, we can compute the degree after 47 more rounds. The degree in all bits of $\mathbf{x}^{(435)}$ after 435 rounds is upper bounded by 22. Using the subspace ν mentioned in (5), after applying the permutation for 544 rounds, i.e., $\mathcal{P}^{(544)}$, we get a zero sum on the following 18 bit positions

$$(x_{17}^{(544)}, x_{16}^{(544)}, \ldots, x_0^{(544)}).$$

Similar distinguishers also works on TinyJAMBU-192. The results are summarized in Table 4.

Table 4. Secret-key zero-sum distinguishers TinyJAMBU permutation. * implies the basic distinguisher of sec Sect. 4.4.

Key size	#rounds	#balanced bits	Complexity
128	480	16	16
128	480	38	18
128	512	9	18
128	544	18	23
128	448	12	39*
192	490	6	16
192	555	8	23
256	490	6	16
256	555	8	23

5.1 Extending to Full Rounds Using Inside-Out Approach

Similar to the basic distinguisher, we can extend this distinguisher to the full round TinyJAMBU-128 without increasing time or data complexity. The advantage we get for forward direction from projecting degree for 47 rounds at the starting, is not possible for the inverse permutation. However, as we discussed earlier, the degree of the inverse permutation already increases slowly.

For the inverse permutation we set the MILP variable as follows $\bar{u}_i = 0$ if $0 \leq i \leq 58$ or $81 \leq i \leq 105$. After 472 rounds we have that the degree in all the bits of $\bar{\mathbf{x}}^{(472)}$ are upper bounded by 22. Thus, after 576 rounds of the inverse permutation we get a zero sum on the following 23 bit positions

$$(\bar{x}_{22}^{(576)}, \bar{x}_{21}^{(576)}, \ldots \bar{x}_{0}^{(576)}).$$

Similar distinguishers also works on TinyJAMBU-192. The results are summarized in Table 5.

Table 5. Secret-key zero-sum distinguishers TinyJAMBU inverse permutation. * implies the basic distinguisher of sec Sect. 4.4.

Key size	#rounds	#balanced bits	Complexity
128	544	6	16
128	544	16	18
128	576	23	23
128	592	7	23
128	608	21	42*
192	544	6	16
192	597	4	23
256	544	6	16
256	597	4	23

To extend the above results to full rounds we set $r_1 = 512$ and $r_2 = 512$, and we have a zero-sum distinguisher on full rounds with 50 balanced bits on the output of the forward direction and 87 balanced bits on the output of the inverse direction. Other combinations of r_1 and r_2 are possible with the trade-off on data/time complexity to number of balanced bits. Our best distinguisher on full TinyJAMBU-128 works with the time/data complexity of 2^{16}. For these we set $r_1 = 480$ and $r_2 = 544$ such that $r_1 + r_2 = 1024$. We get 16 balanced bits in the forward direction and 6 balanced bits in the backward direction. Results on full round attack with various combinations of r_1 and r_2 are given in Table 6. Also see Fig. 8 for a depiction of the attack.

$$\text{zero-sum} \xleftarrow[deg(\mathcal{P}^{-512}) < 23]{} \mathcal{V} \xrightarrow[deg(\mathcal{P}^{512}) < 23]{} \text{zero-sum}$$

Fig. 8. Improved zero-sum distinguisher on full TinyJAMBU-128.

5.2 Attack on TinyJAMBU-192 and TinyJAMBU-256

The improved distinguisher discussed in the previous of this section also works on the full round TinyJAMBU-192 and reduced round TinyJAMBU-256 in a similar manner. The results are summarized in Table 6.

5.3 Experimental Verification

We conducted experiments to verify the existence of the zero-sum distinguishers. All of our distinguishers on TinyJAMBU-128 are verified with the reference implementation of TinyJAMBU. The source codes of the verification can be found in:

https://github.com/ShibamCrS/zeroSumDistinguishersOnTinyJambu.git.

Table 6. Known-key zero-sum distinguishers on full round TinyJAMBU-128 and TinyJAMBU-192 and reduced round TinyJAMBU-256. $*$ implies basic distinguisher of sec Sect. 4.4. † implies reduced rounds.

Key size	#rounds		#balanced bits		Complexity
	Forward	Inverse	After $\mathcal{P}$	after $\bar{\mathcal{P}}$	
128	480	544	16	6	16
128	480	544	38	16	18
128	512	512	50	87	23
128	448	576	11	33	38^*
192	555	597	8	4	23
256	555	597	8	4	$23^\dagger$

6 Conclusion

We discussed full round zero-sum distinguishers for the TinyJAMBU permutation, based on the algebraic properties of the permutation. All the distinguishers have practical complexities, that allowed for complete experimental verification of the attacks.

One important note is that the keyed permutation is the main contribution of the TinyJAMBU submission to the NIST lightweight competition. Our attack is not possible inside the mode. Nevertheless, as we have shown, the keyed permutation used in TinyJAMBU is easily distinguished. Combining this with the fact that the mode is proven to be secure under the assumption that the internal keyed permutation is robust could lead to problems in the future.

A The Algebraic Degree of TinyJAMBU-128 Permutation and Its Inverse

Table 7. Degree of the 127-th bit of TinyJAMBU-128 permutation up to 333 rounds.

Offset	0	1	2	3	4	5	6	7	8	9	10	11	12	13	14	15	16	17	18	19
0+		1	2	2	2	2	2	2	2	2	2	2	2	2	2	2	2	2	2	2
20+	2	2	2	2	2	2	2	2	2	2	2	2	2	2	2	2	2	2	2	2
40+	2	2	2	2	3	3	3	3	3	3	3	3	3	3	3	3	3	3	3	3
60+	3	3	3	3	3	3	3	3	3	3	3	3	3	3	3	3	3	3	3	3
80+	3	4	4	4	4	4	4	4	4	4	4	4	4	4	4	4	5	5	5	5
100+	5	5	5	5	5	5	5	5	5	5	5	5	5	5	5	5	5	5	5	5
120+	5	5	5	5	6	6	6	6	6	6	6	6	6	6	6	6	6	6	6	7
140+	7	7	7	7	7	7	7	7	7	7	7	7	7	7	7	7	7	7	7	7
160+	8	8	8	8	8	8	8	8	8	8	8	8	8	9	9	9	10	10	10	10
180+	10	10	11	11	11	11	11	11	11	11	11	11	11	11	11	11	11	12	12	12
200+	12	12	12	12	12	12	12	12	12	12	13	13	13	13	13	13	14	14	14	15
220+	15	15	15	15	15	16	16	16	16	16	16	16	16	16	16	16	16	16	16	16
240+	17	17	17	17	17	17	17	17	17	17	17	17	17	18	18	18	19	19	19	20
260+	20	20	21	22	22	22	22	22	22	22	23	24	24	24	24	24	24	25	25	25
280+	25	25	25	26	26	26	26	26	26	26	26	27	27	27	27	27	28	28	28	29
300+	29	29	30	30	30	31	31	31	31	31	31	32	33	33	34	34	34	35	35	35
320+	36	36	36	36	36	36	36	36	36	37	37	37	37	38						

Table 8. Degree of the 127-th bit of TinyJAMBU-128 inverse permutation up to 502 rounds.

Offset	0	1	2	3	4	5	6	7	8	9	10	11	12	13	14	15	16	17	18	19
0+		1	2	2	2	2	2	2	2	2	2	2	2	2	2	2	2	2	2	2
20+	2	2	2	2	2	2	2	2	2	2	2	2	2	2	2	2	2	2	2	2
40+	2	2	2	2	2	2	2	2	2	2	2	2	2	2	2	2	2	2	2	2
60+	2	2	2	2	2	2	2	2	2	2	2	3	3	3	3	3	3	3	3	3
80+	3	3	3	3	3	3	3	3	3	3	3	3	3	3	3	3	3	3	3	3
100+	3	3	3	3	3	3	3	3	3	3	3	3	3	3	3	3	3	3	4	4
120+	4	4	4	4	4	4	4	4	4	4	4	4	4	4	4	4	4	4	4	4
140+	4	5	5	5	5	5	5	5	5	5	5	5	5	5	5	5	5	5	5	5
160+	5	5	5	5	5	5	5	5	5	5	5	5	5	5	5	5	5	5	5	5
180+	5	5	5	5	5	5	5	5	6	6	6	6	6	6	6	6	6	6	6	6
200+	6	6	6	6	6	6	6	6	6	6	6	7	7	7	7	7	7	7	7	7
220+	7	7	7	7	7	7	7	7	7	7	7	7	8	8	8	8	8	8	8	8
240+	8	8	8	8	8	8	8	9	9	9	9	9	9	9	9	9	9	9	10	10
260+	10	10	10	10	10	10	10	10	10	10	10	10	10	10	10	10	10	10	10	10
280+	10	11	11	11	11	11	11	11	11	11	11	11	11	11	11	11	12	12	12	12
300+	12	12	12	12	12	12	12	12	12	12	12	13	13	13	13	13	13	14	14	14
320+	14	14	14	14	14	14	14	14	15	15	15	15	15	15	15	15	15	15	15	15
340+	15	15	15	15	15	15	15	15	15	16	16	16	16	16	16	16	16	16	16	16
360+	16	16	16	16	17	17	18	19	19	19	19	19	19	20	20	20	20	20	20	20
380+	20	20	20	20	20	20	20	21	21	21	22	22	22	22	22	22	23	23	24	24
400+	24	24	24	24	24	24	24	24	24	24	24	24	24	25	25	25	25	25	25	25
420+	26	26	26	26	26	26	26	26	26	26	26	26	26	26	27	27	28	29	29	29
440+	29	29	29	30	30	31	31	31	31	31	31	31	31	31	31	31	31	32	32	32
460+	32	32	32	32	33	33	34	34	35	35	35	36	36	36	36	36	36	36	37	37
480+	37	37	37	38	38	38	39	39	39	39	40	40	40	40	40	40	41	41	41	41
500+	41	41	41																	

References

1. Aumasson, J., Dinur, I., Meier, W., Shamir, A.: Cube testers and key recovery attacks on reduced-round MD6 and Trivium. In: Symmetric Cryptography. Dagstuhl Seminar Proceedings, vol. 09031. Schloss Dagstuhl - Leibniz-Zentrum für Informatik, Germany (2009)
2. Aumasson, J.P., Meier, W.: Zero-sum distinguishers for reduced Keccak-f and for the core functions of Luffa and Hamsi. rump session of Cryptographic Hardware and Embedded Systems-CHES 2009, 67 (2009)

3. Boura, C., Canteaut, A.: Zero-sum distinguishers for iterated permutations and application to KECCAK-*f* and Hamsi-256. In: Biryukov, A., Gong, G., Stinson, D.R. (eds.) SAC 2010. LNCS, vol. 6544, pp. 1–17. Springer, Heidelberg (2011). https://doi.org/10.1007/978-3-642-19574-7_1
4. Boura, C., Canteaut, A.: A zero-sum property for the Keccak-f permutation with 18 rounds. In: ISIT, pp. 2488–2492. IEEE (2010)
5. Boura, C., Canteaut, A., De Cannière, C.: Higher-order differential properties of KECCAK and *Luffa*. In: Joux, A. (ed.) FSE 2011. LNCS, vol. 6733, pp. 252–269. Springer, Heidelberg (2011). https://doi.org/10.1007/978-3-642-21702-9_15
6. Canteaut, A., Videau, M.: Degree of composition of highly nonlinear functions and applications to higher order differential cryptanalysis. In: Knudsen, L.R. (ed.) EUROCRYPT 2002. LNCS, vol. 2332, pp. 518–533. Springer, Heidelberg (2002). https://doi.org/10.1007/3-540-46035-7_34
7. Chen, S., Xiang, Z., Zeng, X., Zhang, S.: On the relationships between different methods for degree evaluation. IACR Trans. Symmetric Cryptol. **2021**(1), 411–442 (2021)
8. Daemen, J., Rijmen, V.: AES and the wide trail design strategy. In: Knudsen, L.R. (ed.) EUROCRYPT 2002. LNCS, vol. 2332, pp. 108–109. Springer, Heidelberg (2002). https://doi.org/10.1007/3-540-46035-7_7
9. Dinur, I., Shamir, A.: Cube attacks on tweakable black box polynomials. In: Joux, A. (ed.) EUROCRYPT 2009. LNCS, vol. 5479, pp. 278–299. Springer, Heidelberg (2009). https://doi.org/10.1007/978-3-642-01001-9_16
10. Dobraunig, C., Eichlseder, M., Mendel, F., Schläffer, M.: Cryptanalysis of Ascon. In: Topics in Cryptology - CT-RSA, pp. 371–387 (2015)
11. Dunkelman, O., Lambooij, E., Ghosh, S.: Practical related-key forgery attacks on the full tinyjambu-192/256. Cryptology ePrint Archive, Paper 2022/1122 (2022). https://eprint.iacr.org/2022/1122
12. Dutta, P., Rajas, M., Sarkar, S.: Weak-keys and key-recovery attack for Tiny-JAMBU, May 2022. https://doi.org/10.21203/rs.3.rs-1646044/v1
13. Eichlseder, M., Grassi, L., Lüftenegger, R., Øygarden, M., Rechberger, C., Schofnegger, M., Wang, Q.: An algebraic attack on ciphers with low-degree round functions: application to full MiMC. In: Moriai, S., Wang, H. (eds.) ASIACRYPT 2020. LNCS, vol. 12491, pp. 477–506. Springer, Cham (2020). https://doi.org/10.1007/978-3-030-64837-4_16
14. Hadipour, H., Eichlseder, M.: Integral cryptanalysis of WARP based on monomial prediction. IACR Trans. Symmetric Cryptol. **2022**(2), 92–112 (2022)
15. Hao, Y., Leander, G., Meier, W., Todo, Y., Wang, Q.: Modeling for three-subset division property without unknown subset. In: Canteaut, A., Ishai, Y. (eds.) EUROCRYPT 2020. LNCS, vol. 12105, pp. 466–495. Springer, Cham (2020). https://doi.org/10.1007/978-3-030-45721-1_17
16. Hebborn, P., Lambin, B., Leander, G., Todo, Y.: Lower bounds on the degree of block ciphers. In: Moriai, S., Wang, H. (eds.) ASIACRYPT 2020. LNCS, vol. 12491, pp. 537–566. Springer, Cham (2020). https://doi.org/10.1007/978-3-030-64837-4_18
17. Hu, K., Sun, S., Wang, M., Wang, Q.: An algebraic formulation of the division property: revisiting degree evaluations, cube attacks, and key-independent sums. In: Moriai, S., Wang, H. (eds.) ASIACRYPT 2020. LNCS, vol. 12491, pp. 446–476. Springer, Cham (2020). https://doi.org/10.1007/978-3-030-64837-4_15
18. Knudsen, L., Wagner, D.: Integral cryptanalysis. In: Daemen, J., Rijmen, V. (eds.) FSE 2002. LNCS, vol. 2365, pp. 112–127. Springer, Heidelberg (2002). https://doi.org/10.1007/3-540-45661-9_9

19. Knudsen, L.R.: Truncated and higher order differentials. In: Preneel, B. (ed.) FSE 1994. LNCS, vol. 1008, pp. 196–211. Springer, Heidelberg (1995). https://doi.org/10.1007/3-540-60590-8_16
20. Knudsen, L.R., Rijmen, V.: Known-key distinguishers for some block ciphers. In: Kurosawa, K. (ed.) ASIACRYPT 2007. LNCS, vol. 4833, pp. 315–324. Springer, Heidelberg (2007). https://doi.org/10.1007/978-3-540-76900-2_19
21. Lai, X.: Higher order derivatives and differential cryptanalysis. In: Communications and Cryptography: Two Sides of One Tapestry, pp. 227–233. Springer (1994)
22. Liu, M.: Degree evaluation of NFSR-based cryptosystems. In: Katz, J., Shacham, H. (eds.) CRYPTO 2017. LNCS, vol. 10403, pp. 227–249. Springer, Cham (2017). https://doi.org/10.1007/978-3-319-63697-9_8
23. Saha, D., Sasaki, Y., Shi, D., Sibleyras, F., Sun, S., Zhang, Y.: On the security margin of TinyJAMBU with refined differential and linear cryptanalysis. IACR Trans. Symmetric Cryptol. **2020**(3), 152–174 (2020)
24. Sibleyras, F., Sasaki, Y., Todo, Y., Hosoyamada, A., Yasuda, K.: Birthday-bound slide attacks on TinyJAMBU's keyed permutation for all key sizes. In: Fifth Lightweight Cryptography Workshop (2022)
25. Technology, N.: Report on Lightweight Cryptography: NiSTIR 8114. CreateSpace Independent Publishing Platform (2017)
26. Teng, W.L., Salam, M.I., Yau, W., Pieprzyk, J., Phan, R.C.: Cube attacks on round-reduced TinyJAMBU. IACR Cryptol. ePrint Arch, p. 1164 (2021)
27. Todo, Y.: Structural evaluation by generalized integral property. In: Oswald, E., Fischlin, M. (eds.) EUROCRYPT 2015. LNCS, vol. 9056, pp. 287–314. Springer, Heidelberg (2015). https://doi.org/10.1007/978-3-662-46800-5_12
28. Todo, Y.: Integral cryptanalysis on full MISTY1. J. Cryptol. **30**(3), 920–959 (2017)
29. Todo, Y., Morii, M.: Bit-based division property and application to SIMON family. In: Peyrin, T. (ed.) FSE 2016. LNCS, vol. 9783, pp. 357–377. Springer, Heidelberg (2016). https://doi.org/10.1007/978-3-662-52993-5_18
30. Wang, Q., Grassi, L., Rechberger, C.: Zero-sum partitions of PHOTON permutations. In: Smart, N.P. (ed.) CT-RSA 2018. LNCS, vol. 10808, pp. 279–299. Springer, Cham (2018). https://doi.org/10.1007/978-3-319-76953-0_15
31. Wu, H., Huang, T.: The JAMBU lightweight authentication encryption mode (v2.1). Submission to CAESAR (2016). https://competitions.cr.yp.to/round3/jambuv21.pdf
32. Wu, H., Huang, T.: TinyJAMBU: a family of lightweight authenticated encryption algorithms: submission to NIST LwC (2019). https://csrc.nist.gov/CSRC/media/Projects/lightweight-cryptography/documents/finalist-round/updated-spec-doc/tinyjambu-spec-final.pdf
33. Wu, H., Huang, T.: TinyJAMBU: a family of lightweight authenticated encryption algorithms (version 2) (2021). https://csrc.nist.gov/CSRC/media/Projects/lightweight-cryptography/documents/finalist-round/updated-spec-doc/tinyjambu-spec-final.pdf

Monte Carlo Tree Search for Automatic Differential Characteristics Search: Application to SPECK

Emanuele Bellini[1], David Gerault[1], Matteo Protopapa[2], and Matteo Rossi[2](✉)

[1] Cryptography Research Centre, Technology Innovation Institute, Abu Dhabi, United Arab Emirates
{emanuele.bellini,david.gerault}@tii.ae
[2] Politecnico di Torino, Torino, Italy
matteo.rossi@polito.it

Abstract. The search for differential characteristics on block ciphers is a difficult combinatorial problem. In this paper, we investigate the performances of an AI-originated technique, Single Player Monte-Carlo Tree Search (SP-MCTS), in finding good differential characteristics on ARX ciphers, with an application to the block cipher SPECK. In order to make this approach competitive, we include several heuristics, such as the combination of forward and backward searches, and achieve significantly faster results than state-of-the-art works that are not based on automatic solvers. We reach 9, 11, 13, 13 and 15 rounds for SPECK32, SPECK48, SPECK64, SPECK96 and SPECK128 respectively. In order to build our algorithm, we revisit Lipmaa and Moriai's algorithm for listing all optimal differential transitions through modular addition, and propose a variant to enumerate all transitions with probability close (up to a fixed threshold) to the optimal, while fixing a minor bug in the original algorithm.

Keywords: Monte Carlo Tree Search · Differential cryptanalysis · ARX · Block ciphers · SPECK

1 Introduction

Block ciphers are a major building block for modern communications and everyday applications. Assessing the security of these primitives is a difficult, yet essential task: in particular, thorough theoretical evaluation of block ciphers permits to estimate their security margin, based on the highest number of rounds that can be attacked by classical attacks, such as differential cryptanalysis [Mat94]. Differential cryptanalysis studies the propagation of a perturbation of the plaintext through the cipher, in the form of differential characteristics. This perturbation represents the difference between the evaluation of two plaintexts throughout the rounds of the cipher. The goal is to find differential characteristics with high

T. Isobe and S. Sarkar (Eds.): INDOCRYPT 2022, LNCS 13774, pp. 373–397, 2022.
https://doi.org/10.1007/978-3-031-22912-1_17

probability, since they can be used to attack the cipher. Finding such characteristics rapidly is important, as a fast search enables designers to test vast sets of parameters in a short amount of time when building new primitives.

Two main approaches coexist to find good differential characteristics: one relies on manually implemented specialized graph-based search strategies, in the line of Matsui's algorithm [Mat94], while the other uses automatic tools, such as SAT, CP, or MILP solvers. The main appeal of using solvers is that the user only needs to implement a representation of the problem in a specific paradigm, and the search itself is performed by an optimized solver, using dedicated propagators. Therefore, using a solver often results in a more efficient implementation, and less chances of human error, as the solvers are typically battle-tested. On the other hand, the generality of automatic solvers may come at the cost of performance, as more efficient specialized algorithms may exist. While the two approaches share the same final goal, the solver-based route mostly focuses on finding efficient ways to model the problem, whereas the graph-based route requires finding better ways to explore the search space.

Indeed, the difficulty in finding good differential characteristics stems from the mere size of the search space, and the resulting combinatorial explosion. However, games such as Go have comparably massive search spaces (over 10^{170} possible games), but are being dominated through AI-originated methods. In particular, Monte-Carlo Tree Search (MCTS) [CBSS08] has proven to be a good exploration strategy for multiplayer games. An extension to single-player games, called single-player MCTS [SWvdH+08], enables similar performances for non-adversarial scenarios.

In this paper, we focus on graph-based searches (as opposed to solver-based), and explore new algorithms for the search of differential characteristics. Among the three main families of block ciphers, Substitution Permutation Networks (SPN), Feistel ciphers and Addition Rotation Xor (ARX), we focus on the latter. In ARX ciphers, modular addition is used to provide non-linearity; its differential properties were extensively studied by Lipmaa and Moriai in [LM01]. Building on their work on efficient algorithms for the differential analysis of modular addition, we propose new variations, as well as a minor correction. We then propose a single-player MCTS based approach for finding differential characteristics, exploiting new heuristics, and obtain promising results on the block cipher SPECK.

Our contributions are the following:

1. We show an inaccuracy in Lipmaa-Moriai Alg. 3 for enumerating optimal transitions through modular addition, and propose a fix.
2. We propose an extension to Lipmaa-Moriai Alg. 3, to enumerate not only the transitions with optimal probability 2^{-t}, but also δ-optimal transitions, with probability better than $2^{-t-\delta}$, for a fixed offset δ. Besides being of theoretical interest, this is useful in our techniques.
3. We propose an adaptation of single-player MCTS to the differential characteristic search problem.

4. We propose a specialization of this algorithm for the block cipher SPECK, using new dedicated heuristics. These heuristics allow our tool to be faster than other graph-based techniques on all instances of SPECK, and sometimes even solver-based ones.

1.1 Related Works

The search for good differential characteristics on SPECK has first been tackled using a variant of Matsui's algorithm. Matsui's algorithm [Mat94] is a Depth-First Search (DFS) algorithm which derives A^*-like heuristics from the knowledge of previous rounds information. Initially proposed for Feistel ciphers, Matsui's algorithm was then extended to ARX ciphers in [BV14], using the concept of threshold search. Threshold search relies on a partial Difference Distribution Table (pDDT), containing all differential transitions up to a probability threshold. The same authors later noted that sub-optimal results were returned by threshold search, and proposed a new variant of Matsui's algorithm, that maintains bit-level optimality through the search. In [LLJW21], a different variant of Matsui's algorithm is proposed, where the differential propagation through modular addition is modeled as a chain of connected S-Boxes, using carry-bit-dependent difference distribution tables (CDDT). A similar method is further improved, both in the construction of the CDDT and in the search process, in [HW19].

Finally, in 2018, Dwivedi et al. used for the first time a MCTS-related method to find differential characteristics on the block cipher LEA [DS18] and, subsequently, on SPECK [DMS19]. Their work have some similarities with ours, especially the fact that we are both using single-player variants of MCTS (in their case, the Nested MCTS). The main differences are:

- in [DMS19] the expansion step is missing. Moreover, when a difference is not in the initial table, the XOR between the two words of SPECK is taken deterministically as the output difference of the modular addition.
- A scoring function is missing, so the paths are completely randomized and the results of the previous searches are not used for the new ones.

The results were sub-optimal, due to the fact that this interpretation of the MCTS is equivalent to a search that optimizes the best differential transition only locally rather than globally.

In addition to these Matsui-based approaches, the state-of-the-art solver-based results are presented in Table 1 for completeness, although we do not directly compare to them, as solver-based approaches, to this day, scale better than Matsui-based techniques for the case of SPECK. In particular, the listed results are an SMT model based on the combination of short trails by Song et al. [SHY16], an MILP model by Fu et al. [FWG+16], and an SMT model by Liu *et al.*, integrating Matsui-like heuristics [LLJW21].

Table 1. Comparison between the different techniques found in literature, with timings when reported. Solver-based works are indicated in italic.

SPECK version	Reference of the attack	Technique	Number of rounds reached	Weight	Time
32	[DMS19]	NMCTS	9	31	–
	[FWG+16]	*MILP*	9	30	–
	[SHY16]	*SMT*	9	30	–
	[BRV14]	Matsui-like	9	30	240 m
	[BVLC16]	Matsui-like	9	30	12 m
	[LLJW21]	Matsui-like (CarryDDT)	9	30	0.15 h
	[SWW21]	*Matsui + SAT*	9	30	7 m
	[HW19]	Matsui-like (CombinationalDDT)	9	30	3 m
	This work	**SP-MCTS**	**9**	**30**	**55 s**
48	[BVLC16]	Matsui-like	9	33	7 d
	[DMS19]	NMCTS	10	43	–
	[BRV14]	Matsui-like	11	47	260 m
	[SHY16]	*SMT*	11	46	12.5 d
	[FWG+16]	*MILP*	11	45	–
	[SWW21]	*Matsui + SAT*	11	45	11 h
	[LLJW21]	Matsui-like (CarryDDT)	11	45	4.66 h
	[HW19]	Matsui-like (CombinationalDDT)	11	45	2 h
	This work	**SP-MCTS**	**11**	**45**	**7 m 18 s**
64	[BVLC16]	Matsui-like	8	27	22 h
	[DMS19]	NMCTS	12	63	–
	This work	**SP-MCTS**	**13**	**55**	**48 m 50 s**
	[BRV14]	Matsui-like	14	60	207 m
	[FWG+16]	*MILP*	15	62	–
	[SWW21]	*Matsui + SAT*	15	62	5.3 h
	[HW19]	Matsui-like (CombinationalDDT)	15	62	1 h
	[SHY16]	*SMT*	15	62	0.9 h
	[LLJW21]	Matsui-like (CarryDDT)	15	62	0.24 h
96	[BVLC16]	Matsui-like	7	21	5d
	[HW19]	Matsui-like (CombinationalDDT)	8	30	162 h
	[LLJW21]	Matsui-like (CarryDDT)	8	30	48.3 h
	[SWW21]	*Matsui + SAT*	10	49	515.5 h
	This work	**SP-MCTS**	**10**	**49**	**1 m 23 s**
	[DMS19]	NMCTS	13	89	–
	This work	**SP-MCTS**	**13**	**84**	**14 m 21 s**
	[FWG+16]	*MILP*	16	87	–
	[SHY16]	*SMT*	16	≤87	≤11.3 h
128	[BVLC16]	Matsui-like	7	21	3 h
	[HW19]	Matsui-like (CombinationalDDT)	7	21	2 h
	[LLJW21]	Matsui-like (CarryDDT)	8	30	76.86 h
	[SWW21]	*Matsui + SAT*	9	39	40.1 h
	This work	**SP-MCTS**	**9**	**39**	**1 m 29 s**
	[DMS19]	NMCTS	15	127	–
	This work	**SP-MCTS**	**15**	**115**	**8 m 34 s**
	[FWG+16]	*MILP*	19	119	–
	[SHY16]	*SMT*	19	≤119	≤5.2 h

1.2 Structure of This Work

This work is structured as follows. In Sect. 2, we give reminders on relevant background knowledge. In Sect. 3, we give an overview of Lipmaa and Moriai's algorithm, which we adapted to our needs; moreover, we address an inaccuracy in the original version of the algorithm. In Sect. 4, we propose a general algorithm to address the problem of searching differential characteristics with the Monte Carlo Tree Search technique. In Sect. 5, we explain the weaknesses of the aforementioned algorithm when it is applied specifically to SPECK and we describe the solutions we adopted. We conclude the paper in Sect. 6.

2 Preliminaries

In this section, we present the main concepts on which our work is based. We describe the Monte Carlo Tree Search algorithm, the concept of differential cryptanalysis, the related structure called Difference Distribution Table, the SPECK family of ciphers and, in conclusion, one key recovery attack strategy for SPECK.

2.1 Notation

In the paper, we use the following notation. We consider bit strings of size n, indexed from 0 to $n-1$, where x_i denotes the ith bit of x, with 0 being the least significant bit, *i.e.* $x = \sum_{i=0}^{n-1} x_i \cdot 2^i$.

We respectively use $\boxplus$, $\lll$, $\ggg$ and $\oplus$ to denote addition modulo 2^n, left and right bitwise rotations and bitwise XOR.

2.2 Monte Carlo Tree Search

Monte Carlo inspired methods are a very popular approach for intelligent playing in board games. They usually extend classical tree-search methods in order to solve the problem of not being able to search the full tree for the best move (as in a BFS or a DFS, both described in [Koz92]) because the game is too complex, or not being able to construct an heuristic evaluation function to apply classical algorithms like A* or IDA*, introduced respectively in [HNR68] and in [Kor85]. The general approach of using Monte Carlo methods for tree-search related problems is referred as Monte Carlo Tree Search (MCTS). Monte Carlo Tree Search was first described as such in 2006 by Coulom [Cou06] on two-player games. Similar algorithms were however already known in the 1990s, for example in Abramson's PhD thesis of 1987 [Abr87]. MCTS for single-player games, or SP-MCTS, was introduced in 2008 by Schadd et al. [SWvdH+08], on the SameGame puzzle game.

The classical algorithm of MCTS has four main steps:

- Selection. In the selection phase, the tree representing the game at the current state is traversed until a leaf node is reached. The root of the tree here is the current state of the game (for example, the positions of the pieces in a chess board), while a leaf is a point ahead in the game (not necessarily the end). The tree is explored using the results of previous simulations.
- Simulation. In the simulation phase, the game is played from a leaf node (reached by selection) until the end. Simulation usually uses completely random choices or heuristics not depending on previous simulations or on the game so far. A payout is given when the end is reached, that in two-player games usually is win, draw or lose (represented as $\{1, 0, -1\}$). Usually for the first runs, when there is no information on the goodness of the moves in the selection phase, only the simulation is done.
- Expansion. In the expansion phase, the algorithm decides, based on the payout, if one or more of the states explored in the simulation phase are worth to be added to the tree. For each simulation a small number of nodes (possibly zero) are added to the initial tree.
- Backpropagation. In the backpropagation phase, the results of the simulation are propagated back to the root. In particular, for every node in the path followed in the selection step, some information about the final payout of the simulation is added, in order to make the following simulation phases more accurate.

Single Player Monte Carlo Tree Search. Single Player MCTS [SWvdH+08] (SP-MCTS), is an application of these techniques to single-player games. The structure of the algorithm is the same as the two-player version, with two major differences:

- In the selection phase, there is no uncertainty linked to the opponent's next moves, so that the scores can be set in a more accurate way for each node.
- In the simulation phase, the space of the payout may be way bigger than 3 elements, leading to difficulties in the backpropagation of the final score. In games where there is a theoretical minimum and maximum payout, it is usually rescaled in the interval $[0, 1]$.

The UCT Formula. For the selection phase, Schadd et al. [SWvdH+08] used a modified version of the UCT (Upper Confidence bounds applied to Trees) formula initially proposed by Kocsis and Szepesvári [KS06]. It computes the score of an edge of the search tree as:

$$UCT(N, i) = \overline{X} + C \cdot \sqrt{\frac{ln\ t(N)}{t(N_i)}} + \sqrt{\frac{\sum x_j^2 - t(N_i) \cdot \overline{X}^2 + D}{t(N_i)}}$$

where N is the current node, N_i is the i-th child node of N ($i \in \{1, 2, \dots n\}$ if the node N has n children nodes), the x_j are the scores of the runs started from node N_i, $\overline{X}$ is the average of them, $t(N)$ is the number of visits of the node N, and C, D are constants to be chosen.

2.3 Differential Cryptanalysis

Differential cryptanalysis is a technique introduced by Biham and Shamir in [BS91] and used to analyze the security of cryptographic primitives. The basic element used in this field is a *difference*, which is a perturbation of the input or the output of the studied function. Usually the differences are defined as XOR ones, so, given two plaintexts p_0, p_1 and the corresponding ciphertexts c_0, c_1, we call an *input difference* a value $\Delta p = p_0 \oplus p_1$ coming from the XOR of the two plaintexts, and an *output difference* $\Delta c = c_0 \oplus c_1$ the one coming from the two ciphertexts. The pair of input and output differences $(\Delta p, \Delta c)$ is called a *differential.* For primitives divided in rounds, we call the sequence of differentials for each round a *differential characteristic.*

Differentials and differential characteristics are (usually) not deterministic due to non-linear components in the structure of cryptographic primitives, so the main goal for the cryptanalyst is to calculate their probability for randomly sampled plaintexts. More formally, for a function f we have

$$\mathbb{P}_f(\Delta p \to \Delta c) = \frac{\sum_{p_0 \in P} \mathrm{Id}(f(p_0) \oplus f(p_0 \oplus \Delta p) = \Delta c)}{|P|},$$

where P is the space of possible plaintexts and Id is the identity function, assuming value 1 if the condition is true and 0 otherwise.

For differential characteristics we can usually rely on the Markov assumption, which is formalized in [LMM91], having

$$\begin{aligned} &\mathbb{P}_f(\Delta p \to \Delta_1 \to \Delta_2 \to \cdots \to \Delta_n \to \Delta_C) = \\ &= \mathbb{P}_f(\Delta p \to \Delta_1) \cdot \mathbb{P}_f(\Delta_1 \to \Delta_2) \cdot \ldots \cdot \mathbb{P}_f(\Delta_n \to \Delta c). \end{aligned}$$

This assumption does not hold in general since it relies on particular conditions. In the case of key-alternating ciphers, i.e., the round keys are added by XOR as in SPECK, having independent and uniformly distributed round keys is sufficient. However, the assumption is usually made for practical reasons.

The key point of differential cryptanalysis is usually to find differential characteristics that propagate with a high probability through the largest possible numbers of rounds.

2.4 Modular Addition and (Partial) DDTs

The source of branching in our search is the non-linear component, the modular addition modulo 2^n. Its differential properties were famously studied by Lipmaa et al. [LM01] and Biryukov et al. [BV13].

Given a differential we can define the *XOR differential probability of modular addition* xdp^+ as

$$\mathrm{xdp}^+(\alpha, \beta, \gamma) = \frac{|\{(a, b) : (a \oplus \alpha) \boxplus (b \oplus \beta) = (a \boxplus b) \oplus \gamma\}|}{2^{2n}}.$$

Similarly we can define xdp^- for modular subtraction.

In this paper, we sometimes refer to the inverse base 2 logarithm of a differential probability, *e.g.*, $-\log_2(\text{xdp}^+(\alpha,\beta,\gamma))$, as its *weight*.

Lipmaa et al. showed that $\text{xdp}^+(\alpha,\beta,\gamma) > 0$ if and only if $\alpha_0 \oplus \beta_0 = \gamma_0$ and for every position i such that $\alpha_i = \beta_i = \gamma_i$ we have $\gamma_{i+1} = \alpha_{i+1} \oplus \beta_{i+1} \oplus \beta_i$.

The authors give then a closed formula for this probability, that is

$$\text{xdp}^+(\alpha,\beta,\gamma) = 2^{-(n-1)+w}$$

where w is the number of indices i such that $\alpha_i = \beta_i = \gamma_i$, excluding the most significant bit.

Moreover, they give an efficient algorithm to find all values of γ such that $\text{xdp}^+(\alpha,\beta,\gamma)$ is maximum for fixed α and β. This algorithm is described in the next section.

For some functions, such as SBoxes, a *difference distribution table* (DDT) containing the possible differential transitions and their probabilities can be built.

In the case of modular addition, as n grows, the size of the DDT makes it impractical to compute and store, as it would need to store all $2^{2\cdot n}$ possible input differences, and up to 2^n output differences for each input difference. To address this issue, in [BV13], Biryukov et al. proposed the idea of a *partial DDT* (pDDT), where only differential transitions with probability greater than a fixed threshold are stored. The authors have shown that, for some families of functions, an efficient algorithm to compute pDDT entries exists, and this is the case for modular addition.

The algorithm relies on the fact that, calling

$$p_k = \text{xdp}^+(\alpha_{k-1}...\alpha_0, \beta_{k-1}...\beta_0, \gamma_{k-1}...\gamma_0),$$

it holds $1 := p_0 \geq p_1, \geq ... \geq p_{n-1}$. From this fact, the algorithm constructs the table bit-by-bit. The interested reader can find the details in the original work.

2.5 The SPECK Family of Block Ciphers

SPECK [BSS+15] is a family of ARX block ciphers proposed in 2013 by the National Security Agency (NSA). SPECK comes in five versions, identified by their block sizes (in bits) as SPECK32, SPECK48, SPECK64, SPECK96 and SPECK128; each version has different options for the key size, which, together with the block size, determines the number of rounds.

The state of the cipher is divided in two words of $N/2$ bits, where N is the block size (for example, SPECK32 has two words of 16 bits); calling x_i and y_i the input words at round i, the cipher can be described as

$$x_{i+1} = ((x_i \ggg \alpha) \boxplus y_i) \oplus k_i,$$

$$y_{i+1} = (y_i \lll \beta) \oplus x_{i+1},$$

where α and β are constants depending on the version of SPECK ($(\alpha,\beta) = (7,2)$ for SPECK32 and $(\alpha,\beta) = (8,3)$ otherwise). The term k_i refers to the round key, obtained from the master key through the key schedule algorithm.

2.6 Differential Characteristics and Key Recovery in SPECK

In 2014, Dinur [Din14] proposed an attack on round-reduced versions of all the variants of SPECK. Starting from an r round differential characteristic, the attack recovers the last two subkeys of the $r+2$ rounds cipher working with a guess-and-determine strategy on the last two modular additions of the cipher. The attack can be extended to $r+4$ rounds by bruteforcing two more subkeys, adding a complexity of 2^{2n}.

3 Lipmaa's Algorithms: Known Facts and New Results

In [LM01], Lipmaa and Moriai present a set of algorithms for the study of the differential behaviour of modular addition. The most widely used of these algorithms is Algorithm 2, which, given α, β, γ, returns $\text{xdp}^+(\alpha, \beta \to \gamma)$; it is a cornerstone in the differential cryptanalysis of ARX ciphers. A less known, yet very useful result, is Algorithm 3 (Lipmaa-Moriai Alg. 3), which, given α, β, enumerates all output differences γ such that $\text{xdp}^+(\alpha, \beta \to \gamma)$ is maximal.

In this section, we present a generalization of Lipmaa-Moriai Alg. 3 to find good but not optimal transitions, and a fix for an inaccuracy in the original algorithm, leading to wrong results for some inputs. The final algorithm is reported at the end of the section.

3.1 Overview of Algorithm 2

As a reminder, the output difference γ to a modular addition is equal to $\alpha \oplus \beta \oplus \delta_c$, where δ_c denotes a difference in the carry.

Algorithm 2 first determines whether a transition from (α, β) to γ is valid, before computing its probability. A transition is said to be valid iff

$$eq(\alpha \ll 1, \beta \ll 1, \gamma \ll 1) \wedge (\alpha \oplus \beta \oplus \gamma \oplus (\beta \ll 1)) = 0 \tag{1}$$

where $x \ll 1$ is the left shift, which append a 0 at the rightmost side of x's bit representation, and $\text{eq}(x, y, z)$ is 1 in all positions where $x_i = y_i = z_i$, and 0 elsewhere.

This condition stems from the observation that three carry patterns are deterministic, whereas the other cases all have probability $\frac{1}{2}$:

1. $\gamma_0 = \alpha_0 \oplus \beta_0$
2. If $\alpha_i = \beta_i = \gamma_i = 0$, then $\gamma_{i+1} = \alpha_{i+1} \oplus \beta_{i+1}$ (because it implies that $\delta_{c_{i+1}} = 0$)
3. If $\alpha_i = \beta_i = \gamma_i = 1$, then $\gamma_{i+1} = \alpha_{i+1} \oplus \beta_{i+1} \oplus 1$ (because it implies that $\delta_{c_{i+1}} = 1$)

Any transition violating these conditions is invalid; all other transitions are possible. It is easy to verify that Eq. 1 eliminates the invalid transitions.

The probability of a valid transition is determined by the number of occurrences w of above mentioned deterministic carry propagation cases 2 and 3, excluding the most significant bit, as 2^{-n+1+w}.

3.2 High Level Overview of Lipmaa-Moriai Alg. 3

Following the notations of [LM01], let l_i be the length of the longest common alternating bit chain: $\alpha_i = \beta_i \neq \alpha_{i+1} = \beta_{i+1} \neq ... \neq \alpha_{i+l_i} = \beta_{i+l_i}$, and let the *common alternation parity* $C(\alpha, \beta)$ be a binary string with length n defined as:

- $C(\alpha, \beta)_i = 1$ if l_i is even and non-zero,
- $C(\alpha, \beta)_i = 0$ if l_i is odd,
- unspecified when $l_i = 0$ (can be both 0 and 1, not affecting subsequent algorithms since there is no chain).

The interested reader can find an algorithm to retrieve $C(x, y)$ in $O(\log n)$ in the original work [LM01]. This tool is the main ingredient used by the authors to construct Algorithm 3, an algorithm that, given in input two n-bit values α, β, retrieves all the possible values γ such that the probability of modular addition with respect to xor: $\mathrm{xdp}^+(\alpha, \beta \to \gamma)$ is maximum.

Alternating chains are relevant to Lipmaa-Moriai Alg. 3, because in the case of a chain of length 2, the carry propagation rules force at least one probabilistic transition: if $\gamma_i = \alpha_i = \beta_i$, then we have $\gamma_{i+1} = \alpha_{i+1} \oplus \beta_{i+1} \oplus \beta_i$, and by definition $\gamma_{i+1} \neq \alpha_{i+1}$, so that γ_{i+2} is free. Conversely, if $\gamma_i \neq \alpha_i$, then γ_{i+1} is free; in both cases, a probability is paid. Intuitively, the number if times a probability is paid for an even length chain is $\frac{l_i}{2}$, whereas for an odd length chain, it depends on which value is chosen first.

In Lipmaa-Moriai Alg. 3, the list of optimal γ values is built bit-by-bit, starting from position 1; position 0 is always set to $\alpha_0 \oplus \beta_0$, following rule 1.

For the remaining bits, 3 cases are to be distinguished:

(a) if $\alpha_{i-1} = \beta_{i-1} = \gamma_{i-1}$, then the choice $\gamma_i = \alpha_{i-1} \oplus \alpha_i \oplus \beta_i$ is the only valid option, by transition rule 1.
(b) else if $\alpha_i \neq \beta_i$, then both choices of γ_i incur a probability of $\frac{1}{2}$ (as none of the deterministic transitions are available); this is equivalent to a chain of length 0. Similarly, if $i = n - 1$, then both choices are equivalent, as position $i - 1$ is not part of the total probability. Finally, if $\alpha_i = \beta_i$ but $C(\alpha, \beta)_i = 1$, then both choices are equivalent again; in reality, this last case is not completely true, but we will come back to it at the end of the section.
(c) Finally, when $\alpha_i = \beta_i$ and $C(\alpha, \beta)_i = 0$, choosing $\gamma_i = \alpha_i$ results in a probability cost equal to $2^{-\lfloor \frac{l_i}{2} \rfloor}$ for the next l_i positions, whereas the other choice has cost $2^{-\lfloor \frac{l_i}{2}+1 \rfloor}$, so that the optimal choice is $\gamma_i = \alpha_i$.

For the remainder of this section, we refer to these as case or branch (a), (b), (c) respectively.

3.3 A Fix for the Original Algorithm

Lipmaa-Moriai Alg. 3 presents an inconsistency. Consider for example the input difference $(\alpha, \beta) = (1011_2, 1001_2)$; we have $C(\alpha, \beta) = 0100_2$. Applying Algorithm 3, we find:

- $\gamma_0 = 0$ (initialisation case)
- $\gamma_1 = \{0, 1\}$ (case (b), since $\alpha_1 \neq \beta_1$)
- $\gamma_2 = \{0, 1\}$ (case (b), since $C(\alpha, \beta)_2 = 1$)
- $\gamma_3 = 0$ if $\gamma_2 = 0$, $\{0, 1\}$ otherwise.

Therefore, $\gamma = 1110_2$ is listed as optimal. However, we have $\text{xdp}^+(1011_2, 1001_2 \rightarrow 1110_2) = 2^{-3}$, while the optimal probability is 2^{-2} (reached, for instance, with $\gamma = 0010_2$). The discrepancy occurs when $C(\alpha, \beta)_{n-2}$ is equal to 1, and $\alpha_{n-3} \neq \beta_{n-3}$. The proof given in [LM01] considers both choices of γ_i equivalent in the (b) branch when $C(\alpha, \beta)_i = 1$, because the length of the chain is $\frac{l}{2}$, and choosing 0 or 1 only shifts the probability vector. This is however incorrect when the chain ends at position $n - 1$, as this position does not count in the probability, and can therefore not be counted as *bad*.

However, at position $n-2$, picking $\gamma_{n-2} = \alpha_{n-2}$ implies that no probability is paid (because $\text{eq}(\alpha_{n-2}, \beta_{n-2}, \gamma_{n-2}) = 1$), and position $n-1$ is free by definition. On the other hand, picking $\gamma_{n-2} \neq \alpha_{n-2}$ costs a probability, so that both choices are *not* equivalent in this case.

To fix this issue, the bit string returned by the common alternation parity algorithm can be modified so that all positions that are part of a chain ending at position $n-1$ are set to 0. The new algorithm to compute $C(\alpha, \beta)$ is reported in Algorithm 1.

Algorithm 1. Fix for the computation of $C(\alpha, \beta)$

Require: a bit-size $n \geq 1$, two n-bits input differences α, β.
Ensure: the corrected version of $C(\alpha, \beta)$ to make Lipmaa-Moriai Alg. 3 work.

```
p = C_LM(α, β)                    ▷ original version from Lipmaa and Moriai

for i = 0 to n − 1 do
    j = n − 1 − i
    if α_j = β_j and α_{j−1} = β_{j−1} and α_j ≠ α_{j−1} then
        p_j = 0
    else
        break
return p
```

In addition, Lipmaa-Moriai Alg. 3 describes a solution by the values allowed for γ only (rather than building an explicit list). Consider $\alpha = 0b0010, \beta = 0b1011$: for this example, $C(\alpha, \beta)_1 = 1$, so that the elif branch is chosen for bit 1, allowing both 0 and 1 for γ_1: the possible values for γ_2 depends on the choice made for γ_1. Removing information on this dependency leads to invalid or sub-optimal solutions being enumerated (such as 0b1101). This can be addressed either via building an explicit list, or with a graph representation described further. The final fixed algorithm is Algorithm 2 with $\delta = 0$.

3.4 Finding δ-Optimal Transitions

We propose a generalization of Lipmaa-Moriai Alg. 3 (see Algorithm 2), which takes as input α, β, δ, where δ is an offset, such that the algorithm returns all γ having $\text{xdp}^+(\alpha, \beta \rightarrow \gamma) \geq max_\gamma(\text{xdp}^+(\alpha, \beta \rightarrow \gamma)) \cdot 2^{-\delta}$; *i.e.*, solutions with probability within a distance $2^{-\delta}$ of the optimal. We call such solutions δ-optimal.

Intuitively, the goal is to modify a branch to eliminate at most δ visits of case (a) compared to an optimal difference, paying every time a cost of $\frac{1}{2}$.

Violating case (a) immediately leads to a transition with probability 0, per rules 2 and 3. On the other hand, the values chosen in case (b) have no influence on the final probability. Therefore, we focus on case (c).

Our algorithm works as follows: for at most δ times, when in branch (c), chose $\gamma_i = \neg\alpha_i$. Therefore, at position $i + 1$, branch (a) cannot be chosen anymore. Intuitively, this is equivalent to paying a probability cost at a position that should be free. In order to list all solutions, we go through all $\sum_{i=0}^{\delta} \binom{t}{i}$ possible positions, where t is the number of visits to case (c) in Lipmaa-Moriai Alg. 3.

We now give arguments for the soundness and completeness of our algorithm; *i.e.*, show that our algorithm returns only δ-optimal solutions, and that it returns all δ-optimal solutions.

Soundness. By Lemma 2 of [LM01], $\text{xdp}^+(\alpha, \beta, \gamma) = 2^{-(n-1)+w}$, where w is the number of visits to branch (a). In our algorithm, we change the outcome of branch (c), effectively forbidding one access to branch (a), at most δ times, therefore adding a factor at most $2^{-\delta}$ to the final probability.

Completeness. Assume γ' to be a δ-optimal output difference for a given (α, β), such that it is not found by our algorithm. Let γ'' be a δ-optimal returned by our algorithm for the same (α, β). Compare these differences bit-by-bit: if they differ at an index that (in our difference γ'') originated from case (b), we have it in our list. If the difference originates from case (c), then we also have it since we flipped all the possible combinations of indices originating from case (c). As discussed before, the difference can not be originated from case (a). Notice that we can always choose γ'' since our algorithm (as well as Lipmaa's) always outputs at least one valid solution.

Complexity. Lipmaa-Moriai Alg. 3 is described in the original paper as a linear-time algorithm. This is, however, not direct from the description given by the authors: in particular, if we consider the case $\alpha \oplus \beta = 2^n - 1$, then branch (b) is the only possible choice for all bit positions except 0. This means that, all 2^{n-1} choices for the remaining bits of γ are valid, and the enumeration is exponential.

This enumeration issue can be addressed by using a compact representation of all possible γ in linear time, by representing the solution space as a directed graph $G = (V, E)$, with $2 \cdot n$ vertices, and at most $4 \cdot n$ edges. In this representation, vertices $V_{i,0}$ and $V_{i,1}$ represent the statement *bit i of γ takes value 0 (resp. 1)*, and vertex $V_{i,j}$ is connected to vertex $V_{i+1,k}$ if $(\gamma_i, \gamma_{i+1}) = (j, k)$ is a pair that belongs to the set of all optimal γ values. A γ value is 0−optimal iff

$V_{0,\gamma_0}, V_{1,\gamma_1}, \ldots, V_{n-1,\gamma_{n-1}}$ is a connected path in the graph. Through the loop of Lipmaa-Moriai Alg. 3, each vertex is visited at most once, yielding a time complexity in $O(n)$. Sampling an optimal solution from the graph can then be done in $O(n)$, by following a connected path.

This representation is possible because the choice of a bit value at position i is independent from the choices made before position $i-1$. On the other hand, when further dependencies exist, as in our variant, the situation is more complex.

Our variant introduces additional computations:

1. We add a pass to zero some values of $C(\alpha, \beta)$, according to the fix mentioned previously. The computation becomes worse-case n, rather than logarithmic;
2. In order to enumerate all the solutions, we need to go through $\sum_{i=0}^{\delta} \binom{t}{i}$ (with t the maximum number of visits to the (c) branch) possible positions of flip in the (c) case.

Point 1 is not an issue, as the computation of $C(\alpha, \beta)$ is only done once at the start of the algorithm. On the other hand, point 2 prevents application of the aforementioned graph approach, as the possible choices for bit i now depend on a state defined by the number of times branch (c) was flipped. On the contrary, our graph representation requires bit i to only depend on bit $i-1$, and not on the previous choices.

We therefore propose to have one graph for each combination of flipped bits, effectively multiplying the computation time by $\sum_{i=0}^{\delta} \binom{t}{i}$, resulting in a complexity in $\Theta(n^{\delta})$, with δ a constant. Crucially, the number of visits to branch (c) t is loosely upper bounded by $\frac{n}{2}$ (as it requires a chain of odd length), and we restrict ourselves to δ values lower than 3, so that the computation overhead factor is upper bounded by $\sum_{i=0}^{2} \binom{32}{i} = 528$ for 64 bit words, as in SPECK-128.

Sampling a δ-optimal solution from the graph can be done in linear time, by choosing one of the graphs at random, and following a connected path, while the enumeration can be done, for example, with a DFS. This approach can however lead to duplicate solutions, so that using an explicit list of solutions remains the best way for full enumeration.

4 Differential Characteristic Search with MCTS

In this section, we outline a general strategy to find differential characteristics with MCTS, using Lipmaa's algorithm, for ciphers with a single modular addition per round. This generic algorithm is not sufficient in practice, so that cipher specific optimizations are required, which we address in the next section for SPECK.

Algorithm 2. Generalized Lipmaa-Moriai Alg. 3

Require: a bit-size $n \geq 1$, two n-bits input differences α, β and the offset $0 \leq \delta \leq n-1$.
Ensure: all possible output differences γ such that $\mathrm{xdp}^{+}(\alpha, \beta \to \gamma)$ differs by at most a $2^{-\delta}$ factor from the optimal one in the form of graphs. In order to sample from them, we can use a simple randomized traversal.

```
Class Node:
    lsb = -1
    successors = [[False, False], [False, False]]

graphs = []
p = C(α, β)                                ▷ our fixed version, as stated in Algorithm 1

procedure GenGraph(α, β)
    possibleCPositions = [i for i = 1 to n − 1 if α_i = β_i]
    positionsLists = [combinations(possibleCPositions, i) for i = 0 to δ]
    for positions in positionsLists do
        graph = [new Node() for i = 0 to n − 1]
        graph.lsb = α_0 ⊕ β_0
        for i = 1 to n − 1 do
            for j ∈ {0, 1} do
                if (i = 1 and graph.lsb = j) or (i ≥ 2 and graph[i − 2].successors[0][j]
or graph[i − 2].successors[1][j])) then
                    if α_{i−1} = β_{i−1} = j then
                        graph[i − 1].successors[j][α_i ⊕ β_i ⊕ β_{i−1}] = True
                    else if α_i ≠ β_i or p_i = 1 or i = n − 1 then
                        graph[i − 1].successors[j] = [True, True]
                    else
                        if i is in positions then
                            graph[i − 1].successors[j][1 − α_i] = True
                        else
                            graph[i − 1].successors[j][α_i] = True
        Append graph to graphs
    return graphs
```

4.1 A General Algorithm

The general idea behind our algorithm is to start with a tree that is as small as possible and expand it with the algorithm presented in Algorithm 2.

Building the Initial Tree. The initial plaintext difference is chosen from a pDDT with threshold probability $t = 2^{-\tau}$, built following Biryukov et al.'s [BV14] algorithm. A virtual root node is set to have all entries of the pDDT as its children at the start of the search.

Exploring Paths. We begin our simulation of differential characteristics as runs of a single-player game. We start from the virtual root (that can be seen as the fixed starting position of a game), and select one of the differences in the pDDT as our initial plaintext difference. We use a second threshold k to determine *how* we choose this difference. Suppose for the moment that every node has children:

- if the node has already been visited at least k times, we select the best child according to its score, using the UCT formula from Schadd et al. [SWvdH+08]; at the end of the run, we update the score of each node of the path using the same formula.
- If the node has not been visited k times yet, we choose a child uniformly at random from allowed choices, using again the UCT formula to update the scores at the end of the game. This allows us to have enough information on the node before making choices based on the previous games.

These two cases can be seen respectively as the *selection* and *simulation* steps of the classic MCTS algorithm.

Choosing the Plaintext Difference. We add a tweak to the selection of the plaintext difference: we select it uniformly at random from the pDDT for the first k iterations, then we store the input differences in a sorted list in descending order based on their score, and select them using a geometrical distribution with probability p. This favors exploration over exploitation, by permitting each difference to have some probability to be chosen at every run. Experimentally, we found that this techniques dramatically improve the performance of the initial difference selection.

Tree Expansion. If the node has no children, *i.e.* no corresponding entry in the pDDT, then we need to generate some. For this purpose, we use our modified version of Lipmaa-Moriai Alg. 3 presented in the previous section. This comes from the idea that choosing always the best possible next difference is a very local strategy, that does not allow us to look for long characteristics. In practice, we fix a penalty threshold δ and list all the possible choices differing at most $2^{-\delta}$ from the optimal one, *i.e.*, the δ-optimal transitions. We then add them to the tree and proceed with our exploration strategy. This approach, in the case of SPECK, is explained in more details in the following section.

Scoring the Nodes. To score the nodes, we use the UCT formula, with a custom formula for the payouts. Our choice here is to mix the global weight of the characteristic with a measure of the local one, weighted appropriately. This results in a scoring that is similar to the one used in the α-AMAF heuristic presented in [HPW09]. In formulas, we have that each payout used to compute the UCT score has this form:

$$x = \beta G + (1 - \beta)L,$$

where:

- G is the global score of the characteristic, calculated as $\frac{1}{w}$, with w being the weight of the differential characteristic.
- L is the local score, calculated as $\alpha\frac{1}{w'}$, where w' is the weight of the differential characteristic *from this point to the end*, and α is a normalization constant.
- $0 \leq \beta \leq 1$ is a constant to weight the two parts of our score.

The purpose of this kind of scoring is to measure the choice of a difference relatively to the current round, because some choices can be good at some point of the characteristic (i.e. near the end, if they have a very good probability) but very bad in others (i.e. near the beginning, if they do not generate good successive choices). This score is then used to backpropagate the results to each node of the path up to the root, meaning that the value of x is added to the list of scores (used inside the UCT formula) of each encountered node.

4.2 Limitations of This Approach

We outline here the two main issues that can arise from the application of this method to a real cipher.

The Branch Number. Even with a small value of δ, expanding the tree can lead to nodes with a very high number of children. Intuitively, this is bad for MCTS, because for its score to be precise, a node must be visited at least a few times, and this becomes harder as the tree gets wider. Because of this issue we need to find a way to give a limitation on the expansion without affecting the result of the search.

The Choice of the Plaintext Difference. In our outline, we proposed to choose the initial plaintext difference inside a pDDT. Experimentally, this works very well when looking for short differential characteristics, but not too well for longer ones. The motivation here is similar to the one of the tree expansion: with the exception of pathological ciphers or cyclic characteristics, in general, differential characteristics start with differences that allow a long propagation without increasing the cost too much. This is not guaranteed to happen with a small pDDT, and creating a very big one can make the branching number too high for the search to work.

How to solve these issues and their impact on the actual search is very cipher-dependent. In the following section, we try to address both of them in our application to the SPECK cipher.

5 Application to SPECK

In this section, we apply the previously described method to the search for differential characteristics on the SPECK cipher. The initial discussion is done on the SPECK32 version, but applications and results for all the versions of SPECK are discussed in the last subsections. We stress again that our objective is to show that our algorithm can be competitive against the state-of-the-art Matsui-like approaches. For this reason, we put ourselves in the same settings as them instead of pushing for a very large number of rounds, showing that our implementation finds good characteristics way faster. We leave optimizations, generalizations and the understanding of the limits of this algorithm for future works.

5.1 The Start-in-the-Middle Approach

We start by tackling what, in our opinion, is the biggest limitation of our previous approach: the choice of the initial difference. In order to better explain the problem, and our solution, we used a SAT solver to list all the optimal differential characteristics for 9 rounds on SPECK32. They are reported in Appendix A. We start by noticing that there are only two characteristics that start with a transition with probability 2^{-3}, while most of them start with 2^{-5}. As reported by Biryukov et al., a pDDT containing all the possible differential transitions with probability up to 2^{-5} contains about 2^{30} elements in the case of SPECK32, that is impossible to handle with MCTS.

Another observation from the reported characteristics is that each of them contains a transition with probability 1 or 1/2. Our aim is to start from that point. We start by creating a pDDT with all the transitions with probability at most 1/2. For SPECK32 this table contains 183 transitions, that is a lot more tractable than 2^{30}. Suppose for the moment that we are looking for a differential characteristic on r rounds, and that we know the position s of this "low weight" difference inside the characteristic. We build a *cache* by applying our strategy on $r - s$ rounds for a fixed number of iterations of MCTS. At the end of this procedure we have a table that maps every low weight difference to a characteristic starting with it. Then we simply run MCTS again in the backward direction for s rounds. Notice that we can use the exact same algorithm that we described in Sect. 3 because for every α, β, γ it holds

$$\mathrm{xdp}^{+}(\alpha, \beta, \gamma) = \mathrm{xdp}^{-}(\alpha, \beta, \gamma).$$

To conclude, we can simply drop the assumption of knowledge of s by brute-forcing it: we start r parallel processes to do the search with all possible values of s and we find one or more values that generate optimal characteristics. We call this approach the *start-in-the-middle*, as an analogy with the classic meet-in-the-middle one. The pseudocode for this algorithm is given in Sect. C.

5.2 Branching Number and the Choice of δ

We then address the other issue pointed out in the previous section: the branching number. From now on we will call the *offset* of a differential characteristic the maximum possible deviation of a transition inside the characteristic from an optimal one. For example: if all the transitions in the characteristic are optimal, then its offset is 0. Otherwise, if there is at least a transition that deviates from the optimal by a factor $2^{-\delta}$ and no bigger deviations, we say that the offset of that characteristic is δ. We start again by analyzing our characteristics on SPECK32. We can see that none of them has offset equal to 0, while only three, which are very similar to each other, have offset equal to 1. On the contrary, almost all the other characteristics, which are different from the aforementioned three, have at least one transition that makes their offset equal to 2. For completeness, it has to be said that only one characteristic among those 15 has offset equal to 3, and there are no bigger offsets. Motivated by this we decided to run our expansion step keeping δ between 1 and 3. This is a very crucial part of our algorithm: in fact, we stress again that the MCTS algorithm needs to explore each branch several times in order to assign an accurate score and make better choices. This is also the main reason behind the fact that chess (and other games) are dominated by computers, while Go is a lot harder. If we compare the branching factor of the two games, chess's one is 35, while Go's is very large, with a value of about 200 [BW95]. This implies a huge difference when comparing the sizes of the two corresponding trees. When dealing with differential characteristic search, if not limited, the branching factor could be even bigger than Go's one, having a maximum value of 2^{n-1} when $\alpha \oplus \beta$ is $2^n - 1$.

5.3 Adding Further Heuristics to Improve the Search

With the previous approach we produce, on average, 83 children to each node on SPECK32 when $\delta = 1$. This number is in line with what we mentioned for the game of chess, and in fact it is enough to find an optimal differential characteristic for this version of SPECK; however, the branch number becomes too large for bigger versions of SPECK. This is not feasible anymore, so we need to add further heuristics to reduce these numbers.

Low Hamming weight differences. As it can be observed in all characteristics found for SPECK and for several other ARX ciphers, good differentials have, in general, a low Hamming weight. Intuitively, this makes sense because we want the smallest possible number of carry propagations to have higher probabilities. This heuristic has already been used in literature to improve the performances of algorithms that find differential characteristics on SPECK, e.g. Biryukov et al. in [BRV14].

Specifically, in our work, we use two kinds of filters based on the Hamming weight of α, β and γ: the first one is based on the Hamming weight of each word, while the second one limits the sum of the Hamming weights of the three words.

Based on the known list of characteristics for SPECK32, we have that the maximum value for the Hamming weight of each 16-bit words is 8, while the average is 4.7. The sum of the three Hamming weight has a maximum value of 20 and an average of 13.1. We use these to derive the parameters given in the experimental results section.

The Expansion Threshold. Another optimization that we considered is to *choose to not expand some nodes.* In addition to the bounding done through δ-optimal transitions, we choose to further bound the probability of each transition by a fixed threshold. In practice, we do not allow for transitions with probability lower than 2^{-12} This is because nodes with good optimal transition probability generate on average a small number of δ-optimal transitions, while bad optimal transitions usually explode into very big numbers of δ-optimal transitions. Intuitively, a low optimal probability implies numerous visits to branches (b) and (c) in Lipmaa-Moriai Alg. 3; each visit in branch (b) adds valid solution (as both bit values are allowed), and each visit to branch (c) affects the $\sum_{i=0}^{\delta} \binom{t}{i}$ factor in the enumeration, and thus the number of solutions.

Using these heuristics significantly reduces the size of the search space, and enable better scaling for larger versions of SPECK.

5.4 Experimental Results and Discussion

All experiments are performed on a laptop equipped with an Intel® Core™ i7-11800H 3.6 GHz. The code is implemented in Python and executed with PyPy3.6. The results are presented in Table 1. The parameters used in the search were:

- $C = \frac{1}{4}$ and $D = 100$ for the UCT, for all the versions.
- $\beta = \frac{1}{5}$ to balance the scoring function, for all the versions.
- $p = \frac{1}{4}$ for the geometric distribution, for all versions.
- $\delta = 2$ for all the versions except SPECK32, for which $\delta = 1$ was enough.
- 10^5 forward iterations for each version to build the cache.
- $(t_1, t_2) = (8, 20)$ for the two Hamming weight thresholds on SPECK32, while $(12, 24)$ was used for all the other versions.
- A probability threshold of 2^{-12} was used for SPECK32, while 2^{-16} was used on all the other versions.
- $k = 5$ for the number of visits of a node before starting to use the UCT, for all the versions.

A key difference between MCTS and others is that the approach is not complete; therefore, it is not able to determine when a solution is optimal, and can keep searching until it exhausts all its allowed iterations. Because we let the search in the backwards direction run without an iteration limit, we do not have a stopping time to report; however, we report the time after which a solution is found by our program.

For SPECK32 and SPECK48, the optimal differential characteristics are found significantly faster than for state-of-the-art graph-based search methods, as well as solvers. This is encouraging, even though it is worth noting that solvers may require additional time to prove optimality; in that sense, the methods are not directly comparable.

SPECK64 appears to be more difficult for our algorithm, as we can only find good differential characteristics up to 13 rounds. We assume that the depth of the tree makes the search more difficult for MCTS, as we generally struggle with characteristics longer than 12 rounds.

For SPECK96, we find the optimal solution for 10 rounds in less than one and a half minute, significantly outperforming the 48 h of the closest graph-based approach. We also report a non-optimal result for 13 rounds, found in 12 min, as a comparison with the previous Monte-Carlo based approach. However, solver-based methods remain significantly ahead for this version of SPECK.

A similar analysis holds for SPECK128, where our approach dominates for small number of rounds (up to 9), but, similarly to the other graph-based approaches, does not scale to as many rounds as solver-based methods.

6 Conclusions

In this paper, we studied variations of custom search algorithms for the search of differential characteristics for SPECK, using SP-MCTS. In the process, we revisited Lipmaa-Moriai Alg. 3 to provide an efficient algorithm for the enumeration of δ-optimal differentials. A naive implementation of SP-MCTS proved to be inefficient, so that we derived additional heuristics from the structure of known good characteristics, allowing us to outperform all other graph-based methods for most instances, and sometimes even solver-based ones.

Our approach, on the other hand, seems to struggle with longer characteristics, equivalent to deeper trees. Further performance gains could be achieved by additional heuristics, possibly derived through reinforcement learning, or through parallelization, as well as further parameters tuning, in particular in the scoring function.

This research is very specific to the SPECK cipher, and it would be interesting to evaluate how it can be extended to other ARX constructions, in particular those with more than one modular addition per round, or even to SPN constructions. Our results constitute a new step along the graph-based search route, which, while more challenging than solver-based approaches, has the potential to outperform solvers through specialization.

Appendix A All Optimal Characteristics on SPECK32

See Table 2.

Table 2. A list of all the differential characteristics with weight 30 in SPECK32.

r	Δ_L	Δ_R	$-\log_2 p$	r	Δ_L	Δ_R	$-\log_2 p$	r	Δ_L	Δ_R	$-\log_2 p$
-	0211	0a04	–	–	7448	b0f8	–	–	8054	a900	–
1	2800	0010	4	1	01e0	c202	5	1	0000	a402	3
2	0040	0000	2	2	020f	0a04	5	2	a402	3408	3
3	8000	8000	0	3	2800	0010	5	3	50c0	80e0	8
4	8100	8102	1	4	0040	0000	2	4	0181	0203	4
5	8004	840e	3	5	8000	8000	0	5	000c	0800	5
6	8532	9508	8	6	8100	8102	1	6	2000	0000	3
7	5002	0420	7	7	8000	840a	2	7	0040	0040	1
8	0080	1000	3	8	850a	9520	4	8	8040	8140	1
9	1001	5001	2	9	802a	d4a8	6	9	0040	0542	2
-	1488	1008	–	–	ad40	0012	–	–	a540	0012	–
1	0021	4001	4	1	8148	8100	5	1	8148	8100	5
2	0601	0604	4	2	1002	1400	3	2	1002	1400	3
3	1800	0010	6	3	1060	4060	4	3	1060	4060	4
4	0040	0000	3	4	0180	0001	5	4	0180	0001	5
5	8000	8000	0	5	0004	0000	3	5	0004	0000	3
6	8100	8102	1	6	0800	0800	1	6	0800	0800	1
7	8000	840a	2	7	0810	2810	2	7	0810	2810	2
8	850a	9520	4	8	0800	a840	3	8	0800	a840	3
9	802a	d4a8	6	9	a850	0952	4	9	a850	0952	4
-	a000	0508	–	–	7458	b0f8	–	–	0050	8402	–
1	0448	1068	4	1	01e0	c202	5	1	2402	3408	3
2	80a0	c100	5	2	020f	0a04	5	2	50c0	80e0	7
3	0207	0604	6	3	2800	0010	5	3	0181	0203	4
4	1800	0010	5	4	0040	0000	2	4	000c	0800	5
5	0040	0000	3	5	8000	8000	0	5	2000	0000	3
6	8000	8000	0	6	8100	8102	1	6	0040	0040	1
7	8100	8102	1	7	8000	840a	2	7	8040	8140	1
8	8000	840a	2	8	850a	9520	4	8	0040	0542	2
9	850a	9520	4	9	802a	d4a8	6	9	8542	904a	4
-	052a	9000	–	–	056a	9000	–	–	d40a	0120	–
1	440a	0408	5	1	440a	0408	5	1	1488	1008	6
2	1080	00a0	4	2	1080	00a0	4	2	0021	4001	4
3	0083	0203	4	3	0083	0203	4	3	0601	0604	4
4	000c	0800	6	4	000c	0800	6	4	1800	0010	6
5	2000	0000	3	5	2000	0000	3	5	0040	0000	3
6	0040	0040	1	6	0040	0040	1	6	8000	8000	0
7	8040	8140	1	7	8040	8140	1	7	8100	8102	1
8	0040	0542	2	8	0040	0542	2	8	8000	840a	2
9	8542	904a	4	9	8542	904a	4	9	850a	9520	4
-	7c48	b0f8	–	–	540a	0120	–	–	7c58	b0f8	–
1	01e0	c202	5	1	1488	1008	6	1	01e0	c202	5
2	020f	0a04	5	2	0021	4001	4	2	020f	0a04	5
3	2800	0010	5	3	0601	0604	4	3	2800	0010	5
4	0040	0000	2	4	1800	0010	6	4	0040	0000	2
5	8000	8000	0	5	0040	0000	3	5	8000	8000	0
6	8100	8102	1	6	8000	8000	0	6	8100	8102	1
7	8000	840a	2	7	8100	8102	1	7	8000	840a	2
8	850a	9520	4	8	8000	840a	2	8	850a	9520	4
9	802a	d4a8	6	9	850a	9520	4	9	802a	d4a8	6

Appendix B Best Characteristics Found with Our Method

See Table 3.

Table 3. Differential characteristics related to the results listed in Table 1.

SPECK32				SPECK48				SPECK64			
r	Δ_L	Δ_R	$-\log_2 p$	r	Δ_L	Δ_R	$-\log_2 p$	r	Δ_L	Δ_R	$-\log_2 p$
-	7448	b0f8	-	-	001202	020002	-	-	40104200	00400240	-
1	01e0	c202	5	1	000010	100000	3	1	00001202	02000002	5
2	020f	0a04	5	2	000000	800000	1	2	00000010	10000000	3
3	2800	0010	5	3	800000	800004	0	3	00000000	80000000	1
4	0040	0000	2	4	808004	808020	2	4	80000000	80000004	0
5	8000	8000	0	5	8400a0	8001a4	4	5	80800004	80800020	2
6	8100	8102	1	6	608da4	608080	9	6	84008020	80008124	4
7	8000	840a	2	7	042003	002400	11	7	a08481a4	a0808880	8
8	850a	9520	4	8	012020	000020	5	8	04200401	00244004	9
9	802a	d4a8	6	9	200100	200000	3	9	01202000	00022020	6
				10	202001	202000	3	10	00010000	00100100	4
				11	210020	200021	4	11	00100000	00900800	2
								12	00900800	04104800	4
								13	04104808	24920808	7

SPECK96							
r	Δ_L	Δ_R	$-\log_2 p$	r	Δ_L	Δ_R	$-\log_2 p$
-	00800a080808	0800124a0848	-	-	900f00480001	011003084008	-
1	000092400040	400000104200	10	1	00800a080808	0800124a0848	10
2	000000820200	000000001202	6	2	000092400040	400000104200	10
3	000000009000	000000000010	4	3	000000820200	000000001202	6
4	000000000080	000000000000	2	4	000000009000	000000000010	4
5	800000000000	800000000000	0	5	000000000080	000000000000	2
6	808000000000	808000000004	1	6	800000000000	800000000000	0
7	800080000004	840080000020	3	7	808000000000	808000000004	1
8	808080800020	a08480800124	5	8	800080000004	840080000020	3
9	800400008124	842004008801	9	9	808080800020	a08480800124	5
10	a0a000008880	81a02004c88c	9	10	800400008124	842004008801	9
				11	a0a000008880	81a02004c88c	9
				12	000080044804	0d0180220c60	12
				13	010080a20028	690c81b26328	13

SPECK128							
r	Δ_L	Δ_R	$-\log_2 p$	r	Δ_L	Δ_R	$-\log_2 p$
-	00000000924000c0	4000000000104200	-	-	0000900f00480001	0100001003084008	-
1	0000000000820200	0000000000001202	6	1	000000800a080808	08000000124a0848	10
2	0000000000009000	0000000000000010	4	2	0000000092400040	4000000000104200	10
3	0000000000000080	0000000000000000	2	3	0000000000820200	0000000000001202	6
4	8000000000000000	8000000000000000	0	4	0000000000009000	0000000000000010	4
5	8080000000000000	8080000000000004	1	5	0000000000000080	0000000000000000	2
6	8000800000000004	8400800000000020	3	6	8000000000000000	8000000000000000	0
7	8080808000000020	a084808000000124	5	7	8180000000000000	8180000000000004	2
8	8004000080000124	8420040080000801	9	8	8000800000000004	8c00800000000020	5
9	a0a0000080800800	81a020048080480c	9	9	8080808000000020	e084808000000124	6
				10	0004000080000124	0420040080000803	10
				11	2020000080800800	0120200480804818	9
				12	0100000480004800	08010020840208c0	11
				13	0800002080820808	48080124a0924e08	11
				14	4000012480124000	0040080184803042	17
				15	00000800a0000202	0200480c84018012	12

Appendix C Pseudocode for the Search Algorithm

Algorithm 3. MCTS search for optimal differential characteristics for SPECK

Require: a bit-size $n \geq 1$, the number of forward rounds and backward rounds for the search, all the parameters specified in Section 5.
Ensure: Differential characteristics of decreasing weights.
  Class Node:
    visits, children, payout = 0, [], []
  Class Cached:
    path, path_weights, best_weight = [], [], ∞
  Build the initial tree from the pDDT as a collection of Node
  Initialize cache as a collection of Cached
  procedure MCTS_ITERATION(Δ_L, Δ_R, *num_rounds*)
    path, path_weights = $[(\Delta_L, \Delta_R)]$, []; increment tree$[(\Delta_L, \Delta_R)]$.visits
    for $i = 1$ to num_rounds **do**
      if $(\Delta_L, \Delta_R) \in$ tree **then**
        if tree$[(\Delta_L, \Delta_R)]$.visits $\leq k$ **then**
          $\Delta_{L,new}$, $\Delta_{R,new}$, p = random choice from tree$[(\Delta_L, \Delta_R)]$.children
        else
          $\Delta_{L,new}$, $\Delta_{R,new}$, p = node with max UCT in tree$[(\Delta_L, \Delta_R)]$.children
      else
        possible_children = δ-optimal(Δ_L, Δ_R, δ)   ▷ All δ-optimal transitions
        for child in possible_children **do**
          if xdp$^+$(child) > expand_threshold **then**
            Add child to tree$[(\Delta_L, \Delta_R)]$.children
        $\Delta_{L,new}$, $\Delta_{R,new}$, p = random choice from tree$[(\Delta_L, \Delta_R)]$.children
      Add $(\Delta_{L,new}, \Delta_{R,new})$ to path and $-\log_2 p$ to path_weights
      tree$[(\Delta_{L,new}, \Delta_{R,new})]$.visits = tree$[(\Delta_{L,new}, \Delta_{R,new})]$.visits + 1
      $\Delta_L, \Delta_R = \Delta_{L,new}, \Delta_{R,new}$
    weight = sum(path_weights)
    for $i = 0$ to *num_rounds* **do**
      payout = $\beta \frac{1}{\text{weight}} + (1-\beta) \frac{num\_rounds - i}{num\_rounds} \frac{1}{sum(path\_weights[i, i+1, i+2, \ldots])}$
      Add payout to tree[path[i]].payouts
    return path, path_weights, weight
  procedure MAIN
    for $i = 1$ to forward_iterations **do**
      $\Delta_L, \Delta_R \leftarrow$ sample from the first level of tree
      path, path_weights, weight = MCTS_ITERATION(Δ_L, Δ_R, *fwd_rounds*)
      if weight < cache$[(\Delta_L, \Delta_R)]$.best_weight **then**
        update cache$[(\Delta_L, \Delta_R)]$
    for $i = 1$ to backward_iterations **do**
      $\Delta_L, \Delta_R \leftarrow$ sample from the first level of tree
      path, path_weights, weight = MCTS_ITERATION(Δ_L, Δ_R, *bwd_rounds*)
      weight = weight + cache$[(\Delta_L, \Delta_R)]$.best_weight
      if weight < global_best_weight **then**
        print the full characteristic and update global_best_weight

---

References

[Abr87] Abramson, B.D.: The expected-outcome model of two-player games. Ph.D. thesis, Columbia University, USA (1987). AAI8827528

[BRV14] Biryukov, A., Roy, A., Velichkov, V.: Differential analysis of block ciphers SIMON and SPECK. In: Cid, C., Rechberger, C. (eds.) FSE 2014. LNCS, vol. 8540, pp. 546–570. Springer, Heidelberg (2015). https://doi.org/10.1007/978-3-662-46706-0_28

[BS91] Biham, E., Shamir, A.: Differential cryptanalysis of des-like cryptosystems. J. Cryptol. **4**, 3–72 (1991)

[BSS+15] Beaulieu, R., Shors, D., Smith, J., Treatman-Clark, S., Weeks, B., Wingers, L.: The SIMON and SPECK lightweight block ciphers. In: Proceedings of the 52nd Annual Design Automation Conference, DAC 2015. Association for Computing Machinery, New York (2015)

[BV13] Biryukov, A., Velichkov, V.: Automatic search for differential trails in ARX ciphers (extended version). IACR Cryptology ePrint Archive 2013:853 (2013)

[BV14] Biryukov, A., Velichkov, V.: Automatic search for differential trails in ARX ciphers. In: Benaloh, J. (ed.) CT-RSA 2014. LNCS, vol. 8366, pp. 227–250. Springer, Cham (2014). https://doi.org/10.1007/978-3-319-04852-9_12

[BVLC16] Biryukov, A., Velichkov, V., Le Corre, Y.: Automatic search for the best trails in ARX: application to block cipher Speck. In: Peyrin, T. (ed.) FSE 2016. LNCS, vol. 9783, pp. 289–310. Springer, Heidelberg (2016). https://doi.org/10.1007/978-3-662-52993-5_15

[BW95] Burmeister, J., Wiles, J.: The challenge of go as a domain for AI research: a comparison between go and chess. In: Proceedings of Third Australian and New Zealand Conference on Intelligent Information Systems. ANZIIS-95, pp. 181–186 (1995)

[CBSS08] Chaslot, G., Bakkes, S., Szitai, I., Spronck, P.: Monte-Carlo tree search: a new framework for game AI1. In: Belgian/Netherlands Artificial Intelligence Conference, pp. 389–390 (2008)

[Cou06] Coulom, R.: Efficient selectivity and backup operators in Monte-Carlo tree search. In: van den Herik, H.J., Ciancarini, P., Donkers, H.H.L.M.J. (eds.) CG 2006. LNCS, vol. 4630, pp. 72–83. Springer, Heidelberg (2007). https://doi.org/10.1007/978-3-540-75538-8_7

[Din14] Dinur, I.: Improved differential cryptanalysis of round-reduced speck. In: Joux, A., Youssef, A. (eds.) SAC 2014. LNCS, vol. 8781, pp. 147–164. Springer, Cham (2014). https://doi.org/10.1007/978-3-319-13051-4_9

[DMS19] Dwivedi, A.D., Morawiecki, P., Srivastava, G.: Differential cryptanalysis of round-reduced speck suitable for internet of things devices. IEEE Access **7**, 16476–16486 (2019)

[DS18] Ashutosh Dhar Dwivedi and Gautam Srivastava: Differential cryptanalysis of round-reduced lea. IEEE Access **6**, 79105–79113 (2018)

[FWG+16] Fu, K., Wang, M., Guo, Y., Sun, S., Hu, L.: MILP-based automatic search algorithms for differential and linear trails for speck. In: Peyrin, T. (ed.) FSE 2016. LNCS, vol. 9783, pp. 268–288. Springer, Heidelberg (2016). https://doi.org/10.1007/978-3-662-52993-5_14

[HNR68] Hart, P.E., Nilsson, N.J., Raphael, B.: A formal basis for the heuristic determination of minimum cost paths. IEEE Trans. Syst. Sci. Cybern. **4**(2), 100–107 (1968)

[HPW09] Helmbold, D., Parker-Wood, A.: All-moves-as-first heuristics in Monte-Carlo go. In: Proceedings of the 2009 International Conference on Artificial Intelligence, ICAI 2009, vol. 2, pp. 605–610, January 2009

[HW19] Huang, M., Wang, L.: Automatic tool for searching for differential characteristics in ARX ciphers and applications. In: Hao, F., Ruj, S., Sen Gupta, S. (eds.) INDOCRYPT 2019. LNCS, vol. 11898, pp. 115–138. Springer, Cham (2019). https://doi.org/10.1007/978-3-030-35423-7_6

[Kor85] Korf, R.E.: Depth-first iterative-deepening: an optimal admissible tree search. Artif. Intell. **27**(1), 97–109 (1985)

[Koz92] Kozen, D.C.: Depth-first and breadth-first search. In: Kozen, D.C. (ed.) The Design and Analysis of Algorithms. Texts and Monographs in Computer Science, pp. 19–24. Springer, New York (1992). https://doi.org/10.1007/978-1-4612-4400-4_4

[KS06] Kocsis, L., Szepesvári, C.: Bandit based Monte-Carlo planning. In: Fürnkranz, J., Scheffer, T., Spiliopoulou, M. (eds.) ECML 2006. LNCS (LNAI), vol. 4212, pp. 282–293. Springer, Heidelberg (2006). https://doi.org/10.1007/11871842_29

[LLJW21] Liu, Z., Li, Y., Jiao, L., Wang, M.: A new method for searching optimal differential and linear trails in ARX ciphers. IEEE Trans. Inf. Theory **67**(2), 1054–1068 (2021)

[LM01] Lipmaa, H., Moriai, S.: Efficient algorithms for computing differential properties of addition. In: Matsui, M. (ed.) FSE 2001. LNCS, vol. 2355, pp. 336–350. Springer, Heidelberg (2002). https://doi.org/10.1007/3-540-45473-X_28

[LMM91] Lai, X., Massey, J.L., Murphy, S.: Markov ciphers and differential cryptanalysis. In: Davies, D.W. (ed.) EUROCRYPT 1991. LNCS, vol. 547, pp. 17–38. Springer, Heidelberg (1991). https://doi.org/10.1007/3-540-46416-6_2

[Mat94] Matsui, M.: On correlation between the order of S-boxes and the strength of DES. In: De Santis, A. (ed.) EUROCRYPT 1994. LNCS, vol. 950, pp. 366–375. Springer, Heidelberg (1995). https://doi.org/10.1007/BFb0053451

[SHY16] Song, L., Huang, Z., Yang, Q.: Automatic differential analysis of ARX block ciphers with application to SPECK and LEA. In: Liu, J.K., Steinfeld, R. (eds.) ACISP 2016. LNCS, vol. 9723, pp. 379–394. Springer, Cham (2016). https://doi.org/10.1007/978-3-319-40367-0_24

[SWvdH+08] Schadd, M.P.D., Winands, M.H.M., van den Herik, H.J., Chaslot, G.M.J.-B., Uiterwijk, J.W.H.M.: Single-player Monte-Carlo tree search. In: van den Herik, H.J., Xu, X., Ma, Z., Winands, M.H.M. (eds.) CG 2008. LNCS, vol. 5131, pp. 1–12. Springer, Heidelberg (2008). https://doi.org/10.1007/978-3-540-87608-3_1

[SWW21] Sun, L., Wang, W., Wang, M.: Accelerating the search of differential and linear characteristics with the sat method. IACR Trans. Symmetric Cryptol. 269–315 (2021)

Finding Three-Subset Division Property for Ciphers with Complex Linear Layers

Debasmita Chakraborty(✉)

Applied Statistics Unit, Indian Statistical Institute, Kolkata, India
debasmitachakraborty1@gmail.com

Abstract. Conventional bit-based division property (CBDP) and bit-based division property using three subsets (BDPT) introduced by Todo *et al.* at FSE 2016 are the most effective techniques for finding integral characteristics of symmetric ciphers. At ASIACRYPT 2019, Wang *et al.* proposed the idea of modeling the propagation of BDPT, and recently Liu *et al.* described a model set method that characterized the BDPT propagation. However, the linear layers of the block ciphers which are analyzed using the above two methods of BDPT propagation are restricted to simple bit permutation. Thus the feasibility of the MILP method of BDPT propagation to analyze ciphers with complex linear layers is not settled. In this paper, we focus on constructing an automatic search algorithm that can accurately characterize BDPT propagation for ciphers with complex linear layers. We first introduce BDPT propagation rule for the binary diffusion layer and model that propagation in MILP efficiently. The solutions to these inequalities are exact BDPT trails of the binary diffusion layer. Next, we propose a new algorithm that models Key-Xor operation in BDPT based on MILP technique. Based on these ideas, we construct an automatic search algorithm that accurately characterizes the BDPT propagation and we prove the correctness of our search algorithm. We demonstrate our model for the block ciphers with non-binary diffusion layers by decomposing the non-binary linear layer trivially by the COPY and XOR operations. Therefore, we apply our method to search integral distinguishers based on BDPT of SIMON, SIMON(102), PRINCE, MANTIS, PRIDE, and KLEIN block ciphers. For PRINCE and MANTIS, we find $(2+2)$ and $(3+3)$ round integral distinguishers respectively which are longest to date. We also improve the previous best integral distinguishers of PRIDE and KLEIN. For SIMON, SIMON(102), the integral distinguishers found by our method are consistent with the existing longest distinguishers.

Keywords: BDPT · Complex linear layer · Binary matrix · MILP

1 Introduction

Division Property. At Eurocrypt 2015, Todo [30] introduced Division property which is a novel strategy to discover integral characteristics to search integral distinguishers of block cipher structures (Feistel structure and SPN structure).

T. Isobe and S. Sarkar (Eds.): INDOCRYPT 2022, LNCS 13774, pp. 398–421, 2022.
https://doi.org/10.1007/978-3-031-22912-1_18

Later, Todo and Morii [31] introduced bit-based division property (which is actually called Conventional Bit-based Division Property (CBDP)), which could be treated as an exceptional instance of division property. Actually CBDP classify all vectors $\boldsymbol{u}$ in $\mathbb{F}_2^n$ into two subsets such that the parity of $\bigoplus_{\boldsymbol{x}\in\mathbb{X}} \boldsymbol{x}^{\boldsymbol{u}}$ is 0 or *unknown* (where $\boldsymbol{x}^u$ is defined as $\boldsymbol{x}^u := \prod_{i=1}^{n} x_i{}^{u_i}$). Moreover, at CRYPTO 2016, Boura and Canteaut [9] presented a different perspective on the division property, called 'parity set'.

The intricacy of using CBDP was generally equivalent to 2^n for a n-bit primitives. Henceforth, the gigantic intricacy limited the wide uses of CBDP. To tackle the limitation of the tremendous complexity, Xiang *et al.* [34] applied MILP-strategy to look through integral distinguisher dependent on CBDP and they applied this modeling technique to six lightweight block ciphers. By extending and improving this method, the integral attacks have been applied to many ciphers and many better integral distinguisher has been found [18,19,21,23,28,29,37].

Three-Subset Division Property. Although CBDP can find more precise integral distinguishers than other methods, the accuracy is never perfect. To find more accurate distinguishers, the bit-based division property using three subsets (BDPT) was proposed in [31]. BDPT divides all vectors $\boldsymbol{u}$ in $\mathbb{F}_2^n$ into two subsets such that the parity of $\bigoplus_{\boldsymbol{x}\in\mathbb{X}} \boldsymbol{x}^{\boldsymbol{u}}$ is 0, 1 or *unknown*. Essentially, the set *unknown* in CBDP is divided into 1-subset and *unknown* subset in BDPT. As a result, BDPT can find more precise integral characteristics than CBDP. For example, CBDP demonstrated the existence of SIMON32's 14-round integral distinguisher whereas BDPT discovered SIMON32's 15-round integral distinguisher [30].

Despite of its successful combination of the MILP and the CBDP, the MILP modeling technique does not work quite well with the BDPT. As in case of BDPT we have to track the division property propagation of two sets ($\mathbb{K}$ (the *unknown* subset) and $\mathbb{L}$ (the 1-subset)) as well as the influence of the set $\mathbb{L}$ on the set $\mathbb{K}$ should also be traced which makes the procedure of constructing automatic search algorithm based on BDPT complicated.

First, Hu *et al.* [20] proposed variant three subset division property (VTDP) and applied this method to improve some integral distinguishers although it sacrifices quite some accuracy of BDPT. Therefore, Wang *et al.* [32] proposed the idea of modeling the propagation for the BDPT and recently Liu *et al.* [24] proposed a model set method to search integral distinguishers based on BDPT. Both of these methods have been applied to the block ciphers having simple bit permutation as their linear layer.

1.1 Motivation

The idea of modeling BDPT propagation which is described in [32] is that each node on the breadth-first search algorithm is regarded as the starting point of division trails, and the MILP evaluates whether there is a feasible solution from every node. According to their searching algorithm, we can run this algorithm to any block cipher efficiently only if we can divide the round function into several appropriate parts. Therefore, it is very difficult to model BDPT propagation using this technique for the ciphers with complex linear layers. Next, Liu *et al.* [24] proposed model set method to search BDPT where the authors constructed r different MILP models for r-round block ciphers which is a bit complicated. Moreover, both these methods have been applied to the block ciphers having linear layers as simple bit permutation. Now, the following question arises: *Is MILP method of BDPT propagation efficiently applicable for ciphers with complex linear layers?*

1.2 Our Contributions

To address this question, first we propose an idea to find BDPT propagation through the binary (complex) linear layer accurately and then we construct an automatic search algorithm for BDPT in this paper. The details of our technical contributions are listed as follows:

Model the BDPT Propagation of Binary Linear Layer. We give an idea to find exact BDPT propagation through the binary (complex) linear layer which is a new method that helps us to construct MILP model of BDPT propagation through the binary linear layer accurately. We actually find that the rows of the primitive matrix corresponding to the binary mixcolumn matrix can be divided into some cosets with the property that the rows in different cosets have no common nonzero entries in the same column. Using this interesting property, we can easily find accurate BDPT propagation and can give a description of such propagation by smallest number of inequalities.

Construction of Automatic Search Algorithm for BDPT. To search for BDPT, first we construct the MILP models for key-independent components of the round function of block ciphers. When a Key-Xor operation is applied, new vectors generated from the set $\mathbb{L}$ will be added to the set $\mathbb{K}$. Therefore, how to model Key-Xor operation accurately is a complex problem. To solve this problem, we construct a new efficient algorithm that models each Key-Xor operation based on MILP technique. Finally, by selecting appropriate initial BDPT and stopping rules we construct an automatic search algorithm that accurately characterize BDPT propagation using only two MILP models which is much simpler than the algorithm described in [24]. Moreover, we prove the correctness of our search algorithm.

Applications. As for the application of our methodology, first time we apply BDPT on block ciphers with complex linear layers. We apply our automatic search model to search integral distinguishers of PRINCE [8], MANTIS [6], KLEIN [16], PRIDE [4], SIMON [5], and SIMON(102) [22]. The results are shown in Table 1.

At first, we apply our method on PRINCE and MANTIS which have binary linear layer. We find $2+2$ round integral distinguisher for PRINCE which is one more round than the previous best integral distinguisher [15] and find $3+3$ round integral distinguisher for MANTIS which is also one more round than the previous best integral distinguisher [15] where we denote a are the rounds before the middle layer, and b are the rounds after the middle layer and $a+b$ as total number of rounds.

Table 1. Summarization of integral distinguishers

Cipher	Data	Round	Number of constant bits	Time	References
MANTIS	2^{32}	3+2	16	–	[15]
	$\mathbf{2^{63}}$	**3+3**	**64**	**2 h 8 m**	**Sect.** 5.1
PRINCE	2^{32}	2+1	64	–	[15]
	$\mathbf{2^{63}}$	**2+2**	**64**	**21 h 45 m**	**Sect.** 5.1
PRIDE64*	-	8	–	–	[33]
	$\mathbf{2^{63}}$	**9**	**32**	**2 h 35 m**	**Sect.** 5.2
KLEIN64	2^{32}	5	64	–	[36]
	$\mathbf{2^{62}}$	**6**	**64**	**45 m**	**Sect.** 5.2

* In [33], the authors have only mentioned that PRIDE64 has 8-round integral distinguisher and no other information is available best known to us.

To complete our BDPT analysis on ciphers with complex linear layers, we apply our method to KLEIN and PRIDE which have non-binary linear layers. As there are no known results on them related to CBDP, then we first apply MILP based CBDP on them and find 6-round and 9-round integral distinguishers for KLEIN and PRIDE respectively which are one more rounds to previous best integral distinguishers [33,36]. Therefore, we apply our MILP based BDPT method and the integral distinguishers we find are in accordance with the integral distinguishers we find based on CBDP. Finally, we apply our method to all variants of SIMON, and SIMON(102) block ciphers and the distinguishers we find are in accordance with the previous longest distinguishers [24] but we get these results in better time.

1.3 Organization of the Paper

This paper is organized as follows: In Sect. 2, we briefly recall some background knowledge about the bit-based division property. In Sect. 3, we studies how to model basic operations used in the round function of a block cipher by the MILP technique and introduce exact modelling of complex (binary) linear layer in BDPT. Section 4 studies the initial and stopping rules, and search algorithm. We show some applications of our model in Sect. 5. At last we conclude the paper in Sect. 6.

2 Preliminaries

2.1 Notation

Let $\mathbb{F}_2$ denote the finite field $\{0,1\}$ and $\boldsymbol{a} = (a_0, a_1, \ldots, a_{n-1}) \in \mathbb{F}_2^n$ be an n-bit vector, where a_i denotes the i-th bit of $\boldsymbol{a}$. For n-bit vectors $\boldsymbol{x}$ and $\boldsymbol{u}$, define $\boldsymbol{x}^{\boldsymbol{u}} = \prod_{i=0}^{n-1} x_i^{u_i}$. Then, for any $\boldsymbol{k} \in \mathbb{F}_2^n$ and $\boldsymbol{k}' \in \mathbb{F}_2^n$, define $\boldsymbol{k} \succeq \boldsymbol{k}'$ if $k_i \geq k_i'$ holds for all $i = 0, 1, \ldots, n-1$, and define $\boldsymbol{k} \succ \boldsymbol{k}'$ if $k_i > k_i'$ holds for all $i = 0, 1, \ldots, n-1$. For a subset $\mathcal{I} \subseteq \{0, 1, ..., n-1\}$, $\boldsymbol{u}_{\mathcal{I}}$ denotes an n-dimensional bit vector $(u_0, u_1, \ldots, u_{n-1})$ satisfying $u_i = 1$ if $i \in \mathcal{I}$ and $u_i = 0$ otherwise. We simply write $\mathbb{K} \leftarrow \boldsymbol{k}$ when $\mathbb{K} = \mathbb{K} \cup \{\boldsymbol{k}\}$ and $\mathbb{K} \rightarrow \boldsymbol{k}$ when $\mathbb{K} = \mathbb{K} \setminus \{\boldsymbol{k}\}$. And $|\mathbb{K}|$ denotes the number of elements in the set $\mathbb{K}$. We denote $[n] = \{1, 2, \ldots, n\}$, $\mathbf{1} = 1^n$, and $\mathbf{0} = 0^n$. We denote i-th unit vector in $\mathbb{F}_2^n$ as $\boldsymbol{e}_i$.

2.2 Bit-Based Division Property

Two kinds of bit-based division property (CBDP and BDPT) were introduced by Todo and Morii at FSE 2016 [31]. Their definitions are as follows.

Definition 1 *(**CBDP** [31]). Let $\mathbb{X}$ be a multiset whose elements take a value of $\mathbb{F}_2^n$. Let $\mathbb{K}$ be a set whose elements take an n-dimensional bit vector. When the multiset $\mathbb{X}$ has the division property $D_{\mathbb{K}}^{1^n}$, it fulfills the following conditions:*

$$\bigoplus_{\boldsymbol{x} \in \mathbb{X}} \boldsymbol{x}^{\boldsymbol{u}} = \begin{cases} unknown, & if\ there\ is\ \boldsymbol{k} \in \mathbb{K}\ satisfying\ \boldsymbol{u} \succeq \boldsymbol{k}, \\ 0 & otherwise. \end{cases}$$

Some propagation rules of CBDP are proven in [30,31,34].

Definition 2 *(**BDPT** [31]). Let $\mathbb{X}$ be a multi-set whose elements take a value of F_2^n. Let $\mathbb{K}$ and $\mathbb{L}$ be two sets whose elements take n-dimensional bit vectors. When the multi-set $\mathbb{X}$ has the division property $D_{\mathbb{K},\mathbb{L}}^{1^n}$, it fulfils the following conditions:*

$$\bigoplus_{\boldsymbol{x} \in \mathbb{X}} \boldsymbol{x}^{\boldsymbol{u}} = \begin{cases} unknown, & if\ there\ is \boldsymbol{k} \in \mathbb{K}\ satisfying\ \boldsymbol{u} \succeq \boldsymbol{k}, \\ 1, & else\ if\ there\ is\ \boldsymbol{l} \in \mathbb{L}\ satisfying\ \boldsymbol{u} = \boldsymbol{l}, \\ 0, & otherwise. \end{cases}$$

If there are $\boldsymbol{k} \in \mathbb{K}$ and $\boldsymbol{k}' \in \mathbb{K}$ satisfying $\boldsymbol{k} \succeq \boldsymbol{k}'$ in the CBDP $D_{\mathbb{K}}^{1^n}$, then $\boldsymbol{k}$ can be removed from $\mathbb{K}$ because the vector $\boldsymbol{k}$ is redundant. This progress is denoted as ***Reduce0***$(\mathbb{K})$. Moreover, if there are $\boldsymbol{l} \in \mathbb{L}$ and $\boldsymbol{k} \in \mathbb{K}$, then the vector $\boldsymbol{l}$ is also redundant if $\boldsymbol{l} \succeq \boldsymbol{k}$. This progress is denoted as ***Reduce1***$(\mathbb{K}, \mathbb{L})$. The redundant vectors in $\mathbb{K}$ and $\mathbb{L}$ will not affect the parity of $\boldsymbol{x}^{\boldsymbol{u}}$ for any $\boldsymbol{u}$.

The propagation rules of $\mathbb{K}$ in CBDP are the same with BDPT. So we only introduce the propagation rules of BDPT which are needed in this paper. For further details, please refer to [31,32].

BDPT Rule 1 (Xor with The Secret Key [31].**)** *Let* $\mathbb{K}$ *be the input multiset satisfying* $D^{1^n}_{\mathbb{K},\mathbb{L}}$. *For the input* $\boldsymbol{x} \in \mathbb{X}$, *the output* $y \in \mathbb{Y}$ *is* $\boldsymbol{y} = (x_0, \dots, x_i \oplus r_k, x_{i+1}, \dots, x_{n-1})$, *where* r_k *is the secret key. Then, the output multiset* $\mathbb{Y}$ *has* $D^{1^n}_{\mathbb{K}',\mathbb{L}'}$, *where* $\mathbb{K}'$ *and* $\mathbb{L}'$ *are computed as*

$$\begin{cases} \mathbb{L}^{'} \leftarrow \boldsymbol{l} \ \ for \ \boldsymbol{l} \in \mathbb{L}, \\ \mathbb{K}^{'} \leftarrow \boldsymbol{k} \ \ for \ \boldsymbol{k} \in \mathbb{K}, \\ \mathbb{K}^{'} \leftarrow (l_1, l_2, ..., l_i \vee 1, ..., l_n) \ \ for \ \boldsymbol{l} \in \mathbb{L} \ satisfying \ l_i = 0. \end{cases}$$

BDPT Rule 2 (S-box [32].**)** *For an S-box* : $\mathbb{F}_2^n \rightarrow \mathbb{F}_2^n$, *let* $\boldsymbol{x} = (x_0, \dots, x_{n-1})$ *and* $\boldsymbol{y} = (y_0, \dots, y_{n-1})$ *denote the input and output variables. And every* y_i, $i \in \{0, 1, \dots, n-1\}$ *can be expressed as a boolean function of* $(x_0, x_1, \dots, x_{n-1})$. *If the input BDPT of S-box is* $D^{1^n}_{\mathbb{K},\mathbb{L}=\{l\}}$, *then the output BDPT of S-box can be calculated by* $D^{1^n}_{\boldsymbol{Reduce0}(\underline{\mathbb{K}}),\boldsymbol{Reduce1}(\underline{\mathbb{K}},\,\underline{\mathbb{L}})}$,

$$\begin{cases} \underline{\mathbb{K}} = \{\boldsymbol{u'} \in \mathbb{F}_2^n \,|\, for \ any \ \boldsymbol{u} \in \mathbb{K}, \ if \ \boldsymbol{y}^{\boldsymbol{u}'} \ contains \ any \ term \ \boldsymbol{x}^{\boldsymbol{v}} \ satisfying \ \boldsymbol{v} \succeq \boldsymbol{u}\} \\ \underline{\mathbb{L}} = \{\boldsymbol{u} \in \mathbb{F}_2^n \,|\, \boldsymbol{y}^{\boldsymbol{u}} \ contains \ the \ term \ \boldsymbol{x}^{\boldsymbol{l}}\} \end{cases}$$

Let $D^{1^n}_{\mathbb{K},\,\mathbb{L}=\{l_0,\dots,l_{r-1}\}}$ *and* $D^{1^n}_{\mathbb{K}',\mathbb{L}'}$ *be the input and output BDPT of S-box, respectively. We can get the output BDPT* $D^{1^n}_{\mathbb{K}',\mathbb{L}'_i}$ *from the corresponding input BDPT* $D^{1^n}_{\mathbb{K},\,\mathbb{L}=\{l_i\}}$ *where* $i = 0, 1, \dots, r-1$. *Then,*

$$\mathbb{L}' = \{\boldsymbol{l} \,|\, \boldsymbol{l} \ appears \ odd \ times \ in \ sets \ \mathbb{L}'_0, \dots, \mathbb{L}'_{r-1}\}$$

2.3 The MILP Model for CBDP

At Asiacrypt 2016, Xiang *et al.* [34] applied MILP method to search integral distinguishers based in CBDP, which allowed them to analyze block ciphers with large sizes. Firstly they introduced the concept of CBDP trail as follows:

Definition 3 (CBDP Trail [34]**).** *Consider the propagation of the division property* $\{\boldsymbol{k}\} \equiv \mathbb{K}_0 \xrightarrow{f_1} \mathbb{K}_1 \xrightarrow{f_2} \mathbb{K}_2 \xrightarrow{f_3} ....$ *Moreover, for any vector* $\boldsymbol{k}_i^* \in \mathbb{K}_i (i \geq 1)$, *there must exist an vector* $\boldsymbol{k}_{i-1}^* \in \mathbb{K}_{i-1}$ *such that* $\boldsymbol{k}_{i-1}^*$ *can propagate to* $\boldsymbol{k}_i^*$ *by CBDP propagation rules. Furthermore, for* $(\boldsymbol{k}_0^*, \boldsymbol{k}_1^*, ..., \boldsymbol{k}_r^*) \in \mathbb{K}_0 \times \mathbb{K}_1 \times ... \times \mathbb{K}_r$, *if* $\boldsymbol{k}_{i-1}^*$ *can propagate to* $\boldsymbol{k}_i^*$ *for all* $i \in \{1, 2, \dots, r\}$, *we call* $(\boldsymbol{k}_0^*, \boldsymbol{k}_1^*, \dots, \boldsymbol{k}_r^*)$ *an* r*-round CBDP trail.*

With the help of division trail, finding the CBDP is transformed into a problem of finding a division trail ended at a unit vector. For more details please refer to [34].

3 The MILP Model for BDPT

Suppose E_r is a r-round iterated block cipher whose round function f_i for $i \in [r]$ consists of a non-linear layer, linear layer, and Key-Xor operation. Let f_k^i be the Key-Xor operation, and f_e^i be the rest of the operations in the ith round function f_i i.e.

$$f_i = f_k^i \circ f_e^i$$

Let, the input multiset $\mathbb{X}$ to the block cipher E_r has initial BDPT as $D^{1^n}_{\mathbb{K}_0=\{\boldsymbol{k}\},\mathbb{L}_0=\{\boldsymbol{l}\}}$, and for any $i \in [r]$, we denote the output BDPT as $D^{1^n}_{\mathbb{K}_i,\mathbb{L}_i}$. Now, for the operation f_e^i, we denote the BDPT propagation as

$$f_e^i(\mathbb{K}_{i-1}) = \mathbb{K}^*_{i-1},\ \ f_e^i(\mathbb{L}_{i-1}) = \mathbb{L}^*_{i-1}$$

We can evaluate the BDPT propagation for $\mathbb{K}$ (*unknown* subset) and $\mathbb{L}$ (1 subset) independently as per the BDPT propagation rules for linear and non-linear layers. Now, for the operation f_k^i, according to the BDPT Rule 1 some new vectors which are produced from the vectors in $\mathbb{L}^*_{i-1}$ and those new vectors along with the vectors in $\mathbb{K}^*_{i-1}$ are the vectors in the set $\mathbb{K}_i$, and the set $\mathbb{L}_i$ is same as $\mathbb{L}^*_{i-1}$.

Now, we divide the operation f_k^i into two parts say f_1^i, f_2^i such that f_1^i is the operation where new elements are produced from each elements in $\mathbb{L}^*_{i-1}$ according to BDPT Rule 1, and f_2^i is the operation which includes the new vectors and the vectors from $\mathbb{K}^*_{i-1}$ in $\mathbb{K}_i$ which is as follows:

$$(\mathbb{K}_i, \mathbb{L}_i) = f_k^i(\mathbb{K}^*_{i-1},\ \mathbb{L}^*_{i-1}) = (f_2^i(f_1^i(\mathbb{L}^*_{i-1}),\ \mathbb{K}^*_{i-1}),\ \mathbb{L}^*_{i-1}) \tag{1}$$

Precisely, f_2^i is the union operation i.e. $\mathbb{K}_i = f_1^i(\mathbb{L}^*_{i-1}) \cup \mathbb{K}^*_{i-1}$.

To model the propagation of BDPT for the operations f_e^i and f_k^i for all $i \in [r]$, we reintroduce a notion named ***BDPT trail***.[1]

Definition 4 (BDPT Trail) *Let $\mathbb{X}$ be the input multiset to the block cipher which has initial BDPT $D^{1^n}_{\mathbb{K}_0=\{\boldsymbol{k}\},\ \mathbb{L}_0=\{\boldsymbol{l}\}}$, and denote the BDPT after r-round propagation through f_e^i, f_k^i for all $i \in [r]$ by $D^{1^n}_{\mathbb{K}_r,\ \mathbb{L}_r}$, where $r \geq 1$. Thus we have the following chain of BDPT propagations:*

$$\begin{array}{ccccccccc} \{\boldsymbol{k}\} \triangleq \mathbb{K}_0 \xrightarrow{f_e^1} \mathbb{K}_0^* & & \mathbb{K}_1 \xrightarrow{f_e^2} \mathbb{K}_1^* \cdots\cdots & \mathbb{K}_{r-1} \xrightarrow{f_e^r} \mathbb{K}_{r-1}^* & & \mathbb{K}_r \\ & \overset{f_k^1}{\times} & & & \overset{f_k^r}{\times} & \\ \{\boldsymbol{l}\} \triangleq \mathbb{L}_0 \xrightarrow{f_e^1} \mathbb{L}_0^* & & \mathbb{L}_1 \xrightarrow{f_e^2} \mathbb{L}_1^* \cdots\cdots & \mathbb{L}_{r-1} \xrightarrow{f_e^r} \mathbb{L}_{r-1}^* & & \mathbb{L}_r \end{array}$$

where $\mathbb{K}^*_{i-1} = f_e^i(\mathbb{K}_{i-1})$, $\mathbb{L}^*_{i-1} = f_e^i(\mathbb{L}_{i-1})$, and $(\mathbb{K}_i,\ \mathbb{L}_i) = f_k^i(\mathbb{K}^*_{i-1},\ \mathbb{L}^*_{i-1})$ for all $1 \leq i \leq r$. Moreover, for any vector tuple $(\boldsymbol{k}_i,\ \boldsymbol{l}_i)$, $\boldsymbol{k}_i \in \mathbb{K}_i$, and $\boldsymbol{l}_i \in \mathbb{L}_i$ ($i \in [r]$), there must exist $(\boldsymbol{k}^*_{i-1},\ \boldsymbol{l}^*_{i-1})$, where $\boldsymbol{k}^*_{i-1} \in \mathbb{K}^*_{i-1}$, and $\boldsymbol{l}^*_{i-1} \in \mathbb{L}^*_{i-1}$ such that $\boldsymbol{k}^*_{i-1} \in \mathbb{K}^*_{i-1}$ propagate to $(\boldsymbol{k}_i,\ \boldsymbol{l}_i)$ by BDPT propagation rule of Key-Xor, and

[1] In [24], the authors have defined ***BDPT trail***. We actually rewrite it according to our notations.

there must exist $(\boldsymbol{k}_{i-1}, \boldsymbol{l}_{i-1}) \in \mathbb{K}_{i-1} \times \mathbb{L}_{i-1}$ *such that* $\boldsymbol{k}_{i-1}$ *propagate to* $\boldsymbol{k}^*_{i-1}$, *and* $\boldsymbol{l}_{i-1}$ *propagate to* $\boldsymbol{l}^*_{i-1}$ *by BDPT propagation rules of linear and non-linear layers. Furthermore, for* $(\boldsymbol{k}_0, \boldsymbol{l}_0), \ldots, (\boldsymbol{k}_r, \boldsymbol{l}_r) \in \mathbb{K}_0 \times \mathbb{L}_0 \times \ldots \times \mathbb{K}_r \times \mathbb{L}_r$, *if* $(\boldsymbol{k}_{i-1}, \boldsymbol{l}_{i-1})$ *can propagate to* $(\boldsymbol{k}_i, \boldsymbol{l}_i)$ *for all* $i \in \{1, 2, \ldots, r\}$, *we call*

$$(\boldsymbol{k}_0, \boldsymbol{l}_0) \overset{f_e^1, f_k^1}{\longrightarrow} (\boldsymbol{k}_1, \boldsymbol{l}_1) \overset{f_e^2, f_k^2}{\longrightarrow} \ldots \overset{f_e^r, f_k^r}{\longrightarrow} (\boldsymbol{k}_r, \boldsymbol{l}_r)$$

an r*-round BDPT trail.*

Now, to model BDPT trail, we propose Proposition 1 according to Definition 4.

Proposition 1. *Let the input multiset* $\mathbb{X}$ *has initial BDPT* $D^{1^n}_{\{\boldsymbol{k}\}, \{\boldsymbol{l}\}}$ *and* $D^{1^n}_{\mathbb{K}_r, \mathbb{L}_r}$ *denote the BDPT of the output multiset after* r*-round propagation. Then, the set of first components of the last vectors of all* r*-round BDPT trails which starts with the vector* $(\boldsymbol{k}, \boldsymbol{l})$ *is equal to the set* $\mathbb{K}_r$ *and the set of second components of the last vectors of all* r*-round BDPT trails which starts with the vector* $(\boldsymbol{k}, \boldsymbol{l})$ *is equal to the set* $\mathbb{L}_r$.

Proof of this Proposition 1 directly follows from Definition 4.

3.1 Some Observations on BDPT Propagation Rule for S-box

S-box is an important component of block ciphers. For a lot of block ciphers it is the only non-linear part. Although any Boolean function can be evaluated by using three rules (COPY, XOR, AND), the propagation requires much time and memory complexity when Boolean function is complex. Inspired by the algorithm of calculating CBDP trails of S-box [34], Wang *et al.* proposed a generalized method to calculate BDPT division trails of S-box in [32] and we have mentioned the rule in BDPT Rule 2.

Let, the input BDPT of S-box is $D^{1^n}_{\mathbb{K},\, \mathbb{L}=\{\boldsymbol{l}\}}$, and according to the BDPT Rule 2, we have found the sets $\underline{\mathbb{K}}$, and $\underline{\mathbb{L}}$ from $\mathbb{K}$ and $\mathbb{L}$ respectively as follows:

$$\begin{cases} \underline{\mathbb{K}} = \{\boldsymbol{u}' \in \mathbb{F}_2^n \mid \text{for any } \boldsymbol{u} \in \mathbb{K}, \text{ if } \boldsymbol{y}^{\boldsymbol{u}'} \text{ contains any term } \boldsymbol{x}^{\boldsymbol{v}} \text{ satisfying } \boldsymbol{v} \succeq \boldsymbol{u}\} \\ \underline{\mathbb{L}} = \{\boldsymbol{u} \in \mathbb{F}_2^n \mid \boldsymbol{y}^{\boldsymbol{u}} \text{ contains the term } \boldsymbol{x}^{\boldsymbol{l}}\} \end{cases} \tag{2}$$

Now, according to the BDPT Rule 2, the output BDPT would be $D^{1^n}_{\mathbb{K}',\, \mathbb{L}'}$ which is as follows:

$$\mathbb{K}' = \boldsymbol{Reduce0}(\underline{\mathbb{K}}),\ \mathbb{L}' = \boldsymbol{Reduce1}(\underline{\mathbb{K}},\ \underline{\mathbb{L}})$$

Therefore, it is obvious that, $\mathbb{K}' \subseteq \underline{\mathbb{K}}$, and $\mathbb{L}' \subseteq \underline{\mathbb{L}}$. Here, we come to two observations as follows:

Observation 1. $\mathbb{L}'$ *does not contain* $\mathbf{1}$ *vector.*

According to the BDPT propagation rule of S-box, as $\mathbb{L}' = \boldsymbol{Reduce1}(\underline{\mathbb{K}},\ \underline{\mathbb{L}})$, and for any $\boldsymbol{u} \in \underline{\mathbb{K}}$, $\mathbf{1} \succeq \boldsymbol{u}$, then $\mathbb{L}'$ does not contain $\mathbf{1}$ vector.

Observation 2. *If* $\mathbb{L} = \{\mathbf{0}\}$, *then* $\mathbb{L}' = \{\mathbf{0}\}$.

Whenever, $\mathbb{L} = \{\mathbf{0}\}$, then $\bigoplus_{\boldsymbol{x} \in \mathbb{X}} \boldsymbol{x}^{\mathbf{0}} = 1$ which implies that the input multiset $\mathbb{X}$ contains a constant term. Therefore, for all $\boldsymbol{u} \succ \mathbf{0}$, $\bigoplus_{\boldsymbol{x} \in \mathbb{X}} \boldsymbol{x}^{\boldsymbol{u}} =$ unknown. Hence, trivially $\bigoplus_{\boldsymbol{y} \in \mathbb{Y}} \boldsymbol{y}^{\mathbf{0}} = 1$ and $\mathbb{L}' = \{\mathbf{0}\}$ where $\mathbb{Y}$ is the output multiset.

Therefore, given an n-bit S-box and its input BDPT $D^{1^n}_{\mathbb{K}=\{\boldsymbol{k}\},\ \mathbb{L}=\{\boldsymbol{l}\}}$, BDPT Rule 2 returns the output BDPT $D^{1^n}_{\mathbb{K}',\ \mathbb{L}'}$. Thus for any vector $\boldsymbol{k}' \in \mathbb{K}'$, $(\boldsymbol{k}, \boldsymbol{k}')$ is a valid division trail for $\mathbb{K}'$ of the S-box. Similarly, this holds for $\mathbb{L}'$ as well. We know that, the vector $\boldsymbol{l}$ does not affect the propagation of vector $\boldsymbol{k}$ through the S-box, we will obtain a complete list of the division trail for $\mathbb{K}'$ by traversing $\boldsymbol{k} \in \mathbb{F}_2^n$ [34].

Similarly, for a certain input vector $\boldsymbol{l} \in \mathbb{F}_2^n$, we will obtain a certain set of division trails for $\mathbb{L}$ using Eq. 2 and then using Observation 1, and Observation 2 we will remove some invalid division trails from $\mathbb{L}$ and obtain a set of division trail for $\mathbb{L}'$. Therefore, if we try all the 2^n possible input vector $\boldsymbol{l}$, we will get a complete list of division trails for $\mathbb{L}'$.

In [32], the authors included some invalid BDPT trail for $\mathbb{L}'$ set while obtaining a complete list of division trails for $\mathbb{L}'$. In [24], the authors have removed those invalid BDPT trail from $\mathbb{L}'$ by introducing another algorithm which is actually equivalent to the algorithm of finding BDPT trail of S-box in [32] and by traversing $\boldsymbol{k} \in \mathbb{F}_2^n$. Now, our approach is similar to their idea [24] in a much simplified manner using two observations from BDPT Rule 2 which was introduced in [32].

In the full version of this paper [10], we present the complete lists of all the division trails for $\mathbb{L}$ of PRINCE S-box according to our method which is same if we apply the method the authors described in [24]. Therefore, after getting the BDPT trails for $\mathbb{K}$ and $\mathbb{L}$ of S-box, we construct the linear inequalities using the method described in [34] whose feasible solutions are exactly those BDPT trails which are shown in the full version of this paper [10].

3.2 MILP Model of BDPT for Complex Linear Layer

In this section, we establish the idea to construct MILP model of BDPT for complex linear layer represented by a matrix $M = (m_{i,j})_{s \times s} \in \mathbb{F}_{2^m}^{s \times s}$. Given the irreducible polynomial of the field $\mathbb{F}_2^m$ where the multiplications operate, the representation of the matrix over $\mathbb{F}_2$ is unique, which we call the primitive matrix of M and is denoted by $M' = (m'_{i,j})_{n \times n}$ where $m'_{i,j} \in \mathbb{F}_2$ and $n = m \times s$. Therefore, if each $m_{i,j}$ in M which is a polynomial in the extension field $F_{2^m} \simeq \mathbb{F}[x]/(f)$, where f is the irreducible polynomial over $\mathbb{F}_2$ with degree m, is either 0 or 1 then M is called binary matrix and otherwise M is non-binary matrix.

Therefore, block ciphers with complex linear layer can be partitioned into two parts: (i) Block ciphers with binary linear layer and (ii) Block ciphers with non-binary linear layer, depending on the binary or non-binary matrix as its linear layer. Examples of block ciphers having binary linear layer are MIDORI,

SKINNY, CRAFT, PRINCE, MANTIS etc. and AES, LED, KLEIN, PRIDE etc. have non-binary linear layer.

Now, an obvious way to model the BDPT propagation through any complex linear layer i.e. $\boldsymbol{u}_1 \xrightarrow{M} \boldsymbol{v}_1$ in $\mathbb{K}$ subset, and $\boldsymbol{u}_2 \xrightarrow{M} \boldsymbol{v}_2$ in $\mathbb{L}$ subset is that one can introduce some auxiliary binary variables and decompose it into the COPY and XOR operations. Therefore, by following the BDPT propagation rule of COPY and XOR, BDPT propagation through linear layer can be modelized. The obvious advantage of this model is that using this technique we can model BDPT propagation of any complex linear layer.

In [21,37], the authors have shown that using this technique one may introduce many invalid division trails in $\mathbb{K}$ subset. Now, here we are going to show that if we use this COPY-XOR technique to handle binary linear layer then many invalid division trails may be added to the $\mathbb{L}$ subset as well which we have shown by giving an example in the full version of this paper [10].

Exact BDPT Modelization for Ciphers Having Binary Linear Layer. Given a binary matrix $M = (m_{i,j})_{s\times s} \in \mathbb{F}_{2^m}^{s\times s}$, and denote $n = m \times s$, we can derive an equivalent matrix working at a bit level which is called primitive matrix $M' = (m'_{i,j})_{n\times n} \in \mathbb{F}_2^{n\times n}$. Now, M' has $n = ms$ number of rows which we denote say $R_0, R_1, \ldots, R_{n-1}$, and define a set of all rows $\mathcal{R} = \{R_i \,|\, 0 \le i \le n-1\}$. Therefore, we can construct m disjoint sets $\mathcal{R}_0,\ \mathcal{R}_1, \ldots, \mathcal{R}_{m-1}$ in the following way:

$$\mathcal{R}_i = \{R_{mj+i} \,|\, 0 \le j \le s-1\} \; for \; all \; 0 \le i \le m-1 \tag{3}$$

Now, it is obvious that $\sqcup_{i=0}^{m-1}\mathcal{R}_i = \mathcal{R}$, and $\mathcal{R}_i$ contains exactly a number s of rows from M' where $0 \le i \le m-1$. Here we come to an important property that the rows in different sets have no common nonzero entries in the same column, which is the key feature of a binary matrix. Exploiting this property of a binary matrix, the binary linear layer can actually be seen as the application of m many s-bit S-box with algebraic degree 1 in parallel.

Therefore, if $\boldsymbol{x} = (x_0, x_1, \ldots, x_{n-1})$, and $\boldsymbol{y} = (y_0, y_1, \ldots, y_{n-1})$ are corresponding input and output variables w.r.t the linear layer i.e. $\boldsymbol{y} = M' \cdot \boldsymbol{x}$, then we can write ANF of m many s-bit S-box with algebraic degree 1 as follows:

$$\begin{cases} S_0(\boldsymbol{x_0}) & = (R_0^0 \cdot \boldsymbol{x_0},\ R_m^0 \cdot \boldsymbol{x_0}, \ldots, R_{(s-1)m}^0 \cdot \boldsymbol{x_0}) \\ S_1(\boldsymbol{x_1}) & = (R_1^1 \cdot \boldsymbol{x_1},\ R_{m+1}^1 \cdot \boldsymbol{x_1}, \ldots, R_{(s-1)m+1}^1 \cdot \boldsymbol{x_1}) \\ \quad\vdots & \\ S_{m-1}(\boldsymbol{x_{m-1}}) & = (R_{m-1}^{m-1} \cdot \boldsymbol{x_{m-1}},\ R_{2m-1}^{m-1} \cdot \boldsymbol{x_{m-1}}, \ldots, R_{sm-1}^{m-1} \cdot \boldsymbol{x_{m-1}}) \end{cases}$$

where R^i_{mj+i} is a vector which belongs to the set $\mathbb{F}_2^s$ such that $R^i_{mj+i} = (m'_{mj+i,\,i},\ m'_{mj+i,\,m+i}, \ldots, m'_{mj+i,\,(s-1)m+i})$, and $\boldsymbol{x_i} = (x_i,\, x_{m+i}, \ldots, x_{(s-1)m+i}) \in \mathbb{F}_2^s$ where $i = 0, 1, \ldots, m-1$, and $j = 0, 1, \ldots, s-1$.

An Example of Exact BDPT Modelization of Binary Matrix. The MixColumns matrix M of the block cipher MANTIS which is as follows:

$$M = \begin{pmatrix} 0\,1\,1\,1 \\ 1\,0\,1\,1 \\ 1\,1\,0\,1 \\ 1\,1\,1\,0 \end{pmatrix} \in \mathbb{F}_{2^4}^{4\times 4}$$

Therefore, for this example, $s = 4$, and $m = 4$, and the primitive matrix M' corresponding to the matrix M is a 16×16 matrix where each matrix element is either 0 or 1 i.e. the primitive matrix $M' \in \mathbb{F}_2^{16\times 16}$ is as follows:

$$M' = \begin{pmatrix}
0&0&0&0&1&0&0&0&1&0&0&0&1&0&0&0\\
0&0&0&0&0&1&0&0&0&1&0&0&0&1&0&0\\
0&0&0&0&0&0&1&0&0&0&1&0&0&0&1&0\\
0&0&0&0&0&0&0&1&0&0&0&1&0&0&0&1\\
1&0&0&0&0&0&0&0&1&0&0&0&1&0&0&0\\
0&1&0&0&0&0&0&0&0&1&0&0&0&1&0&0\\
0&0&1&0&0&0&0&0&0&0&1&0&0&0&1&0\\
0&0&0&1&0&0&0&0&0&0&0&1&0&0&0&1\\
1&0&0&0&1&0&0&0&0&0&0&0&1&0&0&0\\
0&1&0&0&0&1&0&0&0&0&0&0&0&1&0&0\\
0&0&1&0&0&0&1&0&0&0&0&0&0&0&1&0\\
0&0&0&1&0&0&0&1&0&0&0&0&0&0&0&1\\
1&0&0&0&1&0&0&0&1&0&0&0&0&0&0&0\\
0&1&0&0&0&1&0&0&0&1&0&0&0&0&0&0\\
0&0&1&0&0&0&1&0&0&0&1&0&0&0&0&0\\
0&0&0&1&0&0&0&1&0&0&0&1&0&0&0&0
\end{pmatrix} \in \mathbb{F}_2^{16\times 16}$$

Now, we can easily conclude that applying the matrix M' to a vector $\boldsymbol{x} = (x_0, x_1, \ldots, x_{15}) \in \mathbb{F}_2^{16}$ is actually equivalent to performing the following 4-bit S-box in parallel:

$$S_i(x_i, x_{i+4}, x_{i+8}, x_{i+12}) = \begin{pmatrix} 0\,1\,1\,1 \\ 1\,0\,1\,1 \\ 1\,1\,0\,1 \\ 1\,1\,1\,0 \end{pmatrix} \begin{pmatrix} x_i \\ x_{i+4} \\ x_{i+8} \\ x_{i+12} \end{pmatrix}, \; i \in \{0, 1, 2, 3\}$$

Therefore, we can construct exact BDPT trail for $\mathbb{K}$ and $\mathbb{L}$ for the mixcolumn operation and the linear inequalities whose feasible solutions are exactly those BDPT trail.

Now, the exact BDPT modelization of S-box we have discussed in the previous section. Applying that approach we can get the exact BDPT trail through the binary linear layer and then we can easily represent the BDPT trails of binary linear layer as linear inequalities following the approach mentioned in [34]. Thus, we give a way to generate a set of inequalities that exactly model the valid BDPT propagations through a binary linear layer. For the ciphers with non-binary linear layer, we decompose its linear layer through the COPY and

XOR operation trivially and generate a set of linear inequalities that model the propagations through the linear layer.

3.3 MILP Model of BDPT for Key-XOR

In this section, we explain how to construct MILP model of BDPT for the Key-Xor operation. As per the notation discussed above E_r is the r-round block cipher where we denote f_i is the ith round function and f_k^i is the ith round Key-Xor operation. Moreover, we denote the initial and output BDPT for the Key-Xor operation as $(\mathbb{K}_{i-1}^*, \mathbb{L}_{i-1}^*)$, and $(\mathbb{K}_i, \mathbb{L}_i)$ respectively. Therefore, as per BDPT Rule 1, we decompose f_k^i into two operations say f_1^i which actually produces some new elements from each elements of $\mathbb{L}_{i-1}^*$ and f_2^i which includes the new vectors and the vectors from $\mathbb{K}_{i-1}^*$ in $\mathbb{K}_i$ which is described in Eqn 1. Hence, we model the operations f_1^i, and f_2^i which jointly present the MILP model for Key-Xor operation.

Table 2. Trails Corresponding to the Function f_1^i

(l_0, l_1, l_2, l_3)	(l_0', l_1', l_2', l_3')
$(0, 0, l_2, l_3)$	$(0, 1, l_2, l_3)$, $(1, 0, l_2, l_3)$, $(1, 1, l_2, l_3)$
$(0, 1, l_2, l_3)$	$(1, 1, l_2, l_3)$,
$(1, 0, l_2, l_3)$	$(1, 1, l_2, l_3)$,
$(1, 1, l_2, l_3)$	X

Modeling f_1^i. In many ciphers, round key is only XORed with a part of block. Without loss of generality, we assume that the round key is XORed with the left s $(0 \leq s \leq n-1)$ bits. Let, $\mathbb{L}_{i-1}^* \subseteq \mathbb{F}_2^4$ and $s = 2$ i.e. round key is XORed with the leftmost 2 bits. Therefore, according to the BDPT rule 1, f_1^i function creates $\boldsymbol{l}' = (l_0', l_1', l_2', l_3')$ from $\boldsymbol{l} = (l_0, l_1, l_2, l_3)$ where for every vector $\boldsymbol{l} \in \mathbb{L}_{i-1}^*$ satisfying $l_i = 0$, $l_i' = l_i \vee 1$ where $i \in \{0, 1\}$ and $l_j' = l_j$ for all $j = 2, 3$. Therefore, we write the propagation table (Table 2) corresponding to the function f_1^i using which we construct linear inequalities whose feasible solutions are exactly those trails. Now, we are ready to give linear inequalities description of these trails listed in Table 2 as follows:

$$\begin{cases} l_j' \geq l_j, & \text{for } j = 0, 1 \\ l_j' = l_j, & \text{for } j = 2, 3 \\ 2\sum_{j=0}^{1} l_j' - \sum_{j=0}^{1} l_j \geq 2 \\ \sum_{j=0}^{3} l_j' - \sum_{j=0}^{3} l_j \geq 1 \end{cases} \tag{4}$$

where l_0', l_1', l_2', l_3', l_0, l_1, l_2, l_3 are binaries.

Apparently, all feasible solutions of the inequalities in Eq. 4 corresponding to $\boldsymbol{l}$, and $\boldsymbol{l}'$ are exactly the trails of f_1^i function described above in Table 2. Similarly, for a n-bit block cipher where $\mathbb{L}_{i-1}^* \subseteq \mathbb{F}_2^n$, and round key is XORed with the leftmost s $(0 \leq s \leq n-1)$ bits, the linear inequalities we get which describe the trails $\boldsymbol{l} \xrightarrow{f_1^i} \boldsymbol{l}'$ as follows:

$$\begin{cases} l'_j & \geq l_j, \quad \text{for } j = 0,\, 1,\, ...,\, s-1 \\ l'_j & = l_j, \quad \text{for } j = s,\, s+1 \ldots, n-1 \\ s\sum_{j=0}^{s-1} l'_j - (s-1)\sum_{j=0}^{s-1} l_j & \geq s \\ \sum_{j=0}^{n-1} l'_j - \sum_{j=1}^{n} l_j & \geq 1 \end{cases} \tag{5}$$

where $l'_0,\, l'_1,\, ...,\, l'_{n-1}, l_0,\, l_1,\, ...,\, l_{n-1}$ are binaries.

Modeling f_2^i. After applying f_1^i on each element of the set $\mathbb{L}_{i-1}^*$, we get the set say $\mathbb{L}'_{i-1}$ as follows:

$$\mathbb{L}'_{i-1} = \{\boldsymbol{l}' \in \mathbb{F}_2^n \mid f_1^i(\boldsymbol{l}) = \boldsymbol{l}',\ \forall\, \boldsymbol{l} \in \mathbb{L}_{i-1}^*\}$$

Now, from BDPT Rule 1 we know that:

$$f_2^i(\mathbb{K}_{i-1}^*,\ \mathbb{L}'_{i-1}) = \mathbb{K}_{i-1}^* \cup \mathbb{L}'_{i-1} = \mathbb{K}_i$$

Therefore, to model f_2^i, we define another function $g : (\mathbb{F}_2^2 \setminus \{(0,0),(1,1)\}) \times \mathbb{K}_{i-1}^* \times \mathbb{L}'_{i-1} \rightarrow \mathbb{K}_i$ such that:

$$g(\lambda_0, \lambda_1, \boldsymbol{k}^*, \boldsymbol{l}') = (\lambda_0 \wedge k_0^*, \ldots, \lambda_0 \wedge k_{n-1}^*) \oplus (\lambda_1 \wedge l'_0, \ldots, \lambda_1 \wedge l'_{n-1}) \tag{6}$$

where $\lambda = (\lambda_0,\, \lambda_1) \in \mathbb{F}_2^2 \setminus \{(0,0),(1,1)\}$, and $\boldsymbol{k}^* = (k_0^*, \ldots, k_{n-1}^*)$, and $\boldsymbol{l}' = (l'_0, \ldots, l'_{n-1})$. Therefore, from the definition of g we can easily conclude that $\mathbb{K}_i$ contain all the elements of $\mathbb{L}'_{i-1}$, and $\mathbb{K}_{i-1}^*$. Hence, modeling g is actually equivalent to modeling f_2^i. Now, we are going to construct the linear inequalities whose feasible solutions are exactly the g function trail. In order to do that first we have to construct the linear inequalities which are sufficient to describe the propagation $(a,\, b) \xrightarrow{\wedge} c$ where $a,\, b,\, c \in \mathbb{F}_2$ which is as follows:

$$\begin{cases} a - c & \geq 0 \\ b - c & \geq 0 \\ a + b - c & \leq 1 \end{cases} \tag{7}$$

where a, b, c are binaries. Therefore, using Eq. 6 and Eq. 7 we can easily conclude that the following inequalities are sufficient to describe the propagation of g function i.e. $(\lambda_0,\, \lambda_1,\, \boldsymbol{k}^*,\, \boldsymbol{l}') \xrightarrow{g} \boldsymbol{k}$:

$$\begin{cases} \lambda_0 - p_j & \geq 0, \quad \text{for } j = 0, 1, ..., n-1 \\ k_j^* - p_j & \geq 0, \quad \text{for } j = 0, 1, ..., n-1 \\ \lambda_0 + k_j^* - p_j & \leq 1, \quad \text{for } i = 0, 1, ..., n-1 \\ \lambda_1 - q_j & \geq 0, \quad \text{for } i = 0, 1, ..., n-1 \\ l_j' - q_j & \geq 0, \quad \text{for } i = 0, 1, ..., n-1 \\ \lambda_1 + l_j' - q_j & \leq 1, \quad \text{for } i = 0, 1, ..., n-1 \\ p_j + q_j - k_j & = 0, \quad \text{for } j = 0, 1, ..., n-1 \\ \lambda_0 + \lambda_1 & = 1 \end{cases} \tag{8}$$

where $p_0, ..., p_{n-1}, q_0, ..., q_{n-1}, l_0', ..., l_{n-1}', k_0, ..., k_{n-1}, k_0^*, ..., k_{n-1}^*, \lambda_0, \lambda_1$ are binaries and $p = (p_0, p_1, ..., p_{n-1})$, $q = (q_0, q_1, ..., q_{n-1})$ are auxiliary variables. Hence Eq. 8 and Eq. 5 describe the complete MILP model of the Key-XOR operation w.r.t BDPT.

3.4 MILP Model Construction of r-Round Function

For all the functions based on these above mentioned operations, we are finally making a set of linear inequalities depicting one round BDPT propagation. In order to construct an MILP model for r round BDPT propagation we have to iterate this above mentioned procedure r times and finally we conclude upon getting a system of linear inequalities $\mathcal{L}$ which we describe in Algorithm 1.

Algorithm 1 constructs a system of linear inequalities which charecterizes all r-round BDPT trails i.e.

$$(\boldsymbol{k}^0 = \boldsymbol{k}, \boldsymbol{l}^0 = \boldsymbol{l}) \xrightarrow{f_1} (\boldsymbol{k}^1, \boldsymbol{l}^1) \xrightarrow{f_2} \ldots \xrightarrow{f_r} (\boldsymbol{k}^r, \boldsymbol{l}^r)$$

Therefore, we have to construct MILP model using $\mathcal{L}$ and appropriate initial and stopping rules and the search algorithm in order to find integral distinguisher.

4 Automatic Search Algorithm for r-Round Integral Distinguisher

In this section, we first study the initial BDPT and stopping rule to use when searching for integral distinguisher based on BDPT. From Algorithm 1 we got the linear inequality system $\mathcal{L}$ with the input vector $\boldsymbol{k}$ and $\boldsymbol{l}$. Now, we convert the stopping rule into an objective function and combining $\mathcal{L}$ and objective function, we construct the MILP model $\mathcal{M}_{\mathbb{K},\mathbb{L}}$. At last we propose an algorithm to search integral distinguisher based on BDPT given the initial BDPT $D^{1^n}_{\{\boldsymbol{k}\},\{\boldsymbol{l}\}}$ for an n-bit block cipher and prove the correctness of the algorithm.

4.1 Initial BDPT

In [31], Todo and Morii set the initial BDPT as ($\mathbb{K} = \{\mathbf{1}\}$, $\mathbb{L} = \{7fffffff\}$) to search the BDPT of SIMON32, where the active bits of the vector $\boldsymbol{l}$ are set as 1 and 0 for constant bits. Hence we do the same. Let the initial input BDPT variables are $\boldsymbol{k}^0 = (k_0^0, k_1^0, ..., k_{n-1}^0)$, and $\boldsymbol{l}^0 = (l_0^0, l_1^0, ..., l_{n-1}^0)$ where n is the block size. The constraints on k_i^0 and l_i^0 are

$$k_i^0 = 1 \quad \text{for } i = 0, 1, ..., n-1$$

$$l_i^0 = \begin{cases} 1, & \text{if } i-th \text{ bit is active} \\ 0, & \text{otherwise} \end{cases}$$

Algorithm 1: Computing A Set of Constraints Characterizing BDPT Propagation

---

Input: The initial input BDPT of an n-bit iterated cipher $D^{1^n}_{\mathbb{K}_0=\{\boldsymbol{k}\},\mathbb{L}_0=\{\boldsymbol{l}\}}$
$\mathcal{L}_k(\mathbb{K}_{i-1}, \mathbb{K}^*_{i-1})$: a constraint set of linear inequalities whose feasible solutions are all division trails from the set $\mathbb{K}_{i-1}$ to set $\mathbb{K}^*_{i-1}$, $\forall i \in [r]$.
$\mathcal{L}_l(\mathbb{L}_{i-1}, \mathbb{L}^*_{i-1})$: a constraint set of linear inequalities whose feasible solutions are all division trails from the set $\mathbb{L}_{i-1}$ to set $\mathbb{L}^*_{i-1}$, $\forall i \in [r]$.
$New_k(\mathbb{L}^*_{i-1}, \mathbb{L}'_{i-1})$: a constraint set of linear inequalities whose feasible solutions are all f_1^i function trails, $\forall i \in [r]$.
$Union_k(\mathbb{L}'_{i-1}, \mathbb{K}^*_{i-1}, \mathbb{K}_i)$: a constraint set of linear inequalities whose feasible solutions are all f_2^i function trails, $\forall i \in [r]$.
Output: A constraint set of linear inequalities $\mathcal{L}$ describing r-round BDPT propagation
begin
  $\mathcal{L} = \emptyset$, $C^i = C^{i,*} = \emptyset$ where $i = 1, 2, \ldots, r$
  Allocate n-bit variables $\boldsymbol{k}^i$, $\boldsymbol{l}^i$ to denote vectors in the set $\mathbb{K}_i$, $\mathbb{L}_i$ respectively where $i = 0, 1, \ldots, r$
  Allocate n-bit variables $\boldsymbol{l}^{i,*}$, $\boldsymbol{p}^i$, and $\boldsymbol{k}^{i,*}$ to denote vectors in the set $\mathbb{L}^*_i$, $\mathbb{L}'_i$, and $\mathbb{K}^*_i$ respectively where $i = 0, 1, \ldots, r-1$
  $\mathcal{L} \leftarrow (\boldsymbol{k}^0 = \boldsymbol{k})$
  $\mathcal{L} \leftarrow (\boldsymbol{l}^0 = \boldsymbol{l})$
  for $(i = 1\,; i \leq r\,; i++)$ **do**
    $C^i \leftarrow \mathcal{L}_l(\mathbb{L}_{i-1}, \mathbb{L}^*_{i-1}) \cup \mathcal{L}_k(\mathbb{K}_{i-1}, \mathbb{K}^*_{i-1})$
    $C^{i,*} \leftarrow New_k(\mathbb{L}^*_{i-1}, \mathbb{L}'_{i-1})$
    $C^{i,*} \leftarrow Union_k(\mathbb{L}'_{i-1}, \mathbb{K}^*_{i-1}, \mathbb{K}_i)$
    $\mathcal{L} \leftarrow (\boldsymbol{l}^{i-1,*} = \boldsymbol{l}^i)$
    $\mathcal{L} \leftarrow (C^i \cup C^{i,*})$
  end
  return $\mathcal{L}$
end

4.2 Stopping Rule

Our automatic search model only focuses on the parity of one output bit. Without loss of generality, we consider the q-th output bit. After r round, the output set has BDPT $D^{1^n}_{\mathbb{K}_r, \mathbb{L}_r}$. Therefore, according to the Proposition 1, we know that the set of the first components of the last vectors of all r-round BDPT trails which start from the vector $(\boldsymbol{k}, \boldsymbol{l})$ is equal to $\mathbb{K}_r$. Hence, to check whether there exist any unit vector in the $\mathbb{K}_r$, the objective function can be set as follows:

$$Obj : Minimize(k_0^r + k_1^r + \ldots, k_{n-1}^r) \tag{9}$$

Similarly, according to the Proposition 1, the set of the second components of the last vectors of all r-round BDPT trails which start from the vector $(\boldsymbol{k}, \boldsymbol{l})$ is equal to $\mathbb{L}_r$. Thus, we can set the objective function as :

$$Obj : Minimize(l_0^r + l_1^r + \ldots, l_{n-1}^r) \tag{10}$$

Now, at first we construct the MILP model $\mathcal{M}_{\mathbb{K},\mathbb{L}}$ using the system of linear inequalities $\mathcal{L}$ we get from Algorithm 1 and the objective function defined in Eq. 9. Moreover, we construct another MILP model $\mathcal{M}_{\mathbb{L}}$ as follows:

$$\mathcal{M}_{\mathbb{L}} = ConstructModel(\mathcal{L}^*, Min(l_0^r + \ldots, l_{n-1}^r))$$

where $\mathcal{L}^*$ is the constraint set of linear inequalities whose feasible solutions are all division trails from the set $\mathbb{L}_0$ to $\mathbb{L}_r$.

Stopping Rule for the MILP Model $\mathcal{M}_{\mathbb{K},\mathbb{L}}$. To check whether $\mathbb{K}_r$ contains the unit vector $\boldsymbol{e}_q$ is equivalent to check whether the MILP model $\mathcal{M}_{\mathbb{K},\mathbb{L}}$ has feasible solution satisfying $\boldsymbol{k}^r = \boldsymbol{e}_q$. Therefore, we can set the stopping rule as:

$$k_j^r = \begin{cases} 1 & if\ j = q \\ 0 & otherwise \end{cases} \tag{11}$$

If $\mathcal{M}_{\mathbb{K},\mathbb{L}}$ has such feasible solutions, then the q-th output bit is unknown.

Stopping Rule for the MILP Model $\mathcal{M}_{\mathbb{L}}$. If $\mathbb{K}_r$ does not contain $\boldsymbol{e}_q$, then to check whether $\mathbb{L}_r$ contains $\boldsymbol{e}_q$ is equivalent to check whether the MILP model $\mathcal{M}_{\mathbb{L}}$ has feasible solution satisfying $\boldsymbol{l}^r = \boldsymbol{e}_q$. Therefore, we can set the stopping rule as :

$$l_j^r = \begin{cases} 1 & if\ j = q \\ 0 & otherwise \end{cases} \tag{12}$$

If both $\mathbb{K}_r$ and $\mathbb{L}_r$ do not contain $\boldsymbol{e}_q$, then q-th output bit is balanced. Otherwise, we need to count the number of feasible solutions satisfying $\boldsymbol{l}^r = \boldsymbol{e}_q$ of the model $\mathcal{M}_{\mathbb{L}}$. Therefore, the parity of q-th output bit is 0 or 1 if the number of solutions are even or odd respectively as $\mathbb{K}_r$ does not contain $\boldsymbol{e}_q$.

Algorithm 2: Deciding Parity of q-th Output Bit

Input: The r-round cipher E_r, the initial input BDPT of an n-bit iterated cipher $D^{1^n}_{\mathbb{K}_0=\{\boldsymbol{k}\},\mathbb{L}_0=\{\boldsymbol{l}\}}$, the number q, and $\mathcal{L}_l(\mathbb{L}_{i-1}, \mathbb{L}_i)$: a constraint set of linear inequalities whose feasible solutions are all division trails from the set $\mathbb{L}_{i-1}$ to set $\mathbb{L}_i$, $\forall i \in [r]$.

Output: The balanced information of the q-th output bit based on BDPT

```
begin
    Allocate all the variables denoting the input and output BDPT
     Obj_1 = Minimize(k_0^r + k_1^r + ..., k_{n-1}^r)
     Obj_2 = Minimize(l_0^r + l_1^r + ..., l_{n-1}^r)
     Call Algorithm 1 and get a constraint set L whose feasible solutions
     are r-round BDPT trail
     M_{K,L} = ConstructModel(L, Obj_1)
     M_{K,L}.AddConstraint(k^r = e_q)
     if the MILP model M_{K,L} has solutions then
         return unknown
    end
    else
        M_L = ConstructModel(∪_{i=1}^r L_l(L_{i-1}, L_i), Obj_2)
         M_L.AddConstraint(l^0 = l)
         if the MILP model M_L has no feasible solution satisfying
         l^r = e_q then
             return 0
        end
        else
            M_L.AddConstraint(l^r = e^q)
             Count the number of solutions in M_L
             if Count is even then
                 return 0
            end
            else
                 return 1
            end
        end
    end
end
```

4.3 Search Algorithm

We present the automatic search algorithm to find integral distinguisher based on BDPT, which decides the parity of the q-th output bit with the given initial BDPT $D^{1^n}_{\mathbb{K}_0=\{\boldsymbol{k}\},\mathbb{L}_0=\{\boldsymbol{l}\}}$ for an n-bit block cipher. Firstly, we allocate all round variables and auxiliary variables. Therefore, we construct a MILP model $\mathcal{M}_{\mathbb{K},\mathbb{L}}$ that describes all r-round BDPT trails, and another MILP model $\mathcal{M}_{\mathbb{L}}$ that describes all r-round division trails for $\mathbb{L}$. Finally, using appropriate initial and stopping rules, we can obtain the parity of q-th output bit based on BDPT. We illustrate the whole framework in Algorithm 2.

4.4 Correctness of Search Algorithm

Let the initial input division property of an n-bit iterated cipher be $D^{1^n}_{\mathbb{K}_0=\{\boldsymbol{k}\},\mathbb{L}_0=\{\boldsymbol{l}\}}$, and after r-round propagation, the output BDPT we denote as $D_{\mathbb{K}_r,\mathbb{L}_r}$. It is obvious that if $\boldsymbol{e}_q \in \mathbb{K}_r$, then the parity of q-th bit is *unknown* and if $\boldsymbol{e}_q$ does not belongs to $\mathbb{K}_r$ as well as $\mathbb{L}_r$, then the parity of q-th bit is 0.

Therefore, to prove correctness of Algorithm 2 we have to prove that if the q-th unit vector does not belong to $\mathbb{K}_r$ and belongs to $\mathbb{L}_r$, then the parity of q-th output bit is 0 or 1 provided the number of division trails from $\boldsymbol{l}$ to $\boldsymbol{e}_q$ is even or odd respectively. We first prove the following Lemma:

Lemma 1. *Let $\mathbb{X} \subseteq \mathbb{F}_2^n$ has division property $D^{1^n}_{\mathbb{K}_0=\{\boldsymbol{k}\},\mathbb{L}_0=\{\boldsymbol{l}\}}$ and after r-round propagation, the output set $\mathbb{Y}_r$ has division property $D^{1^n}_{\mathbb{K}_r,\mathbb{L}_r}$. For any $\boldsymbol{l}' \in \mathbb{L}_r$, if the number of division trail in $\mathbb{L}$ from $\boldsymbol{l}$ to $\boldsymbol{l}'$ is even, then there exist at least one j in $[r]$ s.t $\mathbb{L}_j$ contains at least one element $\boldsymbol{u}$ which is produced even number of times from the elements in $\mathbb{L}_{j-1}$.*

The proof is provided in the full version of this paper [10]. Therefore, using Lemma 1 we prove the final result as follows:

Proposition 2. *Let $\mathbb{X} \subseteq \mathbb{F}_2^n$ has division property $D^{1^n}_{\mathbb{K}_0=\{\boldsymbol{k}\},\mathbb{L}_0=\{\boldsymbol{l}\}}$ and after r-round propagation, the output set $\mathbb{Y}$ has division property $D^{1^n}_{\mathbb{K}_r,\mathbb{L}_r}$. If $\boldsymbol{e}_q$ doesn't belongs to the set $\mathbb{K}_r$, where $q \in [n]$, then we have:*

1. *If the number of division trail from $\boldsymbol{l}$ to $\boldsymbol{e}_q$ is even in $\mathbb{L}$, then $\bigoplus_{\boldsymbol{y}\in\mathbb{Y}} y_q = 0$.*
2. *If the number of division trail from $\boldsymbol{l}$ to $\boldsymbol{e}_q$ is odd in $\mathbb{L}$, then $\bigoplus_{\boldsymbol{y}\in\mathbb{Y}} y_q = 1$.*

Proof. Let $S \subseteq (\mathbb{F}_2^n)^{r+1}$ be the set which contains all the division trail in $\mathbb{L}$ from $\boldsymbol{l}$ to $\boldsymbol{e}_q$ and $|S|$ is even. Now, by using Lemma 1, we can easily conclude that there exist at least one $j \in \{2, 3, ..., r\}$ s.t $\mathbb{L}_j$ contains an element $\boldsymbol{u}$ which is produced even number of times from the elements in $\mathbb{L}_{j-1}$. Without loss of generality we choose smallest such j.

According to the BDPT propagation rule of XOR and S-box, we can see that if an element $\boldsymbol{u}$ is produced even number of times in $\mathbb{L}_j$ from $\mathbb{L}_{j-1}$, then the following holds:

$$\bigoplus_{\boldsymbol{y}\in\mathbb{Y}_j} \boldsymbol{y}^{\boldsymbol{u}} = 0$$

where $\mathbb{Y}_j$ is the multiset whose BDPT is $D^{1^n}_{\mathbb{K}_j,\mathbb{L}_j}$ and that implies $\boldsymbol{u}$ shouldn't be in $\mathbb{L}_j$. Hence, all the division trails from $\boldsymbol{l}$ to $\boldsymbol{e}_q$ which contains the vector $\boldsymbol{u}$ are actually redundant and those number of redundant division trails must be even. Therefore, we can remove these redundant division trails from S and we can call the new set as S_1. It is trivial that either $|S_1|$ is even or $|S_1| = 0$.

Case-I. If $|S_1| = 0$, then it implies that all the division trails from $\boldsymbol{l}$ to $\boldsymbol{e}_q$ contains the element $\boldsymbol{u}$. Therefore, as $\boldsymbol{u}$ shouldn't be in $\mathbb{L}_j$, so $\boldsymbol{e}_q$ also shouldn't be in $\mathbb{L}_r$ and it is given that $\boldsymbol{e}_q$ doesn't belongs to $\mathbb{K}_r$ which means

$$\bigoplus_{\boldsymbol{y}\in\mathbb{Y}} \boldsymbol{y}^{\boldsymbol{e}_q} = \bigoplus_{\boldsymbol{y}\in\mathbb{Y}} y_q = 0.$$

Case-II. If $|S_1|$ is even, then in a similar way we can find even number of redundant division trails from $\boldsymbol{l}$ to $\boldsymbol{e}_q$ in $\mathbb{L}$ and construct S_2 from S_1 where $|S_2|$ is either even or 0 and so on.

As $|S|$ is finite, then after finitely many p steps, we must get some S_p s.t $|S_p| = 0$. Hence, $\boldsymbol{e}_q$ shouldn't be in $\mathbb{L}_r$ and it is given that $\boldsymbol{e}_q$ doesn't belongs to $\mathbb{K}_r$ which means

$$\bigoplus_{\boldsymbol{y}\in\mathbb{Y}} \boldsymbol{y}^{\boldsymbol{e}_q} = \bigoplus_{\boldsymbol{y}\in\mathbb{Y}} y_q = 0$$

which completes the first part of the proof.

Now, it is given that the number of division trail in $\mathbb{L}$ from $\boldsymbol{l}$ to $\boldsymbol{e}_q$ is odd. Similarly we can construct a set S' containing all such division trails. Therefore, there may or may not exist $j \in \{2, 3, ..., r\}$ s.t $\mathbb{L}_j$ contains an element $\boldsymbol{u}$ which is produced even number of times from the elements in $\mathbb{L}_{j-1}$.

Case-A. If there doesn't exist any such j, then by BDPT propagation rules, we can easily conclude that no division trail from $\boldsymbol{l}$ to $\boldsymbol{e}_q$ is redundant. Therefore, it implies that $\boldsymbol{e}_q$ belongs to $\mathbb{L}_r$ which means $\bigoplus_{\boldsymbol{y}\in\mathbb{Y}} y_q = 1$.

Case-B. If there exist some j s.t $\mathbb{L}_j$ contains an element $\boldsymbol{u}$ which is produced even number of times from the elements in $\mathbb{L}_{j-1}$, then similarly by the previous argument we can easily conclude that all the division trails from $\boldsymbol{l}$ to $\boldsymbol{e}_q$ which contains $\boldsymbol{u}$ are actually redundant. Therefore, we can remove these redundant division trails from S' and we can call the new set as S'_1. It is obvious that $|S'_1|$ is odd.

Now, continuing like this way, after finitely many steps we arrive at a situation where the number of remaining division trails from $\boldsymbol{l}$ to $\boldsymbol{e}_q$ is odd and no redundant division trails are left which implies e_q belongs to $\mathbb{L}_r$. Therefore, $\bigoplus_{\boldsymbol{y}\in\mathbb{Y}} \boldsymbol{y}^{\boldsymbol{e}_q} = \bigoplus_{\boldsymbol{y}\in\mathbb{Y}} y_q = 1$ which completes the second part of the proof. □

5 Applications to Block Ciphers

In this section, we apply our automatic search algorithm for BDPT to SIMON, SIMON(102), MANTIS, PRINCE, KLEIN, and PRIDE block ciphers. All the experiments are conducted on the platform Intel Core i5-8250U CPU @ 1.60 GHz, 8 G RAM, 64 bit Ubuntu 18.04.5 LTS. The optimizer we used to solve MILP models is Gurobi 9.1.2 [17]. For the integral distinguishers, '?' denotes the bit whose balanced information is unknown, '0' denotes the bit whose sum is zero, '1' denotes the bit whose sum is 1. The detailed integral distinguishers of PRINCE, MANTIS, KLEIN and PRIDE are listed in supporting material of the full version of this paper [10].

5.1 Applications to PRINCE and MANTIS

In this section we present the application of our BDPT model to the cipher PRINCE and MANTIS which have binary matrices to conduct their mixcolumn operations in the round functions. Hence, we apply our method to model binary linear layer in BDPT and construct the MILP model efficiently. Then, choosing appropriate initial BDPT, we find improved integral distinguisher as follows:

Integral Attack on PRINCE. Block ciphers based on the reflection design strategy, introduced by PRINCE [8], are a popular choice for low-latency designs. PRINCE is the 64-bit block cipher which uses 128-bit key. The PRINCE cipher is the substitution-permutation network composed of 12 rounds. The 64-bit state can be organised as the 4×4 array of nibbles. For a complete specification and design rationale of the cipher, a reader is referred to [8].

We will denote the number of rounds of PRINCE as $a + b$ where a are the rounds before the middle layer, and b are the rounds after the middle layer. There are several attacks (Integral attack, higher order differential attack, boomerang attack) on PRINCE [2,25,27]. Now, in [15], the authors applied CBDP on PRINCE and found $2 + 1$ and $1 + 2$ round integral distinguishers which are best integral distinguisher till date.

For PRINCE, we find a $2+2$ round integral distinguisher which is one more round than the previous best results [15].

Integral Attack on MANTIS. MANTIS is a tweakable block cipher published at CRYPTO 2016 by Beierle *et al.* [6] and the cipher's structure is similar to PRINCE. This block cipher operate on a 64-bit message block and work with a 64-bit tweak and (64+64) bit key and has a SPN structure. For a more detailed description of the MANTIS family, we refer to the design paper [6].

In the light of cryptanalysis, there are several attacks [7,11,14] on MANTIS. For MANTIS, we find a $3+3$ round integral distinguisher based on BDPT which is one more round than the previous best results [15].

5.2 Applications to KLEIN and PRIDE

To complete our BDPT analysis on ciphers with complex linear layers, we apply our automatic search algorithm for BDPT to block ciphers KLEIN and PRIDE which have non-binary linear layers. In order to handle non-binary linear layers we trivially decompose the linear layers as COPY and XOR operations and construct the MILP model accordingly. Then, choosing appropriate initial BDPT, we find integral distinguisher as follows:

Integral Attack on KLEIN. KLEIN [16] is a family of block ciphers, with a fixed 64-bit block size and variable key length-64, 80 or 96-bits. The structure of KLEIN is a typical Substitution Permutation Network. For more details, please refer to [16].

In the light of cryptanalysis, there are several attacks [1,3,26,36] on the block cipher KLEIN, mostly on KLEIN-64 (key length 64 bits). In [36], the authors have presented a 5-round integral distinguisher using the higher-order integral and the higher-order differential properties which is best integral distinguisher known to us. First we apply MILP based CBDP on KLEIN and find a 6-round integral distinguisher which is one more round than the previous best results [36]. Therefore, we apply the MILP based BDPT on KLEIN and the integral distinguishers we find are in accordance with the distinguishers we find based on CBDP.

Integral Attack on PRIDE. PRIDE is a lightweight block cipher designed by Albrecht et al. [4], appears in CRYPTO 2014. PRIDE is an SPN structure block cipher with 64-bit block cipher and 128-bit key. For more details, please refer to [4]. In the light of cryptanalysis, there are several attacks on PRIDE [12,13,35,38].

First we apply MILP based CBDP on PRIDE and find a 9-round integral distinguisher which is one more round than the previous best results [33]. Therefore, we apply the MILP based BDPT on PRIDE and the integral distinguishers we find are in accordance with the distinguishers we find based on CBDP.

5.3 Applications to SIMON, SIMON (102)

We apply our method to all variants of SIMON [5], and SIMON(102) [22] block ciphers and the distinguishers we find are in accordance with the previous longest distinguishers [24] but we get these results in better time which are shown in the full version of this paper [10].

6 Conclusion and Future Work

In this paper, we provide an idea to model BDPT propagation of ciphers with binary (complex) linear layers and furthermore we construct an efficient automatic search algorithm that accurately characterize BDPT propagation. Based on these, more accurate BDPT for ciphers with binary (complex) linear layers such as PRINCE, MANTIS can be obtained.

For ciphers with non-binary linear layers we trivially decompose the linear layer by COPY-XOR technique which may ignore some balanced property. Therefore, how to model BDPT propagation for ciphers with non-binary linear layers accurately and efficiently is an open problem. Moreover, we construct our model using MILP solver whereas SAT/SMT are also very popular and efficient solvers in this domain. How to implement our model using SAT/SMT solvers or similar ones will be a future work.

Acknowledgement. The authors would like to thank the anonymous reviewers for their valuable comments and suggestions.

References

1. Abed, F., Forler, C., List, E., Lucks, S., Wenzel, J.: Biclique cryptanalysis of the PRESENT and LED lightweight ciphers. IACR Cryptology ePrint Archive 2012:591 (2012)
2. Abed, F., List, E., Lucks, S.: On the security of the core of PRINCE against biclique and differential cryptanalysis. IACR Cryptology ePrint Archive, p. 712 (2012)
3. Ahmadian, Z., Salmasizadeh, M., Aref, M.R.: Biclique cryptanalysis of the full-round KLEIN block cipher. IET Inf. Secur. **9**(5), 294–301 (2015)
4. Albrecht, M.R., Driessen, B., Kavun, E.B., Leander, G., Paar, C., Yalçın, T.: Block ciphers – focus on the linear layer (feat. PRIDE). In: Garay, J.A., Gennaro, R. (eds.) CRYPTO 2014. LNCS, vol. 8616, pp. 57–76. Springer, Heidelberg (2014). https://doi.org/10.1007/978-3-662-44371-2_4
5. Beaulieu, R., Shors, D., Smith, J., Treatman-Clark, S., Weeks, B., Wingers, L.: The SIMON and SPECK lightweight block ciphers. In: Proceedings of the 52nd Annual Design Automation Conference, San Francisco, CA, USA, 7–11 June 2015, pp. 175:1–175:6. ACM (2015)
6. Beierle, C., et al.: The SKINNY family of block ciphers and its low-latency variant MANTIS. In: Robshaw, M., Katz, J. (eds.) CRYPTO 2016. LNCS, vol. 9815, pp. 123–153. Springer, Heidelberg (2016). https://doi.org/10.1007/978-3-662-53008-5_5
7. Beyne, T.: Block cipher invariants as eigenvectors of correlation matrices. J. Cryptol. **33**(3), 1156–1183 (2020)
8. Borghoff, J., et al.: PRINCE - a low-latency block cipher for pervasive computing applications (full version). IACR Cryptology ePrint Archive, p. 529 (2012)
9. Boura, C., Canteaut, A.: Another view of the division property. In: Robshaw, M., Katz, J. (eds.) CRYPTO 2016. LNCS, vol. 9814, pp. 654–682. Springer, Heidelberg (2016). https://doi.org/10.1007/978-3-662-53018-4_24
10. Chakraborty, D.: Finding three-subset division property for ciphers with complex linear layers (full version). Cryptology ePrint Archive, Paper 2022/1444 (2022). https://eprint.iacr.org/2022/1444
11. Chen, S., Liu, R., Cui, T., Wang, M.: Automatic search method for multiple differentials and its application on MANTIS. Sci. China Inf. Sci. **62**(3), 32111:1–32111:15 (2019)
12. Dai, Y., Chen, S.: Cryptanalysis of full PRIDE block cipher. Sci. China Inf. Sci. **60**(5), 052108:1–052108:12 (2017)
13. Dinur, I.: Cryptanalytic time-memory-data tradeoffs for FX-constructions with applications to PRINCE and PRIDE. In: Oswald, E., Fischlin, M. (eds.) EUROCRYPT 2015. LNCS, vol. 9056, pp. 231–253. Springer, Heidelberg (2015). https://doi.org/10.1007/978-3-662-46800-5_10
14. Eichlseder, M., Kales, D.: Clustering related-tweak characteristics: application to MANTIS-6. IACR Trans. Symmetric Cryptol. **2018**(2), 111–132 (2018)
15. Eskandari, Z., Kidmose, A.B., Kölbl, S., Tiessen, T.: Finding integral distinguishers with ease. In: Cid, C., Jacobson Jr., M. (eds.) SAC 2018. LNCS, vol. 11349, pp. 115–138. Springer, Cham (2018). https://doi.org/10.1007/978-3-030-10970-7_6

16. Gong, Z., Nikova, S., Law, Y.W.: KLEIN: a new family of lightweight block ciphers. In: Juels, A., Paar, C. (eds.) RFIDSec 2011. LNCS, vol. 7055, pp. 1–18. Springer, Heidelberg (2012). https://doi.org/10.1007/978-3-642-25286-0_1
17. Gurobi Optimization, LLC.: Gurobi Optimizer Reference Manual (2021)
18. Hebborn, P., Lambin, B., Leander, G., Todo, Y.: Lower bounds on the degree of block ciphers. In: Moriai, S., Wang, H. (eds.) ASIACRYPT 2020. LNCS, vol. 12491, pp. 537–566. Springer, Cham (2020). https://doi.org/10.1007/978-3-030-64837-4_18
19. Hebborn, P., Lambin, B., Leander, G., Todo, Y.: Strong and tight security guarantees against integral distinguishers. In: Tibouchi, M., Wang, H. (eds.) ASIACRYPT 2021. LNCS, vol. 13090, pp. 362–391. Springer, Cham (2021). https://doi.org/10.1007/978-3-030-92062-3_13
20. Hu, K., Wang, M.: Automatic search for a variant of division property using three subsets. In: Matsui, M. (ed.) CT-RSA 2019. LNCS, vol. 11405, pp. 412–432. Springer, Cham (2019). https://doi.org/10.1007/978-3-030-12612-4_21
21. Hu, K., Wang, Q., Wang, M.: Finding bit-based division property for ciphers with complex linear layer. IACR Cryptology ePrint Archive, p. 547 (2020)
22. Kölbl, S., Leander, G., Tiessen, T.: Observations on the SIMON block cipher family. In: Gennaro, R., Robshaw, M. (eds.) CRYPTO 2015. LNCS, vol. 9215, pp. 161–185. Springer, Heidelberg (2015). https://doi.org/10.1007/978-3-662-47989-6_8
23. Lambin, B., Derbez, P., Fouque, P.-A.: Linearly equivalent s-boxes and the division property. Des. Codes Cryptogr. **88**(10), 2207–2231 (2020)
24. Liu, H., Wang, Z., Zhang, L.: A model set method to search integral distinguishers based on division property for block ciphers. Cryptology ePrint Archive, Paper 2022/720 (2022). https://eprint.iacr.org/2022/720
25. Morawiecki, P.: Practical attacks on the round-reduced PRINCE. IET Inf. Secur. **11**(3), 146–151 (2017)
26. Nikolic, I., Wang, L., Shuang, W.: The parallel-cut meet-in-the-middle attack. Cryptogr. Commun. **7**(3), 331–345 (2015)
27. Rasoolzadeh, S., Raddum, H.: Cryptanalysis of PRINCE with minimal data. In: Pointcheval, D., Nitaj, A., Rachidi, T. (eds.) AFRICACRYPT 2016. LNCS, vol. 9646, pp. 109–126. Springer, Cham (2016). https://doi.org/10.1007/978-3-319-31517-1_6
28. Sun, L., Wang, W., Wang, M.: Milp-aided bit-based division property for primitives with non-bit-permutation linear layers. IACR Cryptology ePrint Archive, p. 811 (2016)
29. Sun, L., Wang, W., Wang, M.: Automatic search of bit-based division property for ARX ciphers and word-based division property. In: Takagi, T., Peyrin, T. (eds.) ASIACRYPT 2017. LNCS, vol. 10624, pp. 128–157. Springer, Cham (2017). https://doi.org/10.1007/978-3-319-70694-8_5
30. Todo, Y.: Structural evaluation by generalized integral property. In: Oswald, E., Fischlin, M. (eds.) EUROCRYPT 2015. LNCS, vol. 9056, pp. 287–314. Springer, Heidelberg (2015). https://doi.org/10.1007/978-3-662-46800-5_12
31. Todo, Y., Morii, M.: Bit-based division property and application to SIMON family. In: Peyrin, T. (ed.) FSE 2016. LNCS, vol. 9783, pp. 357–377. Springer, Heidelberg (2016). https://doi.org/10.1007/978-3-662-52993-5_18
32. Wang, S., Hu, B., Guan, J., Zhang, K., Shi, T.: MILP-aided method of searching division property using three subsets and applications. In: Galbraith, S.D., Moriai, S. (eds.) ASIACRYPT 2019. LNCS, vol. 11923, pp. 398–427. Springer, Cham (2019). https://doi.org/10.1007/978-3-030-34618-8_14

33. Xiang, Z., Zeng, X., Zhang, S.: On the bit-based division property of s-boxes. Sci. China Inf. Sci. **65**(4), 149101 (2021)
34. Xiang, Z., Zhang, W., Bao, Z., Lin, D.: Applying MILP method to searching integral distinguishers based on division property for 6 lightweight block ciphers. In: Cheon, J.H., Takagi, T. (eds.) ASIACRYPT 2016. LNCS, vol. 10031, pp. 648–678. Springer, Heidelberg (2016). https://doi.org/10.1007/978-3-662-53887-6_24
35. Yang, Q., et al.: Improved differential analysis of block cipher PRIDE. In: Lopez, J., Wu, Y. (eds.) ISPEC 2015. LNCS, vol. 9065, pp. 209–219. Springer, Cham (2015). https://doi.org/10.1007/978-3-319-17533-1_15
36. Yu, X., Wu, W., Li, Y., Zhang, L.: Cryptanalysis of reduced-round KLEIN block cipher. In: Wu, C.-K., Yung, M., Lin, D. (eds.) Inscrypt 2011. LNCS, vol. 7537, pp. 237–250. Springer, Heidelberg (2012). https://doi.org/10.1007/978-3-642-34704-7_18
37. Zhang, W., Rijmen, V.: Division cryptanalysis of block ciphers with a binary diffusion layer. IET Inf. Secur. **13**(2), 87–95 (2019)
38. Zhao, J., Wang, X., Wang, M., Dong, X.: Differential analysis on block cipher PRIDE. IACR Cryptology ePrint Archive 2014:525 (2014)

Improved Truncated Differential Distinguishers of AES with Concrete S-Box

Chengcheng Chang[1,2], Meiqin Wang[1,2,3], Ling Sun[1,2,3], and Wei Wang[1,2,3](✉)

[1] School of Cyber Science and Technology, Shandong University, Qingdao 266237, Shandong, China
chengchengchang@mail.sdu.edu.cn, {mqwang,lingsun,weiwangsdu}@sdu.edu.cn
[2] Key Laboratory of Cryptologic Technology and Information Security, Ministry of Education, Shandong University, Qingdao 266237, Shandong, China
[3] Quancheng Laboratory, Jinan 250103, China

Abstract. The security of Advanced Encryption Standard (AES) is one of the most important issues in cryptanalysis. In ToSC 2020, Bao et al. proposed an open question about the relation between the input-output indices and the probability of truncated differentials. In this work, we try to answer this question, and accomplish a tighter bound for several types of truncated differential distinguishers based on the differential distribution table (DDT) of the S-box of AES.

In order to reduce the computational complexity, we choose the starting point in the middle of the differential instead of the beginning, construct the DDT of 32-bit to 8/16-bit Super-Sboxes adopting an *integrated S-box* technique, and explore the *divide-and-combine* algorithm to perform the accurate calculation. For the 4-round truncated differentials with only one active byte in the input difference and one inactive byte in the output difference, we investigate the concrete probability of all 256 combinations of input-output indices. Moreover, our computation algorithms remove the independence assumption of functions in Bao et al.'s work, and can be generalized to compute the probability of truncated differentials ended with two inactive bytes in one column. To take full advantage of the results, we construct statistical model based on conditional probability, and propose 4/5/6-round truncated differential distinguishers, respectively. Our 6-round distinguisher needs $2^{62.88}$ chosen-plaintexts and $2^{63.42}$ encryptions, which is better than the published 6-round distinguishers in key-independent secret-key setting. For all truncated differentials presented in this work, we perform experimental verifications on Small-AES variants, and the results show our algorithms can provide reliable results. It is noted that the results do not threaten the security of AES.

Keywords: AES · Cryptanalysis · Secret-key attacks · Truncated differential · Concrete S-box

T. Isobe and S. Sarkar (Eds.): INDOCRYPT 2022, LNCS 13774, pp. 422–445, 2022.
https://doi.org/10.1007/978-3-031-22912-1_19

1 Introduction

Advanced Encryption Standard (AES) [1] was adopted by National Institute of Standards and Technology (NIST) in 2000, and has become the most widespread block cipher in the world. The security evaluation of AES has attracted the attention of international cryptographic community, triggered a series of excellent ideas and cryptanalysis techniques since its publication. Besides the biclique cryptanalysis of the full AES [9], the cryptanalysis results can be divided into two kinds, one is the key recovery attack [3,4,10,19], and the other is the distinguishing attack [3,5,6,35], which distinguishes AES from a random permutation. The construction of distinguishers may lead to key recovery attacks. Moreover, a block cipher or its round-reduced version often serves as the pseudo-random permutation, which is the assumption in the security proof of block cipher operation modes [7], hash functions [33,34], and MACs [13], etc. Therefore, the existence of distinguishers may lead to serious consequences, and we focus on distinguishers of AES in this paper.

Because of the *Wide Trail Strategy*, AES seems immune to traditional differential and linear cryptanalysis. The constructions of distinguishers on AES often study a set of plaintexts or ciphertexts in specific form, investigate various aspects of its properties, and may combine different types of non-random characteristics to achieve better results.

For integral distinguisher, Daemen et al. presented a 3-round integral distinguisher and extended it to 4-round by prepending a round at the beginning [12]. Later, the time complexity is improved by Ferguson et al. by *partial-sum* technique [21]. And then came a series of meet-in-the-middle attacks [18,20,30], where [20] took truncated differential to enumerate the input and output differential values instead of state values, which is called *differential enumeration*, reduced the space complexity significantly.

Several novel and powerful distinguishers arose. Grassi et al. [26] introduced subspace trail cryptanalysis which includes techniques based on (truncated) impossible differentials and integrals as special cases, moreover, they observed the *multiple-of-8* property, which led to the first key-independent secret-key distinguisher for 5-round AES. Besides, Grassi [23–25] introduced Mixture Differential Cryptanalysis on round-reduced AES-like ciphers, translated the *multiple-of-8* 5-round distinguisher into a simpler and more convenient one, and showed how to combine the new 4-round distinguisher with a modified version of truncated differential distinguisher in order to set up new 5-round distinguishers. Rnjom et al. [35] proposed *yoyo-distinguishers* for $3 \sim 6$-round AES, which is the first key-independent distinguisher for 6-round AES, and was improved by Bardeh et al. [6]. Their data and time complexity were further reduced by Bardeh [5] through adopting a variant with adaptively chosen ciphertexts. Recently, Bao et al. [3] illustrated how the well-known integral distinguisher on 3-round AES resembles a sum of PRPs and extended it to truncated-differential distinguishers of 4/5/6-round.

Note that there are various settings in the cryptanalysis of AES variants, such as secret-key, known-key, key-(in)dependent, etc., and we only list the key-independent secret-key distinguishers on AES-128 over 4 rounds in Table 1.

Our Contributions. From the results of Table 1, all the previous distinguishers over 4 rounds do not take the details of S-box into consideration. Bao et al. [3] asked an open question: Can we predict a-priori which input-output indices yield particularly strong distinguishers, given the S-box? To answer this question, we calculate a tighter bound for the truncated differentials of AES.

We start with the analysis of 4-round truncated differentials proposed in [3], which begin with one active cell in the input difference and end with one inactive cell in the output difference. In order to compute the concrete probability, we have to list all differential characteristics following the 4-round truncated differential, obtain the probability of each one, and take the mean value of them as the final result. Obviously, we can not afford such a huge calculation. A *divide-and-combine* technique is put forward, and the complexity is reduced by choosing the appropriate starting point and time-memory tradeoffs. Especially, for the differential distribution table (DDT) of 32-bit to 8-bit Super-box, an *integrated* idea is introduced and cuts down the complexity significantly. Three algorithms are proposed to settle this question. We can also deduce the precise probability of the extended 5/6-round truncated differentials in [3] under the assumption of independent and random round keys.

Furthermore, we check some different 4-round truncated differentials, which begin with one active cell and end with two inactive cells. The *divide-and-combine* technique and statistical framework are refined, the DDT of a 32-bit to 16-bit *integrated* S-box is deduced, and the corresponding results are obtained. The new 4-round distinguisher straightforwardly leads to a 5-round one, and can be extended to 6-round, which is the best distinguisher in key-independent secret-key setting till now.

For two kinds of 4-round distinguishers, we compute the probability based on the output of our algorithms, and perform experimental verifications on AES.

Moreover, for all distinguishers, we test on Small-AES, the downscale variants of AES [11] with 64-bit block size and different 4-bit S-boxes[1], and present the comparison of the experimental results and probability deduced from our algorithms[2]. The results show that our algorithms are reasonable and suitable for a concrete block cipher.

Outline. Section 2 reviews some preliminaries and backgrounds. Section 3 and Sect. 4 focus on the probability calculation and experimental verifications of 4-round truncated differentials with one active input cell. Section 3 analyzes the obstacles of direct calculation, introduces the *divide-and-combine* technique and corresponding algorithms for the truncated differential ended with one inactive

[1] The details of the Small-AES and the different 4-bit S-boxes are presented in Appendix A.

[2] The results will be presented in the full paper.

cell, and describes the refined *divide-and-combine* technique and corresponding algorithm for the one ended with two inactive cells. Section 4 introduces statistical framework using conditional probability, and construct the new 4-round truncated differential distinguisher of AES. Section 5 and Sect. 6 extend it to 5/6-round, respectively. Section 7 concludes this work.

2 Preliminaries and Backgrounds

2.1 Notations

Some notations used throughout this paper are given as follows:

X_i means the input state of the i-th round function; X_i^j stands for the j-th state in a δ-set of the i-th round function, where $j = 0, ..., 2^8 - 1$; X_i^{SB}, X_i^{SR}, and X_i^{MC} mean the state X_i after the application of SubBytes, ShiftRows, and Mixcolumns, respectively; $X_i[j]$ is the j-th byte of state X_i, where $j = 0, \cdots, 15$; k_i stands for the subkey of round i, k_0 denotes the whitening key; ΔX means the difference in the state X; $\cdot$ denotes the multiplication over GF(2^8) for AES; the difference distribution table (DDT) of S-box, is denoted as $DDTS$, and $DDTS[i][j]$ stands for the number of input pairs following input difference i and output difference j.

2.2 Short Description of AES

AES [15] is a block cipher with SP-network in which the plaintext is 128-bit, and supports three versions, denoted as AES-128, AES-192, and AES-256, where the corresponding key size is 128, 192, 256 bits and the round number is 10, 12, 14, respectively. The 128-bit plaintext is treated as a 4×4 matrix, and each round performs the following four basic operations:

1) SubBytes (SB): a nonlinear operation that applies the same 8-bit to 8-bit invertible S-box to every state byte. 2) ShiftRows (SR): a linear operation that rotates i bytes on the i-th row to the left. 3) MixColumns (MC): matrix multiplication over the $GF(2^8)$ that applies to each column. 4) AddRoundKey (AK): a linear operation that XORs the state with the round key.

Before the first round of encryption, there is an additional AK operation, which is named XOR whitening key. And in the last round of encryption, the MC operation is omitted.

2.3 4-Round Truncated Differential Distinguisher of AES

Bao et al.'s Work. In 2020, Bao et al. [3] resembled the 3-round integral distinguisher [12] as a sum of PRPs, and extended it to 4/5/6-round truncated differential distinguisher. Adapting Patarin's setting [32], for an input set of which only different at $X_0[1]$, Bao et al. approximated that, for AES,

$$p_{\mathrm{AES}}\{\Delta X_3^{MC}[0] = 0\} \simeq \frac{1}{2^8} + \frac{1}{2^8 \times (2^8 - 1)^3} \simeq 2^{-8} + 2^{-31.983}. \quad (1)$$

Table 1. Comparison of existing Secret-Key Distinguishers for AES. Data complexity is measured in the number of (adaptive) chosen plaintexts/ciphertexts ((A)CP/(A)CC) for $P_S \geq 0.95$. Time complexity is measured in memory accesses (M) and in round-reduced AES encryptions (E) or XOR operations (XOR) – using the common approximation 20 M $\simeq$ 1-round Encryption. * indicates that the details of S-box are considered.

Property	Rounds	Data	Cost	Ref.
Yoyo Game	4	2CP+2ACC	2XOR	[35]
Impossible Differential	4	$2^{16.25}$CP	$2^{22.3}$M$\simeq 2^{16}$E	[21]
Mixture Differential	4	2^{17}CP	$2^{23.1}$M$\simeq 2^{16.75}$E	[23]
Truncated Differential *	**4**	$\mathbf{2^{30.56}}$CP	$\mathbf{2^{30.94}}$E	**Sect.** 4
Integral	4	2^{32}CP	2^{32}XOR	[12]
Multiple-of-8	4	2^{33}CP	2^{40}M$\simeq 2^{33.7}$E	[27]
Truncated Differential	4	$2^{51.4}$CP	-	[3]
Yoyo Game	5	2^{12}CP+$2^{25.8}$ACC	$2^{24.8}$XOR	[35]
Multiple-of-8	5	2^{32}CP	$2^{35.6}$M$\simeq 2^{29}$E	[27]
Truncated Variance Differential	5	2^{34}CP	$2^{37.6}$M$\simeq 2^{31}$E	[25]
Truncated Differential *	**5**	$\mathbf{2^{42.59}}$CP	$\mathbf{2^{42.91}}$E	**Sect.** 5
Truncated Mean Differential	5	$2^{48.96}$CP	$2^{52.6}$M$\simeq 2^{46}$E	[25]
Probabilistic Mixture Differential	5	2^{52}CP	$2^{71.5}$M$\simeq 2^{64.9}$E	[24]
Truncated Differential	5	2^{68}CP	$2^{73.3}$M	[3]
Impossible Mixture Differential	5	2^{82}CP	$2^{97.8}$M$\simeq 2^{91.1}$E	[24]
Threshold Mixture Differential	5	2^{89}CP	$2^{98.1}$M$\simeq 2^{91.5}$E	[24]
Truncated Differential *	**6**	$\mathbf{2^{62.88}}$CP	$\mathbf{2^{63.42}}$E	**Sect.** 6
Exchange	6	2^{83}ACC	2^{83}E	[5]
Exchange	6	$2^{88.2}$CP	$2^{88.2}$E	[6]
Truncated Differential	6	$2^{89.43}$CP	$2^{96.52}$M	[3]
Impossible Yoyo	6	$2^{122.83}$ACC	$2^{121.83}$XOR	[35]

While for a random permutation, the probability is roughly

$$p_{\text{rand}}\{\Delta X_3^{MC}[0] = 0\} \simeq \frac{2^{120}-1}{2^{128}-1} \simeq 2^{-8}. \quad (2)$$

Then, once $\Delta X_3^{MC}[0] = 0$, $\Delta X_5[0] = 0$ with probability 1 since zero difference is unaffected by AK, SB, SR operations. Thus, the difference between the p_{AES} and the p_{rand} can be exploited to build truncated differential distinguishers of AES.

Statistical Framework. We briefly recall the satistical model adopted by Bao et al.'s [3], which follows the framework of Grassi et al. [25].

The plaintext pairs which satisfy the output difference of the truncated differential are called collisions. For N pairs of plaintexts, let X_{rand} be a random variable for the number of collisions of PRP, and X_{AES} be a random variable of r-round AES. Then $X_{rand} \sim \mathcal{N}(N \times p_{rand}, N \times p_{rand}(1 - p_{rand}))$ and $X_{AES} \sim \mathcal{N}(N \times p_{AES}, N \times p_{AES}(1 - p_{AES}))$. If follows that the number of chosen plaintexts pairs N should satisfy

$$N \geq \frac{2 \times [p_{\text{rand}} \times (1 - p_{\text{rand}}) + p_{\text{AES}} \times (1 - p_{\text{AES}})]}{(p_{\text{rand}} - p_{\text{AES}})^2} \times [\text{erfinv}(2 \times P_S - 1)^2] \quad (3)$$

to obtain a success probability of at least P_S. The erfinv(x) is the inverse error function. It is noted that the setting we considered in our work is form pairs from δ-sets that have no non-trivial relation to each other.

3 Divide-and-Combine Technique

Truncated differential cryptanalysis is widely adopted in the security evaluation of block ciphers, and obtain many good results. However, the critical factor affecting the complexity of the attack, which is the probability of truncated differential, is always roughly evaluated in the secret-Sbox setting. We wonder if we can check all differential characteristics belonging to a truncated differential, compute the corresponding probability, and obtain their mean value, i.e., a realistic bound of the truncated differential taking the details of the underlying S-boxes into consideration.

In this section, we first propose a *divide-and-combine* technique to overcome the obstacles of direct computation, accomplish such kind of calculation for 4-round truncated differentials of AES proposed in [3]. Then, we modify the *divide-and-combine* technique and consider the performance of 4-round truncated differentials with one active cell in input and two inactive cells in the same column of the output, which illustrate some new insights on the security of AES. While the setting of Bao et al. is not applicable to such situation because they require independence assumption of functions, and the bytes in one column are not independent.

3.1 Obstacles of Direct Calculation

To calculate the truncated differential probability, one direct way is to list all differentials leading to the specific truncated differential and sum up their probabilities, which is similar to a tree-based breadth-first search algorithm. While for AES, the biggest challenge is that the computation complexity or memory access is too high to calculate since there are too many differentials involved. We take the 3-round truncated differential (0*00000000000000 $\xrightarrow{3r}$ 0???????????????) (Fig. 1), which is the core of distinguishers presented by Bao et al., as an example, where 0 corresponds to byte with zero (inactive) difference, $*$ means non-zero (active), and ? is unknown, and illustrate the possibilities of differential characteristic satisfying it.

As depicted in Fig. 1, for the starting point ΔX_0, there is only one non-zero byte, i.e., 2^8-1 possible input difference values. Since AK does not affect the value of difference, so does ΔX_1. We recall that in the DDTS of AES, for each non-zero

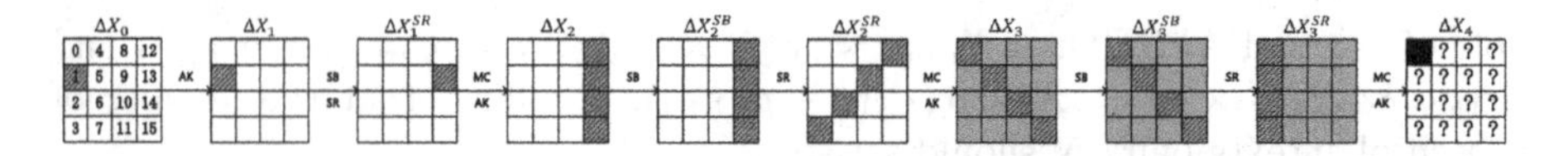

Fig. 1. Truncated difference propagation of $(0*00000000000000 \xrightarrow{3r} \Delta X_4[0]$??????????? ?????). The gray bytes represent the active bytes. The black one in ΔX_4 is the difference we want to predict. The slashed represent bytes related to the black one in ΔX_4. The ? represent bytes that are not sure whether they are active.

input difference, there are $2^7 - 1 = 127$ non-zero output differences.Then each ΔX_1 corresponds to $2^7 - 1$ ΔX_1^{SB}. Similar analysis are performed. It is noted that for each byte in ΔX_4, we just need to consider four related bytes in ΔX_3. For example, $\Delta X_4[0]$ only relates to $\Delta X_3^{MC}[0]$, i.e., $\Delta X_3^{SB}[0, 5, 10, 15]$. We take $\#(\Delta in \to \Delta out)$ to denotes the number of possible differential trails from input difference Δin to output difference Δout. Then the possibilities to calculate the route probability of $(0*00000000000000 \xrightarrow{3r} \Delta X_4[0]$???????????????) is

$$\begin{aligned}
&\#(\Delta X_0[1] \neq 0 \to \Delta X_3^{MC}[0]) \\
=&\#(\Delta X_0[1] \to \Delta X_1^{SB}[1]) \times \#(\Delta X_1^{SB}[1] \to \Delta X_2^{SB}[12, 13, 14, 15]) \\
&\times \#(\Delta X_2^{SB}[12, 13, 14, 15] \to \Delta X_3^{SB}[0, 5, 10, 15]) \\
\simeq& (2^8 - 1) \times 2^7 \times (2^7)^4 \times (2^7)^4 \simeq 2^{71}.
\end{aligned} \tag{4}$$

2^{71} possibilities are infeasible to calculate straightly, therefore, the key issue is how to deal with inner summation in order to reduce complexity.

3.2 Calculate the Probability of 4-Round Truncated Differential with One Active Cell in Input and One Inactive in Output

4-Round Truncated Differentials Ended with One Inactive Cell. As depicted in Fig. 1, if $\Delta X_4[0] = 0$, there must be $\Delta X_4^{SR}[0] = 0$, which leads to a 4-round truncated differential $(0*00000000000000 \xrightarrow{4r} \Delta X_5[0]$???????????????) where the MC operation is omitted in the last round. Therefore, we only need to focus on the first 3-round transformation analyzed in Sect. 3.1.

According to Eq. (4), the computation can be splited into several parts, compute the probability separately, and combine the related ones together. Similarly with the partial sum technique [21], we can divide $\Delta X_1[1] \to \Delta X_4[0]$ into two parts, for example, $\Delta X_1[1] \to \Delta X_1^{SB}[1]$ and $\Delta X_1^{SB}[1] \to \Delta X_3^{SB}[0, 5, 10, 15]$. Then, we compute $Pr(\Delta X_1^{SB}[1] \to \Delta X_3^{SB}[0, 5, 10, 15])$ for each possible $\Delta X_1^{SB}[1]$ and connect the beginner $Pr(\Delta X_1[1] \to \Delta X_1^{SB}[1])$ according to the same $\Delta X_1^{SB}[1]$. The probability of whole truncated differential is just the multiplication of them.

Choose the Appropriate Starting Point. In order to balance the complexity of each part, we need to choose a proper starting point first. Inspired by the meet-in-the-middle idea of [18, 20, 30], we think it is more appropriate to initiate the calculation in the middle of the differential instead of $\Delta X_1[1]$.

For example, we choose the starting point at ΔX_2^{SB} and calculate two round forward and one round backward. Then the possibilities of differential trails, 2^{71} in Eq. (4), becomes about $(2^8-1)\times 2^7 \times (2^7)^4 + (2^8-1)^4 \times (2^7)^4 \simeq 2^{60}$. In addition, due to the different circumstances, the algorithm in Sect. 3.2 places the starting point at ΔX_2^{SB}, while the algorithms in Sect. 3.3 places the starting point at ΔX_2^{MC} for the convenience of code operation.

32-bit to 8-bit Integrated S-box. We develop a new "*integrated*" S-box technique to further cut down the complexity enlighted by Zhang et al.'s work [37] about large-unit linear approximation and Daemen et al.'s Super-box [14,16,17,22]. This technique focuses on the computation of $Pr(\Delta X_2^{SB}[12,13,14,15] \rightarrow \Delta X_4[0])$ which can be accelerated by precomputation.

The mapping from $X_2^{SB}[12,13,14,15]$ to $X_4[0]$ can be taken as a 32-bit to 8-bit Super-Sbox, if we construct its DDT directly, the complexity is $2^{32}\times 2^{32} = 2^{64}$. Instead, we integrate the Super-Sbox into three 16-bit to 8-bit mappings in the following.

According to $X_4[0] = 2\cdot X_3^{SB}[0]\oplus 3\cdot X_3^{SB}[5]\oplus X_3^{SB}[10]\oplus X_3^{SB}[15]$, $X_2^{MC}[0] = X_2^{SB}[15]$, $X_2^{MC}[5] = 3\cdot X_2^{SB}[14]$, $X_2^{MC}[10] = X_2^{SB}[13]$, and $X_2^{MC}[15] = 2\cdot X_2^{SB}[12]$, the function $X_2^{SB}[12,13,14,15] \rightarrow X_4[0]$ is decomposed to:

$$\begin{aligned}
&f_1 : (X_2^{MC}[0], X_2^{MC}[5]) \rightarrow \mathcal{X},\\
&\text{where } \mathcal{X} = 2\cdot S(X_2^{MC}[0]\oplus k_2[0]) \oplus 3\cdot S(X_2^{MC}[5]\oplus k_2[5]),\\
&f_2 : (X_2^{MC}[10], X_2^{MC}[15]) \rightarrow \mathcal{Y},\\
&\text{where } \mathcal{Y} = S(X_2^{MC}[10]\oplus k_2[10]) \oplus S(X_2^{MC}[15]\oplus k_2[15]),\\
&f_3 : (\Delta\mathcal{X}, \Delta\mathcal{Y}) \rightarrow \Delta X_4[0], \text{ where } \Delta X_4[0] = \Delta\mathcal{X}\oplus\Delta\mathcal{Y}.
\end{aligned} \tag{5}$$

In the above definitions, f_1, f_2 are two 16-bit to 8-bit mappings, in fact, both of them are XOR (linear operation) of two 8-bit to 8-bit permutations. According to the distribution function of two discrete random variables sum [28], their DDTs can be calculated efficiently by **MergeTableDDT** in Algorithm 1, and stored as two precomputation tables, denoted as $DDTX$ and $DDTY$, respectively. Since f_3 is a linear function too, its distribution function can be constructed from $DDTX$ and $DDTY$ in a similar way, and the DDT of the whole function can be obtained. However, we only precompute $DDTX$ and $DDTY$, and generate the related part of f_3's DDT for the given $X_2^{SB}[12,13,14,15]$ to reduce memory overhead. Once the DDT is obtained, the corresponding probability can be deduced and the detailed implementation is processed in Algorithm 1.

Algorithms to Calculate the Probability. We present the whole algorithm based on our *divide-and-combine* technique to calculate the probability of the 4-round truncated differential distinguisher introduced in [3]. We divide the 4-round truncated differential into four parts and only list the bytes that affect the probability:

$$\begin{aligned}
&\text{(I) } \Delta X_0[1] \rightarrow \Delta X_1^{SB}[1], && \text{(II) } \Delta X_1^{SB}[1] \rightarrow \Delta X_2^{SB}[12,13,14,15],\\
&\text{(III) } \Delta X_2^{SB}[12,13,14,15] \rightarrow \Delta X_4[0], && \text{(IV) } \Delta X_4[0] \rightarrow \Delta X_5[0].
\end{aligned} \tag{6}$$

Algorithm 1. $Pr(\Delta X_2^{SB}[12,13,14,15] \rightarrow \Delta X_4[0])$ for given $\Delta X_2^{SB}[12,13,14,15]$ and $\Delta X_4[0]$

```
Parameters: ΔX_2^SB[12, 13, 14, 15], ΔX_4[0];
Precomputation:      /* run once for each thread corresponding to ΔX_2^SB[12] */
  DDTX ← MergeTableDDT(2, 3);      /* compute the DDT of f_1 */
  DDTY ← MergeTableDDT(1, 1);      /* compute the DDT of f_2 */
  function MergeTableDDT(me1, me2)
    Initialize the array DDTM of size 2^8 × 2^8 × 2^8 with zeros;
    for u = 0,···,255 and v = 0,···,255 do
      T_u ← DDTS[u];
      T_v ← DDTS[v];
      for i = 0,···,255 and j = 0,···,255 do
        if T_u[i] ≠ 0 and T_v[j] ≠ 0 then
          Δx = me1 × i ⊕ me2 × j;
          DDTM[u][v][Δx] = DDTM[u][v][Δx] + T_u[i] × T_v[j];
    return DDTM

Processing:
  Compute ΔX_2^MC[0, 5, 10, 15] from ΔX_2^SB[12, 13, 14, 15];
  T_x ← DDTX[ΔX_2^MC[0]][ΔX_2^MC[5]];
  T_y ← DDTY[ΔX_2^MC[10]][ΔX_2^MC[15]];
  p3 = 0;
  for i = 0,···,255 do
    if T_x[i] ≠ 0 then
      j = i ⊕ ΔX_{4,0};
      if T_y[j] ≠ 0 then
        p3 = p3 + T_x[i] × T_y[j]/2^32;
  return p3;
```

It is noted that $Pr(\Delta X_4[0] = 0 \rightarrow \Delta X_5[0] = 0) = 1$ if we do not consider the MC operation in the 4-th round. Thus, only the probability of parts (I) $\sim$ (III) needs to be computed, where the probability of part (III) for fixed $(\Delta X_2^{SB}[12,13,14,15], \Delta X_4[0])$ can be obtained from Algorithm 1.

For the connection of part (II) and (III), i.e., $Pr(\Delta X_1^{SB}[1] \rightarrow \Delta X_4[0])$, we perform a multithreading implementation, compute $Pr(\Delta X_1^{SB}[1] \rightarrow \Delta X_2^{SB}[12] \rightarrow \Delta X_4[0])$ for each possible $\Delta X_2^{SB}[12]$ in parallel, and sum them up to obtain $Pr(\Delta X_1^{SB}[1] \rightarrow \Delta X_4[0])$ in Algorithm 2. Algorithm 3 connects the three parts, and outputs the $p[i] = Pr(\Delta X_0 \rightarrow \Delta X_4)$ where $\Delta X_0 = (0i00000000000000)$, $\Delta X_4 = (\Delta X_4[0]???????????????)$ for all non-zero $\Delta X_0[1] = i$ $(i = 1, ..., 255)$ and a fixed $\Delta X_4[0]$. Especially, the probability of the 4-round truncated differential is the mean value of the sum of corresponding $p[i]$, obtained by Algorithm 3 with parameter $\Delta X_4[0] = 0$, i.e., $\frac{1}{255}\sum_{i=1}^{255} p[i]$.

Remark: It is noted that $\Delta X_4[0]$ can take arbitrary value besides 0 in Algorithm 1 $\sim$ 3.

Computational Complexity Analysis. A brief assessment of the computational complexity of Algorithm 3 is as follows. The complexity of Algorithm 3 is dominated by Algorithm 2, while Algorithm 2 is determined by the complexity of calling Algorithm 1 in the function **GetProbability**. As the complexity of Algorithm 1 is approximately 2^8, the computational complexity is approximately $2^8 \times (2^8)^3 \times 2^{-3} \times 2^8 = 2^{37}$.

Algorithm 2. $Pr(\Delta X_1^{SB}[1] \rightarrow \Delta X_4[0])$ for all non-zero $\Delta X_1^{SB}[1]$, and given $\Delta X_4[0]$

Parameters: $\Delta X_4[0]$
Processing:
1: Initial the array p23 of size $2^8 \times 2^8$ with zeros;
2: /* *for* $\Delta X_2^{SB}[12] = 1, ..., 255$, *use 255 threads to perform* ***function*** **GetProbability** *in parallel* */
3: $p23[\Delta X_2^{SB}[12]] \leftarrow$ **GetProbability**$(\Delta X_2^{SB}[12], \Delta X_4[0])$; /* $p23[\Delta X_2^{SB}[12]][i]$ *means* $Pr(\Delta X_1^{SB}[1] = i \rightarrow \Delta X_2^{SB}[12] \rightarrow \Delta X_4[0])$ */
4: Initial the array p1 of size 2^8 with zeros;
5: **for** $\Delta X_2^{SB}[12] = 1, \cdots, 255$ **and** $\Delta X_1^{SB}[1] = 1, \cdots, 255$ **do**
6: $\quad p1[\Delta X_1^{SB}[1]] = p1[\Delta X_1^{SB}[1]] + p23[\Delta X_2^{SB}[12]][\Delta X_1^{SB}[1]]$;
7: **return** $p1$; /* $p1[i] = Pr(\Delta X_1^{SB}[1] \rightarrow \Delta X_4[0])$ *for* $1 \leq i \leq 255$ */
8:
9: **function GetProbability**$(\Delta X_2^{SB}[12], \Delta X_4[0])$
10: $\quad$ **for** $\Delta X_1^{SB}[1] = 1, \cdots, 255$ **do**
11: $\quad\quad$ Compute $\Delta X_1^{MC}[12, 13, 14, 15]$ from $\Delta X_1^{SB}[1]$;
12: $\quad\quad$ **for** 2^{24} values of $\Delta X_2^{SB}[13, 14, 15]$ **do**
13: $\quad\quad\quad p2 = DDTS[\Delta X_1^{MC}[12]][\Delta X_2^{SB}[12]] \times DDTS[\Delta X_1^{MC}[13]][\Delta X_2^{SB}[13]] \times DDTS[\Delta X_1^{MC}[14]][\Delta X_2^{SB}[14]] \times DDTS[\Delta X_1^{MC}[15]][\Delta X_2^{SB}[15]]$;
14: $\quad\quad\quad$ **if** $p2 \neq 0$ **then**
15: $\quad\quad\quad\quad p3 = Algorithm1(\Delta X_2^{SB}[12, 13, 14, 15], \Delta X_4[0])$;
16: $\quad\quad\quad\quad pm[\Delta X_1^{SB}[1]] = pm[\Delta X_1^{SB}[1]] + (p2/2^{32}) \times p3$;
17: **return** pm; /* $pm[i] = Pr(\Delta X_2^{SB}[12] = i \rightarrow \Delta X_4[0])$ *for* $1 \leq i \leq 255$ */

Algorithm 3. $Pr(\Delta X_0[1] \rightarrow \Delta X_4[0])$ for all non-zero $\Delta X_0[1]$ and given $\Delta X_4[0]$

Parameters: $\Delta X_4[0]$
Processing:
1: $p1 \leftarrow Algorithm2(\Delta X_4[0])$;
2: **for** $\Delta X_0[1] = 1, \cdots, 255$ **and** $\Delta X_1^{SB}[1] = 1, \cdots, 255$ **do**
3: $\quad$ **if** $DDTS[\Delta X_0[1]][\Delta X_1^{SB}[1]] \neq 0$ **then**
4: $\quad\quad p[\Delta X_0[1]] = p[\Delta X_0[1]] + (DDTS[\Delta X_0[1]][\Delta X_1^{SB}[1]]/2^8) \times p1[\Delta X_1^{SB}[1]]$;
5: **return** p; /* $p[i] = Pr(\Delta X_0[1] = i \rightarrow \Delta X_4[0])$ *for* $1 \leq i \leq 255$ */

Implemented Results and Experimental Verifications. For DDTS of AES, we run Algorithm 3 with parameter $\Delta X_4[0] = 0$, and derive $Pr(\Delta X_0 \rightarrow \Delta X_4)$ where ΔX_0=(0i00000000000000) and ΔX_4=(0???????????????) for all non-zero i, which equals to $Pr(\Delta X_0 \rightarrow \Delta X_5^{SB})$ where ΔX_5^{SB}=(0???????????????). The average value is $2^{-8} + 2^{-31.816}$.

Index Dependencies. Moreover, we test the index dependencies, as pointed out in [3], i.e., the relation between the probability and the position of the active input cell i_0 and inactive output cell i_5, after three rounds. For all $16 \times 16 = 256$ combinations of (i_0, i_5), we perform Algorithm 3 on Small-AES and AES. The result of Small-AES coincides with Bao et al., and shows that $(i_0, i_5) = (1, 0)$ corresponding to the combination with the highest probability. However, the result of AES is different from that of Small-AES, which is summarized in Table 2. It is noted that the 256 combinations can be divied into 16 equivalent classes because of the structure of AES, and $(i_0, i_5) = (1, 6)$ instead of $(i_0, i_5) = (1, 0)$ seems to be the one of best when $i_0 = 1$, and the probability is $2^{-8} + 2^{-31.654}$.

Data Complexity and Success Probability. We employ the statistical framework in Sect. 2.3 to analyze the amount of data for an acceptable success probability. For the 4-round truncated differential with probability $2^{-8} + 2^{-31.654}$, we need to choose the number of plaintexts pairs $N \simeq 2^{57.738}$ to guarantee a success probability of $P_S = 0.95$, and $N \simeq 2^{58.738}$ with $P_S = 0.99$. Since one δ-set contains 2^8 texts, and can constitute $\binom{2^8}{2}$ pairs. Thus, for $Ps = 0.95$, the distinguisher needs approximately $2^{42.738}$ δ-sets, i.e., $2^{50.738}$ chosen plaintexts. And for $Ps = 0.99$, the number of δ-sets is roughly $2^{43.738}$, i.e., $2^{51.738}$ chosen plaintexts. The complexity is slightly lower than Bao et al. [3].

Table 2. Results of the probability of 4-round truncated differentials of AES in [3] for all combinations of (i_0, i_5). $p_{\text{4r-AES}}$ represents the mean value of Algorithm 3.

(i_0, i_5)	$p_{\text{4r-AES}}$
$(0,0),(1,5),(2,10),(3,15),(4,4),(5,9),(6,14),(7,3),(8,8),(9,13),(10,2),(11,7),(12,12),(13,1),(14,6),(15,11)$	$2^{-8}+2^{-31.982}$
$(0,1),(1,6),(2,11),(3,12),(4,5),(5,10),(6,15),(7,0),(8,9),(9,14),(10,3),(11,4),(12,13),(13,2),(14,7),(15,8)$	$2^{-8}+2^{-31.654}$
$(0,2),(1,7),(2,8),(3,13),(4,6),(5,11),(6,12),(7,1),(8,10),(9,15),(10,0),(11,5),(12,14),(13,3),(14,4),(15,9)$	$2^{-8}+2^{-31.765}$
$(0,3),(1,4),(2,9),(3,14),(4,7),(5,8),(6,13),(7,2),(8,11),(9,12),(10,1),(11,6),(12,15),(13,0),(14,5),(15,10)$	$2^{-8}+2^{-31.924}$
$(0,4),(1,9),(2,14),(3,3),(4,8),(5,13),(6,2),(7,7),(8,12),(9,1),(10,6),(11,11),(12,0),(13,5),(14,10),(15,15)$	$2^{-8}+2^{-31.682}$
$(0,5),(1,10),(2,15),(3,0),(4,9),(5,14),(6,3),(7,4),(8,13),(9,2),(10,7),(11,8),(12,17),(13,6),(14,11),(15,12)$	$2^{-8}+2^{-31.774}$
$(0,6),(1,11),(2,12),(3,1),(4,10),(5,15),(6,0),(7,5),(8,14),(9,3),(10,4),(11,9),(12,2),(13,7),(14,8),(15,13)$	$2^{-8}+2^{-31.924}$
$(0,7),(1,8),(2,13),(3,2),(4,11),(5,12),(6,1),(7,6),(8,15),(9,0),(10,5),(11,10),(12,3),(13,4),(14,9),(15,14)$	$2^{-8}+2^{-32.002}$
$(0,8),(1,13),(2,2),(3,7),(4,12),(5,1),(6,6),(7,11),(8,0),(9,5),(10,10),(11,15),(12,4),(13,9),(14,14),(15,3)$	$2^{-8}+2^{-31.797}$
$(0,9),(1,14),(2,3),(3,4),(4,13),(5,2),(6,7),(7,8),(8,1),(9,6),(10,11),(11,12),(12,5),(13,10),(14,15),(15,0)$	$2^{-8}+2^{-31.674}$
$(0,10),(1,15),(2,0),(3,5),(4,14),(5,3),(6,4),(7,9),(8,2),(9,7),(10,8),(11,13),(12,6),(13,11),(14,12),(15,1)$	$2^{-8}+2^{-31.988}$
$(0,11),(1,12),(2,1),(3,6),(4,15),(5,0),(6,5),(7,10),(8,3),(9,4),(10,9),(11,14),(12,7),(13,8),(14,13),(15,2)$	$2^{-8}+2^{-31.947}$
$(0,12),(1,1),(2,6),(3,11),(4,0),(5,5),(6,10),(7,15),(8,4),(9,9),(10,14),(11,3),(12,8),(13,13),(14,2),(15,7)$	$2^{-8}+2^{-31.687}$
$(0,13),(1,2),(2,7),(3,8),(4,1),(5,6),(6,11),(7,12),(8,5),(9,10),(10,15),(11,0),(12,9),(13,14),(14,3),(15,4)$	$2^{-8}+2^{-31.922}$
$(0,14),(1,3),(2,4),(3,9),(4,2),(5,7),(6,8),(7,13),(8,6),(9,11),(10,12),(11,1),(12,10),(13,15),(14,0),(15,5)$	$2^{-8}+2^{-31.908}$
$(0,15),(1,0),(2,5),(3,10),(4,3),(5,4),(6,9),(7,14),(8,7),(9,8),(10,13),(11,2),(12,11),(13,12),(14,1),(15,6)$	$2^{-8}+2^{-31.816}$

3.3 Calculate the Probability of 4-Round Truncated Differential with One Active Cell in Input and Two Inactive in Output

We check 4-round truncated differentials with one active cell in the input difference and one inactive cell in the output difference in Sect. 3.2, and wonder the performance of distinguishers when there are more than one inactive cells in the output difference. However, the more output cells are involved, the more related cells are considered, so that the possibilities to be considered may be out of control. In addition, the setting of Bao et al. is not applicable to multiple bytes in the same column, since it requires independence between bytes in one column, while considering concrete S-box removes the requirement of independence.

It is noted that the computation of cells in the same column of the output difference, e.g. $\Delta X_4[0,1]$, relate to the same four bytes, $\Delta X_3^{SB}[0,5,10,15]$, so that the increment of possibilities is limited in this situation. Thus, we choose the inactive cells of the output difference in the same column. Based on the relation of cells in one column, we consider the joint distribution, and adjust the *integrated* technique to compatible with the 32-bit to 16-bit Super-Sbox, investigate the *divide-and-combine* technique in Sect. 3.2, and achieve the probability corresponding to the case of two inactive cells in the same column of the output difference, which may lead to new distinguishers.

4-Round Truncated Differentials Ended with Two Inactive Cells. For the convenience of description, we construct the new truncated differential (see Fig. 2(a)) with the same active cell in ΔX_0 and inactive cell in ΔX_5, and only add another inactive cell in ΔX_5, i.e., $\Delta X_0 = (0{*}00000000000000) \xrightarrow{4r} \Delta X_5 = (0???\ ?????????0??)$.

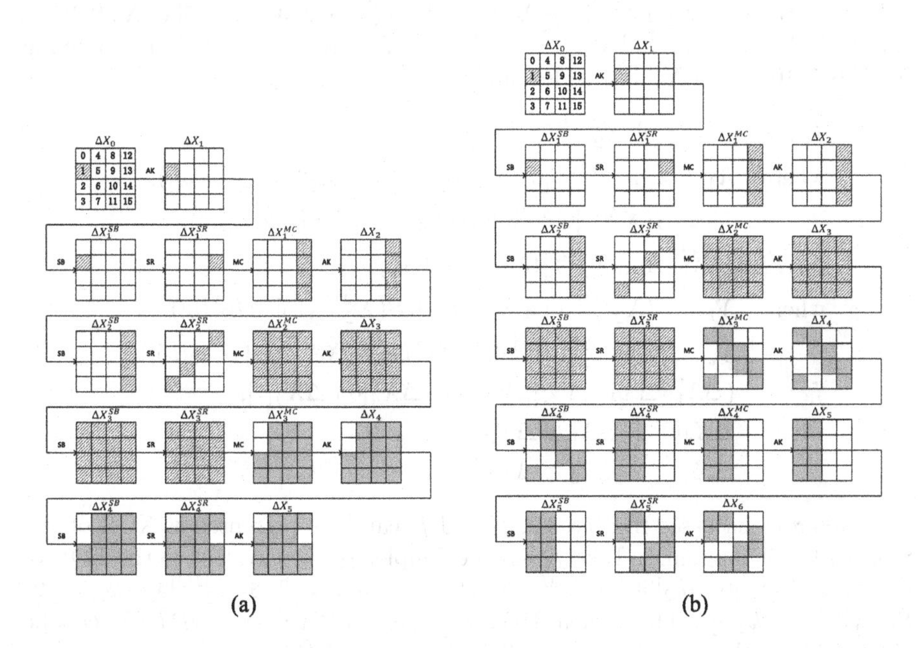

Fig. 2. The figure on the left shows our new 4-round truncated differential distinguisher. The figure on the right shows our new 5-round truncated differential distinguisher. White cells represent bytes with zero difference, while slashed cells represent the active (non-zero) ones, and the gray cells stand for unknown.

Algorithms to Calculate the Probability. Similar with Sect. 3.2, to calculate the probability of new 4-round truncated differential depicted in Fig. 2(a), we divide it into four parts (for the convenience of calculation, we move the connecting point to ΔX_2^{MC} instead of ΔX_2^{SB}):

$$\begin{aligned} &(\text{I})\ \Delta X_0[1] \to \Delta X_1^{SB}[1], &&(\text{II})\ \Delta X_1^{SB}[1] \to \Delta X_2^{MC}[0,5,10,15], \\ &(\text{III})\ \Delta X_2^{MC}[0,5,10,15] \to \Delta X_4[0,1], &&(\text{IV})\ \Delta X_4[0,1] \to \Delta X_5[0,13]. \end{aligned} \tag{7}$$

It is noted that $Pr(\Delta X_4[0,1] = 0 \to \Delta X_5[0,13] = 0) = 1$ if we do not consider the MC operation in the 4-th round. Thus, only the probability of parts (I) $\sim$ (III) needs to be computed, where only the computation of part (III) involves

one more parameters $\Delta X_4[1]$, and the others can be deduced in a similar way as Algorithm 2∼3. Thus, due to the space limination, we just present the corresponding algorithm to compute part (III).
Remark: Note that $\Delta X_4[0,1]$ can take arbitrary values in our algorithms.

32-bit to 16-bit Integrated S-box. To accomplish the computation of part (III), i.e., $Pr(\Delta X_2^{MC}[0,5,10,15]) \to \Delta X_4[0,1]$, we take advantage of a 32-bit to 16-bit integrated S-box, which is the mapping $X_2^{MC}[0,5,10,15] \to X_4[0,1]$. It seems that we can compute $Pr(\Delta X_2^{MC}[0,5,10,15] \to \Delta X_4[0,1])$ in the similar way with Algorithm 1. Since $X_4[0] = 2 \cdot X_3^{SB}[0] \oplus 3 \cdot X_3^{SB}[5] \oplus X_3^{SB}[10] \oplus X_3^{SB}[15]$, $X_4[1] = X_3^{SB}[0] \oplus 2 \cdot X_3^{SB}[5] \oplus 3 \cdot X_3^{SB}[10] \oplus X_3^{SB}[15]$, the mapping $X_2^{MC}[0,5,10,15] \to X_4[0,1]$ is decomposed to:

$$\begin{aligned}
f_1 : \quad & (X_2^{MC}[0], X_2^{MC}[5]) \to \mathcal{X}_1 || \mathcal{X}_2, \\
\text{where} \quad & \mathcal{X}_1 = 2 \cdot S(X_2^{MC}[0] \oplus k_2[0]) \oplus 3 \cdot S(X_2^{MC}[5] \oplus k_2[5]), \\
& \mathcal{X}_2 = S(X_2^{MC}[0] \oplus k_2[0]) \oplus 2 \cdot S(X_2^{MC}[5] \oplus k_2[5]), \\
f_2 : \quad & (X_2^{MC}[10], X_2^{MC}[15]) \to \mathcal{Y}_1 || \mathcal{Y}_2, \\
\text{where} \quad & \mathcal{Y}_1 = S(X_2^{MC}[10] \oplus k_2[10]) \oplus S(X_2^{MC}[15] \oplus k_2[15]), \\
& \mathcal{Y}_2 = 3 \cdot S(X_2^{MC}[10] \oplus k_2[10]) \oplus S(X_2^{MC}[15] \oplus k_2[15]), \\
f_3 : \quad & (\Delta\mathcal{X}_1 || \Delta\mathcal{X}_2,\ \Delta\mathcal{Y}_1 || \Delta\mathcal{Y}_2) \to \Delta X_4[0] || \Delta X_4[1], \\
\text{where} \quad & \Delta X_4[0] = \Delta\mathcal{X}_1 \oplus \Delta\mathcal{Y}_1\ , \\
& \Delta X_4[1] = \Delta\mathcal{X}_2 \oplus \Delta\mathcal{Y}_2.
\end{aligned} \tag{8}$$

Similar as discussed in Sect. 3.2, f_1 and f_2 can be represented as XOR of two 8-bit to 16-bit mappings. Take f_1 as an example. f_1 is equivalent to the XOR of $g_1 = 2 \cdot S(X_2^{MC}[0] \oplus k_2[0]) || S(X_2^{MC}[0] \oplus k_2[0])$ and $g_2 = 3 \cdot S(X_2^{MC}[5] \oplus k_2[5]) || 2 \cdot S(X_2^{MC}[5] \oplus k_2[5])$. Thus, their DDTSs, resp. $DDTXX$ and $DDTYY$ can be obtained from the sum of corresponding elements in DDTs of g_i, and this process is implemented by the **MergeTableDDT** function. Similarly, f_3 can be taken as the XOR of two discrete variables, whose distribution function are $DDTXX$ and $DDTYY$. Thus the distribution function of f_3 can be computed, and turns to $Pr(\Delta X_2^{MC}[0,5,10,15] \to \Delta X_4[0,1])$ directly. The detailed implementation is processed in Algorithm 4[3].

Remark: We take advantage of a kind of time-memory tradeoff, solve this problem by multithreading implementing a cycle calculation. First, in the connection of part (II) and (III), we create 256 threads for each $\Delta X_2^{MC}[0]$ and perform in parallel. In this way, f_1 actually becomes an 8-bit to 16-bit mapping in each thread, whose DDTS has $2^8 \times 2^{16} = 2^{24}$ elements, and is denoted as $DDTXX$. $DDTXX[\Delta X_2^{MC}[5]][\Delta x_1 || \Delta x_2]$ means the number of pairs following input difference $\Delta X_2^{MC}[0] || \Delta X_2^{MC}[5]$ and output difference $\Delta x_1 || \Delta x_2$. While, f_2 is turned to an 8-bit to 16-bit mapping by the loop of $\Delta X_2^{MC}[10]$, and the

[3] The complete algorithms contain Algorithm 4, 5, and 6. Algorithm 5 and 6 are presented in Appendix B.

Algorithm 4. $Pr(\Delta X_2^{MC}[0,5,10,15] \rightarrow \Delta X_4[0,1])$ for given $\Delta X_2^{MC}[0,5,10,15]$ and $\Delta X_4[0,1]$

```
Parameters: DDTYY, ΔX_2^MC[0,5,10,15], ΔX_4[0,1];
Precomputation:      /* run once for each thread corresponding to ΔX_2^MC[0] */
  DDTXX ← MergeTableDDT(ΔX_2^MC[0], 2, 3, 1, 2)      /* compute the DDT of f_1 */;

  function MergeTableDDT(u, me1, me2, me3, me4)
     Initialize the array DDTMM of size 2^8 × 2^16 with zeros;
     T_uu ← DDTS[u];
     for v = 0,···,255 do
        T_vv ← DDTS[v];
        for i = 0,···,255 and j = 0,···,255 do
           if T_uu[i] ≠ 0 and T_vv[j] ≠ 0 then
              Δx1 = me1 × i ⊕ me2 × j;
              Δx2 = me3 × i ⊕ me4 × j;
              DDTMM[v][Δx1||Δx2] = DDTMM[v][Δx1||Δx2] + T_uu[i] × T_vv[j];
     return DDTMM

Processing:
  T_xx ← DDTXX[ΔX_2^MC[5]];
  T_yy ← DDTYY[ΔX_2^MC[15]];
  p3 = 0;
  for i = 0,···,2^16 − 1 do
     if T_xx[i] ≠ 0 then
        j = i ⊕ (ΔX_4[0]||ΔX_4[1])
        if T_yy[j] ≠ 0 then
           p3 = p3 + T_xx[i] × T_yy[j]/2^32;
  return p3;
```

corresponding DDTS is denoted as $DDTYY$. $DDTYY[\Delta X_2^{MC}[15]][\Delta x_1||\Delta x_2]$ means the number of pairs following input difference $\Delta X_2^{MC}[10]||\Delta X_2^{MC}[15]$ and output difference $\Delta x_1||\Delta x_2$.

Furthermore, we calculate the probability for all $16 \times 24 = 384$ combinations of $(i_0, i_5[i,j])$, where $i_5[i]$ and $i_5[j]$ are bytes in the same column and $i \neq j$. The results for $i_5[i,j]$ in the first column are listed in Table 3.

Computational Complexity Analysis. Similar with the assessment in Sect. 3.2, the computational complexity to calculate the probability of new 4-round truncated differential is about $2^{16} \times 2^{16} \times 2^{-2} \times 2^8 \times 2^8 \simeq 2^{46}$.

4 New 4-Round Truncated Differential Distinguisher

In order to further improve the complexity of distinguishing attack, we take advantage of conditional probability instead of joint probability, and illustrate that the distribution of new variable is still applicable to the statistical framework in Sect. 2.3. Moreover, we checkout our results with experimental verifications, which show our model is reasonable.

4.1 Statistical Framework Using Conditional Probability

According to Eq. (3) of the statistical framework in Sect. 2.3, the number of chosen-plaintext pairs N is in reverse proportion to $(p_{\text{rand}} - p_{\text{AES}})^2$ for a given

Table 3. Results of the probability of 4-round truncated differentials of AES in Sect. 3.3 for $i_5[i, j]$ in the first column, where each item means the probability of the corresponding truncated differential.

i_0 \ i_5	0, 1	0, 2	0, 3	1, 2	1, 3	2, 3
0	$2^{-16} - 2^{-30.398}$	$2^{-16} - 2^{-30.414}$	$2^{-16} - 2^{-30.408}$	$2^{-16} - 2^{-29.955}$	$2^{-16} - 2^{-30.407}$	$2^{-16} - 2^{-29.972}$
1	$2^{-16} - 2^{-29.566}$	$2^{-16} - 2^{-29.226}$	$2^{-16} - 2^{-25.863}$	$2^{-16} - 2^{-25.853}$	$2^{-16} - 2^{-30.464}$	$2^{-16} - 2^{-29.558}$
2	$2^{-16} - 2^{-30.409}$	$2^{-16} - 2^{-30.391}$	$2^{-16} - 2^{-30.402}$	$2^{-16} - 2^{-30.388}$	$2^{-16} - 2^{-30.403}$	$2^{-16} - 2^{-30.403}$
3	$2^{-16} - 2^{-23.942}$	$2^{-16} - 2^{-31.479}$	$2^{-16} - 2^{-23.949}$	$2^{-16} - 2^{-23.948}$	$2^{-16} + 2^{-32.966}$	$2^{-16} - 2^{-23.944}$
4	$2^{-16} - 2^{-23.944}$	$2^{-16} - 2^{-31.552}$	$2^{-16} - 2^{-23.950}$	$2^{-16} - 2^{-23.948}$	$2^{-16} + 2^{-33.868}$	$2^{-16} - 2^{-23.943}$
5	$2^{-16} - 2^{-29.955}$	$2^{-16} - 2^{-30.407}$	$2^{-16} - 2^{-30.398}$	$2^{-16} - 2^{-29.972}$	$2^{-16} - 2^{-30.414}$	$2^{-16} - 2^{-30.408}$
6	$2^{-16} - 2^{-29.660}$	$2^{-16} - 2^{-29.126}$	$2^{-16} - 2^{-25.868}$	$2^{-16} - 2^{-25.869}$	$2^{-16} - 2^{-30.301}$	$2^{-16} - 2^{-29.568}$
7	$2^{-16} - 2^{-30.401}$	$2^{-16} - 2^{-30.419}$	$2^{-16} - 2^{-30.190}$	$2^{-16} - 2^{-30.412}$	$2^{-16} - 2^{-30.411}$	$2^{-16} - 2^{-30.279}$
8	$2^{-16} - 2^{-30.403}$	$2^{-16} - 2^{-30.391}$	$2^{-16} - 2^{-30.388}$	$2^{-16} - 2^{-30.402}$	$2^{-16} - 2^{-30.403}$	$2^{-16} - 2^{-30.409}$
9	$2^{-16} - 2^{-23.944}$	$2^{-16} - 2^{-31.479}$	$2^{-16} - 2^{-23.948}$	$2^{-16} - 2^{-23.949}$	$2^{-16} + 2^{-32.966}$	$2^{-16} - 2^{-23.942}$
10	$2^{-16} - 2^{-29.972}$	$2^{-16} - 2^{-30.414}$	$2^{-16} - 2^{-29.955}$	$2^{-16} - 2^{-30.408}$	$2^{-16} - 2^{-30.407}$	$2^{-16} - 2^{-30.398}$
11	$2^{-16} - 2^{-29.558}$	$2^{-16} - 2^{-29.226}$	$2^{-16} - 2^{-25.853}$	$2^{-16} - 2^{-25.863}$	$2^{-16} - 2^{-30.464}$	$2^{-16} - 2^{-29.566}$
12	$2^{-16} - 2^{-29.568}$	$2^{-16} - 2^{-29.126}$	$2^{-16} - 2^{-25.869}$	$2^{-16} - 2^{-25.868}$	$2^{-16} - 2^{-30.301}$	$2^{-16} - 2^{-29.660}$
13	$2^{-16} - 2^{-30.279}$	$2^{-16} - 2^{-30.419}$	$2^{-16} - 2^{-30.412}$	$2^{-16} - 2^{-30.190}$	$2^{-16} - 2^{-30.411}$	$2^{-16} - 2^{-30.401}$
14	$2^{-16} - 2^{-23.943}$	$2^{-16} - 2^{-31.552}$	$2^{-16} - 2^{-23.948}$	$2^{-16} - 2^{-23.950}$	$2^{-16} + 2^{-33.868}$	$2^{-16} - 2^{-23.944}$
15	$2^{-16} - 2^{-30.408}$	$2^{-16} - 2^{-30.407}$	$2^{-16} - 2^{-29.972}$	$2^{-16} - 2^{-30.398}$	$2^{-16} - 2^{-30.414}$	$2^{-16} - 2^{-29.955}$

P_S. To enlarge the absolute value of the bias between p_{AES} and p_{rand}, inspired by a series of works on multiple linear approximations [2,8,29,31], we try to achieve this goal by computing p_{AES} and p_{rand} in conditional probability setting. Without loss of generality, we consider

$$\begin{aligned} Pr_{\text{4r-AES}} &= Pr(\Delta X_0[1] \neq 0 \rightarrow \Delta X_5[13] = 0 | \Delta X_0[1] \neq 0 \rightarrow \Delta X_5[0] = 0) \\ &= \frac{Pr(\Delta X_0[1] \neq 0 \rightarrow \Delta X_5[0,13] = 0)}{Pr(\Delta X_0[1] \neq 0 \rightarrow \Delta X_5[0] = 0)} = \frac{Pr(\Delta X_0[1] \neq 0 \rightarrow \Delta X_4[0,1] = 0)}{Pr(\Delta X_0[1] \neq 0 \rightarrow \Delta X_4[0] = 0)} \end{aligned} \tag{9}$$

instead of $Pr(\Delta X_0[1] \neq 0 \rightarrow \Delta X_5[0,13] = 0)$. In the following, we prove that the distribution of variable under conditional probability is still applicable to the statistical framework in Sect. 2.3.

Denote $X_1|X_0$ as the conditional distribution of X_1 under the condition that $X_0 = x_0$. Suppose two-dimensional random variable (X_0, X_1) follows a two-dimensional normal distribution $\mathcal{N}(\mu_{X_0}, \sigma^2_{X_0}; \mu_{X_1}, \sigma^2_{X_1}; \rho)$. Let $f_{X_0}(x_0)$ and $f_{X_1}(x_1)$ denote the marginal density functions of X_0 and X_1, respectively,

$$f_{X_0}(x_0) = \frac{1}{\sqrt{2\pi} \times \sigma_{X_0}} e^{-\frac{(x_0 - \mu_{X_0})^2}{2\sigma^2_{X_0}}}, \quad f_{X_1}(x_1) = \frac{1}{\sqrt{2\pi} \times \sigma_{X_1}} e^{-\frac{(x_1 - \mu_{X_1})^2}{2\sigma^2_{X_1}}}. \tag{10}$$

Next, we discuss the conditional probability density function. Let $f_{X_1|X_0}(x_1|x_0)$ represents the conditional probability density function of X_1 under the condition that $X_0 = x_0$, then $f_{X_1|X_0}(x_1|x_0)$

$$= \frac{f(x_0, x_1)}{f_{X_0(x_0)}} = \frac{1}{\sqrt{2\pi}\sigma_{X_1}\sqrt{1-\rho^2}} \times e^{-\frac{1}{2(1-\rho^2)\sigma_{X_1}^2}[x_1-(\mu_{X_1}+\rho\frac{\sigma_{X_1}}{\sigma_{X_0}}(x_0-\mu_{X_0}))]^2}. \quad (11)$$

Therefore, the conditional distribution of X_1 under the condition that $X_0 = x_0$ follows a normal distribution, that is, $X_1|X_0 \sim \mathcal{N}(\mu_{X_1}+\rho\frac{\sigma_{X_1}}{\sigma_{X_0}}(x_0-\mu_{X_0}), \sigma_{X_1}^2(1-\rho^2))$. Let $\mu_{X_1|X_0} = \mu_{X_1} + \rho\frac{\sigma_{X_1}}{\sigma_{X_0}}(x_0 - \mu_{X_0})$ and $\sigma_{X_1|X_0}^2 = \sigma_{X_1}^2(1-\rho^2)$, then $X_1|X_0 \sim \mathcal{N}(\mu_{X_1|X_0}, \sigma_{X_1|X_0}^2)$. Based on the Central Limit Theorem, we can use the normal distribution to approximate the binomial distribution $\mathcal{B}(N, p_{X_1|X_0})$. Then the mean $\mu_{X_1|X_0}$ and variance $\sigma_{X_1|X_0}^2$ is given by

$$\mu_{X_1|X_0} = N \times p_{X_1|X_0},\ \sigma_{X_1|X_0}^2 = N \times p_{X_1|X_0} \times (1 - p_{X_1|X_0}). \quad (12)$$

Analogously, we can get another condition distribution $Y_1|Y_0 \backsim \mathcal{N}(\mu_{Y_1|Y_0}, \sigma_{Y_1|Y_0}^2)$. In addition, the difference of two one-dimensional normal distributions is also a normal distribution. That is to say, $X_1|X_0 - Y_1|Y_0 \backsim \mathcal{N}(\mu, \sigma^2)$ with

$$\mu = \mu_{X_1|X_0} - \mu_{Y_1|Y_0} = N \times |p_{X_1|X_0} - p_{Y_1|Y_0}|,$$
$$\sigma^2 = \sigma_{X_1|X_0}^2 + \sigma_{Y_1|Y_0}^2 = N \times [p_{X_1|X_0} \times (1 - p_{X_1|X_0}) + p_{Y_1|Y_0} \times (1 - p_{Y_1|Y_0})].$$

So we can get that

$$P_S = \int_0^{+\infty} \frac{1}{\sqrt{2\pi}} e^{-\frac{(t-\mu)^2}{2\sigma^2}} dt = \int_{-\frac{\mu}{\sigma}}^{+\infty} \frac{1}{\sqrt{2\pi}} e^{-\frac{t^2}{2}} dt = \frac{1}{2}[1 + \text{erf}(\frac{-\mu}{\sqrt{2}\sigma})]. \quad (13)$$

Thus for a success probability of at least P_S, the number N of independent Boolean experiments needs to satisfy

$$N \geq \frac{2 \times [p_{X_1|X_0} \times (1 - p_{X_1|X_0}) + p_{Y_1|Y_0} \times (1 - p_{Y_1|Y_0})]}{(p_{X_1|X_0} - p_{Y_1|Y_0})^2} \times [\text{erfinv}(2 \times P_S - 1)^2]. \quad (14)$$

Where the erfinv(x) is the inverse error function.

4.2 4-Round Truncated Differential Distinguisher Using Conditional Probability

In order to compute the conditional probability $p_{4\text{r-AES}}$ defined in Eq. (9), for example, we obtain the joint probability $Pr(\Delta X_0[1] \neq 0 \rightarrow \Delta X_4[0,1] = 0) \simeq 2^{-16} - 2^{-29.566}$ from Table 3, and find the corresponding probability of the condition $Pr(\Delta X_0[1] \neq 0 \rightarrow \Delta X_4[0] = 0) \simeq 2^{-8} + 2^{-31.816}$ from Table 2, thus, the conditional probability is

$$p_{4\text{r-AES}} = \frac{2^{-16} - 2^{-29.566}}{2^{-8} + 2^{-31.816}} \simeq 2^{-8} - 2^{-21.565}. \quad (15)$$

While for a random permutation, the conditional probability is roughly:

$$p_{\text{rand}} = \left(\frac{2^{112} - 1}{2^{128} - 1}\right) / \left(\frac{2^{120} - 1}{2^{128} - 1}\right) \simeq 2^{-8} - 2^{-120}. \quad (16)$$

Thus, the absolute value of the bias is $|p_{\text{rand}} - p_{4\text{r-AES}}| \simeq 2^{-21.565}$.

Experimental Verifications on 4-Round AES. We follow the statistical framework in Sect. 4.1 to evaluate the data complexity and get that $N \simeq 2^{37.56}$, which is at least $2^{22.56}\delta$-sets, i.e., $2^{30.56}$ chosen plaintexts, for $P_S = 0.95$. For $P_S = 0.99$, we get that $N \simeq 2^{38.56}$, which is at least $2^{23.56}\delta$-sets, i.e., $2^{31.56}$ chosen plaintexts. To illustrate the accuracy of conditional probability in Eq. (15), we perform experimental verifications with different size of samples, which randomly choose 2^{24} and 2^{25} δ-sets, respectively, and encrypt them under 100 random keys to compute the mean value. The experimental results, which are approximately $2^{-8} - 2^{-22.077}$ for 2^{24} δ-sets and $2^{-8} - 2^{-21.474}$ for 2^{25} δ-sets, are consistent with the probability $2^{-8} - 2^{-21.565}$ deduced from Eq. (15), and show that the statistical framework in conditional probability is reasonable.
Remark: Obviously, for different $(i_0, i_5[i,j])$, we can deduce the corresponding $p_{4\text{r-AES}}$ based on Table 2 and 3, and construct 4-round distinguishers similarly.

5 Extend to 5-Round Truncated Differential

Inspired by Bao et al.'s construction, we extend our 4-round truncated differential to 5-round. It is noted that four inactive bytes in one column of ΔX_5 come from four columns of ΔX_4, and transfer to four inactive bytes in ΔX_6 with probability 1 if the MC operation is omitted. Thus, collisions in ΔX_6 relate to collisions in ΔX_4, whose probability can be deduced from our algorithms introduced in Sect. 3. Take the 5-round truncated differential depicted in Fig. 2(b), which is the best one, i.e., the position where the absolute value of the bias is the largest we found, as an example. We can see that if $\Delta X_4[1,2,6,7,8,11,12,13] = 0$, then $\Delta X_6[2,3,5,6,8,9,12,15] = 0$. Suppose the computation between the columns of ΔX_4 are independent of each other, we obtain the conditional probability for 5-round truncated differential:

$$
\begin{aligned}
p_{5\text{r-AES}} &= \frac{Pr(\Delta X_0[1] \neq 0 \rightarrow \Delta X_6[2,3,5,6,8,9,12,15] = 0)}{Pr(\Delta X_0[1] \neq 0 \rightarrow \Delta X_6[2,5,8,15] = 0)} \\
&= \frac{Pr(\Delta X_0[1] \neq 0 \rightarrow \Delta X_4[1,2,6,7,8,11,12,13] = 0)}{Pr(\Delta X_0[1] \neq 0 \rightarrow \Delta X_4[2,7,8,13] = 0)} \\
&= \frac{Pr(\Delta X_0[1] \neq 0 \rightarrow \Delta X_4[1,2] = 0)}{Pr(\Delta X_0[1] \neq 0 \rightarrow \Delta X_4[2] = 0)} \times \frac{Pr(\Delta X_0[1] \neq 0 \rightarrow \Delta X_4[6,7] = 0)}{Pr(\Delta X_0[1] \neq 0 \rightarrow \Delta X_4[7] = 0)} \\
&\times \frac{Pr(\Delta X_0[1] \neq 0 \rightarrow \Delta X_4[8,11] = 0)}{Pr(\Delta X_0[1] \neq 0 \rightarrow \Delta X_4[8] = 0)} \times \frac{Pr(\Delta X_0[1] \neq 0 \rightarrow \Delta X_4[12,13] = 0)}{Pr(\Delta X_0[1] \neq 0 \rightarrow \Delta X_4[13] = 0)} \\
&= \frac{2^{-16} - 2^{-25.853}}{2^{-8} + 2^{31.922}} \times \frac{2^{-16} - 2^{-30.401}}{2^{-8} + 2^{31.765}} \times \frac{2^{-16} - 2^{-23.948}}{2^{-8} + 2^{32.002}} \times \frac{2^{-16} - 2^{-29.955}}{2^{-8} + 2^{31.797}} \\
&\simeq 2^{-32} - 2^{-39.578}.
\end{aligned}
$$

While the conditional probability for a random permutation is $p_{\text{rand}} = \frac{2^{64}-1}{2^{96}-1} \simeq 2^{-32} - 2^{-96}$. Thus, the absolute value of bias is $|p_{\text{rand}} - p_{5\text{r-AES}}| \simeq 2^{-39.578}$.

6 6-Round Truncated Differential Distinguisher

6.1 Extended 6-Round Truncated Differential

Based on the 5-round truncated differential in Sect. 5, we extend it to a 6-round truncated differential by adding one round at the beginning. Our 6-round truncated differential distinguisher is depicted in Fig. 3.

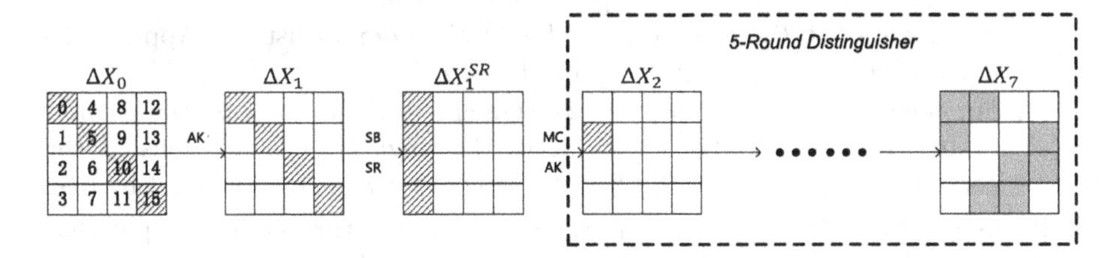

Fig. 3. Extended 6-round truncated differential. White cells represent bytes with zero difference, while slashed cells represent the active ones, and the gray cells stand for unknown.

The 6-round truncated differential is the connection of a 1-round truncated differential and a 5-round truncated differential. So that, the computation of the probability can be discussed in two cases: 1) the output difference of the first round, ΔX_2, coincides with the input of the 5-round distinguisher. In this case, there are four choices for ΔX_2, i.e., $\Delta X_2[i] \neq 0 (i = 0, 1, 2, 3)$, respectively, and the other 15 bytes are inactive, of which the probability of the first round is $\frac{255}{2^{32}-1}$ for $\Delta X_2[i] \neq 0 (i = 0, 1, 2, 3)$, and the probability of the corresponding 5-round truncated differential are $2^{-32} - 2^{-39.878}$, $2^{-32} - 2^{-39.578}$, $2^{-32} - 2^{-39.878}$, and $2^{-32} - 2^{-39.578}$, respectively. 2) ΔX_2 does not satisfy case 1), and we suppose the difference propagates as a random permutation. Then the conditional probability of the 6-round truncated differential for AES is:

$$p_{\text{6r-AES}} \simeq \sum_{i=0}^{3} p_{\text{5r-AES},\Delta X_2[i]\neq 0} \times \frac{255}{2^{32}-1} + p_{\text{rand}} \times (1 - \frac{4 \times 255}{2^{32}-1}) \simeq 2^{-32} - 2^{-61.725}.$$

While the conditional probability for a random permutation is $p_{\text{rand}} \simeq 2^{-32} - 2^{-96}$. Thus, the absolute value of bias is roughly $|p_{\text{rand}} - p_{\text{6r-AES}}| \simeq 2^{-61.725}$.

6.2 Distinguishing Attack on 6-Round AES

We describe an improved distinguishing attack on 6-round AES based on the extended 6-round truncated differential in Sect. 6.1.

Attack Procedure. The details of the attack are as follows:

1. Initialize two counters c_1 and c_2, where c_1 records the number of pairs s.t. $\Delta X_7[2,5,8,15]{=}0$, and c_2 counts the one with $\Delta X_7[2,3,5,6,8,9,12,15] = 0$.
2. For $i = 1, 2, \cdots, 2^s$:
   (a) Generate a structure $\mathcal{S}$ with 2^{32} randomly chosen plaintexts, which iterate over all values in the active byte at positions 0,5,10,15 and set 12 remaining bytes to constants. For each plaintext in $\mathcal{S}$, query the encryption oracle and get the corresponding ciphertext X_7.
   (b) Initialize a list $\mathcal{L}_1$. Denote u as the 32-bit index of list $\mathcal{L}_1$. Append the ciphertext X_7 to $\mathcal{L}_1[u]$ if $X_7[2,5,8,15] = u$.
   (c) For each u, if there are m ($m \geq 2$) elements in $\mathcal{L}_1[u]$, increase the counter c_1 by $m(m-1)/2$. For each one of the $m(m-1)/2$ collisions, if the corresponding $\Delta X_7[3,6,9,12] = 0$, the counter c_2 is incremented by one.
3. If the result $c = 2^s \times \binom{2^{32}}{2} \times c_2/c_1$ is smaller than the threshold t, output that the oracle is the 6-round AES, otherwise, it's a random permutation.

Complexity and Success Probability. We set the threshold $t = N \times (p_{\text{6r-AES}} + p_{\text{rand}})/2$, following the setting in [36]. For $P_S = 0.95$, the number of chosen-plaintext pairs $N \simeq 2^{93.88}$, $p_{\text{6r-AES}} = 2^{-32} - 2^{-61.725}$ and $p_{\text{rand}} = 2^{-32} - 2^{-96}$, so that $t \simeq 2^{61.88}$.

A structure $\mathcal{S}$ contains 2^{32} plaintexts, thus, one structure $\mathcal{S}$ can generate $\binom{2^{32}}{2} \simeq 2^{63}$ pairs, we need to choose $2^{93.88-63} = 2^{30.88}$ structures. The data complexity is $2^{30.88} \times 2^{32} = 2^{62.88}$. The time complexity is dominated by step 2. For step 2.(a), the time complexity consists of $2^{62.88}$ 6-round encryptions. For step 2.(b), it takes $2^{62.88}$ memory accesses. And there are about $\binom{2^{32}}{2} \times 2^{-32} = 2^{31}$ pairs colliding at $\Delta X_7[2,5,8,15]$ for each structure. Thus, step 2.(c) checks about $2^{30.88} \times 2^{31}$ collisions with $2^{61.88}$ memory accesses. To sum up, the total computational complexity can be approximated as $2^{62.88} + 2^{61.88} \simeq 2^{63.46}$ memory accesses and $2^{62.88}$ encryptions, i.e., $2^{63.42}$ encryptions. The memory complexity is about $2^{32} \times 128\text{-bit} = 2^{39}\text{-bit}$.

7 Conclusion and Future Work

In FSE 2020, Bao et al. [3] extended the 3-round integral distinguisher of AES, which is proposed by Daemen et al. [12], to 4/5/6-round truncated differential distinguishers in the secret S-box setting. A natural question is whether we can get a tighter bound when considering the details of the underlying S-boxes.

However, it is infeasible to check all differentials belonging to a truncated differential, and compute the corresponding probability directly, since too many differentials are involved. Thus, we propose a *divide-and-combine* technique, including choosing the appropriate starting point and integrated S-box, to present a whole algorithm to overcome this obstacle, and accomplish the accurate calculation for Bao et al.'s truncated differentials on AES.

In addition, we investigate the relation of cells in one column, modify the parameters of the integrated S-box and related algorithms, and achieve the probability corresponding to the case of two inactive cells in the output difference. And the results lead to new 4/5/6-round distinguishers of AES with concrete S-box, while the setting of Bao et al. is not applicable because the bytes in one column are not independent. In order to further improve the complexity of distinguishing attack, we try to take advantage of the conditional probability to depict the distribution of variables, rather than joint probability, and illustrate the settings are still applicable to the statistical framework proposed by Grassi et al. [25]. We also construct a 6-round truncated differential distinguisher and describe an improved distinguishing attack in conditional probability setting, which has a data complexity of $2^{62.88}$ chosen plaintexts and time complexity of $2^{63.42}$ encryptions.

Meanwhile, as a verification of the 4-round truncated differential, we conduct experimental verifications on AES, and for all 4/5/6-round truncated differentials mentioned in this paper, we check out our results with experimental verifications on Small-AES[4].

One of the future work is whether our results can influence other techniques making use of truncated differentials, such as the *differential enumeration*. The other is to generalize the *integrated* computation of DDTs of Super-Sbox, and obtain tighter bounds for more distinguishing attacks.

Acknowledgments. We sincerely thank the anonymous reviewers for providing valuable comments to help us improve the overall quality of the paper. This work is supported by the National Key Research and Development Program of China (Grant No. 2018YFA0704702 & 2022YFB2701700), the National Natural Science Foundation of China (Grant No. 62032014), the Shandong Provincial Natural Science Foundation (Grant No. ZR2020MF053), the Major Basic Research Project of Natural Science Foundation of Shandong Province (Grant No. ZR202010220025), and the Education Teaching Reform and Research Program of Shandong University (Grant No. 2022Y286).

Appendix

A Brief Description of Small-AES [11].

Small-AES is a 4-bit variant of AES. The differences compared to AES are: 1) Its length is 64-bit of a 4×4 state matrix, where each element is a 4-bit nibble instead of a byte. 2) The operations are performed in the finite field $GF(2^4)$. The modulo polynomial of the MC operation becomes $X^4 + X + 1$. The details of key schedule are omitted.

The details of the S-boxes are defined in Table 4.

[4] The source code of all algorithms to compute the probability and experimental verifications, the supplementary algorithms, and the verified results are provided in the github: https://github.com/ccchang123456/truncated_differential.git.

Table 4. Different 4-bit S-boxes were employed to perform our tests on Small-AES.

	$S(x)$															
x	0	1	2	3	4	5	6	7	8	9	a	b	c	d	e	f
Small-AES	6	b	5	4	2	e	7	a	9	d	f	c	3	1	0	8
Present	c	5	6	b	9	0	a	d	3	e	f	8	4	7	1	2

B Algorithm 5 and Algorithm 6 in the Calculation of 4-Round Truncated Differential with One Active Cell in Input and Two Inactive in Output

Algorithm 4, 5, and 6 make up the complete algorithm for calculating the truncated differential intruduced in Sect. 3.3. Algorithm 5 corresponds to the modified Algorithm 2, and Algorithm 6 corresponds to the modified Algorithm 3.

Algorithm 5. $Pr(\Delta X_1^{SB}[1] \rightarrow \Delta X_4[0,1])$ for all non-zero $\Delta X_1^{SB}[1]$, and given $\Delta X_4[0,1]$

Parameters: $\Delta X_4[0,1]$;
Processing:
1: Initial the array p23 of size $2^8 \times 2^8$ with zeros;
2: /* *for* $\Delta X_2^{MC}[0] = 0, ..., 255$, *use 256 threads to perform* ***function*** **GetProbability** *in parallel* */
3: $p23[\Delta X_2^{MC}[0]] \leftarrow$ **GetProbability**$(\Delta X_2^{MC}[0], \Delta X_4[0,1])$; /* $p23[\Delta X_2^{MC}[0]][i]$ *means* $Pr(\Delta X_1^{SB} = i \rightarrow \Delta X_2^{MC}[0] \rightarrow \Delta X_4[0,1])$ */
4: Initial the array p1 of size 2^8 with zeros;
5: **for** $\Delta X_2^{MC}[0] = 0, \cdots, 255$ and $\Delta X_1^{SB}[1] = 1, \cdots, 255$ **do**
6: $\quad p1[\Delta X_1^{SB}[1]] = p1[\Delta X_1^{SB}[1]] + p23[\Delta X_2^{MC}[0]][\Delta X_1^{SB}[1]]$;
7: **return** $p1[i]$; /* $p1[i] = Pr(\Delta X_1^{SB}[1] \rightarrow \Delta X_4[0,1])$ *for* $1 \leq i \leq 255$ */
8:
9: **function GetProbability**$(\Delta X_2^{MC}[0], \Delta X_4[0,1]))$
10: $\quad$ **for** $\Delta X_2^{MC}[10] = 0, \cdots, 255$ **do**
11: $\quad\quad DDTYY \leftarrow$ **MergeTableDDT**$(\Delta X_2^{MC}[10], 1, 1, 3, 1)$; /* *compute the DDT of* f_2, *run once for each loop corresponding to* $\Delta X_2^{MC}[10]$ */
12: $\quad\quad$ **for** 2^{16} values of $\Delta X_2^{MC}[5, 15]$ **do**
13: $\quad\quad\quad$ Compute $\Delta X_2^{SB}[12, 13, 14, 15]$ from $\Delta X_2^{MC}[0, 5, 10, 15]$;
14: $\quad\quad\quad$ **for** $\Delta X_1^{SB}[1] = 1, \cdots, 255$ **do**
15: $\quad\quad\quad\quad$ Compute $\Delta X_1^{MC}[12, 13, 14, 15]$ from $\Delta X_1^{SB}[1]$;
16: $\quad\quad\quad\quad$ $p2 = DDTS[\Delta X_1^{MC}[12]][\Delta X_2^{SB}[12]] \times DDTS[\Delta X_1^{MC}[13]][\Delta X_2^{SB}[13]] \times DDTS[\Delta X_1^{MC}[14]][\Delta X_2^{SB}[14]] \times DDTS[\Delta X_1^{MC}[15]][\Delta X_2^{SB}[15]]$;
17: $\quad\quad\quad\quad$ **if** $p2 \neq 0$ **then**
18: $\quad\quad\quad\quad\quad$ $p3 = Algorithm4(DDTYY, \Delta X_2^{MC}[0, 5, 10, 15], \Delta X_4[0,1])$;
19: $\quad\quad\quad\quad\quad$ $pm[\Delta X_1^{SB}[1]] = pm[\Delta X_1^{SB}[1]] + (p2/2^{32}) \times p3$;
20: **return** pm; /* $pm[i] = Pr(\Delta X_1^{SB}[1] \rightarrow \Delta X_2^{MC}[0] = i \rightarrow \Delta X_4[0,1])$ *for* $1 \leq i \leq 255$ */

Algorithm 6. $Pr(\Delta X_0[1] \rightarrow \Delta X_4[0,1])$ for all non-zero $\Delta X_0[1]$, and given $\Delta X_4[0,1]$

Parameters: $\Delta X_4[0,1]$
Processing:
1: $p1 \leftarrow Algorithm5(\Delta X_4[0,1])$;
2: **for** $\Delta X_0[1] = 1, \cdots, 255$ **and** $\Delta X_1^{SB}[1] = 1, \cdots, 255$ **do**
3: **if** $DDTS[\Delta X_0[1]][\Delta X_1^{SB}[1]] \neq 0$ **then**
4: $p[\Delta X_0[1]] = p[\Delta X_0[1]] + (DDTS[\Delta X_0[1]][\Delta X_1^{SB}[1]]/2^8) \times p1[\Delta X_1^{SB}[1]]$;
5: **return** p; /* $p[i] = Pr(\Delta X_0[1] = i \rightarrow \Delta X_4[0,1])$ *for* $1 \leq i \leq 255$ */

References

1. National Institute of Standards and Technology: Advanced Encryption Standard, FIPS 197. US Department of Commerce, Washington D.C., November 2001. http://csrc.nist.gov/publications/fips/fips197/fips-197.pdf
2. Baignères, T., Junod, P., Vaudenay, S.: How far can we go beyond linear cryptanalysis? In: Lee, P.J. (ed.) ASIACRYPT 2004. LNCS, vol. 3329, pp. 432–450. Springer, Heidelberg (2004). https://doi.org/10.1007/978-3-540-30539-2_31
3. Bao, Z., Guo, J., List, E.: Extended truncated-differential distinguishers on round-reduced AES. IACR Trans. Symmetric Cryptol. **2020**(3), 197–261 (2020). https://doi.org/10.13154/tosc.v2020.i3.197-261
4. Bar-On, A., Dunkelman, O., Keller, N., Ronen, E., Shamir, A.: Improved key recovery attacks on reduced-round AES with practical data and memory complexities. In: Shacham, H., Boldyreva, A. (eds.) CRYPTO 2018. LNCS, vol. 10992, pp. 185–212. Springer, Cham (2018). https://doi.org/10.1007/978-3-319-96881-0_7
5. Bardeh, N.G.: A Key-Independent Distinguisher for 6-round AES in an Adaptive Setting. IACR Cryptol. ePrint Arch., 2019:945. https://eprint.iacr.org/2019/945
6. Bardeh, N.G., Rønjom, S.: The exchange attack: *How to Distinguish Six Rounds of AES with* $2^{88.2}$ *Chosen Plaintexts*. In: Galbraith, S.D., Moriai, S. (eds.) ASIACRYPT 2019. LNCS, vol. 11923, pp. 347–370. Springer, Cham (2019). https://doi.org/10.1007/978-3-030-34618-8_12
7. Bellare, M., Kilian, J., Rogaway, P.: The security of cipher block chaining. In: Desmedt, Y.G. (ed.) CRYPTO 1994. LNCS, vol. 839, pp. 341–358. Springer, Heidelberg (1994). https://doi.org/10.1007/3-540-48658-5_32
8. Biryukov, A., De Cannière, C., Quisquater, M.: On multiple linear approximations. In: Franklin, M. (ed.) CRYPTO 2004. LNCS, vol. 3152, pp. 1–22. Springer, Heidelberg (2004). https://doi.org/10.1007/978-3-540-28628-8_1
9. Bogdanov, A., Khovratovich, D., Rechberger, C.: Biclique cryptanalysis of the full AES. In: Lee, D.H., Wang, X. (eds.) ASIACRYPT 2011. LNCS, vol. 7073, pp. 344–371. Springer, Heidelberg (2011). https://doi.org/10.1007/978-3-642-25385-0_19
10. Boura, C., Lallemand, V., Naya-Plasencia, M., Suder, V.: Making the Impossible Possible. J. Cryptol. **31**(1), 101–133 (2017). https://doi.org/10.1007/s00145-016-9251-7
11. Cid, C., Murphy, S., Robshaw, M.J.B.: Small scale variants of the AES. In: Gilbert, H., Handschuh, H. (eds.) FSE 2005. LNCS, vol. 3557, pp. 145–162. Springer, Heidelberg (2005). https://doi.org/10.1007/11502760_10
12. Daemen, J., Knudsen, L., Rijmen, V.: The block cipher Square. In: Biham, E. (ed.) FSE 1997. LNCS, vol. 1267, pp. 149–165. Springer, Heidelberg (1997). https://doi.org/10.1007/BFb0052343

13. Daemen, J., Rijmen, V.: The Pelican MAC Function 2.0. IACR Cryptol. ePrint Arch., 2005:88. http://eprint.iacr.org/2005/088
14. Daemen, J., Rijmen, V.: Two-Round AES Differentials. IACR Cryptol. ePrint Arch., 2006:39. http://eprint.iacr.org/2006/039
15. Daemen, J., Rijmen, V.: The Design of Rijndael: AES - The Advanced Encryption Standard. Information Security and Cryptography. Springer (2002). https://doi.org/10.1007/978-3-662-04722-4
16. Daemen, J., Rijmen, V.: Understanding two-round differentials in AES. In: De Prisco, R., Yung, M. (eds.) SCN 2006. LNCS, vol. 4116, pp. 78–94. Springer, Heidelberg (2006). https://doi.org/10.1007/11832072_6
17. Daemen, J., Rijmen, V.: Plateau characteristics. IET Inf. Secur. **1**(1), 11–17 (2007). https://doi.org/10.1049/iet-ifs:20060099
18. Demirci, H., Selçuk, A.A.: A meet-in-the-middle attack on 8-round AES. In: Nyberg, K. (ed.) FSE 2008. LNCS, vol. 5086, pp. 116–126. Springer, Heidelberg (2008). https://doi.org/10.1007/978-3-540-71039-4_7
19. Derbez, P., Fouque, P.-A., Jean, J.: [Improved key recovery attacks on reduced-round, in the single-key setting]. In: Johansson, T., Nguyen, P.Q. (eds.) EUROCRYPT 2013. LNCS, vol. 7881, pp. 371–387. Springer, Heidelberg (2013). https://doi.org/10.1007/978-3-642-38348-9_23
20. Dunkelman, O., Keller, N., Shamir, A.: Improved single-key attacks on 8-round AES-192 and AES-256. In: Abe, M. (ed.) ASIACRYPT 2010. LNCS, vol. 6477, pp. 158–176. Springer, Heidelberg (2010). https://doi.org/10.1007/978-3-642-17373-8_10
21. Ferguson, N., Kelsey, J., Lucks, S., Schneier, B., Stay, M., Wagner, D., Whiting, D.: Improved cryptanalysis of rijndael. In: Goos, G., Hartmanis, J., van Leeuwen, J., Schneier, B. (eds.) FSE 2000. LNCS, vol. 1978, pp. 213–230. Springer, Heidelberg (2001). https://doi.org/10.1007/3-540-44706-7_15
22. Gilbert, H., Peyrin, T.: Super-Sbox cryptanalysis: improved attacks for AES-like permutations. In: Hong, S., Iwata, T. (eds.) FSE 2010. LNCS, vol. 6147, pp. 365–383. Springer, Heidelberg (2010). https://doi.org/10.1007/978-3-642-13858-4_21
23. Grassi, L.: Mixture Differential Cryptanalysis and Structural Truncated Differential Attacks on Round-Reduced AES. IACR Cryptol. ePrint Arch., 2017:832. https://ia.cr/2017/832
24. Grassi, L.: Probabilistic mixture differential cryptanalysis on round-reduced AES. In: Paterson, K.G., Stebila, D. (eds.) SAC 2019. LNCS, vol. 11959, pp. 53–84. Springer, Cham (2020). https://doi.org/10.1007/978-3-030-38471-5_3
25. Grassi, L., Rechberger, C.: Truncated Differential Properties of the Diagonal Set of Inputs for 5-round AES. Accepted by ACISP 2022. IACR Cryptol. ePrint Arch., 2018:182. https://eprint.iacr.org/2018/182
26. Grassi, L., Rechberger, C., Rønjom, S.: Subspace Trail Cryptanalysis and its Applications to AES. IACR Trans. Symmetric Cryptol. **2016**(2), 192–225 (2016). https://doi.org/10.13154/tosc.v2016.i2.192-225
27. Grassi, L., Rechberger, C., Rønjom, S.: A new structural-differential property of 5-round AES. In: Coron, J.-S., Nielsen, J.B. (eds.) EUROCRYPT 2017. LNCS, vol. 10211, pp. 289–317. Springer, Cham (2017). https://doi.org/10.1007/978-3-319-56614-6_10
28. Gupta, B., Guttman, I., Jayalath, K.: Statistics and probability with applications for engineers and scientists using MINITAB. R JMP (2020). https://doi.org/10.1002/9781119516651

29. Kaliski, B.S., Robshaw, M.J.B.: Linear cryptanalysis using multiple approximations. In: Desmedt, Y.G. (ed.) CRYPTO 1994. LNCS, vol. 839, pp. 26–39. Springer, Heidelberg (1994). https://doi.org/10.1007/3-540-48658-5_4
30. Li, L., Jia, K., Wang, X.: Improved single-key attacks on 9-round AES-192/256. In: Cid, C., Rechberger, C. (eds.) FSE 2014. LNCS, vol. 8540, pp. 127–146. Springer, Heidelberg (2015). https://doi.org/10.1007/978-3-662-46706-0_7
31. Matsui, M.: The first experimental cryptanalysis of the data encryption standard. In: Desmedt, Y.G. (ed.) CRYPTO 1994. LNCS, vol. 839, pp. 1–11. Springer, Heidelberg (1994). https://doi.org/10.1007/3-540-48658-5_1
32. Patarin, J.: Generic attacks for the Xor of k random permutations. In: Jacobson, M., Locasto, M., Mohassel, P., Safavi-Naini, R. (eds.) ACNS 2013. LNCS, vol. 7954, pp. 154–169. Springer, Heidelberg (2013). https://doi.org/10.1007/978-3-642-38980-1_10
33. Preneel, B.: Davies-Meyer Hash Function. In: van Tilborg, H.C.A. (ed.) Encyclopedia of Cryptography and Security. Springer (2005). https://doi.org/10.1007/0-387-23483-7_96
34. Preneel, B.: Davies-Meyer. In: van Tilborg, H.C.A., Jajodia, S. (eds.) Encyclopedia of Cryptography and Security, 2nd Ed, pp. 312–313. Springer (2011). https://doi.org/10.1007/978-1-4419-5906-5_569
35. Rønjom, S., Bardeh, N.G., Helleseth, T.: Yoyo tricks with AES. In: Takagi, T., Peyrin, T. (eds.) ASIACRYPT 2017. LNCS, vol. 10624, pp. 217–243. Springer, Cham (2017). https://doi.org/10.1007/978-3-319-70694-8_8
36. Samajder, S., Sarkar, P.: Rigorous upper bounds on data complexities of block cipher cryptanalysis. J. Math. Cryptol. **11**(3), 147–175 (2017). https://doi.org/10.1515/jmc-2016-0026
37. Zhang, B., Xu, C., Meier, W.: Fast correlation attacks over extension fields, large-unit linear approximation and cryptanalysis of SNOW 2.0. In: Gennaro, R., Robshaw, M. (eds.) CRYPTO 2015. LNCS, vol. 9215, pp. 643–662. Springer, Heidelberg (2015). https://doi.org/10.1007/978-3-662-47989-6_31

Boolean Functions

Modifying Bent Functions to Obtain the Balanced Ones with High Nonlinearity

Subhamoy Maitra[1], Bimal Mandal[2(✉)], and Manmatha Roy[1]

[1] Applied Statistics Unit, Indian Statistical Institute, Kolkata 700108, India
subho@isical.ac.in

[2] Department of Mathematics, Indian Institute of Technology Jodhpur, Karwar 342030, India
bimalmandal@iij.ac.in

Abstract. Balanced Boolean functions with high nonlinearity are considered as major cryptographic primitives in the design of symmetric key cryptosystems. Dobbertin, in early nineties, gave an explicit construction for balanced functions on (even) n variables, with nonlinearity $2^{n-1} - 2^{\frac{n}{2}} + nlb(\frac{n}{2})$, where $nlb(t)$ is the maximum nonlinearity of a balanced Boolean functions in t variables and conjectured that $nlb(n) \leq 2^{n-1} - 2^{\frac{n}{2}} + nlb(\frac{n}{2})$. This bound still holds. In this paper we revisit the problem. First we present a detailed combinatorial analysis related to highly nonlinear balanced functions exploiting the inter-related properties like weight, nonlinearity, and Walsh–Hadamard spectrum. Our results provide a general framework to cover the works of Sarkar-Maitra (Crypto 2000), Maity-Johansson (Indocrypt 2002), and Maity-Maitra (FSE 2004) as special cases. In this regard, we revisit the well-known construction methods through modification of bent functions and provide supporting examples for 8, 10, 12, and 14 variables. We believe these results will advance the understanding related to highly nonlinear balanced Boolean functions on even numbers of variables as well as the Dobbertin's conjecture.

Keywords: Boolean functions · Balancedness · Bent functions · Nonlinearity · Walsh–Hadamard transform

1 Introduction

The most important question on nonlinearity of balanced Boolean functions (on even number of input variables) circles around Dobbertin's conjecture [6]. This conjecture is open for around three decades. Towards disproving this conjecture, there are a few studies that considered how an n-variable balanced function can be expected with nonlinearity strictly greater than $2^{n-1} - 2^{\frac{n}{2}} + nlb(\frac{n}{2})$. As such functions are still undiscovered, this remains as one of the most coveted problem to be solved in the domain of Boolean functions.

T. Isobe and S. Sarkar (Eds.): INDOCRYPT 2022, LNCS 13774, pp. 449–470, 2022.
https://doi.org/10.1007/978-3-031-22912-1_20

In symmetric key cryptography, Boolean functions are most of the time the central design primitive. Not to mention that such Boolean functions must not be chosen arbitrarily, but with great care such that they satisfy certain criteria. There are handful of criteria, e.g., nonlinearity, balancedness, strong avalanche criteria, correlation immunity, algebraic degree etc. Each of these cryptographic properties essentially makes the cryptographic design resilient to certain cryptanalytic attempts. For example, high nonlinearity resists affine approximation of the design, good propagation criteria takes care of resistance against differential attacks, and balancedness is desired to eliminate any potential analysis which relies on guessing output of some Boolean function. Therefore, in this line of work, one important task is to construct Boolean functions with good cryptographic properties. What makes this task more difficult is that these properties are sometimes orthogonal to each other, e.g., high algebraic degree implies low correlation immunity. For decades, constructing Boolean functions with good cryptographic properties has been an active area of research. In this paper, we discuss about constructing Boolean functions with high nonlinearity and balancedness, in even number of variables.

It is known that the nonlinearity of any Boolean function in n variables is an integer value less than or equal to $2^{n-1} - 2^{\frac{n}{2}-1}$. This bound is achieved when n is even, and these functions are called bent functions, introduced by Rothaus [19]. Thus, bent functions offer the maximum resistance to affine approximations. Although bent functions are not directly used as cryptographic primitives due to not being balanced. There are some standard techniques of obtaining balanced Boolean functions with good cryptographic properties by modifying bent functions [6,7,13,14,20,22,23]. Research on different classes of bent functions and their relationship with coding theory received very serious attention in literature and one may have a look at [2–5,15,17] and the pointers therein. On the other hand, very little is known for odd n, so far as nonlinearity is concerned. We do not even know the exact upper bound of nonlinearity in this case. It is known that for odd n, one can construct a Boolean function with nonlinearity $2^{n-1} - 2^{\frac{n-1}{2}}$, by concatenating bent functions of $(n-1)$ variables. Construction of an n-variable Boolean function with nonlinearity exceeding this bound (popularly called as bent concatenation bound) for odd n is an interesting problem, and in many papers [8–10,12,18,21] such functions are identified. While most of the results are open here, serious efforts have also been made through years.

Despite optimal nonlinearity of bent functions, which in turn provides optimal resistance to affine approximation attempts, the notable weakness which forbid its cryptographic usage, is lack of balancedness. In [6], Dobbertin presented a construction of balanced Boolean functions in even n with nonlinearity $2^{n-1} - 2^{\frac{n}{2}} + nlb(\frac{n}{2})$, where $nlb(t)$ is the maximum nonlinearity of balanced Boolean functions in t variables. Dobbertin also conjectured that

$$nlb(n) \leq 2^{n-1} - 2^{\frac{n}{2}} + nlb(\frac{n}{2}),$$

which is still standing. Early works related to this conjecture require mention of Sarkar et al. [20], and Maity et al. [14]. More specifically, Sarkar et al. [20]

derived a sufficient condition for a balanced Boolean function in 8 variables having nonlinearity 118 (Dobbertin's construction achieves 116) by concatenating two 7-variable Boolean functions each having nonlinearity 55 and degree 7. Essentially they shifted the question of existence of a Boolean function f in 8 variable with nonlinearity 118 to existence of a pair of Boolean functions (f_1, f_2) each in 7 variable with nonlinearity 55 and degree 7.

Maity et al. [14] took an approach which is more constructive in essence. They attempted to construct balanced functions in 8 variables having nonlinearity 118 by modifying the bent functions. In fact, they focused on the distance between resilient and bent functions, and concluded that for $n = 8$ this distance is 10. They could also construct one resilient Boolean functions in 8 variables with nonlinearity 116. Because resiliency is a stronger criteria than balancedness, they further proceeded to study the existence of balanced Boolean functions in 8 variables with nonlinearity 118, which could have disproved Dobbertin's conjecture for $n = 8$. They ended up by suggesting certain modifications on a suitably chosen bent function in 8 variables which would produce a balanced function in $n = 8$ variables with nonlinearity 118. However, the final result could not be achieved and Dobbertin's conjecture remains open till date.

1.1 Contribution and Organization

Construction of a balanced Boolean function in n variables, $n \geq 8$ is even, having nonlinearity strictly greater than $2^{n-1} - 2^{\frac{n}{n}} + nlb(\frac{n}{2})$ is still an open problem. In this paper we derive necessary and sufficient conditions for existence of such functions in terms of existence of a particular class of linear codes. The rest of the paper is organized as follows.

- In Sect. 3 we derive an equivalent representation of balanced Boolean functions in terms of the existence of certain kind of linear codes and show that the existence of balanced Boolean functions with nonlinearity exceeding Dobbertin's bound can be rephrased as the existence of certain kind of linear codes with required parameters.
- Next, in Sect. 4 we compare our characterization with the earlier attempts [14, 20], and show that the existing results can be subsumed by our more general analysis.
- Further, in Sect. 5, we analyze the method of modifying bent function to get cryptographically significant Boolean function, in light of our new characterization and identify required conditions for construction of balanced Boolean functions with nonlinearity exceeding Dobbertin's bound, in the aforesaid way. We explore such conditions that might produce (we did not get any example yet) balanced Boolean functions having nonlinearity strictly greater than $2^{n-1} - 2^{\frac{n}{n}} + nlb(\frac{n}{2})$ and explain those for $n = 8, 10, 12, 14$.

Sect. 6 concludes the paper. The caveat is, we did not obtain any function to disprove the Dobbertin's conjecture. However, our results shed important insights which may be useful towards further progress in this area of research.
Before proceeding further let us present some background material.

2 Preliminaries

Let $\mathbb{F}_2$ be the prime field of characteristic 2 and $\mathbb{F}_2^n$ be an n-dimensional vector space over $\mathbb{F}_2$. An element of $\mathbb{F}_2^n$ can be expressed as an n-tuple, i.e., $\mathbf{x} = (x_1, x_2, \ldots, x_n)$, where $x_i \in \mathbb{F}_2$, $1 \leq i \leq n$. Addition and scalar product over $\mathbb{F}_2^n$ are defined in a natural way. The inner product of $\mathbf{x}$ and $\mathbf{y}$ in $\mathbb{F}_2^n$ is defined as $\mathbf{x} \cdot \mathbf{y} = \oplus_{i=1}^n x_i y_i$. The weight of an element $\mathbf{x} \in \mathbb{F}_2^n$ is the number of nonzero coordinates, i.e., $wt(\mathbf{x}) = \sum_{i=1}^n x_i$, the sum is over integer. The cardinality of a set A, denoted by $|A|$, is defined by the number of elements in A. Let $\mathbf{0}$ and $\mathbf{1}$ denote the all 0's and all 1's vectors of $\mathbb{F}_2^n$, respectively. Any function from $\mathbb{F}_2^n$ to $\mathbb{F}_2$ is called a Boolean function in n variables, and the set of n-variable Boolean functions is denoted by $\mathcal{B}_n$. Any Boolean function f in n variables can be uniquely written as a multivariate polynomial of the form

$$f(\mathbf{x}) = \bigoplus_{\mathbf{a} \in \mathbb{F}_2^n} \mu_{\mathbf{a}} x_1^{a_1} x_2^{a_2} \cdots x_n^{a_n},$$

where $\mu_{\mathbf{a}} \in \mathbb{F}_2$, for all $\mathbf{a} \in \mathbb{F}_2^n$. This polynomial form of f is called algebraic normal form (ANF). The algebraic degree of $f \in \mathcal{B}_n$ is the degree of highest degree term(s) with nonzero coefficient in its ANF, i.e., $deg(f) = \max_{\mathbf{a} \in \mathbb{F}_2^n} \{wt(\mathbf{a}) : \mu_{\mathbf{a}} \neq 0\}$. If the algebraic degree of a Boolean functions is at most 1, then it is called an affine function. The support of $f \in \mathcal{B}_n$, denoted by $\mathcal{S}_f$, is the set of inputs on which the function is always nonzero, i.e., $\mathcal{S}_f = \{\mathbf{x} \in \mathbb{F}_2^n : f(\mathbf{x}) = 1\}$. The weight of a Boolean function is the cardinality of its support set, i.e., $wt(f) = |\mathcal{S}_f|$. If the weight of an n-variable Boolean function f is 2^{n-1}, then f is called balanced.

The Walsh–Hadamard transform of an n-variable Boolean function f at $\mathbf{a} \in \mathbb{F}_2^n$, denoted by $\mathcal{W}_f(\mathbf{a})$, is expressed as

$$\mathcal{W}_f(\mathbf{a}) = \sum_{\mathbf{x} \in \mathbb{F}_2^n} (-1)^{f(\mathbf{x}) \oplus \mathbf{a} \cdot \mathbf{x}}$$

The multiset $[\mathcal{W}_f(\mathbf{a}) : \mathbf{a} \in \mathbb{F}_2^n]$ is called Walsh–Hadamard spectrum of f. The distance between two n-variable Boolean functions f and g is defined as

$$d(f, g) = |\{\mathbf{x} \in \mathbb{F}_2^n : f(\mathbf{x}) \neq g(\mathbf{x})\}|.$$

The minimum distance of $f \in \mathcal{B}_n$ from the set of all n-variable affine functions is called nonlinearity of f, denoted by $nl(f)$. The relation between nonlinearity and Walsh–Hadamard transform of a Boolean function $f \in \mathcal{B}_n$ is

$$nl(f) = 2^{n-1} - \frac{1}{2} \max_{\mathbf{a} \in \mathbb{F}_2^n} |\mathcal{W}_f(\mathbf{a})|.$$

It is known that nonlinearity of any $f \in \mathcal{B}_n$ is upper bounded by $2^{n-1} - 2^{\frac{n}{2}-1}$ as $\max_{\mathbf{a} \in \mathbb{F}_2^n} |\mathcal{W}_f(\mathbf{a})| \geq 2^{\frac{n}{2}}$ for any Boolean function f in n variables. A Boolean function that achieves this bound is called a bent function. Bent functions are defined for even number of variables and not balanced as $wt(f) = \mathcal{W}_f(\mathbf{0}) \neq 0$. A Boolean function $f \in \mathcal{B}_n$ is said to be correlation immune of order t, $1 \leq t \leq n$, if its values are statistically independent of any subset of t input variables. A function is called t-resilient if it is balanced and correlation immune of order t. In the other word, a function $f \in \mathcal{B}_n$ is called t-resilient if $\mathcal{W}_f(\mathbf{a}) = 0$, for all $\mathbf{a} \in \mathbb{F}_2^n$ with $0 \leq wt(\mathbf{a}) \leq t$.

Let $\mathbb{F}_2^{n\times r}$ be the set of all binary matrices of order $n \times r$. The support matrix of a Boolean function f in n variables with weight r is enumeration of all $\mathbf{x} \in \mathbb{F}_2^n$ from the support set of f in some order (we will consider the lexicographical order from left to right), denoted as $\mathcal{M}_f \in \mathbb{F}_2^{n\times r}$, i.e., a binary matrix of order $n \times r$, where column vectors are the elements of its support set. It is clear that any Boolean function in n variables can be written as an element of $\mathbb{F}_2^{n\times r}$, where $wt(f) = r$. For example let f be an 4-variable Boolean function with support set $\mathcal{S}_f = \{(0,0,1,1),(0,1,1,1),(1,0,1,1),(1,1,0,0),(1,1,0,1),(1,1,1,0)\}$. The support matrix of f, $\mathcal{M}_f \in \mathbb{F}_2^{4\times 6}$, is

$$\mathcal{M}_f = \begin{pmatrix} 0\,0\,1\,1\,1\,1 \\ 0\,1\,0\,1\,1\,1 \\ 1\,1\,1\,0\,0\,1 \\ 1\,1\,1\,0\,1\,0 \end{pmatrix}.$$

with a slight abuse of notation, we denote enumeration of members $\mathbf{x} \in \mathbb{F}_2^n$ from some arbitrary set S with cardinality r in some order as $\mathcal{M}_S \in \mathbb{F}_2^{n\times r}$. Suppose $S = \{(0,0,0,1),(0,1,0,0),(0,1,0,1),(0,1,1,0),(1,0,0,0),(1,1,0,0),(1,1,1,1)\}$ $\subseteq \mathbb{F}_2^4$. The binary matrix corresponding to S, $\mathcal{M}_S \in \mathbb{F}_2^{4\times 7}$, is

$$\mathcal{M}_S = \begin{pmatrix} 0\,0\,0\,0\,1\,1\,1 \\ 0\,1\,1\,1\,0\,1\,1 \\ 0\,0\,0\,1\,0\,0\,1 \\ 1\,0\,1\,0\,0\,0\,1 \end{pmatrix}.$$

A binary linear code [11] of length n and dimension t is a subspace of $\mathbb{F}_2^n$ with dimension t. An element of linear code is called codeword. The weight of a codeword is the number of its coordinates that are nonzero and the distance d of the linear code is the minimum weight of its nonzero codewords (equivalently, the minimum distance between distinct codewords). A binary linear code of length n, dimension t, and distance d is said to be an $[n,t,d]$ code. A generator matrix G of a binary linear code $\mathcal{L} = [n,t,d]$ is an $t \times n$ matrix over $\mathbb{F}_2$ such that its rows a set of basis vectors of $\mathcal{L}$. We now define a subclass of binary linear codes, called Bounded Linear Code, where weights of codewords are bounded from both sides.

Definition 1. *We call a subclass of binary linear code $[n,t,d]$ is a Bounded Linear Code with parameter (n,t,d_{min},d_{max}) if the weight of any nonzero codeword $\mathbf{c}$ is bounded with $d_{min} \leq wt(\mathbf{c}) \leq d_{max}$ and there exist at least one codeword $\mathbf{c}'$ such that either $wt(\mathbf{c}') = d_{min}$ or $wt(\mathbf{c}') = d_{max}$.*

3 Nonlinearity of Balanced Boolean Functions: A Combinatorial Characterization

In this section, we provide a necessary and sufficient characterization for nonlinearity of a Boolean function, in terms of distance of a special kind of linear code, whose generator matrix is the support matrix of the concerned Boolean function. Then we use this characterization to derive more specific conditions towards the existence of balanced Boolean functions with nonlinearity exceeding Dobbertin's bound. Such characterization is fairly robust, in the sense that the similar characterization is possible for other related problems, e.g., the maximum nonlinearity of Boolean functions in odd number of variables. Towards establishing the results, we first observe that every nonzero $\mathbf{a} \in \mathbb{F}_2^n$ defines a partition over the support set of a given Boolean function. Such partition was first defined in [16], and we follow their exposition with minor modifications. Let f be an n-variable Boolean function and $\mathbf{a} \in \mathbb{F}_2^n$. We define two sets as

$$\begin{aligned} \mathcal{S}_f^0(\mathbf{a}) &= \{\mathbf{x} \in \mathcal{S}_f : \mathbf{a} \cdot \mathbf{x} = 0\}, \\ \mathcal{S}_f^1(\mathbf{a}) &= \{\mathbf{x} \in \mathcal{S}_f : \mathbf{a} \cdot \mathbf{x} = 1\}. \end{aligned} \tag{1}$$

By definition, they are mutually disjoint, in particular, $\mathcal{S}_f^0(\mathbf{a}) = \mathcal{S}_f \setminus \mathcal{S}_f^1(\mathbf{a})$, and $\mathcal{S}_f^0(\mathbf{0}) = \mathcal{S}_f$. Now, if $f \in \mathcal{B}_n$ is balanced, then $|\mathcal{S}_f^0(\mathbf{0})| = |\mathcal{S}_f| = 2^{n-1}$.

At this point, we are ready to describe the key insights used in our main result. First observe that if the support matrix of a Boolean function f is considered as generator G of some linear code $\mathcal{L}$, then the cardinality of the set $\mathcal{S}_f^1(\mathbf{a})$ is exactly equal to weight of the codeword corresponding to $\mathbf{a}$, e.g., $|\mathcal{S}_f^1(\mathbf{a})| = wt(\mathbf{a}G)$. On the other hand, the set $\mathcal{S}_f^1(\mathbf{a})$ is basically the intersection between support set of two functions, f and the linear function $\mathbf{a} \cdot \mathbf{x}$. The cardinality of the set $\mathcal{S}_f^1(\mathbf{a})$ is then kind of correlation measure between those two functions. In this way, one can the derive suitable expression relating $|\mathcal{S}_f^1(\mathbf{a})|$ and $\mathcal{W}_f(\mathbf{a})$, Walsh–Hadamard transform of f at $\mathbf{a}$. Here comes the tricky part of the proof. For Boolean functions with high nonlinearity, we expect absolute values of elements of Walsh–Hadamard spectrum of f to be small, and that essentially puts a restriction on the weight of codewords of linear code $\mathcal{L}$. More specifically, we show that for highly nonlinear functions, support matrix of the function, as the generator of some linear code produces codewords with weight, lying within a very short interval around its mean. We use this result to derive specific parameters of such codes for a function with given nonlinearity.

We will present our main result in Theorem 1. In this regard, we first prove two ingredients for the proof. As a first step, towards proving aforesaid result, we derive relationship between cardinality of $\mathcal{S}_f^1(\mathbf{a})$ and Walsh–Hadamard transformation of f at $\mathbf{a} \in \mathbb{F}_2^n \setminus \{0\}$.

$$
\begin{aligned}
\mathcal{W}_f(\mathbf{a}) &= \sum_{\mathbf{x}\in\mathbb{F}_2^n} (-1)^{f(\mathbf{x})\oplus\mathbf{a}\cdot\mathbf{x}} = \sum_{\mathbf{x}\in\mathbb{F}_2^n} (1-2f(\mathbf{x}))(-1)^{\mathbf{a}\cdot\mathbf{x}} \\
&= -2\sum_{\mathbf{x}\in\mathcal{S}_f} (-1)^{\mathbf{a}\cdot\mathbf{x}} = -2\sum_{\mathbf{x}\in\mathcal{S}_f} (1-2\mathbf{a}\cdot\mathbf{x}) \\
&= -2|\mathcal{S}_f| + 4\sum_{\mathbf{x}\in\mathcal{S}_f} \mathbf{a}\cdot\mathbf{x} = 4|\mathcal{S}_f^1(\mathbf{a})| - 2|\mathcal{S}_f|.
\end{aligned} \tag{2}
$$

If $f \in \mathcal{B}_n$ is balanced, Walsh–Hadamard transformation of f at $\mathbf{a}$ evaluates to $4|\mathcal{S}_f^1(\mathbf{a})| - 2^n$ for all nonzero $\mathbf{a} \in \mathbb{F}_2^n$. So, for any balanced function $f \in \mathcal{B}_n$, $n \geq 2$, $\mathcal{W}_f(\mathbf{a}) \equiv 0 \pmod 4$, for all $\mathbf{x} \in \mathbb{F}_2^n$. Suppose the nonlinearity of f is $nl(f) \geq r$. Then

$$
2^{n-1} - \frac{1}{2}\max_{\mathbf{a}\in\mathbb{F}_2^n} |\mathcal{W}_f(\mathbf{a})| \geq r \;\Leftrightarrow\; \max_{\mathbf{a}\in\mathbb{F}_2^n} |\mathcal{W}_f(\mathbf{a})| \leq 2^n - 2r.
$$

Hence, for all $\mathbf{a} \in \mathbb{F}_2^n \setminus \{0\}$

$$
\begin{aligned}
& -2^n + 2r \leq \mathcal{W}_f(\mathbf{a}) \leq 2^n - 2r \\
\Rightarrow\; & -2^n + 2r \leq 4|\mathcal{S}_f^1(\mathbf{a})| - 2^n \leq 2^n - 2r \\
\Rightarrow\; & 2r \leq 4|\mathcal{S}_f^1(\mathbf{a})| \leq 2^{n+1} - 2r \\
\Rightarrow\; & \frac{r}{2} \leq |\mathcal{S}_f^1(\mathbf{a})| \leq \frac{2^n - r}{2}.
\end{aligned}
$$

We formalize above result in following lemma.

Lemma 1. *Let f be a balanced Boolean function in n variables. The nonlinearity of f is $nl(f) \geq r$ if and only if $\frac{r}{2} \leq |\mathcal{S}_f^1(\mathbf{a})| \leq \frac{2^n-r}{2}$, for all nonzero $\mathbf{a} \in \mathbb{F}_2^n$.*

Proof. The "if" direction has been already shown. For the other direction, let r be a positive integer and $f \in \mathcal{B}_n$ is a balanced function such that $\frac{r}{2} \leq |\mathcal{S}_f^1(\mathbf{a})| \leq \frac{2^n-r}{2}$ for all nonzero $\mathbf{a} \in \mathbb{F}_2^n$. Then $\max_{\mathbf{a}\in\mathbb{F}_2^n} |\mathcal{W}_f(\mathbf{a})| \leq 2^n - 2r$, and so, $nl(f) \geq 2^{n-1} - \frac{1}{2}(2^n - 2r)$, i.e., $nl(f) \geq r$. □

In the above result, if there exist an element $\mathbf{a} \in \mathbb{F}_2^n$ such that $|\mathcal{S}_f^1(\mathbf{a})| = \frac{r}{2}$ or $\frac{2^n-r}{2}$, then $\max_{\mathbf{a}\in\mathbb{F}_2^n} |\mathcal{W}_f(\mathbf{a})| = 2^n - 2r$, and so, $nl(f) = r$. Now, we move to deriving relation between distance of the linear code as generated by support matrix of f and cardinality of the set $\mathcal{S}_f^1(\mathbf{a})$ as discussed earlier. Towards that, we first observe that, if all the elements of the support set of f is enumerated in some order and then considered as a binary matrix $\mathcal{M}_f$, following facts are obvious.

- Dimension of the $\mathcal{M}_f$ is $n \times 2^{n-1}$.
- Cardinality of the set $\mathcal{S}_f^1(\mathbf{a})$ is exactly equal to weight of matrix vector product $\mathbf{a}\mathcal{M}_f$, e.g., $|\mathcal{S}_f^1(\mathbf{a})| = wt(\mathbf{a}\mathcal{M}_f)$.

If one considers $\mathcal{M}_f$ as generator of some linear code $\mathcal{L}$, then above properties of $\mathcal{M}_f$ characterizes the linear code to a great extent. We formally put the above observation below.

Lemma 2. *The support matrix $\mathcal{M}_f$ of a balanced function $f \in \mathcal{B}_n$, when considered as a generator of some linear code, generates codewords in the range space $\mathbb{F}_2^{2^{n-1}}$ and weight of codeword corresponding to $\mathbf{a} \in \mathbb{F}_2^n \setminus \{\mathbf{0}\}$, is $|\mathcal{S}_f^1(\mathbf{a})| = wt(\mathbf{a}\mathcal{M}_f)$.*

Now we present our key result below.

Theorem 1. *The support matrix $\mathcal{M}_f$ of a balanced function f with nonlinearity r can be considered as a generator of the Bounded Linear Code with parameter $(2^{n-1}, n, \frac{r}{2}, \frac{2^n-r}{2})$ with all distinct columns.*

Proof. Consider any balanced Boolean function with nonlinearity r. Lemma 1 dictates that $\frac{r}{2} \leq |\mathcal{S}_f^1(\mathbf{a})| \leq \frac{2^n-r}{2}$, for all nonzero $\mathbf{a} \in \mathbb{F}_2^n$. On the other hand, when support matrix of f viewed as the generator matrix of some linear code, Lemma 2 tells that, weight of the codeword corresponding to $\mathbf{a} \in \mathbb{F}_2^n$, is $|\mathcal{S}_f^1(\mathbf{a})|$. These two result together directly implies Theorem 1. □

3.1 Nonlinearity Strictly Greater Than $2^{n-1} - 2^{\frac{n}{2}} + nlb(\frac{n}{2})$

Direct application of Lemma 1 immediately imposes the necessary and sufficient conditions on $|\mathcal{S}_f^1(\mathbf{a})|$ for a balanced function $f \in \mathcal{B}_n$ that have nonlinearity strictly greater than $2^{n-1} - 2^{\frac{n}{2}} + nlb(\frac{n}{2})$, where $nlb(m)$ is the maximum nonlinearity among all balanced Boolean functions in m variables.

Corollary 1. *Let f be a balanced Boolean function in n variables. The nonlinearity of f is $nl(f) > 2^{n-1} - 2^{\frac{n}{2}} + nlb(\frac{n}{2})$ if and only if for any nonzero $\mathbf{a} \in \mathbb{F}_2^n$,*

$$2^{n-2} - 2^{\frac{n}{2}-1} + \frac{nlb(\frac{n}{2})}{2} < |\mathcal{S}_f^1(\mathbf{a})| < 2^{n-2} + 2^{\frac{n}{2}-1} - \frac{nlb(\frac{n}{2})}{2}.$$

From the above result we get the following relation between such balanced functions and linear codes with desired parameters.

Corollary 2. *The nonlinearity of a balanced function $f \in \mathcal{B}_n$ is $nl(f) > 2^{n-1} - 2^{\frac{n}{2}} + nlb(\frac{n}{2})$ if and only if there exists Bounded Linear Code $\mathcal{L}$ with parameters $(n, 2^{n-1}, 2^{n-2} - 2^{\frac{n}{2}-1} + \frac{nlb(\frac{n}{2})}{2}, 2^{n-2} + 2^{\frac{n}{2}-1} - \frac{nlb(\frac{n}{2})}{2})$, with all columns distinct.*

Above result shifts the question of existence of balanced Boolean function having nonlinearity exceeding Dobbertin's bound into the question of existence of some linear codes with desired parameters.

3.2 Deriving Specific Conditions for $n = 8, 10, 12$ and 14

One must know $nlb(\frac{n}{2})$ to derive exact parameters for our desired linear code, for all n. Because maximum nonlinearity of balanced Boolean function in 4, 5, 6 and 7 variables are known, we now proceed to derive exact parameters for required linear code for $n = 8, 10, 12$ and 14. At first we calculate the parameters for $n = 8$ in detail, for rest of the case, we skip the calculation and present the end results in a Table 1.

Table 1. Conditions to obtain highly nonlinear balanced functions for $n = 8, 10, 12$ and 14

n	$2^{n-1} - 2^{\frac{n}{2}-1}$	$nlb(\frac{n}{2})$	$2^{n-1} - 2^{\frac{n}{2}} + nlb(\frac{n}{2})$	$nl(f) > 2^{n-1} - 2^{\frac{n}{2}} + nlb(\frac{n}{2})$	$\frac{nl(f)}{2} \leq \lvert\mathcal{S}_f^1(\mathbf{a})\rvert \leq \frac{2^n - nl(f)}{2}$	$(2^{n-1}, n, \frac{nl(f)}{2}, \frac{2^n - nl(f)}{2})$
8	120	4	116	118	$59 \leq \lvert\mathcal{S}_f^1(\mathbf{a})\rvert \leq 69$	(128, 8, 59, 69)
10	496	12	492	494	$247 \leq \lvert\mathcal{S}_f^1(\mathbf{a})\rvert \leq 265$	(512, 10, 247, 265)
12	2016	26	2010	2012	$1006 \leq \lvert\mathcal{S}_f^1(\mathbf{a})\rvert \leq 1042$	(2048, 12, 1006, 1042)
				2014	$1007 \leq \lvert\mathcal{S}_f^1(\mathbf{a})\rvert \leq 1041$	(2048, 12, 1007, 1041)
14	8128	56	8120	8122	$4061 \leq \lvert\mathcal{S}_f^1(\mathbf{a})\rvert \leq 4131$	(8192, 14, 4061, 4131)
				8124	$4062 \leq \lvert\mathcal{S}_f^1(\mathbf{a})\rvert \leq 4130$	(8192, 14, 4062, 4130)
				8126	$4063 \leq \lvert\mathcal{S}_f^1(\mathbf{a})\rvert \leq 4129$	(8192, 14, 4063, 4129)

For $n = 8$ and $nl(f) = 118$: Since $nlb(4) = 4$. The maximum known nonlinearity of balanced functions in 8 variables is 116. The (non)-existence of a balanced function in 8 variables having nonlinearity 118 still an open question. We now derive the following characterization of such balanced Boolean functions.

Corollary 3. *Let f be a balanced Boolean function in 8 variables. The nonlinearity of f is 118 if and only if $59 \leq |\mathcal{S}_f^1(\mathbf{a})| \leq 69$ for all nonzero $\mathbf{a} \in \mathbb{F}_2^8$, where $\mathcal{S}_f^1(\mathbf{a})$ is defined in* (1).

Proof. Let $f \in \mathcal{B}_8$ be balanced. From Lemma 1, we have $nl(f) = 118$ if and only if $59 \leq |\mathcal{S}_f^1(\mathbf{a})| \leq 69$, for all nonzero $\mathbf{a} \in \mathbb{F}_2^8$. □

Corollary 4. *For each balanced function in 8 variables having nonlinearity 118, there exist a Bounded Linear Code $\mathcal{L}$ with parameters (128, 8, 59, 69) with all distinct columns, and converse is also true.*

We present the similar results in Table 1 for $n = 10, 12$ and 14.

4 Comparison with Existing Results [14,20]

After Dobbertin [6] proposed his conjecture, many attempts were made towards proving or disproving his conjecture. Among them, three attempts [13,14,20] requires special mention. All of them considered the problem for $n = 8$ and proposed characterization of functions with nonlinearity exceeding Dobbertin's bound for $n = 8$. We present their characterizations below and show that all of them can be subsumed by our characterization.

4.1 On Characterization by Maity and Maitra [14]

In [14, Theorem 6], Maity et al. proposed the following construction of balanced function in 8 variables having nonlinearity 118.

Theorem 2. [14, Theorem 6] *Let $f \in \mathcal{B}_8$ be bent such that*

i. $f(\mathbf{x}) = 0$ *for* $wt(\mathbf{x}) \leq 1$, $f(\mathbf{1}) = 1$,
ii. $\mathcal{W}_f(\mathbf{a}) = 16$ *for* $wt(\mathbf{a}) \leq 2$ *and* $\mathcal{W}_f(\mathbf{a}) = -16$ *for* $wt(\mathbf{a}) \geq 6$.

Let $T = \{\mathbf{x} \in \mathbb{F}_2^8 : wt(\mathbf{x}) = 1\}$ *and* $g \in \mathcal{B}_8$ *such that*

$$g(\mathbf{x}) = \begin{cases} f(\mathbf{x}) \oplus 1, & \text{if } \mathbf{x} \in T \\ f(\mathbf{x}), & \text{otherwise} \end{cases},$$

Then g will be balanced with nonlinearity 118.

Here the set T is fixed and then they tried to identify a bent function in 8 variables as mentioned in Theorem [14, Theorem 6], but could not get any such bent function. We see that aforesaid characterization is a special case of our characterization. We first derive a sufficient characterization in our setup and then show that their characterization fall as a special case of ours.

Construction 1. *Let f be a bent function in 8 variables with cardinality* 120 *and* $T \subset \mathbb{F}_2^8 \setminus \mathcal{S}_f$ *such that* $|T| = 8$ *and*

i. if $|\mathcal{S}_T^1(\mathbf{a})| = |\{\mathbf{x} \in T : \mathbf{a} \cdot \mathbf{x} = 1\}| \leq 2$, *then* $\mathcal{W}_f(\mathbf{a}) = 16$,
ii. if $|\mathcal{S}_T^1(\mathbf{a})| = |\{\mathbf{x} \in T : \mathbf{a} \cdot \mathbf{x} = 1\}| \geq 6$, *then* $\mathcal{W}_f(\mathbf{a}) = -16$.

Let us define a Boolean function g in 8 *variables such that*

$$g(\mathbf{x}) = \begin{cases} f(\mathbf{x}) \oplus 1, & \text{if } \mathbf{x} \in T \\ f(\mathbf{x}), & \text{otherwise} \end{cases}.$$

Then g is balanced and its nonlinearity 118.

Proof. It is clear that $\mathcal{W}_g(\mathbf{0}) = 0$, and from Corollary 3, $59 \leq |\mathcal{S}_g^1(\mathbf{a})| \leq 69$, for all nonzero $\mathbf{a} \in \mathbb{F}_2^8$. Then the nonlinearity of g is 118. □

In the above Construction, if $T = \{\mathbf{x} \in \mathbb{F}_2^8 : wt(\mathbf{x}) = 1\}$, then the results given in [14, Theorem 6] directly follows as that case $|\mathcal{S}_T^1(\mathbf{a})| \leq 2$ for all $wt(\mathbf{a}) \leq 2$, and $|\mathcal{S}_T^1(\mathbf{a})| \geq 6$ for all $wt(\mathbf{a}) \geq 6$. That is, in the above construction, we consider the all possible sets T with the initial bent function instead of fixing one. The initial bent function may not have the Walsh–Hadamard transformation values 16 at $\mathbf{a} \in \mathbb{F}_2^8$ when $wt(\mathbf{a}) \leq 2$ and -16 when $wt(\mathbf{a}) \geq 6$. Here, we extend the search domain both by the choice of T and the initial bent function towards constructing a balanced one in 8 variables with nonlinearity 118. Note that we are yet to exactly obtain an example for such functions, and thus the question of 8-variable balanced Boolean function with nonlinearity 118 still remains unsolved. However, our explanation provides a more general class where such functions can be explored.

It is also worth mentioning at this point that, Maity and Johansson [13, Theorem 3] and Maity and Maitra [14, Theorem 4] constructed 1-resilient Boolean functions in 8 variables with nonlinearity 116 by changing 10 outputs of a class of bent functions. In [14], Maity et al. followed the similar kind of technique that is used in [13], and in addition they improved the previous one that is proposed by Maity et al. [13]. Rationale behind their proposed construction can be seen more clearly in our framework. More specifically, they [14, Theorem 4] considered the bent functions as described below.

Theorem 3. [14, Theorem 4] *Let $f \in \mathcal{B}_8$ be a bent function such that*

i. $f(\mathbf{x}) = 0$ *for* $wt(\mathbf{x}) \leq 1$, $f(\mathbf{1}) = 1$,
ii. $\mathcal{W}_f(\mathbf{x}) = 16$ *for* $wt(\mathbf{x}) \leq 1$ *and* $\mathcal{W}_f(\mathbf{1}) = -16$.

Suppose $V = \{\mathbf{x} \in \mathbb{F}_2^8 : wt(\mathbf{x}) = 0, 1, 8\}$. *Let us define a Boolean function* $g \in \mathcal{B}_8$ *such that*

$$g(\mathbf{x}) = \begin{cases} f(\mathbf{x}) \oplus 1, & \text{if } \mathbf{x} \in V \\ f(\mathbf{x}), & \text{otherwise} \end{cases}.$$

Then g is 1*-resilient with nonlinearity* 116.

In the same way, as done earlier, we first derive a sufficient characterization in our framework for 1-resilient Boolean functions in 8 variables with nonlinearity 116, as follows.

Proposition 1. *Let g be a balanced function in* 8 *variables. If g is* 1*-resilient with nonlinearity* 116 *if and only if*

i. $|\mathcal{S}_g^1(\mathbf{a})| = 64$, *for all* $\mathbf{a} \in \mathbb{F}_2^8$ *with* $wt(\mathbf{a}) = 1$, *and*
ii. $58 \leq |\mathcal{S}_g^1(\mathbf{a})| \leq 70$, *for all nonzero* $\mathbf{a} \in \mathbb{F}_2^8$, *there exists at least one* $\mathbf{a} \in \mathbb{F}_2^8$ *such that* $|\mathcal{S}_g^1(\mathbf{a})| = 58$ *or* 70.

Proof. Since g is balanced and $\mathcal{W}_g(\mathbf{a}) = 4|\mathcal{S}_f^1(\mathbf{a})| - 2|\mathcal{S}_f|$, for all nonzero $\mathbf{a} \in \mathbb{F}_2^8$. So, $\mathcal{W}_g(\mathbf{0}) = 0$, and for any nonzero $\mathbf{a} \in \mathbb{F}_2^8$, $\mathcal{W}_g(\mathbf{a}) = 0$ if and only if $|\mathcal{S}_f^1(\mathbf{a})| = 64$. The second part is directly follow from Lemma 1. □

It is clear that the 1-resilient functions constructed in [13,14] satisfy the conditions given in Proposition 1. Now we show that there are many possible ways to construct such functions using our Proposition 1, other than specified in [13,14]. Here is one such example.

Construction 2. *Let* $S = \{(0,0,0,0,0,0,0,0), (1,0,0,0,0,0,0,0), (1,1,0,0,0,0,0,0), (1,0,1,0,0,0,0,0), (1,0,0,1,0,0,0,0), (1,0,0,0,1,0,0,0), (1,0,0,0,0,1,0,0), (1,0,0,0,0,0,1,0), (1,0,0,0,0,0,0,1)\}$ *and* $f \in \mathcal{B}_n$ *be a bent function such that*

i. $f(\mathbf{x}) = 0$ *for all* $\mathbf{x} \in S$ *and* $f(0,1,1,1,1,1,1,1) = 1$,
ii. $\mathcal{W}_f(\mathbf{a}) = 16$ *for* $\mathbf{x} \in \mathbb{F}_2^8$ *such that* $wt(\mathbf{x}) \leq 1$ *except* $\mathbf{u} = (1,0,0,0,0,0,0,0)$, $\mathcal{W}_f(\mathbf{u}) = -16$ *and* $\mathcal{W}_f(\mathbf{1}) = 16$.

Define a Boolean function g as

$$g(\mathbf{x}) = \begin{cases} f(\mathbf{x}) \oplus 1, \text{ if } \mathbf{x} \in S \cup \{(0,1,1,1,1,1,1,1)\} \\ f(\mathbf{x}), \quad \text{otherwise} \end{cases}.$$

Then g is 1*-resilient with nonlinearity greater than or equal to* 116*. In particular, if* $\mathcal{W}_f(\mathbf{a}, 1) = 16$*, where* $\mathbf{a} = (1,1,0,0,0,0,0,0)$*, then* $nl(g) = 116$*.*

Proof. It is clear that g is balanced. Given that $\mathcal{W}_f(\mathbf{a}) = 16$ for $\mathbf{x} \in \mathbb{F}_2^8$ such that $wt(\mathbf{x}) \leq 1$ except $\mathbf{u} = (1,0,0,0,0,0,0,0)$ and $\mathcal{W}_f(\mathbf{u}) = -16$, then $|\mathcal{S}_g^1(\mathbf{a})| = 64$ for all $\mathbf{x} \in \mathbb{F}_2^8$ with $wt(\mathbf{x}) \leq 1$. So, f is 1-resilient. For nonlinearity, from Proposition 1, it is sufficient to prove that $58 \leq |\mathcal{S}_g^1(\mathbf{a})| \leq 70$, for all nonzero $\mathbf{a} \in \mathbb{F}_2^8$. If $wt(\mathbf{a}) \in \{2,3,\ldots,7\}$, then $58 \leq |\mathcal{S}_g^1(\mathbf{a})| \leq 70$ as the modified weights lies between 2 and 6. For $\mathbf{a} = \mathbf{1}$, $|\mathcal{S}_g^1(\mathbf{a})| = 64$, so $nl(f) \geq 116$. If $\mathcal{W}_f(\mathbf{a}, 1) = 16$, where $\mathbf{a} = (1,1,0,0,0,0,0,0)$, then $|\mathcal{S}_g^1(\mathbf{a})| = 70$, and so, $nl(g) = 116$. □

The above construction is based on the choices of an element $\mathbf{a} = (1,1,0,0,0,0,0,0)$ such that $\mathcal{W}_f(\mathbf{a}, 1) = 16$ need not be unique. One can identify other possible choices so that $nl(g) = 116$. For example, let $\mathbf{a} = (1,0,1,0,0,0,0,0)$. If $\mathcal{W}_f(\mathbf{a}, 1) = 16$, then $|\mathcal{S}_g^1(\mathbf{a})| = 70$, and so, $nl(g) = 116$. Now we give an example of our construction.

Example 1. Let $\pi(\mathbf{y}) = \mathbf{y}$, for all $\mathbf{y} \in \mathbb{F}_2^4 \setminus \{(1,0,0,0),(0,1,1,0)\}$, $\pi(1,0,0,0) = (0,1,1,0)$ and $\pi(0,1,1,0) = (1,0,0,0)$. Define $f(\mathbf{x}, \mathbf{y}) = \mathbf{x} \cdot \pi(\mathbf{y}) \oplus h(\mathbf{y})$, for all $\mathbf{x}, \mathbf{y} \in \mathbb{F}_2^4$, where $h \in \mathcal{B}_4$ such that $h(\mathbf{y}) = 0$, for all $\mathbf{y} \in \mathbb{F}_2^4$ except $(0,1,1,0)$, i.e., $h(0,1,1,0) = 1$. Let us define a Boolean function g in 8 variables such that

$$g(\mathbf{x}) = \begin{cases} f(\mathbf{x}) \oplus 1, \text{ if } \mathbf{x} \in S \cup \{((0,1,1,1),(1,1,1,1))\} \\ f(\mathbf{x}), \quad \text{otherwise} \end{cases},$$

where S is defined in Construction 2. Then g is a 1 resilient function having nonlinearity 116.

Proof. Since π is permutation, f is a bent function. We can check that $f(\mathbf{x}, \mathbf{y}) = 0$, for all $(\mathbf{x}, \mathbf{y}) \in S$, and $f((0,1,1,1),(1,1,1,1)) = 1$. Then g is balanced. It is clear that $\mathcal{W}_f(\mathbf{a}, \mathbf{b}) = 16$, for all $(\mathbf{a}, \mathbf{b}) \in \mathbb{F}_2^4 \times \mathbb{F}_2^4$ such that $wt(\mathbf{a}, \mathbf{b}) \leq 1$ except $(\mathbf{u}, \mathbf{v}) = ((1,0,0,0),(0,0,0,0))$, i.e., $\mathcal{W}_f(\mathbf{u}, \mathbf{v}) = -16$, and $\mathcal{W}_f(\mathbf{1}, \mathbf{1}) = 16$. Also $\mathcal{W}_f((1,1,0,0),(0,0,0,0)) = 16$, so from Construction 2 the nonlinearity of g is 116. □

The bent function f in 8 variables given in the above example belongs to Maiorana–McFarland class. Here f can be written as concatenation of 16 linear functions of the form

$$f = l_0||l_1||l_2||l_3||l_4||l_5||\overline{l}_8||l_7||l_6||l_9||l_{10}||l_{11}||l_{12}||l_{13}||l_{14}||l_{15},$$

where $l_i = \mathbf{x} \cdot \mathbf{y}_i \oplus h(\mathbf{y}_i)$, $\mathbf{y}_i$ is the binary representation of integer i, $0 \leq i \leq 15$, and $\overline{l}_j = l_j \oplus \mathbf{1}$, the complement of l_j. In Example 1, we modify 10 points of f

belonging to $S \cup \{((0,1,1,1),(1,1,1,1))\}$, where S is defined in Construction 2, and construct a 1-resilient function having nonlinearity 116.

To explain the functions in more details, the 2^4 many distinct 4-variable linear functions are concatenated concatenated to obtain the bent functions. Then , the functions l_8 and l_6 are swapped in their places and l_8 is complemented. Thus in the truth table, we obtain the bent function as

00005555333366660F0F5A5AFF0069693C3C55AA33CC66990FF05AA53CC36996.

Next we toggle the outputs corresponding to the ten input points as in Construction 2. The points are as $(\mathbf{x},\mathbf{y})$, for example $\mathbf{x}=(0,1,1,1), \mathbf{y}=(1,1,1,1)$. Now $\mathbf{y}=(1,1,1,1)$ will decide the linear function l_{15} and $\mathbf{x}=(0,1,1,1)$ will decide the 7-th point. That is in the truth table, the 247-th point will be toggled. In this manner, the ten points (0, 0, 0, 0, 0, 0, 0, 0), (1, 0, 0, 0, 0, 0, 0, 0), (1, 1, 0, 0, 0, 0, 0, 0), (1, 0, 1, 0, 0, 0, 0, 0), (1, 0, 0, 1, 0, 0, 0, 0), (1, 0, 0, 0, 1, 0, 0, 0), (1, 0, 0, 0, 0, 1, 0, 0), (1, 0, 0, 0, 0, 0, 1, 0), (1, 0, 0, 0, 0, 0, 0, 1), (0,1,1,1,1,1,1,1) will be identified as the decimal points 0, 8, 12, 10, 9, 136, 72, 40, 24, 247 respectively and toggling those in the bent function, we obtain 8-variable 1-resilient function with nonlinearity 116 as

80E855D533B366660F8F5A5AFF0069693CBC55AA33CC66990FF05AA53CC36896.

We would like to reiterate the following points here.

- We already observed that [14, Theorem 4] directly follows from Proposition 1. For any nonzero $\mathbf{a} \in \mathbb{F}_2^8$, $\mathcal{W}_f(\mathbf{a}) = 16$ implies $|\mathcal{S}_f^1(\mathbf{a})| = 64$ and $\mathcal{W}_f(\mathbf{a}) = -16$ implies $|\mathcal{S}_f^1(\mathbf{a})| = 56$. If $V' = \{\mathbf{x} \in \mathbb{F}_2^8 : wt(\mathbf{x}) = 1\}$, then $|\mathcal{S}_{V'}^1(\mathbf{a})| = wt(\mathbf{a})$, for all nonzero $\mathbf{a} \in \mathbb{F}_2^8$. Thus, if one can construct a balanced function g from a bent function f in 8 variables, defined in [14, Theorem 4] using Proposition 1, then g is 1-resilient having nonlinearity 116.
- There are other bent functions which are not used in [14, Theorem 4] to obtain 1-resilient Boolean functions in 8 variables having nonlinearity 116. In our method, such bent functions can be used to construct 1-resilient Boolean functions in 8 variables having nonlinearity 116 from Proposition 1. For example, the bent functions that are used in Construction 2 to construct 1-resilient functions with nonlinearity 116 in 8 variables cannot be used to provide such functions through [14, Theorem 4], as for any bent function f in these class have property that $\mathcal{W}_f(1,0,0,0,0,0,0,0) = -16$. We prove that such bent functions exist in Example 1.

4.2 On Characterization by Sarkar Et Al. [20]

In [20, Theorem 7], Sarkar et al. proved that if $f \in \mathcal{B}_8$ is balanced with nonlinearity 118, then degree of f must be 7 and it can be written as concatenation of two functions in 7 variables as $f(\mathbf{x}) = (1 \oplus x_8)f_1(\mathbf{x}') \oplus x_8 f_2(\mathbf{x}')$, for all $\mathbf{x} = (\mathbf{x}', x_8) \in \mathbb{F}_2^7 \times \mathbb{F}_2$, where f_1 and f_2 are 7-variable Boolean functions each having nonlinearity 55 and degree 7.
We view their characterization in our framework and show that this is a sufficient condition.

Corollary 5. *Let $f = f_1||f_2$, concatenation of two functions f_1 and f_2, be a balanced function in 8 variables having nonlinearity 118. Then $|\mathcal{S}_{f_1}|+|\mathcal{S}_{f_2}| = 128$, and following are true.*

i. $55 \leq |\mathcal{S}_{f_i}| \leq 73$, *with* $|\mathcal{S}_{f_i}| \neq 64$, $i = 1, 2$
ii. For any nonzero $\mathbf{a} \in \mathbb{F}_2^7$, $\frac{|\mathcal{S}_{f_i}|-9}{2} \leq |\mathcal{S}_{f_i}^1(\mathbf{a})| \leq \frac{|\mathcal{S}_{f_i}|+9}{2}$, $i = 1, 2$.

Proof. From [20, Theorem 7], f_1 and f_2 has nonlinearity 55 and degree 7. Thus, f_1 and f_2 are not balanced. We have $\max_{\mathbf{a}\in\mathbb{F}_2^7} |\mathcal{W}_{f_i}(\mathbf{a})| = 18 \Leftrightarrow -18 \leq \mathcal{W}_{f_i}(\mathbf{a}) \leq 18$, for all $\mathbf{a} \in \mathbb{F}_2^7$, where $i = 1, 2$. Let $\mathbf{a} = \mathbf{0}$. Then $-18 \leq \mathcal{W}_{f_i}(\mathbf{0}) \leq 18 \Leftrightarrow \frac{128-18}{2} \leq |\mathcal{S}_{f_i}| \leq \frac{128+18}{2} \Leftrightarrow 55 \leq |\mathcal{S}_{f_i}| \leq 73$. For any nonzero $\mathbf{a} \in \mathbb{F}_2^7$ and $i = 1, 2$, $-18 \leq 4|\mathcal{S}_{f_i}^1(\mathbf{a})| - 2|\mathcal{S}_{f_i}| \leq 18 \Leftrightarrow \frac{|\mathcal{S}_{f_i}|-9}{2} \leq |\mathcal{S}_{f_i}^1(\mathbf{a})| \leq \frac{|\mathcal{S}_{f_i}|+9}{2}$. □

Now we present the conditions of small functions f_1 and f_2 in Table 2 for different possible weights. We used the same notations as in the above corollary. The nonlinearity of f_1 and f_2 are 55, an odd integer. Therefore, the cardinalities of support sets of these functions are also odd numbers. It is clear that the nonlinearity of $f_1||f_2$ is same as $f_2||f_1$. So without loss of generality we present the weights of small functions f_1 and f_2 such that $|\mathcal{S}_{f_1}| < |\mathcal{S}_{f_2}|$.

Table 2. Properties of small functions given in [20, Theorem 7]

$\lvert\mathcal{S}_{f_1}\rvert$	$\lvert\mathcal{S}_{f_2}\rvert$	$s \leq \lvert\mathcal{S}_{f_1}^1(\mathbf{a})\rvert \leq t$	$s' \leq \lvert\mathcal{S}_{f_2}^1(\mathbf{a})\rvert \leq t'$
55	73	$23 \leq \lvert\mathcal{S}_{f_1}^1(\mathbf{a})\rvert \leq 32$	$32 \leq \lvert\mathcal{S}_{f_2}^1(\mathbf{a})\rvert \leq 41$
57	71	$24 \leq \lvert\mathcal{S}_{f_1}^1(\mathbf{a})\rvert \leq 33$	$31 \leq \lvert\mathcal{S}_{f_2}^1(\mathbf{a})\rvert \leq 40$
59	69	$25 \leq \lvert\mathcal{S}_{f_1}^1(\mathbf{a})\rvert \leq 34$	$30 \leq \lvert\mathcal{S}_{f_2}^1(\mathbf{a})\rvert \leq 39$
61	67	$26 \leq \lvert\mathcal{S}_{f_1}^1(\mathbf{a})\rvert \leq 35$	$29 \leq \lvert\mathcal{S}_{f_2}^1(\mathbf{a})\rvert \leq 38$
63	65	$27 \leq \lvert\mathcal{S}_{f_1}^1(\mathbf{a})\rvert \leq 36$	$28 \leq \lvert\mathcal{S}_{f_2}^1(\mathbf{a})\rvert \leq 37$

The sufficient conditions derived by Sarkar et al. [20, Theorem 7] involve the degree and nonlinearity of small functions. Thus, if we experimentally search a balanced function in 8 variables with nonlinearity 118, we first try to identify the small functions f_1 and f_2 in 7 variables such that their degree and nonlinearity are 7 and 55, respectively. Here we further derive the possible weights of $\mathcal{S}_{f_i}^1(\mathbf{a})$ of the small functions as in Table 2. Now, to experimentally discover a balanced function in 8 variables with nonlinearity 118, we first consider the possible weights of small functions f_1 and f_2, and then check the cardinality of $\mathcal{S}_{f_i}^1(\mathbf{a})$, $i = 1, 2$. These additional properties of small functions can help to reduce the search domain. Further, the sufficient conditions in each cases are related with two binary matrices. For example, let $(|\mathcal{S}_{f_1}|, |\mathcal{S}_{f_2}|) = (55, 73)$. Then the order of support matrices of f_1 and f_2, $\mathcal{M}_{f_1}$ and $\mathcal{M}_{f_2}$ are 7×55 and 7×73, respectively. We check the conditions $23 \leq wt(\mathbf{a}\mathcal{M}_{f_1}) \leq 32$ and $32 \leq wt(\mathbf{a}\mathcal{M}_{f_1}) \leq 41$, for all nonzero $\mathbf{a} \in \mathbb{F}_2^7$ so that nonlinearities of f_1 and f_2 are 55.
Now we prove that weights of the small functions f_1 (and f_2) can not be 55 and 57 (73 and 71, respectively).

Proposition 2. *Let $f = f_1||f_2$ be a balanced function in 8 variables having nonlinearity 118. Then the possible cardinality of the support set of f_1 (respectively, f_2) are 59, 61, 63, 65, 67 and 69.*

Proof. From Corollary 5, we have the cardinality of the support set of f_1 and f_2 are $55, 57, 59, 61, 63, 65, 67, 71$ and 73. Since $|\mathcal{S}_{f_1}| + |\mathcal{S}_{f_2}| = 128$, if $|\mathcal{S}_{f_1}| = 55$, then $|\mathcal{S}_{f_2}| = 73$. Therefore, $|\mathcal{S}_f^1(0,0,0,0,0,0,0,1)| = 73 = |\mathcal{S}_{f_2}|$, and from Equation (2), $nl(f) \leq 128 - \frac{36}{2} = 110$. Similarly, we prove that the cardinality of $\mathcal{S}_{f_1}$ can not be $57, 71$ and 73. □

It is observed that if $f = f_1||f_2 \in \mathcal{B}_8$ is balanced, where f_1 and f_2 are two Boolean functions in 7 variables having nonlinearity 55 and degree 7, then the nonlinearity of f may not be 118, in general. For example, let $f_1 \in \mathcal{B}_7$ such that $nl(f_1) = 55$, $\deg(f_1) = 7$, and $f_2 = f_1 \oplus 1$. Then $nl(f_2) = 55$, $\deg(f_2) = 7$, and nonlinearity of $f = f_1||f_2$ is 110 with degree 7. Thus, the question is for which properties of two small functions f_1 and f_2 in 7 variables so that the concatenation of f_1 and f_2, i.e., $f = f_1||f_2$, is a balanced function with nonlinearity 118. We here prove that the support matrix of f_2 needs to satisfy some additional properties for achieving the nonlinearity 118. The support matrix of f can be written as $\mathcal{M}_f = \mathcal{M}'_{f_1}||\mathcal{M}'_{f_2}$, where $\mathcal{M}'_{f_1}$ and $\mathcal{M}'_{f_2}$ are constructed from the support matrices of f_1 and f_2, respectively, with last row of $\mathcal{M}'_{f_1}$, $\mathcal{M}'_{f_2}$ being all-zero and all-one vectors, respectively. Suppose the row vectors of $\mathcal{M}_f$ are $\mathbf{u}_1, \mathbf{u}_2, \ldots, \mathbf{u}_7, \mathbf{u}_8$. Then $\mathbf{u}_i = \mathbf{u}_i^1||\mathbf{u}_i^2$, where $\mathbf{u}_i^1$ and $\mathbf{u}_i^2$ are row vectors of $\mathcal{M}'_{f_1}$ and $\mathcal{M}'_{f_2}$, respectively, $1 \leq i \leq 8$. In particular, $\mathbf{u}_i^1$ and $\mathbf{u}_i^2$ are row vectors of $\mathcal{M}_{f_1}$ and $\mathcal{M}_{f_2}$, respectively, $1 \leq i \leq 7$, and $\mathbf{u}_8^1 = (0, 0, \ldots, 0)$ of length $|\mathcal{S}_{f_1}|$ and $\mathbf{u}_8^1 = (1, 1, \ldots, 1)$ of length $|\mathcal{S}_{f_2}|$.

Theorem 4. *Let $f = f_1||f_2 \in \mathcal{B}_8$ be balanced, where f_1 and f_2 are two Boolean functions in 7 variables having nonlinearity 55. The nonlinearity of f is 118 if the following conditions hold for any nonzero $\mathbf{a} \in \mathbb{F}_2^7$.*

i. If $wt(\mathbf{a})$ is even, then $59 \leq |\mathcal{S}_{f_1}^1(\mathbf{a})| + |\mathcal{S}_{f_2}^1(\mathbf{a})| \leq 69$, and
ii. if $wt(\mathbf{a})$ is odd, then $59 \leq |\mathcal{S}_{f_1}^1(\mathbf{a})| + |\mathcal{S}_{f_2}^0(\mathbf{a})| \leq 69$ and $59 \leq |\mathcal{S}_{f_1}^1(\mathbf{a})| + |\mathcal{S}_{f_2}^1(\mathbf{a})| \leq 69$,

where $\mathcal{S}_{f_1}^1(\mathbf{a})$, $\mathcal{S}_{f_2}^1(\mathbf{a})$ and $\mathcal{S}_{f_2}^0(\mathbf{a})$ are defined in (1).

Proof. Suppose $\mathcal{M}_f = \mathcal{M}'_{f_1}||\mathcal{M}'_{f_2}$ be the support matrix of f, where $\mathcal{M}'_{f_1}$ and $\mathcal{M}'_{f_2}$ are constructed from the support matrix of f_1 and f_2, respectively, as discuss above. Let $\mathbf{u}_1, \mathbf{u}_2, \ldots, \mathbf{u}_7, \mathbf{u}_8$ be row vectors of $\mathcal{M}_f$ such that $\mathbf{u}_i = \mathbf{u}_i^1||\mathbf{u}_i^2$, where $\mathbf{u}_i^1$ and $\mathbf{u}_i^2$ are row vectors of $\mathcal{M}'_{f_1}$ and $\mathcal{M}'_{f_2}$, respectively, $1 \leq i \leq 8$. Suppose $\bar{\mathcal{M}}_{f_2}$ is constructed from $\mathcal{M}_{f_2}$ by complementing each row, i.e., the row vectors of $\bar{\mathcal{M}}_{f_2}$ are $\bar{\mathbf{u}}_1^2, \bar{\mathbf{u}}_2^2, \ldots, \bar{\mathbf{u}}_7^2$. For any nonzero $\mathbf{a} \in \mathbb{F}_2^7$, we have

$$|\mathcal{S}_f^1(\mathbf{a}, \varepsilon)| = \begin{cases} |\mathcal{S}_{f_1}^1(\mathbf{a})| + |\mathcal{S}_{f_2}^1(\mathbf{a})|, & \text{if } \varepsilon = 0, \\ |\mathcal{S}_{f_1}^1(\mathbf{a})| + |\bar{\mathcal{S}}_{f_2}^1(\mathbf{a})|, & \text{if } \varepsilon = 1, \end{cases}$$

where $|\bar{\mathcal{S}}_{f_2}^1(\mathbf{a})| = wt(\mathbf{a}\bar{\mathcal{M}}_{f_2})$.

Case (i): Suppose $wt(\mathbf{a})$ is even. Then $wt(a_1\bar{\mathbf{u}}_1^2 \oplus a_2\bar{\mathbf{u}}_2^2 \oplus \cdots \oplus a_7\bar{\mathbf{u}}_7^2) = wt(a_1\mathbf{u}_1^2 \oplus a_2\mathbf{u}_2^2 \oplus \cdots \oplus a_7\mathbf{u}_7^2)$, so, $|\bar{\mathcal{S}}_{f_2}^1(\mathbf{a})| = wt(\mathbf{a}\bar{\mathcal{M}}_{f_2}) = wt(\mathbf{a}\mathcal{M}_{f_2}) = |\mathcal{S}_{f_2}^1(\mathbf{a})|$.
Case (ii): Suppose $wt(\mathbf{a})$ is odd. Then $wt(a_1\bar{\mathbf{u}}_1^2 \oplus a_2\bar{\mathbf{u}}_2^2 \oplus \cdots \oplus a_7\bar{\mathbf{u}}_7^2) = wt(a_1\mathbf{u}_1^2 \oplus a_2\mathbf{u}_2^2 \oplus \cdots \oplus a_7\mathbf{u}_7^2 \oplus \mathbf{1}) = |\mathcal{S}_{f_2}| - wt(a_1\mathbf{u}_1^2 \oplus a_2\mathbf{u}_2^2 \oplus \cdots \oplus a_7\mathbf{u}_7^2)$, so, $|\bar{\mathcal{S}}_{f_2}^1(\mathbf{a})| = wt(\mathbf{a}\bar{\mathcal{M}}_{f_2}) = |\mathcal{S}_{f_2}| - wt(\mathbf{a}\mathcal{M}_{f_2}) = |\mathcal{S}_{f_2}^0(\mathbf{a})|$.
From the above two cases, we get

$$|\mathcal{S}_f^1(\mathbf{a},\varepsilon)| = \begin{cases} |\mathcal{S}_{f_1}^1(\mathbf{a})| + |\mathcal{S}_{f_2}^1(\mathbf{a})|, & \text{if } \varepsilon = 0, \\ |\mathcal{S}_{f_1}^1(\mathbf{a})| + |\mathcal{S}_{f_2}^1(\mathbf{a})|, & \text{if } \varepsilon = 1 \text{ and } wt(\mathbf{a}) \text{ is even}, \\ |\mathcal{S}_{f_1}^1(\mathbf{a})| + |\mathcal{S}_{f_2}^0(\mathbf{a})|, & \text{if } \varepsilon = 1 \text{ and } wt(\mathbf{a}) \text{ is odd}. \end{cases}$$

□

Using the above result we propose a possible construction method of balanced Boolean functions in 8 variables having nonlinearity 118. This method is based upon identifying two binary matrices of order $7 \times r_1$ and $7 \times r_2$ with $r_1 + r_2 = 128$ that are satisfied certain properties. The justification of the next construction directly follows from Proposition 2 and Theorem 4.

Construction 3. *Let $\mathcal{M}_1$ and $\mathcal{M}_2$ be two binary matrices of order $7 \times r_1$ and $7 \times r_2$ with $r_1 + r_2 = 128$, respectively. Suppose the column vectors of $\mathcal{M}_1$ are distinct, also for $\mathcal{M}_2$, and $r_1 \in \{59, 61, 63, 65, 67, 69\}$. Let us define $\bar{\mathcal{M}}_1$ and $\bar{\mathcal{M}}_2$ such that its first seven row vectors are taken from $\mathcal{M}_1$ and $\mathcal{M}_2$, respectively, and the last rows are all-zero and all-one, respectively. Then $\mathcal{M} = \bar{\mathcal{M}}_1 || \bar{\mathcal{M}}_2$ is the support matrix of a balanced function in 8 variables having nonlinearity 118, if the following conditions hold for any nonzero $\mathbf{a} \in \mathbb{F}_2^7$.*

i. If $wt(\mathbf{a})$ is even, then $59 \leq wt(\mathbf{a}\mathcal{M}_1) + wt(\mathbf{a}\mathcal{M}_2) \leq 69$, and
ii. if $wt(\mathbf{a})$ is odd, then $59 \leq wt(\mathbf{a}\mathcal{M}_1) + wt(\mathbf{a}\mathcal{M}_2) \leq 69$ and $59 \leq r_2 + wt(\mathbf{a}\mathcal{M}_1) - wt(\mathbf{a}\mathcal{M}_2) \leq 69$.

We believe that aforesaid analysis 7-variable functions f_1, f_2 provides additional useful insights towards the construction of balanced function in 8 variable with nonlinearity 118.

5 Construction Method of Highly Nonlinear Balanced Functions from Bent Functions

Though bent functions are directly of little use in cryptography, many constructions of cryptographically significant functions starts with an appropriate bent function and suitable modify it to achieve required properties. Not to mention, Dobbertin's construction for highly nonlinear balanced functions is one notable example of such techniques. We take a similar approach for our very purpose e.g. start with a bent function and modify it in certain way to construct balanced functions with high nonlinearity. It is remembered that $2^{\frac{n}{2}-1}$ many changes (0 to 1) of a bent function in n variables with weight $2^{n-1} - 2^{\frac{n}{2}-1}$ (or 1 to 0 of a bent function in n variables with weight $2^{n-1}+2^{\frac{n}{2}-1}$), we get a balanced function. Nonlinearity of such functions may not exceed Dobbertin's bound (if at all possible). Our attempt here is to characterize the required modification which will produce our desired functions.

Proposition 3. [16, Proposition 3] *Let $f \in \mathcal{B}_n$ have weight $2^{n-1}-2^{\frac{n}{2}-1}$. Then f is bent if and only if, for any nonzero $\mathbf{a} \in \mathbb{F}_2^n$, we have:*

$$(|\mathcal{S}_f^0(\mathbf{a})|, |\mathcal{S}_f^1(\mathbf{a})|) \in \{(2^{n-2}, 2^{n-2}-2^{\frac{n}{2}-1}), (2^{n-2}-2^{\frac{n}{2}-1}, 2^{n-2})\},$$

where $|\mathcal{S}_f^0(\mathbf{a})|$ and $|\mathcal{S}_f^1(\mathbf{a})|$ are defined in (1).

Now we characterize the modification procedure for a given bent function, which will lead to a function with desired nonlinearity.

Theorem 5. *Let $f \in \mathcal{B}_n$ be bent with weight $2^{n-1}-2^{\frac{n}{2}-1}$. Suppose $g \in \mathcal{B}_n$ such that $\mathcal{S}_g = \mathcal{S}_f \cup T$, where $T \subset \mathbb{F}_2^n \setminus \mathcal{S}_f$ with $|T| = 2^{n-1}$. Then g is balanced, and its nonlinearity is $nl(g) > 2^{n-1}-2^{\frac{n}{2}}+nlb(\frac{n}{2})$ if and only if for any nonzero $\mathbf{a} \in \mathbb{F}_2^n$*

i. $|\mathcal{S}_T^1(\mathbf{a})| < 2^{\frac{n}{2}-1}-\frac{nlb(\frac{n}{2})}{2}$, if $\mathcal{W}_f(\mathbf{a}) = 2^{\frac{n}{2}}$, and
ii. $|\mathcal{S}_T^1(\mathbf{a})| > \frac{nlb(\frac{n}{2})}{2}$, if $\mathcal{W}_f(\mathbf{a}) = -2^{\frac{n}{2}}$.

Proof. Since $\mathcal{S}_f \cap T = \emptyset$, $|\mathcal{S}_g| = 2^{n-1}-2^{\frac{n}{2}-1}+2^{\frac{n}{2}-1} = 2^{n-1}$. Thus, g is balanced. From Corollary 1, we have the nonlinearity of f is $nl(g) > 2^{n-1}-2^{\frac{n}{2}}+nlb(\frac{n}{2})$ if and only if for any nonzero $\mathbf{a} \in \mathbb{F}_2^n$,

$$2^{n-2}-2^{\frac{n}{2}-1}+\frac{nlb(\frac{n}{2})}{2} < |\mathcal{S}_g^1(\mathbf{a})| < 2^{n-2}+2^{\frac{n}{2}-1}-\frac{nlb(\frac{n}{2})}{2}$$
$$\Leftrightarrow 2^{n-2}-2^{\frac{n}{2}-1}+\frac{nlb(\frac{n}{2})}{2} < |\mathcal{S}_f^1(\mathbf{a})|+|\mathcal{S}_T^1(\mathbf{a})| < 2^{n-2}+2^{\frac{n}{2}-1}-\frac{nlb(\frac{n}{2})}{2}$$

Since f is bent, from Proposition 3 we have

$$|\mathcal{S}_f^1(\mathbf{a})| \in \{2^{n-2}-2^{\frac{n}{2}-1}, 2^{n-2}\},$$

for all nonzero $\mathbf{a} \in \mathbb{F}_2^n$, and we get the result. □

To sum up we essentially ended up defining the modification procedure as a concatenation of linear code of appropriate dimension and parameters. So the question of finding such a right modification essentially backed by the question of existence of right linear code of special type.

5.1 Studying the Specific Conditions for $n = 8$, and Explaining Some Non-existence Issues

Let f be a bent functions in 8 variables such that $\mathcal{S}_f$ is 120, i.e., the order of support matrix $\mathcal{M}_f$ is 8×120. Thus, the weight of any nonzero linear combinations of row vectors of $\mathcal{M}_f$ is 56 or 64. Let $g \in \mathcal{B}_8$ such that $\mathcal{S}_g = \mathcal{S}_f \cup T$, where $T \subset \mathbb{F}_2^8 \setminus \mathcal{S}_f$ with $|T| = 8$. Observe that $2^{8-1}-2^{\frac{8}{2}}+nlb(\frac{8}{2}) = 116$. The following result is a direct consequence from Theorem 5.

Corollary 6. *Let $f \in \mathcal{B}_8$ be bent such that $|\mathcal{S}_f| = 120$. Suppose $g \in \mathcal{B}_8$ such that $\mathcal{S}_g = \mathcal{S}_f \cup T$, where $T \subset \mathbb{F}_2^8 \setminus \mathcal{S}_f$ with $|T| = 8$. Then g is balanced and its nonlinearity $nl(g) = 118$, if and only if, for any nonzero $\mathbf{a} \in \mathbb{F}_2^8$*

i. $|\mathcal{S}_T^1(\mathbf{a})| \leq 5$, *if* $\mathcal{W}_f(\mathbf{a}) = 16$, *and*
ii. $|\mathcal{S}_T^1(\mathbf{a})| \geq 3$, *if* $\mathcal{W}_f(\mathbf{a}) = -16$.

So, if we could manage to find such a subset T and corresponding bent function in 8 variables, then we could construct a balanced function in 8 variables with nonlinearity 118. Here it is sufficient to identify a set $T \subset \mathbb{F}_2^8 \setminus \mathcal{S}_f$ such that $|\mathcal{S}_T^1(\mathbf{a})| \in \{3, 4, 5\}$ for constructing a balanced function in 8 variables with nonlinearity 118. Unfortunately, no such set can exist as we present below.

Proposition 4. *There does not exist any set $T = \{\mathbf{u}_1, \mathbf{u}_2, \ldots \mathbf{u}_8\} \subset \mathbb{F}_2^8$ such that $|\mathcal{S}_T^1(\mathbf{a})| \in \{3, 4, 5\}$, for all nonzero $\mathbf{a} \in \mathbb{F}_2^8$.*

Proof. If T is a linearly dependent set, then there exist at least one nonzero linear combination of vectors of T such that its weight is 0. Suppose T is linearly independent. Then linear span of T is $\mathbb{F}_2^8$, and so, there are many linear combination of vectors of T such that the weights are not 3 or 4 or 5. □

The above discussion shows non-existence through one particular technique, thus we need to explore different other directions in future. It may be possible to construct a balanced function in 8 variables having nonlinearity 118, where the values of $|\mathcal{S}_T^1(\mathbf{a})|$ may be different from 3, 4 and 5. To achieve this, if $|\mathcal{S}_T^1(\mathbf{a})| \leq 2$ (or ≥ 6) for a nonzero $\mathbf{a} \in \mathbb{F}_2^8$, then the Walsh–Hadamard transform value of initial bent function at $\mathbf{a}$ must be 16 (or -16, respectively) at those points.

Example 2. Let us present a potential example of such construction method considering a simple bent function. Let $f(\mathbf{x}, \mathbf{y}) = \mathbf{x} \cdot \mathbf{y}$, for all $\mathbf{x}, \mathbf{y} \in \mathbb{F}_2^4$. Then f is bent, its Walsh–Hadamard spectrum value is 16 for 136 many times, and is -16 in rest 120 cases. Thus, there are 135 many nonzero elements $(\mathbf{a}, \mathbf{b}) \in \mathbb{F}_2^4 \times \mathbb{F}_2^4$ such that $|\mathcal{S}_f^1(\mathbf{a}, \mathbf{b})| = 64$ and 120 many elements $(\mathbf{a}, \mathbf{b})$ such that $|\mathcal{S}_f^1(\mathbf{a}, \mathbf{b})| = 56$.

Let $f(\mathbf{x}, \mathbf{y}) = \mathbf{x} \cdot \mathbf{y}$, for all $\mathbf{x}, \mathbf{y} \in \mathbb{F}_2^4$. Suppose $T \subset \mathbb{F}_2^4 \times \mathbb{F}_2^4 \setminus \mathcal{S}_f$ with cardinality 8 and satisfies the following conditions.

i. $|\mathcal{S}_T^1(\mathbf{a}, \mathbf{b})| \leq 5$, if $\mathbf{a} \cdot \mathbf{b} = 0$, and
ii. $|\mathcal{S}_T^1(\mathbf{a}, \mathbf{b})| \geq 3$, if $\mathbf{a} \cdot \mathbf{b} = 1$.

Define a Boolean function $g \in \mathcal{B}_8$ as

$$g(\mathbf{x}, \mathbf{y}) = \begin{cases} f(\mathbf{x}, \mathbf{y}) \oplus 1, & \text{if } (\mathbf{x}, \mathbf{y}) \in T \\ f(\mathbf{x}, \mathbf{y}), & \text{otherwise} \end{cases}.$$

Then g is balanced and from Corollary 6 its nonlinearity is 118.

Unfortunately, we are yet to identify any T to achieve such a result. Construction of a balanced nonlinearity 118 function demands whether we can find a proper bent function and obtain a T corresponding to that.

Simply speaking, fixing $f(\mathbf{x}, \mathbf{y}) = \mathbf{x} \cdot \mathbf{y}$, a Maiorana-McFarland bent function is obtained and one may try to toggle 8 output points from 0 to 1 to have a balanced function. This will naturally produce functions with nonlinearities 112, 114, 116 or 118. Given that the conditions above are satisfied as in Example 2, we will obtain 8-variable balanced functions with nonlinearity 118. However, to check that, we need to exhaust $\binom{136}{8}$ options, which is not computationally achievable by the hardware available with us. Obtaining some better filtering strategies in this regard will be useful for further research. For experimental purpose, given f as above, we have chosen 10^8 random T's and obtained nonlinearities 112, 114 and 116 in the proportions 0.398549, 0.594860, 0.006591 respectively.

5.2 Studying the Cases for $n = 10, 12, 14$

Fortunately, for $n = 10$, there is no such obvious impossibility result as in Proposition 4 above. To construct a balanced Boolean function in 10 variables with nonlinearity 494, we need a set of 16 elements outside the support set of f with certain weight properties.

Let $f \in \mathcal{B}_{10}$ such that $\mathcal{S}_f$ is 496, i.e., the order of support matrix $\mathcal{M}_f$ is 10×496. Then $wt(\mathbf{a}\mathcal{M}_f)$ is 240 or 256, for all nonzero $\mathbf{a} \in \mathbb{F}_2^{10}$. Suppose $g \in \mathcal{B}_{10}$ such that $\mathcal{S}_g = \mathcal{S}_f \cup T$, where $T \subset \mathbb{F}_2^{10} \setminus \mathcal{S}_f$ with $|T| = 16$. Since $2^{10-1} - 2^{\frac{10}{2}} + nlb(\frac{10}{2}) = 492$, the question is to identify a balanced Boolean function having nonlinearity 494.

Corollary 7. *Let $f \in \mathcal{B}_{10}$ be bent such that $|\mathcal{S}_f| = 496$. Suppose $g \in \mathcal{B}_{10}$ such that $\mathcal{S}_g = \mathcal{S}_f \cup T$, where $T \subset \mathbb{F}_2^{10} \setminus \mathcal{S}_f$ with $|T| = 16$. Then g is balanced and its nonlinearity $nl(g) = 494$, if and only if, for any nonzero $\mathbf{a} \in \mathbb{F}_2^{10}$*

i. $|\mathcal{S}_T^1(\mathbf{a})| \leq 9$, *if* $\mathcal{W}_f(\mathbf{a}) = 32$, *and*
ii. $|\mathcal{S}_T^1(\mathbf{a})| \geq 7$, *if* $\mathcal{W}_f(\mathbf{a}) = -32$.

Similar as in the case for $n = 8$, if we could manage to find $T \subset \mathbb{F}_2^{10} \setminus \mathcal{S}_f$ and corresponding bent function f in 10 variables, then we could construct a balanced function in 10 variables with nonlinearity 494. It is sufficient to identify a set T such that $|\mathcal{S}_T^1(\mathbf{a})| \in \{7, 8, 9\}$. Here it might be possible to identify such set T. For this one requires to identify 10 binary strings of length 16 such that the weight of any nonzero linear combinations is 7, 8 or 9. Note that, the number of binary strings of length 7, 8 or 9 is $\binom{16}{7} + \binom{16}{8} + \binom{16}{9} = 35750$. For existence of T, we need $2^{10} - 1 = 1023 < 35750$ binary strings of length 16 should have weights 7, 8 or 9. We could not prove an immediate non-existence result as in Proposition 4 for $n = 8$ in this regard.

Similar characterizations hold for for $n = 12$ and 14. We enumerate the observations in Table 3.

Table 3. Construction of highly nonlinear balanced functions for $n = 8, 10, 12$ and 14

n	$\|S_f\|$	$\|S_T\|$	Nonlinearity	$s \leq \|S_T^1(\mathbf{a})\| \leq t$
8	120	8	118	$3 \leq \|S_T^1(\mathbf{a})\| \leq 5$
10	496	16	494	$7 \leq \|S_T^1(\mathbf{a})\| \leq 9$
12	2016	32	2012	$14 \leq \|S_T^1(\mathbf{a})\| \leq 18$
			2014	$15 \leq \|S_T^1(\mathbf{a})\| \leq 17$
14	8128	64	8122	$29 \leq \|S_T^1(\mathbf{a})\| \leq 35$
			8124	$30 \leq \|S_T^1(\mathbf{a})\| \leq 34$
			8126	$31 \leq \|S_T^1(\mathbf{a})\| \leq 33$

Thus, we have two clear directions here towards obtaining balanced Boolean functions on even number of variable beating the Dobbertin's bound.

- A general idea in modifying bent functions and we can have several different approaches to obtain such functions. One such example is to move in the direction of Example 2.
- Putting more specific structures on that. One such idea fails for $n = 8$ as we note in Proposition 4. However, we need further investigation in this regard for even $n \geq 10$.

6 Conclusion

Disproving Dobbertin's Conjecture (if at all possible) is a long standing open problem in Boolean function research. This is related to nonlinearity of n-variable (n even) Boolean functions with balancedness. In this paper, we have studied this problem in a disciplined manner and tried to understand several issues that are related to nonlinearity exceeding the Dobbertin's bound. The necessary and sufficient conditions are enumerated in details and we examine instances with $n = 8, 10, 12, 14$. The weight conditions are studied in details and further relationships with certain linear codes are also explained. While we could not achieve functions having nonlinearity exceeding the Dobbertin's bound, we could add new results that provide more transparent understanding in this domain.

Acknowledgments. We would like to thank the anonymous reviewers of Indocrypt 2022 for their valuable suggestions and comments, which considerably improved the quality of our paper.

References

1. Carlet, C., Guillot, P.: A characterization of binary bent functions. J. Comb. Theor. Ser. A **76**, 328–335 (1996)
2. Carlet, C.: Boolean functions for cryptography and error correcting codes. In: Crama, Y., Hammer, P. (eds.) Boolean Methods and Models, pp. 257–397. Cambridge University Press, Cambridge (2010)
3. Cusick, T.W., Stănică, P.: Cryptographic Boolean Functions and Applications. Elsevier-Academic Press, Cambridge (2009)
4. Dillon J.F.: Elementary Hadamard Difference Sets, PhD Thesis, University of Maryland (1974)
5. Dillon J.F.: Elementary hadamard difference sets, In: proceedings of 6th S. E. Conference of Combinatorics, Graph Theory, and Computing, Utility Mathematics, Winnipeg, pp. 237–249 (1975)
6. Dobbertin H.: Construction of bent functions and balanced Boolean functions with high nonlinearity, Fast Software Encryption 1994 LNCS 1008, pp. 61–74 (1994)
7. Kavut, S., Maitra, S., Tang, D.: Construction and search of balanced Boolean functions on even number of variables towards excellent autocorrelation profile. Des. Codes Crypt. **87**(2–3), 261–276 (2019)
8. Kavut, S., Maitra, S., Yucel, M.D.: Search for Boolean functions with excellent profiles in the rotation symmetric class. IEEE Trans. Inf. Theor. **53**(5), 1743–1751 (2007)
9. Kavut, S., Yucel, M.D.: 9-variable Boolean functions with nonlinearity 242 in the generalized rotation symmetric class. Inf. Comput. **208**(4), 341–350 (2010)
10. Kavut, S., Maitra, S.: Patterson-Wiedemann type functions on 21 variables with nonlinearity greater than bent concatenation bound. IEEE Trans. Inf. Theor. **62**(4), 2277–2282 (2016)
11. MacWilliams, F.J., Sloane, N.J.A.: The Theory of Error-Correcting Codes, Amsterdam, North-Holland, The Netherlands (1977)
12. Maitra, S., Sarkar, P.: Maximum nonlinearity of symmetric Boolean functions on odd number of variables. IEEE Trans. Inf. Theor. **48**(9), 2626–2630 (2002)
13. Maity, S., Johansson, T.: Construction of cryptographically important Boolean functions, INDOCRYPT 2002 LNCS 2551, pp. 234–245 (2002)
14. Maity, S., Maitra, S.: Minimum distance between bent and 1-resilient Boolean functions, FSE 2004 LNCS 3017, pp. 143–160 (2004)
15. McFarland, R.L.: A family of noncyclic difference sets. J. Comb. Theor. Ser. A **15**, 1–10 (1973)
16. Mesnager, S., Mandal, B., Tang, C.: New characterizations and construction methods of bent and hyper-bent Boolean functions. Discrete Math. **343**(11), 112081 (2020)
17. Mesnager, S.: Bent Functions. Springer, Cham (2016). https://doi.org/10.1007/978-3-319-32595-8
18. Patterson, N.J., Wiedemann, D.H.: The covering radius of the $(2^{15}, 16)$ Reed-Muller code is at least 16276. IEEE Trans. Inf. Theor. **29**(3), 354–356 (1983)
19. Rothaus, O.S.: On bent functions. J. Comb. Theor. Ser. A **20**, 300–305 (1976)
20. Sarkar, P., Maitra, S.: Nonlinearity bounds and constructions of resilient Boolean functions, CRYPTO 2000 LNCS 1880, pp. 515–532 (2000)
21. Sun, Y., Zhang, J., Gangopadhyay, S.: Construction of resilient Boolean functions in odd variables with strictly almost optimal nonlinearity. Des. Codes Crypt. **87**(12), 3045–3062 (2019). https://doi.org/10.1007/s10623-019-00662-5

22. Tang, D., Maitra, S.: Constructions of n-variable ($n \equiv 2 \bmod 4$) balanced Boolean functions with maximum absolute value in autocorrelation spectra $< 2^{\frac{n}{2}}$. IEEE Trans. Inf. Theor. **64**(1), 393–402 (2018)
23. Tang, D., Kavut, S., Mandal, B., Maitra, S.: Modifying Maiorana-McFarland type bent functions for good cryptographic properties and efficient implementation. SIAM J. Discrete Math. **33**(1), 238–256 (2019)

Revisiting *BoolTest* – On Randomness Testing Using Boolean Functions

Bikshan Chatterjee[1(✉)], Rachit Parikh[1], Arpita Maitra[2], Subhamoy Maitra[1], and Animesh Roy[1]

[1] Indian Statistical Institute, Kolkata 700108, India
bchatterjee7980@gmail.com, prachit@me.iitr.ac.in, subho@isical.ac.in
[2] TCG Crest, Kolkata 700091, India

Abstract. Pseudo-random number generation is crucial in cryptology and other areas related to information technology. In a broad sense, the security of a protocol relies on the 'randomness' provided by the pseudo-random number generators. It is thus important to examine whether a random-looking stream has some kind of non-randomness in it. Here we consider that a binary stream is divided into blocks of a certain length m and we try to identify an m-bit Boolean function in this regard that is optimal to provide the highest Z-score for the output stream generated by the said function. In this regard, we show certain limitations of the *BoolTest* strategy by Sýs et al (2017) and present combinatorial results related to identifying the most suitable Boolean functions. We show that the existing works related to *BoolTest* identify the Boolean functions that are sub-optimal, constrained by the low degree in the Algebraic Normal Form. Our results find out the best Boolean function in this regard that will produce the maximum Z-score and the complexity is $O(N \log N)$ on the amount of random-looking stream of length N that we read during the evaluation process. We present substantial experimental evidence corresponding to our theoretical ideas. While we solve certain combinatorial problems related to *BoolTest*, the caveat is, this test is not sufficient to conclude on randomness or non-randomness of a given stream of data.

Keywords: Boolean functions · *BoolTest* · Randomness testing · Statistical analysis · Z-score

1 Introduction

Random number generators have wide applications in the broad areas of communication and cryptography. However, classical computers are deterministic and hence, it is not possible to produce any randomness out of them, other than the effect of the initial random seed, if any. The prime development here is in the direction of a Pseudo Random Number Generator (PRNG), where a small seed is used (may be from a random source) as an input and then a deterministic

T. Isobe and S. Sarkar (Eds.): INDOCRYPT 2022, LNCS 13774, pp. 471–491, 2022.
https://doi.org/10.1007/978-3-031-22912-1_21

algorithm generates a stream of random-looking data. This is not random, as the same seed will always generate the same stream of data and the randomness only depends on the initial seed. Only looking at the data, may be computationally or information-theoretically hard to distinguish the data from a truly random source, without knowing the seed. That is the main idea in designing symmetric ciphers.

As there is a need to study the security parameters of a cipher, randomness plays an important role. Towards cryptanalysis of a cipher, an important tool is to design certain distinguishers that can provide information regarding the inappropriate confusion and diffusion properties, i.e., to identify how far the output of the cipher deviates from true randomness. Looking at the algorithm of the cipher to obtain such a distinguisher is naturally a more scientific way. For example, the famous distinguisher [4] against RC4 could be identified from its algorithm. However, a complicated design will always make identifying such distinguishers harder. In this regard, applying statistical tests on the reduced rounds of the cipher might provide a quicker way to identify certain non-randomness, and a more formal design of a distinguisher may be initiated from that observation. For methodologies where the random numbers are generated from physical processes such as quantum mechanics or thermal processes, such statistical tests might be useful. That is, here we will not look at the algorithms to find the weaknesses, but study the data and will try to obtain some statistical measure that will possibly differentiate the available stream at hand from some ideal data generated from a random source. In this direction, one may refer to the well defined statistical test-beds like Diehard [5], Dieharder [1], NIST SP 800-22 [6], Cryp-X [3] and ENT [9].

All these test suites (often called a battery) generally consist of a series of empirical tests of randomness. Each test aims to find a predefined pattern of bits (or block of bits) in the data under consideration and examines the randomness property by certain measures that can be computed from the occurrences of the predefined bit patterns. Each test results in a distribution of a specific feature of bits or the blocks of bits. The distribution is then statistically compared with the expected one from data coming from random data. The data under examination is considered to be non-random if the distributions differ significantly.

In principle, one may design an unlimited number of tests for randomness certification and each of them should have certain merits and demerits. Understanding each statistical test is thus crucial rather than using them as black boxes. In this regard, we concentrate on the *BoolTest* [7], where each block of the bit-stream is applied to a suitably chosen Boolean function and the output bits are studied. In a later work, by Sýs et al [8], a similar technique has been used and many experimental results have been provided. As mentioned in [8], the *BoolTest* is a generalization of the frequency mono-bit test [6]. In this paper, we revisit the techniques presented in [7], identify certain limitations of the test, and then provide some techniques to optimize the method.

1.1 Organization and Contribution

We present certain preliminary ideas in Sect. 1.2.

Then, in Sect. 2, we consider different kinds of data streams and show how the Z-score varies. Our results show that the values from the Z-score might not provide logical characteristics in terms of randomness in certain cases. Section 1.3 presents a brief description of *BoolTest* by Sýs et al [7].

In Sect. 3, we provide a deterministic algorithm to find the best Boolean function that will maximize the Z-score for a given data. The algorithm runs in $O(N \log N)$ time, given N amount of data.

Section 4 presents substantial experimental evidence corresponding to our theoretical ideas. In this direction, we consider RC4 [4], AES [2], and Java rand ? in studying files of various lengths to identify the Z-scores, with properly chosen Boolean functions through our algorithms.

Section 5 concludes the paper. Some of the implementation related codes are available in Appendix.

1.2 Preliminaries

Consider N-bit of data $\mathcal{D}$, whose randomness we would like to test. The data is divided into non-overlapping blocks of m bits. For simplicity, we consider N is divisible by m, i.e., there are $n = \frac{N}{m}$ blocks. Each block of data is applied to a Boolean function of m-input bits to obtain one-bit output. Thus, given N-bit data, we obtain $\frac{N}{m}$ bits out of the Boolean function. Let us call the collection of all m block inputs obtained in such a manner $\mathcal{I}$. The collection $\mathcal{I}$ is a multi-set, not a set since there might be m length blocks that occur more than once.

Naturally, we need to study the frequency distribution from the set $\mathcal{I}$ to identify any non-randomness. By the method described in *BoolTest* [7], one may try to find the best distinguisher function on the frequency distribution obtained. The method used in *BoolTest* to obtain the Boolean function for the best distinguisher involves the metric Z-score. Before defining Z-score in our interpretation, let us introduce some notations. Support of an m-input 1-output Boolean function f is defined as

$$supp(f) = \{x \in \{0,1\}^m : f(x) = 1\}. \tag{1}$$

We also define $W(f)$ as the following set,

$$W(f) = \{x \in \mathcal{I} : f(x) = 1\}. \tag{2}$$

Let q_f be the proportion of inputs for which the f returns 1, i.e.,

$$q_f = \frac{|supp(f)|}{2^m}. \tag{3}$$

If the input distribution had uniformly been random, we would get each input $x \in \{0,1\}^m$ with equal probability. If we had a uniform distribution of m-block inputs $\{0,1\}^m$, then the number of data blocks for which the function f will output 1 is nq_f. Let p_f be the proportion of elements in $\mathcal{I}$ which will output 1 when given as an input to f.

$$p_f = \frac{|x \in \mathcal{I} : f(x) = 1|}{2^m} = \frac{|W(f)|}{2^m}. \tag{4}$$

In other words, the number of inputs for which the function f will output 1 from the collection $\mathcal{I}$ will be np_f.

Definition 1 (*Z*-score). *For a given function f of m-variables, and a collection $\mathcal{I}$, p_f and q_f as defined above, the Z-score defined in [7] is given as,*

$$z_f = \frac{\#1 - nq_f}{\sqrt{nq_f(1-q_f)}}. \tag{5}$$

where $\#1$ is the random variable that describes the number of m-bit blocks in the input data $\mathcal{D}$, that when fed to the function f returns 1. Here, $\#1$ can be written as

$$\#1 = |W(f)| = np_f \tag{6}$$

Thus the Z-score would be

$$z_f = \left| \frac{np_f - nq_f}{\sqrt{nq_f(1-q_f)}} \right|. \tag{7}$$

To provide more intuition to the definition above, if Y is a random variable representing the number of ones obtained as output from the function f over some input distribution, then Z-score is the normalization of binomially distributed random variable Y.

Now that we have defined Z-score, let us proceed to outline a brief introduction to the method presented by [7]. Note that, we are interested in identifying the most optimal Boolean function. Thus, the Z-score we discuss will always be related to a suitable Boolean function f, and thus, as in Definition 1, we always have z_f that is to be studied. Note that the number of distinct m-variable Boolean functions is 2^{2^m} and choosing the optimal Boolean function out of that super-exponential space is the main challenge.

Let $\mathbf{x} = (x_1, \ldots, x_m)$ and $\mathbf{a} = (a_1, \ldots, a_m)$. Any Boolean function f in m variables can be uniquely written as a multivariate polynomial of the form $f(\mathbf{x}) = \bigoplus_{\mathbf{a} \in \{0,1\}^m} \mu_{\mathbf{a}} x_1^{a_1} x_2^{a_2} \cdots x_m^{a_m}$, where $\mu_{\mathbf{a}} \in \{0,1\}$, for all $\mathbf{a} \in \{0,1\}^m$. This polynomial form of f is called the algebraic normal form (ANF). The algebraic degree of f is $deg(f) = \max_{\mathbf{a} \in \{0,1\}^m} \{wt(\mathbf{a}) : \mu_{\mathbf{a}} \neq 0\}$, the degree of highest degree term(s) with nonzero coefficient in its ANF. Here $wt(\mathbf{a})$ is the number of 1's in $\mathbf{a}$.

1.3 Brief Description of *BoolTest* by Sýs et al. [7]

The basic idea is to construct an m-bit Boolean function that will produce the highest Z-score. Given that there are 2^{2^m} Boolean functions, it has been commented in [7] that only a heuristic method in the set of m-variable Boolean functions will be attempted to identify the function. We will later show that this technique [7] is sub-optimal and consequently we will present an optimal algorithm. Now let us explain the strategy of [7].

The *BoolTest* algorithm $\mathcal{B}(deg, m, t, k)$ takes in as input the following parameters:

- deg: Each term in the ANF of the functions searched by *BoolTest* would be of degree deg.
- m : Block size, which is also the number of inputs to the Boolean functions.
- t : Top t monomials of degree deg are chosen and combined by XOR in the next step.
- k : Distinguishers are formed by combining k many monomials of degree deg.

Algorithm 1: BoolTest $\mathcal{B}(deg, m, t, k)$

$M \leftarrow \{1, \ldots, m\}$
$T \leftarrow \{1, \ldots, t\}$

// GET-SUBSETS(j, M) returns all subsets of M of size j
$S_{deg} \leftarrow \texttt{GET-SUBSETS}(deg, M)$

// Set F_{deg} contains all monomials of degree deg
$F_{deg} \leftarrow \{f : f = \prod_{j \in J} x_j, \forall J \in S_{deg}\}$

// Choose top t functions from F_{deg} with highest Z-score
// GET-MAX function takes t monomials with highest Z-score
$F_t \leftarrow \texttt{GET-MAX}(t, F_{deg})$

$S_k \leftarrow \texttt{GET-SUBSETS}(k, T)$

// Take combinations of k monomials from F_t
$F = \{f : f = \bigoplus_{k \in K} f_k, \forall K \in S_k, f_k \in F_t\}$

// Return the max Z-score and the corresponding distinguisher function from F
return $\arg\max_{f \in F} z(f)$

The Algorithm 1 given above provides an algorithmic outline of the *BoolTest* [7]. There are many other details related to optimization to improve the performance of the algorithm. To understand the approach, let us consider an example. If the given parameters are $\mathcal{B}(deg = 2, m = 4, t = 5, k = 3)$, then *BoolTest* first computes Z-score for all the monomials of degree 2 of the form $x_i x_j$ where

$i, j \in \{1, 2, 3, 4\}$(since $m = 4$). There are six such degree 2 monomials in this case. From these monomials, the top $t = 5$ monomials with the highest Z-score will be selected. Now, choose $k = 3$ monomials out of the 5 obtained in the first step (total $\binom{5}{3}$ combinations) and combine them using XOR operation to form a new function with 3 monomials each having degree 2, and obtain the Z-score for each combination. The function with the maximum Z-score out of these combinations is considered to be the function that will provide the best distinguisher through this heuristic.

Let the data $\mathcal{D}$ be a sequence of n random variables $X_1, \ldots, X_n$. The null hypothesis is,

$$H_0 : X_i \sim \text{Uniform}(0, 2^m - 1), \forall i \in \{1, \ldots, n\} \tag{8}$$

If the Boolean function f had been fixed, the number of ones (#1) would be a random variable that follows Binomial distribution $B(n, q_f)$ and z_f would approximately follow standard normal distribution $\mathcal{N}(0, 1)$.

The highest Z-score would be of the form $Z = \max\{z_{f_1}, \ldots, z_{f_{2^{2^m}}}\}$ where each z_{f_i} approximately follows standard normal distribution and, its CDF would be difficult to calculate as the z_{f_i}'s are not independent.

The value [7] calculates is the random variable $\max\{z_{f_1}, \ldots, z_{f_{\binom{t}{k}}}\}$ (it considers only $\binom{t}{k}$ among all possible boolean functions, $f_1, f_2, \ldots f_{\binom{t}{k}}$ are functions constructed from the data) whose CDF would be similarly difficult to calculate. Instead, [7] estimates the acceptance region using a "reference window" created by running the process on (assumed) true random data. If the highest Z-score achieved by the procedure on some sample of data falls within the reference window, the data is assumed to be random; otherwise, it is considered that the data might have non-randomness. However, we see later very high Z-scores do not necessarily imply non-randomness. For example, for large block size m, and the amount of data much smaller than 2^m blocks, very high Z-scores are possible even for truly random data.

2 Critical Evaluations of Z-score

In this section, we analyze a few issues related two the values that we receive from Z-score.

2.1 Z-Score for Data with All and Equal Frequency Inputs

Let us consider $N = nm$ many bits of data, where $n = s2^m$, and each m-bit pattern has frequency s in the data stream, in any order. Denote such a data stream as $D_{all,m,s}$. Then we have the following result.

Lemma 1. *The Z-score for the input stream $D_{all,m,s}$ would be 0 considering any m-bit Boolean function.*

Proof. If we parse through the bit stream and generate a frequency table of all m-bit blocks then we know we will get all the patterns from $\{0,1\}^m$ equal number of times (here s) hence our distribution of inputs is uniform. Z-score for a function f shows how the output bit pattern is different for a particular distribution inferred from the data, from that of a uniform distribution. In this case, the probability of getting any m-length block in $D_{all,m,s}$ is,

$$p(x) = \frac{1}{2^m} = q(x), \; \forall x \in \{0,1\}^m \tag{9}$$

where $q(x)$ is the probability of getting any input in the uniform distribution. Then the Z-score would be,

$$z_f = \frac{n}{\sqrt{nq_f(1-q_f)}} \left| \sum_{x:f(x)=1} (p(x) - q(x)) \right| = 0. \tag{10}$$

Thus, the Z-score is 0 independent of the choice of the function. □

Now, consider a simple counter circuit that generates all the m-bit patterns in a cycle, in the increasing order of decimal digits 0 to $2^m - 1$, and continues from 0 again. This data should not be considered random, but the Z-score will always be zero when we have a multiple of full cycles. The result is the same when the data comes in a cycle, but according to some permutation of 0 to $2^m - 1$. However, there are many other tests, such that linear complexity analysis, that can obtain the simplest LFSRs to distinguish among such streams. This is not possible using Z-score.

2.2 Maximum Z-score for Frequencies s and $s + 1$

Let us consider $N = nm$ many bits of data, where $n = s2^m + u$. There are u many m-bit patterns with frequency $s+1$ and the rest are having the frequency s, in any order. Denote this data stream as $D_{all-two,m,s}$. Let f_i denote the Boolean function with the highest z-score among those functions that output 1 for exactly i inputs. As we have denoted, z_{f_i} is the Z-score of f_i. We prove later in Sect. 3 that the truth table of f_i will contain 1 in the top i most occurring m-bit patterns. We claim that for the stream $D_{all-two,m,s}$, the maximum Z-score would be obtained for $i = u$. Here, #1 would be

$$\#1 = \begin{cases} i(s+1) & i \leq u, \\ u(s+1) + (i-u)s & i > u. \end{cases} \tag{11}$$

When $i \leq u$, we put 1's in the outputs of the u patterns that occur $s+1$ times, so $\#1 = i(s+1)$. Similarly, for the case where $i > u$, the top u patterns will occur $(s+1)$ times whereas rest of the $i - u$ patterns will occur s times making $\#1 = u(s+1) + (i-u)s$. Now, we calculate z_{f_i} as,

$$z_{f_i} = \left| \frac{\#1 - nq_f}{\sqrt{nq_f(1-q_f)}} \right|, \tag{12}$$

where $q_f = \frac{i}{2^m}$. For simplicity of notation let $M = 2^m$. Then z_{f_i} for both cases is

$$z_{f_i} = \begin{cases} \left| \frac{i(s+1) - \frac{ni}{M}}{\sqrt{n(\frac{i}{M})(1-\frac{i}{M})}} \right|, & i \leq u \\ \\ \left| \frac{u(s+1)+s(i-u) - \frac{ni}{M}}{\sqrt{n(\frac{i}{M})(1-\frac{i}{M})}} \right|, & i > u \end{cases} \tag{13}$$

$$= \begin{cases} c_1 \sqrt{\frac{i}{M-i}}, & i \leq u, \\ c_2 \sqrt{\frac{M-i}{i}}, & i > u, \end{cases} \tag{14}$$

where,

$$c_1 = \frac{M-u}{\sqrt{n}} \quad c_2 = \frac{u}{\sqrt{n}}. \tag{15}$$

From the above result, we can see that z_{f_i} is a decreasing function for $i > u$ and it is an increasing function when $i \leq u$. By this, we can conclude that z_{f_i} will be maximum for $i = u$. Thus, maximum Z-score obtained by plugging in $i = u$ will be,

$$Z_{f_u} = \sqrt{\frac{u(M-u)}{n}}. \tag{16}$$

2.3 Maximum *Z*-score When Some of the Patterns Arrive Only, and Only Once

Let us consider $N = nm$ many bits of data, where $n = u < 2^m$. There are u many m-bit patterns with frequency 1 in any order and the rest are not appearing. Denote such a data stream as $D_{some,m,u}$. It can be seen that this data stream is a special case of the $D_{all-two,m,s}$ for $s = 0$. So by plugging in the values from the above equation, we obtain

$$Z = \sqrt{\frac{u(M-u)}{n}} = \sqrt{\frac{u(M-u)}{sM+u}} = \sqrt{M-u}. \tag{17}$$

This is important for large block sizes, as, for large blocks, it is very clear that only a few patterns will arrive, and most of them will arrive only once. For example, if we consider $m = 256$, then in a stream of 2^{38} bits, only 2^{30} blocks will be generated. This is a very small part of 2^{256}, and thus, each of the blocks that appear will appear generally only once, and the rest huge numbers will not appear at all. That is the reason this situation needs to be studied for practical purposes in the cases of larger block sizes.

To provide specific data, we consider 1MB (megabyte) data generated by AES (OFB mode, random IV) with block size $m = 64$ bits, i.e., 8 bytes. Since the number of blocks in data is much less than 2^{64}, the probability of getting all blocks distinct is high. The number of blocks for this data would be $n = \frac{2^{20}}{2^3} =$

131072 (because 1MB $= 2^{20}$ bytes and 64 bits $= 2^3$ bytes). We have checked that the generated data had all the blocks distinct, so $u = 131072$. The maximum Z-score obtained by our implementation (see Appendix) is $4294967295.9999847 = \sqrt{2^{64} - 131072} = \sqrt{M - u}$, that matches the theory. It is important to highlight that the Z-score might be very high in such a scenario.

3 Finding the Best Boolean Function to Have Maximum Z-score

As described in Sect. 1.3, the $BoolTest$ algorithm $\mathcal{B}(deg, m, t, k)$ searches through $\alpha = \binom{m}{deg}$ monomials of degree deg. Now the top t monomials (with high Z-scores) are considered for the second stage where k out of these are added (XORed). That is, $BoolTest$ searches through a very limited space of Boolean functions since all the terms in monomials are of the same degree. On top of that, it has a fixed number of terms k, which makes it at most $\binom{\alpha}{k}$ candidate functions. This limited function search space significantly fails to discover the function with the best Z-score. As described in [7], the Z-score is considered to heuristically find the good Boolean functions.

In this section, we will present that one can devise a deterministic algorithm to discover the Boolean function that will indeed provide the highest possible Z-score. It is computationally elusive to exhaustively search all the 2^{2^m} Boolean functions for $m \geq 6$. We demonstrate an $\mathcal{O}(N \log N)$ algorithm for data size N, to achieve this for any arbitrary block size m. So it would be possible to run this algorithm on almost any size of data that can be stored (and read in reasonable time) in a particular machine.

In this regard, let us first define some notations and prove a few technical results. We define MS for a function f as follow:

$$MS(f) = |p_f - q_f|, \tag{18}$$

where p_f and q_f are same as defined earlier. If there are t inputs for which we get 1 as output then, $q_f = \frac{t}{2^m}$. We first show how using this metric we will obtain the function with the highest Z-score in $\mathcal{O}(m2^m)$ time (an improvement over $\mathcal{O}(2^{2^m})$). Define $\mathcal{F}_t$ to be the set of all m-variable Boolean functions that will output 1 on exactly t of the 2^m possible inputs. Thus, for all functions $f \in \mathcal{F}_t$, we have

$$q_f = \frac{t}{2^m} \tag{19}$$

Since we have fixed the value q_f for a given set $\mathcal{F}_t$, we have essentially fixed the denominator of the Z-score. Thus it would be easy to maximize it among the functions in this set. Further, if $\mathcal{F}$ is a set of all Boolean functions of m variables,

$$\mathcal{F} = \bigcup_{t=0}^{n} F_t \tag{20}$$

Let $f_t \in \mathcal{F}_t$ be the function with maximum Z-score in the set $\mathcal{F}_t$. So the Boolean function with the best Z-score in $\mathcal{F}$ would be,

$$f = \arg\max_{f_t} z(f_t), \tag{21}$$

where, $z(f_t)$ gives the Z-score for the function f_t. We claim that we can find the function with maximum MS in each $\mathcal{F}_t$ (say $f_t \in \mathcal{F}_t$) efficiently. We further claim that the same f_t will provide us with the highest Z-score inside each $\mathcal{F}_t$.

Let us first present a technical result. The complement of a Boolean function f will have a truth table with negated outputs for each input. In other words, if f is a Boolean function and f' is its complement then for input x

$$f'(x) = 1 \oplus f(x) \tag{22}$$

So if a Boolean function f of m-input, outputs 1 for t-inputs then f' will output 1 for the other $2^m - t$ inputs. Now, we prove that the Z-score for both f and f' would be the same.

Proposition 1. *The Z-score for a function f and its complement f' would be the same.*

Proof. Let us assume that f is an m-input Boolean function that will output 1 t times then,

$$z_f = \frac{n}{\sqrt{nq_f(1-q_f)}} \left| \sum_{x:f(x)=1} p(x) - \frac{t}{2^m} \right| \tag{23}$$

where $p(x)$ is the proportion of input x in the input data. Now, $\sum_{x:f(x)=1} p(x) = p_f$. This can also be written as,

$$z_f = \frac{n}{\sqrt{nq_f(1-q_f)}} \left| \left(1 - \sum_{x:f(x)=0} p(x)\right) - \frac{t}{2^m} \right| \tag{24}$$

By definition, if f outputs 0 for some input x, then for the similar inputs f' outputs 1, then

$$z_f = \frac{n}{\sqrt{nq_f(1-q_f)}} \left| \left(1 - \frac{t}{2^m}\right) - \sum_{x:f'(x)=1} p(x) \right| \tag{25}$$

This is the Z-score for f' as,

$$z_f = \frac{n}{\sqrt{nq_f(1-q_f)}} \left| \sum_{x:f'(x)=1} p(x) - \left(\frac{2^m - t}{2^m}\right) \right| \tag{26}$$

$$= \frac{n}{\sqrt{n(1-q_{f'})q_{f'}}} \left| \sum_{x:f'(x)=1} p(x) - \left(\frac{2^m - t}{2^m}\right) \right| = z_{f'} \tag{27}$$

□

Now we present the main result.

Theorem 1. *If f_t is a function in $\mathcal{F}_t$ and $MS(f_t) = \max_{f \in \mathcal{F}_t} MS(f)$ then*

$$f_t(x) = \begin{cases} 1 & \text{if } x \in A_t \\ 0 & \text{otherwise} \end{cases} \tag{28}$$

where A_t is the set of t-most occurring inputs in the data file.

Proof. For any $f \in \mathcal{F}_t$ we have,

$$MS(f) = |p_f - q_f|$$

Our goal is to maximize MS and find the function f for which we get the maximum score. As the function f outputs 1 on exactly t inputs, we have

$$MS(f) = \left| p_f - \frac{t}{2^m} \right| \tag{29}$$

Since we have fixed t, maximizing $MS(f)$ means either maximizing p_f or minimizing p_f.

Following Proposition 1, we show that maximization is enough, i.e., it is not necessary to minimize the sum p_f and this can be seen by the fact that a function and its complement share the same Z-score. Let's say we minimize the value of p_f and f gives 1 for t inputs. Since we are minimizing the summation what we are doing is taking the least frequent t inputs and making the function f output 1 on these inputs. In other words, f outputs 0 for the top $2^m - t$ inputs, so f' (the complement of f) outputs 1 for those $2^m - t$ inputs. Since $f' \in \mathcal{F}_{2^m-t}$, we can calculate the function $g \in \mathcal{F}_{2^m-t}$ that has the best Z-score in $\mathcal{F}_{2^m-t}$. If $g = f'$, then we do not need to minimize p_f because we would anyway find the function f', and calculating the Z-score for f would not be needed. If $g \neq f'$, then also we do not need to minimize p_f, because there is already a function g, that has Z-score greater than that of f. Thus we do not need to minimize the summation separately as we gain information about a function's complement from the function itself.

For a given input file, we calculate the number of occurrences (or probability of occurrence) of each m-bit block $[0 \text{ to } 2^m - 1]$. We will sort an array containing the 2^m m-bit blocks (consider truth table rows) by their proportion in non-decreasing order. For each t, we construct a function f_t that returns 1 on the top t highest probability blocks and 0 for the rest.

For any other function ${f'}_t$ that outputs 1 on exactly t inputs, $p_{{f'}_t} \leq p_{f_t}$ because we are summing over the highest $p(x)$. Note that there might be other functions ${f'}_t$ with $p_{{f'}_t} = p_{f_t}$, but $p_{{f'}_t}$ can never be greater than p_{f_t}. □

Now that we have established a way to find the maximum MS, we use it to calculate the Z-score. The relation between MS and Z-score can be shown as:

$$\begin{aligned} z_f &= \left| \frac{\#1 - nq_f}{\sqrt{nq_f(1-q_f)}} \right| = \left| \frac{np_f - nq_f}{\sqrt{nq_f(1-q_f)}} \right| = \frac{n}{\sqrt{nq_f(1-q_f)}} |p_f - q_f| \\ &= \frac{n}{\sqrt{nq_f(1-q_f)}} MS(f). \end{aligned} \tag{30}$$

Fixing the set F_t from which the function will be chosen, we know that the term $\frac{n}{\sqrt{nq_f(1-q_f)}}$ is constant and so if a function maximizes MS, it maximizes the Z-score. So, we find the highest Z-score for each $\mathcal{F}_t$, $t = \{1, \ldots, 2^m - 1\}$. Note that, for $t = 0$ or $t = 2^m$ only constant functions are possible with undefined Z-scores, that we will not consider. Then we find the maximum among these $2^m - 1$ Z-scores. Based on this, we have the following algorithm.

Algorithm 2: modified booltest (run time: $\mathcal{O}(2^m)$)

```
/* Suppose S is the set of inputs for which the function under
   consideration returns 1 (this set completely defines the
   function)                                                    */

I ← 0 to 2^m − 1 (truth table inputs);
P ← probability of occurance of each m-bit block according to data;
z ← 0 // The Z-score value
t ← 1 // Number of ones as output
p ← 0
SORT (I, P) // Sort array of truth-table-inputs wrt their
probability of occurrence
while t < 2^m do
    // try to maximize |p - q|
    p ← p + P[t − 1] // Adding next most occurring input to S
    q ← t / 2^m
    z_t ← n · (p − q)
    z_t ← z_t / sqrt(nq(1−q))
    if z_t > z then
        t_max ← t
        z ← z_t
    t ← t + 1
end
```

In Algorithm 2 above, to get the actual Boolean function providing the highest Z-score, we can sort the inputs to the truth table using their probabilities and output the truth table with 1's in the t_{max} highest probability inputs.

3.1 Improving the Time and Space Complexity Further

The main drawbacks of the above algorithm are as follows:

- It takes $\mathcal{O}(m2^m + N)$ time, $\mathcal{O}(N)$ for calculating the probabilities, $\mathcal{O}(m2^m)$ for sorting the inputs according to the probabilities, and $\mathcal{O}(2^m)$ iterations in the loop.
- It would also take $\mathcal{O}(2^m)$ space for storing the probability array.

Note that N is the length of the data in bits, and thus, we have to accept that for analysis. On the other hand, if m is large, such as $m = 128$, then the above algorithm cannot be executed with the present computational power. Thus, we now improve the algorithm so that it requires $\mathcal{O}(N \log N)$ time and $\mathcal{O}(N)$ space.

Some Observations. Suppose all the 2^m possible blocks do not appear in the input data. This is natural in the case where the block size is large. If we have data of the order of 2^{40} (say 2^{32} blocks of length 2^8 each), then the block size being 256 bits, it is very clear that at most 2^{32} different patterns may appear. Thus, the algorithm should be redesigned. We already considered that there are n blocks in the data, i.e., $n = \frac{N}{m}$. Now let us consider there are d distinct m bit patterns, i.e., $d \leq n$. That is, there are d data blocks that have non-zero probabilities of occurrences. Now we explain that as in the algorithm above, we should not check all t. Rather, increasing t above d is not required as the Z-score for those values of t would not be more than the Z-score achieved for $t = d$.

Theorem 2. *Let, d be the number of distinct blocks that appear in the data and z_t be the highest possible Z-score for a function $f_t \in \mathcal{F}_t$. Then, $z_{f_d} > z_{f_j}$, $\forall j > d$.*

Proof. For block size m and fixed input data having n many blocks (may not be all distinct), let, f_t be the function with highest Z-score in $\mathcal{F}_t$ and z_{f_t} be the Z-score corresponding to f_t. For $t = d$, consider the truth table of the function f_d providing the highest Z-score z_{f_d}. By Algorithm 2, we know that there would be 1's in the d highest probability blocks in the truth table, which would be all the blocks with non-zero probability. Thus, for every possible data block that appears in the data, there would be a 1 in the corresponding row in f_d's truth table. So, #1, i.e., the number of data blocks in the input which when fed into the function f_d would return 1 is n, the number of blocks. Thus the Z-score will be:

$$z_{f_d} = \left| \frac{n - \frac{d}{2^m}n}{\sqrt{n\frac{d}{2^m}(1 - \frac{d}{2^m})}} \right| = \sqrt{n} \left| \frac{1 - \frac{d}{2^m}}{\sqrt{\frac{d}{2^m}(1 - \frac{d}{2^m})}} \right| \tag{31}$$

When we increase t, by the above algorithm, the truth table corresponding to each f_t would contain 1's in t highest probability blocks. Hence, it would contain 1's in all the blocks with non-zero probability (since there are $d < t$ such blocks), i.e., all blocks that appear in the data will have 1 in the corresponding truth-table row. Thus, #1 will remain fixed at n. Hence the Z-score will be:

$$z_{t>d} = \left| \frac{n - \frac{t}{2^m}n}{\sqrt{n\frac{t}{2^m}(1 - \frac{t}{2^m})}} \right| = \sqrt{n} \left| \frac{1 - \frac{t}{2^m}}{\sqrt{\frac{t}{2^m}(1 - \frac{t}{2^m})}} \right| \tag{32}$$

Now, the function,

$$h(y) = \left| \frac{1-y}{\sqrt{y(1-y)}} \right| \tag{33}$$

is a decreasing function for $y \in (0, 1)$. The graph of h is shown below:

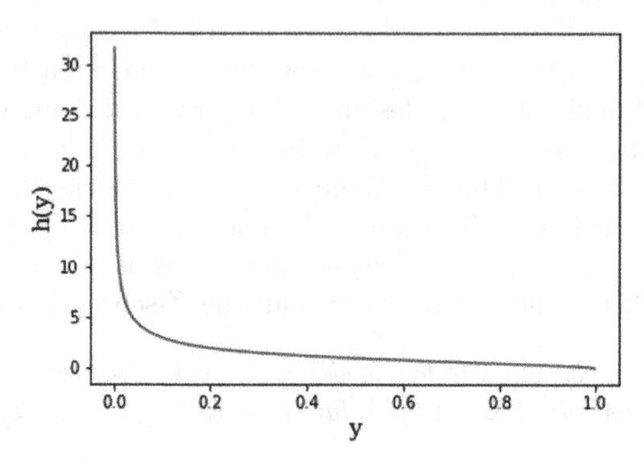

So, for $t > d$, we have: $\sqrt{n}h(\frac{d}{2^m}) > \sqrt{n}h(\frac{t}{2^m})$, i.e.,

$$\sqrt{n} \left| \frac{1 - \frac{d}{2^m}n}{\sqrt{\frac{d}{2^m}(1 - \frac{d}{2^m})}} \right| > \sqrt{n} \left| \frac{1 - \frac{t}{2^m}n}{\sqrt{\frac{t}{2^m}(1 - \frac{t}{2^m})}} \right|,$$

i.e., $z_{f_d} > z_{f_t}$, for $t > d$. Thus the proof. □

Based on the above result, we obtain a more efficient algorithm. Now t will be considered till d which is $\mathcal{O}(N)$ and not up to 2^m like in the previous method. Since we are considering large m now, we cannot use the earlier idea of having 2^m length array for storing the probability of each m-bit block or storing and sorting all the truth table rows. Instead, we sort the m-bit blocks in the input data interpreting their natural decimal values. This brings the same blocks together. Then we can count the number of occurrences of each block by counting the number of consecutive similar blocks, and the number of distinct blocks that are appearing in the data.

Function Encoding. Note that representing an arbitrary m-input Boolean function would require $O(2^m)$ space in the worst case, irrespective of the function encoding scheme such as truth table or ANF, etc. However, in our case, the algorithm ensures that whatever is the highest Z-score Boolean function, it would only output 1 for at most d inputs. So, consider the encoding scheme where we just list the inputs (truth table rows) for which the function returns 1, that is we will work with the support set of the function. This will require $O(N)$ space. Based on all these we present an efficient algorithm.

Algorithm 3: optimized modified booltest (run time: $\mathcal{O}(N \log N)$)

```
// Sort data by value of m bit blocks to bring same blocks together
SORT(Data)
m ← block size
n ← number of blocks in data
count ← 1
idx ← 0
DistinctBlocks ← empty N length array of m-bit blocks
Occurrences ← N length array of integers initialized to 0s
for i ← 1 to N do
    if i = N or Data[i] ≠ Data[i-1] then
        Occurrences[idx] ← count
        DistinctBlocks[idx] ← Data[i − 1]
        count ← 0
        idx ← idx + 1
    count ← count + 1
end
d ← idx // d is the number of distinct blocks in data
SORT-BY-OCCURRENCES(DistinctBlocks, Occurrences)
Δ ← 0
// Δ is the number of occurrences of the blocks in Data, for which
the candidate function being considered returns 1
z_max ← −1
t_max ← −1
for i ← 0 to d − 1 do
    t ← i + 1
    Δ ← Δ + Occurrences[i]
    p ← Δ/N
    q ← t/2^m
    z ← | N(p−q) / sqrt(Nq(1−q)) |
    if z > z_max then
        z_max ← z
        t_max ← i + 1

end
return z_max, list of top t_max inputs of DistinctBlocks
```

4 Results

In this section, we present the experimental results and compare them with the existing works.

4.1 RC4

It is well known [4] that the second byte of the RC4 keystream is biased towards zero with probability almost $\frac{2}{256}$, which is significantly higher than the uniform random value $\frac{1}{256}$. Consider that a long keystream byte sequence is generated with randomly chosen secret keys and then accumulates the second output bytes of RC4 in each case. Since the probability of any other value except zero is slightly less than $\frac{1}{256}$, according to our strategy, the Boolean function that should provide the highest Z-score should have output 1 for all zero input and the rest of the outputs should be 0. This function contains all the terms in ANF.

Based on the bias, it can be shown that the best distinguisher Boolean function for this data, working on 8-bit blocks is the one that returns 1 on the (0, 0, 0, 0, 0, 0, 0, 0) input and 0 on everything else (say f_0). The complement of this function may also be considered. Let us name the input variables as $(x_0, x_1, \ldots, x_7)$ for the eight-bit block. The ANF of the function to maximize the Z-score contains all the terms in ANF, as provided by our Algorithm 3 in all the runs with different sets of data. It is clear that the ANF is quite complicated and such an ANF will never be considered for BoolTest [7]. We note that taking the constraint of degree 3, *BoolTest* [7] provides different functions in different runs towards the sub-optimal efforts in maximizing the Z-score, i.e., cannot provide the correct answer due to sub-optimality.

Table 1. Testing RC4 2nd byte samples.

File	B1		B2		Bool-Test-2	
	Highest Z-score	Best-distinguisher	Highest Z-score	Best-distinguisher	Highest Z-score	Best-distinguisher
RC4 1MB	4.55	$x_3x_6 + x_3x_4$	4.47	$x_1x_3x_5 + x_3x_6x_7 + x_3x_4x_7$	64.16	f_0
RC4 10MB	11.56	$x_2x_6 + x_3x_7$	9.90	$x_3x_4x_7 + x_0x_2x_7 + x_2x_4x_6$	204.36	f_0
RC4 100MB	31.05	$x_6x_7 + x_4x_5$	23.64	$x_2x_6x_7 + x_3x_4x_5 + x_1x_4x_6$	643.14	f_0

In Table 1, B1 is BoolTest [7] with parameters (degree = 2, combine-degree = 2) and B2 represents BoolTest [7] run with parameters (degree = 3, combine-degree = 3). Our results are presented with Bool-Test-2, where it could be seen that the Z-score is much higher.

4.2 Comparison with Java Rand and AES

In this section, we show a comparison of the Z-scores that we have obtained from Java Random and AES. We use 10MB files for both AES and Java Random and apply *BoolTest* [7] as well as our Algorithm 3.

Table 2. Results

File	Block-size	B_1	B_2	Best Z-score
Java 10MB	8	2.668	3.8144	11.7884
Java 10MB	32	4.2611	5.8122	65526.278
Java 10MB	256	5.6797	9.3637	3.4×10^{38}
AES 10MB	8	3.4817	4.7543	12.1083
AES 10MB	32	4.322	6.289	65526.10
AES 10MB	256	5.557	8.4509	3.4×10^{38}

In the Table 2 above, B_1 is the *BoolTest* algorithm with parameters $\mathcal{B}(deg = 2, m = \text{Block-size}, t = 128, k = 2)$ and B_2 is the BoolTest algorithm with parameters $\mathcal{B}(deg = 3, m = \text{Block-size}, t = 128, k = 3)$. As explained in this initiative, since *BoolTest* [7] searches in limited function space, the Z-score obtained with those constraints is sub-optimal. On the other hand, we obtain very high scores in this regard. Our analysis in Sect. 2.3 theoretically explains why such large values in Z-score are possible.

4.3 Cross-testing by the Generated Polynomials, i.e., Functions

The motivation is to generate the Boolean function for which the Z-score will be maximized so that one can interpret a high value outside some interval as non-randomness. One interesting methodology to evaluate this *BoolTest* is to generate the best function from one data set and to use that function to evaluate the Z-score of another random-looking data set. First, let us consider the *BoolTest* [7] heuristics in this regard.

Table 3. Cross-testing with *BoolTest* [7].

Function generated by	Z-score				
	Java 1MB	Java 10MB	Java 100MB	AES 1MB	AES 10MB
Java 1MB (m=8)	3.22	0.8	1.0	0.156	1.124
Java 1MB (m=256)	6.08	0.22	0.80	0.148	1.977
Java 100MB (m=256)	2.6	8.5	36.75	1.1	2.0
AES 10MB (m=256)	0.56	0.30	1.34	0.02	8.45

In the Table 3, for Java 1MB ($m = 8$), we have used the parameters $\mathcal{B}(deg = 2, m = 8, t = 128, k = 2)$. For Java 1MB ($m = 256$), we have used the parameters $\mathcal{B}(deg = 2, m = 256, t = 128, k = 2)$ and for the Java 100MB ($m = 256$) we have used the parameters $\mathcal{B}(deg = 3, m = 256, t = 128, k = 3)$.

Table 4. Cross-testing for our algorithm

Function generated by	Z-score				
	Java 1MB	Java 10MB	Java 100MB	AES 1MB	AES 10MB
Java 1MB (m=8)	13	0.26	0.87	0.12	0.38
Java 1MB (m=8)	0.58	11.8	0.50	0.20	1.0
Java 100MB(m=256)	10^{-33}	3×10^{-33}	3.4×10^{38}	9.6×10^{-34}	3×10^{-33}
AES 10MB(m=256)	3×10^{-34}	9.6×10^{-34}	3×10^{-33}	3×10^{-34}	3.4×10^{-38}

Using our Algorithm 3 (with implementation in Appendix) we performed cross-testing too and obtained the following results.

The small values that we obtained for $m = 256$ are probably because most of the 256-bit blocks for which the generated function outputs 1 will never arrive in other samples of the data. The expected number of 1 s will also be small.

To summarize, we generate the Boolean function for a distinguisher based on a particular sample of data. Then with this function, we run the distinguisher for a different set of random-looking data and observe the Z-score. Generally, the lower values related to AES provide the understanding that it demonstrates more randomness than the Java random number generator. This is a natural conclusion, but these kinds of cross-testing require further investigation and more concrete theoretical support.

5 Conclusion

In this paper, we have presented several critical points related to *BoolTest* [7], which is considered to be a method to evaluate the randomness of a stream of data. In this connection, we present combinatorial results related to identifying the most suitable Boolean functions in maximizing the Z-score that could not be achieved in the heuristic presented for *BoolTest* [7]. Our Algorithm 3 finds the best Boolean function having the maximum Z-Score in $O(N \log N)$ time, given N amount of data. While we solve certain combinatorial problems related to *BoolTest*, the caveat is, this test is not sufficient to conclude on randomness or non-randomness of a given stream of data. Certain statistical interpretations have been discussed in [7, Section 5], but we believe that this tool needs further evaluation. For example, one may consider cross-testing based on the generated polynomials from *BoolTest* [7] or by our method that we have discussed in Sect. 4.3. Further analysis in this regard might provide a better understanding of this domain, which we put forward for future research.

Acknowledgment. The authors like to acknowledge the anonymous reviewers for the detailed comments that improved the editorial as well as the technical presentation of the paper.

Appendix : Implementation Details

For large block sizes, the Z-score would be very large and it would not be possible to store the results accurately in 64 bits data elements of C programming compilers. For example, the highest Z-score for a block size of 256 might be of the order of 10^{38}. It would require $\sim$ 126 bits to represent such integers up to 10^{38}. To maintain accuracy, we instead use the GNU multi-precision library (GMP) for the calculations [10].

Listing 1.1. C code for final algorithm

```
// data = address (of first byte) of nm bit data
// len = length in bytes of data, i.e, nm/8
// m = block size
void generate(unsigned char* data, int len, int m)
{
    unsigned long n = (len*8)/m;

    // merge sort m-bit blocks at address 'data'
    // by the value of m-bit blocks
    sort_large_block(data, 0, n - 1, m);


    if(n == 0)
    {
        printf("no data\n\n");
        return;
    }

    // need length(occurences) = num_distinct_blocks ( <= n )
    int* occurences = malloc(n * sizeof(int));
    // need length(distinct_blocks) = num_distinct_blocks * size_of_block
        (<= data size)
    unsigned char* distinct_blocks = malloc(len * sizeof(unsigned char));

    unsigned long num_distinct_blocks = 0;
    unsigned long curr_count = 1, idx = 0;
    unsigned long sum_occurences = 0;
    for(unsigned long i=1; i<=n; i++)
    {
        if(i==n || !same_block(data, i, data, i-1))
        {
            copy_block(distinct_blocks, idx, data, i-1, m);
            occurences[idx] = curr_count;
            sum_occurences += occurences[idx];
            curr_count = 0;
            idx++;
        }
        curr_count++;
    }
    num_distinct_blocks = idx;

    // merge sort m-bit blocks at address 'distinct_blocks'
    // by their number of occurrences in the data
    sort_by_occurences(distinct_blocks, occurrences, 0, num_distinct_blocks
        - 1, m);

    mpf_t z_max;
    mpf_init(z_max); mpf_set_ui(z_max, (unsigned long) 0);
    mpf_t t, q, p, MS, z, d1, d2;
    mpf_init(t);
    mpf_init(q);
    mpf_init(p);
    mpf_init(MS);
    mpf_init(z);
```

```
    mpf_init(d1);
    mpf_init(d2);

    int t_max = -1; int num_one = 0;
    sum_occurences = 0;

    for(unsigned long i=0; i<num_distinct_blocks; i++)
    {
        sum_occurences += occurences[i];

        mpf_set_ui(t, (unsigned long)(i + (unsigned long)1));
        mpf_set_ui(d1, 2); mpf_pow_ui(d1, d1, m);

        if(mpf_cmp(t, d1) == 0)
        {
            printf("t = 2^m: break\n\n");
            break;
        }

        // MS
        mpf_div(q, t, d1);
        mpf_set_ui(p, sum_occurences);
        mpf_div_ui(p, p, (unsigned long)n);
        mpf_sub(MS, p, q);
        mpf_abs(MS, MS);

        // z-score
        mpf_mul_ui(MS, MS, (unsigned long) n);
        mpf_set_ui(d2, (unsigned long) 1);
        mpf_sub(d2, d2, q);
        mpf_mul(d2, d2, q);
        mpf_mul_ui(d2, d2, (unsigned long) n);
        mpf_sqrt(d2, d2);
        mpf_div(z, MS, d2);

        if(mpf_cmp(z, z_max) > 0)
        {
            mpf_set(z_max, z);
            t_max = i+1;
            num_one = sum_occurences;
        }

    }

    // get anf
    if(m <= 16)
    {
        int x = m - 3; if(x < 0) x = 0;
        unsigned char* truth_table = (unsigned char*)malloc((1<<x)*sizeof(
             unsigned char));

        for(int i=0; i<(1<<x); i++)
            truth_table[i] = 0;

        for(int i=0; i<t_max; i++)
            set_bit(truth_table, get_block_as_int(distinct_blocks, i));
        anf_from_truth_table(truth_table, m);
        free(truth_table);
    }

    printf("highest z-score --- \n");
    mpf_out_str(stdout, 10, 0, z_max);
    printf("\n t_max=%d, num_one = %d\n\n", t_max, num_one);

    save_bool_function(distinct_blocks, num_distinct_blocks, t_max, m);\\
    free(occurences);
    free(distinct_blocks);
}
```

References

1. Brown, R.G., Eddelbuettel, D., Bauer, D.: Dieharder: a random number test suite (Version 3.31.1) (2014). https://webhome.phy.duke.edu/rgb/General/dieharder.php
2. Daemen, J., Rijmen, V.: AES proposal: Rijndael (1998). https://csrc.nist.gov/csrc/media/projects/cryptographic-standards-and-guidelines/documents/aes-development/rijndael-ammended.pdf
3. Gustafson, H., Dawson, E., Nielsen, L., Caelli, W.: A computer package for measuring the strength of encryption algorithms. Comput. Secur. **13**(8), 687–697 (1994)
4. Mantin, I., Shamir, A.: A practical attack on broadcast RC4. In: Matsui, M. (ed.) FSE 2001. LNCS, vol. 2355, pp. 152–164. Springer, Heidelberg (2002). https://doi.org/10.1007/3-540-45473-X_13
5. Marsaglia, G.: The Marsaglia random number CDROM including the diehard battery of tests of randomness; National Science Foundation: Alexandria, VA, USA (1995). https://en.wikipedia.org/wiki/Diehard_tests, https://web.archive.org/web/20160125103112/,http://stat.fsu.edu/pub/diehard/
6. Rukhin, A., et al.: A statistical test suite for random and pseudorandom number generators for cryptographic applications. https://nvlpubs.nist.gov/nistpubs/legacy/sp/nistspecialpublication800-22r1a.pdf (2010). Random Bit Generation, NIST, https://csrc.nist.gov/projects/random-bit-generation
7. Sýs, M., Klinec, D., Svenda, P.: The efficient randomness testing using Boolean functions. In: 14th International Conference on Security and Cryptography (Secrypt 2017), pp. 92–103. SciTePress (2017). https://www.scitepress.org/papers/2017/64251/64251.pdf
8. Sýs, M., Klinec, D., Kubicek, K., Svenda, P.: BoolTest: the fast randomness testing strategy based on Boolean functions with application to DES, 3-DES, MD5, MD6, and SHA-256. E-Business and Telecommunications (2019). https://crocs.fi.muni.cz/public/papers/secrypt2017selected
9. Walker, J.: Pseudorandom number sequence test program (2018). https://www.fourmilab.ch/random/
10. The GNU MP Bignum Library. Available at: https://gmplib.org/. Accessed 6 Sept 2022

Weightwise Almost Perfectly Balanced Functions: Secondary Constructions for All n and Better Weightwise Nonlinearities

Agnese Gini(✉) and Pierrick Méaux

University of Luxembourg, Esch-sur-Alzette, Luxembourg
{agnese.gini,pierrick.meaux}@uni.lu

Abstract. The design of FLIP stream cipher presented at Eurocrypt 2016 motivates the study of Boolean function with good cryptographic criteria when restricted to subsets of $\mathbb{F}_2^n$. Since the security of FLIP relies on properties of functions restricted to subsets of constant Hamming weight, called slices, several studies investigate functions with good properties on the slices, *i.e.* weightwise properties. A major challenge is to build functions balanced on each slice, from which we get the notion of Weightwise Almost Perfectly Balanced (WAPB) functions. Although various constructions of WAPB functions have been exhibited since 2017, building WAPB functions with high weightwise nonlinearities remains a difficult task. Lower bounds on the weightwise nonlinearities of WAPB functions are known for very few families, and exact values were computed only for functions in at most 16 variables.

In this article, we introduce and study two new secondary constructions of WAPB functions. This new strategy allows us to bound the weightwise nonlinearities from those of the parent functions enabling us to produce WAPB functions with high weightwise nonlinearities. As a practical application, we build several novel WAPB functions in up to 16 variables by taking parent functions from two different known families. Moreover, combining these outputs, we also produce the 16-variable WAPB function with the highest weightwise nonlinearities known so far.

Keywords: FLIP cipher · Boolean functions · Weightwise (almost) perfectly balanced function · Weightwise nonlinearity

1 Introduction

The study of Boolean functions with good cryptographic criteria when restricted to subsets of $\mathbb{F}_2^n$ became recently relevant due to their role in the security of FLIP stream cipher introduced by Méaux, Journault, Standaert, and Carlet at Eurocrypt 2016 [MJSC16]. FLIP's filter function is evaluated on a set of vectors of $\mathbb{F}_2^n$ having constant Hamming weight, as a consequence of design choices to make the cipher homomorphic-friendly. Hence, the security of FLIP family relates to certain properties of Boolean functions when they are restricted to some input

T. Isobe and S. Sarkar (Eds.): INDOCRYPT 2022, LNCS 13774, pp. 492–514, 2022.
https://doi.org/10.1007/978-3-031-22912-1_22

subsets, *e.g.* slices $\mathsf{E}_{k,n} = \{x \in \mathbb{F}_2^n \mid \mathsf{w}_\mathsf{H}(x) = k\}$ of the hypercube $\mathbb{F}_2^n$. In [CMR17], the Boolean cryptographic criteria on restricted sets such as balancedness, nonlinearity and algebraic immunity were first studied. In particular, the concept of balancedness for a Boolean function $f\colon \mathbb{F}_2^n \to \mathbb{F}_2$, *i.e.* the preimages of 0 and 1 under f have the same cardinality, is extended to *weightwise perfectly balancedness*, *i.e.* all the restrictions of f to the slices $\mathsf{E}_{k,n}$ are balanced. As balanced functions are generally suitable for avoiding constructions with statistical biases, we expect the same for Weightwise Perfectly Balanced (WPB) functions in the context of inputs with fixed Hamming weight. More precisely, WPB functions are functions balanced on each slice with $1 \leq k \leq n-1$, equal to 0 in 0_n and to 1 in 1_n. However, WPB functions only exist for n a power of 2, since the balancedness on each slice requires the cardinality of each one of these sets to be even. Thus, the authors also introduced the notion of *weightwise almost perfectly balancedness* allowing a tolerance for slices of odd cardinality sufficiently small to preserve the reliability of these functions. Namely, for Weightwise Almost Perfectly Balanced (WAPB) functions we allow the cardinalities of the preimages of 0 and 1 to differ of 1 when the slice $\mathsf{E}_{k,n}$ has an odd cardinality.

Carlet *et al.* also provided in [CMR17] a recursive construction of WAPB functions for all n and a secondary construction of WPB functions. Afterwards, several other constructions have been proposed [LM19, TL19, LS20, MS21, ZS21, MSL21, GS22, ZS22, MPJ+22, GM22]. Being WAPB function relevant in a cryptographic context, all these works aim to produce W(A)PB functions having good parameters relatively to the other cryptographic criteria such as restricted and global nonlinearity, algebraic immunity and degree. For instance, the functions proposed in [TL19] have optimal algebraic immunity, while the family described in [LM19] has good nonlinearity on all the slices, also called weightwise nonlinearities. In fact, the weightwise nonlinearity is the criterion that got the most attention in these constructions, often used to compare the different families. It is also the criterion with more open problems; differently from $\mathbb{F}_2^n$ (and the associated concept of bent functions), the maximum nonlinearity that can take a function restricted to a slice is unknown, and bounds on this maximum are studied in different works [CMR17, MZD19, GM22]. Furthermore, a notion of restricted Walsh transform has been introduced [MMM+18] to study better the weightwise nonlinearity. Except for the exact weightwise nonlinearities obtained experimentally on functions up to 16 variables, in very few cases, this parameter is known or even bounded for a construction. There are lower bounds known for two families of WPB functions, the recursive construction of [CMR17], whose weightwise nonlinearities are studied in [Su21], and one construction from [LM19].

In this article, we present two novel secondary constructions of WAPB functions for all n with proven bound on their weightwise nonlinearities, and we use them to build a 16-variable WPB function with the highest weightwise nonlinearities exhibited so far. More precisely, our contributions are the following. First, we study the impact of the addition of symmetric functions and of Siegenthaler's construction on the restricted Walsh transform. Secondly, we introduce the notion of Special WAPB (SWAPB) functions, a sub-family where we fix the support size on the slices of odd cardinality. Then, we give two secondary constructions of SWAPB functions, first from an n-variable SWAPB function and an n-variable WAPB function to an $(n+1)$-variable WAPB function, and then from an n-variable SWAPB function to a $n+t$-variable SWAPB function. Very differently from the precedent constructions, these functions are obtained combining Siegenthaler's construction and addition of symmetric functions, which allows to derive a lower bound on the weightwise nonlinearities of the child function from the parameters of the parent functions. Furthermore, we prove that the recursive construction of [CMR17] gives WAPB functions that are inherently special.

Finally, we provide an experimental part, where we determine the exact parameters of functions in 8 and 16 variables. Specifically, we first build 8 and 16-variable WPB functions from our second construction seeded with CMR functions and with LM functions, *i.e.* functions from [CMR17] and [LM19], respectively. Thereafter, we combine (slice by slice) these functions in 16 variables to obtain the 16-variable function with the highest weightwise nonlinearities exhibited so far.

Organization: In Sect. 2 we give the necessary preliminaries on Boolean functions and (weightwise) cryptograhic criteria, and properties on the parity of binary coefficients. In Sect. 3 we introduce and study special WAPB functions, we give two secondary constructions and prove a lower bound on their weightwise nonlinearities. We prove that CMR WAPB functions are special functions in Sect. 4. Then, We give concrete functions in 8 and 16 variables, they are obtained from one of our new construction seeded by CMR functions, by LM functions, of mixing such functions to obtain higher weightwise nonlinearities. Finally, we conclude briefly the article in Sect. 5.

2 Preliminaries

In addition to classic notations we use $[a, b]$ to denote the subset of all integers between a and b: $\{a, a+1, \dots, b\}$. For readability we use the notation $+$ instead of $\oplus$ to denote the addition in $\mathbb{F}_2$ and $\sum$ instead of $\bigoplus$. For a vector $v \in \mathbb{F}_2^n$ we denote $\mathsf{w_H}(v)$ its Hamming weight $\mathsf{w_H}(v) = |\{i \in [1, n] \mid v_i = 1\}|$.

2.1 Boolean Functions and Weightwise Considerations

In this subsection we recall the main concepts on Boolean functions and their weightwise properties we will use in this article. We refer to *e.g.* [Car21] for Boolean functions and cryptographic parameters and to [CMR17] for the weightwise properties, also called properties on the slices. For $k \in [0, n]$ we call slice of the Boolean hypercube (of dimension n) the set $\mathsf{E}_{k,n} = \{x \in \mathbb{F}_2^n \mid \mathsf{w_H}(x) = k\}$. Accordingly, the Boolean hypercube is partitioned into $n+1$ slices where the elements have the same Hamming weight.

Definition 1 (Boolean Function). *A Boolean function f in n variables is a function from $\mathbb{F}_2^n$ to $\mathbb{F}_2$. The set of all n-variable Boolean functions is denoted $\mathcal{B}_n$.*

Definition 2 (Algebraic Normal Form (ANF) and degree). *We call Algebraic Normal Form of a Boolean function f its n-variable polynomial representation over $\mathbb{F}_2$ (i.e. belonging to $\mathbb{F}_2[x_1, \dots, x_n]/(x_1^2 + x_1, \dots, x_n^2 + x_n)$):*

$$f(x_1, \dots, x_n) = \sum_{I \subseteq [1,n]} a_I \left(\prod_{i \in I} x_i \right)$$

where $a_I \in \mathbb{F}_2$. The (algebraic) degree of f, denoted $\mathsf{deg}(f)$ is:

$$\mathsf{deg}(f) = \max_{I \subseteq [1,n]} \{|I| \mid a_I = 1\} \text{ if } f \text{ is not null}, 0 \text{ otherwise.}$$

To denote when a definition or a property is restricted to a slice we will use the subscript k. For example, for a n-variable Boolean function f we denote its support $\mathsf{supp}(f) = \{x \in \mathbb{F}_2^n \mid f(x) = 1\}$ and we refer to $\mathsf{supp}_k(f)$ for its support restricted to a slice, that is $\mathsf{supp}(f) \cap \mathsf{E}_{k,n}$.

Definition 3 (Balancedness). *A Boolean function $f \in \mathcal{B}_n$ is called balanced if $|\mathsf{supp}(f)| = 2^{n-1} = |\mathsf{supp}(f+1)|$.*

For $k \in [0, n]$, f is said balanced on the slice k if $||\mathsf{supp}_k(f)| - |\mathsf{supp}_k(f+1)|| \leq 1$. In particular when $|\mathsf{E}_{k,n}|$ is even $|\mathsf{supp}_k(f)| = |\mathsf{supp}_k(f+1)| = |\mathsf{E}_{k,n}|/2$.

Definition 4 (Weightwise (Almost) Perfectly Balanced Function (WPB and WAPB)). *Let $m \in \mathbb{N}^*$ and f be a Boolean function in $n = 2^m$ variables. It will be called weightwise perfectly balanced (WPB) if, for every $k \in [1, n-1]$, f is balanced on the slice k, that is $\forall k \in [1, n-1]$, $|\mathsf{supp}_k(f)| = \binom{n}{k}/2$, and:*

$$f(0, \cdots, 0) = 0, \quad \text{and } f(1, \cdots, 1) = 1.$$

The set of WPB functions in 2^m variables is denoted $\mathcal{WPB}_m$.

When n is not a power of 2, other weights than $k = 0$ and n give slices of odd cardinality, in this case we call $f \in \mathcal{B}_n$ weightwise almost perfectly balanced (WAPB) if:

$$|\mathsf{supp}_k(f)| = \begin{cases} |\mathsf{E}_{k,n}|/2 & \text{if } |\mathsf{E}_{k,n}| \text{ is even,} \\ (|\mathsf{E}_{k,n}| \pm 1)/2 & \text{if } |\mathsf{E}_{k,n}| \text{ is odd.} \end{cases}$$

The set of WAPB functions in n variables is denoted $\mathcal{WAPB}_n$.

Note that the definition of WAPB functions above (as introduced in [CMR17]) is more general than the one of WPB functions, for $n = 2^m$ the WPB functions are a subset of the WAPB functions since the value in 0_n and 1_n can be taken freely for the latter. Alternatively, $\mathcal{WAPB}_n$ corresponds to the set of functions at maximal distance from the set of n-variable symmetric functions $\mathcal{SYM}_n$, that is $\mathcal{WAPB}_n$ is metrically regular for the Hamming distance and $\mathcal{SYM}_n$ is its metric complement. We refer to [Tok12] for the notion of metrically regular sets and the survey [Obl20]. In [SSB18] various metrically regular sets are considered, WAPB functions are presented under the name of maximally asymmetric functions, and the authors provide the cardinality of $\mathcal{WAPB}_n$ (also given in [IMM13]) and the number of balanced WAPB functions.

Definition 5 (Nonlinearity and weightwise nonlinearity). *The nonlinearity* $\mathsf{NL}(f)$ *of a Boolean function* $f \in \mathcal{B}_n$*, where* n *is a positive integer, is the minimum Hamming distance between* f *and all the affine functions in* $\mathcal{B}_n$*:*

$$\mathsf{NL}(f) = \min_{g,\, \mathsf{deg}(g) \le 1} \{d_H(f,g)\},$$

where $g(x) = a \cdot x + \varepsilon$*,* $a \in \mathbb{F}_2^n, \varepsilon \in \mathbb{F}_2$ *(where* $\cdot$ *is some inner product in* $\mathbb{F}_2^n$*; any choice of an inner product will give the same value of* $\mathsf{NL}(f)$*).*

For $k \in [0,n]$ *we denote* NL_k *the nonlinearity on the slice* k*, the minimum Hamming distance between* f *restricted to* $\mathsf{E}_{k,n}$ *and the restrictions to* $\mathsf{E}_{k,n}$ *of affine functions over* $\mathbb{F}_2^n$*. Accordingly:*

$$\mathsf{NL}_k(f) = \min_{g,\, \mathsf{deg}(g) \le 1} |\mathsf{supp}_k(f+g)|.$$

We also recall the concept of Walsh transform, and restricted Walsh transform [MMM+18], which are of particular interest to study the (restricted) nonlinearity or balancedness.

Definition 6 (Walsh transform and restricted Walsh transform). *Let* $f \in \mathcal{B}_n$ *be a Boolean function, its Walsh transform* W_f *at* $a \in \mathbb{F}_2^n$ *is defined as:*

$$W_f(a) := \sum_{x \in \mathbb{F}_2^n} (-1)^{f(x)+a\cdot x}.$$

Let $f \in \mathcal{B}_n$*,* $S \subset \mathbb{F}_2^n$*, its Walsh transform restricted to* S *at* $a \in \mathbb{F}_2^n$ *is defined as:*

$$W_{f,S}(a) := \sum_{x \in S} (-1)^{f(x)+ax}.$$

For $S = \mathsf{E}_{k,n}$ *we denote* $W_{f,\mathsf{E}_{k,n}}(a)$ *by* $\mathcal{W}_{f,k}(a)$*.*

Property 1 (Nonlinearity on the slice, adapted from [CMR17], Proposition 6). *Let* $n \in \mathbb{N}^*$*,* $k \in [0,n]$*, for every* n*-variable Boolean function* f *over* $\mathsf{E}_{k,n}$*:*

$$\mathsf{NL}_k(f) = \frac{|\mathsf{E}_{k,n}|}{2} - \frac{\max_{a \in \mathbb{F}_2^n} |\mathcal{W}_{f,k}(a)|}{2}.$$

Property 2 (Balancedness on the slice and restricted Walsh transform). *Let* $n \in \mathbb{N}^*$*,* $k \in [0,n]$*,* $f \in \mathcal{B}_n$ *is balanced over* $\mathsf{E}_{k,n}$ *if and only if:*

$$\mathcal{W}_{f,k}(0_{|\mathsf{E}_{k,n}|}) = \begin{cases} 0 & \text{if } |\mathsf{E}_{k,n}| \text{ is even,} \\ \pm 1 & \text{if } |\mathsf{E}_{k,n}| \text{ is odd.} \end{cases}$$

2.2 Siegenthaler's Construction, Symmetric Functions

In the following we recall the Siegenthaler construction, a common secondary construction which combines two n-variable functions to obtain an $(n+1)$-variable function:

Definition 7 (Siegenthaler's Construction). *Let $n \in \mathbb{N}$, $f_0, f_1 \in \mathcal{B}_n$, we call Siegenthaler's construction f from components f_0 and f_1:*

$$f \in \mathcal{B}_{n+1}, \quad \forall x \in \mathbb{F}_2^n, \forall y \in \mathbb{F}_2, \; f(x, y) = (1 + y) \cdot f_0(x) + y \cdot f_1(x).$$

We recall definitions and properties on symmetric functions since they will be used for the main secondary construction we present in the article. Symmetric functions are functions such that changing the order of the inputs does not change the output. They have been the focus of many works for their cryptographic parameters such as [Car04, CV05, BP05, DMS06, QLF07, SM07, QFLW09, CL11], or more recently [TLD16, CM19, CZGC19, Méa19, Méa21, CM22].

Definition 8 (Symmetric Functions). *Let $n \in \mathbb{N}^*$, the Boolean symmetric functions are the functions which are constant on each $\mathsf{E}_{k,n}$ for $k \in [0, n]$. The set of n variable symmetric functions is denoted $\mathcal{SYM}_n$ and $|\mathcal{SYM}_n| = 2^{n+1}$. We distinguish families of symmetric functions:*

- *Elementary symmetric functions. Let $i \in [0, n]$, the elementary symmetric function of degree i in n variables, denoted $\sigma_{i,n}$, is the function which ANF contains all monomials of degree i and no monomials of other degrees.*
- *Threshold Functions. Let $d \in [0, n]$, the threshold function of threshold d is defined as:*

$$\forall x \in \mathbb{F}_2^n, \quad \mathsf{T}_{d,n}(x) = 1 \text{ if and only if } \mathsf{w_H}(x) \geq d.$$

- *Slice indicator functions. Let $k \in [0, n]$, the indicator function of the slice of weight k is defined as:*

$$\forall x \in \mathbb{F}_2^n, \quad \varphi_{k,n}(x) = 1 \text{ if and only if } \mathsf{w_H}(x) = k.$$

The $n+1$ n-variable symmetric functions of each family form a basis of $\mathcal{SYM}_n$ (that is every element of $\mathcal{SYM}_n$ can be written as a linear combination of these $n + 1$ functions). Now, we precise on how to express $\varphi_{k,n}$ as a sum of symmetric elementary function. To do so, we use the expression of threshold functions in term of symmetric elementary functions from [Méa19], since $\varphi_{k,n}$ is the sum of two consecutive threshold functions.

Property 3 (Algebraic normal form of threshold functions (adapted from [Méa19], Theorem 1)). *Let $n, d \in \mathbb{N}^*$ such that $0 < d \leq n + 1$, let $D = 2^{\lceil \log d \rceil}$. For $v \in \mathbb{F}_2^n$ we denote $\overline{v}$ the complementary of $v \in \mathbb{F}_2^n$: $\forall i \in [1, n]$, $\overline{v_i} = 1 - v_i$. We denote $\preceq$ the partial order on $\mathbb{F}_2^n$ defined as $a \preceq b \Leftrightarrow \forall i \in [1, n], a_i \leq b_i$, where $\leq$ denotes the usual order on $\mathbb{Z}$ and the elements a_i and b_i of $\mathbb{F}_2$ are identified to 0 or 1 in $\mathbb{Z}$. We denote the set:*

$$A_d = \{v \in [0, D - 1] \,|\, v \preceq D - d\} = \{v \in \mathbb{F}_2^{\lceil \log d \rceil} \,|\, v \preceq \overline{d - 1}\},$$

where $d - 1$ is considered over $\log D - 1$ bits. We also denote:

$$B_{d,n} = \{kD + d + v \,|\, k \in \mathbb{N}, v \in A_d\} \cap [1, n] = \{kD - v \,|\, k \in \mathbb{N}^*, v \in A_d\} \cap [1, n].$$

The ANF of the threshold function is given by: $\mathsf{T}_{d,n} = \sum_{i \in B_{d,n}} \sigma_{i,n}$.

Since $\varphi_{k,n} = \mathsf{T}_{k,n} + \mathsf{T}_{k+1,n}$ its ANF is given by $B_{k,n} \Delta B_{k+1,n}$, where Δ denotes the symmetric difference of sets (*i.e.* $A \Delta B = (A \cup B) \setminus (A \cap B)$).

2.3 Parity of Binomial Coefficients

This section contains results about binomial coefficients that will be used in this article. As a convention we set $\binom{a}{b} = 0$ if $b < 0$ and $b > a$.

Property 4 (Pascal's formula). *Let $a, b \in \mathbb{N}$. Then*

$$\binom{a}{b} = \binom{a-1}{b} + \binom{a-1}{b-1}.$$

Property 5 (Vandermonde Convolution). *Let $a, b, c \in \mathbb{N}$. Then*

$$\binom{a+c}{b} = \sum_{j=0}^{b} \binom{c}{b-j}\binom{a}{j}.$$

Property 6 Lucas' Theorem, *e.g.* [Fin47]). *Let $a, b, p \in \mathbb{N}$ be integers such that $a > b$ and p is a prime. Consider their p-adic expansions $a = \sum_{j=0}^{q} a_j p^j$ and $b = \sum_{j=0}^{q} b_j p^j$ such that $0 \le a_j < p$ and $0 \le b_j < p$ for each $j \in [0, q]$ and $a_q \neq 0$. Then*

$$\binom{a}{b} \equiv \prod_{j=0}^{q} \binom{a_j}{b_j} \pmod{p}.$$

Proposition 1. *Let $a, b \in \mathbb{N}$ and their binary decomposition be $a = \sum_{j=0}^{q_a} a_j 2^j$ and $b = \sum_{j=0}^{q_b} b_j 2^j$ such that $0 \le a_j < 2$ and $0 \le b_j < 2$ for each j, and $a_{q_a}, b_{q_b} \neq 0$.*

1. *$\binom{2^a}{b}$ is even for $0 < b < 2^a$.*
2. *If $a \equiv 0 \mod 2$ and $b \equiv 1 \mod 2$, then $\binom{a}{b} \equiv 0 \mod 2$.*
3. *If $a \equiv 1 \mod 2$ and $b \equiv 0 \mod 2$, then $\binom{a}{b} \equiv \binom{a-1}{b} \mod 2$.*
4. *$\binom{a}{b} \equiv 1 \mod 2$ if and only if for all $j \in [0, q_b]$ it holds $a_j \ge b_j$.*

Proof. 1. If $0 < b < 2^a$, there exists at least a coefficient $b_j = 1$ in the binary expansion of b for $j < a$. Then by Property 6 $\binom{2^a}{b} \equiv 0 \mod 2$ since $\binom{0}{b_j} \equiv 0$.
2. If $a \equiv 0 \mod 2$, then $0 \equiv a\binom{a-1}{b-1} \equiv b\binom{a}{b} \equiv \binom{a}{b} \mod 2$.
3. This comes from Property 4 and point 2.
4. From Lucas' theorem we have that $\binom{a}{b} \equiv 1 \mod 2$ if and only if $\binom{a_j}{b_j} \equiv 1 \mod 2$ for each $j \in [0, q_b]$ if and only if $a_j \ge b_j$ for each $j \in [0, q_b]$.

□

We prove the following fact, illustrated by Fig. 1 for $n < 16$.

Lemma 1. *Let $u \ge 2$ and $t \in [1, 2^{u-2}]$, for all $k \in [2^{u-1} - 2t + 1, 2^{u-1} - 1]$ the binomial coefficient $\binom{2^u - 2t}{k}$ is even.*

Proof. We write $2^u - 2t = 2^{u-1} + (2^{u-1} - 2t)$, then using Property 5 we obtain

$$\binom{2^u - 2t}{k} = \sum_{j=0}^{k} \binom{2^{u-1} - 2t}{k-j}\binom{2^{u-1}}{j}$$

Since the coefficients $\binom{2^{u-1}}{j}$ are even for $0 < j < 2^{u-1}$ by Proposition 1.1, reducing the convolution modulo 2 we obtain (recall that $\binom{a}{b} = 0$ if $b < 0$ and $b > a$, therefore for certain values of k some addenda can be zero by default):

$$\binom{2^u - 2t}{k} \equiv \binom{2^{u-1} - 2t}{k} + \binom{2^{u-1} - 2t}{k - 2^{u-1}} \mod 2.$$

Therefore, $\binom{2^u - 2t}{k}$ is even if $k \in [2^{u-1} - 2t + 1, 2^{u-1} - 1]$. □

Fig. 1. Binomial coefficients and parity for $n \in [0, 15]$. The square labeled with k at level n corresponds to the binomial coefficient $\binom{n}{k}$ and it is colored in yellow if the coefficient is even and teal if the coefficient is odd. (Color figure online)

3 Special WAPB Functions and Secondary Constructions

In this section, we begin with properties of the restricted Walsh transform relatively to Siegenthaler's construction and addition of symmetric functions. Then, we define a subset of balanced WAPB functions and give a construction to transform any WAPB function into a function in this subclass. Finally, we provide and study a secondary construction of $(n+1)$-variable WAPB function from two n-variable WAPB functions.

3.1 Restricted Walsh Transform and Properties

First, we study the weightwise restricted Walsh transform of functions obtained through Siegenthaler's construction.

Proposition 2 (Weightwise restricted Walsh transform and Siegenthaler's construction). *Let $n \in \mathbb{N}$, $f_0, f_1 \in \mathcal{B}_n$, f obtained through Siegenthaler's construction with components f_0 and f_1 has the following property:*

$$\forall k \in [0, n], \forall (a, b) \in \mathbb{F}_2^n \times \mathbb{F}_2, \quad \mathcal{W}_{f,k}(a, b) = \mathcal{W}_{f_0,k}(a) + (-1)^b \mathcal{W}_{f_1,k-1}(a).$$

Proof. We rewrite $\mathcal{W}_{f,k}(a,b)$:

$$\begin{aligned}\mathcal{W}_{f,k}(a,b) &= \sum_{(x,y)\in \mathsf{E}_{k,n+1}} (-1)^{f(x,y)+(a,b)\cdot(x,y)} \\ &= \sum_{x\in\mathsf{E}_{k,n}} (-1)^{f(x,0)+(a,b)\cdot(x,0)} + \sum_{x\in\mathsf{E}_{k-1,n}} (-1)^{f(x,1)+(a,b)\cdot(x,1)} \\ &= \sum_{x\in\mathsf{E}_{k,n}} (-1)^{f_0(x)+a\cdot x} + \sum_{x\in\mathsf{E}_{k-1,n}} (-1)^{f_1(x)+a\cdot x+b} \\ &= \mathcal{W}_{f_0,k}(a) + (-1)^b \mathcal{W}_{f_1,k-1}(a).\end{aligned}$$

□

Proposition 3 (Weightwise nonlinearity bound on Siegenthaler's construction). *Let $n \in \mathbb{N}$, $f_0, f_1 \in \mathcal{B}_n$, f obtained through Siegenthaler's construction with components f_0 and f_1 has the following property:*

$$\forall k \in [0,n], \quad \mathsf{NL}_k(f) \geq \mathsf{NL}_k(f_0) + \mathsf{NL}_{k-1}(f_1).$$

Proof. First, we bound $\max_{(a,b)\in\mathbb{F}_2^n\times\mathbb{F}_2} |\mathcal{W}_{f,k}(a,b)|$ using Proposition 2:

$$\begin{aligned}\max_{(a,b)\in\mathbb{F}_2^n\times\mathbb{F}_2} |\mathcal{W}_{f,k}(a,b)| &= \max_{(a,b)\in\mathbb{F}_2^n\times\mathbb{F}_2} |\mathcal{W}_{f_0,k}(a) + (-1)^b\mathcal{W}_{f_1,k-1}(a)| \\ &= \max_{a\in\mathbb{F}_2^n} \left(|\mathcal{W}_{f_0,k}(a)| + |\mathcal{W}_{f_1,k-1}(a)|\right).\end{aligned}$$

Then, we use Property 1:

$$\begin{aligned}\max_{a\in\mathbb{F}_2^n} \left(|\mathcal{W}_{f_0,k}(a)| + |\mathcal{W}_{f_1,k-1}(a)|\right) &\leq \max_{a\in\mathbb{F}_2^n} |\mathcal{W}_{f_0,k}(a)| + \max_{a\in\mathbb{F}_2^n} |\mathcal{W}_{f_1,k-1}(a)| \\ &\leq |\mathsf{E}_{k,n}| - 2\mathsf{NL}_k(f_0) + |\mathsf{E}_{k-1,n}| - 2\mathsf{NL}_{k-1}(f_1) \\ &\leq |\mathsf{E}_{k,n+1}| - 2(\mathsf{NL}_k(f_0) + \mathsf{NL}_{k-1}(f_1)).\end{aligned}$$

Finally, using again Property 1: $\mathsf{NL}_k(f) \geq \mathsf{NL}_k(f_0) + \mathsf{NL}_{k-1}(f_1)$. □

In the following we consider the impact on the weightwise restricted Walsh transform of adding a symmetric function.

Proposition 4 (Weightwise restricted Walsh transform and addition of symmetric function). *Let $n \in \mathbb{N}^*$, $k \in [0,n]$ and $f \in \mathcal{B}_n$, the following holds on $f + \varphi_{k,n}$*

$\forall a \in \mathbb{F}_2^n, \forall i \in [0,n]\backslash\{k\}, \mathcal{W}_{f+\varphi_{k,n},i}(a) = \mathcal{W}_{f,i}(a),$ *and* $\mathcal{W}_{f+\varphi_{k,n},k}(a) = -\mathcal{W}_{f,i}(a).$

Proof. Rewriting $\mathcal{W}_{f+\varphi_{k,n},i}(a)$ we obtain:

$$\mathcal{W}_{f+\varphi_{k,n},i}(a) = \sum_{x\in\mathsf{E}_{i,n}} (-1)^{(f+\varphi_{k,n})(x)+a\cdot x} = \begin{cases} \mathcal{W}_{f,i}(a) & \text{if } i \neq k, \\ -\mathcal{W}_{f,i}(a) & \text{if } i = k. \end{cases}$$

□

Consequently, Proposition 4 directly implies that adding symmetric functions do not alter the weightwise balancedness nor the weightwise nonlinearity of a function.

3.2 Special WAPB Functions

In the following we specify a sub-part of balanced WAPB functions called special WAPB. To do so we use the characterization of WAPB through the weightwise restricted Walsh transform.

Definition 9 (Special Weightwise Almost Perfectly Balanced functions (SWAPB)). *Let $n \in \mathbb{N}^*$, f is a WAPB function if:*

$$\mathcal{W}_{f,k}(0_n) = \begin{cases} 0 & \textit{if } |\mathsf{E}_{k,n}| \textit{ is even,} \\ \pm 1 & \textit{if } |\mathsf{E}_{k,n}| \textit{ is odd.} \end{cases}$$

Additionally, the function is called special WAPB (SWAPB) if:

$$\mathcal{W}_{f,k}(0_n) = \begin{cases} 0 & \textit{if } |\mathsf{E}_{k,n}| \textit{ is even,} \\ 1 & \textit{if } |\mathsf{E}_{k,n}| \textit{ is odd and } k < n/2, \\ -1 & \textit{if } |\mathsf{E}_{k,n}| \textit{ is odd and } k > n/2. \end{cases}$$

The set of SWAPB functions in n variables is denoted $\mathcal{SWAPB}_n$.

Property 7 (Basic properties of SWAPB functions). *Let $n \in \mathbb{N}^*$, the following hold for $\mathcal{SWAPB}_n$:*

- $\mathcal{SWAPB}_n \subset \mathcal{WAPB}_n$,
- *if* $n = 2^m$ *then* $\mathcal{SWAPB}_n = \mathcal{WPB}_m$,
- $|\mathcal{SWAPB}_n| = \prod_{k=0}^{n} \binom{\nu}{\lfloor \nu/2 \rfloor}$ *for* $\nu = \binom{n}{k}$.

The next proposition allows to build a SWAPB function from a WAPB function.

Proposition 5 (From WAPB to SWAPB). *Let $n \in \mathbb{N}^*$ and $f \in \mathcal{WAPB}_n$. Let $S_f \subset [0, n]$ the set defined as $S_f = \{k \in [0, n/2[, \mid \mathcal{W}_{f,k}(0_n) = -1\} \cup \{k \in]n/2, n], \mid \mathcal{W}_{f,k}(0_n) = 1\}$, the function $f' = f + \sum_{k \in S_f} \varphi_{k,n}$ belongs to $\mathcal{SWAPB}_n$.*

Proof. Using the characterization through the restricted Walsh transform and the definition of S_f we get:

$$\mathcal{W}_{f,k}(0_n) = \begin{cases} 0 & \text{if } |\mathsf{E}_{k,n}| \text{ is even,} \\ 1 & \text{if } |\mathsf{E}_{k,n}| \text{ is odd, } k < n/2, \text{ and } k \notin S_f, \\ -1 & \text{if } |\mathsf{E}_{k,n}| \text{ is odd, } k < n/2, \text{ and } k \in S_f, \\ -1 & \text{if } |\mathsf{E}_{k,n}| \text{ is odd, } k > n/2, \text{ and } k \notin S_f, \\ 1 & \text{if } |\mathsf{E}_{k,n}| \text{ is odd, } k > n/2, \text{ and } k \in S_f. \end{cases}$$

Applying Proposition 4, the value of $\mathcal{W}_{f',k}(0_n)$ is flipped for all $k \in S_f$ and unchanged for the other weights (relatively to f). Thereafter, f' is SWAPB. ☐

3.3 Secondary Constructions of WAPB Functions

We introduce a secondary construction from two n-variables SWAPB functions to one $n+1$ SWAPB function. Repetitively using this construction allows us to build WAPB functions for all n.

Construction 1

Input: Let $n \in \mathbb{N}^*$ f_0, f_1 two n-variable SWAPB functions.
Output: f an $n+1$-variable SWAPB function.
1: Define S_n as $S_n = \{k \in [1, n/2[\mid \binom{n}{k-1} \equiv \binom{n}{k} \equiv 1 \mod 2\}$.
2: **for** $k \in S_n$ **do**
3: $\quad f_1 \leftarrow f_1 + \varphi_{k-1,n} + \varphi_{n-k,n}$,
4: **end for**
5: Compute $f = (1 + x_{n+1})f_0 + x_{n+1}f_1$.
6: **return** f.

Theorem 1 (Special weightwise almost perfectly balancedness of Construction 1). *Let $n \in \mathbb{N}^*$, $f_0 \in \mathcal{SWAPB}_n$, and $f_1 \in \mathcal{WAPB}_n$, the function f given by Construction 1 belongs to $\mathcal{SWAPB}_{n+1}$.*

Proof. By construction f is obtained from Siegenthaler's construction with components f_0 and $f_1' = f_1 + \sum_{k \in S_n}(\varphi_{k-1,n} + \varphi_{n-k,n})$ where f_0 and f_1 are SWAPB functions. Accordingly, the restricted Walsh transform values of f can be obtained from the ones of f_0 and f_1 using Proposition 2. The values of the restricted Walsh transform of f_0 and f_1 are given by Definition 9 since these two functions are SWAPB. Then, $\mathcal{W}_{f_1',k}(0_n)$ can be determined by using Proposition 4.

We do a disjunction of cases to determine $\mathcal{W}_{f,k}(0_{n+1})$, considering the parity of $\binom{n}{k-1}$ and $\binom{n}{k}$, for $k \in [0, n/2[$:

- Case $\binom{n}{k-1} \equiv \binom{n}{k} \equiv 0 \mod 2$. In this case:

$$\mathcal{W}_{f,k}(0_{n+1}) = \mathcal{W}_{f_0,k}(0_n) + \mathcal{W}_{f_1',k-1}(0_n) = 0 + \mathcal{W}_{f_1,k-1}(0_n) = 0,$$

and

$$\mathcal{W}_{f,n+1-k}(0_{n+1}) = \mathcal{W}_{f_0,n-k+1}(0_n) + \mathcal{W}_{f_1',n-k}(0_n) = 0 + \mathcal{W}_{f_1,n-k}(0_n) = 0.$$

- Case $\binom{n}{k-1} \not\equiv \binom{n}{k} \mod 2$. In this case:

$$\mathcal{W}_{f,k}(0_{n+1}) = \mathcal{W}_{f_0,k}(0_n) + \mathcal{W}_{f_1',k-1}(0_n) = \mathcal{W}_{f_0,k}(0_n) + \mathcal{W}_{f_1,k-1}(0_n) = 1,$$

and $\mathcal{W}_{f,n+1-k}(0_{n+1}) = \mathcal{W}_{f_0,n-k+1}(0_n) + \mathcal{W}_{f_1,n-k}(0_n) = -1$.
- Case $\binom{n}{k-1} \equiv \binom{n}{k} \equiv 1 \mod 2$. In this case:

$$\mathcal{W}_{f,k}(0_{n+1}) = \mathcal{W}_{f_0,k}(0_n) + \mathcal{W}_{f_1',k-1}(0_n) = 1 + \mathcal{W}_{f_1+\varphi_{k,n},k-1}(0_n) = 1 - 1 = 0,$$

and

$$\begin{aligned}\mathcal{W}_{f,n+1-k}(0_{n+1}) &= \mathcal{W}_{f_0,n-k+1}(0_n) + \mathcal{W}_{f_1',n-k}(0_n)\\ &= -1 + \mathcal{W}_{f_1+\varphi_{n-k,n},n-k}(0_n) = -1+1 = 0.\end{aligned}$$

Using Pascal's formula $|\mathsf{E}_{k,n+1}|$ is even if and only if $\binom{n}{k-1} \equiv \binom{n}{k} \mod 2$, and regrouping the different cases we obtain:

$$\mathcal{W}_{f,k}(0_{n+1}) = \begin{cases} 0 & \text{if } |\mathsf{E}_{k,n+1}| \text{ is even,}\\ 1 & \text{if } |\mathsf{E}_{k,n+1}| \text{ is odd and } k < (n+1)/2,\\ -1 & \text{if } |\mathsf{E}_{k,n+1}| \text{ is odd and } k > (n+1)/2.\end{cases}$$

Therefore, $f \in \mathcal{SWAPB}_n$. □

Remark 1. From Proposition 1 we have that for each $n \in \mathbb{N}$ $S_n = \emptyset$ if $n \equiv 0 \mod 2$. Therefore, if n is even, the input function f_1 of Construction 1 is not modified by Step 1 to 4. Thus, one can output directly $f = (1 + x_{n+1})f_0 + x_{n+1}f_1$.

Combining Proposition 5 and Theorem 1 enables us to obtain a SWAPB function in $n+1$ variable from any n variable WAPB function. Since the obtained function is SWAPB, the theorem can be reapplied with twice this function. Thus, repeating this procedure allows us to build SWAPB functions for all $n' > n$. Moreover, the weightwise nonlinearity of such built functions can be bounded using Proposition 3. Thereafter, we describe the construction obtained by using t times the same SWAPB function, *i.e.* Construction 2.

Theorem 2 (Special weightwise almost perfectly balancedness and weightwise nonlinearity bound of Construction 2). *Let $n, t \in \mathbb{N}^*$ and $f \in \mathcal{SWAPB}_n$, the function g generated by Construction 2 is such that:*

$$g \in \mathcal{SWAPB}_{n+t}, \quad and\, \forall k \in [0, n+t],\ \mathsf{NL}_k(g) \geq \sum_{i=0}^{\min\{k,t\}} \binom{t}{i} \mathsf{NL}_{k-i}(f).$$

Proof. Note that for $t = 1$, it corresponds to:

$$\begin{aligned} g =& f + x_{n+1}\left(\sum_{k\in S_n} \varphi_{k-1,n} + \varphi_{n-k,n}\right)\\ =&(1+x_{n+1})f + x_{n+1}\left(f + \sum_{k\in S_n} \varphi_{k-1,n} + \varphi_{n-k,n}\right),\end{aligned}$$

i.e. the function obtained by Construction 1 from $f_0 = f_1 = f$. Therefore, using Theorem 1, g is SWABP, and Proposition 3 gives the bound on $\mathsf{NL}_k(g)$. The results for $t > 1$ are obtained by immediate recursion. □

Construction 2

Input: Let $n, t \in \mathbb{N}^*$ f a n-variable SWAPB functions.
Output: g an $(n+t)$-variable SWAPB function.

1: Initialize g, $g \leftarrow f$.
2: **for** $i \in [1, t]$ **do**
3: $h = 0$
4: **if** $n + i - 1 \equiv 0 \mod 2$ **then**
5: $S_{n+i-1} \leftarrow \{k \in [1, (n+i-1)/2[\mid \binom{n+i-1}{k-1} \equiv \binom{n+i-1}{k} \equiv 1 \mod 2\}$,
6: **for** $k \in S_{n+i-1}$ **do**
7: $h \leftarrow h + \varphi_{k-1,n+i-1} + \varphi_{n+i-1-k,n+i-1}$,
8: **end for**
9: **end if**
10: $g \leftarrow g + x_{n+i}h$,
11: **end for**
12: **return** g.

4 Concrete Constructions and Parameters

In the first part of this section we recall the CMR construction from [CMR17] of WAPB functions for all n, and we prove that CMR functions are SWAPB. This implies that we can use functions from this family as seeds for Construction 2 to obtain other SWAPB functions. Hence, we collect some relevant cryptographic parameters of new WPB functions in 8 and 16 variables computed by using this strategy. Finally, we also apply Construction 2 with some LM functions from [LM19] as input, and we explain how to combine all these functions to get another function in $\mathcal{WPB}_4$ having high weightwise nonlinearity on every slice.

The methods that we applied to explicitly determine the functions and the value of their cryptographic parameters are discussed in Sect. 4.4

4.1 Building SWAPB Functions from CMR Construction

Definition 10 (CMR WAPB construction (adapted from [CMR17], Proposition 5)). *Let $n \in \mathbb{N}, n \geq 2$, the WAPB function f_n is recursively defined by $f_2(x_1, x_2) = x_1$ and for $n \geq 3$:*

$$f_n(x_1,\ldots,x_n) = \begin{cases} f_{n-1}(x_1,\ldots,x_{n-1}) & \text{if } n \text{ odd,} \\ f_{n-1}(x_1,\ldots,x_{n-1}) + x_{n-2} + \prod_{i=1}^{2^{d-1}} x_{n-i} & \text{if } n = 2^d; d > 1, \\ f_{n-1}(x_1,\ldots,x_{n-1}) + x_{n-2} + \prod_{i=1}^{2^d} x_{n-i} & \text{if } n = p \cdot 2^d; p \text{ odd.} \end{cases}$$

For example, the 16-variable function from this construction is:

$$f_{16} = x_1 + x_2 + x_2x_3 + x_4 + x_4x_5 + x_6 + x_4x_5x_6x_7$$
$$+ x_8 + x_8x_9 + x_{10} + x_8x_9x_{10}x_{11} + x_{12} + x_{12}x_{13} + x_{14} + x_8x_9x_{10}x_{11}x_{12}x_{13}x_{14}x_{15},$$

and the function f_i for $i \in [2,15]$ is given by the ANF of f_{16} reduced to the variables with index smaller than i for i even and $i-1$ for i odd.

We prove that functions from CMR WAPB construction are SWAPB.

Theorem 3. *Let $n \in \mathbb{N}, n \geq 2$ and f_n be the n-variable WAPB function from CMR construction (Definition 10). Then, $f_n \in \mathcal{SWAPB}_n$.*

Proof. If $n = 2^d$ for $d > 1$ we have that f_n is WPB by [CMR17, Proposition 5], hence it is special by Property 7. If $n = 3$, explicit computations show that $|\mathsf{supp}_0(f_3)| = 0 = (|\mathsf{E}_{0,3}| - 1)/2$, $|\mathsf{supp}_1(f_3)| = 1 = (|\mathsf{E}_{1,3}| - 1)/2$, $|\mathsf{supp}_2(f_3)| = 2 = (|\mathsf{E}_{2,3}| + 1)/2$ and $|\mathsf{supp}_3(f_3)| = 1 = (|\mathsf{E}_{0,3}| + 1)/2$. This implies that $f_3 \in \mathcal{SWAPB}_3$, too.

Now, we prove that $f_n \in \mathcal{SWAPB}_n$ by induction on n for the missing values. Since our results extends [CMR17, Proposition 5], for the sake of simplicity, we recall here some facts from its proof denoting them by $(\star)$, and we refer to the original paper for details. Specifically, let us assume that for $n \geq 5$ for $2 \leq i < n$ f_i is SWAPB.

- If $n \equiv 1 \mod 2$, we can write it as $2\ell + 1$. For any $k \in [1, n-1]$ it holds $|\mathsf{supp}_k(f_n)| = |\mathsf{supp}_{k-1}(f_{n-1})| + |\mathsf{supp}_k(f_{n-1})|$. Namely, $\mathcal{W}_{f_n,k}(0_n) = \mathcal{W}_{f_{n-1},k}(0_{n-1}) + \mathcal{W}_{f_{n-1},k-1}(0_{n-1})$. From Proposition 1, we get that at least one cardinality between $|\mathsf{E}_{k-1,n-1}|$ and $|\mathsf{E}_{k,n-1}|$ is even. If both are even, $|\mathsf{E}_{k,n-1}| + |\mathsf{E}_{k-1,n-1}| = |\mathsf{E}_{k,n}|$ is even and $\mathcal{W}_{f_n,k}(0_n) = 0$.
  If one is odd, then $|\mathsf{E}_{k,n}|$ is also odd and we have the following cases:
  - Suppose $k < \ell$. Then $|\mathsf{supp}_k(f_n)| = |\mathsf{E}_{k,n-1}|/2 + |\mathsf{E}_{k-1,n-1}|/2 - 1/2 = (|\mathsf{E}_{k,n}| - 1)/2$, *i.e.* $\mathcal{W}_{f_n,k}(0_n) = 1$, since f_{n-1} is SWABP.
  - Suppose $k > \ell + 1$. Then $|\mathsf{supp}_k(f_n)| = |\mathsf{E}_{k,n-1}|/2 + |\mathsf{E}_{k-1,n-1}|/2 + 1/2 = (|\mathsf{E}_{k,n}| + 1)/2$, *i.e.* $\mathcal{W}_{f_n,k}(0_n) = -1$, since f_{n-1} is SWABP.
  - The central binomial $\binom{2\ell}{\ell}$ is always even for $\ell > 1$, since by Pascal's formula (Property 4) $\binom{2\ell}{\ell} \equiv 2^{2\ell} - 2\sum_{j=0}^{\ell-1}\binom{2\ell}{j} \equiv 0 \mod 2$. Being $n - 1 = 2\ell$, we have $|\mathsf{E}_{\ell,n-1}| \equiv 0 \mod 2$. Then, by Pascal's formula we obtain that $|\mathsf{E}_{\ell,n}| \equiv |\mathsf{E}_{\ell-1,n-1}| \mod 2$ and $|\mathsf{E}_{\ell+1,n}| \equiv |\mathsf{E}_{\ell+1,n-1}| \mod 2$.
    Therefore, since f_{n-1} is SWABP

$$\mathcal{W}_{f_n,\ell}(0_n) = \mathcal{W}_{f_{n-1},\ell}(0_{n-1}) + \mathcal{W}_{f_{n-1},\ell-1}(0_{n-1}) = \mathcal{W}_{f_{n-1},\ell-1}(0_{n-1}) = 1,$$

$$\mathcal{W}_{f_n,\ell+1}(0_n) = \mathcal{W}_{f_{n-1},\ell+1}(0_{n-1}) + \mathcal{W}_{f_{n-1},\ell}(0_{n-1}) = \mathcal{W}_{f_{n-1},\ell+1}(0_{n-1}) = -1.$$

  Moreover, $|\mathsf{supp}_0(f_n)| = |\mathsf{supp}_0(f_{n-1})| = 0$ and $|\mathsf{supp}_n(f_n)| = |\mathsf{supp}_n(f_{n-1})| = 1$. Therefore, f_n is SWAPB if $n \equiv 1 \mod 2$.
- Suppose $n = p \cdot 2^d$ and $p > 1$ odd. Let us denote $n_d = n - 2^d$. We have the following cases:
  - If $k = 0$, $|\mathsf{supp}_0(f_n)| = |\mathsf{supp}_0(f_n)| = 0$ $(\star)$.
  - If $k \in [1, 2^d - 1]$, it holds

$$|\mathsf{supp}_k(f_n)| = |\mathsf{supp}_k(f_{n_d})| + \frac{1}{2}\left(\binom{n}{k} - \binom{n_d}{k}\right) (\star),$$

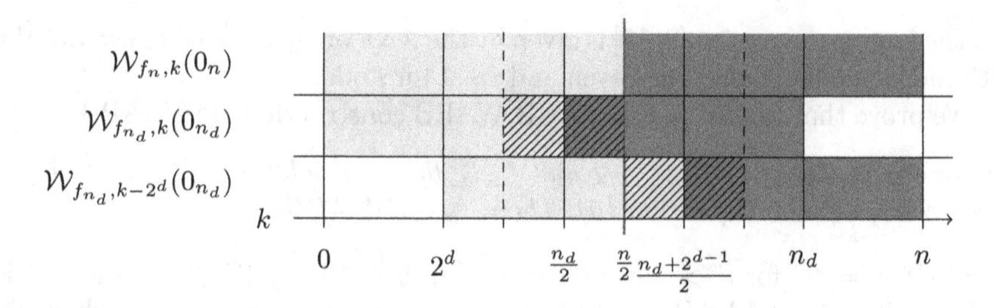

Fig. 2. Light orange and light blue areas correspond to intervals of k where the restricted Walsh transform of the corresponding CMR function is either in $\{0, 1\}$ or $\{0, -1\}$, respectively. While, dashed areas correspond to intervals of k where we prove it to be zero. For the studied k we have $\mathcal{W}_{f_n,k}(0_n) = \mathcal{W}_{f_{n_d},k}(0_{n_d}) + \mathcal{W}_{f_{n_d},s}(0_{n_d})$. Therefore, the coloring of the top row is fully determined by those of the rows below. (Color figure online)

that is $\mathcal{W}_{f_n,k}(0_n) = \mathcal{W}_{f_{n_d},k}(0_{n_d})$.
If $|\mathsf{E}_{k,n_d}| \equiv 0 \mod 2$, then $\mathcal{W}_{f_n,k}(0_n) = 0$. Conversely, since $(n_d)/2 = 2^{d-1}(p-1) > 2^d$ for $p > 3$, $\mathcal{W}_{f_n,k}(0_n) = \mathcal{W}_{f_{n_d},k}(0_{n_d}) = 1$. If $p = 3$, $\mathcal{W}_{f_n,k}(0_n) = 0$ for each $k \in [1, 2^d - 1]$, since f_{n_d} is WPB.

- If $k \in [2^d, n-1]$, setting $s = k - 2^d$ it holds

$$|\mathsf{supp}_k(f_n)| = |\mathsf{supp}_k(f_{n_d})| + |\mathsf{supp}_s(f_{n_d})| + \frac{1}{2}\left(\binom{n}{k} - \binom{n_d}{k} - \binom{n_d}{s}\right) (\star).$$

This is equivalent to

$$\mathcal{W}_{f_n,k}(0_n) = \mathcal{W}_{f_{n_d},k}(0_{n_d}) + \mathcal{W}_{f_{n_d},s}(0_{n_d})$$

Depending on the value of k by induction we know that

$$\mathcal{W}_{f_{n_d},k}(0_{n_d}) \in \begin{cases} \{1,0\} & \text{if } k < n_d/2, \\ \{-1,0\} & \text{if } k \geq n_d/2. \end{cases}$$

$$\mathcal{W}_{f_{n_d},s}(0_{n_d}) \in \begin{cases} \{1,0\} & \text{if } k < n_d/2 + 2^d, \\ \{-1,0\} & \text{if } k \geq n_d/2 + 2^d. \end{cases}$$

Notice that, at least one between $|\mathsf{E}_{k,n_d}|$ and $|\mathsf{E}_{s,n_d}|$ is even. Indeed, consider the binary decomposition $n_d = \sum_{j=0}^{q} a_j 2^j$, $k = \sum_{j=0}^{q} k_j 2^j$ and $s = k - 2^d = \sum_{j=0}^{q} s_j 2^j$ (where $q = \lfloor \log_2(n) \rfloor$). If $\binom{n_d}{k}$ is odd, from Proposition 1 we have that $a_j \geq b_j$ for each j. In particular, since $n_d = 2^d(p-1)$, $a_d = 0$ and consequently $k_d = 0$. This implies that $s_d = 1$, hence $\binom{a_d}{s_d} = 0$. Thus, by Lucas' theorem $\binom{n_d}{s}$ is even if $\binom{n_d}{k}$ is odd.
This implies that for $k \in [2^d, n_d/2]$ we have that $\mathcal{W}_{f_n,k}(0_n) \in \{1, 0\}$, while that $\mathcal{W}_{f_n,k}(0_n) \in \{-1, 0\}$ for $k \in [n_d/2 + 2^d, n-1]$. See Fig. 2.
In order to conclude, it is sufficient to show that $\binom{n_d}{k}$ is even for $k \in [n_d/2 - 2^{d-1}, n_d/2]$. Indeed, by using the symmetries of binomial coefficient

this implies

$$\mathcal{W}_{f_n,k}(0_n) = \begin{cases} \mathcal{W}_{f_{n_d},s}(0_{n_d}) \in \{1,0\} & \text{if } k \in [n_d/2, n/2], \\ \mathcal{W}_{f_{n_d},k}(0_{n_d}) \in \{-1,0\} & \text{if } k \in [n/2, n_d/2 + 2^d]. \end{cases} \quad (1)$$

Recall that $n_d = 2^d(p-1)$. If $p = 3$, f_{n_d} is WPB and all the $\binom{n_d}{k}$ are even for $k \in [1, n_d - 1]$.
Hence, suppose $p > 3$. Setting $L = \lfloor \log_2(p) \rfloor + 1$, since p is odd, we can write $p = 2^{L-1} + \sum_{j=1}^{L-2} p_j 2^j + 1$ with $p_j \in \{0,1\}$. Let $u = d + L$, then $2^d < n_d < n < 2^u$ and

$$\begin{aligned} n_d =& 2^d(p-1) = 2^{d+L-1} + \sum_{j=1}^{L-2} p_j 2^{j+d} = 2^{u-1} + \sum_{j=d+1}^{u-2} p_{j-d} 2^j \\ =& 2^u - 2^{u-1} + \sum_{j=d+1}^{u-2} p_{j-d} 2^j = 2^u - 2(2^{u-2} - \sum_{i=d}^{u-3} p_{i-d+1} 2^i) \end{aligned}$$

This implies that we can write $n_d = 2^u - 2t$ with $t \in [1, 2^{u-2}]$. Therefore, applying Lemma 1 we obtain that for $k \in [2^{u-1} - 2t + 1, 2^{u-1} - t] = [n_d/2 - t + 1, n_d/2]$ the binomial coefficients $\binom{n_d}{k}$ are even.
Furthermore, since $n_d + 2^d = n < 2^{d+L} = 2^u$, we must have $t > 2^{d-1}$, *i.e.* $t \in]2^{d-1}, 2^{u-2}]$. Then, $[n_d/2 - 2^{d-1}, n_d/2] \subseteq [n_d/2 - t + 1, n_d/2]$.
This implies that for $k \in [n_d/2 - 2^{d-1}, n_d/2]$ the coefficients $\binom{n_d}{k}$ are even and consequently (1) holds true.

- If $k = n$, $|\mathsf{supp}_n(f_n)| = 1$ $(\star)$.

Therefore, f_n is SWAPB.

□

Now, we can define a novel family of functions obtained by Construction 2 seeded by the SWAPB CMR functions.

Definition 11 (SWAPB functions $g_{\ell,n}$). *Let $n, \ell \in \mathbb{N}$ with $\ell \in [2, n-1]$, we call $g_{\ell,n}$ the SWAPB function obtained by applying Proposition 5 and Construction 2 with $t = n - \ell$ and f_ℓ, the ℓ-variable WAPB function from CMR construction. We set $g_{n,n} = f_n$.*

In Tables 1 and 2 we report degree, algebraic immunity, nonlinearity and NL_k for $k = 2, \ldots, n-2$ of the functions $g_{\ell,n}$ for $n = 8$ and $n = 16$, respectively. Studying only $g_{\ell,n}$ for ℓ even is sufficient, since the following fact holds:

Proposition 6. *Let $n, s \in \mathbb{N}$ and $s \in [1, (n-1)/2]$. Then $g_{2s,n} = g_{2s+1,n}$.*

Proof. Following Definition 11, the function $g_{2s,n}$ is obtained by applying Construction 2 with f_{2s} as input. Therefore, we have

$$g_{2s,n} = f_{2s} + \sum_{i=1}^{n-2s} x_{2s+i} h_{2s+i-1}$$

where $h_j = \sum_{k \in S_j} \varphi_{k-1,j} + \varphi_{j-k,j}$. By Remark 1 we have $S_{2s} = \emptyset$ and consequently $h_{2s} = 0$. Moreover, we have that $f_{2s+1}(x_1, \dots, x_{2s+1}) = f_{2s}(x_1, \dots, x_{2s})$ from Definition 10. This implies that

$$
\begin{aligned}
g_{2s,n} =& f_{2s} + \sum_{i=1}^{n-2s} x_{2s+i} h_{2s+i-1} = f_{2s} + x_{2s+1} h_{2s} + x_{2s+2} h_{2s+1} + \cdots + x_n h_{n-1} \\
=& f_{2s+1} + x_{2s+2} h_{2s+1} + \cdots + x_n h_{n-1} = g_{2s+1,n}.
\end{aligned}
$$

□

Table 1. Cryptographic parameters of the SWAPB functions $g_{\ell,8}$.

	Degree	AI	NL	NL_2	NL_3	NL_4	NL_5	NL_6
$g_{2,8}$	4	3	88	5	10	16	12	5
$g_{4,8}$	4	3	88	3	7	15	11	3
$g_{6,8}$	4	3	96	2	12	18	12	2
$g_{8,8}$	4	3	88	2	12	19	12	6

Table 2. Cryptographic parameters of the SWAPB functions $g_{\ell,16}$.

	Deg	AI	NL	NL_2	NL_3	NL_4	NL_5	NL_6	NL_7	NL_8	NL_9	NL_{10}	NL_{11}	NL_{12}	NL_{13}	NL_{14}
$g_{2,16}$	8	6	28576	16	97	459	1508	3078	4209	4699	4441	3157	1674	671	170	26
$g_{4,16}$	8	6	28032	14	75	383	1343	2879	4010	4534	4354	3126	1555	627	168	24
$g_{6,16}$	8	6	29792	10	44	344	1458	3110	4502	4947	4321	2897	1326	580	157	20
$g_{8,16}$	8	6	27712	10	44	328	1326	2818	3815	4083	4105	3047	1534	656	144	16
$g_{10,16}$	8	6	29840	5	43	377	1595	3279	4446	5066	4714	3320	1655	507	105	11
$g_{12,16}$	8	5	29152	5	43	265	1397	3148	4439	4971	4803	3396	1712	627	151	13
$g_{14,16}$	8	5	29824	4	56	350	1288	3108	4774	5540	4902	3228	1664	638	152	12
$g_{16,16}$	8	4	29488	4	56	350	1288	3108	4774	5539	4902	3236	1672	654	152	28

4.2 Building Other WPB Functions from LM Construction

In this subsection we study the output of Construction 2 seeded by WPB functions introduced in [LM19]. We recall the definition of these LM functions, referring to the original paper and to [Car21] for the notions of coset leaders of the cyclotomic classes and trace form of a Boolean function.

Definition 12 (LM WPB construction (adapted from [LM19], Corollary 3.5)). *Let $n \in \mathbb{N}, n \geq 2$, we denote by Γ_n the set of all the coset leaders of the cyclotomic classes of 2 modulo $2^n - 1$ and by $o(j)$ the cardinality of the cyclotomic class of 2 modulo $2^n - 1$ containing j. Define $T_j \colon \mathbb{F}_{2^{o(j)}} \to \mathbb{F}_2$ the function $y \mapsto \sum_{i=0}^{o(j)-1} y^{2^i}$. For any fixed β primitive element of $\mathbb{F}_{2^2}$ and any given any function $\iota \colon \Gamma_n \setminus \{0\} \to \{1, 2\}$, the LM WPB function associate to ι is*

$$
LM_\iota(x) = \sum_{j \in \Gamma_n \setminus \{0\}} T_j(\beta^{\iota(j)} x^j).
$$

These functions are proven to be WPB functions defined in 2^m variables, hence SWAPB. Therefore, they can be used to generate other SWAPB by using Construction 2 for all n. We observed that when we apply Construction 2 exhaustively to all LM functions in 4 variables to construct new 8-variable WPB functions we obtain functions having two possible configurations of degree, algebraic immunity, nonlinearity and NL_k for $k = 2, \ldots, n-2$, summarized by Table 3.

Table 3. Profiles of WPB functions in 8 variables returned by Construction 2 applied to the LM family in 4 variables.

	Degree	AI	NL	NL_2	NL_3	NL_4	NL_5	NL_6
Profile 1	4	4	96	5	13	19	17	5
Profile 2	4	4	96	5	16	20	17	5

In order to get new 16-variable WPB functions, we considered in practice two functions as a seed for Construction 2 derived from LM construction having good cryptographic properties. See Table 4.

Table 4. Cryptographic parameters of two WPB functions in 8 variables derived from LM construction.

	Degree	AI	NL	NL_2	NL_3	NL_4	NL_5	NL_6
l	7	4	108	6	21	27	22	9
l_0	7	4	104	9	22	27	22	9

Specifically, we took l as a LM WPB function optimizing NL_4, NL_5,and NL_6 for LM construction (see [LM19, Table 1]), while we obtained l_0 as $\varphi_{0,n}(x)l(x) + \sum_{k=1}^{3} \varphi_{k,n}(x)\bar{l}(x) + \sum_{k=4}^{n} \varphi_{k,n}(x)l(x)$, where for any $f \in \mathcal{B}_n$ we denote by $\bar{f}(x)$ the Boolean function $f(x + 1_n)$ obtained by the composition of the bit-wise negation of x and f. Applying Construction 2 for $n = t = 8$ and as a input either l or l_0, we get two distinct functions g and g_0, respectively. We collect in Table 5 their degree, algebraic immunity, nonlinearity and NL_k for $k = 2, \ldots, n-2$.

Table 5. Cryptographic parameters of the WPB functions g and g_0.

	Deg	AI	NL	NL_2	NL_3	NL_4	NL_5	NL_6	NL_7	NL_8	NL_9	NL_{10}	NL_{11}	NL_{12}	NL_{13}	NL_{14}
g	8	7	30720	22	160	672	1878	3570	4983	5567	5103	3629	1884	688	172	24
g_0	8	7	30592	22	160	672	1865	3581	4951	5455	5071	3603	1880	688	172	24

4.3 Hybrid Function with High Weightwise Nonlinearity in $\mathcal{WPB}_4$

In the previous subsections we described the properties of some WPB in 16 variables obtained by Construction 2 seeded both with CMR and LM functions. Namely, we computed some functions in $\mathcal{WPB}_4$ having high weightwise nonlinearity on certain slices. In Table 2 and 5 the maximal realised values are in red. Therefore, by combining these functions we can obtain the following *hybrid function*:

$$h_{16}(x) = \sum_{k\in\{1,2\}}^{2} \varphi_{k,n}(x)\bar{f}_{16}(x) + \sum_{k\in\{3,4,5,6,7\}} \varphi_{k,n}(x)\bar{g}(x) + \\ + \sum_{k\in\{8,9,10,11,12,13\}} \varphi_{k,n}(x)g(x) + \sum_{k\in\{14,15,16,0\}} \varphi_{k,n}(x)f_{16}(x) \in \mathcal{WPB}_4$$

Table 6 contains the degree, algebraic immunity, nonlinearity and NL_k for $k = 2, \ldots, n-2$ of h_{16}.

Table 6. Cryptographic parameters of h_{16}. By construction $\mathsf{NL}_k(h_{16}) = \mathsf{NL}_{n-k}(h_{16})$.

	Deg	AI	NL	NL_2	NL_3	NL_4	NL_5	NL_6	NL_7	NL_8
h_{16}	14	8	30704	28	172	688	1884	3629	5103	5567

Table 7. Comparison with known lower bound [GM22, Proposition 9] and upper bound [GM22, Proposition 10] for $\mathsf{M}_{k,16}$, *i.e.* the maximum weightwise nonlinearity of $\mathcal{WPB}_4$ over $\mathsf{E}_{k,16}$.

	NL_2	NL_3	NL_4	NL_5	NL_6	NL_7	NL_8
h_{16}	28	172	688	1884	3629	5103	5567
Lower bound	34	222	803	2016	3774	5443	6141
Upper bound	54	268	888	2150	3959	5666	6378

Table 7 shows that the values $\mathsf{NL}_k(h_{16})$ are below the known lower bound of $\mathsf{M}_{k,16}$, the maximum weightwise nonlinearity of $\mathcal{WPB}_4$ over $\mathsf{E}_{k,16}$. Nevertheless, according to [GM22, Table 5], h_{16} is the currently known (explicitly constructed) function with the best weightwise nonlinearity on the slices.

4.4 Computational Aspects

We provided the exact value of cryptographic parameters of the WPB functions that we analyzed, both in 8 and 16 variables. We retrieved them by concrete computations via `sagemath` [The22]. Specifically, we used `BooleanFunction` class from the module `sage.crypto.boolean_function` to encode the functions, and we applied the built-in methods for computing degree and algebraic immunity. Then, we computed the weightwise nonlinearity on the slices NL_k for $k = 2, \ldots, n-2$ by adapting the strategy from [GM22]. See Algorithm 1. For Construction 2 we built the $\varphi_{k,n}$ functions via truth tables for compatibility. Another possible approach can be via ANF using Property 3.

Data parallelism and iterators allowed us to obtain these values in less then one hour by using 128 cores, by 2xAMD Epyc ROME 7H12 @ 2.6 GHz [64c/280W], *i.e.* one regular node of the UL Aion supercomputer https://hpc.uni.lu/ [VBCG14]. Our code is available at https://github.com/agnesegini/WAPB_pub.

Algorithm 1

Input: Let $n, k \in \mathbb{N}^*$ with $0 < k < n$, and $f \in \mathcal{B}_n$.
Output: $\mathsf{NL}_k(f)$
1: Compute v_f the vector of evaluations of f over the $\mathsf{E}_{k,n}$.
2: Generate $\mathsf{P}_{k,n}$ the spherically punctured Reed Muller code of order 1 of length $\nu = \binom{n}{k}$.
3: Compute δ the distance between v_f and $\mathsf{P}_{k,n}$. ▷ This can be performed in parallel.
4: **return** δ

---

5 Conclusion

In this article we introduced two secondary constructions of weightwise almost perfectly balanced functions and provided examples up to 16 variables. While former approaches focused on modifying the support of a low degree functions to make it W(A)PB, our technique is based on an iterative application of Siegenthaler's construction and addition of symmetric functions. This directly provides us a theoretical lower bound on the weightwise nonlinearities based on the parameters of the parent function (Theorem 2). Moreover, via this construction, we explicitly built SWAPB functions up to 16 variables and determined exactly their main cryptographic parameters. Finally, we combined these functions by taking for each slice k the one from the function obtaining the highest NL_k, which gave us the function h_{16} with the highest weightwise nonlinearities exhibited so far.

Open Questions:

- *Higher weightwise nonlinearities.* The function h_{16} is obtained by combining the functions with highest NL_k built with Construction 2 from CMR of LM functions. One natural next step would be to use other WPB families as seed for Construction 2 and possibly combine those functions with best NL_k. Moreover, it would be interesting to try to reach (or overcome) the non-constructive lower bound from [GM22]. See Table 7.
- *Parameters of equivalent WAPB functions.* Considering W(A)PB functions relatively to classes equivalent up to addition of symmetric functions is a good start to build more constructions, and it has the advantage to group WAPB

functions having exactly the same NL_k. As a matter of fact, using special WAPB functions rather than WAPB functions has been useful in this article to exhibit a secondary construction. Taking a special WAPB function is not restrictive since any WAPB function is equivalent to a special one up to the addition of symmetric functions. Major questions relatively to these classes would be to determine the variation of cryptographic parameters inside the same class, and find a criterion to choose the best representative.

Acknowledgments. The two authors were supported by the ERC Advanced Grant no. 787390.

References

[BP05] Braeken, A., Preneel, B.: On the algebraic immunity of symmetric Boolean functions. In: Progress in Cryptology - INDOCRYPT 2005, 6th International Conference on Cryptology in India, Bangalore, India, December 10–12, 2005, pp. 35–48. Proceedings (2005)

[Car04] Carlet, C.: On the degree, nonlinearity, algebraic thickness, and nonnormality of Boolean functions, with developments on symmetric functions. IEEE Trans. Inf. Theor. **50**, 2178–2185 (2004)

[Car21] Carlet, C.: Boolean Functions for Cryptography and Coding Theory. Cambridge University Press, Cambridge (2021)

[CL11] Chen, Y., Lu, P.: Two classes of symmetric Boolean functions with optimum algebraic immunity: construction and analysis. IEEE Trans. Info. Theor. **57**(4), 2522–2538 (2011)

[CM19] Carlet, C., Méaux, P.: Boolean functions for homomorphic-friendly stream ciphers. In: Gueye, C.T., Persichetti, E., Cayrel, P.-L., Buchmann, J. (eds.) A2C 2019. CCIS, vol. 1133, pp. 166–182. Springer, Cham (2019). https://doi.org/10.1007/978-3-030-36237-9_10

[CM22] Carlet, C., Méaux, P.: A complete study of two classes of Boolean functions: direct sums of monomials and threshold functions. IEEE Trans. Inf. Theor. **68**(5), 3404–3425 (2022)

[CMR17] Carlet, C., Méaux, P., Rotella, Y.: Boolean functions with restricted input and their robustness; application to the FLIP cipher. IACR Trans. Symmetric Cryptol. **3**, 2017 (2017)

[CV05] Canteaut, A., Videau, M.: Symmetric Boolean functions. IEEE Trans. Inf. Theor. **51**, 2791–2811 (2005)

[CZGC19] Chen, Y., Zhang, L., Guo, F., Cai, W.: Fast algebraic immunity of 2^{m+2} and 2^{m+3} variables majority function. IEEE Access **7**, 80733–80736 (2019)

[DMS06] Dalai, D.K., Maitra, S., Sarkar, S.: Basic theory in construction of Boolean functions with maximum possible annihilator immunity. Des. Codes Crypt. **40**, 41–58 (2006)

[Fin47] Fine, N.J.: Binomial coefficients modulo a prime. Am. Math. Monthly **54**(10), 589–592 (1947)

[GM22] Gini, A., Méaux, P.: On the weightwise nonlinearity of weightwise perfectly balanced functions. Discrete Appl. Math. **322**, 320–341 (2022)

[GS22] Guo, X., Sihong, S.: Construction of weightwise almost perfectly balanced Boolean functions on an arbitrary number of variables. Discrete Appl. Math. **307**, 102–114 (2022)

[IMM13] Ivchenko, I., Medvedev, Y.I., Mironova, V.A.: Symmetric Boolean functions and their metric properties matrices of transitions of differences when using some modular groups. Mat. Vopr. Kriptogr. **4**, 49–63 (2013)

[LM19] Liu, J., Mesnager, S.: Weightwise perfectly balanced functions with high weightwise nonlinearity profile. Des. Codes Crypt. **87**(8), 1797–1813 (2019)

[LS20] Li, J., Sihong, S.: Construction of weightwise perfectly balanced Boolean functions with high weightwise nonlinearity. Discrete Appl. Math. **279**, 218–227 (2020)

[Méa19] Méaux, P.: On the fast algebraic immunity of majority functions. In: Schwabe, P., Thériault, N. (eds.) LATINCRYPT 2019. LNCS, vol. 11774, pp. 86–105. Springer, Cham (2019). https://doi.org/10.1007/978-3-030-30530-7_5

[Méa21] Méaux, P.: On the fast algebraic immunity of threshold functions. Crypt. Commun. **13**(5), 741–762 (2021). https://doi.org/10.1007/s12095-021-00505-y

[MJSC16] Méaux, P., Journault, A., Standaert, F.-X., Carlet, C.: Towards stream ciphers for efficient FHE with low-noise ciphertexts. In: Fischlin, M., Coron, J.-S. (eds.) EUROCRYPT 2016. Part I, volume 9665 of LNCS, pp. 311–343. Springer, Heidelberg (2016). https://doi.org/10.1007/978-3-662-49890-3_13

[MMM+18] Maitra, S., Mandal, B., Martinsen, T., Roy, D., Stănică, P.: Tools in analyzing linear approximation for Boolean functions related to FLIP. In: Chakraborty, D., Iwata, T. (eds.) INDOCRYPT 2018. LNCS, vol. 11356, pp. 282–303. Springer, Cham (2018). https://doi.org/10.1007/978-3-030-05378-9_16

[MPJ+22] Mario, L., Picek, S., Jakobovic, D., Djurasevic, M., Leporati, A.: Evolutionary construction of perfectly balanced Boolean functions (2022)

[MS21] Mesnager, S., Su, S.: On constructions of weightwise perfectly balanced Boolean functions. Crypt. Commun. **13**(6), 951–979 (2021). https://doi.org/10.1007/s12095-021-00481-3

[MSL21] Mesnager, S., Su, S., Li, J.: On concrete constructions of weightwise perfectly balanced functions with optimal algebraic immunity and high weightwise nonlinearity. Boolean Functions and Applications (2021)

[MZD19] Mesnager, S., Zhou, Z., Ding, C.: On the nonlinearity of Boolean functions with restricted input. Crypt. Commun. **11**(1), 63–76 (2019)

[Obl20] Oblaukhov, A.K.: On metric complements and metric regularity in finite metric spaces. Prikl. Diskr. Mat. **49**, 35–45 (2020)

[QFLW09] Qu, L., Feng, K., Liu, F., Wang, L.: Constructing symmetric Boolean functions with maximum algebraic immunity. IEEE Trans. Inf. Theor. **55**, 2406–2412 (2009)

[QLF07] Qu, L., Li, C., Feng, K.: A note on symmetric Boolean functions with maximum algebraic immunity in odd number of variables. IEEE Trans. Inf. Theor. **53**, 2908–2910 (2007)

[SM07] Sarkar, P., Maitra, S.: Balancedness and correlation immunity of symmetric Boolean functions. Discrete Math. **307**, 2351–2358 (2007)

[SSB18] Stanica, P., Sasao, T., Butler, J.T.: Distance duality on some classes of Boolean functions. J. Comb. Math. Comb. Comput. **107**, 181–198 (2018)

[Su21] Sihong, S.: The lower bound of the weightwise nonlinearity profile of a class of weightwise perfectly balanced functions. Discrete Appl. Math. **297**, 60–70 (2021)

[The22] The Sage Developers. SageMath, the Sage Mathematics Software System (Version 9.5) (2022). https://www.sagemath.org

[TL19] Tang, D., Liu, J.: A family of weightwise (almost) perfectly balanced Boolean functions with optimal algebraic immunity. Crypt. Commun. **11**(6), 1185–1197 (2019). https://doi.org/10.1007/s12095-019-00374-6

[TLD16] Tang, D., Luo, R., Du, X.: The exact fast algebraic immunity of two subclasses of the majority function. IEICE Trans. Fund. Electron. Commun. Comput. Sci. **E99.A**, 2084–2088 (2016)

[Tok12] Tokareva, N.N.: Duality between bent functions and affine functions. Discrete Math. **312**(3), 666–670 (2012)

[VBCG14] Varrette, S., Bouvry, P., Cartiaux, H., Georgatos, F.: Management of an academic HPC cluster: The UL experience. In: 2014 International Conference on High Performance Computing & Simulation (HPCS), pp. 959–967 (2014)

[ZS21] Zhang, R., Su, S.: A new construction of weightwise perfectly balanced Boolean functions. Adv. Math. Commun. (2021)

[ZS22] Zhu, L., Sihong, S.: A systematic method of constructing weightwise almost perfectly balanced Boolean functions on an arbitrary number of variables. Discrete Appl. Math. **314**, 181–190 (2022)

Quantum Cryptography and Cryptanalysis

Improved Quantum Analysis of SPECK and LowMC

Kyungbae Jang[1(✉)], Anubhab Baksi[2], Hyunji Kim[1], Hwajeong Seo[1], and Anupam Chattopadhyay[2]

[1] Division of IT Convergence Engineering, Hansung University, Seoul, South Korea
starj1023@gmail.com

[2] Temasek Laboratories, Nanyang Technological University, Singapore, Singapore
{anubhab001,anupam}@ntu.edu.sg

Abstract. As the prevalence of quantum computing is growing in leaps and bounds over the past few years, there is an ever-growing need to analyze the symmetric-key ciphers against the upcoming threat. Indeed, we have seen a number of research works dedicated to this. Our work delves into this aspect of block ciphers, with respect to the SPECK family and LowMC family.

The SPECK family received two quantum analysis till date (Jang et al., Applied Sciences, 2020; Anand et al., Indocrypt, 2020). We revisit these two works, and present improved benchmarks SPECK (all 10 variants). Our implementations incur lower full depth compared to the previous works.

On the other hand, the quantum circuit of LowMC was explored earlier in Jaques et al.'s Eurocrypt 2020 paper. However, there is an already known bug in their paper, which we patch. On top of that, we present two versions of LowMC (on L1, L3 and L5 variants) in quantum, both of which incur significantly less full depth than the bug-fixed implementation.

Keywords: Quantum implementation · Grover's search · SPECK · LowMC

Hyunji Kim and Hwajeong Seo were supported by the Institute for Information & Communications Technology Planning & Evaluation (IITP) grant funded by the Korean government (MSIT) (⟨Q|Crypton, number 2019-0-00033, Study on Quantum Security Evaluation of Cryptography based on Computational Quantum Complexity); and Kyungbae Jang was supported by the Basic Science Research Program through the National Research Foundation of Korea (NRF) funded by the Ministry of Education (2022R1A6A3A13062701) of the Korean government. Anupam Chattopadhyay was partly supported by the NRF Grant Award, number NRF2021-QEP2-02-P05 by the Singaporean government. Further, we thank Da Lin (Hubei University, Wuhan, PR China) for the kind support during preparation of the manuscript.

T. Isobe and S. Sarkar (Eds.): INDOCRYPT 2022, LNCS 13774, pp. 517–540, 2022.
https://doi.org/10.1007/978-3-031-22912-1_23

1 Introduction

Among the major progress in the computational science in recent times, the quantum computing is certainly included in the topmost contenders. While a massive race of research is underway to build a functional quantum computer, it stands to reason that we should investigate how such a device can undermine the current security notions. As a matter of fact, it is well-known that certain public-key systems would face major problem [10,16,19,20,34,38] against an adversary equipped with a quantum computer. Going further, one may also notice that the symmetric-key counterpart would also be affected, mostly due to the so-called Grover's search algorithm [18].

Due to the power of the quantum properties of matter (namely, superposition and entanglement), quantum algorithms can find (with a high probability) the solution to certain types of problems faster than the best-known classical algorithms. In this case, the Grover's search algorithm can find the secret key of a symmetric-key cipher with about square-root search of what would be required for a classical computer, roughly speaking.

Therefore, it is not surprising that the research community in the symmetric-key cryptography as well would take interest in figuring out the possible impact a functional quantum computer can have—see Sect. 2.2 for a collection of related works. This work, too, makes a humble attempt to evaluate the quantum security of the block cipher families, SPECK [12] and LowMC [1].

Contribution

In brief, we present the followings in this work:

1. **SPECK Family** (10 variants; Sect. 4). We improve the quantum implementations of the SPECK family from the Indocrypt 2020 paper by Anand, Maitra and Mukhopadhyay, [6]; and the same from the Applied Sciences paper by Jang, Choi, Kwon, Kim, Park and Seo [23] in terms lower depth (though the X gate count is higher in our case). By improving the quantum adders and parallelization in the architecture, we show noticeable reduction of depth[1].
2. **LowMC Family** (L1, L3 and L5; Sect. 5). We observe that the implementation (LowMC) by [31] contains some programming related issue, which probably resulted in underestimating the resources for non-linear components (similar issue with respect to AES was reported by the Asiacrypt'20 authors [43]; and later in [22,26]); although the linear components (Sects. 5.2, 5.3) were not affected. We patch the issues (❋, such as impossible parallelism and omitting initialization of ancilla qubits) and estimate the correct quantum gates and depth from the number of qubits they reported in Sect. 5.5.

[1] However the reduction of full depth is less prominent (ranging from 10% to 12% depending on the variant of SPECK), still our implementation takes less quantum resource. See Table 4 for the benchmark.

Independent to that, we present two versions of three LowMC variants, which we refer to as, the *regular* (⌑) and the *shallow* (⋈) versions. Both the regular and the shallow versions provide high parallelism as the linear layer and key schedule work simultaneously. The regular (respectively, shallow) version uses the S-box implementation that has the Toffoli depth of 3 (respectively, 1), as described in Fig. 4. Further, we show some improvement in the implementation of the linear layer, key schedule, and also in the parallelization of both.

Table 1 shows the benchmarks for the SPECK cipher family, including the results from [6,23]. The proposed SPECK quantum circuits require a higher number of X gates than previous works. This is due to the nature of the quantum adder used in our implementation (detailed in Sect. 4.1). Similarly, a summary of results of on LowMC can be found in Table 2, where we consolidate results from the bug-fixed implementation of [31]. The T-depth of the shallow version of LowMC is higher than the bug-fixed implementation of [31], but actually, this is derived from the difference in the decomposition method of the Toffoli gate although the Toffoli depth is the same (see Sect. 2.3 for details).

When the basic implementation of the ciphers is available, in Sect. 6, we elaborate the estimated cost of running the Grover's search algorithm. We estimate only the cost of oracle in the Grover's search algorithm with the proposed quantum circuits. There is a module called diffusion operator that amplifies the amplitude of the solution returned by oracle, but the overhead is negligible, so it is excluded from the cost estimation. Lastly, the parallel operation of the Grover's search algorithm required according to the block and key size of the cipher is reflected in the cost estimation. We also comment on the quantum security level proposed in [33].

Our source codes are written in ProjectQ[2]. Developed by the researchers from ETH Zurich, it is a Python-based open-source framework for quantum computing, and offers a support for IBM's quantum chips. The variable `resource_check` is set to 0 in `ClassicalSimulator` to check the test vectors and set to 1 in `ResourceCounter` to decompose Toffoli gates in our codes. All relevant codes, along with a toy version of SPECK (where it is possible to simulate the Grover's search), are released in public[3]. For more information, one may refer to the full version of the paper, which can be found in [25].

2 Prerequisite

2.1 Backdrop and Motivation

The Grover's search algorithm is a quantum algorithm that can find a solution in an n-qubit search space with $\lfloor \frac{\pi}{4}\sqrt{2^n} \rfloor$ (about $\sqrt{2^n}$) serial application. Theoretically, this algorithm can reduce symmetric-key ciphers (having an n-bit key)

[2] Homepage: https://projectq.ch/. Code: https://github.com/ProjectQ-Framework/ProjectQ. Documentation: https://projectq.readthedocs.io/en/latest/.

[3] https://github.com/starj1023/SPECK_LowMC_QC.

Table 1. Comparison of quantum resources required for variants of SPECK.

SPECK		#Toffoli ☆	#CNOT ✲	#NOT ✱	#qubits ❂	Depth	Full depth ❉
	JCKKPS [23]	1,290	3,706	42	97	3,313	N/A
32/64	AMM [6]	1,290	4,222	42	96	1,694	5,873
	This work	1,247	4,179	1,160	98	814	5,258
	JCKKPS [23]	1,982	5,606	42	121	4,969	N/A
48/72	AMM [6]	1,978	6,462	42	120	2,574	9,153
	This work	1,935	6,419	1,848	122	1,166	8,075
	JCKKPS [23]	2,074	5,866	45	145	5,203	N/A
48/96	AMM [6]	2,070	6,762	45	144	2,691	9,541
	This work	2,025	6,717	1,935	146	1,219	8,441
	JCKKPS [23]	3,162	8,890	54	161	8,009	N/A
64/96	AMM [6]	3,162	10,318	54	160	4,082	14,563
	This work	3,111	10,267	3,012	162	1,794	12,870
	JCKKPS [23]	3,286	9,238	57	193	8,323	N/A
64/128	AMM [6]	3,286	10,722	57	192	4,239	15,181
	This work	3,233	10,669	3,131	194	1,863	13,365
	JCKKPS [23]	5,172	14,436	60	193	12,923	N/A
96/96	AMM [6]	5,170	16,854	60	192	6,636	23,657
	This work	5,115	16,799	5,010	194	2,828	21,028
	JCKKPS [23]	5,360	14,960	64	241	13,397	N/A
96/144	AMM [6]	5,358	17,466	64	240	6,873	23,657
	This work	5,301	17,409	5,194	242	2,929	21,779
	JCKKPS [23]	7,942	22,086	75	257	19,797	N/A
128/128	AMM [6]	7,938	25,862	75	256	10,144	36,358
	This work	7,875	25,799	7,761	256	4,256	32,224
	JCKKPS [23]	8,192	22,784	80	321	20,427	N/A
128/192	AMM [6]	8,190	26,682	80	320	10,461	37,381
	This work	8,125	26,617	8,010	322	4,389	33,231
	JCKKPS [23]	8,444	23,484	81	385	21,061	N/A
128/256	AMM [6]	8,442	27,502	81	384	10,778	38,431
	This work	8,375	27,435	8,255	386	4,522	34,238

with n-bit security on a classical computer to $n/2$-bit security on a quantum computer.

An abridged description of the algorithm is given as follows. The Grover's search algorithm operates on n-qubits in the superposition state and finds a solution by iterating the set of oracle and diffusion operators about n times. First, n Hadamard gates are used to prepare n-qubits in superposition state. This causes 2^n queries to coexist as probabilities in n-qubit. In the oracle, the logic to find a solution is implemented as quantum gates. For the quantum key search, quantum encryption of the target cipher must be implemented as logic in the oracle. The oracle finds a solution (i.e., the secret key), but the measurement probabilities with non-solutions are still the same. So, the diffusion operator amplifies the amplitude of the solution returned by the oracle. After increasing the amplitude of the solution sufficiently by repeating the oracle and diffusion operators, n-qubits are finally measured.

However, the catch is that the quantum attack using the Grover's algorithm on the symmetric-key cipher requires a lot of quantum resources. Despite much

Table 2. Comparison of quantum resources required for variants of LowMC.

LowMC		#CNOT ✲	#1qCliff ❁	#T ✢	T-depth ✤	#qubits ✪	Full depth ✻
L1	❋	344,972	2,466	4,200	20	1,006	49,350
	⌑	498,208	2,466	4,200	240	3,200	4,708
	⋈	500,208	2,466	4,200	80	3,830	4,708
L3	❋	1,135,935	4,699	6,300	30	1,434	159,659
	⌑	1,669,456	4,699	6,300	360	6,720	10,571
	⋈	1,672,456	4,699	6,300	120	7,650	10,571
L5	❋	2,535,162	7,137	7,980	38	1,802	346,736
	⌑	3,754,484	7,137	7,980	456	11,008	17,789
	⋈	3,758,284	7,137	7,980	152	12,178	17,789

❋: Bug-fixed JNRV [31]

⌑: Regular version.

⋈: Shallow version.

progress though, the state-of-the-art quantum computers have only very limited resources, and consequently cannot afford to run the Grover's algorithm.

If the quantum cost required to attack the cipher is high, it can be expected to provide the desired security (i.e., n-bit security) even in the post-quantum era (without increasing the key size). Thus, it is important to estimate and analyze the cost of quantum attacks on various ciphers.

2.2 Related Works

Estimating the quantum resources required for key recovery using the Grover search algorithm was probably first presented for AES by Grassl, Langenberg, Roetteler, and Steinwand [17]. This work has been followed up by the research community with various implementations of AES [2,22,26,31,32,43]. These papers all focus on the efficient implementation of quantum circuits, thereby reducing the cost for running the Grover's search algorithm with increasingly low resource. Apart from AES, a large number of other ciphers have also received the quantum analysis, SIMON [7], SPECK [6,23], SKINNY [13], PRESENT and GIFT [28], SHA-2 and SHA-3 [3], FSR-based ciphers [5], ChaCha [11], SM3 [39,42], RECTANGLE and KNOT [9], KATAN [35], DEFAULT [24], GIFT–SKINNY–SATURNIN [13], PIPO [30], to name some of those.

2.3 Quantum Gates

There are several commonly used quantum gates to implement ciphers into quantum circuits, such as X (NOT), CNOT, and Toffoli (CCNOT) gates. The X gate inverts the value of a qubit, which can replace the classical NOT operation (i.e., X $(a) = \sim a$). The CNOT gate operates on two qubits, and the value of the target qubit is determined according to the value of the control qubit. If the value of the control qubit is 1, the target qubit is inverted, and if the value of the control qubit is 0, it is maintained (i.e., CNOT $(a, b) = (a, a \oplus b)$. Since this is equivalent to XORing the value of the control qubit to the target qubit, the CNOT gate can replace the classic XOR operation. Toffoli gates operate on three qubits, with two control qubits and one target qubit. The value of the target qubit is reversed only when the values of both control qubits are 1 (i.e., Toffoli $(a, b, c) = (a, b, c \oplus ab)$). Since this is equivalent to XORing the ANDed value of control qubits to the target qubit, Toffoli gate can replace the classic AND operation. We can implement cipher encryption in quantum using these quantum gates, which can replace the classic NOT, XOR, and AND operations.

Among these gates, it is important from an optimization point of view that we need to reduce the number of Toffoli gates. Because the Toffoli gate is implemented as a combination of T gates (determine the T-depth) and Clifford gates (i.e., CNOT, H, or X gate), the cost is relatively high. There are several ways to decompose the Toffoli gate [4,21,36], and the full depth means the depth when the Toffoli gates are decomposed. In our work, when estimating decomposed resources, we adopt the decomposition method of 7 T gates + 8 Clifford gates, T-depth of 4, and full depth of 8 for one Toffoli gate [4].

2.4 NIST Security Levels

In order to describe the security of cipher against a quantum adversary, NIST stated the following security margins for a cipher [33]:

- Level 1: Cipher is at least as hard to break as AES-128.
- Level 2: Cipher is at least as hard to break as SHA-256.
- Level 3: Cipher is at least as hard to break as AES-192.
- Level 4: Cipher is at least as hard to break as SHA-384.
- Level 5: Cipher is at least as hard to break as AES-256.

NIST recommended that ciphers should achieve at least Levels 1, 2 and/or 3, to provide sufficient security in the post-quantum era. The estimates used in [33] were based on the results of AES circuits were taken from that of [17], and are as listed as follows: Level 1: 2^{170}, Level 3: 2^{233}, Level 5: 2^{298}. These figures were calculated as total number of gates $\times$ full depth of the quantum key search (as estimated in [17]) respectively for AES-128, 192, and 256 under the Grover's algorithm.

3 Target Ciphers

3.1 SPECK Family (32/64, 48/72, 48/96, 64/96, 64/128, 96/96, 96/144, 128/128, 128/192, 128/256)

SPECK [12] is a family of lightweight block ciphers that was developed by the National Security Agency (NSA) in 2013. The SPECK family adopts a Feistel-like structure and contains 10 variants. The parameters for each variant are specified in Table 3.

Table 3. Parameters for SPECK variants.

Word size (n)	Key words (m)	Block size ($2n$)	Key size (nm)	α	β	Rounds (T)
16	4	32	64	7	2	22
24	3	48	72	8	3	22
	4		96			23
32	3	64	96	8	3	26
	4		128			27
48	2	96	96	8	3	28
	3		144			29
64	2	128	128	8	3	32
	3		192			33
	4		256			34

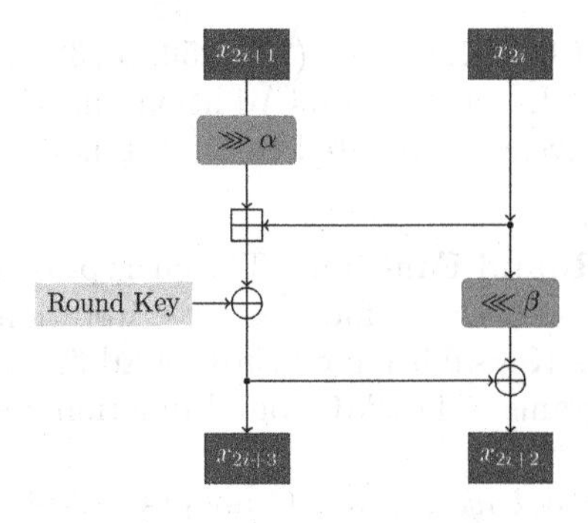

Fig. 1. Round function of SPECK.

Round Function. The round function of SPECK consists of modular addition, bit-wise rotation and exclusive-OR (XOR) as shown in Fig. 1. Let (x_{2i+1}, x_{2i}) be the $2n$-bit input of the ith round, where x_{2i+1} and x_{2i} are both n-bit words. In each round, the state is updated as follows:

1. Updating x_{2i+1} by cyclically shifting its bits to the right by α bits, and then performing the addition modulo 2^n on x_{2i+1} and x_{2i} via $x_{2i+1} = x_{2i+1} + x_{2i}$.
2. XORing the n-bit round key to x_{2i+1}, and cyclically shifting the bits in x_{2i} to the left by β bits, simultaneously.
3. XORing x_{2i+1} to x_{2i} and finishing the update of round function.

Key Schedule. The sub-keys of SPECK are expanded in a similar way as the state in each round. Denote $l_0, l_1, \cdots, l_{m-2}$ the variables for producing the sub-keys of SPECK family. In order to generate the $(i+1)^{\text{th}}$-round sub-key, where $i \in \{0, T\}$, take (l_i, k_i) as the input of round function as shown in Fig. 1 with the number i served as round key in key addition step. Denote the output (l_{i+m-1}, k_{i+1}), k_{i+1} is the generated sub-key.

3.2 LowMC Family (L1, L3, L5)

LowMC [1] is a family of SPN-based block ciphers. Motivated by the fact that non-linear gates are costly compared to the linear gates in applications such as Multi-party Computation (MPC), Fully Homomorphic Encryption (FHE) and Zero Knowledge (ZK), the ciphers specific to these niches are designed to have a low small AND gate/depth count. LowMC is flexible in design (some components of it can be determined randomly), the recommended instance of [1] can be characterized by the block size n, the key size k, the number of S-boxes m in the non-linear layer, the allowed data complexity d of attacks and the round r, where $(n, k, m, d, r) \in \{(256, 80, 49, 64, 11), (256, 128, 63, 128, 12)\}$. Note that in the post-quantum digital signature Picnic[4] [41], the adopted variants of LowMC can be characterized by $(n, k, m, r) \in \{(128, 128, 10, 20),$

[4] Apart from LowMC, Picnic also uses SHA-3 in some form.

(192, 192, 10, 30), (256, 256, 10, 38)}. LowMC round consists of SboxLayer, LinearLayer, ConstantAddition and KeyAddition; and in the Key Schedule, round keys are generated through LinearLayer[5].

Round Function. The encryption of LowMC starts with a whitening key addition over $\mathbb{F}_2$, followed by r iterations of the round function which is composed as KeyAddition ∘ ConstantAddition ∘ LinearLayer ∘ SboxLayer. Schematic diagrams of LowMC round function and key schedule can be found in Fig. 2.

SboxLayer. LowMC adopts a 3-bit S-box (in the look-up form, it is given by `01367452`) with the coordinate function representation (in ANF) as $(a \oplus bc, a \oplus b \oplus ac, a \oplus b \oplus c \oplus ab)$ in its substitution layer, where a, b, c are the input bits. For a specific instance of LowMC, only the first $3m$ bits of the state will go through the S-box.

LinearLayer. The linear layer of LowMC is matrix multiplication in $\mathbb{F}_2$.

ConstantAddition. Round constants are XORed to the sate by the operation of addition in $\mathbb{F}_2$.

KeyAddition. The n-bit round keys generated by key schedule are XORed to the state after each round. Also, the encryption with LowMC starts with a key whitening.

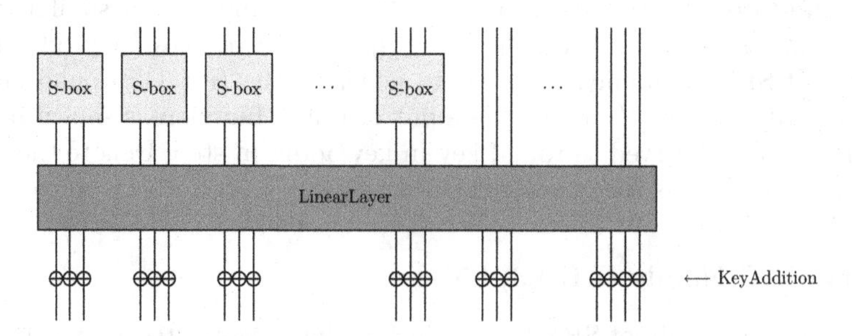

Fig. 2. Round function of LowMC.

Key Schedule. The round keys are derived from the master key via multiplication with a random matrix with full rank.

[5] As the exact specification is generated at random, it is suggested in [8] to call LowMC as a 'meta-cipher' (instead of a 'cipher').

4 SPECK in Quantum

For implementation of SPECK in quantum, we present a parallel addition implementation for a quantum circuit. We design a parallel addition structure by allocating one more carry qubit. We take an on-the-fly approach to perform round functions and key schedules together. Then the additions of the round function and key schedule are performed in parallel. As a result, compared to the implementation in [6], we save one Toffoli gate per addition and provide a 56% performance improvement in terms of depth.

4.1 Quantum Adder for SPECK

A quantum adder is implemented as a combination of quantum gates. Previous implementations of SPECK [27] used a ripple carry-based quantum adder [15]. The quantum adder in the previous work uses one ancilla qubit, $(2n-2)$ Tofffoli gates, $(4n-2)$ CNOT gates, with a depth of $(5n-3)$. Later, Anand et al. improved performance in terms of depth and saved one ancilla qubit by adopting a different quantum adder (from [40]) in their SPECK quantum circuit implementation [6]. The quantum adder used in their work uses a $(2n-2)$ Tofffoli gate, $(5n-6)$ CNOT gates, with a depth of $(5n-5)$ where no ancilla qubits are used. For this reason, it saves 1 qubit compared to the quantum adder used in [27].

We use an improved quantum adder based on the ripple-carry approach, which is referred to as the improved Cuccaro–Draper–Kutin–Moulton (CDKM) adder [15]. This quantum adder uses one ancilla and more X gates, but reduce the Toffoli gates and circuit depth, significantly. When the condition is $n \geq 4$ for n-bit addition, an improved quantum adder is available. Since the 16-bit addition operator is the smallest unit in SPECK, the improved quantum adder can be applied to all variants of SPECK. In modular addition, one ancilla qubit can be saved (generic addition uses two ancilla), and the quantum gates and circuit depth can also be reduced. Finally, the quantum adder we used requires one ancilla qubit, $(2n-3)$ Toffoli gates, $(5n-7)$ CNOT gates, $(2n-6)$ X gates, and the circuit depth is $(2n+3)$. We do not know exactly why, but when we implemented in ProjectQ, a depth of $(2n+3)$ was estimated. Details of the implementation can be found in [15]. Algorithm 1 describes the improved CDKM adder used in our implementation of SPECK.

4.2 Quantum Circuit for SPECK Using Parallel Addition

We briefly reiterate the round function of SPECK and the key schedule process (refer to Sect. 3.1) for better clarity. The round function of SPECK uses an n-bit round key (k) for a $2n$-bit (x, y) block, and the process is shown in Eq. (1). Notations $\lll$ and $\ggg$ mean left and right rotation, respectively.

$$R_k(x, y) = ((x \lll \alpha) + y) \oplus k, (y \ggg \beta) \oplus ((x \lll \alpha) + y) \oplus k) \quad (1)$$

Algorithm 1. Quantum circuit for improved n-bit CDKM adder ($n \geq 6$).

Input: n-qubit operands a, b, carry qubit c $(= 0)$.
Output: $a = a, b = a + b, c = 0$.

1: **for** $i = 0$ to $n - 3$ **do**
2: $\quad b[i + 1] \leftarrow \text{CNOT}(a[i + 1], b[i + 1])$
3: **end for**

4: $c \leftarrow \text{CNOT}(a[1], c)$
5: $c \leftarrow \text{Toffoli}(a[0], b[0], c)$
6: $a[1] \leftarrow \text{CNOT}(a[2], a[1])$
7: $a[1] \leftarrow \text{Toffoli}(c, b[1], a[1])$
8: $a[2] \leftarrow \text{CNOT}(a[3], a[2])$

9: **for** $i = 0$ to $n - 6$ **do**
10: $\quad a[i + 2] \leftarrow \text{Toffoli}(a[i + 1], b[i + 2], a[i + 2])$
11: $\quad a[i + 3] \leftarrow \text{CNOT}(a[i + 4], a[i + 3])$
12: **end for**

13: $a[n - 3] \leftarrow \text{Toffoli}(a[n - 4], b[n - 3], a[n - 3])$
14: $b[n - 1] \leftarrow \text{CNOT}(a[n - 2], b[n - 1])$
15: $b[n - 1] \leftarrow \text{CNOT}(a[n - 1], b[n - 1])$
16: $b[n - 1] \leftarrow \text{Toffoli}(a[n - 3], b[n - 2], b[n - 1])$

17: **for** $i = 0$ to $n - 4$ **do**
18: $\quad b[i + 1] \leftarrow \text{X}(b[i + 1])$
19: **end for**
20: $b[1] \leftarrow \text{CNOT}(c, b[1])$

21: **for** $i = 0$ to $n - 4$ **do**
22: $\quad b[i + 2] \leftarrow \text{CNOT}(a[i + 1], b[i + 2])$
23: **end for**

24: $a[n - 3] \leftarrow \text{Toffoli}(a[n - 4], b[n - 3], a[n - 3])$

25: **for** $i = 0$ to $n - 6$ **do**
26: $\quad a[n - 4 - i] \leftarrow \text{Toffoli}(a[n - 5 - i], b[n - 4 - i], a[n - 4 - i])$
27: $\quad a[n - 3 - i] \leftarrow \text{CNOT}(a[n - 2 - i], a[n - 3 - i])$
28: $\quad b[n - 3 - i] \leftarrow \text{X}(b[n - 3 - i])$
29: **end for**

30: $a[1] \leftarrow \text{Toffoli}(c, b[1], a[1])$
31: $a[2] \leftarrow \text{CNOT}(a[3], a[2])$
32: $b[2] \leftarrow \text{X}(b[2])$
33: $c \leftarrow \text{Toffoli}(a[0], b[0], c)$
34: $a[1] \leftarrow \text{CNOT}(a[2], a[1])$
35: $b[1] \leftarrow \text{X}(b[1])$
36: $c \leftarrow \text{CNOT}(a[1], c)$

37: **for** $i = 0$ to $n - 2$ **do**
38: $\quad b[i] \leftarrow \text{CNOT}(a[i], b[i])$
39: **end for**

40: **return** a, b, c

The initial key is $K = k_0, l_0, \ldots, l_{m-2}$, and the generated $RK_i = k_0, k_1, \ldots, k_{r-1}$ are used as the ith round key ($0 \leq i \leq r - 1$, r being the total number of rounds). The key schedule process is given in Eq. (2).

$$l_{i+m-1} = (k_i + (l_i \lll \alpha)) \oplus i, k_{i+1} = (k_i \ggg \beta) \oplus l_{i+m-1}. \tag{2}$$

In this part, we explore where the parallel addition is available in the implementation of SPECK as a quantum circuit. We use the initial k_0 in the first round, then update k_0 to k_i to use it as the round key in the ith round ($0 \leq i \leq r - 1$). By taking this on-the-fly approach, there is no need to allocate qubits for the key schedule. For each round, the round function and key schedule are executed together. Due to this, addition $(x \lll \alpha) + y$ in the round function and addition $k_i + (l_i \lll \alpha)$ in the key schedule can be performed in parallel.

In the previous implementation, the key schedule is performed after the round function in the i-th round by adopting the same on-the-fly approach. And only one carry qubit c_0 for addition is allocated. We take two different approaches for parallel addition.

First, k should not be updated in the key schedule until the round key k is used in the round function. In general, parallel addition is impossible because the round function and the key schedule are performed sequentially. We present the procedure for each round as round function (1/2) $\rightarrow$ key schedule (1/2) $\rightarrow$ round function (2/2) $\rightarrow$ key schedule (2/2) instead of round function (1/1) $\rightarrow$ key schedule (1/1).

Then, Round function (1/2) is $x = (x \lll \alpha) + y$ and key schedule (1/2) is $l_i = k_i + (l_i \lll \alpha)$, which are parallel addition targets. Since k_i should not be updated (required in round function (2/2)), the result of addition in key schedule (1/2) is stored in l_i. Round function (2/2) is $x = x \oplus k_i$, $y = (y \ggg \beta) \oplus x$ and key schedule (2/2) is $k_i = (k_i \ggg \beta) \oplus (l_i \oplus i)$.

Algorithm 2. Quantum circuit implementation of SPECK-32/64.

Input: 32-qubit block (x, y), 64-qubit keywords (k_0, l_0, l_1, l_2),

carry qubits c_0 $(= 0)$, c_1 $(= 0)$.

Output: 32-qubit ciphertext (x, y).

```
1: for i = 0 to r − 2 do
2:     Round function (1/2) :
3:         x ← x ⋘ 7
4:         x ← ADD(y, x, c0)
5:     Key schedule (1/2) :
6:         l_{i%3} ← l_{i%3} ⋘ 7
7:         l_{i%3} ← ADD(k0, l_{i%3}, c1)
8:     Round function (2/2) :
9:         x ← CNOT16(k0, x)
10:        y ← y ⋙ 2
11:        y ← CNOT16(x, y)
12:    Key schedule (2/2) :
13:        for j = 0 to 5 do                ▷ Constant XOR
14:            if (i ≫ j)&1 then
15:                l_{i%3}[j] ← X(l_{i%3}[j])
16:            end if
17:        end for
18:        k0 ← k0 ⋙ 2
19:        k0 ← CNOT16(l_{i%3}, k0)
20: end for
21: Round function (1/2) :              ▷ Last round
22:     x ← x ⋘ 7
23:     x ← ADD(y, x, c0)
24: Round function (2/2) :
25:     x ← CNOT16(k0, x)
26:     y ← y ⋙ 2
27:     y ← CNOT16(x, y)
28: return (x, y)
```

Second, now that parallel addition is possible, we need one more carry qubit for this. A ripple-carry quantum adder requires a carry qubit with an initial value of 0, and when the addition is completed, the carry qubit is reset to 0 again. The previous implementation takes advantage of this to allocate only one carry qubit c_0 and reuse it in all additions. However, in order to reuse c_0, the next addition cannot be performed until the addition is finished. Therefore, since we will perform two additions in parallel, we allocate two carry qubits c_0 and c_1 and use them in each addition.

Finally, the proposed quantum circuit implementation provides a 56% performance improvement in terms of depth. Algorithm 2 describes the quantum circuit implementation for SPECK-32/64.

This technique is applied to all SPECK versions, only the parameters are changed. Implementations for other versions can be found in our code. Rotations (i.e., $\lll, \ggg$) can be implemented with the swap gates, but we do not use quantum resources by implementing a logical swap that changes the index of the qubits. CNOT16 means CNOT gate operation of a 16-qubit array. Figure 3 shows the quantum circuit of SPECK-32/64 operating for 3 rounds.

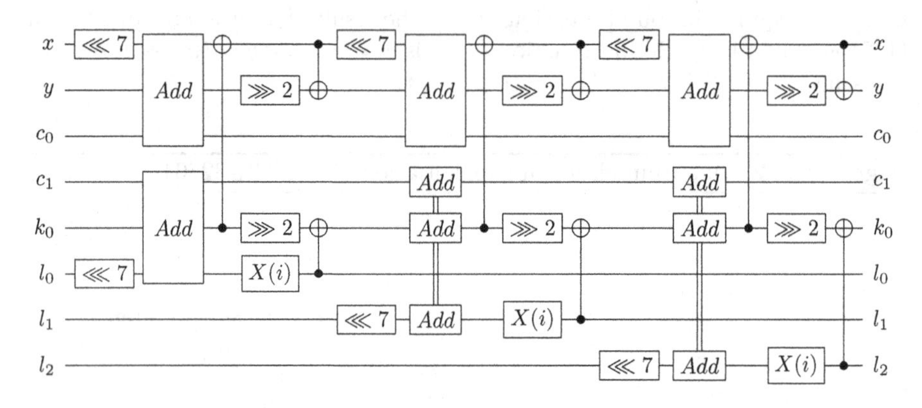

Fig. 3. Quantum circuit for SPECK-32/64 (3 rounds only).

4.3 Architecture and Resource Requirement

As shown in Table 1, the quantum resources required to implement our SPECK quantum circuits are much cheaper compared to the previous SPECK quantum circuits. In [23,27], Jang et al. used a ripple carry-based quantum adder and did not take into account the room for parallel addition. In [6], Anand et al. improved the performance by using a different quantum adder than that of the previous implementation. The quantum circuit they implemented does not use additional qubits and offers performance improvements in terms of depth. We use two more carry qubits and X gates, but parallel addition using an improved ripple carry quantum adder provides performance improvement in terms of circuit depth and a reduction in the number of Toffoli gates.

In the case of SPECK, which is based on an ARX structure, it is important which quantum adder is used. In this work, a quantum circuit is designed so that the additions of the round function of SPECK and the key schedule are performed in parallel, and a few ancilla qubits are allocated accordingly. Also, this approach is expandable because it works even if it is changed to another quantum adder.

In Table 1, quantum resources are reported when the Toffoli gates are not decomposed for simplicity of comparison. However, the Toffoli gate is decomposed into several quantum gates. For detailed resource estimation in this paper, we follow the Toffoli gate decomposition in [4]. One Toffoli gate is decomposed into 7 T gates + 8 Clifford gates (T-depth is 4 and full depth is 8). Table 4 shows the detailed quantum resources required for our SPECK quantum circuits.

Table 4. Quantum resources (decomposed gates) required for variants of SPECK (this work).

SPECK	#CNOT ✲	#1qCliff ❁	#T ✢	T-depth ✤	#qubits ❂	Full depth ❊
32/64	11,661	3,654	8,729	2,552	98	5,258
48/72	18,029	5,718	13,545	3,960	122	8,074
48/96	18,867	5,985	14,175	4,140	146	8,441
64/96	28,933	9,234	21,777	6,344	162	12,870
64/128	30,067	9,597	22,631	6,588	194	13,365
96/96	47,489	15,240	35,805	10,416	194	21,028
96/144	49,215	15,796	37,107	10,788	242	21,779
128/128	73,049	23,511	55,125	16,000	258	32,224
128/192	75,367	24,260	56,875	16,500	322	33,231
128/256	77,685	25,005	58,625	17,000	386	34,238

5 LowMC in Quantum

Regular and Shallow Versions

As mentioned earlier, our quantum circuits of LowMC are divided into regular (⌑) and shallow (⋈) versions. The regular version offers high parallelism while taking into account the trade-off of qubit-depth. Both the regular and the shallow versions provide high parallelism as the linear layer and key schedule work simultaneously. The difference is that the regular version of the S-box has a Toffoli depth of 3 and the shallow version of the S-box has a Toffoli depth of 1, as detailed in Sect. 5.1.

5.1 Implementation of S-box

In [31], two quantum circuit implementation for the 3-bit S-box of LowMC were described as shown in Fig. 4. The in-place S-box (Fig. 4(a)) stores the output value in the input, and the shallow S-box (Fig. 4(b)) additionally uses 3 output qubits and 3 ancilla qubits, but the Toffoli depth can be reduced and the shallow S-box is adopted in their implementation. Notice that the 3 garbage lines are reset, this is because those are reused in the next S-box (save for the last one). When the Toffoli gate is decomposed in the case of the in-place S-box, the full depth is 23, and the shallow S-box is lower at 12. Table 5(a) shows the quantum resources required for the two implementations of the 3-bit S-box.

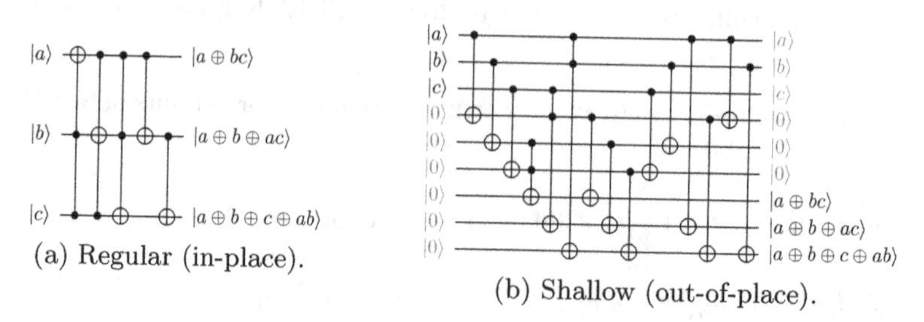

(a) Regular (in-place).

(b) Shallow (out-of-place).

Fig. 4. Quantum circuit for LowMC S-box.

Several trade-offs are to be considered when choosing the quantum S-box implementation. The Toffoli depth of the in-place S-box is 3 and that of the shallow S-box is 1. This is definitely an advantage for the shallow S-box. However, we found that the full depth of the S-box does not affect the full depth of the LowMC when using 10 S-boxes. This is because the depth for S-box is covered by the key schedule and the linear layer. One thing to note is that in-place S-box can be operated in parallel without additional cost, but shallow S-box requires additional ancilla qubits for parallel operation, and qubits for output are allocated every round. Considering these trade-offs, we adopt and compare both S-boxes in our implementations. The regular version of LowMC (⌑) uses the regular/in-place S-box implementation and the shallow version (⋈) uses the shallow/out-of-place S-box implementation.

5.2 Implementation of Linear Layer and Key Schedule

In the linear layer, the pseudo-randomly generated matrix over GF(2) of dimension $n \times n$ in LowMC instantiation is multiplied by an n-bit block. In [31], it is possible to implement an in-place implementation in which CNOT gates are used only in an n-qubit block due to PLU factorization (i.e., internal mixing). In contrast, in our quantum circuit implementation, CNOT gate is performed depending on where the bit value of the matrix is 1. In the CNOT gate, the n-qubit block acts as a control, and a newly allocated n-qubit acts as a target. Finally, the matrix product is stored in the newly allocated n-qubit. Although n-qubit to store the output of the linear layer is newly allocated every round, our approach can obtain a compact quantum circuit. Because we allocate new n-qubits for matrix multiplication, it frees up space and allows for parallelism. Table 5(b) shows the quantum resources required to implement quantum circuits for the linear layer. Since the CNOT gates and depths required for a round are slightly different according to the pseudo-randomly generated matrices, our results in Table 5(b) show the average for all rounds.

In the key schedule, round keys are generated by multiplying the k-bit input key with the matrix of dimension $k \times k$ of each round in the same way. Unlike the linear layer, we can save qubits by using the reverse operation in the key schedule. Only in the first key schedule, a new k-qubit for storing the round key is allocated. After KeyAddition, the reverse operation of the key schedule is performed to return the round key (k-qubit) to a clean state, and it is reused in the next key schedule. Due to the reverse operation, the CNOT gates are doubled. However, in terms of depth, we perform the reverse operation of the key schedule in parallel with the linear layers for the n-qubit block by using two input keys and round key qubits. Figure 5 shows our LowMC quantum circuit operating fully in parallel by operating two input keys (reverse operation of Key Schedule is denoted as Key schedule†[6]). We initially allocate additional $2 \cdot k$ qubits (k_1 and rk_1) and use them alternately in rounds. Although it is omitted in Fig. 5, the input key k_0 is copied to k_1 through the CNOT gates and then the circuit is executed. Through this, the key schedule and the reverse operation of the key schedule can be executed simultaneously with the linear layer. Table 5(c) shows the quantum resources required to implement quantum circuits for the key schedule. It should be pointed out that in Table 5(c), our result excludes the initially allocated $3 \cdot k$-qubit (rk_0, k_1 and rk_1 in Fig. 5). The regular version of Fig. 5(a) and the shallow version of Fig. 5(b) differ in whether the output qubits for the S-boxes are allocated or not, and the Toffoli depth.

5.3 Implementation of KeyAddition and ConstantAddition

KeyAddition and ConstantAddition are implemented the same as in SPECK. KeyAddition is simply implemented using k CNOT gates. In ConstantAddition, since the constants are already known, the X gates are performed where the bit value of the constant is 1.

5.4 Architecture and Resource Requirement

As already presented in Table 2, one may find the quantum resources required to implement our LowMC quantum circuits. In LowMC quantum circuits, the most quantum resources are used for matrix multiplication in the key schedule and linear layer. In [31], an in-place implementation was presented through matrix multiplication using the PLU factorization. On the other hand, we design with a general structure, using more qubits, but more compact quantum circuits are obtained. Lastly, our quantum circuit design using two input keys simultaneously executes the linear layer, key schedule, and reverse operation of the key schedule.

[6] Key Schedule in quantum (of LowMC) denotes the product of the matrix of the round and the input key, and the product is stored in qubits for the round key. The reverse operation (i.e., uncompute) of Key Schedule is defined as Key Schedule†, and cleans the qubits for the round key.

Table 5. Comparison of quantum resources (decomposed gates) required for LowMC.

(a) S-box.

Method	#CNOT ✲	#1qCliff ❁	#T ✢	Toffoli depth ✦	#qubits ❂	Full depth ❈
⌑ S-box [31]	20	6	21	3	3	26
⋈ S-box [31]	30	6	21	1	9	12

⌑: Regular version.

⋈: Shallow version.

(b) Linear layer.

Method	#CNOT ✲	#1qCliff ❁	#qubits ❂	Full depth ❈
Linear layer L1 [31]	8,093	60	128	2,365
Linear layer L3 [31]	18,080	90	192	5,301
Linear layer L5 [31]	32,714	137	256	8,603
Linear layer L1	8,205	0	256	225
Linear layer L3	18,418	0	384	339
Linear layer L5	32,793	0	512	455

(c) Key schedule.

Method	#CNOT ✲	#1qCliff ❁	#qubits ❂	Full depth ❈
Key schedule L1 [31]	8,104	0	128	2,438
Key schedule L3 [31]	18,242	0	192	4,896
Key schedule L5 [31]	32,525	0	256	9,358
Key schedule L1	8,183	0	128	224
Key schedule L3	18,418	0	192	340
Key schedule L5	32,772	0	256	456

5.5 Corrected LowMC Implementation from Eurocrypt'20 (JNRV)

For a clearer context, here we give a brief description of the situations where Q#'s `ResourcesEstimator` issues arise and how those issues affect the quantum benchmarks given in the Eurocrypt'20 paper [31]. This was discovered when we tried to cross-check their publicly available source codes[7]. Indeed, this was also noted in [43] as a bug; and this apparently led to underestimation of gate count, qubit count and depth reported in [31] for the non-linear components (and also the S-box of LowMC).

To our understanding, some problems arise if the qubits are allocated by the `using` command in Q# (and it affects the non-linear components). However

[7] https://github.com/microsoft/grover-blocks.

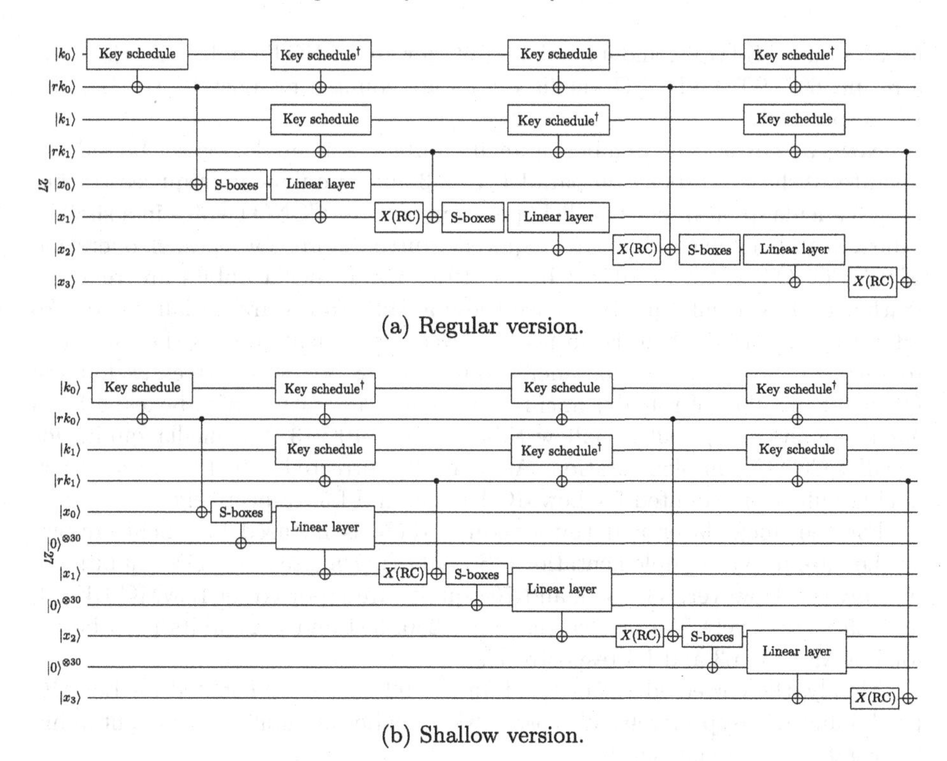

(a) Regular version.

(b) Shallow version.

Fig. 5. Architecture of LowMC quantum circuit.

more experiments are to be carried out in order to be completely certain about it. The `using` command automatically disposes when the function ends. If ancilla qubits to implement LowMC S-box are allocated with the `using` command, the consistency between depth and qubits is lost. When 10 S-boxes are executed in SubBytes, the ancilla qubits allocated by the `using` are counted only for the first S-box and not after. Also counts the depth for executing 10 S-boxes simultaneously. In order to derive the correct result, the number of qubits or depth must be increased. To be modified, the number of qubits must be increased or the depth must be increased. Q#'s `ResourcesEstimator` tries to find its own lower bound for depth and qubit. That is, to achieve the qubits of the lower bound, the depth may have to be increased, and to achieve the depth of the lower bound, the qubits may have to be increased.

For LowMC quantum circuits in [31], the key schedule and the linear layer are in-place implementations, so only the shallow S-box is reported as lower-bound. We correct the number of qubits so that $10 \times$ S-boxes can be operated in parallel. In LowMC, CCNOT implementation with T-depth of one in [37] is adopted rather than AND gate. This CCNOT implementation requires 4 ancilla qubits (see [37] for details). We correct the number of qubits while keeping the CCNOT implementation they adopted.

They count the qubits for ($10 \times$ S-boxes $\times$ number of rounds) as follows: ($10 \times 3 \times$ number of rounds) ancilla qubits for the output of S-boxes, 3 ancilla qubits for all shallow S-boxes, and 4 ancilla qubits for all CCNOT implementations. As a result, 607, 907 and 1,147 ancilla qubits are counted for LowMC L1, L3, and L5, respectively[8].

Now, we correct the number of ancilla qubits estimated as lower-bound. To operate 10 shallow S-boxes in parallel, 10×3 ancilla qubits are required (rather than 3 ancilla qubits). For parallel operation of three CCNOT gates in a shallow S-box, 3×4 ancilla qubits are required. Furthermore, for parallel operation of all CCNOT gates in $10 \times$ S-boxes, $10 \times 3 \times 4$ ancilla qubits are required (rather than 4 ancilla qubits). Since these ancilla qubits are initialized to zero after the operation, there is no need to clean the ancilla qubits (i.e., no need to reverse). So the count for ancilla qubits for S-boxes is correct as follows: ($10 \times 3 \times$ number of rounds) ancilla qubits for the output of S-boxes, 10×3 ancilla qubits for parallel shallow S-boxes, and $10 \times 3 \times 4$ ancilla qubits for parallel CCNOT implementation. As a result, corrected 750, 1,050, and 1,290 ancilla qubits are counted for LowMC L1, L3, and L5, respectively.

For the linear layer and key schedule, there is no need for ancilla qubits as they are in-place implementations. So only (block size + key size) qubits are initially set. However, 384, 576, and 768 qubits are reported for LowMC L1, L3, and L5 respectively. We believe that only 256, 384 and 512 qubits need be set for LowMC L1, L3 and L5 respectively.

Finally, the corrected 1,056, 1,434 and 1,802 qubits are counted for LowMC L1, L3 and L5, respectively. We correct the number of qubits while maintaining their gates and circuit depth.

6 Estimating Cost of Grover's Key Search

In this part, we evaluate the performance (quantum resources required) of the proposed quantum circuits (i.e., SPECK and LowMC). Our quantum implementation results from a quantum simulator on a classical computer, not on a real quantum computer. Due to the difficulty in accessing real quantum computers (and there is also no large-scale quantum computer), most studies report quantum implementations and resource analysis on quantum simulators [6,7,26,29,31,35]. In our work, we use the quantum programming tool ProjectQ to implement and simulate quantum circuits. We use two internal libraries (`ClassicalSimulator` and `ResourceCounter`) of ProjectQ to verify the test vector and then estimate the required quantum resources. `ClassicalSimulator` can simulate large-scale quantum circuits by limiting only quantum gates with Boolean functions (i.e., that have analogy with classical gates) such as X, CNOT, and CCNOT gates. We use `ResourceCounter` to check the number of qubits, the number of quantum gates, and the depth of our quantum circuits. Tables 2 and 4 show the quantum resources required to implement our SPECK and LowMC quantum circuits.

[8] https://github.com/microsoft/grover-blocks/blob/master/numbers/lowmc.csv.

We estimate the cost of the Grover's key search for SPECK and LowMC based on the proposed quantum circuits. The Grover's search algorithm operates by iteration of oracle and diffusion operator. Commonly, the cost of the diffusion operator is ignored in the estimation [17,24,26,31]. The diffusion operator operates on key qubits, and has very little overhead compared to oracle. For this reason, in most studies, the cost of iterating the oracle is estimated as the final cost of the Grover's key search.

In the Grover's oracle, the quantum circuit for the target cipher is operated twice. The first operation encrypts the known plaintext using the key in superposition. Then, we need to check that the (n-bit) ciphertext in the superposition state matches the ciphertext we know. An n-Controlled X gate is used for this. This single gate (i.e., n-Controlled X gate) is also excluded from resource estimation for simplicity because it is a negligible overhead in oracle. Therefore, the cost of the oracle is calculated as the quantum resources required for the encryption circuit (Table 2 or 4) to operate twice sequentially.

As mentioned earlier, the Grover's search algorithm operates as an iteration of oracle and diffusion operator, and we exclude the cost of diffusion operator from resource estimation. Then, the final cost of the Grover's key search is calculated as (oracle $\times$ number of iterations). The number of times the Grover's oracle is applied is in turn determined by the key size. For a k-bit key (i.e., k-bit search space), the number of iterations to get the solution key is $\lfloor \frac{\pi}{4}\sqrt{2^k} \rfloor$ [14] (about $\sqrt{2^k}$). That is, the Grover's search algorithm reduces the security (by the square root) of symmetric key ciphers. Lastly, the Grover's key search on r (plaintext, ciphertext) pairs must be performed (which can be done in parallel) to exclude spurious keys. In [31,32], $r = \lceil k/n \rceil$ (plaintext, ciphertext) pairs are used for Grover's key search for ciphers using n-bit blocks and k-bit keys, and we also follow this structure.

Table 6 shows the Grover's key search cost for SPECK variants. According to the block and key size of SPECK variants, r (plaintext, ciphertext) pairs are required. However, since r (plaintext, ciphertext) pairs are operated in parallel, the depth is not affected. Table 6 and 7 show the Grover's key search cost for SPECK and LowMC variants, respectively. Table 6 and 7 are calculated as (Table 4 and 2, respectively) $\times$ 2 $\times$ $\lfloor \frac{\pi}{4}\sqrt{2^k} \rfloor \times r$ (the number of qubits is not needed in the calculation, owing to the sequential nature of the quantum circuits).

Now, we evaluate the post-quantum security levels of SPECK and LowMC based on NIST's post-quantum security requirements [33]. NIST defined the post-quantum security level as the Grover's's key search cost of AES variants calculated in [17], as stated already in Sect. 2.4. For instance, if the complexity to mount a quantum attack for a given cipher is comparable to or more difficult to that of AES-128 (i.e., 2^{170}), then Level 1 is said to be achieved; since the estimate of Level 1 was taken as 2^{170} in [33]. It may be stated that, the count of qubit was not directly included in computing the security levels (i.e., high full depth was allowed).

Following the security levels stated in [33], the cost of running the Grover's's key search on SPECK and LowMC, for the variants of $\leq$ 128-bit sized keys, none

Table 6. Quantum resources required for key search for SPECK (this work).

SPECK	r	#qubits ❂	Total gates ✳	Full depth ✵	Cost ✳ × ✵	Level of security NIST [33]	G$^+$ [17]	J$^+$ [26]
32/64	2	133	$1.749 \cdot 2^{47}$	$1.008 \cdot 2^{45}$	$1.747 \cdot 2^{92}$	Not achieved ($< 2^{170}$)	Not achieved ($< 2^{169}$)	Not achieved ($< 2^{157}$)
48/72	2	173	$1.357 \cdot 2^{52}$	$1.548 \cdot 2^{49}$	$1.05 \cdot 2^{102}$			
48/96	2	197	$1.419 \cdot 2^{64}$	$1.619 \cdot 2^{61}$	$1.149 \cdot 2^{126}$			
64/96	2	229	$1.089 \cdot 2^{65}$	$1.234 \cdot 2^{62}$	$1.344 \cdot 2^{127}$			
96/96	1	195	$1.181 \cdot 2^{65}$	$1.008 \cdot 2^{63}$	$1.19 \cdot 2^{128}$			
64/128	2	261	$1.132 \cdot 2^{81}$	$1.281 \cdot 2^{78}$	$1.45 \cdot 2^{159}$			Level 1 ($> 2^{157}$)
96/144	2	341	$1.854 \cdot 2^{89}$	$1.044 \cdot 2^{87}$	$1.936 \cdot 2^{176}$	Level 1 ($< 2^{233}$)	Level 1 ($< 2^{233}$)	Level 1 ($< 2^{222}$)
128/128	1	259	$1.818 \cdot 2^{81}$	$1.545 \cdot 2^{79}$	$1.404 \cdot 2^{161}$	Not achieved ($< 2^{170}$)	Not achieved ($< 2^{169}$)	Level 1 ($> 2^{157}$)
128/192	2	453	$1.42 \cdot 2^{114}$	$1.593 \cdot 2^{111}$	$1.131 \cdot 2^{226}$	Level 1 ($< 2^{233}$)	Level 1 ($< 2^{233}$)	Level 3 ($> 2^{222}$)
128/256	2	517	$1.463 \cdot 2^{146}$	$1.641 \cdot 2^{143}$	$1.201 \cdot 2^{290}$	Level 3 ($< 2^{298}$)	Level 3 ($< 2^{298}$)	Level 5 ($> 2^{286}$)

Table 7. Quantum resources required for key search for LowMC (this work).

LowMC		r	#qubits ❂	Total gates ✳	Full depth ✵	Cost ✳ × ✵	Level of security NIST [33]	G$^+$ [17]	J$^+$ [26]
L1	⌑	1	3,201	$1.513 \cdot 2^{83}$	$1.806 \cdot 2^{76}$	$1.366 \cdot 2^{160}$	Not achieved ($< 2^{170}$)	Not achieved ($< 2^{169}$)	Level 1 ($> 2^{157}$)
	⋈		3,831	$1.519 \cdot 2^{83}$	$1.806 \cdot 2^{76}$	$1.371 \cdot 2^{160}$			
L3	⌑	1	6,721	$1.259 \cdot 2^{117}$	$1.013 \cdot 2^{110}$	$1.276 \cdot 2^{227}$	Level 1 ($< 2^{233}$)	Level 1 ($< 2^{233}$)	Level 3 ($> 2^{222}$)
	⋈		7,651	$1.261 \cdot 2^{117}$	$1.013 \cdot 2^{110}$	$1.278 \cdot 2^{227}$			
L5	⌑	1	11,009	$1.412 \cdot 2^{150}$	$1.706 \cdot 2^{142}$	$1.204 \cdot 2^{293}$	Level 3 ($< 2^{298}$)	Level 3 ($< 2^{298}$)	Level 5 ($> 2^{286}$)
	⋈		12,179	$1.413 \cdot 2^{150}$	$1.706 \cdot 2^{142}$	$1.205 \cdot 2^{293}$			

⌑: Regular version.
⋈: Shallow version.

achieves Level 1 security. When the key size is increased, SPECK using 144-bit key achieves Level 1 security; similarly the variants with 192-bit and 256-bit sized keys respectively achieve Level 1 and Level 3 security. On the other hand, the bounds that were actually computed based on the circuits presented in [17] are quite close, but not exactly the same as that of [33] for Level 1 ($< 2^{169}$ from [17], but 2^{170} in [33]).

That said, one may note that the bounds stated in [33] or [17], in some sense overestimated the cost for the respective levels. With each newer implementation, the quantum costs is reduced. In other words, as the quantum costs for the AES variants are reduced, the security levels are to be adjusted accordingly. As far as we know, the best-known implementation (i.e., with the least cost) of AES-128, 192 and 256 as quantum circuits were presented in [26]; were calculated as Level 1: $\approx 2^{157}$, Level 3: $\approx 2^{222}$, Level 5: $\approx 2^{286}$. When adjusted with these newly computed figures, we observe that SPECK and LowMC achieve Level 1 for 128-bit keys, Level 3 for 192-bit keys, and Level 5 (highest) for 256-bit keys.

Apart from the cost itself, there is another requirement from NIST in terms of full depth. The quantum circuits should have less full depth than the so-called "MAXDEPTH" limit [33]. No clear boundary for MAXDEPTH was specified; instead 2^{40}, 2^{64} and 2^{96} are to be considered as landmarks. However, as discussed in [26, Section 2.3], this limit is not always respected in the literature. Looking at Tables 6 and 7 that, one may notice that, our implementations overtook the MAXDEPTH boundaries, particularly those with larger key size. As a follow-up work, one may be interested in adopting a proper procedure (see [26, Section 2.3] for three possible options), as those are out-of-scope for this work.

7 Conclusion

In this work, we follow the previous works [6,23,31] where the quantum analysis of the SPECK and LowMC cipher families was conducted. As a synopsis of our work, it can be mentioned that, we manage to find a reduced depth implementation of the 10 SPECK variants (thereby improving from [6,23]) and 3 LowMC variants, on top of bug-fixing the LowMC implementation from [31] (and benchmark those). All in all, our implementations achieve these security bounds (which are defined in terms of the quantum cost of the AES family [33]):

- Variants of SPECK that use $\leq$ 96-bit key: Not achieved ($< 2^{157}$), SPECK-64/128: 2^{159} (Level 1), SPECK-96/144: 2^{176} (Level 1), SPECK-128/128: 2^{161} (Level 1), SPECK-128/192: 2^{226} (Level 3), SPECK-128/256: 2^{290} (Level 5);
- LowMC L1: 2^{160} (Level 1), LowMC L3: 2^{227} (Level 3), LowMC L5: 2^{293} (Level 5);

when the results from [26] are taken into account. We anticipate our work would be useful to the broader community when analyzing the quantum security of ciphers in the coming future. In particular, we anticipate future researcher will take interest in implementing other ARX ciphers (for instance, by utilizing the quantum adder, see Sect. 4.1) as well as SHA-256 and SHA-384 (those are important milestones to figure out the quantum security levels, see Sect. 2.4).

References

1. Albrecht, M.R., Rechberger, C., Schneider, T., Tiessen, T., Zohner, M.: Ciphers for MPC and FHE. In: Oswald, E., Fischlin, M. (eds.) EUROCRYPT 2015. LNCS, vol. 9056, pp. 430–454. Springer, Heidelberg (2015). https://doi.org/10.1007/978-3-662-46800-5_17
2. Almazrooie, M., Samsudin, A., Abdullah, R., Mutter, K.N.: Quantum reversible circuit of AES-128. Quant. Inf. Process. **17**(5), 1–30 (2018). https://doi.org/10.1007/s11128-018-1864-3

3. Amy, M., Di Matteo, O., Gheorghiu, V., Mosca, M., Parent, A., Schanck, J.: Estimating the cost of generic quantum pre-image attacks on SHA-2 and SHA-3. In: Avanzi, R., Heys, H. (eds.) SAC 2016. LNCS, vol. 10532, pp. 317–337. Springer, Cham (2017). https://doi.org/10.1007/978-3-319-69453-5_18
4. Amy, M., Maslov, D., Mosca, M., Roetteler, M., Roetteler, M.: A meet-in-the-middle algorithm for fast synthesis of depth-optimal quantum circuits. IEEE Trans. Comput. Aided Des. Integr. Circuits Syst. **32**(6), 818–830 (2013). https://doi.org/10.1109/tcad.2013.2244643
5. Anand, R., Maitra, A., Maitra, S., Mukherjee, C.S., Mukhopadhyay, S.: Quantum resource estimation for FSR based symmetric ciphers and related Grover's attacks. In: Adhikari, A., Küsters, R., Preneel, B. (eds.) INDOCRYPT 2021. LNCS, vol. 13143, pp. 179–198. Springer, Cham (2021). https://doi.org/10.1007/978-3-030-92518-5_9
6. Anand, R., Maitra, A., Mukhopadhyay, S.: Evaluation of quantum cryptanalysis on SPECK. In: Bhargavan, K., Oswald, E., Prabhakaran, M. (eds.) INDOCRYPT 2020. LNCS, vol. 12578, pp. 395–413. Springer, Cham (2020). https://doi.org/10.1007/978-3-030-65277-7_18
7. Anand, R., Maitra, A., Mukhopadhyay, S.: Grover on *SIMON*. Quant. Inf. Process. **19**(9), 1–17 (2020). https://doi.org/10.1007/s11128-020-02844-w
8. Baksi, A., Bhattacharjee, A., Breier, J., Isobe, T., Nandi, M.: Big brother is watching you: a closer look at backdoor construction. Cryptology ePrint Archive, Paper 2022/953 (2022). https://eprint.iacr.org/2022/953
9. Baksi, A., Jang, K., Song, G., Seo, H., Xiang, Z.: Quantum implementation and resource estimates for Rectangle and Knot. Quant. Inf. Process. **20**(12), 1–24 (2021). https://doi.org/10.1007/s11128-021-03307-6
10. Banegas, G., Bernstein, D.J., Van Hoof, I., Lange, T.: Concrete quantum cryptanalysis of binary elliptic curves. Cryptology ePrint Archive (2020)
11. Bathe, B., Anand, R., Dutta, S.: Evaluation of Grover's algorithm toward quantum cryptanalysis on ChaCha. Quant. Inf. Process. **20**(12), 1–19 (2021). https://doi.org/10.1007/s11128-021-03322-7
12. Beaulieu, R., Shors, D., Smith, J., Treatman-Clark, S., Weeks, B., Wingers, L.: The SIMON and SPECK families of lightweight block ciphers. Cryptology ePrint Archive, Report 2013/404 (2013). https://eprint.iacr.org/2013/404
13. Bijwe, S., Chauhan, A.K., Sanadhya, S.K.: Quantum search for lightweight block ciphers: gift, skinny, saturnin. Cryptology ePrint Archive, Paper 2020/1485 (2020). https://eprint.iacr.org/2020/1485
14. Boyer, M., Brassard, G., Høyer, P., Tapp, A.: Tight bounds on quantum searching. Fortschritte der Physik **46**(4–5), 493–505 (1998). https://doi.org/10.1002/(SICI)1521-3978(199806)46:4/5⟨493::AID-PROP493⟩3.0.CO;2-P
15. Cuccaro, S., Draper, T., Kutin, S., Moulton, D.: A new quantum ripple-carry addition circuit. arXiv (2008). https://arxiv.org/pdf/quant-ph/0410184.pdf
16. Gidney, C.: Factoring with n+2 clean qubits and n-1 dirty qubits. arXiv preprint arXiv:1706.07884 (2017)
17. Grassl, M., Langenberg, B., Roetteler, M., Steinwandt, R.: Applying Grover's algorithm to AES: quantum resource estimates. In: Takagi, T. (ed.) PQCrypto 2016. LNCS, vol. 9606, pp. 29–43. Springer, Cham (2016). https://doi.org/10.1007/978-3-319-29360-8_3
18. Grover, L.K.: A fast quantum mechanical algorithm for database search. In: Proceedings of the Twenty-Eighth Annual ACM Symposium on Theory of Computing, pp. 212–219 (1996)

19. Häner, T., Jaques, S., Naehrig, M., Roetteler, M., Soeken, M.: Improved quantum circuits for elliptic curve discrete logarithms. In: Ding, J., Tillich, J.-P. (eds.) PQCrypto 2020. LNCS, vol. 12100, pp. 425–444. Springer, Cham (2020). https://doi.org/10.1007/978-3-030-44223-1_23
20. Häner, T., Roetteler, M., Svore, K.M.: Factoring using 2n+ 2 qubits with toffoli based modular multiplication. arXiv preprint arXiv:1611.07995 (2016)
21. He, Y., Luo, M.X., Zhang, E., Wang, H.K., Wang, X.F.: Decompositions of n-qubit toffoli gates with linear circuit complexity. Int. J. Theor. Phys. **56**(7), 2350–2361 (2017)
22. Huang, Z., Sun, S.: Synthesizing quantum circuits of AES with lower t-depth and less qubits. Cryptology ePrint Archive, Report 2022/620 (2022). https://eprint.iacr.org/2022/620
23. Jang, K., Choi, S., Kwon, H., Kim, H., Park, J., Seo, H.: Grover on Korean block ciphers. Appl. Sci. **10**(18) (2020). https://doi.org/10.3390/app10186407
24. Jang, K., Baksi, A., Breier, J., Seo, H., Chattopadhyay, A.: Quantum implementation and analysis of default. Cryptology ePrint Archive, Paper 2022/647 (2022). https://eprint.iacr.org/2022/647
25. Jang, K., Baksi, A., Kim, H., Seo, H., Chattopadhyay, A.: Improved quantum analysis of speck and LOWMC (full version). Cryptology ePrint Archive, Paper 2022/1427 (2022). https://eprint.iacr.org/2022/1427
26. Jang, K., Baksi, A., Kim, H., Song, G., Seo, H., Chattopadhyay, A.: Quantum analysis of AES. Cryptology ePrint Archive, Paper 2022/683 (2022). https://eprint.iacr.org/2022/683
27. Jang, K., Choi, S., Kwon, H., Seo, H.: Grover on SPECK: quantum resource estimates. Cryptology ePrint Archive, Report 2020/640 (2020). https://eprint.iacr.org/2020/640
28. Jang, K., Song, G., Kim, H., Kwon, H., Kim, H., Seo, H.: Efficient implementation of PRESENT and GIFT on quantum computers. Appl. Sci. **11**(11) (2021). https://www.mdpi.com/2076-3417/11/11/4776
29. Jang, K., Song, G., Kim, H., Kwon, H., Kim, H., Seo, H.: Parallel quantum addition for Korean block cipher. IACR Cryptology ePrint Archive, p. 1507 (2021). https://eprint.iacr.org/2021/1507
30. Jang, K., et al.: Grover on PIPO. Electronics **10**(10), 1194 (2021)
31. Jaques, S., Naehrig, M., Roetteler, M., Virdia, F.: Implementing Grover oracles for quantum key search on AES and LowMC. In: Canteaut, A., Ishai, Y. (eds.) EUROCRYPT 2020. LNCS, vol. 12106, pp. 280–310. Springer, Cham (2020). https://doi.org/10.1007/978-3-030-45724-2_10
32. Langenberg, B., Pham, H., Steinwandt, R.: Reducing the cost of implementing the advanced encryption standard as a quantum circuit. IEEE Trans. Quant. Eng. **1**, 1–12 (2020). https://doi.org/10.1109/TQE.2020.2965697
33. NIST.: Submission requirements and evaluation criteria for the post-quantum cryptography standardization process (2016). https://csrc.nist.gov/CSRC/media/Projects/Post-Quantum-Cryptography/documents/call-for-proposals-final-dec-2016.pdf
34. Putranto, D.S.C., Wardhani, R.W., Larasati, H.T., Kim, H.: Another concrete quantum cryptanalysis of binary elliptic curves. Cryptology ePrint Archive (2022)
35. Rahman, M., Paul, G.: Grover on katan: quantum resource estimation. IEEE Trans. Quant. Eng. **3**, 1–9 (2022)
36. Selinger, P.: Quantum circuits of t-depth one. Phys. Rev. A **87**(4), 042302 (2013)
37. Selinger, P.: Quantum circuits of *t*-depth one. Phys. Rev. A **87**, 042302 (2013). https://doi.org/10.1103/PhysRevA.87.042302

38. Shor, P.W.: Algorithms for quantum computation: discrete logarithms and factoring. In: Proceedings 35th Annual Symposium on Foundations of Computer Science, pp. 124–134. IEEE (1994)
39. Song, G., Jang, K., Kim, H., Lee, W., Hu, Z., Seo, H.: Grover on SM3. IACR Cryptology ePrint Archive (2021). https://eprint.iacr.org/2021/668
40. Takahashi, Y., Tani, S., Kunihiro, N.: Quantum addition circuits and unbounded fan-out (2009). https://arxiv.org/abs/0910.2530
41. Zaverucha, G., et al.: The Picnic signature algorithm. Submission to PQC Third Round (2020). https://github.com/microsoft/Picnic/blob/master/spec/spec-v3.0.pdf
42. Zou, J., Li, L., Wei, Z., Luo, Y., Liu, Q., Wu, W.: New quantum circuit implementations of SM4 and sm3. Quant. Inf. Process. **21**(5), 1–38 (2022)
43. Zou, J., Wei, Z., Sun, S., Liu, X., Wu, W.: Quantum circuit implementations of AES with fewer qubits. In: Moriai, S., Wang, H. (eds.) ASIACRYPT 2020. LNCS, vol. 12492, pp. 697–726. Springer, Cham (2020). https://doi.org/10.1007/978-3-030-64834-3_24

A Proposal for Device Independent Probabilistic Quantum Oblivious Transfer

Jyotirmoy Basak[1](✉), Kaushik Chakraborty[2], Arpita Maitra[3], and Subhamoy Maitra[1]

[1] Applied Statistics Unit, Indian Statistical Institute, Kolkata, India
bjyotirmoy.93@gmail.com
[2] School of Informatics, The University of Edinburgh, Edinburgh, UK
[3] TCG Centre for Research and Education in Science and Technology, Kolkata, India

Abstract. In this paper, we propose a novel Probabilistic Quantum Oblivious Transfer (also known as Quantum Private Query or QPQ) scheme with full Device-Independent (DI) certification. To the best of our knowledge, this is the first time we provide such a full DI-QPQ scheme using EPR pairs. Our proposed scheme exploits the self-testing of shared EPR pairs along with the self-testing of projective measurement operators in a setting where the client and the server do not trust each other. To certify full device independence, we exploit a strategy to self-test a particular class of POVM elements that are used in the protocol. Further, we provide formal security analysis and obtain an upper bound on the maximum cheating probabilities for both the dishonest client as well as the dishonest server.

1 Introduction

Since the very first proposal by Chor et al. [6], both Private Information Retrieval (PIR), and Symmetric PIR have attracted extensive attention from the classical cryptography domain. [5,7,11,12,22,28]. SPIR is a two-party (say Server, and Client) mistrustful crypto primitive. Informally, in SPIR one party, Client would like to retrieve some information from a database that is stored at the other party, i.e., Server's side without revealing any information about the retrieved data bits to the Server. The Server's goal is not to reveal any information about the rest of the database. The task of SPIR is similar to the 1 out of N oblivious transfer. Similar to most of the secure two-party cryptographic primitives, designing a secure SPIR scheme is a difficult task. Since the client's privacy and the database security appear to be conflicting, it is elusive to design information-theoretically secure SPIR schemes both in classical and in quantum domain [12,23]. This paper focuses on a more weaker version of SPIR, called Private Query (PQ), where the client is allowed to gain more information about the database than SPIR or 1 out of N oblivious transfer. On the other hand, the client's privacy is ensured in the sense of cheat sensitivity i.e., if the server tries to gain the information about the client's queries then the client can detect that.

T. Isobe and S. Sarkar (Eds.): INDOCRYPT 2022, LNCS 13774, pp. 541–565, 2022.
https://doi.org/10.1007/978-3-031-22912-1_24

The PQ primitive is weaker than SPIR but stronger than PIR. However, this type of primitive suffers from the same limitation as in the PIR schemes. For example, in order to respond client's query, the server must process the entire database. Otherwise, the server will gain information regarding the indices corresponding to the client's query. Moreover, the server needs to send the encrypted version of the entire database; otherwise, it would get an estimate about the number of records that match the query.

In Quantum Private Query (QPQ), the client issues queries to a database and obtains the values of the data bits corresponding to the queried indices such that the client can learn a small amount of extra information about the database bits that are not intended to know by her (known as database security), whereas the server can gain a small amount of information about the query indices of the client (known as user privacy) in a cheat-sensitive way. The functionality of this QPQ primitive can be explained as a probabilistic n-out-of-N Oblivious Transfer (here we consider $n = 1$) where the client has probabilistic knowledge about the other (the bits that are not intended to know by her) database bits.

The first protocol in this domain had been proposed by Giovannetti et al. [13], followed by [14] and [27]. However, all these protocols used quantum memories and none of these are practically implementable at this point. For implementation purpose, Jakobi et al. [18] presented an idea which was based on a Quantum Key Distribution (QKD) protocol [32]. This is the first QPQ protocol based on a QKD scheme. In 2012, Gao et al. [9] proposed a flexible generalization of [18]. Later, Rao et al. [30] suggested two more efficient modifications of classical post-processing in the protocol of Jakobi et al. In 2013, Zhang et al. [37] proposed a QPQ protocol based on the counterfactual QKD scheme [26]. Then, in 2014, Yang et al. came up with a flexible QPQ protocol [36] which was based on the B92 QKD scheme [3]. This domain is still developing, as evident from the number of recent publications [10,34]. Some of these protocols exploit entangled states to generate a shared key between the server (Bob) and the client (Alice). In some other protocols, a single qubit is sent to the client. The qubit is prepared in certain states based on the value of the key and the client has to perform certain measurements on this encoded qubit to extract the key bit. Although these protocols differ in the process of key generation, the basic ideas are the same. The security of all these protocols is defined based on the following facts.

- The server (Bob) knows the whole key which would be used for the encryption of the database.
- The client (Alice) knows a fraction of bits of the key.
- Bob does not get any information about the position of the bits which are known to Alice.

It is natural to consider that one of the legitimate parties may play the role of an adversary. Alice tries to extract more information about the raw key bits (which implies additional information about data bits), whereas Bob tries to know the position of the bits that are known to Alice. For this reason, QPQ can be viewed as a two-party mistrustful cryptographic primitive. Despite its cheat-sensitive property, the server Bob and the client Alice are allowed to violate user privacy and data security, respectively, with a negligible probability based on the security requirements. In practice, the exact primitive that one tries to achieve is as follows-

- Malicious Alice can only know a small amount of additional data bits than that is intended to know by her. Here the aim is to minimise Alice's of extra information about the database.
- Malicious Bob can only gain a small amount of information about the query indices of Alice. Here Alice tries to hides her query indices from Bob.

Very recently, Maitra et al. [24] identified that the securities of all the existing protocols are based on the fact that the communicating parties rely on their devices, i.e., the source that supplies the qubits and the detectors that measure the qubits. Thus, similar to the QKD protocols, the trustworthiness of the devices are implicit in the security proofs of the QPQ protocols. However, in Device Independent (DI) scenario, these trustful assumptions over the devices are removed and the security is guaranteed even after removing these assumptions. But unlike QKD, it is hard to prove DI security in the case of QPQ because of its mistrustful property.

To remove the trustful assumptions and enhance the overall security, recently a DI-QPQ protocol has been described in [24] and it's finite sample analysis has been discussed in [1]. In [24], the authors introduced a testing phase at the server-side and proposed a semi-device independent version of the Yang et al. [36] QPQ scheme.

In this QKD based QPQ scheme [36] (and also in the other QKD based QPQ schemes), the main idea of partial key generation at the client's side relies on the distinction between non-orthogonal states. For the QPQ scheme [36], the server Bob and the client Alice share non maximally entangled states and Alice performs projective measurements at her side on some specified basis randomly to guess the raw key bits (chosen by Bob) with certainty.

It is well-known that contrary to the non maximally entangled states, maximally entangled states are easy to prepare in practice and are also more robust in the case of DI certification. Moreover, it is also known that POVM measurement provides optimal distinction [17,29] between non-orthogonal quantum states.

Keeping these in mind, here we propose a novel QPQ scheme using shared EPR pairs (between the server and the client) and POVM measurement (at the client's side to retrieve the maximum number of raw key bits with certainty). Our proposed scheme provides full DI certification exploiting self-testing of EPR pairs along with self-testing of POVM measurement (at the client's side) and projective measurement (at server's side). We further provide formal security proofs (considering all the strategies that preserve the correctness condition) and obtain an upper bound on the maximum cheating probabilities for both dishonest server and dishonest client.

1.1 Our Contribution

In a distrustful cryptographic primitive like QPQ, it is much harder to prove Device Independence (DI). Keeping this in mind, here we try to achieve data privacy and user security (so that no significant information is leaked to any of the parties) and also try to maintain the cheat-sensitive property. Our main contributions in this paper are threefold which we enumerate below.

1. We propose a novel QPQ scheme and remove the trustworthiness from the devices (source as well as measurement devices) using the self-testing of EPR pairs, self-testing of projective measurements (mentioned in [19]) and self-testing of POVM measurements. Recently, Maitra et al. [24] proposed a semi DI version of the QPQ scheme [36]. However, the QPQ scheme [36] uses non maximally entangled states which are difficult to prepare in practice and are also less robust in the case of DI certification as compared to the maximally entangled states. Keeping this in mind, here we propose a QPQ scheme using EPR pairs and a proper self-testing mechanism that guarantees full DI security of our protocol. To the best of our knowledge, this is the first time we provide such a full DI-QPQ scheme.
2. We replace the usual projective measurement at client Alice's side with optimal POVM measurement so that (on average) Alice can obtain maximum raw key bits with certainty and (possibly) retrieve the maximum number of data bits in a single query. We also show that our proposed scheme provides (on average) the maximum number of raw key bits with certainty for Alice.
3. Contrary to all the existing QPQ protocols, in the present effort, we provide a general security analysis (considering all the attacks that preserve the correctness condition) and provide an upper bound on the cheating probabilities (i.e., a lower bound on the amount of information leakage in terms of entropy) for both the parties (server as well as the client).

1.2 Notations and Definitions

Atfirst, we list down a few notations that we use throughout this paper to describe our scheme.

- $\mathcal{K}$: Initial number of states for the QPQ protocol. Here, we assume that $\mathcal{K}$ is asymptotically large for our scheme.
- $|\psi\rangle_{\mathcal{B}_i\mathcal{A}_i}$: the i-th copy of the shared state where the first qubit corresponds to Bob (subscript $\mathcal{B}_i$ denotes the subsystem corresponds to Bob) and the second qubit corresponds to Alice (subscript $\mathcal{A}_i$ denotes the subsystem corresponds to Alice).
- $\rho_{\mathcal{B}_i\mathcal{A}_i}$: the density matrix representation for the i-th shared state.
- $\rho_{\mathcal{A}_i}$ ($\rho_{\mathcal{B}_i}$): the reduced density matrix at Alice's (Bob's) side for the i-th shared state.
- X: the N-bit database which corresponds to server Bob.
- X_i: i-th bit of the database.
- $R(R_{\mathcal{A}})$: the entire raw key corresponding to Bob (Alice) of size kN bits for some integer $k > 1$.
- $F(F_{\mathcal{A}})$: the entire final key corresponding to Bob (Alice) of size N bits.
- $R_i(R_{\mathcal{A}_i})$: the i-th raw key bit at Bob's (Alice's) side.

- $F_i(F_{\mathcal{A}_i})$: the i-th final key bit at Bob's (Alice's) side.
- $\mathcal{I}_l$: the index set of the elements which are quaried by the client Alice.
- M: POVM device used at Alice's side.
- $A(B)$: measurement outcome at Alice's (Bob's) side.
- $\mathcal{A}(\mathcal{A}^*)$: honest (dishonest) client Alice.
- $\mathcal{B}(\mathcal{B}^*)$: honest (dishonest) server Bob.
- $|\phi_0\rangle = \cos\frac{\theta}{2}|0\rangle + \sin\frac{\theta}{2}|1\rangle$.
- $|\phi_1\rangle = \cos\frac{\theta}{2}|0\rangle - \sin\frac{\theta}{2}|1\rangle$.

Now we list down a few definitions that are required to understand the security related issues of our scheme.

- **Trace Distance:** The trace distance allows us to compare two probability distributions $\{p_i\}$ and $\{q_i\}$ over the same index set which can be defined as

$$Dist(p_i, q_i) = \frac{1}{2}\sum_i |p_i - q_i|.$$

- In quantum paradigm, the trace distance is a measure of closeness of two quantum states ρ and σ. The trace norm of an operator M is defined as,

$$||M||_1 = Tr|M|,$$

where $|M| = \sqrt{M^\dagger M}$. The trace distance between quantum states ρ and σ is given by,

$$\begin{aligned} Dist(\rho, \sigma) &= Tr|\rho - \sigma| \\ &= ||\rho - \sigma||_1, \end{aligned}$$

where $|A| = \sqrt{A^\dagger A}$ is the positive square root of $\sqrt{A^\dagger A}$.

- **Fidelity:** Like trace distance, fidelity is an alternative measure of closeness. In terms of fidelity, the similarity between the two probability distributions $\{p_i\}$ and $\{q_i\}$ can be defined as,

$$F(p_i, q_i) = \left(\sum_i \sqrt{p_i q_i}\right)^2.$$

- The fidelity of two quantum states ρ and σ is defined as

$$F(\rho, \sigma) = \left[Tr(\sqrt{\rho^{1/2}\sigma\rho^{1/2}})\right]^2.$$

- In case of pure states, the fidelity is a squared overlap of the states $|\psi\rangle$ and $|\phi\rangle$, i.e.,

$$F(\rho, \sigma) = |\langle\psi|\phi\rangle|^2,$$

where $\rho = |\psi\rangle\langle\psi|$ and $\sigma = |\phi\rangle\langle\phi|$ are corresponding density matrix representation of the pure states $|\psi\rangle$ and $|\phi\rangle$ respectively.

- The two measures of closeness of quantum states, trace distance and fidelity, are related by the following inequality [8],

$$1-\sqrt{F(\rho,\sigma)} \leq \frac{1}{2}Tr|\rho-\sigma| \leq \sqrt{1-F(\rho,\sigma)}.$$

- Trace distance has a relation with the distinguishability of two quantum states. Suppose, one referee prepares two quantum states ρ and σ for another party (say Alice) to distinguish. The referee prepares each of the states with probability $\frac{1}{2}$. Let p_{correct} denotes the optimal guessing probability for Alice and it is related to trace distance by the following expression,

$$p_{correct} = \frac{1}{2}\left(1+\frac{1}{2}Tr|\rho-\sigma|\right).$$

 It implies that trace distance is linearly dependent to the maximum success probability in distinguishing two quantum states ρ and σ. For further details one may refer to [15].

- **Conditional Minimum Entropy:** Let $\rho = \rho_{AB}$ be the density matrix representation of a bipartite quantum state. Then the conditional minimum entropy of subsystem A conditioned on subsystem B is defined by [21]

$$H_{min}(A|B)_\rho = -\inf_{\sigma_B} D_\infty(\rho_{AB}||\mathbb{I}_A \otimes \sigma_B),$$

 where $\mathbb{I}_A$ denotes the identity matrix of the dimension of system A and the infimum ranges over all normalized density operators σ_B on subsystem B and also for any two density operators T, T' we define,

$$D_\infty(T||T') = \inf\{\lambda \in \mathbb{R} : T \leq 2^\lambda T'\}.$$

- Let ρ_{XB} be a bipartite quantum state where the X subsystem is classical. For the given state ρ_{XB} if $p_{\text{guess}}(X|B)_{\rho_{XB}}$ denotes the maximum probability of guessing X given the subsystem B, then from [21] we have,

$$p_{\text{guess}}(X|B)_{\rho_{XB}} = 2^{-H_{min}(X|B)_\rho}. \tag{1}$$

1.3 Adversarial Model

In the distrustful primitive QPQ, each of the parties has different security goals. The requirement of the entire protocol for the honest case is termed as *Protocol Correctness* whereas the security of the server (Bob) is termed as *Data Privacy* and the security requirement for the client (Alice) is termed as *User privacy*. Formally, these terms are defined below.

Definition 1. *Protocol Correctness:*

If the user (i.e., the client) Alice and the database owner (i.e., the server) Bob both are honest, then after the key establishment phase, the probability that

Alice correctly retrieves the expected number of data bits in a single database query is very high. This implies that in case of honest implementation of the protocol, if X denotes the actual number of data bits known by Alice and $E[X]$ denotes the expected number of data bits that are supposed to be known by Alice then, after the key establishment phase,

$$\Pr(|X - E[X]| \leq \delta_t \wedge \textit{the protocol does not abort}) \geq P_c \quad (2)$$

where δ_t denotes the amount of deviation allowed by Bob and P_c denotes the probability with which the value of X lies within the interval $[E[X]-\delta_t, E[X]+\delta_t]$ (ideally, the value of P_c should be high).

Definition 2. *Protocol Robustness:*

If the user (i.e., the client) Alice and the database owner (i.e., the server) Bob both are honest, then after the key establishment phase of our proposed scheme, the probability that Alice will know none of the final key bits (as well as the database bits) and the protocol has to be restarted is very low. More formally,

$$\Pr(\textit{the protocol aborts in honest scenario}) \leq P_a \quad (3)$$

where P_a denotes the probability that Alice knows none of the final key bits and aborts the protocol (ideally, the value of P_a should be small).

Definition 3. *Data Privacy:*

A QPQ protocol satisfies the data privacy property if either the protocol aborts with high probability in the asymptotic limit, or in a single database query, dishonest Alice ($\mathcal{A}^*$) can correctly extract (on average) at most τ fraction of bits of the N-bit database X where $\tau (0 < \tau < 1)$ is very small compared to the size of the entire database i.e., N. This implies that if $D_{\mathcal{A}^*}$ denotes the number of data bits that dishonest Alice can extract (on average) in a particular query then,

$$E_R(D_{\mathcal{A}^*}) \leq \tau N \quad (4)$$

where the expectation is taken all over the random coins R that are used in the protocol.

Definition 4. *User Privacy:*

Let $\mathcal{I}_l = \{i_1, \ldots, i_l\}$ denotes the indices of the data bits that Alice wants to know from the database by performing l many queries. Then for a QPQ protocol, after l many queries, either the protocol aborts with high probability in the asymptotic limit, or the dishonest Bob ($\mathcal{B}^*$) can correctly guess (on average) at most δ fraction of indices from the index set $\mathcal{I}_l$ where δ $(0 < \delta < 1)$ is very small compared to the size of the index set i.e., l. This implies that after l many queries to the database by Alice, if $\mathcal{I}_{\mathcal{B}^*}$ denotes the number of correctly guessed indices by dishonest Bob then,

$$E_{R'}(\mathcal{I}_{\mathcal{B}^*}) \leq \delta l \quad (5)$$

where the expectation is taken all over the random coins R' that are used in the protocol.

1.4 Assumptions for Our Device Independent Proposal

The list of assumptions for the security of our proposed QPQ-protocol can be summarized as follows.

1. Devices follow the laws of quantum mechanics i.e., the quantum states and the measurement operators involved in this scheme lead to the observed outcomes via the Born rule.
2. Like the recent DI proposal for oblivious transfer from the bounded-quantum-storage-model and computational assumptions in [4], here also we assume that the state generation device and the measurement devices (both at honest and dishonest party's end) are described by a tensor product of Hilbert spaces, one for each device. That means for this proposal, we assume that the devices follow the *i.i.d.* assumptions such that each use of a device is independent of the previous use and they behave the same in all trials. This also implies that the statistics of all the rounds are independent and identically distributed (i.i.d.) and the devices are memoryless. We also assume that the honest party chooses the inputs randomly and independently for each rounds.
   Note : As QPQ is a distrustful primitive, to detect the fraudulent behavior (if any) of the dishonest party, the *i.i.d.* assumption on the inputs chosen by the honest party seems justified here.
3. The honest party can interact with the unknown devices at his end only by querying the devices with the inputs and getting the corresponding outputs whereas the dishonest party can manipulate all the devices before the start of the protocol. However, we assume that after the protocol starts, the dishonest party can no longer change this behavior - s/he cannot manipulate any devices held by the honest party, and also cannot "open up" any devices s/he possesses at her/his end (i.e. the dishonest party is also restricted to only supplying the inputs and getting the corresponding outputs from those devices after the start of the protocol). We also assume that the dishonest party processes their data in an *i.i.d.* fashion.
4. Generally, in the Device Independent (DI) scenario, it is assumed that Alice's and Bob's laboratories are perfectly secured, i.e., there is no communication between the laboratories. As QPQ is a distrustful scheme, here we assume that each party's aim is not only to retrieve as much additional information as possible from the other party but also to leak as little additional information as possible from his side. For this reason, while testing the cheating of a dishonest party in a particular testing phase, the other party must act honestly in that test to detect the fraudulent behavior (if any) of the dishonest party. If both the parties act deceitfully in any testing phase, then none of them can detect the cheating of the other party. So, one party must act honestly in every testing phase.
   In the local tests, the honest party chooses the input bits randomly for the devices at his end (on behalf of the referee). So, there is no communication between the laboratories. But for distributed tests (i.e., the tests performed by both of them with the shared states), we assume that the honest party

chooses the input bits for both the parties on behalf of the referee and then dishonest party announces the measurement outcomes. That means for the distributed tests, we allow communication regarding the input and output bits from the honest party's and the dishonest party's end respectively.
We also assume that the honest party can somehow "shield" his devices such that no information (regarding the inputs and the outputs) is leaked from his laboratory until he chooses to announce something.
Note: Here, one may think that in case of distributed test, the dishonest party may not measure his qubits according to the values of the input bits chosen by the honest party. In that case, how the honest party can detect this dishonest behaviour in the corresponding testing phase is clearly mentioned later in the analysis of *device independent security.*
5. The inputs for self-tests are chosen freely and independently i.e., the device used to generate input bits for one party does not have any correlations (classical or quantum) with the devices of the other party.

2 Our Proposed DI-QPQ Scheme

Here we describe the step by step procedures of the protocol. Note that we haven't considered the channel noise here. So, here we assume that all the operations are perfect.

Algorithm 1: LoaclCHSHtest($\mathcal{S}, \mathcal{P}$)

- For each $i \in \mathcal{S}$, $\mathcal{P}$ does the following-
  1. For the inputs $s_i = 0$ and $s_i = 1$, $\mathcal{P}$'s device performs a measurement in the first qubit of the i-th state and outputs $c_i = 0$ or $c_i = 1$.
  2. For the inputs $r_i = 0$ and $r_i = 1$, $\mathcal{P}$'s device performs a measurement in the second qubit of the i-th state and outputs $b_i = 0$ or $b_i = 1$.
- From the inputs s_i, r_i and corresponding outputs c_i, b_i, $\mathcal{P}$ estimates the following quantity,

$$\mathcal{C} = \frac{1}{|\mathcal{S}|} \sum_{i \in \mathcal{S}} \mathcal{C}_i$$

where the parameter $\mathcal{C}_i$ is defined as follows ,

$$\mathcal{C}_i := \begin{cases} 1 & \text{If } s_i r_i = c_i \oplus b_i \\ 0 & \text{otherwise.} \end{cases}$$

- If $\mathcal{C} = \cos^2 \frac{\pi}{8}$ then the protocol continues.
- Otherwise, the protocol aborts.

1. **Entanglement Distribution Phase:**
   (a) A third party distributes $\mathcal{K}$ copies of two qubit states $|\phi\rangle_{\mathcal{AB}}$ between Alice and Bob such that Alice (Bob) receives $\mathcal{A}$ ($\mathcal{B}$) subsystem of $|\phi\rangle_{\mathcal{AB}}$.

Algorithm 2: OBStest($\mathcal{S}$)

- For each $i \in \mathcal{S}$, Alice and Bob does the following-
  (a) Bob generates a random bit $s_i \in_R \{0,1\}$ as an input of Alice's device and declares the input publicly.
  (b) For the inputs $s_i = 0$ and $s_i = 1$, Alice's device performs a measurement in her part of the i-th copy of the shared states and outputs $c_i = 0$ or $c_i = 1$.
  (c) Bob already generates the input bits $r_i = 0$ or $r_i = 1$ randomly for his measurement device in the i-th instance and obtains the outcome $b_i = 0$ or $b_i = 1$.
  (d) Alice and Bob declare their inputs s_i, r_i and corresponding outputs c_i, b_i.
- From the declared outcomes Alice and Bob estimate the following quantity,

$$\beta = \frac{1}{4} \sum_{s,r,c,b \in \{0,1\}} (-1)^{d_{srcb}} \alpha^{1 \oplus s} \langle \phi_{\mathcal{AB}} | A_c^s \otimes B_b^r | \phi_{\mathcal{AB}} \rangle$$

where $\alpha = \frac{(\cos\theta + \sin\theta)}{|(\cos\theta - \sin\theta)|}$ and d_{srcb} is defined as follows ,

$$d_{srcb} := \begin{cases} 0 & \text{If } sr = c \oplus b \\ 1 & \text{otherwise.} \end{cases}$$

- If $\beta = \frac{1}{\sqrt{2}|(\cos\theta - \sin\theta)|}$ then the protocol continues.
- Otherwise, the protocol aborts.

---

2. **Source Device Verification Phase:**
   This phase comprises of two subphases. Atfirst Bob acts as a referee and performs a local test and then Alice does the same at her end. The different steps are as follows.
   (a) Bob randomly chooses $\frac{\gamma_1 \mathcal{K}}{2}$ instances from the shared $\mathcal{K}$ instances, declares the instances publicly and constructs a set $\Gamma_{\text{CHSH}}^{\mathcal{B}}$ with these instances.
   (b) Alice sends her qubits for all the instances in $\Gamma_{\text{CHSH}}^{\mathcal{B}}$ to Bob.
   (c) For every i-th sample in $\Gamma_{\text{CHSH}}^{\mathcal{B}}$, Bob generates random bits $r_i \in_R \{0,1\}$ and $s_i \in_R \{0,1\}$ as the inputs of his two measurement devices (these devices act as the devices of two different parties).

(d) Bob performs LocalCHSHtest($\Gamma^{\mathcal{B}}_{\text{CHSH}}$, Bob), mentioned in Algorithm 1 for the set $\Gamma^{\mathcal{B}}_{\text{CHSH}}$.
(e) If Bob passes the LocalCHSHtest($\Gamma^{\mathcal{B}}_{\text{CHSH}}$, Bob) game then they proceed further, otherwise they abort.
(f) From the rest $\left(\mathcal{K} - \frac{\gamma_1 \mathcal{K}}{2}\right)$ instances, Alice randomly chooses $\frac{\gamma_1 \mathcal{K}}{2}$ instances, declares the instances publicly and constructs a set $\Gamma^{\mathcal{A}}_{\text{CHSH}}$ with these instances.
(g) Bob sends her qubits for all the instances in $\Gamma^{\mathcal{A}}_{\text{CHSH}}$ to Alice.
(h) For every i-th sample in $\Gamma^{\mathcal{A}}_{\text{CHSH}}$, Alice generates random bits $r_i \in_R \{0,1\}$ and $s_i \in_R \{0,1\}$ as the inputs of her two measurement devices (these devices act as the devices of two different parties).
(i) Alice performs LocalCHSHtest($\Gamma^{\mathcal{A}}_{\text{CHSH}}$, Alice), mentioned in Algorithm 1 for the set $\Gamma^{\mathcal{A}}_{\text{CHSH}}$.
(j) If Alice passes the LocalCHSHtest($\Gamma^{\mathcal{A}}_{\text{CHSH}}$, Alice) game then they proceed to the next part of the protocol where Bob self-tests his observables, otherwise they abort.

Algorithm 3: KEYgen($\mathcal{S}$)

- For each $i \in \mathcal{S}$, Alice does the following-
  (a) If Bob declared $a_i = 0$, Alice measures her qubit of the i-th shared state using the measurement device $M^0 = \{M^0_0, M^0_1, M^0_2\}$.
  (b) If Bob declared $a_i = 1$, Alice measures her qubit of the i-th shared state using the measurement device $M^1 = \{M^1_0, M^1_1, M^1_2\}$.

3. **DI Testing for Bob's Measurement Device:**
   (a) Bob and Alice consider the rest $(\mathcal{K}-\gamma_1\mathcal{K})$ states and for $1 \leq i \leq (\mathcal{K}-\gamma_1\mathcal{K})$, Bob does the following-
   - Bob first generates a random bit r_i for the i-th instance (i.e., bit $r_i \in_R \{0,1\}$).
   - If $r_i = 0$, Bob's device applies measurement operator $\{B^0_0, B^0_1\}$, and generates the output $b_i = 0$ and $b_i = 1$ respectively.
   - If $r_i = 1$, Bob's device applies measurement operator $\{B^1_0, B^1_1\}$, and generates the output $b_i = 0$ and $b_i = 1$ respectively.
   - Bob declares $a_i = 0$ whenever his device outputs $b_i = 0$ (i.e., the device applies measurement operator B^0_0 or B^1_0 for the i-th instance).
   - Bob declares $a_i = 1$ whenever his device outputs $b_i = 1$ (i.e., the device applies measurement operator B^0_1 or B^1_1 for the i-th instance).

(b) From these $(\mathcal{K} - \gamma_1\mathcal{K})$ instances, Bob randomly chooses $\frac{\gamma_2(\mathcal{K}-\gamma_1\mathcal{K})}{2}$ instances, declares the instances publicly and constructs a set $\Gamma_{\text{obs}}^{\mathcal{B}}$ with these instances.
(c) Alice then randomly chooses $\frac{\gamma_2(\mathcal{K}-\gamma_1\mathcal{K})}{2}$ instances from the rest $(\mathcal{K}-\gamma_1\mathcal{K}-\frac{\gamma_2(\mathcal{K}-\gamma_1\mathcal{K})}{2})$ instances, declares the instances publicly and make a set $\Gamma_{\text{obs}}^{\mathcal{A}}$ with these instances.
(d) Alice and Bob construct a set Γ_{obs} with all their chosen instances i.e., $\Gamma_{\text{obs}} = \Gamma_{\text{obs}}^{\mathcal{A}} \cup \Gamma_{\text{obs}}^{\mathcal{B}}$.
(e) Alice and Bob perform OBStest(Γ_{obs}), mentioned in Algorithm 2, for the set Γ_{obs}.

Algorithm 4: POVMtest$(\mathcal{S})$

- Alice considers all those instances of the set $\mathcal{S}$ where Bob declared $a_i = 0$ and creates a set $\mathcal{S}^0$ with those instances.
- Similarly, with the rest of the instances (i.e., the instances where Bob declared $a_i = 1$), Alice creates a set $\mathcal{S}^1$.
- Let us assume that y denotes the value of a_i and for the set $\mathcal{S}^y$, the states at Alice's side are either ρ_x^y or $\rho_{x\oplus 1}^y$ (for input $x \in_R \{0,1\}$ at Bob's side).
- For each set $\mathcal{S}^y$, Alice calculates the value of the parameter

$$\Omega^y = \sum_{b,x\in\{0,1\}} (-1)^{b\oplus x}\text{Tr}[M_b^y \rho_x^y]$$

where M_b^y is the measurement outcome at Alice's side in KEYgen().
- If for every $\mathcal{S}^y$ $(y \in \{0,1\})$,

$$\Omega^y = \frac{2\sin^2\theta}{(1+\cos\theta)}$$

then the protocol continues.
- Otherwise, the protocol aborts.

4. **DI Testing for Alice's POVM Elements:**
(a) After the DI testing phase for Bob's measurement device, Alice and Bob proceed to this phase with the rest $(\mathcal{K} - |\Gamma_{\text{CHSH}}| - |\Gamma_{\text{obs}}|)$ shared states. Let us denote this set as Γ_{POVM}.
(b) Alice randomly chooses $\gamma_3|\Gamma_{\text{POVM}}|$ samples from the rest shared $|\Gamma_{\text{POVM}}|$ states. We call this set as $\Gamma_{\text{POVM}}^{\text{test}}$. Alice performs KEYgen$(\Gamma_{\text{POVM}}^{\text{test}})$, mentioned in Algorithm 3, for the set $\Gamma_{\text{POVM}}^{\text{test}}$.

   (c) Alice then performs POVMtest($\Gamma_{\text{POVM}}^{\text{test}}$), mentioned in Algorithm 4, for the set $\Gamma_{\text{POVM}}^{\text{test}}$.
5. **Key Establishment Phase:**
   (a) After the DI testing phase for POVM elements, Alice proceeds to this phase with the rest ($|\Gamma_{\text{POVM}}| - \gamma_3|\Gamma_{\text{POVM}}|$) shared states. Let us denote this set as Γ_{Key}.
   (b) For the shared states of the set Γ_{Key}, Alice performs KEYgen(Γ_{Key}).
   (c) After KEYgen(Γ_{Key}),
      - If Alice gets $M_0^0(M_1^0)$ for $a_i = 0$, she concludes that the original raw key bit for i-th instance is 0(1). Whenever Alice gets M_2^0, she ignores that outcome.
      - Similarly, if Alice obtains $M_0^1(M_1^1)$ for $a_i = 1$, she concludes that the original raw key bit for i-th instance is 0(1). Whenever Alice gets M_2^1, she ignores that outcome.
   (d) After these key generation, Alice and Bob proceed to private query phase with this $|\Gamma_{\text{Key}}|$ shared states. Note that $|\Gamma_{\text{Key}}| = kN$ for some positive integer $k > 1$ where N is the number of bits in the database and k is exponentially smaller than N.
   (e) Alice and Bob use the raw key bits obtained from these kN many states for the next phase.
6. **Private Query Phase:**
   (a) After the *key establishment phase*, Alice and Bob share a raw key of length kN bits where Bob knows every bit value and Alice knows partially (and Bob does not know the indices of the bits known by Alice).
   (b) Alice and Bob then cut the raw key into k substrings of length N, and add these k strings bitwise to obtain the final key of length N.
   (c) Now suppose that Alice knows only the i-th bit F_i of Bob's final key F and wants to know the j-th bit m_j of the database. Alice then announces a permutation P_A such that after applying the permutation, the i-th bit of the final key goes to j-th position.
   (d) Consequently, Bob applies this permutation P_A on the final key F at his side, encrypt the database with this modified key using one time pad and send the encrypted database to Alice.
   (e) As, the data bit m_j will be encrypted by the final key bit F_i, Alice can correctly recover the intended data bit after decrypting the encrypted database with her key.

Proposed QPQ Scheme (In Case of Honest Implementation):

- The server Bob and the client Alice share $\mathcal{K}$ EPR pairs among themselves such that the first qubit of every shared EPR state corresponds to Alice and the second qubit corresponds to Bob.
- For each of these $\mathcal{K}$ shared EPR pairs, Bob and Alice generate raw key bits in the following way-
  - Bob randomly chooses the value of the i-th raw key bit r_i (i.e., $r_i \in_R \{0,1\}$).
  - If $r_i = 0$, Bob measures his qubit of the i-th shared state in $\{|0\rangle, |1\rangle\}$ basis, otherwise (i.e., for $r_i = 1$) he measures in $\{|0'\rangle, |1'\rangle\}$ basis where $|0'\rangle = \cos\theta |0\rangle + \sin\theta |1\rangle$ and $|1'\rangle = \sin\theta |0\rangle - \cos\theta |1\rangle$ (here Bob chooses the value of θ according to the relation as mentioned in equation 8).
  - Bob declares a classical bit $a_i = 0 (a_i = 1)$ whenever the measurement outcome at his side for the i-th instance is either $|0\rangle$ ($|1\rangle$) or $|0'\rangle$ ($|1'\rangle$).
  - Whenever Bob declared $a_i = 0$, Alice measures her qubit of the i-th EPR pair using the POVM $M^0 = \{M_0^0, M_1^0, M_2^0\}$ where

$$M_0^0 \equiv \frac{(\sin\theta |0\rangle - \cos\theta |1\rangle)(\sin\theta \langle 0| - \cos\theta \langle 1|)}{1 + \cos\theta}$$
$$M_1^0 \equiv \frac{1}{1 + \cos\theta} |1\rangle \langle 1|$$
$$M_2^0 \equiv I - M_0^0 - M_1^0$$

  - Similarly, whenever Bob declared $a_i = 1$, Alice measures her qubit of the i-th EPR pair using the POVM $M^1 = \{M_0^1, M_1^1, M_2^1\}$ where

$$M_0^1 \equiv \frac{(\cos\theta |0\rangle + \sin\theta |1\rangle)(\cos\theta \langle 0| + \sin\theta \langle 1|)}{1 + \cos\theta}$$
$$M_1^1 \equiv \frac{1}{1 + \cos\theta} |0\rangle \langle 0|$$
$$M_2^1 \equiv I - M_0^1 - M_1^1$$

  - If Alice gets $M_0^0 (M_1^0)$ for $a_i = 0$, she concludes that the original raw key bit for i-th instance is $0(1)$. Whenever Alice gets M_2^0, her measurement outcome remains uncertain.
  - Similarly, if Alice obtains $M_0^1 (M_1^1)$ for $a_i = 1$, she concludes that the original raw key bit for i-th instance is $0(1)$. Whenever Alice gets M_2^1, her measurement outcome remains uncertain.
  - After this raw key generation, Alice and Bob first perform some postprocessing on their raw key bits to generate the final key and then perform database query according to the strategy mentioned in the *private query phase* of the proposed scheme.

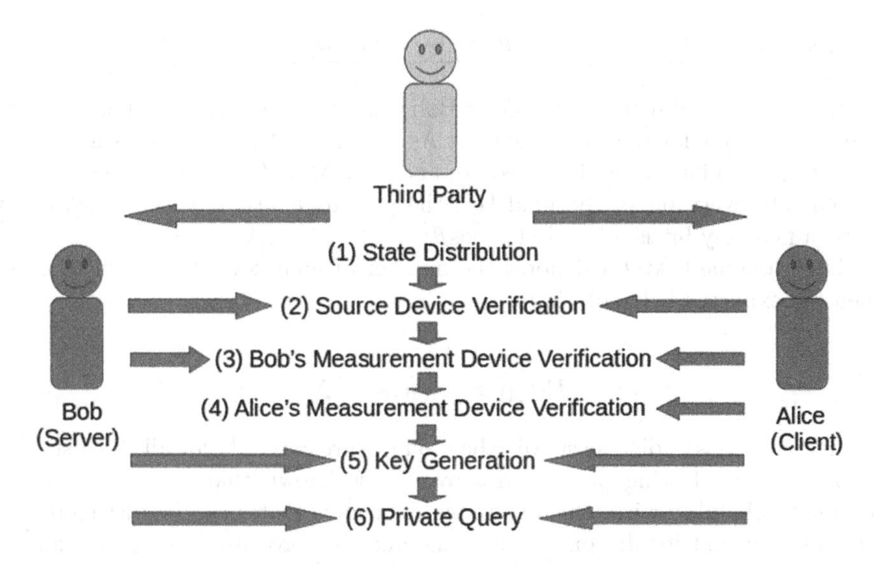

Fig. 1. Schematic diagram of our proposed DI-QPQ scheme.

3 Analysis of Our Protocol

In this section we discuss the functionality of our proposed scheme. At first, we discuss the correctness of our protocol in Subsect. 3.1. Next, we estimate (in Subsect. 3.2) the values of different parameters involved in our scheme for security purpose. At last, we discuss the security related issues of our proposed scheme in Subsect. 3.3 (Fig. 1).

3.1 Correctness of the Protocol

Theorem 1. *For honest Alice and honest Bob scenario, at the end of key establishment phase, Alice can correctly guess (on average) only* $(1 - \cos\theta)$ *fraction of bits of the entire raw key* R.

The proof of this theorem directly follows from our proposed scheme. However, a detailed proof of this theorem can be found in [2].

3.2 Parameter Estimation for Private Query Phase

Here we estimate the values of different parameters considering the honest implementation of our proposed scheme such that the protocol preserves both user privacy and data privacy.

Estimation of the parameter θ for security purpose:

In our proposed scheme, the server Bob wants the client Alice to know atmost one final key bit for database security. As the client Alice can guess a raw key bit with probability around $(1 - \cos\theta)$, and they XOR k number of raw key bits to generate every bit of the final key, the probability that Alice can correctly guess a final key bit is around $(1 - \cos\theta)^k$.

If we assume that $F_{\mathcal{A}}$ denotes the number of final key bits known by Alice then the expected value of $F_{\mathcal{A}}$ is,

$$E[F_{\mathcal{A}}] \approx (1 - \cos\theta)^k N$$

In our proposal, dishonest Alice has to measure correctly for all the instances to pass the DI testing phases. Moreover, it is known that $(1 - \cos\theta)$ is the optimal probability with which two non orthogonal states can be distinguished. This implies that for dishonest Alice, the guessing probability of a raw key bit and a final key bit such that the protocol does not abort is upper bounded by $(1 - \cos\theta)$ and $(1 - \cos\theta)^k$ respectively.

So, the expected value of $F_{\mathcal{A}^*}$ will also be upper bounded by,

$$E[F_{\mathcal{A}^*}] \leq (1 - \cos\theta)^k N \tag{6}$$

As the database is encrypted by performing bitwise XOR with the final key, the expected number of data bits that dishonest Alice can guess correctly in a single query is also upper bounded by $(1 - \cos\theta)^k N$ i.e.,

$$E[D_{\mathcal{A}^*}] \leq (1 - \cos\theta)^k N \tag{7}$$

For security purpose, we want dishonest Alice to know at least one and less than two final key bits. That means,

$$1 \leq E[F_{\mathcal{A}}] < 2$$

$$\text{i.e.,} \quad \frac{1}{N} \leq (1 - \cos\theta)^k < \frac{2}{N} \tag{8}$$

From these results, one can conclude the following,

Corollary 1. *If the server Bob wants the client Alice to know at least one final key bit but less than two final key bits, then Bob needs to choose the parameter k and the value of θ such that,*

$$\frac{1}{N} \leq (1 - \cos\theta)^k < \frac{2}{N}$$

Estimation of the security parameter P_a and P_c :

In our scheme, the probability that Alice can successfully guess a final key bit is equal to $(1-\cos\theta)^k$. So, the probabilty that Alice can not guess a final key bit is equal to $\left[1-(1-\cos\theta)^k\right]$. That means the probability that Alice knows none of the N final key bits is equal to

$$\left[1-(1-\cos\theta)^k\right]^N \approx e^{-(1-\cos\theta)^k N} \tag{9}$$

This implies that for our proposed scheme, we get the following bound on the value of the parameter P_a

$$P_a \leq e^{-(1-\cos\theta)^k N} \tag{10}$$

From the Eq. 8, we get that $\frac{1}{N} \leq (1-\cos\theta)^k < \frac{2}{N}$. If we consider that Bob chooses the value of θ such that $(1-\cos\theta)^k = \frac{1}{N}$ then replacing this value in Eq. 10, we can get,

$$\boxed{P_a \leq e^{-1}} \tag{11}$$

That means the value of P_a is small for our proposed scheme. Similarly, the probability that the protocol does not abort in honest scenario (i.e., Alice knows atleast one final key bit) is equal to

$$\Pr(\text{protocol doesn't abort in honest scenario}) \geq \left[1-e^{-1}\right] \tag{12}$$

Hence, for our scheme, the probability that the protocol does not abort is high. Now, we recall the Chernoff-Hoeffding [16] inequality.

Proposition 1. *(Chernoff-Hoeffding Inequality) Let $X = \frac{1}{m}\sum_{1\leq i\leq m} X_i$ be the average of m independent random variables $X_1, X_2, \cdots, X_m$ with values $(0,1)$, and let $\mathbb{E}[X] = \frac{1}{m}\sum_{1\leq i\leq m}\mathbb{E}[X_i]$ be the expected value of X. Then for any $\delta_{CH} > 0$, we have $\Pr\left[|X-\mathbb{E}[X]| \geq \delta_{CH}\right] \leq \exp(-2\delta_{CH}^2 m)$.*

Here, after the *key establishment phase* of our scheme, we consider $X_i = 1$ whenever Alice knows the value of the i-th final key bit (i.e., for all the raw key bits corresponding to the i-th final key bit, Alice gets the conclusive outcomes i.e., either $M_0^0(M_0^1)$ or $M_1^0(M_1^1)$) or equivalently the corresponding data bit (i.e., the data bit which is encrypted with the i-th final key bit after permutation on the final key) and the protocol does not abort and $X_i = 0$ otherwise. As there are total N number of final key bits, we consider the value of the random variable X as $X = \sum_{i=1}^{N} X_i$.

Whenever the protocol does not abort, the expected number of final key bits that Alice should know after the *key establishment phase* is $E[X] =$

$(1-\cos\theta)^k N$ and there are total $m = N$ number of final key bits. Now, we want that the value of X lie within the error margin $\delta_{CH} = \epsilon\,(1-\cos\theta)^k N$ from the expected value. Here, we can calculate the corresponding probability using the Chernoff-Hoeffding inequality as the final key bits at Alice's side are all independent. For our proposed scheme, the value of the random variable X and also the expected value $E[X]$ is calculated considering the scenario that the protocol does not abort. So, from the expression of Chernoff-Hoeffding bound in Proposition 1, we can write that,

$$\begin{aligned}&\Pr\left[|X - \mathbb{E}[X]| < \delta_{CH} \wedge \text{protocol doesn't abort}\right]\\ &\geq 1 - \exp(-2\delta_{CH}^2\, m)\end{aligned} \tag{13}$$

We want that the number of final key bits known by Alice lie within the interval $[p - \epsilon p, p + \epsilon p]$ where $p = (1-\cos\theta)^k N$ i.e., the deviation is $\delta_{CH} = \epsilon\,(1-\cos\theta)^k N$. From the expression 13, replacing the value of δ_{CH} and m, we get that,

$$\boxed{\begin{aligned}&\Pr\left[|X - \mathbb{E}[X]| < \delta_{CH} \wedge \text{protocol doesn't abort}\right]\\ &\geq 1 - \exp(-2\delta_{CH}^2\, N)\\ &\text{where}\ \ \delta_{CH} = \epsilon\,(1-\cos\theta)^k N\end{aligned}} \tag{14}$$

Now, if we consider that Bob chooses the value of θ such that $(1-\cos\theta)^k = \frac{1}{N}$ then replacing this value in Eq. 14, we can get,

$$\boxed{\begin{aligned}&\Pr\left[|X - \mathbb{E}[X]| < \epsilon \wedge \text{protocol doesn't abort}\right]\\ &\geq 1 - \exp(-2\epsilon^2\, N)\end{aligned}}$$

As the correct guessing of a final key bit implies the correct guessing of the corresponding data bit, from Definition 1, we can say that for honest Alice and honest Bob, the value of the parameter P_c for our scheme is lower bounded by,

$$\boxed{P_c \geq \left[1 - \exp(-2\epsilon^2\, N)\right]} \tag{15}$$

In practice, this probability is high as the value of N is very large.

From the condition in Eq. 8, we can get the following bound on the value of δ_{CH}.

$$\epsilon \leq \delta_{CH} < 2\epsilon \tag{16}$$

This implies that for security purpose, the upper bound on the value of ϵ will be $\epsilon \leq \frac{1}{2}$.

To evaluate the performance, here we consider our scheme as 1 out of 2 probabilistic oblivious transfer (i.e., $N = 2$ and $k = 1$). From Eq. 6, we can say that if Bob chooses the value of θ such that $(1-\cos\theta) = \frac{1}{2}$, then the expected number of final key bit (or data bit) that Alice can guess in a single round of our scheme is $\frac{1}{2} \times 2 = 1$. From Eq. 15, one can conclude that for our scheme, the probability of getting this final key bit is lower bounded by

$$P_c \geq 1 - \exp(-1) \approx 0.632 \tag{17}$$

We now discuss the security related issues of this modified scheme.

3.3 Security of Our Protocol

In this section, we discuss the security related issues of our proposed scheme.

Device Independent Security

In our proposal, the device independent (DI) testing has been done in three phases. The first two DI testing are done in *source device verification phase* and *DI testing for Bob's measurement device.* The third DI testing occurs in *DI testing phase for Alice's POVM elements.* From the rigidity of CHSH game [31, Lemma 4.2], one can conclude the following for the LocalCHSHtest.

Corollary 2 (DI testing of shared states). *In the LocalCHSH test of the source device verification phase of our proposed scheme, either the devices achieve $C = \cos^2 \frac{\pi}{8}$ for both Alice and Bob (i.e., the shared states are EPR pairs), or the protocol aborts with high probability in the asymptotic limit.*

In the next phase, Bob checks the functionality of his measurement devices. Here we assume that Bob will act honestly in this phase to check the functionality of his devices because from the result in Lemma 2, it is clear that if dishonest Bob wants to guess Alice's query indices with more certain probability then he should allow Alice to know more data bits in a single query which violates our assumption 4 that none of the parties reveal additional information from his side to get more information from the other party. From OBStest, one can conclude the following about the functionality of Bob's measurement device.

Theorem 2 (DI testing of Bob's measurement devices). *In OBStest, either Bob's measurement devices achieve $\beta = \frac{1}{\sqrt{2}|(\cos\theta - \sin\theta)|}$ (i.e., his devices measure correctly in $\{|0\rangle, |1\rangle\}$ and $\{|0'\rangle, |1'\rangle\}$ basis where $|0'\rangle = (\cos\theta|0\rangle + \sin\theta|1\rangle)$, $|1'\rangle = (\sin\theta|0\rangle - \cos\theta|1\rangle)$) or the protocol aborts with high probability in the asymptotic limit.*

The proof of this theorem follows exactly the same approach that is mentioned in [19] for certifying non-maximally incompatible observables. For space limitation, the detailed proof is omitted here. However, one can refer to [2] for the proof of this theorem.

The third DI testing is done in *DI testing for Alice's POVM elements.* In this phase, Alice measures the chosen states using the device $M^0 = \{M_0^0, M_1^0, M_2^0\}$ or $M^1 = \{M_0^1, M_1^1, M_2^1\}$ based on the declared a_i values for each of the instances. From the measurement outcomes, one can conclude the following.

Theorem 3 (DI testing of Alice's measurement device M_0). *In POVMtest, for the instances where Bob declares $a_i = 0$, either the protocol aborts with high probability in the asymptotic limit or Alice's measurement devices achieve $\Omega^0 = \frac{2\sin^2\theta}{1+\cos\theta}$ i.e., the devices are of the following form (up to a local unitary),*

$$M_0^0 = \frac{1}{(1+\cos\theta)}(|1'\rangle\langle 1'|) \tag{18}$$

$$M_1^0 = \frac{1}{(1+\cos\theta)}(|1\rangle\langle 1|) \tag{19}$$

$$M_2^0 = \mathbb{I} - M_0^0 - M_1^0, \tag{20}$$

where $|1'\rangle = \sin\theta|0\rangle - \cos\theta|1\rangle$.

Theorem 4 (DI testing of Alice's measurement device M_1). *In POVMtest, for the instances where Bob declares $a_i = 1$, either the protocol aborts with high probability in the asymptotic limit or Alice's measurement devices achieve $\Omega^1 = \frac{2\sin^2\theta}{1+\cos\theta}$, i.e., the devices are of the following form (up to a local unitary),*

$$M_0^1 = \frac{1}{(1+\cos\theta)}(|0'\rangle\langle 0'|) \tag{21}$$

$$M_1^1 = \frac{1}{(1+\cos\theta)}(|0\rangle\langle 0|) \tag{22}$$

$$M_2^1 = \mathbb{I} - M_0^1 - M_1^1, \tag{23}$$

where $|0'\rangle = \cos\theta|0\rangle + \sin\theta|1\rangle$.

The proofs of these two theorems are omitted here because of the space limitation. However, one can refer to Appendix C (entitled *DI Testing of POVM Elements*) in the full version of this paper in [2] for the detailed proof.

All these discussions lead us to the following conclusion.

Corollary 3. *Either our DI proposal aborts with high probability in the asymptotic limit, or it certifies that the devices involved in our QPQ scheme achieve the intended values of $\mathcal{C}, \beta$ and Ω^0 (or Ω^1) in LocalCHSHtest, OBStest and POVMtest respectively.*

Database Security Against Dishonest Alice

In this subsection, we establish an upper bound on the cheating probability of dishonest Alice during the *key establishment phase.*

Theorem 5. *After OBStest, if Alice's measurement device is not tested then in the key establishment phase, dishonest Alice can inconclusively (i.e., can not know the positions of the correctly guessed bits with certainty) retrieve (on average) at most $\left(\frac{1}{2} + \frac{1}{2}\sin\theta\right)$ fraction of bits of the entire raw key.*

Lemma 1. *In the key establishment phase of our scheme, either the protocol aborts with high probability in the asymptotic limit, or dishonest Alice's strategy can retrieve (on average)* $(1-\cos\theta)$ *fraction of bits of the entire raw key.*

The result in Theorem 5 actually follows from [35]. The detailed proof of Theorem 5 and Lemma 1 can be found in [2].

For our QPQ proposal, the database contains N number of data bits. Now relating Eq. 7 with the expression in Definition 3, we can derive the following bound on the value of τ for this proposed scheme.

Corollary 4. *In our DI-QPQ proposal, for dishonest Alice and honest Bob, either the protocol aborts with high probability in the asymptotic limit, or dishonest Alice can guess on average* τ *fraction of bits of the final key, where*

$$\tau \leq (1-\cos\theta)^k \tag{24}$$

Replacing the value of $(1-\cos\theta)^k$ *with the upper bound mentioned in Eq. 8, we can get the following upper bound on the value of* τ.

$$\boxed{\tau < \frac{2}{N}} \tag{25}$$

This relation shows that for this proposed scheme, τ *is small compared to* N.

The comparative study between maximum inconclusive (i.e., the positions of the correctly guessed bits can't be known with certainty) success probability and maximum conclusive (i.e., the positions can be known with certainty) success probability is shown in Fig. 2. From the figure, it is clear that the maximum inconclusive success probability outperforms maximum conclusive success probability for small values of θ.

User Security Against Dishonest Bob

In this section, we establish an upper bound on the guessing probability of dishonest Bob in guessing an index correctly from Alice's query index set.

Lemma 2. *In the proposed QPQ protocol, after* l *many queries to the* N*-bit database by Alice, dishonest Bob* ($\mathcal{B}^*$) *can successfully guess whether a particular index* i *belongs to Alice's query index set* $\mathcal{I}_l$ *(i.e.,* $i \in \mathcal{I}_l$*) with probability atmost* $\frac{l}{N}$*, i.e.,*

$$\Pr(\textit{Bob guesses } i \in \mathcal{I}_l) \leq \frac{l}{N}$$

The proof of Lemma 2 is omitted here because of space limitation and can be found in the full version of this paper in [2].

This result implies that whenever Bob guesses a particular index from the data bits, the chosen index will be in Alice's query index set with probability around $\frac{l}{N}$. Here we assume that after l many queries, Alice's query index set $\mathcal{I}_l$ has l many data bits and Alice chooses these l bits independently. That means

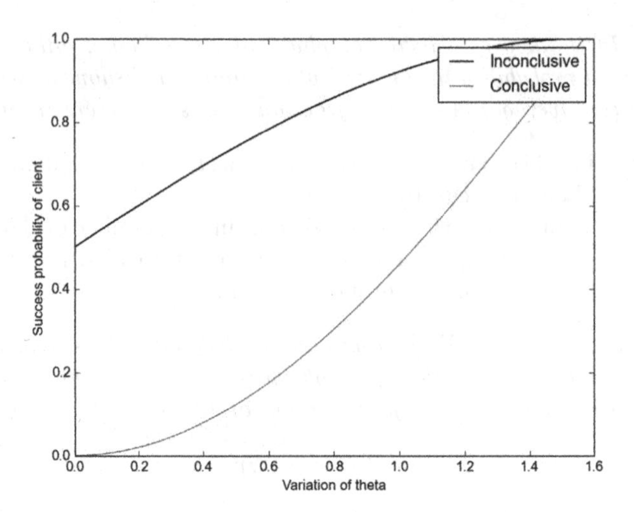

Fig. 2. Comparison between maximum inconclusive and conclusive success probabilities

if Bob guesses l many indices after the query phases then the expected value of the number of indices ($\mathcal{I}_{\mathcal{B}^*}$) that dishonest Bob guesses correctly from the index set $\mathcal{I}_l$ will be,

$$\begin{aligned} E[\mathcal{I}_{\mathcal{B}}] &= \Pr(\text{Bob guesses } i \in \mathcal{I}_l).l \\ &\leq \frac{l^2}{N} \end{aligned} \tag{26}$$

Note that, due to the statistical fluctuations, the dishonest Bob can pass all the tests in our DI-QPQ proposal, and can learn much more than negligible (in l) fraction of query indices from Alice's query index set. However, from the result in Corollary 3, we can say that in the asymptotic limit, dishonest Bob's probability of passing all such tests is very low. Moreover, it is also clear from the result of Lemma 1 that if dishonest Bob wants to guess a query index with more certain probability then he has to allow Alice to know more data bits in a particular query which violates the assumption 4.

Now, comparing Eq. 26 with the expression in the Definition 4, we can derive the following upper bound on the value of δ for our proposed scheme.

Corollary 5. *In our proposed QPQ scheme, for dishonest Bob and honest Alice, either the protocol aborts with high probability in the asymptotic limit, or dishonest Bob can guess on average δ fraction of indices from Alice's query index set $\mathcal{I}_l$ where,*

$$\delta \leq \left(\frac{l}{N}\right) \tag{27}$$

Usually, in practice, the size of the database (i.e., N) is exponentially larger as compared to the size of Alice's query index set (i.e., l). More precisely, $N \approx l^n$, *for some positive integer n. Now, replacing this value of N in Eq. 27, we can get the following bound on the value of δ.*

$$\boxed{\delta \leq \frac{1}{l^{(n-1)}}} \tag{28}$$

This relation shows that for this proposed scheme, δ is small compared to l.

4 Discussion and Conclusion

Initially, all the QPQ schemes were proposed considering that the devices involved are trusted. Thus, a significant portion of the security issues depends on the functionality of the underlying devices. To remove such assumptions, Maitra et al. [24] first initiated the idea of DI in the QPQ domain by suggesting a semi-DI version of the QPQ scheme [36]. In this present effort, we move one step further and propose a novel QPQ scheme considering maximally entangled states with full DI certification to improve the robustness. We also discuss the optimality of the number of raw key bits that a dishonest client Alice can retrieve at her side for these kinds of QKD based QPQ schemes and show that for this proposal, Alice retrieves the optimal number of raw key bits. Contrary to all the existing QPQ schemes which analyze the security issues considering certain eavesdropping strategies, here in this effort, we analyze the security of our scheme in the most general way considering all the attacks that preserve the correctness condition. We further manage to get upper bounds on the cheating probabilities for both the dishonest client and the dishonest server. As the recent QPQ schemes incorporate the idea of QKD, along with the other applications of oblivious transfer, QPQ may soon become a crucial near-term application of quantum internet.

References

1. Basak, J., Maitra, S.: Clauser-Horne-Shimony-Holt versus three-party pseudo-telepathy: on the optimal number of samples in device-independent quantum private query. Quantum Inf. Process. **17**, 77 (2018)
2. Basak, J., Chakraborty, K., Maitra, A., Maitra, S.: Improved and formal proposal for device independent quantum private query (2022). https://arxiv.org/abs/1901.03042
3. Bennett, C.H.: Quantum cryptography using any two nonorthogonal states. Phys. Rev. Lett. **68**(21), 3121–3124 (1992)
4. Broadbent, A., Yuen, P.: Device-independent oblivious transfer from the bounded-quantum-storage-model and computational assumptions. arxiv.org/abs/2111.08595 (2021)
5. Cachin, C., Micali, S., Stadler, M.: Computationally private information retrieval with polylogarithmic communication. In: Stern, J. (ed.) EUROCRYPT 1999. LNCS, vol. 1592, pp. 402–414. Springer, Heidelberg (1999). https://doi.org/10.1007/3-540-48910-X_28

6. Chor, B., Goldreich, O., Kushilevitz, E., Sudan, M.: Private information retrieval. In: Proceedings of the 36th Annual Symposium on Foundations of Computer Science, pp. 41–50 (1995)
7. Di Crescenzo, G., Malkin, T., Ostrovsky, R.: Single database private information retrieval implies oblivious transfer. In: Preneel, B. (ed.) EUROCRYPT 2000. LNCS, vol. 1807, pp. 122–138. Springer, Heidelberg (2000). https://doi.org/10.1007/3-540-45539-6_10
8. Fuchs, C.A., de Graaf, J.V.: Cryptographic distinguishability measures for quantum-mechanical states. IEEE Trans. Inf. Theory **45**, 1216 (1999)
9. Gao, F., Liu, B., Wen, Q.Y., Chen, H.: Flexible quantum private queries based on quantum key distribution. Opt. Express **20**, 17411–17420 (2012)
10. Liu, B., Gao, F., Huang, W., Wen, Q.Y.: QKD-based quantum private query without a failure probability. Sci. China Physics, Mech. Astron. **58**(10), 1–6 (2015). https://doi.org/10.1007/s11433-015-5714-3
11. Gentry, C., Ramzan, Z.: Single-database private information retrieval with constant communication rate. In: Caires, L., Italiano, G.F., Monteiro, L., Palamidessi, C., Yung, M. (eds.) ICALP 2005. LNCS, vol. 3580, pp. 803–815. Springer, Heidelberg (2005). https://doi.org/10.1007/11523468_65
12. Gertner, Y., Ishai, Y., Kushilevitz, E., Malkin, T.: Protecting data privacy in private information retrieval schemes. In: Proceedings of the Thirtieth Annual ACM Symposium on Theory of Computing, pp. 151–160 (1998)
13. Giovannetti, V., Lloyd, S., Maccone, L.: Quantum random access memory. Phys. Rev. Lett. **100**(23), 230502 (2008)
14. Giovannetti, V., Lloyd, S., Maccone, L.: Quantum private queries: security analysis. IEEE Trans. Info. Theory **56**(7), 3465–3477 (2010)
15. Helstrom, C.W.: Quantum Detection and Estimation Theory. Mathematics in Science and Engineering, vol. 123. Academic Press, New York (1976)
16. Hoeffding, W.: Probability inequalities for sums of bounded random variables. J. Am. Stat. Assoc. **58**(301), 13–30 (1963)
17. Ivanovic, I.D.: How to differentiate between non-orthogonal states. Physics Lett. A **123**(6), 257–259 (1987)
18. Jakobi, M., et al.: Practical private database queries based on a quantum-key-distribution protocol. Phys. Rev. A **83**(2), 022301 (2011)
19. Kaniewski, J.: Self-testing of binary observables based on commutation. Phys. Rev. A **95**(6), 062323 (2017)
20. Kon, W.Y., Lim, C.C.W.: Provably-secure symmetric private information retrieval with quantum cryptography. https://arxiv.org/abs/2004.13921 (2020)
21. Konig, R., Renner, R., Schaffner, C.: The operational meaning of min- and max-entropy. IEEE Trans. Info. Theory **55**(9), 4337–4347 (2009)
22. Kushilevitz, E., Ostrovsky, R.: Replication is not needed: single database, computationally-private information retrieval. In: Proceedings of the 38th Annual Symposium on Foundations of Computer Science, pp. 364–373, 1997
23. Lo, H.K.: Insecurity of quantum secure computations. Phys. Rev. A **56**(2), 1154 (1997)
24. Maitra, A., Paul, G., Roy, S.: Device-independent quantum private query. Phys. Rev. A **95**(4), 042344 (2017)
25. Kumar Mishra, S., Sarkar, P.: Symmetrically private information retrieval. In: Roy, B., Okamoto, E. (eds.) INDOCRYPT 2000. LNCS, vol. 1977, pp. 225–236. Springer, Heidelberg (2000). https://doi.org/10.1007/3-540-44495-5_20
26. Noh, T.G.: Counterfactual Quantum Cryptography. Phys. Rev. Lett. **103**(23), 230501 (2009)

27. Olejnik, L.: Secure quantum private information retrieval using phase-encoded queries. Phys. Rev. A **84**(2), 022313 (2011)
28. Ostrovsky, R., Skeith III, W.E.: A survey of single-database private information retrieval: techniques and applications. In: Proceedings of the 10th International Conference on Practice and Theory in Public-Key Cryptography, pp. 393–411 (2007)
29. Peres, A., Terno, D.R.: Optimal distinction between non-orthogonal quantum states. J. Phys. A: Math. Gen. **31**, 7105 (1998)
30. Rao, M.V.P., Jakobi, M.: Towards communication-efficient quantum oblivious key distribution. Phys. Rev. A **87**(1), 012331 (2013)
31. Reichardt, B., Unger, F., Vazirani, U.: A classical leash for a quantum system: command of quantum systems via rigidity of CHSH games. Nature **496**(7446), 456 (2013)
32. Scarani, V., Acín, A., Ribordy, G., Gisin, N.: Quantum cryptography protocols robust against photon number splitting attacks for weak laser pulse implementations. Phys. Rev. Lett. **92**, 057901 (2004)
33. P.W. Shor, Algorithms for Quantum Computation: Discrete Logarithms and Factoring. In: Foundations of Computer Science (FOCS) 1994, pp. 124–134. IEEE Computer Society Press (1994)
34. Wei, C.Y., Gao, F., Wen, Q.Y., Wang, T.Y.: Practical quantum private query of blocks based on unbalanced-state Bennett-Brassard-1984 quantum-key-distribution protocol. Sci. Rep. **4**, 7537 (2014)
35. Wiesner, S.: Conjugate coding. SIGACT News **15**(1), 78–88 (1983)
36. Yang, Y.G., Sun, S.J., Xu, P., Tiang, J.: Flexible protocol for quantum private query based on B92 protocol. Quant. Info. Proc. **13**, 805 (2014)
37. Zhang, J.L., Guo, F.Z., Gao, F., Liu, B., Wen, Q.Y.: Private database queries based on counterfactual quantum key distribution. Phys. Rev. A **88**(2), 022334 (2013)

Quantum Attacks on PRFs Based on Public Random Permutations

Tingting Guo[1,2], Peng Wang[1,2(✉)], Lei Hu[1,2], and Dingfeng Ye[1,2]

[1] SKLOIS, Institute of Information Engineering, CAS, Beijing, China
w.rocking@gmail.com, {guotingting,hulei,yedingfeng}@iie.ac.cn
[2] School of Cyber Security, University of Chinese Academy of Sciences, Beijing, China

Abstract. Plenty of permutation-based pseudorandom functions (PRFs) were proposed. In order to analyze their quantum security uniformly, we proposed three general frameworks $F1, F2$, and $F3$ for n-to-n-bit PRFs with one, two parallel, and two serial public permutation calls respectively, where every permutation is preceded and followed by any bitwise linear mappings. We analyze them in the $Q2$ model where attackers have quantum-query access to PRFs and permutations. Our results show $F1$ is not secure with $\mathcal{O}(n)$ quantum queries while its PRFs achieve $n/2$-bit security in the classical setting, and $F2, F3$ are not secure with $\mathcal{O}(2^{n/2}n)$ quantum queries while their PRFs, such as SoEM, PDMMAC, and pEDM, achieve $2n/3$-bit security in the classical setting. Besides, we attack three general instantiations XopEM, EDMEM, and EDMDEM of $F2, F3$ with at most $\mathcal{O}(2^{n/2}n)$ quantum queries, which derive from replacing the two PRPs in Xop, EDM, and EDMD with two independent EM constructions. We also attack pre-existing concrete PRF instantiations of $F2, F3$: DS-SoEM, PDMMAC, pEDM, and SoKAC21, with at most $\mathcal{O}(2^{n/2}n)$ quantum queries.

Keywords: PRF · Permutation · Quantum attack

1 Introduction

Symmetric-key Schemes Based on PRFs. A Message Authentication Code (MAC) is a symmetric-key primitive that ensures message integrity. For a popular nonce-based MAC, the Wegman-Carter (WC) scheme [5,10,34,36], it offers better security when replacing the underlying Pseudorandom Permutation (PRP) with Pseudorandom Function (PRF). For other cryptographic designs, such as encryption mode CTR [1] and authenticated encryption mode GCM [27], it also offers better security when replacing the underlying PRPs (block ciphers) with PRFs. Thus it is of great necessity to design pseudorandom functions (PRFs) even with fixed-input length. Unfortunately, dedicated fixed input length PRF designs are scarce. The well-known PRP/PRF switching lemma [4,23] suggests simply viewing the PRP as a PRF. However, it makes the cryptographic designs be limited to only birthday bound security, i.e., $n/2$-bit security (We say a design m-bit security if it is secure up to $\mathcal{O}(2^m)$ queries) assuming

T. Isobe and S. Sarkar (Eds.): INDOCRYPT 2022, LNCS 13774, pp. 566–591, 2022.
https://doi.org/10.1007/978-3-031-22912-1_25

the size of the output of PRP is n bits. Thus, plenty of researchers make a great effort to transform PRPs to PRFs with high quality.

PRP-to-PRF Conversion Methods with BBB Security. Fortunately, there have existed four main PRP-to-PRF transformation methods in achieving security beyond the birthday bound: Trunc, Xop, EDM, and EDMD. Let block ciphers be modeled as PRPs. Trunc [22] truncates the output of an n-bit block cipher by $m < n$ bits, resulting $(m+n)/2$-bit security [2,19]. Let E_1, E_2 be two independent block ciphers. Xop, EDM, and EDMD based on E_1, E_2 all provide n-bit security [28,30–32]. Xop [3] is the XOR of PRPs for input M:

$$\mathrm{XoP}_{E_1,E_2}(M) = E_1(M) \oplus E_2(M).$$

Encrypted Davies-Meyer (EDM) [16] and Encrypted Davies-Meyer Dual (EDMD) [28] EDMD serially perform two block ciphers:

$$\mathrm{EDM}_{E_1,E_2}(M) = E_2(E_1(M) \oplus M),$$
$$\mathrm{EDMD}_{E_1,E_2}(M) = E_2(E_1(M)) \oplus E_1(M).$$

In fact, at ASIACRYPT 2021, Chen et al. [15] have proved XoP, EDM, and EDMD are the only constructions with Beyond-Birthday-Bound (BBB) security ($> n/2$-bit security) of all n-to-n-bit PRFs based on two block cipher calls.

Advantages of Permutation-Based Designs. It is well known that designing a block cipher is more complex than a keyless public permutation, as the former involves evaluating the underlying key scheduling algorithm. Besides, we do not need to store the round keys in public permutation-based designs. In addition, the theory of analyzing the security of any cryptographic design based on public permutations is full-fledged. Therefore, it has been an extraordinarily popular approach to design cryptographic schemes based on public permutations straightforwardly.

Even-Mansour Constructions. We can view PRPs as PRFs directly. One of the most famous public permutation-based PRPs is Even-Mansour (EM) cipher [18]: $\mathrm{EM}(M) = \pi(M \oplus K_1) \oplus K_2$, where π is a public random permutation and K_1, K_2 are two independent keys. Later, Bogdanov et al. [7] introduced a more general PRP KAC by iterating EM for multiple rounds. However, they both only provide birthday bound security with respect to the block size by PRP/PRF switching lemma.

n-to- n -Bit PRFs with One or Two Permutation Calls. Researchers try to design public permutation-based n-to-n-bit PRFs that provide BBB security with one or two permutations calls. At CRYPTO 2019 Chen et al. [14] firstly delved into the methods of designing such PRFs. They presented the general design of a PRF with only one public permutation call and whose permutation is preceded and followed by any linear mappings consisting of bitwise exclusive-OR and scalar multiplication (see Fig. 1). They showed such construction cannot be secure beyond the birthday bound for any linear mapping in the classical setting.

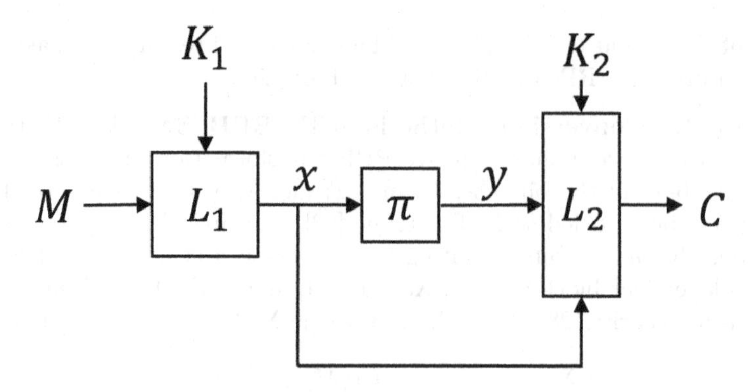

Fig. 1. Function proposed by Chen et al. [14] using two keys K_1 and K_2, and making one public random permutations evaluation π, where L_1, L_2 are two blockwise linear mappings.

So they try to design PRFs with higher security by making two public permutations calls. In the same paper [14], they try to get such PRFs by instantiating generic BBB secure PRP-to-PRF conversion functions with EMs or their variants, i.e., replacing the two PRPs in Xop and EDMD with two independent EMs or their variants. They firstly proposed SoKAC by instantiating with the variants of EM for EDMD construction, which is similar to 2-round KAC [7], with two public permutations and two keys. They named it SoKAC1 if the two permutations are the same, which only provides birthday bound security [14]. And they named it SoKAC21 if the two keys are the same, which provides BBB security [14] but unfortunately attacked by Nandi [29] at EUROCRYPT 2020 with only birthday bound complexity. In addition to SoKAC, they also put forward SoEM by instantiating with EM for Xop construction. SoEM is based on two public permutations and two keys as well. They called it SoEM1 if the two permutations are the same and SoEM21 if the two keys are the same, which are both only birthday bound securities [14]. Delightfully, they proved that SoEM22, with two independent permutations and two independent keys, is secure up to $2n/3$ bits [14].

Following their design method, plentiful fantabulous PRFs have been put forward. Quickly, at CRYPTO 2020 Chakraborti et al. [13] introduced PDMMAC, which is based on only single public permutation and its reverse and only takes a single key, by instantiating with the EM appropriately for EDM construction. It also provides $2n/3$-bit security [13]. Next to PDMMAC, in 2020, Bhattacharjee et al. [6] designed DS-SoEM, which is based on only one public permutation and even doesn't need the inverse of the permutation like PDMMAC. It is a Xop construction instantiated with EM with two same public permutation calls and two independent keys and still maintains $2n/3$-bit security. Another preeminent PRF based on only one public permutation and two keys is pEDM, which is introduced by Dutta et al. [17] in 2021. It is also an EDM construction instantiated with EMs with $2n/3$-bit security.

Previous Quantum Attacks on PRFs with One or Two Permutation Calls. There have existed attacks for permutation-based PRFs in the $Q2$ model, which means attackers can make superposition queries to a quantum oracle of $U_F : |x, y\rangle \mapsto |x, y \oplus F(x)\rangle$, where F is a classical primitive implemented on a quantum computer and attacker has quantum access to it. Kuwakado et al. [25] and Kaplan et al. [24] firstly recovered the keys of EM cipher by constructing periodic function based on this cipher and applying Simon's algorithm with $O(n)$ queries to recover the secret period which is useful for key recovery. For PRFs based on two public permutation calls, recently in 2022 Shinagawa et al. [33] presented key recovery attacks against SoEM. They successfully attacked SoEM1 and SoEM21 with quantum queries for polynomial times by applying Simon's algorithm, and SoEM22 with $\mathcal{O}(2^{n/2}n)$ quantum queries by applying Grover-meet-Simon algorithm. For SoEM variants with linear key schedules, Zhang [37] showed they are also vulnerable to Simon's algorithm and Grover-meet-Simon algorithm.

Motivations. There are still plenty of PRFs based on permutations haven't been analyzed in the $Q2$ model, such as SoKAC, PDMMAC, DS-SoEM, pEDM and so on. What about the security of such PRFs in the $Q2$ model? How to propose general frameworks and analyze their securities?

Our Contributions. We assume all permutations in all PRFs we analyzed are on n bits. And the following functions all are n-to-n-bit functions except for DS-SoEM.

1. The first contribution is to systematically tackle the security of a PRF with one random permutation call whose permutation is preceded and followed by linear mappings from a generalized perspective in the $Q2$ model. The general function we considered (See Fig. 2) is more universal than Chen et al. [14] (See Fig. 1):
   1) First, we change the value from the first linear mapping to the permutation (i.e. x) and the value from the first linear mapping to the second linear mapping (i.e. z) from same to be independent;
   2) Second we extend blockwise linear mappings to bitwise linear mappings.

   We name our generalized function as $F1$. It actually generalize the pre-existing constructions with one permutation call. We considering different types of linear mappings and prove that, whatever the linear mappings are, such construction is not secure with quantum queries for polynomial times in the $Q2$ model in spite of its birthday bound security [14] in the classical setting.

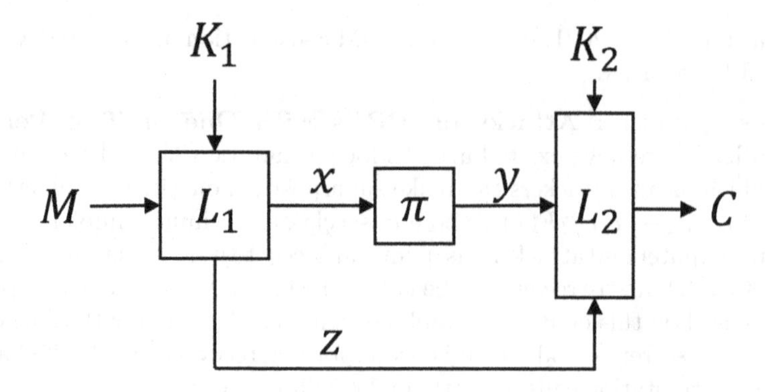

Fig. 2. Function $F1$ using two keys K_1 and K_2, and making one public random permutations evaluation π, where L_1, L_2 are two bitwise linear mappings.

2. The second contribution is to systematically tackle the security of a PRF with two public random permutations calls and both permutations are preceded and followed by bitwise linear mappings from a generalized perspective. We show that all such PRFs can be divided into two kinds: one's two permutation calls are parallel and the other's are serial. We call the general design of the former as $F2$ as pictured in Fig. 3(a) and the latter as $F3$ as pictured in Fig. 3(b). They actually generalize the pre-existing constructions with two permutation calls. We find that, whatever the linear mappings are, the both two constructions cannot be secure beyond $\mathcal{O}(2^{n/2}n)$ quantum queries in the $Q2$ model in spite of BBB security of their concrete instantiations in the classical setting [13,14,17].

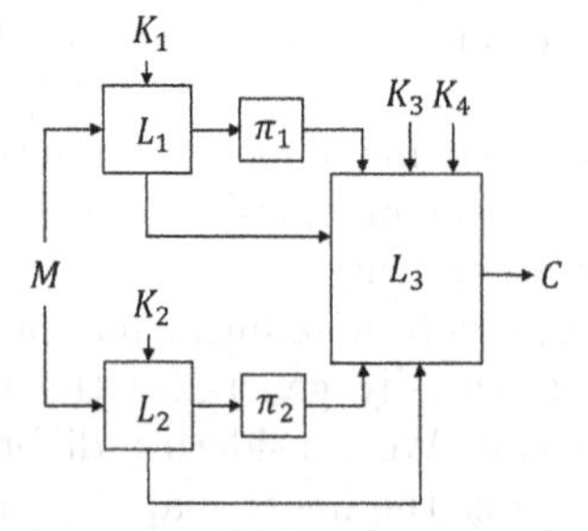

(a) Function $F2$ based on four keys K_1, K_2, K_3 and K_4, and making two parallel public random permutations evaluation π_1 and π_2, where L_1, L_2, L_3 are three bitwise linear mappings.

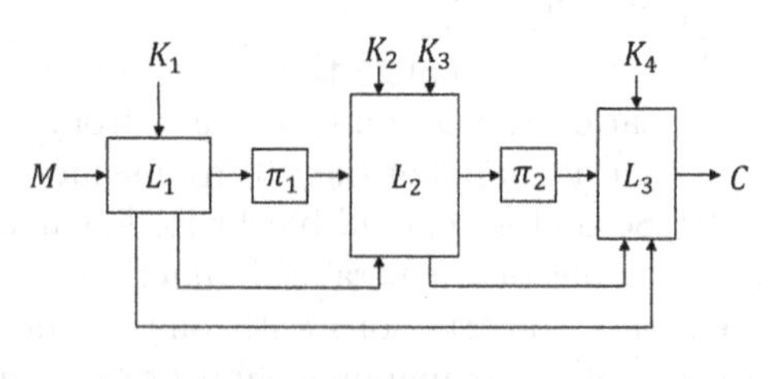

(b) Function $F3$ using four keys K_1, K_2, K_3 and K_4, and making two serial public random permutations evaluation π_1 and π_2, where L_1, L_2, L_3 are three bitwise linear mappings.

Fig. 3. Functions based on two public permutations calls.

3. Our third contribution is to present the quantum security of general and cencrete instantiations of $F2, F3$. We show the hierarchy of all PRFs based on two public permutations in Fig. 4.

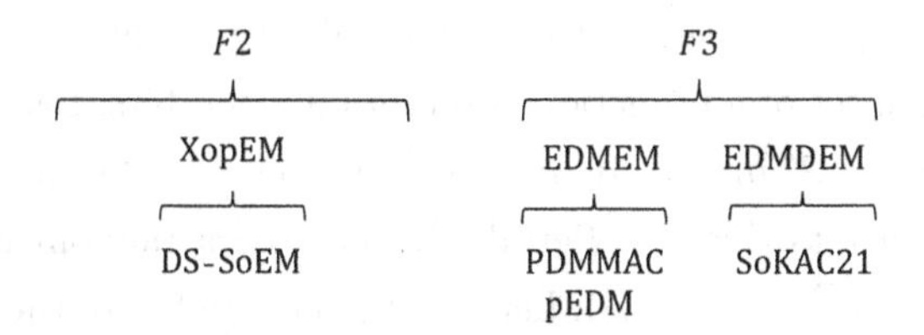

Fig. 4. The hierarchy of all PRFs based on two public permutations calls.

1) By replacing the two PRPs in Xop, EDM, and EDMD with two independent EMs respectively, we get three constructions and name them XopEM, EDMEM, and EDMDEM. It is a general and natural idea to get PRFs with two permutation calls. The attack for EM cipher [24,25] is inapplicable to these combined constructions for their more complex constructions which are difficult to construct periodic functions. So the combined constructions may be safer in the $Q2$ model. We show they are not secure with at most $\mathcal{O}(2^{n/2}n)$ quantum queries in the $Q2$ model in spite of their concrete instantiations (such as PDMMAC, pEDM, and SoEM22) are secure up to $2n/3$ bits.
2) We show the security of pre-existing concrete PRF designs instantiated with EM or its variants for Xop, EDM, and EDMD which make two permutation calls. Our results show $2n/3$-bit secure DS-SoEM, PDMMAC and pEDM in the classical can be broken with at most $\mathcal{O}(2^{n/2}n)$ queries in the $Q2$ model. We also show that SoKAC21 breaks with $\mathcal{O}(2^{n/3})$ queries in the $Q2$ model.

We summarize our main results in Table 1.

Table 1. Summary of the our main results. n is the size of permutation. d is a truncation parameter.

	Functions	The number of calls of public permutations	The number of public permutations	The number of keys	The query complexity of our quantum attack
Generic functions	$F1$	1	1	2	$\mathcal{O}(n)$
	$F2$	2	2	4	$\mathcal{O}(2^{n/2}n)$
	$F3$	2	2	4	$\mathcal{O}(2^{n/2}n)$
General Instantiations	XopEM	2	2	4	$\mathcal{O}(2^{n/2}n)$
	EDMEM	2	2	4	$\mathcal{O}(2^{n/2}n)$
	EDMDEM	2	2	4	$\mathcal{O}(2^{n/2}n)$
Special Instantiations	DS-SoEM [6]	2	1	2	$\mathcal{O}(2^{(n-d)/2}(n-d))$
	PDMMAC [13]	2	1	1	$\mathcal{O}(2^{n/2})$
	pEDM [17]	2	1	2	$\mathcal{O}(2^{n/2}n)$
	SoKAC21 [14]	2	2	1	$\mathcal{O}(2^{n/3})$

2 Preliminaries

2.1 Notations

Let $\mathbb{N}$ be the set of positive integers. For $n \in \mathbb{N}$, let $\{0,1\}^n$ be the set of all n-bit binary strings. Let $\text{Perm}(n)$ be the set of all permutations on n bits and $\text{Func}(m,n)$ be the set of all functions from m bits to n bits. Let $x \xleftarrow{\$} \mathcal{X}$ indicate choosing x from set $\mathcal{X}$ uniformly and random. Let $\pi \xleftarrow{\$} \text{Perm}(n)$ be a random permutation on n bits (i.e. $\pi \xleftarrow{\$}\text{Perm}(n)$). Let ρ be a random function from n bits to n bits (i.e. $\rho \xleftarrow{\$}\text{Func}(n,n)$). Let $\#\mathcal{X}$ be the number of the elements in set $\mathcal{X}$. Let $\hat{0}$ indicate the zero linear mappings which maps all values in $\{0,1\}^n$ to 0^n. Let $x := y$ indicate defining x as y.

2.2 Decomposition of Linear Mappings

Chen et al. [14] considered blockwise linear mappings. That is to say, the linear mappings consisting of bitwise exclusive-OR '$\oplus$' and scalar multiplication '$\times$' in $\{0,1\}^n$. We give an example. Let matrix

$$\mathbf{M} = \begin{bmatrix} a_1 & a_2 \\ a_3 & a_4 \end{bmatrix},$$

where $a_1, a_2, a_3, a_4 \in \{0,1\}^n$. For any $x_1, x_2 \in \{0,1\}^n$,

$$\mathbf{M}\begin{bmatrix} x_1 \\ x_2 \end{bmatrix} = \begin{bmatrix} a_1 & a_2 \\ a_3 & a_4 \end{bmatrix}\begin{bmatrix} x_1 \\ x_2 \end{bmatrix} = \begin{bmatrix} (a_1 \times x_1) \oplus (a_2 \times x_2) \\ (a_3 \times x_1) \oplus (a_4 \times x_2) \end{bmatrix}. \quad (1)$$

Let linear mapping $L(x_1, x_2) =$ Eq. (1) and branch linear mappings $l_1(x_1) = a_1 \times x_1, l_2(x_2) = a_2 \times x_2, l_3(x_1) = a_3 \times x_1, l_4(x_2) = a_4 \times x_2$. Then

$$L(x_1, x_2) = \begin{bmatrix} l_1(x_1) \oplus l_2(x_2) \\ l_3(x_1) \oplus l_4(x_2) \end{bmatrix}.$$

So they decomposed L to l_1, l_2, l_3, l_4.

In this paper, we considered bitwise linear mappings. That is to say, the linear mappings consisting of bitwise exclusive-OR '$\oplus$' and dot product '$\cdot$' with constants in $\{0,1\}$ (i.e., $b' \cdot y' = 1$ iff $b' = y' = 1$; otherwise, $b' \cdot y' = 0$). We give an example. Let matrix

$$\mathbf{M} = \begin{bmatrix} b_{1,1} & \dots & b_{1,2n} \\ \vdots & \ddots & \vdots \\ b_{2n,1} & \dots & b_{2n,2n} \end{bmatrix}.$$

where $b_{i,j} \in \{0,1\}$ for $i,j = 1,\dots,2n$. For any $x_1 = y_1 y_2 \dots y_n \in \{0,1\}^n$ and $x_2 = y_{n+1} y_{n+2} \dots y_{2n} \in \{0,1\}^n$, we interpret them to

$$\mathbf{x}_1 = \begin{bmatrix} y_1 \\ \vdots \\ y_n \end{bmatrix}, \mathbf{x}_2 = \begin{bmatrix} y_{n+1} \\ \vdots \\ y_{2n} \end{bmatrix}.$$

Then

$$\mathbf{M}\begin{bmatrix}\mathbf{x}_1\\ \mathbf{x}_2\end{bmatrix} = \begin{bmatrix} b_{1,1} & \dots & b_{1,2n} \\ \vdots & \ddots & \vdots \\ b_{2n,1} & \dots & b_{2n,2n}\end{bmatrix}\begin{bmatrix} y_1 \\ \vdots \\ y_{2n}\end{bmatrix} = \begin{bmatrix} b_{1,1}\cdot y_1 \oplus \dots \oplus b_{1,2n}\cdot y_{2n} \\ \vdots \\ b_{2n,1}\cdot y_1 \oplus \dots \oplus b_{2n,2n}\cdot y_{2n}\end{bmatrix} \quad (2)$$

We partitioning matrix $\mathbf{M}$ to four $n \times n$ partitioned matrices $\mathbf{M}_1, \mathbf{M}_2, \mathbf{M}_3, \mathbf{M}_4$:

$$\mathbf{M} = \begin{bmatrix}\mathbf{M}_1 \, \mathbf{M}_2 \\ \mathbf{M}_3 \, \mathbf{M}_4\end{bmatrix}.$$

It also holds that

$$\mathbf{M}\begin{bmatrix}\mathbf{x}_1\\ \mathbf{x}_2\end{bmatrix} = \begin{bmatrix}\mathbf{M}_1\mathbf{x}_1 \oplus \mathbf{M}_2\mathbf{x}_2 \\ \mathbf{M}_3\mathbf{x}_1 \oplus \mathbf{M}_4\mathbf{x}_2\end{bmatrix}.$$

Let linear mapping $L(x_1, x_2) =$ Eq. (2) and branch linear mappings $l_1(x_1) = \mathbf{M}_1\mathbf{x}_1$,$l_2(x_2) = \mathbf{M}_2\mathbf{x}_2, l_3(x_1) = \mathbf{M}_3\mathbf{x}_1, l_4(x_2) = \mathbf{M}_4\mathbf{x}_2$. Then

$$L(x_1, x_2) = \begin{bmatrix} l_1(x_1) \oplus l_2(x_2) \\ l_3(x_1) \oplus l_4(x_2)\end{bmatrix}.$$

So we can decomposed L to l_1, l_2, l_3, l_4. For simplicity, we write $L = (l_1, l_2, l_3, l_4)$.

Multiplication by a in $\{0,1\}^n$ is a linear transformation over the n column vectors space with every element in $\{0,1\}$, and can therefore be represented by a $n \times n$ matrix $\mathbf{M}_a$ with every element in $\{0,1\}$. Thus, for any element $x \in \{0,1\}^n$ and its corresponding n column vector $\mathbf{x}$ whose every element in $\{0,1\}$, we can write

$$a \times x = \mathbf{M}_a\mathbf{x}.$$

So the bitwise linear mappings are more general than blockwise linear mappings.

2.3 The Security of qPRF Based on Public Random Permutations

Let $\pi_1, \dots, \pi_\ell$ be public random permutations. Let F be a keyed function that may depend on $\pi_1, \dots, \pi_\ell$ and ρ be a random function that is independent of $\pi_1, \dots, \pi_\ell$. Given the quantum oracle of $\pi_1^{\pm}, \dots, \pi_\ell^{\pm}$ and function F or ρ, where the superscript $\pm$ for π_i indicates the distinguishers has bi-directional access. The security of quantum pseudorandom function (qPRF) of F is defined by the minimum number of quantum queries of all distinguishers to distinguish $(F, \pi_1^{\pm}, \dots, \pi_\ell^{\pm})$ from $(\rho, \pi_1^{\pm}, \dots, \pi_\ell^{\pm})$. We call the queries to F or ρ construction queries and the queries to $\pi_1^{\pm}, \dots, \pi_\ell^{\pm}$ primitive queries.

2.4 Quantum Algorithms

1) Simon's Algorithm. Simon's algorithm [35] finds the period of a periodic function with quantum queries for polynomial times and polynomial qubits. It solves the Simon's problem.

Definition 1 (Simon's Problem). *Let n be a positive integer. Given a Boolean function $f : \{0,1\}^n \to \{0,1\}^n$, and the promise that there exists $s \in \{0,1\}^n \backslash \{0^n\}$ such that for any $x, y \in \{0,1\}^n$, $[f(x) = f(y)] \Leftrightarrow [x \oplus y \in \{0^n, s\}]$. Find s.*

Classically, we can find s by searching collisions with $\mathcal{O}(2^{n/2})$ queries. However, in the $Q2$ model Simon's algorithm [35] can reduce the queries rapidly to only polynomial times. Recall that the Hadamard transform $H^{\otimes n}$ applied on an n-qubit state $|x\rangle$ for some $x \in \{0,1\}^n$ gives $H^{\otimes n}|x\rangle = \frac{1}{\sqrt{2^n}} \sum_{y \in \{0,1\}^n} (-1)^{x \cdot y} |y\rangle$, where $x \cdot y := x_1 y_1 \oplus \cdots \oplus x_n y_n$.

The Steps of Simon's Algorithm [35]:

1. Initialize the state of $2n$ qubits to $|0\rangle^{\otimes n}|0\rangle^{\otimes n}$;
2. Apply Hadamard transformation $H^{\otimes n}$ to the first n qubits to obtain quantum superposition $\frac{1}{\sqrt{2^n}} \sum_{x \in \{0,1\}^n} |x\rangle |0\rangle^{\otimes n}$;
3. A quantum query to the function f maps this to the state: $\frac{1}{\sqrt{2^n}} \sum_{x \in \{0,1\}^n} |x\rangle |f(x)\rangle$;
4. Measure the last n qubits to get the output of $f(z)$, and the first n qubits collapse to $\frac{1}{\sqrt{2}}(|z\rangle + |z \oplus s\rangle)$;
5. Apply the Hadamard transform $H^{\otimes n}$ to the first n quantum qubits again, we can get $\frac{1}{\sqrt{2}} \frac{1}{\sqrt{2^n}} \sum_{y \in \{0,1\}^n} (-1)^{y \cdot z} \left(1 + (-1)^{y \cdot s}\right) |y\rangle$. If $y \cdot s = 1$ then the amplitude of $|y\rangle$ is 0. So measuring the state in the computational basis yields a random vector y such that $y \cdot s = 0$, which means that y must be orthogonal to s.

By repeating these steps $O(n)$ times, $n-1$ independent vectors y orthogonal to s can be obtained with high probability, then we can recover s with high probability by using linear algebra.

At CRYPTO 2016, Kaplan et al. [24] relaxed the promise in Simon's problem. They defined $\varepsilon(f)$ to quantify how far the function is from satisfying Simon's promise, where

$$\varepsilon(f) := \max_{t \in \{0,1\}^n \backslash \{0^n, s\}} \Pr_x[f(x) = f(x \oplus t)].$$

$\varepsilon(f) = 0$ means satisfying the promise in Simon's problem. They proved Simon's algorithm still can recover the period s provided bounded $\varepsilon(f)$ [24].

Theorem 1 (Simon's Algorithm with Approximate Promise *[24]***).** *If $\varepsilon(f) \leq p_0 < 1$, then Simon's algorithm returns s with cn queries to f using $\mathcal{O}(n)$ qubits, with probability at least $1 - \left(2\left(\frac{1+p_0}{2}\right)^c\right)^n$.*

Choosing $c \geq \frac{3}{(1-p_0)}$ ensures that the error decreases exponentially with n.

2) Grover's Algorithm. Grover's algorithm [20] can find a target from a set. It solves the Grover's problem.

Definition 2 (Grover's Problem). *Let m be a positive integer, and $test : \{0,1\}^m \to \{0,1\}$ be a Boolean function. Find an u such that $test(u) = 1$.*

Classically, we can find an u such that $test(u) = 1$ with $\mathcal{O}(\frac{2^m}{\#\{u:test(u)=1\}})$ queries to $test(\cdot)$. However, in the $Q2$ model Grover's algorithm [20] can speed up the search by square root [11].

The Steps of Grover's Algorithm [20]:

1. Initializing a n-bit register $|0\rangle^{\otimes m}$.
2. Apply Hadamard transformation $H^{\otimes m}$ to the first register to obtain quantum superposition $H^{\otimes m}|0\rangle = \frac{1}{\sqrt{2^m}} \sum_{x\in\{0,1\}^m} |u\rangle = |\varphi\rangle$.
3. Construct an oracle $O : |u\rangle \mapsto (-1)^{test(u)}|u\rangle$.
4. Apply Grover iteration for $R \approx \frac{\pi}{4}\sqrt{\frac{2^m}{\#\{u:test(u)=1\}}}$ times to amplify the amplitudes of goal elements in $\mathcal{U}$: $[(2|\varphi\rangle\langle\varphi| - I)O]^R|\varphi\rangle$.
5. Measure the register to get an u such that $test(u) = 1$.

More generally, the *test* function can't describe the target set so precisely. That is to say, $test(u)$ always outputs 1 for elements in the target set, but for elements not in the target set that $test(u)$ also output 1 with some probability.

Definition 3 (Grover's Problem with Biased *test* Function). *Let m be a positive integer, $\mathcal{U}$ be a subset in $\{0,1\}^m$, $test : \{0,1\}^m \to \{0,1\}$ be a Boolean function who satisfies*

$$\begin{cases} \Pr[test(u) = 1] = 1, & u \in \mathcal{U}, \\ \Pr[test(u) = 1] \leq p_1, & u \notin \mathcal{U}. \end{cases}$$

Find an $u \in \mathcal{U}$.

Luckily, the Grover's algorithm with $\mathcal{O}(2^{m/2})$ quantum queries to $test(\cdot)$ using $\mathcal{O}(m)$ qubits can find a $u \in \mathcal{U}$ as well assuming $\#\mathcal{U} \leq 2, p_1 \leq \frac{1}{2^{2m}}$ [8,21].

3) Grover-meet-Simon Algorithm. At ASIACRYPT 2017 Leander and May [26] combined Grover's algorithm with Simon's algorithm to recover the keys of FX construction. They named their technique as Grover-meet-Simon algorithm. Paper [8,21] considered the general algorithm to solve the general Grover-meet-Simon problem.

Definition 4 (Grover-meet-Simon Problem). *Let m, n be two positive integers, set $\mathcal{U} \subseteq \{0,1\}^m$ and $f : \{0,1\}^m \times \{0,1\}^n \to \{0,1\}^n$ be a function who satisfies*

$$\begin{cases} f(u,\cdot) \text{ is a periodic function with period } s_u, & u \in \mathcal{U}, \\ f(u,\cdot) \text{ is an aperiodic function}, & u \notin \mathcal{U}. \end{cases}$$

Set $\mathcal{U}_s := \{(u, s_u) : u \in \mathcal{U}, s_u \text{ is the period of } f(u,\cdot)\}$. Find any tuple $(u, s_u) \in \mathcal{U}_s$.

The main idea of the Grover-meet-Simon algorithm is to search $u \in \mathcal{U}$ by Grover's algorithm and check whether or not $u \in \mathcal{U}$ by whether $f(u, \cdot)$ is periodic or not, which can be implemented by Simon's algorithm. Let $\varepsilon(f)$ to quantify how far the function is from satisfying $[f(u,x) = f(u,y)] \Leftrightarrow [u \in \mathcal{U}, x \oplus y \in \{0^n, s_u\}]$.

$$\varepsilon(f) := \max_{(u,t)\in\{0,1\}^m\times\{0,1\}^n\setminus(\mathcal{U}_s\cup\{0,1\}^m\times\{0^n\})} \Pr_x[f(u,x) = f(u, x\oplus t)].$$

Then Grover-meet-Simon algorithm with $\mathcal{O}(2^{m/2}n)$ quantum queries to f using $\mathcal{O}(m+n^2)$ qubits will output a tuple $(u, s_u) \in \mathcal{U}_s$ assuming $\varepsilon(f) \le 7/8, \#\mathcal{U} \le 2$ [21].

At ASIACRYPT 2019 Bonnetain et al. [9] improved the Grover-meet-Simon algorithm in the Q2 model for the function $f(u, \cdot)$ in Definition 4 which can be constructed from the sum of two functions, i.e.,

$$f(u,\cdot) = h(u,\cdot) \oplus g(\cdot).$$

Given the quantum oracle of $h(u,\cdot)$ and $g(\cdot)$, it is easy to construct the quantum oracle of $f(u,\cdot)$. They also used Grover's algorithm to search $u \in \mathcal{U}$ and check whether $f(u,\cdot)$ is periodic or not by Simon's algorithm. Each time a new u is tested, a new function $h(u,\cdot)$ is queried. But, in contrast, the function g is always the same. So they firstly got a superposition quantum state about g by making $\mathcal{O}(n)$ queries to the quantum oracle of g'. Then they reused it every time the algorithm queries about $f(u,\cdot)$. They showed after making $\mathcal{O}(2^{m/2}n)$ queries to $h(u,\cdot)$ and $\mathcal{O}(n)$ queries to g using $\mathcal{O}(m+n^2)$ qubits, the algorithm can recover the unique $u \in \mathcal{U}$ and its corresponding period assuming $\varepsilon(f) \le 1/2, \#\mathcal{U} = 1$ [9]. It is easy to prove it is also suitable for $\varepsilon(f) \le 7/8, \#\mathcal{U} = 2$. This improved algorithm greatly reduces the query complexity from $\mathcal{O}(2^{m/2}n)$ to $\mathcal{O}(n)$ for $f(u,\cdot)$ whose $h(u,\cdot)$ is public computable by the adversary itself, such as $f(u,\cdot)$ based on FX construction.

3 Attack on Function with One Permutation Call

We will show that any function that makes only one public random permutation call and has linear pre- and post-processing functions of the permutation only is not secure with queries for polynomial times in the $Q2$ model. Let $M, C \in \{0,1\}^n$ and K_1, K_2 be two independent keys in $\{0,1\}^n$. Let π be a public random permutation, $L_1 : (\{0,1\}^n)^2 \to (\{0,1\}^n)^2$ and $L_2 : (\{0,1\}^n)^3 \to \{0,1\}^n$ be any two linear mappings. Then we let $F1 : \{0,1\}^{2n} \times \{0,1\}^n \to \{0,1\}^n$ be the general function using keys K_1, K_2 with input M and output C. And it makes one call to public random permutation π and has the pre- and post-linear mapping L_1, L_2. See $F1$ in Fig. 2.

Theorem 2. *Let $n \in \mathbb{N}$, and consider the function $F1 : \{0,1\}^{2n} \times \{0,1\}^n \to \{0,1\}^n$ of Fig. 2 based on a public random permutation π with block length of n bits and using two keys $K_1, K_2 \xleftarrow{\$} \{0,1\}^n$, for any linear mappings L_1, L_2. There exists a distinguisher $\mathcal{D}$ making at most $\mathcal{O}(n)$ construction queries and at most $\mathcal{O}(n)$ primitive queries to distinguish $F1$ from random function.*

Proof. The linear mappings L_1, L_2 are public. So firstly, we decompose

$$L_1 = (l_{11}, l_{12}, l_{13}, l_{14}),$$
$$L_2 = (l_{21}, l_{22}, l_{23}),$$

in the classical setting such that

$$L_1(K_1, M) = \begin{bmatrix} l_{11}(K_1) \oplus l_{12}(M) \\ l_{13}(K_1) \oplus l_{14}(M) \end{bmatrix},$$
$$L_2(K_2, y, z) = \begin{bmatrix} l_{21}(K_2) \oplus l_{22}(y) \oplus l_{23}(z) \end{bmatrix},$$

where every branch linear mapping l_{ij} maps $\{0,1\}^n \to \{0,1\}^n$. The function $F1$ after decomposition is in Fig. 5(a). Then we can distinguish $(F1, \pi)$ from (ρ, π) by considering these branch linear mappings in three cases, which cover all scenarios. In the case 1), 2), and 3.1), we refer to the attack in [14] to distinguish $F1$ from random function just by $\mathcal{O}(1)$ classical queries. The subcase 3.2) is a bit more complicated. However, we can still attack it by constructing a periodic function and applying Simon's algorithm to recover the secret period of $F1$, which leads to distinguishing attack as well. Let e denote a value only related to keys. And $h(M)$ denotes a function which can been calculated by public functions with M. For simplicity, then we can write function $F1$ as:

$$F1(M) = l_{22}\pi(l_{12}(M) \oplus l_{11}(K_1)) \oplus h(M) \oplus e,$$

where

$$e = l_{23}l_{13}(K_1) \oplus l_{21}(K_2),$$
$$h(M) = l_{23}l_{14}(M).$$

See Fig. 5(b).

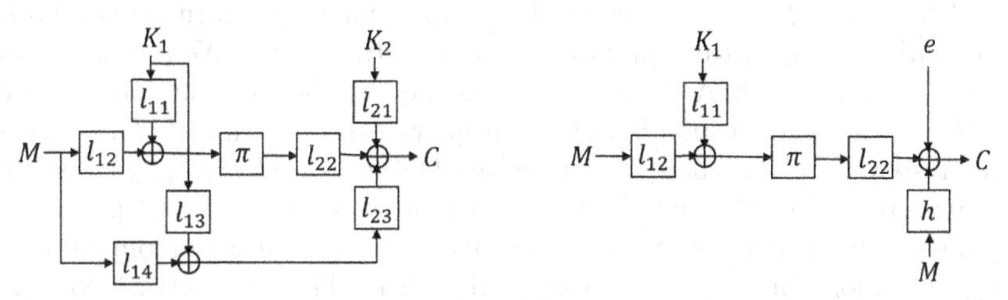

(a) The decomposition of function $F1$ by $L_1 = (l_{11}, l_{12}, l_{13}, l_{14})$ and $L_2 = (l_{21}, l_{22}, l_{23})$.

(b) The simple form of function $F1$ after decomposition.

Fig. 5. The decomposition of function $F1$.

Case 1) $\boldsymbol{l_{22} = \hat{0}}$. When $l_{22} = \hat{0}$, the output of the permutation π is not related to C. That is to say, $F1(M) = h(M) \oplus e$. We select arbitrary two different messages

M and M' and query the construction oracle with them to get answers C and C'. If the function is $F1$, then $C' \oplus C = h(M) \oplus h(M')$. However, for random function it holds with negligible probability. So we distinguish them.

Case 2) $l_{11}(K_1) = 0^n$**.** In this case, the input of the function π is independent of the key K_1. We select arbitrary two different messages M and M' and query the construction oracle with them to get answers C and C'. Then we distinguish $F1$ from random function by whether or not $C' \oplus C = l_{22}\pi(l_{12}(M)) \oplus l_{22}\pi(l_{12}(M')) \oplus h(M) \oplus h(M')$.

Case 3) $l_{22} \neq \hat{0}, l_{11}(K_1) \neq 0^n$**.**

Subcase 3.1) l_{12} **is not invertible.** Firstly, we find two different M and M' who satisfies $l_{12}(M) = l_{12}(M')$. It is easy to achieve by using basic linear algebra to find the kernel of l_{12} classically. We name the kernel $\mathsf{kel}(l_{12})$ and for any $a \in \mathsf{kel}(l_{12}), l_{12}(a) = 0^n$. We choose arbitrary $a \in \mathsf{kel}(l_{12})\backslash\{0^n\}$. Then for any $M \in \{0,1\}^n$ and $M' = M \oplus a$ that $l_{12}(M) = l_{12}(M')$. After that we query the construction oracle with M and M' to obtain C and C'. And we can distinguish $F1$ from random function by whether or not $C' \oplus C = h(M) \oplus h(M')$.

Subcase 3.2) l_{12} **is invertible.** We let

$$\begin{aligned} f : \{0,1\}^n \to& \{0,1\}^n \\ M \mapsto& F1(M) \oplus h(M) \oplus l_{22}\pi(l_{12}(M)) \\ f(M) =& l_{22}\pi(l_{12}(M) \oplus l_{11}(K_1)) \oplus l_{22}\pi(l_{12}(M)) \oplus e. \end{aligned}$$

Public classical primitives h, l_{22}, l_{12} can be implemented on quantum computers by adversaries using at most $\mathcal{O}(n^2)$ qubits. So with the addition of given the quantum oracle of $F1$ and π, we can construct the quantum oracle of f using at most $\mathcal{O}(n^2)$ qubits similar to paper [24]. It is easily to obtain $f(M) = f(M \oplus l_{12}^{-1}l_{11}(K_1))$ for all M. That is to say, f is a periodic function with period $s := l_{12}^{-1}(l_{11}(K_1))$. If $\varepsilon(f) \leq 1/2$, then by Theorem 1, Simon's algorithm can find the period with $\mathcal{O}(n)$ quantum queries to f using $\mathcal{O}(n)$ qubits. We put the proof of $\varepsilon(f) \leq 1/2$ in Appendix A. After recovering s, query the construction oracle with any $M, M \oplus s$ to get responds C, C' and query $l_{22}\pi(\cdot)$ with $l_{12}(M), l_{12}(M \oplus s)$ to get responses y, y'. Then $C' \oplus C = h(M) \oplus y \oplus h(M \oplus s) \oplus y'$. Instead, if the adversary is given quantum access to random function ρ and permutation π, it doesn't hold. Because Simon's algorithm will output a random value after querying random function. So we distinguish them. The whole attack costs $\mathcal{O}(n)$ queries to $F1$ and π with at most $\mathcal{O}(n^2)$ qubits. This method can be applied to EM construction. □

4 Pseudorandom Function with Two Permutation Calls

We will show that any pseudorandom function that makes two serial (see Fig. 3(b)) or parallel (see Fig. 3(a)) public permutation calls and every permutation has linear pre- and post-processing functions is not secure with $\mathcal{O}(2^{n/2}n)$

queries in the $Q2$ model by applying the improved Grover-meet-Simon algorithm [9]. In Sect. 5, the method applies to EDM [16], EDMD [28] and Xop [3] constructions instantiated with EM construction [18], and concrete schemes DS-SoEM [6], PDMMAC [13] and pEDM [17].

4.1 Attack on Pseudorandom Function with Two Parallel Permutation Calls

Let $\pi_1, \pi_2 \in \mathrm{Perm}(n)$ and K_1, K_2, K_3, K_4 are four independent keys in $\{0,1\}^n$. Let $L_1 : (\{0,1\}^n)^2 \to (\{0,1\}^n)^2, L_2 : (\{0,1\}^n)^2 \to (\{0,1\}^n)^2, L_3 : (\{0,1\}^n)^6 \to \{0,1\}^n$ be any three linear mappings. Then let the general function $F2 : \{0,1\}^{4n} \times \{0,1\}^n \to \{0,1\}^n$ based on two parallel public permutation calls be defined as Fig. 3(a).

Theorem 3. *Let $n \in \mathbb{N}$, and consider the function $F2 : \{0,1\}^{4n} \times \{0,1\}^n \to \{0,1\}^n$ of Fig. 3(a) based on public random permutations π_1 and π_2 with block length of n bits and using four keys $K_1, K_2, K_3, K_4 \xleftarrow{\$} \{0,1\}^n$, for any linear mapplings L_1, L_2, L_3. There exists a distinguisher $\mathcal{D}$ making at most $\mathcal{O}(n)$ construction queries and at most $\mathcal{O}(2^{n/2}n)$ primitive queries to distinguish $F2$ from random function.*

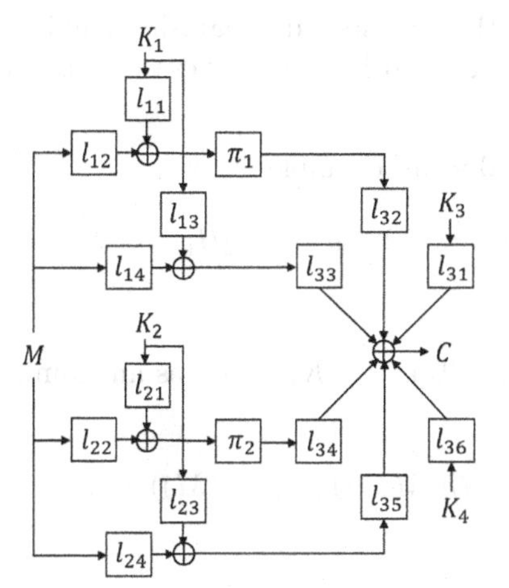

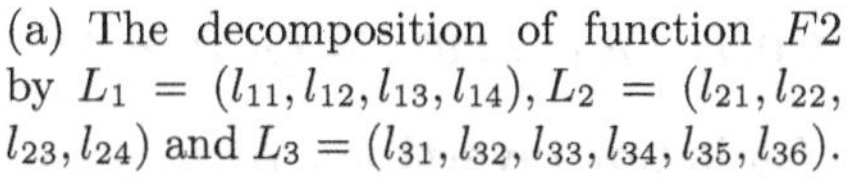
(a) The decomposition of function $F2$ by $L_1 = (l_{11}, l_{12}, l_{13}, l_{14}), L_2 = (l_{21}, l_{22}, l_{23}, l_{24})$ and $L_3 = (l_{31}, l_{32}, l_{33}, l_{34}, l_{35}, l_{36})$.

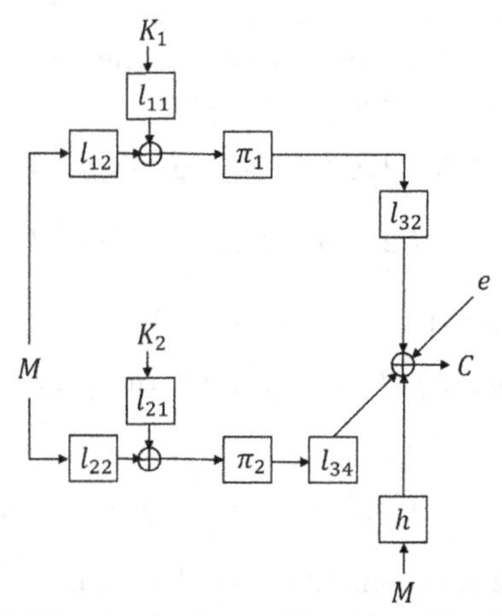

(b) The simple form of function $F2$ after decomposition.

Fig. 6. The decomposition of function $F2$.

Proof. Firstly, we decompose L_1, L_2, L_3 into

$$\begin{aligned} L_1 &= (l_{11}, l_{12}, l_{13}, l_{14}), \\ L_2 &= (l_{21}, l_{22}, l_{23}, l_{24}), \\ L_3 &= (l_{31}, l_{32}, l_{33}, l_{34}, l_{35}, l_{36}) \end{aligned}$$

as in Fig. 6(a) in the classical setting, where every branch linear mapping $l_{ij} : \{0,1\}^n \rightarrow \{0,1\}^n$. Then we will attack the decomposition form of $F2$. We consider four cases as follows, which cover all scenarios. Let e denote a value only related to keys. And $h(M)$ denotes a function which can been calculated by public functions with M. For simplicity, then we can write

$$F2(M) = l_{32}\pi_1(l_{12}(M) \oplus l_{11}(K_1)) \oplus l_{34}\pi_2(l_{22}(M) \oplus l_{21}(K_2)) \oplus h(M) \oplus e,$$

where

$$\begin{aligned} h(M) &= l_{33}l_{14}(M) \oplus l_{35}l_{24}(M), \\ e &= l_{31}(K_1) \oplus l_{36}(K_4) \oplus l_{35}l_{23}(K_2) \oplus l_{33}l_{13}(K_1). \end{aligned}$$

See Fig. 6(b).

Case 1) $\boldsymbol{l_{32} = \hat{0}}$ **or** $\boldsymbol{l_{34} = \hat{0}}$. Take $l_{32} = \hat{0}$ as an example. Now

$$F2(M) = l_{34}\pi_2(l_{22}(M) \oplus l_{21}(K_2)) \oplus h(M) \oplus e,$$

which degenerates into $F1$. By Theorem 2 there exists a distinguisher making at most $\mathcal{O}(n)$ construction queries and at most $\mathcal{O}(n)$ primitive queries to distinguish it from random function.

Case 2) $\boldsymbol{l_{12} = \hat{0}}$ **or** $\boldsymbol{l_{22} = \hat{0}}$. Take $l_{12} = \hat{0}$ as an example. Now

$$F2(M) = l_{34}\pi_2(l_{22}(M) \oplus l_{21}(K_2)) \oplus h(M) \oplus e \oplus l_{32}\pi_1(l_{11}(K_1)),$$

which degenerates into $F1$, too.

Case 3) $\boldsymbol{l_{11}(K_1) = 0^n}$ **or** $\boldsymbol{l_{21}(K_2) = 0^n}$. Take $l_{11}(K_1) = 0^n$ as an example. Now

$$F2(M) = l_{34}\pi_2(l_{22}(M) \oplus l_{21}(K_2)) \oplus h(M) \oplus l_{32}\pi_1(l_{12}(M)) \oplus e,$$

which degenerates into $F1$, too.

Case 4) $\boldsymbol{l_{32} \neq \hat{0}, l_{12} \neq \hat{0}, l_{34} \neq \hat{0}, l_{22} \neq \hat{0}, l_{11}(K_1) \neq 0^n, l_{21}(K_2) \neq 0^n}$.

Subcase 4.1) $\boldsymbol{l_{12}}$ **is not invertible or** $\boldsymbol{l_{22}}$ **is not invertible.** We take l_{12} is not invertible as an example.

If there are two different M and M' such that $l_{12}(M) = l_{12}(M')$ and $l_{22}(M) = l_{22}(M')$, we query the construction oracle with M and M' to obtain C and C'. Then we can distinguish $F2$ from random function by whether or not $C \oplus C' = h(M) \oplus h(M')$.

If there are no two different M and M' such that $l_{12}(M) = l_{12}(M')$ and $l_{22}(M) = l_{22}(M')$, no nonzero element of the kernel of linear mapping l_{22} (resp.l_{12}) belongs to the kernel of l_{12} (resp.l_{22}). Fix an arbitrary nonzero element a of the kernel of l_{12} and any M. Then

$$\begin{cases} l_{12}(M) = l_{12}(M \oplus a), \\ l_{22}(M) \oplus l_{22}(M \oplus a) = l_{22}(a)(\neq 0^n) \end{cases}$$

Assume the size of the kernel of l_{12} (resp. l_{22}) is r (resp. x), then linear mapping l_{12} (resp. l_{22}) has $2^n/r$ (resp. $2^n/x$) different images and every image of l_{12} (resp. l_{22}) has r (resp. x) pre-images. By there being no two different M and M' such that $l_{12}(M) = l_{12}(M')$ and $l_{22}(M) = l_{22}(M')$, we get every different pre-images corresponding to the same image of l_{12} (resp. l_{22}) correspond to different images of l_{22} (resp. l_{12}), which leads $r \leq 2^n/x$ (resp. $x \leq 2^n/r$). Thus $\max\{2^n/x, 2^n/r\} \geq 2^{n/2}$, which means the larger size of the images of l_{12}, l_{22} is at least $2^{n/2}$. Assume the image size of l_{22} is larger than l_{12}. Under this assumption there exist the following attack, or there exist another similar attack as well. Let

$$\begin{aligned} f : \{0,1\}^n \to & \{0,1\}^n \\ u \mapsto & F2(M) \oplus h(M) \oplus F2(M \oplus a) \oplus h(M \oplus a) \oplus \\ & l_{34}\pi_2(l_{22}(M) \oplus u) \oplus l_{34}\pi_2(l_{22}(M \oplus a) \oplus u) \\ f(u) = & l_{34}\pi_2(l_{22}(M) \oplus l_{21}(K_2)) \oplus l_{34}\pi_2(l_{22}(M \oplus a) \oplus l_{21}(K_2)) \oplus \\ & l_{34}\pi_2(l_{22}(M) \oplus u) \oplus l_{34}\pi_2(l_{22}(M \oplus a) \oplus u). \end{aligned}$$

Let $\mathcal{U} := \{l_{21}(K_2), l_{21}(K_2) \oplus l_{22}(a)\}$. It is easy to obtain when $u \in \mathcal{U}$, $f(u) = 0^n$ for all $M \in \{0,1\}^n$. So we try to search an $u \in \mathcal{U}$ by Grover's algorithm through defining a *test* function, which filters $u \in \mathcal{U}$ from all u's by whether or not $f(u) = 0^n$. Firstly, fix $\mathcal{M} := \{M_1, M_2, \dots, M_q\}$ which satisfies for any M_i that $l_{22}(M_i), l_{22}(M_i \oplus a) \notin \{l_{22}(M_j), l_{22}(M_j \oplus a) | M_j \in \mathcal{M} \backslash \{M_i\}\}$. Secondly, calculate $b_i := F2(M_i) \oplus h(M_i) \oplus F2(M_i \oplus a) \oplus h(M_i \oplus a)$ for $i = 1, \dots, q$ through querying $F2$. Then let $test : \{0,1\}^n \to \{0,1\}$ be

$$test(u) = \begin{cases} 1, \text{ if } b_i = l_{34}\pi_2(l_{22}(M_i) \oplus u) \oplus l_{34}\pi_2(l_{22}(M_i \oplus a) \oplus u) & i = 1, \dots, q, \\ 0, \text{ otherwise.} \end{cases}$$

Given the quantum oracle of $F2$ and π_2, we can construct the quantum oracle of $test(\cdot)$ using at most $\mathcal{O}(n^2)$ qubits. It is easy to obtain that $test(u) = 1$ for any $u \in \mathcal{U}$. If $\Pr[test(u) = 1] \leq \frac{1}{2^{2n}}$ holds for any $u \notin \mathcal{U}$, then we can recovery an $u \in \mathcal{U}$ by Grover's algorithm with at most $\mathcal{O}(2^{n/2})$ queries to $test(\cdot)$ using $\mathcal{O}(n)$ qubits. We prove $\Pr[test(u) = 1] \leq \frac{1}{2^{2n}}$ for any $u \notin \mathcal{U}$ when $q \geq 4n$ in Appendix B. After recovering a $u \in \mathcal{U}$, for a fixed $M \in \{0,1\}^n \backslash \mathcal{M}$ we check whether $F2(M) \oplus h(M) \oplus F2(M \oplus a) \oplus h(M \oplus a) \oplus l_{34}\pi_2(l_{22}(M) \oplus u) \oplus l_{34}\pi_2(l_{22}(M \oplus a) \oplus u) = 0^n$ or not by $\mathcal{O}(1)$ classical queries to $F2$ and π_2. It holds beyond doubt. However, if we replace the construction function from $F2$ to a random function, it happens with negligible probability. Thus we distinguish $F2$ from the random function. The whole attack costs $\mathcal{O}(n)$ queries to $F2$ and $\mathcal{O}(2^{n/2}n)$ queries to π_2 with at most $\mathcal{O}(n^2)$ qubits.

Subcase 4.2) l_{12}, l_{22} are invertible. Because π_1 and π_2 are two independent random permutations, so $\pi_1 = \pi_2$ with negligible probability. We only consider $\pi_1 \neq \pi_2$. We let

$$\begin{aligned} f : \{0,1\}^n \times \{0,1\}^n \to & \{0,1\}^n \\ (u, M) \mapsto & F2(M) \oplus h(M) \oplus l_{34}\pi_2(l_{22}(M) \oplus u) \oplus l_{32}\pi_1(l_{12}(M)) \\ f(u, M) = & l_{34}\pi_2(l_{22}(M) \oplus l_{21}(K_2)) \oplus l_{32}\pi_1(l_{12}(M) \oplus l_{11}(K_1)) \oplus \\ & l_{34}\pi_2(l_{22}(M) \oplus u) \oplus l_{32}\pi_1(l_{12}(M)) \oplus e. \end{aligned}$$

Let $\mathcal{U} := \{l_{21}(K_2), l_{22}l_{12}^{-1}l_{11}(K_1) \oplus l_{21}(K_2)\}$ and $s := l_{12}^{-1}l_{11}(K_1)$. It is easy to get when $u \in \mathcal{U}$, $f(u, M) = f(u, M \oplus s)$ holds for all M. Thus if $\varepsilon(f) \leq 7/8$, the improved Grover-meet-Simon algorithm [9] can recover an $u \in \mathcal{U}$ and s with $\mathcal{O}(n)$ quantum queries to $F2$ and π_1 and $\mathcal{O}(2^{n/2}n)$ quantum queries to π_2 using at most $\mathcal{O}(n^2)$ qubits. After that, we can distinguish $F2$ from random function. We put the proof of $\varepsilon(f) \leq 7/8$ in Appendix C. □

4.2 Attack on Pseudorandom Function with Two Serial Permutation Calls

Let $\pi_1, \pi_2 \xleftarrow{\$} \text{Perm}(n)$ and K_1, K_2, K_3, K_4 are four independent keys in $\{0,1\}^n$. Let $L_1 : (\{0,1\}^n)^2 \to (\{0,1\}^n)^3, L_2 : (\{0,1\}^n)^4 \to (\{0,1\}^n)^2, L_3 : (\{0,1\}^n)^4 \to \{0,1\}^n$ be any three linear mappings. And let the general function $F3 : \{0,1\}^{4n} \times \{0,1\}^n \to \{0,1\}^n$ based on two serial public permutation calls be defined as in Fig. 3(b). Similar to $F1$ and $F2$, we can decompose L_1, L_2, L_3 into

$$\begin{aligned} L_1 &= (l_{11}, l_{12}, l_{13}, l_{14}, l_{15}, l_{16}), \\ L_2 &= (l_{21}, l_{22}, l_{23}, l_{24}, l_{25}, l_{26}, l_{27}, l_{28}), \\ L_3 &= (l_{31}, l_{32}, l_{33}, l_{34}) \end{aligned}$$

as in Fig. 7(a) in the classical setting, where every branch linear mapping $l_{ij} : \{0,1\}^n \to \{0,1\}^n$. For whether general or concrete instantiations of $F3$ in Sect. 5, l_{12} is identical mappling. Thus for $l_{12} \neq \hat{0}$, we only consider l_{12} is invertible.

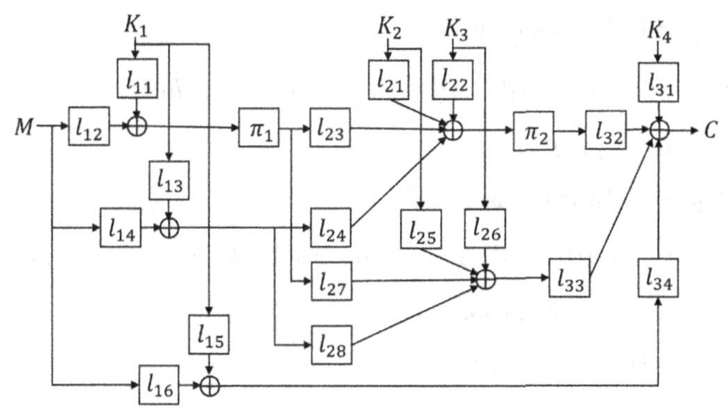

(a) The decomposition of function $F3$ by $L_1 = (l_{11}, l_{12}, l_{13}, l_{14}, l_{15}, l_{16}), L_2 = (l_{21}, l_{22}, l_{23}, l_{24}, l_{25}, l_{26}, l_{27}, l_{28})$ and $L_3 = (l_{31}, l_{32}, l_{33}, l_{34})$.

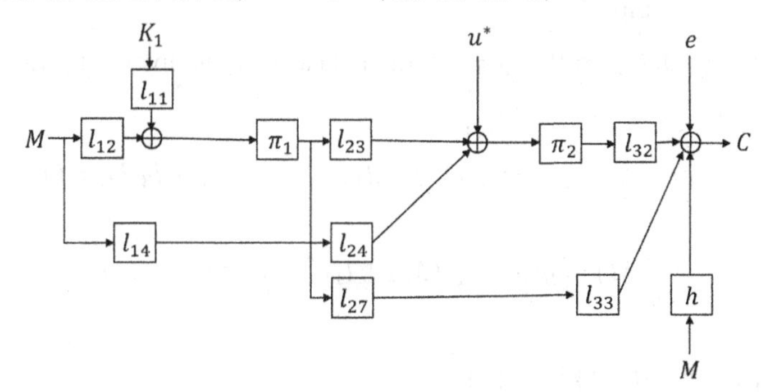

(b) The simple form of function $F3$ after decomposition.

Fig. 7. The decomposition of function $F3$.

Theorem 4. *Let $n \in \mathbb{N}$, and consider the function $F3 : \{0,1\}^{4n} \times \{0,1\}^n \to \{0,1\}^n$ of Fig. 3(b) based on two public random permutation π_1 and π_2 with block length of n bits and using four keys $K_1, K_2, K_3, K_4 \xleftarrow{\$} \{0,1\}^n$, for any linear mapplings L_1, L_2, L_3 where $l_{12} = \hat{0}$ or invertible. There exists a distinguisher $\mathcal{D}$ making at most $\mathcal{O}(n)$ construction queries and at most $\mathcal{O}(2^{n/2}n)$ primitive queries to distinguish $F3$ from random function.*

Proof. For simplicity, we let

$$h(M) := l_{33}l_{28}l_{14}(M) \oplus l_{34}l_{16}(M),$$
$$e := l_{33}(l_{28}l_{13}(K_1) \oplus l_{25}(K_2) \oplus l_{26}(K_3)) \oplus l_{31}(K_4) \oplus l_{34}l_{15}(K_1),$$
$$u^* := l_{21}(K_2) \oplus l_{22}(K_3) \oplus l_{24}l_{13}(K_1).$$

Then

$$F3(M) = l_{32}\pi_2(l_{23}\pi_1(l_{12}(M) \oplus l_{11}(K_1)) \oplus l_{24}l_{14}(M) \oplus u^*) \oplus l_{33}l_{27}\pi_1(l_{12}(M) \oplus l_{11}(K_1)) \oplus h(M) \oplus e.$$

See Fig. 7(b). We will attack $F3$ by attacking four cases as follows.

Case 1) $l_{32} = \hat{\mathbf{0}}$. In this case,

$$F3(M) = l_{33}l_{27}\pi_1(l_{12}(M) \oplus l_{11}(K_1)) \oplus h(M) \oplus e,$$

which degenerates into $F1$.

Case 2) $l_{12} = \hat{\mathbf{0}}$. In this case,

$$F3(M) = l_{32}\pi_2(l_{23}\pi_1(l_{11}(K_1)) \oplus l_{24}l_{14}(M) \oplus u^*) \oplus h(M) \oplus e \oplus l_{33}l_{27}\pi_1(l_{11}(K_1)),$$

which degenerates into $F1$, too.

Case 3) $l_{23} = \hat{\mathbf{0}}$. In this case,

$$F3(M) = l_{32}\pi_2(l_{24}l_{14}(M) \oplus u^*) \oplus l_{33}l_{27}\pi_1(l_{12}(M) \oplus l_{11}(K_1)) \oplus h(M) \oplus e,$$

which degenerates into $F2$.

Case 4) $l_{32} \neq \hat{\mathbf{0}}, l_{12} \neq \hat{\mathbf{0}}, l_{23} \neq \hat{\mathbf{0}}$. In this case, l_{12} is invertible. Let

$$u^{**} := u^* \oplus l_{24}l_{14}l_{12}^{-1}l_{11}(K_1),$$
$$g(u, x) := l_{32}\pi_2\left(l_{23}\pi_1(x) \oplus l_{24}l_{14}l_{12}^{-1}(x) \oplus u\right) \oplus l_{33}l_{27}\pi_1(x).$$

Then

$$F3(M) = g(u^{**}, l_{12}(M) \oplus l_{11}(K_1)) \oplus h(M) \oplus e.$$

Let

$$\begin{aligned} f : \{0,1\}^n \times \{0,1\}^n \to& \{0,1\}^n \\ (u, M) \mapsto& F3(M) \oplus h(M) \oplus g(u, l_{12}(M)) \\ f(u, M) =& g(u^{**}, l_{12}(M) \oplus l_{11}(K_1)) \oplus g(u, l_{12}(M)) \oplus e. \end{aligned}$$

Then it is easy to get when $u = u^{**}$, $f(u^{**}, M) = f(u^{**}, M \oplus s)$ holds for all $M \in \{0,1\}^n$ where $s := l_{12}^{-1}l_{11}(K_1)$. Thus if $\varepsilon(f) \leq 7/8$, the improved Grover-meet-Simon algorithm [9] can recover an u^{**} and s with $\mathcal{O}(n)$ quantum queries to $F3$ and $\mathcal{O}(2^{n/2}n)$ quantum queries to π_1, π_2 using at most $\mathcal{O}(n^2)$ qubits. We put the proof of $\varepsilon(f) \leq 7/8$ in Appendix D. After that, we can distinguish $F3$ from random function. □

5 Instantiations of Some PRFs

In this section, we show the security of general and some concrete instantiations of $F2$ and $F3$. In the following, we always assume $K_1, K_2, K_3, K_4 \xleftarrow{\$} \{0,1\}^n$ and $\pi_1, \pi_2 \xleftarrow{\$} \mathrm{Perm}(n)$. For the reason that $K_i = 0^n$ happens with negligible probability, we assume $K_i \neq 0^n$ for $i = 1, 2, 3, 4$. These instantiations are simpler than $F2$ and $F3$. So it is easier to constructing the Grover-meet-Simon function $f(u, .)$ for them. In the following, we only put the key recovery methods of PRFs. After recovery the distinguishing attacks from random function are similar as $F2, F3$, so we omit them.

5.1 Xop Construction Instantiated with EM Construction

We instantiate Xop construction by replacing two block ciphers with two EM constructions $\mathrm{EM}(x) = \pi_1(x \oplus K_1) \oplus K_2$ and $\mathrm{EM}(x) = \pi_2(x \oplus K_3) \oplus K_4$, and get

$$\mathrm{XopEM}(M) = \pi_1(M \oplus K_1) \oplus \pi_2(M \oplus K_2) \oplus K_3 \oplus K_4.$$

It is a general instantiation of $F2$. Thus we can recover K_1, K_2 by applying the improved Grover-meet-Simon algorithm [9] with $\mathcal{O}(2^{n/2}n)$ queries using $\mathcal{O}(n^2)$ qubits when considering function

$$f(u, M) = \mathrm{XopEM}(M) \oplus \pi_1(M) \oplus \pi_2(M \oplus u),$$

which has a period K_1 in its second component when $u = K_1 \oplus K_2$ or K_2.

DS-SoEM. For message $M \in \{0,1\}^{n-d}$, 'msb_{n-d}' means the truncation of key masks at the input to their $n-b$ most significant bits. Bhattarcharjee et al. [6] defined

$$\begin{aligned}\text{DS-SoEM}(M) =&\pi_1((M \oplus \mathrm{msb}_{n-d}(K_1))\|0^d)\oplus \\ &\pi_1((M \oplus \mathrm{msb}_{n-d}(K_2))\|1^d) \oplus K_1 \oplus K_2.\end{aligned}$$

It is a concrete variant of the instantiation of Xop. We can recover $\mathrm{msb}_{n-d}(K_1)$, $\mathrm{msb}_{n-d}(K_2)$ by applying the improved Grover-meet-Simon algorithm [9] with $\mathcal{O}(2^{\frac{n-d}{2}}(n-d))$ queries using $\mathcal{O}(n(n-d))$ qubits when considering function

$$f(u, M) = \text{DS-SoEM}(M) \oplus \pi_1(M\|0^d) \oplus \pi_1((M \oplus u)\|1^d),$$

which has a period $\mathrm{msb}_{n-d}(K_1)$ in its second component when $u = \mathrm{msb}_{n-d}(K_1 \oplus K_2)$ or $\mathrm{msb}_{n-d}(K_2)$.

5.2 EDM Construction Instantiated with EM Construction

We can instantiate EDM construction with two EM construction and get

$$\mathrm{EDMEM}(M) = \pi_2(\pi_1(M \oplus K_1) \oplus M \oplus K_2 \oplus K_3) \oplus K_4.$$

It is a general instantiation of $F3$. We can recover $K_1, K_2 \oplus K_3$ by applying the improved Grover-meet-Simon algorithm [9] with $\mathcal{O}(2^{n/2}n)$ queries using $\mathcal{O}(n^2)$ qubits when considering function

$$f(u, M) = \mathrm{EDMEM}(M) \oplus \pi_2(\pi_1(M) \oplus M \oplus u),$$

which has a period K_1 in its second component when $u = K_1 \oplus K_2 \oplus K_3$.

PDMMAC. Chakraborti et al. [13] defined

$$\mathrm{PDMMAC}(M) = \pi_1^{-1}(\pi_1(M \oplus K_1) \oplus M \oplus K_1 \oplus 2K_1) \oplus 2K_1.$$

It is a concrete instantiation of EDM. We search K_1 straightforwardly by Grover's search, which costs $\mathcal{O}(2^{n/2})$ queries and $\mathcal{O}(n)$ qubits.

pEDM. Dutta et al. [17] defined

$$\mathrm{pEDM}(M) = \pi_1\left(\pi_1\left(M \oplus K_1\right) \oplus M \oplus K_1 \oplus K_2\right) \oplus K_1.$$

It is a concrete instantiation of EDM. We apply the improved Grover-meet-Simon algorithm [9] to attack it with $\mathcal{O}\left(2^{n/2}n\right)$ queries using $\mathcal{O}\left(n^2\right)$ qubits when considering function

$$f(u, M) = \mathrm{pEDM}(M) \oplus \pi_1\left(\pi_1(M) \oplus M \oplus u\right),$$

which has a period K_1 in its second component when $u = K_2$.

5.3 EDMD Construction Instantiated with EM Construction

We instantiate EDMD construction with EM construction and get

$$\mathrm{EDMDEM}(M) = \pi_2(\pi_1(M \oplus K_1) \oplus K_2 \oplus K_3) \oplus \pi_1(M \oplus K_1) \oplus K_2 \oplus K_4.$$

It is a general instantiation of $F3$. We can recover $K_1, K_2 \oplus K_3$ by applying the improved Grover-meet-Simon algorithm [9] with $\mathcal{O}(2^{n/2}n)$ queries using $\mathcal{O}(n^2)$ qubits when considering function

$$f(u, M) = \mathrm{EDMDEM}(M) \oplus \pi_2(\pi_1(M) \oplus u) \oplus \pi_1(M),$$

which has a period K_1 in its second component when $u = K_2 \oplus K_3$.

SoKAC21. SoKAC21 [14] is as follows:

$$\mathrm{SoKAC21}(M) = \pi_2(\pi_1(M \oplus K_1) \oplus K_1) \oplus \pi_1(M \oplus K_1) \oplus K_1.$$

It is a concrete instantiation of EDMD. It is well known that BHT algorithm [12] is a time-memory trade-off algorithm of Grover's algorithm. By applying this algorithm to speed up the birthday bound classical attack [29] by Nandi, we can distinguish it from random function with $\mathcal{O}(2^{n/3})$ quantum queries with at most $\mathcal{O}(2^{n/3})$ qubits.

6 Conclusion

In this paper, we systematically analyze the securities of PRFs based on one or two public random permutation calls with pre- and post-linear processes of each permutation in the $Q2$ model. Besides, we present the security of some popular instantiations: contain general instantiations (XopEM, EDMEM, EDMDEM) and concrete PRFs (DS-SoEM, PDMMAC, SoKAC21). Section 4.2 does not include that case l_{12} not invertible. We find it is more complicated to find attack when l_{12} is not invertible for whatever other branch linear mapplings be. We leave it an open problem. Generally, it is sufficient to consider $l_{12} = O$ or invertible with respect to pre-existing instantiations. The further question is if there is provable security in the $Q2$ model to show the tightness of the bound.

Acknowledgments. The authors thank the anonymous reviewers for many helpful comments. This paper was supported by the NSFC of China (61732021 and 62202460) and the National Key R&D Program of China (2018YFA0704704).

A Proof of $\varepsilon(f) \leq 1/2$ in Subcase 3.2) in Sect. 3

In fact, we can prove $\varepsilon(f)$ is at most $\frac{1}{2}$, i.e., for any $t \in \{0,1\}^n \backslash \{0^n, s\}$ that

$$\Pr_M \left[\begin{array}{c} l_{22}\pi(l_{12}(M) \oplus l_{11}(K_1)) \oplus l_{22}\pi(l_{12}(M)) \\ l_{22}\pi(l_{12}(M \oplus t) \oplus l_{11}(K_1)) \oplus l_{22}\pi(l_{12}(M \oplus t)) = 0^n \end{array} \right] \leq 1/2. \quad (3)$$

By $t \notin \{0^n, s\}$ we know the four inputs of $l_{22}\pi$, i.e., $l_{12}(M) \oplus l_{11}(K_1), l_{12}(M)$, $l_{12}(M \oplus t) \oplus l_{11}(K_1)$, and $l_{12}(M \oplus t)$, are different from each other. Then by the randomness of π, the four inputs of $l_{22}(\cdot)$ are four distinct random values in $\{0,1\}^n$. By $l_{22} \neq \hat{0}$, we obtain the range of $l_{22}(\cdot)$ has at least two elements and the probability of $l_{22}(x) = y$ for any random $x \in \{0,1\}^n$ and y in the range is at most $\frac{1}{2}$. Thus the Eq. (3) happens with probability no more than $\frac{1}{2}$.

B Proof of $\Pr[test(u) = 1] \leq \frac{1}{2^{2n}}$ for Any $u \notin \mathcal{U}$ in Subcase 4.1) in Sect. 4.1

$$\begin{aligned} \text{Let } f_i(u) :=& F2(M_i) \oplus h(M_i) \oplus F2(M_i \oplus a) \oplus h(M_i \oplus a) \oplus l_{34}\pi_2(l_{22}(M_i) \oplus u) \oplus \\ & l_{34}\pi_2(l_{22}(M_i \oplus a) \oplus u) \\ =& l_{34}\pi_2(l_{22}(M_i) \oplus l_{21}(K_2)) \oplus l_{34}\pi_2(l_{22}(M_i \oplus a) \oplus l_{21}(K_2)) \oplus \\ & l_{34}\pi_2(l_{22}(M_i) \oplus u) \oplus l_{34}\pi_2(l_{22}(M_i \oplus a) \oplus u), \end{aligned}$$

and $y_i^1 := l_{22}(M_i) \oplus l_{21}(K_2), y_i^2 := l_{22}(M_i \oplus a) \oplus l_{21}(K_2), y_i^3 := l_{22}(M_i) \oplus u, y_i^4 := l_{22}(M_i \oplus a) \oplus u$, for $i = 1, 2, \ldots, q$. By $l_{22}(a) \neq 0^n, u \notin \mathcal{U}$ we get for any function f_i, the y_i^1, y_i^2, y_i^3, and y_i^4 are different from each other. To calculate the probability of these q equations $f_i(u) = 0^n$ where $u \notin \mathcal{U}$, we consider sampling about π_2. If y_i^1, y_i^2, y_i^3, and y_i^4, who are the inputs of π_2 in ith equation, all have appeared in the other $q-1$ equations, then we don't sample in the ith equation. By any M_i that $l_{22}(M_i), l_{22}(M_i \oplus a) \notin \{l_{22}(M_j), l_{22}(M_j \oplus a) : M_j \in \mathcal{M} \setminus \{M_i\}\}$, we get $y_i^1, y_i^2 \notin \{y_j^1, y_j^2 : j \in \{1, 2, \ldots, q\} \setminus \{i\}\}$. However, if $u = l_{22}(M_i) \oplus l_{22}(M_j) \oplus l_{21}(K_2)$ then $y_i^1 = y_j^3, y_i^2 = y_j^4, y_i^3 = y_j^1, y_i^4 = y_j^2$. Or if $u = l_{22}(M_i) \oplus l_{22}(M_j) \oplus l_{21}(K_2) \oplus l_{22}(a)$ then $y_i^1 = y_j^4, y_i^2 = y_j^3, y_i^3 = y_j^2, y_i^4 = y_j^1$. Therefore, even in the worst case we have to sample π_2 in at least $\lfloor \frac{q}{2} \rfloor$ equations among q. For every equation needing sample, by the randomness of π_2, it holds with probability at most $\frac{1}{2}$. Therefore, for any $u \notin \mathcal{U}$, we have $\Pr[test(u) = 1] \leq (\frac{1}{2})^{\lfloor \frac{q}{2} \rfloor}$. We have $\Pr[test(u) = 1] \leq 1/2^{2n}$ for $q \geq 4n$. We notice that this attack requires l_{22} with at least $4n$ different images. When $4n \leq 2^{n/2}$, that is to say, $n \geq 6$, it works.

C Proof of $\varepsilon(f) \leq 7/8$ in Subcase 4.2) in Sect. 4.1

Let $\mathcal{U}_t = \{0,1\}^n \times \{0,1\}^n \backslash (\{(l_{21}(K_2), s), (l_{22}l_{12}^{-1}l_{11}(K_1) \oplus l_{21}(K_2), s)\} \cup \{0,1\}^n \times \{0^n\})$. In this case, $\varepsilon(f) = \max\limits_{(u,t) \in \mathcal{U}_t} \Pr_M[f(u, M) = f(u, M \oplus t)]$. The function

$f(u, M) = f(u, M \oplus t)$ equals

$$\begin{aligned} & l_{34}\pi_2(l_{22}(M) \oplus l_{21}(K_2)) \oplus l_{34}\pi_2(l_{22}(M) \oplus u) \oplus \\ & l_{34}\pi_2(l_{22}(M \oplus t) \oplus l_{21}(K_2)) \oplus l_{34}\pi_2(l_{22}(M \oplus t) \oplus u) \oplus \\ & l_{32}\pi_1(l_{12}(M) \oplus l_{11}(K_1)) \oplus l_{32}\pi_1(l_{12}(M)) \oplus \\ & l_{32}\pi_1(l_{12}(M \oplus t) \oplus l_{11}(K_1)) \oplus l_{32}\pi_1(l_{12}(M \oplus t)) = 0^n \end{aligned} \quad (4)$$

1) $\boldsymbol{u \in \mathcal{U}, t \notin \{0^n, s\}}$. By $l_{11}(K_1) \neq 0^n, t \notin \{0^n, s\}$ we get the four inputs of $l_{32}\pi_1$ in Eq. (4) are different. By the randomness of π_1 the Eq. (4) holds with probability at most 1/2.

2) $\boldsymbol{u \notin \mathcal{U}, t = s}$. Now the Eq. (4) equals

$$\begin{aligned} & l_{34}\pi_2(l_{22}(M) \oplus l_{21}(K_2)) \oplus l_{34}\pi_2(l_{22}(M) \oplus u) \oplus \\ & l_{34}\pi_2(l_{22}(M \oplus l_{12}^{-1}l_{11}(K_1)) \oplus l_{21}(K_2)) \oplus l_{34}\pi_2(l_{22}(M \oplus l_{12}^{-1}l_{11}(K_1)) \oplus u) = 0^n \end{aligned}$$

By $u \notin \mathcal{U}, l_{22}l_{12}^{-1}l_{11}(K_1) \neq 0^n$, we get the four inputs of $l_{34}\pi_2$ in Eq. (4) are different. By the randomness of π_2 the Eq. (4) holds with probability at most 1/2.

3) $\boldsymbol{u \notin \mathcal{U}, t \notin \{0^n, s\}}$. We can prove the Eq. (4) holds with probability at most 1/2 the same as 1), so we omit it.

D Proof of $\varepsilon(f) \leq 7/8$ in Case 4) of Sect. 4.2

Let $\mathcal{U}_t = \{0,1\}^n \times \{0,1\}^n \backslash (\{(u^{**}, s)\} \cup \{0,1\}^n \times \{0^n\})$. In this case, $\varepsilon(f) = \max_{(u,t) \in \mathcal{U}_t} \Pr_M[f(u, M) = f(u, M \oplus t)]$. we take $l_{33}l_{27} = l_{24}l_{14} = O$ as an example. The other cases when $l_{33}l_{27} \neq O, l_{24}l_{14} \neq O$ are similar. We divide $(u, t) \in \mathcal{U}_t$ into the following cases, which cover all sceneries.

1) $\boldsymbol{u = u^{**}, t \notin \{0^n, s\}}$. Now the equation $f(u, M) = f(u, M \oplus t)$ equals

$$l_{32}\pi_2(y_1) \oplus l_{32}\pi_2(y_2) \oplus l_{32}\pi_2(y_3) \oplus l_{32}\pi_2(y_4) = 0^n, \quad (5)$$

wherec $y_1 = l_{23}\pi_1(l_{12}(M) \oplus l_{11}(K_1)) \oplus u^{**}, y_2 = l_{23}\pi_1(l_{12}(M)) \oplus u^{**}$, $y_3 = l_{23}\pi_1(l_{12}(M \oplus t) \oplus l_{11}(K_1)) \oplus u^{**}, y_4 = l_{23}\pi_1(l_{12}(M \oplus t)) \oplus u^{**}$. If $y_1 = y_2, y_3 = y_4$ or $y_1 = y_3, y_2 = y_4$ or $y_1 = y_4, y_2 = y_3$, then Eq. (5) holds. We observe that four inputs of $l_{23}\pi_1$: y_1, y_2, y_3, and y_4 are distinct from each other by $l_{11}(K_1) \neq 0^n$ and $t \notin \{0^n, s\}$. So this case happens with probability at most 3/4 by the randomness of π_1. Otherwise, there is at least one $y_i (i \in \{1, 2, 3, 4\})$ is different from the other three. In this case, by the randomness of π_2, the Eq. (5) holds with probability at most 1/2. So the Eq. (5) holds with a bound $3/4 + 1/4 \cdot 1/2 = 7/8$.

2) $\boldsymbol{u \neq u^{**}, t = s}$. Now the equation $f(u, M) = f(u, M \oplus t)$ is equal to

$$l_{32}\pi_2(y_1) \oplus l_{32}\pi_2(y_2) \oplus l_{32}\pi_2(y_3) \oplus l_{32}\pi_2(l_{23}\pi_1(y_4) = 0^n, \quad (6)$$

where $y_1 = l_{23}\pi_1(l_{12}(M) \oplus l_{11}(K_1)) \oplus u^{**}, y_2 = l_{23}\pi_1(l_{12}(M)) \oplus u, y_3 = l_{23}\pi_1(l_{12}(M)) \oplus u^{**}, y_4 = l_{23}\pi_1(l_{12}(M) \oplus l_{11}(K_1)) \oplus u$. By $u \neq u^{**}$, we get

$y_1 \neq y_4$. And we observe that $[y_1 = y_2 \Leftrightarrow y_3 = y_4]$ (resp. $[y_1 = y_3 \Leftrightarrow y_2 = y_4]$). So $y_1 = y_2$ and $y_1 = y_3$ don't hold simultaneously, or it leads to $y_1 = y_4$. If $y_1 = y_2$, the Eq. (6) holds. This case holds with probability at most 1/2 by the randomness of π_1. Otherwise, if $y_1 \neq y_2$ and $y_1 = y_3$, the Eq. (6) holds as well. This case holds with probability at most $1/2 \cdot 1/2 = 1/4$ by the randomness of π_1. At last, if $y_1 \neq y_2$ and $y_1 \neq y_3$, then y_1, y_2, y_3, and y_4 are different from each other, the Eq. (6) holds with probability of $1/2 \cdot 1/2 \cdot 1/2 = 1/8$ by the randomness of π_2. So the Eq. (6) holds with a bound 7/8.

3) $\boldsymbol{u \neq u^{**}, t \notin \{0^n, s\}}$. This case is similar to 1), so we omit it.

References

1. Bellare, M., Desai, A., Jokipii, E., Rogaway, P.: A concrete security treatment of symmetric encryption. In: FOCS 1997, pp. 394–403. IEEE Computer Society (1997). https://doi.org/10.1109/SFCS.1997.646128
2. Bellare, M., Impagliazzo, R.: A tool for obtaining tighter security analyses of pseudorandom function based constructions, with applications to PRP to PRF conversion. IACR Cryptol. ePrint Arch. 1999, 24 (1999). https://eprint.iacr.org/1999/024
3. Bellare, M., Krovetz, T., Rogaway, P.: Luby-Rackoff backwards: increasing security by making block ciphers non-invertible. In: Nyberg, K. (ed.) EUROCRYPT 1998. LNCS, vol. 1403, pp. 266–280. Springer, Heidelberg (1998). https://doi.org/10.1007/BFb0054132
4. Bellare, M., Rogaway, P.: The security of triple encryption and a framework for code-based game-playing proofs. In: Vaudenay, S. (ed.) EUROCRYPT 2006. LNCS, vol. 4004, pp. 409–426. Springer, Heidelberg (2006). https://doi.org/10.1007/11761679_25
5. Bernstein, D.J.: Stronger security bounds for wegman-carter-shoup authenticators. In: Cramer, R. (ed.) EUROCRYPT 2005. LNCS, vol. 3494, pp. 164–180. Springer, Heidelberg (2005). https://doi.org/10.1007/11426639_10
6. Bhattacharjee, A., List, E., Nandi, M.: CENCPP - beyond-birthday-secure encryption from public permutations. IACR Cryptol. ePrint Arch. 2020, 602 (2020). https://eprint.iacr.org/2020/602
7. Bogdanov, A., Knudsen, L.R., Leander, G., Standaert, F.-X., Steinberger, J., Tischhauser, E.: Key-alternating ciphers in a provable setting: encryption using a small number of public permutations. In: Pointcheval, D., Johansson, T. (eds.) EUROCRYPT 2012. LNCS, vol. 7237, pp. 45–62. Springer, Heidelberg (2012). https://doi.org/10.1007/978-3-642-29011-4_5
8. Bonnetain, X.: Tight Bounds for Simon's algorithm. In: Longa, P., Ràfols, C. (eds.) LATINCRYPT 2021. LNCS, vol. 12912, pp. 3–23. Springer, Cham (2021). https://doi.org/10.1007/978-3-030-88238-9_1
9. Bonnetain, X., Hosoyamada, A., Naya-Plasencia, M., Sasaki, Yu., Schrottenloher, A.: Quantum attacks without superposition queries: the offline Simon's algorithm. In: Galbraith, S.D., Moriai, S. (eds.) ASIACRYPT 2019. LNCS, vol. 11921, pp. 552–583. Springer, Cham (2019). https://doi.org/10.1007/978-3-030-34578-5_20
10. Brassard, Gilles: On computationally secure authentication tags requiring short secret shared keys. In: Chaum, David, Rivest, Ronald L.., Sherman, Alan T.. (eds.) Advances in Cryptology, pp. 79–86. Springer, Boston, MA (1983). https://doi.org/10.1007/978-1-4757-0602-4_7

11. Brassard, G., Hoyer, P., Mosca, M., Tapp, A.: Quantum amplitude amplification and estimation. Contemp. Math. **305**, 53–74 (2002)
12. Brassard, G., Hoyer, P., Tapp, A.: Quantum algorithm for the collision problem. arXiv preprint quant-ph/9705002 (1997)
13. Chakraborti, A., Nandi, M., Talnikar, S., Yasuda, K.: On the composition of single-keyed tweakable Even-Mansour for achieving BBB security. IACR Trans. Symmetric Cryptol. **2020**(2), 1–39 (2020). https://doi.org/10.13154/tosc.v2020.i2.1-39
14. Chen, Y.L., Lambooij, E., Mennink, B.: How to build pseudorandom functions from public random permutations. In: Boldyreva, A., Micciancio, D. (eds.) CRYPTO 2019. LNCS, vol. 11692, pp. 266–293. Springer, Cham (2019). https://doi.org/10.1007/978-3-030-26948-7_10
15. Chen, Y.L., Mennink, B., Preneel, B.: Categorization of faulty nonce misuse resistant message authentication. In: Tibouchi, M., Wang, H. (eds.) ASIACRYPT 2021. LNCS, vol. 13092, pp. 520–550. Springer, Cham (2021). https://doi.org/10.1007/978-3-030-92078-4_18
16. Cogliati, B., Seurin, Y.: EWCDM: an efficient, beyond-birthday secure, nonce-misuse resistant MAC. In: Robshaw, M., Katz, J. (eds.) CRYPTO 2016. LNCS, vol. 9814, pp. 121–149. Springer, Heidelberg (2016). https://doi.org/10.1007/978-3-662-53018-4_5
17. Dutta, A., Nandi, M., Talnikar, S.: Permutation based EDM: an inverse free BBB secure PRF. IACR Trans. Symmetric Cryptol. **2021**(2), 31–70 (2021). https://doi.org/10.46586/tosc.v2021.i2.31-70
18. Even, S., Mansour, Y.: A construction of a cipher from a single pseudorandom permutation. J. Cryptol. **10**(3), 151–162 (1997). https://doi.org/10.1007/s001459900025
19. Gilboa, S., Gueron, S.: The advantage of truncated permutations. CoRR abs/1610.02518 (2016). https://arxiv.org/abs/1610.02518
20. Grover, L.K.: A fast quantum mechanical algorithm for database search. In: Proceedings of the Twenty-Eighth Annual ACM Symposium on the Theory of Computing, 1996. pp. 212–219 (1996). https://doi.org/10.1145/237814.237866
21. Guo, T., Wang, P., Hu, L., Ye, D.: Attacks on beyond-birthday-bound MACs in the quantum setting. In: Cheon, J.H., Tillich, J.-P. (eds.) PQCrypto 2021 2021. LNCS, vol. 12841, pp. 421–441. Springer, Cham (2021). https://doi.org/10.1007/978-3-030-81293-5_22
22. Hall, C., Wagner, D., Kelsey, J., Schneier, B.: Building PRFs from PRPs. In: Krawczyk, H. (ed.) CRYPTO 1998. LNCS, vol. 1462, pp. 370–389. Springer, Heidelberg (1998). https://doi.org/10.1007/BFb0055742
23. Impagliazzo, R., Rudich, S.: Limits on the provable consequences of one-way permutations. In: Goldwasser, S. (ed.) CRYPTO 1988. LNCS, vol. 403, pp. 8–26. Springer, New York (1990). https://doi.org/10.1007/0-387-34799-2_2
24. Kaplan, M., Leurent, G., Leverrier, A., Naya-Plasencia, M.: Breaking symmetric cryptosystems using quantum period finding. In: Robshaw, M., Katz, J. (eds.) CRYPTO 2016. LNCS, vol. 9815, pp. 207–237. Springer, Heidelberg (2016). https://doi.org/10.1007/978-3-662-53008-5_8
25. Kuwakado, H., Morii, M.: Security on the quantum-type Even-Mansour cipher. In: ISITA 2012, pp. 312–316. IEEE (2012). https://ieeexplore.ieee.org/document/6400943/
26. Leander, G., May, A.: Grover meets Simon – quantumly attacking the FX-construction. In: Takagi, T., Peyrin, T. (eds.) ASIACRYPT 2017. LNCS, vol. 10625, pp. 161–178. Springer, Cham (2017). https://doi.org/10.1007/978-3-319-70697-9_6

27. McGrew, D.A., Viega, J.: The security and performance of the Galois/Counter Mode (GCM) of operation. In: Canteaut, A., Viswanathan, K. (eds.) INDOCRYPT 2004. LNCS, vol. 3348, pp. 343–355. Springer, Heidelberg (2004). https://doi.org/10.1007/978-3-540-30556-9_27
28. Mennink, B., Neves, S.: Encrypted Davies-Meyer and its dual: towards optimal security using mirror theory. In: Katz, J., Shacham, H. (eds.) CRYPTO 2017. LNCS, vol. 10403, pp. 556–583. Springer, Cham (2017). https://doi.org/10.1007/978-3-319-63697-9_19
29. Nandi, M.: Mind the composition: birthday bound attacks on EWCDMD and SoKAC21. In: Canteaut, A., Ishai, Y. (eds.) EUROCRYPT 2020. LNCS, vol. 12105, pp. 203–220. Springer, Cham (2020). https://doi.org/10.1007/978-3-030-45721-1_8
30. Patarin, J.: A proof of security in $O(2^n)$ for the Xor of Two Random Permutations. In: Safavi-Naini, R. (ed.) ICITS 2008. LNCS, vol. 5155, pp. 232–248. Springer, Heidelberg (2008). https://doi.org/10.1007/978-3-540-85093-9_22
31. Patarin, J.: Introduction to mirror theory: Analysis of systems of linear equalities and linear non equalities for cryptography. IACR Cryptol. ePrint Arch. 2010, 287 (2010). https://eprint.iacr.org/2010/287
32. Patarin, J.: Generic attacks for the XOR of k random permutations. In: Jacobson, M., Locasto, M., Mohassel, P., Safavi-Naini, R. (eds.) ACNS 2013. LNCS, vol. 7954, pp. 154–169. Springer, Heidelberg (2013). https://doi.org/10.1007/978-3-642-38980-1_10
33. Shinagawa, K., Iwata, T.: Quantum attacks on sum of Even-Mansour pseudorandom functions. Inf. Process. Lett. **173**, 106172 (2022). https://doi.org/10.1016/j.ipl.2021.106172
34. Shoup, V.: On fast and provably secure message authentication based on universal hashing. In: Koblitz, N. (ed.) CRYPTO 1996. LNCS, vol. 1109, pp. 313–328. Springer, Heidelberg (1996). https://doi.org/10.1007/3-540-68697-5_24
35. Simon, D.R.: On the power of quantum computation. SIAM J. Comput. **26**(5), 1474–1483 (1997). https://doi.org/10.1137/S0097539796298637
36. Wegman, M.N., Carter, L.: New hash functions and their use in authentication and set equality. J. Comput. Syst. Sci. **22**(3), 265–279 (1981). https://doi.org/10.1016/0022-0000(81)90033-7
37. Zhang, P.: Quantum attacks on sum of Even-Mansour construction with linear key schedules. Entropy **24**(2), 153 (2022)

On Security Notions for Encryption in a Quantum World

Céline Chevalier[1], Ehsan Ebrahimi[2], and Quoc-Huy Vu[1](✉)

[1] CRED, Université Panthéon-Assas, Paris, France
{celine.chevalier,quoc.huy.vu}@ens.fr
[2] SnT, University of Luxembourg, Esch-sur-Alzette, Luxembourg
ehsan.ebrahimi@uni.lu

Abstract. Indistinguishability against adaptive chosen-ciphertext attacks (IND-CCA2) is usually considered the most desirable security notion for classical encryption. In this work, we investigate its adaptation in the quantum world, when an adversary can perform superposition queries. The security of quantum-secure classical encryption has first been studied by Boneh and Zhandry (CRYPTO'13), but they restricted the adversary to classical challenge queries, which makes the indistinguishability only hold for classical messages (IND-qCCA2). We extend their work by giving the first security notions for fully quantum indistinguishability under quantum adaptive chosen-ciphertext attacks, where the indistinguishability holds for superposition of plaintexts (qIND-qCCA2).

1 Introduction

Recent advances in quantum computing show the possible emergence of new kinds of attacks due to quantum adversaries. The first type of attacks would be due to adversaries owning a quantum computer and using it to break computational assumptions (thus attacking classical cryptographic cryptosystems). This has been made possible by the invention of quantum algorithms that solve factoring and discrete logarithm problems in polynomial time [20] and consequently, break the security of many classical public-key encryption schemes based on these assumptions. This threat has led to the emergence of so-called *post-quantum cryptography*, based on arguably quantum-resistant assumptions. But this change of assumptions may not be sufficient, and symmetric cryptosystems may also be impacted, in case we allow a quantum adversary, not only to perform computation on a quantum computer it may own, but also to carry out a second type of attacks, by interacting with the target in superposition. Quantum algorithms for unstructured search [13] or period finding [21] could then be applied to attack classical constructions using superposition queries [10,15]. Cryptosystems secure against this type of attacks would be called *quantum secure*.

As we approach the quantum era, it thus becomes necessary to construct new public-key cryptosystems based on quantum-resistant assumptions, and to

E. Ebrahimi—Work done while at École Normale Supérieure.

T. Isobe and S. Sarkar (Eds.): INDOCRYPT 2022, LNCS 13774, pp. 592–613, 2022.
https://doi.org/10.1007/978-3-031-22912-1_26

investigate the security of both symmetric and public-key cryptosystems against an attacker allowed to interact with honest parties using quantum communication. Recently, there has been towards this goal extensive research works that consider this scenario of quantum superposition attacks for different classical cryptographic constructions such as random oracles, pseudorandom functions, encryption and signature schemes [1,4–6,11,22] and give corresponding new security definitions. Furthermore, this new field of research is also motivated by the existence of concrete attacks against classical constructions using superposition queries (e.g., see [10,15] and their follow up works). In this paper, we continue this line of work and focus on the security for classical encryption schemes against quantum adversaries allowed to make quantum encryption and decryption queries.

1.1 Defining Security for Encryption Against Quantum Adversaries

Classical Security Notions. Indistinguishability-based security definitions are modeled as a game between a challenger and an adversary $\mathcal{A}$. In the Find-Then-Guess style, the game starts with a first learning phase (with access to some oracles), followed by a challenge phase where $\mathcal{A}$ sends a challenge query (two messages x_0 and x_1 to be encrypted) and receives a challenge ciphertext (encryption of x_b). Afterwards, a second learning phase follows, and finally, $\mathcal{A}$ outputs a solution (its guess for the bit b). The security reduction consists in constructing a new adversary which simulates $\mathcal{A}$ and solves some hard underlying problem. The learning phases define the type of attacks: chosen-plaintext attacks (CPA) if the adversary has access to an encryption oracle in both learning phases, and chosen-ciphertext attacks (CCA) in case it also has access to a decryption oracle in the learning phases (non-adaptive or CCA1 if it is restricted to the first learning phase, and adaptive or CCA2 otherwise).

Indistinguishability against adaptive chosen-ciphertext attack (IND-CCA2) is usually considered the most desirable security notion for encryption. In the CCA2 games, the adversary is restricted not to ask for decryption of the challenge ciphertext, otherwise, this would lead to a trivial guess of the bit b. It is the role of the challenger to ensure that the adversary obeys this rule, which intrinsically requires the ability to copy, store and compare classical strings.

Quantum Attacks on Encryption. With recent advances in quantum computing, a quantum adversary may become a tangible threat in not so long. Switching to post-quantum computational assumptions is a beginning but may not be enough in case the adversary gains quantum access to honest parties and protocols. Consider for instance the well-known construction of CCA2 secure encryption schemes from lossy trapdoor functions [19]: if the construction is instantiated with lattice-based problems, it is arguably post-quantum secure. But we show later that, the insecurity may arise from the use of a one-time pad inside the construction. Furthermore, [10,15] and their follow up works show that the security of several classical constructions can be compromised if the adversary can perform superposition attacks.

Boneh-Zhandry's Security Notions [6]. Boneh and Zhandry propose the first definition of IND-CCA for both symmetric and public-key encryption schemes against quantum adversaries allowed to make quantum encryption and decryption queries. But they show that the natural translation of the classical Find-then-Guess paradigm to the quantum setting is unachievable, even for IND-CPA security. To overcome this impossibility, they resort to considering quantum queries during the learning phases only, and classical queries during the challenge phase. In addition to looking artificial, this inconsistency between the learning phases and the challenge phase may lead to a cryptographic construction that fulfills this security notion (IND-qCPA or IND-qCCA) while being subject to an attack.

For instance, in [2], the authors verify IND-qCPA security of XTS mode of operation (with quantum learning queries and classical challenge queries). They design a block cipher such that an encryption scheme in XTS mode, instantiated with that block cipher, can be attacked during the learning phase using quantum learning queries. However, this attack cannot be used to violate the IND-qCPA security definition. The explanation for this inconsistency is that this attack cannot be implemented in the challenge phase due to the classical restriction imposed on the adversary. This example supports our claim that the inconsistency between the learning phases and the challenge phase can be problematic and should be overcome.

IND-CCA2 Security Notions. To date, defining the CCA2 security with quantum challenge queries remains unsolved. In [11], the authors address the inconsistency described above for the case of symmetric encryption, but only for IND-CPA, and leave as an open problem the IND-CCA definitions.

The main obstacle is to define how the challenger should reply to the quantum decryption queries after the adversary has made the quantum challenge queries. When the challenge queries are classical, they can be stored and later the challenger can return $\perp$ if the adversary submits one of them as a decryption query. Although it is trivial and inherent to store the challenge ciphertext in the classical setting, it is highly non-trivial to store ciphertexts in the quantum world, due to a number of technical obstacles, all of which can be traced to quantum no-cloning and the destructiveness of quantum measurements.

In this paper, we manage to overcome this recording barrier by using Zhandry's compressed oracle technique [23] (an overview is given in Sect. 1.2) and we propose the first quantum version for IND-CCA security notion. We justify our definitions in Sect. 1.3. Due to the space limitations, we defer discussions on our security notions and related work to the full version of our paper [8].

1.2 Our Approach

Towards resolution, we start from a recent groundbreaking technique that allows for on-the-fly simulation of random oracles in the quantum setting: Zhandry's compressed oracles [23]. The goal of his work is to overcome the recording barrier, by allowing the reduction to record information about the adversary's queries, which is a key feature of many classical ROM proofs.

Zhandry's key observations are threefold. First, instead of considering a random function h being chosen beforehand, one can purify the adversary's mixed state by putting h in uniform superposition $\sum_h |h\rangle$. This observation is a technicality that allows us to fulfill the two next points. Then, the next observation is that, by doing the queries in the Fourier basis, the data will be written to the oracle's registers instead of writing to the opposite direction. This enables the simulator to get some information about the adversary's queries. Finally, the last and most important one is that the simulator needs to be ready to forget some point it simulated previously, by performing a particular test on the database after answering the query. In particular, Zhandry defines a test computation that maps $|+\rangle \mapsto |+\rangle|1\rangle$ and $|\phi\rangle \mapsto |\phi\rangle|0\rangle$ for any $|\phi\rangle$ orthogonal to $|+\rangle$, where $|+\rangle = \sum_x |x\rangle$ is the uniform superposition state. The "test-and-forget" procedure can be implemented by first performing the query in the Fourier basis and then doing the test operation on the output registers (of the simulator). This test determines whether the adversary has any information from the oracle at some input. If not, that pair will be removed from the database so that the adversary cannot detect that it is interacting with a simulated oracle.

This technique has been extended from random oracles to lazy-sampling of non-uniform random functions in [9]. The intuition is almost the same, except that now one starts from the all-zero state, performs an *efficient* sampling operation that computes the function $f(x)$ according to some non-uniform distribution – it is the quantum Fourier transform (QFT) operation in the uniform setting. One then performs the query in the Fourier basis, transforms back to the computational basis and applies the "test-and-forget" operation (which is defined similarly as in the uniform setting). For this to work, the two important requirements are that: i) the sampling operation must be efficient; ii) the function distribution must be independent for every input.

To define security for encryption, we choose the real-or-random paradigm to work with. This is because partially, the real-or-random paradigm does not suffer from Boneh-Zhandry's impossibility (discussion below). Furthermore, it is actually possible to define quantum chosen-ciphertext security for this paradigm using the quantum lazy-sampling technique we just described. In what follows, let us focus on the random world of the paradigm. For each challenge query in the random world, the challenger applies a random function to the plaintext registers before encrypting, all aforementioned requirements are met: the encryption of each submitted plaintext is actually an encryption of another uniformly random plaintext, and since the encryption algorithm is efficient, the sampling operation can also be efficiently constructed.

The above idea gives us a reasonable way to define adaptive chosen ciphertext security against quantum challenge queries: by instantiating the encryption oracle with this lazy-sampling technique, we are able to keep track of the information needed to formulate the CCA2 notions, namely the challenge queries the adversary has made, and the challenge ciphertexts it has received. However, applying Zhandry's framework directly to our setting does not work, and more efforts are needed. For example, one main difference is that in our setting, when making queries to the random oracle, there is no response register (from the adversary). In Zhandry's framework, this response register is essential for the technique, as the "test-and-forget" procedure works based on the value of this register. Another problem is how to implement the oracle with an *one-shot* call to the encryption algorithm: this is necessary when defining "one-time" security, or when doing security reductions. We refer the reader to Sect. 3 for technical details.

1.3 Our Contributions

New Notions of Quantum Indistinguishability and Their Achievability. We define novel security notions for encryption in both the symmetric (Definition 2 in Sect. 4) and public-key settings (Defnition 3 in Sect. 5). Our main contribution is to propose the first definitions for adaptive chosen ciphertext security that support *fully quantum indistinguishability*, resolving an outstanding open problem posed by Gagliardoni *et al.* [11]. Furthermore, to justify our formalization, we show that our notions

- are achievable (see Theorem 2 and Theorem 4);
- are all closed under composition (see Theorem 1 and Theorem 3);
- are strictly stronger than previous notions with classical challenge queries. In particular, this shows the quantum (in)security of various symmetric encryption schemes including stream cipher and some block cipher modes of operation such as CFB, OFB, CTR. This even extends to authenticated encryption, in which some most widely used encryption modes like GCM are also resulting in an insecure scheme.
- (when restricted to classical challenge queries) are equivalent to Boneh-Zhandry's notions [6].

In this work, we adopt the Real-or-Random security definition. Informally, in the real game, the adversary has no restrictions on the use of the decryption oracle Dec. Only in the random game, the challenge encryption oracle is implemented as a compressed oracle: it applies a *random function* h[1] to the plaintext register before doing the encryption. For each decryption query, the challenger looks for the query's basis state in the database (in superposition) and if found, it reasonably guesses that the adversary is trying to decrypt the challenge ciphertext, and so it returns the adversary's original message (which is what is stored in the database). Otherwise, it decrypts normally. Intuitively, the security is established by the distinguishing probability of the adversary between whether its message is encrypted with Enc or $\mathsf{Enc} \circ h$.

[1] We note that previous works [7,17] use *random permutations* instead of random functions in the random world. It is arguable which security definition is the right adaptation of the classical Real-or-Random security definition to the quantum setting. However, the two notions are equivalent if the message space has size super-polynomial. This is because in this case, random functions and random permutations are indistinguishable.

We then provide constructions satisfying these security notions in Sect. 4.2 and Sect. 5.2. For the symmetric-key setting, our construction follows the classical Encrypt-then-Mac paradigm, in which we use a pseudorandom function in the role of the MAC scheme (see Theorem 2). Concerning the public-key setting, we propose a compiler that lifts any secure encryption scheme in the sense of [6] to an encryption scheme secure in the sense of our notions in Sect. 5.2 (Theorem 4). The compiler follows the classical hybrid encryption paradigm, where we encrypt the message with a one-time symmetric encryption which can be constructed from pseudorandom functions, and then encrypt the symmetric key with a secure public-key scheme (in the sense of [6]).

Due to the space limitations, we defer formal proofs of some theorems stated in the paper, as well as other results to the full version [8].

2 Preliminaries

2.1 Notations

Let $\lambda \in \mathbb{N}$ be the security parameter. The notation $\mathsf{negl}(\lambda)$ denotes any function f such that $f(\lambda) = \lambda^{-\omega(1)}$. When sampling uniformly at random a value a from a set $\mathcal{U}$, we employ the notation $a \xleftarrow{\$} \mathcal{U}$. When sampling a value a from a probabilistic algorithm $\mathcal{A}$, we employ the notation $a \leftarrow \mathcal{A}$. For $a \in \mathbb{N}$, $[a] = \{x \in \mathbb{N} \mid x \leq a\}$ will denote the closed integer interval with endpoints 0 and a. Let $|\cdot|$ denote either the length of a string, or the cardinal of a finite set, or the absolute value. By PPT we mean a polynomial-time non-uniform family of probabilistic circuits, and by QPT we mean a polynomial-time non-uniform family of quantum circuits. Let $\delta_{x,x'}$ denote the Kronecker delta function of x and x'.

2.2 Quantum Computing

For notation and conventions regarding quantum information, we refer the reader to [18]. We recall a few basics here. We let $|\phi\rangle$ denote an arbitrary pure quantum state, let $|x\rangle$ denote an element of the standard (computational) basis. A mixed state will be denoted by lowercase Greek letters, e.g., ρ. We let $|+\rangle$ denote the uniform superposition, that is $|+\rangle := \sum_x |x\rangle$.

A pure state $|\phi\rangle$ can be manipulated by performing a unitary transformation U to the state $|\phi\rangle$, which we denote $U|\phi\rangle$. The identity on a n-bit quantum system is denoted $\mathcal{I}_n$. Given two quantum systems A, B, with corresponding Hilbert spaces $\mathcal{H}_A, \mathcal{H}_B$, let $|\phi\rangle = |\phi_0, \phi_1\rangle$ be a state of the joint system. We write $U^A|\phi\rangle$ to denote that we act with U on register A, and with identity $\mathcal{I}$ on register B, and we write U^{AB} to denote that we act with U on both registers A, B simultaneously, that is $U^{AB} = U^A \otimes U^B$.

Quantum Computations. Let Q be a n-bit quantum system over $\mathbb{Z}_q$ for some integer q. The Quantum Fourier Transform (QFT) performs the following operation efficiently:

$$\mathsf{QFT}|x\rangle := \frac{1}{\sqrt{q^n}} \sum_{y \in \{0,1\}^n} \omega_q^{x \cdot y} |y\rangle,$$

where $\omega_q := \exp(\frac{2\pi i}{q})$, and $x \cdot y$ denotes the dot product. In this paper, we usually consider $q = 2$, so that $\omega_q = (-1)$.

Given a function $f : \mathcal{X} \to \mathcal{Y}$, we model a quantum-accessible oracle $\mathcal{O}$ for f as a unitary transformation $\mathcal{O}_f$ acting on three registers X, Y, Z with the property that $\mathcal{O}_f : |x, y, 0\rangle \mapsto |x, y \oplus f(x), 0\rangle$, where $\oplus$ is some involutive group operation (so-called quantum query model). Given an algorithm $\mathcal{A}$, we sometimes write $y \leftarrow \mathcal{A}^{\mathcal{O}_1, \mathcal{O}_2, \ldots}(x)$ for the event that a quantum adversary $\mathcal{A}$ takes x as input, makes quantum queries to $\mathcal{O}_1, \mathcal{O}_2, \ldots$, and finally outputs y.

2.3 Cryptosystems and Notions of Security

Here we briefly recall standard notations of classical cryptosystems [12].

Symmetric-key Encryption. A symmetric-key cryptosystem $\mathcal{SE}$ consists of three PPT algorithms $\mathcal{SE} = (\mathcal{K}, \mathsf{SymEnc}, \mathsf{SymDec})$.

The standard correctness requirement is that for any key $\mathbf{k} \leftarrow \mathcal{K}()$, any random coin r of SymEnc and any $x \in \mathcal{X}$, we have $\mathsf{SymDec}_{\mathbf{k}}(\mathsf{SymEnc}_{\mathbf{k}}(x; r)) = x$. We sometimes omit the randomness r in SymEnc.

Public-key Encryption. A public-key cryptosystem $\mathcal{E}$ consists of three PPT algorithms $\mathcal{E} = (\mathsf{KeyGen}, \mathsf{Enc}, \mathsf{Dec})$.

The following correctness definition is taken from [14]. We call a public-key encryption scheme $\mathcal{E}$ *δ-correct* if

$$\mathbb{E}\left[\max_{x \in \mathcal{X}} \Pr_{r \in \mathcal{R}}[\mathsf{Dec}_{\mathbf{sk}}(\mathsf{Enc}_{\mathbf{pk}}(x; r)) \neq x]\right] \leq \delta,$$

where the expectation is taken over $(\mathbf{pk}, \mathbf{sk}) \leftarrow \mathsf{KeyGen}(\lambda)$.

Game-Based Definitions. Previously, quantum indistinguishability for adaptive chosen-ciphertext security has been defined in the work of Boneh and Zhandry [6]. At a high level, their notions allow quantum encryption and decryption queries, but require challenge queries to be *classical*. Regarding the attack models, the following security notions are then defined: IND-qCPA, IND-qCCA1, IND-qCCA2.

3 How to Record Encryption Queries in the Random World?

The starting point towards our goal of defining indistinguishability-based security notions for encryption is to explain how the challenger should reply to quantum decryption queries in the second learning phase after the adversary has made the quantum encryption queries in the challenge phase. This implies explaining how it could record these quantum challenge queries. In this section, we show how this can be done in the random world.

3.1 Ciphertext Decomposition

For simplicity, let we denote the encryption algorithm as a function f that takes as input a plaintext $x \in \mathcal{X}$, a randomness $r \in \mathcal{R}$ and outputs a ciphertext $y \leftarrow f(x;r) \in \mathcal{Y}$. We also assume that the domain of f is $\mathcal{X} = \{0,1\}^m$, its range is $\mathcal{Y} = \{0,1\}^n$, and the randomness space $\mathcal{R} = \{0,1\}^\ell$. We make a convention that $f(\perp) = 0$, where $\perp$ denotes some symbol outside the domain $\mathcal{X}$ and the range $\mathcal{Y}$. We define ciphertext decomposition as follows.

Definition 1. *For a function f, for all messages $x \in \mathcal{X}$, we write $y := (y_1\|y_2) \leftarrow f(x;r)$ and define:*

- **Message-independent:** *y_1 is message-independent if for all randomness r, there exists a function g such that $y_1 := g(r)$. In other words, the message-independent component of the ciphertext can be computed solely from the randomness r, independent of the message x. Furthermore, we require that $0 \leq |y_1| \leq |y|$.*
- **Message-dependent:** *y_2 is message-dependent if for all randomness r, there exists no function g such that $y_2 := g(r)$. In other words, the message-dependent component of the ciphertext can not be computed solely from the randomness r. Furthermore, we require that $1 \leq |y_2| \leq |y|$.*

We will also write $f := f_2 \circ f_1$, where f_1 acts only on the randomness, and f_2 acts on both the randomness and the plaintext.

Remark 1. Our definition above can be defined for any encryption scheme, without losing of generality. Furthermore, it also does not exclude some artificial encryption scheme such that the encryption is deterministic when the plaintext x is some special value (for example, the secret key), that is, there exists a function g such that $y_2 := g(x)$.

Remark 2. The definition of ciphertext decomposition is merely served as a technical step towards constructing the compressed encryption oracle in the random world in subsequent sections. We note that in an actual proof of security of an encryption scheme, one usually needs not to pay attention to this decomposition definition.

3.2 Oracle Variations

Here, we describe some oracle variations which will be used later in subsequent sections, the so-called *standard oracle* and *Fourier oracle*. These oracles and their equivalence are proven in much of literature on quantum-accessible oracles (e.g., see [9,16,23]).

Standard Oracles. For any function f with domain $\mathcal{X} = \{0,1\}^m$ and range $\mathcal{Y} = \{0,1\}^n$, the standard oracle for f is a unitary defined as

$$\mathsf{StdO}_f \sum_{x,y} \alpha_{x,y}|x,y\rangle_{XY} \mapsto \sum_{x,y} \alpha_{x,y}|x, y \oplus f(x)\rangle_{XY}.$$

The standard oracle can also be implemented in the truth table form: for each query, the oracle's internal state consists of $n2^m$-qubit F registers containing the truth table of the function. For short, we write $|f(0)\| \ldots \|f(2^m-1)\rangle$ as $|D\rangle$. Then, StdO_f performs the following map (on the adversary's basis states):

$$\begin{aligned}\mathsf{StdO}_f|x,y\rangle_{XY} \otimes |D\rangle_F \mapsto & |x, y \oplus D(x)\rangle_{XY}|D\rangle_F \\ = & |x, y \oplus f(x)\rangle_{XY}|D\rangle_F\end{aligned}$$

The equivalence of these two oracle variations follows directly from the fact that for each query, if we trace out the oracle's internal registers, the mixed state of the adversary in both cases will be identical.

Fourier Oracles. The Fourier oracle model $\mathsf{FourierO}_f$, while technically provides a different interface to the adversary, can be mapped to the standard oracle by QFT operations. The initial state of $\mathsf{FourierO}_f$ is

$$\mathsf{QFT}^F|D\rangle_F = \frac{1}{\sqrt{2^{n2^m}}}\sum_E(-1)^{E\cdot F}|E\rangle_F.$$

On the basis states, the Fourier oracle $\mathsf{FourierO}_f$ is defined as follows.

$$\begin{aligned}&\mathsf{FourierO}_f|x,z\rangle_{XY} \otimes \frac{1}{\sqrt{2^{n2^m}}}\sum_E(-1)^{E\cdot D}|E\rangle_F \\ &\mapsto \frac{1}{\sqrt{2^{n2^m}}}\sum_E(-1)^{E\cdot D}|x,z\rangle_{XY}|E \oplus P_{x,z}\rangle_F.\end{aligned}$$

where $P_{x,z}$ is the point function that outputs z on x and 0 everywhere else. Intuitively, with the Fourier oracle, instead of adding data from the oracle's registers to the adversary's registers, it adds in the opposite direction.

Lemma 1 [16,23]. *For any adversary $\mathcal{A}$ making queries to* StdO_f, *let $\mathcal{B}$ be the adversary that is identical to $\mathcal{A}$, except it performs the Fourier transformation to the response registers before and after each query. Then* $\Pr\left[\mathcal{A}^{\mathsf{StdO}_f}()=1\right] = \Pr\left[\mathcal{B}^{\mathsf{FourierO}_f}()=1\right]$.

Proof. Each oracle can be constructed by an f-independent quantum circuit containing just one copy of the other, that is

$$\begin{aligned}\mathsf{QFT}^{YF} \circ \mathsf{StdO}_f \circ \mathsf{QFT}^{\dagger YF} &= \mathsf{FourierO}_f, \\ \mathsf{QFT}^{\dagger YF} \circ \mathsf{FourierO}_f \circ \mathsf{QFT}^{YF} &= \mathsf{StdO}_f.\end{aligned}$$

□

3.3 Recording Queries in the Random World

As we have explained in Sect. 1, to define chosen-ciphertext security, we follow the real-or-random paradigm. In this section, we show how to process queries and record them in the random world, in which before applying the encryption algorithm f, the challenger chooses a random function h and applies it to the plaintext registers. As such, we also denote the encryption procedure in the random world as $f \circ h$. In what follows, we abuse the notation and write $f \circ h$ in the subscript of the oracle's notation with this meaning: for each query, a random function h is chosen uniformly by the oracle, so that h is not a pre-defined function. We note that the function f is known to the adversary though.

Single-Query Setting. We first start describing the oracle operations handling a single query and describe the general case later.

Without loss of generality, we assume that the query's response register Y can be decomposed into two parts Y_1, Y_2, in which the first part corresponds to the message-independent component, and the second part corresponds to the message-dependent component. Let $|Y_1| := n_1$ and $|Y_2| := n_2$ where $n_1 + n_2 = n$.

In the standard oracle model, the encryption oracle is implemented by first sampling a randomness r, a function $h : \mathcal{X} \to \mathcal{X}$ uniformly at random, and then applying the encryption algorithm f on the input $(h(x); r)$. From the adversary's point of view, this is equivalent to h being in uniform superposition $\sum_h |h\rangle$ and performing the following map:

$$|x, y\rangle_{XY} \otimes |r\rangle_R \sum_h |h\rangle_H \mapsto \sum_h |x, y \oplus f((h(x)); r)\rangle_{XY} |r\rangle_R |h\rangle_H. \tag{1}$$

Augmenting the joint system with a uniform superposition register H is a *purification* of the adversary's mixed state, and tracing out H (i.e., projecting onto the one-dimensional subspace spanned by $|h\rangle$) recovers the original mixed state. Moreover, this projection, which is outside of the adversary's view, is undetectable by any adversary $\mathcal{A}$.

Using ciphertext decomposition definition, we can write Eq. (1) as follows.

$$\begin{aligned}|x, y_1 \| y_2\rangle_{XY_1Y_2} \otimes |r\rangle_R \sum_h |h\rangle_H &\mapsto \sum_h |x, (y_1 \| y_2) \oplus f(h(x); r)\rangle_{XY_1Y_2} \otimes |r\rangle_R |h\rangle_H \\ &= \sum_h |x, y_1 \oplus f_1(r), y_2 \oplus f_2(h(x); r)\rangle_{XY_1Y_2} \otimes |r\rangle_R |h\rangle_H.\end{aligned}$$

We further note that, since the same randomness r is used for all "slots" in superposition, $f_1(r)$ is also the same for all "slots". In other words, $f_1(r)$ is just a classical value, which can be computed independently from the adversary's query. As a result, only the message-dependent registers are needed for recording queries. From now on to the rest of this section, we only consider the message-dependent parts in the adversary's response registers as well as the oracle's registers. These parts are denoted with index 2 in subscript (e.g., $y_2, z_2, f_2, \ldots$).

Now we describe our compressed encryption oracles. We first introduce some local procedures acting on the oracle's side, possibly controlled by the adversary's registers. Let Decomp_x be the identity operator except for

$$\mathsf{Decomp}_x \left(|r\rangle |x\rangle \frac{1}{\sqrt{2^m}} \sum_{u \in \{0,1\}^m} |u\rangle \frac{1}{\sqrt{2^{n_2}}} \sum_v (-1)^{f_2(u;r) \cdot v} |v\rangle \right) = |r\rangle |\perp\rangle |0\rangle |0\rangle,$$

and

$$\mathsf{Decomp}_x\left(|r\rangle|\bot\rangle|0\rangle|0\rangle\right) = |r\rangle|x\rangle\frac{1}{\sqrt{2^m}}\sum_{u\in\{0,1\}^m}|u\rangle\frac{1}{\sqrt{2^{n_2}}}\sum_v(-1)^{f_2(u;r)\cdot v}|v\rangle.$$

It is clear that Decomp_x is a unitary operator. Furthermore, applying it twice results in the identity, thus Decomp_x is an involution.

Using the notion similar to the description of Zhandry's compressed random oracle in [23], we introduce the notion of a database D that is maintained by the oracle as follows. A database D will be a collection of tuples $(x, (x', y))$, where $(x, (x', y)) \in D$ corresponds to $D(x) = (x', y)$. We say $D(x) = \bot$ if there is no such pair for an input x. For a database D with $D(x) \neq \bot$, we also write $D = \{x, u, v\} \cup D'$ where $D'(x) = \bot$. D consists of all the oracle's registers, except the randomness registers R. Decomp is then defined as the related unitary acting on the joint quantum system as follows.

$$\mathsf{Decomp}|x, z_2\rangle \otimes |r\rangle|D\rangle = |x, z_2\rangle \otimes \mathsf{Decomp}_x|r\rangle|D\rangle.$$

Let Init be the procedure that samples a random r uniformly and initializes a new register $|r\rangle|\bot, 0, 0\rangle$. Let $\mathsf{FourierO}'$ be unitary defined on the adversary's basis states as:

$$\begin{aligned}
&\mathsf{FourierO}'|x, z_2\rangle \otimes |r\rangle|D\rangle \\
&= \mathsf{FourierO}'|x, z_2\rangle \otimes |r\rangle\frac{1}{\sqrt{2^m}}\frac{1}{\sqrt{2^{n_2}}}\sum_{u,v}(-1)^{v\cdot f_2(u;r)}|\{x, u, v\} \cup D'\rangle \\
&= |x, z_2\rangle \otimes |r\rangle\frac{1}{\sqrt{2^m}}\frac{1}{\sqrt{2^{n_2}}}\sum_{u,v}(-1)^{v\cdot f_2(u;r)}|\{x, u, v \oplus z_2\} \cup D'\rangle.
\end{aligned}$$

Finally, we define the $\mathsf{CFourierO}_{f_2\circ h}$ oracle[2]:

$$\mathsf{CFourierO}_{f_2\circ h} := \mathsf{Decomp} \circ \mathsf{FourierO}' \circ \mathsf{Decomp} \circ \mathsf{Init}.$$

We state the following lemma:

Lemma 2. *In the single-query setting, the compressed Fourier oracle* $\mathsf{CFourierO}_{f_2\circ h}$ *acts on a basis state* $|x, z_2\rangle$ *where* $x \in \mathcal{X}$ *and* $z_2 \in \{0,1\}^{n_2}$*, as follows.*

- *If* $z_2 = 0$*, then* $\mathsf{CFourierO}_{f_2\circ h}|x, z_2\rangle \mapsto |x, z_2\rangle \otimes |r\rangle|\bot, 0, 0\rangle$.
- *If* $z_2 \neq 0$*, then* $\mathsf{CFourierO}_{f_2\circ h}|x, z_2\rangle \mapsto |x, z_2\rangle \otimes |\phi_{x,z_2}\rangle$*, where*

$$|\phi_{x,z_2}\rangle := |r\rangle\frac{1}{\sqrt{2^{m+n_2}}}\sum_u\sum_v(-1)^{f_2(u;r)\cdot v}|x, u, v \oplus z_2\rangle.$$

[2] For notation consistency, we use the same subscript in compressed oracles as for standard oracles. However, we note that there is no real function h in the implementation of $\mathsf{CFourierO}$ and its variants.

Furthermore, for any adversary $\mathcal{A}$ making a single query to $\mathsf{StdO}_{f_2 \circ h}$*, let* $\mathcal{B}$ *be the adversary that is identical to* $\mathcal{A}$*, except it performs the Hadamard transformation* $\mathsf{H}^{\otimes n}$ *to the response registers before and after the query. Then* $\Pr\left[\mathcal{A}^{\mathsf{StdO}_{f_2 \circ h}}() = 1\right] = \Pr\left[\mathcal{B}^{\mathsf{CFourierO}_{f_2 \circ h}}() = 1\right]$.

Proof. To prove the lemma, it is enough to show that $\mathsf{CFourierO}_{f_2 \circ h}$ and $\mathsf{FourierO}_{f_2 \circ h}$ are perfectly indistinguishable.

We prove this through a sequence of games. In what follows, we ambiguously denote $\mathsf{QFT}|f_2(x;r)\rangle$ by $|\eta_x\rangle$ for each $x \in \{0,1\}^m$. We will also take $y \oplus \perp = y, y \cdot \perp = 0$. When the adversary's response register is $|+\rangle$ (which corresponds to $|0\rangle$ in the Fourier basis), we can write, on the truth table of the oracle (for both $\mathsf{FourierO}_{f_2 \circ h}$ and $\mathsf{StdO}_{f_2 \circ h}$), the column with index x where x is the query's input as $\perp$.

Game G_0: In this game, the adversary interacts with the Fourier oracle $\mathsf{FourierO}_{f_2 \circ h}$, whose initial state is $|r\rangle \frac{1}{\sqrt{2^{m 2^m}}} \sum_h (h(0), \eta_{h(0)}) \| \cdots \| (h(2^{m-1}), \eta_{h(2^m-1)} |)\rangle$.

Game G_1: In this game, we represent the oracle in the form:

$$|r\rangle \frac{1}{\sqrt{2^{m 2^m}}} \sum_h |(0, h(0), \eta_{h(0)}) \| \cdots \| (2^m - 1, h(2^{m-1}), \eta_{h(2^m-1)})\rangle.$$

The update procedure for a query is then simply $\mathsf{FourierO}'$. G_1 is identical to G_0, since we have inserted the input points $0, \ldots, 2^m - 1$ into the oracle's state, which is independent from the adversary's state.

Game G_2: In this game, the oracle starts out as the "zero" database:

$$|r\rangle |(\perp, 0, 0) \| \cdots \| (\perp, 0, 0)\rangle.$$

Then a query is implemented as $\mathsf{Decomp}'^{\dagger} \circ \mathsf{FourierO}' \circ \mathsf{Decomp}'$, where $\mathsf{Decomp}' := \bigotimes_{i=0}^{2^m-1} \mathsf{Decomp}_i$. At the beginning, Decomp' is applied to the "zero" database, which maps it to the complete database

$$|r\rangle \frac{1}{\sqrt{2^{m 2^m}}} \sum_h |(0, h(0), \eta_{h(0)}) \| \cdots \| (2^m - 1, h(2^{m-1}), \eta_{h(2^m-1)})\rangle.$$

Then $\mathsf{FourierO}'$ is applied and the output state of G_2 in this stage will be exactly the output state of G_1. Since $\mathsf{Decomp}'^{\dagger}$ is a unitary that only operates on the oracle's register, its applications is undetectable to the adversary. So G_2 is perfectly indistinguishable from G_1.

Game G_3: In this final game, we use the compressed oracle $\mathsf{CFourierO}_{f_2 \circ h}$. Let x be the query's input. We note that $\mathsf{FourierO}'$ and $\mathsf{Decomp}_{x'}$ commute for any $x' \neq x$. Thus, we can move the computation of $\mathsf{Decomp}_{x'}$ to come after $\mathsf{FourierO}'$, consequently, its applications cancel out. We then have:

$$\begin{aligned}
&\mathsf{Decomp}'^{\dagger} \circ \mathsf{FourierO}' \circ \mathsf{Decomp}'(|x, z\rangle \otimes |r\rangle |D\rangle) \\
&= \mathsf{Decomp}_x^{\dagger} \circ \mathsf{FourierO}' \circ \mathsf{Decomp}_x(|x, z\rangle \otimes |r\rangle |D\rangle) \\
&= \mathsf{Decomp}^{\dagger} \circ \mathsf{FourierO}' \circ \mathsf{Decomp}(|x, z\rangle \otimes |r\rangle |D\rangle).
\end{aligned}$$

We are left with a database D whose support has at most 1 defined point after the query in G_2. The remaining $\geq 2^m - 1$ points are all $(\perp, 0, 0)$. So we may end up with a superposition of databases that have at most one defined point. We then can move this defined point in the database to the first register (this is a unitary operator and is undetectable to the adversary) and obtain a superposition of databases that have a defined point only in the first register. Therefore we can discard all but the first register, without affecting the adversary's state. This shows that G_3 and G_2 are identical. □

The compressed Fourier encryption oracle in the random world $\mathsf{CFourierO}_{f \circ h}$ is straightforwardly obtained by running the message-independent function f_1 on the randomness r, transforming it to the Fourier basis and then composing it with $\mathsf{CFourierO}_{f_2 \circ h}$. Formally, $\mathsf{CFourierO}_{f \circ h} := (\mathsf{QFT}^{F_1} U^R_{f_1}) \circ \mathsf{CFourierO}_{f_2 \circ h}$. We then have

Lemma 3. *For any adversary $\mathcal{A}$ making a single query to $\mathsf{StdO}_{f \circ h}$, let $\mathcal{B}$ be the adversary that is identical to $\mathcal{A}$, except it performs the Hadamard transformation $\mathsf{H}^{\otimes n}$ to the response registers before and after the query. Then* $\Pr\left[\mathcal{A}^{\mathsf{StdO}_{f \circ h}}() = 1\right] = \Pr\left[\mathcal{B}^{\mathsf{CFourierO}_{f \circ h}}() = 1\right]$.

Compressed Standard Encryption Oracles. By applying Hadamard to the adversary's response registers before and after the query, and to the oracle's register F after the query, we also obtain the compressed standard encryption oracle $\mathsf{CStO}_{f \circ h}$. The oracle's state after the query is (in superposition of) $|r, x, u, f(u;r)\rangle$. Formally, $\mathsf{CStO}_{f \circ h} := \mathsf{QFT}^{YF} \circ \mathsf{CFourierO}_{f \circ h} \circ \mathsf{QFT}^{Y}$. By applying the same argument as in Lemma 1 to $\mathsf{CFourierO}_{f \circ h}$ and $\mathsf{CStO}_{f \circ h}$, and combining with Lemma 3, the following lemma follows:

Lemma 4. *$\mathsf{CStO}_{f \circ h}$ and $\mathsf{StdO}_{f \circ h}$ are perfectly indistinguishable. That is, for any adversary $\mathcal{A}$, we have that* $\Pr\left[\mathcal{A}^{\mathsf{StdO}_{f \circ h}}() = 1\right] = \Pr\left[\mathcal{A}^{\mathsf{CStO}_{f \circ h}}() = 1\right]$.

Many-Query Setting. We denote $\mathsf{CStO}_{f \circ H}$ as the following oracle: for each query, $\mathsf{CStO}_{f \circ H}$ invokes a new instance of $\mathsf{CStO}_{f \circ h}$ with uniformly and independently randomness r. Similarly, $\mathsf{StdO}_{f \circ H}$ denote the following oracle: for each query, $\mathsf{StdO}_{f \circ H}$ samples uniformly and independently a randomness r and a random function h, and then answers that query using $\mathsf{StdO}_{f \circ h}$. By the standard hybrid argument, it is easy to verify that:

Lemma 5. *$\mathsf{CStO}_{f \circ H}$ and $\mathsf{StdO}_{f \circ H}$ are perfectly indistinguishable, in the many-query setting.*

For each i-th query, its oracle's database is $|D_i\rangle := |x_i, u_i, f(u_i; r_i)\rangle$. Overall, the oracle's database D will be a collection of many tuples $(x, (x', y))$ where $(x, (x', y)) \in D$ means $f(x'; r) = y$ and $h(x) = x'$ for different random functions h.

3.4 A Technical Observation

Notice that from the proof of Lemma 2 above, we implement this compressed encryption oracle with at least two computations of f_2 (and so f) via two applications of Decomp. However, as we will see in later sections, it is crucial for our security reductions to simulate $\mathsf{CFourierO}_{f \circ h}$ with only one computation of f, which allows us to "outsource" f computations to other oracles. We now give an intuition why we can reduce many computations of f to one computation. Let's consider the following cases.

- The z_2 registers are all-zero. Note that since the initial state of the oracle database D is also all-zero, applying the first Decomp and then XORing the adversary's registers to the oracle's (i.e., the application of $\mathsf{FourierO}'$) does not change the database's state. Finally, the second application of Decomp brings it back to all-zero state, which can be discarded. At the end of this step, D is empty. In this case, we see that we can skip $\mathsf{FourierO}'$, and two applications of Decomp cancel out, leaving us no applications of f.
- The z_2 registers are not zero. By a similar argument, we have that the second application of Decomp has no effects on the joint system, leaving us only one application of f in the first application of Decomp.

We describe a quantum circuit in Fig. 1, which applies a single computation of f_2 (denoted as a unitary U_{f_2}), implementing our compressed encryption oracle in the random world. Let Test be the unitary defined as $\mathsf{Test}|0\rangle|b\rangle \mapsto |0\rangle|b\rangle$ and $\mathsf{Test}|\phi\rangle|b\rangle \mapsto |\phi\rangle|b \oplus 1\rangle$ for any $|\phi\rangle$ orthogonal to $|0^{n_2}\rangle$ and $b \in \{0,1\}$. A concrete computation reveals that this circuit outputs the same quantum state as stated in Lemma 2.

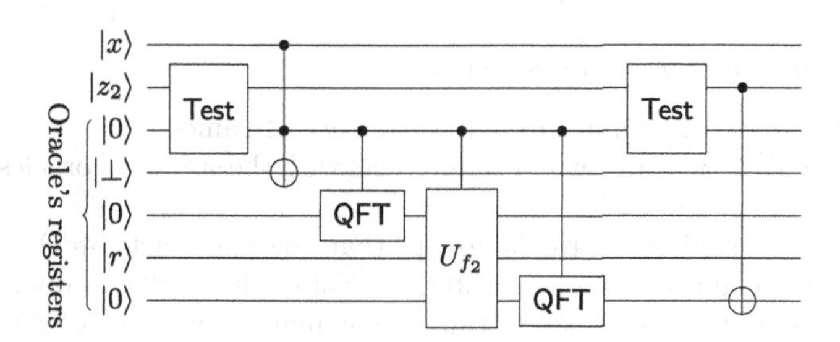

Fig. 1. A quantum circuit implementing our $\mathsf{CFourierO}_{f_2 \circ h}$ oracle. Depending on the control bit b which is the output of Test, if $b = 1$, we apply U_{f_2}, otherwise, we apply the identity. The bit b will be discarded after the computation.

3.5 How to Answer Decryption Queries?

We now describe how to answer decryption queries in the random world using the database constructed above. Generally, we will consider any δ-correct encryption scheme (see Definition in Sect. 2.3).

We will start with a technical lemma, in which the decryption will answer "naively", that is if the ciphertext is $f(x'; r)$ for some x', the decryption oracle is expected to return x', even if x' was the output of a random function. (Roughly speaking, this decryption oracle mimics a standard decryption oracle with no restrictions on the adversary.) We call this decryption oracle the *naive decryption oracle*.

In the following, we abuse the notation and denote f^{-1} as the decryption algorithm. We then give the adversary access to a new oracle denoted $\mathsf{CInvO}_{f^{-1}}$ (this is our naive decryption oracle) which acts on the database, instead of $\mathsf{StdO}_{f^{-1}}$. Given access to $\mathsf{CInvO}_{f^{-1}}$, the bound on the distinguishing probability of the adversary when interacting with the compressed oracle $\mathsf{CStO}_{f\circ H}$ is stated in Lemma 6.

We define a classical procedure $\mathsf{FindImage}'$ which takes as input a ciphertext $y \in \mathcal{Y}$, and a database D. Then, it looks for a tuple $(x,(x',y)) \in D$. If found, it outputs $(b=1, w=x')$, otherwise, it outputs $(b=0, w=0)$. Notice that there may be many tuples with the same y in D, but since an encryption scheme must be injective (for decryption to work), these pairs must have the same x'.

We define the unitary operation $\mathsf{CInvO}_{f^{-1}}$ for the inverse queries which maps the basis state $|y,z\rangle \otimes |D\rangle$ to:

$$\begin{cases} U_{f^{-1}}|y,z\rangle \otimes |D\rangle = |y, z \oplus f^{-1}(y)\rangle \otimes |D\rangle & \text{if } \mathsf{FindImage}'(y,D) = (0,0), \\ |y, z \oplus w\rangle \otimes |D\rangle & \text{if } \mathsf{FindImage}'(y,D) = (1,w). \end{cases}$$

This unitary is implemented by a single call to f^{-1}, controlled by the output bit b of $\mathsf{FindImage}'$ recorded in some ancilla registers[3].

Lemma 6. *For any (unbounded) oracle algorithm $\mathcal{A}$, and any δ-correct encryption scheme:*

$$\left|\Pr\left[\mathcal{A}^{\mathsf{StdO}_{f\circ H},\mathsf{StdO}_{f^{-1}}}() = 1\right] - \Pr\left[\mathcal{A}^{\mathsf{CStO}_{f\circ H},\mathsf{CInvO}_{f^{-1}}}() = 1\right]\right| \leq \mathcal{O}(q_i \cdot \delta),$$

where q_i is the number of inverse queries.

Proof. We prove this lemma through a sequence of games.
Game G_0: This is the game where $\mathcal{A}$ interacts with the standard oracles $\mathsf{StdO}_{f\circ H}$ and $\mathsf{StdO}_{f^{-1}}$.
Game G_1: This is identical to G_0, except that now the oracle $\mathsf{StdO}_{f\circ H}$ is simulated using the compressed oracle $\mathsf{CStO}_{f\circ H}$. Notice that $\mathsf{StdO}_{f^{-1}}$ operation does not touch the database registers, thus it commutes with any $\mathsf{CStO}_{f\circ h}$ operation. Since $\mathsf{CStO}_{f\circ H}$ is equivalent to the standard oracle $\mathsf{StdO}_{f\circ H}$, $\mathcal{A}$ cannot distinguish G_1 and G_0.
Game G_2: This is identical to G_1, except that now the oracle $\mathsf{StdO}_{f^{-1}}$ is replaced by the oracle $\mathsf{CInvO}_{f^{-1}}$.

Let $|\Psi\rangle$ be the joint system state of the adversary and the oracle before making any inverse query. Denote $\Delta = \mathsf{StdO}_{f^{-1}} - \mathsf{CInvO}_{f^{-1}}$. For each query $|y,z\rangle$ to the inverse oracle, we consider the registers y, z, D. We now examine three cases.

(a) Let D be such that $y \notin D$, that is, $\mathsf{FindImage}(y,D) = (0,0)$. Let P_1 be the projection onto the registers y, D such that $y \notin D$. In this case, the inverse oracle in both games applies the unitary mapping $|y,z\rangle \otimes |D\rangle \mapsto |y, z\oplus f^{-1}(y)\rangle \otimes |D\rangle$. Thus, $\Delta P_1 |\Psi\rangle = 0$.

[3] The oracle first computes $\mathsf{FindImage}'$, records the output in some ancilla register, performs the CNOT operation controlled on the output and finally un-compute $\mathsf{FindImage}'$.

(b) Let D be such that $y \in D$, that is, $\mathsf{FindImage}(y, D) = (1, w)$. Let P_2 be the projection onto the registers y, D such that $y \in D$ and $f^{-1}(y) = w$. In this case, we also have $\Delta P_2 |\Psi\rangle = 0$.

(c) Let D be such that $y \in D$. Let P_3 be the projection onto the registers y, D such that $y \in D$ but $f^{-1}(y) \neq w$. Thus $\|P_3|\Psi\rangle\|^2$ is the probability of measuring y, D and get $y \in D$ such that $f^{-1}(y = f(x)) \neq x$ for some pre-image x of y. In this case, we have $\|\Delta P_3|\Psi\rangle\|^2 \leq \delta$, by the definition that the encryption scheme is δ-correct.

Notice that $P_1 + P_2 + P_3 = \mathcal{I}$. Therefore, we have $\|\Delta|\Psi\rangle\|^2 = \left\|\sum_{i=1}^3 \Delta P_i |\Psi\rangle\right\|^2 \overset{(*)}{\leq} \sum_{i=1}^3 \|\Delta P_i|\Psi\rangle\|^2 \leq \delta$, where $(*)$ uses triangle inequality. Then the same holds true for any mixed state since any mixed state is in the convex hull of pure states. If $\mathcal{A}$ makes at most q_i inverse queries, the trace distance of the mixed state of the adversary in games G_2 and G_1 is at most $\mathcal{O}(q_i \cdot \delta)$. This completes the proof. □

Now we describe our actual decryption oracle in the random world. Instead of using $\mathsf{FindImage}'$ which returns $(1, x')$, we use an identical $\mathsf{FindImage}$ except that it returns $(b = 1, w = x)$ when $(x, (x', y)) \in D$. The oracle $\mathsf{CInvO}_{f^{-1}}$ is redefined using $\mathsf{FindImage}$ as follows. It maps the basis state $|y, z\rangle \otimes |D\rangle$ to:

$$\begin{cases} U_{f^{-1}}|y, z\rangle \otimes |D\rangle = |y, z \oplus f^{-1}(y)\rangle \otimes |D\rangle & \text{if } \mathsf{FindImage}(y, D) = (0, 0), \\ |y, z \oplus w\rangle \otimes |D\rangle & \text{if } \mathsf{FindImage}(y, D) = (1, w). \end{cases}$$

3.6 Notation

From now on to the rest of the paper, we will use the following notation:

- $\mathcal{O}$ to denote the standard encryption and decryption oracles StdO (which are distinguished by subscript, e.g., $\mathcal{O}_{\mathsf{SymEnc}}$ for encryption and $\mathcal{O}_{\mathsf{SymDec}}$ for decryption) in the real world.
- $\mathcal{R}$ to denote the compressed encryption and decryption oracles (which are distinguished by subscript, e.g., $\mathcal{R}_{\mathsf{SymEnc}}$ for encryption and $\mathcal{R}_{\mathsf{SymDec}}$ for decryption) in the random world. In particular, the encryption one will be implemented using CStO, and the decryption one using CInvO.

4 Quantum-Secure Symmetric Encryption

4.1 Definitions of Security

In this section, we use the compressed oracle technique defined above to define quantum real-or-random indistinguishability security notions.

High-Level View. During the learning phases, $\mathcal{A}$ has access to the encryption standard oracle $\mathcal{O}_{\mathsf{SymEnc}_k}$. In the CCA case, it also has access to $\mathcal{O}_{\mathsf{SymDec}_k}$ in the first learning phase. We describe informally how we handle the challenge phase and the decryption queries in the second learning phase. The goal is to mimic the (purely) classical CCA security game in which: $\mathcal{A}$ gives a challenge plaintext and receives either encryption of it or encryption of a random message; during the second learning phase, if $\mathcal{A}$ makes a decryption query on the challenge ciphertext, it is given back the challenge plaintext in both games.

In the real-world ($b = 1$), the adversary has no restrictions on the use of the decryption oracle (in particular, $\mathcal{A}$ can freely decrypt the challenge ciphertext – getting back the challenge plaintext, as in the classical case), so that the encryption oracle is simply implemented as the standard encryption oracle $\mathcal{O}_{\mathsf{SymEnc_k}}$ and the decryption oracle as the standard decryption oracle $\mathcal{O}_{\mathsf{SymDec_k}}$.

In the random-world ($b = 0$), the challenger implements the challenge encryption oracle using a compressed encryption oracle $\mathcal{R}_{\mathsf{SymEnc_k}}$, and the decryption oracle in the second phase $\mathcal{R}_{\mathsf{SymDec_k}}$ as described in Sect. 3.5. As in the real-world, this decryption oracle always returns the original plaintext (x) if the query is a challenge one, using the database. Otherwise, it just decrypts normally.

Definitions. Formally, denote $\mathcal{A} = (\mathcal{A}_1, \mathcal{A}_2)$. In both games, $\mathcal{A}_1$ outputs an internal state $|\Phi\rangle$ after the first phase (i.e., the first learning phase), which will be given to $\mathcal{A}_2$ in the second phase (including the challenge and the second learning phase). We define a "real-or-random" oracle $\mathcal{RR}$ allowing $\mathcal{A}_2$ to make quantum challenge queries. For learning queries, $\mathcal{A}_2$ has access to $\mathcal{O}_{\mathsf{SymEnc_k}}$ and potentially a decryption oracle $\mathcal{DEC}$ defined as follows.

$$\mathcal{RR}(b) = \begin{cases} \mathcal{O}_{\mathsf{SymEnc_k}} & \text{if } b = 1, \\ \mathcal{R}_{\mathsf{SymEnc_k}} & \text{if } b = 0, \end{cases} \quad \mathcal{DEC}(b) = \begin{cases} \mathcal{O}_{\mathsf{SymDec_k}} & \text{if } b = 1, \\ \mathcal{R}_{\mathsf{SymDec_k}} & \text{if } b = 0. \end{cases}$$

Definition 2 (Indistinguishability notions for symmetric encryption (qIND-qCPA, qIND-qCCA1, qIND-qCCA2)).
Let $\mathcal{SE} = (\mathcal{K}, \mathsf{SymEnc}, \mathsf{SymDec})$ be a symmetric encryption scheme and let $\mathcal{A} = (\mathcal{A}_1, \mathcal{A}_2)$ be a quantum adversary. For $qatk \in [qcpa, qcca1, qcca2]$, we define the following game, where the oracles $\mathcal{O}_1, \mathcal{O}_2$ are defined according to qatk:

Experiment $\mathsf{Expt}^{qind\text{-}qatk-b}_{\mathcal{SE}}(\lambda, \mathcal{A})$:
1 : $\mathbf{k} \xleftarrow{\$} \mathcal{K}$
2 : $\lvert\Phi\rangle \leftarrow \mathcal{A}_1^{\mathcal{O}_{\mathsf{SymEnc_k}}, \mathcal{O}_1}(\lambda)$
3 : $b' \leftarrow \mathcal{A}_2^{\mathcal{RR}(b), \mathcal{O}_{\mathsf{SymEnc_k}}, \mathcal{O}_2}(\lvert\Phi\rangle)$
4 : **return** b'

qatk	*Oracle* $\mathcal{O}_1$	*Oracle* $\mathcal{O}_2$
qcpa	$\varnothing$	$\varnothing$
qcca1	$\mathcal{O}_{\mathsf{SymDec_k}}$	$\varnothing$
qcca2	$\mathcal{O}_{\mathsf{SymDec_k}}$	$\mathcal{DEC}(b)$

We define $\mathcal{A}$'s advantage by

$$\mathsf{Adv}^{qind\text{-}qatk}_{\mathcal{A},\mathcal{SE}}(\lambda) := \left| \Pr\left[\mathsf{Expt}^{qind\text{-}qatk-1}_{\mathcal{SE}}(\lambda, \mathcal{A}) = 1\right] - \Pr\left[\mathsf{Expt}^{qind\text{-}qatk-0}_{\mathcal{SE}}(\lambda, \mathcal{A}) = 1\right] \right|.$$

We say $\mathcal{SE}$ is secure in the sense of qIND-qATK *if $\mathcal{A}$ being* QPT *implies that $\mathsf{Adv}^{qind\text{-}qatk}_{\mathcal{A},\mathcal{SE}}(\lambda)$ is negligible.*

Comparison with Boneh-Zhandry's Notions. To justify our notions, in the full version of the paper [8], we show that when restricting our definitions to classical challenge queries, they are equivalent to Boneh-Zhandry's notions (IND-qATK). If we denote our restricted notions by IND-qATK', a scheme $\mathcal{SE}$ is IND-qATK' secure iff it is IND-qATK secure.

Furthermore, we also show that upgrading from classical challenge queries to quantum challenge queries gives the adversary more power. In particular, we show that the IND-qCCA2 secure symmetric encryption scheme given by Boneh and Zhandry [6] is insecure once the adversary can make even a single quantum challenge query in the sense of chosen plaintext security (qIND-qCPA). Our attack can be considered as an impossibility to achieve quantum indistinguishability for encryption schemes which follow the stream cipher-like paradigm (such as stream ciphers, block cipher modes of operation including CFB, OFB, CTR, or even some most widely used modes like GCM for authenticated encryptions).

Single-Message versus Many-Message Security. We have presented definitions which allow the adversary to make $q(\lambda)$-many challenge queries to the real-or-random oracle. A scheme satisfying the definitions in the case when $q(\lambda) = 1$ is said to be *single-message* secure. The question of whether single-message security implies many-message security is the question of composability of the definitions, which is answered affirmatively below.

Theorem 1. *A symmetric encryption scheme $\mathcal{SE}$ is many-message* qIND-qATK *secure iff it is single-message* qIND-qATK *secure.*

The proof follows the classical hybrid argument; we give it in the full version of our paper [8].

4.2 Feasibility of Quantum CCA2 Security

The classical Encrypt-then-MAC paradigm [3] shows that an IND-CPA secure symmetric encryption scheme can be made IND-CCA2 secure if combined with an EUF-CMA MAC scheme. However, it is not obvious how to prove security in the quantum setting, as the reduction algorithm has no way to tell which ciphertexts the adversary received as the result of an encryption query in the learning phases, and no way to decrypt the ciphertexts if it has received them. To remedy these problems, we choose a specific type of MAC scheme in the construction (that is, any quantum-secure PRF) and leave the general security proof as an open question. The encryption scheme can be instantiated with any qIND-qCPA encryption scheme. In the proof, we simulate the MAC with random oracle and use Zhandry's compressed oracles technique to efficiently check if the adversary has seen a particular ciphertext as a result of an encryption query, and to decrypt in this case. Due to space limitations, the proof of Theorem 2 is given in the full version of our paper [8].

Construction 1. *Let* $\mathcal{SE} = (\mathcal{K}_{\mathcal{SE}}, \mathsf{SymEnc}, \mathsf{SymDec})$ *be a symmetric encryption scheme and* $\mathsf{qPRF} = \{\mathsf{qPRF}_k\}_{k \in \mathbb{N}}$ *be a family of quantum-secure pseudorandom functions. A composition of base schemes* $\mathcal{SE}$ *and* qPRF *is the symmetric* *encryption scheme* $\mathcal{SE}' = (\mathcal{K}', \mathsf{SymEnc}', \mathsf{SymDec}')$ *whose constituent algorithms are defined as follows.*

$\mathcal{K}'(\lambda)$:	$\mathsf{SymEnc}'_{\mathrm{k}_1\|\mathrm{k}_2}(x)$:	$\mathsf{SymDec}'_{\mathrm{k}_1\|\mathrm{k}_2}(c \parallel \tau)$:
1 : $\mathrm{k}_1 \xleftarrow{\$} \mathcal{K}_{\mathcal{SE}}()$	*1* : $c \leftarrow \mathsf{SymEnc}_{\mathrm{k}_1}(x)$	*1* : $x \leftarrow \mathsf{SymDec}_{\mathrm{k}_1}(c)$
2 : $\mathrm{k}_2 \xleftarrow{\$} \{0,1\}^\lambda$	*2* : $\tau \leftarrow \mathsf{qPRFk2}(c\|x)$	*2* : **if** $\mathsf{qPRF}_{\mathrm{k}_2}(c\|x) \neq \tau$ **then**
3 : **return** $\mathrm{k}_1 \parallel \mathrm{k}_2$	*3* : **return** $c \parallel \tau$	*3* : **return** $\perp$
		4 : **return** x

Theorem 2. *Let* $\mathcal{SE}$ *be an* qIND-qCPA *secure symmetric encryption scheme. Let* qPRF *be a family of quantum-secure pseudorandom functions. Then the encryption scheme* $\mathcal{SE}'$ *defined in Construction 1 is* qIND-qCCA2 *secure.*

Remark 3. As shown in [22], quantum-secure PRFs can be constructed from quantum-secure one-way functions. In addition, [7,11] shows how to construct qIND-qCPA secure encryption schemes from quantum-secure pseudorandom permutations.

5 Quantum-Secure Public-Key Encryption

5.1 Definitions of Security

Indistinguishability Security. The indistinguishability notions can be defined analogously to the ones given in Sect. 4. We define a real-or-random oracle allowing quantum queries and the decryption oracle in the second learning phase as follows.

$$\mathcal{RR}(b) = \begin{cases} \mathcal{O}_{\mathsf{Enc}_{\mathsf{pk}}} & \text{if } b = 1, \\ \mathcal{R}_{\mathsf{Enc}_{\mathsf{pk}}} & \text{if } b = 0, \end{cases} \quad \mathcal{DEC}(b) = \begin{cases} \mathcal{O}_{\mathsf{Dec}_{\mathsf{sk}}} & \text{if } b = 1, \\ \mathcal{R}_{\mathsf{Dec}_{\mathsf{sk}}} & \text{if } b = 0. \end{cases}$$

Definition 3 (qIND-qCPA, qIND-qCCA1, qIND-qCCA2). *Let* $\mathcal{E} = (\mathsf{KeyGen}, \mathsf{Enc}, \mathsf{Dec})$ *be a public-key encryption scheme and let* $\mathcal{A} = (\mathcal{A}_1, \mathcal{A}_2)$ *be a quantum adversary. For qatk* $\in$ *[qcpa, qcca1, qcca2], we define the following game, where the oracles* $\mathcal{O}_1, \mathcal{O}_2$ *are defined according to qatk:*

Experiment $\mathsf{Expt}_{\mathcal{E}}^{qind\text{-}qatk-b}(\lambda, \mathcal{A})$:	*qatk*	*Oracle* $\mathcal{O}_1$	*Oracle* $\mathcal{O}_2$
1 : $(\mathsf{pk}, \mathsf{sk}) \leftarrow \mathsf{KeyGen}(\lambda)$	*qcpa*	$\varnothing$	$\varnothing$
2 : $\|\Phi\rangle \leftarrow \mathcal{A}_1^{\mathcal{O}_1}(\mathsf{pk})$	*qcca1*	$\mathcal{O}_{\mathsf{Dec}_{\mathsf{sk}}}$	$\varnothing$
3 : $b' \leftarrow \mathcal{A}_2^{\mathcal{RR}(b), \mathcal{O}_2}(\|\Phi\rangle)$	*qcca2*	$\mathcal{O}_{\mathsf{Dec}_{\mathsf{sk}}}$	$\mathcal{DEC}(b)$
4 : **return** b'			

We define $\mathcal{A}$'s advantage by

$$\mathsf{Adv}^{qind\text{-}qatk}_{\mathcal{A},\mathcal{E}}(\lambda) := \left|\Pr\left[\mathsf{Expt}^{qind\text{-}qatk-1}_{\mathcal{E}}(\lambda,\mathcal{A})=1\right] - \Pr\left[\mathsf{Expt}^{qind\text{-}qatk-0}_{\mathcal{E}}(\lambda,\mathcal{A})=1\right]\right|.$$

We say $\mathcal{E}$ is secure in the sense of qIND-qATK *if $\mathcal{A}$ being* QPT *implies that* $\mathsf{Adv}^{qind\text{-}qatk}_{\mathcal{A},\mathcal{E}}(\lambda)$ *is negligible.*

Similarly as in Sect. 4, our definitions, restricted to classical challenge queries, are equivalent to Boneh-Zhandry's notions (IND-qATK). Furthermore, the following theorem shows that our notions are closed under composition.

Theorem 3. *An encryption scheme $\mathcal{E}$ is many-message* qIND-qATK *secure iff it is single-message* qIND-qATK *secure.*

5.2 A Lifting Theorem: From IND-qCCA2 to qIND-qCCA2

We present a compiler transforming IND-qATK security to qIND-qATK security. Our compiler follows the classical hybrid encryption paradigm. The message is encrypted under a random symmetric key each time, and the key is encrypted by the public-key encryption scheme. Since the same randomness is used for each query in superposition, we can use the same random symmetric key in superposition each time. This means that the adversary never has quantum access to the encryption algorithm of the public-key scheme, only the symmetric encryption needs to be secure against quantum queries, which we know how to construct from one-way functions (Theorem 2).

Construction 2. *Let $\mathcal{E} = (\mathsf{KeyGen}, \mathsf{Enc}, \mathsf{Dec})$ be a public-key encryption scheme which is* IND-qATK *secure and δ-correct. Let $\mathcal{SE} = (\mathsf{SymEnc}, \mathsf{SymDec})$ be a one-time* qIND-qATK *secure symmetric-key encryption scheme. We construct a new public-key encryption scheme $\mathcal{E}' = (\mathsf{KeyGen}', \mathsf{Enc}', \mathsf{Dec}')$ as follows.*

$\mathsf{KeyGen}'(\lambda)$:	$\mathsf{Enc}'_{\mathsf{pk}}(x)$:	$\mathsf{Dec}'_{\mathsf{sk}}(c_1\|c_2)$:
1 : $(\mathsf{pk}, \mathsf{sk}) \xleftarrow{\$} \mathsf{KeyGen}(\lambda)$	1 : $\mathsf{k} \xleftarrow{\$} \mathcal{K}()$	1 : $\mathsf{k} \leftarrow \mathsf{Dec}_{\mathsf{sk}}(c_1)$
2 : **return** $(\mathsf{pk}, \mathsf{sk})$	2 : $c_1 \leftarrow \mathsf{Enc}_{\mathsf{pk}}(\mathsf{k})$	2 : $x \leftarrow \mathsf{SymDec}_{\mathsf{k}}(c_2)$
	3 : $c_2 \leftarrow \mathsf{SymEnc}_{\mathsf{k}}(x)$	3 : **return** x
	4 : **return** $c_1\|c_2$	

Remark 4. In this construction, we make no extra assumptions. We know that the existence of IND-qATK secure encryption implies the existence of quantum-secure one-way functions. IND-qATK secure public-key encryption can be constructed based on quantum-resistant assumptions (e.g., Learning With Errors) [6].

Theorem 4. *The encryption scheme $\mathcal{E}'$ defined in Construction 2 is* qIND-qCCA2 *secure, if $\mathcal{E}$ is* IND-qCCA2 *secure, and $\mathcal{SE}$ is one-time* qIND-qCCA2 *secure. In particular, for any* QPT *adversary $\mathcal{A}$, there exist* QPT *adversaries $\mathcal{B}, \mathcal{C}$ such that*

$$\mathsf{Adv}^{qind\text{-}qcca2}_{\mathcal{A},\mathcal{E}'}(\lambda) \leq \mathcal{O}(q_d \cdot \delta) + 2 \cdot \mathsf{Adv}^{ind\text{-}qcca2}_{\mathcal{B},\mathcal{E}}(\lambda) + \mathsf{Adv}^{qind\text{-}qcca2}_{\mathcal{C},\mathcal{SE}}(\lambda),$$

where q_d is the number of decryption queries in the second phase.

Acknowledgments. This work was supported in part by the French ANR project CryptiQ (ANR-18-CE39-0015) and the French *Programme d'Investissement d'Avenir* under national project RISQ P141580. The authors want to thank Damien Vergnaud, David Pointcheval and Christian Majenz for fruitful discussions, as well as the anonymous reviewers for useful comments.

References

1. Alagic, G., Majenz, C., Russell, A., Song, F.: Quantum-access-secure message authentication via blind-unforgeability. In: Canteaut, A., Ishai, Y. (eds.) EUROCRYPT 2020. LNCS, vol. 12107, pp. 788–817. Springer, Cham (2020). https://doi.org/10.1007/978-3-030-45727-3_27
2. Anand, M.V., Targhi, E.E., Tabia, G.N., Unruh, D.: Post-quantum security of the CBC, CFB, OFB, CTR, and XTS modes of operation. In: Takagi, T. (ed.) PQCrypto 2016. LNCS, vol. 9606, pp. 44–63. Springer, Cham (2016). https://doi.org/10.1007/978-3-319-29360-8_4
3. Bellare, M., Namprempre, C.: Authenticated encryption: relations among notions and analysis of the generic composition paradigm. J. Cryptol. **21**(4), 469–491 (2008)
4. Boneh, D., Dagdelen, Ö., Fischlin, M., Lehmann, A., Schaffner, C., Zhandry, M.: Random oracles in a quantum world. In: Lee, D.H., Wang, X. (eds.) ASIACRYPT 2011. LNCS, vol. 7073, pp. 41–69. Springer, Heidelberg (2011). https://doi.org/10.1007/978-3-642-25385-0_3
5. Boneh, D., Zhandry, M.: Quantum-secure message authentication codes. In: Johansson, T., Nguyen, P.Q. (eds.) EUROCRYPT 2013. LNCS, vol. 7881, pp. 592–608. Springer, Heidelberg (2013). https://doi.org/10.1007/978-3-642-38348-9_35
6. Boneh, D., Zhandry, M.: Secure signatures and chosen ciphertext security in a quantum computing world. In: Canetti, R., Garay, J.A. (eds.) CRYPTO 2013. LNCS, vol. 8043, pp. 361–379. Springer, Heidelberg (2013). https://doi.org/10.1007/978-3-642-40084-1_21
7. Carstens, T.V., Ebrahimi, E., Tabia, G.N., Unruh, D.: Relationships between quantum IND-CPA notions. In: Nissim, K., Waters, B. (eds.) TCC 2021. LNCS, vol. 13042, pp. 240–272. Springer, Cham (2021). https://doi.org/10.1007/978-3-030-90459-3_9
8. Chevalier, C., Ebrahimi, E., Vu, Q.-H.: On security notions for encryption in a quantum world. Cryptology ePrint Archive, Report 2020/237 (2020)
9. Czajkowski, J., Majenz, C., Schaffner, C., Zur, S.: Quantum lazy sampling and game-playing proofs for quantum indifferentiability. Cryptology ePrint Archive, Report 2019/428 (2019)

10. Damgård, I., Funder, J., Nielsen, J.B., Salvail, L.: Superposition attacks on cryptographic protocols. In: Padró, C. (ed.) ICITS 2013. LNCS, vol. 8317, pp. 142–161. Springer, Cham (2014). https://doi.org/10.1007/978-3-319-04268-8_9
11. Gagliardoni, T., Hülsing, A., Schaffner, C.: Semantic security and indistinguishability in the quantum world. In: Robshaw, M., Katz, J. (eds.) CRYPTO 2016. LNCS, vol. 9816, pp. 60–89. Springer, Heidelberg (2016). https://doi.org/10.1007/978-3-662-53015-3_3
12. Goldreich, O.: Foundations of Cryptography: Basic Applications, vol. 2. Cambridge University Press, Cambridge (2004)
13. Grover, L.K.: A fast quantum mechanical algorithm for database search. In: 28th ACM STOC, pp. 212–219. ACM Press, May 1996
14. Hofheinz, D., Hövelmanns, K., Kiltz, E.: A modular analysis of the Fujisaki-Okamoto transformation. In: Kalai, Y., Reyzin, L. (eds.) TCC 2017. LNCS, vol. 10677, pp. 341–371. Springer, Cham (2017). https://doi.org/10.1007/978-3-319-70500-2_12
15. Kaplan, M., Leurent, G., Leverrier, A., Naya-Plasencia, M.: Breaking symmetric cryptosystems using quantum period finding. In: Robshaw, M., Katz, J. (eds.) CRYPTO 2016. LNCS, vol. 9815, pp. 207–237. Springer, Heidelberg (2016). https://doi.org/10.1007/978-3-662-53008-5_8
16. Kashefi, E., Kent, A., Vedral, V., Banaszek, K.: Comparison of quantum oracles. Phys. Rev. A **65**(5), 050304 (2002)
17. Mossayebi, S., Schack, R.: Concrete security against adversaries with quantum superposition access to encryption and decryption oracles. arXiv preprint arXiv:1609.03780 (2016)
18. Nielsen, M.A., Chuang, I.L.: Quantum Computation and Quantum Information: 10th Anniversary Edition, 10th edn. Cambridge University Press, Cambridge (2011)
19. Peikert, C., Waters, B.: Lossy trapdoor functions and their applications. In: 40th ACM STOC, pp. 187–196. ACM Press, May 2008
20. Shor, P.W.: Polynomial-time algorithms for prime factorization and discrete logarithms on a quantum computer. SIAM Rev. **41**(2), 303–332 (1999)
21. Simon, D.R.: On the power of quantum computation. In: 35th FOCS, pp. 116–123. IEEE Computer Society Press, November 1994
22. Zhandry, M.: How to construct quantum random functions. In: 53rd FOCS, pp. 679–687. IEEE Computer Society Press, October 2012
23. Zhandry, M.: How to record quantum queries, and applications to quantum indifferentiability. In: Boldyreva, A., Micciancio, D. (eds.) CRYPTO 2019. LNCS, vol. 11693, pp. 239–268. Springer, Cham (2019). https://doi.org/10.1007/978-3-030-26951-7_9

Post Quantum Cryptography

A One-Time Single-bit Fault Leaks All Previous NTRU-HRSS Session Keys to a Chosen-Ciphertext Attack

Daniel J. Bernstein[1,2](✉)

[1] Department of Computer Science, University of Illinois at Chicago, Chicago, USA
[2] Horst Görtz Institute for IT Security, Ruhr University Bochum, Bochum, Germany
djb@cr.yp.to

Abstract. This paper presents an efficient attack that, in the standard IND-CCA2 attack model plus a one-time single-bit fault, recovers the NTRU-HRSS session key. This type of fault is expected to occur for many users through natural DRAM bit flips. In a multi-target IND-CCA2 attack model plus a one-time single-bit fault, the attack recovers every NTRU-HRSS session key that was encapsulated to the targeted public key before the fault. Software carrying out the full multi-target attack, using a simulated fault, is provided for verification. This paper also explains how a change in NTRU-HRSS in 2019 enabled this attack.

Keywords: Chosen-ciphertext attacks · Natural faults · Implicit rejection

1 Introduction

In 2016, the call for submissions for the NIST Post-Quantum Cryptography Standardization Project [78] said that NIST intends to standardize "one or more schemes that enable existentially unforgeable digital signatures with respect to an adaptive chosen message attack" and "one or more schemes that enable 'semantically secure' encryption or key encapsulation with respect to adaptive chosen ciphertext attack"—in other words, signature systems providing EUF-CMA security, and PKEs or KEMs providing IND-CCA2 security.

The EUF-CMA game allows the attacker to call an oracle that signs arbitrary messages; the only restriction is that the attacker does not win the game if

This work was funded by the Intel Crypto Frontiers Research Center; by the Deutsche Forschungsgemeinschaft (DFG, German Research Foundation) as part of the Excellence Strategy of the German Federal and State Governments—EXC 2092 CASA—390781972 "Cyber Security in the Age of Large-Scale Adversaries"; by the U.S. National Science Foundation under grant 1913167; by the Taiwan's Executive Yuan Data Safety and Talent Cultivation Project (AS-KPQ-109-DSTCP); and by the Cisco University Research Program. "Any opinions, findings, and conclusions or recommendations expressed in this material are those of the author(s) and do not necessarily reflect the views of the National Science Foundation" (or other funding agencies). Permanent ID of this document: 662cf4ad8f5bff33ae4d71d56051a656d8a62e48. Date: 2022.10.24.

T. Isobe and S. Sarkar (Eds.): INDOCRYPT 2022, LNCS 13774, pp. 617–643, 2022.
https://doi.org/10.1007/978-3-031-22912-1_27

the attacker's forged message was specifically provided as input to the oracle. Similarly, the IND-CCA2 game for KEMs allows the attacker to call an oracle that decapsulates arbitrary messages, although the attacker does not win the game if the oracle was used specifically to decapsulate the target message.

An application providing such powerful oracles is thoroughly flawed and should not be used. But applications do sign and decapsulate *some* messages, providing *some* of the same information. Aiming merely for security in the absence of such oracles would then be a disaster, as illustrated by Bleichenbacher's million-message attack [24], which was demonstrated against real HTTPS servers and played an important role in ensuring attention to chosen-ciphertext attacks. See also [25] demonstrating continued exploitability of essentially the same attack against some servers two decades later.

Sometimes the literature suggests that it suffices to aim for security against *the oracles provided by applications.*[1] But this would be an evaluation nightmare. One would have to check all the different ways that applications handle signatures and decapsulations, consider how this can change in the future, and then evaluate whether a cryptographic system is secure in all of these contexts. So the community asks for EUF-CMA signature systems and for IND-CCA2 KEMs, rather than for something weaker.[2]

The literature often presents a simpler justification for stronger security models: namely, the blanket statement that it is always better (e.g., "more conservative") to ask for security in stronger models.[3] This blanket statement goes far beyond saying that it is better to ask for IND-CCA2 than for IND-CPA: it also implies that any proposal to replace IND-CCA2 with stronger model M_1 should be accepted, and then any proposal to replace M_1 with a stronger model M_2 should be accepted, and so on. This is its own form of evaluation nightmare.

The critical question to ask is how to manage the risk of real-world security failures so as to best protect real users from attack. The answer cannot be to devote more and more security-analysis resources to more and more obscure risks: time taken chasing a neverending series of academic targets is time taken away from ensuring more important security properties. This does not imply, however, that the right answer is to stop with EUF-CMA and IND-CCA2.

---

[1] See, e.g., [77]: "We conclude that the CNS attack is a concern for the ISO 9796-2 signature scheme with partial message recovery in environments where the attacker is capable of obtaining the signatures of a significant number (e.g., one million) of chosen messages. In environments where the attacker is not capable of obtaining these signatures, the CNS attack is not a concern.".

[2] Exception: In the context of protocols that use the cryptosystem key just once, such as the SIGMA approach to secure sessions, the literature often encourages targeting merely IND-CPA. See [60] for a recent example. On the other hand, it is a mistake from a systems-security perspective to give users (1) a cryptosystem designed for IND-CCA2 and (2) a non-IND-CCA2 cryptosystem designed merely for IND-CPA. As [71] put it: "CPA vs CCA security is a subtle and dangerous distinction, and if we're going to invest in a post-quantum primitive, better it not be fragile.".

[3] Occasionally exceptions are made for security notions *proven* to be unachievable.

1.1. Fragility. Beyond EUF-CMA security and IND-CCA2 security, NIST's call for submissions said that "additional security properties ... would be desirable". Let's focus on the last item in NIST's list:

> A final desirable, although ill-defined, property is resistance to misuse. Schemes should ideally not fail catastrophically due to isolated coding errors, random number generator malfunctions, nonce reuse, keypair reuse (for ephemeral-only encryption/key establishment) etc.

In 2018, a catastrophic failure was reported in Dilithium because of an isolated coding error in the official Dilithium software. Specifically, the software generated random values incorrectly, reusing randomness at a place where the specification instead generated new randomness; [75] announced that this "reuse of randomness can easily be exploited to recover the secret key". Evidently Dilithium fails to provide "resistance to misuse".

On the other hand, it is difficult to imagine how *any* scheme could prevent "isolated coding errors" from causing catastrophic failures,[4] never mind all the other forms of potential "misuse". Did NIST have some reason to think that "resistance to misuse" could be achieved?

Perhaps the intent was not to ask the yes-no question of whether one can construct a misuse scenario, but rather the tricky risk-assessment question of *how likely* it is for people to make mistakes that will cause a scheme to fail. It could be that other cryptographic systems are *more* failure-prone than Dilithium, and that the official Dilithium software was simply unlucky.

It is not easy to evaluate such a complicated, open-ended security "property". The lack of a clear definition violates the following Katz–Lindell [68] statement: "One of the key intellectual contributions of modern cryptography has been the realization that formal definitions of security are *essential* prerequisites for the design, usage, or study of any cryptographic primitive or protocol." It is also easy to see how an attacker can use this "property" as a tool to attack cryptosystem-selection processes, promoting weaker cryptosystems by selectively objecting to stronger cryptosystems.[5]

---

[4] A standard could insist that implementors take a majority vote of three independent implementations, but experience shows that there are correlations among errors from different implementors. Furthermore, a coding error could replace the majority vote with taking just the result of the first implementation, or an implementor could "misuse" the scheme by taking just one implementation; either way, a coding error in that implementation could cause disaster even if other implementations are perfect.

[5] In its latest report [2], NIST criticized Classic McEliece for a "misuse scenario" where "reusing the same error vector when encapsulating for multiple public keys" would damage security—even though (1) there have been no examples of this scenario occurring for Classic McEliece, (2) the official Classic McEliece software has always used RNGs correctly, and (3) no encapsulation mechanism is safe against external RNG failures. Meanwhile none of NIST's reports criticized Dilithium for the "misuse scenario" of reusing randomness inside a single signature—even though (1) this scenario occurred in the official Dilithium software, (2) this destroyed the security of that software, and (3) the problem was in that software, not in an external RNG.

The literature nevertheless provides clear reasons to believe that some cryptographic systems are more failure-prone than others. For example, for ECDH systems that transmit curve points in affine coordinates (x, y), there are endless reports (e.g., [21]) of implementations that fail to check whether the incoming point is on the curve, and that are easily breakable as a result. This attack is structurally eliminated by ECDH systems that (as in [8] and [9]) choose twist-secure curves and transmit merely x.

Presumably there are also ways to adjust post-quantum design decisions to reduce the chance of implementation failures. It is important to keep in mind here that there is far less evidence available today regarding post-quantum implementation failures than regarding pre-quantum implementation failures, and the general difficulty of evaluating implementation security means that claims of security improvements need to be investigated carefully before they are used for making decisions. This is not a reason to avoid study of the topic.

1.2. Natural DRAM Faults. In 2009, Schroeder, Pinheiro, and Weber [93] reported the results of a 2006–2008 study of failure rates in the DRAM in "the majority of machines in Google's fleet". The observed failure rates were "25,000 to 70,000 errors per billion device hours per Mbit".

Conventional SECDED ECC DRAM encodes 64 bits of logical data in 72 bits of physical DRAM, using a distance-4 linear error-correcting code.[6] "SECDED" here means "single-error correcting, double-error detecting", and "ECC" means "error-correcting code". In particular, SECDED ECC DRAM corrects any single bit flip, while reporting the correction to the operating system. Some computer buyers make sure to buy SECDED ECC DRAM; this is also how the study from [93] collected data.

However, most computing devices today simply store 64 bits of logical data in 64 bits of physical DRAM. Any single physical bit flip is then a logical bit flip, directly corrupting data, with no warning to the user. For example, flipping a single bit in DRAM can silently convert the ASCII letters "NTRU" to "NTRW".

Consider a reasonably popular cryptosystem that, worldwide, has a billion active 256-bit keys stored in DRAM without SECDED. An extrapolation from the error rates reported in [93] suggests that between 50000 and 140000 of those keys will have a bit flipped each year.[7] This is frequent enough to mandate investigation of the security consequences.

---

[6] This 12.5% overhead is not the best that can be done. The overhead of a distance-4 error-correcting code, such as an extended Hamming code, drops as the dimension increases. DRAM today is normally accessed in 512-bit blocks ("lines"), larger than the 64-bit blocks conventionally used for SECDED. A 512-bit line encoded as 528 bits can be stored as 16 bits on each chip in a 33-chip module, which in principle should cost just 3.125% more than a 32-chip module; and 523 bits are enough to encode 512 bits with SECDED, as noted in, e.g., [104].

[7] Presumably this is an underestimate of the error rate: one would not expect average user devices to be as reliable as Google's air-conditioned, systematically monitored, frequently replaced servers.

1.3. Contributions of This Paper. This paper shows that, in the IND-CCA2 attack model augmented to include a one-time flip of one bit stored by the legitimate user, NTRU-HRSS is devastatingly insecure: there is an efficient attack that recovers the NTRU-HRSS session key. In a multi-target IND-CCA2 attack model similarly augmented to include a one-time single-bit fault, the same attack efficiently recovers all of the NTRU-HRSS session keys that were encapsulated to the targeted public key before the fault.

Section 4.2 presents the full multi-target attack. For verification, as a supplement to this paper, attack software is provided that carries out the multi-target attack against the official NTRU-HRSS software, using a simulation of the required fault. See Sect. 2 for a comparison to previous fault attacks.

Section 4.3 formulates analogous fault attacks against Streamlined NTRU Prime and Classic McEliece, and explains why both of those attacks are blocked by plaintext confirmation, a CCA defense already built into the CCA conversions inside those cryptosystems. (This should not be interpreted as a claim that Streamlined NTRU Prime and Classic McEliece are immune to *all* fault attacks.) See Sect. 3 for a survey of chosen-ciphertext attacks and defenses.

Interestingly, NTRU-HRSS had included the same CCA defense in its original design, but then *removed the defense on the basis of papers claiming to have proven that the defense was not necessary.* See Sect. 4.4. Those papers were considering a more limited attack model.

2 Fault Attacks

This section explains how this paper's fault attack fits into the broader literature on fault attacks.

A fault is like a software bug or a hardware bug in that it complicates analyses of computer behavior: it violates the implicit assumption that each computation is being carried out correctly. As a further complication, a fault is like a physical side channel in that it comes from physical effects whose boundaries are hard to formalize and analyze. Even if a system is secure in the absence of faults, the attacker can hope that the system becomes breakable when faults occur.

2.1. A Generic Fault Attack. If one wants to skip the complications of analyzing physical effects—or if one believes the blanket statement that it is better to ask for security in stronger models; see Sect. 1—then one might hypothesize that the attacker has the power to induce arbitrary faults in computations. Under this hypothesis, the following generic fault attack extracts the internal secrets from *any* computation whose output is visible to the attacker.

View the computation as an unrolled circuit consisting of NAND gates, and consider a NAND gate $a, b \mapsto 1 - ab$ producing output at the end of the computation. If the output is 0 then $a = b = 1$. Otherwise the attack deduces

a, b by re-running the computation with a bit-flip fault on a and then with a bit-flip fault on b. The attack now knows the inputs to the NAND gate.

The attack then targets the inputs to an earlier NAND gate that produced a, while using a set-to-1 fault to force $b = 1$ so that changes in a are visible as changes in the output $1 - ab$. Set-to-1 faults can also be used in place of the bit-flip faults in the previous paragraph.

The attack proceeds upwards in the same way through each gate to extract the entire internal state of the computation. The number of runs of the computation is $\Theta(n)$ where n is the circuit size. Each run uses $O(d)$ faults to ensure that the targeted bit is visible in the output, where $d \leq n$ is the circuit depth.

Internal checks in the computation, such as verifying signatures before releasing them, do nothing to stop this attack: checks are just like any other computation in succumbing to faults. Randomizing the computation simply requires the attacker to apply further faults to zero the randomness. Destroying the device after 1000 computations requires keeping track of the number of computations; the attacker can apply faults to zero that number. Destroying the device after *one* computation does not need a counter but still requires triggering a self-destruct mechanism; the attacker can apply faults to clear the trigger.

2.2. Specializing, Optimizing, and Demonstrating the Generic Fault Attack. A typical fault attack in the literature can be viewed as (1) specializing the generic attack from Sect. 2.1 to a particular target and (2) optimizing the specialized attack so that the attacker does not need to induce as many faults. The resulting attacks vary in how many faults they use and in how precisely targeted those faults are.

Sometimes fault-attack papers include real-world demonstrations that one can produce the necessary faults by, e.g., heating a circuit, firing lasers at the circuit, etc.; see, e.g., [34]. Sometimes faults can be induced by software; see, e.g., [94].

For most attacks, one cannot reasonably expect the requisite faults to occur naturally. One can try to stop these attacks by cutting off data flow that the attacker might be able to use to induce faults in the legitimate user's computation. This includes keeping the attacker physically away from the device, and constraining software behavior so as to avoid faults.

2.3. Natural-Fault Attacks. Occasionally a fault attack relies on such a small number of faults that one *can* expect naturally occurring physical effects to produce the requisite faults. Eliminating the attacker's ability to induce faults does nothing to stop an attack of this type. The classic example, pointed out by Boneh, Demillo, and Lipton [27], is as follows.[8]

[8] As a different example of using just one fault, consider the IND-CCA2 game for KEMs. The attacker is free to send a ciphertext with one bit flipped, and to inspect the resulting session key; now simply hypothesize that a fault flips the bit back at the beginning of decapsulation. One reason that this is a less satisfactory example than [27] is that it requires a specific fault to occur during a narrow window of time, while a fault in a stored secret key at any moment—something more likely to occur naturally—opens up the attack of [27].

The job of an RSA signer is to compute an eth root s of h modulo pq, where (pq, e) is the public key and h is a hash of the message being signed. This is the same as computing $s = h^d \bmod pq$ for a suitable decryption exponent d. "RSA-CRT", the usual speed-oriented choice of RSA signature algorithm, computes an eth root s_p of h modulo p as $h^{d_p} \bmod p$ where $d_p = d \bmod (p-1)$, computes an eth root s_q of h modulo q, and combines s_p with s_q to obtain s.

Now say the signer signs the same message again, but this time there is a fault in the computation of s_p—anything that changes the output; e.g., a bit flip in d_p. The resulting signature S will then be the same as the correct signature s modulo q but not modulo p, and the attacker can compute q as $\gcd\{S - s, pq\}$.

A variant by Lenstra [72] is to compute q as $\gcd\{S^e - h, pq\}$. This variant assumes that the attacker also sees the message m being signed, but avoids the need for multiple signatures of m, so the attack works with passive observation of objects that are normally sent in the clear, namely messages and signatures.

One of the countermeasures suggested in [27] is to check signatures before releasing them. In real-world RSA, the exponent e is chosen to be small, so the check adds very little to the cost of signing. But typical RSA descriptions do not include this check, and typical tests of RSA software do not detect the check, so it is easy to imagine RSA software being deployed without the check.

Sullivan–Sippe–Heninger–Wustrow [96] announced in 2022 that they had exploited faults to extract "private RSA keys associated with a top-10 Alexa site" and "browser-trusted wildcard certificates for organizations that used a popular VPN product". [96, Section 5.3] found some hosts producing bad signatures for months, suggesting that faults "are persistent: disk corruption or memory corruption affecting the private key." Other faults were transient; perhaps a secret key was copied from disk to DRAM, then a bit flipped in DRAM, and then the same DRAM was reused for other data, wiping out the flipped bit. On the other hand, [96] reported unsuccessfully trying some possibilities for flipped bits. Another hypothesis noted in [96] is "failing hardware".

2.4. Algorithm Dependence in Natural-Fault Attacks. At the time of [27], the primary RSA specification was PKCS #1 v1.5, released in 1993. Secret keys were specified to have the following components (see [66, Section 7.2]): the public key n; the encryption exponent e; the decryption exponent d; the secret primes p and q; the integers d_p and d_q; and the inverse of q modulo p. There are many ways to double-check these secret keys so as to detect flipped bits: check whether n matches pq, check whether d_p matches $d \bmod (p-1)$, check whether ed is 1 modulo $p-1$, etc. With more work one can *correct* flipped bits (and also correct any errors that might occur inside the signing computation).

Consequently, the fault attack from [27] was an attack against *some* algorithms computing the specified signing function. Stopping the attack required changing commonly used algorithms (for example, to check signatures as mentioned above), but did not require a new specification of the signing function,[9] new test vectors, etc.

[9] Perhaps the signing function could have been changed to reduce the chance of problems—see Sect. 1.1—but this is a separate issue.

As another example of algorithm dependence in natural-fault attacks, consider the following three versions of the Ed25519 signature system:

- In standard Ed25519 (see [63]), the secret key is a 32-byte string that is hashed to obtain (1) a secret scalar and (2) another secret that is hashed together with the message to obtain a nonce. Any bit flip in the stored secret key will produce completely different hash output, leading to garbage signatures of no evident value for the attacker.
- In the most commonly used variant of Ed25519, the secret key is 64 bytes: the same 32-byte string as above, plus a copy of the 32-byte public key. With the simplest signing algorithm, a fault in these 64 bytes will leak the secret key. This is an algorithm-dependent attack; a signing algorithm that double-checks the secret scalar against the public key will detect the fault.
- A more efficient fault-attack countermeasure incorporates another 32 bytes of randomness into the input to the hash producing the nonce, without the cost of checking the public key. This variant of Ed25519 was considered in, e.g., [11] and (as a fault-attack countermeasure) [85, Section 8].

To summarize, the availability of fault attacks is sensitive to details of (1) the cryptosystem at hand and (2) the algorithms used for that cryptosystem.

2.5. Comparison. Like the attack from [27] against RSA-CRT, this paper's attack against NTRU-HRSS works if a single bit is flipped in a stored secret key, an event that will occur naturally for some users.

Unlike the attack from [27], this paper's attack has the further feature of being algorithm-independent: it works against *any* algorithm that computes the specified function of the secret key. The NTRU-HRSS secret key does not contain any data that a decapsulation algorithm can use to detect the fault exploited in this paper's attack. This paper's attack against NTRU-HRSS is thus a decapsulation-algorithm-independent natural-fault attack.

A disadvantage of this paper's attack (compared to the attack from [27] with the improvement of [72]) is that it is active. The attack takes full advantage of the flexibility of the attack model: for each target ciphertext, the attack sends some modified versions of the ciphertext before and after the fault occurs, and sees some information about the resulting session keys. Hopefully the application does not actually provide so much flexibility to the attacker. On the other hand, the rationale for asking for IND-CCA2 security (see Sect. 1), rather than investigating whether IND-CCA2 security is overkill for applications, applies with equal force when one extends the IND-CCA2 attack model to include a natural fault. It is interesting that the IND-CCA2 security of NTRU-HRSS is so fragile in the presence of natural faults.

Another disadvantage of this paper's attack is that it is recovering only session keys, not Alice's secret key. On the other hand, the reason an attacker wants to recover Alice's secret key is to be able to recover all session keys; this attack recovers all session keys that were communicated before the fault.

2.6. The Cold-Boot Argument Against Error Correction. The literature on cold-boot DRAM attacks often uses redundancy in stored data to correct flipped bits; see, e.g., [49, Section 5]. This is occasionally used as an argument that secret data should be stored in maximally compressed format; see, e.g., [49, Section 8, "suggested countermeasures", including "avoiding precomputation"]. The same argument implies that users should *not* include redundancy in data to detect and correct errors, and in particular should *not* use SECDED ECC DRAM; [49, Section 3.4] says "ECC memory could turn out to *help* the attacker".

However, users who avoid SECDED ECC DRAM are exposed to a large class of hard-to-analyze correctness risks and security risks that they would otherwise have avoided. Meanwhile it is clear that well-executed cold-boot DRAM attacks rarely encounter errors in the first place (see, e.g., [49, Table 2, "no errors" entries]) and are thus *not* stopped by attempts to avoid redundancy.

Encrypting DRAM, using a key stored in better-protected hardware, is a simpler and much more convincing defense to cold-boot DRAM attacks. Encrypting DRAM is also compatible with SECDED ECC DRAM and other protections against faults.

3 Chosen-Ciphertext Attacks and Defenses

This section surveys the general structure of chosen-ciphertext attacks against code-based and lattice-based systems, and of various cryptosystem features that *seem to* interfere with these attacks. Beware that the literature often overstates the extent to which (some of) these features are *known* to interfere with these attacks; see Sect. 4.4.

The round-3 versions of NTRU-HRSS, Streamlined NTRU Prime, and Classic McEliece are used as running examples, abbreviated `ntruhrss`, `sntrup`, and `mceliece` respectively. Table 3.1 summarizes the features of these cryptosystems.

3.2. Ciphertext Structure. Throughout this section, Bob's ciphertext has the form $B = bG + d$, where G is Alice's public key and b, d are secrets, in particular with d chosen to be small. The choice of letters here is as in [12, Section 8], unifying notation between ECDH, "noisy DH" lattice-based and code-based systems, and further lattice-based and code-based systems.

The `mceliece` description uses an optimized ciphertext structure due to Niederreiter: simply He, where H is the public key and e is small. However, H internally consists of two parts, an identity matrix and another matrix Q, so He can be written as $e_1 + Qe_2$. This is, modulo transposition and relabeling, again a ciphertext of the form $bG + d$.

3.3. Decryption. Alice uses her private key to recover b and d. Let's assume at the outset that this recovery process is labeled as a PKE returning plaintext (b, d).

The original McEliece system [76] instead viewed b as the plaintext—not required to be small—and d as something chosen randomly in encryption. The original NTRU system [52] instead viewed d as the plaintext and b as something

Table 3.1. Cryptosystem features that *seem to* (but do not necessarily) interfere with chosen-ciphertext attacks. The `mceliece`, `sntrup`, and `ntruhrss` columns indicate whether the features appear in Classic McEliece, Streamlined NTRU Prime, and NTRU-HRSS respectively. All entries are for the round-3 versions of `mceliece`, `sntrup`, and `ntruhrss`; implicit rejection appeared in `sntrup` and `ntruhrss` in 2019, while plaintext confirmation was removed from `ntruhrss` in 2019.

feature (see main body for definitions)	`mceliece`	`sntrup`	`ntruhrss`
hashing the plaintext	yes	yes	yes
rigidity	yes	yes	yes
no decryption failures	yes	yes	yes
plaintext confirmation	yes	yes	no
implicit rejection	yes	yes	yes
hashing the ciphertext	yes	yes	no
limited ciphertext space beyond small plaintext	no	yes	no
limited plaintext space beyond small plaintext	no	no	no
no derandomization	yes	yes	yes

chosen randomly in encryption. A 1996 NTRU handout [53, Section 4.2] had also considered a deterministic PKE with (b, d) as the plaintext—although this handout was not put online until 2016, after deterministic NTRU PKEs had already been recommended in, e.g., [10].

Linear algebra easily recovers b from bG (assuming G is public and injective), but recovering b from a noisy multiple $bG+d$ is conjectured to be hard (for appropriate choices of parameters). This conjecture is often described as conjectured hardness of the "LPN", "LWE", "Ring-LPN", "Ring-LWE", "Module-LPN", or "Module-LWE" problems (again for appropriate choices of parameters), where the choice of name depends on various details of the algebraic structure containing G. These problems, in turn, are typically claimed to have been introduced in various 21st-century papers. However, the original McEliece [76] and NTRU [52, Section 3] papers had already analyzed the cost of various algorithms for the cases of LPN and Ring-LWE that matter for those cryptosystems, so it is wrong to credit those problems to subsequent papers. There is some value in generalizing the problems (for example, to study other cryptosystems), but credit for the general problems has to include credit to the cases considered earlier.

The rest of this paper ignores the possibility of recovering (b, d) purely from $(G, bG + d)$, and instead focuses on the extra power of chosen-ciphertext attacks.

3.4. Exploiting Linearity for Chosen-Ciphertext Attacks. Given the linear structure of a ciphertext $B = bG+d$ and the definition of IND-CCA2 security,[10] the obvious attack sends a modified $B' = B + \delta = bG + d + \delta$ for some small nonzero δ. The attacker hopes that the decryption process successfully returns

[10] Beware that there are several slightly different definitions of IND-CCA2 security for PKEs. See generally [7].

$(b, d + \delta)$, at which point the attacker simply subtracts δ and wins. The attacker chooses δ to be small because decryption does not work for arbitrarily large d.

For example, a `mceliece` decoder requires (b, d) to have a specific Hamming weight. The attacker chooses a random weight-2 vector δ, a vector of the form $(0, \ldots, 0, 1, 0, \ldots, 0, 1, 0, \ldots, 0)$. There is then a good chance that $(b, d + \delta)$ has the right weight, meaning that decryption returns $(b, d + \delta)$. Various features described below are included in `mceliece` to stop this attack.

In the same example, the attacker can, more generally, choose $B' = B + \beta G + \delta$ where β, δ have total weight 2. To simplify notation, the comments below focus mainly on $B' = B + \delta$, but similar comments apply to $B' = B + \beta G + \delta$.

3.5. Feature 0: Hashing the Plaintext. As a preliminary step in limiting the information provided to chosen-ciphertext attacks, let's switch from a PKE to a KEM that hashes the plaintext.

Specifically, let's define encapsulation to choose the input (b, d) randomly (not necessarily uniformly; other distributions can be more convenient), and let's define decapsulation to return a hash $H(b, d)$. The attacker sending $B + \delta$ and receiving $H(b, d + \delta)$ has no obvious way to reconstruct or otherwise recognize $H(b, d)$, unless the hash function H is remarkably weak.

Let's assume from now on that the goal is to build a KEM that resists chosen-ciphertext attacks. This was the target for most encryption submissions to the NIST Post-Quantum Cryptography Standardization Project, and in particular is the target for `ntruhrss`, `sntrup`, and `mceliece`. Internally, each of these KEMs is built from a PKE that produces ciphertext $bG + d$ and recovers (b, d), or something equivalent to (b, d), during decryption.

Generic transformations convert any KEM into various other cryptographic objects. For example, in the paper [95] that introduced the KEM abstraction (and specifically KEMs that hash the plaintext), Shoup built a PKE handling variable-length user messages by using a KEM to encapsulate a session key and then using symmetric cryptography to encrypt user data under that key.

There are arguments against using KEMs. For example, the literature explains how to build a variable-length PKE with smaller ciphertexts by encoding some of the user data inside the input to a fixed-length PKE: in particular, encoding some user data inside (b, d). The usual approach is to take some randomness and some user data, apply an "all-or-nothing transform" (see generally [91]), and encode the result as (b, d); decryption reverses these steps. However, the space savings seems less important than the simplification of independently analyzing a KEM layer. All-or-nothing transforms might still be useful inside KEM designs; see Sect. 3.14 below.

3.6. Probing the Boundaries of Successful Decryption. Hashing by itself does not stop chosen-ciphertext attacks. The main issue is that the attacker sending $B + \delta$ does not always receive a hash of $(b, d + \delta)$. Sometimes $d + \delta$ is too large to be decoded successfully, and then decapsulation returns a failure report instead of a hash.

The pattern of successes and failures is valuable information for the attacker. For example, consider again the `mceliece` decoder, which works exactly when (b, d) has a specific weight. If adding $\delta = (0, \ldots, 0, 1, 0, \ldots, 0, 1, 0, \ldots, 0) \in \mathbb{F}_2^n$ to d preserves weight then exactly one of the two 1 positions must match a position set in d. Seeing enough such δ quickly reveals all of the positions in d. One can try to accelerate this by using each failing δ as a statistical indication that both 1 positions are likely to be unset in d, but the attack works quickly in any case.

This attack against the original McEliece system was introduced by Hall, Goldberg, and Schneier in [50] and by Verheul, Doumen, and van Tilborg in [101] (which says it was submitted in 1998, before [50] appeared). To be more precise, this is essentially the attack in [101, Section 4]; the attacks in [50, Section 2] and [101, Section 3] are variants that assume that the decoder works when d has *at most* a specific weight.

As another example, `ntruhrss` chooses $d \in \mathbb{Z}[x]/(x^n - 1)$ as $x - 1$ times a polynomial T with coefficients in $\{-1, 0, 1\}$, and checks the same condition during decapsulation. Adding $\delta = 2(x-1)$ changes T to $T+2$, which works when the constant coefficient of T is -1 and seems very unlikely to work otherwise; adding $\delta = -2(x - 1)$ works when the constant coefficient of T is 1; adding $\delta = 2x(x - 1)$ works when the next coefficient of T is -1; etc.

3.7. Probing as an Attack Against the Secret Key. Failure patterns have further consequences for PKEs that are not *rigid*. Non-rigidity means that the specified decryption function can successfully decrypt multiple ciphertexts to the same plaintext.

For example, recall that the original NTRU system has just d as a plaintext, with b chosen randomly in encryption. A closer look at the system reveals that decrypting $B + \beta G$ for small β has a good chance of producing d—there are multiple ciphertexts that produce the same plaintext—and then the resulting session key is exactly the legitimate user's session key, breaking IND-CCA2.

Even worse, the pattern of successes and failures for small β reveals the secret key. Here the attacker does not need to see any information about the session keys except for knowing which $B + \beta G$ succeeded and which failed. This paper suppresses details of this attack, aside from noting that it is easiest for the attacker to begin with a known (b, d). Attacks of this type against NTRU were published by Hoffstein–Silverman [54] and Jaulmes–Joux [62]; variants include [45], [38], [40], [17], [4], [39], [59], [81], [22], [88], [105], and [89].

An analogous problem occurs for PKEs that have *decryption failures*, meaning that the specified decryption function will sometimes fail to decrypt a legitimate ciphertext to the original plaintext. For example, the original NTRU system had a noticeable frequency of decryption failures, and this was exploited by Howgrave-Graham, Nguyen, Pointcheval, Proos, Silverman, Singer, and Whyte [56] to recover the secret key.

3.8. Feature 1: Rigidity. The first step in limiting the power of probing is to choose a rigid PKE, so that multiple ciphertexts cannot produce the same plaintext. It is easy to convert any deterministic PKE into a rigid PKE by modifying decryption to reencrypt the plaintext and to check the result against the ciphertext. This is the Fujisaki–Okamoto [46] transform in the case of deterministic PKEs.

All of `ntruhrss`, `sntrup`, and `mceliece` are designed as rigid PKEs starting from deterministic PKEs, although not always with an obvious step of reencrypting via the encryption procedure:

- Simple facts about error-correcting codes are used inside `mceliece` to accelerate the reencryption procedure. The resulting algorithm uses, asymptotically, an essentially linear number of operations, and avoids storage of the public key inside the private key.[11]
- For `ntruhrss`, the reencryption procedure is optimized to share a multiplication with the original decryption algorithm.
- For `sntrup`, the original decryption algorithm automatically avoids the analogous multiplication (since d is chosen by rounding), and reencryption simply calls the same procedure as encryption.

What matters for this feature is not whether there is a visible reencryption step, but whether the resulting PKE is rigid; this is why Table 3.1 lists "rigidity" rather than "reencryption".

3.9. Feature 2: No Decryption Failures. The second step in limiting the power of probing is to choose a PKE where the specified decryption function always recovers the original plaintext from the corresponding ciphertext. There are no decryption failures in `ntruhrss`, `sntrup`, and `mceliece`.

Note that "no decryption failures" refers to decryption failures for ciphertexts obtained from the encryption algorithm. Decryption can still fail for other ciphertexts created by the attacker.

If a rigid PKE has no decryption failures then it decrypts exactly the ciphertexts $bG + d$ for a key-independent set of valid plaintexts (b, d). An attacker replacing $B = bG + d$ with $B' = B + \beta G + \delta$ will obtain a valid ciphertext if $(b+\beta, d+\delta)$ is in the same key-independent set, and presumably will not obtain a valid ciphertext if $(b+\beta, d+\delta)$ is not in this key-independent set. Otherwise some valid $(b', d') \neq (b+\beta, d+\delta)$ has $B' = b'G+d'$, i.e., $(b+\beta-b')G+(d+\delta-d') = 0$; but it is supposed to be hard for the attacker to find small nonzero s, t such that $sG + t = 0$. Taking large β or large δ seems even less useful.

In short, there is no obvious way for the attacker to find (β, δ) where failures will provide any information about the secret key. In the absence of such information, the secret key is protected against the attack of Sect. 3.7.

However, the attacker can still target the legitimate user's plaintext (b, d) via the attack from Sect. 3.6. This is addressed in Sect. 3.10.

[11] See generally [14, Section 8]. Even better, the usual decoding algorithm inside `mceliece` is shown in [14, Section 7] to be rigid even without reencryption. However, [14, Section 8.4] recommends reencryption for robustness.

3.10. Feature 3: Plaintext Confirmation. The third step in limiting the power of probing is to replace the ciphertext B with $(B, H'(b, d))$, where H' is another hash function, and to check $H'(b, d)$ on decryption. This transformation was published by Dent [37, Table 4] and is now known as plaintext confirmation.

The point of plaintext confirmation is to prevent the attacker from modifying a ciphertext for the legitimate user's secret (b, d) into a ciphertext for $(b, d + \delta)$. The attacker can replace B with $B+\delta$, but has no obvious way to replace $H'(b, d)$ with $H'(b, d + \delta)$ without first finding (b, d). If the attacker knew (b, d) then the attacker could compute the session key $H(b, d)$ without bothering to carry out a chosen-ciphertext attack. An attacker can still choose (b, d) and modify the resulting ciphertext to try to attack the secret key, but this is addressed by a rigid PKE without decryption failures; see Sect. 3.9.

Typically H and H' are both chosen as a cryptographic hash function applied to separate input spaces: $H(b, d) = F(1, b, d)$ and $H'(b, d) = F(2, b, d)$. An alternative is to choose H and H' as the left and right halves of the output of a cryptographic hash function: $F(b, d) = (H(b, d), H'(b, d))$. Obviously one must not select H' as H, or as any other function whose outputs reveal the H outputs on the same inputs; see [6] for examples of attacks against real proposals where H and H' were not adequately separated.

3.11. Feature 4: Implicit Rejection. An alternative to plaintext confirmation is "implicit rejection". This means replacing any failure output for a ciphertext B with a string $H(r, B)$, where r is a random string, part of Alice's secret key.

The idea is that replacing the failures with random garbage hides the pattern of successfully modified ciphertexts. The attacker sees $H(b, d+\delta)$ in success cases and $H(r, B + \delta)$ in failure cases, and—without knowing (b, d) in advance—has no way to distinguish these situations.

For comparison, plaintext confirmation stops the attacker's $B + \delta$ from being a valid ciphertext. These features are compatible: one can use implicit rejection to hide the pattern of successes, and use plaintext confirmation to limit the attacker's ability to create a pattern of successes in the first place.

With implicit rejection, care is required to avoid leaking the pattern of failures through timing. A typical approach starts with B, computes (b, d) in constant time along with a bit indicating failure, computes $H(r, B)$, computes $H(b, d)$, and uses the bit to select either $H(r, B)$ or $H(b, d)$ in constant time.

More generally, one can replace any failure output with $R(B)$, where R is a secretly keyed function producing output of the same length as the normal hash outputs $H(b, d)$. Well-studied message-authentication codes are faster than general-purpose hash functions.

Implicit rejection was introduced by Persichetti [86] in the McEliece context, and generalized by Hofheinz–Hövelmanns–Kiltz [55].

3.12. Feature 5: Hashing the Ciphertext. Instead of choosing the session key as $H(b, d)$, one can choose it as $H(b, d, B)$ where B is the ciphertext. If an attacker-chosen $B + \delta$ decrypts to the same (b, d) then the resulting session key $H(b, d, B + \delta)$ will be different from $H(b, d, B)$.

This extra hash input hides any collisions produced by decryption. For comparison, reencryption creates rigidity, preventing any collisions from appearing in the first place. These features are compatible. Note the analogy to implicit rejection hiding the pattern of successfully modified ciphertexts while plaintext confirmation eliminates those ciphertexts.

For implementors, a convenient feature of using $H(b, d, B)$ for a valid session key and $H(r, B)$ for implicit rejection is that one can easily merge the hash calls if r has the same length as (b, d). Security analysis is slightly easier if a valid session key uses $H(1, b, d, B)$ and implicit rejection uses $H(0, r, B)$; this still allows the same merging.

Hofheinz–Hövelmanns–Kiltz [55] observed that ciphertext hashing changed what they could prove regarding security. See [19, Appendix A.5] for an example of a broken cryptosystem that seems to be rescued by ciphertext hashing.

3.13. Feature 6: Limited Ciphertext Space. Another way to reduce the attacker's ability to modify ciphertexts is to force legitimate ciphertexts $bG + d$ to be in a constrained set checked by Alice.

For example, `sntrup` chooses b randomly, and then rounds each entry of bG to the nearest multiple of 3 to obtain $B = bG + d$; each entry of d is -1 or 0 or 1. The ciphertext format enforces the multiple-of-3 rule, so an attacker's modified ciphertexts also have to follow this rule.

An advantage of constraining ciphertexts via the ciphertext format is that this constraint does not rely on Alice's decapsulation algorithm. This does not mean that the constraint is as effective as other defenses. Typically such ciphertext constraints are presented as a way to reduce the use of randomness and reduce ciphertext sizes; there is very little cryptanalytic literature considering the extent to which these constraints interfere with chosen-ciphertext attacks.[12]

Here is another example of constraining the set of ciphertexts. Recall that McEliece's original cryptosystem has ciphertexts $bG + d$ where b is arbitrary and d is small. To constrain $bG + d$ to a linear subspace V, first choose a small d and then find, by linear algebra, b for which $bG + d \in V$. It is easy to select V so that b always exists and is unique, and it is easy to show that these constrained ciphertexts $bG + d$ are equivalent to Niederreiter's ciphertexts.

As noted above, Niederreiter's ciphertexts can also be viewed as having the form $bG + d$, where different variables are now labeled as b, d, G, and where (b, d) is required to be small. One could further constrain $bG + d$ to a limited subspace by choosing b randomly and then finding a small d for which $bG + d$ is in that subspace; this means solving a decoding problem for that subspace.

In Table 3.1, "limited ciphertext space beyond small plaintext" means that $bG + d$ is constrained *beyond* requiring small (b, d), so `mceliece`'s use of

[12] Given recent misinformation regarding rounding, it seems necessary to emphasize that the cryptanalytic question here is whether rounding is *stronger* than adding random errors: this attack avenue *obviously* works against random errors, whereas analysis is required of the extent to which the attack avenue is blocked by rounding. See also [90], which finds that rounding complicates side-channel-assisted chosen-ciphertext attacks.

Niederreiter's ciphertexts does not qualify, whereas further constraining $bG + d$ as in the previous paragraph would qualify.

3.14. Feature 7: Limited Plaintext Space. One last way to reduce the attacker's ability to modify ciphertexts is to limit the space of plaintexts (b, d).

In the standard attacks, the attacker is choosing (β, δ) so that the target plaintext (b, d) has a noticeable chance of $(b + \beta, d + \delta)$ also being a plaintext. Constraining the plaintext space can reduce this chance to something negligible.

Typically there is a reasonably efficient way to compress (b, d) into an s-bit string where the number N of choices of (b, d) is not far below 2^s. Normally N, and therefore 2^s, is much larger than 2^{256}. A standard way to sample from a "structureless" set of s-bit strings is as follows: start with a 256-bit string, zero-pad to s bits, and then apply an all-or-nothing transform.[13] One can then try decompressing the resulting s-bit string to (b, d); if this fails then one can try again with a new 256-bit string. Unless there is some surprising interaction between the all-or-nothing transform and the compression mechanism, each try will succeed with probability approximately $N/2^s$, and one can statistically check this with experiments.

Alice, upon decrypting a ciphertext to obtain (b, d), compresses (b, d) to s bits, inverts the all-or-nothing transform, and checks for the zero-padding. Defining hashes in terms of the 256-bit string instead of (b, d) forces implementations to invert the all-or-nothing transform, although one still has to worry that implementations will skip the zero-padding check.

An alternative way to limit the plaintext space is as follows. Take any algorithm to randomly generate (b, d), and compose it with any cryptographic random-number generator producing the necessary bits of randomness from a 256-bit seed. This is generally hard to invert, but one can transmit, as part of the ciphertext, the seed encrypted under a hash of (b, d).

Because of various patent issues that remain unresolved at the time of this writing,[14] I'm currently limiting time spent investigating Kyber [3] and other cryptosystems in the GAM/LPR family. However, it is interesting to note that this family relies on the seed approach for another reason, namely "derandomization". Care is required here regarding security: my paper [13] gives examples of cryptosystems where derandomization loses about 100 bits of security, and the impact of derandomization on GAM/LPR systems requires cryptanalysis. None of `ntruhrss`, `sntrup`, and `mceliece` have this issue. This is reported in the "no derandomization" line in Table 3.1.

---

[13] Presumably an all-or-nothing transform is overkill here, since most of the structure in the plaintext (b, d) is not easy to see in ciphertexts $bG + d$. It would be interesting to identify the relevant security properties of plaintext sets, and to optimize construction algorithms and recognition algorithms for secure sets.

[14] See, e.g., [2, page 18]: "If the agreements are not executed by the end of 2022, NIST may consider selecting NTRU instead of Kyber." There are also various relevant patents that do not seem to be considered in [2], such as CN107566121A.

4 The NTRU-HRSS Attack

This section presents this paper's attack against `ntruhrss`, and describes an accompanying software package `attackntrw` [16] that successfully carries out the attack against existing `ntruhrss` software with a simulated fault.

This section also presents analogous attacks against `sntrup` and `mceliece`, and explains why these attacks are blocked by the plaintext confirmation built into `sntrup` and `mceliece`. This section continues by reviewing how "provable security" led `ntruhrss` to remove plaintext confirmation, and concludes by evaluating possible countermeasures to protect `ntruhrss`.

4.1. Attack Model. The model considered here is the standard IND-CCA2 attack model for KEMs, plus a one-time bit flip at a uniform random position inside Alice's stored secret key. "One time" means that there is a time at which a bit flips—and then the bit *stays* flipped, not magically returning to its previous value. The attacker can carry out many chosen-ciphertext queries to the original secret key before the bit flip and to the new secret key after the bit flip.

The attack below requires the bit flip to occur within 256 specific bits inside Alice's secret key. This does not occur with probability 1, but it does occur with noticeable probability, namely $256/z$, where the secret key has z bits. In the real world, one expects a fault in these 256 bits to naturally occur for the fraction of users described in Sect. 1.2. Note that padding the secret key, increasing z, would not reduce the number of users affected, although it would reduce $256/z$.

It is easy to see that one can achieve security in this model (unlike the more general fault-attack models reviewed in Sect. 2) if and only if one can achieve standard IND-CCA2 security without faults: simply change the secret-key format to include error correction, for example with a distance-3 Hamming code or a distance-4 extended Hamming code, and apply an error-correcting decoder inside the decapsulation algorithm. However, a KEM that lacks this feature in its secret-key format might be breakable in this model whether or not it is IND-CCA2. The attack below shows that `ntruhrss` is breakable in this model.

The attack is actually stated for multi-target IND-CCA2 (plus a one-time bit flip), but readers not familiar with multi-target IND-CCA2 can freely focus on the case of a single target ciphertext.

A weaker starting attack model than IND-CCA2 would suffice for this attack. What the attack needs to see is simply whether specified pairs of session keys are identical within the attacker-chosen ciphertexts.

4.2. Attack Details. The available `ntruhrss` software supports one parameter set, namely `ntruhrss701`. The following description focuses on `ntruhrss701`.

Eve sees Alice's `ntruhrss701` public key G and any number of legitimate ciphertexts $B_1, B_2, \ldots$. These are elements of the ring $(\mathbb{Z}/8192)[x]/(x^{701} - 1)$, encoded as strings. The `attackntrw` software uses the official `nturhrss701` software[15] to generate a public key and 10 target ciphertexts.

[15] Officially, NTRU-HRSS has three software releases and a development repository. Software release 1, via PQClean, was eliminated by PQClean in July 2022 [67] since

For each j, Eve sends Alice various modified versions (described below) of the legitimate ciphertext B_j, and observes the resulting session keys, as allowed by the (multi-target) IND-CCA2 attack model. To ensure that there is no cheating, the `attackntrw` software carries out decapsulation via an `alice_oracle` function that (1) aborts if the input matches any of $B_1, B_2, \ldots$ and otherwise (2) calls the official `ntruhrss701` software.

A fault then occurs, flipping a bit anywhere inside the implicit-rejection key from Sect. 3.11, the random string r stored inside the secret key.[16] The `attackntrw` software simulates such a fault by flipping the next-to-bottom bit of the last byte of Alice's secret key; this bit happens to be inside r, and flipping any other bit inside r would also work.

Eve then sends the same modified ciphertexts to Alice, observes the resulting session keys, and performs a simple calculation (described below) to extract the secrets b_j, d_j inside each ciphertext $B_j = b_j G + d_j$. The `attackntrw` software performs this calculation and verifies that the session keys computed by the attack match the session keys obtained from the official `ntruhrss701` software.

Eve's modified versions of B_j have the form $B_j + 2(x-1)x^e$ and $B_j - 2(x-1)x^e$ for $0 \le e < 701$, so overall there are 1402 modifications of each ciphertext. One could try to improve this—for example, just 701 modifications would identify about 1/3 of the coefficients of the relevant secret and limit the other 2/3 to just two values, presumably enough information to make a lattice attack feasible—but `attackntrw` is already very fast with 1402 modifications.

The point of these modifications is that, as noted in Sect. 3.6, d_j has the form $(x-1)T_j$ where T_j has coefficients in $\{-1, 0, 1\}$, and the modified ciphertext $B_j \pm 2(x-1)x^e = b_j G + d_j \pm 2(x-1)x^e$ will decrypt successfully[17] when $T_j \pm 2x^e$ has coefficients in $\{-1, 0, 1\}$, i.e., when the coefficient of x^e in T_j is ∓ 1, whereas it cannot be expected to decrypt successfully otherwise.

Without the fault, the pattern of decryption failures would be hidden by implicit rejection. However, with the fault, a decryption failure is immediately visible as a ciphertext producing a different session key before and after the fault: it would be astonishing if changing a bit in r produced a hash collision! Eve sees these mismatches, reconstructs T_j and thus d_j, and follows the relevant steps in

NTRU is "no longer under consideration by NIST", even though, as noted above, [2] says "NIST may consider selecting NTRU instead of Kyber". Software release 2, via BoringSSL, is of the `ntruhrss` variant used in the CECPQ2 post-quantum deployment experiments in Google Chrome; this is "not compatible" with the NTRU-HRSS specification, although the reported reason for this—a different choice of hash function—should not matter for this paper. Software release 3, via the SUPERCOP [18] benchmarking framework, is what `attackntrw` uses.

[16] Faults could also flip other bits of the secret key, or—in a broader model—bits of code, intermediate bits in computations, etc. This paper is analyzing the impact of faults in r; again, this should not be interpreted as making security claims regarding arbitrary fault attacks.

[17] Exception: The multi-target IND-CCA2 attack model will also prevent successful decryption if a modified ciphertext happens to collide with another legitimate ciphertext. However, such collisions are so rare that they can safely be ignored.

the decapsulation algorithm to reconstruct b_j and the corresponding session key, completely breaking `ntruhrss` in this attack model.

To recap: `ntruhrss` relies critically on implicit rejection for (conjecturally) achieving IND-CCA2, but implicit rejection is fragile, losing security when a natural fault occurs.

4.3. How Plaintext Confirmation Stops Analogous `mceliece` and `sntrup` Attacks. A valid `mceliece` ciphertext has the form $B = bG + d$ where (b, d) has a specific Hamming weight. An analogous chosen-ciphertext attack replaces B with $B + \beta G + \delta$, where (β, δ) is chosen by the attacker to have a good chance of having the right Hamming weight of $(b + \beta, d + \delta)$, as in Sect. 3.4. The attacker again detects whether implicit rejection has occurred by checking whether a session key is the same before and after a fault.

Similarly, a valid `sntrup` ciphertext has the form $bG + d$ where (b, d) has coefficients in $\{-1, 0, 1\}$ and b has a specific Hamming weight. An analogous chosen-ciphertext attack replaces B with $B + \beta G + \delta$ where (β, δ) are chosen by the attacker to have a good chance of still having coefficients in $\{-1, 0, 1\}$ in $(b + \beta, d + \delta)$ and the right Hamming weight for $b + \beta$; e.g., take $\beta = 0$ and set exactly one coefficient in δ to 1 to detect whether that coefficient of d is 1.

However, for both `mceliece` and `sntrup`, the ciphertext also includes plaintext confirmation, another hash of (b, d). As in Sect. 3.10, the attacker has no way to replace this with a hash of $(b+\beta, d+\delta)$. So *all* of the modified ciphertexts are (implicitly) rejected, eliminating the information that the attack needs.

For `sntrup`, there is an independent reason that the attack does not work as stated: see Sect. 3.13. However, there could be workarounds for the attacker. Plaintext confirmation makes much more obvious that the attack fails.

4.4. How Proofs Led `ntruhrss` to Remove Plaintext Confirmation. The original version of `ntruhrss` in 2017 included plaintext confirmation, as did the `ntruhrss` submission to round 1 of the NIST competition: see [57, Algorithm 6, "e_2"] and [58, Section 1.10.4, "*qrom_hash*"]. However, the `ntruhrss` submission to round 2 of the NIST competition in 2019 removed plaintext confirmation. It is interesting to look at why.

The reason for original `ntruhrss` and round-1 `ntruhrss` to include plaintext confirmation was not that *plaintext confirmation interferes with attacks*, but rather that *plaintext confirmation seemed necessary for certain types of proofs.* This distinction became important later.

Saito, Xagawa, and Yamakawa [92] proposed a modification of round-1 `ntruhrss`, writing in [92, Section 1.2] that "the obtained KEM is CCA secure in the QROM" under a specific assumption. The modification was designed to be as simple as possible to support the underlying QROM proof; the proof relied on implicit rejection but not on plaintext confirmation; consequently, the modification did not include plaintext confirmation.

The round-2 `ntruhrss` submission [35, page 24] said that the KEM from [92] "has a tight security reduction in the ROM and avoids the plaintext-confirmation hash", along with having "a tight reduction in the QROM". The round-2 `ntruhrss` KEM is the KEM from [92] plus some further changes that are not relevant here. For comparison, previous versions of `ntruhrss` had appealed to the QROM proofs from [98], which assumed plaintext confirmation.

To summarize: Why did `ntruhrss` end up deciding that it was not useful to spend ciphertext space on plaintext confirmation? Answer: because plaintext confirmation turned out to be unnecessary for various types of proofs. But this paper shows that even a one-time single-bit fault is enough to break the proofs!

The practice of eliminating any cryptosystem features not needed for proofs is common in cryptography—but not universal. The possibility of plaintext confirmation stopping attacks not stopped by implicit rejection was noted in [19, Section 17]: implicit rejection and plaintext confirmation "target different aspects of attacks", so it is "difficult to justify a recommendation against the dual-defense construction". More broadly, Koblitz wrote the following in [69, page 977]: "Anyone working in cryptography should think very carefully before dropping a validation step that had been put in to prevent security problems. Certainly someone with Krawczyk's experience and expertise would never have made such a blunder if he hadn't been over-confident because of his 'proof' of security." The "proof" critiqued in [69] was erroneous, but the same danger appears when a correct proof is in a model too narrow to capture real-world attacks.

4.5. Countermeasures for NTRU-HRSS. Any algorithm computing the specified `ntruhrss` decapsulation function will be vulnerable to the same attack. There is nothing in the secret-key format that the algorithm can use to detect that r has had a fault: r is simply 256 bits of randomness generated independently of the rest of the secret key.[18] The fault converts a valid secret key into another valid secret key.

Consequently, to stop this attack, implementors have to use a cryptosystem that is not the currently specified `ntruhrss` cryptosystem. Perhaps the simplest approach is to switch to another secret-key format that makes bit flips detectable or even correctable; see, e.g., the generic use of Hamming codes in Sect. 4.1.

Implementors can also replace the specified decapsulation function with a more complicated *stateful* function that tries to detect attack patterns and to limit the exposure of each ciphertext. One approach is to maintain a database of previously seen values of d and reject nearby values, and similarly for b; but this could be a serious performance problem if "nearby" is too generous, and could allow attacks if "nearby" is too strict. An alternative is to maintain a database of ciphertexts and reject any repeated ciphertexts (modulo any "benign malleability" allowed by the cryptosystem), if this is suitable for the application. See [54, Section 2] for further stateful approaches. All of these approaches complicate the data flow and raise denial-of-service questions.

More options are available for implementors willing to break interoperability with `ntruhrss` ciphertexts; see Sect. 3. Plaintext confirmation is an obvious

[18] For comparison, the specified `mceliece` secret-key format already includes a 256-bit seed that can be double-checked against the rest of the secret key. This seed was specified to allow compression, but implementors can reuse it for double-checks of whether various faults have occurred.

choice. Limiting the ciphertext space or plaintext space could help, but this needs analysis. Hashing the ciphertext does not help: the attack detects failing ciphertexts by seeing that a fault changes the results for the same ciphertext.

4.6. Whose Responsibility Is Error Correction? Let's assume that there's an objective of changing the secret-key format, specifically encoding the secret key using a distance-4 extended Hamming code. This fixes natural bit flips anywhere in the secret key, not just in r, so it is attractive whether or not there is plaintext confirmation.

There's still a question of *who* should encode the secret key. Should the `ntruhrss` specification be updated to specify an encoded secret-key format? Or should applications encode secret keys, and much more data, to protect all of that data against bit flips? Or should the operating system build error correction into paging mechanisms, and continually sweep through pages to check for errors? Or should the hardware apply error correction to all data stored in DRAM?

The attack relies on all of these layers failing to act. Note that the fact that there are multiple layers that can act gives each layer an excuse not to act, especially when nobody is responsible for the security of the system as a whole.

One could respond that any layer that *can* take action should do so: the `ntruhrss` designers can specify error correction, so they should; applications can correct errors, so they should; the operating system can correct errors, so it should; and the hardware can correct errors, so it should. These layers can share specifications, and to some extent implementations, of the error-correction mechanisms. But this nevertheless means added complications at each layer. Surely a simpler, more easily reviewed system can address the problem at hand, the same way that twist-security and x-coordinates address the ECDH security problem mentioned in Sect. 1.1 without the complications of implementations having to check point validity.

SECDED ECC DRAM handles DRAM bit flips in a way that is measurable and seems robust. Unfortunately, computer manufacturers appear to have used the minor costs of SECDED ECC DRAM for market segmentation, in much the same way that 19th-century railroad companies installed a roof on *some* train cars for market segmentation; see generally [80, Section 3]. Perhaps DDR5 "on-die ECC"—which tries to catch DRAM errors, although it does not protect data in transit to the CPU—will eventually put an end to the non-ECC era, but non-ECC equipment will continue to be in use for many years.

It is clear that many options require software for error correction. As another supplement to this paper, I have released a `libsecded` software library [15] that encodes arrays in RAM using a distance-4 Hamming code. However, this paper does not draw conclusions regarding the optimal way forward.

Acknowledgments. This paper is inspired by a series of discussions with Tanja Lange regarding IND-CCA2 attacks and defenses. In particular, Lange pointed out plaintext confirmation as a countermeasure to fault attacks.

References

1. — (no editor), *IEEE international conference on communications, ICC 2017*, IEEE, 2017. See [38]
2. Gorjan Alagic, Daniel Apon, David Cooper, Quynh Dang, Thinh Dang, John Kelsey, Jacob Lichtinger, Carl Miller, Dustin Moody, Rene Peralta, Ray Perlner, Angela Robinson, Daniel Smith-Tone, Yi-Kai Liu, *Status report on the third round of the NIST Post-Quantum Cryptography Standardization Process* (2022). NISTIR 8413. Cited in §1.1, §3.14, §3.14, §4.2
3. Roberto Avanzi, Joppe Bos, Léo Ducas, Eike Kiltz, Tancrede Lepoint, Vadim Lyubashevsky, John M. Schanck, Peter Schwabe, Gregor Seiler, Damien Stehlé, *CRYSTALS-Kyber: Algorithm specifications and supporting documentation* (2020). Cited in §3.14
4. Ciprian Baetu, F. Betül Durak, Loïs Huguenin-Dumittan, Abdullah Talayhan, Serge Vaudenay, *Misuse attacks on post-quantum cryptosystems*, in Eurocrypt 2019 [61] (2019), 747–776. Cited in §3.7
5. Mihir Bellare (editor), *Advances in cryptology—CRYPTO 2000*, LNCS, 1880, Springer, 2000. See [62]
6. Mihir Bellare, Hannah Davis, Felix Günther, *Separate your domains: NIST PQC KEMs, oracle cloning and read-only indifferentiability*, in Eurocrypt 2020 [32] (2020), 3–32. Cited in §3.10
7. Mihir Bellare, Dennis Hofheinz, Eike Kiltz, *Subtleties in the definition of IND-CCA: when and how should challenge decryption be disallowed?*, Journal of Cryptology **28** (2015), 29–48. Cited in §3.4
8. Daniel J. Bernstein, *Re: Current consensus on ECC* (2001). Cited in §1.1
9. Daniel J. Bernstein, *Curve25519: new Diffie-Hellman speed records*, in PKC 2006 [103] (2006), 207–228. Cited in §1.1
10. Daniel J. Bernstein, *A subfield-logarithm attack against ideal lattices* (2014). Cited in §3.3
11. Daniel J. Bernstein, *How to design an elliptic-curve signature system* (2014). Cited in §2.4
12. Daniel J. Bernstein, *Comparing proofs of security for lattice-based encryption* (2019). Second PQC Standardization Conference. Cited in §3.2
13. Daniel J. Bernstein, *On the looseness of FO derandomization* (2021). Cited in §3.14
14. Daniel J. Bernstein, *Understanding binary-Goppa decoding* (2022). Cited in §3.8, §3.8, §3.8
15. Daniel J. Bernstein, *libsecded (software package)* (2022). Cited in §4.5
16. Daniel J. Bernstein, *attackntrw (software package)* (2022). Cited in §4
17. Daniel J. Bernstein, Leon Groot Bruinderink, Tanja Lange, Lorenz Panny, *HILA5 Pindakaas: On the CCA security of lattice-based encryption with error correction*, in Africacrypt 2018 [64] (2018), 203–216. Cited in §3.7
18. Daniel J. Bernstein, Tanja Lange (editors), *eBACS: ECRYPT Benchmarking of Cryptographic Systems* (2022). Accessed 25 August 2022. Cited in §4.2
19. Daniel J. Bernstein, Edoardo Persichetti, *Towards KEM unification* (2018). Cited in §3.12, §4.4
20. Eli Biham (editor), *Fast software encryption, 4th international workshop, FSE '97*, LNCS, 1267, Springer, 1997. See [91]
21. Eli Biham, Lior Neumann, *Breaking the Bluetooth pairing—the fixed coordinate invalid curve attack*, in SAC 2019 [84] (2019), 250–273. Cited in §1.1

22. Nina Bindel, Douglas Stebila, Shannon Veitch, *Improved attacks against key reuse in learning with errors key exchange*, in Latincrypt 2021 [74] (2021), 168–188. Cited in §3.7
23. Mario Blaum, Patrick G. Farrell, Henk C. A. van Tilborg (editors), *Information, coding and mathematics*, Kluwer International Series in Engineering and Computer Science, 687, Kluwer, 2002. MR 2005a:94003. See [101]
24. Daniel Bleichenbacher, *Chosen ciphertext attacks against protocols based on the RSA encryption standard PKCS #1*, in Crypto 1998 [70] (1998), 1–12. Cited in §1
25. Hanno Böck, Juraj Somorovsky, Craig Young, *Return of Bleichenbacher's oracle threat (ROBOT)*, in [43] (2018), 817–849. Cited in §1
26. Dan Boneh (editor), *Advances in cryptology—CRYPTO 2003*, LNCS, 2729, Springer, 2003. See [56]
27. Dan Boneh, Richard A. DeMillo, Richard J. Lipton, *On the importance of checking cryptographic protocols for faults (extended abstract)*, in Eurocrypt 1997 [47] (1997), 37–51; see also newer version [28]. Cited in §2.3, §2.3, §2.3, §2.3, §2.4, §2.4, §2.5, §2.5, §2.5
28. Dan Boneh, Richard A. DeMillo, Richard J. Lipton, *On the importance of eliminating errors in cryptographic computations*, Journal of Cryptology 14 (2001), 101–119; see also older version [27]
29. Joe P. Buhler (editor), *Algorithmic number theory, third international symposium, ANTS-III*, LNCS, 1423, Springer, 1998. See [52]
30. Kevin Butler, Kurt Thomas (editors), *31st USENIX Security Symposium*, USENIX Association, 2022. See [96]
31. L. Jean Camp, Stephen Lewis (editors), *Economics of information security*, Advances in Information Security, 12, Springer, 2004. See [80]
32. Anne Canteaut, Yuval Ishai (editors), *Advances in cryptology—EUROCRYPT 2020*, LNCS, 12106, Springer, 2020. See [6]
33. Anne Canteaut, François-Xavier Standaert (editors), *Advances in cryptology—EUROCRYPT 2021*, LNCS, 12697, Springer, 2021. See [34]
34. Pierre-Louis Cayrel, Brice Colombier, Vlad-Florin Dragoi, Alexandre Menu, Lilian Bossuet, *Message-recovery laser fault injection attack on the Classic McEliece cryptosystem*, in [33] (2021), 438–467. Cited in §2.2
35. Cong Chen, Oussama Danba, Jeffrey Hoffstein, Andreas Hulsing, Joost Rijneveld, John M. Schanck, Peter Schwabe, William Whyte, Zhenfei Zhang, *NTRU: algorithm specifications and supporting documentation* (2019). Cited in §4.4
36. Mauro Conti, Jianying Zhou, Emiliano Casalicchio, Angelo Spognardi (editors), *Applied cryptography and network security—18th international conference, ACNS 2020*, LNCS, 12146, Springer, 2020. See [59]
37. Alexander W. Dent, *A designer's guide to KEMs*, in Cirencester 2003 [83] (2003), 133–151. Cited in §3.10
38. Jintai Ding, Saed Alsayigh, R. V. Saraswathy, Scott R. Fluhrer, Xiaodong Lin, *Leakage of signal function with reused keys in RLWE key exchange*, in ICC 2017 [1] (2017), 1–6. Cited in §3.7
39. Jintai Ding, Joshua Deaton, Kurt Schmidt, Vishakha, Zheng Zhang, *A simple and efficient key reuse attack on NTRU cryptosystem* (2019). Cited in §3.7
40. Jintai Ding, Scott R. Fluhrer, Saraswathy RV, *Complete attack on RLWE key exchange with reused keys, without signal leakage*, in ACISP 2018 [97] (2018), 467–486. Cited in §3.7

41. John R. Douceur, Albert G. Greenberg, Thomas Bonald, Jason Nieh (editors), *Proceedings of the eleventh international joint conference on measurement and modeling of computer systems, SIGMETRICS/Performance 2009*, ACM, 2009. See [93]
42. Orr Dunkelman, Stefan Dziembowski (editors), *Advances in cryptology—EUROCRYPT 2022*, LNCS, 13277, Springer, 2022. See [60]
43. William Enck, Adrienne Porter Felt (editors), *27th USENIX security symposium, USENIX Security 2018, Baltimore, MD, USA, August 15-17, 2018*, USENIX Association, 2018. See [25]
44. Wieland Fischer, Naofumi Homma (editors), *Cryptographic hardware and embedded systems—CHES 2017*, LNCS, 10529, Springer, 2017. See [57]
45. Scott R. Fluhrer, *Cryptanalysis of ring-LWE based key exchange with key share reuse* (2016). Cited in §3.7
46. Eiichiro Fujisaki, Tatsuaki Okamoto, *Secure integration of asymmetric and symmetric encryption schemes*, in Crypto 1999 [102] (1999), 537–554. Cited in §3.8
47. Walter Fumy (editor), *Advances in cryptology—EUROCRYPT '97*, LNCS, 1233, Springer, 1997. See [27]
48. Debin Gao, Qi Li, Xiaohong Guan, Xiaofeng Liao (editors), *Information and communications security-23rd international conference, ICICS 2021*, LNCS, 12919, Springer, 2021. See [105]
49. J. Alex Halderman, Seth D. Schoen, Nadia Heninger, William Clarkson, William Paul, Joseph A. Calandrino, Ariel J. Feldman, Jacob Appelbaum, Edward W. Felten, *Lest we remember: cold boot attacks on encryption keys*, in USENIX Security 2008 [82] (2008), 45–60. Cited in §2.6, §2.6, §2.6, §2.6
50. Chris Hall, Ian Goldberg, Bruce Schneier, *Reaction attacks against several public-key cryptosystems*, in ICICS 1999 [100] (1999), 2–12. Cited in §3.6, §3.6, §3.6
51. Martin Hirt, Adam D. Smith (editors), *Theory of cryptography—14th international conference, TCC 2016-B*, LNCS, 9986, 2016. See [98]
52. Jeffrey Hoffstein, Jill Pipher, Joseph H. Silverman, *NTRU: a ring-based public key cryptosystem*, in ANTS III [29] (1998), 267–288. Cited in §3.3, §3.3
53. Jeffrey Hoffstein, Jill Pipher, Joseph H. Silverman, *NTRU: a new high speed public key cryptosystem* (2016). Circulated at Crypto 1996, put online in 2016. Cited in §3.3
54. Jeffrey Hoffstein, Joseph H. Silverman, *Reaction attacks against the NTRU public key cryptosystem* (2000). Cited in §3.7, §4.5
55. Dennis Hofheinz, Kathrin Hövelmanns, Eike Kiltz, *A modular analysis of the Fujisaki-Okamoto transformation*, in TCC 2017-1 [65] (2017), 341–371. Cited in §3.11, §3.12
56. Nick Howgrave-Graham, Phong Q. Nguyen, David Pointcheval, John Proos, Joseph H. Silverman, Ari Singer, William Whyte, *The impact of decryption failures on the security of NTRU encryption*, in Crypto 2003 [26] (2003), 226–246. Cited in §3.7
57. Andreas Hülsing, Joost Rijneveld, John M. Schanck, Peter Schwabe, *High-speed key encapsulation from NTRU*, in [44] (2017), 232–252. Cited in §4.4
58. Andreas Hülsing, Joost Rijneveld, John M. Schanck, Peter Schwabe, *NTRU-HRSS-KEM: algorithm specifications and supporting documentation* (2017). Cited in §4.4

59. Loïs Huguenin-Dumittan, Serge Vaudenay, *Classical misuse attacks on NIST round 2 PQC—the power of rank-based schemes*, in ACNS 2020 [36] (2020), 208–227. Cited in §3.7
60. Loïs Huguenin-Dumittan, Serge Vaudenay, *On IND-qCCA security in the ROM and its applications: CPA security is sufficient for TLS 1.3*, in Eurocrypt 2022 [42] (2022), 613–642. Cited in §1
61. Yuval Ishai, Vincent Rijmen (editors), *Advances in cryptology—EUROCRYPT 2019*, LNCS, 11477, Springer, 2019. See [4]
62. Éliane Jaulmes, Antoine Joux, *A chosen-ciphertext attack against NTRU*, in Crypto 2000 [5] (2000), 20–35. Cited in §3.7
63. Simon Josefsson, Ilari Liusvaara, *Edwards-curve digital signature algorithm (EdDSA)* (2017). Cited in §2.4
64. Antoine Joux, Abderrahmane Nitaj, Tajjeeddine Rachidi (editors), *Progress in cryptology—AFRICACRYPT 2018*, LNCS, 10831, Springer, 2018. See [17]
65. Yael Kalai, Leonid Reyzin (editors), *Theory of cryptography—15th international conference, TCC 2017*, LNCS, 10677, Springer, 2017. See [55]
66. Burt Kaliski, *PKCS #1: RSA encryption version 1.5* (1998). Cited in §2.4
67. Matthias Kannwischer, *Remove schemes that are no longer under consideration by NIST* (2022). Cited in §4.2
68. Jonathan Katz, Yehuda Lindell, *Introduction to modern cryptography: principles and protocols*, Chapman & Hall/CRC, 2007. Cited in §1.1
69. Neal Koblitz, *The uneasy relationship between mathematics and cryptography*, Notices of the American Mathematical Society **54** (2007), 972–979. Cited in §4.4, §4.4
70. Hugo Krawczyk (editor), *Advances in cryptology—CRYPTO '98*, LNCS, 1462, Springer, 1998. See [24]
71. Adam Langley, *CECPQ2* (2018). Cited in §1
72. Arjen K. Lenstra, *Memo on RSA signature generation in the presence of faults* (1996). Cited in §2.3, §2.5
73. Joseph K. Liu, Hui Cui (editors), *Information security and privacy—25th Australasian conference, ACISP 2020*, LNCS, 12248, Springer, 2020. See [80]
74. Patrick Longa, Carla Ràfols (editors), *Progress in cryptology—LATINCRYPT 2021*, LNCS, 12912, Springer, 2021. See [22]
75. Vadim Lyubashevsky, *OFFICIAL COMMENT: CRYSTALS-DILITHIUM* (2018). Cited in §1.1
76. Robert J. McEliece, *A public-key cryptosystem based on algebraic coding theory* (1978), 114–116. JPL DSN Progress Report. Cited in §3.3, §3.3
77. Alfred Menezes, *Evaluation of security level of cryptography: RSA signature schemes (PKCS#1 v1.5, ANSI X9.31, ISO 9796)* (2002). Cited in §1
78. National Institute of Standards and Technology, *Submission requirements and evaluation criteria for the post-quantum cryptography standardization process* (2016). Cited in §1
79. Jesper Buus Nielsen, Vincent Rijmen (editors), *Advances in cryptology—EUROCRYPT 2018*, LNCS, 10822, Springer, 2018. See [92]
80. Andrew M. Odlyzko, *Privacy, economics, and price discrimination on the internet*, in [31] (2004), 187–211. Cited in §4.5
81. Satoshi Okada, Yuntao Wang, Tsuyoshi Takagi, *Improving key mismatch attack on NewHope with fewer queries*, in ACISP 2020 [73] (2020), 505–524. Cited in §3.7
82. Paul C. van Oorschot (editor), *Proceedings of the 17th USENIX security symposium*, USENIX Association, 2008. See [49]

83. Kenneth G. Paterson (editor), *Cryptography and coding, 9th IMA international conference*, LNCS, 2898, Springer, 2003. See [37]
84. Kenneth G. Paterson, Douglas Stebila (editors), *Selected areas in cryptography—SAC 2019*, LNCS, 11959, Springer, 2020. See [21]
85. Trevor Perrin, *The XEdDSA and VXEdDSA signature schemes* (2016). Cited in §2.4
86. Edoardo Persichetti, *Improving the efficiency of code-based cryptography*, Ph.D. thesis, 2012. Cited in §3.11
87. Bart Preneel (editor), *Advances in cryptology—EUROCRYPT 2000*, LNCS, 1807, Springer, 2000. See [95]
88. Yue Qin, Chi Cheng, Xiaohan Zhang, Yanbin Pan, Lei Hu, Jintai Ding, *A systematic approach and analysis of key mismatch attacks on lattice-based NIST candidate KEMs*, in Asiacrypt 2021 [99] (2021), 92–121. Cited in §3.7
89. Yue Qin, Ruoyu Ding, Chi Cheng, Nina Bindel, Yanbin Pan, Jintai Ding, *Light the signal: optimization of signal leakage attacks against LWE-based key exchange* (2022). Cited in §3.7
90. Prasanna Ravi, Martianus Frederic Ezerman, Shivam Bhasin, Anupam Chattopadhyay, Sujoy Sinha Roy, *Will you cross the threshold for me? Generic side-channel assisted chosen-ciphertext attacks on NTRU-based KEMs*, IACR Transactions on Cryptographic Hardware and Embedded Systems **2022.1** (2022), 722–761. Cited in §3.13
91. Ronald L. Rivest, *All-or-nothing encryption and the package transform*, in FSE 1997 [20] (1997), 210–218. Cited in §3.5
92. Tsunekazu Saito, Keita Xagawa, Takashi Yamakawa, *Tightly-secure key-encapsulation mechanism in the quantum random oracle model*, in Eurocrypt 2018 [79] (2018), 520–551. Cited in §4.4, §4.4, §4.4, §4.4
93. Bianca Schroeder, Eduardo Pinheiro, Wolf-Dietrich Weber, *DRAM errors in the wild: a large-scale field study*, in [41] (2009), 193–204. Cited in §1.2, §1.2, §1.2
94. Mark Seaborn, Thomas Dullien, *Exploiting the DRAM rowhammer bug to gain kernel privileges* (2015). Cited in §2.2
95. Victor Shoup, *Using hash functions as a hedge against chosen ciphertext attack*, in Eurocrypt 2000 [87] (2000), 275–288. Cited in §3.5
96. George Arnold Sullivan, Jackson Sippe, Nadia Heninger, Eric Wustrow, *Open to a fault: On the passive compromise of TLS keys via transient errors*, in USENIX Security 2022 [30] (2022), 233–250. Cited in §2.3, §2.3, §2.3, §2.3
97. Willy Susilo, Guomin Yang (editors), *Information security and privacy—23rd Australasian conference*, ACISP 2018, LNCS, 10946, Springer, 2018. See [40]
98. Ehsan Ebrahimi Targhi, Dominique Unruh, *Post-quantum security of the Fujisaki-Okamoto and OAEP transforms*, in [51] (2016), 192–216. Cited in §4.4
99. Mehdi Tibouchi, Huaxiong Wang (editors), *Advances in cryptology—ASIACRYPT 2021*, LNCS, 13093, Springer, 2021. See [88]
100. Vijay Varadharajan, Yi Mu (editors), *Information and communication security, second international conference, ICICS'99*, Springer, 1999. See [50]
101. Eric R. Verheul, Jeroen M. Doumen, Henk C. A. van Tilborg, *Sloppy Alice attacks! Adaptive chosen ciphertext attacks on the McEliece public-key cryptosystem*, in [23] (2002), 99–119. MR 2005b:94041. Cited in §3.6, §3.6, §3.6
102. Michael J. Wiener (editor), *Advances in cryptology—CRYPTO '99*, LNCS, 1666, Springer, 1999. See [46]
103. Moti Yung, Yevgeniy Dodis, Aggelos Kiayias, Tal Malkin (editors), *Public key cryptography—9th international conference on theory and practice in public-key cryptography*, LNCS, 3958, Springer, 2006. See [9]

104. Meilin Zhang, Vladimir M. Stojanovic, Paul Ampadu, *Reliable ultra-low-voltage cache design for many-core systems*, IEEE Transactions on Circuits and Systems II: Express Briefs **59** (2012), 858–862. Cited in §1.2
105. Xiaohan Zhang, Chi Cheng, Ruoyu Ding, *Small leaks sink a great ship: an evaluation of key reuse resilience of PQC third round finalist NTRU-HRSS*, in ICICS 2021 [48] (2021), 283–300. Cited in §3.7

An Efficient Key Recovery Attack Against NTRUReEncrypt from AsiaCCS 2015

Zijian Song[1,2], Jun Xu[1,2(✉)], Zhiwei Li[1,2], and Dingfeng Ye[1,2]

[1] State Key Laboratory of Information Security, Institute of Information Engineering, Chinese Academy of Sciences, Beijing 100093, China
xujun@iie.ac.cn

[2] School of Cyber Security, University of Chinese Academy of Sciences, Beijing 100093, China

Abstract. At AsiaCCS 2015, Nuñez et al. proposed a NTRU-based proxy re-encryption (PRE) scheme, called NTRUReEncrypt. A complete PRE scheme permits the sender to encrypt messages to the proxy, and allows the receiver to decrypt the ciphertexts re-encrypted by the proxy. At PQCrypto 2019, Liu et al. provided cryptanalysis of the scheme based on decryption failures and statistical analysis, both of which need huge amount of ciphertexts. For instance, for ees1171ep1 parameter set, the number of ciphertexts required are $4.68 \cdot 10^{17}$ and $4.83 \cdot 10^{17}$ respectively. In this paper we point out that the security of NTRUReEncrypt would be impacted by an efficient key recovery attack based on linearization technique, it can reduce the number of required ciphertexts drastically. To be specific, two parties sending and receiving messages can recover the other's private key by communicating $O(N + [\frac{N}{2}])$ times, where N is an odd prime in the ring $\mathcal{R} = \mathbb{Z}[x]/\left(x^N - 1\right)$. For specific scheme on parameter sets ees1087ep1, ees1171ep1, ees1499ep1, where N equals 1087, 1171 and 1499 respectively, the amount of ciphertexts used in our attack is only on the order of 10^3, and our experiments are all completed within one hour on PC. Moreover, we discuss the NTRUReEncrypt instantiated with the NTRU parameter sets in the third round of NIST-PQC competition and give the theoretical analysis.

Keywords: NTRUReEncrypt · NTRU · Linearization technique · Key recovery attack

1 Introduction

In 1998, Blaze, Bleumer and Strauss [3] proposed a new type of public-key cryptographic scheme, namely proxy re-encryption (PRE) scheme. A complete PRE scheme consists of three parties: the sender Alice, the receiver Bob, and the proxy. It permits Alice to encrypt messages to the proxy, and allows Bob to decrypt the ciphertexts re-encrypted by the proxy. Further, in the communication process, the proxy only provides the re-encryption operation without knowing any information about messages. In fact, proxy re-encryption scheme is a variant of

T. Isobe and S. Sarkar (Eds.): INDOCRYPT 2022, LNCS 13774, pp. 644–657, 2022.
https://doi.org/10.1007/978-3-031-22912-1_28

the traditional public key cryptosystem. Its basic algorithm is the same as that of public key encryption scheme, except that there are two more steps: generating proxy re-encryption key and re-encrypting ciphertexts. After a period of development, many PRE schemes have been constructed, but the vast majority of these are based on traditional number theoretic problems, such as discrete logarithm problem [4]. However, these problems suffered the impact after Shor's algorithm [14,15] was put forward. Therefore, the attention turned to the field of post-quantum cryptography, such as lattice-based schemes [1].

At AsiaCCS 2015 [13], Nuñez et al. proposed a NTRU-based proxy re-encryption (PRE) scheme, called NTRUReEncrypt. It only has one more re-encryption step, and the rest is the same as NTRU scheme including parameter sets. In 1996, Hoffstein, Pipher and Silverman proposed a cryptosystem called NTRU [7], which has the advantages of high efficiency and low memory usage. Due to these properties, it becomes an indispensable part of post-quantum cryptography, and has been standardized by IEEE P1363.1 [2]. Recently, it was also submitted to the third round of NIST-PQC competition, e.g. NTRU-HPS, NTRU-HRSS [5]. One reason for the efficiency of NTRU is that, some of polynomials in NTRU have small coefficients, which we will call "small" polynomials for ease of description. The brief process of NTRUReEncrypt scheme is as follows: (1) Alice encrypts the message m as $C_A = h_A * r + m$ by selecting a small polynomial r, where Alice's private key is (f_A, g_A) and public key is $h_A = p * g_A * f_A^{-1}$. (2) The proxy selects a small polynomial e, encrypts C_A sent by Alice as $C_B = C_A * rk_{A \to B} + p * e$, where $rk_{A \to B} = f_A * f_B^{-1} \bmod q$ is the re-encrypted key of the proxy and f_B is Bob's private key. (3) Bob decrypts C_B sent by the proxy as $(C_B * f_B \bmod q) \bmod p$ to obtain the message m.

There are many attacks against NTRU, such as decryption failure attack [8], broadcast attack [6,10]. The method of latter uses the linearization technique, whose main idea is to generate a linear system by linearizing monomials into new variables. At PQCrypto 2019 [11], Liu et al. proposed cryptanalysis of NTRUReEncrypt, whose strategy is based on two points: one is decryption failure, the other is statistical analysis. Due to the huge amount of data required, these two attacks are hard to implement in practice. For instance, for ees1171ep1 parameter set, the number of required ciphertexts are $4.68 \cdot 10^{17}$ and $4.83 \cdot 10^{17}$ respectively.

Our Contribution. We present a key recovery attack based on the linearization technique against NTRUReEncrypt, where the parameter sets are from those in AsiaCCS 2015 and PQCrypto 2019. To implement an attack, $O(N + [\frac{N}{2}])$ legal communications are needed to collect ciphertexts C_{A_i} and C_{B_i}, where N is an odd prime in the ring $\mathcal{R} = \mathbb{Z}[x]/\left(x^N - 1\right)$. The comparison of PQCrypto 2019 and our work is shown in Table 1.

The technical overview is as follows. First, we focus on the following relation from the proxy's re-encryption stage:

$$C_B = C_A * rk_{A \to B} + p * e \bmod q.$$

Here, $C_A, C_B, p, q = 2^\gamma$ are known, where γ is an integer, and $rk_{A \to B}, e$ are unknown. Our goal is to recover the re-encryption key $rk_{A \to B}$, and then obtain

Table 1. Number of ciphertexts needed

	ees1087ep1	ees1171ep1	ees1499ep1
PQCrypto 2019	$4.06 \cdot 10^{17}$	$4.83 \cdot 10^{17}$	$9.67 \cdot 10^{17}$
Our work	$3.17 \cdot 10^3$	$3.58 \cdot 10^3$	$4.45 \cdot 10^3$

the private key f_A or f_B based on $rk_{A\to B} = f_A * f_B^{-1} \bmod q$. For the sake of efficiency, we first choose to recover $rk_{A\to B} \bmod 2$ instead of $rk_{A\to B} \bmod q$. Due to the special structure of coefficients in the polynomial e, i.e., its coefficients have certain numbers of $+1$, -1, and 0. Hence, the inner product of the coefficient vector of e is fixed. Thus, we can establish a system of linear congruence equations by using inner product calculation, and then obtain $rk_{A\to B} \bmod 2$ by using the linearization technique. According to $rk_{A\to B} = f_A * f_B^{-1} \bmod q$ and q is a power of 2, we get that $f_B * (rk_{A\to B} \bmod 2) = f_A \bmod 2$. Without loss of generality, suppose that Bob is an attacker, where f_B is the private key of Bob. Based on the above equation, Bob can determine the position of 0 bits of Alice's private key f_A. Notice that the private key pair (f_A, g_A) of Alice satisfies $h_A = p*g_A*f_A^{-1} \bmod q$, where h_A is the public key. It implies $f_A * h_A = p * g_A \bmod 2$. Furthermore, the attacker Bob can also deduce the position of 0 bits of g_A. Finally, combining with the position of 0 bits about f_A and g_A, we get a new system of linear congruence equations derived from $f_A * h_A = p * g_A \bmod q$, and compute the remaining bits of f_A and g_A using Gaussian elimination. Theoretically, the algorithm overhead is divided into two main parts: constructing linear equations from the proxy's re-encryption stage and solving linear equations. Since we choose to work on $\mathbb{F}_2$ rather than $\mathbb{Z}_{2048}$, the cost of the latter is greatly reduced to negligible. This means that the time required to implement an attack is almost dependent on constructing a system of equations, which could be completed within one hour on PC.

Our another contribution is to discuss the NTRUReEncrypt instantiated with the NTRU parameter sets in the third round of NIST-PQC competition. Unlike the parameter sets from AsiaCCS 2015 and PQCrypto 2019, the parameter sets in the third round of NIST-PQC competition, e.g., NTRU-HPS and NTRU-HRSS [5], no longer determine the certain numbers of $+1, -1, 0$ in the coefficients of the secret polynomials. It means that the inner product of e is not fixed. However, we can still take advantage of another property of ternary polynomials e. Denote the coefficient vector of e as $\mathbf{e}$, hence each component $\mathbf{e}_i \in \{-1, 0, 1\}$ satisfies $\mathbf{e}_i = (\mathbf{e}_i)^3$. The remaining operations are the same as the previous attack, except that the number of communications is increased to $O(N^2)$.

Organization. The rest of this paper is organized as follows: In Sect. 2, we provide some basic preliminaries for the linear form and parameter sets of NTRU. In Sect. 3, we briefly describe NTRU and NTRUReEncrypt schemes, provide the specific parameter sets used in this paper. In Sect. 4, we present our attack in detail and give a comparison with PQCrypto 2019 [11]. In Sect. 5, we discuss the NTRUReEncrypt instantiated with the NTRU parameter sets in the third round of NIST-PQC competition, and also compare with previous parameter sets used in Sect. 4. In Sect. 6, we present the experimental results, whose parameter sets are ees1087ep1, ees1171ep1, ees1499ep1 respectively. In Sect. 7, we conclude the paper.

2 Preliminaries

In this section, we provide some basic preliminaries of NTRU and NTRUReEncrypt scheme. The operations of both schemes are defined over the quotient ring $\mathcal{R} = \mathbb{Z}_q[x]/\left(x^N - 1\right)$, where N is an odd prime. Other parameters p, q are integers, where p is much smaller than q and $\gcd(p, q) = 1$.

The polynomials are selected from four subset of $\mathcal{R}$, denote as $\mathcal{L}_f = \mathcal{T}_{(d_f, d_f-1)}$, $\mathcal{L}_g = \mathcal{T}_{(d_g, d_g)}$, $\mathcal{L}_r = \mathcal{T}_{(d_r, d_r)}$,

$$\mathcal{L}_m = \left\{ m \in \mathcal{R} : \text{ every coefficient of } m \text{ lies between } -\frac{p-1}{2} \text{ and } \frac{p-1}{2} \right\}.$$

In addition, elements in $\mathcal{L}_f$, $\mathcal{L}_g$, $\mathcal{L}_r$ are ternary polynomials. We introduce the definition of ternary polynomial from PQCrypto 2019 [11].

Definition 1. *A ternary polynomial $\mathcal{T}$ with positive integers d_1, d_2 is defined as:*

$$\mathcal{T}_{(d_1,d_2)} = \left\{ \begin{array}{c} \textit{trinary polynomials of } \mathcal{R} \textit{ with } d_1 \textit{ entries} \\ \textit{equal to } 1 \textit{ and } d_2 \textit{ entries equal to } -1 \end{array} \right\}.$$

2.1 Vector and Matrix Forms of NTRU

A polynomial $f \in \mathcal{R}$ in NTRU can be presented as $f = \sum_{i=0}^{N-1} f_i x^i$. Its vector form can be presented as $\mathbf{f} = \left(f_0, f_1, \cdots, f_{N-1}\right)^T$. The polynomial f can be written in the form of a circular matrix $\mathbf{F}$ in $\mathbb{Z}_q^{N \times N}$:

$$\mathbf{F} = \begin{pmatrix} f_0 & f_{N-1} & \cdots & f_1 \\ f_1 & f_0 & \cdots & f_2 \\ \vdots & \vdots & \ddots & \vdots \\ f_{N-1} & f_{N-2} & \cdots & f_0 \end{pmatrix}$$

Further, the matrix form of multiplication of two polynomials $f, g \in \mathcal{R}$ can be presented as:

$$\begin{pmatrix} f_0 & f_{N-1} & \cdots & f_1 \\ f_1 & f_0 & \cdots & f_2 \\ \vdots & \vdots & \ddots & \vdots \\ f_{N-1} & f_{N-2} & \cdots & f_0 \end{pmatrix} \begin{pmatrix} g_0 \\ g_1 \\ \vdots \\ g_{N-1} \end{pmatrix}.$$

As needed, there are the following fundamental lemmas [9]:

Lemma 1. *If* $\mathbf{H} \in \mathbb{Z}_q^{N\times N}$ *is a circular matrix over* $\mathbb{Z}_q^{N\times N}$*, then* $\mathbf{H}^T$ *is also a circular matrix over* $\mathbb{Z}_q^{N\times N}$.

Lemma 2. *If* $\mathbf{G}, \mathbf{H} \in \mathbb{Z}_q^{N\times N}$ *are circular matrices, then* $\mathbf{GH}$ *is also a circular matrix. In particular,* $\mathbf{H}^T\mathbf{H}$ *is a symmetric circular matrix.*

3 NTRU and Its Proxy Re-encryption Scheme

In this section, we overview the NTRU and NTRU-based proxy re-encryption scheme, called NTRUReEncrypt. Parameters sets are shown in the following table, related to version 3.3 of the EESS#1 specification [2], from IEEE P1363.1 standard. For ees1087ep1, ees1171ep1, ees1499ep1, the private keys are f, g selected from $\mathcal{L}_f = \mathcal{T}_{(d_f, d_f-1)}$ and $\mathcal{L}_g = \mathcal{T}_{(d_g, d_g)}$ respectively, the set of small polynomials is $\mathcal{L}_r = \mathcal{T}_{(d_r, d_r)}$, where small means that the coefficients of the polynomials are small.

Table 2. Instance of polynomial sets

Instance	N	p	q	d_g	$d_f = d_r$
ees1087ep1	1087	3	2048	362	120
ees1171ep1	1171	3	2048	390	106
ees1499ep1	1499	3	2048	499	79

In practice, some variants of NTRU take the following approach to generating key f for efficiency: f have the form of $1 + p * F$ with $F \in \mathcal{T}_{(d_f, d_f)}$, first generate $F \in \mathcal{T}_{(d_f, d_f)}$, and then calculate $1 + p * F$ to obtain f. We would use this form throughout the rest of the paper.

3.1 NTRU Cryptosystem

The brief description of NTRU cryptosystem is as follows, see [7] for more details.

- **Key Generation:** Randomly chooses $F \in \mathcal{T}_{(d_f, d_f)}$, then calculate $1 + p * F$ to obtain f, where f has inverse f_p^{-1}, f_q^{-1} in R_p, R_q, then randomly chooses $g \in \mathcal{T}_{(d_g, d_g)}$. Outputs public key $pk = h = p * g * f_q^{-1} \bmod q$, private key $sk = (f, g)$.
- **Encryption:** To encrypt a plaintext $m \in \mathcal{L}_m$, randomly chooses $r \in \mathcal{L}_r$. Outputs ciphertext $c = h * r + m \bmod q$.
- **Decryption:** To decrypt a ciphertext c, receiver uses private key f and computes $a = f * c \bmod q$ such that coefficients of a are all lie between $(-q/2, q/2]$. Outputs plaintext $m = a * f_p^{-1} \bmod p$.

Note that f, g, s, m are small, i.e. each of its coefficients is small, then all coefficients of $a = c * f = p * g * s + m * f \bmod q$ lie in $(-q/2, q/2]$ with high probability. Thus, one computes $a = c * f \bmod q$ turns to $a = c * f$ over $\mathbb{Z}$. Then can decrypt the message:

$$a * f_p^{-1} = p * g * s * f_p^{-1} + m * f * f_p^{-1} = m \mod p.$$

3.2 NTRUReEncrypt

NTRUReEncrypt is a NTRU-based proxy re-encryption scheme, all parameter sets are related to formal NTRU scheme. Its initial key generation and first encryption stage are consistent with NTRU encryption, at second re-encryption stage, algorithm selects the same set of polynomials as NTRU. NTRUReEncrypt has a unique re-encrypt key generation, which ensures that Bob can decrypt the re-encrypted ciphertext sent from proxy.

The flow of the algorithm is as follows:

- **Key Generation:** Key generation algorithm is the same as that in NTRU. Outputs a pair of public and secret keys (pk_A, sk_A) for Alice, where $sk_A = (f_A, g_A)$ and $pk_A = h_A$, and Bob also obtains a public-private key pair in the same way.
- **Re-encrypt Key Generation:** The algorithm requires two private keys sk_A and sk_B, from sender Alice and receiver Bob respectively. Outputs re-encrypt key $rk_{A \to B} = f_A * f_B^{-1} \bmod q$. The re-encryption key can be computed by a simple three-party protocol below:
  1. Alice selects $t \in \mathcal{R}_q$, sends $t * f_A \bmod q$ to Bob and t to proxy;
  2. Bob sends $t * f_A * f_B^{-1} \bmod q$ to proxy;
  3. Proxy computes $rk_{A \to B} = f_A * f_B^{-1} \bmod q$.
- **Encryption:** Alice encrypts a plaintext $m \in \mathcal{L}_m$, randomly chooses $r \in \mathcal{L}_r$. Outputs ciphertext $C_A = h_A * r + m \bmod q$.
- **Re-encryption:** Proxy encrypts a ciphertext C_A sent by Alice, randomly chooses $e \in \mathcal{L}_r$. Outputs ciphertext $C_B = C_A * rk_{A \to B} + p * e \bmod q$.
- **Decryption:** Bob decrypts a ciphertext C_B, uses private key f_B and compute

  $$C_B * f_B = p * g_A * r + m * f_A + p * e * f_B \mod q$$

  such that coefficients of $C_B * f_B$ are all lie between $(-q/2, q/2]$. Outputs plaintext $m = C_B * f_B \bmod p$.

Decryption stage is similar to previous NTRU decryption, see [13] for more details.

4 Key Recovery Attack Against NTRUReEncrypt

In this section, we propose an efficient key recovery attack by only collecting ciphertexts C_A and C_B based on the algorithm of Li et al. [10]. They proposed a broadcast attack against NTRU only to recover messages at AsiaCCS 2015, however in NTRUReEncrypt, we find out that the re-encryption key $rk_{A \to B}$ can be recovered from the proxy's re-encryption stage, then a malicious receiver (sender) can directly recover the private key of the other one.

4.1 Construction of Equations

We now recall the re-encryption stage, proxy encrypts a ciphertext C_A sent by Alice, randomly choosese $\in \mathcal{L}_r$. Outputs ciphertext

$$C_B = C_A * rk_{A \to B} + p * e \mod q. \tag{4.1}$$

For convenience, we denote $\mathbf{e}$, $\mathbf{c_B}$, λ as their vector form in lowercase, and denote $\mathbf{C_A}$ as its matrix form in uppercase, then write Eq. (4.1) in linear form:

$$p\mathbf{e} = \mathbf{c_B} - \mathbf{C_A}\lambda \mod q,$$

where λ is the vector form of re-encryption key $rk_{A \to B}$.

Then, do the inner product of both sides of the equation:

$$(p\mathbf{e})^T(p\mathbf{e}) = (\mathbf{c_B} - \mathbf{C_A}\lambda)^T(\mathbf{c_B} - \mathbf{C_A}\lambda) \mod q.$$

Note that $p = 3$ and secret polynomial e selected in set $\mathcal{L}_r$, the numbers of $+1$ and -1 in their coefficients are d_r, thus $(p\mathbf{e})^T(p\mathbf{e}) = 2d_r p^2$ is a constant, denote as d.

We can get

$$d - \mathbf{c_B}^T\mathbf{c_B} = \lambda^T\mathbf{C_A}^T\mathbf{C_A}\lambda - 2\mathbf{c_B}^T\mathbf{C_A}\lambda \mod q. \tag{4.2}$$

4.2 Linearization

For convenience, let $d - \mathbf{c_B}^T\mathbf{c_B} = u$, $\mathbf{c_B}^T\mathbf{C_A} = (k_0, k_1, \cdots, k_{N-1})$, and

$$\mathbf{C_A}^T\mathbf{C_A} = \begin{pmatrix} c_0 & c_{N-1} & \cdots & c_1 \\ c_1 & c_0 & \cdots & c_2 \\ \vdots & \vdots & \ddots & \vdots \\ c_{N-1} & c_{N-2} & \cdots & c_0 \end{pmatrix}.$$

From **Lemma 2.2**, $\mathbf{C_A}^T\mathbf{C_A}$ is a symmetric circular matrix, where $c_i = c_{N-i}$, for $i \in \{0, 1, \cdots, N-1\}$. Then expanding Eq. (4.2), we can get

$$\begin{aligned} u = {} & c_0\left(\lambda_0^2 + \lambda_1^2 + \cdots + \lambda_{N-1}^2\right) \\ & + c_1\left(\lambda_1\lambda_0 + \lambda_2\lambda_1 + \cdots + \lambda_0\lambda_{N-1}\right) \\ & + \cdots\cdots \\ & + c_{N-1}\left(\lambda_{N-1}\lambda_0 + \lambda_0\lambda_1 + \cdots + \lambda_{N-2}\lambda_{N-1}\right) \\ & - 2k_0\lambda_0 - 2k_1\lambda_1 - \cdots - 2k_{N-1}\lambda_{N-1} \mod q \end{aligned} \tag{4.3}$$

Note that when choosing a specific parameter N, vector $\lambda = (\lambda_0, \lambda_1, \cdots, \lambda_{N-1})$ has N unknown components. After the inner product operation, it generates $O(N^2)$ new monomials $\lambda_i\lambda_j$, for $0 \le i \le j \le N-1$.

A trivial idea is to linearize these variables to $O(N^2)$ one-order monomials, denoted as $\mathbf{x} = (x_0, x_1, \cdots, x_{O(N^2)-1})$. Then Eq. (4.3) turns to a congruence equation with $O(N^2+N)$ variables, thus λ_i can be recovered by collecting $O(N^2)$ equations in time $O(N^6)$ by Gaussian elimination. In certain parameter sets defined by NTRUReEncrypt, the size of N generally amounts to 10^3, which means the system of linear equations with around 10^6 variables and it is hard to implement in practice.

To reduce the number of variables, let

$$x_i = \lambda_i\lambda_0 + \lambda_{i+1}\lambda_1 + \cdots + \lambda_{N-1}\lambda_{N-i-1} + \lambda_0\lambda_{N-i} + \cdots + \lambda_{i-1}\lambda_{N-1},$$

for $i = 0, 1, \cdots, N-1$. In the parameter sets we attacked, N is an odd prime. Note that $c_i = c_{N-i}$ and $x_i = x_{N-i}$ for $i = 0, 1, \cdots, N-1$, the Eq. (4.3) is equivalent to

$$\begin{aligned} u = {} & c_0x_0 + 2c_1x_1 + \cdots + 2c_{[\frac{N}{2}]}x_{[\frac{N}{2}]} \\ & - 2k_0\lambda_0 - 2k_1\lambda_1 - \cdots - 2k_{N-1}\lambda_{N-1} \mod q, \end{aligned} \tag{4.4}$$

where q is a power of 2 denoted as $q = 2^\gamma$, γ is a positive integer. Further, assuming that c_0, u are even, the equation could be converted to

$$\begin{aligned} \frac{1}{2}u = {} & \frac{1}{2}c_0x_0 + c_1x_1 + \cdots + c_{[\frac{N}{2}]}x_{[\frac{N}{2}]} \\ & - k_0\lambda_0 - k_1\lambda_1 - \cdots - k_{N-1}\lambda_{N-1} \mod 2^{\gamma-1}. \end{aligned} \tag{4.5}$$

Notice that we can get one congruence Eq. (4.4) with $(N+[\frac{N}{2}]+1)$ variables by collecting C_A and C_B through one legal communication, so we could collect a series of samples by communicating relevant times. In fact through experiment, we could always select enough equations in the form of (4.5) by choosing these samples, and the number of samples is $O(N + [\frac{N}{2}])$, which is related to the number of variables.

4.3 Solving the System of Linear Congruence Equations

Denote n as the number of variables and $n = N + [\frac{N}{2}] + 1$, then we build a linear system $\mathbf{L} \times \mathbf{X} = \mathbf{S} \mod 2^{\gamma-1}$ by collecting $n+l$ equations from Eq. (4.5), where l is a positive integer, the vector $\mathbf{X} = (x_0, x_1, \cdots, x_{[\frac{N}{2}]}, \lambda_0, \lambda_1, \cdots, \lambda_{N-1})^T$, the row of the matrix $\mathbf{L}$ corresponds to (4.5) equals

$$(\frac{1}{2}c_0, c_1, \cdots, c_{[\frac{N}{2}]}, -k_0, -k_1, \cdots, -k_{N-1})^T,$$

and $\mathbf{S}$ is the column vector related to $\frac{1}{2}u$. For the sake of efficiency, we choose to apply our algorithm to work over the finite field $\mathbb{F}_2$ but not the ring $\mathbb{Z}_{2^{\gamma-1}}$, which means that we turn to solve the system of equations $\mathbf{L} \times \mathbf{X} = \mathbf{S} \mod 2$. That is, our goal is to find $rk_{A\to B} \mod 2$ not $rk_{A\to B} \mod 2^{\gamma-1}$, and we would show that it is enough for recovering the private key in the next subsection.

Note that the vector $\mathbf{S} \in \mathbb{F}_2^n$, the matrix $\mathbf{L} \in \mathbb{F}_2^{(n+l)\times n}$, we aim to find $rk_{A\to B} \mod 2 = (\lambda_0, \lambda_1, \cdots, \lambda_{N-1})^T$ by selecting last N bits of $\mathbf{X} \in \mathbb{F}_2^n$. It is obvious that there is a unique solution is equivalent to the matrix $\mathbf{L}$ is invertible, which means that the rank of $\mathbf{L}$ equals to n. The problem turns to figure out the proportion of the matrices of rank n in $\mathbf{L} \in \mathbb{F}_2^{(n+l)\times n}$. Li et al. [10] gave the following result estimating the proportion of invertible matrices in finite field among all matrices:

Theorem 1. *Let $\mathbb{F}_q$ be the finite field with q elements, where q is a prime power. The proportion of matrices of rank n in the set of $(n+l)\times n$ matrices with entries in $\mathbb{F}_q$ is equal to:*

$$\prod_{k=l+1}^{n+l} \left(1 - q^{-k}\right), \quad l = 0, 1, 2, \cdots .$$

According to the theorem above, we give the proportion of the matrices of rank n in $\mathbb{F}_q$ in Table 3 blow. It implies that if l grows, the probability that the random matrix $\mathbf{L}$ is invertible is also increasing. In the case of our attack, $q = 2$, $l = 4$, and the random matrix $\mathbf{L}$ is invertible with high probability.

Table 3. The proportion of the matrices of rank n in $\mathbf{L} \in \mathbb{F}_q^{(n+l)\times n}$

q	$l=0$	$l=1$	$l=2$	$l=3$	$l=4$
2	0.2889	0.5776	0.7701	0.8801	0.9388
3	0.5601	0.8402	0.9452	0.9816	0.9938
7	0.8368	0.9763	0.9966	0.9995	0.9999

For any ciphertext pair (C_A, C_B) in Eq. (4.1), we could always get $C_B(1) = C_A(1) rk_{A\to B}(1) \bmod q$, which also holds on $\mathbb{F}_2$. Specifically, we could obtain a new equation:

$$C_B(1) = C_A(1)(\lambda_0 + \lambda_1 + \cdots + \lambda_{N-1}) \mod 2,$$

where $C_A(1), C_B(1)$ are fixed number. Adding this equation to the system of linear equations that we seek to solve, and now we can take $l = 3$ to implement our attack. Since the number of variables is $n = N + \left[\frac{N}{2}\right] + 1$, $l = 3$, thus we can construct a system of linear equations with the number of equations $n + l + 1 = N + \left[\frac{N}{2}\right] + 5$, which could be solved to obtain a unique solution in time $O(N^3)$ using Gaussian elimination. Compared to running on $\mathbb{Z}_{1024}$, our algorithm requires significantly less time to run on $\mathbb{F}_2$, just a few seconds.

4.4 Recovering Private Keys

In Sect. 4.3, we have obtained the re-encryption key $rk_{A\to B} \bmod 2$. Now, we discuss how to recover the private key in this subsection. First, we recover the position of 0 bits of the private key pair (f, g) by means of the obtained

$rk_{A\to B}$ mod 2, and then reveal the remaining bits of the private key f by solving a system of linear equations.

Since $rk_{A\to B} = f_A * f_B^{-1} \bmod q$, we have that $rk_{A\to B} = f_A * f_B^{-1} \bmod 2$. If one party to the communication obtains $rk_{A\to B}$ mod 2, then can immediately calculate the other party's private key in the sense of modulo 2. Now we design a roadmap to show how to recover the private keys. For the sake of description, we assume that Bob is the malicious party, who knows $rk_{A\to B}$ mod 2 and the private key f_B:

Step 1. Considering $f_A = f_B * rk_{A\to B} \bmod 2$. Since $f_A = 1 + p * F$ with $F \in \mathcal{L}_{(d_f,d_f)}$ and $p = 3$, we get $p * F = f_B * rk_{A\to B} - 1 \bmod 2$. Note that there are d_f $+1$'s, d_f -1's and $(N - 2d_f)$ 0's in the coefficients of F, so the position of 0 bits of F can be determined by counting the position of the 0 bits of $f_B * rk_{A\to B} - 1 \bmod 2$, where the number of 0 bits of F is $N - 2d_f$. It means that we can also get the position of the 0 bits of f_A.

Step 2. Since the public key $h_A = p * g_A * f_A^{-1} \bmod q$ with $g_A \in \mathcal{L}_{(d_g,d_g)}$ holds, $h_A = p * g_A * f_A^{-1} \bmod 2$ is also satisfied, where the coefficients of g_A have d_g $+1$'s and d_g -1's, $(N - 2d_g)$ 0's. Based on $p * g_A = h_A * f_A \bmod 2$, the position of the 0 bits of g_A can be determined by counting the position of 0 bits of $h_A * f_A \bmod 2$, where the number of 0 bits is $N - 2d_g$.

Step 3. Plugging $f_A = 1 + p * F$ into $h_A * f_A = p * g_A \bmod q$, we get $h_A * (1 + p * F) = p * g_A \bmod q$, which is equivalent to the equation

$$p * h_A * F = p * g_A - h_A \bmod q. \tag{4.6}$$

For convenience, we denote $\mathbf{f}$, $\mathbf{g}$, $\mathbf{h}$ as the vector form of F, g_A, h_A, and $\mathbf{H_A}$ as the matrix form of h_A. The Eq. (4.6) can be rewritten as the following linear form:

$$p\mathbf{H_A}\mathbf{f} = p\mathbf{g} - \mathbf{h} \mod q.$$

That is,

$$p \cdot \mathbf{H_A} \begin{pmatrix} f_0 \\ f_1 \\ \vdots \\ f_{N-1} \end{pmatrix} = p \cdot \begin{pmatrix} g_0 \\ g_1 \\ \vdots \\ g_{N-1} \end{pmatrix} - \begin{pmatrix} h_0 \\ h_1 \\ \vdots \\ h_{N-1} \end{pmatrix} \mod q, \tag{4.7}$$

where $\mathbf{H_A}$ is a $N \times N$ matrix. Considering the $2N$ variables ($\mathbf{f}$ and $\mathbf{g}$) of Eq. (4.7), there are $N - 2d_f$ and $N - 2d_g$ known in $\mathbf{f}$ and $\mathbf{g}$ respectively. Hence, the number of unknown variables is $2d_f + 2d_g$, whereas the number of equations is N. According to Table 2, N is larger than $2d_f + 2d_g$ (e.g. in ees1171ep1, $N = 1171$, $d_g = 390$, $d_f = 106$, the number of equations $N = 1171$ is larger than the number of variables $2d_f + 2d_g = 992$). Hence we can recover the remaining bits of $\mathbf{f}$ by solving the system of linear equations using Gaussian elimination, then recover all bits of f_A which is Alice's private key.

At PQCrypto 2019, Liu, Pan, and Zhang [11] proposed a key recovery attack based on statistical methods, malicious receiver Bob needed huge amount of ciphertexts C_{B_i} encrypted by the same plaintext m, which is illegal and hard to implement. Here are the approximate number of ciphertexts in Table 4.

Table 4. Comparison of our work with PQCrypto 2019

	ees1087ep1	ees1171ep1	ees1499ep1
PQCrypto 2019	$4.06 \cdot 10^{17}$	$4.83 \cdot 10^{17}$	$9.67 \cdot 10^{17}$
Our work	$3.17 \cdot 10^{3}$	$3.58 \cdot 10^{3}$	$4.45 \cdot 10^{3}$

Remark. The cryptanalysis proposed by PQCrypto 2019 is based on decryption failure and statistical analysis, both require huge amount of ciphertexts and the chosen plaintexts. Moreover, the ciphertexts in latter case should be encrypted by the same plaintext. Unlike the previous ones, our attack has two advantages: (1) The amount of ciphertext required is greatly reduced. (2) There are no restrictions on plaintext, our attack only needs to be done in legal communication.

5 Case of NTRU Scheme with Different Parameter Sets

In this section, we discuss other schemes of NTRU with different parameter sets instantiated to the NTRUReEncrypt. We divide them into two cases, one with a constant number of $+1$, -1 (if any) coefficients of the secret polynomial selected in $\mathcal{L}_r$, in which case we can still attack with the same method as in the previous section, and the other with the NTRU schemes in the third round of NIST-PQC competition, which have a variable number of $+1$, -1 (if any) coefficients of the secret polynomial selected in $\mathcal{L}_r$, and we analyze this case by a new trick.

5.1 Case of Certain Secret Polynomial Coefficients

For the case of certain secret polynomial coefficients, [12] summarised some instantiations of NTRU, and their specific parameter sets are listed in the table below, where $\mathcal{B}$ denotes the set of all polynomials with binary coefficients, $\mathcal{B}(d)$ denotes a subset of $\mathcal{B}$ with exactly d coefficients equal 1, $\mathcal{L}_m$ denotes the polynomial set whose coefficients lying between $-\frac{1}{2}(p-1)$ and $\frac{1}{2}(p-1)$ (Table 5).

One can check that, as for the secret polynomial e selected from $\mathcal{L}_r$ in these schemes, the inner product of its coefficient vectors is a constant. Then we can use the method proposed in Sect. 4 to recover the private keys.

Table 5. Some instantiations of NTRU

Parameter Sets	q	p	$\mathcal{L}_f$	$\mathcal{L}_g$	$\mathcal{L}_m$	$\mathcal{L}_r$
NTRU-1998	$2^k \in [\frac{N}{2}, N]$	3	$\mathcal{L}_{(d_f, d_f-1)}$	$\mathcal{L}_{(d_g, d_g)}$	$\mathcal{L}_m$	$\mathcal{L}_{(d_r, d_r)}$
NTRU-2001	$2^k \in [\frac{N}{2}, N]$	$x+2$	$1+p*F$	$\mathcal{B}(d_g)$	$\mathcal{B}$	$\mathcal{B}(d_r)$
NTRU-2005	prime	2	$1+p*F$	$\mathcal{B}(d_g)$	$\mathcal{B}$	$\mathcal{B}(d_r)$

5.2 Case of Uncertain Secret Polynomial Coefficients

We now discuss the case in the third round of NIST-PQC competition, such as NTRU-HPS, NTRU-HRSS [5], whose parameter sets are instantiated to the NTRUReEncrypt. For specific parameter sets in ees1087ep1, ees1171ep1, ees1499-ep1, our attack's point is that the secret polynomial e selected in set $\mathcal{L}_r$, whose coefficients have a certain number of $+1$, -1, and 0.

However, in NTRU-HPS and NTRU-HRSS, polynomial set $\mathcal{L}_r = \mathcal{T}$ and $\mathcal{T}$ is the set of non-zero ternary polynomials of degree at most $N-2$. It indicates that we no longer have information on the number of coefficients in the secret polynomial e, thus the inner product calculation would fail. Ding et al. [6] used the property $\mathbf{e}_i = {\mathbf{e}_i}^3$, for $i \in \{0, 1, \cdots, N-1\}$ in the broadcast attack against NTRU to recover plantexts, it could also be used in this case to recover the secret keys.

Separating p from Eq. (4.1) and write it in linear form, we can get

$$\mathbf{e} = (\mathbf{C_B} - \mathbf{C_A}\mathbf{r}) * p^{-1} \mod q.$$

Since $\mathbf{e}_i = {\mathbf{e}_i}^3$, so we can get equations that eliminates $\mathbf{e}$:

$$[(\mathbf{C_B} - \mathbf{C_A}\mathbf{r}) * p^{-1}]_i = [(\mathbf{C_B} - \mathbf{C_A}\mathbf{r}) * p^{-1}]_i^3 \mod q, \tag{5.1}$$

for $i \in \{0, 1, \cdots, N-1\}$. Note that in Eq. (5.1) only $\mathbf{r}$ is the unknown variable, cubic computation generates $O(N^3)$ new monomials, and we can also linearize these monomials into new variables. Since one legal communication produces N equations, the system of linear congruence equations can be built by communicating N^2 times, thus recover $\mathbf{r}$ in time $O(N^9)$. The following table is the comparison of parameter sets between EESS#1 and NTRU-Round3, where NTRU-HPS is the same as NTRU-HRSS (Table 6).

Table 6. Comparison of EESS#1 with NTRU-Round3

Instance	Number of communications	Variables	Gaussian elimination
EESS#1	$O(N)$	$O(N)$	$O(N^3)$
NTRU-HPS	$O(N^2)$	$O(N^3)$	$O(N^9)$

6 Experiments

In this section, we present experimental results on the assumption that Bob is a malicious receiver. Due to ciphertexts C_A could be collected on the public channel and C_B could be received normally by Bob, we assumed in our experiment that the attacker could collect enough ciphertext pairs (C_A, C_B). All experiments were performed in SageMath 9.6 on a macOS Monterey 12.5.1 system with Apple M1 CPU @ 3.2GHz, 8GB RAM, and our implement was available at https://github.com/s4lTea/NTRUReEncrypt_Attack. We implemented our attack against NTRUReEncrypt scheme, whose parameter sets defined by EESS#1 are the same as those from AsiaCCS 2015 [13] and PQCrypto 2019 [11]. We performed our attack 50 times for each instance, and gave the average number of communications and running time required by the algorithm. In our experimental results, let $n = N + [\frac{N}{2}] + 1$, we could always find a matrix $\mathbf{L}$ of rank n. We splited the algorithm into 3 steps:

1) Focusing on the proxy's re-encryption stage, then generate a system of linear congruence equations with $n+4$ equations and n variables by communicating enough times.
2) Solving it on $\mathbb{F}_2$ using Gaussian elimination to obtain re-encryption key $rk_{A\rightarrow B}$ mod 2.
3) Building another system of linear congruence equations with N equations and $2d_f + 2d_g$ variables to solve, finally obtain Alice's private key.

Table 7. Experimental Results with different parameter sets

Instance	N	p	q	Rank($\mathbf{L}$)	Number of communications	Total time(min)
ees1087ep1	1087	3	2048	1634	3174	17.4
ees1171ep1	1171	3	2048	1757	3579	22.8
ees1499ep1	1499	3	2048	2249	4454	41.9

Step 1 takes some time (minutes) due to matrix multiplication operations. As it works on $\mathbb{F}_2$, so step 2 takes only a few seconds and the running time could be negligible. There are small number of variables related to the equations in step 3, so the time required to either construct or solve the equations is negligible. The experimental results are shown in Table 7. For ease of description, we take the cost of step 1 as the total time of our algorithm.

7 Conclusion

In this paper, we presented an efficient key recovery attack against NTRUReEncrypt scheme, whose parameter sets are defined by EESS#1 specification [2] from IEEE P1363.1 standard. The attack is based on a special structure of secret polynomials from the set $\mathcal{L}_r$. In addition, the key recovery attack could be extended to the NTRUReEncrypt instantiated with the NTRU parameter sets in the third round of NIST-PQC competition.

Acknowledgments. The authors would like to thank anonymous reviewers for their helpful comments and suggestions. The work of this paper was supported in part by the National Natural Science Foundation of China (Grants 61732021, 62272454).

References

1. Aono, Y., Boyen, X., Phong, L.T., Wang, L.: Key-private proxy re-encryption under LWE. In: Paul, G., Vaudenay, S. (eds.) INDOCRYPT 2013. LNCS, vol. 8250, pp. 1–18. Springer, Cham (2013). https://doi.org/10.1007/978-3-319-03515-4_1
2. Key Cryptographic Techniques Based. IEEE p1363. 1^{TM}/d1211 (2008)
3. Blaze, M., Bleumer, G., Strauss, M.: Divertible protocols and atomic proxy cryptography. In: Nyberg, K. (ed.) EUROCRYPT 1998. LNCS, vol. 1403, pp. 127–144. Springer, Heidelberg (1998). https://doi.org/10.1007/BFb0054122
4. Canetti, R., Hohenberger, S.: Chosen-ciphertext secure proxy re-encryption. In: Proceedings of the 14th ACM Conference on Computer and Communications Security, pp. 185–194 (2007)
5. Cong, C., Oussama, D., Jerey, H.: NTRU: the round 3 NIST submission (2020). https://ntru.org/release/NIST-PQ-Submission-NTRU-20201016
6. Ding, J., Pan, Y., Deng, Y.: An algebraic broadcast attack against NTRU. In: Susilo, W., Mu, Y., Seberry, J. (eds.) ACISP 2012. LNCS, vol. 7372, pp. 124–137. Springer, Heidelberg (2012). https://doi.org/10.1007/978-3-642-31448-3_10
7. Hoffstein, J., Pipher, J., Silverman, J.H.: NTRU: a ring-based public key cryptosystem. In: Buhler, J.P. (ed.) ANTS 1998. LNCS, vol. 1423, pp. 267–288. Springer, Heidelberg (1998). https://doi.org/10.1007/BFb0054868
8. Howgrave-Graham, N., et al.: The impact of decryption failures on the security of NTRU encryption. In: Boneh, D. (ed.) CRYPTO 2003. LNCS, vol. 2729, pp. 226–246. Springer, Heidelberg (2003). https://doi.org/10.1007/978-3-540-45146-4_14
9. Kra, I., Simanca, S.R.: On circulant matrices. Notices AMS **59**(3), 368–377 (2012)
10. Li, J., Pan, Y., Liu, M., Zhu, G.: An efficient broadcast attack against NTRU. In: Proceedings of the 7th ACM Symposium on Information, Computer and Communications Security, pp. 22–23 (2012)
11. Liu, Z., Pan, Y., Zhang, Z.: Cryptanalysis of an NTRU-based proxy encryption scheme from ASIACCS'15. In: Ding, J., Steinwandt, R. (eds.) PQCrypto 2019. LNCS, vol. 11505, pp. 153–166. Springer, Cham (2019). https://doi.org/10.1007/978-3-030-25510-7_9
12. Mol, P., Yung, M.: Recovering NTRU secret key from inversion oracles. In: Cramer, R. (ed.) PKC 2008. LNCS, vol. 4939, pp. 18–36. Springer, Heidelberg (2008). https://doi.org/10.1007/978-3-540-78440-1_2
13. Nuñez, D., Agudo, I., Lopez, J.: NTRUReEncrypt: an efficient proxy re-encryption scheme based on NTRU. In Proceedings of the 10th ACM Symposium on Information, Computer and Communications Security, pp. 179–189 (2015)
14. Shor, P.W.: Algorithms for quantum computation: discrete logarithms and factoring. In: Proceedings 35th Annual Symposium on Foundations of Computer Science, pp. 124–134. IEEE (1994)
15. Shor, P.W.: Polynomial-time algorithms for prime factorization and discrete logarithms on a quantum computer. SIAM Rev. **41**(2), 303–332 (1999)

Two Remarks on the Vectorization Problem

Wouter Castryck(✉) and Natan Vander Meeren

imec-COSIC, KU Leuven, Leuven, Belgium
wonter.castryck@esat.kuleuven.be

Abstract. We share two small but general observations on the vectorization problem for group actions, which appear to have been missed by the existing literature. The first observation is pre-quantum: explicit examples show that, for classical adversaries, the vectorization problem cannot in general be reduced to the parallelization problem. The second observation is post-quantum: by combining a method for solving systems of linear disequations due to Ivanyos with a Kuperberg-style sieve, one can solve the hidden shift problem, and therefore the vectorization problem, for any finite abelian $2^t p^k$-torsion group in polynomial time and predominantly relying on classical work; here t, k are any fixed non-negative integers and p is any fixed prime number.

Keywords: Group actions · Vectorization problem · Linear disequations

1 Introduction

This paper discusses two unrelated aspects of the *vectorization problem* for abelian group actions, which specializes to the classical discrete logarithm problem in the case of exponentiation in finite cyclic groups.

The first formal study of cryptographic group actions was carried out in 1990 by Brassard and Yung [8], but non-discrete-logarithm-based examples go back, at least, to the work of Brassard and Crépeau from 1986 [6]. However, none of the early concrete instances were genuinely novel, perhaps with the exception of finite symmetric groups (or abelian subgroups thereof) acting on sets of graphs, whose vectorization problem is just the graph isomorphism problem, famously solved by Babai in 2017 [4,20]. This situation changed with the independent works of Couveignes [13] and Rostovtsev–Stolbunov [31,36], who proposed to use ideal-class groups acting on sets of elliptic curves over finite fields through isogenies. Also CSIDH [10] fits within this framework. It is Couveignes who coined the term *vectorization*. The isogeny-based construction attracted a lot of attention, lately, because the corresponding vectorization problem is supposed to be hard even in the presence of quantum adversaries. At the same time, being an abelian group action, it inherits many of the features of the celebrated exponentiation map.

To date, the list of cryptographically interesting group actions remains rather limited, but since it concerns such a basic and flexible concept, it is well imaginable that new constructions remain to be discovered, both for use in a classical

T. Isobe and S. Sarkar (Eds.): INDOCRYPT 2022, LNCS 13774, pp. 658–678, 2022.
https://doi.org/10.1007/978-3-031-22912-1_29

and in a (post-)quantum context, e.g., see [23] for a candidate based on tensors. General statements on the hardness of the vectorization problem help in understanding the fundamental features and limitations of group-action-based cryptography. We present two such statements, which are small addenda to the existing literature, including surveys such as [1,17,35], but which appear to have been missed and therefore seem worth reporting.

A Pre-quantum Observation. Our first statement is classical and negative in nature: very simple constructions show that, classically, one cannot expect in general that the vectorization problem for an abelian group action reduces in polynomial time to the *parallelization problem.*[1] This contrasts with the post-quantum setting, where the vectorization and parallelization problems become computationally equivalent [16,27]. Our conclusion also contrasts with the discrete logarithm problem, which is believed to be no harder than the computational Diffie–Hellman problem in view of the Maurer–Wolf reduction [26]. It had already been pointed out, e.g. by Smith [35, §11] and Gnilke–Zumbrägel [17, p3], that Maurer–Wolf does not extend to the group action framework. But, as far as we are aware, the existence of alternative classical reductions was not ruled out yet. To the contrary: some researchers have suggested that such a reduction should exist, see e.g. [11, §1.2]. The current work refutes this.

A Post-quantum Observation. Our second observation revisits [9, §3], where it was shown how to combine a classical (= pre-quantum) method due to Friedl et al. [15, §3] for solving systems of linear disequations modulo p with a Kuperberg-style sieve [24]. This led to an easy polynomial-time quantum algorithm which solves the *hidden shift problem*, and therefore the vectorization problem, for groups of the form

$$(\mathbb{Z}_{2^{t_1}} \times \mathbb{Z}_{2^{t_2}} \times \cdots \times \mathbb{Z}_{2^{t_m}} \times \mathbb{Z}_p^n, +) \tag{1}$$

while relying mainly on classical computations; most notably, the requirements in terms of quantum memory are very limited. Here, p is a fixed prime number and the exponents t_i are bounded by a fixed integer t, but n and m can vary freely. For $t = 1$ and $n = 0$ the algorithm specializes to Simon's method [34].

In [9] it was left unnoticed that a generalization of the method of Friedl et al. due to Ivanyos [22], capable of solving systems of linear disequations modulo *powers* of p, is equally compatible with Kuperberg's sieve. This allows one to extend the algorithm from (1) to groups of the form

$$(\mathbb{Z}_{2^{t_1}} \times \mathbb{Z}_{2^{t_2}} \times \cdots \times \mathbb{Z}_{2^{t_m}} \times \mathbb{Z}_{p^{k_1}} \times \mathbb{Z}_{p^{k_2}} \times \cdots \times \mathbb{Z}_{p^{k_n}}, +) \tag{2}$$

for any fixed prime number p and any number of exponents t_i, resp. k_i, that are bounded by fixed integers t, resp. k. Without affecting the polynomial runtime and the memory-efficiency, that is.

[1] Except under cataclysmic hypotheses such as P=NP, in which case hard instances of the vectorization problem do not exist.

Moreover, as in the case of [9], this extended algorithm can be combined with Kuperberg's collimation sieve [25,29], yielding the following refinement of [9, Thm. 1.2]:

Theorem 1. *For any fixed prime number* p *and non-negative integers* t, k*, there exists a quantum algorithm for solving the hidden shift problem in any finite abelian group* $(G, +)$ *with time, query and QROM-complexity*

$$\text{poly}(\log |G|) \cdot 2^{\mathcal{O}(\sqrt{\log |2^t p^k G|})}$$

and requiring storage of $\text{poly}(\log |G|)$ *qubits.*

Here QROM stands for quantum read-only memory; this is also known as quantum random-access classical memory (QRACM), see [25, §2]. Let us also clarify that the group

$$2^t p^k G = \{ 2^t p^k g \mid g \in G \}$$

is the group obtained from G by annihilating its $2^t p^k$-torsion.

Paper Organization. In Sect. 2 we quickly recall the vectorization and parallelization problems as well as their connection to the abelian hidden shift problem. In Sect. 3 we present examples of group actions proving the non-equivalence between vectorization and parallelization in a classical context. We devote a separate Sect. 4 to solving systems of linear disequations, because a secondary aim of our paper is to make this interesting problem (which is open for moduli as small as 6) more widespread in the cryptographic community; indeed, perhaps naively, we hope that this problem will find other cryptographic applications. In Sect. 5 we describe our method for finding hidden shifts in finite abelian $2^t p^k$-torsion groups, while spending time on recalling the details of its most important plug-in: Ivanyos' algorithm from [22]. We take the opportunity to correct a minor error and to considerably sharpen the estimated runtime. The final Sect. 6 gives some concluding remarks.

2 Vectorization, Parallelization and Hidden Shift

Let $(G, +)$ be an abelian group. An *action* of G on a finite set X is a map

$$\star : G \times X \to X : (g, x) \mapsto g \star x$$

satisfying $0 \star x = x$ and $g_1 \star (g_2 \star x) = (g_1 + g_2) \star x$ for all $g_1, g_2 \in G$ and all $x \in X$. Throughout, we make the implicit assumption that the action is only ever evaluated in elements of G and X that admit an efficient representation, and that computing this evaluation is efficient as well. The *stabilizer* of an element $x \in X$ is the subgroup $\text{St}(x) = \{ g \mid g \star x = x \} \subseteq G$. The *orbit* of $x \in X$ is the subset $\text{Or}(x) = \{ g \star x \mid g \in G \} \subseteq X$ and as soon as G is finite we have $|\text{Or}(x)| \cdot |\text{St}(x)| = |G|$ for all $x \in X$. Two orbits either coincide or are disjoint, and together the orbits partition X. All elements within one orbit have the same stabilizer. The action is called *free* if all stabilizers are trivial. It is called *transitive* if there is one orbit, only.

Definition 2. *The* vectorization problem *for* $\star$ *is about explicitly determining* $g \bmod \mathrm{St}(x)$ *upon input of* $x, g \star x \in \mathrm{Or}(x)$.

One basic example of a group action is the exponentiation map

$$\mathbb{Z}_n^* \times X : (g, x) \mapsto x^g$$

in a finite cyclic group X of order n. Here, the vectorization problem specializes to the discrete logarithm problem. Note that the generators of X form one orbit, and when restricting the action to this orbit it becomes free and transitive.

The classical Diffie–Hellman key exchange protocol naturally generalizes from exponentiation in cyclic groups to arbitrary abelian group actions. Indeed, after Alice and Bob agree on a base element $x \in X$, Alice acts with a secret $g_0 \in G$ on x and sends the result $g_0 \star x$ to Bob, and likewise Bob acts with a secret $g_1 \in G$ on x and sends $g_1 \star x$ to Alice. Both parties can now compute

$$(g_0 + g_1) \star x = g_1 \star (g_0 \star x) = g_0 \star (g_1 \star x), \tag{3}$$

which can be fed to a key derivation function; note that (3) uses the assumption that G is abelian. Breaking this protocol directly relates to:

Definition 3. *The* parallelization problem *for* $\star$ *is about explicitly determining* $(g_0 + g_1) \star x$ *upon input of* $x, g_0 \star x, g_1 \star x \in \mathrm{Or}(x)$.

The parallelization problem straightforwardly reduces to the vectorization problem but the converse reduction, as we will see in Sect. 3, does not apply in general. We recall that this story changes in the presence of quantum adversaries, where the converse reduction does apply [16,27].

When studying the hardness of vectorization and parallelization, one can assume that the action is free and transitive. Indeed, it clearly suffices to assume transitivity because the vectorization problem and the parallelization problem are formulated within one orbit. But then all $x \in X$ have the same stabilizer S, so we can assume freeness by acting with G/S rather than with G. Free and transitive actions necessarily satisfy $|G| = |X|$.

Remark 4. The explicit determination of the stabilizer S can be viewed as an instance of the *hidden subgroup problem* in the abelian group G. Quantumly, this is easy using Shor's algorithm [33], but classically this may be a hard problem. Nevertheless, it is possible to compute in G/S without knowing S explicitly, because testing equivalence mod S is easy: $g_0 - g_1 \in S$ if and only if $g_0 \star x = g_1 \star x$ for whatever $x \in X$ (assuming transitivity).

For free actions, the vectorization problem can be viewed as an instance of:

Definition 5. *Given oracle access to injective functions* $f_0, f_1 : G \to X$ *such that there exists an* $s \in G$ *such that for all* $g \in G$ *we have* $f_0(g) = f_1(g + s)$, *the (abelian)* hidden shift problem *is about finding* s.

Indeed, from an input $x, s \star x$ to the vectorization problem we can construct the functions $f_i : G \to X$ as

$$f_0 : g \mapsto g \star (s \star x),$$
$$f_1 : g \mapsto g \star x,$$

which hide the shift s. Assuming access to an oracle for evaluating the functions f_0, f_1 on arbitrary superpositions over elements of G, there exist quantum algorithms due to Kuperberg [24,25] for solving the hidden shift problem in subexponential time

$$2^{\mathcal{O}(\sqrt{\log |G|})} \tag{4}$$

as well as subexponential quantum space; more precisely the algorithm from [25] requires storage of poly(log $|G|$) qubits and an amount (4) of QROM. Kuperberg studied this in the context of the hidden subgroup problem in the associated dihedral group Dih(G), which turns out to be equivalent with the hidden shift problem in G, see [24, §6].

Remark 6. There exist non-injective versions of the abelian hidden shift problem, where the problem of breaking the Legendre pseudo-random function is arguably the best-known instance in cryptography. Such versions may be easier to tackle, quantumly, and will not be considered here; see [19, Ch. 7].

While Kuperberg's algorithm admits variants with different time-memory trade-offs, see e.g. [12,30], none of them breaks through the $L_{|G|}(1/2)$-barrier in general. This does not mean that better quantum algorithms are not possible for special classes of G. Famously, this is true for 2-torsion groups, which can be handled in polynomial time using Simon's method [34]. This generalizes to 2^t-torsion for any fixed t using Kuperberg's sieve, see [5]. In a different direction, it generalizes to p-torsion for any fixed prime p using the aforementioned method due to Friedl et al. [15]. The latter authors also present a self-reducibility tool, allowing for a polynomial-time quantum solution to the abelian hidden shift problem in finite abelian groups of *any* fixed exponent r.[2] However, this requires a quantization of otherwise classical post-processing steps, resulting in more complicated quantum algorithms with more restrictive memory requirements; in particular the self-reducibility does not seem suitable for obtaining memory-friendly statements like Theorem 1. As mentioned, in [9, §3] it was shown that for $r = 2^t p$ there exists an easy workaround; we revisit this in Sect. 5, where we generalize it to $r = 2^t p^k$.

3 Non-equivalence of Vectorization and Parallelization

We claim that the vectorization problem and the parallelization problem are *not* equivalent as soon as one believes in the existence of one-way group homomorphisms, see e.g. [7, §5]. This does not contradict the results from [16,27] because,

[2] Recall: the *exponent* of a finite group is the least common multiple of the orders of its elements.

in the presence of quantum adversaries, no such one-way homomorphisms exist. But pre-quantumly we have several very classical candidates.

The construction is really simple: consider two finite abelian groups $(G_0, +)$ and $(G_1, +)$ along with an easy-to-compute but hard-to-invert group homomorphism $f : G_0 \to G_1$. Then the map

$$\star : G_0 \times G_1 \to G_1 : (g, x) \mapsto g \star x := x + f(g)$$

is a well-defined action of G_0 on G_1. The vectorization problem amounts to extracting g from a pair $x, x + f(g)$, which is of course equivalent to extracting g from $f(g)$: this is hard by assumption. On the other hand, the parallelization problem is about computing $x + f(g_0 + g_1) = x + f(g_0) + f(g_1)$ from x, $x + f(g_0)$ and $x + f(g_1)$, which is trivial.

Example 7. One classical example of a one-way group homomorphism is the squaring map

$$(\mathbb{Z}_N^*, \cdot) \;\to\; (\mathbb{Z}_N^*, \cdot) \;:\; x \mapsto x^2$$

in the unit group of the ring of integers modulo an RSA modulus N. So the vectorization problem for the corresponding group action $(\mathbb{Z}_N^* \times \mathbb{Z}_N^*) \to \mathbb{Z}_N^* : (g, x) \mapsto g^2 x$ is hard, while parallelization is trivial.

Example 8. A free and transitive example can be obtained from exponentiation

$$(\mathbb{Z}_n, +) \;\to\; (X, \cdot) \;:\; g \mapsto \alpha^g$$

in a cyclic order-n group $X = \langle \alpha \rangle$ in which the discrete logarithm problem is believed to be hard. The vectorization problem for the corresponding group action $(\mathbb{Z}_n \times X) \to X : (g, x) \mapsto x\alpha^g$ is hard, and parallelization is straightforward.

Interestingly, Example 7 may have been the first non-exponentiation based group action that saw study in the context of cryptography [6], yet for the purpose of bit commitment rather than key exchange.

4 Systems of Linear Disequations and the Standard Approach to Hidden Shift Finding

A system of *linear disequations* over an integer residue ring $\mathbb{Z}_r$, for some $r > 1$, is a system of the form

$$\begin{cases} a_{11}s_1 + a_{12}s_2 + \ldots + a_{1n}s_n \neq b_1, \\ a_{21}s_1 + a_{22}s_2 + \ldots + a_{2n}s_n \neq b_2, \\ \qquad \vdots \\ a_{m1}s_1 + a_{m2}s_2 + \ldots + a_{mn}s_n \neq b_m, \end{cases}$$

with known $a_{ij}, b_i \in \mathbb{Z}_r$ $(1 \leq i \leq m,\ 1 \leq j \leq n)$, where one wants to solve for $s_1, \ldots, s_n$. It is an intriguing (and not very widespread) open problem how to do this in general. Of course, for $r = 2$ one just faces a system of linear equations in disguise. More generally, for $r = p$ a prime number, one can re-express every disequation as

$$(a_{i1}s_1 + a_{i2}s_2 + \ldots + a_{in}s_n - b_i)^{p-1} = 1, \tag{5}$$

thus obtaining a system of non-linear (as soon as $p > 2$) equations, which can be fed to a Gröbner basis calculation. Alternatively, if we have unlimited access to random disequations then we can solve this by linearization: this is the approach from [15] and it runs in polynomial time for fixed p. This can be generalized to $r = p^k$ for any $k \geq 1$, following Ivanyos [22], but away from prime powers we are clueless about how to approach this problem. Even seemingly harmless rings such as $\mathbb{Z}_6 \cong \mathbb{Z}_2 \times \mathbb{Z}_3$ remain unsolved. Let us stress that the algorithms from [15,22] are pre-quantum. This being said, we do not know of quantum algorithms that perform significantly better than their pre-quantum counterparts (apart from speed-ups of Grover type [18] in search steps).

Systems of linear disequations naturally show up in the "standard" quantum approach to solving the hidden shift problem in a finite abelian group $(G, +)$, which can always be assumed to be of the form

$$\mathbb{Z}_{r_1} \times \mathbb{Z}_{r_2} \times \cdots \times \mathbb{Z}_{r_n}, +$$

for integers r_i. This standard approach works by generating many *phase vectors*:

Definition 9. *Given a finite abelian group G, let*

$$G^\vee = \{\, \textit{group homomorphisms } (G, +) \to (\mathbb{C}^*, \cdot) \,\}$$

denote the dual group, equipped with point-wise multiplication. Then for any $\chi \in G^\vee$ and $s \in G$ the quantum state

$$|\Psi_s(\chi)\rangle = \frac{1}{\sqrt{2}}(|0\rangle + \chi(s)|1\rangle)$$

is called a phase vector *over G.*

Within our context, the value of $s = (s_1, s_2, \ldots, s_n)$ will always be the hidden shift we are looking for: therefore we drop the subscript and just write $|\Psi(\chi)\rangle$. Creating such a phase vector for some uniformly random $\chi \in G^\vee$ is standard practice and comes at the cost of two quantum Fourier transforms, one call to f_0 and one call to f_1 [24,30]. We treat this as a black box and assume throughout that we have oracle access to phase vectors. We stress that the result of an oracle call is $|\Psi(\chi)\rangle$ with χ a uniformly random, *known* element of $G^\vee$. The amplitude $\chi(s)$ is unknown, though.

Phase vectors serve as input to the hidden shift finding algorithms due to Kuperberg and others [24,25,29,30]. These algorithms proceed by gradually converting the phase vectors into more interesting ones through a process of combination and measurement; a basic version of Kuperberg's sieve will appear as a subroutine in Sect. 5.

For now, we just note that when measuring $|\Psi(\chi)\rangle$ in the $|\pm\rangle$-basis, where as usual

$$|\pm\rangle = \frac{|0\rangle \pm |1\rangle}{\sqrt{2}},$$

we measure '$-$' with probability $|1-\chi(s)|^2/4$. Upon such a measurement we can conclude that $\chi(s) \neq 1$. Writing

$$\chi : (g_1, g_2, \ldots, g_n) \mapsto \exp\left(2\pi \mathbf{i}\left(\frac{a_1 g_1}{r_1} + \frac{a_2 g_2}{r_2} + \ldots + \frac{a_n g_n}{r_n}\right)\right)$$

for known a_i, this translates into a disequation

$$\frac{r}{r_1} a_1 s_1 + \frac{r}{r_2} a_2 s_2 + \cdots + \frac{r}{r_n} a_n s_n \not\equiv 0 \mod r \tag{6}$$

where $r = \operatorname{lcm}(r_1, r_2, \ldots, r_n)$ denotes the exponent of G. Querying many phase vectors leads to a large system of linear disequations, unless $s = (0, 0, \ldots, 0)$ in which case one never measures '$-$'; but this will be noticed quickly (or it can be tested beforehand). This means that we have effectively reduced the hidden shift problem over G to the problem of solving a system of linear disequations. A more formal discussion will be given in Sect. 5.2.

Remark 10. Clearly, disequations of the form (6) are not arbitrary. The presence of the cofactors r/r_i is totally natural, since we can only expect to determine s_i modulo r_i. But we also see that each disequation is homogeneous, i.e., all constants b_i are zero. Consequently, this approach will only allow to determine $(s_1, s_2, \ldots, s_n)$ up to multiplication with an unknown scalar $\lambda \in \mathbb{Z}_r^*$. This means that, after solving the system, one is still left with the task of determining this scalar, e.g., by exhaustive search.

Unfortunately, as mentioned before, the only moduli r for which we have a solution with polynomial run-time (for fixed r) are prime powers. Our objective however lies in solving the hidden shift problem and, as shown in [9, §3], it is possible to get rid of powers of 2 using a Kuperberg-style sieve prior to running the above reduction. This is recalled, in a generalized setting, in the next section.

5 Finding Hidden Shifts in $2^t p^k$-torsion Groups

This section covers our algorithm for solving the hidden shift problem in finite abelian $2^t p^k$-torsion groups. It is an adaptation of [9, §3], where we aim for an incorporation of Ivanyos' algorithm rather than that of Friedl et al. We can assume that our group $(G, +)$ is of the form (2) with $t = t_1 \geq \ldots \geq t_m \geq 1, k = k_1 \geq \ldots \geq k_n \geq 1$ for integers $m, n \geq 0$, and p an odd prime. The hidden shift s is written as $s = (s'_1, \ldots s'_m, s_1, \ldots, s_n)$ with $s'_i \in \mathbb{Z}_{2^{t_i}}$ and $s_i \in \mathbb{Z}_{p^{k_i}}$.

5.1 Kuperberg Sieve

The goal of this first part of the algorithm is to turn phase vectors over G into phase vectors over the subgroup $H = \mathbb{Z}_{p^{k_1}} \times \ldots \times \mathbb{Z}_{p^{k_n}}$. This is done through Kuperberg's process of *combining phase vectors,* which is about merging $|\Psi(\chi_1)\rangle$ and $|\Psi(\chi_2)\rangle$ into $|\Psi(\chi_1\chi_2^{\pm})\rangle$, as follows:

1. Tensor the two phase vectors together:
   $|\Psi(\chi_1)\rangle|\Psi(\chi_2)\rangle = \frac{1}{2}(|00\rangle + \chi_2(s)|01\rangle + \chi_1(s)|10\rangle + \chi_1(s)\chi_2(s)|11\rangle.$
2. Perform a CNOT gate on the second qubit:
   $\frac{1}{2}(|00\rangle + \chi_2(s)|01\rangle + \chi_1(s)|11\rangle + \chi_1(s)\chi_2(s)|10\rangle.$
3. Measure the second qubit:
   $|\Psi(\chi_1\chi_2^{\pm})\rangle = \frac{1}{\sqrt{2}}(|0\rangle + \chi_1(s)\chi_2^{\pm}(s)|1\rangle).$

More generally, one can combine q phase vectors $|\Psi(\chi_1)\rangle, |\Psi(\chi_2)\rangle \ldots, |\Psi(\chi_q)\rangle$ into one phase vector $|\Psi(\chi_1\chi_2^{\pm}\cdots\chi_q^{\pm})\rangle$ by repeating this procedure $q-1$ times.

We can use this to obtain phase vectors that are ℓ-*divisible* for increasing values of ℓ, in the following sense:

Definition 11. *If the character* $\chi \in G^\vee$ *satisfies*

$$\chi^{2^{t-\ell}p^k} = 1$$

for some $0 \le \ell \le t$, *then the phase vector* $|\Psi(\chi)\rangle$ *is said to be* ℓ-divisible.

More precisely, if we let r_ℓ denote the largest positive integer for which $t_{r_\ell} \ge t-\ell$, then one can combine $r_\ell + 1$ ℓ-divisible phase vectors

$$|\Psi(\chi_1)\rangle, |\Psi(\chi_2)\rangle, \ldots, |\Psi(\chi_{r_\ell+1})\rangle$$

into a single $(\ell+1)$-divisible phase vector. Indeed, write every χ_i as

$$(g_1, \ldots, g_m, h_1, \ldots, h_n) \mapsto \exp\left(2\pi\mathbf{i}(\frac{a_{i,1}g_1}{2^{t_1}} + \ldots + \frac{a_{i,m}g_m}{2^{t_m}} + \frac{b_{i,1}h_1}{p^{k_1}} + \ldots + \frac{b_{i,n}h_n}{p^{k_n}})\right)$$

By assumption, for all $1 \le j \le r_\ell$ we have $2^{t_j-t+\ell} \mid a_{i,j}$. Setting

$$c_{i,j} := a_{i,j}/2^{t_j-t+\ell} \bmod 2$$

thus yields $r_\ell + 1$ vectors of the form $(c_{i,1}, \ldots, c_{i,r_\ell})$ for $1 \le i \le r_\ell + 1$. Furthermore, these vectors are linearly dependent in $\mathbb{Z}_2$, which means that there are coefficients $d_1, \ldots, d_{r_\ell+1} \in \mathbb{Z}_2$ such that

$$d_1c_{1,j} + \ldots + d_{r_\ell+1}c_{r_\ell+1,j} = 0 \bmod 2$$

for all $1 \le j \le r_\ell$. We can calculate these coefficients classically, and combine the phase vectors $|\Psi(\chi_i)\rangle$ for which $d_i = 1$, in the sense of Kuperberg. The result is a phase vector $|\Psi(\chi)\rangle$ for which

$$\chi : (g_1, \ldots, g_m, h_1, \ldots, h_n) \mapsto \exp\left(2\pi\mathbf{i}(\frac{a_1g_1}{2^{t_1}} + \ldots + \frac{a_mg_m}{2^{t_m}} + \frac{b_1h_1}{p^{k_1}} + \ldots + \frac{b_nh_n}{p^{k_n}})\right)$$

is such that the coefficients a_j satisfy $2 \mid \frac{a_j}{2^{t_j}} 2^{t-\ell}$ for $1 \leq j \leq r_\ell$. This implies that the phase vector is in fact $(\ell+1)$-divisible. Note that in the procedure above, the phase vectors $|\Psi(\chi_i)\rangle$ for which $d_i = 0$ need not be discarded: they can be kept aside for possible later use.

Pipelining this procedure for $\ell = 0, 1, \ldots, t-1$ eventually yields a phase vector $|\Psi(\chi)\rangle$ where $\chi \in G^\vee$ is such that all the coefficients $a_1, \ldots, a_m$ are zero. This means that χ depends only on $h_1, \ldots, h_n$. We can therefore interpret this phase vector as a phase vector over H.

5.2 Disequations

Now that we have a procedure returning phase vectors over $H = \mathbb{Z}_{p^{k_1}} \times \ldots \times \mathbb{Z}_{p^{k_n}}$, we can use these for generating linear disequations over $\mathbb{Z}_{p^k}$ along the lines of Sect. 4. Here we discuss this more formally, while explaining how these disequations can be solved for $(s_1, \ldots, s_n)$. This follows Ivanyos [22], but we take the opportunity to fix a small error in step (a) and to provide a sharper degree bound in step (d), leading to an improved complexity estimate. We stress that these steps are entirely classical. We need the notion of *near uniformity*:

Definition 12. *Given a probability distribution over a finite set A along with a subset $A' \subseteq A$, we say that the distribution is* nearly uniform over A' with tolerance $c \geq 1$ *if* $\Pr(a) = 0$ *when* $a \in A \setminus A'$ *and*

$$\frac{1}{c|A'|} \leq \Pr(a) \leq \frac{c}{|A'|}$$

when $a \in A'$.

For any finite abelian group $(G, +)$ and tolerance $c \geq 1$, we formally define the problems RLD-s(G,c) and RLD-d(G, c), which are the search and decision versions of the homogeneous random linear disequations problem:

Definition 13. RLD-s(G, c) *is about finding any generator of a secret cyclic subgroup $\langle s \rangle \subseteq G$, given access to samples from a nearly uniform distribution with tolerance c over the subset $\{\, \chi \in G^\vee \mid \chi(s) \neq 1 \,\} \subseteq G^\vee$.*

It should be clear from the definition that, indeed, one can only hope to find a generator of $\langle s \rangle$ rather than s itself. This directly relates to the fact that the corresponding linear disequations are homogeneous, see Remark 10.

Definition 14. *Given unlimited access to characters $\chi \in G^\vee$ which are consistently sampled from either*

- *a nearly uniform distribution with tolerance c over $\{\, \chi \in G^\vee \mid \chi(s) \neq 1 \,\}$ for a fixed $s \in G \setminus \{0\}$, or*
- *a nearly uniform distribution with tolerance c over the entirety of $G^\vee$,*

the RLD-d(G, c) *problem is about deciding which is the case.*

Of course, in our case, we will apply these definitions to the group

$$H = \mathbb{Z}_{p^{k_1}} \times \cdots \times \mathbb{Z}_{p^{k_n}},$$

and the element s in the above problems will take the value of the corresponding component $(s_1, \ldots, s_n)$ of our hidden shift.

(a) From finding $(s_1,\ldots,s_n)$ to RLD-s(H, 3). To sample from

$$H^\vee_{s_1,\ldots,s_n} = \{\chi \in H^\vee \mid \chi(s_1, \ldots, s_n) \neq 1\},$$

we use the following method. First, using the Kuperberg sieve from Sect. 5.1, we generate a phase vector $|\Psi(\chi)\rangle$ over H, where it is easy to check that $\chi \in H^\vee$ is uniformly random. We then measure this phase vector in the $|\pm\rangle$-basis. When measuring '+' we reject the sample and start over. When measuring '−', we return χ^j for some uniformly random $j \in \{0, 1, \ldots, p^k - 1\}$ that is coprime to p.

Note that the overall probability of measuring '−' is

$$\frac{1}{|H|} \sum_{\chi \in H^\vee} \frac{|1 - \chi(s_1, \ldots, s_n)|^2}{4} = \frac{1 - \delta_{(s_1,\ldots,s_n),(0,\ldots,0)}}{2},$$

where $\delta_{\cdot,\cdot}$ denotes the Kronecker delta. If we fail to measure '−' for (say) 128 consecutive times then with overwhelming probability $(s_1, \ldots s_n) = (0, \ldots, 0)$ and we are done. Else, it follows from Bayes' theorem that the above procedure samples $\chi \in H^\vee$ with probability

$$\frac{1}{2\varphi(p^k)|H|} \sum_{\substack{j=0 \\ \gcd(j,p)=1}}^{p^k-1} |1 - \chi^j(s_1, \ldots, s_n)|^2$$

which equals 0 if $\chi(s_1, \ldots, s_n) = 1$, i.e., if $\chi \notin H^\vee_{s_1,\ldots,s_n}$, and is contained in the interval $[1/2|H|, 2/|H|]$ in the other case; see [22, Lem. 2] (here φ denotes Euler's totient function). Therefore the resulting distribution is nearly uniform over $H^\vee_{s_1,\ldots,s_n}$ with tolerance $2|H|/|H^\vee_{s_1,\ldots,s_n}| \leq 2p/(p-1) \leq 3$. Thus, by solving RLD-s(H, 3) we can find a generator of the cyclic group $\langle(s_1, \ldots, s_n)\rangle$; note that there is a small error in the corresponding statement in Ivanyos' paper [22, Prop. 1], who reduces to RLD-s(H, 2) instead. Finding the exact value of $(s_1, \ldots, s_n)$ then amounts to exhaustive search over a set of size $\langle(s_1, \ldots, s_n)\rangle \leq p^k$.

Remark 15. Testing whether a guess $(\tilde{s}_1, \ldots, \tilde{s}_n)$ is correct can be done as explained in [9, §3], by transforming phase vectors $|\Psi(\chi)\rangle$ into

$$\frac{1}{\sqrt{2}}(|0\rangle + \chi(\tilde{s}_1, \ldots, \tilde{s}_n)^{-1}\chi(s_1, \ldots, s_n)|1\rangle)$$

before measuring it in the $|\pm\rangle$-basis. As soon as we measure '−', the guess is wrong. If we fail to measure '−' for (say) 128 consecutive times then the guess was correct with overwhelming probability.

(b) From RLD-s(H, 3) to RLD-d(S, 6). For any subgroup $S \subseteq H$, we obtain a distribution on $S^\vee$ by restricting the domain of the characters from H to S. Depending on whether $(s_1, \ldots, s_n) \in S$ or not, this distribution is nearly uniform over

$$S^\vee_{s_1,\ldots,s_n} = \{\, \chi \in S^\vee \mid \chi(s_1, \ldots, s_n) \neq 1 \,\} \quad \text{or} \quad \text{the entirety of } S^\vee$$

where the tolerance doubles at worst; see [22, Lem. 3]. This can be used to reduce RLD-s$(H, 3)$ to $O(p(k_1 + \ldots + k_n))$ instances of RLD-d$(S, 6)$ for varying subgroups $S \subseteq H$, as follows. The first goal is to find a cyclic subgroup containing $(s_1, \ldots, s_n)$. To this end, we will assume that H is non-cyclic; if H is already cyclic, we can skip the next paragraph.

We start by setting $S = H$ and repeat the following procedure. Choose an isomorphism

$$\iota : S \xrightarrow{\cong} \mathbb{Z}_{p^{k'_1}} \times \ldots \times \mathbb{Z}_{p^{k'_r}} \tag{7}$$

with $r \in \{2, \ldots, n\}$ and all k'_i positive. Pick any two indices $i, j \in \{1, \ldots, r\}$ and consider the $p + 1$ index-p subgroups

$$S_{(\lambda_i:\lambda_j)} = \iota^{-1}\{\, (x_1, \ldots, x_r) \in S \mid \lambda_i x_i + \lambda_j x_j \equiv 0 \bmod p \,\}$$

with $(\lambda_i : \lambda_j) \in \mathbb{P}^1(\mathbb{Z}_p) = \{\, (a : 1) \mid a \in \mathbb{Z}_p \,\} \cup \{\, (1 : 0) \,\}$. We distinguish between two cases:

(i) if $\iota(s_1, \ldots, s_n)$ has zero components at indices i and j then $(s_1, \ldots, s_n) \in S_{(\lambda_i:\lambda_j)}$ for all $(\lambda_i : \lambda_j)$,
(ii) if not, then $(s_1, \ldots, s_n) \in S_{(\lambda_i:\lambda_j)}$ for exactly one $(\lambda_i : \lambda_j)$.

Using at most 2 calls of the form RLD-d$(S_{(\lambda_i:\lambda_j)}, 6)$ we can figure out whether we are in case (i) or (ii), and in the latter case at most $p-2$ further calls identify the unique $(\lambda_i : \lambda_j)$ for which $(s_1, \ldots, s_n) \in S_{(\lambda_i:\lambda_j)}$. In the former case we know

$$(s_1, \ldots, s_n) \in \bigcap_{(\lambda_i:\lambda_j) \in \mathbb{P}^1(\mathbb{F}_p)} S_{(\lambda_i:\lambda_j)} = \iota^{-1}\{\, (x_1, \ldots, x_r) \mid x_i \equiv x_j \equiv 0 \bmod p \,\}.$$

Thus we have replaced S by a subgroup of index p or p^2.

Repeating this process eventually leads to a cyclic subgroup

$$S \cong \mathbb{Z}_{p^{k'}}$$

of H that contains $(s_1, \ldots, s_n)$. This means that $\langle (s_1, \ldots, s_n) \rangle = p^{i-1} S$, where $i \in \{1, \ldots, k'\}$ is minimal such that a call to RLD-d$(p^i S, 6)$ reveals near uniformity over the entirety of $p^i S$.

(c) Reduction to the case of free modules. Reconsider the isomorphism ι from (7) and write $k' = \max_i k'_i$. Through composition of ι with

$$\varepsilon : \mathbb{Z}_{p^{k'_1}} \times \cdots \times \mathbb{Z}_{p^{k'_r}} \hookrightarrow \mathbb{Z}^r_{p^{k'}} : (x_1, \ldots, x_n) \mapsto \left(x_1 p^{k'-k'_1}, \ldots, x_r p^{k'-k'_r} \right) \tag{8}$$

we can embed S in the free $\mathbb{Z}_{p^{k'}}$-module $\mathbb{Z}^r_{p^{k'}}$. We turn our distribution on $S^\vee$ into a distribution on $\mathbb{Z}^{r\vee}_{p^{k'}}$ as follows: for any sample χ we can write

$$\chi \circ \iota^{-1} : (x_1, \ldots, x_r) \mapsto e^{2\pi \mathbf{i}\left(\frac{a_1 x_1}{p^{k'_1}} + \ldots + \frac{a_r x_r}{p^{k'_r}}\right)}$$

and we lift a_i to $\tilde{a}_i = a_i + f p^{k'_i}$ for some uniformly random $f \in \{0, \ldots, p^{k'-k'_i} - 1\}$ in order to end up with a character

$$\tilde{\chi} : (x_1, \ldots, x_r) \mapsto e^{2\pi \mathbf{i}\left(\frac{\tilde{a}_1 x_1 + \ldots + \tilde{a}_r x_r}{p^{k'}}\right)}. \tag{9}$$

The resulting distribution is nearly uniform over either

$$\left\{ \psi \in \mathbb{Z}^{r\vee}_{p^{k'}} \;\middle|\; \psi(\varepsilon(\iota(s_1, \ldots, s_n))) \neq 1 \right\} \qquad \text{or} \qquad \text{all of} \mathbb{Z}^{r\vee}_{p^{k'}}$$

depending on whether the distribution on $S^\vee$ was nearly uniform over $S^\vee_{s_1,\ldots,s_n}$ or all of $S^\vee$. The tolerance is not affected. Thus the calls to RLD-d$(S_{(\lambda_i:\lambda_j)}, 6)$ from above can be replaced with calls to RLD-d$(\mathbb{Z}^r_{p^{k'}}, 6)$.

(d) Solving RLD-d for free modules. From now on we simply assume that

$$H = \mathbb{Z}^n_{p^k} \quad \text{and} \quad s = (s_1, \ldots, s_n)$$

and we recall Ivanyos' method for solving RLD-d$(H, 6)$; in order to use this method in the above reduction, one needs to replace $k \leftarrow k'$, $n \leftarrow r$, $s \leftarrow \varepsilon(\iota(s))$. Along the way, we reduce the value $D = (p-1)((2p-2)^k - 1)/(2p-3)$ from Ivanyos' paper by roughly a factor 2^{k-1}.

Concretely, we let $D = p^k - 1$ and consider the space

$$V = \mathbb{Z}_p^D[x_{1,0}, \ldots, x_{1,k-1}, \ldots, x_{n,0}, \ldots, x_{n,k-1}]$$

of polynomials in nk variables of total degree at most D, where each variable occurs in degree at most $p-1$; we can assume $D \leq nk(p-1)$. The dimension of V admits the crude estimate

$$\dim V \leq \binom{nk + D}{nk} = O(n^D)$$

(remember that p and k are treated as fixed constants), although for concrete parameter sets it is more convenient to work with the precise formula

$$\dim V = \sum_{i=0}^{nk} (-1)^i \binom{nk}{i} \binom{nk + D - ip}{D - ip}, \qquad D = p^k - 1 \tag{10}$$

from [3, Thm. 5.5]. We refer to [14] for alternative upper bounds obtained from Cramér's theorem. For every character

$$\chi : (x_1, \ldots, x_n) \mapsto e^{2\pi \mathbf{i}\left(\frac{a_1 x_1 + \ldots + a_n x_n}{p^k}\right)} \tag{11}$$

that we sample, we add a new row to a matrix M with entries in $\mathbb{Z}_p$ having $\dim V$ columns, as follows. By applying the base-p expansion map

$$\delta : \mathbb{Z}_{p^k} \to \mathbb{Z}_p^k : x_0 + x_1 p + \ldots + x_{k-1} p^{k-1} \mapsto (x_0, x_1, \ldots, x_{k-1})$$

component-wise to $(a_1, \ldots, a_n)$ we end up with a vector of length nk: the corresponding row then consists of the evaluations of the monomials in V at this vector. After sampling N characters, we have $M \in \mathbb{Z}_p^{N \times \dim V}$.

If our distribution is nearly uniform over the entirety of $H^\vee$, then the kernel of M describes polynomials in V that vanish at N nearly uniformly randomly sampled points of $\mathbb{Z}_p^{nk}$. Since V does not contain non-zero polynomials that vanish everywhere, this kernel must eventually become trivial as N grows. To estimate how large N must be taken, Ivanyos makes the following beautiful (but crude) reasoning: if $\ker M$ contains a non-zero polynomial P then this polynomial is non-vanishing in at least about $p^{nk-D/(p-1)}$ points, in view of [3, Cor. 5.26]. Therefore the probability that P gets removed when passing to the next sample, and therefore the probability that $\dim \ker M$ drops, is at least roughly

$$\frac{p^{nk-D/(p-1)}}{p^{nk}} = p^{-D/(p-1)}.$$

So, incorporating our tolerance $c = 6$, we can expect about $6p^{D/(p-1)} \dim V = O(n^D)$ samples to be sufficient for revealing that $\ker M = \{0\}$.

On the other hand, if all characters χ are non-vanishing at $(s_1, \ldots, s_n)$ then the kernel is never empty. We quickly recall the argument, while highlighting the source of the improved value of D: this comes from a sharp estimate on the degree of the *carry-polynomial*

$$C(x, y) = \sum_{i=1}^{p-1} (1 - (x - i)^{p-1}) \sum_{j=p-i}^{p-1} (1 - (y - j)^{p-1}) \in \mathbb{Z}_p[x, y]$$

which for all $a, b \in \mathbb{Z}_p$ satisfies $C(a, b) = 1$ if $a + b \geq p$ and $C(a, b) = 0$ if $a + b < p$, thereby explaining its name. Ivanyos, who describes $C(x, y)$ using Langrange basis polynomials, provided the naive bound $\deg C(x, y) \leq 2p - 2$, but from [21, Thm. 1] applied to $C(x + p - 1, y)$ it follows that the degree is actually p. Through a repeated use of this carry-polynomial, for any positive integer T it is easy to construct polynomials $Q_i \in \mathbb{Z}_p[x_{1,0}, \ldots, x_{1,k-1}, \ldots, x_{T,0}, \ldots, x_{T,k-1}]$ of degree at most $(\deg C)^i = p^i$ such that

$$\delta(a_1 + \ldots + a_T) = \Big(Q_0(\delta(a_1), \ldots, \delta(a_T)), \ldots, Q_{k-1}(\delta(a_1), \ldots, \delta(a_T))\Big) \quad (12)$$

for all $a_1, \ldots, a_T \in \mathbb{Z}_{p^k}$, see [22, Lem. 5]. Choosing $T = (p^k - 1)n$, for every tuple $(a_1, \ldots, a_n)$ coming from a character χ as in (11), we can use (12) to view

$$\delta(a_1 s_1 + \ldots + a_n s_n) = \delta(\underbrace{a_1 + \ldots + a_1}_{\times s_1} + \ldots + \underbrace{a_n + \ldots + a_n}_{\times s_n} + \underbrace{0 + \ldots + 0}_{\times (T - s_1 - \ldots - s_n)})$$

as the evaluation in $(\delta(a_1), \ldots, \delta(a_n))$ of a tuple of fixed but unknown polynomials

$$P_0, \ldots, P_{k-1} \in \mathbb{Z}_p[x_{1,0}, \ldots, x_{1,k-1}, \ldots, x_{n,0}, \ldots, x_{n,k-1}],$$

of degrees satisfying $\deg P_i \leq p^i$. So we know that $\chi(s_1, \ldots, s_n) \neq 1$ if and only if the polynomial P obtained from

$$\prod_{j=0}^{k-1} (P_j^{p-1} - 1)$$

by reduction mod $x_{1,0}^p - x_{1,0}, \ldots, x_{n,k-1}^p - x_{n,k-1}$ vanishes at $(\delta(a_1), \ldots, \delta(a_n))$. This is the desired non-zero element of V.

Remark 16. We can also view

$$\delta(a_1 s_1 + \ldots + a_n s_n) = \delta(\underbrace{s_1 + \ldots + s_1}_{\times a_1} + \ldots + \underbrace{s_n + \ldots + s_n}_{\times a_n} + \underbrace{0 + \ldots + 0}_{\times (T - a_1 - \ldots - a_n)})$$

as a tuple $P'_0, \ldots, P'_{k-1}$ of *known* polynomials evaluated in the unknown entries of $(\delta(s_1), \ldots, \delta(s_n))$. The polynomial

$$\prod_{j=0}^{k-1} (P_j'^{p-1} - 1)$$

then serves as an analogue of (5): gathering enough such polynomials should allow one the recover the hidden shift (or rather the cyclic subgroup it generates) using Gröbner bases, or via linearization. We expect this to run in time $O(n^D)$, although a precise runtime analysis of this direct search approach seems hard.

Remark 17. As was suggested to us by Frederik Vercauteren, instead of using base-p expansions it may be enlightening to work with Witt vector expansions [32, §II.6], for which formulae for addition (i.e., analogues of the above polynomials Q_i) and multiplication have seen more systematic study. But we will not pursue this track here.

5.3 Kuperberg Sieve, Again

Once $s_1, \ldots, s_n$ are found, we can define $f'_0, f'_1 : \mathbb{Z}_{2^{t_1}} \times \cdots \times \mathbb{Z}_{2^{t_m}} \to X$ by letting

$$f'_0(g_1, \ldots, g_m) = f_0(g_1, \ldots, g_m, 0, \ldots, 0)$$

and

$$f'_1(g_1, \ldots, g_m) = f_1(g_1, \ldots, g_m, s_1, \ldots, s_n).$$

This gives a new hidden shift problem with hidden shift $(s'_1, \ldots, s'_m)$. We solve this by rerunning Kuperberg's sieve from Sect. 5.1. Concretely, we sieve until we obtain $(t-1)$-divisible phase vectors. Measuring these in the $|\pm\rangle$-basis results in linear disequations mod 2 in the least significant bits of $s'_1, \ldots, s'_m$. Of course, these disequations can be seen as exact equations; also note that both '+' and '−' give rise to an equation. After solving this system of linear equations, we repeat this process for $(t-2)$-divisible phase vectors, obtaining the second most significant bits. We continue until we have found all of $(s'_1, \ldots, s'_m)$.

5.4 Algorithm Summary and Complexity

The method is summarized in Algorithm 3 and determines the hidden shift $s = (s_1', \ldots, s_m', s_1, \ldots, s_n)$ in two stages. In the pseudo-code, the unspecified parameter ϵ can be increased to reduce the probability of false positives, i.e., to decrease the likeliness that a nearly uniform distribution over the entirety of $H^\vee$ is not recognized as such (leading one to conclude that $s \in S$ in Step 7 of Algorithm 2 while in fact $s \notin S$, and likewise for Step 14). The quantity $\dim V$ refers to the formula from (10).

The cost of determining $(s_1, \ldots, s_n)$ is dominated by the runs of the decision algorithm from Step 5.2(d) on the (free module versions of the) groups $S_{(\lambda_i:\lambda_j)}$ from Step 5.2(b). There are $O(n)$ such groups to be considered. In order to run the decision algorithm, we need to prepare

$$O(n^D) \text{ characters } \chi \in H_{s_1,\ldots,s_n}^\vee \tag{13}$$

and transform them into characters

$$\tilde{\chi} \in \mathbb{Z}_{p^{k'}}^{r\vee}, \qquad r \leq n, k' \leq k$$

Algorithm 1: Ivanyos' decision algorithm

Input : $H \cong \mathbb{Z}_{p^{k_1}} \times \cdots \times \mathbb{Z}_{p^{k_n}}$, p odd prime, $k = k_1 \geq k_2 \geq \cdots \geq k_n \geq 1$
$(6+\epsilon)p^{(p^k-1)/(p-1)} \dim V$ characters $\chi \in H^\vee$ which are either
(i) nearly uniform over $\{\chi \in H^\vee | \chi(s) \neq 1\}$ for a fixed $s \in H \setminus \{0\}$,
(ii) or nearly uniform over the entirety of $H^\vee$,
with tolerance 6 and where $\dim V$ is as in (10)

Output: the correct distribution: *(i)* or *(ii)*

1 Choose an isomorphism $\iota : H \to \mathbb{Z}_{p^{k_1}} \times \cdots \times \mathbb{Z}_{p^{k_n}}$;
2 Initialize a matrix M over $\mathbb{Z}_p$ with $\dim V$ columns ;
3 **for** *χ in given list of characters* **do**
4 | Lift $\chi \circ \iota^{-1}$ to a character

$$\tilde{\chi} : \mathbb{Z}_{p^k}^n \to \mathbb{C}^* : (x_1, \ldots, x_n) \mapsto e^{2\pi \mathrm{i}\left(\frac{\tilde{a}_1 x_1 + \ldots + \tilde{a}_n x_n}{p^k}\right)}$$

| as in (9) ;
5 | Add row to M consisting of the evaluations of the $\dim V$ monomials of degree at most $p-1$ in each variable and total degree at most $p^k - 1$ in

$$(\delta(\tilde{a}_1), \ldots, \delta(\tilde{a}_n)) \in \mathbb{Z}_p^{nk}$$

| (component-wise base-p expansion) ;
6 **end**
7 **if** $\ker M \neq \{0\}$ **then**
8 | **return** "distribution *(i)*" ;
9 **else**
10 | **return** "distribution *(ii)*" ;
11 **end**

---

via the methods described in Steps 5.2(b-c); the costs of these transformations are largely dominated by the estimates that follow.

Once we have the characters $\tilde{\chi}$ at our disposal, we can build the $O(n^D) \times O(n^D)$ matrix M and compute its kernel, requiring $O(n^{D\omega})$ time and $O(n^{2D})$ space, where $\omega \approx 2.373$ denotes the Alman–Williams constant [2]. Note that the same characters (13) can be reused during each run of the decision algorithm. Each character (13) is obtained by generating $O(1)$ phase vectors over H via the Kuperberg sieve from Sect. 5.1 and proceeding as in Step 5.2(a). In turn, each phase vector over H requires us to combine $O(m^t)$ phase vectors over G, and this combination takes $O(m^t)$ quantum gates, $O(m)$ quantum space, $O(m^{t-1+\omega})$ classical work and $O(m^2)$ classical space. Finally, generating a phase vector over G costs two quantum Fourier transforms over G and one call to f_0, f_1 each.

If we measure the cost of the quantum Fourier transform by $O(\log^2 |G|)$ time and $O(\log |G|)$ space [28, §5.1], we arrive at the following overall estimates for retrieving $(s_1, \ldots, s_n)$:

- $O(n^D m^t (m+n)^2)$ quantum gates, $O(m+n)$ qubits and $O(n^D m^t)$ oracle calls to f_0, f_1,
- $O(n^D m^{t-1+\omega} + n n^{D\omega})$ classical time and $O(m^2 + n^{2D})$ classical space.

The cost of determining $(s'_1, \ldots, s'_m)$ once $(s_1, \ldots, s_n)$ is found again amounts to the combination of $O(m^t)$ phase vectors via Kuperberg's sieve, but now over the smaller group

$$\mathbb{Z}_{2^{t_1}} \times \cdots \times \mathbb{Z}_{2^{t_m}}.$$

Algorithm 2: Ivanyos' search algorithm

Input : $H \cong \mathbb{Z}_{p^{k_1}} \times \cdots \times \mathbb{Z}_{p^{k_n}}$, p odd prime, $k = k_1 \geq k_2 \geq \cdots \geq k_n \geq 1$
$(6+\epsilon)p^{(p^k-1)/(p-1)} \dim V$ nearly uniform $\chi \in H^\vee$ satisfying $\chi(s) \neq 1$ for a fixed $s \in H \setminus \{0\}$, with tolerance 3 and with $\dim V$ as in (10)

Output: a generator of $\langle s \rangle$

1 Choose an isomorphism $\iota : H \to \mathbb{Z}_{p^{k_1}} \times \cdots \times \mathbb{Z}_{p^{k_n}}$;
2 **if** $n > 1$ **then**
3   Pick $i, j \in \{1, \ldots, n\}$ with k_i, k_j maximal ;
4   **repeat**
5   Take new point $(\lambda_i : \lambda_j) \in \mathbb{P}^1(\mathbb{Z}_p)$;
6         $S \leftarrow \iota^{-1}\{(x_1, \ldots, x_n) \mid \lambda_i x_i + \lambda_j x_j \equiv 0 \bmod p\}$;
7   **until** $s \in S$ *(decide by running Algorithm 1 on group S and chars $\chi|_S$)* ;
8   Find generator of $\langle s \rangle$ by running Algorithm 2 on group S and chars $\chi|_S$;
9 **else**
10   Pick generator s_0 of H ;
11     $i \leftarrow 0$;
12   **repeat**
13         $i \leftarrow i + 1$;
14   **until** $s \notin p^i H$ *(decide using Algorithm 1 on group $p^i H$ and chars $\chi|_{p^i H}$)* ;
15   **return** $p^{i-1} s_0$;
16 **end**

Algorithm 3: Finding hidden shifts in finite abelian $2^t p^k$-torsion groups

Input : $G = \mathbb{Z}_{2^{t_1}} \times \cdots \times \mathbb{Z}_{2^{t_m}} \times \mathbb{Z}_{p^{k_1}} \times \cdots \times \mathbb{Z}_{p^{k_n}}$, with p an odd prime, $t = t_1 \geq \cdots \geq t_m \geq 1$, $k = k_1 \geq \cdots \geq k_n \geq 1$
Oracle access to phase vectors $(|0\rangle + \chi(s)|1\rangle)/\sqrt{2}$ for known but uniformly random $\chi \in G^\vee$ and unknown but fixed $s \in G$

Output: $s = (s'_1, \ldots, s'_m, s_1, \ldots, s_n)$

```
1  for ℓ from 0 to t do
2  |   Determine r_ℓ maximal such that t_{r_ℓ} ≥ t − ℓ
3  end
4  if n > 0 then
5  |   L ← {} ; (will contain characters of H = Z_{p^{k_1}} × ··· Z_{p^{k_n}})
6  |   repeat
7  |   |   Call for (r_0 + 1)(r_1 + 1)···(r_{t−1} + 1) phase vectors
8  |   |   for j from 0 to t − 1 do
9  |   |   |   Divide the phase vectors into groups of size r_j + 1
10 |   |   |   Create a (j + 1)-divisible phase vector from every such group
11 |   |   end
12 |   |   sign ← measurement of resulting t-divisible |Ψ(χ)⟩ w.r.t. |±⟩ ;
13 |   |   if sign = − then
14 |   |   |   sample j from Z*_{p^k} uniformly at random ;
15 |   |   |   L ← L ∪ {χ^j} ;
16 |   |   end
17 |   |   ((if failure for 128 consecutive times ⇒ (s_1, ..., s_n) = (0, ..., 0)))
18 |   until |L| = (6 + ϵ)p^{(p^k − 1)/(p − 1)} dim V;
19 |   Find generator of ⟨(s_1, ..., s_n)⟩ by running Algorithm 2 on L ;
20 |   Find (s_1, ..., s_n) as in Remark 15 (requires few extra t-divisible |Ψ(χ)⟩) ;
21 else
22 |   Call for (r_0 + 1)(r_1 + 1)···(r_{t−2} + 1) phase vectors
23 |   for j from 0 to t − 2 do
24 |   |   Divide the phase vectors into groups of size r_j + 1
25 |   |   Create a (j + 1)-divisible phase vector from every such group
26 |   end
27 |   Apply Simon's method to obtain s mod 2G
28 |   Apply Algorithm 3 on 2G ≅ Z_{2^{t_1 − 1}} × ··· × Z_{2^{t_m − 1}}
29 end
```

So this cost is dominated by the above estimates.

We stress that the implicit constants in the O-notations above strongly depend on p, k, t, which are treated as fixed values. Finally, revisiting Remark 16, we expect that a direct search variant would reduce the classical runtime from $O(n^D m^{t-1+\omega} + nn^{D\omega})$ to $O(n^D m^{t-1+\omega} + n^{D\omega})$.

5.5 Hidden Shift Finding in Groups with Large $2^t p^k$-torsion

An almost word-by-word copy of the discussion from [9, §5] shows that the algorithm described in Sect. 5 naturally merges into Peikert's "least-significant bit first" variant [29] of Kuperberg's collimation sieve [25]. More concretely, all one needs to do is make the following adjustments to [9, Alg. 3] and [9, Alg. 4]: replace each occurrence of p with p^k, and at the point where [9, Alg. 1] is invoked, call the algorithm from Sect. 5 instead. This yields Theorem 1.

6 Conclusion

In this paper, we have presented two unrelated addenda to the existing literature on cryptographic group actions.

The first addendum is the observation that, classically, the vectorization problem does not in general admit a polynomial-time (or even sub-exponential time) reduction to the parallelization problem. This contrasts with the quantum setting, where both problems were shown to be computationally equivalent [16,27]. It also contrasts with the special case of exponentiation in finite cyclic groups, where convincing arguments in favour of the existence of a classical polynomial-time reduction were provided by Maurer and Wolf [26].

The second addendum is the remark that an algorithm due to Ivanyos [22] for solving systems of linear disequations over $\mathbb{Z}_{p^k}$ (p prime, k a positive integer) can be combined with a Kuperberg-style sieve in order to obtain a polynomial-time quantum algorithm for solving the hidden shift problem in finite abelian $2^t p^k$-torsion groups (t a positive integer, p, k, t fixed) that involves mostly classical work; in particular, the requirements in terms of quantum memory are very limited. This extends the observation from [9, §3] from $k = 1$ to arbitrary fixed values of k. Along the way, we fixed a small error in Ivanyos' reduction and we provided a sharper complexity estimate. More importantly, we hope that this paper succeeds in bringing the intriguing problem of solving systems of linear disequations to the attention of a wider audience.

As in [9], our algorithm can be merged with Kuperberg's collimation sieve into a single quantum algorithm for solving the hidden shift problem in any finite abelian group G in time

$$\mathrm{poly}(\log |G|) \cdot 2^{O(\sqrt{\log |2^t p^k G|})},$$

where the main memory requirements are in terms of quantum read-only memory: only polynomially many qubits are needed. The consequences for group-action based cryptography are as discussed in [9, Ex. 2.3]: the vectorization problem is weakened by the presence of a large $2^t p^k$-torsion group, however ideal-class groups, as used by Couveignes [13], Rostovtsev–Stolbunov [31] and in CSIDH [10], are well-protected against this, in view of the Cohen–Lenstra heuristics. Nevertheless, wariness of this potential weakness is advisable.

Acknowledgements. The author names are in alphabetical order: see https://www.ams.org//profession/leaders/CultureStatement04.pdf. The paper was written in the context of the second-listed author's participation in the Honours @ KU Leuven Programme and was supported in part by the European Research Council (ERC) under the European Union's Horizon 2020 research and innovation programme (grant agreement ISOCRYPT - No. 101020788), by CyberSecurity Research Flanders with reference number VR20192203, and by the Research Council KU Leuven under grant number C14/18/067. Both authors would like to thank the anonymous reviewers for helpful comments.

References

1. Alamati, N., De Feo, L., Montgomery, H., Patranabis, S.: Cryptographic group actions and applications. In: Moriai, S., Wang, H. (eds.) ASIACRYPT 2020. LNCS, vol. 12492, pp. 411–439. Springer, Cham (2020). https://doi.org/10.1007/978-3-030-64834-3_14
2. Alman, J., Williams, V.V.: A refined laser method and faster matrix multiplication. In: SODA 2020, pp. 522–539. SIAM (2021)
3. Assmus, E.F., Key, J.D.: Polynomial codes and finite geometries. In: Handbook of Coding Theory, vol. I, II, pp. 1269–1343 (1998)
4. Babai, L.: Graph isomorphism in quasipolynomial time [extended abstract]. In: STOC'16, pp. 684–697. ACM, New York (2016)
5. Bonnetain, X., Naya-Plasencia, M.: Hidden shift quantum cryptanalysis and implications. In: Peyrin, T., Galbraith, S. (eds.) ASIACRYPT 2018. LNCS, vol. 11272, pp. 560–592. Springer, Cham (2018). https://doi.org/10.1007/978-3-030-03326-2_19
6. Brassard, G., Crépeau, C.: Non-transitive transfer of confidence: a perfect zero-knowledge interactive protocol for SAT and beyond. In: Proceedings of the 27th IEEE Symposium on Foundations of Computer Science, pp. 188–195 (1986)
7. Brassard, G., Crépeau, C., Yung, M.: Everything in NP can be argued in *perfect* zero-knowledge in a *bounded* number of rounds. In: Ausiello, G., Dezani-Ciancaglini, M., Della Rocca, S.R. (eds.) ICALP 1989. LNCS, vol. 372, pp. 123–136. Springer, Heidelberg (1989). https://doi.org/10.1007/BFb0035756
8. Brassard, G., Yung, M.: One-way group actions. In: Menezes, A.J., Vanstone, S.A. (eds.) CRYPTO 1990. LNCS, vol. 537, pp. 94–107. Springer, Heidelberg (1991). https://doi.org/10.1007/3-540-38424-3_7
9. Castryck, W., Dooms, A., Emerencia, C., Lemmens, A.: A fusion algorithm for solving the hidden shift problem in finite abelian groups. In: Cheon, J.H., Tillich, J.-P. (eds.) PQCrypto 2021 2021. LNCS, vol. 12841, pp. 133–153. Springer, Cham (2021). https://doi.org/10.1007/978-3-030-81293-5_8
10. Castryck, W., Lange, T., Martindale, C., Panny, L., Renes, J.: CSIDH: an efficient post-quantum commutative group action. In: Peyrin, T., Galbraith, S. (eds.) ASIACRYPT 2018. LNCS, vol. 11274, pp. 395–427. Springer, Cham (2018). https://doi.org/10.1007/978-3-030-03332-3_15
11. Chenu, M., de La Morinerie: Supersingular group actions and post-quantum key exchange. PhD thesis, Institut Polytechnique de Paris (2021)
12. Childs, A.M., Jao, D., Soukharev, V.: Constructing elliptic curve isogenies in quantum subexponential time. J. Math. Crypt. **8**(1), 1–29 (2014)
13. Couveignes, J.-M.: Hard homogeneous spaces (1997). IACR Cryptology ePrint Archive, Report 2006/291. https://ia.cr/2006/291
14. Ellenberg, J.S., Gijswijt, D.: On large subsets of $\mathbb{F}_q^n$ with no three-term arithmetic progression. Ann. Math. **185**(1), 339–343 (2017)
15. Friedl, K., Ivanyos, G., Magniez, F., Santha, M., Sen, P.: Hidden translation and Orbit Coset in quantum computing. In: STOC'03, pp. 1–9. ACM, New York (2003)

16. Galbraith, S., Panny, L., Smith, B., Vercauteren, F.: Quantum equivalence of the DLP and CDHP for group actions. Math. Crypt. **1**(1), 40–44 (2021)
17. Gnilke, O.W., Zumbrägel, J.: Cryptographic group and semigroup actions. In: *WCC 2022*, Designs, Codes and Cryptography. Springer (2022)
18. Grover, L.K.: A fast quantum mechanical algorithm for database search. In: STOC'96, pp. 212–219. ACM (1996)
19. Hallgren, S.: Quantum Fourier sampling, the hidden subgroup problem, and beyond. PhD thesis, University of California, Berkeley (2000)
20. Helfgott, H.A.: Isomorphismes de graphes en temps quasi-polynomial [d'après Babai et Luks, Weisfeiler-Leman,...]. Astérisque, (407), pp.135–182 (2019). Séminaire Bourbaki (exp. no. 1125)
21. Iliashenko, I., Zucca, V.: Faster homomorphic comparison operations for BGV and BFV. In: PETS 2021 (3), pp. 246–264. De Gruyter (2021)
22. Ivanyos, G.: On solving systems of random linear disequations. Quantum Inf. Comput. **8**(6–7), 579–594 (2008)
23. Ji, Z., Qiao, Y., Song, F., Yun, A.: General linear group action on tensors: a candidate for post-quantum cryptography. In: Hofheinz, D., Rosen, A. (eds.) TCC 2019. LNCS, vol. 11891, pp. 251–281. Springer, Cham (2019). https://doi.org/10.1007/978-3-030-36030-6_11
24. Kuperberg, G.: A subexponential-time quantum algorithm for the dihedral hidden subgroup problem. SIAM J. Comput. **35**(1), 170–188 (2005)
25. Kuperberg, G.: Another subexponential-time quantum algorithm for the dihedral hidden subgroup problem. In: TQC 2013, volume 22 of Leibniz International Proceedings in Informatics (LIPIcs), pp. 20–34 (2013)
26. Maurer, U., Wolf, S.: The relationship between breaking the Diffie-Hellman protocol and computing discrete logarithms. SIAM J. Comput. **28**(5), 1689–1721 (1999)
27. Montgomery, H., Zhandry, M.: Full quantum equivalence of group action DLog and CDH, and more. In: Asiacrypt, Lecture Notes in Computer Science. Springer (2022)
28. Nielsen, M.A., Chuang, I.L.: Quantum Computation and Quantum Information. Cambridge University Press, Cambridge (2000)
29. Peikert, C.: He gives C-sieves on the CSIDH. In: Eurocrypt 2, volume 12106 of Lecture Notes in Computer Science, pp. 463–492 (2020)
30. Regev, O.: A subexponential time algorithm for the dihedral hidden subgroup problem with polynomial space (2004). Cornell University arXiv https://arxiv.org/abs/quant-ph/0406151
31. Rostovtsev, A., Stolbunov, A.: Public-key cryptosystem based on isogenies (2006). IACR Cryptology ePrint Archive, Report 2006/145. https://ia.cr/2006/145/
32. Jean-Pierre Serre. Local fields, volume 67 of Graduate Texts in Mathematics. Springer-Verlag, 1979. Translated from the French by Marvin Jay Greenberg
33. Shor, P.W.: Polynomial-time algorithms for prime factorization and discrete logarithms on a quantum computer. SIAM J. Comput. **26**(5), 1484–1509 (1997)
34. Simon, D.R.: On the power of quantum computation. SIAM J. Comput. **26**(5), 1474–1483 (1997)
35. Smith, B.: Pre- and post-quantum Diffie-Hellman from groups, actions, and isogenies. In: WAIFI 2018, volume 11321 of Lecture Notes in Computer Science, pp. 3–40 (2018)
36. Stolbunov, A.: Cryptographic schemes based on isogenies. PhD thesis, Norwegian University of Science and Technology (2012)

Efficient IBS from a New Assumption in the Multivariate-Quadratic Setting

Sanjit Chatterjee and Tapas Pandit(✉)

Department of Computer Science and Automation, Indian Institute of Science Bangalore, Bengaluru, India
{sanjit,tapas}@iisc.ac.in

Abstract. Since its introduction in 1984, identity-based signature (IBS) schemes have been studied in different settings. But, there are very few constructions available in the multivariate quadratic polynomials (MQ) setting. The existing IBS schemes in the MQ-setting are either less efficient or do not have any formal security reduction. In this paper, we investigate the problem of constructing an efficient and provably secure IBS scheme in the MQ-setting. Our starting point is the recent IBS scheme of Chen et al. which is very efficient but has some issues related to correctness and lacks a formal justification of security. We propose a modified construction that addresses the limitations of the Chen et al. proposal while retaining its efficiency. For the security reduction, we introduce a new cryptographic parameterized assumption in the MQ-setting. Our modified proposal allows any arbitrary bit string to be an identity and the size of the public parameters does not depend on the size of the universe of identities in contrast to the original proposal. Therefore, our modified scheme works as an *unbounded* IBS. Finally, we provide some justification towards the intractability of the newly introduced assumption.

Keywords: Identity-based signature · Multivariate cryptography · Post-quantum security · Provable security

1 Introduction

Shamir [Sha84] introduced the concept of identity-based cryptography to simplify certificate management process of the traditional public key cryptosystems [DH76]. In the same paper, he illustrated an identity-based signature (IBS) scheme in the RSA-setting. Since then, several IBS schemes [CC03,BLMQ05] have been proposed based on different cryptographic assumptions.

It is widely known that all the deployed public-key cryptosystems will be insecure, once a full-scale quantum computer is ready. Therefore, the research community is now involved in the race of designing cryptosystems in the post-quantum setting. Multivariate quadratic polynomials (MQ) setting is an attractive choice for quantum-safe signature schemes. In fact, four multivariate signatures Rainbow, LUOV, GeMSS and MQDSS were shortlisted for the 2nd round

T. Isobe and S. Sarkar (Eds.): INDOCRYPT 2022, LNCS 13774, pp. 679–696, 2022.
https://doi.org/10.1007/978-3-031-22912-1_30

of the NIST PQC standardization competition [NIS19], and among them Rainbow [DS05] and GeMSS were selected as a finalist and alternative candidate respectively for the 3rd round of evaluation [NIS20].

In this paper, we are interested to study post-quantum identity-based signature in the MQ-setting. In the literature, only a handful of IBS schemes [STX13, Luy19, CLND19, CDP21] have been designed in the MQ-setting. The first two proposals hardly have any practical relevance as both the signatures contain the UOV/Rainbow public-key which is rather large. On the other hand, the proposal in [CLND19] is very efficient. In fact, when instantiated using UOV as the underlying primitive, signature size as well as signing and verification time for the IBS is comparable to that of UOV. However, the proposal in [CLND19] neither ensures correctness (see Sect. 3) nor provides any formal security argument. Most recently, Chatterjee et al. [CDP21] proposed an identity-based signature scheme in the MQ-setting following the style of MQDSS signature [CHR+16]. Although the authors provided a formal security reduction of their scheme, the size of the signature is proportional to the size of the 3-pass version of MQDSS which is considerably less efficient than UOV and Rainbow signatures. So, all these limitations in the existing proposals indicate the relative difficulty of constructing a secure IBS in the MQ-setting with efficiency comparable to that of UOV.

Our Result. In this paper, we make some progress on this problem. Basically, we revisit the proposal of Chen at al. [CLND19] and modify it to resolve the issues (Sect. 3) related to the correctness. In contrast to the original proposal, our modification (Sect. 4) allows any arbitrary bit string to be an identity and the size of the public parameters does not depend on the size of the universe. Therefore, our construction essentially works as an *unbounded* identity-based signature. Further, in this modified proposal, we make some improvement in terms of efficiency as well.

As mentioned earlier in this section, the original proposal of Chen et al. does not have any formal security analysis. We provide a formal treatment (Sect. 4.1) of security of our modified proposal in the random oracle model. For the security reduction, we introduce a new parameterized assumption (Sect. 2.2) in the MQ-setting. We show that a special case[1] of the assumption is equivalent (Proposition 1) to the WMQ-problem [SSH11] and also give some justification towards the intractability of the general case (Sect. 5).

2 Preliminaries

The basic notations, like oil-and-vinegar polynomials, different classes of polynomials and a special UOV-map [CLND19] are defined in this section. The hardness assumptions, and the syntax and security of identity-based signature scheme are also provided in this section.

[1] Security of our proposal relies on this special case, if the attacker is not given any access to the key extraction oracle.

2.1 Notations and Background

Notations. For a set X, the notation $x \xleftarrow{\text{U}} X$ denotes that x is drawn uniformly at random from X. For $a, b \in \mathbb{N} \cup \{0\}$, define $[a, b] = \{x \in \mathbb{N} \cup \{0\} : a \leq x \leq b\}$ and when $b \in \mathbb{N}$, define $[b] = [1, b]$.

Oil-and-Vinegar Polynomials. Suppose v, m and n are three positive integers such that $n = v+m$ and they would be called the number of vinegar variables, oil variables and total variables respectively. Without loss of generality, we assume that out of n variables, the first v variables are vinegar variables and the rest m variables are oil variables. If $\boldsymbol{x} = (x_1, \ldots, x_v, x_{v+1}, \ldots, x_{v+m})$, then we write $\boldsymbol{x} = (\boldsymbol{x}_v, \boldsymbol{x}_m)$, where $\boldsymbol{x}_v = (x_1, \ldots, x_v)$ and $\boldsymbol{x}_m = (x_{v+1}, \ldots, x_{v+m})$. The same integer m also represents the number of polynomials in the context of system of equations.

By a quadratic polynomial map $\mathcal{F} : \mathbb{F}^n \to \mathbb{F}^m$ of oil-and-vinegar type, we mean $\mathcal{F} = (f^{(1)}, \ldots, f^{(m)})$ and each $f^{(k)} : \mathbb{F}^n \to \mathbb{F}$ is a quadratic polynomial of oil-and-vinegar type:

$$f^{(k)}(\boldsymbol{x}) = \sum_{i=1}^{v} \sum_{j=v+1}^{n} \alpha_{i,j}^{(k)} \cdot x_i x_j + \sum_{i=1}^{v} \sum_{j=1}^{v} \beta_{i,j}^{(k)} \cdot x_i x_j + \sum_{i=1}^{n} \gamma_i^{(k)} \cdot x_i + \delta^{(k)} \quad (1)$$

where $\mathbb{F}$ is a field and $\alpha_{i,j}^{(k)}, \beta_{i,j}^{(k)}, \gamma_i^{(k)}, \delta^{(k)} \in \mathbb{F}$ for $k \in [m]$.

For a quadratic polynomial map $\mathcal{F} : \mathbb{F}^n \to \mathbb{F}^m$ and a fixed $\boldsymbol{x}_v \in \mathbb{F}^v$, define a map $\mathcal{F}_{\boldsymbol{x}_v} : \mathbb{F}^m \to \mathbb{F}^m$ by $\mathcal{F}_{\boldsymbol{x}_v}(\boldsymbol{x}_m) = \mathcal{F}(\boldsymbol{x}_v, \boldsymbol{x}_m)$ for all $\boldsymbol{x}_m \in \mathbb{F}^m$. So, for $\boldsymbol{y} \in \mathbb{F}^m$, its preimage set under $\mathcal{F}_{\boldsymbol{x}_v}$ is given by:

$$\begin{aligned} \mathcal{F}_{\boldsymbol{x}_v}^{-1}(\boldsymbol{y}) &= \{\boldsymbol{z}_m \in \mathbb{F}^m : \mathcal{F}_{\boldsymbol{x}_v}(\boldsymbol{z}_m) = \boldsymbol{y}\} \\ &= \{\boldsymbol{z}_m \in \mathbb{F}^m : \mathcal{F}(\boldsymbol{x}_v, \boldsymbol{z}_m) = \boldsymbol{y}\}. \end{aligned}$$

Non-singular Matrices. The notation, $\mathsf{GL}_n(\mathbb{F})$ denotes the collection of all $n \times n$ non-singular matrices over $\mathbb{F}$. When $n = v + m$, we can write an $n \times n$ matrix B over $\mathbb{F}$ through its block representation as follows

$$B = \begin{bmatrix} B_{11} & B_{12} \\ B_{21} & 0 \end{bmatrix} \quad (2)$$

where B_{11}, B_{12} and B_{21} are $v \times v$, $v \times m$ and $m \times v$ matrices respectively, and 0 represents $m \times m$ zero matrix. Note that if $\mathcal{F} : \mathbb{F}^n \to \mathbb{F}^m$ is an oil-and-vinegar map, then $\mathcal{F} \circ B : \mathbb{F}^n \to \mathbb{F}^m$ will be of oil-and-vinegar type. Let $\widetilde{\mathsf{GL}}_n(\mathbb{F}) = \{B \in \mathsf{GL}_n(\mathbb{F}) : B$ can be written as in Eq. (2)$\}$.

Invertible Affine Maps. By invertible affine map, we mean $\mathcal{T} = (A, \boldsymbol{a}) \in \mathsf{GL}_n(\mathbb{F}) \times \mathbb{F}^n$. For $\boldsymbol{x} \in \mathbb{F}^n$, define $\mathcal{T}(\boldsymbol{x}) := A\boldsymbol{x} + \boldsymbol{a}$. So, $\mathcal{T}$ is a map from $\mathbb{F}^n$ onto $\mathbb{F}^n$. We use the notation $\mathsf{invAff}(\mathbb{F}^n, \mathbb{F}^n)$ to denote the set of all invertible affine maps from $\mathbb{F}^n$ onto $\mathbb{F}^n$.

Classes of Polynomial Maps. Let $\mathcal{F}_{\mathsf{uov}}(\mathbb{F}^n, \mathbb{F}^m)$ be the collection of all quadratic polynomial maps $\mathcal{F} : \mathbb{F}^n \to \mathbb{F}^m$ of oil-and-vinegar type. Define

$\mathcal{P}_{\mathsf{uov}}(\mathbb{F}^n, \mathbb{F}^m) = \{\mathcal{F} \circ \mathcal{T} : \mathcal{F} \in \mathcal{F}_{\mathsf{uov}}(\mathbb{F}^n, \mathbb{F}^m) \wedge \mathcal{T} \in \mathsf{invAff}(\mathbb{F}^n, \mathbb{F}^n)\}$. Let $\mathcal{P}(\mathbb{F}^n, \mathbb{F}^m)$ be the collection of all quadratic polynomial maps $\mathcal{P} : \mathbb{F}^n \to \mathbb{F}^m$. Obviously, $\mathcal{P}_{\mathsf{uov}}(\mathbb{F}^n, \mathbb{F}^m)$ is a subset of $\mathcal{P}(\mathbb{F}^n, \mathbb{F}^m)$.

Special UOV-Maps [CLND19]**.** Usually, the key-pair of a UOV-system consists of a public UOV-map $\mathcal{P} = \mathcal{F} \circ \mathcal{T} \in \mathcal{P}_{\mathsf{uov}}(\mathbb{F}^n, \mathbb{F}^m)$ and a secret key $(\mathcal{F}, \mathcal{T}) \in \mathcal{F}_{\mathsf{uov}}(\mathbb{F}^n, \mathbb{F}^m) \times \mathsf{invAff}(\mathbb{F}^n, \mathbb{F}^n)$, where the coefficients of each map $f^{(k)}$ and the entries in $\mathcal{T}$ are constant. In this work, we consider a special type of central map and affine map which are described next. In the following, we first define $\widetilde{\mathcal{F}} = (f^{(1)}, \ldots, f^{(m)})$ and $\widetilde{\mathcal{T}} = (A, \boldsymbol{a})$ respectively, where the coefficients of each $f^{(k)}$ and the entries of the matrix A and the vector $\boldsymbol{a}$ involved in $\mathcal{T}$ are not constant, rather linear maps in the variables in $\boldsymbol{z} := (z_1, \ldots, z_d)$ for some positive integer d. In fact, for $\boldsymbol{x} = (x_1, \ldots, x_n) \in \mathbb{F}^n$, $f^{(k)}$ for $k \in [m]$ and $\widetilde{\mathcal{T}} = (A, \boldsymbol{a})$ are given by

$$f^{(k)}(\boldsymbol{x}) = \sum_{i=1}^{v} \sum_{j=v+1}^{n} \alpha_{i,j}^{(k)}(\boldsymbol{z}) \cdot x_i x_j + \sum_{i=1}^{v} \sum_{j=1}^{v} \beta_{i,j}^{(k)}(\boldsymbol{z}) \cdot x_i x_j + \sum_{i=1}^{n} \gamma_i^{(k)}(\boldsymbol{z}) \cdot x_i + \delta^{(k)}(\boldsymbol{z}) \tag{3}$$

$$(\widetilde{\mathcal{T}}(\boldsymbol{x}))_i := \sum_{j=1}^{n} A_{i,j}(\boldsymbol{z}) \cdot x_j + a_i(\boldsymbol{z}) \text{ for } i \in [n] \tag{4}$$

where $\alpha_{i,j}^{(k)}, \beta_{i,j}^{(k)}, \gamma_i^{(k)}, \delta^{(k)}$, $A_{i,j}$ and a_i are random linear maps in $z_1, \ldots, z_d$. When $\alpha_{i,j}^{(k)}, \beta_{i,j}^{(k)}, \gamma_i^{(k)}$ and $\delta^{(k)}$ are evaluated at a fixed $\boldsymbol{z}$, then $f^{(k)} : \mathbb{F}^n \to \mathbb{F}$ will become a quadratic oil-and-vinegar polynomial and hence $\widetilde{\mathcal{F}} : \mathbb{F}^n \to \mathbb{F}^m$ will be a central oil-and-vinegar map. Similarly, when $A_{i,j}$ and a_i are evaluated at a fixed $\boldsymbol{z}$, then $\widetilde{\mathcal{T}} : \mathbb{F}^n \to \mathbb{F}^n$ will become an affine map. The public parameter $\widetilde{\mathcal{P}} = (p^{(1)}, \ldots, p^{(m)})$ is defined by

$$\widetilde{\mathcal{P}} := \widetilde{\mathcal{F}} \circ \widetilde{\mathcal{T}} = (f^{(1)} \circ \widetilde{\mathcal{T}}, \ldots, f^{(m)} \circ \widetilde{\mathcal{T}}). \tag{5}$$

Note that the coefficients of 2-degree monomials, 1-degree monomials and constant term in each $p^{(k)}$ are polynomials in $z_1, \ldots, z_d$ of degree 3, 2 and 1 respectively. Similarly as above, when the coefficients of each $p^{(k)}$ are evaluated at a fixed $\boldsymbol{z}$, then $\widetilde{\mathcal{P}} : \mathbb{F}^n \to \mathbb{F}^m$ will be a public oil-and-vinegar map. To summarize, for a fixed $\boldsymbol{z}$, the maps $\widetilde{\mathcal{F}} : \mathbb{F}^n \to \mathbb{F}^m$, $\widetilde{\mathcal{T}} : \mathbb{F}^n \to \mathbb{F}^n$ and $\widetilde{\mathcal{P}} : \mathbb{F}^n \to \mathbb{F}^m$ constitute an instantiation of UOV-maps, which will be denoted by $\widetilde{\mathcal{F}}_{\boldsymbol{z}} : \mathbb{F}^n \to \mathbb{F}^m$, $\widetilde{\mathcal{T}}_{\boldsymbol{z}} : \mathbb{F}^n \to \mathbb{F}^n$ and $\widetilde{\mathcal{P}}_{\boldsymbol{z}} : \mathbb{F}^n \to \mathbb{F}^m$ respectively in the rest of this paper.

2.2 Hardness Assumption

In this section, we recall the MQ-problem and some variant of it. We also introduce a new parameterized problem based on which we provide a security reduction of our proposed IBS construction in Sect. 4.

Definition 1 (MQ-Problem). *Given* $(\mathcal{P}, \boldsymbol{y}^*) \in \mathcal{P}(\mathbb{F}^n, \mathbb{F}^m) \times \mathbb{F}^m$, *find an* $\boldsymbol{x}^* \in \mathbb{F}^n$ *such that* $\boldsymbol{y}^* = \mathcal{P}(\boldsymbol{x}^*)$. *The advantage of an algorithm* $\mathcal{A}$ *in breaking the MQ-Problem is defined by*

$$\mathsf{Adv}_{\mathcal{A}}^{\mathrm{MQ}}(\kappa) = \Pr\left[\mathcal{P}(\boldsymbol{x}^*) = \boldsymbol{y}^* : (\mathcal{P}, \boldsymbol{y}^*) \xleftarrow{\mathrm{U}} \mathcal{P}(\mathbb{F}^n, \mathbb{F}^m) \times \mathbb{F}^m;\ \boldsymbol{x}^* \leftarrow \mathcal{A}(\mathcal{P}, \boldsymbol{y}^*)\right].$$

We say the MQ-Problem is intractable, if for every quantum PPT algorithm $\mathcal{A}$, *the advantage* $\mathsf{Adv}_{\mathcal{A}}^{\mathrm{MQ}}(\kappa)$ *is a negligible function in* κ.

Now, consider a special case of the MQ-Problem which we call the WMQ-Problem. The authors [SSH11] considered this problem for proving security of their salted version of UOV-signature.

Definition 2 (WMQ-Problem[SSH11]). *Given* $(\mathcal{P}, \boldsymbol{y}^*) \in \mathcal{P}_{\mathsf{uov}}(\mathbb{F}^n, \mathbb{F}^m) \times \mathbb{F}^m$, *find an* $\boldsymbol{x}^* \in \mathbb{F}^n$ *such that* $\boldsymbol{y}^* = \mathcal{P}(\boldsymbol{x}^*)$. *The advantage* $\mathsf{Adv}_{\mathcal{A}}^{\mathrm{WMQ}}(\kappa)$ *of an algorithm* $\mathcal{A}$ *in breaking the WMQ-Problem is defined by*

$$\Pr\left[\mathcal{P}(\boldsymbol{x}^*) = \boldsymbol{y}^* : (\mathcal{P}, \boldsymbol{y}^*) \xleftarrow{\mathrm{U}} \mathcal{P}_{\mathsf{uov}}(\mathbb{F}^n, \mathbb{F}^m) \times \mathbb{F}^m;\ \boldsymbol{x}^* \leftarrow \mathcal{A}(\mathcal{P}, \boldsymbol{y}^*)\right].$$

We say the WMQ-Problem is intractable, if for every quantum PPT algorithm $\mathcal{A}$, *the advantage* $\mathsf{Adv}_{\mathcal{A}}^{\mathrm{WMQ}}(\kappa)$ *is a negligible function in* κ.

The following newly introduced problem is no harder than the WMQ problem.

Definition 3 (ν-Parameterized Weak Multivariate Quadratic (PWMQ) Problem). *For a non-negative integer* ν, *we define the problem instance as follows: Let* d *be a positive integer.*

1. *Consider* $\widetilde{\mathcal{F}} = (f^{(1)}, \ldots, f^{(m)})$, $\widetilde{\mathcal{T}} = (A, \boldsymbol{a})$ *and* $\widetilde{\mathcal{P}} = (p^{(1)}, \ldots, p^{(m)})$ *as defined in Eqs. (3), (4) and (5) respectively.*
2. *Pick a random subset* $V \subset \mathbb{F}^d$ *with* $|V| = \nu$ *such that* $\widetilde{\mathcal{T}}_{\boldsymbol{z}}$ *is non-singular*[2] *for all* $\boldsymbol{z} \in V$.
3. *Pick* $\boldsymbol{z}^* \xleftarrow{\mathrm{U}} \mathbb{F}^d \setminus V$ *such that* $\widetilde{\mathcal{T}}_{\boldsymbol{z}^*}$ *is non-singular.*
4. *For each* $\boldsymbol{z} \in V$, *choose* $R_{\boldsymbol{z}} \xleftarrow{\mathrm{U}} \widetilde{\mathsf{GL}}_n(\mathbb{F})$ *and compute* $\widetilde{\mathcal{F}}_{\boldsymbol{z}}$ *and* $\widetilde{\mathcal{T}}_{\boldsymbol{z}}$. *Then, set* $\mathcal{F}_{\boldsymbol{z}} := \widetilde{\mathcal{F}}_{\boldsymbol{z}} \circ R_{\boldsymbol{z}}$, $\mathcal{T}_{\boldsymbol{z}} := (R_{\boldsymbol{z}})^{-1} \circ \widetilde{\mathcal{T}}_{\boldsymbol{z}}$ *and* $\mathcal{SK}_{\boldsymbol{z}} := (\mathcal{F}_{\boldsymbol{z}}, \mathcal{T}_{\boldsymbol{z}})$.
5. *Pick* $\boldsymbol{y}^* \xleftarrow{\mathrm{U}} \mathbb{F}^m$ *and set* $\mathcal{D} := \left(\widetilde{\mathcal{P}}, (\boldsymbol{z}, \mathcal{SK}_{\boldsymbol{z}})_{\boldsymbol{z} \in V}, \boldsymbol{z}^*, \boldsymbol{y}^*\right)$.

Given $\mathcal{D}$, *find* $\boldsymbol{x}^* \in \mathbb{F}^n$ *such that* $\widetilde{\mathcal{P}}_{\boldsymbol{z}^*}(\boldsymbol{x}^*) = \boldsymbol{y}^*$. *The advantage of an algorithm* $\mathcal{A}$ *in breaking the* ν*-PWMQ problem is defined by*

$$\mathsf{Adv}_{\mathcal{A}}^{\nu\text{-PWMQ}}(\kappa) = \Pr\left[\widetilde{\mathcal{P}}_{\boldsymbol{z}^*}(\boldsymbol{x}^*) = \boldsymbol{y}^* : \boldsymbol{x}^* \leftarrow \mathcal{A}(\mathcal{D})\right].$$

We say the ν*-PWMQ problem is intractable, if for every quantum PPT algorithm* $\mathcal{A}$, *the advantage* $\mathsf{Adv}_{\mathcal{A}}^{\nu\text{-PWMQ}}(\kappa)$ *is a negligible function in* κ.

A special case of this problem is for $\nu = 0$. It is useful for arguing security of the proposed construction against an attacker who does not have access to the key-extract oracle. In Sect. 5, we discuss about the plausible intractability of this new parameterized problem.

---

[2] If we write $\widetilde{\mathcal{T}}_{\boldsymbol{z}} = (A_{\boldsymbol{z}}, \boldsymbol{a}_{\boldsymbol{z}})$, then $\widetilde{\mathcal{T}}_{\boldsymbol{z}}$ is singular if and only if $A_{\boldsymbol{z}}$ is singular.

2.3 Identity-Based Signature

Definition 4 (IBS Scheme). *It consists of four PPT algorithms -* IBS.Setup, IBS.Extract, IBS.Sign *and* IBS.Ver.

- IBS.Setup: *It takes as input a security parameter κ and outputs public parameters and master secret key pair $(\mathcal{PP}, \mathcal{MSK})$.*
- IBS.Extract: *It takes as input public parameters $\mathcal{PP}$, master secret key $\mathcal{MSK}$ and an identity $\mathsf{id} \in \mathcal{U}$, where $\mathcal{U}$ is the universe of identities, and outputs a signing key $\mathcal{SK}_{\mathsf{id}}$.*
- IBS.Sign: *It takes as input public parameters $\mathcal{PP}$, a message $\mathrm{m} \in \mathcal{M}$, where $\mathcal{M}$ is the message space, and a secret key $\mathcal{SK}_{\mathsf{id}}$ and outputs a signature σ.*
- IBS.Ver: *It takes as input public parameters $\mathcal{PP}$, a message m, a signature σ and an identity* id. *It outputs a value 1, if σ is a valid signature for $(\mathrm{m}, \mathsf{id})$, else it outputs 0.*

Correctness: *For all $(\mathcal{PP}, \mathcal{MSK}) \xleftarrow{\text{U}} \mathsf{IBS.Setup}(1^\kappa)$, for all $\mathsf{id} \in \mathcal{U}$, $\mathcal{SK}_{\mathsf{id}} \xleftarrow{\text{U}} \mathsf{IBS.Extract}(\mathcal{PP}, \mathcal{MSK}, \mathsf{id})$ and for all $\mathrm{m} \in \mathcal{M}$, it is required that*

$$\mathsf{IBS.Ver}(\mathcal{PP}, \mathrm{m}, \mathsf{IBS.Sign}(\mathcal{PP}, \mathrm{m}, \mathcal{SK}_{\mathsf{id}}), \mathsf{id}) = 1.$$

Next, we define the standard security notion of IBS, called existentially unforgeable under chosen-key and chosen-message attack (EUF-ID-CMA).

Definition 5 (EUF-ID-CMA). *An IBS scheme is said to be EUF-ID-CMA secure, if for all quantum PPT algorithms $\mathcal{A}$, the advantage*

$$\mathsf{Adv}_{\mathcal{A}}^{\text{EUF-ID-CMA}}(\kappa) := \Pr\left[\mathsf{Exp}_{\mathcal{A}}^{\text{EUF-ID-CMA}}(\kappa) = 1\right]$$

in $\mathsf{Exp}_{\mathcal{A}}^{\text{EUF-ID-CMA}}(\kappa)$ (defined in Fig. 1) is a negligible function in κ, where $\mathcal{A}$ is provided access to signature oracle $\mathcal{O}_{\mathsf{Sign}}$ and key-extract oracle $\mathcal{O}_{\mathsf{Extr}}$ at most polynomial number of times, and $\mathsf{Q}_{\mathsf{extr}}$ is the set of identities on which key-extract queries were made and $\mathsf{Q}_{\mathsf{sign}}$ is the set of message-identity pairs on which signature queries were made.

$\mathsf{Exp}_{\mathcal{A}}^{\text{EUF-ID-CMA}}(\kappa)$:

1. $(\mathcal{PP}, \mathcal{MSK}) \xleftarrow{\text{U}} \mathsf{IBS.Setup}(1^\kappa)$
2. $(\mathrm{m}^*, \mathsf{id}^*, \sigma^*) \longleftarrow \mathcal{A}^{\{\mathcal{O}_{\mathsf{Extr}}, \mathcal{O}_{\mathsf{Sign}}\}}(\mathcal{PP})$
3. return 0, if $\mathsf{id}^* \in \mathsf{Q}_{\mathsf{extr}}$ or $(\mathrm{m}^*, \mathsf{id}^*) \in \mathsf{Q}_{\mathsf{sign}}$ or $\mathsf{IBS.Ver}(\mathcal{PP}, \mathrm{m}^*, \sigma^*, \mathsf{id}^*) = 0$
4. return 1

Fig. 1. Experiment for EUF-ID-CMA of IBS scheme

EUF-nID-CMA. The security notion, existentially unforgeable under no-key and chosen-message attack (EUF-nID-CMA) can be defined similarly as EUF-ID-CMA, except that no key-extract oracle is provided to the adversary.

3 Revisiting the IBS of Chen et al.

In [CLND19], the authors proposed two IBS constructions: the first construction, called ID-UOV, uses UOV-style signature generation [KPG99], whereas the second one, called ID-Rainbow, uses the Rainbow-style signature generation [DS05]. The novelty of their constructions lies in the key-extraction for different identities. Here, we mainly discuss ID-UOV as ID-Rainbow has a similar structure. The universe of identities is considered to be $\mathcal{U} = \{U_1, \ldots, U_N\}$ for some positive integer N. For each user U_i, a unique identifier $\boldsymbol{z} = (z_1, \ldots, z_d) \in \mathbb{F}^d$ is assigned, where d is some positive but fixed integer. [CLND19] did not provide any details on the derivation of these public identifiers but from the scheme description one can surmise that the size of $\mathcal{U}$ can be at most a polynomial in the security parameter.

The master secret key $\mathcal{MSK}$ consists of the random linear maps $\alpha_{i,j}^{(k)}(\boldsymbol{z})$, $\beta_{i,j}^{(k)}(\boldsymbol{z}), \gamma_i^{(k)}(\boldsymbol{z}), \delta^{(k)}(\boldsymbol{z}), A_{i,j}(\boldsymbol{z})$ and $a_i(\boldsymbol{z})$ as defined in Eqs. (3) and (4). These random linear maps basically represent the expressions for the central map $\widetilde{\mathcal{F}}$ and the affine map $\widetilde{\mathcal{T}}$. For each user with unique identifier $\boldsymbol{z}$, a non-singular linear transformation[3] $R_{\boldsymbol{z}}$ (having the form as defined in Eq. (2)) is also considered as part of the $\mathcal{MSK}$. This linear transformation works as a randomizer while extracting the secret key for a user having the identifier $\boldsymbol{z}$. The public map $\widetilde{\mathcal{P}}$ defined in Eq. (5) is included as part of the public parameters $\mathcal{PP}$.

For a user U with the unique identifier $\boldsymbol{z}$, the secret key $\mathcal{SK}_U$ is extracted as follows. Evaluate each linear map appearing in $\mathcal{MSK}$ at $\boldsymbol{z}$ to obtain the concrete secret maps $\widetilde{\mathcal{F}}_{\boldsymbol{z}}$ and $\widetilde{\mathcal{T}}_{\boldsymbol{z}}$. Then compute $\mathcal{F}_{\boldsymbol{z}} = \widetilde{\mathcal{F}}_{\boldsymbol{z}} \circ R_{\boldsymbol{z}}$ and $\mathcal{T}_{\boldsymbol{z}} = (R_{\boldsymbol{z}})^{-1} \circ \widetilde{\mathcal{T}}_{\boldsymbol{z}}$, and assign $\mathcal{SK}_U = (\mathcal{F}_{\boldsymbol{z}}, \mathcal{T}_{\boldsymbol{z}})$ to the user U. This key basically plays the role of a particular instance of UOV, where the corresponding public key is $\mathcal{PK}_U = \widetilde{\mathcal{P}}_{\boldsymbol{z}} = \widetilde{\mathcal{F}}_{\boldsymbol{z}} \circ \widetilde{\mathcal{T}}_{\boldsymbol{z}}$. Note that $\widetilde{\mathcal{P}}_{\boldsymbol{z}}$ can be computed by evaluating each coefficient of the polynomials involved in the public parameter $\widetilde{\mathcal{P}}$ at $\boldsymbol{z}$. The signature generation and verification process are the same as that of the UOV scheme using $(\mathcal{SK}_U, \mathcal{PK}_U)$ as the corresponding UOV key-pair.

Problem of Non-singularity. We point out an issue related to the non-singularity of $\widetilde{\mathcal{T}}_{\boldsymbol{z}}$ that has a bearing on the correctness of the scheme. A necessary condition for the scheme to function properly is that $(\mathcal{SK}_U, \mathcal{PK}_U)$ must be valid for all the registered users $U \in \mathcal{U}$. This means, the linear transformation $\mathcal{T}_{\boldsymbol{z}}$ must be non-singular, which essentially implies that $\widetilde{\mathcal{T}}_{\boldsymbol{z}}$ must be non-singular. Note that for $\boldsymbol{z} \in \mathbb{F}^d$, with non-negligible probability[4], say, μ, $\widetilde{\mathcal{T}}_{\boldsymbol{z}}$ will be non-singular. If all the users $U \in \mathcal{U}$ get registered, then the corresponding key-pairs

[3] Although the authors considered a non-singular randomizer for each user, they, perhaps erroneously also mentioned that KDC will compute the randomizer via ID without detailing how, see [CLND19, pages 4 and 6].

[4] The formula in [Lev05] says that the probability of $\widetilde{\mathcal{T}}_{\boldsymbol{z}}$ being non-singular is roughly $(1-1/q)$, where q is the size of the underlying field $\mathbb{F}$. The formula needs the entries of $\widetilde{\mathcal{T}}_{\boldsymbol{z}}$ to be uniformly and independently distributed over $\mathbb{F}$, which is assumed to be the case in practice. A similar assumption was also considered in [SSH11, Beu21, Beu22]..

$(\mathcal{SK}_U, \mathcal{PK}_U)$ will all be valid UOV key-pairs with probability approximately μ^N. This, in turn, implies that for all users $U \in \mathcal{U}$ with unique identifier $\boldsymbol{z}$, all the corresponding $\widetilde{\mathcal{T}}_{\boldsymbol{z}}$'s will be non-singular with probability roughly μ^N.

For the correctness of ID-UOV, one has to ensure that μ^N is non-negligible in κ. Thus, the total number of identities N cannot be exponential[5]. Further, for a given N (polynomial in κ), whether μ^N is non-negligible or not will depend upon the size of $\mathbb{F}$. So, it cannot be conclusively said that the universe considered in [CLND19] supports even a polynomial (in κ) number of identities. However, for real-world applications, one will prefer an explicit large universe construction. Finally and most importantly, the authors did not provide any formal security argument for their proposal.

4 Modified Construction and Its Security

In this section, we propose an IBS scheme by modifying the original construction of Chen et al. [CLND19]. This modified IBS scheme not only resolves the issues of non-singularity discussed above, but also provides some additional advantages such as registering new arbitrary identities to the existing IBS system. Further, our IBS proposal can handle any binary string as identity, i.e., the universe of identities is considered to be $\{0,1\}^*$. We use a cryptographic hash function to assign a unique value from $\mathbb{F}^d$ to each identity. Following the style of salted UOV [SSH11], we introduce random token and random salt in key-extraction and signature-generation respectively. Formally, the description of the scheme is given below.

$\mathsf{IBS.Setup}(\kappa)$. Let $\mathcal{U} = \{0,1\}^*$ be the universe of identities. Let $\mathcal{H} : \mathcal{U} \times \mathsf{TokSpac} \to \mathbb{F}^d$ and $\mathcal{H}_{\mathsf{uov}} : \mathcal{M} \times \mathsf{SaltSpac} \to \mathbb{F}^m$ be cryptographic hash functions, where $\mathcal{M}$, $\mathsf{TokSpac} = \{0,1\}^{\ell_t}$ and $\mathsf{SaltSpac} = \{0,1\}^{\ell_s}$ denote the message space, token space and salt space respectively. Note that to avoid the birthday attack, $|\mathbb{F}|^d$ should be chosen sufficiently large. In fact, for 128-bit security, we want $|\mathbb{F}|^d$ to be at least 2^{256}. Pick $\widetilde{\mathcal{F}} = (f^{(1)}, \ldots, f^{(m)})$, $\widetilde{\mathcal{T}} = (A, \boldsymbol{a})$ and $\widetilde{\mathcal{P}} = (p^{(1)}, \ldots, p^{(m)})$ as defined in Eqs. (3), (4) and (5) respectively. The public parameters and master secret key are given as follows:

$$\mathcal{PP} = \left(\widetilde{\mathcal{P}}, \mathcal{H}, \mathcal{H}_{\mathsf{uov}}, \ell_t, \ell_s\right)$$

$$\mathcal{MSK} = \Big(\{\alpha_{i,j}^{(k)}(\boldsymbol{z}), \beta_{i,\iota}^{(k)}(\boldsymbol{z})\}_{(k,i,j,\iota)\in[m]\times[v]\times[v+1,n]\times[v]}, \\ \{\gamma_i^{(k)}(\boldsymbol{z}), \delta^{(k)}(\boldsymbol{z})\}_{(k,i)\in[m]\times[n]}, \{A_{i,j}(\boldsymbol{z}), a_i(\boldsymbol{z})\}_{i,j\in[n]}\Big).$$

Note that each of the components involved in $\mathcal{MSK}$ is a polynomial in $\boldsymbol{z}$ of degree 1 as mentioned in Sect. 2.1.

[5] This can also be viewed from the fact that $\mathcal{MSK}$ has to contain the randomizer $R_{\boldsymbol{z}}$ for each user with unique identifier $\boldsymbol{z}$.

$\mathsf{IBS.Extract}(\mathcal{PP}, \mathcal{MSK}, \mathsf{id} \in \mathcal{U})$. The following is the procedure for generating the secret key $\mathcal{SK}_{\mathsf{id}}$:

1. choose $\mathsf{tok}_{\mathsf{id}} \xleftarrow{U} \mathsf{TokSpac}$
2. compute $\boldsymbol{z} = \mathcal{H}(\mathsf{id}, \mathsf{tok}_{\mathsf{id}})$
3. if $\widetilde{\mathcal{T}}_{\boldsymbol{z}}$ is singular, then go to step 1
4. pick a matrix $R_{\boldsymbol{z}} \xleftarrow{U} \widetilde{\mathsf{GL}}_n(\mathbb{F})$
5. set $\mathcal{F}_{\boldsymbol{z}} := \widetilde{\mathcal{F}}_{\boldsymbol{z}} \circ R_{\boldsymbol{z}}$ and $\mathcal{T}_{\boldsymbol{z}} := (R_{\boldsymbol{z}})^{-1} \circ \widetilde{\mathcal{T}}_{\boldsymbol{z}}$
6. return $\mathcal{SK}_{\mathsf{id}} := (\mathsf{tok}_{\mathsf{id}}, \mathcal{F}_{\boldsymbol{z}}, \mathcal{T}_{\boldsymbol{z}})$.

$\mathsf{IBS.Sign}(\mathcal{PP}, \mathrm{m}, \mathcal{SK}_{\mathsf{id}})$. Set $\mathcal{F} := \mathcal{F}_{\boldsymbol{z}}$ and $\mathcal{T} := \mathcal{T}_{\boldsymbol{z}}$ and then perform the following steps:

1. choose $\boldsymbol{x}'_v \xleftarrow{U} \mathbb{F}^v$
2. $\boldsymbol{r} \xleftarrow{U} \mathsf{SaltSpac}$
3. $\boldsymbol{y} = \mathcal{H}_{\mathsf{uov}}(\mathrm{m}||\boldsymbol{r})$
4. if $\mathcal{F}^{-1}_{\boldsymbol{x}'_v}(\boldsymbol{y}) = \emptyset$, go to step 2
5. $\boldsymbol{x}'_m \xleftarrow{U} \mathcal{F}^{-1}_{\boldsymbol{x}'_v}(\boldsymbol{y})$
6. $\boldsymbol{x} = \mathcal{T}^{-1}(\boldsymbol{x}'_v, \boldsymbol{x}'_m)$
7. set $\sigma = (\boldsymbol{x}, \boldsymbol{r})$
8. return $\sigma_{\mathrm{m},\mathsf{id}} = (\mathsf{tok}_{\mathsf{id}}, \sigma)$.

$\mathsf{IBS.Ver}(\mathcal{PP}, \mathrm{m}, \mathsf{id}, \sigma_{\mathrm{m},\mathsf{id}})$. Here $\sigma_{\mathrm{m},\mathsf{id}} = (\mathsf{tok}_{\mathsf{id}}, \sigma)$ and $\sigma = (\boldsymbol{x}, \boldsymbol{r})$. Then, compute $\boldsymbol{z} = \mathcal{H}(\mathsf{id}, \mathsf{tok}_{\mathsf{id}})$ and $\widetilde{\mathcal{P}}_{\boldsymbol{z}}$. The signature is accepted if $\widetilde{\mathcal{P}}_{\boldsymbol{z}}(\boldsymbol{x}) = \mathcal{H}_{\mathsf{uov}}(\mathrm{m}||\boldsymbol{r})$, else rejected.

Here we point out some features of our proposed IBS.

1. The matrix $A_{\boldsymbol{z}}$ (as part of $\widetilde{\mathcal{T}}_{\boldsymbol{z}}$) is non-singular with probability roughly $(1 - 1/q)$ (see footnote 4), where $q = |\mathbb{F}|$. So, the expected number of executions of the loop in $\mathsf{IBS.Extract}$ will be roughly $q/(q-1)$. Note that the additional random tokens involved in $\mathsf{IBS.Extract}$ ensures the non-singularity of $\widetilde{\mathcal{T}}_{\boldsymbol{z}}$ for each id with $\boldsymbol{z} = \mathcal{H}(\mathsf{id}, \mathsf{tok}_{\mathsf{id}})$ unlike the proposal of [CLND19].
2. Further, in the original proposal of Chen et al., the randomizer $R_{\boldsymbol{z}}$ for each user with unique identifier $\boldsymbol{z}$ is part of the $\mathcal{MSK}$. This means, their proposal has the drawback of storing additional $n^2 \cdot N$ field elements corresponding to the randomizers in $\mathcal{MSK}$, where $N = |\mathcal{U}|$. In contrast for key extraction, we sample every time a fresh non-singular matrix $R_{\boldsymbol{z}}$ with $\boldsymbol{z} = \mathcal{H}(\mathsf{id}, \mathsf{tok}_{\mathsf{id}})$. So, we do not need to store the $R_{\boldsymbol{z}}$ for each user with identifier $\boldsymbol{z}$. Hence, our scheme improves upon the efficiency of the original proposal. Note that, our scheme allows the creation of multiple secret keys for the same identity.
3. Unlike [CLND19], our proposal supports any arbitrary bit string to be identity and the size of public parameters does not depend on the size of the universe. Therefore, our proposal gives an *unbounded* identity-based signature scheme.
4. Note that the size of the signatures and the computations involved in signature generation and verification of our proposal are more or less the same as that of the UOV (or Rainbow). Hence, the efficiency of the proposed construction is comparable to that of the underlying primitive.

Correctness. By the construction, $\widetilde{\mathcal{P}} = \widetilde{\mathcal{F}} \circ \widetilde{\mathcal{T}}$. Therefore, for all $\boldsymbol{z} \in \mathbb{F}^d$, we have $\widetilde{\mathcal{P}}_{\boldsymbol{z}} = \widetilde{\mathcal{F}}_{\boldsymbol{z}} \circ \widetilde{\mathcal{T}}_{\boldsymbol{z}}$. Since $R_{\boldsymbol{z}}$ is non-singular, we have $\widetilde{\mathcal{P}}_{\boldsymbol{z}} = \widetilde{\mathcal{F}}_{\boldsymbol{z}} \circ \widetilde{\mathcal{T}}_{\boldsymbol{z}} = \mathcal{F}_{\boldsymbol{z}} \circ \mathcal{T}_{\boldsymbol{z}}$. By the choice of $R_{\boldsymbol{z}}$, $\mathcal{F}_{\boldsymbol{z}}$ will be of oil-and-vinegar type. Also note that for all identities $\mathsf{id} \in \mathcal{U}$ with $\boldsymbol{z} = \mathcal{H}(\mathsf{id}, \mathsf{tok}_{\mathsf{id}})$, $\widetilde{\mathcal{T}}_{\boldsymbol{z}}$ is always non-singular (thanks to point (1.)). Therefore, $(\widetilde{\mathcal{P}}_{\boldsymbol{z}}, (\mathcal{F}_{\boldsymbol{z}}, \mathcal{T}_{\boldsymbol{z}}))$ is a valid UOV key-pair. Hence, correctness follows from that of the salted UOV-signature [SSH11].

4.1 Security Argument

In this section, we argue security of the proposed IBS construction under the ν-PWMQ problem in the random oracle model. Although we closely follow the proof strategy of Sakumoto et al. [SSH11], ours is a bit involved as we handle random oracle for tokens and key-extract oracle additionally. In the following, we state and prove the security of the proposed IBS construction.

Theorem 1. *If the ν_{ext}-PWMQ problem is intractable, then the proposed IBS is EUF-ID-CMA secure in the classical random oracle model, where ν_{ext} is the number of key-extract queries.*

Proof. Let $\mathcal{A}$ be a quantum PPT adversary which can break the EUF-ID-CMA security of the signature scheme. An instance $\mathcal{D} := \left(\widetilde{\mathcal{P}}, (\boldsymbol{z}, \mathcal{SK}_{\boldsymbol{z}})_{\boldsymbol{z} \in V}, \boldsymbol{z}^*, \boldsymbol{y}^*\right)$ of the ν_{ext}-PWMQ problem is given to a simulator S and the goal of S is to find $\boldsymbol{x}^* \in \mathbb{F}^n$ such that $\widetilde{\mathcal{P}}_{\boldsymbol{z}^*}(\boldsymbol{x}^*) = \boldsymbol{y}^*$. We treat both the hash functions $\mathcal{H}$ and $\mathcal{H}_{\mathsf{uov}}$ as random oracles. Let the numbers of oracle queries made by $\mathcal{A}$ to $\mathcal{H}$ and $\mathcal{H}_{\mathsf{uov}}$ be ν_{id} and ν_{uov} respectively. Let $\mathcal{Q}_{\mathsf{ext}}$ be the set for keeping identity-token pairs involved in the key-extract oracle. Let $\mathcal{Q}_{\mathsf{tok}}$ be the set for keeping the information of token for each identity involved in the sign oracle. Initially, $\mathcal{Q}_{\mathsf{ext}} = \emptyset$ and $\mathcal{Q}_{\mathsf{tok}} = \emptyset$. Let ν_s be the number of signature queries made by $\mathcal{A}$. The simulator maintains two lists $\mathsf{List}_{\mathsf{id}}$ and $\mathsf{List}_{\mathsf{uov}}$ for keeping records of the forms: $(\mathsf{id}||\mathsf{tok}, \mathcal{H}(\mathsf{id}||\mathsf{tok}))$ and $(\mathrm{m}||\boldsymbol{r}, \mathcal{H}_{\mathsf{uov}}(\mathrm{m}||\boldsymbol{r}))$ respectively. The simulator S picks $(i^*, j^*) \xleftarrow{\mathrm{U}} [\nu_{\mathsf{id}} + \nu_s] \times [\nu_{\mathsf{uov}} + \nu_s]$ as a guess for the forgery identity id^* and message m^* respectively.

The identity space $\mathcal{U}$ is partitioned into two disjoint sets $\mathcal{U}_\lambda$ and $\mathcal{U} \backslash \mathcal{U}_\lambda$ based on a biased coin having the probability of head λ following Coron's partitioning technique [Cor00, CDP21]. The exact value of λ will be set later. The decision whether an identity id belongs to $\mathcal{U}_\lambda$ is taken as follows. First, toss the biased coin and then id is assigned to $\mathcal{U}_\lambda$ if and only if the outcome is head. In the simulation, the following strategy is used to answer key-extract and sign oracle query.

1. For a key-extract query on id, if $\mathsf{id} \in \mathcal{U}_\lambda$, then abort.
2. For a sign query on $(\mathrm{m}, \mathsf{id})$, do the following.
   (a) If $\mathsf{id} \in \mathcal{U}_\lambda$, the query is answered by manipulating $\mathcal{H}_{\mathsf{uov}}$ oracle as done in the proof of salted UOV [SSH11].
   (b) Otherwise, the query is answered by running the original sign algorithm using the corresponding secret key.

The simulator S answers all the oracle queries (that may appear in arbitrary order) in the following manner.

- **Hash-oracle**($\mathcal{H}$). The simulator handles the i-th query on $\mathsf{id}_i||\mathsf{tok}_i$ as follows. It checks whether $(\mathsf{id}_i||\mathsf{tok}_i, *) \in \mathsf{List}_{\mathsf{id}}$.
  1. If yes, then returns $\mathcal{H}(\mathsf{id}_i||\mathsf{tok}_i)$.
  2. Otherwise.
     (a) If $i = i^*$, it sets $\mathsf{List}_{\mathsf{id}} := \mathsf{List}_{\mathsf{id}} \cup \{(\mathsf{id}_i||\mathsf{tok}_i, \boldsymbol{z}^*)\}$ and returns $\boldsymbol{z}^*$.
     (b) Else, it picks $\boldsymbol{z}_i \xleftarrow{\text{U}} \mathbb{F}^d$, sets $\mathsf{List}_{\mathsf{id}} := \mathsf{List}_{\mathsf{id}} \cup \{(\mathsf{id}_i||\mathsf{tok}_i, \boldsymbol{z}_i)\}$ and returns $\boldsymbol{z}_i$.
- **Key-Extract-oracle**. Let id be the query identity. Then S answers key-extract query as follows. It checks whether $\mathsf{id} \notin \mathcal{U}_\lambda$.
  1. If yes, it performs the following steps.
     (a) It chooses $\mathsf{tok} \xleftarrow{\text{U}} \mathsf{TokSpac}$.
     (b) If $(\mathsf{id}||\mathsf{tok}, *) \in \mathsf{List}_{\mathsf{id}}$, it **aborts**.
     (c) Else, it chooses $\boldsymbol{z}$ from V and then sets $\mathsf{List}_{\mathsf{id}} := \mathsf{List}_{\mathsf{id}} \cup \{(\mathsf{id}||\mathsf{tok}, \boldsymbol{z})\}$, $V := V \setminus \{\boldsymbol{z}\}$ and $Q_{\mathsf{ext}} := Q_{\mathsf{ext}} \cup \{\mathsf{id}||\mathsf{tok}\}$.
     (d) Returns the key $\mathcal{SK}_{\mathsf{id}} := (\mathsf{tok}, \mathcal{SK}_{\boldsymbol{z}})$[6].
  2. Otherwise, it **aborts**.
- **Hash-oracle**($\mathcal{H}_{\mathsf{uov}}$). The simulator handles the j-th query on $\mathrm{m}_j||\boldsymbol{r}_j$ as follows. It checks whether $(\mathrm{m}_j||\boldsymbol{r}_j, *) \in \mathsf{List}_{\mathsf{uov}}$.
  1. If yes, then returns $\mathcal{H}_{\mathsf{uov}}(\mathrm{m}_j||\boldsymbol{r}_j)$.
  2. Otherwise.
     (a) If $j = j^*$, sets $\mathsf{List}_{\mathsf{uov}} := \mathsf{List}_{\mathsf{uov}} \cup \{(\mathrm{m}_j||\boldsymbol{r}_j, \boldsymbol{y}^*)\}$ and returns $\boldsymbol{y}^*$.
     (b) Else, it picks $\boldsymbol{y}_j \xleftarrow{\text{U}} \mathbb{F}^m$, sets $\mathsf{List}_{\mathsf{uov}} := \mathsf{List}_{\mathsf{uov}} \cup \{(\mathrm{m}_j||\boldsymbol{r}_j, \boldsymbol{y}_j)\}$ and returns $\boldsymbol{y}_j$.
- **Sign-oracle**. The simulator answers the signature oracle query on $\mathrm{m}||\mathsf{id}$ as follows. It checks whether $\mathsf{id} \in \mathcal{U}_\lambda$.
  - If yes, it performs the following steps.
    1. If $\mathsf{id}||* \notin Q_{\mathsf{tok}}$, then it chooses $\mathsf{tok} \xleftarrow{\text{U}} \mathsf{TokSpac}$ and sets $Q_{\mathsf{tok}} := Q_{\mathsf{tok}} \cup \{\mathsf{id}||\mathsf{tok}\}$.
    2. Now, there exists a token $\mathsf{tok} \in \mathsf{TokSpac}$ such that $\mathsf{id}||\mathsf{tok} \in Q_{\mathsf{tok}}$.
    3. It then calls its internal oracle $\mathcal{H}$ on $\mathsf{id}||\mathsf{tok}$ and let $\boldsymbol{z}$ be the answer.
    4. Picks $\sigma = (\boldsymbol{x}, \boldsymbol{r}) \xleftarrow{\text{U}} \mathbb{F}^n \times \mathsf{SaltSpac}$.
    5. If $(\mathrm{m}||\boldsymbol{r}, *) \in \mathsf{List}_{\mathsf{uov}}$, it **aborts**, otherwise updates $\mathsf{List}_{\mathsf{uov}}$ with $(\mathrm{m}||\boldsymbol{r}, \widetilde{\mathcal{P}}_{\boldsymbol{z}}(\boldsymbol{x}))$.
    6. Returns $\sigma_{\mathrm{m},\mathsf{id}} = (\mathsf{tok}, \sigma)$.
  - Otherwise, the simulator checks whether there exists a token tok such that $\mathsf{id}||\mathsf{tok} \in Q_{\mathsf{ext}}$.
    1. If yes, then answers by running the original sign algorithm using the corresponding $\mathcal{SK}_{\mathsf{id}} = (\mathsf{tok}, \mathcal{SK}_{\boldsymbol{z}})$, where $\boldsymbol{z} = \mathcal{H}(\mathsf{id}||\mathsf{tok})$[7].

[6] Note that in the construction, tok is not distributed uniformly, but it can be made negligibly-close to uniform distribution, if q is chosen sufficiently large. Sakumoto et al. [SSH11] also faced a similar issue while simulating salts in their reduction and they implicitly assumed that $\mathcal{A}$ cannot distinguish the difference. Nonetheless, we assume that $\mathcal{A}$ cannot distinguish between a uniform token and the token involved in the actual key-extraction.

[7] If for an identity id, there are many tokens in Q_{ext} then consider any one of them.

2. Else, it calls its internal key-extract oracle on id and let $\mathcal{SK}_{\mathsf{id}}$ be the reply. Then answers by running the original sign algorithm using the secret key $\mathcal{SK}_{\mathsf{id}}$.

- **Forgery**. $\mathcal{A}$ submits the forgery $(\mathrm{m}^*, \mathsf{id}^*, \sigma_{\mathrm{m}^*,\mathsf{id}^*} = (\mathsf{tok}_{\mathsf{id}^*}, \boldsymbol{x}^*, \boldsymbol{r}^*))$ to S. If $\mathsf{id}^*||\mathsf{tok}_{\mathsf{id}^*}$ is not the i^*-th argument of $\mathcal{H}$ oracle or $\mathrm{m}^*||\boldsymbol{r}^*$ is not the j^*-th argument of $\mathcal{H}_{\mathsf{uov}}$ oracle, then it **aborts**. It returns $\boldsymbol{x}^*$ to its challenger, if $\widetilde{\mathcal{P}}_{\boldsymbol{z}^*}(\boldsymbol{x}^*) = \boldsymbol{y}^*$, else aborts.
- **Analysis**. Let us assume that 'abort' in step 2 of key-extract oracle does not occur. Then the probability that S aborts in answering a single key-extract query is at most $\zeta_1 = (\nu_{\mathsf{id}} + \nu_{\mathsf{ext}} + \nu_s) \cdot 2^{-\ell_t}$. The probability that S aborts in answering a single sign-query is at most $\zeta_2 = (\nu_{\mathsf{uov}} + \nu_s) \cdot 2^{-\ell_s} + (\nu_{\mathsf{id}} + \nu_{\mathsf{ext}} + \nu_s) \cdot 2^{-\ell_t}$. So, if 'abort' in step 2 of key-extract oracle does not occur, then S answers all the queries successfully with probability at least $\rho = 1 - \zeta_1 \cdot \nu_{\mathsf{ext}} - \zeta_2 \cdot \nu_s$. Note that ρ can be made negligibly-close to 1 by choosing ℓ_t and ℓ_s sufficiently large. Now, the probability that for all the key-extract queries 'abort' in step 2 of key-extract oracle never occurs is $(1 - \lambda)^{\nu_{\mathsf{ext}}}$, which attains the maximum value of $1/e$, when λ is set to $1/(1 + \nu_{\mathsf{ext}})$. Therefore, S answers all the queries successfully (i.e., without any abort) with probability at least ρ/e. With probability $1/\left((\nu_{\mathsf{id}} + \nu_s) \cdot (\nu_{\mathsf{uov}} + \nu_s)\right)$, S correctly guesses the indices i^* and j^* such that $\mathsf{id}_{i^*}||\mathsf{tok}_{i^*} = \mathsf{id}^*||\mathsf{tok}_{\mathsf{id}^*}$ and $\mathrm{m}_{j^*}||\boldsymbol{r}_{j^*} = \mathrm{m}^*||\boldsymbol{r}^*$. This implies that $\mathcal{H}(\mathsf{id}^*||\mathsf{tok}_{\mathsf{id}^*}) = \boldsymbol{z}^*$ and $\mathcal{H}_{\mathsf{uov}}(\mathrm{m}^*||\boldsymbol{r}^*) = \boldsymbol{y}^*$. So, we have $\widetilde{\mathcal{P}}_{\boldsymbol{z}^*}(\boldsymbol{x}^*) = \boldsymbol{y}^*$, if the forgery is a valid one. The advantage of S in solving the given problem instance is

$$\mathsf{Adv}_{\mathsf{S}}^{\nu_{\mathsf{ext}}\text{-PWMQ}}(\kappa) \leq \frac{\rho}{e} \cdot \frac{\mathsf{Adv}_{\mathcal{A}}^{\text{EUF-ID-CMA}}(\kappa)}{(\nu_{\mathsf{id}} + \nu_s) \cdot (\nu_{\mathsf{uov}} + \nu_s)}.$$

□

Remark 1. Note that the components $\mathcal{SK}_{\boldsymbol{z}}$ involved in the instance of the ν-PWMQ are utilized only when the key-extract oracle is invoked. Therefore, the above argument can be easily modified to establish the EUF-nID-CMA security of the proposed IBS scheme based on the 0-PWMQ problem where one has to essentially consider $\mathcal{U}_\lambda$ to be the whole identity space $\mathcal{U}$.

5 On the Hardness Assumption

In this section, we focus on the intractability of the ν-PWMQ problem. As mentioned above, the special case, 0-PWMQ problem can be used to argue EUF-nID-CMA security of the proposed IBS scheme. We first show that the 0-PWMQ problem is equivalent to the WMQ-problem.

Proposition 1. *The WMQ-problem is no harder than the* 0*-PWMQ problem.*

Proof. We will establish a solver S for the WMQ-problem using a solver $\mathcal{A}$ for the 0-PWMQ problem. Let $(\mathcal{P}, \boldsymbol{y}^*) \in \mathcal{P}(\mathbb{F}^n, \mathbb{F}^m) \times \mathbb{F}^m$ be the given WMQ-problem instance. Let $\mathcal{P} = (\mathrm{p}^{(1)}, \ldots, \mathrm{p}^{(m)})$. Note that each $\mathrm{p}^{(k)}$ is a quadratic polynomial in n variables, i.e.,

$$\mathrm{p}^{(k)}(\boldsymbol{x}) = \sum_{i=1}^{n}\sum_{j=1}^{n} \mathrm{p}_{i,j}^{(k)} \cdot x_i x_j + \sum_{i=1}^{n} \mathrm{p}_i^{(k)} \cdot x_i + \mathrm{p}_0^{(k)}.$$

We then write $\mathrm{p}^{(k)}$ as $\left(\{\mathrm{p}_{i,j}^{(k)}\}_{i,j\in[n]}, \{\mathrm{p}_i^{(k)}\}_{i\in[n]}, \mathrm{p}_0^{(k)}\right)$, where $\{\mathrm{p}_{i,j}^{(k)}\}_{i,j\in[n]}$ represents the quadratic part, $\{\mathrm{p}_i^{(k)}\}_{i\in[n]}$ denotes the linear part and $\mathrm{p}_0^{(k)}$ is the constant term.

Let $\boldsymbol{u} = (u_1, \ldots, u_d) \xleftarrow{\mathrm{U}} \mathbb{F}^d$. We will prepare $\widetilde{\mathcal{P}} = \widetilde{\mathcal{F}} \circ \widetilde{\mathcal{T}}$ for random expressions $\widetilde{\mathcal{F}}$ and $\widetilde{\mathcal{T}}$ as defined in Sect. 2.1 such that

$$\widetilde{\mathcal{P}}_{\boldsymbol{u}} = \mathcal{P}. \tag{6}$$

Let us write $\widetilde{\mathcal{F}} = (\mathrm{f}^{(1)}, \ldots, \mathrm{f}^{(m)})$, where each $\mathrm{f}^{(k)}$ is of the form as defined in Eq. (3). Also, note that $\widetilde{\mathcal{T}}$ has the form as defined in Eq. (4). Then we can write $\widetilde{\mathcal{P}} = (\mathrm{f}^{(1)} \circ \widetilde{\mathcal{T}}, \ldots, \mathrm{f}^{(m)} \circ \widetilde{\mathcal{T}})$. By the requirement (as given in Eq. (6)), we have $\mathrm{f}_{\boldsymbol{u}}^{(k)} \circ \widetilde{\mathcal{T}}_{\boldsymbol{u}} = \mathrm{p}^{(k)}$ for all $k \in [m]$, which in turn implies that

$$\mathcal{Q}(\mathrm{f}_{\boldsymbol{u}}^{(k)} \circ \widetilde{\mathcal{T}}_{\boldsymbol{u}}) = \mathcal{Q}(\mathrm{p}^{(k)}) \tag{7}$$

where $\mathcal{Q}(\mathrm{p}^{(k)})$ and $\mathcal{Q}(\mathrm{f}_{\boldsymbol{u}}^{(k)} \circ \widetilde{\mathcal{T}}_{\boldsymbol{u}})$ denote the matrix representations of the quadratic part of $\mathrm{p}^{(k)}$ and $\mathrm{f}_{\boldsymbol{u}}^{(k)} \circ \widetilde{\mathcal{T}}_{\boldsymbol{u}}$ respectively. Note that we can express $\mathcal{Q}(\mathrm{f}^{(k)})$ as a block matrix B_k as given below:

$$\mathcal{Q}(\mathrm{f}^{(k)}) = B_k = \left[\begin{array}{c|c} B_{11}^{(k)} & B_{12}^{(k)} \\ \hline B_{21}^{(k)} & 0 \end{array}\right]$$

$$= \left[\begin{array}{cccc|ccc} \alpha_{11}^{(k)}(\boldsymbol{z}) & \ldots & \alpha_{1v}^{(k)}(\boldsymbol{z}) & & \alpha_{1v+1}^{(k)}(\boldsymbol{z}) & \ldots & \alpha_{1n}^{(k)}(\boldsymbol{z}) \\ \vdots & \ldots & \vdots & & \vdots & \ldots & \vdots \\ \alpha_{v1}^{(k)}(\boldsymbol{z}) & \ldots & \alpha_{vv}^{(k)}(\boldsymbol{z}) & & \alpha_{vv+1}^{(k)}(\boldsymbol{z}) & \ldots & \alpha_{vn}^{(k)}(\boldsymbol{z}) \\ \hline \alpha_{v+11}^{(k)}(\boldsymbol{z}) & \ldots & \alpha_{v+1v}^{(k)}(\boldsymbol{z}) & & 0 & \ldots & 0 \\ \vdots & \ldots & \vdots & & \vdots & \ldots & \vdots \\ \alpha_{n1}^{(k)}(\boldsymbol{z}) & \ldots & \alpha_{nv}^{(k)}(\boldsymbol{z}) & & 0 & \ldots & 0 \end{array}\right] \tag{8}$$

where $\alpha_{\iota\zeta}^{(k)}(\boldsymbol{z}) = b_1^{(k\iota\zeta)} z_1 + \cdots + b_d^{(k\iota\zeta)} z_d$ which is a linear map in $z_1, \ldots, z_d$.

We now discuss how to find each expression $\mathrm{f}^{(k)}$ and $\widetilde{\mathcal{T}}$. Here we only illustrate the steps in finding the coefficients (as expressions) of the quadratic part of $\mathrm{f}^{(k)}$ as the expressions for the linear terms and constant term can be computed similarly. For finding quadratic terms of $\mathrm{f}^{(k)}$, we only use $\{\mathrm{p}_{i,j}^{(k)}\}_{i,j\in[n]}$. The idea is to choose $\widetilde{\mathcal{T}}$ and the blocks $B_{11}^{(k)}, B_{21}^{(k)}$ randomly, and then find[8] B_{12} using Eq. (7) for $k \in [m]$. If we denote $C = \widetilde{\mathcal{T}}_{\boldsymbol{u}} = (\mathrm{c}_{ij})_{i,j\in[n]}$, then we can write $\mathcal{Q}(\mathrm{f}_{\boldsymbol{u}}^{(k)} \circ \widetilde{\mathcal{T}}_{\boldsymbol{u}}) = C^{\top} \cdot \mathcal{Q}(\mathrm{f}_{\boldsymbol{u}}^{(k)}) \cdot C$, where $C^{\top}$ denotes the transpose of C. Therefore, we have the following relation using Eqs. (6) and (8):

$$\begin{bmatrix} \sum_{\iota=1}^{n} \mathrm{c}_{\iota 1} \sum_{\varsigma=1}^{v} \mathrm{c}_{\varsigma 1} \alpha_{\iota\varsigma}^{(k)}(\boldsymbol{u}) + & & \sum_{\iota=1}^{n} \mathrm{c}_{\iota 1} \sum_{\varsigma=1}^{v} \mathrm{c}_{\varsigma n} \alpha_{\iota\varsigma}^{(k)}(\boldsymbol{u}) + \\ \quad \sum_{\iota=1}^{v} \mathrm{c}_{\iota 1} \sum_{\varsigma=v+1}^{n} \mathrm{c}_{\varsigma 1} \alpha_{\iota\varsigma}^{(k)}(\boldsymbol{u}) & \cdots & \quad \sum_{\iota=1}^{v} \mathrm{c}_{\iota 1} \sum_{\varsigma=v+1}^{n} \mathrm{c}_{\varsigma n} \alpha_{\iota\varsigma}^{(k)}(\boldsymbol{u}) \\ \vdots & \vdots & \vdots \\ \sum_{\iota=1}^{n} \mathrm{c}_{\iota n} \sum_{\varsigma=1}^{v} \mathrm{c}_{\varsigma 1} \alpha_{\iota\varsigma}^{(k)}(\boldsymbol{u}) + & & \sum_{\iota=1}^{n} \mathrm{c}_{\iota n} \sum_{\varsigma=1}^{v} \mathrm{c}_{\varsigma n} \alpha_{\iota\varsigma}^{(k)}(\boldsymbol{u}) + \\ \quad \sum_{\iota=1}^{v} \mathrm{c}_{\iota n} \sum_{\varsigma=v+1}^{n} \mathrm{c}_{\varsigma 1} \alpha_{\iota\varsigma}^{(k)}(\boldsymbol{u}) & \cdots & \quad \sum_{\iota=1}^{v} \mathrm{c}_{\iota n} \sum_{\varsigma=v+1}^{n} \mathrm{c}_{\varsigma n} \alpha_{\iota\varsigma}^{(k)}(\boldsymbol{u}) \end{bmatrix} = \mathcal{Q}(p^{(k)}).$$

Note that all $\alpha_{\iota\varsigma}^{(k)}$'s (as polynomials) involved in the above matrix are known, except the ones that are highlighted in gray color. It boils down to find $\alpha_{\iota\varsigma}^{(k)}(\boldsymbol{z}) = b_1^{(k\iota\varsigma)} z_1 + \cdots + b_d^{(k\iota\varsigma)} z_d$ for $(\iota, \varsigma) \in [v] \times [v+1, n]$ involved in the following system:

$$\left.\begin{array}{r} \sum_{\iota=1}^{n} \mathrm{c}_{\iota 1} \sum_{\varsigma=1}^{v} \mathrm{c}_{\varsigma 1} \alpha_{\iota\varsigma}^{(k)}(\boldsymbol{u}) + \sum_{\iota=1}^{v} \mathrm{c}_{\iota 1} \sum_{\varsigma=v+1}^{n} \mathrm{c}_{\varsigma 1} \alpha_{\iota\varsigma}^{(k)}(\boldsymbol{u}) = p_{11}^{(k)} \\ \vdots \\ \sum_{\iota=1}^{n} \mathrm{c}_{\iota 1} \sum_{\varsigma=1}^{v} \mathrm{c}_{\varsigma n} \alpha_{\iota\varsigma}^{(k)}(\boldsymbol{u}) + \sum_{\iota=1}^{v} \mathrm{c}_{\iota 1} \sum_{\varsigma=v+1}^{n} \mathrm{c}_{\varsigma n} \alpha_{\iota\varsigma}^{(k)}(\boldsymbol{u}) = p_{1n}^{(k)} \\ \vdots \\ \sum_{\iota=1}^{n} \mathrm{c}_{\iota i} \sum_{\varsigma=1}^{v} \mathrm{c}_{\varsigma j} \alpha_{\iota\varsigma}^{(k)}(\boldsymbol{u}) + \sum_{\iota=1}^{v} \mathrm{c}_{\iota i} \sum_{\varsigma=v+1}^{n} \mathrm{c}_{\varsigma j} \alpha_{\iota\varsigma}^{(k)}(\boldsymbol{u}) = p_{ij}^{(k)} \\ \vdots \\ \sum_{\iota=1}^{n} \mathrm{c}_{\iota n} \sum_{\varsigma=1}^{v} \mathrm{c}_{\varsigma 1} \alpha_{\iota\varsigma}^{(k)}(\boldsymbol{u}) + \sum_{\iota=1}^{v} \mathrm{c}_{\iota n} \sum_{\varsigma=v+1}^{n} \mathrm{c}_{\varsigma 1} \alpha_{\iota\varsigma}^{(k)}(\boldsymbol{u}) = p_{n1}^{(k)} \\ \vdots \\ \sum_{\iota=1}^{n} \mathrm{c}_{\iota n} \sum_{\varsigma=1}^{v} \mathrm{c}_{\varsigma n} \alpha_{\iota\varsigma}^{(k)}(\boldsymbol{u}) + \sum_{\iota=1}^{v} \mathrm{c}_{\iota n} \sum_{\varsigma=v+1}^{n} \mathrm{c}_{\varsigma n} \alpha_{\iota\varsigma}^{(k)}(\boldsymbol{u}) = p_{nn}^{(k)} \end{array}\right\} \quad (9)$$

[8] One can alternate the choice of the blocks. It is also possible to randomly choose some of the entries of B_k (not necessarily block-wise) and find the expressions of the remaining entries.

where (i,j)-th element of $\mathcal{Q}(p^{(k)})$ is $p_{ij}^{(k)}$. The system in Eq. (9) can be further written as

$$\left.\begin{array}{r}
\sum_{\iota=1}^{v}\sum_{\zeta=v+1}^{n} c_{\iota 1}c_{\zeta 1}u_1 b_1^{(k\iota\zeta)} + \cdots + \sum_{\iota=1}^{v}\sum_{\zeta=v+1}^{n} c_{\iota 1}c_{\zeta 1}u_d b_d^{(k\iota\zeta)} = \theta_{11}^{(k)} \\
\vdots \\
\sum_{\iota=1}^{v}\sum_{\zeta=v+1}^{n} c_{\iota 1}c_{\zeta n}u_1 b_1^{(k\iota\zeta)} + \cdots + \sum_{\iota=1}^{v}\sum_{\zeta=v+1}^{n} c_{\iota 1}c_{\zeta n}u_d b_d^{(k\iota\zeta)} = \theta_{1n}^{(k)} \\
\vdots \\
\sum_{\iota=1}^{v}\sum_{\zeta=v+1}^{n} c_{\iota i}c_{\zeta j}u_1 b_1^{(k\iota\zeta)} + \cdots + \sum_{\iota=1}^{v}\sum_{\zeta=v+1}^{n} c_{\iota i}c_{\zeta j}u_d b_d^{(k\iota\zeta)} = \theta_{ij}^{(k)} \\
\vdots \\
\sum_{\iota=1}^{v}\sum_{\zeta=v+1}^{n} c_{\iota n}c_{\zeta 1}u_1 b_1^{(k\iota\zeta)} + \cdots + \sum_{\iota=1}^{v}\sum_{\zeta=v+1}^{n} c_{\iota n}c_{\zeta 1}u_d b_d^{(k\iota\zeta)} = \theta_{n1}^{(k)} \\
\vdots \\
\sum_{\iota=1}^{v}\sum_{\zeta=v+1}^{n} c_{\iota n}c_{\zeta n}u_1 b_1^{(k\iota\zeta)} + \cdots + \sum_{\iota=1}^{v}\sum_{\zeta=v+1}^{n} c_{\iota n}c_{\zeta n}u_d b_d^{(k\iota\zeta)} = \theta_{nn}^{(k)}
\end{array}\right\} \quad (10)$$

where $\theta_{ij}^{(k)} = p_{ij}^{(k)} - \sum_{\iota=1}^{n}\sum_{\zeta=1}^{v} c_{\iota i}c_{\zeta j}\alpha_{\iota\zeta}^{(k)}(\boldsymbol{u})$. Therefore, the above system is a linear system of equations in the variables $b_\ell^{(k\iota\zeta)}$'s involved in $\alpha_{ij}^{(k)}$ (highlighted in gray color) for $(\iota,\zeta) \in [v] \times [v+1,n]$ and $\ell \in [d]$. Note that $C = (c_{ij})_{i,j\in[n]}$ is obtained by evaluating $\widetilde{\mathcal{T}}$ at $\boldsymbol{u}$. Since each entry of $\widetilde{\mathcal{T}}$ is a random linear map in $\boldsymbol{z}$, w.l.o.g[9] we can assume that all c_{ij}'s are distributed uniformly and independently over $\mathbb{F}$. Further, the coefficients of the polynomials given in Eq. (10) are random variables of the form $c_{\iota i}c_{\zeta j}u_\ell$ for $(\iota,\zeta) \in [v]\times[v+1,n]$, $\ell \in [d]$ and $i,j \in [n]$, where the variables $c_{\iota i}$'s, $c_{\zeta j}$'s and u_ℓ's are uniformly and independently distributed over $\mathbb{F}$. It is easy to check that the number of random variables involved in Eq. (10) is $n^2 \cdot v \cdot m \cdot d$. That is, we derive $n^2 \cdot v \cdot m \cdot d$ new variables of the form $c_{\iota i}c_{\zeta j}u_\ell$ from the same number of original variables, i.e., from $c_{\iota i}$'s, $c_{\zeta j}$'s and u_ℓ. Hence, the new variables $c_{\iota i}c_{\zeta j}u_\ell$'s are expected to behave as if they are chosen uniformly and independently from $\mathbb{F}$. Therefore, the above system looks like a random system. Here the number of equations is $n^2 \approx 9m^2$ and the number of variables is $v \cdot m \cdot d \approx 2m^2 \cdot d$. By choosing d sufficiently large (in fact, choose d to be at least $\lceil 9/2 \rceil$), we can make the above system to be underdetermined and therefore, the system will have a solution with high probability. Assuming $\theta_{ij}^{(k)}$'s are uniform over $\mathbb{F}$, each component of the solution will be uniform over $\mathbb{F}$ (thanks

[9] More precisely, all c_{ij}'s will be distributed uniformly and independently over $\mathbb{F}$, if each entry of $\widetilde{\mathcal{T}}$ is a random affine map in $\boldsymbol{z}$. This is due to the fact that each entry will now have an independently and uniformly chosen constant term.

to the linearity of the system). Further, the system can be solved efficiently. In fact, let the rank of the system be τ ($\leq 9m^2$). So, the degree of freedom will be $2m^2 \cdot d - \tau$ ($\geq m^2(2d-9)$). That is, we can choose $(2m^2 \cdot d - \tau)$ many variables randomly. Then, the remaining τ many variables will be a linear combination of the chosen $(2m^2 \cdot d - \tau)$ random values and n^2 many given random values $\{\theta_{ij}^{(k)}\}_{i.j \in [n]}$. Therefore, $\widetilde{\mathcal{P}}$ will have a close-to-proper distribution from $\mathcal{A}$'s point of view.

Now, S supplies $\mathcal{D} = \left(\widetilde{\mathcal{P}}, \boldsymbol{z}^* := \boldsymbol{u}, \boldsymbol{y}^*\right)$ to $\mathcal{A}$ which then returns $\boldsymbol{x}^* \in \mathbb{F}^n$ such that $\widetilde{\mathcal{P}}_{\boldsymbol{u}}(\boldsymbol{x}^*) = \boldsymbol{y}^*$. By construction of $\widetilde{\mathcal{P}}$, $\widetilde{\mathcal{P}}_{\boldsymbol{u}} = \mathcal{P}$, and hence $\mathcal{P}(\boldsymbol{x}^*) = \boldsymbol{y}^*$. This completes the argument. □

Unlike the 0-PWMQ problem, we do not have any formal reduction or separation result regarding the ν-PWMQ problem. However, we provide some heuristic arguments justifying why this new problem is expected to be computationally intractable. Here the goal is to solve $\widetilde{\mathcal{P}}_{\boldsymbol{z}^*}(\boldsymbol{x}) = \boldsymbol{y}^*$ for random $\boldsymbol{z}^* \in \mathbb{F}^d$ and $\boldsymbol{y}^* \in \mathbb{F}^m$. Note that the UOV-public map $\widetilde{\mathcal{P}}_{\boldsymbol{z}^*}$ is efficiently computable by evaluating each coefficient of the public expression $\widetilde{\mathcal{P}}$ at $\boldsymbol{z}^*$. Since $\widetilde{\mathcal{P}}$ and $\boldsymbol{z}^*$ are chosen uniformly, $\widetilde{\mathcal{P}}_{\boldsymbol{z}^*}$ behaves as a random UOV-public map whose corresponding secret key is $\mathcal{SK}_{\boldsymbol{z}^*} = (\widetilde{\mathcal{F}}_{\boldsymbol{z}^*}, \widetilde{\mathcal{T}}_{\boldsymbol{z}^*})$. First assume that we do not know anything other than the target UOV-public key $\widetilde{\mathcal{P}}_{\boldsymbol{z}^*}$. Then the only available known attacks are the generic attacks, e.g., direct attack [Fau99,Fau02,BFP09], band separation [DYC+08] and UOV-attack [KS98], etc. Under suitable parameter choice, one can compensate for those generic attacks.

Note that $\mathcal{SK}_{\boldsymbol{z}^*}$ is easily computable just by evaluating $\widetilde{\mathcal{F}}$ and $\widetilde{\mathcal{T}}$ at $\boldsymbol{z}^*$. Since $\widetilde{\mathcal{F}}$ and $\widetilde{\mathcal{T}}$ are not publicly available, so the question is whether there is any alternative way to compute $\mathcal{SK}_{\boldsymbol{z}^*}$. Given the structure of the problem instance (Definition 3), this essentially boils down to the question of whether one can efficiently compute the expression of $\widetilde{\mathcal{F}}$ or $\widetilde{\mathcal{T}}$ using the additional information $\mathcal{SK}_{\boldsymbol{z}} = (\widetilde{\mathcal{F}}_{\boldsymbol{z}} \circ R_{\boldsymbol{z}}, (R_{\boldsymbol{z}})^{-1} \circ \widetilde{\mathcal{T}}_{\boldsymbol{z}})$ for $\boldsymbol{z} \in V \setminus \{\boldsymbol{z}^*\}$. For simplicity, let us assume that $\widetilde{\mathcal{T}} = A$ (i.e., without translation by $\boldsymbol{a}$). Since each entry of A is a linear map in d variables, we can find the expression for $\widetilde{\mathcal{T}}$ using Gaussian elimination, if we know $\widetilde{\mathcal{T}}_{\boldsymbol{z}}$ for at least d many $\boldsymbol{z}$ (i.e., $|V| \geq d$). Similarly, we can find the expression of $\widetilde{\mathcal{F}}$, if we know $\widetilde{\mathcal{F}}_{\boldsymbol{z}}$ for at least d many $\boldsymbol{z}$. However, neither $\widetilde{\mathcal{F}}_{\boldsymbol{z}}$ nor $\widetilde{\mathcal{T}}_{\boldsymbol{z}}$ is given directly as part of the secret key $\mathcal{SK}_{\boldsymbol{z}}$ for $\boldsymbol{z} \in V \setminus \{\boldsymbol{z}^*\}$. In fact, they are concealed by a random matrix $R_{\boldsymbol{z}}$ and its inverse respectively. So, the components of $\mathcal{SK}_{\boldsymbol{z}}$ do not reveal any valuable information about $\widetilde{\mathcal{F}}_{\boldsymbol{z}}$ and $\widetilde{\mathcal{T}}_{\boldsymbol{z}}$. Therefore, it is reasonable to conclude that finding neither $\widetilde{\mathcal{T}}_{\boldsymbol{z}}$ nor $\widetilde{\mathcal{F}}_{\boldsymbol{z}}$ is computationally feasible.

6 Concluding Remark

We have proposed an efficient IBS scheme in the MQ-setting by modifying the IBS proposal of Chen et al. The modified proposal ensures that any arbitrary bit string can be considered as an identity without affecting the size of the public parameters. The IBS can be instantiated in the UOV/Rainbow setting and retains the efficiency of that underlying signature primitive. For the security reduction, we have introduced a new parameterized assumption in the MQ-setting and provided some justification towards its plausible intractability. A more rigorous cryptanalysis of the newly introduced assumption is left as an interesting open problem.

Acknowledgement. We would like to thank Dr. M. Prem Laxman Das and the anonymous reviewers of Indocrypt 2022 for their comments and suggestions that helped us in polishing the technical and editorial content of this paper. This work is supported by the Ministry of Electronics and Information Technology, Government of India through its grants for the Center of Excellence in Quantum Technology at IISc Bangalore, India.

References

Beu21. Beullens, W.: Improved cryptanalysis of UOV and rainbow. In: Canteaut, A., Standaert, F.-X. (eds.) EUROCRYPT 2021. LNCS, vol. 12696, pp. 348–373. Springer, Cham (2021). https://doi.org/10.1007/978-3-030-77870-5_13

Beu22. Beullens, W.: Breaking rainbow takes a weekend on a laptop. In: Dodis, Y., Shrimpton, T. (eds.) CRYPTO 2022. LNCS, vol. 13508, pp. 464–479. Springer, Cham (2022). https://doi.org/10.1007/978-3-031-15979-4_16

BFP09. Bettale, L., Faugére, J.-C., Perret, L.: Hybrid approach for solving multivariate systems over finite fields. J. Math. Cryptol. **3**(3), 177–197 (2009)

BLMQ05. Barreto, P.S.L.M., Libert, B., McCullagh, N., Quisquater, J.-J.: Efficient and provably-secure identity-based signatures and signcryption from bilinear maps. In: Roy, B. (ed.) ASIACRYPT 2005. LNCS, vol. 3788, pp. 515–532. Springer, Heidelberg (2005). https://doi.org/10.1007/11593447_28

CC03. Choon, J.C., Hee Cheon, J.: An identity-based signature from gap Diffie-Hellman groups. In: Desmedt, Y.G. (ed.) PKC 2003. LNCS, vol. 2567, pp. 18–30. Springer, Heidelberg (2003). https://doi.org/10.1007/3-540-36288-6_2

CDP21. Chatterjee, S., Dimri, A., Pandit, T.: Identity-based signature and extended forking algorithm in the multivariate quadratic setting. In: Adhikari, A., Küsters, R., Preneel, B. (eds.) INDOCRYPT 2021. LNCS, vol. 13143, pp. 387–412. Springer, Cham (2021). https://doi.org/10.1007/978-3-030-92518-5_18

CHR+16. Chen, M.-S., Hülsing, A., Rijneveld, J., Samardjiska, S., Schwabe, P.: From 5-pass $\mathcal{MQ}$-based identification to $\mathcal{MQ}$-based signatures. In: Cheon, J.H., Takagi, T. (eds.) ASIACRYPT 2016. LNCS, vol. 10032, pp. 135–165. Springer, Heidelberg (2016). https://doi.org/10.1007/978-3-662-53890-6_5

CLND19. Chen, J., Ling, J., Ning, J., Ding, J.: Identity-based signature schemes for multivariate public key cryptosystems. Comput. J. **62**(8), 1132–1147 (2019)

Cor00. Coron, J.-S.: On the exact security of full domain hash. In: Bellare, M. (ed.) CRYPTO 2000. LNCS, vol. 1880, pp. 229–235. Springer, Heidelberg (2000). https://doi.org/10.1007/3-540-44598-6_14

DH76. Diffie, W., Hellman, M.: New directions in cryptography. IEEE Trans. Inf. Theory **22**(6), 644–654 (1976)

DS05. Ding, J., Schmidt, D.: Rainbow, a new multivariable polynomial signature scheme. In: Ioannidis, J., Keromytis, A., Yung, M. (eds.) ACNS 2005. LNCS, vol. 3531, pp. 164–175. Springer, Heidelberg (2005). https://doi.org/10.1007/11496137_12

DYC+08. Ding, J., Yang, B.-Y., Chen, C.-H.O., Chen, M.-S., Cheng, C.-M.: New differential-algebraic attacks and reparametrization of rainbow. In: Bellovin, S.M., Gennaro, R., Keromytis, A., Yung, M. (eds.) ACNS 2008. LNCS, vol. 5037, pp. 242–257. Springer, Heidelberg (2008). https://doi.org/10.1007/978-3-540-68914-0_15

Fau99. Faugére, J.C.: A new efficient algorithm for computing Gröbner bases (F4). J. Pure Appl. Algebra **139**(1–3), 61–88 (1999)

Fau02. Faugére, J.C.: A new efficient algorithm for computing Gröbner bases without reduction to zero (F5). In: Proceedings of the 2002 International Symposium on Symbolic and Algebraic Computation, pp. 75–83. Springer, Cham (2002). https://doi.org/10.1145/780506.780516

KPG99. Kipnis, A., Patarin, J., Goubin, L.: Unbalanced oil and vinegar signature schemes. In: Stern, J. (ed.) EUROCRYPT 1999. LNCS, vol. 1592, pp. 206–222. Springer, Heidelberg (1999). https://doi.org/10.1007/3-540-48910-X_15

KS98. Kipnis, A., Shamir, A.: Cryptanalysis of the oil and vinegar signature scheme. In: Krawczyk, H. (ed.) CRYPTO 1998. LNCS, vol. 1462, pp. 257–266. Springer, Heidelberg (1998). https://doi.org/10.1007/BFb0055733

Lev05. Levitskaya, A.A.: Systems of random equations over finite algebraic structures. Cybernetics and Sys. Anal **41**(1), 67–93 (2005)

Luy19. Van Luyen, L.: An improved identity-based multivariate signature scheme based on rainbow. Cryptography **3**(1) (2019)

NIS19. National Institute of Standards and Technology: Post-quantum crypto project (Second Round Submission) (2019). https://csrc.nist.gov/Projects/post-quantum-cryptography/post-quantum-cryptography-standardization/round-2-submissions. Accessed 16 Aug 2022

NIS20. National Institute of Standards and Technology: Post-quantum crypto project (Third Round Submission) (2020). https://csrc.nist.gov/Projects/post-quantum-cryptography/post-quantum-cryptography-standardization/round-3-submissions. Accessed 16 Aug 2022

Sha84. Shamir, A.: Identity-based cryptosystems and signature schemes. In: Blakley, G.R., Chaum, D. (eds.) CRYPTO 1984. LNCS, vol. 196, pp. 47–53. Springer, Heidelberg (1985). https://doi.org/10.1007/3-540-39568-7_5

SSH11. Sakumoto, K., Shirai, T., Hiwatari, H.: On provable security of UOV and HFE signature schemes against chosen-message attack. In: Yang, B.-Y. (ed.) PQCrypto 2011. LNCS, vol. 7071, pp. 68–82. Springer, Heidelberg (2011). https://doi.org/10.1007/978-3-642-25405-5_5

STX13. Shen, W., Tang, S., Xu, L.: IBUOV, a provably secure identity-based UOV signature scheme. In: IEEE 16th International Conference on Computational Science and Engineering, LNCS, pp. 388–395. IEEE (2013)

Revisiting the Security of Salted UOV Signature

Sanjit Chatterjee[1], M. Prem Laxman Das[2], and Tapas Pandit[1(✉)]

[1] Department of Computer Science and Automation, Indian Institute of Science Bangalore,Bangalore, India
{sanjit,tapas}@iisc.ac.in
[2] Society for Electronic Transactions and Security, Chennai, India

Abstract. Due to the recent attack of Beullens on Rainbow, the crypto community looks back again at the unbalanced oil-and-vinegar (UOV) signature. The original UOV does not have any formal security reduction. It was Sakumoto et al. who added a random salt to the original UOV signature to give a reduction under the UOV-inversion (UOVI) problem in the classical random oracle model (CROM).

In this paper, we revisit the security of salted UOV signature. We start by identifying some issues related to programming the random oracle and the distribution of the salt. Then provide a security reduction of the salted UOV signature in the CROM that clearly addresses these issues. One crucial requirement of our reduction is that the field size needs to be asymptotically superpolynomial in the security parameter. We also give a security reduction of the salted UOV under the UOVI problem in the quantum random oracle model. This work is hoped to aid further concrete security analysis and thereby guide parameter choice of UOV-based schemes in the context of future standardization of post-quantum signature.

Keywords: Digital signature · Multivariate cryptography · UOV · Post-quantum security · QROM

1 Introduction

Multivariate quadratic polynomials (MQ) based signatures [DS05,PCY+15, CHR+16] are attractive candidates for post-quantum cryptography due to their fast verification and short signature. One of these is Rainbow which was a finalist in the recently concluded third round of NIST PQC Standardization competition. Rainbow [DS05] is a multilayered version of unbalanced oil-and-vinegar (UOV) signature scheme [KPG99]. Several variants of signatures, e.g., identity-based signature [CLND19,CDP21], blind signature [PSM17] and ring signature [MP17] have been designed in the MQ-setting using Rainbow (or UOV) as primary building block. Note that Rainbow as a multilayered extension of UOV [KPG99] was solely introduced to gain efficiency. For practical applications,

T. Isobe and S. Sarkar (Eds.): INDOCRYPT 2022, LNCS 13774, pp. 697–719, 2022.
https://doi.org/10.1007/978-3-031-22912-1_31

two-layered version of Rainbow is mainly recommended as further increasing the number of layers does not significantly improve its efficiency. However, the recent attack [Beu22] on the two-layered Rainbow basically works by peeling off the 2nd layer followed by an existing UOV attack in [KPG99] on the 1st layer with much smaller parameter size compared to the original UOV. The attack motivates the research community to look back at the UOV signature with renewed interest. Thus a rigorous security analysis would be useful in designing UOV-based signatures, which could be candidates for future standardization efforts.

Note that the original UOV proposal of [KPG99] does not have any formal security proof. It was Sakumoto et al. [SSH11] who first formally studied security of the UOV signature. They introduced a random salt to make the output signature somewhat uniform and then argued security in the classical random oracle model (CROM) from UOV-inversion (UOVI) problem[1] using the FDH-technique [BR93]. In [SSH11], the authors consider the hash function to be $\mathcal{H} : \mathcal{M} \times \mathsf{SaltSpac} \to \mathbb{F}^m$, where $\mathbb{F}$ is the underlying field. That is, the hash arguments have the form: (m, s), where m is the message and s is the salt (a binary string). A valid signature for m under the salted UOV is of the form: $\sigma = (\boldsymbol{x}, s)$ such that $\mathcal{P}(\boldsymbol{x}) = \mathcal{H}(\mathsf{m}, s)$, where $\mathcal{P} : \mathbb{F}^n \to \mathbb{F}^m$ is the UOV public map. In the security reduction, $\mathcal{H}$ is treated as a random oracle.

Our Result. In this paper, we revisit the security reduction of the salted UOV [SSH11] and identify some gaps pertaining to programming the random oracle and distribution of the salt (Sect. 3). In particular, when queried with (m, s) the random oracle involved in the signature oracle is programmed to return $\mathcal{P}(\boldsymbol{x})$, where $\boldsymbol{x} \in \mathbb{F}^n$ is randomly chosen. That is, the authors implicitly assumed that for a random $\boldsymbol{x} \in \mathbb{F}^n$, $\mathcal{P}(\boldsymbol{x})$ is uniform. The paper also assumed that the salt part of the output signature is uniform, although the distribution of the salt actually depends on the size of the underlying field.

We then provide (Sect. 4) a security reduction of the salted UOV signature in the CROM that clearly addresses these issues. Here we consider the salted homogeneous UOV scheme, but through the subspace description [Beu21] of the UOV-trapdoor (Sect. 4.1). The main reason for using [Beu21] is that it improves secret key sizes (Sect. 4.2). For the reduction, we assume neither the uniformity of $\mathcal{P}(\boldsymbol{x})$ nor the uniformity of the salt involved in the output signature. We essentially show that both distributions deviate from the respective uniform distributions by at most $1/q$ (Proposition 1 and Corollary 2), where q is the size of the underlying field. One crucial implication of our result is that the field size q needs to be asymptotically superpolynomial in the security parameter. Suppose the upper bound on the numbers of signature queries and random oracle queries in practice are 2^{20} and 2^{60} respectively. Then, from a back-of-the-envelope calculation based on Theorem 1, one can see that the underlying field has to be chosen of size roughly 2^{88} for 128-bit security[2]. This will surely impact the efficiency of

[1] Given a random UOV public map $\mathcal{P} : \mathbb{F}^n \to \mathbb{F}^m$ and a random element $\boldsymbol{y} \in \mathbb{F}^m$, find an $\boldsymbol{x} \in \mathbb{F}^n$ such that $\mathcal{P}(\boldsymbol{x}) = \boldsymbol{y}$.

[2] Note that whatever the parameter choice of UOV, the unavoidable degradation due to the total number of random oracle queries will always be there.

the scheme. Thus deriving the parameter sizes and the consequent implication on efficiency based on a concrete analysis of our security reduction of salted UOV could be an interesting future work.

In principle, it is desirable to have security proof in the quantum random oracle model (QROM), rather than just in CROM. We achieve this for salted UOV by providing a security reduction (Sect. 5) from the UOVI problem in the QROM. Again based on this reduction, we do not provide any parameter choice, other than pointing out the fact that q needs to be asymptotically superpolynomial in the security parameter (Theorem 2).

2 Preliminaries

For $a \in \mathbb{N}\backslash\{0\}$, define $[a] = \{x \in \mathbb{N}\backslash\{0\} : x \leq a\}$. For a set X, we write $x \xleftarrow{\$} X$ to mean that x is drawn uniformly at random from X. For an algorithm A and its input x, the notation $y \leftarrow A(x)$ denotes that when A is run on x, it outputs y. We use bold-face lower case letters, e.g., $\boldsymbol{x}$ to denote column vectors. The i-th entry of $\boldsymbol{x}$ is denoted by $\boldsymbol{x}_i$. For a matrix A, the notation $A^\top$ is used to denote its transpose. The fixed finite field on which all the operations take place is denoted by $\mathbb{F}$. The notation q will denote the size of the field $\mathbb{F}$. We make no assumptions about the characteristic of $\mathbb{F}$.

We shall consider only homogeneous quadratic polynomials in n variables over $\mathbb{F}$. While discussing UOV scheme, the number of polynomials in the secret system $\mathcal{F}$ and that in the public system $\mathcal{P}$ will be m. This number would match the number of oil variables, as per the usual description of UOV scheme. The oil (vinegar) variables will be last m (respectively, the first $v = n - m$) of them among $\{X_1, \ldots, X_n\}$. With the secret and public systems one can associate polynomial maps $\mathcal{F} : \mathbb{F}^n \to \mathbb{F}^m$ and $\mathcal{P} : \mathbb{F}^n \to \mathbb{F}^m$, respectively, which will just denote evaluation. All the polynomials which appear in this work are homogeneous. It is well-known in literature that the security of MQ-based systems depends mainly the quadratic part of the polynomials involved. The transformation used for mixing the variables in UOV is assumed to be an invertible matrix. Hence the public key obtained will be homogeneous whenever the secret key is so.

2.1 Quadratic Polynomials and Their Matrix Representation

We shall consider homogeneous quadratic polynomials in m variables over a finite field $\mathbb{F}$. Any such $\mathfrak{f}$ has a associated *polar form* $\mathfrak{f}'$, which is symmetric bilinear, satisfying

$$\mathfrak{f}'(X, Y) = \mathfrak{f}(X + Y) - \mathfrak{f}(X) - \mathfrak{f}(Y).$$

With every homogeneous quadratic polynomial one can associate a matrix. The matrix representing the polynomial is defined as follows.

Definition 1. *Let $\mathfrak{f}$ be a homogeneous quadratic polynomial over $\mathbb{F}$ in n variables. An $n \times n$ matrix $M_\mathfrak{f}$ is said to represent $\mathfrak{f}$ if*

$$\mathfrak{f}(X) = X^\top M_\mathfrak{f} X,$$

where $X = (X_1, \ldots, X_m)^\top$ *is a column vector of variables.*

Remark 1. The polar form of the quadratic form is bilinear. There is an obvious way (see [Beu21]) for obtaining the matrix representing the polar form, depending on the characteristic of the underlying field. If M'_f denotes this matrix, then $\mathfrak{f}'(X, Y) = X^\top M'_\mathfrak{f} Y$.

2.2 (Unbalanced) Oil-Vinegar Signature Schemes

We will be following the treatment of Kipnis *et al.* [KPG99]. Let $\mathbb{F}$ be a fixed finite field. As usual, let n and m be positive integers, and set $v = n - m$. Let $\{X_1, \ldots, X_v\}$ denote the (ordered) set of vinegar variables and $\{X_{v+1}, \ldots, X_n\}$ that of oil variables. The message (digest) space is $\mathbb{F}^m$ and the signature space is $\mathbb{F}^n$ (for plain-UOV scheme).

The central object in such schemes is a polynomial of the following special form. *The oil-vinegar type polynomial* is a quadratic polynomial over $\mathbb{F}$ in the variables described above, but without any quadratic terms involving only the oil variables. In other words, a oil variable is not allowed to mix with another oil variable in such a polynomial. The general form of such a polynomial is as follows:

$$\phi(X_1, \ldots, X_m) = \sum_{j=1}^{v} \sum_{k=j}^{n} \alpha_{jk} X_j X_k + \sum_{j=1}^{n} \beta_j X_j + \gamma. \tag{1}$$

For a given polynomial of the form in Eq. (1), if values are assigned to all vinegar variables, the resulting polynomial is linear in oil variables. This feature is the central theme of the trapdoor.

Let $\mathcal{T} : \mathbb{F}^n \to \mathbb{F}^n$ denote an invertible linear map. Such a transformation is used for mixing the input variables. The oil-vinegar trapdoor is described as follows.

Definition 2. *(Oil-Vinegar Trapdoor) Let* $\mathbb{F}$ *be a system of* m *polynomials of the form given in Eq. (1). Let* $\mathcal{T}$ *be a invertible linear transformation on* $\mathbb{F}^n$*. Let* $\mathcal{P} = \mathcal{F} \circ \mathcal{T}$ *be the polynomial system obtained by composing each polynomial in* $\mathbb{F}$ *with* $\mathcal{T}$ *(i.e., transformed polynomials when* $\mathcal{T}$ *acts on the vector of variables). Given* $\mathcal{P}$ *and* $\boldsymbol{\tau} \in \mathbb{F}^m$*, the challenge is to solve* $\mathcal{P}(\cdot) = \boldsymbol{\tau}$*.*

Remark 2. Solving $\mathcal{P}(\cdot) = \boldsymbol{\tau}$ directly is assumed to be hard. But the knowledge of the trapdoor information, namely $\mathbb{F}$ and $\mathcal{T}$, can be used for solving such a system [KPG99]. Solving $\mathcal{F}(\cdot) = \boldsymbol{\tau}$ is easy. The strategy is to assign random values for vinegar variables and solving the linear system involving only the oil variables. The resulting assignment is then inverted under the affine transformation $\mathcal{T}$. The process of assigning values to vinegar variables and solving the resulting system of linear equations is repeated until one valid assignment for all variables is found.

Matrix Description of Homogeneous Quadratic System. Studying the matrices representing the $\mathcal{F}$ and $\mathcal{P}$ systems becomes useful from the analysis point of view. Let us consider the component polynomials $\mathfrak{f}$ involved in the system $\mathcal{F}$ to be homogeneous quadratic polynomials. Since every polynomial $\mathfrak{f}$ in the system $\mathcal{F}$ is devoid of oil-oil term and every polynomial $\mathfrak{g}$ in $\mathcal{P}$ is constructed as $\mathfrak{g} = \mathfrak{f} \circ \mathcal{T}$, the (block) form of their corresponding matrices will be

$$M_{\mathfrak{f}} = \begin{bmatrix} A & B \\ 0 & 0 \end{bmatrix} \text{ and } M_{\mathfrak{g}} = \mathcal{T}^{\top} M_{\mathfrak{f}} \mathcal{T}, \tag{2}$$

where A is a $v \times v$ upper triangular matrix, B is a $v \times m$ matrix, and 0's are all zero matrices of suitable orders such that $M_{\mathfrak{f}}$ is an $n \times n$ matrix.

2.3 Linear Subspace Interpretation of Oil-Vinegar Trapdoor

Beullens [Beu21] takes a subspace approach to the oil-vinegar trapdoor description. The public key is a MQ system $\mathcal{P}$ (m MQ polynomials in n variables) which vanishes on a secret subspace $\mathsf{O} \subset \mathbb{F}^n$ of dimension m. The trapdoor is set as follows. First, the subspace O is chosen at random. Then a system $\mathcal{P}$, consisting of m multivariate quadratic polynomials in n variables, vanishing at this subspace O, is chosen uniformly at random. The trapdoor information is the description of O. For a target $\boldsymbol{\tau} \in \mathbb{F}^m$, solving $\mathcal{P}(\cdot) = \boldsymbol{\tau}$ is easy. Notice that

$$\mathcal{P}(\boldsymbol{v} + \boldsymbol{o}) = \mathcal{P}(\boldsymbol{v}) + \mathcal{P}(\boldsymbol{o}) + \mathcal{P}'(\boldsymbol{v}, \boldsymbol{o}) \tag{3}$$

holds for any $\boldsymbol{o}$ coming from the subspace O and any $\boldsymbol{v}$ coming from $\mathbb{F}^n$. Thus $\mathcal{P}(\cdot) = \boldsymbol{\tau}$ can be solved by solving

$$\mathcal{P}'(\boldsymbol{v}, \boldsymbol{o}) = \boldsymbol{\tau} - \mathcal{P}(\boldsymbol{v}),$$

where $\boldsymbol{o} \in \mathsf{O}$. The above system is linear in variable $\boldsymbol{o}$. For, the first term in the right hand side of Eq. (3) is fixed once $\boldsymbol{v}$ is fixed, the second term is zero since $\boldsymbol{o}$ is from the distinguished subspace O and the third term is linear in oil variables.

On the other hand, solving the MQ system $\mathcal{P}$, without the knowledge of the trapdoor information is assumed to be hard.

2.4 Syntax and Security of Signature Scheme

Definition 3 (Signature Scheme). *It consists of three PPT algorithms -* KeyGen, Sign *and* Ver.

- KeyGen*: It takes as input a security parameter κ and outputs a public and private key pair $(\mathcal{PK}, \mathcal{SK})$.*
- Sign*: It takes as input a message $\mathsf{m} \in \mathcal{M}$, where $\mathcal{M}$ is the message space, and the secret key $\mathcal{SK}$ and outputs a signature σ.*
- Ver*: It takes as input a message-signature pair (m, σ) and the public key $\mathcal{PK}$. It outputs a value 1 if (m, σ) is a valid message-signature pair else it outputs 0.*

Correctness: *For all* $(\mathcal{PK}, \mathcal{SK}) \leftarrow \mathsf{KeyGen}(1^\kappa)$ *and for all messages* $\mathsf{m} \in \mathcal{M}$, *it is required that*

$$\mathsf{Ver}(\mathsf{m}, \mathsf{Sign}(\mathsf{m}, \mathcal{SK}), \mathcal{PK}) = 1.$$

Next we define security model of the signature scheme. A security notion very useful in practice is called existentially unforgeable under chosen message attack (EUF-CMA).

Definition 4 (EUF-CMA). *A signature scheme is said to be EUF-CMA secure if for all quantum PPT algorithms* $\mathcal{A}$, *the advantage*

$$\mathsf{Adv}_{\mathcal{A}}^{EUF\text{-}CMA}(\kappa) = \Pr\left[\mathsf{Ver}(\mathsf{m}^*, \sigma^*, \mathcal{PK}) = 1 \;\middle|\; \begin{matrix}(\mathcal{PK}, \mathcal{SK}) \leftarrow \mathsf{KeyGen}(1^\kappa);\\ (\mathsf{m}^*, \sigma^*) \leftarrow \mathcal{A}^{\mathcal{O}_{\mathsf{Sign}}}(\mathcal{PK})\end{matrix}\right]$$

is a negligible function in κ, *where* $\mathcal{A}$ *is provided access to the sign oracle* $\mathcal{O}_{\mathsf{Sign}}$ *with a natural restriction that* $\mathsf{m}^* \neq \mathsf{m}$ *for all messages* m *queried to* $\mathcal{O}_{\mathsf{Sign}}$.

3 Revisiting the Security Reduction of Salted UOV

The unbalanced oil-and-vinegar (UOV) signature was proposed in [KPG99] to protect from the attack of [KS98] on the balanced oil-and-vinegar signature of Patarin [Pat97]. However, the authors of UOV-signature did not provide any formal security proof of their construction. The distribution of the signatures generated in the original UOV-signature [KPG99] is not uniform, even if the underlying hash function is treated as random oracle. Therefore, the FDH-technique [BR93] is not directly applicable in arguing security of the UOV-signature.

The signature scheme, salted UOV was presented in [SSH11, Section 4.1] (see Appendix B). The salt is appended to the message and hashed, thereby a system $\mathcal{P}(\cdot) = \mathcal{H}(\mathsf{m}||s)$ is set up. A solution is obtained by first putting values for the vinegar variables and solving (a linear system) for the oil variables. If the system does not have a solution, a fresh salt is chosen. The authors point out that, this way, the distribution of the signature will be uniform, and hence, the FDH-technique can be used to argue the security of the salted UOV signature.

3.1 On the Simulation of Random Oracle and Salt

First, we informally describe the FDH-style security reduction [BR93]. Let $(f : \mathfrak{D} \to \mathfrak{R}, y^* \in \mathfrak{R})$ be the given problem instance, where f is a trapdoor permutation and the goal is to find an $x^* \in \mathfrak{D}$ such that $f(x^*) = y^*$. Recall that the FDH-signature for a message m is of the form $\sigma = x$, where $y = \mathcal{H}(\mathsf{m})$, $x = f^{-1}(y)$ and $\mathcal{H} : \mathcal{M} \to \mathfrak{R}$ is the underlying hash function. A message-signature pair (m, σ) is valid, if $f(\sigma) = \mathcal{H}(\mathsf{m})$.

If an adversary can produce a valid forgery (m^*, σ^*) for this signature scheme, then a solver for the above problem can be constructed. Here the underlying hash function $\mathcal{H} : \mathcal{M} \to \mathfrak{R}$ is treated as a random oracle. That means, the adversary must have queried the corresponding message m^* to the random oracle $\mathcal{H}$. In the reduction, an index is guessed where the forgery message m^* could appear as a random oracle query and the corresponding random oracle value is appropriately programmed. In other words, pick an index i^* randomly as a guess and set $\mathcal{H}(\mathsf{m}_{i^*}) = y^*$. Note that for a correct guess, we have $\mathsf{m}^* = \mathsf{m}_{i^*}$. For a query on message $\mathsf{m} \in \mathcal{M}$ other than m_{i^*}, first pick a signature $\sigma \xleftarrow{\$} \mathfrak{D}$, then program the random oracle at m as $\mathcal{H}(\mathsf{m}) = y = f(\sigma)$ and store the tuple (m, σ, y) in a list List. So using the list, all the oracle queries can be answered. Note that $f(\sigma)$ will be uniform over $\mathfrak{R}$ as f is bijective[3]. If i^* is correctly guessed and (m^*, σ^*) is a valid forgery, then we have $f(\sigma^*) = \mathcal{H}(\mathsf{m}^*) = y^*$, which implies that $\boldsymbol{x}^* = \sigma^*$ is the required solution of the given problem instance.

In [SSH11], the authors showed a security reduction of salted UOV in the CROM under the hardness of UOVI-problem. Since a salt is involved as part of the signature, the FDH-style proof will be slightly different here. We summarize their security proof as follows. In the game between a simulator S and an adversary $\mathcal{A}$, S maintains a list $\mathsf{List}_{\mathsf{uov}}$ of three tuples $(\mathsf{m}, s, \boldsymbol{y})$, where $\boldsymbol{y}$ is the hash of $\mathsf{m}||s$. The random oracle query on challenge message is answered in a similar way as done above. The other queries are answered as follows. For an incoming random oracle query $\mathsf{m}||s$, if $(\mathsf{m}, s, \cdot) \in \mathsf{List}_{\mathsf{uov}}$, then the stored value is returned. Else a random value $\boldsymbol{y}$ is returned and $(\mathsf{m}, s, \boldsymbol{y})$ is appended to the list $\mathsf{List}_{\mathsf{uov}}$. If m is a signing oracle query, the simulator chooses a salt s at random. If $(\mathsf{m}, s, \cdot)$ is in the list, it aborts. Else, it chooses $\boldsymbol{x} \in \mathbb{F}^n$ uniformly at random and returns $(\boldsymbol{x}, s)$ as signature corresponding to m after appending $(\mathsf{m}, s, \boldsymbol{y})$, with $\boldsymbol{y} = \mathcal{P}(\boldsymbol{x})$, to the list $\mathsf{List}_{\mathsf{uov}}$. Similarly as above, when the index i^* is correctly guessed and $(\boldsymbol{x}^*, s^*)$ is a valid forgery for m^*, then $\boldsymbol{x}^*$ will be a solution of the given UOVI-problem instance.

Issue in Random Oracle Programming. Note that while answering the sign-queries, the random oracle $\mathcal{H}$ is programmed by assigning $\mathcal{P}(\boldsymbol{x})$ for random choice of $\boldsymbol{x}$ from $\mathbb{F}^n$. Since $\mathcal{P}$ is neither bijective nor known to be regular, it cannot be definitely said that $\mathcal{P}(\boldsymbol{x})$ is uniform over $\mathbb{F}^m$. Hence, $\mathcal{H}$ is treated as a random oracle in [SSH11] without any proper justification. This, in turn, opens up the possibility of a potential gap in the security claim.

Issue in Salt Distribution. A signature in the salted UOV [SSH11] consists of a salt and an element from $\mathbb{F}^n$. Note that only the salt generated in the last iteration of the loop in the sign algorithm (Algorithm 3) contributes to the final output signature. In other words, the salt in the output signature follows a distribution that samples a couple of salts in a row *without replacement* and outputs the final salt. More precisely, the distribution of the salt in the output signature depends on the rank of an $m \times m$ matrix, which further depends

[3] Note that the reduction also works, if f is considered to be a regular function. Here regular means the preimage sets of all the points in $\mathcal{M}$ under f are of same size.

on $q = |\mathbb{F}|$ (for details, see [SSH11, Section 3.1]). As described earlier, while answering the sign-queries in the reduction, the salts are always chosen uniformly at random. Essentially, this creates a difference between the distributions of salts, one involved in the actual signatures and the other in the simulated signatures. It seems the authors implicitly assume that a computational adversary cannot detect the difference.

4 A Clean Security Reduction of Salted UOV

For addressing the issues raised in the previous section, we consider the underlying maps involved in the public key $\mathcal{P}$ to be homogeneous quadratic polynomials. For general quadratic polynomial maps, closing the above gaps still remains an interesting research problem. Nonetheless, restricting to homogeneous quadratic maps does not weaken the security of the signature as the intractability of the underlying MQ-problem mainly relies on the quadratic part of the MQ-system. Using the result on the distribution of $\mathcal{P}(\boldsymbol{x})$ that we describe in Sect. 4.3, one can derive a clean security proof of the salted homogeneous UOV signature. However, in this paper, we argue the security of an alternative salted UOV signature (see Sect. 4.2) which is based on the subspace approach to UOV trapdoor [Beu21]. The reason for considering this alternative construction is that it improves upon the key sizes a bit. The remainder of this section is organized as follows. We start with the (plain) homogeneous UOV signature based on Beullens' subspace approach in Sect. 4.1. Then, present its salted version in Sect. 4.2. We analyze the distribution of $\mathcal{P}(\boldsymbol{x})$ in Sect. 4.3. Finally, provide a clean proof of the salted homogeneous UOV in Sect. 4.4.

4.1 Homogeneous UOV Signature Scheme Using the Subspace Interpretation

The trapdoor described in Sect. 2.3 can be used to design a signature scheme. We discuss the efficiency aspects of the key generation and signing modules (without salt) in this section. The public key system is an MQ-system, which vanishes on a subspace. The trapdoor information is the description of the subspace. The two major questions are the following. How does one sample a random subspace of $\mathbb{F}^n$ and a uniformly random MQ system which vanishes on this subspace? How does one represent the trapdoor information so that the MQ system can be solved, efficiently, using the trapdoor information? Next we discuss these two points based on [Beu21].

Efficient Setup for the UOV Trapdoor. The trapdoor can be efficiently setup using the matrix representation of the quadratic form. Let $\mathcal{F} = (\mathfrak{f}_1, \ldots, \mathfrak{f}_m)$, be the collection of secret UOV (homogeneous) polynomials. The matrix corresponding to $\mathfrak{f}_i$ has the form

$$M_{\mathfrak{f}_i} = \begin{array}{c} \\ v \\ m \end{array} \begin{array}{c} v \quad m \\ \begin{bmatrix} A_i & B_i \\ 0 & 0 \end{bmatrix} \end{array} \tag{4}$$

where A_i is a random $v \times v$ upper triangular matrix, B_i is a random $v \times m$ matrix, and 0's are all zero matrices of suitable orders such that $M_{\mathfrak{f}_i}$ is an $n \times n$ matrix. The following subspace

$$\mathsf{O}' = \left\{ (x_1, \ldots, x_n)^\top \in \mathbb{F}^n : x_1 = \cdots = x_{n-m} = 0 \right\}$$

is called *oil subspace* of dimension m. Notice that every $\mathfrak{f}_i$ vanishes on O'. If $\mathcal{P} = (\mathfrak{g}_1, \ldots, \mathfrak{g}_m)$ is the public key system, where $\mathfrak{g}_i = \mathfrak{f}_i \circ \mathcal{T}$, then every quadratic form in $\mathcal{P}$ vanishes on $\mathsf{O} = \mathcal{T}^{-1}(\mathsf{O}')$, where $\mathcal{T}$ is an invertible matrix. Since $\mathcal{T}$ is invertible, the dimension of O is equal to m. So, O will be a random subspace when $\mathcal{T}$ is a random invertible matrix.

Note that the distribution of the public keys generated in both ways, one as discussed above and the traditional one are the same. Beullens also pointed out in [Beu21] that the public key generated using the subspace description (discussed in Sect. 2.3) and using the traditional description have the identical distribution.

Solving the MQ System Using Trapdoor, Efficiently. We discuss how a solution for $\mathcal{P}(\cdot) = \boldsymbol{\tau}$ can be obtained, efficiently, using the trapdoor information. Recall that the trapdoor information is a description of the subspace on which this system $\mathcal{P}$ vanishes. From Eq. (3), solving this system amounts to solving $\mathcal{P}'(\boldsymbol{v}, \boldsymbol{o}) = \boldsymbol{\tau}'$ for $\boldsymbol{o} \in \mathsf{O}$, where $\boldsymbol{\tau}' = \boldsymbol{\tau} - \mathcal{P}(\boldsymbol{v})$ and $\boldsymbol{v} \in \mathbb{F}^n$ is fixed. For a homogeneous quadratic polynomial $\mathfrak{g}$, the polar form is given by $\mathfrak{g}'(\boldsymbol{x}, \boldsymbol{y}) = \boldsymbol{x}^\top M'_{\mathfrak{g}} \boldsymbol{y}$ for all $\boldsymbol{x}, \boldsymbol{y} \in \mathbb{F}^n$ (see Remark 1).

We can describe the subspace $\mathsf{O} \subset \mathbb{F}^n$ of dimension m using column-span of a full-rank $n \times m$ matrix $\overline{M}$. For, if O is generated by a linearly independent set of vectors $\{\boldsymbol{w}_1, \ldots, \boldsymbol{w}_m\}$ from $\mathbb{F}^n$, then i-th column of $\overline{M}$ will be $\boldsymbol{w}_i$. Thus $\overline{M}$ is a full-rank matrix and any element of the subspace O can be written as $\boldsymbol{o} = \overline{M}\boldsymbol{y}$ for some $\boldsymbol{y} \in \mathbb{F}^m$.

We now describe an effective procedure for solving the public key system. For each public key polynomial $\mathfrak{g}$, a row vector $\boldsymbol{c}_{\mathfrak{g}}$ is computed as $\boldsymbol{c}_{\mathfrak{g}} = \boldsymbol{v}^\top M'_{\mathfrak{g}} \overline{M}$, where $\boldsymbol{v}$ is a random vector chosen from $\mathbb{F}^n$. We then consider the following linear system: $C\boldsymbol{y} = \boldsymbol{\tau}'$ (recall that $\boldsymbol{\tau}' = \boldsymbol{\tau} - \mathcal{P}(\boldsymbol{v})$), where the $\mathfrak{g}$-th row of the matrix C is the row vector $\boldsymbol{c}_{\mathfrak{g}}$. If a solution $\boldsymbol{y} \in \mathbb{F}^m$ to the system exists, an element $\boldsymbol{o} \in \mathsf{O}$ can hence be obtained as $\boldsymbol{o} = \overline{M}\boldsymbol{y}$. The quantity $\boldsymbol{v} + \boldsymbol{o}$ is a solution for $\mathcal{P}(\cdot) = \boldsymbol{\tau}$.

Remark 3. Note that the matrix C can also be written as $C = C' \cdot \overline{M}$, where the row of C' corresponding to the public polynomial $\mathfrak{g}$ is given by $\boldsymbol{c}'_{\mathfrak{g}} = \boldsymbol{v}^\top M'_{\mathfrak{g}}$. When the map $\mathcal{P}'(\boldsymbol{v}, .) : \mathsf{O} \to \mathbb{F}^m$ is non-singular, then the rank of C' will be m, which in turn implies that the matrix C will be non-singular as $\overline{M}$ is an $n \times m$ full-rank matrix. In [Beu21], Beullens uses the following fact:

$$\mathsf{Pr}[\mathcal{P}'(\boldsymbol{v}, .) : \mathsf{O} \rightarrow \mathbb{F}^m \text{ is non-singular} : \boldsymbol{v} \xleftarrow{\$} \mathbb{F}^n] \approx (1 - 1/q). \tag{5}$$

So, the procedure described above is expected to terminate after a few trials.

The algorithmic version of the above description is given in Appendix A. The signature scheme derived from the trapdoor is also described there.

4.2 Salted Homogeneous UOV

In this section, we illustrate a signature scheme, designed using the subspace description of the UOV trapdoor [Beu21]. The approach is similar to that used in [SSH11]. A salt is used for making the security reduction to go through in the random oracle model. Let us refer to this signature as salted homogeneous UOV (SHUOV) signature.

KeyGen. This takes the security parameter 1^κ as input and outputs the public and secret keys. The secret key $\mathcal{SK}$ is a description of the subspace $\mathsf{O} \subset \mathbb{F}^n$, that is, an $n \times m$ full-rank matrix $\overline{M}$ (as mentioned in Sect. 4.1). The public key $\mathcal{PK}$ is the system $\mathcal{P}$ consisting of m MQ-polynomials in n variables which vanishes at O. See Sect. 4.1 for a description. A hash function $\mathcal{H} : \mathcal{M} \times \mathsf{SaltSpac} \rightarrow \mathbb{F}^m$ for converting message into a fixed-length digest is known publicly, where $\mathcal{M}$ and $\mathsf{SaltSpac} = \{0, 1\}^{\ell_s}$ are respectively the message space and the salt space. Note that the signature space of SHUOV-signature is $\Sigma = \mathbb{F}^n \times \mathsf{SaltSpac}$.

Sign. This takes message m and the secret key $\mathcal{SK}$ as input and outputs a signature σ. The procedure for computing the signature is described in Algorithm 1. The tuple $(\boldsymbol{z}, s)$ is returned as signature σ.

Ver. This module takes message m, the signature σ and the public key $\mathcal{PK}$ as input and outputs accept or reject. The steps are described below:
- Parse the signature as $(\boldsymbol{z}, s)$
- Compute $\boldsymbol{\tau} = \mathcal{H}(\mathsf{m}||s)$
- Accept the signature if $\mathcal{P}(\boldsymbol{z}) = \boldsymbol{\tau}$ holds; otherwise reject.

Correctness. The signature scheme is correct. If a message m, public key $\mathcal{P}$ and a signature $(\boldsymbol{z}, s)$, where $\boldsymbol{z}$ is obtained according to Algorithm 1 are given, then, we need to verify that $\mathcal{P}(\boldsymbol{z}) = \mathcal{H}(\mathsf{m}||s)$ holds. Let $\mathfrak{g}$ be any MQ-polynomial in the public key system $\mathcal{P}$. The following can be easily verified for each such $\mathfrak{g}$:

$$\begin{aligned}
\mathfrak{g}(\boldsymbol{z}) &= \mathfrak{g}(\boldsymbol{v} + \boldsymbol{o}) \\
&= \mathfrak{g}(\boldsymbol{v}) + \mathfrak{g}(\boldsymbol{o}) + \mathfrak{g}'(\boldsymbol{v}, \boldsymbol{o}) \\
&= \mathfrak{g}(\boldsymbol{v}) + \mathfrak{g}(\boldsymbol{o}) + \mathfrak{g}'(\boldsymbol{v}, \overline{M}\boldsymbol{u}) \\
&= \mathfrak{g}(\boldsymbol{v}) + \mathfrak{g}(\boldsymbol{o}) + \boldsymbol{v}^\top M'_{\mathfrak{g}} \overline{M} \boldsymbol{u}
\end{aligned} \tag{6}$$

Algorithm 1. Signing Module for Salted UOV

Require: The message m, secret key $\mathcal{SK}$ and the description of the salt space
Ensure: A signature $(\boldsymbol{z}, s)$ on m

1: Sample a vector $\boldsymbol{v} \xleftarrow{\$} \mathbb{F}^n$
2: Compute $\boldsymbol{c}_{\mathfrak{g}} = \boldsymbol{v}^\top M'_{\mathfrak{g}}\overline{M}$ ▷ $M'_{\mathfrak{g}}\overline{M}$ can be precomputed
3: Construct the $m \times m$ matrix C with $\boldsymbol{c}_{\mathfrak{g}}$ as rows
4: **repeat**
5: Sample $s \xleftarrow{\$} \{0,1\}^{\ell_s}$ ▷ a salt s is sampled
6: Compute $\boldsymbol{\tau} = \mathcal{H}(\mathsf{m}||s)$
7: Compute $\boldsymbol{\tau}' = \boldsymbol{\tau} - \mathcal{P}(\boldsymbol{v})$
8: **until** $\{\boldsymbol{y} \in \mathbb{F}^m : C \cdot \boldsymbol{y} = \boldsymbol{\tau}'\} \neq \emptyset$
9: Sample $\boldsymbol{u} \xleftarrow{\$} \{\boldsymbol{y} \in \mathbb{F}^m : C \cdot \boldsymbol{y} = \boldsymbol{\tau}'\}$
10: Compute $\boldsymbol{o} = \overline{M}\boldsymbol{u}$ ▷ column-span corresponding to $\boldsymbol{u}$
11: Compute $\boldsymbol{z} = \boldsymbol{v} + \boldsymbol{o}$
12: Output $\sigma = (\boldsymbol{z}, s)$

where $\mathfrak{g}'$ is the polar form of $\mathfrak{g}$. Since $\boldsymbol{o} \in \mathsf{O}$, the second term on the RHS of Eq. (6) is zero and the third term is equal to the $\mathfrak{g}$-th coordinate of the vector $\boldsymbol{\tau}' = \boldsymbol{\tau} - \mathcal{P}(\boldsymbol{v})$. Thus, combining the above observation for every such $\mathfrak{g}$, we obtain $\mathcal{P}(\boldsymbol{z}) = \mathcal{H}(\mathsf{m}||s)$. This proves that $(\boldsymbol{z}, s)$ is a valid signature on m.

Efficiency Comparison. One can easily check that both the versions, based on traditional approach [SSH11] and subspace approach (presented above) entertain more or less the same signing and verification time. However, the key sizes are improved in the subspace approach as only the basis information for the secret hidden subspace is required to store. In fact, the number of field elements required to store for both the approaches are presented in Table 1.

Table 1. Public and secret key sizes for UOV signature

Approach	Public key (# of field elements)	Secret key (# of field elements)
Traditional	$mn(n+1)/2$	$m(v(v+1)/2 + vm) + n^2$
Subspace	$mn(n+1)/2$	mn

We now discuss the distribution of the output signatures. Note that the output signature has two components $\boldsymbol{z}$ and the salt s. In the following, we first establish (in Proposition 1) that the statistical distance between the salt part of the output signature and the uniform distribution over $\{0,1\}^{\ell_s}$ is bounded by $1/q$. Then, we show (in Corollary 1) that the distribution of the signature deviates from the uniform distribution over $\mathbb{F}^n \times \{0,1\}^{\ell_s}$ by at most $1/q$. Let us define a good set and a bad set as follows:

$$\mathsf{Good} = \{\boldsymbol{v} \in \mathbb{F}^n : \mathcal{P}'(\boldsymbol{v}, .) : \mathsf{O} \to \mathbb{F}^m \text{ is non-singular}\}$$
$$\mathsf{Bad} = \{\boldsymbol{v} \in \mathbb{F}^n : \mathcal{P}'(\boldsymbol{v}, .) : \mathsf{O} \to \mathbb{F}^m \text{ is singular}\}.$$

Following Eq. (5), we have $|\mathsf{Good}| \approx q^n(1-1/q)$ and $|\mathsf{Bad}| \approx q^n \cdot \frac{1}{q}$, where $q = |\mathbb{F}|$. Sometimes, we refer to an element of Good (resp. Bad) as good (resp. bad) element. Let χ denote the random variable corresponding to the salt part of the output signature. Note that the distribution of χ depends on that of random variables $\boldsymbol{v}$ and s (involved in steps 1 and 5 respectively). Let U denote the uniform distribution over $\{0,1\}^{\ell_s}$. Then, the following proposition gives a bound on their statistical distance.

Proposition 1. *The statistical distance between χ and U is bounded by $1/q$.*

Proof. First observe that for any $\boldsymbol{a} \in \{0,1\}^{\ell_s}$, we have $\mathsf{Pr}[\chi = \boldsymbol{a} \mid \boldsymbol{v} \in \mathsf{Good}] = 1/2^{\ell_s}$, where the probability is taken over the random choice of $\boldsymbol{v} \in \mathbb{F}^n$ and $s \in \{0,1\}^{\ell_s}$. Then, calculate the following probability for any $\boldsymbol{a} \in \{0,1\}^{\ell_s}$.

$$\begin{aligned}\mathsf{Pr}[\chi = \boldsymbol{a}] &= \sum_{S \in \{\mathsf{Good},\mathsf{Bad}\}} \mathsf{Pr}[\chi = \boldsymbol{a} \mid \boldsymbol{v} \in S] \cdot \mathsf{Pr}[\boldsymbol{v} \in S] \\ &\approx \frac{1}{2^{\ell_s}} \cdot \left(1 - \frac{1}{q}\right) + p_{\boldsymbol{a}} \cdot \frac{1}{q} \qquad (7)\end{aligned}$$

where $p_{\boldsymbol{a}} = \mathsf{Pr}[\chi = \boldsymbol{a} \mid \boldsymbol{v} \in \mathsf{Bad}]$. Then, the statistical distance between χ and U is given by

$$\begin{aligned}\Delta(\chi, U) &= \frac{1}{2} \cdot \sum_{\boldsymbol{a} \in \{0,1\}^{\ell_s}} |\mathsf{Pr}[\chi = \boldsymbol{a}] - \mathsf{Pr}[U = \boldsymbol{a}]| \\ &\approx \frac{1}{2} \cdot \sum_{\boldsymbol{a} \in \{0,1\}^{\ell_s}} \left| \frac{1}{2^{\ell_s}} \cdot \left(1 - \frac{1}{q}\right) + p_{\boldsymbol{a}} \cdot \frac{1}{q} - \frac{1}{2^{\ell_s}} \right| \qquad \text{[using Eq. (7)]} \\ &\leq \frac{1}{2} \cdot \sum_{\boldsymbol{a} \in \{0,1\}^{\ell_s}} \left(\frac{1}{2^{\ell_s}} \cdot \frac{1}{q} + p_{\boldsymbol{a}} \cdot \frac{1}{q} \right) \\ &= \frac{1}{2 \cdot q} \cdot \left(1 + \sum_{\boldsymbol{a} \in \{0,1\}^{\ell_s}} p_{\boldsymbol{a}} \right) \\ &= \frac{1}{2 \cdot q} \cdot (1+1) = \frac{1}{q}.\end{aligned}$$

This completes the proof.

Corollary 1. *The distribution of the output signature deviates from the uniform distribution over Σ by at most $1/q$.*

Proof. Since $\boldsymbol{v}$ is chosen uniformly at random from $\mathbb{F}^n$, the $\boldsymbol{z}$-part of the signature is uniform over $\mathbb{F}^n$. Hence, the corollary follows from Proposition 1.

4.3 Uniformity of MQ-Systems

We now analyze the distribution of $\mathcal{P}(\boldsymbol{x})$, when $\boldsymbol{x} \in \mathbb{F}^n$ is chosen uniformly at random. In particular, we quantify the gap between this distribution and the uniform distribution. This is essentially required for giving a concrete security reduction of the salted homogeneous UOV signature. Recall that $\mathcal{P}$ is a random MQ-system which vanishes on a random subspace O. We show (see Corollary 2) that the statistical distance between the distribution of $\mathcal{P}(\boldsymbol{x})$ and the uniform distribution over $\mathbb{F}^m$ is at most $1/q$, where $q = |\mathbb{F}|$. Since $|\mathbb{F}^n/\mathsf{O}| = q^{n-m}$, $\mathbb{F}^n$ can be written as a union of q^{n-m} disjoint cosets of O in $\mathbb{F}^n$, i.e.,

$$\mathbb{F}^n = \bigcup_{i=1}^{q^{n-m}} \mathsf{Coset}_i$$

where $\mathsf{Coset}_i = \boldsymbol{v}_i + \mathsf{O}$, $\boldsymbol{v}_i$ is called a coset representative and $\mathsf{Coset}_j \cap \mathsf{Coset}_k = \emptyset$ for distinct $j, k \in [q^{n-m}]$. We now study the behavior (basically, bijectivity) of $\mathcal{P}$ on each coset Coset_i which is independent of the choice of the representative.

Proposition 2. *When $\boldsymbol{v}_i \in \mathsf{Good}$, then the restricted map $\mathcal{P} : \mathsf{Coset}_i \to \mathbb{F}^m$ is bijective.*

Proof. It suffices to show $\mathcal{P} : \mathsf{Coset}_i \to \mathbb{F}^m$ is injective. Let $\boldsymbol{x}_1', \boldsymbol{x}_2' \in \mathsf{O}$ be two arbitrary distinct elements. Since $\mathcal{P}'(\boldsymbol{v}_i, .) : \mathsf{O} \to \mathbb{F}^m$ is injective, $\mathcal{P}'(\boldsymbol{v}_i, \boldsymbol{x}_1') \neq \mathcal{P}'(\boldsymbol{v}_i, \boldsymbol{x}_2')$, that means $\mathcal{P}(\boldsymbol{v}_i + \boldsymbol{x}_1') \neq \mathcal{P}(\boldsymbol{v}_i + \boldsymbol{x}_2')$.

When $\mathcal{P}$ is bijective on a coset, then we would refer to this coset as 'good coset', otherwise 'bad coset'. Note that given a coset, any element of it can be a representative. So, if a coset contains at least one good element, then $\mathcal{P}$ will be bijective on that coset. We now ask the following question. What is the probability that a randomly picked coset is good? To answer the question let us take a look at the worst case situation, although the likelihood of this is very low: Out of the total q^{n-m} cosets, roughly $\frac{1}{q} \cdot q^{n-m}$ many cosets contain only the bad elements. Therefore, if we pick up any coset randomly, then it will be good with probability roughly $(1 - 1/q)$ in the worst case. Let GSet be the union of all good cosets and BSet be the union of all bad cosets (i.e., $\mathsf{BSet} = \mathbb{F}^n \setminus \mathsf{GSet}$). So, $\Pr[\boldsymbol{x} \in \mathsf{GSet}] \approx 1 - 1/q$ and $\Pr[\boldsymbol{x} \in \mathsf{BSet}] \approx 1/q$, where the probability is taken over the uniform choice of $\boldsymbol{x} \in \mathbb{F}^n$. Note that the statistical distance between the distribution of $\mathcal{P}(\boldsymbol{x})$ and the uniform distribution over $\mathbb{F}^m$ will be maximum in the worst case situation mentioned above. The following corollary quantifies the gap of the two distributions.

Corollary 2. *Let $\mathcal{P} : \mathbb{F}^n \to \mathbb{F}^m$ be a homogeneous UOV public map. When $\boldsymbol{x} \xleftarrow{\$} \mathbb{F}^n$, let χ denote the distribution of $\mathcal{P}(\boldsymbol{x})$ over $\mathbb{F}^m$. Let U be the uniform distribution over $\mathbb{F}^m$. Then $\Delta(\chi, U) \leq \frac{1}{q}$.*

Proof. When $\boldsymbol{x} \xleftarrow{\$} \mathsf{GSet}$, then $\boldsymbol{x}$ belongs to a random good coset; let us call it Coset. Then, $\boldsymbol{x}$ will be uniform over Coset. So, $\mathcal{P}(\boldsymbol{x})$ will be uniform over $\mathbb{F}^m$ thanks to Proposition 2. That is, for any $\boldsymbol{a} \in \mathbb{F}^m$, we have $\Pr[\chi = \boldsymbol{a} \mid \boldsymbol{x} \in \mathsf{GSet}] = 1/q^m$, where the probability is taken over the random choice of $\boldsymbol{x} \in \mathbb{F}^n$. Therefore, for any $\boldsymbol{a} \in \mathbb{F}^m$, we have

$$\begin{aligned}\mathsf{Pr}[\chi = \boldsymbol{a}] &= \sum_{S \in \{\mathsf{GSet},\mathsf{BSet}\}} \mathsf{Pr}[\chi = \boldsymbol{a} \mid \boldsymbol{x} \in S] \cdot \mathsf{Pr}[\boldsymbol{x} \in S] \\ &\approx \frac{1}{q^m} \cdot \left(1 - \frac{1}{q}\right) + p_{\boldsymbol{a}} \cdot \frac{1}{q} \end{aligned} \tag{8}$$

where $p_{\boldsymbol{a}} = \mathsf{Pr}[\chi = \boldsymbol{a} \mid \boldsymbol{x} \in \mathsf{BSet}]$. Then, the statistical distance between χ and U is given by

$$\begin{aligned}\Delta(\chi, U) &= \frac{1}{2} \cdot \sum_{\boldsymbol{a} \in \mathbb{F}^m} |\mathsf{Pr}[X = \boldsymbol{a}] - \mathsf{Pr}[U = \boldsymbol{a}]| \\ &\approx \frac{1}{2} \cdot \sum_{\boldsymbol{a} \in \mathbb{F}^m} \left| \frac{1}{q^m} \cdot \left(1 - \frac{1}{q}\right) + p_{\boldsymbol{a}} \cdot \frac{1}{q} - \frac{1}{q^m} \right| && \text{[using Eq. (8)]} \\ &\leq \frac{1}{2} \cdot \sum_{\boldsymbol{a} \in \mathbb{F}^m} \left(\frac{1}{q^m} \cdot \frac{1}{q} + p_{\boldsymbol{a}} \cdot \frac{1}{q} \right) \\ &= \frac{1}{2 \cdot q} \cdot \left(1 + \sum_{\boldsymbol{a} \in \mathbb{F}^m} p_{\boldsymbol{a}} \right) \\ &= \frac{1}{2 \cdot q} \cdot (1 + 1) = \frac{1}{q}. \end{aligned}$$

This completes the proof.

4.4 Security of Salted Homogeneous UOV Signature in CROM

In this section, we argue the security of SHUOV signature (presented in Sect. 4.2) in the classical random oracle model. Following the proof-style of [SSH11] and Corollaries 1 and 2, a security reduction can be easily shown from the UOVI-problem. For the shake of completeness, we give a proof-sketch in the CROM thereby resolving the issues raised in Sect. 3.1. Similarly, a security reduction for the traditional salted homogeneous UOV of [SSH11] can be derived.

Theorem 1. *If the UOVI-problem is intractable and q is superpolynomial in the security parameter κ, then the SHUOV-Signature is EUF-CMA secure in the CROM.*

Proof-sketch in CROM. The proof uses a hybrid argument over the following games.

0. Game_0. This is exactly the original EUF-CMA security game, where the hash function $\mathcal{H} : \mathcal{M} \times \mathsf{SaltSpac} \to \mathbb{F}^m$ is treated as random oracle. Note that the non-salt part of the output signature is distributed uniformly over $\mathbb{F}^n$. Let q_{uov} and q_{sign} be the number of hash queries and the number of sign-queries respectively. Let δ be the advantage of an adversary $\mathcal{A}_0$ in Game_0, i.e., $\mathsf{Adv}_{\mathcal{A}_0}^{\text{EUF-CMA}}(\kappa) = \delta$.

1. Game_1. This is same as Game_0, except[4] the salts involved in the answers of sign queries are chosen uniformly at random. That is, the output signature in Game_1 is distributed uniformly over Σ. Then, by Corollary 1, the advantage of an adversary $\mathcal{A}_1$ in Game_1 is given by $\mathsf{Adv}_{\mathcal{A}_1}^{\text{EUF-CMA}}(\kappa) \geq \mathsf{Adv}_{\mathcal{A}_0}^{\text{EUF-CMA}}(\kappa) - q_{\mathsf{sign}} \cdot \frac{1}{q} = \delta - q_{\mathsf{sign}} \cdot \frac{1}{q}$.
2. Game_2. This is same as Game_1, except the q_{sign}-many random oracle queries are answered by $\mathcal{P}(\boldsymbol{x})$, where $\boldsymbol{x} \xleftarrow{\$} \mathbb{F}^m$. Then, by Corollary 2, the advantage of an adversary $\mathcal{A}_2$ in Game_2 is given by $\mathsf{Adv}_{\mathcal{A}_2}^{\text{EUF-CMA}}(\kappa) \geq \mathsf{Adv}_{\mathcal{A}_1}^{\text{EUF-CMA}}(\kappa) - q_{\mathsf{sign}} \cdot \frac{1}{q} \geq \delta - 2 \cdot q_{\mathsf{sign}} \cdot \frac{1}{q}$.

We now show that using $\mathcal{A}_2$ in Game_2, we can break the UOVI-problem. An instance $(\mathcal{P}, \boldsymbol{y}^*) \in \mathcal{P}_{\mathsf{uov}}(\mathbb{F}^n, \mathbb{F}^m) \times \mathbb{F}^m$ of the UOVI-problem is given to a simulator S and the goal of S is to find $\boldsymbol{x}^* \in \mathbb{F}^n$ such that $\mathcal{P}(\boldsymbol{x}^*) = \boldsymbol{y}^*$. The simulator maintains a list $\mathsf{List}_{\mathsf{uov}}$ for keeping records of the form: $(\mathsf{m}, s, \mathcal{H}(\mathsf{m}||s))$. The adversary $\mathcal{A}_2$ may ask queries to hash oracle and sign-oracle in any order. The simulator S picks $i^* \xleftarrow{\$} [q_{\mathsf{uov}}]$ as a guess for the forgery message.

- **Hash-oracle**. When $\mathcal{A}_2$ asks the i-th $\mathcal{H}$-query on $\mathsf{m}_i||s_i$, it returns $\mathcal{H}(\mathsf{m}_i||s_i)$ if $(\mathsf{m}_i, s_i, \cdot) \in \mathsf{List}_{\mathsf{uov}}$. Otherwise, if $i = i^*$, then S updates $\mathsf{List}_{\mathsf{uov}}$ with the entry $(\mathsf{m}_i, s_i, \boldsymbol{y}^*)$ and returns $\boldsymbol{y}^*$, else it picks $\boldsymbol{y}_i \xleftarrow{\$} \mathbb{F}^m$, updates $\mathsf{List}_{\mathsf{uov}}$ with $(\mathsf{m}_i, s_i, \boldsymbol{y}_i)$ and returns $\boldsymbol{y}_i$.
- **Sign-oracle**. On the i-th query on message m_i, S picks $(\boldsymbol{x}_i, s_i) \xleftarrow{\$} \Sigma$. If $(\mathsf{m}_i, s_i, \cdot) \in \mathsf{List}_{\mathsf{uov}}$, it aborts, otherwise updates $\mathsf{List}_{\mathsf{uov}}$ with $(\mathsf{m}_i, s_i, \mathcal{P}(\boldsymbol{x}_i))$[5] and returns $\sigma_i = (\boldsymbol{x}_i, s_i)$.
- **Forgery**. When $\mathcal{A}_2$ produces a message-signature pair $(\mathsf{m}^*, \sigma^* = (\boldsymbol{x}^*, s^*))$, S submits $\boldsymbol{x}^*$ as a solution of the given instance of the UOVI-problem.

Note that all the queries of $\mathcal{A}_2$ are answered according to the description in Game_2. With probability $1/q_{\mathsf{uov}}$, S correctly guesses the message $\mathsf{m}^* = \mathsf{m}_{i^*}$, and $\boldsymbol{x}^*$ is a correct solution of the given problem instance if $(\mathsf{m}^*, \sigma^* = (\boldsymbol{x}^*, s^*))$ is a valid pair. So, the advantage of breaking the UOVI-problem is given by

$$\begin{aligned}
\mathsf{Adv}_{\mathsf{S}}^{\text{UOVI}}(\kappa) &\geq \frac{1}{q_{\mathsf{uov}}} \cdot \mathsf{Adv}_{\mathcal{A}_2}^{\text{EUF-CMA}}(\kappa) \\
&= \frac{1}{q_{\mathsf{uov}}} \cdot \left(\delta - 2 \cdot q_{\mathsf{sign}} \cdot \frac{1}{q}\right) \qquad (9)\\
&\approx \frac{1}{q_{\mathsf{uov}}} \cdot \delta \qquad [\text{as } q \text{ is superpolynomial in } \kappa\ ]
\end{aligned}$$

This ends the proof-sketch. □

[4] As mentioned earlier in Sect. 3.1, there is a gap between the distribution of salts involved in the construction and the security reduction of [SSH11]. That gap essentially depends on the size of the underlying field. But the authors implicitly assumed that a computational adversary cannot distinguish the difference. Unlike [SSH11], our security treatment takes into account this difference.

[5] Note that $\mathcal{H}(\mathsf{m}_i||s_i)$ is programmed by the value $\mathcal{P}(\boldsymbol{x}_i)$, instead of uniformly random value of $\mathbb{F}^m$ and this change is already captured in Game_2.

Remark 4. As mentioned earlier, we are able to resolve the issues in the security argument of [SSH11] (raised in Sect. 3) for the case of homogeneous salted UOV signature. While we utilize the subspace description of the scheme, one can easily check that the same strategy works for the case of conventional description (thanks to the identical distribution of keys in both the approaches). However, for general (not necessarily homogeneous) salted UOV signature, it is not known whether the corresponding key can be expressed through the subspace structure. Hence, one cannot directly apply Proposition 2 in this case.

Remark 5. Note that the above reduction makes sense, if q (that is, the size of the underlying field) involved in Eq. (9) is a superpolynomial in the security parameter. This q appears in Eq. (9) due to the bounds involved in Corollaries 1 and 2. Improving these bounds is an interesting future research problem as they have a direct bearing on the size of the underlying field.

5 Security of Salted Homogeneous UOV in QROM

In this section, we prove the security of SHUOV-signature in the quantum random oracle model. We start by recalling some notations and important results required for the security reduction. For two sets $\mathcal{X}$ and $\mathcal{Y}$, the notation $\mathcal{Y}^{\mathcal{X}}$ denotes the set of all functions from $\mathcal{X}$ to $\mathcal{Y}$. For a distribution D on $\mathcal{Y}$, the notation $g \longleftarrow D^{\mathcal{X}}$ denotes sampling a function $g : \mathcal{X} \to \mathcal{Y}$ as follows: for $x \in \mathcal{X}$, $g(x)$ is sampled according to the distribution D. For a given function $f : \mathcal{X} \to \mathcal{Y}$, we can always handle on-the-fly simulation of the function by the following unitary (see [NC00]):

$$\begin{aligned} \mathcal{O}_f : \mathcal{X} \times \mathcal{Y} &\to \mathcal{X} \times \mathcal{Y} \\ |x, y\rangle &\mapsto |x, y \oplus f(x)\rangle \end{aligned} \tag{10}$$

So, for handling superposition queries to the random oracle $\mathcal{H}$, it suffices to give a function description of the oracle. Here, we will use the fact [Zha12b] that the advantage of a quantum algorithm in distinguishing a randomly chosen $2k$-wise independent function from a truly random function is 0, where the number of quantum queries is at most k. This means a quantum-accessible random oracle can be implemented by choosing a random $2k$-wise independent function.

We show a reduction in the QROM based on small-range distributions [Zha12a]. Here, we first give the definition and related results of small-range distribution.

Definition 5 (Small-range distributions [Zha12a]**).** *Given an integer $r \in \mathbb{N}$, two sets $\mathcal{X}$ and $\mathcal{Y}$, and a distribution D on $\mathcal{Y}$, a small-range distribution, denoted by $\mathsf{SR}_r^D(\mathcal{X})$, is defined to be the following distribution on $\mathcal{Y}^{\mathcal{X}}$:*

1. *For each $i \in [r]$, choose a random value y_i from $\mathcal{Y}$ according to the distribution D, i.e., sample a function, say, $g : [r] \to \mathcal{Y}$ according to $D^{[r]}$.*
2. *For each $x \in \mathcal{X}$, pick $i \xleftarrow{\$} [r]$ and set $\mathcal{O}(x) = y_i$.*

This distribution can be alternatively viewed as follows: choose $g \leftarrow D^{[r]}$ and $f \xleftarrow{\$} [r]^{\mathcal{X}}$ and return the composition $\mathcal{O} = g \circ f$. Now, we state a result which is very important for arguing security of public-key schemes in the QROM. It essentially says that the difference between the output distributions of a quantum algorithm making k quantum queries to an oracle sampled either according to $\mathsf{SR}_r^D(\mathcal{X})$ or randomly from $\mathcal{Y}^{\mathcal{X}}$ is at most $27k^3/r$. The result is stated below.

Lemma 1 (**[Zha12a, Corollary 7.5]**). *Suppose a quantum algorithm asks k many quantum queries to an oracle either drawn from $\mathsf{SR}_r^D(\mathcal{X})$ or drawn randomly from $\mathcal{Y}^{\mathcal{X}}$. Then, the output distributions of the algorithm are $\ell(k)/r$-close, where $\ell(k) = \pi^2(2k)^3/3 < 27k^3$.*

Next, we describe another important result that can be used for programming random oracles. In particular, the result is useful in a situation, where oracle values were supposed to be assigned uniformly, but are assigned by sampling according to a distribution which is ϵ (negligible) distance apart from uniform distribution. Then, any quantum algorithm making k many queries to one of them can distinguishing them with probability at most $\mathcal{O}(k^{3/2}) \cdot \epsilon^{1/2}$. The result is stated below.

Lemma 2 (**[BZ13, Lemma 2.5]**). *Let $\mathcal{X}$ and $\mathcal{Y}$ be two sets. Suppose for each $x \in \mathcal{X}$, there are two distributions D_x and D'_x on $\mathcal{Y}$ with $|D_x - D'_x| \leq \epsilon$. Let two functions $\mathcal{O} : \mathcal{X} \to \mathcal{Y}$ and $\mathcal{O}' : \mathcal{X} \to \mathcal{Y}$ be defined as follows: for each $x \in \mathcal{X}$, $\mathcal{O}(x)$ and $\mathcal{O}'(x)$ are set by sampling from $\mathcal{Y}$ according to the distributions D_x and D'_x respectively. Then, any quantum algorithm making at most k quantum queries to $\mathcal{O}$ or $\mathcal{O}'$ can not distinguishing them, except with probability at most $\sqrt{8C_0k^3\epsilon}$, where $C_0 = 27$ (a universal constant).*

Let $\widetilde{\mathcal{M}} = \mathcal{M} \times \mathsf{SaltSpac}$. The sets $\mathcal{X}$ and $\mathcal{Y}$ that appear in the above lemma are considered to be $\widetilde{\mathcal{M}}$ and $\mathbb{F}^m$ in our context respectively. Further, we consider for all $x \in \mathcal{X}$, the distributions D_x (resp. D'_x) are to be the same and let us call it D (resp. D'). Now, we set D to be the uniform distribution over $\mathbb{F}^m$ and define the distribution D' over $\mathbb{F}^m$ as follows: Pick $\boldsymbol{x} \xleftarrow{\$} \mathbb{F}^n$ and output $\mathcal{P}(\boldsymbol{x})$, where $\mathcal{P} : \mathbb{F}^n \to \mathbb{F}^m$ is a random public key of SHUOV-scheme. Note that the statistical distance between D and D' is at most ϵ (thanks to Corollary 2), where $\epsilon = 1/q$ and $q = |\mathbb{F}|$. In the reduction, we use the following corollary.

Corollary 3. *Let $\widetilde{\mathcal{O}} : \widetilde{\mathcal{M}} \to \mathbb{F}^n$ and $\mathcal{O} : \widetilde{\mathcal{M}} \to \mathbb{F}^m$ be two quantum-accessible random oracles. Let $\mathcal{O}' : \widetilde{\mathcal{M}} \to \mathbb{F}^m$ be a quantum-accessible oracle defined as follows: for $\mathsf{m}||s \in \widetilde{\mathcal{M}}$, $\mathcal{O}'(\mathsf{m}||s) = \mathcal{P}(\widetilde{\mathcal{O}}(\mathsf{m}||s))$. Then, any quantum algorithm making at most k queries to $\mathcal{O}$ or $\mathcal{O}'$ can not distinguishing them, except with probability at most $\sqrt{8C_0k^3\epsilon}$.*

Proof. Let D be the uniform distribution over $\mathbb{F}^m$. Then, we define $D_{\mathsf{m}||s} = D$ for all $\mathsf{m}||s \in \widetilde{\mathcal{M}}$ and the computation of $\mathcal{O} : \widetilde{\mathcal{M}} \to \mathbb{F}^m$ can be thought of via sampling from $\mathbb{F}^m$ according to distribution D. The distribution D' picks $\boldsymbol{x} \xleftarrow{\$} \mathbb{F}^n$ and returns $\mathcal{P}(\boldsymbol{x})$. For each $\mathsf{m}||s \in \widetilde{\mathcal{M}}$, the distribution $D'_{\mathsf{m}||s}$ samples $\mathcal{P}(\boldsymbol{x})$, where $\boldsymbol{x} = \widetilde{\mathcal{O}}(\mathsf{m}||s)$ is uniform over $\mathbb{F}^n$. Basically, D' and $D'_{\mathsf{m}||s}$ have identical distribution. The remainder of the proof immediately follows from Lemma 2 and Corollary 2.

Theorem 2. *If the UOVI-problem is intractable and q is exponential in the security parameter κ, then the SHUOV-Signature is EUF-CMA secure in the QROM.*

Proof. We adopt the proof strategy [Zha15] of signature from trapdoor permutations. The proof essentially follows from a hybrid argument over the following games.

0. Game_0. This is exactly the original EUF-CMA security game, where the hash function $\mathcal{H} : \widetilde{\mathcal{M}} \to \mathbb{F}^m$ is treated as random oracle. Note that the non-salt part of the output signature is distributed uniformly over $\mathbb{F}^n$. Let q_{uov} and q_{sign} be the number of hash queries and the number of sign-queries respectively. Let $q_{\mathsf{tot}} = q_{\mathsf{uov}} + q_{\mathsf{sign}} + 1$. Let δ be the advantage of an adversary $\mathcal{A}_0$ in Game_0, i.e., $\mathsf{Adv}^{\text{EUF-CMA}}_{\mathcal{A}_0}(\kappa) = \delta$.
1. Game_1. This is same as Game_0, except the random oracle is programmed as follows: Pick a quantum-accessible random oracle $\widetilde{\mathcal{O}} : \widetilde{\mathcal{M}} \to \mathbb{F}^n$. Then, for each $\mathsf{m}||s \in \widetilde{\mathcal{M}}$, define $\mathcal{H}(\mathsf{m}||s) = \mathcal{P}(\widetilde{\mathcal{O}}(\mathsf{m}||s))$. By Corollary 3, the advantage of an adversary $\mathcal{A}_1$ in Game_1 is
$$\mathsf{Adv}^{\text{EUF-CMA}}_{\mathcal{A}_1}(\kappa) \geq \mathsf{Adv}^{\text{EUF-CMA}}_{\mathcal{A}_0}(\kappa) - \sqrt{8 \cdot C_0 \cdot q^3_{\mathsf{tot}} \cdot \epsilon} = \delta - \sqrt{8 \cdot C_0 \cdot q^3_{\mathsf{tot}} \cdot \epsilon}.$$
2. Game_2. This is same as Game_1, except the function $\widetilde{\mathcal{O}} : \widetilde{\mathcal{M}} \to \mathbb{F}^n$ is sampled according to the small-range distribution $\mathsf{SR}^D_r(\widetilde{\mathcal{M}})$, where D is the uniform distribution over $\mathbb{F}^n$, $r = \lceil 2 \cdot \ell(q_{\mathsf{tot}})/\delta \rceil$ and $\ell(q_{\mathsf{tot}}) = \pi^2 \cdot (2q_{\mathsf{tot}})^3/3 < 27 \cdot q^3_{\mathsf{tot}}$. Note that as mentioned earlier, $\widetilde{\mathcal{O}}$ can be viewed as $\widetilde{\mathcal{O}} = g \circ f$, where $g : [r] \to \mathbb{F}^n$ is described by the elements $\boldsymbol{x}_1, \ldots, \boldsymbol{x}_r \xleftarrow{\$} \mathbb{F}^n$ and $f \xleftarrow{\$} [r]^{\widetilde{\mathcal{M}}}$, i.e., for $\mathsf{m}||s \in \widetilde{\mathcal{M}}$, $\widetilde{\mathcal{O}}(\mathsf{m}||s) = \boldsymbol{x}_i$, where $f(\mathsf{m}||s) = i$. Then, by Lemma 1, the advantage of an adversary $\mathcal{A}_2$ in Game_2 is
$$\mathsf{Adv}^{\text{EUF-CMA}}_{\mathcal{A}_2}(\kappa) \geq \mathsf{Adv}^{\text{EUF-CMA}}_{\mathcal{A}_1}(\kappa) - \ell(q_{\mathsf{tot}})/r \geq \delta/2 - \sqrt{8 \cdot C_0 \cdot q^3_{\mathsf{tot}} \cdot \epsilon}.$$
3. Game_3. This is same as Game_2, except the salt computation in sign-oracle which is handled as follows: Let $\mathcal{O}_{\mathsf{salt}} : \widetilde{\mathcal{M}} \times [q_{\mathsf{sign}}] \to \mathsf{SaltSpac}$ be a classical[6] random oracle. A counter ctr (initially, set to 0) is maintained to keep track the index[7] of the current message queried to the sign-oracle. For a query message m, $\mathsf{ctr} \leftarrow \mathsf{ctr} + 1$ and the salt value for m is computed as $\mathcal{O}_{\mathsf{salt}}(\mathsf{m}||\mathsf{ctr})$. That is, the output signature in Game_3 is distributed uniformly over Σ. By Corollary 1, the advantage of an adversary $\mathcal{A}_3$ in Game_3 is
$$\mathsf{Adv}^{\text{EUF-CMA}}_{\mathcal{A}_3}(\kappa) \geq \mathsf{Adv}^{\text{EUF-CMA}}_{\mathcal{A}_2}(\kappa) - q_{\mathsf{sign}} \cdot \epsilon \geq \delta/2 - \sqrt{8 \cdot C_0 \cdot q^3_{\mathsf{tot}} \cdot \epsilon} - q_{\mathsf{sign}} \cdot \epsilon.$$

[6] Since the salt generation in the security game is involved only in answering sign-oracle (classically), it is sufficient to have a salt generation random oracle $\mathcal{O}_{\mathsf{salt}}$ which is classical.

[7] The whole purpose of this counter is to generate different salts even for the same message queried multiple times to the sign-oracle.

4. Game_4. This is same as Game_3, except the following:
   (a) At the beginning of the game, pick $i^* \xleftarrow{\$} [r]$. (This is the guess where the forged message-salt appears to the oracle f, i.e., $f(\mathsf{m}^*||s^*) = i^*$.)
   (b) Abort, if $f(\mathsf{m}^*||s^*) \neq i^*$ or if for any sign-query on m, $f(\mathsf{m}||s) = i^*$, where s is computed as described in Game_3.

   The probability of not abort is

$$\Pr[\neg\mathsf{abort}] = \frac{1}{r} \cdot \left(1 - \frac{1}{r}\right)^{q_{\mathsf{sign}}} \geq \frac{1}{r} - \frac{q_{\mathsf{sign}}}{r^2} \geq \frac{1}{2 \cdot r} \text{ (as } r \geq 2 \cdot q_{\mathsf{sign}}).$$

   Then, the advantage of an adversary $\mathcal{A}_4$ in Game_4 is

$$\begin{aligned}\mathsf{Adv}^{\text{EUF-CMA}}_{\mathcal{A}_4}(\kappa) &\geq \frac{1}{2 \cdot r} \cdot \mathsf{Adv}^{\text{EUF-CMA}}_{\mathcal{A}_4}(\kappa) \\ &\geq \frac{1}{2 \cdot r} \cdot \left(\frac{\delta}{2} - \sqrt{8 \cdot C_0 \cdot q^3_{\mathsf{tot}} \cdot \epsilon} - q_{\mathsf{sign}} \cdot \epsilon\right).\end{aligned}$$

5. Game_5. This is same as Game_4, except the following change in answering hash queries: Pick $\boldsymbol{y} \xleftarrow{\$} \mathbb{F}^m$ and set $\mathcal{H}(\mathsf{m}||s) = \boldsymbol{y}$ (instead of defining $\mathcal{H}(\mathsf{m}||s) = \mathcal{P}(\boldsymbol{x}_{i^*})$) for all $\mathsf{m}||s \in \widetilde{\mathcal{M}}$ such that $f(\mathsf{m}||s) = i^*$. Then, the advantage of an adversary $\mathcal{A}_5$ in Game_5 (using Corollary 2) is

$$\begin{aligned}\mathsf{Adv}^{\text{EUF-CMA}}_{\mathcal{A}_5}(\kappa) &\geq \mathsf{Adv}^{\text{EUF-CMA}}_{\mathcal{A}_4}(\kappa) - \epsilon \\ &\geq \frac{1}{2 \cdot r}\left(\frac{\delta}{2} - \sqrt{8 \cdot C_0 \cdot q^3_{\mathsf{tot}} \cdot \epsilon} - q_{\mathsf{sign}} \cdot \epsilon\right) - \epsilon.\end{aligned}$$

Now, we create a solver for the UOVI-problem using the adversary $\mathcal{A}_5$ (in Game_5). An instance $(\mathcal{P}, \boldsymbol{y}^*) \in \mathcal{P}_{\mathsf{uov}}(\mathbb{F}^n, \mathbb{F}^m) \times \mathbb{F}^m$ of the UOVI-problem is given to a simulator S and the goal of S is to find $\boldsymbol{x}^* \in \mathbb{F}^n$ such that $\mathcal{P}(\boldsymbol{x}^*) = \boldsymbol{y}^*$. The simulator will use $\mathcal{A}_5$ in the environment of Game_5 for breaking the problem instance. S picks $i^* \xleftarrow{\$} [r]$ and answers the following queries that may appear in any order:

- **Hash-oracle.** For answering quantum queries to hash-oracle, it suffices to describe only the classical description of the oracle function (without using any history) thanks to the on-the-fly simulation due to the unitary given in Eq. (10). For an input $\mathsf{m}||s \in \widetilde{\mathcal{M}}$, the function is defined as follows:

$$\mathcal{H}(\mathsf{m}||s) = \begin{cases} \boldsymbol{y}^* & \text{if } i = i^* \\ \mathcal{P}(\boldsymbol{x}_i) & \text{otherwise} \end{cases}$$

  where $f(\mathsf{m}||s) = i$. As mentioned earlier the quantum random oracle $f : \widetilde{\mathcal{M}} \to [r]$ can be implemented using random $2 \cdot q_{\mathsf{tot}}$-wise independent function.
- **Sign-oracle.** For a sign-query on m, S sets $\mathsf{ctr} \leftarrow \mathsf{ctr} + 1$ and computes $s = \mathcal{O}_{\mathsf{salt}}(\mathsf{m}||\mathsf{ctr})$ and $i = f(\mathsf{m}||s)$. If $i = i^*$, it aborts, otherwise returns the signature $\sigma = (\boldsymbol{x}, s)$, where $\boldsymbol{x} = \widetilde{\mathcal{O}}(\mathsf{m}||s)$.

- **Forgery**. When $\mathcal{A}$ produces a message-signature pair $(\mathsf{m}^*, \sigma^* = (\boldsymbol{x}^*, s^*))$, S checks whether $f(\mathsf{m}^*||s^*) = i^*$. If not, S aborts, otherwise it submits $\boldsymbol{x}^*$ as a solution of the given problem instance.

Note that when $(\mathsf{m}^*, \sigma^* = (\boldsymbol{x}^*, s^*))$ is a valid forgery, we have $\mathcal{P}(\boldsymbol{x}^*) = \mathcal{H}(\mathsf{m}^*||s^*) = \boldsymbol{y}^*$ and hence, $\boldsymbol{x}^*$ is a valid solution to the instance of the UOVI-problem. Therefore, we have

$$\begin{aligned}
\mathsf{Adv}_{\mathsf{S}}^{\mathrm{UOVI}}(\kappa) &= \mathsf{Adv}_{\mathcal{A}_5}^{\mathrm{EUF\text{-}CMA}}(\kappa) \\
&\geq \frac{1}{2\cdot r}\left(\frac{\delta}{2} - \sqrt{8\cdot C_0\cdot q_{\mathsf{tot}}^3\cdot\epsilon} - q_{\mathsf{sign}}\cdot\epsilon\right) - \epsilon \\
&\geq \frac{\delta}{4\cdot 27\cdot q_{\mathsf{tot}}^3}\left(\frac{\delta}{2} - \sqrt{8\cdot C_0\cdot q_{\mathsf{tot}}^3\cdot\epsilon} - q_{\mathsf{sign}}\cdot\epsilon\right) - \epsilon \\
&= \frac{\delta^2}{216\cdot q_{\mathsf{tot}}^3} - \epsilon\cdot\left(1 + \frac{\delta\cdot q_{\mathsf{sign}}}{108\cdot q_{\mathsf{tot}}^3}\right) - \sqrt{\epsilon}\cdot\sqrt{\frac{C_0}{54}}\cdot\frac{\delta}{\sqrt{q_{\mathsf{tot}}^3}} \\
&\approx \frac{\delta}{216\cdot q_{\mathsf{tot}}^3}
\end{aligned} \tag{11}$$

where the 2nd and the 3rd terms involved in Eq. (11) are ignored as $\epsilon = 1/q$ is negligible in κ. When δ is non-negligible in κ, then $\mathsf{Adv}_{\mathsf{S}}^{\mathrm{UOVI}}(\kappa)$ is non-negligible – a contradiction.

6 Concluding Remark

In this paper, we have identified some issues related to the security reduction of the salted UOV signature in the CROM [SSH11] and then addressed these issues through the subspace description [Beu21] of the scheme. This alternative construction of salted UOV improves the signing key size a bit. We also have provided a security reduction of the same scheme in the QROM. Our security treatment is applicable only to the homogeneous salted UOV signature. A clean security reduction for general salted UOV signature remains an interesting research problem.

Acknowledgement. We would like to thank the anonymous reviewers of Indocrypt 2022 for their comments and suggestions that helped us in polishing the technical and editorial content of this paper. This work is supported by the Ministry of Electronics and Information Technology, Government of India through its grants for the Center of Excellence in Quantum Technology at IISc Bangalore, India.

A Signature Using Trapdoor Information

A.1 Algorithm for Solving the Public Key System Using Trapdoor Information

In this section, we give the method for solving the public key system using the trapdoor information as an algorithm. The procedure was described in Sect. 4.1.

Algorithm 2. Inverting Public Key System Using Trapdoor

Require: The matrices $M'_{\mathfrak{g}}$ for each public key polynomial $\mathfrak{g}$, the hidden subspace O and an image point $\boldsymbol{\tau} \in \mathbb{F}^m$, where $\mathcal{P}(\mathsf{O}) = \{\mathbf{0}\}$ and O is described as column-space of an $n \times m$ matrix $\overline{M}$.

Ensure: A solution $\boldsymbol{z} \in \mathbb{F}^n$ such that $\mathcal{P}(\boldsymbol{z}) = \boldsymbol{\tau}$.

1: **repeat**
2: Sample a vector $\boldsymbol{v} \xleftarrow{\$} \mathbb{F}^n$
3: Compute $\boldsymbol{\tau}' = \boldsymbol{\tau} - \mathcal{P}(\boldsymbol{v})$
4: Compute $\boldsymbol{c}_{\mathfrak{g}} = \boldsymbol{v}^\top M'_{\mathfrak{g}}\overline{M}$
5: Construct $m \times m$ matrix C with $\boldsymbol{c}_{\mathfrak{g}}$ as rows
6: **until** $\{\boldsymbol{y} \in \mathbb{F}^m : C \cdot \boldsymbol{y} = \boldsymbol{\tau}'\} \neq \emptyset$
7: Sample $\boldsymbol{u} \xleftarrow{\$} \{\boldsymbol{y} \in \mathbb{F}^m : C \cdot \boldsymbol{y} = \boldsymbol{\tau}'\}$
8: Compute $\boldsymbol{o} = \overline{M}\boldsymbol{u}$ ▷ column-span corresponding to $\boldsymbol{u}$
9: Compute $\boldsymbol{z} = \boldsymbol{v} + \boldsymbol{o}$
10: Output $\sigma = \boldsymbol{z}$

A.2 Signature Scheme

Let us write down the complete signature scheme based on this trapdoor.

KeyGen. This takes the security parameter 1^κ as input and outputs the public and secret keys. The secret key is a description of the subspace $\mathsf{O} \subset \mathbb{F}^n$ and the public key is the system $\mathcal{P}$ consisting of m MQ-polynomials in n variables which vanish at O. Note that O can be represented by an $n \times m$ matrix as described in Sect. 4.1. Thus $\mathcal{SK} = \mathsf{O}$ and $\mathcal{PK} = \mathcal{P}$. A hash function $\mathcal{H} : \mathcal{M} \to \mathbb{F}^m$ for converting message into a fixed-length digest is known publicly.

Sign. This takes message m and the secret key $\mathcal{SK}$ as input and outputs a signature σ. The signature σ is obtained by solving $\mathcal{P}(\cdot) = \mathcal{H}(\mathsf{m})$ using Algorithm 2.

Ver. This takes the message m, the signature σ and the public key $\mathcal{PK}$ as input and outputs accept or reject. If $\mathcal{P}(\sigma) = \mathcal{H}(\mathsf{m})$, holds, the signature is accepted. Otherwise, the signature is rejected.

B Signature of Sakumoto et al.

We reproduce the salted version of UOV signature given in [SSH11, Section 4.1]. The secret key is a UOV type MQ system $\mathcal{F}$ of m polynomials in n variables. The authors consider non-homogeneous polynomials. Then, as usual, an affine invertible transformation $\mathcal{T}$ is used for mixing the variables. The public key is obtained in the obvious way as $\mathcal{P} = \mathcal{F} \circ \mathcal{T}$. The scheme uses a salt of length ℓ_s, which is a polynomial in the security parameter κ. The public and the secret keys contain a description of the salt space.

The verification follows the obvious procedure. We describe the signing algorithm in Algorithm 3. The variable list is parsed as $(\boldsymbol{x}_v, \boldsymbol{x}_m)$, where $\boldsymbol{x}_v$ denotes the vector of vinegar variables and $\boldsymbol{x}_m$ that of oil variables. There are v vinegar variables and m oil variables. The notation $\mathcal{F}(\boldsymbol{x}'_v, \boldsymbol{x}_m)$ is used to denote the linear system in oil variables which is obtained after the vinegar variables have been specialized to the vector $\boldsymbol{x}'_v$.

Algorithm 3. Signing Algorithm of Sakumoto, Shirai and Hiwatari

Require: $\mathcal{F}$, $\mathcal{T}$ and the message m
Ensure: A signature σ on m such that $\mathcal{P}(\sigma) = \mathcal{H}(\mathsf{m}||s)$

1: Sample $\boldsymbol{x}'_v \xleftarrow{\$} \mathbb{F}^v$ ▷ uniform assignment for vinegar variables
2: **repeat**
3: Sample salt $s \xleftarrow{\$} \{0,1\}^{\ell_s}$ ▷ sampling random salt
4: Compute $\boldsymbol{y} = \mathcal{H}(\mathsf{m}||s)$
5: **until** $\{\boldsymbol{x}_m \in \mathbb{F}^m : \mathcal{F}(x'_v, \boldsymbol{x}_m) = y\} \neq \emptyset$
6: Sample $\boldsymbol{x}'_m \xleftarrow{\$} \{\boldsymbol{x}_m \in \mathbb{F}^m : \mathcal{F}(\boldsymbol{x}'_v, \boldsymbol{x}_m) = \boldsymbol{y}\}$
7: Compute $\boldsymbol{x} = \mathcal{T}^{-1}(\boldsymbol{x}'_v, \boldsymbol{x}'_m)$ ▷ applying $\mathcal{T}^{-1}$ on a length n vector
8: Output $(\boldsymbol{x}, s)$ as signature

References

[Beu21] Beullens, W.: Improved cryptanalysis of UOV and rainbow. In: Canteaut, A., Standaert, F.-X. (eds.) EUROCRYPT 2021. LNCS, vol. 12696, pp. 348–373. Springer, Cham (2021). https://doi.org/10.1007/978-3-030-77870-5_13

[Beu22] Beullens, W.: Breaking rainbow takes a weekend on a laptop. In: Dodis, Y., Shrimpton, T. (eds.) CRYPTO 2022. LNCS, vol. 13508, pp. 464–479. Springer, Cham (2022). https://doi.org/10.1007/978-3-031-15979-4_16

[BR93] Bellare, M., Rogaway, P.: Random oracles are practical: a paradigm for designing efficient protocols. In: 1st ACM conference on Computer and communications security, pp. 62–73. SIAM (1993)

[BZ13] Boneh, D., Zhandry, M.: Secure signatures and chosen ciphertext security in a quantum computing world. In: Canetti, R., Garay, J.A. (eds.) CRYPTO 2013. LNCS, vol. 8043, pp. 361–379. Springer, Heidelberg (2013). https://doi.org/10.1007/978-3-642-40084-1_21

[CDP21] Chatterjee, S., Dimri, A., Pandit, T.: Identity-based signature and extended forking algorithm in the multivariate quadratic setting. In: Adhikari, A., Küsters, R., Preneel, B. (eds.) INDOCRYPT 2021. LNCS, vol. 13143, pp. 387–412. Springer, Cham (2021). https://doi.org/10.1007/978-3-030-92518-5_18

[CHR+16] Chen, M.-S., Hülsing, A., Rijneveld, J., Samardjiska, S., Schwabe, P.: From 5-pass $\mathcal{MQ}$-based identification to $\mathcal{MQ}$-based signatures. In: Cheon, J.H., Takagi, T. (eds.) ASIACRYPT 2016. LNCS, vol. 10032, pp. 135–165. Springer, Heidelberg (2016). https://doi.org/10.1007/978-3-662-53890-6_5

[CLND19] Chen, J., Ling, J., Ning, J., Ding, J.: Identity-based signature schemes for multivariate public key cryptosystems. Comput. J. **62**(8), 1132–1147 (2019)

[DS05] Ding, J., Schmidt, D.: Rainbow, a new multivariable polynomial signature scheme. In: Ioannidis, J., Keromytis, A., Yung, M. (eds.) ACNS 2005. LNCS, vol. 3531, pp. 164–175. Springer, Heidelberg (2005). https://doi.org/10.1007/11496137_12

[KPG99] Kipnis, A., Patarin, J., Goubin, L.: Unbalanced oil and vinegar signature schemes. In: Stern, J. (ed.) EUROCRYPT 1999. LNCS, vol. 1592, pp. 206–222. Springer, Heidelberg (1999). https://doi.org/10.1007/3-540-48910-X_15

[KS98] Kipnis, A., Shamir, A.: Cryptanalysis of the oil and vinegar signature scheme. In: Krawczyk, H. (ed.) CRYPTO 1998. LNCS, vol. 1462, pp. 257–266. Springer, Heidelberg (1998). https://doi.org/10.1007/BFb0055733

[MP17] Mohamed, M.S.E., Petzoldt, A.: RingRainbow – an efficient multivariate ring signature scheme. In: Joye, M., Nitaj, A. (eds.) AFRICACRYPT 2017. LNCS, vol. 10239, pp. 3–20. Springer, Cham (2017). https://doi.org/10.1007/978-3-319-57339-7_1

[NC00] Nielsen, M.A., Chuang, I.L.: Quantum Computation and Quantum Information. Cambridge University Press, New York (2000)

[Pat97] Patarin, J.: The oil and vinegar algorithm for signatures. In: Dagstuhl Workshop on Cryptography (1997)

[PCY+15] Petzoldt, A., Chen, M.-S., Yang, B.-Y., Tao, C., Ding, J.: Design principles for HFEv- based multivariate signature schemes. In: Iwata, T., Cheon, J.H. (eds.) ASIACRYPT 2015. LNCS, vol. 9452, pp. 311–334. Springer, Heidelberg (2015). https://doi.org/10.1007/978-3-662-48797-6_14

[PSM17] Petzoldt, A., Szepieniec, A., Mohamed, M.S.E.: A practical multivariate blind signature scheme. In: Kiayias, A. (ed.) FC 2017. LNCS, vol. 10322, pp. 437–454. Springer, Cham (2017). https://doi.org/10.1007/978-3-319-70972-7_25

[SSH11] Sakumoto, K., Shirai, T., Hiwatari, H.: On provable security of UOV and HFE signature schemes against chosen-message attack. In: Yang, B.-Y. (ed.) PQCrypto 2011. LNCS, vol. 7071, pp. 68–82. Springer, Heidelberg (2011). https://doi.org/10.1007/978-3-642-25405-5_5

[Zha12a] Zhandry, M.: How to construct quantum random functions. In: FOCS, pp. 679–687. IEEE Computer Society (2012)

[Zha12b] Zhandry, M.: Secure identity-based encryption in the quantum random oracle model. In: Safavi-Naini, R., Canetti, R. (eds.) CRYPTO 2012. LNCS, vol. 7417, pp. 758–775. Springer, Heidelberg (2012). https://doi.org/10.1007/978-3-642-32009-5_44

[Zha15] Zhandry, M.: Cryptography in the age of quantum computers. Ph.D. thesis, Stanford University (2015)

Author Index

Printed in the United States
by Baker & Taylor Publisher Services